Handbook of Spectrum Auction Design

Following the successful PCS Auction conducted by the US Federal Communications Commission in 1994, auctions have replaced traditional ways of allocating valuable radio spectrum, a key resource for the telecom sector. Spectrum auctions have raised billions of dollars worldwide and have become a role model for market-based approaches in the public and private sectors.

The design of spectrum auctions is a central application of game theory due to its economic relevance and the theoretical challenges it presents. This comprehensive handbook features classic papers and new contributions on all aspects of spectrum auction design, including the pros and cons of different auctions and lessons learned from theory, experiments, and the field, providing a valuable resource for regulators, telecommunications professionals, consultants, and researchers.

Martin Bichler is Professor of Informatics at the Technical University of Munich (TUM), and a faculty member at the TUM School of Management. He is known for his academic work on market design, and he has acted as a consultant for private and public organizations including regulators, telecoms, and procurement organizations. Projects in which he is involved include the design of auctions for industrial procurement, logistics, advertising, fishery access rights, and spectrum sales. His research addresses algorithmic, game-theoretical, and behavioral questions and has appeared in leading journals in computer science, economics, operations research, and management science. He is currently Editor of *Business and Information Systems Engineering* and serves on the editorial boards of several academic journals.

Jacob K. Goeree is Scienta Professor and Director of the AGORA Centre for Market Design at UNSW Australia. Jacob is a world-leading experimenter, theorist, and market designer. His research has influenced the design of spectrum auctions and markets for pollution permits and fishery access rights. Jacob was awarded a Research Fellowship by the Alfred P. Sloan Foundation and is Fellow of the Econometric Society, the Royal Netherlands Academy of Arts and Sciences, and the Royal Netherlands Society for the Sciences. He served as President of the Economic Science Association and as Editor of *Experimental Economics*. He is currently an Associate Editor of the *Journal of Economic Theory* and *Games and Economic Behavior*, and an Advisory Editor of *Experimental Economics*.

Handbook of Spectrum Auction Design

Edited by

Martin Bichler
Technical University of Munich

Jacob K. Goeree
AGORA Center for Market Design, University of New South Wales, Sydney

CAMBRIDGE
UNIVERSITY PRESS

University Printing House, Cambridge CB2 8BS, United Kingdom

One Liberty Plaza, 20th Floor, New York, NY 10006, USA

477 Williamstown Road, Port Melbourne, VIC 3207, Australia

314–321, 3rd Floor, Plot 3, Splendor Forum, Jasola District Centre, New Delhi – 110025, India

79 Anson Road, #06-04/06, Singapore 079906

Cambridge University Press is part of the University of Cambridge.

It furthers the University's mission by disseminating knowledge in the pursuit of education, learning and research at the highest international levels of excellence.

www.cambridge.org
Information on this title: www.cambridge.org/9781107135345
DOI: 10.1017/9781316471609

© Cambridge University Press 2017

This publication is in copyright. Subject to statutory exception and to the provisions of relevant collective licensing agreements, no reproduction of any part may take place without the written permission of Cambridge University Press.

First published 2017

Printed in the United States of America by Sheridan Books, Inc.

A catalogue record for this publication is available from the British Library

ISBN 978-1-107-13534-5 Hardback
ISBN 978-1-316-50114-6 Paperback

Cambridge University Press has no responsibility for the persistence or accuracy of URLs for external or third-party internet websites referred to in this publication and does not guarantee that any content on such websites is, or will remain, accurate or appropriate.

Contents

List of Contributors	page ix
Preface	xiii
Martin Bichler and Jacob K. Goeree	
List of Papers	xix

Part I   The Simultaneous Multiple-Round Auction

1   **Putting Auction Theory to Work: The Simultaneous Ascending Auction**   3
    Paul R. Milgrom

2   **An Equilibrium Analysis of the Simultaneous Ascending Auction**   26
    Jacob K. Goeree and Yuanchuan Lien

3   **The Efficiency of the FCC Spectrum Auctions**   54
    Peter Cramton

4   **Measuring the Efficiency of an FCC Spectrum Auction**   62
    Jeremy T. Fox and Patrick Bajari

Part II   The Combinatorial Clock Auction Designs

5   **Combinatorial Auction Design**   111
    David P. Porter, Stephen J. Rassenti, Anil Roopnarine, and Vernon L. Smith

6   **The Clock-Proxy Auction: A Practical Combinatorial Auction Design**   120
    Lawrence M. Ausubel, Peter Cramton, and Paul R. Milgrom

7   **Spectrum Auction Design**   141
    Peter Cramton

8	**A Practical Guide to the Combinatorial Clock Auction** Lawrence M. Ausubel and Oleg Baranov	170
9	**Market Design and the Evolution of the Combinatorial Clock Auction** Lawrence M. Ausubel and Oleg Baranov	187
10	**Quadratic Core-Selecting Payment Rules for Combinatorial Auctions** Robert Day and Peter Cramton	195
11	**Core-Selecting Package Auctions** Robert Day and Paul R. Milgrom	226
12	**A New Payment Rule for Core-Selecting Package Auctions** Aytek Erdil and Paul Klemperer	241
13	**On the Impossibility of Core-Selecting Auctions** Jacob K. Goeree and Yuanchuan Lien	253
14	**Ascending Combinatorial Auctions with Risk Averse Bidders** Kemal Guler, Martin Bichler, and Ioannis Petrakis	264
15	**Properties of the Combinatorial Clock Auction** Jonathan Levin and Andrzej Skrzypacz	294
16	**Budget Constraints in Combinatorial Clock Auctions** Maarten Janssen, Vladimir A. Karamychev, and Bernhard Kasberger	318
17	**(Un)expected Bidder Behavior in Spectrum Auctions: About Inconsistent Bidding and its Impact on Efficiency in the Combinatorial Clock Auction** Christian Kroemer, Martin Bichler, and Andor Goetzendorff	338

Part III Alternative Auction Designs

18	**A Combinatorial Auction Mechanism for Airport Time Slot Allocation** Stephen J. Rassenti, Vernon L. Smith, and Robert L. Bulfin	373
19	**A New and Improved Design for Multi-Object Iterative Auctions** Anthony M. Kwasnica, John O. Ledyard, David P. Porter, and Christine DeMartini	391
20	**Hierarchical Package Bidding: A Paper & Pencil Combinatorial Auction** Jacob K. Goeree and Charles A. Holt	418
21	**Assignment Messages and Exchanges** Paul R. Milgrom	443
22	**The Product-Mix Auction: A New Auction Design for Differentiated Goods** Paul Klemperer	462

23	**The Continuous Combinatorial Auction Architecture**	**477**
	Charles R. Plott, Hsing Yang Lee, and Travis Maron	
24	**Coalition-based Pricing in Ascending Combinatorial Auctions**	**493**
	Martin Bichler, Zhen Hao, and Gediminas Adomavicius	

Part IV Experimental Comparisons of Auction Designs

25	**Experiments Testing Multiobject Allocation Mechanisms**	**531**
	John O. Ledyard, David P. Porter, and Antonio Rangel	
26	**Laboratory Experimental Testbeds: Application to the PCS Auction**	**561**
	Charles R. Plott	
27	**An Experimental Test of Flexible Combinatorial Spectrum Auction Formats**	**588**
	Christoph Brunner, Jacob K. Goeree, Charles A. Holt, and John O. Ledyard	
28	**On the Impact of Package Selection in Combinatorial Auctions: An Experimental Study in the Context of Spectrum Auction Design**	**607**
	Tobias Scheffel, Georg Ziegler, and Martin Bichler	
29	**Do Core-Selecting Combinatorial Clock Auctions Lead to High Efficiency? An Experimental Analysis of Spectrum Auction Designs**	**640**
	Martin Bichler, Pasha Shabalin, and Jürgen Wolf	
30	**Spectrum Auction Design: Simple Auctions For Complex Sales**	**672**
	Martin Bichler, Jacob K. Goeree, Stefan Mayer, and Pasha Shabalin	

Part V The Bidders' Perspective

31	**Winning Play in Spectrum Auctions**	**689**
	Jeremy I. Bulow, Jonathan Levin, and Paul R. Milgrom	
32	**Up in the Air: GTE's Experience in the MTA Auction for Personal Communication Services Licenses**	**713**
	David J. Salant	
33	**Bidding Complexities in the Combinatorial Clock Auction**	**731**
	Vitali Gretschko, Stephan Knapek, and Achim Wambach	
34	**Strategic Bidding in Combinatorial Clock Auctions – A Bidder Perspective**	**748**
	Richard Marsden and Soren T. Sorensen	
35	**Impact of Budget Constraints on the Efficiency of Multi-lot Spectrum Auctions**	**764**
	Nicholas Fookes and Scott McKenzie	

Part VI Secondary Markets and Exchanges

36 **Spectrum Markets: Motivation, Challenges, and Implications** 785
    Randall Berry, Michael L. Honig, and Rakesh V. Vohra

37 **Designing the US Incentive Auction** 803
    Paul R. Milgrom and Ilya R. Segal

38 **Solving the Station Repacking Problem** 813
    Alexandre Fréchette, Neil Newman, and Kevin Leyton-Brown

39 **ICE: An Expressive Iterative Combinatorial Exchange** 828
    Benjamin Lubin, Adam I. Juda, Ruggiero Cavallo, Sébastien Lahaie, Jeffrey Shneidman, and David C. Parkes

40 **ACE: A Combinatorial Market Mechanism** 874
    Leslie Fine, Jacob K. Goeree, Tak Ishikida, and John O. Ledyard

Outlook 903
    Martin Bichler and Jacob K. Goeree

Contributors

Gediminas Adomavicius Carlson School of Management, University of Minnesota

Lawrence M. Ausubel Department of Economics, University of Maryland

Patrick Bajari Department of Economics, University of Washington

Oleg Baranov Department of Economics, University of Colorado Boulder

Randall Berry Department of Electrical Engineering and Computer Science, Northwestern University

Martin Bichler Department of Informatics, Technical University of Munich

Christoph Brunner Alfred-Weber-Institut, University of Heidelberg

Robert L. Bulfin Department of Industrial and Systems Engineering, Auburn University

Jeremy I. Bulow Graduate School of Business, Stanford University

Ruggiero Cavallo Yahoo Research

Peter Cramton Department of Economics, University of Maryland

Robert Day School of Business, University of Connecticut

Christine DeMartini Division of the Humanities and Social Sciences, California Institute of Technology

Aytek Erdil Faculty of Economics, University of Cambridge

Leslie Fine Salesforce

Nicholas Fookes Communications Chambers

Jeremy T. Fox Department of Economics, Rice University

Alexandre Fréchette Department of Computer Science, University of British Columbia

Jacob K. Goeree School of Economics, UNSW Business School

Andor Goetzendorff Department of Informatics, Technical University of Munich

Vitali Gretschko Center for European Economic Research (ZEW)

Kemal Guler Department of Economics, Anadolu University

Zhen Hao Department of Informatics, Technical University of Munich

Charles A. Holt Department of Economics, University of Virginia

Michael L. Honig Department of Electrical Engineering and Computer Science, Northwestern University

Takashi Ishikida OR Consultant

Maarten Janssen Department of Economics, University of Vienna and Higher School of Economics, Moscow

Adam Juda Google

Vladimir A. Karamychev Department of Economics, Erasmus University and Tinbergen Institute

Bernhard Kasberger Department of Economics, University of Vienna

Paul Klemperer Department of Economics, University of Oxford

Stephan Knapek TWS Partners

Christian Kroemer Department of Informatics, Technical University of Munich

Anthony M. Kwasnica Smeal College of Business, The Pennsylvania State University

Sébastien Lahaie Google

John O. Ledyard Division of the Humanities and Social Sciences, California Institute of Technology

Hsing Yang Lee Division of the Humanities and Social Sciences, California Institute of Technology

Jonathan Levin Graduate School of Business, Stanford University

Kevin Leyton-Brown Department of Computer Science, University of British Columbia

Yuanchuan Lien Department of Economics, Stanford University

Benjamin Lubin Information Systems Department, Questrom School of Business, Boston University

LIST OF CONTRIBUTORS

Travis Maron Division of the Humanities and Social Sciences, California Institute of Technology

Richard Marsden NERA Economic Consulting

Stefan Mayer Department of Informatics, Technical University of Munich

Scott McKenzie Coleago Consulting

Paul R. Milgrom Department of Economics, Stanford University

Neil Newman Department of Computer Science, University of British Columbia

David C. Parkes Paulson School of Engineering and Applied Sciences, Harvard University

Jannis Petrakis Department of Informatics, Technical University of Munich

Charles R. Plott Division of the Humanities and Social Sciences, California Institute of Technology

David P. Porter Economic Science Institute, Chapman University

Antonio Rangel Division of the Humanities and Social Sciences, California Institute of Technology

Stephen J. Rassenti Economic Science Institute, Chapman University

Anil Roopnarine Cybernomics Inc.

David J. Salant Auction Technologies, and Toulouse School of Economics

Tobias Scheffel Department of Informatics, Technical University of Munich

Ilya R. Segal Department of Economics, Stanford University

Pasha Shabalin Astradi

Jeffrey Shneidman Fish & Richardson

Andrzej Skrzypacz Graduate School of Business, Stanford University

Vernon L. Smith Economic Science Institute, Chapman University

Soren T. Sorensen NERA Economic Consulting

Rakesh V. Vohra Department of Economics, University of Pennsylvania

Achim Wambach Center for European Economic Research (ZEW)

Jürgen Wolf Department of Informatics, Technical University of Munich

Georg Ziegler Department of Informatics, Technical University of Munich

Preface

The 1994 sales of spectrum for "personal communication services" (PCS) marked a sharp change in policy by the US Federal Communications Commission (FCC), which had allocated spectrum for free until then. The PCS auction was designed by Stanford professors Paul Milgrom and Robert Wilson.[1] The July auction with just ten licenses raised over $600 million while the December auction raised more than $7 billion, breaking all records for the sale of public assets in the US and leading the *New York Times* to hail it as "the greatest auction ever." Substantial revenues are an obvious benefit but even more important is the fact that the PCS auction allocated this valuable public asset efficiently. By forcing firms "to put their money where their mouths are" the PCS auction selected firms that could utilize spectrum the best – to the greater benefit of society. Milgrom and Wilson's design has since been adopted by many regulators and has generated hundreds of billions of dollars for treasuries worldwide.

The PCS auction was organized as a simultaneous multiple-round auction (SMRA), which is also known as a simultaneous ascending auction (SAA). The SMRA is a simple but flexible format to sell multiple licenses in parallel. Despite the simplicity of its rules, the SMRA may create strategic difficulties for bidders interested in acquiring combinations of licenses. Since licenses have to be won one-by-one in the SMRA, bidders who compete aggressively for a desired combination risk winning an inferior subset at high prices. This is known as the *exposure problem*. Foreseeing the possibility of being exposed, bidders may act cautiously with adverse effects for revenue and efficiency.

Combinatorial auctions solve the exposure problem by allowing for bids on combinations of licenses. While this feature has the potential to improve efficiency, it also leads to new design challenges such as the computational hardness of the allocation problem, or the combinatorial growth in the number of package bids in some auction formats. Spectrum auction design has seen several recent innovations such as the single-stage and two-stage combinatorial clock auction (CCA), hierarchical package bidding (HPB), or sealed-bid combinatorial auctions.

---

[1] A closely related format that used a different stopping rule was proposed by Preston McAfee.

The motivation for this edited volume came from discussions with regulators, consultants, and telecom operators who were asking for literature on recent trends in spectrum auction design. High-stakes spectrum auctions are being conducted regularly across the world, and just like academics, practitioners would like to get an overview of the various developments in this field. The need for an edited volume that takes stock of the rapid developments in this field is illustrated, for instance, by the "Market-Design Experiments" chapter of the recent *Second Handbook of Experimental Economics* (2016). In this chapter, Al Roth, who received the 2012 Nobel Prize for his work on improving matching institutions, writes:

> In summary, if I had written this section on FCC auctions in early 2008, it would have been tempting to conclude on a triumphant note: after years of experiments promoting package bidding, the FCC had finally implemented a limited version of it. In view of the subsequent return to auctions without package bidding, a more sober assessment may be called for.

It just goes to show how *the times they are a-changin'*. In retrospect the year 2008 was a turning point for combinatorial auction design, in terms of practical application and fundamental research. That year, the FCC successfully applied the hierarchical package bidding (HPB) auction to sell their 700 MHz spectrum, resulting in record revenues of close to $19 billion. At the same time, regulators around the globe decided to substitute away from the SMRA and employ a combinatorial clock auction (CCA) or other combinatorial formats (e.g. Australia, Austria, Canada, Denmark, Montenegro, the Netherlands, Ireland, Slovenia, Switzerland, and the UK). Depending on the type of spectrum being sold the SMRA is still being used, as Roth notes, but the vast majority of spectrum auctions conducted since 2008 have allowed for combinatorial bidding. Moreover, combinatorial spectrum auction design has since blossomed into a prime example of innovative, impactful, and interdisciplinary research as attested by the contributions to this *Handbook*.

Besides spectrum auctions, the contributions to this book deal with resource allocation problems involving hard computational allocation problems and strategic market participants. These questions are fundamental to computer science, economics, operations research, and the management sciences alike. Actually, combinatorial auctions are only possible nowadays due to the substantial advances in combinatorial optimization in the past decades. While we focus on spectrum sales, the questions raised are clearly not restricted to this application. Multi-object markets of this sort can be found in industrial procurement, logistics, the sale of pollution permits, in day-ahead energy markets, or the sale of TV ad slots, to name just a few. Successful auction designs for spectrum markets are a likely role model for other domains as well.

The volume is organized in six parts. Part I focuses on the Simultaneous Multi-Round Auction, Part II on the Combinatorial Clock Auction, and Part III on alternative auction formats. Part IV summarizes experimental comparisons of different auction formats in the lab. Part V provides experiences and strategies of bidders in different auction designs, and Part VI includes contributions on secondary spectrum markets and exchanges.

Part I: The Simultaneous Multi-Round Auction

The Simultaneous Multi-Round Auction (SMRA) is a beautifully simple generalization of the English auction to multiple licenses. All licenses are sold at the same time, each with a price associated with it, and the bidders can bid on any one of the licenses. The auction proceeds in rounds, which is a specific period of time in which all bidders can submit bids. After the round is closed, the auctioneer discloses provisional winners and current license prices, which equal the highest bids submitted for the licenses. The bidding continues until no bidder is willing to raise the bid on any of the licenses any more. In other words, if in one round no new bids are placed then the auction ends with each bidder winning the licenses on which he has the high bid, and pays the bid for any license won.

The SMRA has successfully been used to allocate spectrum for more than two decades raising hundreds of billions of dollars for Treasuries worldwide. Part I summarizes key contributions on the SMRA. It includes the seminal overview paper by Paul Milgrom (Chapter 1), a game-theoretical analysis by Goeree and Lien (Chapter 2), as well as empirical analyses by Cramton (Chapter 3) and Fox and Bajari (Chapter 4).

Part II: The Combinatorial Clock Auction Designs

The Combinatorial Clock Auction (CCA) refers to a family of different but related designs, which were used world-wide since 2008. Chapter 5 describes a one-stage ascending clock auction in a paper by Porter, Rassenti, Roopnarine, and Smith. The first two-stage CCA design is outlined in Chapter 6 by Ausubel, Cramton, and Milgrom. Cramton discusses various properties and the design rationale in Chapter 7, while in Chapter 8 Ausubel and Baranov provide an accessible guide to the CCA as it is used world-wide nowadays. In Chapter 9 Ausubel and Baranov summarize differences in successive versions of the CCA. Day and Cramton treat computational issues about the quadratic core-selecting payment rule used in the two-stage CCA in Chapter 10, and Day and Milgrom analyze the core-selecting payment rule game-theoretically in Chapter 11. In Chapter 12 Erdil and Klemperer provide alternatives to the quadratic core-selecting payment rule.

Chapter 13 provides a Bayesian Nash equilibrium analysis of the broader class of core-selecting auctions assuming risk-neutral bidders and shows that no core-selecting auction can be in the core with respect to the true valuations if the Vickrey auction is not in the core. Guler, Bichler, and Petrakis (Chapter 14) show that the result extends to arbitrarily risk-averse bidders, although risk aversion can reduce the scope of inefficient equilibria. Levin and Skrzypacz provide a game-theoretical analysis of the specifics of the CCA and show that there are multiple equilibria in Chapter 15. Janssen, Karamychev, and Kasberger analyze the impact of budget constraints on the CCA in Chapter 16. Finally, Kroemer, Bichler, and Goetzendorff analyze bidding behavior in the CCA based on bid data from the field in Chapter 17.

Part III: Alternative Auction Designs

SMRA and the CCA both exhibit advantages and disadvantages, but they are not the only auction formats for single-sided multi-object auctions. Part III of this edited volume summarizes alternative auction designs. Some of them have been evaluated by regulators, some also been used for spectrum sales or other applications. The original design of a combinatorial auction for the allocation of airport time slot by Rassenti, Smith, and Bulfin is described in Chapter 18. In Chapter 19 Kwasnica, Ledyard, Porter, and DeMartini introduce ascending combinatorial auctions with pseudo-dual linear and anonymous prices. Such designs have been analyzed for the sale of spectrum licenses and used in industrial procurement. Chapter 20 by Goeree and Holt describes Hierarchical Package Bidding (HPB), an auction format which has been used by the US Federal Communications Commission to sell spectrum licenses.

Milgrom describes an auction format for substitutable preferences in Chapter 21, which is related to a design outlined by Klemperer in Chapter 22. This product-mix auction has been used to auction loans of funds secured against different varieties of collateral. Plott, Lee, and Maron introduce a continuous (not round-based) combinatorial auction format which has been used in field applications in Chapter 23, and Bichler, Hao, and Adomavicius introduce a pricing rule to address the coordination problem that bidders face in larger ascending combinatorial auctions with exponentially many possible package bids in Chapter 24.

Part IV: Experimental Comparisons of Auction Designs

Laboratory experiments have been recognized as important complements to game-theoretical analyses of auctions. They are particularly important for multi-object auctions, because game-theoretical models often need to make simplifying assumptions and human bidder behavior can deviate significantly from normative theoretical models. The chapters in Part IV provide results of experiments which aimed at a comparison of different auction formats.

Ledyard, Porter, and Rangel (Chapter 25) as well as Plott (Chapter 26) describe initial experiments to compare SMRA against sequential and combinatorial auctions. Brunner, Goeree, Holt, and Ledyard compare SMRA to auction formats using pseudo-dual linear and anonymous prices and a single-stage CCA in Chapter 27. With high complementarities in the valuations, the combinatorial auction formats achieved higher efficiency than SMRA. Scheffel, Ziegler, and Bichler compare HPB with the single-stage CCA, and an auction with pseudo-dual linear prices in Chapter 28. Both, the CCA and HPB achieved high efficiency and revenue, but the package selection heuristics of bidders had a negative impact on efficiency in all combinatorial auction formats.

Chapter 29 by Bichler, Shabalin, and Wolf reports the first experiments to compare the two-stage CCA with SMRA in larger auctions with more licenses based on realistic spectrum band plans. The two-stage CCA achieved lower efficiency than SMRA in particular in multi-band auctions, which is partly due to the fully enumerative bid language used in the two-stage CCA and the fact that bidders can only submit subsets of the exponentially many packages with positive value. Chapter 30 by Bichler, Goeree, Mayer, and Shabalin then addresses the problem with compact bid languages and shows that combinatorial auctions with compact bid languages, where bidders can specify their preferences succinctly, have a significant positive impact on efficiency.

Part V: The Bidders' Perspective

Analytical models and lab experiments typically require some simplifications. The strategic challenges and problems of bidders are often beyond what can be modeled or analyzed in the lab experimentally. Part V covers reports of colleagues, who consulted in spectrum auctions shedding light on additional aspects which are important in the field.

In Chapter 31 Bulow, Levin, and Milgrom discuss bidding strategies in a simultaneous ascending auction organized by the US Federal Communications Commission leveraging information about other market participants and their budget revealed throughout the auction. Also Salant discusses bidding strategies in an SMRA with regional licenses in Chapter 32. Chapter 33 by Gretschko, Knapek, and Wambach, Chapter 34 by Marsden and Sorensen, and Chapter 35 by Fookes and McKenzie focus on various strategic problems in the two-stage CCA.

Part VI: Secondary Markets and Exchanges

The move of regulators to use markets to allocate spectrum rights through auctions in the mid-1990s is widely considered a success. Yet regardless of how efficiently initial rights are allocated, changing supply and demand conditions mean that initial allocations can quickly become inefficient. Well-functioning secondary markets can ensure that spectrum can shift to new, more efficient uses. Such secondary markets might also be organized as a centralized market, which allows to better address technical or strategic allocation constraints. Part VI discusses related ideas.

Berry, Honig, and Vohra provide a discussion of challenges and implications of secondary spectrum markets in Chapter 36. Chapter 37 by Milgrom and Segal describes the remarkable design of the US incentive auction in 2016–2017, a large two-sided spectrum auction market allowing TV broadcasters to sell and telecoms to buy spectrum licenses. The allocation problem in this auction is a computationally hard problem discussed in a contribution by Newman and Fréchette in Chapter 38. Spectrum auction markets among telecommunication providers will require some support for package bids. Combinatorial exchanges are in their infancy, but we provide two examples of exchange designs in Chapters 39 and 40. Lubin, Juda, Cavallo, Lahaie, Shneidman, and Parkes propose an expressive iterative combinatorial exchange design in Chapter 39. Fine, Ishikida, Goeree, and Ledyard describe a combinatorial call market for pollution permits in Chapter 40. Such designs provide valuable ideas for future spectrum markets.

The book ends with an outlook chapter discussing current challenges in the design of spectrum auctions.

Acknowledgements

We would like to thank all contributors for their support in organizing this edited volume. We also appreciate the support of the various academic organizations and publishers that have allowed us to reprint papers that have already appeared in journals. Special thanks to Marianne Thanner, Lauren Cowles, and Esther Miguéliz Obanos for their help in organizing the edited volume, and to Yuqi Goeree for suggesting the cover. Finally, we would like to thank MSM staff for their hospitality during much of the editing process and the AGORA center for market design at UNSW, see www.agora.group/, for generous financial support.

Papers

This edited volume intends to provide key contributions on spectrum auction design and make them accessible in a single book. The following list provides references to the original articles.

Part I The Simultaneous Multiple-Round Auction

- Chapter 1: Paul R. Milgrom. Putting Auction Theory to Work: The Simultaneous Ascending Auction. *Journal of Political Economy*, 108(2):245–272, 2000. © The University of Chicago Press
- Chapter 2: Jacob K. Goeree and Yuanchuan Lien. An Equilibrium Analysis of the Simultaneous Ascending Auction. Reprint from *Journal of Economic Theory*, 153:506–533, © Elsevier 2014. With permission of Elsevier
- Chapter 3: Peter Cramton. The Efficiency of the FCC Spectrum Auctions. *The Journal of Law and Economics*, 41(S2):727–736, 1998. © The University of Chicago Press
- Chapter 4: Jeremy T. Fox and Patrick Bajari. Measuring the Efficiency of an FCC Spectrum Auction. *American Economic Journal: Microeconomics*, 5(1):100–146, 2013. © American Economic Association

Part II The Combinatorial Clock Auction Designs

- Chapter 5: David P. Porter, Stephen J. Rassenti, Anil Roopnarine, and Vernon L. Smith. Combinatorial Auction Design. *Proceedings of the National Academy of Sciences*, 100(19):11153–11157, 2003. © (2003) National Academy of Sciences, U.S.A
- Chapter 6: Lawrence M. Ausubel, Peter Cramton, and Paul R. Milgrom. The Clock-Proxy Auction: A Practical Combinatorial Auction Design. In *Combinatorial Auctions*, pages 115–138. 2006. © 2005 Massachusetts Institute of Technology, by permission of the MIT Press

- Chapter 7: Peter Cramton. Spectrum Auction Design. *Review of Industrial Organization*, 42(2):161–190, 2013. © Springer Science+Business Media New York 2013. With permission of Springer
- Chapter 8: Lawrence M. Ausubel and Oleg Baranov. A Practical Guide to the Combinatorial Clock Auction. *Economic Journal*, (to appear), 2017. © 2017 Royal Economic Society
- Chapter 9: Lawrence M. Ausubel and Oleg Baranov. Market Design and the Evolution of the Combinatorial Clock Auction. *The American Economic Review*, 104(5):446–451, 2014. © American Economic Association
- Chapter 10: Robert Day and Peter Cramton. Quadratic Core-Selecting Payment Rules for Combinatorial Auctions. *Operations Research*, 60(3):588–603, 2012. Reproduced with permission. © INFORMS, www.informs.org
- Chapter 11: Robert Day and Paul R. Milgrom. Core-Selecting Package Auctions. *International Journal of Game Theory*, 36(3–4):393–407, 2008. © Springer-Verlag 2007. With permission of Springer
- Chapter 12: Aytek Erdil and Paul Klemperer. A New Payment Rule for Core-Selecting Package Auctions. *Journal of the European Economic Association*, 8(2–3):537–547, 2010. © 2010 by the European Economic Associations
- Chapter 13: Jacob K. Goeree and Yuanchuan Lien. On the Impossibility of Core-Selecting Auctions. *Theoretical Economics*, 11(1), 2016. Econometric Society
- Chapter 14: Kemal Guler, Martin Bichler, and Ioannis Petrakis. Ascending Combinatorial Auctions with Risk Averse Bidders. *Group Decision and Negotiation*, 25(3):609–639, 2016. © Springer Science+Business Media Dordrecht 2015. With permission of Springer
- Chapter 15: Jonathan Levin and Andrzej Skrzypacz. Properties of the Combinatorial Clock Auction. *The American Economic Review*, (to appear), 2017. © American Economic Association
- Chapter 16: Maarten Janssen, Vladimir A. Karamychev, and Bernhard Kasberger. Budget Constraints in Combinatorial Clock Auctions, 2017
- Chapter 17: Christian Kroemer, Martin Bichler, and Andor Goetzendorff. (Un)expected Bidder Behavior in Spectrum Auctions: About Inconsistent Bidding and its Impact on Efficiency in the Combinatorial Clock Auction. *Group Decision and Negotiation*, 25(1):31–63, 2016. © Springer Science+Business Media Dordrecht 2015. With permission of Springer

Part III Alternative Auction Designs

- Chapter 18: Stephen J. Rassenti, Vernon L. Smith, and Robert L. Bulfin. A Combinatorial Auction Mechanism for Airport Time Slot Allocation. *The Bell Journal of Economics*, pages 402–417, 1982. © The RAND Corporation
- Chapter 19: Anthony M. Kwasnica, John O. Ledyard, David P. Porter, and Christine DeMartini. A New and Improved Design for Multi-Object Iterative Auctions. *Management Science*, 51(3):419–434, 2005. Reproduced with permission. © INFORMS, www.informs.org
- Chapter 20: Jacob K. Goeree and Charles A. Holt. Hierarchical Package Bidding: A Paper & Pencil Combinatorial Auction. Reprint from *Games and Economic Behavior*, 70(1):146–169, 2010. © Elsevier 2010. With permission of Elsevier

- Chapter 21: Paul R. Milgrom. Assignment Messages and Exchanges. *American Economic Journal: Microeconomics*, 1(2):95–113, 2009. © American Economic Association
- Chapter 22: Paul Klemperer. The Product-Mix Auction: A New Auction Design for Differentiated Goods . *Journal of the European Economic Association*, 8(2–3):526–536, 2010. © 2010 by the European Economic Association
- Chapter 23: Charles Plott, Hsing Yang Lee, and Travis Maron. The Continuous Combinatorial Auction Architecture. *The American Economic Review*, 104(5):452–456, 2014. © American Economic Association
- Chapter 24: Martin Bichler, Zhen Hao, and Gediminas Adomavicius. Coalition-Based Pricing in Ascending Combinatorial Auctions. *Information Systems Research*, 28(1), 159–179, 2017. Reproduced with permission. © INFORMS, www.informs.org

Part IV Experimental Comparisons of Auction Designs

- Chapter 25: John O. Ledyard, David P. Porter, and Antonio Rangel. Experiments Testing Multi-Object Allocation Mechanisms. *Journal of Economics & Management Strategy*, 6(3):639–675, 1997. © 1997 Massachusetts Institute of Technology
- Chapter 26: Charles Plott. Laboratory Experimental Testbeds: Application to the PCS Auction. *Journal of Economics & Management Strategy*, 6(3):605–638, 1997. © 1997 Massachusetts Institute of Technology
- Chapter 27: Christoph Brunner, Jacob K. Goeree, Charles A. Holt, and John O. Ledyard. An Experimental Test of Flexible Combinatorial Spectrum Auction Formats. *American Economic Journal: Microeconomics*, 2(1):39–57, 2010. © American Economic Association
- Chapter 28: Tobias Scheffel, Georg Ziegler, and Martin Bichler. On the Impact of Package Selection in Combinatorial Auctions: An Experimental Study in the Context of Spectrum Auction Design. *Experimental Economics*, 15(4):667–692, 2012. © Economic Science Association 2012. With permission of Springer
- Chapter 29: Martin Bichler, Pasha Shabalin, and Jürgen Wolf. Do Core-Selecting Combinatorial Clock Auctions Always Lead to High Efficiency? An Experimental Analysis of Spectrum Auction Designs. *Experimental Economics*, 16(4):511–545, 2013. © Economic Science Association 2013. With permission of Springer
- Chapter 30: Martin Bichler, Jacob K. Goeree, Stefan Mayer, and Pasha Shabalin. Spectrum Auction Design: Simple Auctions for Complex Sales. Reprint from *Telecommunications Policy*, 38(7):613–622, 2014. © Economic Science Association 2013. With permission of Elsevier

I. Part V The Bidders' Perspective

- Chapter 31: Jeremy I. Bulow, Jonathan Levin, and Paul R. Milgrom. Winning Play in Spectrum Auctions, 2017
- Chapter 32: David J. Salant. Up in the Air: GTE's Experience in the MTA Auction for Personal Communication Services Licenses. *Journal of Economics & Management Strategy*, 6(3):549–572, 1997. © 1997 Massachusetts Institute of Technology

- Chapter 33: Vitali Gretschko, Stephan Knapek, and Achim Wambach. Bidding Complexities in the Combinatorial Clock Auction, 2017
- Chapter 34: Richard Marsden and Soren T. Sorensen. Strategic Bidding in Combinatorial Clock Auctions – A Bidder Perspective, 2017
- Chapter 35: Nicholas Fookes and Scott McKenzie. Impact of Budget Constraints on the Efficiency of Multi-Lot Spectrum Auctions, 2017

Part VI Secondary Markets and Exchanges

- Chapter 36: Randall Berry, Michael L. Honig, and Rakesh V. Vohra. Spectrum Markets: Motivation, Challenges, and Implications. *Communications Magazine, IEEE*, 48(11):146–155, 2010. © 2010 IEEE. Reprinted, with permission
- Chapter 37: Paul R. Milgrom and Ilya R. Segal. Designing the US Incentive Auction, 2017
- Chapter 38: Alexandre Fréchette, Neil Newman, and Kevin Leyton-Brown. Solving the Station Repacking Problem. *Proc. of the Thirtieth AAAI Conference on Artificial Intelligence*, pages 702–709, 2016. Association for the Advancement of Artificial Intelligence, Palo Alto, CA. © 2016 Association for the Advancement of Artificial Intelligence
- Chapter 39: Benjamin Lubin, Adam Juda, Ruggiero Cavallo, Sébastien Lahaie, Jeffrey Shneidman, and David C. Parkes. ICE: An Expressive Iterative Combinatorial Exchange. *Journal of Artificial Intelligence Research*, 33:33–77, 2008. AAAI Press
- Chapter 40: Leslie Fine, Jacob K. Goeree, Tak Ishikida, and John O. Ledyard. ACE: A Combinatorial Market Mechanism, 2017

PART I
The Simultaneous Multiple-Round Auction

CHAPTER 1

Putting Auction Theory to Work: The Simultaneous Ascending Auction[*]

Paul R. Milgrom

I. Introduction

The "simultaneous ascending auction" was first introduced in 1994 to sell licenses to use bands of radio spectrum in the United States. Much of the attention devoted to the auction came from its role in reducing federal regulation of the radio spectrum and allowing market values, rather than administrative fiat, to determine who would use the spectrum resource. Many observers were also fascinated by the then-novel use of weblike interfaces for bidders. The large amounts of money involved were yet another source of interest. The very first use of the auction rules was a $617 million sale of 10 paging licenses in July 1994. In the broadband personal communications services (PCS) auction, which began in December 1994, 99 licenses were sold for a total price of approximately $7 billion. Once the auctions had been conducted, it became much harder to ignore the tremendous value of the large amounts of spectrum allocated to uses such as high-definition television, for which Congress had demanded no compensation at all. Moreover, the perceived successes with the new rules inspired imitators to conduct similar spectrum auctions in various countries around the world and to recommend similar auctions for other applications.

Among academic economists, attention was also piqued because the auction design made detailed use of the ideas of economic theory and the recommendations of economic theorists. Indeed, the U.S. communications regulator adopted nearly all its important rules[1] from two detailed proposals for a simultaneous ascending auction: one by Preston McAfee and the other by Robert Wilson and me. Economic analysis dictated nearly all the rule choices in the first few auctions. Various reviews suggest that the new auction design realized at least some of the theoretical advantages that had been claimed for it.[2]

---

[*] My thanks go to Peter Cramton, Paul Klemperer, Yoav Shoham, and Padmanabhan Srinagesh, as well as seminar participants at Stanford, the University of Pittsburgh, and Yale for comments on an earlier draft.

[1] The sole exceptions were the financing rules, which were devised to encourage participation in the auctions by financially weak smaller businesses and those owned by women and minorities.

[2] See Cramton (1995), Milgrom (1995), and McAfee and McMillan (1996) for accounts of the auction and the run-up to it.

Several parts of economic theory proved helpful in designing the rules for the simultaneous ascending auction and in thinking about how the design might be improved and adapted for new applications. After briefly reviewing the major rules of the auction in Section II, I turn in Section III to an analysis based on *tatonnement* theory, which regards the auction as a mechanism for discovering an efficient allocation and its supporting prices. The analysis reveals a fundamental difference between situations in which the licenses are mutual substitutes and others in which the same licenses are sometimes substitutes and sometimes complements. When the licenses are mutual substitutes for all bidders, not only is it true that equilibrium prices exist, but straightforward, "myopic" bidding in the auction leads bidders to prices and an allocation that are close to competitive equilibrium. This happens even though, in contrast to traditional *tatonnement* processes, prices in the auction process can never fall and can rise only by fixed increments. However, if even one bidder has demand in which licenses are not mutual substitutes, then there is a profile of demands for the other bidders, all of which specify that licenses are mutual substitutes, such that no competitive equilibrium prices exist. There is an inherent limitation in the very conception of the auction as a process for discovering a competitive allocation and competitive prices in that case.

Section IV is a selective account of some applications of game theory to evaluating the design of the simultaneous ascending auction for spectrum sales. Game-theoretic arguments were among those that convinced regulators to adopt my suggestion of an "activity rule," which helps ensure that auctions end in a reasonable amount of time. Game theory also provided the decisive argument against the first "combinatorial bidding" proposals and has also been employed to evaluate various other suggested rule changes.

Results like those reported in Section III have led to renewed interest in auctions in which bids for license packages are permitted. In Section V, I use game theory to analyze the biases in a leading proposal for dynamic combinatorial bidding. Section VI briefly answers two additional questions that economists often ask about auction design: If trading of licenses after the auction is allowed, why does the auction form matter at all for promoting efficient license assignments? If the number of licenses to be sold is held fixed, how sharp is the conflict between the objectives of assigning licenses efficiently and obtaining maximum revenue? Section VII presents a conclusion.

II. Simultaneous Ascending Auction Rules in Brief

A simultaneous ascending auction is an auction for multiple items in which bidding occurs in rounds. At each round, bidders simultaneously make sealed bids for any items in which they are interested. After the bidding, round results are posted. For each item, these results consist of the identities of the new bids and bidders[3] as well as the "standing high bid" and the corresponding bidder. The initial standing high bid for each item is given (it may be zero), and the "corresponding bidder" is the auctioneer. As the

---

[3] The first trial of the simultaneous ascending auction did not include announcements of bidder identities, but the larger bidders were often able to infer identities anyway, leading to a change in the rules to remove that advantage. The practice of identifying the bidders continues to be controversial.

auction progresses, the new standing high bid at the end of a round for an item is the larger of the previous standing high bid or the highest new bid, and the corresponding bidder is the one who made that bid. In addition to the round results, the minimum bids for the next round are also posted. These bids are computed from the "standing high bid" by adding a predetermined bid increment. For spectrum licenses, the increments are typically the larger of some fixed amount or a fixed percentage of the standing high bid.[4]

A bid represents a real commitment of resources by the bidder. In the most common version of the rules, a bidder is permitted to withdraw bids, but there is a penalty for doing so: if the selling price of the item is less than the withdrawn bid, the withdrawing bidder must pay the difference. In other applications, bid withdrawals are simply not permitted.

A bidder's eligibility to make new bids during the auction is controlled by the "activity rule." Formally, the rule is based on a "quantity" index, such as spectrum bandwidth or population covered by a license, that roughly corresponds to the value of the license. During the auction, a bidder may not have active bids on licenses that exceed its *eligibility*, measured in terms of the index.

At the outset of the auction, each bidder establishes its *initial eligibility* for bidding by making deposits covering the quantity of spectrum for which it wishes to be eligible. During the auction, a bidder is considered *active* for a license at a round if it makes an eligible new bid for the license or if it owns the standing high bid from the previous round. At each round, a bidder's activity is constrained not to exceed its eligibility. If a bid is submitted that exceeds the bidder's eligibility, the bid is simply rejected.

The auction is conducted in a sequence of three stages, each consisting of multiple rounds. The auction begins in stage 1, and the administrator advances the auction to stage 2 and later to stage 3 when there are two or more consecutive rounds with little new bidding. In each round during stage j, a bidder that wishes to maintain its eligibility must be active on licenses covering a fraction f_j of its eligibility. If a bidder with eligibility x is active on a license quantity $y < f_j x$ during stage j, then its eligibility is reduced at the next round to y/f_j.[5]

The activity rules have two functions. First, they create pressure on bidders to bid actively, which increases the pace of the auction. Second, they increase the information available to bidders during the auction, particularly late in the auction. For example, in stage 3, bidders know that the remaining eligible demand for licenses at the current prices is just $1/f_3$ times the current activity level, which can be rather informative when f_3 is close to one.[6]

The auction also provides five "waivers" of the activity rule for each bidder, which can be used at any time during the auction, that allow the bidder to avoid reduction in its eligibility in a given round. The waivers were included to prevent errors in the bid

---

[4] In the spectrum auctions, the percentage has usually been 5 percent or 10 percent (and in recent auctions has been dependent on the level of bidding in the auction). The appropriate size of the increment has also been subjected to economic analysis that takes into account the cost of adding rounds to the auction and the extent and type of the uncertainty about bidder values.

[5] In the 1998 auction of licenses to spectrum in the 220 MHz range, the fractions used were $(f_1, f_2, f_3) = (.8, .90, .98)$.

[6] See Cramton (1997) for evidence on the informational content of bids.

submission process from causing unintended reductions in a bidder's eligibility, but they also have some strategic uses.

There are several different options for rules to close the bidding that were filed with the regulator. One proposal, made by Preston McAfee, specified that when a license had received no new bids for a fixed number of rounds, bidding on that license would close. That proposal was coupled with a suggestion that the bid increments for licenses should reflect the bidding activity on a license. A second proposal, made by Robert Wilson and me, specified that bidding on all licenses should close simultaneously when there is no new bidding on any license. To date, the latter rule is the only one that has been used.[7]

When the auction closes, the licenses are sold at prices equal to the standing high bids to the corresponding bidders. The rules that govern deposits, payment terms, and so on are quite important to the success of the auction,[8] but they are mostly separable from the other auction rule issues and receive no further comment here.

III. Auctions and *Tatonnement* Theory

The simultaneous ascending auction is a process that, on its surface, bears a strong resemblance to the *tatonnement* process of classical economics. Like the *tatonnement* process, the objective of the auction is to identify allocations (which the spectrum regulators call "assignments") and supporting prices to approximate economic efficiency. Yet there are differences as well. First, bids in the auction represent real commitments of resources, and not tentatively proposed trades. Consequently, bidders are reluctant to commit themselves to purchases that may become unattractive when the prices of related licenses change. Second, in the auction, prices can never decrease. That is an important limitation because the ability of prices to adjust both upward and downward is a fundamental requirement in theoretical analyses of the *tatonnement*. Third, in the initial version of the simultaneous ascending auction, the bidders themselves name the prices. That contrasts with the Walrasian *tatonnement*, in which some fictitious auctioneer names the prices. Other differences arise from the nature of the application. The licenses sold in the auction are indivisible. This fact means that the set of allocations cannot be convex, so the usual theorems about the existence of competitive equilibrium do not apply. My analysis focuses on all these issues: the risk that bidders take when they commit resources in early rounds of the auction, the existence of competitive equilibrium, and whether the simultaneous auction process in which prices increase monotonically can converge to the equilibrium.

---

[7] The activity rule and the closing rule make this auction perform very differently from the otherwise similar "silent auction" commonly used in charity sales. In a silent auction, the items being sold are typically set on tables in a room and bidders walk around the room, entering their bids and bidder identification on a paper sheet in front of the items. Bidding closes at a predetermined time. It is a common experience that bidders in silent auctions often delay placing their bids until the final moment, completing their entry on the paper just as the bidding closes. With its closing and activity rules, the simultaneous ascending auction eliminates both the "final moment" that bidders exploit in silent auctions and also bidders' ability to wait until late in the auction before making any serious bids.

[8] Failure to establish these rules properly led to billions of dollars of bidder defaults in the U.S. "C-block auction." Similar problems on a smaller scale occurred in some Australian spectrum auctions.

Let $L = \{1, \ldots, L\}$ be the set of *indivisible* licenses to be offered for sale. Denote a typical subset of L by S. In describing license demand, I also use S to represent the vector $\mathbf{1}_S$.

Assume that a typical bidder i who acquires the set of licenses S and pays an amount of money m for the privilege enjoys utility of $v_i(S) - m$. Given a vector of prices $\mathbf{p} \in \mathbb{R}_+^L$, $\mathbf{p} \cdot S$ denotes the total price of the licenses composing S. The demand correspondence for i is defined by $D_i(\mathbf{p}) = \mathrm{argmax}_S\{v_i(S) - \mathbf{p} \cdot S\}$. Assume that there is *free disposal*, so $S \subset S'$ implies that $v_i(S) \leq v_i(S')$. I sometimes omit the subscript from demand functions, relying on the context to make the meaning clear.

During an auction, it often happens that a bidder is the high bidder on a subset of the licenses it would wish to acquire at the current bid prices. To describe such situations, I introduce the following definition: An individual bidder demands the set of licenses T at price vector $\mathbf{p}$, written $T \in X(\mathbf{p})$, if there exists $S \in D(\mathbf{p})$ such that $S \supset T$.

The usual definition of substitutes needs to be generalized slightly to deal with the case of demand correspondences. The idea is still the same, though: raising the prices of licenses not in any set S cannot reduce the demand for licenses in the set S.

Definition. Licenses are *mutual substitutes* if for every pair of price vectors $\mathbf{p}' \geq \mathbf{p}$, $S \in X(\mathbf{p})$ implies that $S \in (X(p_S, p'_{L\setminus S}))$.

After any round of bidding, the minimum bids for the next round are given by the rule described in Section II. If the standing high bids at a round are given by the vector $\mathbf{p} \in \mathbb{R}_+^L$, then the minimum bid at the next round for the lth license is $p_l + \epsilon \max(p_l, \hat{p}_l)$ for some $\epsilon > 0$. The vector of minimum bids is then $\mathbf{p} + \epsilon(\mathbf{p} \vee \hat{\mathbf{p}})$, where $\hat{\mathbf{p}} \in \mathbb{R}_{++}^L$ is a parameter of the auction design, and the "join" $\mathbf{p} \vee \hat{\mathbf{p}}$ denotes the price vector that is the component-wise maximum of $\mathbf{p}$ and $\hat{\mathbf{p}}$.

During a simultaneous ascending auction, the minimum bid increment drives a wedge between the prices faced by different individual bidders. To analyze the progress of the auction, it is useful to define the *personalized* price vector $\mathbf{p}^j$ facing bidder j at the end of a round to be $\mathbf{p}^j = (p_{S_j}, (\mathbf{p} + \epsilon(\mathbf{p} \vee \hat{\mathbf{p}}))_{L\setminus S_j})$. That is, j's prices for the licenses S_j that it has been assigned are j's own standing high bids, but its prices for the other licenses are the standing high bids plus the minimum bid increment. This reflects the fact that under the rules of the auction, j can no longer purchase those other licenses at their current standing high bids.

My analysis of the *tatonnement* process consists of a study of what happens to bidder j when it (possibly) alone bids in a "straightforward" (nonstrategic) manner, and what happens when all bidders bid in a straightforward manner. When I say that j bids "straightforwardly," I mean that if, at the end of some round n, bidder j demands the licenses assigned to it (formally, if $S_j \in X_j(\mathbf{p}^j)$), then j makes the minimum bid at round $n + 1$ on a maximal set of licenses T such that $S_j \cup T \subset D_j(\mathbf{p}^j)$. Of course, bidders that wish to acquire multiple licenses commonly have an incentive to withhold some of their demand in order to reduce prices.[9] Consequently, the part of the analysis employing straightforward bidding must be understood as only a partial analysis,

---

[9] Ausubel and Cramton (1996) argue that an incentive of this sort is unavoidable in a wide class of auctions, including all those that establish uniform prices for identical objects.

which ignores strategic incentives to highlight important nonstrategic properties of the auction design.

Intuitively, whenever the auction allows, the straightforward bidder bids to acquire the set of licenses that it demands at its personalized prices. Notice that the antecedent condition is automatically satisfied at the beginning of the auction because no bidder has yet been assigned any licenses.

Straightforward bidding often leads to ties at some rounds of the auctions. For the analysis of this section, any tie-breaking rule that selects a winner from among the high bidders will work.

My first theorem says that if j bids straightforwardly from the beginning of the auction and if licenses are mutual substitutes for j, then the antecedent condition for straightforward bidding continues to be satisfied round after round.

Theorem 1. *Assume that all the licenses are mutual substitutes for bidder j. Suppose that, at the end of round n, bidder j's assignment $S_j \in X_j(\mathbf{p}^j)$. If, at round $n + 1$, bidder j bids straightforwardly, then, regardless of the bids made by others, j's assignment S'_j at the end of round $n + 1$ satisfies $S'_j \in X_j(\mathbf{p}^{j\prime})$, where $\mathbf{p}^{j\prime}$ is j's personalized price at the end of round $n + 1$. Moreover, j's tentative profit after any round—what it would earn if the auction were terminated after that round at the then-current prices and allocation—is always non-negative.*

Proof. Let T be the set of licenses on which j bids at round $n + 1$. Then, by the rules of the auction, $S'_j \subset S_j \cup T$. Also, by the hypothesis of straightforward bidding, $S_j \cup T \in D_j(\mathbf{p}^j)$. So, by definition, $S'_j \in X_j(\mathbf{p}^j)$.

The rules also imply that j's personalized prices $\mathbf{p}^{j\prime}$ for the licenses in S'_j coincide with the prices of those licenses according to $\mathbf{p}^j$. Moreover, j's personalized prices cannot fall from round to round: $\mathbf{p}^{j\prime} \geq \mathbf{p}^j$. Hence, by the definition of mutual substitutes, $S'_j \in X_j(\mathbf{p}^{j\prime})$.

Finally, j's tentative payoff after any round is independent of the prices of items outside the set S_j. Hence, without affecting j's payoff, we may replace those other prices by prices $\mathbf{p}'$ so high that j would prefer not to acquire any of those other items at these prices. By mutual substitutes, $S_j \in X_j(p^j_{S_j}, p'_{L \setminus S_j})$, and hence $S_j \in D_j(p^j_{S_j}, p'_{L \setminus S_j})$. This implies that j's tentative profits are indeed nonnegative. □

The next issue is what happens when all bidders bid in a straightforward way. Theorem 2 provides an answer.

Theorem 2. *Assume that the licenses are mutual substitutes for all bidders and that all bidders bid straightforwardly. Then the auction ends with no new bids after a finite number of rounds. Let $(\mathbf{p}^*, S^*)$ be the final standing high bids and license assignment. Then $(\mathbf{p}^*, S^*)$ is a competitive equilibrium for the economy with modified valuation functions defined by $\hat{v}_j(T) = v_j(T) - \epsilon(\mathbf{p} \vee \hat{\mathbf{p}}) \cdot (T \setminus S_{j^*})$ for each bidder j. The final assignment maximizes total value to within a single bid increment:*

$$\max_{\{S_j\}} \sum_j v_j(S_j) - \sum_j v_j(S^*_j) \leq \epsilon(\mathbf{p}^* \vee \hat{\mathbf{p}}) \cdot L.$$

Proof. In view of theorem 1, at the end of every round, every bidder's tentative profit is positive. This implies that the total price of the licenses assigned to the bidders after any round of the auction is bounded above by the maximum total value of the licenses. Given the positive lower bounds on the bid increments, it follows that the auction ends after a finite number of rounds.

By construction, bidder j's demand at the final price vector $\mathbf{p}^*$ with j's modified valuation is the same as its demand at the corresponding personalized price vector $\mathbf{p}^j$ for the original valuation. Since there are no new bids by j at the final round, we may conclude from the condition of straightforward bidding and theorem 1 that $S_j^* \in \hat{D}(\mathbf{p}^*)$. Since this holds for all j, $(\mathbf{p}^*, S^*)$ is a competitive equilibrium with the modified valuations.

For the second statement of the theorem, we can make the following calculation:

$$\max_S \sum_j v_j(S_j) = \max_S \sum_j [\hat{v}_j(S_j) + \epsilon(\mathbf{p}^* \vee \hat{\mathbf{p}}) \cdot (S_j \setminus S_j^*)]$$

$$\leq \max_S \sum_j [\hat{v}_j(S_j) + \epsilon(\mathbf{p}^* \vee \hat{\mathbf{p}}) \cdot S_j]$$

$$= \max_S \sum_j \hat{v}_j(S_j) + \epsilon(\mathbf{p}^* \vee \hat{\mathbf{p}}) \cdot L$$

$$= \sum_j \hat{v}_j(S_j^*) + \epsilon(\mathbf{p}^* \vee \hat{\mathbf{p}}) \cdot L$$

$$= \sum_j v_j(S_j^*) + \epsilon(\mathbf{p}^* \vee \hat{\mathbf{p}}) \cdot L.$$

The first equality follows from the definition of the modified valuations, the inequality from the restriction that all prices are nonnegative, and the following equality from the fact that S partitions L. The fourth step follows from the already proven fact that $(\mathbf{p}^*, S^*)$ is a competitive equilibrium for the modified valuations combined with the first welfare theorem and the fact that, with quasi-linear payoffs, a license assignment is efficient if and only if it maximizes the total value to all the bidders. Finally, the last equality follows by the definition of $\hat{v}_j(\cdot)$, which coincides with $v_j(\cdot)$ when evaluated at S_j^*. □

If the coefficient ϵ varies during the auction, then the most relevant values of ϵ for this analysis are ones that apply when bidders are last eligible to make new bids, which is normally near the end of the auction. (The activity rule is what makes this statement inexact.) This suggests that very high levels of *tatonnement* efficiency might be obtained by using small increments near the end of the auction. It was with this in mind that the Milgrom-Wilson rules originally adopted in the United States by the Federal Communications Commission (FCC) called for using smaller minimum bid increments in the final stage of the auction.[10]

---

[10] That rule was later changed for reasons of transaction costs: smaller increments late in the auction led to large numbers of costly rounds with relatively little bidding activity.

The final questions in this section are, What relation does the auction outcome have to the competitive equilibrium outcome? Does a competitive equilibrium even exist in this setting with indivisible licenses? Theorem 3 provides answers.

Theorem 3. *Suppose that the licenses are mutual substitutes in demand for every bidder. Then a competitive equilibrium exists. For ϵ sufficiently small, the final license assignment $S^*(\epsilon)$ is a competitive equilibrium assignment.*[11]

Proof. Let $\epsilon_n \to 0$ and let $S^*(\epsilon_n)$ and $\mathbf{p}^*(\epsilon_n)$ be corresponding sequences of final license assignments and prices. Since there are only finitely many possible license assignments, some assignment S^{**} must occur infinitely often along the sequence. Also, each license price is bounded above by the maximum value of a license package. So there exists a subsequence $n(k)$ along which $S^*(\epsilon_{n(k)}) = S^{**}$ and such that $\mathbf{p}^*(n(k))$ converges to some $\mathbf{p}^{**}$. By theorem 2, for all k, $S_j^{**} \in D_j(\mathbf{p}^*(n(k)), \epsilon_{n(k)})$, where the second argument of D_j identifies the relevant perturbed preferences. By the standard closed graph property of the demand correspondence, $S_j^{**} \in D_j(\mathbf{p}^{**})$, so $(S^{**}, \mathbf{p}^{**})$ is a competitive equilibrium. □

Intuitively, because the number of possible allocations is finite, a value-maximizing allocation generates a greater total value than the best nonmaximizing allocation by some amount $\delta > 0$. If the bid increment ϵ is sufficiently much smaller than δ, then, according to theorem 2, only an efficient allocation can result from straightforward bidding. The auction prices that support that allocation approximate the competitive equilibrium prices.

Thus, when all licenses are mutual substitutes for all bidders, the simultaneous ascending auction with straightforward bidding is an effective *tatonnement*. First, a bidder that bids straightforwardly during the auction is "safe": it is sure to acquire a set of licenses that is nearly optimal relative to its valuation and the final license prices, and it never risks actually losing money. If every bidder bids straightforwardly, then the auction eventually ends with an assignment that approximately maximizes the total value. Indeed, if the bid increment is small, then the final assignment exactly maximizes the total value and is a competitive equilibrium assignment. The final bids "approximately support" the solution, in the sense that they are close to the personalized prices that support the solution for each bidder. A number proportional to the bid increment bounds the error in each of these approximations.

The first three theorems were developed only for the case of licenses that are mutual substitutes. In practice, the status of spectrum licenses as substitutes or complements

---

[11] Milgrom and Roberts (1991) show the existence of a competitive equilibrium with mutual substitutes using a lattice-theoretic argument that does not require all goods to be divisible. They proceed to show that a wide variety of discrete and continuous "adaptive" and "sophisticated" price adjustment processes converge to the competitive equilibrium price vector. Unlike the present analysis, however, their analysis assumes that demand is given by a function, rather than by a correspondence, and they do not address the monotonicity of the auction process. Kelso and Crawford (1982) obtain results analogous to theorems 1 and 3 in a model of job matching. Gul and Stacchetti (1997) characterize utility functions that display "no complementarities," which is an alternative formulation of the idea of mutual substitutes. They also introduce a new auction process in which an auctioneer announces price vectors $\mathbf{p}$ and the bidders report their corresponding sets of demands $D_j(\mathbf{p})$. The auctioneer uses the reported information to control a continuous process of price increases. For the case of no complementarities in bidder utility, they demonstrate that their new auction process converges monotonically up to a competitive equilibrium.

Table 1.

	A	B	AB
1	a	b	$a+b+c$
2	$a+d$	$b+d$	$a+b+d$

may often depend on how the licenses are defined. For example, in the DCS-1800 spectrum auction conducted in the Netherlands in February 1998, some of the offered licenses permitted use of only very small amounts of bandwidth relative to the efficient scale. A bidder that sought to establish an efficiently scaled mobile wireless telephone system would find that the value of, say, two small licenses is more than two times the value of a single license. Formally, that scale economy creates a complementarity among licenses: the value of a pair of licenses is more than the sum of the individual values. A similar complementarity from economies of scale and scope would be created by recent proposals in Australia to establish licenses covering small geographic areas with small amounts of bandwidth.

While there may be positive results available for some environments in which some of the goods are complements, my next result establishes a sharp limit. It shows that introducing into the previous model a single bidder for which licenses are not mutual substitutes leads to drastic changes in the conclusions.

Theorem 4. *Suppose that the set of possible individual valuation functions includes all the ones for which licenses are mutual substitutes in individual demand. Suppose that, in addition, the set includes at least one other valuation function. Then if there are at least three bidders, there is a profile of possible individual valuation functions such that no competitive equilibrium exists.*

Intuition for theorem 4 is given in a two-license, two-bidder example, summarized in table 1. In the table, the licenses are denoted by A and B and the bidders by 1 and 2. Bidder 1 is the bidder for which licenses are not substitutes. This requires that the value of the pair AB exceed the sum of the individual values, that is, $c > 0$. Now let us introduce another bidder for which the same two licenses are substitutes. Let us take $c/2 < d < c$. In this case, the unique value-maximizing license allocation is for bidder 1 to acquire both licenses. In order to arrange for bidder 2 not to demand licenses, the prices must be $p_A \geq a+d$ and $p_B \geq b+d$, but at these prices bidder 1 is unwilling to buy the licenses. Consequently, there exist no equilibrium prices.

Proof of Theorem 4. Suppose that there is a bidder in the auction with valuation function v for which licenses are not mutual substitutes. Then there is some price vector $\mathbf{p}$, real number $\epsilon > 0$, and licenses j and k such that $\{j, k\} \in X(\mathbf{p})$, but $j \notin X(\mathbf{p}\backslash(\mathbf{p}_j + \epsilon))$ and $k \notin X(\mathbf{p}\backslash(\mathbf{p}_j + \epsilon))$. For this bidder, define an indirect valuation function w on the set of licenses $\{j, k\}$ by

$$w(S) = \max_{T \subset L \backslash \{j,k\}} v(T \cup S) - \mathbf{p} \cdot T.$$

The bidder's demand for licenses in the set $\{j, k\}$ given the established prices $p_{L\backslash\{j,k\}}$ for the licenses besides j and k is determined by w. Set $a = w(j), b = w(k)$, and

$c = w(jk) - a - b$. From our assumptions about the bidder's demand, it follows that $c > 0$ and that $p_j + p_k < a + b + c < p_j + p_k + \epsilon$. Let us now introduce two new bidders whose values are given by the following valuation function:

$$\hat{v}(S) = \mathbf{p}(S \backslash \{j, k\}) + (a + d)\mathbf{1}_{j \in S} + (b + d)\mathbf{1}_{k \in S} - d\mathbf{1}_{j \in S, k \in S},$$

where $c/2 < d < c$. For the new bidders, the various licenses are mutual substitutes. (Indeed, the bidders' demands for each license in $L \backslash \{j, k\}$ are independent of all prices except the license's own price. For the two licenses j and k, the verification is routine.) By construction, the competitive equilibrium prices, if they exist, of licenses in $L \backslash \{j, k\}$ are given by p_{-jk}. But then the problem of finding market-clearing prices for j and k is reduced to the example analyzed above, in which nonexistence of equilibrium prices has already been established. □

This nonexistence is related as well to a problem sometimes called the "exposure problem" that is faced by participants in a simultaneous ascending auction. This refers to the phenomenon that a bidder that bids straightforwardly according to its demand schedule is exposed to the possibility that it may wind up winning a collection of licenses that it does not want at the prices it has bid, because the complementary licenses have become too expensive. If the bidders in the example in the table were to adopt only undominated strategies in every subgame of the simultaneous ascending auction game, then it is not possible that at the end of the auction bidder 1 will acquire both licenses unless the prices are at least $a + d$ and $b + d$ minus one increment. The reason is that bidder 2 always does at least as well (and could do better) in that subgame of the auction by placing one more bid. Whenever bidder 1 wins both licenses, it loses money, and at equilibrium it will anticipate that. To avoid "exposure" completely, bidder 1 must bid no more than a for license A and no more than b for license B. If it does so, then the outcome will be inefficient and the prices, a and b, will not reflect any of the potential "synergy" between the licenses.

One puzzle raised by the preceding analysis is that there have been spectrum auctions involving complements that appeared to function quite satisfactorily. The U.S. regional narrowband auction in 1994 was an auction in which several bidders successfully assembled collections of regional paging licenses in single spectrum bands to create the package needed for a nationwide paging service. In Mexico, the 1997 sale of licenses to manage point-to-point microwave transmissions in various geographic areas exhibited a similar pattern. What appears to be special about these auctions is that licenses covering different regions in the same spectrum band that were complementary for bidders planning nationwide paging or microwave transmission networks were not substitutes for any other bidders. The nonexistence theorem given above depended on the idea that licenses that are complements for one bidder are substitutes for another.[12]

---

[12] The following table presents an example of nonexistence even when licenses are mutual complements for all bidders, but the degrees of complementarity vary. Tabulated are the values of three bidders (labeled 1, 2, and 3) for three licenses (A, B, and C). If a competitive equilibrium did exist, its assignment would be efficient, assigning licenses A and C to bidder 3 and B to bidder 1 or 2. For bidders 1 and 2 to demand their equilibrium assignments, the prices must satisfy $p_B \leq 1$, $p_A + p_B \geq 3$, and $p_B + p_C \geq 3$. However, these together imply

The potential importance of the exposure problem is illustrated by the Netherlands auction mentioned earlier: the DCS-1800 auction in February 1998 in which 18 spectrum licenses were offered for sale. In that auction, two of the lots—designated A and B—were believed to be efficiently scaled; the remaining 16 lots were too small to be valuable alone and needed to be combined in groups of perhaps six licenses to be useful for a mobile telephone business. In this auction, the smaller licenses would naturally be complements for one another for bidders with no other licensed spectrum, but they would be substitutes for one another for bidders that were merely seeking to increase their amounts of licensed spectrum.

The outcome of the auction involved final prices per unit of bandwidth in lots A and B that were more than twice as high as those for any of the 16 smaller lots. It might seem that the bidders on lots A and B behaved foolishly since they might have acquired as much spectrum for less by bidding on smaller lots. However, bidders may have been deterred from aggressive bidding for the smaller lots for fear that that would drive up the prices of those lots while still leaving some of the winning bidders with too little bandwidth for an efficiently scaled business. This may simply be an instance in which, as suggested by theory, the prices fail to reflect the potential synergies among the licenses.

The problem of bidding for complements has inspired continuing research both to clarify the scope of the problem and to devise practical auction designs that overcome the exposure problem.

IV. Auctions and Game Theory

Another part of economic theory that has proved useful for evaluating alternative auction designs is game theory. Here I consider two such applications. The first model formalizes the ideas that motivated the introduction of the activity rule. The second is a study of how the auction closing rules affect the likelihood of collusive outcomes.

The Need for Activity Rules

In the design of the auction, one of the concerns was to estimate how long the auction would take to complete. This, in turn, depended on forecasting how aggressively bidders would behave. Could one count on the bidders to move the auction along, perhaps

that $p_A + p_C \geq 4$, which is inconsistent with bidder 3's demand for the pair AC. So no competitive equilibrium exists.

Bidder	A	B	C	AB	AC	BC	ABC
1	1	1	0	3	1	1	3
2	0	1	1	1	1	3	3
3	1	0	1	1	3.5	1	3.5

Licenses and Combinations

Table 2.

	LICENSE	
BIDDER	A	B
1	15	30
2	10	Not eligible
3	Not eligible	5 with probability .9 15 with probability .1

to economize on their own costs of participating? Or would the bidders sometimes have a strategic incentive to hold back, slowing the pace of the auction substantially?

There were several reasons to be skeptical that the bidders themselves could be relied on to enforce a quick pace. In the mutual substitutes model analyzed earlier, there is no affirmative gain to a bidder from bidding aggressively early in the auction, since all naive bidding paths lead to the same competitive equilibrium outcome. So bidders with a positive motive to delay might find little reason not to do so. In some of the spectrum auctions, the major bidders included established competitors in the wireless industry that stood to profit from delays in new entry caused by delays in the auction process.

There can also be a variety of strategic motives for delay in the auction itself. Here I shall use a model to investigate one that is so common as to be decisive for planning the auction design. The model is based on the notion that the bidders are, or may be, budget-constrained.[13] (A large measure of strategic behavior in the actual spectrum auctions seemed to be motivated by this possibility.) If a bidder's competitor for a particular license is budget-constrained and its values or budget or both are private information, then the bidder may gain by concealing its ability or willingness to pay a high price until its competitor has already committed most of its budget to acquiring other licenses. The budget-constrained competitor may respond with its own delay, hoping to learn something about the prices of its highest-valued licenses before committing resources to other licenses. These behaviors delay the completion of the auction. What follows is a sample bidding game verifying that such behaviors are possible equilibrium phenomena.

Suppose that there are three bidders—1, 2, and 3—and two licenses—A and B. Each bidder has a total budget of 20, and its total payments cannot exceed this limit. A bidder's payoff is its value for the licenses it acquires minus the total amount it pays. The values of the three bidders for the two licenses are listed in table 2.

The rules of the game are as follows. Initially, the prices are zero, and both items are assigned to the auctioneer. At any round, a bidder can raise the bid by one unit on any license for which it is eligible to bid. Ties are broken at random. After a round with no new bids, the auction ends. Payoffs are determined as described above.

My question is, Does there exist a (sequential) equilibrium in which bidders 2 and 3 bid "straightforwardly," that is, in which each raises the bid on a license whenever

---

[13] Budget constraints can have profound effects on bidding behavior and equilibrium strategies. Pitchik and Schotter (1988) initiated research into the effects of budget constraints; see also Che and Gale (1996, 1998). For some of the other effects of budget constraints on actual bidder behavior in the spectrum auctions, see Milgrom (1995, chap. 1).

it is not assigned the license and its value strictly exceeds the current highest bid? If bidder 3's value is common knowledge among the bidders, then one can routinely verify that the answer is affirmative. Bidder 1's corresponding strategy depends on bidder 3's value for license B. If that value is 5, then at the equilibrium, bidder 1 bids in the same straightforward manner as the other two bidders. If, however, bidder 3's value is 15, then bidder 1's best reply is different. At one equilibrium, 1 bids straightforwardly on license B and limits its bids on license A to ensure that it will win license B with its limited budget.

If 3's value is private information, however, then the answer changes. For suppose that bidders 2 and 3 bid straightforwardly. Then 1 could learn 3's value by bidding on license B until it was assured of acquiring that license, then devoting its remaining budget in an attempt to win license A. In particular, 1 would always win license B. It would also win license A at a price of 10 or 11 when 3's value for B was low. There can be no equilibrium with these properties, however. For if there were, then when bidder 3 has the high value, it could wait until 1 bids 10 or 11 on license A before bidding more than 5 on license B. Then 3 would win license B and earn a positive profit.

Theorem 5. *There is no sequential equilibrium of the private information game in table 2 in which bidders 2 and 3 each bid "straightforwardly," as described above.*

Both bidders 2 and 3 may have an incentive to slow their bidding in this auction, each hoping that bidder 1 will become unable to compete effectively for one license because it has spent its budget on another license. What the equilibrium in this example does not show is a delay induced by bidder 1, since it avoids committing resources until after bidder 3 has shown its hand. I conjecture that the example can be extended to incorporate that feature, so that all bidders have a tendency to delay.

In the actual spectrum auctions, the activity rule limited such wait-and-see strategies by specifying that a bidder that remained inactive in the early rounds of the auction would be ineligible to bid in later rounds. However, the first auctions cast doubt on the necessity of the rule. In the national and regional narrowband auctions, there was far more bidding activity than required by the activity rule, leading some to propose that the auction be simplified by dropping the rule. However, the AB block PCS auction, which was the third simultaneous ascending auction, followed quite a different pattern.

For the AB auction, the volume of activity associated with each license is measured by the population in the region covered by the licenses according to the 1990 U.S. census ("POPs"). The average license in this auction covered a region with approximately 5 million POPs. The auction generated 3,333 data points, each consisting of a vector of bids made by a bidder at a round.[14] Only 30 of the 3,333 observations reveal activity that exceeds the required level by at least one average size license, that is, 5 million POPs, and only 140 observations reveal activity exceeding required activity by more than 1 million POPs. Thus bidders most often bid only slightly more than was minimally necessary to maintain their current bidding eligibility.

---

[14] Observations in which bidders take a "waiver" are excluded for two reasons. First, the required activity does not apply at rounds with waivers, so there is no natural x variable. Second, each bidder that ceases bidding before the end of the auction automatically exercises five waivers according to the FCC rules, so those observations contain no information about bidder decision making.

Table 3.

	A	B	AB	Budget
1	4	...	...	3
2	...	4	...	3
3	$1+\epsilon$	$1+\epsilon$	$2+\epsilon$	2

Free Riding

One of the main issues in the early debates about the spectrum auction was whether all bidding should apply to individual licenses or whether, instead, bids for combinations of licenses should be allowed. According to one combinatorial bidding proposal, bids would first be accepted for certain predetermined packages of licenses, such as a nationwide collection of licenses, and then bidding on individual licenses would ensue. After all bidding had ceased, the collection of bids that maximize total revenues would be the winning bids, and licenses would be assigned accordingly.[15] The model of this auction below assumes that in the event of ties, package bids are selected in preference to bids on individual licenses and that bids must be entered as whole numbers.

The primary economic argument against allowing combination bids is that such bids can give rise to a free-rider problem among bidders on the individual licenses, leading to avoidable inefficiencies. Table 3 provides a simplified version of an example I presented during the deliberations to show how that can happen. In this example, there are three bidders—labeled 1, 2, and 3—and two licenses—A and B. Bidders 1 and 2 are willing to pay up to 4 for licenses A and B, respectively, and neither is eligible to acquire the other license.[16] With ϵ small and positive, bidder 3 has the lowest values for the licenses but is distinguished by its desire to acquire both. To keep the strategy spaces small and ease the analysis, I impose economically insignificant budget constraints on the bidders, as shown in table 3.

With the specified values, the sole efficient license assignment has bidders 1 and 2 acquiring licenses A and B, respectively. With bids restricted to be whole numbers, that corresponds to a subgame-perfect equilibrium of the simultaneous ascending auction. At the equilibrium, bidders 1 and 2 make minimum bids at each round as necessary to acquire their respective licenses of interest, whereas bidder 3 bids 1 for each license and then gives up.

If the proposed combinatorial auction is used, bidder 3 can refrain from bidding for licenses A and B directly, bidding instead for the pair AB. This strategy creates a free-rider problem for bidders 1 and 2. A high bid by bidder 1 on license A helps bidder 2

---

[15] Depending on what combinations are allowed, there may also need to be rules specifying the winner when there are overlapping combinations. Generally, the recommendation was that the winning set of bids should be the set that maximizes the total bid price.

[16] In the actual auctions, bidders were ineligible to acquire additional wireless telephone licenses for areas they already served. This restriction was motivated by competition policy.

Table 4.

	Raise Bid	Don't Raise
Raise bid	2, 3	2, 3
Don't raise	3, 2	0, 0

to acquire license B. A symmetric observation applies to bidder 2. Each would prefer that the other raise the total of the individual bids sufficiently to beat 3's bid.

Even in the complete information case shown here, this free-rider problem can lead to inefficient mixed-strategy equilibria. The corresponding equilibrium strategies are as follows. In the combination bidding round, bidder 3 bids 2 for the license combination AB. Bidder 1 raises the price of license A by 1 whenever it does not own the standing high bid for that license. Otherwise, if at any time during the auction the license prices are 1 for A, 1 for B, and 2 for AB, then bidder 1 raises its high bid on license A with probability two-thirds. Bidder 2's strategy is symmetrical to bidder 1's but is focused on license B instead of license A.

The key to understanding this equilibrium is to recognize the payoffs in the subgame after the prices are 1 for A, 1 for B, and 2 for package AB. The payoff matrix for bidders 1 and 2 in that subgame is shown in table 4.

This subgame has a symmetric equilibrium in which each bidder raises the bid with probability two-thirds. Backward induction from there supports the equilibrium strategies described above. At the equilibrium, there is a one-ninth probability that 3 acquires both licenses, even though its value for those licenses is just one-fourth of the total of the competitors' values. This example is representative of a robust set of examples, including especially ones with asymmetric information that make the free-rider problem even harder to resolve.

The following theorem summarizes this discussion.

Theorem 6. *The proposed two-stage auction (in which combinatorial bidding is followed by a simultaneous ascending auction for individual licenses) can introduce inefficient equilibrium outcomes that would be avoided in the simultaneous ascending auction without combinatorial bidding.*

It bears emphasis that this defect applies to the particular combinatorial rule that was proposed and is not a general criticism of all combinatorial bidding.

Collusion and Closing Rules

Motivated by the idea of the *tatonnement*, the rules of the spectrum auction specified that bidding would close on no licenses until there were no new bids on any license. In that way, if a license that changed hands at some round were a substitute or complement for another license, the losing bidder could react by bidding for the substitute or withdrawing a bid for a complement, and the winner could react in the reverse way.

Strategically, however, simultaneous closings create opportunities for collusion that can be mitigated by other closing rules.[17] To illustrate this in a simple model, suppose that there are two bidders, 1 and 2, and two licenses, A and B. Each bidder has a value for each license of 10. The auction rules are the same as in the preceding subsection, with a simultaneous close of bidding on all licenses when there is no bidding on any license. The next two theorems, the proofs of which are straightforward, show that both "competitive" and "collusive" outcomes are consistent with equilibrium in this game.

Theorem 7. *The following strategy, adopted by both bidders, constitutes a sequential equilibrium of the game with simultaneous closes of bidding: if the price of either license is below 10, bid again on that license.*

This is the "competitive" outcome and results in prices of 10 for both licenses and zero profits for the bidders. However, other outcomes are also possible.

Theorem 8. *The following strategies constitute a sequential equilibrium of the game with simultaneous closes of bidding. (1) For bidder 1, if 2 has never bid on license A, then if license A has received no bids, bid $1 on license A; otherwise, do not bid. If 2 has ever bid on license A, then bid according to the strategy described in theorem 7. (2) Bidder 2 bids symmetrically.*

This is the most collusive equilibrium, resulting in prices of just 1 for each license and total profits of 18 for the two bidders, which are the lowest prices possible if the licenses are to be sold. The collusive outcome is supported by the threat, inherent in the strategies, to shift to competitive behavior if the other party to the arrangement does not refrain from bidding on a particular license.

An extreme alternative is to close bidding on a license after any round in which there is no new bid on *that license*. This rule excludes the possibility that bidders can each retaliate if the other cheats on the arrangement. For example, suppose that the auction is supposed to end after round n with a bid price of $b \leq 8$ on license A, won by bidder 1. Then bidder 2 has nothing to lose and, in the trembling-hand logic of equilibrium, something to gain by raising the price at round $n + 1$. Consequently, we have the following result.

Theorem 9. *In the game with license-by-license closes of bidding, at every (trembling-hand) perfect equilibrium, the price of each license is at least 9.*

Similar results can be obtained from a rule that arranges for bidding to close on a license if there has been no new bid in the past three rounds. Alternatively, bidding may close on a license when there has been no new bid for three rounds and the total number of new bids on all licenses for the past five rounds is less than some trigger value. Rules along these lines can allow for substitution among licenses until late in the auction while still deterring some of the most obvious opportunities for collusion.

---

[17] An unpublished paper by Rob Gertner (1995) inspired my analysis of closing rules. His presentation analyzed the vulnerability to collusion of the simultaneous ascending auction with simultaneous closings and showed that the same form of collusion is not consistent with equilibrium in the traditional auctions in which items are sold one at a time, in sequence.

V. Dynamic Bidding for Combinations of Licenses

The considerations raised in the *tatonnement* analysis suggest the need to use a mechanism that does not rely simply on prices for individual licenses and instead allows bidding for license packages. An auction design that, in theory, uses combination bidding to good effect is the generalized Vickrey auction, also called the Groves-Clarke "pivot mechanism" (Vickrey 1961; Clarke 1971; Groves and Loeb 1975). Since that will serve as our standard of comparison, I review it briefly here.

Let L denote the set of available licenses and let P be the set of license assignments; these are indexed partitions of L. For any assignment $S \in P$, partition element S_i represents the set of licenses assigned to bidder i.

The rules of the generalized Vickrey auction are as follows. Each bidder submits a bid that specifies a value for every nonempty subset of L. For any set of licenses T, let $v_i(T)$ denote i's bid for that set. The auctioneer chooses the license assignment S^* that maximizes $v_1(S_1^*) + \ldots + v_N(S_N^*)$. Each bidder i pays a price p_i for its licenses according to the formula

$$p_i = \max_{S \in P} \sum_{j \neq i} v_j(S_j) - \sum_{j \neq i} v_j(S_j^*).$$

It is well known that, subject to certain assumptions,[18] the bidders in a generalized Vickrey auction have a dominant strategy, which is to set their bids for each license package equal to its actual value. When each bidder uses its dominant strategy, licenses are assigned efficiently. Moreover, if the bidder types have independent, atomless[19] distributions, then any other auction design that leads to efficient outcomes must involve the same expected payments by all the types of all the bidders (see, e.g., Engelbrecht-Wiggans 1988).

The generalized Vickrey auction itself is not practical for use in spectrum sales. If there were no restrictions on feasible license combinations, the number of combinations would be $2^{|L|} - 1$. Most of the sales being conducted presently involve hundreds of licenses, and even though in practice most of the combinations can be ruled out as infeasible or irrelevant, the number of potentially important combinations is still infeasibly large.[20] I seek to use the Vickrey auction here as a benchmark, in much the same way that the competitive equilibrium benchmark is used in market welfare analyses.

---

[18] Among the important assumptions are the following. First, the bidders know their own values; i.e., this is a pure private-value model with no common-value elements. (See Milgrom and Weber [1982] for a discussion of this assumption.) Second, bidders must care only about the sets of licenses they acquire and the prices they pay, and not about the identities of the other license acquirers and the prices they pay (although extensions of the Vickrey auction can accommodate bidders that care about the entire license allocation). Third, budget constraints must never be binding. Each of these assumptions is a strong one. None precisely fits the facts about the U.S. spectrum auctions. In addition, there is the relatively more innocuous assumption that bidder preferences are quasi-linear. This means that a bidder's utility is representable as the value of the licenses assigned to it minus the price that it pays.

[19] I am indebted to Paul Klemperer for pointing out the necessity of the atomless type distribution condition. In this application, a "type" is a vector of values for licenses and combinations of licenses.

[20] An additional objection to Vickrey auctions is that they require bidders to reveal their value estimates. Bidders have been reluctant to do that, possibly because they fear that reporting their values would reveal information to competitors about how they form estimates, what discount rates they use, what financing they have available, or what their business plans are.

Table 5.

	License ID			
Bidder	A	B	AB	Vickrey Price
1	V_1	V_1	V_1	$V_3 - V_2$
2	V_2	V_2	V_2	$V_3 - V_1$
3	0	0	V_3	$V_1 + V_2$

Given that it is infeasible to specify all relevant combinations in advance, one idea to economize on computing power is to specify combinations as the auction progresses. The leading such proposal is based on a procedure called the "adaptive user selection mechanism" (AUSM) that was developed in experimental economics laboratories for solving what the experimenters regarded as "difficult" resource allocation problems (Banks, Ledyard, and Porter 1989; Ledyard, Noussair, and Porter 1996).

The AUSM differs from the simultaneous ascending auction in a number of respects, and many of its features have been proposed for adoption in the spectrum auctions. Among the proposed changes are the following: First, allow bidding to take place continuously in time, rather than force bidders to bid simultaneously in discrete rounds. Second, in place of an activity rule, follow the experimenters' technique of using random closing times, which motivate bidders to be active before the end of the auction. Third, permit bids for combinations of licenses rather than just for individual licenses. When a new combinatorial bid is accepted, it displaces all previous standing high bids for individual licenses or combinations of licenses that overlap the licenses in the new bid. The new bid should be accepted if the amount of the bid is greater than the sum of the displaced bids. Fourth, allow the use of a "standby queue" on which bidders may post bids that cannot, by themselves, displace existing bids but become available for use in new combinations. For example, suppose that bidder 1 owns the standing high bid of 20 for license combination ABCD. Bidder 2 is interested in acquiring AB for a price of up to 15 but has no interest in CD. It may post a bid of 12 for AB on the standby queue. Suppose that it does so and that bidder 3 is willing to pay up to 15 for CD. Then bidder 3 may "lift" 2's bid from the standby queue and submit that together with its bid of 10 for license combination CD, thereby creating a bid of 22 for the combination ABCD. Under the rules, bidders 2 and 3 become the new owners of the standing high bids.

We can begin to analyze this proposal using a simple example, represented in table 5. There are three bidders, labeled 1–3, and two licenses. The first two bidders each want to acquire a single license; the third bidder is interested only in the pair. The final column shows what price the bidder would pay in a generalized Vickrey auction in which it is a license winner.

The bidders' values are drawn from continuous distributions. For the first two bidders, the distribution has support on $[a, b]$, and for the third bidder, it has support on $[c, d]$. We assume that $2a < d$ and that $2b > c \geq b$. These inequalities mean that (1) there is a priori uncertainty about the efficient license assignment, and (2) the two single-license bidders need to coordinate to be able to outbid bidder 3.

Since there are many different implementations of AUSM, I regard it as a class of games. I limit attention to implementations in which bidding takes place in rounds and does not end after a round in which there are new bids. I look for properties of equilibrium in undominated strategies of any such AUSM game in which no bidder makes jump bids. Three general properties hold. First, no bidder j bids more than its own actual value V_j, for to do so would entail using a weakly dominated strategy. With no jump bids, this implies that bidder 3 never pays more than $V_1 + V_2$. Second, since bidder 3 always has an opportunity to respond to the bids by 1 and 2, equilibrium entails that bidder 3 wins a license when $V_3 > V_1 + V_2$. Free-riding among the individual bidders may mean that bidder 3's AUSM equilibrium price is strictly less than the Vickrey price $V_1 + V_2$. Third, when the single-license bidders 1 and 2 win licenses in an AUSM game, the total price they pay is V_3. They win only when $V_1 + V_2 > V_3$, and given the free-rider problem, they may not always win even when that inequality holds.[21] From the preceding inequality, the total price V_3 that the bidders pay when they win is strictly greater than the total Vickrey price of $2V_3 - V_1 - V_2$. This leads to the following conclusion.

Theorem 10. *In the example analyzed here, the total equilibrium prices under AUSM for the single-license bidders are always at least as high as and sometimes higher than the Vickrey prices, whereas the price paid by the combination bidder is never more and sometimes less than the Vickrey price. The combination bidder wins (weakly) more often than it would at an efficient auction, and the single-license bidders win (weakly) less often than they would at such an auction.*

Experiments have established that AUSM performs well in some environments with significant complementarities. The questions for auction designers are, Which kinds? And how can their disadvantages be minimized? Identifying biases is a first step toward answering such questions.

VI. Two Additional Questions

One of the most frequently expressed doubts about the spectrum auctions is the doubt that the form of the auction matters at all. After all, the argument goes, one should expect that if the initial assignment resulting from the auction is inefficient and if licenses are tradable, the license owners will be motivated after the auction to buy, sell, and swap licenses until an efficient assignment is achieved.

There are both theoretical and empirical grounds for rejecting this argument. The theoretical argument is developed at length in Milgrom (1995). Briefly, the argument combines two theoretical observations from the theory of resource allocation under incomplete information in private-values environments. The first observation is that, once property rights have been assigned, ex post bargaining cannot generally achieve efficient rearrangement of the rights. The older theoretical literature shows this for the

---

[21] Notice that a solution to the free-rider problem may require that one bidder pay more for its license than another bidder pays for a perfectly substitutable license. One may guess that such a solution would be particularly difficult to achieve if the bidders are ex ante identically situated.

case in which there are just two parties to the bargain and the efficient allocation of the license is uncertain. Recent work by Cai (1997) suggests that the efficient outcomes become even less likely when there are multiple parties involved, as is the case when a bidder needs to assemble a collection of spectrum licenses from multiple owners to offer the most valuable mobile telephone service.[22] The years of delay in developing nationwide mobile telephone services in the United States, despite the value that customers reportedly assign to the ability to "roam" widely with their phones, testify to the practical importance of this theoretical effect. An inefficient initial assignment cannot, in general, be quickly corrected by trading in licenses after the auction is complete.

In contrast, the generalized Vickrey auction applied to the initial assignment of rights in the same environment can achieve an efficient license assignment—at least in theory. There are significant practical difficulties in implementing a Vickrey auction in the spectrum sales environment, but the theoretical possibility of an auction that always yields an efficient assignment establishes the possibility that a good initial design can accomplish objectives that ex post bargaining cannot.

A second common question concerns the trade-off between the goals of allocational efficiency and revenue. The primary goal of the spectrum auctions was set by the 1993 budget legislation as one of promoting the "efficient and intensive use" of the radio spectrum. However, the simultaneous ascending auction is now also being touted for other applications, such as the sale of stranded utility assets (Cameron, Cramton, and Wilson 1997) in which revenue is regarded as an important objective. Such applications call for putting more emphasis both on how the auction rules affect revenue and on the extent of the conflict between the goals of efficiency and revenue in multiobject auctions.

Particularly when the number of bidders is small, the goals of efficiency and revenue can come into substantial conflict. A particularly crisp example of this is found in the decision about how to package groups of objects when there are only two bidders.[23] Using the spectrum sale as an example, suppose that the available bands of spectrum are denoted $\{1, \ldots, B\}$ and that these are packaged in licenses $\mathbf{L} = \{1, \ldots, L\}$. The jth license consists of a set of bands $S_j \subseteq \{1, \ldots, B\}$, and a "band plan" is a partition $S = \{S_1, \ldots, S_N\}$ of the L bands into $N \leq L$ licenses.

Next, I introduce a special assumption. Suppose that each bidder i's valuation for any license is given by $X_i(S_j) = \sum_{k \in S_j} x_{ik}$. This assumption abstracts from some potential interactions between efficiency and revenue and isolates the one effect on which I wish to focus.

Let $R(S)$ denote the revenue from the license sales corresponding to the band plan S, and let $V(S)$ be the total value of the licenses to the winning bidders when the licenses are sold individually in simultaneous second-price auctions and each bidder adopts its dominant strategy.

The conflict between efficiency and revenue in this context is very sharp. When one is choosing band plans in this setting, there is a *dollar-for-dollar trade-off* between the

---

[22] The same theoretical analysis applies to attempts to resolve the problem by contracting: ex post bargaining under incomplete information after property rights have already been assigned does not generally lead to efficient outcomes.

[23] See Palfrey (1983) for a related analysis, showing that bundling can increase revenue even when it reduces efficiency in various kinds of auctions.

seller's revenue $R(S)$ and the value $V(S)$ of the final license assignment: any change in the band plan S that increases the value of the assignment reduces the seller's revenue by an equal amount!

Theorem 11. *The sum of the value created and the revenue generated by the auction is a constant, independent of the band plan S: $R(S) + V(S) = X_1(\mathbf{L}) + X_2(\mathbf{L})$. Coarser band plans generate higher revenues and create less value.*

Proof. For the first statement, it suffices to show that, for any license S_j, the value created by the auction plus the license price is equal to $X_1(S_j) + X_2(S_j)$, for the result then follows by summing over licenses.

Suppose (without loss of generality) that bidder 1 has the higher value for the license. Then in an English auction, bidder 1 will win; the winner's value will be $X_1(S_j)$; and the price will be the second-highest value, $X_2(S_j)$.

For the second statement, recall that the outcome of the ascending auction is to assign each license to the bidder that values it most highly. Given two band plans S and S', with S coarser than S', the associated values are

$$V(S) = \sum_{T \in S} \max \left(\sum_{k \in T} x_{1k}, \sum_{k \in T} x_{2k} \right)$$

$$\leq \sum_{T \in S} \sum_{\substack{T \subseteq T \\ T \in S}} \max \left(\sum_{k \in T} x_{1k}, \sum_{k \in T} x_{2k} \right) = V(S).$$

The inequality applies term by term to the maxima over sets $T \in S$. □

To illustrate the theorem, suppose that there are two bands with $x_{11} > x_{21}$ but $x_{12} < x_{22}$, and suppose in addition that $x_{11} + x_{12} > x_{21} + x_{22}$. There are two possible band plans according to whether the bands are sold as one license or two. When the bands are sold separately, bidder 1 wins band 1 at price x_{21} and bidder 2 wins band 2 at price x_{12}, creating a total value of $x_{11} + x_{22}$ and revenue of $x_{21} + x_{12}$. When the bands are sold together, bidder 1 acquires both at price $x_{21} + x_{22}$, creating a total value of $x_{11} + x_{12}$. The loss of value from adopting this plan is $x_{22} - x_{12}$, which is precisely the same as the increase in revenue from the same change.

In the analysis of Cameron et al. (1997), the items being sold are electrical generating plants or other "stranded utility assets" associated with deregulation. In that case, revenue (which reduces the burden on ratepayers) and efficiency are both typically among the goals of the public authority. In that case, if the number of serious bidders is sufficiently small, then the effect identified in this suggestion contributes to a trade-off in the public decision process between the goals of revenue and efficiency.

VII. Conclusion

In the last few years, theoretical analyses have clearly proved their worth in the practical business of auction design. Drawing on both traditional and new elements of auction theory, theorists have been able to analyze proposed designs, detect biases, predict shortcomings, identify trade-offs, and recommend solutions.

It is equally clear that designing real auctions raises important practical questions for which current theory offers no answers. The "bounded rationality" constraints that limit the effectiveness of the generalized Vickrey auction are important ones and have so far proved particularly resistant to simple analysis. Because of such limits to our knowledge, auction design is a kind of engineering activity. It entails practical judgments, guided by theory and all available evidence, but it also uses ad hoc methods to resolve issues about which theory is silent. As with other engineering activities, the practical difficulties of designing effective, real auctions themselves inspire new theoretical analyses, which appears to be leading to new, more efficient and more robust designs.

References

Ausubel, Lawrence M., and Cramton, Peter. "Demand Reduction and Inefficiency in Multi-unit Auctions." Working Paper no. 96–07. College Park: Univ. Maryland, 1996.

Banks, Jeffrey S.; Ledyard, John O.; and Porter, David P. "Allocating Uncertain and Unresponsive Resources: An Experimental Approach." *Rand J. Econ.* 20 (Spring 1989): 1–25.

Cai, Hong-Bin. "Delay in Multilateral Bargaining under Complete Information." Working paper. Los Angeles: Univ. California, 1997.

Cameron, Lisa; Cramton, Peter; and Wilson, Robert. "Using Auctions to Divest Generating Assets." *Electricity J.* 10, no. 10 (1997): 22–31.

Che, Yeon-Koo, and Gale, Ian. "Financial Constraints in Auctions: Effects and Antidotes." In *Advances in Applied Microeconomics*, vol. 6, edited by Michael R. Baye. Greenwich, Conn.: JAI, 1996.

———. "Standard Auctions with Financially Constrained Bidders." *Rev. Econ. Studies* 65 (January 1998): 1–21.

Clarke, Edward H. "Multipart Pricing of Public Goods." *Public Choice* 11 (Fall 1971): 17–33.

Cramton, Peter C. "Money Out of Thin Air: The Nationwide Narrowband PCS Auction." *J. Econ. and Management Strategy* 4 (Summer 1995): 267–343.

———. "The FCC Spectrum Auctions: An Early Assessment." *J. Econ. and Management Strategy* 6 (Fall 1997): 431–95.

Engelbrecht-Wiggans, Richard. "Revenue Equivalence in Multi-object Auctions." *Econ. Letters* 26, no. 1 (1988): 15–19.

Gertner, Robert. "Revenue and Efficiency Differences between Sequential and Simultaneous Multiple-Unit Auctions with Limited Competition." Manuscript. Chicago: Univ. Chicago, Grad. School Bus., 1995.

Groves, Theodore, and Loeb, Martin. "Incentives and Public Inputs." *J. Public Econ.* 4 (August 1975): 211–26.

Gul, Faruk, and Stacchetti, Ennio. "Walrasian Equilibrium without Complementarities." Working Paper no. 97–01. Ann Arbor: Univ. Michigan, Dept. Econ., 1997.

Kelso, Alexander S., Jr., and Crawford, Vincent P. "Job Matching, Coalition Formation and Gross Substitutes." *Econometrica* 50 (November 1982): 1483–1504.

Ledyard, John O.; Noussair, Charles; and Porter, David. "The Allocation of Shared Resources within an Organization." *Econ. Design* 2 (November 1996): 163–92.

McAfee, R. Preston, and McMillan, John. "Analyzing the Airwaves Auction." *J. Econ. Perspectives* 10 (Winter 1996): 159–75.

Milgrom, Paul. "Auction Theory for Privatization." Book manuscript. Stanford, Calif.: Stanford Univ., 1995.

Milgrom, Paul, and Roberts, John. "Adaptive and Sophisticated Learning in Normal Form Games." *Games and Econ. Behavior* 3 (February 1991): 82–100.

Milgrom, Paul, and Weber, Robert J. "A Theory of Auctions and Competitive Bidding." *Econometrica* 50 (September 1982): 1089–1122.

Palfrey, Thomas R. "Bundling Decisions by a Multiproduct Monopolist with Incomplete Information." *Econometrica* 51 (March 1983): 463–83.

Pitchik, Carolyn, and Schotter, Andrew. "Perfect Equilibria in Budget-Constrained Sequential Auctions: An Experimental Study." *Rand J. Econ.* 19 (Autumn 1988): 363–88.

Vickrey, William. "Counterspeculation, Auctions, and Competitive Sealed Tenders." *J. Finance* 16 (March 1961): 8–37.

CHAPTER 2

An Equilibrium Analysis of the Simultaneous Ascending Auction

Jacob K. Goeree and Yuanchuan Lien

1. Introduction

In recent years, governments around the world have employed auctions to award licenses for the rights to operate in certain markets. The spectrum auctions conducted by the US Federal Communications Commission (FCC) provide a particularly prominent example. In the FCC auctions, telecom firms compete for blocks of frequencies (typically on the order of 10-20 MHz) defined over certain geographic areas.[1] The format pioneered by the FCC is the simultaneous ascending auction (SAA), which is a dynamic, multi-round format in which the items are put up for sale simultaneously and the auction closes only when bidding on all items has stopped. The SAA has become the standard to conduct large-scale, large-stakes spectrum auctions and has generated close to $80 billion for the US Treasury and hundreds of billions worldwide.

An important property of the SAA is that when items are substitutes and bidding is "straightforward," i.e. in each round of the auction bidders place minimum acceptable bids on those licenses that provide the highest current profits, then prices converge to competitive equilibrium prices and a fully efficient outcome results (Milgrom [18]; Gul and Stacchetti [14]). However, in many of the FCC auctions there are synergies between licenses for adjacent geographic regions, and bidders' values for combinations of licenses exceed the sum of individual license values. For example, the bid regressions reported by Ausubel, Cramton, McAfee, and McMillan [2] show that the highest losing bid on a license is higher if the bidder who placed the bid has won or eventually wins a license. Bajari and Fox [5] apply a structural econometrics model to data from FCC auction #5 and find evidence for substantial value complementarities: they estimate that the value of a nationwide package is 69% more than the sum of underlying values.[2] Value complementarities were considered even more important in

---

[1] The different frequency bands that have been put up for sale in the 73 FCC auctions since 1994 accommodate different usages, including wireless and cellular phone applications, mobile television broadcasting, and air-to-ground communication. See http://wireless.fcc.gov/auctions/default.htm?job=auctions_all.
[2] FCC auction #5 (the "C-block auction") was conducted in 1995 and generated over $10 billion in revenue.

the recently conducted FCC auction #73, where potential entrants, e.g. Google, competed against established incumbents such as Verizon and AT&T for highly valuable 700 MHz spectrum.[3] Most experts believed that an entrant could have a viable business plan only if it would acquire a "national footprint," i.e. a set of licenses covering the entire United States.

In this paper, we consider an environment where one or more global bidders (entrants) have super-additive values for the licenses, i.e. for global bidders licenses are *complements* rather than substitutes. For this environment an often cited problem of the item-by-item competition that occurs under SAA is that global bidders face an *exposure problem* – when competing aggressively for a package, global bidders may incur a loss when winning only an inferior subset. Foreseeing the possibility of being "exposed," global bidders may decide to bid cautiously and drop out early, which could adversely affect the auction's revenue and efficiency.[4] For example, the counterfactual experiments conducted by Bajari and Fox [5] demonstrate that in FCC auction #5 only 50% of the total available surplus was captured.[5] In addition, a substantial body of laboratory evidence documents the negative impact of the exposure problem on the SAA's performance (see, e.g., Brunner et al. [6], and references therein).

Despite the potential shortcomings of the item-by-item competition underlying SAA, it has been the preferred choice for most spectrum auctions. Alternatives allowing for package bids were either considered too complex or thought to be prone to "free riding" (Milgrom [18]).[6] Furthermore, the familiar Vickrey-Clarke-Groves (VCG) mechanism, which guarantees full efficiency even in the presence of value complementarities, is generally dismissed because of its perverse revenue properties. In particular, the VCG mechanism can lead to non-core outcomes that result in high bidder profits and low seller revenue. Moreover, seller revenue can decrease when more bidders participate. The following three-bidder, two-item example provided by Ausubel and Milgrom [4] illustrates these shortcomings. Suppose local bidder 1 is interested only in item A, local bidder 2 is interested only in item B, the global bidder 3 is interested only in the package AB, and all bidders' values (for individual items or the package) are $1 billion. The VCG mechanism assigns the items efficiently to bidders 1 and 2, but at zero prices![7] Besides generating the lowest possible revenue, this outcome is outside the core as the seller and global bidder can form a blocking coalition. Moreover, excluding one of the local bidders, raises the seller's revenue to $1 billion. These perverse revenue

---

[3] FCC auction #73 was conducted in 2008 and generated a record $19 billion in revenue. It was the first combinatorial auction conducted by the FCC, based on hierarchically structured packages (Rothkopf, Pekeč, and Harstad [21]) and a novel pricing rule (Goeree and Holt [13]).

[4] Milgrom [18] argues that the different per-unit-of-bandwidth prices observed for small and large licenses in the Dutch DCS-1800 auction reflect the exposure problem. A similar observation applies to the recent FCC auction #66, where 12 large (F-band) licenses providing 20 MHz of nationwide coverage sold for $4.2 billion while 734 small (A-band) licenses also providing 20 MHz of nationwide coverage went for $2.3 billion.

[5] Bajari and Fox [5] show that surplus could have been doubled had the FCC offered large regional licenses or a nationwide package in addition to individual licenses.

[6] In package auctions, local bidders who drop out early ("free ride") may earn windfall profits if other local bidders remain active and outbid the global bidders. After all, a local bidder's concern is simply whether *as a group* they meet the threshold set by a global bidder's package bid. Of course, if all local bidders free ride this threshold may not be met with adverse effects for revenue and efficiency – this is known as the *threshold problem*.

[7] Ausubel and Milgrom [4] develop this example further in a theorem that shows that bidders' Vickrey payoffs are the highest payoffs over all points in the core.

properties, shown here in a complete-information setting, carry over to the Bayesian framework studied in this paper where bidders' values are private information (see Example 1 in Section 2.1).

This paper compares the SAA and VCG mechanisms in settings with value complementarities.[8] Our approach differs from previous studies of the SAA, which assume that bidders have complete information and that the items for sale are substitutes (e.g. Milgrom [18]). Either assumption, however, precludes the possibility of an exposure problem for global bidders interested in aggregating many items. One contribution of this paper is the introduction of a tractable model that allows for a Bayes-Nash equilibrium analysis of the exposure problem in the SAA.

We first consider an environment where global bidders with super-additive (convex) valuations compete with local bidders who each value a different item, e.g. when local bidders are interested only in the license for the region where they have local monopoly power. The setup is general in that it allows for arbitrary numbers of local and global bidders, arbitrary distributions of local and global bidders' values over one-dimensional types, and a general convex valuation function to capture global bidders' value complementarities. We provide a recursive characterization of equilibrium bidding in this setting, which enables us to quantify the effects of the exposure problem on efficiency and revenue. In particular, due to the exposure problem the SAA is not efficient and the efficiency loss is borne by the global bidder. The SAA shares the poor revenue-generating features of the VCG mechanism, e.g. its revenues may be low and may decline with the number of bidders. Finally, the similarities between the SAA and VCG mechanisms in this environment become even stronger as the number of items grows: the two mechanisms are revenue and efficiency equivalent in the limit.

The intuition why the SAA is able to produce full efficiency in the limit is that global bidders face no exposure with respect to each other because of the convexity assumption and they face no exposure with respect to the local bidders whose behavior becomes deterministic by the law of large numbers. Importantly, however, the global bidders can avoid the exposure problem only because of the dynamic nature of the SAA: it allows the global bidder with the second-highest value to drop out before any of the local bidders that are part of the efficient allocation do. In contrast, in a sealed-bid version of the SAA, global bidders face an exposure problem and, in equilibrium, the highest-value global bidder wins fewer items than is socially optimal.

Next, we relax the assumptions that local bidders value a specific item or that global bidders' valuations are convex, and demonstrate that under these more realistic conditions the SAA generally underperforms vis-à-vis the VCG mechanism. We extend our recursive characterization of equilibrium bidding in the SAA to the case where items are perfect substitutes for local bidders so that price arbitrage will occur (as observed in many spectrum auctions). We demonstrate that for this environment, which combines complements for the global bidders with substitutes for the local bidders,

---

[8] We compare the SAA to VCG not because the latter is a serious contender for practical applications, see e.g. Rothkopf [20] for (thirteen) reasons why not. But despite its shortcomings in practice, VCG is the relevant benchmark because of its broad theoretical applicability. Since it is dominant-strategy implementable it can readily be used to study incomplete-information environments with multiple units and value complementarities.

the exposure problem is much worse and the performance of the SAA suffers as a result. In particular, when the number of items grows large, it may be optimal for the global bidder to drop out immediately with dramatic consequences for efficiency and revenue.

Finally, we study a setting with two "medium-sized" global bidders with non-convex valuation functions. When the global bidders each need less than half the items, they compete head-to-head only at very low prices after which they compete only with the local bidders. In other words, the global bidders follow a strategy of mutual forbearance and divide the market at low prices. In contrast, when the global bidders each need more than half the items, the equilibrium involves mutually destructive bidding: global bidders may bid above their values and even the winning bidder may incur a loss.[9] We show that the SAA again performs worse than the VCG mechanism, whether or not market sharing is feasible.

1.1. Related Literature

Auctions in which bidders have synergistic values have often been analyzed within a complete-information setting, see, for instance, Szentes and Rosenthal [22, 23]. There are relatively few theoretical papers that apply the standard Bayesian framework of incomplete information. An early exception is Krishna and Rosenthal [15] who study the simultaneous sealed-bid second-price auction (SSA). Similar to our bidding environment, local bidders in their setup are interested in only one object while global bidders are interested in multiple objects for which they have synergistic values. Krishna and Rosenthal derive an explicit solution for the case of two items and show how it varies with the synergy level. They also discuss the extension to more than two items and provide a numerical comparison of revenue in alternative formats. Other papers that study the SSA include Rosenthal and Wang [19], who allow for common values and partially overlapping bidder interests, and a more recent paper by Chernomaz and Levin [10] who use theory and experiments to analyze the SSA and a package bidding variant when local bidders have identical values.

Ascending formats have been analyzed either assuming a clock price that rises in response to excess demand or assuming that bidders can name their own bids (i.e. submit any bid they want). In the latter category, Brusco and Lopomo [7] demonstrate the possibility of collusive demand-reduction equilibria in the SAA. They find that increasing the number of bidders and objects narrows the scope for collusion. Brusco and Lopomo [8] analyze the effects of budget constraints. Zheng [24] shows that jump bidding may serve as a signaling device to alleviate the inefficiencies that result from the exposure problem. Albano, Germano, and Lovo [1] analyze a ("Japanese style") clock version of the ascending auction for a setting with only two items. They note the equivalence between the SAA and a "survival auction" and point out that many of the collusive or signaling equilibria that occur when bidders can name their bids do not arise for the clock variant of the SAA.

---

[9] For a complete-information environment, Bykowsky, Cull, and Ledyard [9] have argued the possibility of such behavior using the concept of a "local Nash equilibrium."

1.2. Organization

Section 2 provides an equilibrium analysis of the SAA when each local bidder is interested in a different item and global bidders' values are convex. We start by considering a single global bidder (Section 2.1) and determine the effects of the exposure problem for efficiency (Section 2.2). In Section 2.3 we extend the result to multiple global bidders. In Section 3 we establish the equivalence of the SAA and VCG mechanisms as the number of items grows large. In Section 4 we demonstrate the poor performance of the SAA in an environment that combines substitutes and complements (Section 4.1) and in an environment with two medium-sized global bidders with non-convex valuation functions (Sections 4.2 and 4.3). Section 5 concludes and Appendix A contains all proofs.

2. The Simultaneous Ascending Auction

Consider an environment with $n \geq 1$ local bidders and $K \geq 1$ global bidders who compete for n items labeled $1, \ldots, n$. We assume that local bidder i values only item i, e.g. the license for the region where she has monopoly power.

Assumption 1 (non-substitutability). *Local bidder $i = 1, \ldots, n$ is interested only in item i.*

Local bidder i's value for item i is denoted v_i. The local bidders' values are identically and independently distributed according to $F(\cdot)$, with support $[0,1]$.[10]

Global bidder j's value for winning k items is $\alpha(k)V^j$, where the V^j are identically and independently distributed according to $G(\cdot)$ with support $[0, n]$,[11] and $\alpha(k)$ is increasing in k with $\alpha(0) = 0$ and $\alpha(n) = 1$. We define the "marginal values" $V_k^j = (\alpha(n + 1 - k) - \alpha(n - k))V^j$ for $k = 1, \ldots, n$ so that global bidder j's marginal value of the first item is V_n^j, of the second item is $V_{n-1}^j, \ldots$, and of the n-th item is V_1^j.[12]

Assumption 2 (convexity). *For global bidder $j = 1, \ldots, K$, the marginal values form a non-decreasing sequence $V_n^j \leq V_{n-1}^j \leq \ldots \leq V_1^j$. Moreover, we say that*

(i) *values are "additive" when $V_n^j = \ldots = V_1^j = V^j/n$.*
(ii) *values exhibit "complementarities" when $V_n^j < \ldots < V_2^j < V_1^j$.*
(iii) *values exhibit "extreme complementarities" when $V_n^j = \ldots = V_2^j = 0$ and $V_1^j = V^j$.*

---

[10] The assumption of a [0,1] support is without loss of generality since we can rescale local bidders' values. The symmetry assumption can be relaxed at the expense of more cumbersome notation as can the assumption of a single local bidder per item.

[11] We restrict the support of the global bidders' values to [0, n] because global bidders with values higher than n face no exposure problem and always win all items. The assumption that global bidders' values are identically distributed can be relaxed see Remark 1 below.

[12] Alternative ways to capture synergies include adding a known constant to the sum of individual item values (e.g. Albano, Germano, Lovo [1]; Rosenthal and Wang [19]) or multiplying the sum of individual item values by a known multiplier that is bigger than 1 (e.g. Chernomaz and Levin [10]). A more general formulation would assign values to each item and package separately. This multi-dimensional type approach has so far only been tried for the case of two items (e.g. Brusco and Lopomo [7]; Zheng [24]). Since we are interested in auctions with many items we retain the one-dimensional type assumption for tractability.

The SAA is modeled using n price clocks that tick upward (at equal and constant pace) when two or more bidders accept the current price levels.[13] If only one bidder accepts the new price for an item then this bidder becomes the provisional winner for the item and its price clock pauses.[14] If, at a later point, others decide to accept the new price for the item then its price clock resumes and the current provisional winner is unassigned. In other words, bidders can switch back and forth between items as long as the auction has not ended, i.e. as long as some new price is accepted by more than one bidder or some provisional winner is outbid. Once the auction ends, provisional winners become final winners and pay the final prices they accepted for the items they win. To expedite the auction, a simplified activity rule is imposed: the total number of items bidders compete for (i.e. bid for or provisionally win) can never rise.[15]

We assume that local bidders compete only for the item they are interested in so that they have a dominant strategy to bid up to their values.[16] In contrast, global bidders are interested in and compete for all items. As a result, prices rise uniformly and a global bidder's optimal strategy is characterized by a single drop-out level. A global bidder's strategy is complicated by the fact that when competing aggressively for a package, the global bidder may suffer a loss when she is able to win only an inferior subset. Foreseeing the possibility of being "exposed" and incurring a loss, the global bidder may decide to bid cautiously and drop out early, which could adversely affect the auction's revenue and efficiency — this is known as the *exposure problem*.

2.1. Single Global Bidder

We write $B_k^K(V)$ to denote a global bidder's bidding function when K global bidders and k out of n local bidders are active. A global bidder's optimal drop-out level is then given by $\max\{p, B_k^K(V)\}$, where p denotes the current price level. For ease of notation we simply refer to $B_k^K(V)$ as the optimal drop-out level, where it is implicitly understood that the global bidder finds it unprofitable to stay in the auction and prefers to drop out immediately when $B_k^K(V) < p$. It will prove useful to introduce the notation $F_k(v|p) = 1 - ((1 - F(v))/(1 - F(p)))^k$, which is the conditional probability that the minimum of k active local bidders' values is less than v given that the minimum is no less than p. In this section we focus on the case of a single global bidder ($K = 1$).

Suppose first that there is only one item for sale ($k = 1$). If the current price level is p and the global bidder chooses a drop-out price level $B_1^1(V)$, her expected profits are

$$\Pi_1^1(V, p) = \int_p^{B_1^1(V)} (V_1 - v_1) dF_1(v_1|p),$$

with v_1 the local bidder's value. The integrand $\pi_1^1(V, v_1) = V_1 - v_1$ is the global bidder's profit if the local bidder drops out at price v_1. Clearly, the global bidder's expected

---

[13] This modeling approach is common in the literature on ascending auctions. In practice, the number of bid increments is often restricted – when it is limited to 1 the auction is essentially a clock auction. Most recent spectrum auctions have used a clock format.

[14] If no bidder accepts the new price, then the provisional winner is chosen randomly among the bidders that accepted the previous price. Such "ties" occur with probability 0 in the Bayes-Nash equilibria described below.

[15] Activity rules of this type have been used in all (FCC) spectrum auctions.

[16] In many FCC auction, bidders have to state on which items they will bid prior to the auction.

profit is maximized by choosing the drop-out price $B_1^1(V)$ such that $\pi_1^1(V, B_1^1(V)) = 0$, or $B_1^1(V) = V_1$.

Next, consider the case $k = 2$. Suppose the current price level is p and the global bidder chooses a drop-out price level $B_2^1(V)$ (for both items), her expected profit is nontrivial only when the local bidder with the lower value drops out before $B_2^1(V)$. Once a local bidder drops out, the global bidder faces competition only in a single market and she is willing to bid up to V_1 in this market. The reason is that her profit for the item on which bidding stopped is sunk (i.e. independent of whether or not she wins an additional item). The global bidder's expected profit can be written as

$$\Pi_2^1(V, p) = \int_p^{B_2^1(V)} \left\{ \int_{v_2}^{V_1} (V_1 - v_1) dF_1(v_1|v_2) + (V_2 - v_2) \right\} dF_2(v_2|p),$$

where v_2 (v_1) denotes the lower (higher) of the local bidders' values. The integrand $\pi_2^1(V, v_2) = \int_{v_2}^{V_1} (V_1 - v_1) dF_1(v_1|v_2) + (V_2 - v_2)$ is the global bidder's expected profit conditional on the local bidder with the lower value dropping out at v_2. The first term arises when the global bidder wins the remaining item, i.e. when $v_1 \geq v_2$ is less than V_1 (since the global bidder bids up to V_1 for the remaining item) in which case the global bidder wins the additional item and pays v_1 for it. The second term indicates that the global bidder profits $V_2 - v_2$ from the item for which bidding stopped first, irrespective of whether she wins the additional item.

Again, the global bidder's optimal drop-out level follows from $\pi_2^1(V, B_2^1(V)) = 0$, which yields $B_2^1(V) = V_2 + \Pi_1^1(V, B_2^1(V))$. Note that the global's profit can be recursively expressed using the profit for the single local-bidder case: $\Pi_2^1(V, p) = \int_p^{B_2^1(V)} \{\Pi_1^1(V, v_2) + (V_2 - v_2)\} dF_2(v_2|p)$. This recursive relation can be generalized to the case of more than two items.

Proposition 1. *The global bidder's optimal drop-out level solves* $B_k^1(V) = V_k + \Pi_{k-1}^1(V, B_k^1(V))$, *where the payoffs satisfy the recursive relation*

$$\Pi_k^1(V, p) = \int_p^{B_k^1(V)} \{\Pi_{k-1}^1(V, v_k) + (V_k - v_k)\} dF_k(v_k|p), \quad (2.1)$$

with $\Pi_0^1(V, p) = 0$.

The proposition implies a set of fixed-point equations from which the optimal bids can be solved recursively: $B_1^1(V) = V_1$, $B_2^1(V) = V_2 + \int_{B_1^1(V)}^{B_2^1(V)} (V_1 - v_1) dF_1(v_1|B_2^1(V))$, etc. The intuition behind the bidding functions in Proposition 1 stems form a familiar break-even condition: when a license is marginally won at price $B_n^1(V)$, its value plus the expected payoffs from continuing (knowing all remaining local bidders' values exceed $B_n^1(V)$) must balance this cost.

Example 1. To illustrate, suppose $n = 2$ and the local bidders' values are uniformly distributed on [0,1]. We have $B_1^1(V) = V_1$ and a simple calculation shows that

$$B_2^1(V) = \frac{1}{3}\left(1 + V_1 + V_2 - \sqrt{(2 - V_1 - V_2)^2 - 3(1 - \min(1, V_1))^2}\right) \quad (2.2)$$

For the case of extreme complementarities ($V_2 = 0$, $V_1 = V$), these bidding functions are illustrated by the two left-most lines in the left panel of Figure 1 (the right panel will

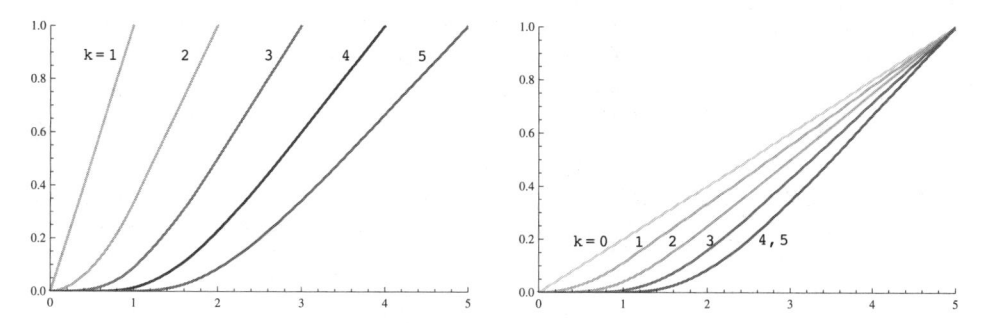

Figure 1. The Global Bidder's Optimal Drop-Out Level When Complementarities are Extreme: $B_k^1(V)$ (left panel) for $k = 1, \ldots, 5$ and $B_k^2(V)$ (right panel) for $k = 0, \ldots, 5$ (with $B_5^2(V) = B_4^2(V) = B_5^1(V)$).

be discussed in Section 2.3). The left panel also shows the optimal bidding functions $B_n^1(V)$ for higher values of n, which can be used to illustrate some of the perverse revenue properties of the SAA. For instance, suppose that also the global bidder's value is uniformly distributed on [0,1], then revenues of the SAA are (0.33, 0.27, 0.087) with $n = 1, 2, 3$ local bidders respectively. The corresponding revenue numbers for the VCG mechanism are ($\frac{1}{3}, \frac{1}{4}, \frac{1}{10}$). To summarize, the revenue of the SAA can fall as more bidders enter the auction and can be *lower* than that of the VCG mechanism.[17]

2.2. Constrained Efficiency

The optimal drop-out levels of the global bidder shown in the left panel of Figure 1 illustrate the effects of the exposure problem in equilibrium. Consider, for instance, the case of five local bidders and suppose the global bidder is equally strong in expectation, i.e. the global bidder's value for the package is 2.5. When all five local bidders are active, the global bidder drops out when the price for each item is 0.2 (see the lowest curve in the left panel of Figure 1), which means that the global bidder drops out at 40% of the package value! This does not necessarily mean, however, that efficiency is negatively affected. The lowest curve in the left panel of Figure 1 only applies when all five local bidders are active, and if this occurs at an item price of 0.2 then the sum of the local bidders' expected values is 3 (not 2.5). Hence, efficiency may be improved when the global bidder drops out (especially when complementarities are extreme, as in Figure 1, and the global bidder derives no value from winning less than five items).

Compared to the fully efficient VCG mechanism, there are two potential sources of inefficiencies in the SAA. First, when global bidders drop out, not all local bidders' values have been revealed. Due to this residual uncertainty there are necessarily some ex post inefficiencies in the SAA (unlike the VCG mechanism where the allocation is based on ex post information about bidders' values). Second, global bidders' profit-maximizing drop-out levels may differ from welfare-maximizing drop-out levels. Distinguishing these two sources of inefficiency is useful in that the first one becomes

---

[17] When the global bidder's value is uniformly distributed on $[0, n]$, SAA revenues are (0.33, 0.60, 0.85) for $n = 1, 2, 3$ compared to (0.33, 0.58, 0.83) for the VCG mechanism.

irrelevant with many items since then the uncertainty about local bidders' values vanishes because of the law of large numbers. To quantify the second source of inefficiency, we say that the global bidder's drop-out level is *constrained* efficient when it maximizes expected welfare, which is based on a mixture of ex post information (values of bidders that have dropped out) and ex ante information (values of active bidders).

To derive the constrained efficient drop-out level for the global bidder consider first the case of a single item ($k = 1$). A social planner would choose $B_1^1(V)$ to maximize

$$W_1^1(V, p) = \int_p^{B_1^1(V)} V_1 dF_1(v_1|p) + \int_{B_1^1(V)}^1 v_1 dF_1(v_1|p)$$

where the first (second) term corresponds to the global (local) bidder winning the item. Comparing the expression for welfare to the global's profit $\Pi_1^1(V, p)$ in Section 2.1 shows that $W_1^1(V, p) = \Pi_1^1(V, p) + E(v|v > p)$. In other words, welfare and the global bidder's profit differ only by a constant independent of $B_1^1(V)$. Hence, the profit-maximizing drop-out levels also maximize welfare. We next generalize this to an arbitrary number of items.

Proposition 2. *The global bidder's drop-out level is constrained efficient.*

As in the benchmark VCG mechanism, individual and social incentives are aligned in the SAA. The difference is that in the VCG mechanism all values are revealed at once, which allows for a fully efficient outcome. In contrast, the SAA is only constrained efficient because of the residual uncertainty the global bidder faces about the values of active local bidders.

The efficiency gain in the VCG mechanism does not benefit the seller, however, but only the global bidder. Let $W^{SAA}(V)$, $R^{SAA}(V)$, $\Pi^{SAA}(V)$, and $\pi^{SAA}(V)$ denote the expected welfare, expected revenue, expected global bidder's profit, and expected local bidders' total profit under the SAA mechanism, where the (ex ante) expectation is taken over local bidders' values only. Similar definitions apply with respect to the VCG mechanism.

Proposition 3. *The efficiency gain of the VCG mechanism accrues to the global bidder*

$$W^{VCG}(V) - W^{SAA}(V) = \Pi^{VCG}(V) - \Pi^{SAA}(V) \qquad (2.3)$$

while the difference in seller's revenue accrues to the local bidders

$$R^{SAA}(V) - R^{VCG}(V) = \pi^{VCG}(V) - \pi^{SAA}(V) \qquad (2.4)$$

2.3. Multiple Global Bidders

A global bidder's optimal bidding function when there are multiple global bidders follows from the same 'break even' logic that underlies the result of Proposition 1. First, consider the case of $K = 2$ global bidders and suppose k out the n local bidders are still active: the optimal bid $B_k^2(V)$ is determined by requiring that at this price level the marginal costs and benefits of staying in a little longer (i.e. by bidding as of type $V + \epsilon$) cancel. There are two possible marginal events: one occurs when a local drops out, in which case the other global bidder has a value no less than V. Hence, the continuation profits for a global bidder with value V are zero in this case: $\Pi_{k-1}^2(V, B_k^2(V)) = 0$.

Alternatively, the other global bidder drops out, in which case the continuation profits are given by $\Pi_k^1(V, B_k^2(V))$. Furthermore, the global bidder now wins all the $(n-k)$ items for which the local bidders had already dropped out at a price of $B_k^2(V)$ for each item. Finally, when there are $K \geq 3$ global bidders, the only non-vanishing marginal term results from $K-1$ global bidders dropping out at the same time, which produces the same marginal equation as when $K=2$.

Proposition 4. *The global bidder's optimal drop-out level satisfies $B_k^K(V) = B_k^2(V)$ for $K \geq 2$ and*

$$\Pi_k^1(V, B_k^2(V)) + \sum_{\ell=k+1}^{n} (V_\ell - B_k^2(V)) = 0, \tag{2.5}$$

where the $\Pi_k^1(V, p)$ satisfy the recursion relations of Proposition 1.

It is worthwhile pointing out a few cases: $k=0$ occurs when all local bidders have dropped out and two (or more) global bidders are active. We then have $B_0^2(V) = V/n$ since at a price of V/n the global bidder is indifferent between winning nothing and winning everything at that price. For $k=n$ we have $\Pi_n^1(V, B_n^2(V)) = 0$ so $B_n^2(V) = B_n^1(V)$. Likewise, for $k=n-1$ we have $\Pi_{n-1}^1(V, B_{n-1}^2(V)) + V_n = B_{n-1}^2(V)$ so $B_{n-1}^2(V) = B_n^1(V)$ (see Proposition 1). In other words,

$$B_n^2(V) = B_{n-1}^2(V) = B_n^1(V). \tag{2.6}$$

This result may seem surprising given that $B_{n-1}^2(V)$ is determined by the marginal event when the other global bidder drops out, while $B_n^2(V)$ is determined by the marginal event when the other global bidder and a local bidder drop out. Nevertheless, in both scenarios the optimal drop-out level follows from considering the cost and benefit of winning the first item, which yields $B_n^1(V)$ as the optimal drop-out level.

Example 1 (continued). The right panel of Figure 1 shows the global bidder's optimal drop-out levels $B_n^2(V)$ for $0 \leq n \leq 5$ when local bidders' values are uniform and complementarities are extreme. Note that there are five (not 6) lines since $B_5^2(V) = B_4^2(V)$, see (2.6), and that $B_k^2(V) \leq B_{k+1}^1(V) \leq B_k^1(V)$ for all k.

As we show next, the ranking of the global bidders' optimal drop-out levels holds more generally. Competition from other global bidders aggravates the exposure problem and *lowers* a global bidder's optimal drop-out level: $B_k^2(V) \leq B_{k+1}^2(V)$ for all k (see the proof of Proposition 2 continued). The fact that global bidders are more cautious when facing competition from other global bidders does not hurt constrained efficiency. On the contrary, it implies that global bidders who do not have the highest value, and who should therefore not win any items in the optimal allocation, drop out before local bidders that should win items in the optimal allocation do.

Proposition 2 (continued). *Global bidders' drop-out levels are constrained efficient for $K \geq 1$.*

As we show in the next section, the differences between the VCG and SAA mechanisms vanish when the number of items grows large. Full limit efficiency of the SAA can be understood from the aforementioned distinction between the two sources of inefficiency. In the limit, the residual uncertainty about active local bidders' values disappears, which turns the constrained efficiency result of Proposition 2 into a full efficiency result.

3. Large Auctions

In many applications of the SAA the number of items is very large, e.g. in some of the FCC spectrum auctions more than a thousand items are sold. In this section, we show how our approach extends to large auctions. In particular, we derive the optimal drop-out levels when multiple global bidders are active, $B^K(V)$ for $K \geq 2$, and show that they are less than $B^1(V)$, which applies with only a single global bidder. Since in large auctions there is no residual uncertainty about active local bidders' values, this result implies that the SAA becomes fully efficient in the limit. In addition, we show that in large auctions the SAA generates the same profits for the seller and the bidders as the VCG mechanism.

Let V denote the highest of the global bidders' values and $v_n \leq \ldots \leq v_1$ denote the (ordered) local bidders' values. When the highest-value global bidder wins k of n items, welfare is

$$W(k, V) = \sum_{\ell=n-k+1}^{n} V_\ell + \sum_{\ell=1}^{n-k} v_\ell.$$

In the limit when $n \to \infty$ the sum of local bidders' values will diverge, and we assume that the highest of the global bidders' values diverges as well, i.e. $V = n\hat{V}$ where $\hat{V}$ is distributed according to $\hat{G}(\hat{V}) = G(V)$ with support $[0,1]$. We can then normalize welfare and profits on a per-item basis. Suppose the highest-value global bidder wins a fraction κ of all items then normalized welfare is $W(\kappa, \hat{V}) = \lim_{n \to \infty} W(\kappa n, n\hat{V})/n$ and the global bidder's normalized value of winning a fraction κ of all items is $\mathcal{V}(\kappa) = \lim_{n \to \infty} \alpha(\kappa n)\hat{V}$. Assumption 2 implies that $\alpha(\cdot)$ is convex, and, hence, so is $\mathcal{V}(\cdot)$. To simplify notation, below we write V and $G(\cdot)$ instead of $\hat{V}$ and $\hat{G}(\cdot)$ to indicate the normalized value and its distribution.

The welfare maximizing fraction of items assigned to the highest-value global bidder now follows from $W(V) \equiv \max_\kappa W(\kappa, V)$, or, equivalently,

$$W(V) = \max_{0 \leq \kappa \leq 1} \mathcal{V}(\kappa) + \int_{F^{(-1)}(\kappa)}^{1} v dF(v), \tag{3.1}$$

where we used that in the limit when n grows large, $v_{(1-\kappa)n}$ is asymptotically normally distributed with mean $F^{(-1)}(\kappa)$ and variance of order $1/n$ (David and Nagajara [12]). The solution to (3.1) is denoted $\kappa^*(V) = \text{argmax}(W(\kappa, V))$ so that $W(V) = W(\kappa^*(V), V)$.

In the SAA, local bidders drop out at a known rate, e.g. at price level p a total of $F(p)$ local bidders have dropped out. Suppose there is only one global bidder ($K = 1$). The global bidder's optimal strategy is to bid up to a level $B^1(V)$ that maximizes her per-item profit:

$$\Pi^1(V) = \mathcal{V}(F(B^1(V))) - \int_0^{B^1(V)} v dF(v). \qquad (3.2)$$

Note that $\Pi^1(V) = W(F(B^1(V)), V) - E(v)$ so the global bidder's optimal drop-out level is simply $B^1(V) = F^{(-1)}(\kappa^*(V))$.

Next consider the case of multiple global bidders. First, let $K = 2$. The optimal drop-out level $B^2(V)$ follows by requiring that the marginal benefits and costs of staying in a little longer (by bidding as of type $V + \epsilon$) cancel. This deviation affects the outcome only when the rival global bidder drops out in between (with probability $\epsilon g(V)$), in which case the net benefit is

$$\mathcal{V}(F(B^2(V))) - B^2(V)F(B^2(V)) + \int_{B^2(V)}^{B^1(V)} (\mathcal{V}'(F(v)) - v)dF(v) = 0.$$

Here the first term reflects the value of the $F(B^2(V))$ items the global bidder wins when her rival drops out, the second term is how much she pays for them, and the third term is her continuation profit when she proceeds to win additional items by bidding up to $B^1(V)$. Integrating this last term and using the definition of $W(V)$ shows that $B^2(V)$ solves

$$B^2(V)F(B^2(V)) + \int_{B^2(V)}^1 v dF(v) - W(V) = 0. \qquad (3.3)$$

The left side of (3.3) is strictly increasing in $B^2(V)$ so the solution is unique. We next show that $B^2(V) \leq B^1(V)$. Evaluating the left side of (3.3) at $B^2(V) = B^1(V)$ yields

$$B^1(V)F(B^1(V)) - \mathcal{V}(F(B^1(V))) = \kappa^*(V)F^{(-1)}(\kappa^*(V)) - \mathcal{V}(\kappa^*(V))$$
$$\geq \kappa^*(V)(F^{(-1)}(\kappa^*(V)) - \mathcal{V}'(\kappa^*(V)))$$
$$= 0,$$

where the equality in the first line follows from the definition of $B^1(V)$, the weak inequality in the second line follows from convexity of $\mathcal{V}$, and the equality in the third line follows from the first-order condition for $\kappa^*(V)$, see (3.1).[18] Since the left side of (3.3) is strictly increasing in $B^2(V)$ this implies that $B^2(V) \leq B^1(V)$, with strict inequality when there are complementarities and $\mathcal{V}(\cdot)$ is strictly convex.

Finally, when $K \geq 3$, the marginal equation that determines $B^K(V)$ follows by requiring that the marginal benefits and costs of staying in a little longer (e.g. by bidding as of type $V + \epsilon$) cancel. This deviation affects the outcome only when all rival global bidders drop out in between, and the resulting marginal equation is the same as when $K = 2$.

Example 2. Suppose local bidders' values are uniformly distributed and a global bidder's value of winning a fraction κ of the items is given by $\mathcal{V}(\kappa) = \kappa^\rho V$.

---

[18] To be precise, the equality in the third line holds only for interior solutions. To account for possible boundary solutions, note that for $\kappa^*(V) = 0$ the expression in the second line vanishes. Furthermore, if $\kappa^*(V) = F(B^1(V)) = 1$, then the global bidder's optimal drop-out level is determined solely by the event when other global bidders drop out and $B^1(V) = B^2(V) = \mathcal{V}(1)$.

When $1 \leq \rho < 2$, the optimal drop-out levels are $B^1(V) = \min(1, (\rho V)^{1/(2-\rho)})$ and $B^K(V) = \min(\sqrt{2V-1}, \sqrt{2/\rho - 1}(\rho V)^{1/(2-\rho)})$ for $K \geq 2$. When $\rho \geq 2$, $B^1(V) = \mathbf{1}_{V \geq \frac{1}{2}}$ and $B^K(V) = \sqrt{2V-1}\,\mathbf{1}_{V \geq \frac{1}{2}}$ for $K \geq 2$, where $\mathbf{1}$ denotes an indicator function. With extreme complementarities ($\rho = \infty$), these results are readily extended to general distributions of the local bidders' values: $B^1(V) = \mathbf{1}_{V \geq E(v)}$ and $B^2(V)$ solves $\int_{B^2(V)}^1 F(y)dy = 1 - V$ for $V \geq E(v)$ and is zero otherwise.

Since $B^1(V)$ maximizes welfare the SAA is fully efficient in the limit when the number of items grows large. Surprisingly, the SAA also generates the same revenue as the VCG mechanism in this limit. To glean some insight for this result, note that (3.3) implies

$$B^2(Z)F(B^2(Z)) + \int_{B^2(Z)}^{B^1(V)} y\, dF(y) = W(Z) - \int_{B^1(V)}^1 y\, dF(y).$$

On the left side is the SAA payment of the global bidder with the highest value V when the second-highest of the global bidders' values is Z, and on the right side is the corresponding Vickrey payment.

Proposition 5. *Under Assumptions 1 & 2, the SAA becomes fully efficient and yields the same seller revenue and bidder profits as the VCG mechanism when the number of items grows large.*

Remark 1. Since the global bidders' optimal drop-out levels are independent of the value distribution $G(\cdot)$, Propositions 1 and 4 directly apply when there are asymmetries among the global bidders' value distributions. Also the equivalence result of Proposition 5 extends to the asymmetric case (seller's revenue and the winning global bidder's profit are affected by asymmetries but not efficiency or the local bidders' profits, see the expressions in the proof of Proposition 5).

In the next section we show that Assumptions 1 and 2 are necessary for the SAA to be efficient in the limit. The intuition for this limit result is that the global bidders face no exposure with respect to each other because of the convexity assumption and they face no exposure with respect to the local bidders whose behavior becomes deterministic by the law of large numbers. It is important to point out, however, that it is the dynamic nature of the SAA that allows the global bidders to avoid the exposure problem. Since $B^2(V) \leq B^1(V)$, the global bidder with the second-highest value drops out before any of the local bidders that are part of the efficient allocation do. In contrast, in a sealed-bid version of the SAA (Krishna and Rosenthal [15]) global bidders suffer from the exposure problem and generally obtain less than the socially-optimal number of items, even in large auctions.[19] The advantage of the dynamic SAA over sealed-bid formats in the setting characterized by Assumptions 1 and 2 is an important new insight

---

[19] An efficient outcome of the simultaneous sealed-bid second-price auction (SSA) dictates that global bidders place identical "uniform" bids for all items. Suppose not and some global bidder's bid B_i for item i is less than her bid B_j for item $j \neq i$. Then with positive probability $B_i < v_i < v_j < B_j$ and other global bidders' bids for items i and j are lower than B_i and B_j respectively. In this case, the lower-value local bidder i wins an item and the higher-value local bidder j does not, which is inefficient. Hence, a necessary condition for full efficiency is that global bidders place uniform bids. In the Supplemental Material we show that while it is an equilibrium for global bidders to bid uniformly in the SSA, such uniform bids are not sufficient to ensure efficiency with two or more global bidders – due to the exposure problem, global bidders win fewer items than is socially optimal.

that complements the usual motivation for the SAA when common-value elements are present.[20]

4. More General Environments

An important virtue of the SAA is that it facilitates substitution: the items are put up for sale simultaneously and the auction does not close until bidding on all items has stopped. As a result, bidders can switch back and forth between items in response to developing prices without having to worry that certain items are gone because they did not bid on them. This virtue played no role in the analysis above where each local bidder had a specific (regional) interest and after the local bidder dropped out, the market for an item remained inactive regardless of the prices of other items. One extension studied in this section is to assume that items are substitutes for local bidders so that price arbitrage will result in similar prices for the items.

Another extension is to consider market sharing among global bidders. In the analysis above, global bidders have an interest in all items and the value of each item increases the more items are won. Due to this convexity assumption only the highest-value global bidder obtains any items. In this section, we study the case where each global bidder needs only a subset of the items, which implies that global bidders' valuation functions are no longer convex. We consider the case when market sharing is feasible and the fitting problems that arise when it is not.

As we demonstrate below, these extensions exacerbate the exposure problem. With price arbitrage, the highest-value local bidder will drive up the prices for all items, making it difficult for the global bidder to acquire a large collection of items. When there is a fitting problem, global bidders run the risk of "mutually destructive bidding," i.e. when global bidders compete fiercely for their desired combination and losers end up paying for inferior subsets. Finally, when market sharing is feasible, global bidders may follow a strategy of mutual forbearance in order to keep prices low, at the expense of efficiency and the local bidders. As we will show, the aggravated exposure problem causes the SAA to perform dramatically worse than the VCG mechanism.

4.1. Substitutes and Complements

In this section we assume local bidders want at most one item and value all items the same (perfect substitutes).[21] The global bidder has super-additive values as before (complements).

---

[20] This positive aspect is somewhat reminiscent of results by Compte and Jehiel [11] who show that a dynamic (single-unit) auction raises more revenue than a sealed-bid format when bidders have to acquire costly information to determine the object's value. The intuition is that even when ex ante competition is strong, the possibility to observe that few competitors are left in the auction can create strong incentives for information acquisition.

[21] It is useful to illustrate the SAA rules of Section 2 for the substitutes environment. For simplicity, suppose there is one local and one global bidder and prices go up by at most Δ in each round that lasts Δ seconds. In the first round, prices are 0 and the global bidder bids for all items and the local bidder bids for, say, item 1. Then the price for item 1 is raised by Δ while all other prices stay at 0. The global bidder is the provisional winner on items $2, \ldots, n$. In the second round, the global bidder again bids for all items while the local bidder switches to, say, item 2. Now the global bidder becomes the provisional winner on items 1 and $3, \ldots, n$. This process continues until either the local or the global bidder drops out. If the local bidder drops out, the global bidder wins all items. If the global bidder drops out, the local bidder only wins the item for which she bids in the round that the global bidder drops out. For example, if the global bidder drops out in round 3, the global bidder is the

Assumption 1' (substitutability). *Local bidder $i = 1, \ldots, n$ wants a single item and values all items the same.*

Intuitively, the exposure problem is worse under this setup since local bidders will switch licenses[22] and drive up their prices uniformly until they drop out[23] – the global bidder will thus have to pay the value of the last local bidder that dropped out for all the items she wins.[24]

Proposition 6. *The global bidder's optimal drop-out level $B_k^1(V)$ maximizes $\Pi_k^1(V, p)$, where the payoffs satisfy the recursive relation*

$$\Pi_k^1(V, p) = \int_p^{B_k^1(V)} \{\Pi_{k-1}^1(V, v_k) + (V_k - v_k)\} dF_k(v_k|p)$$

$$- (n - k) \int_p^1 \{\min(v_k, B_k^1(V)) - p\} dF_k(v_k|p), \qquad (4.1)$$

with $\Pi_0^1(V, p) = 0$.

Note that if the global bidder is successful in winning an additional item (i.e. when $v_k < B_k^1(V)$) then she gains V_k and pays v_k, and the price increase of the $(n - k)$ items the global bidder would anyhow win is $(v_k - p)$. Otherwise, the price increase for these items is $(B_k^1(V) - p)$.

The profit functions in Proposition 6 are not necessarily concave unlike their counterparts in Proposition 1, which do not have the additional term in the second line of (4.1). As a result, the drop-out levels that maximize profits are not necessarily interior solutions. An exception is when all local bidders are active ($k = n$) since then the second line of (4.1) disappears.

Example 1 (continued). Suppose $n = 2$ and local bidders' values are uniformly distributed. For $k = 1$, Proposition 6 implies $\Pi_1^1(V, p) = (B_1^1(V) - p)(V_1 - 1)/(1 - p)$, which is maximized at $B_1^1(V) = \mathbf{1}_{V_1 \geq 1}$ (recall that the bidder drops out at $\max\{p, B_1^1(V)\}$, i.e. the bidder drops out immediately when $B_1^1(V) < p$). Therefore, the continuation profit is $\Pi_1^1(V, p) = (V_1 - 1)\mathbf{1}_{V_1 \geq 1}$ and the first-order condition for $k = 2$

$$B_2^1(V) = V_2 + \Pi_1^1(V, B_2^1(V)),$$

yields $B_2^1(V) = V_2 + (V_1 - 1)\mathbf{1}_{V_1 \geq 1}$. Note that the optimal drop-out levels are lower than those in the original Example 1, which reflects the aggravated exposure problem in this setup. In particular, the global bidder's drop-out levels no longer maximize expected welfare. This example can be extended to more than two local bidders when complementarities are extreme.

---

provisional winner for items 1 and 2 (ties are settled in favor of global bidders) as well as items $4, \ldots, n$, so the local bidder is assigned only item 3. The continuous format studied here corresponds to the limit $\Delta \to 0$.

[22] The activity rule for local bidders is modified so that they can compete for any item.

[23] In FCC auction #4, for instance, a single bidder drove up the prices on many items before dropping out of the auction and winning nothing.

[24] The Bayes-Nash equilibria derived in this section are supported by beliefs that global bidders' drop-out levels are based on the lowest current price. As a result, local bidders have no incentive to deviate from driving up the prices in a uniform manner (perfect arbitrage).

Proposition 7. *When $n \geq 2$ local bidders' values are uniformly distributed and complementarities are extreme, the global bidder's optimal drop-out level is given by*

$$B_n^1(V) = \begin{cases} 0 & \text{if } V_1 < n-1 \\ V - (n-1) & \text{if } V_1 \geq n-1 \end{cases}$$

and $B_k^1(V) = 1$ for $1 \leq k < n$ and $V \geq n-1$.

To summarize, the global bidder drops out of the auction immediately unless her value exceeds $n-1$.[25] Clearly, with many items this outcome is very inefficient since the expected value of all the items to the local bidders is only $n/2$. We next generalize this inefficiency result to more general distributions and valuation functions.

Proposition 8. *Under Assumptions 1' & 2, the global bidder wins fewer items than is socially optimal when there are complementarities. Compared to the VCG mechanism, the SAA is (i) inefficient, (ii) may yield less revenue, (iii) benefits local bidders, and (iv) hurts the global bidder.*

Example 2 (continued). Recall that, for $1 \leq \rho < 2$, the socially optimal fraction of items won by the global bidder is $\min(1, (\rho V)^{1/(2-\rho)})$, while in equilibrium, it is only $(\frac{1}{2}\rho V)^{1/(2-\rho)}$.[26] So the global bidder wins fewer items than is socially optimal even with linear valuations (demand reduction). For $\rho \geq 2$, the socially optimal fraction is $\mathbf{1}_{V \geq \frac{1}{2}}$ while in equilibrium the global bidder wins no items *irrespective of her value*.[27] Due to the aggravated exposure problem in this environment with complements and substitutes, the global bidder drops out immediately.

4.2. Fitting Problems

In this section and the next, we keep Assumption 1 (i.e. local bidders' interests are fixed) but we drop the convexity Assumption 2. In particular, we consider large auctions with two "medium-sized" global bidders who each demand fewer than all items. Complementarities are extreme, i.e. a global bidder has no value for less than α of the items, nor does a global bidder value additional items beyond an α fraction. Because of the latter, global bidders' valuations are no longer convex.[28]

Assumption 2' (regional complementarities). *For global bidder $j = 1, 2$, the value of obtaining a fraction κ of the items is*

$$\mathcal{V}_j(\kappa) = \begin{cases} V_j & \text{if } \kappa \geq \alpha \\ 0 & \text{if } \kappa < \alpha \end{cases}$$

for some $0 < \alpha < 1$.

---

[25] Even if one of the local bidders drops out at a price of 0, the global bidder still expects to pay n times the expected value of the expected value of the highest of $n-1$ local bidders' values. So the expected payment would be $n(n-1)/n$, which explains why $n-1$ is the threshold value for the global bidder to start bidding.

[26] The global bidder's objective is to maximize $V\kappa^\rho - \kappa^2$, which yields $(\frac{1}{2}\rho V)^{1/(2-\rho)}$ for $1 \leq \rho < 2$.

[27] For $\rho \geq 2$, the interior solution is a local minimum, while $\kappa = 0$ yields 0 and $\kappa = 1$ yields $V - 1 \leq 0$.

[28] Global bidders can compete for at most an α fraction of the items. As in Section 2, the prices for active items rise uniformly and a global bidder's strategy is characterized by a single drop-out level.

We first study the case $\alpha > 1/2$ so that there is a "fitting problem," i.e. it is not possible for both global bidders to win their desired fraction of items. As a result, global bidders that drop out have to pay for (worthless) items if fewer than $1 - \alpha$ local bidders are active.[29] This lowers the value of dropping out and may cause global bidders to stay in at prices that exceed their values. Such behavior can be mutually destructive since prices may become so high that even the winning global bidder incurs a loss.

Consider the case where only one global bidder is active and suppose this global bidder's value is low enough so that, in equilibrium, the global bidder drops out before a fraction α of the local bidders has dropped out. In this case, the optimal drop-out level is akin to that of Example 2 in Section 3. Define $V^*(\alpha) = \alpha E(v|F(v) \le \alpha)$ then $B^1(V) = F^{(-1)}(\alpha) \mathbf{1}_{V \ge V^*(\alpha)}$. Similarly, from (3.3), $B^2(V)$ is zero when $V < V^*(\alpha)$ and for $V \ge V^*(\alpha)$ it follows from

$$\int_{B^2(V)}^{F^{(-1)}(\alpha)} F(y)dy = \alpha F^{(-1)}(\alpha) - V$$

cf. Example 2. This solution, however, only applies when at least a fraction $1 - \alpha$ of the local bidders are active, which requires $B^2(V) \le F^{(-1)}(\alpha)$ or $V \le V^{**}(\alpha) = \alpha F^{(-1)}(\alpha)$.

We next determine $B^2(V)$ for $V \ge V^{**}(\alpha)$. Consider a global bidder with value V who bids as if her value is $V + \epsilon$. The effect of this deviation is twofold. First, with probability $\epsilon g(V)$ the other global bidder drops out in between and the deviation results in a winning payoff of $V - \alpha B^2(V)$ instead of a losing payoff of $-B^2(V)(F(B^2(V)) - \alpha)$. Second, with probability $(1 - G(V))$ the other global bidder has a higher value and the deviation simply raises the costs of the items that are won. Equating the marginal cost and benefit of the deviation yields

$$g(V)\Big(V - B^2(V)(2\alpha - F(B^2(V)))\Big) = (1 - G(V))\Big(B^2(V)(F(B^2(V)) - \alpha)\Big)'$$

The right side vanishes at $V = 1$, which implies $B^2(1) = 1/(2\alpha - 1) > 1$. The intuition is that when the price exceeds 1, local bidders are no longer active so the global bidder is assigned a fraction $1 - \alpha$ of the items when she drops out or a fraction α of the items when she wins. Hence, a global bidder with $V = 1$ continues to bid as long as the cost of winning an additional $2\alpha - 1$ of the items is less than 1. The global bidders' aggressive bids can cause losses, even for the winning bidder. For example, if global bidders have values close to 1 then the winning bidder's profit is $1 - \alpha/(2\alpha - 1) = -(1 - \alpha)/(2\alpha - 1) < 0$, which is also the profit for the losing bidder. The global bidders' aggressive bids do not necessarily translate into higher seller revenue.

Proposition 9. *Suppose Assumptions 1 & 2' hold and $1/2 < \alpha < 1$ so that there is a fitting problem. Compared to the VCG mechanism, the SAA is (i) inefficient, (ii) may*

---

[29] Due to the SAA's activity rule, global bidders will compete for exactly an α fraction of the items. The Bayes-Nash equilibrium considered here involves perfect arbitrage, i.e. global bidders switch among items and drive up prices uniformly (this equilibrium is supported by beliefs that global bidders' drop-out levels depend on the lowest current price). With $2\alpha > 1$, the losing global bidder, who fails to win α items at the end of the auction, will necessarily win some of the items once fewer than $1 - \alpha$ local bidders are active. We do not claim uniqueness, i.e. other (asymmetric) equilibria may exist.

yield less revenue (and yields less revenue when α is high enough), (iii) hurts local bidders, and (iv) creates a positive probability of losses for both losing and winning global bidders.

4.3. Market Sharing

When $\alpha \leq 1/2$, market sharing is possible and global bidders have an alternative to dropping out: they can "yield" by not bidding for the same items as the other global bidder and instead compete with the remaining local bidders to obtain a fraction α of the items. Compared to dropping out, yielding results in an extra payoff of

$$V - \int_{B^2(V)}^{F^{(-1)}(2\alpha)} v dF(v) \tag{4.2}$$

i.e. the value of winning the additional items minus their cost. Note that the additional payoff (4.2) is strictly increasing in V. Furthermore, (4.2) is negative when evaluated at $V = V^{**}(\alpha)$ and positive when evaluated at $V = 1$.[30] So there is a unique $V^{***}(\alpha)$, which satisfies $V^{**}(\alpha) < V^{***}(\alpha) < 1$, where the global switches from dropping out to yielding.

To determine the first-order condition for $V \geq V^{***}(\alpha)$, consider a global bidder with value V who bids as if her value is $V + \epsilon$. With probability $\epsilon g(V)$ the other global bidder drops out in between and the deviation results in a winning payoff of $V - \alpha B^2(V)$ instead of a yielding payoff of $V - \int_{B^2(V)}^{F^{(-1)}(2\alpha)} v dF(v) - B^2(V)(F(B^2(V)) - \alpha)$. Second, with probability $(1 - G(V))$ the other global bidder has a higher value and the deviation simply raises the costs of the $(F(B^2(V)) - \alpha)$ items that are eventually won. Equating the marginal cost and benefit yields

$$g(V) \left(\int_{B^2(V)}^{F^{(-1)}(2\alpha)} v dF(v) - B^2(V)(2\alpha - F(B^2(V))) \right)$$
$$= (1 - G(V))(F(B^2(V)) - \alpha)(B^2(V))'$$

The right side vanishes at $V = 1$, which implies that $B^2(1) = F^{(-1)}(2\alpha) < 1$. The intuition is that at a price of $F^{(-1)}(2\alpha)$ a fraction of 2α of the local bidders has dropped out, so global bidders of all types prefer to yield and divide the market rather than to compete further.

To summarize, in equilibrium, the price at which the global bidders divide the market is less than $F^{(-1)}(2\alpha)$. This low price contrasts with the price that would have resulted in a single-price clock auction, for instance, where both global bidders would have to pay $F^{(-1)}(2\alpha)$ to drive 2α local bidders out of the market. We next demonstrate that the price at which global bidders yield is also too low from a welfare viewpoint. In the VCG mechanism, the global bidder with the lowest value, Z, would obtain a fraction α of the items when

$$Z \geq V^o(\alpha) \equiv \int_{F^{(-1)}(\alpha)}^{F^{(-1)}(2\alpha)} t dF(t)$$

---

[30] Recall that $B^2(V^{**}(\alpha)) = F^{(-1)}(\alpha)$ so (4.2) evaluated at $V^{**}(\alpha)$ becomes $\int_{F^{(-1)}(\alpha)}^{F^{(-1)}(2\alpha)} (F(v) - 2\alpha) dv < 0$.

In equilibrium, the lowest-value global bidder obtains a fraction α of the items when $Z \geq V^{***}(\alpha)$, where $V^{***}(\alpha) < V^o(\alpha)$.[31] Hence, there are two value ranges where the lowest-value global bidder obtains too many items. When $V^{**}(\alpha) < Z < V^{***}(\alpha)$ she obtains a strictly positive fraction (although strictly less than α) while it is optimal that she wins nothing. When $V^{***}(\alpha) < Z < V^o(\alpha)$ she obtains an α fraction of the items, while it is optimal that she wins nothing.

Proposition 10. *Suppose Assumptions 1 & 2' hold and $0 < \alpha \leq 1/2$ so that market sharing is feasible. Compared to the VCG mechanism, the SAA is (i) inefficient, (ii) may yield less revenue, (iii) hurts local bidders, and (iv) benefits the global bidders who divide the market at low prices.*

Example 3. Suppose the local and global bidders' values are uniformly distributed. The top-right panel of Figure 2 shows the optimal drop-out level, $B^2(V)$, for $\alpha = 3/4$ and $0 \leq V \leq 1$, and illustrates the degree to which mutually destructive bidding can occur. For instance, a bidder with a value of 1 is willing to stay in until the price levels reaches 2, at which point the bidder wins but her profit is $1 - 2(3/4) < 0$. The top-left panel of Figure 2 shows the global bidder's optimal drop-out levels for $\alpha = 2/5$ and $0 \leq V \leq 1$, and illustrates the extent to which global bidders divide the market at low prices. For instance, a global bidder with a value of $V = 1/2$ yields at a per-item price of approximately 0.6, at which point the cost of the 2/5 fraction of items the global bidder is competing for is roughly half the value.

The middle and bottom panels of Figure 2 illustrate the performance differences of the SAA and VCG mechanisms for $0 \leq \alpha \leq 1$. The middle-left panel shows welfare differences, the middle-right panel shows revenue differences, the lower-left panel shows differences in the global bidders' profits, and the lower-right panel shows differences in local bidders' profits. In each of these four panels, the relevant difference is normalized by the (maximal possible) welfare of the VCG mechanism. Note that both welfare and revenue are lower in the SAA for all values of α, i.e. whether or not market sharing is feasible. Furthermore, the local bidders are always worse off while the global bidders benefit only when market sharing is feasible.

5. Conclusions

The simultaneous ascending auction (SAA) is generally considered one of the most successful applications of game theory. Since its initial use in the Personal Communication Services (PCS) auction conducted by the FCC in 1994, the SAA has become the dominant format to conduct large-scale spectrum auctions and it has raised hundreds of billions worldwide. In their review of the process that shaped the first PCS auction, McAfee, McMillan, and Wilkie [16] recall that while the economic literature generally favored an ascending format, "the case was far from transparent." Indeed,

---

[31] Since $V^{***}(\alpha) = \int_{B^2(V^{***}(\alpha))}^{F^{(-1)}(2\alpha)} t dF(t) < \int_{B^2(V^{**}(\alpha))}^{F^{(-1)}(2\alpha)} t dF(t) = V^o(\alpha)$ where the first equality follows from the definition of $V^{***}(\alpha)$, the inequality follows since $V^{***}(\alpha) > V^{**}(\alpha)$, and the second equality follows since $B^2(V^{**}(\alpha)) = F^{(-1)}(\alpha)$.

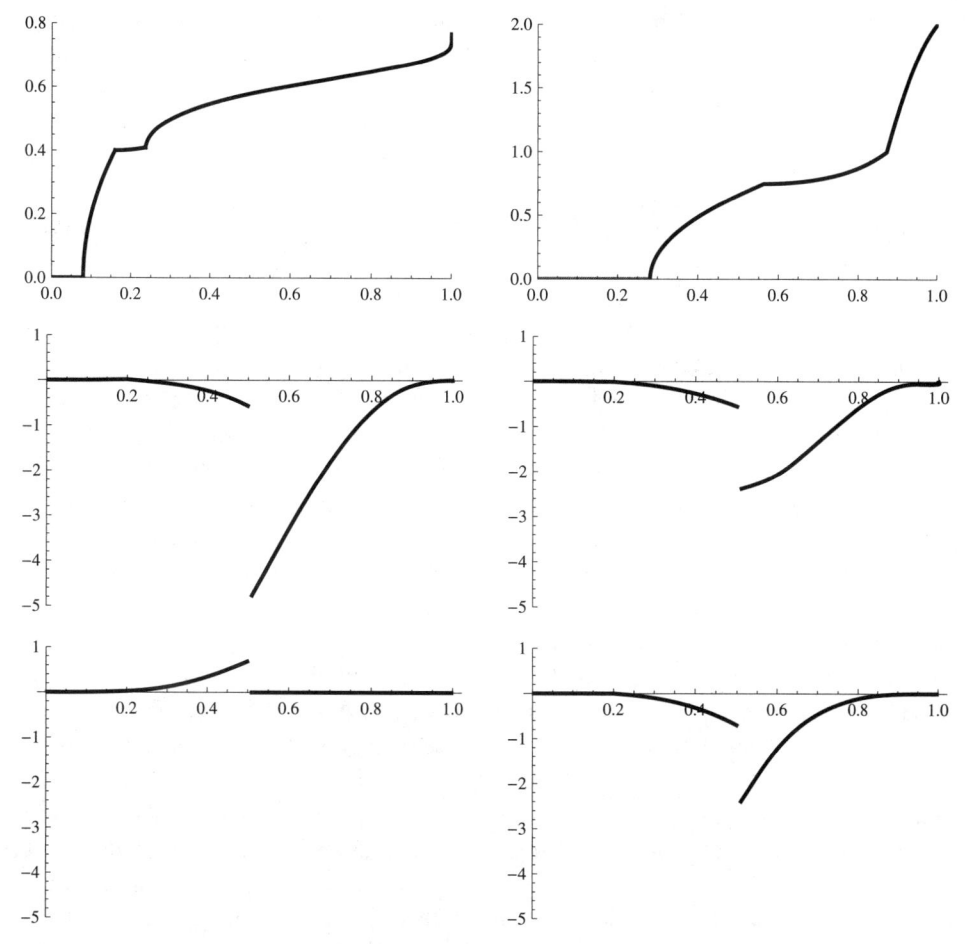

Figure 2. The Optimal Drop-Out Level, $B^2(V)$, for $\alpha = 2/5$ (top left) and $\alpha = 3/4$ (top right). Differences in Welfare (middle left), Revenue (middle right), Global Bidders' Profits (bottom left), and Local Bidders' Profits (bottom right) as Percentages of Maximal Welfare for $0 \leq \alpha \leq 1$.

the typical motivation for dynamic formats is that in common-value environments they allow bidders to refine their value estimates by observing others' bids (Milgrom and Weber [17]). However, a major concern in many of the spectrum auctions conducted to date is that due to the item-by-item competition that underlies the SAA, global bidders interested in acquiring large combinations face an *exposure problem* – when competing aggressively for a combination of items, a global bidder may incur a loss when winning an inferior subset.

This paper introduces a tractable model of local/global competition that allows for a general Bayes-Nash equilibrium analysis of the exposure problem. We first assume each local bidder values a different item and global bidders values are convex. Our setup is general in that it allows for arbitrary numbers of local and global bidders, arbitrary distributions of local and global bidders' values over one-dimensional types, and a general convex valuation function to capture global bidders' value complementarities. For this environment, we demonstrate that the exposure problem results in perverse revenue properties of the SAA. In particular, the SAA may yield non-core outcomes in which local bidders obtain the items at low prices – prices that may fall short of Vickrey prices. Moreover, the seller's revenue may decline as more bidders enter the auction. These shortcomings, which are well known for the benchmark Vickrey-Clark-Groves (VCG) mechanism (e.g. Ausubel and Milgrom [4]), were hitherto not known for the SAA simply because a general equilibrium analysis did not exist.

The similarity between the SAA and VCG mechanisms becomes even more pronounced in auctions with a large number of items, as is the case in many FCC spectrum auctions. In the limit, the SAA is fully efficient and yields identical profits for the bidders and the seller as the VCG mechanism. Importantly, it is the dynamic nature of the SAA that enables the global bidders to avoid the exposure problem as it allows the global bidder with the second-highest value to drop out before any of the efficient local bidders do. This is not possible in a sealed-bid version of the SAA (Krishna and Rosenthal [15]) where global bidders suffer from the exposure problem and generally obtain fewer items than is socially optimal. The advantage of the dynamic SAA over sealed-bid formats in this setting is an important new insight that complements the usual motivation for the SAA when common-value elements are present.

We also consider an environment where the items are perfect substitutes for local bidders. In this case, local bidders switch licenses and drive up prices in a uniform manner until they drop out. Such behavior is commonly observed in actual spectrum auctions, e.g. Ausubel and Cramton [3] note "... in the SAA used for spectrum licenses, there is a strong tendency toward arbitrage of the prices for identical items. Indeed, in the PCS auction of July 1994, similar licenses were on average priced within 0.3 percent of the mean price for that category of license." We demonstrate that the exposure problem is much more severe in this case and that the global bidder always wins fewer items than is socially optimal. Indeed, in large auctions it can be optimal for the global bidder to drop out right away, with dramatic effects for revenue and efficiency of the SAA. Hence, for this arguably more realistic setting with substitutes and complements, it is the dynamic feature of the SAA that creates a severe exposure problem for global bidders because it allows high-value local bidders to drive the prices up on all licenses. In contrast, the impact of one or more high-value local bidders would be limited in a sealed-bid format.

Finally, we relax the assumption that global bidders' values are convex and consider an environment with medium-sized global bidders who demand fewer than all items. We show that the SAA is generally inefficient whether or not market sharing is feasible. When it is, high-value global bidders follow a strategy of mutual forbearance to divide the market at low prices. When it is not, high-value global bidders engage in mutually destructive bidding, which drives out high-value local bidders and may cause even the winning global bidder to incur a loss. In both cases, the SAA yields lower welfare and typically less seller revenue than the VCG mechanism.

To summarize, our approach enables us to quantify the adverse effects of the exposure problem in the SAA across a broad array of bidding environments. We find that the SAA generally performs worse than the benchmark VCG mechanism, e.g. when both substitutes and complements play a role, when there are fitting problems, or when global bidders can divide the market. Our findings contrast with the superior performance of the SAA in a substitutes-only environment (e.g. Milgrom [18]), and reinforce the interest of policy makers in more flexible auction institutions that accommodate bidders' synergistic preferences.

Acknowledgments

We gratefully acknowledge financial support from the European Research Council (ERC Advanced grant, ESEI-249433). We thank Oleg Baranov, Alexey Kushnir, participants at the "Rothkopf Memorial Conference" (Penn State, June 2009), the "International Game Theory Meetings" (Stony Brook, July 2009), the "Wilson Scholar Lecture" (University of Arizona, November 2009), and the "Industrial Organization: Theory, Empirics, and Experiments," (Lecce, July 2011), as well as seminar participants at the University of Edinburgh (February 2010), Carlos III Madrid (March 2010), UCL (May 2010), Bocconi University (January, 2011), University of Munich (February 2011), Paris School of Economics (March, 2011), Collegio Alberto (April, 2011), University of Cologne (May, 2011), University of Basel (May, 2011), GREQAM Marseille (May, 2011), UPF Barcelona (October 2011) for helpful comments.

A. Appendix: Proofs

Proof of Proposition 1. Recall that

$$B_k^1(V) = \text{argmax}_b \left\{ \int_p^b (\Pi_{k-1}^1(V, v_k) + (V_k - v_k))dF_k(v_k|p) \right\} \quad (A.1)$$

The necessary first-order condition is $B_k^1(V) = V_k + \Pi_{k-1}^1(V, B_k^1(V))$, which has a unique solution since the right side is decreasing in $B_k^1(V)$ and the left side is increasing. Moreover, the solution corresponds to a maximum since the second derivative of the objective function in (A.1) evaluated at $b = B_k^1(V)$ is negative. *Q.E.D.*

Proof of Proposition 2. We prove, by induction, that $W_k^1(V, p) = \Pi_k^1(V, p) + kE(v|v > p)$ for all $k \geq 1$. In the main text we have shown it is true for $k = 1$. For $k \geq 2$ we have:

$$W_k^1(V, p) = \int_p^{B_k^1(V)} (W_{k-1}^1(V, v_k) + V_k)dF_k(v_k|p) + \int_{B_k^1(V)}^1 \sum_{i=1}^k v_i dF_k(v_k|p)$$

$$= \int_p^{B_k^1(V)} (\Pi_{k-1}^1(V, v_k) + V_k - v_k)dF_k(v_k|p)$$

$$+ \int_p^{B_k^1(V)} ((k-1)E(v|v > v_k) + v_k)dF_k(v_k|p) + \int_{B_k^1(V)}^1 \sum_{i=1}^k v_i dF_k(v_k|p)$$

$$= \Pi_k^1(V, p) + kE(v|v > p)$$

In the first line, the second term on the right side occurs when the global bidder drops out before the lowest-value local bidder (among the k active local bidders), in which case all remaining items are awarded to the local bidders. The first term corresponds to the case where the local bidder drops out first (at price level v_k), in which case the social planner optimizes the continuation welfare $W^1_{k-1}(V, v_k)$ with one fewer local bidder. In going from the first to the second line we used the induction hypothesis, and in going from the second to the third line we used the recursive property of the global bidder's profit, see Proposition 1. Since welfare and the global bidder's profit differ only by a constant, $B^1_k(V)$ is chosen in a socially optimal manner. Q.E.D.

Proof of Proposition 3. Recall that $W^{SAA}(V) = W^1_n(V, 0)$ and $\Pi^{SAA}(V) = \Pi^1_n(V, 0)$ differ by $nE(v)$, see the proof of Proposition 2. Suppose the VCG mechanism assigns k of the n licenses to the global bidder for which she pays the opportunity cost, which is the sum of the k lowest local bidders' values. Let $\hat{\Pi}^{VCG}$ and $\hat{W}^{VCG}$ denote the global's profit and welfare respectively as a function of the entire profile of bidders' valuations: $\hat{\Pi}^{VCG} = \sum_{\ell=1}^k (V_{n-\ell+1} - v_{n-\ell+1})$ and $\hat{W}^{VCG} = \sum_{\ell=1}^k V_{n-\ell+1} + \sum_{\ell=k+1}^n v_{n-\ell+1}$, so $\hat{W}^{VCG} = \hat{\Pi}^{VCG}(V) + \sum_{k=1}^n v_k$. Taking expectations with respect to local bidders' value shows that the global's expected profit $\Pi^{VCG}(V)$ and expected welfare $W^{VCG}(V)$ differ by $nE(v)$. This establishes (2.3). The equality in (2.4) now follows from the "accounting identity" $R = W - \Pi - \pi$. Q.E.D.

Proof of Proposition 4. When $(n - k)$ local bidders have dropped out and k are still active, a global bidder's optimal drop-out level is determined by

$$B^2_k(V) = \operatorname{argmax}_b \left\{ \int_{(B^2_k)^{(-1)}(p)}^{(B^2_k)^{(-1)}(b)} \left(\Pi^1_k(V, B^2_k(W)) + \sum_{\ell=k+1}^n (V_\ell - B^2_k(W)) \right) dG(W)^{K-1} \right\}$$

(A.2)

The necessary first-order condition yields (2.5), which has a unique solution since the left side is strictly decreasing in $B^2_k(V)$. Moreover, the solution corresponds to a maximum since the second derivative of the objective function in (A.2) evaluated at $b = B^2_k(V)$ is negative. Q.E.D.

Proof of Proposition 2 (continued). We first prove that competition among global bidders lowers drop-out levels.

Lemma A1. $B^2_k(V) \leq B^1_{k+1}(V)$ for $k = 0, 1, \ldots, n-1$.

Proof. For $k = n - 1$, this follows from (2.6). To prove the lemma for $k \leq n - 2$ note that when $B^2_k(V) = B^1_{k+1}(V)$ the left-side of (2.5) is equal to

$$\Pi^1_k(V, B^1_{k+1}(V)) + \sum_{\ell=k+1}^n (V_\ell - B^1_{k+1}(V)) = \sum_{\ell=k+2}^n (V_\ell - B^1_{k+1}(V))$$

$$\leq \sum_{\ell=k+2}^n (V_\ell - V_{k+1}) \leq 0$$

where the first equality follows since $\Pi^1_k(V, B^1_{k+1}(V)) = B^1_{k+1}(V) - V_{k+1}$ (see Proposition 1), the first inequality follows since $B^1_{k+1}(V) \geq V_{k+1}$, and the second inequality follows from Assumption 2. Since the left side of (2.5) is strictly decreasing in $B_k^2(V)$ the above inequality implies that $B_k^2(V) \leq B_{k+1}^1(V)$ for $k = 0, 1, \ldots, n - 1$. Q.E.D.

Next, consider the global bidder with the highest value, V, and suppose in the optimal allocation this global bidder is assigned k^* items. Once other global bidders have dropped out, social optimality follows from Proposition 2, i.e. $B_k^1(V) > v_k$ for $k = n - k^* + 1, \ldots, n$ and $B_k^1(V) < v_k$ for $k = 1, \ldots, n - k^*$. We need to show that all other global bidders drop out before $B_{n-k^*+1}^1$. This follows since for all $V' < V$ and $k = n - k^* + 1, \ldots, n$, we have $B_k^K(V') = B_k^2(V') \leq B_{k+1}^1(V') \leq B_k^1(V') < B_k^1(V) \leq B_{n-k^*+1}^1(V)$. Q.E.D.

Proof of Proposition 5. The best global bidder wins $F(B^1(V)) = \kappa^*(V)$ items so welfare is maximized: $W^{SAA}(V) = W(V)$. To determine the best global bidder's profit note that she wins an optimal fraction of items $F(B^1(V))$, which she values at $\mathcal{V}(F(B^1(V)))$, and for which she pays

$$\int_0^V \left\{ B^2(Z)F(B^2(Z)) + \int_{B^2(Z)}^{B^1(V)} y \, dF(y) \right\} dG(Z|V)^{K-1}$$

where $G(Z|V) = G(Z)/G(V)$. Here the first term in the integral corresponds to the items the best global bidder wins (all at once) when the second-best global bidder drops out at $B^2(Z)$, and the second term corresponds to the items she wins when local bidders subsequently drop out between $B^2(Z)$ and $B^1(V)$. Using (3.1) and (3.3), we can rewrite the global bidder's profit as

$$\Pi^{SAA}(V) = \int_0^V (W(V) - W(Z)) dG(Z|V)^{K-1}$$

(Note that for $K = 1$, the expression reduces to $W(V) - W(0) = W(V) - E(v)$.) Local bidders with values higher than $B^1(V)$ win an item at price $B^1(V)$ and the total profits for the local bidders as a group therefore are

$$\pi^{SAA}(V) = \int_{B^1(V)}^1 (v - B^1(V)) dF(v) = \int_{B^1(V)}^1 (1 - F(v)) dv$$

Finally, revenue follows from $R = W - \Pi - \pi$. It is standard to verify that the VCG expressions for welfare, revenue, and profits are identical to those for the SAA. Q.E.D.

Proof of Proposition 6. To establish the recursive relation, note that if at price level p there are k active local bidders then the global bidder will at least win $n - k$ items since $n - k$ local bidders have already dropped out. If the lowest of the active local bidders' values, v_k, is less than $B_k^1(V)$, then the global bidder's gain is the continuation profit $\Pi_{k-1}^1(V, v_k)$ plus $V_k - v_k$ for the additional item she wins, and she has to pay an extra $(n - k)(v_k - p)$ for the items she would anyhow have won. If v_k is greater than $B_k^1(V)$ then the global bidder wins no additional items but the price of the $(n - k)$ items she does win is raised from p to $B_k^1(V)$. Q.E.D.

Proof of Proposition 7. Recall that

$$B_k^1(V) = \text{argmax}_b \left\{ \int_p^b \left(\Pi_{k-1}^1(V, v_k) + (V_k - v_k) \right) dF_k(v_k|p) \right.$$
$$\left. - (n-k) \int_p^1 (\min(v_k, b) - p) dF_k(v_k|p) \right\}$$

The derivative of the term between curly brackets, evaluated at $b = B_k^1(V)$, is proportional to

$$\Pi_{k-1}^1(V, B_k^1(V)) + V_k - B_k^1(V) - \frac{n-k}{k}(1 - B_k^1(V)) \qquad (A.3)$$

For $k = 1$ we have $V_1 = V$ and $\Pi_0^1 = 0$ so the slope is $V - (n-1) + (n-2)B_1^1(V)$, which is positive for $V \geq n-1$. So profits are increasing in $B_1^1(V)$ and, hence, maximized at $B_1^1(V) = 1$. We next prove, by induction, that for $1 \leq k < n$ and $V \geq n-1$, the global bidder bids up to 1. Suppose with $k-1$ active local bidders, the global bidder bids up to 1. To derive $\Pi_{k-1}^1(V, B_k^1(V))$ note that the global bidder pays $B_k^1(V) + (1 - B_k^1(V))(k-1)/k$ for each of the $k-1$ additional items she wins plus an additional $(1 - B_k^1(V))(k-1)/k$ for each of the $n - (k-1)$ items she is already winning:

$$\Pi_{k-1}^1(V, B_k^1(V)) = V - (n-1) + (n-k)B_k^1(V) + \frac{n-k}{k}(1 - B_k^1(V))$$

For $k > 1$, (A.3) becomes (recall that complementarities are extreme so $V_{k>1} = 0$)

$$V - (n-1) + (n-k-1)B_k^1(V)$$

So the slope is positive and the profit increasing in $B_k^1(V)$ when $V \geq n-1$ and $k \leq n-1$. Hence, $B_k^1(V) = 1$ for $1 \leq k < n$ and $V \geq n-1$. Furthermore, for $k = n$ and $V \geq n-1$, the optimal bid follows from the first-order condition that the slope is zero: $B_n^1(V) = V - (n-1)$. Q.E.D.

Proof of Proposition 8. When local bidders treat the items as substitutes, the global bidder's optimal drop-out level satisfies $B(V) = \text{argmax}_b(\mathcal{V}(F(b)) - bF(b))$, compared to $B^1(V) = \text{argmax}_b(\mathcal{V}(F(b)) - \int_0^b v dF(v))$, when each local bidder is interested in a different item. Note that the objective functions in the maximization problems differ by $\int_0^b F(v)dv$ so the solutions differ unless $B(V) = B^1(V) = 0$, which can hold only for low values of V. When $B(V) \neq B^1(V)$, the first optimization problem implies

$$\mathcal{V}(F(B(V))) - B(V)F(B(V)) > \mathcal{V}(F(B^1(V))) - B^1(V)F(B^1(V))$$

and the second optimization problem implies

$$\mathcal{V}(F(B^1(V))) - \int_0^{B^1(V)} v dF(v) > \mathcal{V}(F(B(V))) - \int_0^{B(V)} v dF(v)$$

Hence, we have

$$B^1(V)F(B^1(V)) - B(V)F(B(V)) > \mathcal{V}(F(B^1(V))) - \mathcal{V}(F(B(V)))$$

$$> \int_{B(V)}^{B^1(V)} v dF(v)$$

$$= B^1(V)F(B^1(V)) - B(V)F(B(V)) - \int_{B(V)}^{B^1(V)} F(v) dv$$

which implies $B(V) < B^1(V)$. Hence, the global bidder wins fewer items ($F(B(V))$) than is socially optimal ($F(B^1(V))$). Also, the lower bid of the global bidder implies higher profits for the local bidders. An envelope theorem argument implies that the global bidder's profit is lower since the fraction of items the global bidder wins is less. Finally, revenue may be lower (see, e.g., Example 2 continued) or higher depending on distributional assumptions. Q.E.D.

Proof of Proposition 9. The SAA and VCG yield the same allocations, and the same profits, when the lowest of the two global bidders' values, Z, is less than $V^{**}(\alpha)$. When $Z \geq V^{**}(\alpha)$, the lowest-value global bidder bids up to $B^2(Z) > F^{(-1)}(\alpha)$ thereby driving local bidders with values between $F^{(-1)}(\alpha)$ and $B^2(Z)$ out of the market. Hence, the welfare difference is

$$W^{VCG} - W^{SAA} = \int_{V^{**}(\alpha)}^{1} \int_{F^{(-1)}(\alpha)}^{B^2(Z)} v dF(v) dG_{\min}(Z)$$

where $G_{\min}(Z) = 1 - (1 - G(Z))^2$. Note that the right side is strictly positive. Furthermore, if the lowest-value global bidder had dropped out at $F^{(-1)}(\alpha)$, the local bidders' profit would have been $\int_{F^{(-1)}(\alpha)}^{1}(y - F^{(-1)}(\alpha))dF(y) = \int_{F^{(-1)}(\alpha)}^{1}(1 - F(y))/f(y)dF(y)$. Instead, the local bidders' profit is $\int_{B^2(Z)}^{1}(y - B^2(Z))dF(y) = \int_{B^2(Z)}^{1}(1 - F(y))/f(y)dF(y)$. The difference

$$\pi^{VCG} - \pi^{SAA} = \int_{V^{**}(\alpha)}^{1} \int_{F^{(-1)}(\alpha)}^{B^2(Z)} \frac{1 - F(v)}{f(v)} dF(v) dG_{\min}(Z)$$

is strictly positive. In contrast, equality of the global bidders' profits, $\Pi^{VCG} = \Pi^{SAA}$, follows from an envelope argument since the probability of obtaining α items is the same under the VCG and SAA mechanisms (i.e. the probability that a global bidder with value $V \geq V^*(\alpha)$ obtains α items is $G(V)$). Finally, the revenue difference follows from the identity $R = W - \Pi - \pi$:

$$R^{VCG} - R^{SAA} = \int_{V^{**}(\alpha)}^{1} \int_{F^{(-1)}(\alpha)}^{B^2(Z)} \left(v - \frac{1 - F(v)}{f(v)}\right) dF(v) dG_{\min}(Z)$$

The revenue difference is intuitive: under SAA, the seller no longer collects the marginal revenues, $MR(V) \equiv v - (1 - F(v))/f(v)$, for local bidders with values between $F^{(-1)}(\alpha)$ and $B^2(Z)$. These marginal revenues can be negative for low values of v but are positive for high values of v. For example, for uniformly distributed values, the marginal revenues are $MR(v) = 2v - 1 > 0$ for $v > 1/2$. Hence, in the uniform case, the seller's revenue under SAA is always lower than under VCG since

$F^{(-1)}(\alpha) = \alpha > 1/2$. More generally, a sufficient (but not necessary) condition for revenues to be higher under VCG is that $\alpha > \max(1/2, F(MR^{(-1)}(0)))$. Q.E.D.

Proof of Proposition 10. The SAA and VCG yield the same allocations, and the same profits, when the lowest of the two global bidders' values, Z, is less than $V^{**}(\alpha)$. The difference in welfare when $Z \geq V^{**}(\alpha)$ follows from the discussion in the main text (see the proof of Proposition 9):

$$W^{VCG} - W^{SAA} = \int_{V^{**}(\alpha)}^{V^{***}(\alpha)} \int_{F^{(-1)}(\alpha)}^{B^2(Z)} v dF(v) dG_{\min}(Z) + \int_{V^{***}(\alpha)}^{V^o(\alpha)} (V^o(\alpha) - Z) dG_{\min}(Z)$$

which is strictly positive. Similarly, the difference in local bidders' profits is

$$\pi^{VCG} - \pi^{SAA} = \int_{V^{**}(\alpha)}^{V^{***}(\alpha)} \int_{F^{(-1)}(\alpha)}^{B^2(Z)} \frac{1 - F(v)}{f(v)} dF(v) dG_{\min}(Z)$$
$$+ \int_{V^{***}(\alpha)}^{V^o(\alpha)} \int_{F^{(-1)}(\alpha)}^{F^{(-1)}(2\alpha)} \frac{1 - F(v)}{f(v)} dF(v) dG_{\min}(Z)$$

which is strictly positive. To understand the difference in the global bidders' profit, note that the SAA results in a higher probability of obtaining α items than the VCG mechanism. In particular, under the SAA mechanism this probability is $G(V)$ for $V < V^{***}(\alpha)$ and 1 for $V \geq V^{***}(\alpha)$, compared to $G(V)$ for $V < V^o(\alpha)$ and 1 for $V \geq V^o(\alpha)$ under the VCG mechanism. The global bidders' profit difference is therefore

$$\Pi^{SAA} - \Pi^{VCG} = 2 \int_{V^{***}(\alpha)}^{1} \int_{V^{***}(\alpha)}^{\min(V, V^o(\alpha))} (1 - G(Z)) dZ dG(V)$$
$$= \int_{V^{***}(\alpha)}^{V^o(\alpha)} \frac{1 - G(Z)}{g(Z)} dG_{\min}(Z)$$

which is strictly positive. Finally, the expression for the revenue difference follows from $R = W - \Pi - \pi$:

$$R^{VCG} - R^{SAA} = \int_{V^{**}(\alpha)}^{V^{***}(\alpha)} \int_{F^{(-1)}(\alpha)}^{B^2(Z)} \left(v - \frac{1 - F(v)}{f(v)} \right) dF(v) dG_{\min}(Z)$$
$$+ \int_{V^{***}(\alpha)}^{V^o(\alpha)} \int_{F^{(-1)}(\alpha)}^{F^{(-1)}(2\alpha)} \left(v - \frac{1 - F(v)}{f(v)} \right) dF(v) dG_{\min}(Z)$$

which can be either positive (see, e.g., Example 3) or negative depending on distributional assumptions. Q.E.D.

References

[1] G. L. Albano, F. Germano, S. Lovo, Ascending Auctions for Multiple Objects: the Case for the Japanese Design, Econ. Theory 28 (2006) 331–355.

[2] L. M. Ausubel, P. Cramton, R. P. McAfee, J. McMillan, Synergies in Wireless Telephony: Evidence from the Broadband PCS Auctions, J. Econ. Manage. Strat. 6 (1997) 497–527.

[3] L. M. Ausubel, P. Cramton, Demand Reduction and Inefficiency in Multi-Unit Auctions, working paper (2002).
[4] L. M. Ausubel, P. Milgrom, The Lovely But Lonely Vickrey Auction, in P. Cramton, Y. Shoham, and R. Steinberg, eds., Combinatorial Auctions (2006), MIT Press.
[5] P. Bajari, J. Fox, Measuring the Efficiency of an FCC Spectrum Auction, Amer. Econ. J.: Microeconomics 5(1) (2013) 100–147.
[6] C. Brunner, J. K. Goeree, C. A. Holt, J. O. Ledyard, An Experimental Test of Flexible Combinatorial Spectrum Auction Formats, Amer. Econ. J.: Microeconomics 2 (2010) 39–57.
[7] S. Brusco, G. Lopomo, Collusion via Signaling in Simultaneous Ascending Bid Auctions with Heterogeneous Objects, with and without Complementarities, Rev. Econ. Stud. 69 (2002) 407–463.
[8] S. Brusco, G. Lopomo, Simultaneous Ascending Auctions with Complementarities and Budget Constraints, Econ. Theory 38 (2009) 105–124.
[9] M. M. Bykowsky, R. J. Hull, J. O. Ledyard, Mutually Destructive Bidding: The FCC Auction Design Problem, J. Regul. Econ. 17 (2000) 205–228.
[10] K. Chernomaz, D. Levin, Efficiency and Synergy in a Multi-Unit Auction with and without Package Bidding: an Experimental Study, Games Econ. Behav. 76 (2012) 611–635.
[11] O. Compte, P. Jehiel, Auctions and Information Acquisition: Sealed Bid or Dynamic Formats?, RAND J. Econ. 38 (2007) 355–372.
[12] H. A. David, H. N. Nagajara, Order Statistics, Wiley-Interscience (2003).
[13] J. K. Goeree, C. A. Holt, Hierarchical Package Bidding: A 'Paper & Pencil' Combinatorial Auction, Games Econ. Behav. 70 (2010) 146–169.
[14] F. Gul, E. Stacchetti, Walrasian Equilibrium without Complementarities, J. Econ. Theory, 87 (2000) 95–124.
[15] V. Krishna, R. W. Rosenthal, Simultaneous Auctions with Synergies, Games Econ. Behav. 17 (1996) 1–31.
[16] R. P. McAfee, J. McMillan, S. Wilkie, The Greatest Auction in History, in Better Living Through Economics, ed: John Siegfried, Harvard University Press (2010) 168–184.
[17] P. R. Milgrom, R. J. Weber, A Theory of Auctions and Competitive Bidding, Econometrica 50 (1982) 1089–1122.
[18] P. R. Milgrom, Putting Auction Theory to Work: the Simultaneous Ascending Auction, J. Polit. Econ. 108 (2000) 245–272.
[19] R. W. Rosenthal, R. Wang, Simultaneous Auctions with Synergies and Common Values, Games Econ. Behav. 17 (1996) 32–55.
[20] M. H. Rothkopf, Thirteen Reasons Why the Vickrey-Clarke-Groves Process Is Not Practical, Oper. Res. 55 (2007) 191–197.
[21] M. Rothkopf, A. Pekeč, R. Harstad, Computationally Manageable Combinatorial Auctions, Manage. Sci. 44 (1998) 1131–1147.
[22] B. Szentes, R. W. Rosenthal, Three-Object, Two-Bidder Simultaneous Auctions: Chopsticks and Tetrahedra, Games and Econ. Behav. 44 (2003) 114–133.
[23] B. Szentes, R. W. Rosenthal, Beyond Chopsticks: Symmetric Equilibria in Majority Auction Games, Games and Econ. Behav. 45 (2003) 278–295.
[24] C. Z. Zheng, Jump Bidding and Overconcentration in Decentralized Simultaneous Ascending Auctions, Games and Econ. Behav. 76 (2012) 648–664.

CHAPTER 3

The Efficiency of the FCC Spectrum Auctions

Peter Cramton[*]

From July 1994 to July 1996, the Federal Communications Commission (FCC) conducted nine spectrum auctions, raising about $20 billion for the U.S. Treasury. The auctions assigned thousands of licenses to hundreds of firms. These firms are now in the process of creating the next generation of wireless communication services. The questions addressed in this note are: Were the auctions efficient? Did they award the licenses to the firms best able to turn the spectrum into valuable services for consumers?

In addressing these questions, I focus on the narrow question of license assignment. Assignment is the second step in the process of utilizing spectrum. The first step is the allocation of the spectrum for licensing. The allocation defines the license (the frequency band, the geographic area, the time period, and the restrictions on use). I focus on assignment, since that is what the FCC spectrum auctions were asked to do. More general auctions that determine aspects of the allocation, such as band plans, have yet to be implemented.

Why should we care about auction efficiency? If resale is allowed, will not postauction transactions fix any assignment inefficiencies? The answer is "yes" in a Coasean world without transaction costs. However, transaction costs are not zero. Postauction transactions are often made difficult by strategic behavior between parties with private information and market power. The experience with the cellular lotteries is a case in point. It took a decade of negotiations and private auctions for the eventual service providers to acquire desirable packages of licenses from the lottery winners. Efficient auctions are possible before assignments are made but may become impossible after an initial assignment. The problem is that the license holder exercises its substantial market power in the resale of the license.[1] For this reason, it is important to get the assignment right the first time.

---

[*] I am grateful to the National Science Foundation for funding. I served as an auction advisor to the Federal Communications Commission and several firms (PageNet, MCI, Pocket Communications, and CD Radio) in various auctions. The views expressed are my own.
[1] Peter Cramton, Robert Gibbons & Paul Klemperer, Dissolving a Partnership Efficiently, 55 *Econometrica* 615 (1987).

All but two of the FCC auctions have used a simultaneous ascending design in which groups of related licenses are auctioned simultaneously over many rounds of bidding. In each round, bidders submit new higher bids on any of the licenses they desire, bumping the standing high bidder. The auction ends when a round passes without any bidding; that is, no bidder is willing to raise the price on any license. This design, proposed by Preston McAfee, Paul Milgrom, and Robert Wilson, is the natural extension of the English auction to multiple related goods.[2] Its advantage over a sequence of English auctions is that it gives the bidders more flexibility in moving among license packages as prices change. As one license gets bid up, a bidder can shift to an alternative that represents a better value. In this way, bidders are able to arbitrage across substitutable licenses. Moreover, they can build packages of complementary licenses using the information revealed in the process of bidding. Here I examine whether these potential advantages were realized.

The analysis is speculative. Since I do not observe the bidders actual valuations, it is impossible to say exactly how efficient the auctions were. Nonetheless, there is substantial evidence that the auctions were successful. I present this evidence and then identify the problems inherent in the auctions that suggest possible inefficiencies.

1. Evidence of Success

Revenue is a first sign of success. Auction revenues have been substantial, breaking $20 billion in the first 2 years.[3] Revenues have exceeded industry and government estimates.[4] The simultaneous ascending auction may be partially responsible for the large revenues. By revealing information in the auction process, the winner's curse is reduced, and the bidders can bid more aggressively.[5] Also, revenues may increase to the extent the design enables bidders to piece together more efficient packages of licenses.

Revenue maximization and efficiency are closely aligned goals. Indeed, in ex ante symmetric settings, the seller's expected revenue is maximized by assigning the goods to those with the highest values.[6] High prices are consistent with an efficient auction since only bidders with high values are willing to pay high prices. Moreover, efficiency-minded governments should care about the revenues raised at auction since auction revenues are less distortionary than the principal source of government revenues—taxation. Economists estimate that the welfare loss from increasing taxes in the United States is in the range of 17–56 cents per dollar of extra revenue raised.[7] Hence, in

---

[2] See McAfee, R. Preston, Auction Design of Personal Communication Services, attached to Comments of PacTel Corporation in PP Docket No. 93-253 (1993); and Milgrom, Paul & Robert Wilson, Affidavit of Paul R. Milgrom and Robert B. Wilson, attached to Comments of Pacific Bell and Nevada Bell in PP Docket No. 93-253 (1993).

[3] All auction data are available from the FCC's web site at www.fcc.gov.

[4] An exception is the Wireless Communications Services auction, which began April 15, 1997.

[5] Paul R. Milgrom & Robert J. Weber, A Theory of Auctions and Competitive Bidding, 50 *Econometrica* 1089 (1982).

[6] Lawrence M. Ausubel & Peter Cramton, Demand Reduction and Inefficiency in Multi-Unit Auctions (Working paper, Univ. Maryland 1996).

[7] Charles L. Ballard, John B. Shoven, and John Whalley, General Equilibrium Computations of the Marginal Welfare Costs of Taxes in the United States, 75 *Am. Econ. Rev.* 128 (1985).

designing the auction, the government should be willing to accept some assignment inefficiency if the gain in revenues is sufficiently large.

A second indicator of success is that the auctions tended to generate market prices. Similar items sold for similar prices. In the narrowband auctions, the price differences among similar licenses were at most a few percent and often zero. In the first broadband auction, where two licenses were sold in each market, the prices differed by less than one minimum bid increment in 42 of the 48 markets.

A third indicator of success is the formation of efficient license aggregations. Bidders did appear to piece together sensible license aggregations. This is clearest in the narrowband auctions. In the nationwide narrowband auction, bidders buying multiple bands preferred adjacent bands. The adjacency means that the buffer between bands can be used for transmission, thus increasing capacity. The two bidders that won multiple licenses were successful in buying adjacent bands. In the regional narrowband auction, the aggregation problem was more complicated. Several bidders had nationwide interests, and these bidders would have to piece together a license in each of the five regions, preferably all on the same band, in order to achieve a nationwide market. The bidders were remarkably successful in achieving these aggregations. Four of the six bands sold as nationwide aggregations. Bidders were able to win all five regions within the same band. Even in the two bands that were not sold as nationwide aggregations, bidders winning multiple licenses won geographically adjacent licenses within the same band.

Large aggregations were also formed in the Major Trading Area (MTA) broadband auction. Bidders tended to win the same band when acquiring adjacent licenses. The three bidders with nationwide interests appear to have efficient geographic coverage when one includes their cellular holdings. The footprints of smaller bidders also seem consistent with the bidders' existing infrastructures. In the C-block auction, bidders were able to piece together contiguous footprints, although many bidders were interested in stand-alone markets.

Two studies analyze the MTA and Basic Trading Area (BTA) auction data to see if there is evidence of local synergies.[8] Consistent with local synergies, these studies find that bidders did pay more when competing with a bidder holding neighboring licenses. Hence, bidders did bid for synergistic gains and, judging by the final footprints, often obtained them.

The two essential features of the FCC auction design are (1) the use of multiple rounds, rather than a single sealed bid, and (2) simultaneous, rather than sequential sales. The goal of both of these features is to reveal information and then give the bidders the flexibility to respond to the information. There is substantial evidence that the auction was successful in revealing extensive information. Bidders had good information about both prices and assignments at a point in the auction where they had the flexibility to act on the information.[9] The probability that a high bidder would eventually win the market was high at the midpoint of each auction. Also the correlation between midauction and final prices was high in each auction. Information about prices

---

[8] Lawrence M. Ausubel, Peter Cramton, R. Preston McAfee & John McMillan, Synergies in Wireless Telephony: Evidence from the Broadband PCS Auctions 6 *J. Econ. & Mgmt. Strategy* 497 (1997); and Patrick S. Moreton & Pablo T. Spiller, What's in the Air: Interlicense Synergies and Their Impact on the FCC's Broadband PCS License Auctions, 41 *J. Law & Econ.* 677–716 (1998).

[9] Peter Cramton, The FCC Spectrum Auctions: An Early Assessment, 6 *J. Econ. & Mgmt. Strategy* 431 (1997).

and assignments improved throughout each auction and was of high quality before bidders lost the flexibility to move to alternative packages.

The absence of resale also suggests that the auctions were highly efficient. In the first two years, there has been little resale. GTE is the one exception. Shortly after the MTA auction ended, GTE sold its MTA winnings for about what it paid for the licenses. Apparently there was a shift in corporate strategy away from Personal Communication Services (PCS) and toward cellular.

2. Potential Problems

Despite the apparent success of these auctions there are several potential problems that stand in the way of an efficient assignment.

Standard auctions at best ensure that the bidder with the highest private value wins, rather than the bidder with the highest social value. Private and social values can diverge in these auctions because the winners will be competing in a marketplace. One collection of winners may lead to a more collusive industry structure. For example, a license may be worth more to an incumbent than a new entrant, simply because of the greater market power the incumbent would enjoy without the new entrant. Recognizing this, the FCC limits the amount of spectrum any one firm can hold in any geographic area. Indeed, the FCC forbids incumbent cellular firms from bidding in their markets. Another example comes from the battle over technology standards in broadband PCS. Supporters of one standard may value a license more highly if it creates a hole in the footprint of a competing standard, putting the competing standard at a competitive disadvantage. This may have been an issue in the fight over Chicago in the C-block auction. Chicago was a major hole in the Global System for Mobile communications (GSM) footprint, but was already covered by the Code Division Multiple Access (CDMA) footprint. However, a GSM bidder only got the license after a long fight with the largest CDMA bidder.

A second issue stems from the fact that these are multiple-item auctions. The efficiency results from single-item auctions do not carry forward to the multiple-item setting. In an ascending auction for a single item, each bidder has a dominant strategy of bidding up to its private valuation. Hence, the item always goes to the bidder with the highest value. If, instead, two identical items are being sold in a simultaneous ascending auction, then a bidder has an incentive to stop bidding for the second item before its marginal valuation is reached. Continuing to bid for two items raises the price paid for the first.[10] As a result, the bidder with the highest value for the second item may be beaten by a bidder demanding just a single unit.

This logic is quite general. In multiunit uniform-price auctions, every equilibrium is inefficient.[11] Bidders have an incentive to shade their bids for multiple units, and the incentive to shade increases with the quantity being demanded. Hence, large bidders will shade more than small bidders. This differential shading creates an inefficiency. The small bidders will tend to inefficiently win licenses that should be won by the

---

[10] For settings where this effect is strong, see Richard Engelbrecht-Wiggans & Charles M. Kahn, Multiunit Auctions with Uniform Prices (Working paper, Univ. Illinois 1995).

[11] Ausubel & Cramton, supra note 6.

large bidders. The intuition for this result is analogous to why a monopolist's marginal revenue curve lies below its demand curve: bringing more units to market reduces the price paid on all units. In the auction, demanding more units raises the price paid on all units. Hence, the incentive to reduce demand.

To a large extent, the FCC spectrum auctions can be viewed as a uniform-price auction. Certainly, for licenses that are close substitutes, the simultaneous ascending auction has generated near uniform prices for similar items. Hence, large bidders in the spectrum auctions had an incentive to make room for smaller rivals.

Direct evidence of demand reduction was seen in the nationwide narrowband auction. The largest bidder, PageNet, reduced its demand from three of the large licenses to two, at a point when prices were still well below its marginal valuation for the third unit.[12] PageNet felt that, if it continued to demand a third license, it would drive up the prices on all the others to disadvantageously high levels.

An examination of the bidding in the MTA broadband auction is suggestive that the largest bidders did drop out of certain markets at prices well below plausible values. Although this could be tacit collusion, demand reduction is another explanation. Individual maximizing behavior would cause large bidders to make room for rivals to keep prices down.

Further evidence of demand reduction comes from the C-block auction. One large bidder defaulted on the down payment, so the FCC reauctioned the licenses. Interestingly, the licenses sold for 3 percent more than in the original auction. Consistent with demand reduction, NextWave, the largest winner in the C-block auction, bought 60 percent of the reauctioned spectrum. This occurred despite the fact that NextWave was not the second-highest bidder on any of these licenses in the original auction. NextWave was able to bid aggressively in the reauction, knowing that its bidding would have no affect on prices in the original auction.

In auctions for identical items, the inefficiencies of demand reduction can be eliminated with a Vickrey auction. Alternatively, one can use Lawrence Ausubel's ascending implementation of the static Vickrey auction, which has the additional advantages of an ascending-bid design.[13] However, the spectrum auctions were not for identical items, so Vickrey-type mechanisms were not practical.

There are good reasons to think that any inefficiencies caused by demand reduction in the FCC spectrum auctions are overstated. Demand reduction favors small bidders. Hence, small bidders, having less of an incentive to reduce demand, are able to win licenses they might otherwise not get. Demand reduction, then, fosters competition in the auction by encouraging the participation of small bidders. Perhaps more important, demand reduction may increase competition in the market for wireless services by increasing the number of competing firms.

The FCC, as mandated by Congress, has taken more direct steps to increase the number and diversity of winning firms. One-third of the broadband PCS spectrum has been set aside to small businesses. Preferences, both installment payments and bidding

---

[12] I was a member of the PageNet bidding team. See Peter Cramton, Money Out of Thin Air: The Nationwide Narrowband PCS Auction, 4 *J. Econ. & Mgmt. Strategy* 267 (1995), for a detailed analysis of this auction.

[13] Lawrence M. Ausubel, An Efficient Ascending-Bid Auction for Multiple Objects (Working paper, Univ. Maryland 1997).

credits, also have been given to designated bidders in the other auctions. One might think that these set asides and preferences would be a prominent source of inefficiency. They let small firms win licenses that might otherwise go to large firms with higher values. However, the auction experience of the first 2 years suggests that the inefficiencies from preferences are small or even negative. In the regional narrowband auction, there is strong evidence that preferences to firms controlled by women or minorities raised revenues.[14] The preferences stimulated competition, forcing the large firms to pay more than they otherwise would. In the C-block broadband auction, which was a set aside for small businesses, competition was so intense that it resulted in prices that were about 80 percent higher than in the earlier MTA auction.[15] Revenues were certainly stimulated by the preferences to the small firms. Given the dramatic increase in revenues, it is hard to imagine that the assignment of these licenses to small firms involved substantial inefficiencies. These small firms expressed valuations well in excess of what the big firms paid in the prior auction. Moreover, even if the small firms have lower values, the entrance of these small firms is likely to stimulate competition in the market for wireless services.

Another source of inefficiency in the spectrum auctions comes from the difficulties firms may have in piecing together efficient sets of licenses. The ability to form efficient aggregations is greatly enhanced by the excellent information about prices and assignments that is revealed in the auction process. Nonetheless, bidders may be hesitant to bid for synergistic gains they are unlikely to achieve. This exposure problem may lead to a failure to obtain efficient synergies. Similarly, bidders may bid for a synergistic gain, only to find they are inefficiently stuck with some individual licenses that do not make sense without others. Mark Bykowsky, Robert Cull, and John Ledyard emphasize this potential problem and recommend package bidding—being able to bid on a collection of licenses, rather than just on individual licenses.[16] Although I agree that package bidding may be a good idea in settings where synergies are both strong and varied among the bidders, I do not think that the early spectrum auctions fit this case. Bidders in the narrowband auctions had little difficulty in forming efficient aggregations. In the MTA broadband auction, it appears that the individual markets were sufficiently large to capture most local synergies.

The C-block broadband auction provided the greatest challenge to bidders since the BTA licenses were only about one-tenth the size of the MTA licenses and competition was much more intense. BTA-level synergies were certainly more important than the MTA-level synergies. However, I do not believe that the exposure problem stifled bidding or prevented firms from forming efficient aggregations. Early in the auction, competition was sufficient that it was easy to move from one package to another. When this was difficult, bidders would focus their bidding in the major markets that were key to synergistic gains. For example, Chicago is a key market in obtaining a strong Midwest presence. Hence, the fight over Chicago was resolved before the winning firm

---

[14] Ian Ayres & Peter Cramton, Pursuing Deficit Reduction through Diversity: A Case Study of How Affirmative Action at the FCC Increased Auction Competition, 48 *Stanford L. Rev.* 401 (1996).

[15] The 80 percent figure comes after netting out the 25 percent bidding credit and an additional credit of about 30 percent derived from the value of the favorable installment payments.

[16] Mark M. Bykowsky, Robert J. Cull, and John O. Ledyard, Mutually Destructive Bidding: The FCC Auction Design Problem (Working paper, California Inst. Technology 1995).

would bid seriously for the smaller complementary licenses neighboring Chicago. As a result of this strategy, major markets tended to receive final bids before the smaller markets.

Several firms did acquire clusters of adjacent licenses in the C-block auction. However, the bidding of the largest bidder, NextWave, suggests that local synergies were not large. NextWave pursued a strategy of acquiring major markets around the United States. Spending nearly $5 billion, it had the resources to instead acquire large contiguous clusters in a few parts of the country but chose not to do so. NextWave's strategy would not make sense if local synergies were large at the BTA level.

A potentially serious inefficiency in the C-block auction was speculative bidding caused by overly attractive installment payment terms. Bidders only had to put down 5 percent of their bids at the end of the auction, a second 5 percent at the time of license award, and then quarterly installment payments at the 10-year Treasury rate with interest-only payments for the first 6 years. These attractive terms favor bidders that are speculating in spectrum. If prices go up, the speculators do well; if prices fall, the speculators can walk away from their down payments. Indeed, spectrum prices did fall after the C-block auction, and most of the large bidders in the C-block auction defaulted on the payments. As a result of this experience, the FCC no longer offers installment payments. Bids must be paid in full when the licenses are awarded.

A final source of inefficiency comes, not from bad assignments, but from delayed assignments. Each month of delay means a loss of consumer surplus. The simultaneous ascending auctions took a significant amount of time to conduct. Relatively simple auctions, like the nationwide narrowband auction, were done in a week. However, the more complex auctions with hundreds of bidders and licenses (the C-block, Multipoint Distribution Service (MDS), and Specialized Mobile Radio (SMR) auctions) took about 80 bidding days.

The simultaneous ascending auction has a number of parameters (minimum bid increments, activity requirements, and rounds per day) that let the FCC control the pace of the auction. The parameters are adjusted during the auction to balance the goals of a timely and desirable assignment. The bidders need time to adjust strategies in light of information revealed in the bidding. Too much haste may lead to bidder error and inefficient assignments. Time also may be needed for bidders to line up additional capital if prices are higher than expected. Certainly, these spectrum auctions could have been conducted more quickly, but probably not without reducing the efficiency of the assignment.

3. Conclusion

Any auction would look good relative to the FCC's past experience with comparative hearings and lotteries. Hence, it is remarkable that the FCC chose an innovative and untested design to auction the spectrum. Fortunately, there is now substantial evidence that the simultaneous ascending auction worked well. It raised large revenues. It revealed critical information in the process of bidding and gave bidders the flexibility to adjust strategies in response to new information. As a result, similar licenses sold for similar prices, and bidders were able to piece together sensible sets of licenses.

The setting of the spectrum auctions is too complex to guarantee full efficiency. Bidders with the highest private values may not have the highest social values. To keep prices low, large bidders may reduce demand, inefficiently making room for smaller rivals. Preferences for designated bidders may distort assignments. And bidders may hesitate to bid for synergistic combinations for fear of not obtaining the synergies. Nonetheless, an examination of the bidding suggests that these problems, although present, probably did not lead to large inefficiencies. Moreover, the cures to these problems have side-effects that may be worse than the disease.

The spectrum auctions are a major step toward creating a market for spectrum. The greatest room for improvement lies, not in the assignment of licenses, but in the allocation process. Some allocations, like PCS, allow flexible use, but others, such as broadcasting, do not. Further steps need to be taken to assure that market forces, not political lobbying, determine spectrum use.

References

Ausubel, Lawrence M., An Efficient Ascending-Bid Auction for Multiple Objects (Working paper, Univ. Maryland 1997).

Ausubel, Lawrence M. & Peter Cramton, Demand Reduction and Inefficiency in Multi-Unit Auctions (Working paper, Univ. Maryland 1996).

Ausubel, Lawrence M., Peter Cramton, R. Preston McAfee & John McMillan, Synergies in Wireless Telephony: Evidence from the Broadband PCS Auctions, 6 *J. Econ. & Mgmt. Strategy* 497–527 (1997).

Ayres, Ian & Peter Cramton, Deficit Reduction through Diversity: A Case Study of How Affirmative Action at the FCC Increased Auction Competition, 48 *Stanford L.Rev.* 761–815 (1996).

Ballard, Charles L., John B. Shoven & John Whalley, General Equilibrium Computations of the Marginal Welfare Costs of Taxes in the United States, 75 *Am. Econ. Rev.* 128–138 (1985).

Bykowsky, Mark M., Robert J. Cull & John O. Ledyard, Mutually Destructive Bidding: The FCC Auction Design Problem, (Working paper, California Inst. Technology 1995).

Cramton, Peter, Money Out of Thin Air: The Nationwide Narrowband PCS Auction, 4 *J. Econ. & Mgmt. Strategy* 267–343 (1995).

Cramton, Peter, The FCC Spectrum Auctions: An Early Assessment, 6 *J. Econ. & Mgmt. Strategy* 431–495 (1997).

Cramton, Peter, Robert Gibbons & Paul Klemperer, Dissolving a Partnership Efficiently, 55 *Econometrica* 615–632 (1987).

Engelbrecht-Wiggans, Richard & Charles M. Kahn, Multi-Unit Auctions with Uniform Prices (Working paper, Univ. Illinois 1995).

McAfee, R. Preston, Auction Design of Personal Communication Services, attached to Comments of PacTel Corporation in PP Docket No. 93–253 (1993).

Milgrom, Paul & Robert J. Weber, A Theory of Auctions and Competitive Bidding, 50 *Econometrica* 1089–1122 (1982).

Milgrom, Paul & Robert Wilson, Affidavit of Paul R. Milgrom and Robert B. Wilson, attached to Comments of Pacific Bell and Nevada Bell in PP Docket No. 93–253 (1993).

Moreton, Patrick S. & Pablo T. Spiller, What's in the Air? Interlicense Synergies and Their Impact on the FCC's Broadband PCS License Auctions, 41 *J. Law & Econ.* 677–716 (1998).

CHAPTER 4

Measuring the Efficiency of an FCC Spectrum Auction[*]

Jeremy T. Fox and Patrick Bajari

1. Introduction

The US Federal Communications Commission (FCC) auctions licenses of radio spectrum for mobile phone service, employing an innovative simultaneous ascending auction. We study data from the 1995–1996 auction of licenses for the C Block of the 1900 MHz PCS spectrum band. The C block divided the continental United States into 480 small, geographically distinct licenses. A mobile phone carrier that holds two geographically adjacent licenses can offer mobile phone users a greater contiguous coverage area. One intent of auctioning small licenses is to allow bidders, rather the FCC, to decide where geographic complementarities lie. Bidders can assemble packages of licenses that maximize the benefits from geographic complementarities. The US practice of dividing the country into small geographic territories differs markedly from European practice, where nationwide licenses are often issued. These nationwide licenses ensure that the same provider will operate in all markets, so that all geographic complementarities are realized.

Economic theory suggests that the allocation of licenses in a simultaneous ascending auction need not be allocatively efficient. Brusco and Lopomo (2002) and Engelbrecht-Wiggans and Kahn (2005) demonstrate that bidders may implicitly collude through the threat of bidding wars. For example, a bidder might not add an additional license to a package to take advantage of complementarities because of threats of higher,

---

[*] Fox would like to thank the NET Institute, the Olin Foundation and the NSF, grant SES-0721036, for financial support. Bajari thanks the National Science Foundation, grants SES-0112106 and SES-0122747 for financial support. Thanks to helpful comments from Christopher Adams, Susan Athey, Lawrence Ausubel, Timothy Conley, Peter Cramton, Nicholas Economides, Philippe Fevrier, Matthew Gentzkow, Philip Haile, Ali Hortacsu, Robert Jacques, Jonathan Levin, Paul Milgrom, Harry Paarsch, Ariel Pakes, Robert Porter, Philip Reny, Bill Rogerson, Gregory Rosston, John Rust, Andrew Sweeting, Chad Syverson and Daniel Vincent. Thanks also to seminar participants at many conferences and universities. Thanks to Peter Cramton for sharing data on license characteristics, to David Porter for sharing data on experimental auctions, to Todd Schuble for help with GIS software, and to Chad Syverson for sharing data on airline travel. Excellent research assistance has been provided by Luis Andres, Wai-Ping Chim, Stephanie Houghton, Dionysios Kaltis, Ali Manning, Denis Nekipelov, David Santiago and Connan Snider. Our email addresses are jeremyfox@gmail.com and bajari@uw.edu.

retaliatory bids on the bidder's other licenses. For auctions of multiple homogeneous items, Ausubel and Cramton (2002) demonstrate that bidders may find it profitable to unilaterally reduce demand for licenses, similarly to a monopolist raising prices and profits by reducing supply. The concern about intimidatory collusion and demand reduction in FCC spectrum auctions is well founded. Cramton and Schwartz (2000, 2002) show that bidders in the AB block did not aggressively compete for licenses and in the later DEF block auction used the trailing digits of their bids to signal rivals not to bid on other licenses.

We provide the first structural estimate of a valuation function in an FCC spectrum auctions, apart from Hong and Shum (2003). They model bidding for each license as a single-unit auction and therefore do not measure complementarities. Our estimator is based on the assumption that the allocation of licenses is pairwise stable in matches, that is, an exchange of two licenses by winning bidders must not raise the sum of the valuations of the two bidders. In our econometric model, bidder valuations are a parametric function of license characteristics, bidder characteristics, and bidder private values. We use the maximum score or maximum rank correlation estimator for matching games introduced in Fox (2010a), where the objective function is the number of inequalities that satisfy pairwise stability. Such estimators for single-agent choice problems were introduced in Manski (1975) and Han (1987). We estimate the influence of various bidder and license characteristics on bidder valuations. Finally, we compare the efficiency of the observed and counterfactual allocations of licenses and discuss the implications of our estimates for alternative auction designs.

Our estimator is consistent under an econometric version of pairwise stability in matches, which we call the rank order property. We first justify the non-econometric version of pairwise stability in matches only with references to the experimental and theoretical literatures on simultaneous ascending auctions. We then state the econometric version of pairwise stability in matches only, the rank order property, as an assumption, with the non-econometric version of pairwise stability as informal motivation.

The are three justifications of the non-econometric version of pairwise stability. In terms of the experimental literature, we use data from experimental simultaneous ascending auctions by Banks et al. (2003), where bidder valuations are known, and show that the outcomes come close to satisfying pairwise stability in matches only.

Second, we analyze the outcomes generated by the equilibria in a simultaneous ascending auction discussed by Brusco and Lopomo (2002) and Engelbrecht-Wiggans and Kahn (2005). In the cases they study, but allowing asymmetric bidders and licenses to some extent (as described in our appendix), the equilibrium outcomes satisfy pairwise stability in matches only. In addition, a version of the demand reduction model of Ausubel and Cramton (2002) satisfies pairwise stability in matches only. The latter result requires straightforward bidding and no complementarities, as in Milgrom (2000).

Finally, there were few or no swaps of licenses between bidders immediately after the auction, even though such swaps were legally permissible and presumably had low transaction costs compared to the license values. Pairwise stability rules out bidders finding it profitable to trade licenses (perhaps with monetary transfers), and any swapping would be direct evidence against pairwise stability. We note that pairwise stability is a weaker condition than allocative efficiency: efficiency implies pairwise stability but not the reverse.

We contribute to the literature on spectrum auctions and the empirical analysis of multiple unit auctions in several ways. First, we structurally estimate bidder valuation functions in a spectrum auction. The existing empirical literature on FCC spectrum auctions is primarily descriptive. McAfee and McMillan (1996) provide an early analysis of the AB auction results. Cramton and Schwartz (2000, 2002) report evidence of attempts at coordination through bid signaling. Ausubel et al. (1997) and Moreton and Spiller (1998) present bid regressions showing evidence for complementarities. The structural approach is useful because it allows the researcher to quantitatively measure components of bidder valuations and the efficiency of the allocation of licenses, given the identifying assumptions.

Second, our estimator contributes to the literature on the structural estimation of multiple-unit auctions. Hortacsu and McAdams (2010), Fevrier, Préget and Visser (2004), Wolak (2007), Chapman, McAdams and Paarsch (2007), and Kastl (2011) study divisible good auctions, like those for electricity and treasury bills. To our knowledge, Cantillon and Pesendorfer (2006), who study sealed-bid auctions for bus routes under package bidding, is the only other structural paper to study auctions of multiple heterogeneous items. In contrast to Cantillon and Pesendorfer, we study an ascending auction without package bidding, and we allow for implicit collusion.

All of the above papers specify a model of equilibrium behavior and invert a bidder's first-order condition to recover its valuation. Athey and Haile (2008) and Paarsch and Hong (2006) survey studies of single-unit auctions that use this strategy. This first-order-condition approach and other approaches using bid data (such as Haile and Tamer, 2003) are not possible in our application because bids may be poor reflections of valuations under intimidatory collusion. None of the above estimators are consistent in the presence of implicit collusion.

Third, our paper contributes to the literature on structural estimation by allowing for a fixed effects model of unobserved heterogeneity in bidder valuations. In previous research, a maintained assumption is that the econometrician observes all publicly available information. We expect FCC bidders to have access to better information than we do. Our approach allows for license-specific fixed effects in valuations. When the first draft of this paper was circulated, the only paper that allowed for unobserved heterogeneity was Krasnokutskaya (2011). However, her approach and subsequent research rely on bid data, which is not our approach as we now explain.

Fourth, previous methods for structural estimation in auctions identify bidder valuations from final bids submitted in the auction. Theorists such as Crawford and Knoer (1981), Kelso and Crawford (1982), Leonard (1983), Demange, Gale and Sotomayor (1986), Hatfield and Milgrom (2005), Day and Milgrom (2007), and Edelman, Ostrovsky and Schwarz (2007), among others, have pointed out that a one-to-many, two-sided matching game is a generalization of an auction of multiple heterogeneous items. We are the first to use this insight in empirical work. We estimate the deterministic component of valuations as a function of recorded license and bidder characteristics, up to a normalization, based on the match between bidder characteristics and license characteristics. We do not use bid data in our preferred estimator. We demonstrate that a closely-related estimator that uses bid data does not yield reasonable estimates of bidder valuations. In part because we do not use bid data, we focus on estimating deterministic components of payoffs, namely how valuations relate to observed bidder and license characteristics, including the gains from geographic complementarities. We do not estimate the distribution of license- and bidder-specific private values.

Fifth, the effective size of the choice set for bidders in our application is very large. In our application, there are 480 licenses and, as a result, any estimator that relies on a direct comparison of the discrete choice between all potential packages will be computationally infeasible. Our estimator, based on pairwise stability, circumvents this computational difficulty.

Sixth, the true data generating process in a simultaneous ascending auction is a noncooperative, dynamic game. This game has multiple equilibria, including implicitly collusive and competitive equilibria. We base estimation on pairwise stability, a condition that holds across a set of equilibria in the situations studied in the theoretical literature on simultaneous ascending auctions. Pairwise stability may not hold across all equilibria and in other contexts, but it facilitates structural estimation for an otherwise intractable dynamic game. Estimation based on this arguably weak condition avoids solving for the equilibrium to the dynamic game. Computing equilibria would not be possible, given the indeterminacy of the equilibrium, the huge state space in a simultaneous ascending auction, and the massive choice set of bidders.

Finally, we estimate a two-sided, non-search matching game with transferable utility. Dagsvik (2000), Choo and Siow (2006), and Chiappori et al. (2012) work with logit-based specifications applied mostly to one-to-one matching, or marriage. We use the matching estimator of Fox (2010a).[1] An FCC bidder can win more than one license, and we focus on complementarities. We are the first paper to estimate a many-to-one matching game where the payoffs of bidders are not additively separable across licenses (unlike, say, Sørensen (2007)).

We find mixed evidence concerning the efficiency of the observed allocation of the licenses. At least since Coase (1959), the use of spectrum auctions has been justified on efficiency grounds. We find that bidders strongly value complementarities between licenses and that bidders with larger initial eligibilities value licenses more. We also find that awarding each license to a distinct bidder would reduce allocative efficiency, justifying spectrum auctions as efficiency enhancing in comparison with the prior lotteries regime. However, we find evidence that the observed packages of licenses were too small for an efficient allocation given the complementarities between licenses. Indeed, we estimate that dividing the continental United States into four, large regional licenses, assortatively matched to the four largest winning bidders, would have raised the allocative efficiency of the C block by 48%, compared to the actual outcome. Our findings suggest that small license territories, together with the possibility of intimidatory collusion, can generate an inefficient allocation of licenses. We briefly discuss more specific policy implications.

---

[1] The empirical application in Fox (2010a) was added after the paper was initially circulated. Subsequent uses of the estimator in Fox also postdate early versions of our paper.

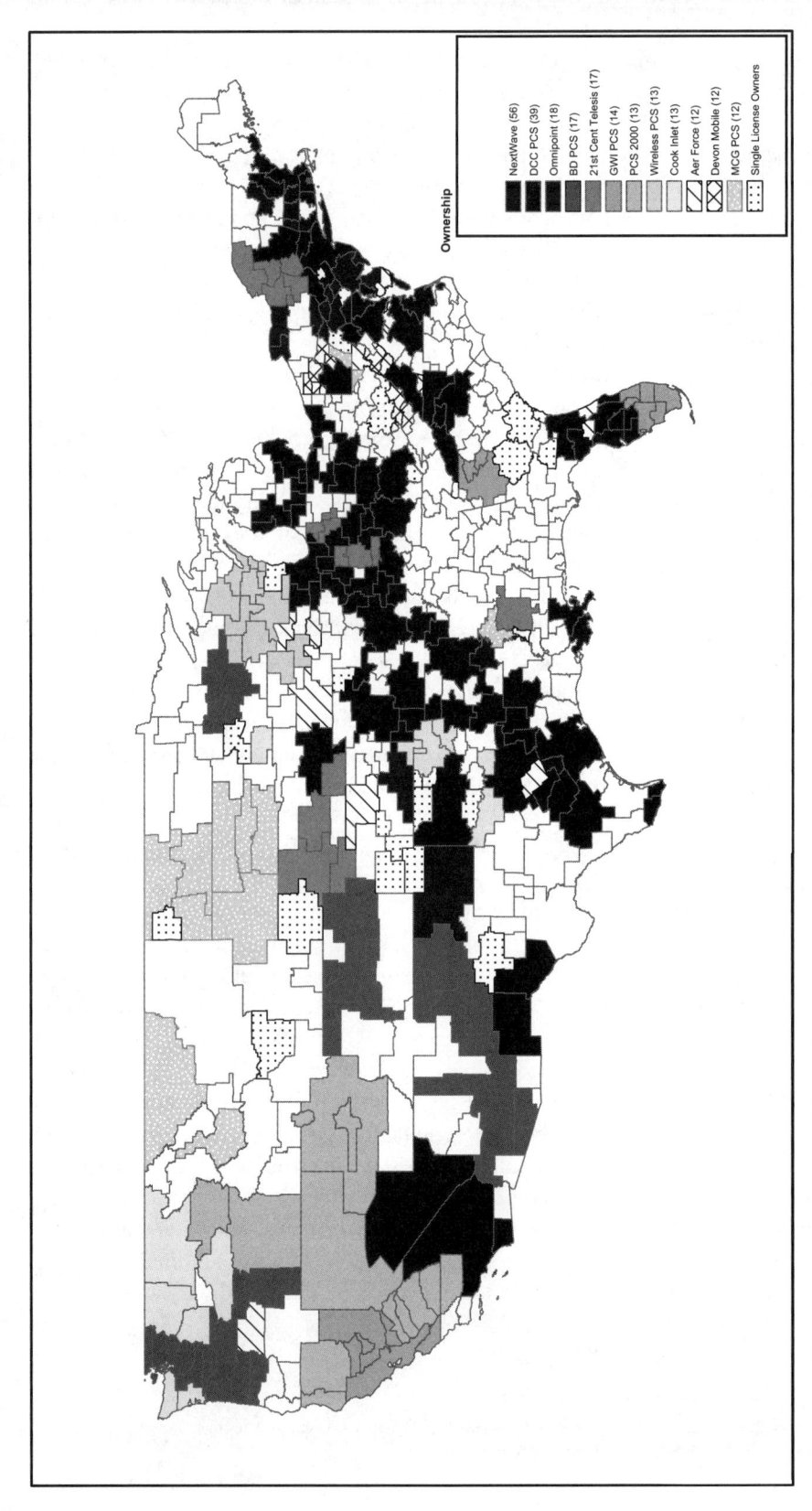

Figure 1. Map of the Licenses Won By the Top 12 Winning Bidders and Bidders Who Won Only One License.

2. Background for the C Block Auction

2.1. FCC Spectrum Auctions for Mobile Phones

Wireless phones transmit on the publicly-owned radio spectrum. In order to prevent interference from multiple radio transmissions on the same frequency, the FCC issues spectrum users licenses to transmit on specified frequencies.

There were three initial auctions of mobile phone spectrum between 1995 and 1997. The first auction (the AB blocks) sold 99 licenses for 30 MHz of spectrum for 51 large geographic regions and raised $7.0 billion for the US Treasury. The second auction (the C block) sold 493 30 MHz licenses in more narrowly-defined geographic regions to smaller bidders that met certain eligibility criteria. The C block auction closed with winning bids totaling $10.1 billion, although some bidders were unable to make payments, and their licenses were later re-auctioned. The third auction (the DEF blocks) sold three licenses for 10 MHz in each of the same 493 markets as the C block. The bids totaled $2.5 billion in the DEF blocks.

There are a number of reasons to prefer to use data from the C block auction instead of the AB or DEF blocks. First, the number of observations is much larger in the C block: there are 255 bidders in the C block compared to only 30 in the AB blocks and 155 in the DEF blocks. Furthermore, there were two licenses for sale for every geographic region in the AB blocks, and three licenses for every geographic region in the DEF blocks. An AB or DEF block winning bidder was thus guaranteed to be competing directly against at least one other winning carrier after the auction ended. This direct externality in the valuations of bidders complicates the analysis of bidding behavior. In the C block, each geographic region had only one license for sale.

The C block auction took 184 rounds, lasting from December 1995 to April 1996. Incumbent carriers did not participate in the C block because of discounts offered to small businesses. Figure 1 is a map of the licenses won by the top 12 winning bidders. The largest winner in the C block auction was NextWave, whose winning bids totaled $4.2 billion for 56 licenses, including close to $1 billion for the New York City license.

2.2. After the Auction: Mergers

C block bidders were given an extended payment plan of ten years. Many of the bidders planned to secure outside funding for both their license bids and other carrier startup costs after the auction. Many C block winners were unable to meet their financial obligations to the FCC. These new carriers were unable to secure enough outside funding to both operate a mobile phone company and pay back the FCC. Many C block winners returned their licenses to the FCC, where they were re-auctioned. Others companies merged with larger carriers (forming a large part of the licenses held by T-Mobile USA, for example) or were able to protect their licenses in bankruptcy court. NextWave is the most famous case of bankruptcy protection; later it settled with the FCC and sold some of its licenses to other carriers for billions of dollars. Ex-post, the C block bidders, who were accused of bidding too aggressively at the time, underpredicted the eventual market value of the licenses. However, much of this value was to larger carriers, not small-business entrants who could not secure the financing to operate as a mobile phone carrier. In 2011, only a few C block winners, such as GWI/MetroPCS, remain true independent carriers marketing service under their own brand.

The merger activity suggests that a bidder's post-auction value for winning licenses was not only a function of the package of territories it planned to serve as a mobile phone carrier. Valuations might be a function of the bidder's beliefs about the expected value from resale of its licenses, from mergers after the auction and from the risk of bankruptcy. Valuations also likely reflect the ability to serve traveling customers through roaming agreements as well as to sign up new subscribers directly. Therefore, we favor an interpretation of the estimates from our structural model that encompasses all these possibilities.

2.3. Auction Rules and Bidder Characteristics

FCC spectrum auctions are simultaneous, ascending-bid, multiple-round auctions that can take more than a hundred days to complete. A simultaneous ascending auction is a dynamic game with incomplete information. Each auction lasts multiple rounds, where in each round all licenses are available for bidding. A bidder can remain silent or enter bids to raise the standing high bids on one or several licenses. During a round, bidding on all licenses closes at the same time. The auction ends when no more bids are placed on any item; bidding on all items remains possible until the end. These auction rules were designed to allow bidders to assemble packages exhibiting complementarities, while letting the bidders themselves and not the FCC determine where the true complementarities lie. If bidders have finite valuations, they will cease bidding after a finite number of rounds, although the length of the auction is not known at the start. Package bidding is not allowed; bidders place bids on each license separately.

Each bidder makes a payment before the auction begins for initial eligibility. A bidder's eligibility is expressed in units of total population. A bidder cannot bid on a package of licenses that exceeds the bidder's eligibility. For example, a bidder who pays to be eligible for 100 million people cannot bid on licenses that together contain more than 100 million residents. Eligibility cannot be increased after the auction starts. During the auction, the eligibility of bidders that do not make enough bids is reduced. By the close of the auction, many bidders are only eligible for a population equal to the population of their winning licenses.

The eligibility payments were 1.5 cents per MHz-individual in a hypothetical license for the C block. These payments are trivial compared to the closing auction prices. We use eligibility to control for a bidder's willingness to devote financial resources towards winning spectrum. This paper does not model strategic motives (such as intimidating rivals) for choosing eligibility levels. Such motives could break our assumed monotone relationship between a bidder's true valuation for licenses and its eligibility, which will make our estimates inconsistent.

Table 1 lists characteristics of the 85 winning and 170 non-winning bidders in the continental United States. The average winning bidder paid fees to be eligible to bid on licenses covering 11 million people, while the average losing bidder was eligible to bid on licenses covering only 5 million people. Bidders also had to submit financial disclosure forms (the FCC's Form 175) in order to qualify as entrepreneurs for the C block, which was limited to new entrants. Table 1 shows that the financial characteristics of

Table 1. *Characteristics of Winners and Non-Winners of Packages in the Continental United States*

	Winners		non-Winners	
Characteristic	Mean	Stand. dev.	Mean	Stand. dev.
initial eligibility (millions of residents)	10.7	28.5	4.69	17.4
revenues ($ millions)	12.8	21.7	12.4	18.8
assets ($ millions)	39.6	67.7	40.4	72.4
# of licenses won	5.65	7.95	0	0
# of licenses ever bid on	40.2	73.9	13.9	41.0
# of bidders	85		170	

winners and non-winners were similar, which leads us to believe that these disclosure forms did not represent the true resources of bidders. Hence, in our structural estimator, we use initial eligibility as an individual bidder characteristic instead of assets or revenues.

Table 1 lists the mean number of licenses bid on and won by winners and non-winners. The mean winning bidder won 6 licenses and entered at least one bid on 40 licenses, compared to bidding on 14 licenses for non-winners. Although not listed in the table, the top 15 winning bidders were active bidders on many licenses. The top 15 winners won an average of 18 licenses and bid on an average of 118 (out of 480) licenses. Most of the major winners and some of the non-winners were investors operating on a national scale.

2.4. Prices and Winning Packages

The C block auction generated closing bids where the underlying characteristics of licenses explain much of the variation in prices across licenses. The most important characteristic of a license is the number of people living in it, who represent potential subscribers to mobile phone service. The population-weighted mean of the winning prices per resident is $40. The second most important characteristic in determining the closing prices is population density. Spectrum capacity is more likely to be binding in more densely populated areas. A regression of a license's winning price divided by its population on its population density gives an R^2 of 0.33.[2] However, prices per resident varied widely across the AB, C and DEF auctions. It is difficult to reconcile this across-auction variation with a view that the final bids closely reflect bidder valuations (Ausubel et al., 1997).

In the C block, the average winning bidder agreed to pay $116 million and won a package covering 2.9 million people. The largest winner, NextWave, bid $4.2 billion for a package covering 94 million people.

Figure 2 plots the log of a bidder's initial eligibility on the horizontal axis and the log of the package's winning population on the vertical axis. A 45 degree line is also included; all observations lie beneath the line because a bidder cannot win more than

---

[2] Ausubel et al. (1997) use proprietary consulting data on the population density of the expected build-out areas for C block mobile phone service. They have provided us the same data, which we use here.

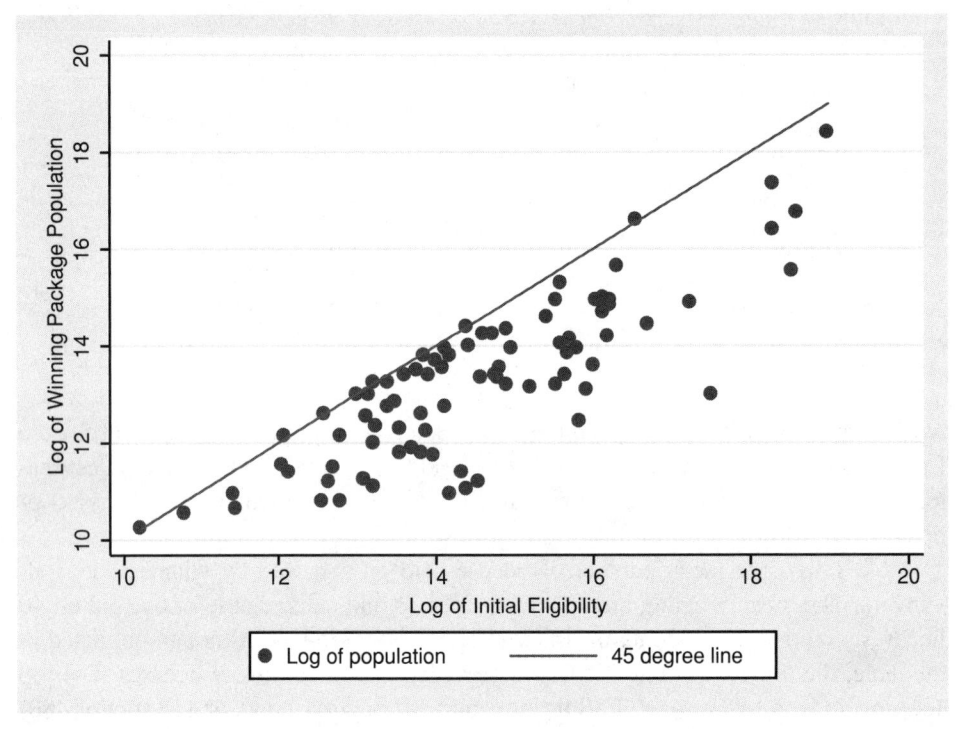

Figure 2. Log of a Winning Package's Population and the Log of the Winning Bidder's Initial Eligibility.

its initial eligibility. Eight winning bidders appear to be constrained; we will impose such eligibility constraints in our estimator later in the paper. The R^2 of a quadratic fit of the log of winning population to the log of initial eligibility is quite high, at 0.70. Initial eligibility is predictive of acquired spectrum.

2.5. Suggestive Evidence on Complementarities

A major justification for the simultaneous ascending auction is that it allows bidders to assemble packages of nearby licenses. Such adjacent licenses are said to exhibit complementarities or synergies. Bajari, Fox and Ryan (2008) use data on calling plan choice to estimate that consumers do have high willingnesses to pay to avoid roaming surcharges while traveling. So there is evidence that economic primitives do justify complementarities in bidders' structural valuation functions.

One's prior might be that complementarities are not important in the spectrum auctions. The FCC chose market boundaries to be in sparsely settled areas in order to minimize complementarities across markets. Furthermore, 1900 MHz PCS wireless phone service is mainly deployed in urban areas and along major highways, so there might not even be PCS service along the boundaries of two markets. Finally, companies can coordinate with contracts (roaming agreements) if the same company does not own the adjacent licenses.

However, an initial inspection of our data is compatible with the existence of geographic complementarities. The map of the top 12 winners in Figure 1 shows several bidders win licenses in markets adjacent to each other. For example, NextWave, the largest winner, purchases clumps of adjacent licenses in different areas of the country. GWI/MetroPCS fits the cluster pattern well, winning licenses in the greater San Francisco, Atlanta and Miami areas.

On the other hand, the majority of winning bidders win only a few licenses. Figure 1 emphasizes this by also plotting the 26 licenses in the continental United States that were the only license won by their winning bidders. Only 20 out of 85 C block winning bidders won packages of licenses where the population in adjacent licenses within the package was more than 1 million. Aer Force is the prime example of a top 12 bidder that did not seem overly concerned with complementarities. Figure 1 shows that Aer Force won 12 licenses, but that none of them are adjacent to each other. From the maps alone, it appears some winning bidders cared more about geographic complementarities than others.

Previous researchers have generally concluded that complementarities were important. Ausubel et al. (1997) and Moreton and Spiller (1998) examine whether adjacent licenses exhibited complementarities by regressing the log of winning bids on market and bidder characteristics. Ausubel et al. study the AB and C block auctions and find that the log of winning bids are positively related to whether the runner-up bidders won adjacent licenses, as one might expect in an ascending-bid auction. However, the coefficient in the C block auction is economically small, meaning that prices do not seem to strongly reflect any value of complementarities. Moreton and Spiller have better measures of incumbency and also find that winning bids are positively related to the runner-up bidder's measures of complementarities.

The previous authors also discuss scale economies, the notion that a wireless network involves fixed costs that can be spread out among more customers in a larger carrier. Scale economies can be represented by valuations convex in the total population of a package. However, because bidders with higher valuations (empirically, higher initial eligibilities) win packages with higher populations, it may be hard to empirically distinguish operating scale economies from heterogeneities in bidder valuations.

Figure 1 suggests that the clusters of nearby licenses in winning packages are possibly too small. If bidder valuations were primarily a function of complementarities, we might expect to see the entire southeast won by one bidder, for example.

The fact that many bidders win clusters of licenses suggests that complementarities matter to some degree. An alternative explanation is that a bidder has correlated license-specific values across licenses in a geographic cluster. There seems to be little scope for distinguishing between the two explanations in a spectrum auction setting. Gentzkow (2007) discusses the difficulties of distinguishing true complementarities from correlated preferences in a consumer demand setting. We assume away spatially correlated, license-specific private values and focus on complementarities because the evidence suggests that the largest winners were not local businessmen with special attachments to particular, large regions. Many of the the largest winners, such as NextWave, Omnipoint and GWI/MetroPCS, won small clusters in many regions of the country. MetroPCS has its headquarters in Dallas, but won licenses only near Atlanta, Miami and San Francisco. DCR/Pocket won licenses stretching from Detroit to Dallas, an oddly-shaped region to be a regional specialist in. PCS2000 won mainly a cluster of licenses in the West, but its headquarters was far away in Puerto Rico. This discussion does not rule out that these bidders have spatially correlated license-specific values, but it suggests that such an explanation is not more likely than complementarities.

We do not view the price regressions of Ausubel et al. (1997) and Moreton and Spiller (1998) as a consistent estimator of bidder valuations, for at least two reasons. First, the auction induces an econometric selection problem in the final allocation of licenses to bidders. Winning packages have high payoffs for observed or unobserved reasons; otherwise they would not win. As both bidder- and license-specific valuations and complementarities across licenses contribute to total payoffs, those packages with relatively low complementarities will have relatively high bidder- and license-specific valuations. As the bidder- and license-specific valuations are typically not observed and are related to the error term in the price regression, there will be correlation between the complementary proxies and the error terms in the price regression. Linear regression will thus be inconsistent.

Even if winning packages' complementarities were somehow uncorrelated with winning packages' bidder- and license-specific valuations, the estimator would still be inconsistent. Under intimidatory collusion as discussed in the next subsection, prices will not reflect valuations and so price regressions will not identify structural parameters. In order to interpret price regressions as estimates of structural parameters, one would need to assume that the outcome to the auction is equivalent to a competitive equilibrium to the underlying economy.

2.6. Suggestive Evidence About Intimidatory Collusion

Milgrom (2000, Theorems 2,3) proves that a simultaneous ascending auction is equivalent to a tatonnement process that finds a competitive equilibrium of the economy, under two assumptions: 1) the licenses are mutual substitutes for all bidders, and 2) all bidders bid straightforwardly. Unfortunately, neither of the assumptions needed to prove that a simultaneous ascending auction finds a competitive equilibrium appear to hold in the C block data. We have already discussed evidence that there may be complementarities in bidders' valuations.

Bidding straightforwardly means that a bidder submits new bids each period in order to maximize its structural profit function, rather than some other continuation value in a dynamic game. One violation of straightforward bidding is jump bidding. When making a jump bid, a bidder enters a bid that exceeds the FCC's minimum bid for that license and round. We define a jump bid to be any bid that is 2.5% greater than the minimum bid. Figure 3 shows that there was a non-trivial level of jump bidding during the C block auction.

When jump bidding, a bidder risks the chance that the jump bid will exceed the valuation of rival bidders and be the final price. A jump bidder therefore has a nonzero probability of overpaying for a license. However, there are possible strategic advantages from jump bidding. In a single unit, affiliated values model, Avery (1998) demonstrates that jump bidding may signal the jump bidder's intentions to bid aggressively throughout the auction. Because other bidders fear the winner's curse or if bidding is costly, they may stop bidding in order to avoid overpaying conditional on winning the item.

Figure 3 shows jump bidding was prevalent towards the beginning of the auction, where the risk of overpaying is much lower. Jump bids might represent signals that

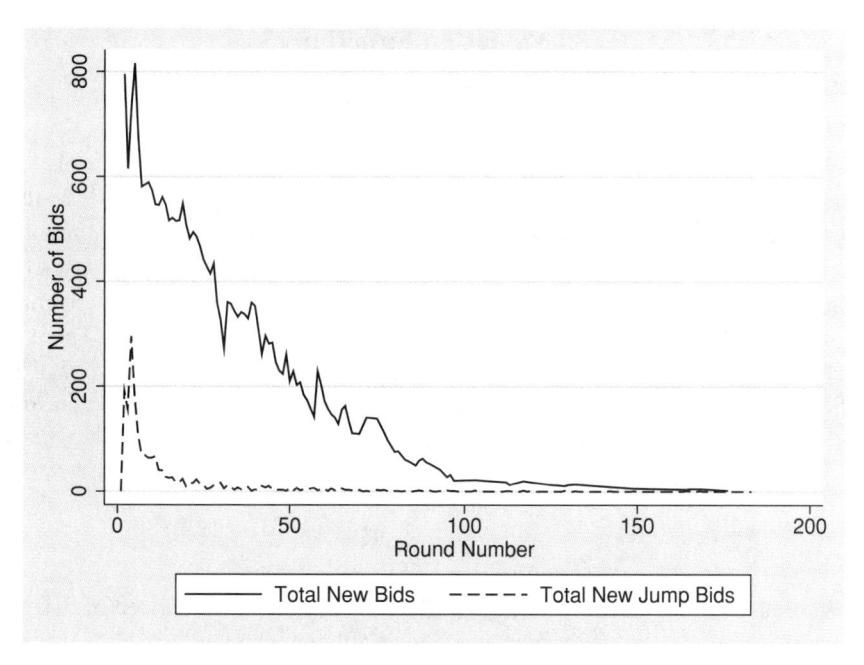

Figure 3. The Number of Jump Bids per Round.

are attempts at intimidation, but jump bids are not evidence the signals caused other bidders to withdraw. There are anecdotes of actual retaliation. In round 3, Pocket (DCR) placed a large jump bid of 60% more than the minimum for Las Vegas. In round 70, MetroPCS (GWI) outbid Pocket for Las Vegas and PCS2000 for Reno. In round 71, Pocket outbid MetroPCS on Reno and Salt Lake City, the only time Pocket bid on either of those licenses. Further, PCS2000 outbid MetroPCS on Las Vegas, the only time since round 12 PCS2000 had bid on Las Vegas. In round 72, after seeming to retaliate against MetroPCS, Pocket enters the winning bid for Las Vegas, meaning the bid stands until the end of the auction, round 184.

There are other instances of intimidation that do not involve jump bids. Towards the end of the auction, NextWave and AerForce were competing for Fredericksburg, Virginia. NextWave needed Fredericksburg to complete a regional cluster around Washington, DC. In round 162, NextWave outbid AerForce for Fredericksburg. In round 163, AerForce responded not only by bidding on Fredericksburg but also by bidding on Lakeland, Florida. Lakeland is a small population territory that AerForce had not bid on in a long while and that NextWave had been winning. In round 164, NextWave bid again and retook Lakeland, but never bid again on Fredericksburg. By challenging AerForce on Fredericksburg, NextWave only succeeded in paying 10% (two bid increments) more to win Lakeland.

Cramton and Schwartz (2000, 2002) provide examples of signaling and implicit collusion through intimidation in the auctions for the AB and DEF blocks. We feel the evidence is strong enough that any estimation method for simultaneous ascending auction data must be based on conditions that hold in the presence of this type of implicitly collusive behavior.

3. Valuation Functions

3.1. Bidders' Valuation Functions

We now introduce the components of a bidder's profit function. There are $a = 1, \ldots, N$ bidders and $j = 1, \ldots, L$ licenses for sale. We will abuse notation and let N be the set of all bidders and L the set of all licenses. Our environment is a multiple-unit auction where bidders may win a package of licenses. We let $J \subset L$ denote such a package of licenses. In the C block, the licenses are permits to transmit mobile phone signals in specified geographic territories and there is only one license per territory. There were $N = 255$ registered bidders in the C block and 493 licenses for sale. We will limit attention to the $L = 480$ licenses for sale in the continental United States and mostly to the $H = 85$ winning bidders in the continental US.

Bidder a maximizes its profit

$$\pi_a(J) - \sum_{j \in J} p_j$$

from winning package J at prices $(p_j)_{j \in J}$. Bidder a's profit is comprised of two parts. The term $\pi_a(J)$ is a's **valuation** for the package of licenses J and $\sum_{j \in J} p_j$ is the price that a pays for this package. In our application, we will parameterize the valuation $\pi_a(J)$ as

$$\pi_a(J) = \bar{\pi}_\beta(w_a, x_J) + \sum_{j \in J} \xi_j + \sum_{j \in J} \epsilon_{a,j}. \tag{1}$$

The function $\bar{\pi}_\beta(w_a, x_J)$ takes as arguments the characteristics w_a of bidder a and the characteristics x_J of the package of licenses J. The function $\bar{\pi}_\beta$ is parameterized by a finite vector of parameters β. Later β will be the object of estimation. The term ξ_j is a fixed effect for license j and $\epsilon_{a,j}$ is a private value specific to license j and bidder a. The fixed effect ξ_j captures the common element to the valuation of the license, such as the base contribution of population, the fact that spectrum is more scarce in more densely-populated territories and the fact that competition from incumbent carriers may be stronger in some territories than others. Let y_j be the vector of observed characteristics of license j. The characteristics x_J of a package J are formed by $x_J = \zeta(Y)$ from the $J < \infty$ license characteristics in $Y = (y_1, \ldots, y_J)$. The function ζ is known to the researcher.

In our application, we let $w_a = \text{elig}_a$ be the initial (before the auction begins) eligibility of bidder a. We treat w_a as economically exogenous, in that firms do not choose w_a for strategic reasons, such as intimidating rivals. We also assume that w_a is strictly monotone in the true preferences of bidders for spectrum; we do not allow unobserved bidder characteristics that affect the valuations of all licenses.

For license characteristics, let

$$x_J = \left((\text{pop}_j)_{j=1}^J, \text{complem.}_J \right)$$

be equal to the population of all licenses in the package J as well as a vector complem.$_J$ of proxies for the complementarities in the package. Our choice of $\bar{\pi}_\beta(w_a, x_J)$ is

$$\bar{\pi}_\beta(w_a, x_J) = \pm 1 \cdot \text{elig}_a \cdot \left(\sum_{j \in J} \text{pop}_j \right) + \beta' \text{complem.}_J. \tag{2}$$

The interaction $\text{elig}_a \cdot (\sum_{j \in J} \text{pop}_j)$ captures the fact in Table 1 and Figure 2 that bidders with more initial eligibility win more licenses. We use $w_a = \text{elig}_a$ as our main measure of bidder characteristics, given that Table 1 shows financial measures are uncorrelated with winning a license. The coefficient on $\text{elig}_a(\sum_{j \in J} \text{pop}_j)$ has been normalized to ± 1 because dividing both sides of an inequality that is used in estimation by a positive constant will not change the inequality. Overall, the term $\text{elig}_a \cdot (\sum_{j \in J} \text{pop}_j)$ captures assortative matching between bidders with higher values and packages of licenses with more population.

The term $\beta' \text{complem.}_J$ provides the total contribution of the several complementarity measures in the vector complem._J. Each element of complem._J is a nonlinear construction from the characteristics of the underlying licenses in the package J. The parameters β describe the relative importance of each complementarity measure in terms the units of $\text{elig}_a \cdot (\sum_{j \in J} \text{pop}_j)$. Overall, the term $\beta' \text{complem.}_J$ captures one-sided matching between related licenses into the same packages.

In terms of units, eligibility is the initial eligibility of a bidder. Population is just the number of residents (in the 1990 census) of the license. To aid interpretation, we divide both measures by the population of the continental United States, so that an eligibility or population of 1 corresponds to a true value of 253 million people. With this normalization, the mean population $\sum_{j \in J} \text{pop}_j$ among the 85 winning packages is 0.012 (standard deviation of 0.044), and the mean $\text{elig}_a(\sum_{j \in J} \text{pop}_j)$ is 0.0046 (standard deviation 0.030). We discuss the measures of geographic complementarities below.

We choose a simple functional form for $\bar{\pi}_\beta(w_a, x_J)$ in order to demonstrate that a parsimonious model can satisfy a high percentage of the inequalities introduced below. A more complicated functional form would have little benefit in terms of the overall fraction of inequalities that are satisfied and would obscure the interpretation of the parameters. As we do not have profit, cost, pricing, sales and merger profits data for firms operating mobile phone carriers as a function of their coverage areas, we cannot decompose $\bar{\pi}_\beta(w_a, x_J)$ into the present discounted values of sales, marginal costs, per-period fixed costs, one-time fixed costs and profits from merger activity.

3.2. Assumptions

We now list a series of assumptions. These assumptions are made to clarify the informational structure of the simultaneous ascending auction. They will be referenced in a series of remarks about the robustness of the theoretical results of Brusco and Lopomo (2002) and Engelbrecht-Wiggans and Kahn (2005), in Appendix A. We first present assumptions about the bidder characteristics w_a.

Assumption 1. *The scalar bidder heterogeneity w_a is public information. Further, $\bar{\pi}_\beta(w_a, x_J) = h_\beta(w_a) \cdot \bar{\pi}_\beta^1(x_J) + \bar{\pi}_\beta^2(x_J)$, where $h_\beta(\cdot)$ is a monotone function, and $\bar{\pi}_\beta^1$ and $\bar{\pi}_\beta^2$ are unrestricted functions of x_J. Each w_a is in the data.*

Making w_a private information simplifies some of the analysis below, as Remark 2 in the appendix argues. A private w_a induces ex-post asymmetry in the values of licenses; a bidder with a high preference for license j_1 will often prefer license j_2 more as well. Privately observed values is the natural case to start with in developing an estimator for simultaneous ascending auctions of multiple heterogeneous items under implicit collusion. A private w_a also tracks the theoretical assumptions in the papers Brusco and Lopomo (2002) and Engelbrecht-Wiggans and Kahn (2005).

The assumption of private information is inaccurate for the C block because w_a is in the data and was disclosed prior to the auction. Instead, we allow w_a to be commonly observed information. A model where w_a is public information is a model with bidders with known asymmetries. See Remarks 2, 6 and 7 in the appendix for discussion about the need for Assumption 1, which relates to implicit collusion in particular models. Note that regardless of whether w_a is private information or public information, w_a is a private value in the sense of Milgrom and Weber (1982) given that rival bidders would not update their own valuations if they learned the value w_a. We next turn to the ξ_j terms.

Assumption 2. *The term ξ_j is a license j fixed effect, which we assume is publicly observed by the bidders. ξ_j may be statistically dependent with y_j and hence x_J, for $j \in J$. Each ξ_j is not in the data.*

The fixed effect enters bidders' valuations additively and is meant to capture the characteristics of license j that are observed by the bidders, but not by the econometrician. For example, we lack controls for the incumbent mobile phone companies as well as the winners of the earlier AB auctions and potential merger and roaming partners. As is standard in fixed effect models, we cannot identify the effects of elements of x_J that are collinear with the fixed effects ξ_j. We can, however, identify β in $\tilde{\pi}_\beta(w_a, x_J)$, which captures the interaction between the bidder and license characteristics observed by the econometrician. We will not estimate the fixed effects or their distribution, but our estimator is consistent in their presence. Our estimator will be inconsistent if the bidder heterogeneity w_a interacts with the fixed effects ξ_j.

The unobservables, $\epsilon_{a,j}$, reflect bidder a's private information about license j. We use the framework of independent private values for the $\epsilon_{a,j}$ term (any correlation occurs through w_a).

Assumption 3. *The $\epsilon_{a,j}$ are i.i.d. across bidders and licenses and are independent of all w's, x's and ξ's. Each $\epsilon_{a,j}$ is privately observed by bidder a, but the distribution of $\epsilon_{a,j}$ is common knowledge among the bidders. Each $\epsilon_{a,j}$ is not in the data.*

These reflect bidder-specific costs and benefits from operating in a particular territory. As we discuss, our maximum rank correlation estimation approach will not allow us to identify the distribution of the $\epsilon_{a,j}$'s. For the C block, the trade press and the number of licenses bid on by each bidder suggest that many winning bidders were willing to operate in any region of the country. This suggests the variance of $\epsilon_{a,j}$ is small. A small variance of $\epsilon_{a,j}$ contrasts with the AB blocks, where many bidders were incumbents trying to win territories near their existing service areas. In Section 2.5, we acknowledge that there is no obvious way to use data from a simultaneous ascending auction to distinguish true complementarities from spatially correlated $\epsilon_{a,j}$'s. Other sources of differential values across bidders and across licenses occur through the observable w's and x's and the unobservable ξ's.

The theoretical literature on the simultaneous ascending auction uses the assumption of private values. Under private values, bidders would not revise their own valuations

Table 2. *Winning Packages: Sample Statistics and Correlation Matrix for Geographic Complementarity Proxies*

Characteristic	Mean	Standard Deviation	Min	Max
Population/distance two markets in a package	0.0055	0.024	0	0.20
Trips between markets in a package in the American Travel Survey	0.0032	0.020	0	0.182
Total trips between airports in markets in a package (thousands)	0.0023	0.017	0	0.150
Correlations	Geo dist.	ATS Trips		
Population/distance two markets in a package	1			
Trips between markets in a package in the American Travel Survey	0.97	1		
Total trips between airports in markets in a package (thousands)	0.95	0.99		

The sample is the 85 winning packages in the continental United States. The formulas for these measures are equations (3) and (4).

if they were to observe the private information of rivals. For the C block, there is some evidence that some bidders stuck to their private evaluations of the value of wireless service and did not update their valuations. The bidder with the second-highest initial eligibility won no licenses because the prices exceeded that bidder's evaluation of the profit potential from wireless services.

In a common values model, ξ_j might be unobserved to the bidders as well. If a bidder was able to learn ξ_j, it would revise its valuations. Again, common values are usually not part of formal models of spectrum auctions because of technical complexity. As Hong and Shum (2003) argue empirically for the AB blocks, at the end of the auction a lot of information about ξ_j has been disclosed, possibly mitigating any winner's curse. However, this conclusion is less obvious under implicit collusion, where the link between bids and true values is imperfect.

3.3. Three Proxies for Potential Complementarities

We construct proxies for geographic economies of scope and use them as our measure of complementarities in (2). Table 2 presents descriptive statistics as well as the correlation matrix for the three measures. The measures are highly but not perfectly correlated with each other. For all geographic complementarity proxies, some fraction of the winning packages has a value of 0. For example, 26 out of the 85 winning packages contain only one license in the continental United States.

3.3.1. Geographic Distance

Our first proxy for geographic scope is based on the geographic distance between pairs of licenses within a package. We measure distance between two licenses using the population-weighted centroid of each license.[3] For a package J in the set L of all licenses, potential complementarities are

$$\text{geocomplem.}_J = \sum_{i \in J} \text{pop}_i \frac{\left(\sum_{j \in J, j \neq i} \frac{\text{pop}_i \text{pop}_j}{\text{dist}_{i,j}^{\delta}}\right)}{\left(\sum_{j \in L, j \neq i} \frac{\text{pop}_i \text{pop}_j}{\text{dist}_{i,j}^{\delta}}\right)}, \qquad (3)$$

where population is measured in fractions of the US total population and distance is measured in kilometers.[4] The distance, $\text{dist}_{i,j}$, between licenses i and j is, in our first set of estimates, raised to a power $\delta = 4$ to make this measure overweight nearby territories. The choice of $\delta = 4$ is arbitrary and was chosen to make the clusters of licenses seen in Figure 1 have non-trivial levels of complementarities. We also estimate the model with the choice of $\delta = 2$. The measure geocomplem.$_J$ proxies for short-distance travel and cost and marketing synergies across nearby territories. Also, the measure is similar to the well-known gravity equation in international trade. The measure also has the desirable feature that any firm's complementarities cannot decrease by adding licenses to a package.

3.3.2. Two Travel Measures

Geographic measures of distance may not capture the returns to scope that concern carriers. Mobile phone customers may travel by means other than ground transportation. For example, many business users travel by air between Los Angeles and New York. In fact, the C block bidder NextWave won both the Los Angeles and New York licenses. We have two complementarity proxies based upon travel between two licenses. The first measure, from the 1995 American Travel Survey (ATS), is proportionate to the number of trips longer than 100 km between major cities. All forms of transportation are covered. The downside of this measure is that for privacy reasons the ATS does not provide enough information about rural origins and destinations to tie rural areas to particular mobile phone licenses. Our second measure, from the Airline Origin and Destination Survey for the calendar year 1994, is the projected number of passengers flying between two mobile phone license areas.[5] The drawback of the air travel measure is that it assumes all passengers stay in the mobile phone license area where their destination airport is located. We effectively code that there are zero potential complementarities between rural licenses for both travel measures. Both travel measures for a

---

[3] The population-weighted centroid is calculated using a rasterized smoothing procedure using county-level population data from the US Census Bureau.

[4] This geographic complementarity proxy can be motivated as follows. Consider a mobile phone user in a home market i. That mobile phone user potentially wants to use his phone in all other markets. He is more likely to use his phone if there are more people to visit, so his visit rate is increasing in the population of the other license, j. The user is less likely to visit j if j is far from his home market i, so we divide by the distance between i and j. We care about all users equally, so we multiply the representative user in i's travel experience by the population of i.

[5] Intermediate stops are not counted for either dataset. For both datasets, geographic information software (GIS) was used to match origins and destinations with mobile phone licenses. For airports, the origin and destination license areas are easy to calculate. For the MSAs (Metropolitan Statistical Areas) used in the ATS, the equivalent C block license area was found using the centroid of the origin or destination MSA. The C block license boundaries for urban areas roughly follow MSAs.

package J are population-weighted means across licenses, and take the form

$$\text{travelcomplem.}_J = \sum_{i \in J} \text{pop}_i \frac{\sum_{j \in J, j \neq i} \text{trips(origin is } i, \text{ destination is } j)}{\sum_{j \in L, j \neq i} \text{trips(origin is } i, \text{ destination is } j)}. \quad (4)$$

Our ATS measure uses the count of raw trips in the survey, and the air travel count is inflated to approximate the total number of trips during 1994.[6] As with geographic distance, if $J = L$, travelcomplem.$_J = \sum_{i \in L} \text{pop}_i = 1$. Here again, adding a license to a package cannot take away complementarities between other licenses, so travelcomplem.$_J$ weakly increases as licenses are added to J.

4. Pairwise Stability

4.1. Pairwise Stability and Other Properties of Auction Outcomes

A spectrum auction is a data generating process that produces a vector of prices for each license $p^L = (p_1, \ldots p_L)$ and an allocation of licenses $A = \{J_1, \ldots, J_N\}$. In this subsection, we define pairwise stability in matches only as well as several alternatives for comparison. This subsection does not present an equilibrium definition, instead listing properties that outcomes to auctions may or may not satisfy.

Definition 1. An allocation of bidders to licenses $A = \{J_1, \ldots, J_N\}$ satisfying $\cup_{a \in N} J_a \subseteq L$ and $J_a \cap J_b = \emptyset$ for all bidders $a \in N$ and $b \in N$ is a **pairwise stable outcome in matches only** if, for each pair of winning bidders $a \in N$ and $b \in N$, corresponding winning packages J_a and J_b, as well as licenses $i_a \in J_a$ and $i_b \in J_b$,

$$\pi_a(J_a) + \pi_b(J_b) \geq \pi_a((J_a \setminus \{i_a\}) \cup \{i_b\}) + \pi_b((J_b \setminus \{i_b\}) \cup \{i_a\}). \quad (5)$$

Keep in mind that private values $\epsilon_{a,j}$ are included in the definition of $\pi_a(J_a)$. Pairwise stability in matches only considers swapping licenses: the total valuations of two bidders must not be increased by an exchange of one license each.[7] One way to motivate pairwise stability in matches only is to say that bidders would not want to exchange licenses along with side payments, at the end of the auction. This is a true mathematical interpretation of pairwise stability in matches only, and we will use the lack of swapping licenses after the auction to suggest that the outcome may have been pairwise stable in matches. However, when examining the output of theoretical models of simultaneous ascending auctions, we will not rely on bidders exchanging licenses with side payments. Rather, certain noncooperative equilibria to dynamic games will end up being pairwise stable in matches only.

---

[6] Our airline passenger measure does not distinguish between origins and destinations, so we simply divide the formula for the complementarity proxy by 2. If all airline trips are round trips during the same calendar year, this measure should be exactly correct.

[7] One could strengthen Definition 1 to consider exchanges of bundles of two or more licenses between each of two bidders. This notion could be called "two bidders, two bundles stability." This is a stronger condition as it implies Definition 1. The rest of the section focuses on motivating Definition 1 and not "two bidders, two bundles stability." Given the lack of motivation and the desire to use weaker rather than stronger assumptions in estimation, we focus on Definition 1. In a previous draft with a slightly different specification for the valuation function, we did estimate a model using inequalities derived from exchanges of bundles of two licenses for each bidder, and found that the point estimates were quite similar to the estimates based on Definition 1.

Pairwise stability in matches only will lead to a matching approach to estimation. The results from Section 2 suggest there is important information about valuations that is contained in which bidders win which licenses. For example, the clustering of licenses in Figure 1 suggests that complementarities in licenses may be important. Table 1 and Figure 2 show that bidders with higher initial eligibilities win more licenses. This is consistent with bidders with higher eligibilities having higher valuations for licenses.

Definition 2. The outcome $(p^L, A) = (p^L, \{J_1, \ldots, J_N\})$ satisfying $\cup_{a \in N} J_a \subseteq L$ and $J_a \cap J_b = \emptyset$ for all bidders a and b is a **pairwise stable outcome in both prices and matches** if, for each bidder $a = 1, \ldots, N$, corresponding winning package $J_a \subset L$, and licenses $i \in J_a$ and $j \notin J_a$, $j \in L$,

$$\pi_a(J_a) - p_i \geq \pi_a((J_a \setminus \{i\}) \cup \{j\}) - p_j. \tag{6}$$

In the above definition, at the closing prices p^L, bidder a must not want to swap one of its winning licenses i for some other bidder's license j. Note that pairwise stability in both prices and matches implies pairwise stability in matches only. Adding the inequality

$$\pi_b(J_b) - p_j \geq \pi_b((J_b \setminus \{j\}) \cup \{i\}) - p_i$$

to (6) cancels the license prices and gives (5). We present estimates from estimators based on both conditions, but we focus on the weaker of the two conditions.

Because our paper seeks to structurally measure efficiency, it is important to distinguish pairwise stability in matches only from efficiency.

Definition 3. An allocation of bidders to licenses $A = \{J_1, \ldots, J_N\}$ is **efficient** whenever

$$\sum_{a \in N} \pi_a(J_a) \geq \sum_{a \in N} \pi_a(J'_a)$$

for all other partitions $\{J'_1, \ldots, J'_N\}$ of L, where a partition satisfies $\bigcup_{a=1}^N J'_a \subseteq L$ and $J_a \cap J_b = \emptyset \; \forall \; a, b \in N$.

Efficiency is a stronger condition than pairwise stability in matches only. It may be efficient for one of two bidders to win all the licenses. Pairwise stability in matches only simply says an equal exchange of one license each does not raise the sum of valuations for the two bidders.

Intuitively, our estimator will measure the importance of clustering patterns and other patterns on the map in Figure 1. One insight is that this way of looking at the map can yield a consistent estimator under pairwise stability in matches only, which is weaker than efficiency. Indeed, Fox (2010b) proves that nonparametric identification of features of $\bar{\pi}(w_a, x_J)$ can occur equally as well with the conditions from pairwise stability as with the conditions from efficiency (also known as full stability). We return to nonparametric identification in Section 5.3.

The definition of efficiency uses knowledge of the private values $\epsilon_{a,j}$ and, if some licenses are not allocated to bidders, fixed effects ξ_j. Our estimation strategy will not recover estimates of the distributions of these unobservables, as we have discussed.

When we turn to measuring efficiency at the end of the paper, we will use the following measure of efficiency, which focuses on the contribution to valuations arising from observed license (package) and bidder characteristics.

Definition 4. An allocation of bidders to licenses $A = \{J_1, \ldots, J_N\}$ is **deterministically efficient** whenever

$$\sum_{a \in N} \bar{\pi}_\beta(a, J_a) \geq \sum_{a \in N} \bar{\pi}_\beta(a, J'_a)$$

for all other partitions $\{J'_1, \ldots, J'_N\}$ of L, where a partition satisfies $\bigcup_{a=1}^{N} J'_a \subseteq L$ and $J_a \cap J_b = \emptyset \ \forall \ a, b \in N$. Likewise, $\sum_{a \in N} \bar{\pi}_\beta(a, J'_a)$ for some partition $\{J'_1, \ldots, J'_N\}$ is a cardinal (non-ordinal) **measure of deterministic efficiency**.

4.2. Experimental Evidence on Pairwise Stability

The rest of Section 4 motivates why the spectrum auction outcome satisfies pairwise stability in matches only. Banks et al. (2003) conducted experimental evaluations of the FCC simultaneous ascending auction. The authors assigned valuation functions to subject bidders and let the winning subjects keep their profits. A key advantage of experimental data is that the valuation functions of bidders are experimentally induced and hence observed in the data (Bajari and Hortaçsu, 2005). We can test directly whether the auction outcome satisfied pairwise stability in matches only.

Banks et al. consider 52 auctions, each with 10 licenses for sale and between 6 and 8 bidders. In some cases, bidder valuation functions exhibited complementarities between some subset of the 10 licenses, and other times bidder valuations were additive across licenses. Within each auction, we analyzed each pair of licenses won by different bidders. We checked whether Definition 1, pairwise stability in matches only, holds for each pair of licenses. We calculate the percentage of the inequalities that are satisfied within each auction. The mean auction had 95.1% of its inequalities formed by the exchange of licenses between two winning bidders satisfy Definition 1. We feel that the approximation of Definition 1 to outcomes to these experimental auctions is high. Of course, the real C block auction has many more bidders and licenses than these experiments, and so the experiments cannot easily be extrapolated to the C block setting.

More ambitiously, one might be interested in the fraction of auctions where the restrictions fit the data perfectly: 100% of theoretically valid inequalities are satisfied. 29 out of the 52 auctions satisfy pairwise stability in matches only. In more than half of the experiments, the restrictions of pairwise stability in matches only are completely satisfied. We repeat the same exercises for pairwise stability in both matches and prices, which is Definition 2. The mean percentage of satisfied inequalities across the 52 auctions is lower than before, at 88%. Also, only 9.6% (5 out of 52) of the auctions satisfy Definition 2 perfectly: prices are such that bidders would prefer the licenses they won over alternative licenses. Thus, pairwise stability in matches only has more experimental evidence in its favor than pairwise stability in both matches and prices.

4.3. Lack of Swapping Licenses After the Auction

After the auction, reports in the trade media and government records indicate there was very little swapping of licenses. Swaps would have been legal: the FCC's unjust enrichment rules penalized transfers only to bidders that were not qualified for the C block, not swaps between C block bidders. Swapping licenses (perhaps with side payments) would have been direct evidence against the outcome being pairwise stable in matches only. While any negotiation is costly, the total bids in the C block auction were more than $10 billion, suggesting that negotiation time would be a small cost to incur in order to improve the profitability of winning bidders.

Cramton (2006) interprets the lack of immediate, post-auction resale as evidence that the C block auction's outcome is efficient, a stronger condition than pairwise stability in matches only. We think Cramton's interpretation is too strong. During the ten year period after the auction, many of the C block bidders were involved in mergers to create the large, national mobile phone carriers of today. Most of these mergers were with companies that did not directly bid in the C block auction. Fox and Perez-Saiz (2006) describe some of these mergers and show that they were primarily designed to expand the geographic coverage area of providers. The revealed preference of C block bidders to participate in mergers to increase scale is evidence that the winning packages may have been too small. Mergers are a costlier form of license adjustment than exchanges, and it is possible an outcome could be pairwise stable in matches only but inefficient, due to an inefficiently small scale for most winning bidders. Consolidation may increase valuations, but swapping licenses may not.

4.4. Results of Brusco and Lopomo (2002) and Engelbrecht-Wiggans and Kahn (2005)

Section 2.6 presented suggestive evidence that bidders might have been implicitly colluding through the auction mechanism. We are not ready to conclude that there was definitely collusion. However, we believe the evidence in favor of implicit collusion is strong enough that any structural estimator for spectrum auction data should be consistent under the models of implicit collusion in simultaneous ascending auctions in the literature.

Brusco and Lopomo (2002), or BL, and Engelbrecht-Wiggans and Kahn (2005), or EK, present models of simultaneous ascending auctions that in many cases have equilibria where implicit collusion between bidders occurs. A common theme will be that BL's and EK's examples often satisfy pairwise stability in matches only. Note that finding symmetric, perfect Bayesian equilibrium to complex dynamic games can be challenging, as consistent sets of beliefs for all players must be found. BL and EK primarily prove theorems about what might be considered simple examples. To our knowledge, there are no general theorems about perfect Bayesian equilibria to spectrum auctions with arbitrary sets of players, licenses and payoff structures. However, Kwasnica and Sherstyuk (2007) conduct experiments with the simultaneous ascending auction. They find that bidders' behavior shares many of the features of the BL and EK equilibria.

We consider two examples from BL. There are two bidders and two licenses. Each bidder has a (privately observed) payoff π^1 for license 1, π^2 for license 2, and

$\pi^{1,2} = \pi^1 + \pi^2 + k$ for licenses 1 and 2 for some $k > 0$. The vector (π^1, π^2, k) is drawn independently across the two bidders from the support $[0, 1]^2 \times [\underline{k}, \overline{k}]$.

BL first study the case of $\underline{k} = \overline{k} = 0$, or no complementarities. In BL's Proposition 2, they find an equilibrium where the two bidders open with bids of 0 on the item with the higher private value. If the two bidders open with bids on different items, they split the items at a price of 0 and the auction ends. If the two bidders bid on the same item, bidding continues until the bids reach $\Delta\pi = \pi^1 - \pi^2$ for one of the two bidders. At that point, the bidder whose value of $\Delta\pi$ has been reached switches to the second item at a price of 0. The auction ends. Although not emphasized by BL, the outcome of this equilibrium satisfies pairwise stability in matches only.

Lemma 1. *In the BL equilibrium in their Proposition 2, the outcome always satisfies pairwise stability in matches only.*

Proof. There are two sets of outcomes. First, the bidders a and b may split the licenses after the first round. Without loss of generality, this happens when $\pi_a^1 \geq \pi_a^2$ and $\pi_b^1 < \pi_b^2$. Addition gives

$$\pi_a^1 + \pi_b^2 > \pi_a^2 + \pi_b^1,$$

and Definition 1 is satisfied. Second and again without loss of generality, bidder a may win license 1 after bidder b deviates to win license 2 when the price of license 1 exceeds $\Delta\pi_b = \pi_b^1 - \pi_b^2$. We thus know $\Delta\pi_b < \Delta\pi_a$. Rearranging the inequality gives

$$\pi_a^1 + \pi_b^2 > \pi_a^2 + \pi_b^1,$$

and again Definition 1 is satisfied. □

It is the use of $\Delta\pi$ to decide when to switch that ensures that this implicitly collusive outcome satisfies pairwise stability in matches only. There are two reasons why this use of $\Delta\pi$ is not arbitrary. First, a bidder switching to the non-preferred license before $\Delta\pi$ would be leaving money on the table: the other bidder might drop out at $\Delta\pi - \eta$ for $\eta > 0$. The second reason is the notion of interim efficiency in Remark 5 in Appendix A. This appendix contains a series of remarks about the robustness of the equilibrium in BL's example. Using references to explicit results in BL and EK, Appendix A suggests that the existence of implicitly collusive equilibria (and to a lesser degree, outcomes that satisfy pairwise stability in matches only) may be relatively robust to the number of bidders, the number of licenses, correlation in private values for each bidder (ex post high and low types), commonly observed correlation in private values for each bidder (ex ante high and low types), ex ante asymmetries in the distribution of private values for each license, and the concern about unstudied equilibria to the model. Further, we believe pairwise stability in matches, by focusing on exchanges that keep the number of licenses won by each bidder the same, satisfies some of the spirit behind budget constraints.

In a second example, BL study the case with large complementarities, or $\underline{k} > 1$. In BL's equilibrium in their Proposition 7, bidders are split into three groups. The first group has a low valuation for each of the two licenses if won separately and will never be intimidated to implicitly collude. The second group will settle for winning license 1 if the other bidder will settle for license 2, here at prices of 0. The third group will settle for license 2 if the other bidder will settle for license 1. This implicitly collusive outcome is inefficient because complementarities are large, $\underline{k} > 1$, and the outcome assigns the licenses to different bidders. BL discuss that if $\underline{k} = \overline{k}$, so that both bidders have the same value for the complementarities, then the value of the complementarities will always be competed away in competitive bidding, so that there will be no first group of bidders that refuse to implicitly collude. In our empirical specification for the valuation function (2), all complementarities will arise from the x_J term and the complementarities between different licenses for the same bidder will not be interacted with the bidder characteristic w_a.

Lemma 2. *In the BL equilibrium in their Proposition 7, the outcome always satisfies pairwise stability in matches only.*

Proof. There are two sets of outcomes. First, competitive bidding may be triggered and the bidder with the highest value, say a, for the package of both licenses will win both licenses. In this case, there are no licenses to exchange and pairwise stability in matches only has no bite. In the other outcome, bidders a and b may split the items so that, without loss of generality, a wins 1 and b wins 2. In BL's equilibrium, this happens only when $\pi_a^1 > \pi_a^2$ and $\pi_b^1 \leq \pi_b^2$. So, (5) becomes, for $i_a = 1$ and $i_b = 2$,

$$\pi_a^1 + \pi_b^2 > \pi_a^2 + \pi_b^1.$$

□

The equilibria BL find are natural. Having a high private-value realization for a license tells the bidder little about the valuations of its rivals. There is little to gain from bidding on a subset of licenses that are not the highest private-value realizations of the bidder. Because agents have private information, they must signal through bids to find sustainable, implicitly collusive equilibria.[8]

4.5. Demand Reduction

Demand reduction is when bidders unilaterally choose to not compete for all units they have positive valuations for. This can be profitable if they know rival bidders have decreasing returns to scale in their valuations, which can include the case of constant marginal valuations for a finite number of homogeneous items that is lower than the number of items for sale (Ausubel and Cramton, 2002).

Pairwise stability is an implication of the tatonnement conditions for the spectrum auction model of Milgrom (2000). In Appendix B, we use the tatonnement conditions of

---

[8] The two BL examples also satisfy pairwise stability in both matches and prices. In the example in Lemma 1, if the bidders split the items at prices of 0, each bidder gets the item that gives the bidder the highest value. If instead they bid up the price on a single item, the bidder who uses $\Delta\pi$ to deviate to the second item also has a higher post-auction profit for the second item at the closing prices. Likewise, the interesting outcome in Lemma 2 has the bidders splitting the items for sale at prices of 0. Other examples may have outcomes that violate pairwise stability in matches and prices while satisfying pairwise stability in matches. In a world where licenses have asymmetric marginal distributions so that both bidders often prefer license 1 to 2, the bidders may still implicitly collude by splitting the items for sale at a price of 0 (see Remark 4 in the appendix). This outcome is not pairwise stable in prices and matches because both bidders prefer license 1 at a price of 0. If the opportunity cost $\Delta\pi$ is used to govern which bidder coordinates on bidding on license 1 (say bid on license 1 when $\Delta\pi > c$ for some constant $c > 0$), the outcome of splitting the licenses will be pairwise stable in matches only.

Milgrom to demonstrate that both Definitions 1 and 2 are satisfied in a model of demand reduction in simultaneous ascending auctions, without complementarities. The analysis of Milgrom requires straightforward bidding; strategic bidding during the auction itself is not allowed. Both BL and EK prove that competitive bidding is a Bayesian Nash equilibrium to the simultaneous ascending auction.

4.6. Existence of a Pairwise Stable Allocation Under Complementarities

While the true data generating process is likely a dynamic Nash game, we rely on the conditions of pairwise stability in matches only for estimation. Milgrom (2000) and Hatfield and Milgrom (2005) give a key condition under which a competitive equilibrium, and so, a pairwise stable in matches only allocation, Definition 1, is guaranteed to exist in a many-to-one matching environment like a spectrum auction, where one bidder matches to many licenses but each license is matched to only one bidder. The key condition is that the valuation functions of bidders exhibit substitutes, not complementarities, across multiple licenses in the same package. Therefore, there is no general existence theorem for a pairwise stable allocation in a many-to-one matching environment with complementarities across multiple licenses in the same package.

Even if a model lacks a general existence theorem, it is certainly possible that the actual data are generated from a valid pairwise stable allocation. This is the maintained assumption for this paper. Not surprisingly there exists a continuum of private-value realizations where the C block satisfies pairwise stability in matches only, when $\beta = \hat{\beta}$, the estimated parameters. Looking ahead to the structural estimates, column 2 of Table 3 will indicate that 95% of the potential inequalities from the estimation analog of Definition 1, pairwise stability in matches only, are satisfied at the point estimates. The estimation analog does not use license-specific private values $\epsilon_{a,j}$ to fit inequalities. So the 95% of satisfied inequalities comes without relying on private values at all. By making private values $\epsilon_{a,j}$ for the observed matches between bidders a and licenses j high, and keeping $\epsilon_{a,j} = 0$ for matches that are not part of the final allocation, the fraction of satisfied inequalities can increase to 100%.

5. The Estimator

5.1. Estimator

Fox (2010a) introduces a semiparametric maximum score or maximum rank correlation estimator for many-to-many matching games with transferable utility. Maximum score was first introduced by Manski (1975) and maximum rank correlation was introduced by Han (1987). In matching, the objective function is the same for the two estimators and the difference is whether the sample grows large in the number of markets (maximum score) or in the number of agents observed in a single market (maximum rank correlation). Our application fits into the one large market asymptotic argument. The estimator is semiparametric as no parametric distributions for the unobservables $\epsilon_{a,j}$

and the fixed effects ξ_j are imposed. The estimator is based on forming the empirical analog of the inequalities in Definition 1, which uses data on matches but not prices.

We will estimate the parameters β in (1), the valuation function. To make the econometric objective functions more readable, we will sometimes write $\bar{\pi}_\beta(a, J) \equiv \bar{\pi}_\beta(w_a, x_J)$ for bidder a and package J. Let H be the number of winning bidders. First consider a simple auction with two bidders a and b and two licenses 1 and 2. In the data, a wins 1 and b wins 2. The estimator $\hat{\beta}$ is any vector that satisfies the inequality

$$\bar{\pi}_\beta(a, \{1\}) + \bar{\pi}_\beta(b, \{2\}) \geq \bar{\pi}_\beta(a, \{2\}) + \bar{\pi}_\beta(b, \{1\}).$$

The inequality is satisfied whenever the sum of the deterministic parts of bidder valuations is not increased by an exchange of licenses. With only two bidders and two licenses, typically many such parameters β will satisfy the inequality. Any one of those parameters is a valid point estimate. Further, the confidence interval for β will be large. We need to use all of the data to produce an estimate of β with a smaller confidence interval.

For the full sample, the estimator $\hat{\beta}$ is any vector that maximizes the objective function

$$Q^{\text{match}}(\beta) = \frac{2}{H(H-1)} \sum_{a=1}^{H-1} \sum_{b=a+1}^{H} \sum_{i=1}^{|J_a|} \sum_{j=1}^{|J_b|}$$

$$\times 1[\text{pop}((J_a \backslash \{i\}) \cup \{j\}) \leq w_a, \text{pop}((J_b \backslash \{j\}) \cup \{i\}) \leq w_b] \cdot$$

$$1[\bar{\pi}_\beta(a, J_a) + \bar{\pi}_\beta(b, J_b) \geq \bar{\pi}_\beta(a, (J_a \backslash \{i\}) \cup \{j\}) + \bar{\pi}_\beta(b, (J_b \backslash \{j\}) \cup \{i\})], \quad (7)$$

where pop(J) gives the population of the package J: $\sum_{k \in J} \text{pop}_k$. The objective function $Q^{\text{match}}(\beta)$ considers all combinations of two licenses won by different bidders, a and b. Only inequalities involving counterfactual packages with populations under the initial eligibility constraints for both bidders are included. If an inequality is satisfied, the count or score of correct predictions increases by 1. The estimator's inequalities include only the deterministic portion of valuations, $\bar{\pi}_\beta(w_a, x_J)$. Many inequalities will remain unsatisfied, even at the true parameter vector, because of the unobserved realizations of private values $\epsilon_{a,j}$, which also affect matches. Because not all inequalities can be satisfied, changing the score objective to squaring the deviations from deterministic pairwise stability makes the estimator inconsistent. $Q^{\text{match}}(\beta)$ is a step function and as a result, in a finite sample there can be a continuum (or multiple continua) of parameters that maximize $Q^{\text{match}}(\beta)$. Any maximizer is a consistent estimator. Reporting a 95% confidence region for each element of β provides a description of the estimates that encompasses the finite sample ambiguity in the point estimates.

The complete valuation function (1) includes fixed effects ξ_j. If instead we worked with $\bar{\pi}_\beta(a, J_a) + \sum_{j \in J} \xi_j$ in (7), the ξ_j's would enter into both sides of (5) and difference out. Therefore, we do not need to estimate these fixed effects.

The maximum rank correlation approach only estimates the parameters β in $\bar{\pi}_\beta(w_a, x_J)$, not the distribution of any error terms. Parameters in a function of observables have always been the object of interest in maximum score (Manski, 1975; Han, 1987; Horowitz, 1992; Matzkin, 1993). We could instead write down a likelihood as the outcome to a dynamic game and attempt to estimate the distribution of unobservables. This would be difficult. 1) There are $N \cdot L = 255 \cdot 480 = 122,400$ private values $\epsilon_{a,j}$ and the likelihood would be an integral over them. 2) The simultaneous ascending auction has multiple equilibria, including both competitive and implicitly collusive equilibria. Estimation would have to impose one equilibrium is selected. 3) An implicitly collusive equilibria is sustained by threats of punishment not typically seen on the path of play. 4) Each ξ_j would be treated as a random effect as it cannot be differenced out of the likelihood. 5) Computing a likelihood would require evaluating all possible packages. Several papers in the collection Cramton, Shoham and Steinberg (2006) explore how even computing a winning bid in an alternative combinatorial auction is an active area of research in computer science. Cramton (2006) argues that a major motivation for using the simultaneous ascending auction over a package-bidding combinatorial auction is the computational challenge in evaluating all packages. Evaluating all possible packages is not a tractable estimation strategy in the C block environment.

5.2. Consistency and Inference

The asymptotics in Fox (2010a), for our application, are in the number $H \leq N$ of winning bidders observed in one large market. We assume the econometrician observes some finite number of recorded agents from an aggregately deterministic auction. Indeed, we introduce the fiction that the real-life matching market or auction with H winning bidders is a subset of some very large auction. As H gets larger in the asymptotic approximation, the researcher collects more data on a single auction.[9] The fiction is not to be taken literally; there were only 85 winning bidders in the C block.

We repeat the notion of pairwise stability in matches only from Fox (2010a) so readers understand the assumptions we make. Let each winning bidder have characteristics w. Likewise, each license has characteristics y. The function $\zeta(Y)$ take a set $Y = \{y_1, \ldots, y_J\}$ of $J < \infty$ license characteristics and forms a package characteristic $x = \zeta(Y)$. The function ζ is known to the researcher. The exogenous features of the matching market include $g_w(w)$, the density of bidder characteristics w, as well as $g_y(y)$, the density of license characteristics y. The terms w, x and y can all be vectors. The equilibrium outcome in this auction includes a density $g_{x,w}^{\beta,S}(\langle w, x \rangle)$, which gives the frequency of the ordered pair $\langle w, x \rangle$, representing a winning package with characteristics x for a bidder with characteristics w. The density $g_{x,w}^{\beta,S}(\langle w, x \rangle)$ is an endogenous outcome, and so it is a function of the vector of the unknown parameters β and the unknown densities $S = (g_\epsilon(\epsilon), g_\xi(\xi \mid y))$ of both the license- and bidder-specific private values and the license fixed effects.

Assumption 4. *Let $\langle w_1, x_1 \rangle$ and $\langle w_2, x_2 \rangle$ be two hypothetical winning packages and let $x_1 = \zeta(Y_1)$ for $Y_1 = \{y_{1,1}, \ldots, y_{1,J_1}\}$ and $x_2 = \zeta(Y_2)$ for $Y_2 = \{y_{2,1}, \ldots, y_{2,J_2}\}$. Let $y_{1,i_1} \in Y_1$ and $y_{2,i_2} \in Y_2$. Let $x_3 = \zeta((Y_1 \setminus \{y_{1,i_1}\}) \cup \{y_{2,i_2}\})$ and $x_4 = \zeta((Y_2 \setminus \{y_{2,i_2}\}) \cup$*

---

[9] Parameter values may affect the rate at which H increases. If complementarities are large, winning packages will likely be large and H might grow slowly compared to the number of licenses. The asymptotics are in H and the speed at which H itself increases is not directly playing a role in our arguments.

$\{y_{1,i_1}\}$). Assume, for any β and S,

$$\bar{\pi}_\beta(w_1, x_1) + \bar{\pi}_\beta(w_2, x_2) > \bar{\pi}_\beta(w_1, x_3) + \bar{\pi}_\beta(w_2, x_4)$$

if and only if

$$g_{x,w}^{\beta,S}(\langle w_1, x_1\rangle) \cdot g_{x,w}^{\beta,S}(\langle w_2, x_2\rangle) > g_{x,w}^{\beta,S}(\langle w_1, x_3\rangle) \cdot g_{x,w}^{\beta,S}(\langle w_2, x_4\rangle).$$

This assumption is also called a rank order property. The econometric version of pairwise stability in matches is a condition on the equilibrium sorting pattern. It says that if an exchange of licenses produces a lower sum of deterministic valuations, then the frequency of observing winning packages with the same characteristics as the exchange of licenses must be lower than observing winning packages with characteristics that give higher valuations. Note that the same number of licenses is won by a bidder on both sides of the inequalities in Assumption 4. We do not ask why a single bidder did not win more licenses, because bidders may split the licenses among them because of intimidatory collusion, not efficiency. Assumption 4 rules out estimation challenges involving multiple equilibria. By using data from only one auction, we condition on the equilibrium being played in that market. Also, the assumption implicitly assumes an equilibrium allocation $g_{x,w}^{\beta,S}(\langle w, x\rangle)$ exists. Fox (2010a) discusses Assumption 4 in more detail.

Given additional assumptions on the support of β, w and x, Fox (2010a) shows that the estimator is consistent, following the original arguments of Han (1987). One part of the usual way of showing that an extremum estimator is consistent is proving that its population objective function is uniquely globally maximized at the true parameter value (Newey and McFadden, 1994). For a simpler example that gives intuition for this type of estimator, consider the binary choice maximum score estimator (Manski, 1975). In the binary choice model, a consumer chooses to buy a product with characteristics c whenever $c'\gamma + \mu > 0$, where c are regressors, γ are parameters to estimate, and μ is an error term. Manski gives conditions on the error term such that a rank order property holds: $c'\gamma > 0$ if and only if $\Pr(\text{buy} \mid c) > \frac{1}{2}$. The finite sample objective function with n observations on consumers indexed by i is

$$Q_n(\gamma) = \frac{1}{n}\sum_{i=1}^{n}(1[i\text{ buys in data}] \cdot 1[c'\gamma > 0] + 1[i\text{ does not buy in data}] \cdot 1[c'\gamma < 0]).$$

Using the law of iterated expectations and the law of large numbers, the population objective function is then

$$\text{plim} Q_n(\gamma) = E_c(\Pr(\text{buy} \mid c) \cdot 1[c'\gamma > 0] + (1 - \Pr(\text{buy} \mid c)) \cdot 1[c'\gamma < 0]).$$

At the true value of γ, the rank order property ensures that $\Pr(\text{buy} \mid c) > \frac{1}{2}$ whenever $c'\gamma > 0$, so that the true γ globally maximizes the population objective function if some element of c has continuous support so that ties have measure zero. Continuous support also ensures that the maximum is unique; we omit the argument.

Sherman (1993) shows that the maximum rank correlation estimator is $\sqrt{H}$-consistent and asymptotically normal. The asymptotic variance matrix derived in Sherman (1993) is complex to use in that it requires additional nonparametric estimates of components that appear in the variance matrix. To avoid this complexity we use a resampling procedure known as subsampling, which is consistent under fairly weak conditions. As Politis and Romano (1994) state, essentially the only assumption needed for the validity of subsampling is that the estimator has a limiting distribution.

5.3. Nonparametric Identification of Features of the Valuation Function

Fox (2010b) proves a sequence of theorems about the nonparametric identification of features of $\bar{\pi}(w_a, x_J)$. Functional form assumptions are not required for $\bar{\pi}$; it is not known up to a finite vector of parameters β.

In this paper, we wish to identify the allocative efficiency of the observed C block allocation and counterfactual allocations, following Definition 4. Fox (2010b) does not prove that the total sum $\sum_{a \in N} \bar{\pi}(a, J_a)$ of an allocation of licenses to bidders is nonparametrically identified up to scale. Some parametric assumptions will be needed for our measurements.

The nonparametric identification theorems do provide some insights. Expression (2) says that the parametric valuation function can be decomposed into one term involving the sorting of bidders with higher valuations to packages with more population and several terms involving the geographic complementarities among a set of licenses in the same package. Theorem 5.2 in Fox (2010b) states that we can identify the sign of the first term nonparametrically, i.e. whether bidder heterogeneity is a complement to package population. We see whether assortative matching between heterogeneous bidders and package populations is more likely than anti-assortative matching in the observed allocation. Likewise, Theorem 5.6 in Fox states that we can identify nonparametrically whether each proxy for complementarities really has a positive sign in valuations. More importantly, Theorems 5.3 and 5.7 in Fox state that we can nonparametrically identify the relative magnitudes of each of the complementarity measures. We can identify, without any functional form assumptions, the ratio of the complementarities between bidder heterogeneity and population to the geographic complementarities between licenses. In effect, we identify the ratio of the two-sided complementarities between bidders and licenses to the one-sided complementarities between licenses in the same package.

The parametric functional form in (2) specifies the valuation function to be known up to parameters that, by the above arguments, are known to be nonparametrically identified. Thus, the extrapolation out of sample to examine the efficiency of counterfactual allocations relies on a parametric functional form whose parameters represent objects that are nonparametrically identified within the sample.

6. Main Estimates of Valuation Functions

Table 3 lists estimates of β in the valuation function, (2), from the maximum rank correlation estimator. The numbers in parentheses are 95% confidence intervals. Computational details are discussed in the footnote to the table. Columns 1 and 2 report baseline estimates. The number of inequalities is 13,428. As in (2), because matches

Table 3. *Maximum Rank Correlation Estimates of Valuation Parameters*

Column	(1)	(2)	(3)	(4)
Distance parameter δ	4	4	2	2
Population * bidder eligibility	+1	+1	+1	+1
		Superconsistent		
Population/distance two markets in a package	0.32 (0.31,0.50)	0.32 (0.30,0.47)	1.06 (0.87,1.56)	0.86 (0.58,1.06)
Trips between markets in a package in the American Travel Survey		0.03 (−0.08,0.40)		−0.62 (−0.96,−0.27)
Total trips between airports in markets in a package (thousands)		−0.16 (−0.37,0.34)		−0.26 (−0.51,0.51)
# possible inequalities		13,428		
% inequalities correct	0.944	0.945	0.956	0.960

The objective function was numerically maximized using differential evolution (Storn and Price, 1997). More than ten runs were performed for all specifications. The reported point estimates are the best found maxima.

The parentheses are 95% confidence intervals computed using subsampling. Subsampling uses 200 replications and 25 packages per replication (sampled without replacement). For each 25 packages, we use only the inequalities where all licenses are from the sampled packages. Subsampled confidence regions are not necessarily symmetric around the point estimate. In unreported results, we take subsets of the data by using only the inequalities corresponding to 120 out of the 480 licenses in the United States. For each license, we evaluate the valuation functions using the full winning package, whether all of the package's licenses are among the subset of 120 or not. The confidence regions from drawing licenses are similar to the regions found by drawing packages. Subsampling has not been extended to allow for spatial autocorrelation, so we do not adjust for such correlation. Parameters that can take on only a finite number of values (here ±1) converge at an arbitrarily fast rate; they are superconsistent.

are qualitative outcomes, we normalize the coefficient on $\text{elig}_a(\sum_{j\in J}\text{pop}_j)$ to be ±1. We estimate the other parameters β separately for the +1 and −1 normalizations and pick the vector with the highest number of satisfied inequalities. The results show that +1 is the correct point estimate. This fits the fact in Figure 2 that bidders with more initial eligibility win packages with more total population.

Column 1 includes only one proxy for geographic complementarities: geographic distance, (3). The coefficient of $\beta_{\text{geo.}} = 0.32$ means, at the furthest extrapolation, that if one bidder with the maximum eligibility of 1 were to win the entire United States (population of 1), then the also maximized complementarities (value of $1 \cdot \beta_{\text{geo.}}$) would give a total package value of $1 \cdot 1 + 0.32 \cdot 1 = 1.32$. The value from complementarities corresponds to $0.32/1.32 = 24\%$ of the total package value. Across the 85 winning packages, the standard deviation of $\text{elig}_a(\sum_{j\in J}\text{pop}_j)$ is 0.029 and the standard deviation of geocomplem.$_j$ is 0.024. Because the standard deviations are roughly the same, the coefficient estimate $\beta_{\text{geo.}} = 0.32$ implies that variation in the geographic location of licenses, geocomplem.$_j$, is roughly $0.32/1 = 32\%$ as important in explaining the valuation of winning bidders as variation in the match between bidders with more eligibility and packages with more population, $\text{elig}_a(\sum_{j\in J}\text{pop}_j)$.[10]

---

[10] The estimates ignore the fact that the bidder OmniPoint was, outside of the auction, given (by the FCC) a special pioneer license for the highly populated market of New York City. We do not include OmniPoint's license so we do not need to make assumptions about its disposition in the counterfactual allocations in Table 5. Failing to account for OmniPoint's total package might induce small biases in the parameter estimates.

Column 2 adds the two travel based complementarity measures to the specification. Now, not only do we measure the relative importance of $\text{elig}_a(\sum_{j\in J}\text{pop}_j)$ and complementarities in sorting, we see which measure of complementarities is most important. Total trips using all forms of travel has a statistically insignificant coefficient of 0.03, while the coefficient on geocomplem.$_J$, 0.32, is similar to the coefficient in column 1. One interpretation is that the geographic pattern of clustering reflects more than just customers wishing to make calls while traveling. Other forms of complementarities include marketing and cost-of-service synergies. The second travel measure, air travel, has a negative point estimate with a wide confidence interval that includes 0. The wide confidence regions are not surprising given the high correlation in Table 2 between the two travel measures among winning packages.[11] The upper bound of its 95% confidence interval of 0.34 does allow for important role for air travel synergies.[12] The standard deviation of air travel complementarities is 0.017, which is only a little smaller than, say, the standard deviation of geographic distance complementarities of 0.024. Given the similar standard deviations, the point estimates show air travel substantially reduces valuations compared to geographic distance or the composite measure of travel.

Table 3 also lists the percentage of satisfied inequalities at the point estimates, which is a measure of statistical fit. 95% of the inequalities are satisfied. Vertical differences in bidder valuations for licenses and complementarities across licenses in the same package can explain most of the sorting patterns at the pair of licenses level.

Columns 3 and 4 use a different δ parameter in the geographic complementarities measure based on distance and population, (3). In column 3, the coefficient on geographic distance complementarities is 1.06, higher than the point estimate of 0.32 in column 1. This is not because of a change in units; the standard deviation of geographic distance complementarities for $\delta = 2$ is similar to the standard deviation for $\delta = 4$. A similar increase of the point estimate happens when comparing the point estimate on geographic complementarities in column 4 to the point estimate in column 2. The coefficients on the two travel complementarities are negative with wide confidence intervals on air travel.[13]

7. Estimators Using Other Inequalities

This section explores two alternative estimators that use inequalities based on different theoretical conditions for estimation. One estimator is explicitly incompatible with intimidatory equilibria where agents split the items for sale and the other estimator uses closing prices data to explain why one license is preferred to another license by

---

[11] In a previous draft, we reported results using inequalities whether or not they violated the initial eligibility constraints. Using more inequalities substantially reduces the width of the confidence regions on some of the estimates.

[12] The point estimate on air travel is a lower bound on the complementarities from air travel, as air travel also appears in the ATS survey and is being double counted. Roughly 75% of trips in the ATS are by car; the fraction by air increases with distance.

[13] If estimation does not drop inequalities where counterfactual package populations violate initial eligibility constraints, the point estimates on the complementarity measures are positive with smaller confidence intervals.

Table 4. *Estimators Using Other Inequalities*

Type of inequalities	Transfer of 1 license		Swaps of 1 license w/prices	
	(1)	(2)	(3)	(4)
Population * bidder eligibility	+1 Superconsistent	+1	0.36 (−0.13,0.41)	0.36 (−0.15,0.42)
Population/distance two markets in a package	6.7 (−3.0,9.2)	9.8 (−12,14)	0.12 (−0.23,0.15)	0.12 (−4.82,0.15)
Trips between markets in a package in the American Travel Survey		−0.37 (−0.49,1.2)		0.03 (−0.81,0.19)
Total trips between airports in markets in a package (thousands)		−0.1 (−0.39,0.06)		−0.09 (−0.22,0.04)
Price (in *trillions*)			−1 Superconsistent	−1
# possible inequalities	16,084		73,409	
% inequalities correct	0.950	0.953	0.913	0.914

All estimates use $\delta = 4$. See Table 3 for computational details.

a bidder. We show that the alternative estimators generate bizarre estimates of bidder valuations.

7.1. Estimates with Forced Transfers of Licenses

Columns 1 and 2 of Table 4 consider a variant of the estimator where bidder a adds a license j to its package J without swapping the license for another. Let $\eta(j)$ be the bidder who wins license j. A corresponding inequality for a's decision not to win j involves an increase in the number of a's licenses by 1 and a decrease in the number of $\eta(j)$'s licenses by 1. Let H be the set of 85 winning bidders. The estimator is any parameter value that maximizes

$$Q^{\text{addmatch}}(\beta) = \sum_{a=1}^{H} \sum_{j=1}^{L} 1[a \neq \eta(j)] \cdot 1[\text{pop}(J_a \cup \{j\}) \leq w_a] \cdot$$
$$1[\bar{\pi}_\beta(a, J_a) + \bar{\pi}_\beta(\eta(j), J_{\eta(j)}) \geq \bar{\pi}_\beta(a, J_a \cup \{j\}) + \bar{\pi}_\beta(\eta(j), J_{\eta(j)} \setminus \{j\})],$$

where $J_{\eta(j)}$ is the complete package won by the bidder that won license j. The estimator imposes the condition that a did not increase its package by one license because the sum of valuations of a and $\eta(j)$ would go down from doing so: it would be less efficient. This condition may be untenable because a may instead have not added the license j to a's package because of a fear of suffering retaliation from bidder $\eta(j)$. Therefore, maximizing $Q^{\text{addmatch}}(\beta)$ produces an inconsistent estimator under the intimidatory equilibria in Brusco and Lopomo (2002) and Engelbrecht-Wiggans and Kahn (2005).

Columns 1 and 2 report a priori unreasonable estimates. In column 1, the coefficient on complementarities is implausibly large (although with a wide confidence interval).

The point estimate of 6.7 shows the contribution to valuations from complementarities is roughly 7 times the valuation from winning an equivalent amount of population (times eligibility). The coefficient in column 2 is an even larger 9.8.[14]

7.2. Estimates with Prices

Columns 3 and 4 of Table 4 report estimates using both matches and prices data. The maximum score objective function is based on Definition 2, pairwise stability in both matches and prices. When using price in addition to matches data, the estimator $\hat{\beta}$ is any vector that maximizes the objective function

$$Q^{\text{price}}(\beta) = \frac{1}{L^2} \sum_{i=1}^{L} \sum_{j=1}^{L} 1[\eta(i) \neq \eta(j)] \cdot 1[\text{pop}((J_{\eta(i)} \setminus \{i\}) \cup \{j\}) \leq w_{\eta(i)}]$$

$$1[\tilde{\pi}_\beta(\eta(i), J_{\eta(i)}) - \tilde{\pi}_\beta(\eta(i), (J_{\eta(i)} \setminus \{i\}) \cup \{j\}) \geq p_i - p_j], \quad (8)$$

where p_i is the final, closing price of license i and $\eta(i)$ is defined above. Here, we impose the restriction that bidder $\eta(i)$ prefers to win its package $J_{\eta(i)}$ instead of winning $(J_{\eta(i)} \setminus \{i\}) \cup \{j\}$, or license j instead of i, at the closing prices to the auction. In other words, we impose the condition that the closing prices explain why bidder $\eta(i)$ won license i instead of j. Rearranging the inequality gives the inequality in Definition 2, except that like the other estimators, the private-value terms $\epsilon_{a,j}$ are not included, as is standard for maximum rank correlation estimators. Fixed effects ξ_j cannot be allowed in this type of estimator; for consistency using prices, we must assume the fixed effects are always zero.[15]

Akkus et al. (2012) were the first to use the estimator with prices and perform a Monte Carlo study. In all of our Monte Carlo experiments (see Appendix C for some) with i.i.d. private-value terms $\epsilon_{a,j}$, the estimator performs extremely well.

In columns 3–4 of Table 4, we have included price, measured in *trillions* of dollars. The coefficient on price is normalized to be ±1 and estimated to be −1. Taken literally, the coefficient on 0.36 on $\text{elig}_a(\sum_{j \in J} \text{pop}_j)$ in column 3 says that the value of a bidder with eligibility equal to the entire US's population winning the entire US is $360 billion (although it is not statistically distinct from zero). Likewise, the value of complementarities from a nationwide license is $120 billion. These estimates are absurdly high, given that the bids for the C block totaled $10.1 billion. Indeed, the annual revenue for the wireless phone industry in 2006, with nine or more active licenses per territory (not just the C block), was $113 billion. It is unlikely that bidders in 1996 felt the C block

---

[14] A previous draft included point estimates from inequalities that violate the initial eligibility constraint. The confidence intervals are smaller, typically exclude zero, and the point estimates are even larger in magnitude.
[15] In Definition 2 and the objective function (8), ξ_j does not difference out of the inequality, like it does in Definition 1. Therefore, we have also estimated specifications including population and population density (we used all inequalities, not just those under initial eligibility). The point estimates on the covariates that affect the efficiency of alternative allocations of licenses to bidders are then $886 billion for winning the entire US's population for $\text{elig}_a(\sum_{j \in J} \text{pop}_j)$ and $743 billion for the geographic complementarities geocomplem_j for winning the entire US. These estimates dramatically reinforce the finding that, under an alternative scale normalization, the coefficient β_{price} is estimated to be economically small.

had 7–8 times the stock of *profit* potential as the yearly flow of *revenue* from all blocks combined 10 years later.

How is the model fitting the outcome data? Only the ratio of two parameters that enter structural payoffs linearly, say $\beta_{\text{geo}}/\beta_{\text{price}}$, is identified from an inequality. A high dollar value for non-price package and bidder characteristics is equivalent to saying the estimated coefficient on license price β_{price} would be economically small in magnitude if some other characteristic's coefficient was normalized to ± 1. A small coefficient on price is consistent with the finding in Section 2.4 that population and population density, characteristics mostly subsumed into ξ_j, explain most price variation.

As we discussed in Section 2.5, Ausubel et al. (1997) included measures of the runner-up bidder's potential complementarities in a license-level price regression, and found a nonzero but economically small coefficient. Together, the estimates from (8) and the price regressions suggest that prices may not clear the market in the sense of sorting price taking bidders to different packages in a competitive market. Pairwise stability in prices and matches, Definition 2, may not be satisfied.

8. Counterfactual Efficiencies and Policy Implications

8.1. Actual and Counterfactual Deterministic Efficiencies

We compare the efficiency of the observed allocation of licenses to that of several counterfactual license allocations. The parameter estimates from the previous section suggest that some of the various measures of complementarities are important determinants of bidder valuations. However, the auction allocated licenses to 85 different bidders, which suggests that an improvement in deterministic efficiency is possible by grouping licenses into larger winning packages. Furthermore, our earlier results suggest that demand reduction and intimidatory collusion may be present in the auction, which causes or exacerbates this inefficiency.

We did not include any bidder or license characteristics in the deterministic valuation function (2) that (ex ante) would seemingly make smaller licenses optimal. Nor can we think of obvious measures in the US mobile phone industry, as this industry does seem to benefit from geographic scope, if only because of the demand side preference for larger calling areas that we estimate in Bajari, Fox and Ryan (2008). It is more or less given that geographically larger licenses will improve deterministic allocative efficiency. The purpose of this section is to quantitatively measure exactly how much more geographically larger licenses will improve allocative efficiency. If the efficiency gain is small, there might not be much scope for policy in improving the allocation. We use the point estimates from column 2 of Table 3 and the definition of deterministic efficiency in Definition 4. The results are in Table 5.

For a given allocation of licenses, Table 5 reports the value of $\sum_{a \in N} \bar{\pi}_\beta(a, J_a)$. It is easiest to look at the last row of the table first. The last row considers the largest winner (and bidder with the highest initial eligibility), NextWave, winning all 480 licenses in the continental United States. NextWave was initially eligible for 176 million people, or 71% of the 1990 population. Therefore, the contribution to total value from NextWave's differential use for licenses is 0.71. NextWave winning all licenses would maximize the

Table 5. *Counterfactual Deterministic Efficiency From Five Allocations, Point Estimates Imposing Eligibility Constraints*

Allocation	$\text{elig}_a(\sum_{j \in J} \text{pop}_j)$	Geographic Distance	Air Travel	ATS Trips	Total
C block: 85 winning packages	$1 \cdot 0.39 =$ 0.39	$0.32 \cdot 0.47 =$ 0.15	$-0.16 \cdot 0.20 =$ -0.03	$0.03 \cdot 0.27 =$ 0.01	0.52
All 480 licenses won by different bidders	$1 \cdot 0.17 =$ 0.17	$0.32 \cdot 0 =$ 0	$-0.16 \cdot 0 =$ 0	$0.03 \cdot 0 =$ 0	0.17
Each 47 MTAs separate package	$1 \cdot 0.20 =$ 0.20	$0.32 \cdot 0.72 =$ 0.23	$-0.16 \cdot 0.04 =$ -0.01	$0.03 \cdot 0.17 =$ ~0	0.43
Four large, regional licenses (top four of the 85 actual winners win)	$1 \cdot 0.50 =$ 0.50	$0.32 \cdot 0.96 =$ 0.31	$-0.16 \cdot 0.37 =$ -0.06	$0.03 \cdot 0.58 =$ 0.02	0.77
Nationwide license for entire United States (NextWave wins)	$1 \cdot 0.71 =$ 0.71	$0.32 \cdot 1 =$ 0.32	$-0.16 \cdot 1 =$ -0.16	$0.03 \cdot 1 =$ 0.03	0.90

Eligibility, population and all three complementarity proxies range from 0 to 1. These counterfactuals use the point estimates from column 2 of Table 3. Only licenses in the continental United States are considered. For the 47 MTAs in the continental United States as well as the four large regions, the top winners in the actual auction are assortatively matched to the counterfactual packages in order of population. For example, NextWave always wins the package with the highest population.

three geographic-complementarity proxies, at values of 1 each. So the total differential value (excluding the ξ_j's) of a nationwide license is $1 \cdot 0.71 + \beta_1 \cdot 1 + \beta_2 \cdot 1 + \beta_3 \cdot 1$, where the three β's are the complementarity parameters estimated in column 2 of Table 3. The total value of a nationwide license is then 0.90.

Now consider the other four efficiency evaluations. The first row considers the actual allocation of bidders to licenses in the C block auction. The total surplus generated by the C block is 0.52, less than the 0.90 from the nationwide license. The terms in three of the four columns (excepting the column using the negative point estimate) are smaller than in the bottom column, suggesting that the C block failed to maximize the potential benefits from complementarities.

The second row considers an extreme where all 480 licenses are won by separate bidders. There can be no across-license complementarities. We impose that the licenses auctioned in the C block are the lowest level of disaggregation possible. There are 255 C block bidders (losers and winners). We assortatively match bidders to licenses by initial eligibility for bidders and population for licenses, so that NextWave wins New York, for example. For the $480 - 255 = 225$ licenses with the smallest populations, we say they are won by bidders with the lowest (255th) level of initial eligibility. The results show that the contribution from the $\text{elig}_a(\sum_{j \in J} \text{pop}_j)$ term is 0.17, smaller than the actual C block allocation's value of 0.39 by about half. This reflects bidders with lower valuations winning licenses.

The third row considers grouping the 480 C block licenses into 47 packages reflecting the 47 Major Trading Areas (MTAs) in the continental United States used for the 1995 AB spectrum auction. No C block license belongs to more than one MTA. The MTAs are natural groupings centered around large metropolitan areas, but including lots of rural territory as well. Again, we assortatively match winning bidders to licenses based on initial eligibility and population, so again NextWave wins New York. However, in the C block auction NextWave won New York and a lot more, so here the contribution from assortative matching between heterogeneous bidders and package population is low, at 0.20. However, the design of the MTA boundaries ensures that most local, geographic distance complementarities are captured. The measure of geographic distance complementarities rises from 0.47 to 0.72. On the other hand, the MTAs are only local areas, and so a great deal of travel between regions occurs across MTAs. The values of the travel geographic complementarity measures are small under the MTA scenario. The total value of this allocation is 0.43, lower than the actual C block allocation.

The fourth row considers splitting the United States into four large regions: the Northeast, Midwest, South and West. We assign each of the 47 MTAs to one of these groupings. The Midwest is roughly from Pittsburgh to Wichita, Washington, DC is in the north, and Oklahoma and Texas (other than El Paso) are in the south. We take the four largest winners by initial eligibility and assortatively match them to the four regions by population. NextWave's package is the Midwest; it is still slightly smaller in population than the package NextWave won in the C block. The fourth row shows that the contribution from differential bidder valuations is now higher, the measure of geographic distance complementarities is close to 1, and the two travel measures are about twice as high as the C block values. Thus, a system of four large regions raises the value from complementarities compared to the C block and significantly raises the amount of the US population won by high-value bidders. The United States is much bigger than a typical Western European nation; auctioning four licenses is workable plan that captures a large fraction of the maximum possible deterministic efficiency, 0.77 out of 0.90. These point estimates indicate that the efficiency from four large regional licenses is 0.77, which is 48% higher than the efficiency of 0.52 from the C block allocation. The figure of 48% is a lower bound on the improvement in deterministic efficiency, because the same, high value bidder could win two or more of the large, regional licenses in an actual simultaneous ascending auction with four licenses.

8.2. Policy Implications for Bidder Anonymity

In 2006, the FCC changed policies so that the bidder identities of submitted bids are now anonymous. The intention of this rule change was to limit intimidation and signaling. A previous draft of this paper addressed one mechanism of signaling other bidders (jump bids) more explicitly. Here, our policy counterfactuals suggest that the simultaneous ascending auction produced inefficiently small winning packages. If bidder anonymity is one way of reducing the scope of intimidation, then it may make the final allocation of licenses to bidders more efficient.

8.3. Bidders with Overly Optimistic Beliefs

One limitation of our revealed preferences approach to estimate bidder valuation functions is that the winning bidders may have overstated the short-term value of the licenses. All the bidders bid less than the long-term license value (compared to the high valuations for 30 MHz of 1900 MHz spectrum in the modern mobile phone industry) but the C block winners might have been overly optimistic about the short-term prospects. In this case, overly optimistic beliefs would lead to a disjunction between the estimated valuation function consistent with bidder behavior and the function a social planner focused on a short horizon might use to evaluate the efficiency of the allocation. Optimistic beliefs may have led the large winners to devote more money to initial eligibility than losers and small winners, thus raising the the estimated economic importance of $\text{elig}_a \cdot (\sum_{j \in J} \text{pop}_j)$ in the valuation function, (2). Likewise, bidders may have overstated complementarities, meaning that the parameters β in (2) are too high. As we impose that the coefficient on $\text{elig}_a \cdot (\sum_{j \in J} \text{pop}_j)$ is normalized to 1, these competing biases may have over or underestimated the parameters of β relative to those a social planner would prefer.

8.4. Competitive Scale-Reducing Economic Forces

Intimidation and demand reduction reduce the size of winning packages and make the resulting mobile phone industry lack true national players. At least three other economic forces that are compatible with competitive bidding work in the same direction. First, bidders may have monetary budget constraints, so that financial constraints from outside of the auction make the auction outcome inefficient. See also Remark 9 in the appendix. Second, a bidder may run down eligibility by focusing on a smaller license and be unable to switch to a license with a larger population once the price of the smaller license becomes too expensive. Path dependence may lock a bidder into considering only substitute licenses with relatively small populations. See also Remark 10. Third, the FCC's rules prevented one bidder from winning more than 98 licenses in the C and F auctions. Only the largest winner, NextWave, was anywhere close to bumping up against this constraint.

The previous descriptive literature and our bidding anecdotes in Section 2.6 show that bid signaling did go on during the C block auction (Cramton and Schwartz, 2000). However, measuring the extent or effectiveness of signaling seems difficult when these other factors also reduce winning package sizes. We note all three of these competitive reasons for inefficiently-small winning packages are consistent with larger licenses raising efficiency.

8.5. Producer vs Consumer Surplus

By "efficiency", we mean the efficiency of the allocation of licenses to bidders, from the viewpoint of bidders and not the consumers of mobile phones. As in all papers on auctions adopting a revealed preferences approach, we cannot use the outcome of the auction to identify a social planner's welfare function separate from the valuations of the bidders. In Bajari, Fox and Ryan (2008), we did measure the preferences of consumers; we estimated a substantial willingness to pay for larger coverage areas. Here, we consider a spectrum auction with only one license per territory and only new entrant bidders, so the auction will increase the number of competitors by one in each territory regardless of which bidder wins each territory. Therefore, an outcome of the auction that results in carriers with geographically large coverage areas will not directly allow such entrants to exercise more market power than entrants with smaller coverage areas, except through offering the higher quality product of more coverage. Still, there might be other reasons why a company with a large geographic coverage area is bad for consumers, such as the exercise of market power in vertical markets (say handset provision) that might deter technological innovation. Our approach is based on bidder revealed preference and will not detect such consumer welfare losses, although we are unaware of any empirical evidence showing negative effects of geographically large coverage areas in the mobile phone industry.

9. Conclusions

We measure the efficiency of the outcome of an FCC spectrum auction using a structural model of the deterministic portion of bidder valuations. A spectrum auction is a complex dynamic game, with many bidders and many items for sale. The simultaneous ascending auction is theoretically susceptible to intimidatory collusion. Intimidation may result in winning packages that are inefficiently small, as bidders split the market to coordinate on paying less to the seller.

Our approach to estimation uses an econometric version of pairwise stability in matches. Pairwise stability says the sum of valuations from two winning bidders must not be increased by swapping licenses. There are four pieces of evidence suggesting that pairwise stability is likely to hold in simultaneous ascending auctions: experimental evidence, the lack of post-auction swapping, theoretical analysis of implicit collusion, and theoretical analysis of demand reduction. Intimidatory collusion is sustained by bid signaling and threats of retaliation by reverting to straightforward bidding.

We employ a matching maximum rank correlation estimator, which maximizes the number of inequalities that satisfy pairwise stability. The estimator is computationally simple as it avoids evaluating all possible counterfactual packages. Also, the estimator controls for additive, license-specific fixed effects. We estimate valuations using data on only the matches between bidders and licenses, not data on the closing prices. Indeed, we show that two alternative estimators, including one using price data, produce bizarre estimates using the C block data.

Our estimates empirically validate the FCC's focus on complementarities when designing the mechanism for allocating radio spectrum. Also, the spectrum auction itself produces a much higher surplus than awarding licenses through the FCC's prior practices, such as lotteries. However, we find that the final allocation of licenses was allocatively inefficient before considering private values. Deterministic efficiency would increase by 48% by awarding four large, regional licenses to the four highest-value bidders. A nationwide license would capture even more of the total deterministic efficiency. To a rough degree, our finding that splitting the United States into four large chunks raises deterministic efficiency validates the European approach of offering nationwide licenses and hence capturing all geographic complementarities, as the largest Western European countries (France, Germany, the United Kingdom) are, in terms of population, on the scale of about a fourth of the United States.

A. Remarks About Generalizations to the Main BL Example

This appendix discusses extensions to the main BL example, where there are two bidders, two items for sale, and no complementarities. The main BL example is Proposition 2 in their paper. We follow the expositional style of the theory papers BL and EK, where formal theorems are proved for simple examples and extensions are discussed less formally.

Remark 1. In the remarks that follow, often the analysis of BL and EK is worried that implicit collusion, as in the BL examples in Section 4.4, is not sustainable. In these cases, bidders may find the expected value (over the private values of rivals) for competing for all the licenses to be higher than implicitly colluding. Competitive bidding does not provide a concern for the estimator. BL and EK show competitive bidding is a perfect Bayesian equilibrium that will result in an efficient outcome. Efficient outcomes are automatically pairwise stable in matches only.

Remark 2. The main BL examples requires an assumption on the marginal distribution of π^1 and (because they are identically distributed) π^2. However, little is assumed about the joint distribution. Thus, the BL examples allow a bidder's private values to be correlated across licenses. There can be ex-post high-value or low-value bidders. In this case, the identities of the high-value or low-value bidders are not common knowledge. Also see Theorem 4 in EK, which also studies the case of joint dependence between π^1 and π^2, or ex-post high and low private-value bidders. Note that if our bidder heterogeneity measure w was private information and entered π^1 and π^2, it would just induce correlation between π^1 and π^2. So a private w is nested in the above analysis. A privately observed w is convenient theoretically, but does not apply to the actual C block auction, where our measure of w was disclosed to rival bidders before the auction.

Remark 3. The main BL example studies the case of two bidders and two licenses. The discussion following Theorem 4 in EK states that simple implicitly collusive equilibria are possible whenever the number of licenses exceeds the number of bidders. Further, Proposition 4 in BL finds a collusive equilibrium where there are more bidders than licenses. In this equilibrium, high-value bidders raise the price to weed out weak bidders, before attempting to signal and implicitly collude. All implicitly-colluding bidders must win an item for collusion to be successful. Because of the need to weed out the weak bidders, we would not necessarily expect to see very low prices in intimidatory-collusive equilibria. Indeed, the prices in the C block were not particularly low.

Remark 4. The main BL example studies the case where each license has an identical marginal distribution. Remark 3 in BL explores the case where each of the private values for licenses 1 and 2 has a known, bidder-invariant marginal distribution and $E[\pi^1] > E[\pi^2]$: license 2 on average has a lower private value π^2. This is the case in our valuation specification: package observables (to the bidders) x_J and $\sum_{j=1}^{J} \xi_j$ shift around the mean of valuations. A formal statement of BL's Proposition 2 refers to BL's Condition A, which is a condition on the marginal distribution that ensures that even a high-value type would find it profitable to implicitly collude and win one item for a low price rather than competing and winning both items. As a high-value type a does not know the privately observed values of a rival b, this is a condition on the rival b's mean private value. Remark 3 in BL states collusion can take place under different assumptions about the private value distribution, i.e. when π^1 and π^2 have different marginal distributions and $E[\pi^1] > E[\pi^2]$. A stronger condition on the distribution of each π^j is needed because the bidder who bids on the item with a lower mean private value must be induced to stick with that item and not also compete for the other license in competitive bidding. If collusion is sustainable, it follows roughly the form in the main BL example. However, the specific equilibrium outcome described in Remark 3 of BL does not necessarily satisfy pairwise stability in matches only because bidders use a multiplicative constant (arising from the particular support conditions on the private values in the example used in BL's Remark 3) to modify the value of π^1 for comparison with π^2 when deciding whether to bid on item 2 instead. If instead the bidders used the opportunity cost $\Delta \pi = \pi^1 - \pi^2$ to decide whether to open bidding on items 1 or 2, pairwise stability in matches will occur. If BL's conditions for implicit collusion fail to hold, Remark 1 states competition ensues, under which the pairwise stability in matches only condition still holds.

Remark 5. The main BL example presents just one symmetric, perfect Bayesian equilibrium to the game in question. Straightforward bidding is always a symmetric equilibrium, as Remark 1 discusses, for example. Neither BL or EK claim to find all possible symmetric equilibria to the game. It is possible that symmetric equilibria that do not satisfy pairwise stability in matches only exist (without changing the assumptions of the main BL example), although they are currently unknown. However, BL use an equilibrium property known as *interim efficiency*. Proposition 3 in BL suggests that, under additional conditions, that the outcome in the main example maximizes a "weighted sum of all types' expected surplus", where the maximization is taken over all incentive compatible allocations such that each bidder always receives one object. Thus, Proposition 3 in BL uses the property of interim efficiency to suggest that, if possible, bidders would want to coordinate on the equilibrium in the main example, rather than some arbitrary, undiscovered equilibrium. Thus, the existence of other symmetric, perfect Bayesian equilibria that have yet to be found should not dissuade us from considering the equilibrium in the main example.

Remark 6. Consider a case where there are ex ante, commonly observed high and low-type bidders. For simplicity, say the payoff to bidder a from winning license j is $\pi_a^j = w_a \cdot y_j + \epsilon_{a,j}$, where here the scalar y_j is a characteristic that raises valuations, such as the population of the license. Let the standard deviation and support of the mean-zero, i.i.d. private values $\epsilon_{a,j}$ be small and let $w_a \gg w_b$ and $y_1 \gg y_2$, so that $\pi_a^1 + \pi_b^2 > \pi_a^2 + \pi_b^1$ for all realizations of $\epsilon_{a,j}$. Implicit collusion without signaling could occur: the high-type bidder a is allocated license 1 and the low-type bidder b is allocated license 2, both at prices of 0. This equilibrium could be sustained through threats of competitive bidding if π_a^1 is always sufficiently higher than π_a^2 and the loss to a from competitive bidding, the valuations of the rival $\pi_b^1 + \pi_b^2$, is sufficiently large. This type of equilibrium does not involve signaling and is based on public information (here w_a, w_b, y_1 and y_2) rather than private information, so the equilibrium strategies are dissimilar to those in the main example, although the outcomes are quite similar. The equilibrium outcome still satisfies pairwise stability in matches only, as the bidders assortatively match to licenses: high w with high y and $\pi_a^1 + \pi_b^2 > \pi_a^2 + \pi_b^1$. Thus, ex ante, commonly observed high- and low-type bidders can be compatible with pairwise stability in matches only, even without signaling. Altogether, this discussion explains Assumption 1, which says that if valuations are monotone in the scalar w_a, w can be publicly observed.[16]

Remark 7. Another case is when there are known asymmetries in the values for bidders a and b for licenses 1 and 2. For example, bidder a may be known to on average have a high private value for license 1, while bidder b may be known to on average have a high private value for license 2. Notationally, bidder-and-license-specific private values $\pi_j^a = \epsilon_{a,j}$ have commonly observed, bidder and license specific distributions $F_{a,j}$. This general notation encompasses Remark 6 as a special case. Under bidder and license specific distributions, a lot of the information on bidder idiosyncratic valuations is public and, definitionally, no longer privately observed information. Indeed, Theorem 5 in EK allows the distribution of private values to vary across bidder/item pairs. In that case, there is less need to signal using the bidding mechanism to coordinate and pairwise stability in matches only may not occur. Intuitively, this type of result just applies a folk-theorem-like result to the publicly observed part of payoffs. For an equivalent of our Lemma 1 to hold, valuations must be private or valuations must be monotone in the observed heterogeneity, as in the example in Remark 6. Many conditions under which Lemma 1 does not hold, in our experience, involve aspects of valuations that are asymmetric across licenses and are not privately observed. However, Section 2.5 argued based on institutional details that this type of extreme asymmetry was not common in the C block auction. As we have discussed, there is little evidence that the major winning bidders were local businessmen with pre-announced, bidder- and license-specific valuations for particular licenses. Further, colluding based on ex ante known bidder-and-license-specific asymmetries would not require signaling via the bidding mechanism, as in the main BL example. Section 2.6 and the previous, descriptive empirical literature argue that there is evidence of bidders signaling each other using the bidding mechanism, rather than relying on known asymmetries in bidder valuations.

Remark 8. Related to the previous remark is the possibility of *asymmetric*, perfect Bayesian equilibria.[17] Asymmetric equilibria are not discussed in BL and EK. Consider a case where $E[\pi^j]$ is high, π^j has a small, bounded support relative to $E[\pi^j]$, and π^1 and π^2 are independent. Then the equilibrium outcome where bidder a wins license 1 at a price of 0 and bidder b wins license 2 at a price of 0 is sustainable by the threat of resorting to competitive bidding. The cost of competition so high that

---

[16] Of course, one could construct similar examples (without one of the conditions $w_a \gg w_b$ and $y_1 \gg y_2$, perhaps) where the high-type bidder is assigned the low-type item. These outcomes strike us as unnatural: we cannot imagine the pre-game coordination that would lead to a high-value bidder accepting a low-value item.

[17] Note the two uses of the word "asymmetric": asymmetric equilibria here and asymmetric bidder valuations in Remark 7.

the expected value of colluding is high for both bidders, even if a has a higher private value for 2 and b has a higher private value for license 1. Here the equilibrium is asymmetric because bidder a takes an action regardless of its license values. Note that this equilibrium requires no signaling: bidders divide up the items before private values are realized. The empirical evidence in the previous literature and in Section 2.6 is strongly suggestive that signaling took place.[18] Thus, relying on the equilibrium refinement of symmetry as in symmetric, perfect Bayesian equilibria, seems logical for a first estimator for simultaneous ascending auctions with complementarities given the empirical evidence. By restricting attention to symmetric equilibria, we follow the theory on the simultaneous ascending auction.

Remark 9. Bulow, Levin and Milgrom (2009) have emphasized the role of budget constraints in a much more recent spectrum auction than the C block. Pairwise stability in matches only respects one version of a monetary budget constraint: the number of matches of each bidder is the same on the left and right sides of the inequality. Pairwise stability in matches only does not ask why one bidder did not win more licenses at the expense of a rival, only why license j_1 was won by bidder 1 and license j_2 was won by bidder 2 and not the reverse. Thus, the inequalities in pairwise stability in matches only capture some of the spirit of budget constraints. We note that almost all other estimators for auctions of a single item do not respect budget constraints at all; bids are suggested to be informative of valuations, not budget constraints.[19]

Remark 10. FCC spectrum auctions have eligibility rules. At the end of the C block auction, all but two bidders, who were competing for a single license, had settled on their final, winning packages. Their final-round eligibilities were only slightly above the populations of their winning packages. The condition of pairwise stability in matches only is not a statement about the behavior of bidders at the auction's final round, when they had no free eligibility. Rather, pairwise stability in matches only is a condition on the entire data generating process (all the rounds of the auction) and the final allocation that results from the data generating process. The interesting signaling behavior in the BL and EK models arise at the start of the auction, when bidders' eligibilities are above the populations of their final winning packages.

Remark 11. An additional concern in simultaneous ascending auctions is the exposure problem, where a bidder fails to secure additional licenses to complete a package and therefore prefers to not to win a license it did win at the end of the auction. Cramton (2006) argues that the price discovery advantages of and the withdrawal options in the FCC's simultaneous ascending auction design mitigate any exposure problem. Pairwise stability in matches will still hold under an exposure problem if valuations would not be increased by swapping licenses. Given the exposure problem, pairwise stability holds if the bidders are exposed on the "best of a bad menu" of licenses.

---

[18] One could possibly write down an asymmetric equilibrium with signaling. In that case, the empirical evidence of signaling would not be evidence in favor of the symmetry refinements used in BL and EK.

[19] One estimation approach would be to impose pairwise stability in matches only for exchanges of licenses with similar closing prices. Our experiments show that this reduces the empirical power (increases the width of the confidence intervals) of the estimator considerably.

Table 6. *Valuations for Two-Bidder Examples of Demand Reduction*

	Bidder a	Bidder b, case 1	Bidder b, case 2
License 1	$\pi_a^1 \geq \pi_a^2$	$\pi_b^1 \leq \pi_a^1, \pi_b^1 \leq \pi_b^2$	$\pi_b^1 \leq \pi_a^1, \pi_b^1 \geq \pi_b^2$
License 2	$\pi_a^2 \leq \pi_a^1$	$\pi_b^2 \leq \pi_a^2, \pi_b^2 \geq \pi_b^1$	$\pi_b^2 \leq \pi_a^2, \pi_b^2 \leq \pi_b^1$
Both 1 & 2	$\pi_a^{1,2} = \pi_a^1 + \pi_a^2$	$\pi_b^{1,2} = \max\{\pi_b^1, \pi_b^2\}$	$\pi_b^{1,2} = \max\{\pi_b^1, \pi_b^2\}$

B. Demand Reduction and Pairwise Stability, Without Complementarities

Demand reduction is studied by Ausubel and Cramton (2002) for the case of sealed bid auctions of multiple homogeneous items. In a simultaneous ascending auction, demand reduction is consistent with straightforward bidding by forward-looking agents. Kagel and Levin (2001) and List and Lucking-Reiley (2000) find substantial demand reduction in experiments. This section considers demand reduction, but in a market without complementarities because of a need to refer to a Milgrom (2000) theorem. Because complementarities are the focus of our empirical work, we place this material in an appendix, although we feel the results are interesting for the estimation method. Also, Milgrom requires bidders to bid straightforwardly, so this analysis does not distinguish between publicly and privately observed information and so does not work with Bayesian Nash equilibria. On the other hand, Brusco and Lopomo (2002) and Engelbrecht-Wiggans and Kahn (2005) show that competitive bidding is a Bayesian Nash equilibrium to simultaneous ascending auctions.

Consider bidders a and b competing for two licenses 1 and 2. Use the shorthand notation $\pi_a^{1,2}$ for $\pi_a(\{1, 2\})$. Let the valuations of bidders a and b for the three possible packages be as listed in Table 6, case 1. Bidder a has a higher value for all packages. Bidder b has decreasing returns to scale: there is no incremental value from winning both licenses.

If both bidders bid straightforwardly in a simultaneous ascending auction, and ignoring minimum bid increments, a will win both licenses at prices equal to b's values: $p_1 = \pi_b^1$ and $p_2 = \pi_b^2$. However, if a reduces its demand and lets b win item 2 at $p_2 = 0$, a can win item 1 at $p_1 = 0$. Bidder b accepts this because it has a demand for only one license and prefers 2 to 1. The demand reduction outcome is inefficient: valuations are maximized by having a win both items. However, when a wins 1 and b wins 2, $\pi_a^1 + \pi_b^2 > \pi_a^2 + \pi_b^1$, so that the sum of valuations cannot be increased with license swaps. Definition 1 is satisfied as the bidders disagree on the valuation ranking of the licenses. One can use the zero prices to show Definition 2 is satisfied as well.

Now we will argue that the example does not rely on bidder disagreement over the valuation ranking. Case 2 in Table 6 changes b's valuations so that a and b agree on the valuation ranking of licenses 1 and 2: $\pi_b^1 \geq \pi_b^2$. At the beginning of the auction, with $p_1 = p_2 = 0$, bidder b will bid on item 1 as b prefers 1 and has a demand for only one item. Only at a price $p_1^\star$ such that $\pi_b^1 - p_1^\star = \pi_b^2$ will b accept winning license 2 instead of 1. If $\pi_a^1 - p_1^\star > \pi_a^2$, then substituting in $p_1^\star = \pi_b^1 - \pi_b^2$ to $\pi_a^1 - p_1^\star > \pi_a^2$ again gives $\pi_a^1 + \pi_b^2 > \pi_a^2 + \pi_b^1$. Definitions 1 and 2 are satisfied.

What if in case 2, $\pi_a^1 - p_1^\star = \pi_a^1 - (\pi_b^1 - \pi_b^2) < \pi_a^2$? If a finds it profitable to reduce its demand, a will reduce its demand on license 1 and win 2, leaving $\pi_a^2 + \pi_b^1 > \pi_a^1 + \pi_b^2$. Again, $p_1^\star$ is set, by straightforward bidding, to make a and b coordinate on a pairwise stable outcome. Definitions 1 and 2 are satisfied. The points made in this example are more general.[20]

Lemma 3. *Consider straightforward bidding in a simultaneous ascending auction with demand reduction. Under the tatonnement conditions of Milgrom (2000), the outcome is a pairwise stable outcome to a matching game where the maximum number, or quota, of licenses that a bidder can win is the number of licenses the bidder won in the outcome. Both Definitions 1 and 2 are satisfied.*

Proof. Let the allocation portion of the demand reduction outcome be A, and let bidder a's wining package be J_a. For all bidders a, redefine a's valuation for a package J to be negative infinity if J has more licenses than J_a. $\pi_a(J) = -\infty$ for $|J| > |J_a|$. Then Milgrom's tatonnement process theorems (Theorems 2 and 3 in Milgrom) show that the simultaneous ascending auction will find a competitive equilibrium (core outcome) of the economy with the truncated valuation functions. Pairwise stability, Definition 2, is implied by being in the core of the economy with truncated valuation functions. As the swaps considered in Definition 2 do not change the number of licenses won by any bidder, the valuations under the swaps are the same as under the pre-truncated valuation functions. So the outcome is pairwise stable under a matching game where bidders cannot add additional licenses to their package. □

Under demand reduction, the outcome may not be efficient, but there is no reason to believe that there exist swaps of licenses that would raise sums of valuations. The lemma does not explain how much demand reduction will go on: the unilateral incentive to reduce demand requires knowledge that another bidder has strong decreasing returns to scale.[21]

C. Monte Carlo for Estimator with Both Matches and Price Data

Fox (2010a) presents Monte Carlo studies showing that the finite-sample performance of the maximum score estimator using matches only is reasonable. However, for a small number of bidders and licenses and a high variance of the error term, the estimator uses data only on matches can have high bias and root mean squared error (RMSE) in a finite sample, as random noise from the $\epsilon_{a,j}$ terms dominates the matching, leaving little signal in the sorting pattern seen in the data. Like similar results in Akkus et al. (2012), Table 7 reports results from a Monte Carlo study from a one-to-one, two-sided matching market. Each bidder a matches to at most one license j, and the payoff of a bidder is

---

[20] The conditions for Milgrom's tatonnement process theorem rule out complementarities, in part to avoid the exposure problem. Definition 1 requires only that sum of valuations not be raised by swapping licenses. It is compatible with many forms of the exposure problem. See Remark 11.

[21] The initial eligibilities of other bidders are known before bidding starts. Therefore, some forms of decreasing returns are public knowledge. Further, Cramton (2006) interprets the purchase of spectrum in a small, quick, post-auction sale (a bidder did not make its payments) by NextWave as evidence that NextWave was reducing its demand during the auction.

Table 7. *Maximum Score Monte Carlo: Comparing Using Data On Only Matches To Data On Both Matches and Prices Under Tatonnement Assumptions with Noise Dominating Matches, True Value Is 1.5*

# bidders	# licenses per auction	# spectrum auctions	error std. dev.	Matches bias	Matches RMSE	Matches + Prices bias	Matches + Prices RMSE
30	30	1	1	0.587	1.93	0.005	0.03
10	10	10	1	0.330	1.05	0.009	0.07
30	30	1	5	1.22	4.22	0.02	0.09
10	10	10	5	1.69	7.36	−0.02	0.446

$\bar{\pi}_\beta(a, j) + \epsilon_{a,j} = x_{1,a}x_{1,j} + \beta x_{2,a}x_{2,j}$. There are two characteristics for bidders and two for licenses, with characteristics for each side distributed as a bivariate normal with means (10, 10), variances (1, 1) and covariance 0.5. The errors are i.i.d. normal with standard deviations listed in the table. For each auction we draw observable characteristics and unobservable error terms and compute an equilibrium assignment and vector of prices using the primal and dual linear programs for two-sided matching (Koopmans and Beckmann, 1957; Shapley and Shubik, 1972). The true β is 1.5. The example is chosen to make using only matches look bad: there is not much signal about $\bar{\pi}_\beta$ in the sorting patterns if the realized matches are visually plotted in characteristic space, especially in the second half of the table where the standard deviation of $\epsilon_{a,j}$ is five times higher than in the upper part of the table. Note that for the C block the map in Figure 1 shows that there are clear sorting patterns; this Monte Carlo study makes using match data bad to show the potential advantages of using price data. The finite-sample bias and RMSE are always much lower with continuous transfer data, even though the data on matches alone are uninformative. For all four cases the absolute value of the bias is small for small samples, and for three of the four cases the RMSE is low compared to the true value of 1.5.

Table 7 shows a major advantage of using price data: the finite-sample performance is much better if prices are generated from a tatonnement process. There are several advantages to using only match data, even if the prices are generated by a tatonnement process. This first is transparency: there is only one type of dependent variable, so inferring parameters from the US map of winning bidders is straightforward. With two types of dependent variables, it is not as clear where identification arises from. The second is robustness. In this paper, we review models where prices are not generated by a tatonnement process, but the matches are still robust to pairwise swaps.

References

Akkus, Oktay, J. Anthony Cookson, and Ali Hortacsu, "The Determinants of Bank Mergers: A Revealed Preference Analysis," 2012. University of Chicago working paper.

Athey, Susan and Philip A. Haile, "Nonparametric Approaches to Auctions," *Handbook of Econometrics*, 2008, 6.

Ausubel, Lawrence M. and Peter Cramton, "Demand Reduction and Inefficiency in Multi-Unit Auctions," July 2002. University of Maryland working paper.

—, **R. Preston McAfee, and John McMillan**, "Synergies in Wireless Telephony: Evidence from the Broadband PCS Auctions," *Journal of Economics and Management Strategy*, Fall 1997, 6 (3), 497–527.

Avery, Christopher, "Strategic Jump Bidding in English Auctions," *Review of Economic Studies*, 1998, 65, 185–210.

Bajari, Patrick and Ali Hortaçsu, "Are structural estimates of auction models reasonable? Evidence from experimental data," *Journal of Political Economy*, 2005, 113 (4), 703–741.

—, Jeremy T. Fox, and Stephen Ryan, "Evaluating Wireless Carrier Consolidation Using Semiparametric Demand Estimation," *Quantitative Marketing and Economics*, 2008, *6* (4), 299–338.

Banks, Jeffrey, Mark Olson, David Porter, Stephen Rassenti, and Vernon Smith, "Theory, experiment and the federal communications commission spectrum auctions," *Journal of Economic Behavior and Organization*, 2003, *51*, 303–350.

Brusco, Sandro and Giuseppe Lopomo, "Collusion via Signalling in Simultaneous Ascending Bid Auctions with Heterogenous Objects, with and without Complementarities," *Review of Economic Studies*, April 2002, *69* (2), 407–436.

Bulow, Jeremy, Jonathan Levin, and Paul Milgrom, "Winning Play in Spectrum Auctions," February 2009. Stanford University working paper.

Cantillon, Estelle and Martin Pesendorfer, "Combination Bidding in Multiple Unit Auctions," 2006. Université Libre de Bruxelles working paper.

Chapman, James T.E., David McAdams, and Harry J. Paarsch, "Bounding Best-Response Violations in Discriminatory Auctions with Private Values," *American Economic Review*, May 2007, *97* (2), 455–458.

Chiappori, Pierre-André, Bernard Salanié, and Yoram Weiss, "Partner Choice and the Marital College Premium," 2012. Columbia University working paper.

Choo, Eugene and Aloysius Siow, "Who Marries Whom and Why," *The Journal of Political Economy*, November 2006, *114* (1), 175–201.

Coase, Ronald H, "The Federal Communications Commission," *Journal of Law and Economics*, 1959, *2*, 1–40.

Cramton, Peter, "Simultaneous Asending Auctions," in Peter Cramton, Yoav Shoham, and Richard Steinberg, eds., *Combinatorial Auctions*, MIT Press, 2006.

— and Jesse A. Schwartz, "Collusive Bidding: Lessons from the FCC Spectrum Auctions," *Journal of Regulatory Economics*, May 2000, *17* (3), 229–252.

— and —, "Collusive Bidding in the FCC Spectrum Auctions," *Contributions to Economic Analysis & Policy*, 2002, *1* (1), Article 11.

—, Yoav Shoham, and Richard Steinberg, *Combinatorial Auctions*, MIT Press, 2006.

Crawford, Vincent P. and Elsie Marie Knoer, "Job Matching with Heterogeneous Firms and Workers," *Econometrica*, 1981, *49* (2), 437–450.

Dagsvik, John K., "Aggregation in Matching Markets," *International Economic Review*, February 2000, *41* (1), 27–57.

Day, Robert and Paul Milgrom, "Core-Selecting Package Auctions," *International Journal of Game Theory*, June 2007.

Demange, Gabrielle, David Gale, and Marilda A. Oliveira Sotomayor, "Multi-Item Auctions," *The Journal of Political Economy*, August 1986, *94* (4), 863–872.

Edelman, Benjamin, Michael Ostrovsky, and Michael Schwarz, "Internet Advertising and the Generalized Second Price Auction: Selling Billions of Dollars Worth of Keywords," *American Economic Review*, March 2007, *97* (1).

Engelbrecht-Wiggans, Richard and Charles M. Kahn, "Low revenue equilibria in simultaneous auctions," *Management Science*, 2005, *51* (3), 508–518.

Fevrier, Philippe, Raphaelle Préget, and Michael Visser, "Econometrics of Share Auctions," 2004. CREST working paper.

Fox, Jeremy T., "Estimating Matching Games with Transfers," 2010. University of Michigan working paper.

—, "Identification in Matching Games," *Quantitative Economics*, 2010, *1* (2), 203–254.

— and Hector Perez-Saiz, "Mobile Phone Mergers and Market Shares: Short Term Losses and Long Term Gains," September 2006. NET Institute Working Paper 06-16.

Gentzkow, Matthew, "Valuing New Goods in a Model with Complementarity: Online Newspapers," *The American Economic Review*, 2007, *97* (3), 713–744.

Haile, Philip A. and Elie Tamer, "Inference with an Incomplete Model of English Auctions," *Journal of Political Economy*, February 2003, *111* (1), 1–51.

Han, A.K., "Non-parametric analysis of a generalized regression model," *Journal of Econometrics*, 1987, *35*, 303–316.

Hatfield, John William and Paul R. Milgrom, "Matching with Contracts," *American Economic Review*, September 2005, *95* (4), 913–935.

Hong, Han and Matthew Shum, "Econometric models of asymmetric ascending auctions," *Journal of Econometrics*, 2003, *112*, 327–358.

Horowitz, Joel L., "A Smoothed Maximum Score Estimator for the Binary Response Model," *Econometrica*, May 1992, *60* (3), 505–551.

Hortacsu, Ali and David McAdams, "Mechanism Choice and Strategic Bidding in Divisible Good Auctions: An Empirical Analysis of the Turkish Treasury Auction Market," *Journal of Political Economy*, 2010, *118* (5), 833–865.

Kagel, John H. and Dan Levin, "Behavior in Multi-Unit Demand Auctions: Experiments with Uniform Price and Dynamic Vickrey Auctions," *Econometrica*, 2001, *69* (2), 413–454.

Kastl, Jakub, "Discrete Bids and Empirical Inference in Divisible Good Auctions," *Review of Economic Studies*, 2011, *78* (3), 974–1014.

Kelso, Alexander S. and Vincent P. Crawford, "Job Matching, Coalition Formation, and Gross Substitutes," *Econometrica*, November 1982, *50* (6), 1483–1504.

Koopmans, Tjalling C. and Martin Beckmann, "Assignment Problems and the Location of Economic Activities," *Econometrica*, January 1957, *25* (1), 53–76.

Krasnokutskaya, Elena, "Identification and Estimation of Auction Models with Unobserved Heterogeneity," *The Review of Economic Studies*, 2011, *78* (3), 293–327.

Kwasnica, Anthony M. and Katerina Sherstyuk, "Collusion and Equilibrium Selection in Auctions," *The Economic Journal*, 2007, *117* (516), 120–145.

Leonard, Herman B., "Elicitation of Honest Preferences for the Assignment of Individuals to Positions," *Journal of Political Economy*, June 1983, *91* (3), 461–479.

List, John A. and David Lucking-Reiley, "Demand Reduction in Multiunit Auctions: Evidence from a Sportscard Field Experiment," *American Economic Review*, 2000, *90* (4), 961–972.

Manski, Charles F., "Maximum Score Estimation of the Stochastic Utility Model of Choice," *Journal of Econometrics*, 1975, *3* (3), 205–228.

Matzkin, Rosa L., "Nonparametric identification and estimation of polychotomous choice models," *Journal of Econometrics*, 1993, *58*, 137–168.

McAfee, R. Preston and John McMillan, "Analyzing the Airwaves Auction," *Journal of Economic Perspectives*, Winter 1996, *10* (1), 159–175.

Milgrom, Paul, "Putting Auction Theory to Work: The Simultaneous Ascending Auction," *Journal of Political Economy*, 2000, *108* (2), 245–272.

Milgrom, Paul R. and Robert J. Weber, "A Theory of Auctions and Competitive Bidding," *Econometrica*, 1982, *50* (5), 1089–1122.

Moreton, Patrick S. and Pablo T. Spiller, "What's In the Air: Interlicense Synergies in the Federal Communications Commission's Broadband Personal Communication Service Spectrum Auctions," *Journal of Law and Economics*, October 1998, *41* (2, Part 2), 677–725.

Newey, W.K. and D. McFadden, "Large Sample Estimation and Hypothesis Testing," in "Handbook of Econometrics," Vol. 4, Elsevier, 1994, pp. 2111–2245.

Paarsch, Harry J. and Han Hong, *An Introduction to the Structural Econometrics of Auction Data*, MIT Press, 2006.

Politis, Dimitris N. and Joseph P. Romano, "Large Sample Confidence Regions Based on Subsamples under Minimal Assumptions," *The Annals of Statistics*, December 1994, *22* (4), 2031–2050.

Shapley, Lloyd S. and Martin Shubik, "The assignment game I: the core," *International Journal of Game Theory*, 1972, *1*, 111–130.

Sherman, Robert P., "The Limiting Distribution of the Maximum Rank Correlation Estimation," *Econometrica*, January 1993, *61* (1), 123–137.

Sørensen, Morten, "How Smart is Smart Money? A Two-Sided Matching Model of Venture Capital," *Journal of Finance*, December 2007, *LXII* (6), 2725–2762.

Storn, Rainer and Kenneth Price, "Differential Evolution – A Simple and Efficient Heuristic for Global Optimization over Continuous Spaces," *Journal of Global Optimization*, 1997, *115*, 341–359.

Wolak, Frank A., "Quantifying the Supply-Side Benefits from Forward Contracting in Wholesale Electricity Markets," *Journal of Applied Econometrics*, November 2007, *22*, 1179–1209.

PART II
The Combinatorial Clock Auction Designs

CHAPTER 5

Combinatorial Auction Design

David P. Porter, Stephen J. Rassenti, Anil Roopnarine, and
Vernon L. Smith

Introduction

Combinatorial auctions enhance our ability to efficiently allocate multiple resources in complex economic environments. They explicitly allow buyers and sellers of goods and services to bid on packages of items with related values or costs. For example, "I bid $10 to buy 1 unit of item A and 2 units of item B, but I won't pay anything unless I get everything." They also allow buyers, sellers and the auctioneer to impose logical constraints that limit the feasible set of auction allocations. For example, "I bid $12 to buy 2 units of item C OR $15 to buy 3 units of item D, but I don't want both." Finally, they can handle functional relationships amongst bids or allocations, such as budget constraints or aggregation limits that allow many bids to be connected together. For example, "I won't spend more than a total of $35 on all my bids" or "This auction will allocate no more than a total of 7 units of items F, G and H."

There are several reasons to prefer to have the bidding message space expanded beyond the simple space used for traditional single commodity auctions. As Bykowsky et al. (2000) point out, when values have strong complementarities, there is a danger of 'financial exposure' that results in losses to bidders if combinatorial bidding is not allowed. For example, in the case of complementary items such as airport take-off and landing times, the ability to reduce uncertainty to the bidder by allowing him to precisely declare his object of value, a cycle of slots for an entire daily flight pattern, is obvious: one component slot not acquired ruins the value of the flight cycle. In the same situation substitution possibilities would also be important to consider: if flight cycle A is not won, cycle B may be an appropriate though less valuable substitute for the crew and equipment available. Allocation inefficiencies due to financial exposure in non-combinatorial auctions have been frequently demonstrated in experiments beginning with Rassenti et al. (1982) (see also Porter (1999), Banks et al. (1989), Ledyard et al. (2002) and Kwasnika et al. (1998)).

This paper is concerned with designing a transparent, efficient and practical combinatorial buyers' auction for multiple distinct items with perhaps multiple units of each available. We focused this endeavor around the past auctions, debate, proposals, and tests precipitated by the FCC's quest for the design of the perfect spectrum auction. Many of the design principles ultimately implemented by us are easily extensible to more complex two sided negotiations with environmental constraints. We begin by mimicking the original FCC debate: describing several combinatorial auction designs and their potential problems. We next present some pertinent laboratory results of auction design already tried. We then describe a new combinatorial auction design that we invented to eliminate many of the problems found in previous designs. We finish with the description and testing of three auctions previously untested in the laboratory, the one currently used by the FCC in the field, a combinatorial auction design solicited by the FCC but not yet implemented, and our own new design.

FCC Auction Design Process

The FCC conducted several high priced auctions for communications' spectra during the 1990s. The Simultaneous Multi-Round Auction (SMR) design employed by the FCC has several notable features:

- Simultaneous auction of the various licenses offered
- Iterative procedure allowing round by round update of bids
- Activity rules demanding continuous participation at minimum increments

After much discussion and debate, but no controlled testing, the FCC chose to implement the SMR. The FCC did not originally opt for a combinatorial form of auction for three stated reasons:

1. <u>Computational uncertainty</u>. At any point during a combinatorial auction, the selection of the winning bids and what it would cost for competition to displace them typically requires the solution of integer programming (IP) problems. These problems are notorious for being computational burdensome: technically described as NP complete or hard problems. At issue here is that there is no guarantee that the solution can be found for such a problem in a "reasonable" amount of time when there are M bidders and N items being auctioned, and M^N is large.[1]
2. <u>Bidding complexity</u>. Combinatorial auctions would be burdensome and difficult for participants and the auctioneer for at least two reasons: there are inconceivably many packages on which a bidder might want to place bids, and selecting any subset may be strategically awkward and provide the auctioneer with incomplete information.[2]
3. <u>Threshold problem</u>. Combinatorial auctions present the following strategic impediment to efficient outcomes. Suppose each of two small bidders is bidding on a separate item, but a third bidder is bidding on a package that contains both items. Then the two small bidders must implicitly negotiate through their bidding to ascertain what price each will pay in order that the sum of both bids exceeds the package bid.

---

[1] This was referred to at various FCC discussions as the M^N boogie-man since there are that many ways N items can be allocated amongst M bidders. For example, with M = 20 and N = 20, which are very small numbers by FCC auction standards, M^N already represents more than a trillion trillion possible ways of handing out the licenses.

[2] This is the supermarket problem. You have $100 in your pocket and you are standing at the entrance to the supermarket. You despondently realize that you may never come out as you must first consider every possible way you might fill your shopping cart.

The FCC chose to implement the SMR design in all of its spectrum auctions rather than confront the above issues. Bidders in an SMR auction who had superadditive values for a particular package of licenses were subject to financial exposure if they miss acquiring at least one of those licenses. The FCC auctions implemented a withdrawal rule to "reduce" the potential financial exposure of the bidders: that is bidders had the right to withdraw bids on particular licenses subject to the fact that they were obligated to pay the difference between their own and the final winning bid if it was less. This and other SMR rules led to results in various FCC auctions that revealed some interesting perverse strategies. That prompted several auction rule adjustments. Two notable rule changes implemented were restrictions on jump bidding (the allowable increment when raising your bid) and restrictions on the number of withdrawals allowed.

Based on the previous problems associated with the SMR design and spurred by the FCC sponsored debate, several new designs have emerged.

Auction Designs for Complex Environments

It is quite natural for an auction participant to wish to reveal as little as possible concerning his interest and values for particular items or packages in an auction that simultaneously sells multiple items. In general, auction systems that provide feedback and allow bidders to revise their bids seem to produce more efficient outcomes. This feedback feature is the cornerstone of the recent FCC SMR spectrum auctions, and many others worldwide. Experience in both field and laboratory suggests that in complex economic environments iterative auctions, which enhance the ability of the participant to detect keen competition and learn when and how high to bid, produce better results than sealed bid auctions.

An iterative combinatorial auction could allow bidders to explore the bid space without having to place bids for all possible items. Amongst the iterative designs so far tried, two specific timing rules have been examined:

1. Continuous Auction (Banks et al. (1989)). A timer is started and bids can be submitted in real time. The best bid combination that fits within the logistic constraints of the auction is posted. New bids can be tendered at any time and can either be placed on a standby list to be "combined" with others or used immediately to directly replace tentatively winning bids. The auction ends if no allocation changing bids are submitted during a fixed time period.
2. Multi-round (Kwasnika et al. (1998), CRA (1998)). A round begins and sealed bids are submitted. An integer program (IP) is solved to find the highest valued combination of winning bids. The winners are posted and a new round is started. New sealed bids can be submitted and the IP is run again to see if a new solution, with some new winners, is found. When there are no new winners or no new bids the auction is over.

Another design issue that must be tackled in multiple item auctions is to provide bidders incentives to solve the threshold problem. The most obvious approach is to use a one-shot sealed bid auction with Vickrey type pricing, in which it is theoretically in the interest of the participant to reveal his true values. In particular, Isaac and James (2000) use the standard Vickrey auction which requires the running of an IP for each winner to determine their price. Rassenti et al. (1982) eliminate the need to run individual IPs for each winner and instead use pseudo competitive prices by running a specialized Linear Program to get prices for each item.

Since the iterative and Vickrey sealed bid approaches are contrary, natural evolution led to an auction design based on combining an iterative process with some form of value-inducing pricing. This design change was spurred by the fact that in the above-mentioned iterative processes, the information content of feedback to the participants was not transparent. In particular, participants had to look at previous bids to compute what it would take to combine with others to replace current winning bids: this required solving another IP problem. Without prices to guide bidders the iterative process can become extremely burdensome.

Banks et al. (1989) implemented a computationally intensive iterative process where each round Vickrey prices are computed and sent back to each bidder along with required bids on rejected packages to displace bidders, given that no one else would change their bid. Following the psuedo-competitive pricing of Rassenti et al. (1982), Kwasnika et al. (1998) use a procedure to calculate prices as feedback information to bidders. They find that the provision of this price information to bidders allows the auction to outperform previous designs by enhancing competitive bid formulation.

In an effort to deal with the computational burden cited by the FCC two approaches have been offered. Rothkof et al. (1998) would simply limit the type and number of bids that can be submitted. Banks et al. (1989) would eliminate the computational burden of the auctioneer through decentralization: requiring the bidders to execute any non-essential computations.

More recently, in response to an FCC RFP for a new combinatorial auction design, CRA (1998) proposed, and Cybernomics tested, a hybrid auction system which combines multi-round and continuous bidding periods. It also implements various activity rules that mimic those in the original SMR auction along with some new variable and more complicated activity rules.

The systematic results so far derived from the body of laboratory-controlled studies of the above mechanisms are:

1. There is little support that the threshold problem is of great concern.
2. Iterative processes seem to be of assistance to bidders.
3. Price feedback information provides for bidder transparency.
4. Computational burden must be taken seriously.

Relying on these results, its own experience with previous real auctions, and various consulting contracts, The FCC, for its upcoming 700 MHz auction, has designed a new multi-round process that limits the number of bids that each bidder can submit[3] and the prices at which he can submit them, provides computationally intensive feedback prices described as short-falls[4], and includes a typical cluster of activity rules. These proposed FCC auction rules for the 700 MHz auction have not been tested in the laboratory.

---

[3] 12 items and 12 package bids.
[4] The shortfall for a particular unsuccessful bid is how much it would need to be increased to become a winning bid given that no others in the auction would change their bids. Thus, every unaccepted bid in every round requires the solution of an equally complicated time-consuming integer programming.

The Combinatorial Clock Auction

During February through October of 1999, while Cybernomics was testing versions of the FCC's SMR and the CRA proposal, we had the chance to carefully observe hundreds of auction participants' behavior and the outcomes their strategies generated. Taking seriously the problem of computational burden associated with generating prices in moderately sized combinatorial auctions, yet not willing to impose activity constraints that interfere with bidding transparency by limiting types and numbers of bid revisions, we developed a simple new auction design: the Cybernomics Combinatorial Clock (CCC) auction.

McCabe et al. (1988) demonstrated that occasionally simple ascending English auctions can be become inefficient because of overstated bids intended to forestall or signal competition. As the FCC has found in the field (and Cybernomics in the laboratory), the threat of financial exposure aggravates this behavior during an SMR auction for multiple items. The McCabe et al. (1988) solution is a simple upward ticking clock controlled by the auctioneer to remove active bidder price control. The CCC follows this lesson and creates a unique price for each item that is controlled by that item's clock.[5] The price ticks upward only when there is excess demand for an item.

The process we devised is quite simple and transparent. The clocks, one for each item, are started at low prices. Each round bidders are given a fixed amount of time to submit which packages or individual items they would like to purchase at the current clock prices. A simple algorithm then counts up the demand for each item by each bidder, without double counting for mutually exclusive packages. The item demands are then aggregated across participants. For items that have more than one bidder demanding more units than are available, the clock price is raised. A new round is started and new bids are requested.

The CCC is transparent because no bidder has to submit a price, and she is free to indicate the contents of any packages that she will buy at the given clock prices. There are no activity rules binding her bidding sequence. Her previous bids for packages at previous prices are remembered and eligible for consideration by the auctioneer unless they are purposefully withdrawn.

The auction continues as long as there is excess demand for one of the items being offered. If the auction reaches the point where there is exactly one bid left for each unit of each item available, then it ends and the standing bidders are awarded the items at the current clock prices: no computation is required!

But it can happen that after a particular clock price increases, the demand for that item becomes less than is available. In the case where there is excess supply for at least one item, and demand exactly equals supply for every other item available, then the auctioneer must compute the solution to an IP to find the allocation of items that would maximize his revenue. He includes all bids at current and previous clock prices that have not been withdrawn. The solution will tend to use old bids to allocate units for which there is excess supply at the current clock prices. If the overall solution does not seek to displace those holding the standing winning bids on items where supply equals

---

[5] If there is more than one unit of an item available then the clock price for that item applies identically to all the units available.

demand, the auction is over and bidders pay prices bid. If the solution does seek to displace at least one bidder holding the standing winning bid on an item where supply equals demand, then that item is considered to have excess demand, the clock price is ticked upward, and the auction continues.

Therefore, the final allocation is one in which all standing bids win items at the final clock prices, and items with excess supply at the final clock prices are awarded at previous clock prices. The auction process that generates this allocation is simply a greedy algorithm to discover psuedo-dual upper bound prices[6]: the lowest prices (final clock prices) at which everyone who submitted a bid is definitely declared a winner. In complex environments these prices are often not unique, and this allows the auction mechanism some flexibility in achieving an efficient outcome.

The beauty of the CCC auction is in its simplicity to the participant, and its minimal computational requirements of the auctioneer. It trivially accommodates the sale of multiple units of multiple items. It provides bidders complete freedom to move in and out of the auction bidding on whichever packages at will. It allows the bidder to impose logical constraints without increasing the computational burden during the auction. For example, a bidder may submit mutually exclusive bids and 'if and only if' bids: the auction simply computes his demand for an item as the maximum number of units he could possibly win. The bidder is also free to blend current and previous clock prices in a current compound bid as long as part of her bid is at current clock prices. The CCC auction is the most flexible combinatorial auction we know of.

Auction Tests

After having conducted a series of experiments that tested versions of the CRA combinatorial auction and the FCC's SMR auction, in the fall of 1999 we continued by independently testing our own invention: the Cybernomics' Combinatorial Clock (CCC) auction. All of the test auctions offered only one unit of each of ten items for sale. Ten bidders each had different values for various packages of items. The value distributions that we created used variations on the following conditions to test the efficacy of the auction designs:

Condition 1. The *Join Factor*. Join measures the relative difference between the value of the optimal allocation (V^*), which was constructed to include several smaller bidder packages, and the next highest value allocation ($¥$), which was constructed to include a single bidder's package covering the optimal set of smaller packages. We define the Join Factor as the ratio $J = (¥/V^*)$. As J increases, we create a stronger necessity for the several small bidders to bid aggressively in order to overcome the large bidder and win the auction at 100% efficiency.

Condition 1 creates the circumstances for the threshold problem. As J gets larger it aggravates the behavioral dilemma of everyone wanting everyone else to pay the final increment to overcome the big package bidder. But note that as J approaches 1, the relative loss in efficiency from having the big bidder win approaches 0.

---

[6] See Rassenti et al. (1982) for a thorough description of pseudo-dual price computation.

Table 1. *Case1 Values generating Join Factors and Own Effects*

Optimal Allocation of 10 Licenses											
A	B	C	D	E	F	G	H	I	J	$ Value	Bidder ID
♦	♦		♦							100	1
		♦						♦		80	2
				♦			♦			80	3
					♦	♦				120	4 or 6
									♦	50	5
2nd Best Allocation of 10 Licenses											
♦	♦	♦	♦	♦	♦	♦	♦	♦	♦	350 or 301	6

Condition 2. The *Own Effect*. This effect is one that is coupled with the join factor. It occurs when b is the large package bidder who demands ¥, but b is also one of the small package bidders included in V*. To achieve the optimal allocation b must forego his large package to be included in the optimal allocation of smaller winning packages. Efficiency may be hurt because b may not collaborate in his role as a small package bidder for two reasons: first, he may feel he is in a stronger negotiating position (since he owns the large package) and may demand more of the surplus than other small package bidders; or second, he may think that displacing himself, even if it is apparently profitable, may create unpredictable dynamics in the subsequent bidding.

Tables 1 and 2 show optimal and second best allocations for the various boundary experiments we conducted to sell ten licenses (A through J) with different values to different bidders. Complete value parameters for these experiments can be found at http://economic.gmu.edu/FCC_Parameters.

In Case 1 (Table 1) bidder 6 has value for the super-package and is either included in the optimal allocation or not depending on whether he **owns** the package (F,G). The Join is either **high** .81 (350/430) or **low** .70 (301/430), depending on bidder 6's assigned value for the super-package.

Table 2. *Cases 2a and 2b, Values generating Join Factors and Own Effects*

Optimal Allocation of 10 Licenses											
A	B	C	D	E	F	G	H	I	J	$ Value	Bidder ID
♦										17.37	4
	♦									36.27	5
		♦	♦	♦						88.59	1
					♦					24.00	2
						♦				30.00	4
							♦			36.00	3
								♦		48.00	1
									♦	54.00	5 or 6
2nd Best Allocations of 10 Licenses											
					♦	♦	♦	♦	♦	180 or 153	5
♦	♦	♦	♦	♦						114	1

Table 3. *CCC vs CRA vs SMR*

Case	Gain	Own	Auction	% Allocation Efficiency
1	.81	Yes	CCC	100, 100, 100
			CRA	78, 79, 78
			SMR	59
1	.81	No	CCC	100, 100, 100
			CRA	97, 79
			SMR	63
1	.70	Yes	CCC	100, 100, 100, 100
			CRA	100, 100
			SMR	70
2a	.80	Yes	CCC	100, 100, 99, 100, 99, 100
			CRA	99, 99, 99, 95, 94, 95, 95
			SMR	100, 99, 95, 95
2b	.94	Yes	CCC	100, 100, 100
			CRA	91, 94, 94
			SMR	100
2b	.94	No	CCC	100, 100, 100
			CRA	95, 95
			SMR	100
2b	.80	Yes	CCC	100, 100, 100
			CRA	100, 91
			SMR	100

In Case 2 (Table 2) the licenses were divided into two separable groups of five, 2a and 2b, with separate join factors and own effects within each group and with no bidders valuing packages containing licenses in both groups. For licenses A-E several bidders had values for the entire license group from A to E, while some of them had values for packages that were part of the optimal allocation. In the second group of licenses F-J, each bidder had values for single licenses except bidder 5 who had value for the entire group of licenses F-J. We divided the separable outcomes of the two groups into Cases 2a and 2b respectively.

In total 55 test auctions were executed. Student participants, bidding anonymously through a local area network of computers, earned a total of approximately $40,000 for their bidding successes. Table 3 shows the allocation efficiencies for all of the CRA, SMR and CCC auctions.

When comparing the outcomes of the non-combinatorial SMR (used until now by the FCC) against the outcomes of the combinatorial auction originally solicited by the FCC from CRA we find:

Result 1: As the join factor increases, efficiency falls. The own effect also reduces efficiency. Even though efficiencies are low for the CRA auction in Case 1, they are higher than for the SMR auction. In Case 2b in which bidders only have values for single license packages, except for one bidder who has a value for all the licenses, the SMR auction outperforms the CRA auction.

But if we compare the more transparent and simpler Cybernomics' Combinatorial Clock auction use we find:

Result 2: The combinatorial clock auction was uniformly more efficient than both the other auction mechanisms across all the test environments. In addition, it yielded 100% efficiency in all but two of the test auctions.

The consistency of Result 2 is remarkable given the typically somewhat noisy results of laboratory experiments with human subjects. As simple as it is, the CCC is extremely promising.

References

Banks, J., Ledyard, J., Porter, D., 1989. Allocating uncertain and unresponsive resources: an experimental approach. *Rand Journal of Economics* 20, 1–25.

Bykowsky M., Cull, R., Ledyard, J., 2000. Mutually destructive bidding: the Federal Communications Commission auction design problem. *Journal of Regulatory Economics* 17(3), 205–228.

CRA – Charles River and Associates Inc. and Market Design Inc., 1998. Report 2: Simultaneous ascending auctions with package bidding. Charles River and Associates No. 1351-00.

Isaac, M, James, D. 2000. Robustness of the incentive compatible combinatorial auction *Experimental Economics* 3(1): 31–53.

Kwasnica, A., John O. Ledyard, David Porter and Christina DeMartini, 1999 "A New and Improved Design for Multi-Object Iterative Auctions," *Caltech Social Science Working Paper No. 1054.*

Ledyard, J., Olson M., Porter D., Swanson J., Torma D. 2002. "The Design of an Auction for Logistics Services," *Informs.*

McCabe, K., Rassenti, S., Smith, V., 1988. Testing Vickrey's and other simultaneous multiple unit versions of the English auction, revised 1989. In: R. M. Isaac (Ed). *Research in Experimental Economics*, Vol. 4, 1991. Greenwich, CT: JAI Press.

Porter, D., 1999. An experimental examination of bid withdrawal in a multi-object auction," *Review of Economic Design.*

Rassenti, S., Smith, V., Bulfin, R., 1982. A combinatorial auction mechanism for airport time slot allocation. *Bell Journal of Economics* 13, 402–417.

Rothkopf, M. Pekec, A., Harstad, R. 1998. Computationally manageable combinational auctions *Management Science* Volume 44, Issue 8.

CHAPTER 6

The Clock-Proxy Auction: A Practical Combinatorial Auction Design

Lawrence M. Ausubel, Peter Cramton, and Paul R. Milgrom

1. Introduction

In this chapter we propose a method for auctioning many related items. A typical application is a spectrum sale in which different bidders combine licenses in different ways. Some pairs of licenses may be substitutes and others may be complements. Indeed, a given pair of licenses may be substitutes for one bidder but complements for another, and may change between substitutes and complements for a single bidder as the prices of the other licenses vary. Our proposed method combines two auction formats—the clock auction and the proxy auction—to produce a hybrid with the benefits of both.

The *clock auction* is an iterative auction procedure in which the auctioneer announces prices, one for each of the items being sold. The bidders then indicate the quantities of each item desired at the current prices. Prices for items with excess demand then increase, and the bidders again express quantities at the new prices. This process is repeated until there are no items with excess demand.

The *ascending proxy auction* is a particular package bidding procedure with desirable properties (see Ausubel and Milgrom 2002, 2006). The bidders report values to their respective proxy agents. The proxy agents iteratively submit package bids on behalf of the bidders, selecting the best profit opportunity for a bidder given the bidder's inputted values. The auctioneer then selects the provisionally winning bids that maximize revenues. This process continues until the proxy agents have no new bids to submit.

The clock-proxy auction is a hybrid auction format that begins with a clock phase and ends with a final proxy round. First, bidders directly submit bids in a clock auction, until there is no excess demand for any item. Then bidders have a single opportunity to input proxy values. The proxy round concludes the auction. All bids are kept live throughout the auction. There are no bid withdrawals. The bids of a particular bidder are mutually exclusive. There is an activity rule throughout the clock phase and between the clock phase and the proxy round.

There are three principal motivations behind our clock-proxy auction proposal. First, Porter et al. (2003) precede us in proposing a particular version of a "combinatorial" clock auction for spectrum auctions, and they provide experimental evidence in its support. Second, the recent innovation of the proxy auction provides a combinatorial auction format suitable for related items such as spectrum. Unlike pure clock auctions, whose anonymous linear prices are not generally rich enough to yield efficient outcomes even with straightforward bidding, the proxy auction leads to efficient outcomes and it yields competitive revenues when bidding is straightforward. It also has some desirable individual and group incentive properties. However, the theoretical development of the proxy auction treats only a sealed-bid procedure, omitting opportunities for bidder feedback and price discovery. Third, our own version of a clock auction has been implemented in the field for products such as electricity in recent years with considerable success (see Ausubel and Cramton 2004). This empirical success in the field suggests that the clock phase would be a simple and effective device for providing essential price discovery in advance of a final proxy round. During the clock phase, bidders learn approximate prices for individual items as well as packages (summing the individual prices). This price information helps bidders focus their valuation analysis on packages that are most relevant.

An important benchmark for comparison is the simultaneous ascending auction (see Cramton, 2006; Milgrom 2000, 2004). This auction form performs well when items are substitutes and competition is strong. The clock phase by itself also does well in this simple setting and, in particular, the outcome is similar to that of a simultaneous ascending auction. However, the addition of the proxy auction round should be expected to handle complications, such as complements, collusion, and market power, much better than the simultaneous ascending auction. In environments—including many spectrum auctions—where such complications are present, the clock-proxy auction is likely to outperform the simultaneous ascending auction both on efficiency and revenues.

We begin by motivating and describing the clock phase. Then we examine the proxy phase. Finally we combine the two together in the clock-proxy auction, describing the important role played by both phases, comparing the auction with the simultaneous ascending auction, and discussing implementation issues. Some aspects of the auction technology are further described by Ausubel and Milgrom (2001), Ausubel, Cramton, and Jones (2002), and Milgrom (2004).

2. Clock Phase

The simultaneous clock auction is a practical implementation of the fictitious "Walrasian auctioneer." The auctioneer announces anonymous linear prices. The bidders respond with quantities desired at the specified prices. Then the prices are increased for items in excess demand, while other prices remain unchanged. This process is repeated until there is no excess demand for any item.

The clock phase has several important benefits. First, it is simple for the bidders. At each round, the bidder simply expresses the quantities desired at the current prices. Linear pricing means that it is trivial to evaluate the cost of any package—it is just the inner product of the prices and quantities. Limiting the bidders' information to a reporting of the excess demand for each item removes much strategizing. Complex bid signaling and collusive strategies are eliminated, as the bidders cannot see individual bids, but only aggregate information. Second, unlike the original Walrasian auction, it is monotonic. This monotonicity contributes to the simplicity of the auction and ensures that it will eventually terminate. Finally, the clock phase produces highly useable price discovery, because of the item prices (linear pricing). With each bidding round, the bidders get a better understanding of the likely prices for relevant packages. This is essential information in guiding the bidders' decision making. Bidders are able to focus their valuation efforts on the most relevant portion of the price space. As a result, the valuation efforts are more productive. Bidder participation costs fall and efficiency improves.

The weakness of the clock auction is its use of linear pricing at the end of the auction. This means that, to the extent that there is market power, bidders will have an incentive to engage in demand reduction to favorably impact prices. This demand reduction implies that the auction outcome will not be fully efficient (Ausubel and Cramton 2002). When goods are substitutes, the clock auction can restore efficiency by utilizing a "clinching" rule instead of linear pricing (Ausubel 2004, 2006). However, in environments with complementary goods, a clock auction with a separate price quoted for each individual item cannot by itself generally avoid inefficiency. The proxy phase will eliminate this inefficiency.

There are several design choices that will improve the performance of the clock phase. Good choices can avoid the exposure problem, improve price discovery, and handle discrete rounds.

2.1. Avoiding the Exposure Problem

One important issue in clock auctions is how to treat quantity changes that, if accepted, would make aggregate demand less than supply. For example, for a particular item, demand may equal supply, so the price of the item does not increase, but the increased price of a complementary item may lead the bidder to reduce the quantity it demands. In both clock auctions and the related simultaneous ascending auctions, the usual rule has been to prohibit quantity reductions on items for which the price does not increase, but this creates an exposure problem when some items are complements. Our design allows a bidder to reduce quantity for any item so long as the price has increased on some item the bidder had demanded. This rule eliminates the exposure problem. The bidder is given the flexibility to drop quantity on items for which there is no price increase.

Another case arises when demand is greater than supply for a particular item so the price increases, and one or more bidders attempt to reduce their demands, making demand less than supply. The common approach in this case is to ration the bidders' reductions so that supply equals demand. However, this again creates an exposure problem when some items are complements. Our approach is not to ration the bidders. All reductions are accepted in full.

The reason for the common restrictions on quantity reductions is to avoid undersell (ending the auction at a point where demand is less than supply). However, these restrictions create an exposure problem. Bidders may be forced to purchase quantities that do not make sense given the final price vector. We eliminate these restrictions and avoid the exposure problem. The consequence is the possibility of undersell in the clock phase, but this is of little importance, as the proxy round can resolve any undersell.

We have conducted over twenty high-stake clock auctions using this rule for electricity products, some of which are substitutes and some of which are complements. These are clock-only auctions without a proxy round. However, because the auctions are conducted quarterly, any undersell in the current auction is added to the quantities in the next auction. Our experience has been that undersell typically is slight (only a few percent of the total). The one exception was an auction in which there was a large negative market price shock near the end of the auction, which resulted in undersell of about fifty percent.

With our rule, the clock auction becomes a package auction. For each price vector, the bidder expresses the package of items desired without committing itself to demanding any smaller package.

All bids in the clock phase are kept live in the proxy round. Including these bids has two effects. It potentially increases revenues after the proxy phase by expanding choices in the winner determination problem, and it encourages sincere bidding in the clock phase, because bidders are on the hook for all earlier bids.

2.2. Improving Price Discovery

In auctions with more than a few items, the sheer number of packages that a bidder might buy makes it impossible for bidders to determine all their values in advance. Bidders adapt to this problem by focusing most of their attention on the packages that are likely to be valuable relative to their forecast prices. A common heuristic to forecast package prices is to estimate the prices of individual items and to take an inner product with quantities to estimate the likely package price. Clock auctions with individual prices assist bidders in this *price discovery* process.

Several recent proposed combinatorial auction procedures, such as the RAD procedure studied by Kwasnica et al. (2005), produce approximate shadow prices on individual items to help guide bidders. The clock auction just does this directly.

Price discovery is undermined to the extent that bidders misrepresent their demands early in the auction. One possibility is that bidders will choose to underbid in the clock phase, hiding as a "snake in the grass" to conceal their true interests from their opponents. To limit this form of insincere bidding, the U.S. Federal Communications Commission (FCC) introduced the Milgrom-Wilson activity rule, and similar activity rules have since become standard in both clock auctions and simultaneous ascending auctions. In its most typical form, a bidder desiring large quantities at the end of the auction must bid for quantities at least as large early in the auction, when prices are lower.

Some clock auctions have performed well in the laboratory without any activity rule (Porter et al. 2003). We suspect that this is because of the limited information that the bidders have about the preferences and plans of the other bidders. This lack of information makes it difficult for participants to know how best to deviate from the straightforward strategy of bidding to maximize profits, ignoring one's impact on prices. In practice, activity rules appear to be important, because of the more detailed knowledge bidders have about the preferences of others and hence a better sense of the benefits of deviating from straightforward bidding. The first U.S. broadband auction is a good example of an auction where the activity rule played an important role (McAfee and McMillan 1996; Cramton 1997).

The most common activity rule in clock auctions is monotonicity in quantity. As prices rise, quantities cannot increase. Bidders must bid in a way that is consistent with a weakly downward sloping demand curve. This works well when auctioning identical items, but is overly restrictive when there are many different products. If the products are substitutes, it is natural for a bidder to want to shift quantity from one product to another as prices change, effectively arbitraging the price differences between substitute products.

A weaker activity requirement is a monotonicity of aggregate quantity across a group of products. This allows full flexibility in shifting quantity among products in the group. This is the basis for the FCC's activity rule. Each license has a number of bidding units associated with it, based on the size of the license. A bidder's activity in a round is the sum of the bidding units of the licenses on which the bidder is active—either the high bidder in the prior round or placing a valid bid in the current round. This aggregate activity level must exceed or equal a specified percentage (the activity requirement) of the bidder's current eligibility (typically, 60 percent in the first stage, 80 percent in the second, and 100 percent in the third stage). Otherwise, the bidder's eligibility in all future rounds is reduced to its activity divided by the activity requirement. Additionally, a bidder has five waivers. A bidder can use a waiver in a round to prevent its eligibility from being reduced in the round.

A weakness of the rule based on monotonicity of aggregate quantities is that it assumes that quantities are readily comparable. For example, in the FCC auctions, the quantity associated with a license is the bandwidth of the license times the population covered (MHz-pop). If prices on a per MHz-pop basis vary widely across licenses, as often is the case, bidders may have an incentive to bid on cheap licenses to satisfy the activity rule. This distortion in bidding compromises price discovery.

We propose an alternative activity rule based on revealed preference that does not require any aggregate quantity measure. The rule is derived from standard consumer theory. Consider any two times, denoted s and t ($s < t$). Let p^s and p^t be the price vectors at these times, let x^s and x^t be the associated demands of some bidder, and let $v(x)$ be that bidder's value of the package x. A sincere bidder prefers x^s to x^t when prices are p^s:

$$v(x^s) - p^s \cdot x^s \geq v(x^t) - p^s \cdot x^t$$

and prefers x^t to x^s when prices are p^t:

$$v(x^t) - p^t \cdot x^t \geq v(x^s) - p^t \cdot x^s.$$

Adding these two inequalities yields the *revealed preference activity rule*:

(RP) $$(p^t - p^s) \cdot (x^t - x^s) \leq 0.$$

At every time t, the bidder's demand x^t must satisfy RP for all times $s < t$.

For the case of a single good, RP is equivalent to the condition that as price goes up, quantity cannot increase; that is, bids must be consistent with a weakly downward-sloping demand curve.

Now suppose there are many goods, but all the goods are perfect substitutes in some fixed proportion. For example, the FCC is auctioning 2-MHz licenses and 20-MHz licenses. Ten 2-MHz blocks substitute perfectly for one 20-MHz block. In this simple case, we would want RP to do the same thing it does when the perfect substitutes are auctioned as a single good, and it does so.

First suppose that all prices are consistent with the rate of substitution (e.g., the 20-MHz block is ten times as expensive as the 2-MHz block) and all are increasing by the same percentage. The bidder then only cares about the total quantity in MHz and does not care about which goods are purchased. In this case, RP allows the bidder to substitute arbitrarily across goods. RP is satisfied with equality so long as the bidder maintains the same total MHz in response to the higher prices, and inequality if the bidder reduces total MHz.

Second suppose that the prices are not consistent with the rate of substitution. Say the price on the 2-MHz block increases too fast relative to the 20-MHz block. The bidder then wants to shift all its quantity to the 20-MHz block, and RP allows this: because the 20 MHz is relatively cheaper, RP gives the bidder more credit for dropping quantity on the 2-MHz blocks than the bidder is debited for the increase in the 20-MHz block. It might seem that the mispricing allows the bidder to expand quantity somewhat, but this is not the case. Because RP is required with respect to all previous bids, the bidder would be constrained by its maximum quantity the last time the 20-MHz block was the best value.

We conclude that RP does just the right thing in the case of perfect substitutes. The activity rule is neither strengthened nor weakened by alternative product definitions.

Now suppose some goods are perfect complements in fixed proportion. For example, in an electricity auction, the bidder wants to maintain a 2-to-1 ratio between baseload product and peakload product. If there are just these two products, then the bidder just cares about the weighted sum of the product prices. As prices increase, the bidder certainly satisfies RP by maintaining the same quantities or by reducing the quantities in the desired ratio; however, the bidder is unable to increase quantities. RP does just the right thing in the case of perfect complements.

If we combine the two cases above so that some goods are perfect substitutes and some are perfect complements, then RP still does the right thing. Bidders will want to shift quantity to the cheapest substitute in building the package of complements. Shifting away from substitute products for which price is increasing too quickly yields a credit that exceeds the debit from shifting toward the relatively cheap product. Hence, this is allowed under RP. Moreover, RP prevents a bidder who always bids on the cheapest among substitutes goods from expanding its quantity of complementary goods as prices rise.

It is useful to compare RP with the current FCC activity rule, which ignores prices and simply looks at aggregate quantity in MHz-pop. "Parking" is the main problem created by the current rule: to maintain flexibility, a bidder has an incentive to bid on underpriced products or low-value products with high quantity, rather than to bid on products that it actually wants to buy. The bidder does this for two reasons: 1) to keep the prices on desired products from increasing too quickly, while maintaining the flexibility to expand demand on products for which competitor demands fall off faster than expected; and 2) to maintain the flexibility to punish a rival by shifting bidding for the rival's desired markets if the rival bids for the bidder's desired markets. Thus, parking is motivated by demand reduction and tacit collusion. But the clock implementation mitigates collusion, because bidders see only excess demand; they do not have the information to know when retaliation is needed, where the retaliation should occur, or how to avoid retaliation. And the final proxy round mitigates demand reduction. Hence, we should expect parking to be much less of a problem in the clock implementation.

The greatest damage from parking comes from price distortions that exclude the high-value bidder from winning an item. Under the FCC rule, bidders are most tempted to park on low-price, high-quantity licenses. These prices may get bid up to the point where the efficient winner drops out, because they enable the parking bidder to bid later on other licenses. In contrast, the RP rule does not allow a bidder to increase its quantity for another license unless there is excess demand for the parking license. Thus, parking is only effective when bidding on underpriced goods. But parking on underpriced goods does no harm; it simply serves to increase the price of the underpriced good. Hence, the revealed-preference activity rule has important advantages over the current FCC activity rule.

The revealed-preference activity rule may appear more complex than the FCC rule based on aggregate quantity. However, it still can be displayed in the same simple way on the bidder's bid entry screen. As the bid is entered, an activity cell indicates the amount of slack in the tightest RP constraint, and changes to red when the constraint is violated. Moreover, to the extent that the revealed preference activity rule eliminates complex parking strategies, the rule may be simpler for bidders.

2.3. Handling Discrete Rounds

Although in theory one can imagine implementing an ascending auction in continuous time, this is hardly ever done in practice. Real clock auctions use discrete rounds for two important reasons. First, communication is rarely so reliable that bidders would be willing to be exposed to a continuous clock. A bidder would find it unsatisfactory if the price clock swept past the bidder's willingness to pay because of a brief communication lapse. Discrete rounds are robust to communication problems. Discrete rounds have a bidding window of significant duration, rarely less than ten minutes and sometimes more than one hour. This window gives bidders time to correct any communication problems, to resort to back-up systems, or to contact the auctioneer and have the round extended. Second, a discrete round auction may improve price discovery by giving the bidders an opportunity to reflect between rounds. Bidders need time to incorporate information from prior rounds into a revised bidding strategy. This updating is precisely the source of price discovery and its associated benefits.

An important issue in discrete-round auctions is the size of the bid increments. Larger bid increments enable the auction to conclude in fewer rounds, but the coarse price grid potentially introduces inefficiencies. Large increments also introduce incentives for gaming as a result of the expanded importance of ties. But using small increments, especially in an auction with many clocks, can greatly increase the number of rounds and, hence, the time required to complete the auction. Bidders generally prefer a shorter auction, which reduces participation costs. It also reduces exposure to market price movements during the auction. This is especially relevant in securities and energy auctions for which there are active secondary markets of close substitutes, and for which underlying price movements could easily exceed the bid increments.

Fortunately, it is possible to capture nearly all of the benefits of a continuous auction and still conduct the auction in a limited number of rounds, using the technique of intra-round bids. With intra-round bids, the auctioneer proposes tentative end-of-round prices. Bidders then express their quantity demands in each auction round at all price vectors along the line segment from the start-of-round prices to the proposed end-of-round prices. If, at any time during the round, the prices reach a point at which there is excess supply for some good, then the round ends with those prices. Otherwise, the round ends with the initially proposed end-of-round prices.

Consider an example with two products. The start-of-round prices are 90, 180 and end-of-round prices are 100, 200. The bidder decides to reduce quantity at two price points (40 percent and 60 percent) between the start-of-round and end-of-round prices, as shown below:

	Product 1		Product 2	
Price Point	Price	Quantity	Price	Quantity
0%	90	8	180	4
40%	94	5	188	4
60%	96	5	192	2
100%	100	5	200	2

The auctioneer aggregates all the bids and determines whether any products clear at price points of up to 100 percent. If not, then the process repeats with new end-of-round prices based on excess demand. If one or more products clear, then we find the first product to clear. Suppose the bidder's drop from 8 to 5 at the 40 percent price point causes product 1 to clear, but product 2 has not yet cleared at the 40 percent price point. Then the current round would post at the 40 percent price point. The next round would have start-of-round prices of 94, 188 (the prices at the 40 percent price point) and, perhaps, end-of-round prices of 94, 208. The price of product 1 stops increasing, as there is no longer excess demand.

Following this exact approach means that the clock phase will typically have more rounds than products. This works fine in an environment where there are multiple units of a relatively limited number of products (all of which are assigned the same price). However, this could be an issue in FCC auctions with hundreds of unique licenses requiring independent prices. In that event, the auctioneer may wish to adopt an approach of settling for approximate clearing in the clock phase in order to economize on the number of rounds.

This use of intra-round bids avoids the inefficiency associated with a coarser price grid. It also avoids the gaming behavior that arises from the increased importance of ties with coarser prices. The only thing that is lost is the within-round price discovery. However, within-round price discovery is much less important than the price discovery that occurs between rounds.

The experience from a number of high-stakes clock auctions indicates that intra-round bidding lets the auctioneer conduct auctions with several products in about ten rounds, with little or no loss from the discreteness of rounds (Ausubel and Cramton 2004). These auctions can be completed in a single day. By way of contrast, early spectrum auctions and some electricity auctions without intra-round bids took weeks or even months to conclude. In a few instances, the longer duration was warranted due to the enormous uncertainty and extremely high stakes, but generally speaking, intra-round bids would have reduced the bidding costs without any meaningful loss in price discovery.

2.4. End of the Clock Phase

The clock phase concludes when there is no excess demand on any item. The result of the clock phase is much more than this final assignment and prices. The result includes all packages and associated prices that were bid throughout the clock phase. Due to complementarities, the clock phase may end with substantial excess supply for many items. If this is the case, the final assignment and prices may not provide a good starting point for the proxy phase. Rather, bids from an earlier round may yield an assignment with higher revenue. (When calculating revenues excess supply should be priced at the reserve price, which presumably represents the seller's opportunity cost of selling the item.)

A sensible approach is to find the revenue maximizing assignment and prices from all the bids in the clock phase. This point is found by backing up the clock to the price point where revenue is at its maximum. The revenue maximizing prices from the clock phase can serve as reasonable lower bounds on prices in the proxy phase. That is, the minimum bid on each package is calculated as the inner product of the revenue maximizing prices and the quantities of items in the package.

In some cases the auctioneer may decide to end the clock phase early—with some excess demand on one or more items. This would be done when the total revenue ceases to increase or when revenue improvements from successive clock rounds are sufficiently small. With the proxy phase to follow, there is little loss in either revenues or efficiency from stopping, say when revenue improvements are less than one-half percent for two consecutive rounds. At this point price discovery is largely over on all but the smallest items. Giving the auctioneer the discretion to end the clock phase early also enables the auction to follow a more predictable schedule.

3. Proxy Phase

Like the clock auction, the proxy auction is based on package bids. However, the incentives are quite different. The main difference is the absence of anonymous linear prices on individual items. Only packages are priced—and the prices may be bidder specific. This weakens price discovery, but the proxy phase is not about price discovery. It is about providing the incentives for efficient assignment. All the price discovery occurs in the clock phase. The second main difference is that the bidders do not bid directly in the proxy phase. Rather, they submit values to the proxy agents, who then bid on their behalf using a specific bidding rule. The proxy agents bid straightforwardly to maximize profits. The proxy phase is a last-and-final opportunity to bid.

The proxy auction works as follows (see Ausubel and Milgrom 2002, 2006). Each bidder reports his values to a proxy agent for all packages that the bidder is interested in. Budget constraints can also be reported. The proxy agent then bids in an ascending package auction on behalf of the real bidder, iteratively submitting the allowable bid that, if accepted, would maximize the real bidder's profit (value minus price), based on the reported values. The auction in theory is conducted with negligibly small bid increments. After each round, provisionally winning bids are determined that maximize seller revenue from compatible bids. All of a bidder's bids are kept live throughout the auction and are treated as mutually exclusive. The auction ends after a round with no new bids (see Hoffman et al., 2006 and Day and Raghavan 2004 for practical methods to implement the proxy phase).

The advantage of this format is that it ends at a core allocation for the reported preferences. Denote the coalition form game (L, w) where L is the set of players ($l = 0$ is the seller and the rest are the bidders) and $w(S)$ is the value of coalition S. Let X denote the set of feasible allocations $(x_l)_{l \in L}$. If S excludes the seller, then $w(S) = 0$; if S includes the seller, then

$$w(S) = \max_{x \in X} \sum_{l \in S} v_l(x_l).$$

The Core(L, w) is the set of all imputations π (payoffs imputed to the players based on the allocation) that are feasible for the coalition of the whole and cannot be blocked by any coalition S; that is, for each coalition S, $\sum_{l \in S} \pi_l(x_l) \geq w(S)$.

Theorem (Ausubel and Milgrom 2002, Parkes and Ungar 2000). *The payoff vector π resulting from the proxy auction is a core imputation relative to the reported preferences*: $\pi \in Core(L, w)$.

Core outcomes exhibit a number of desirable properties, including: 1) efficiency, and 2) competitive revenues for the seller. Thus, the theorem shows that the proxy auction is not subject to the inefficiency of demand reduction: no bidder can ever reduce the price it pays for the package it wins by withholding some of its losing bids for other packages. The theorem also includes the idea that the seller earns competitive revenues: no bidder or coalition of bidders is willing to bid more for the seller's goods. Ausubel and Milgrom (2002, Theorems 2 and 14) establish the core outcome result, whereas Parkes and Ungar (2000, Theorem 1) independently demonstrate the efficiency of outcomes of an ascending proxy auction without addressing the issue of the core.

A payoff vector in the core is said to be *bidder optimal* if there is no other core allocation that all bidders prefer. If the items are substitutes, then the outcome of the proxy auction coincides with the outcome of the Vickrey auction and with the unique bidder-optimal point in the core. If the goods are not substitutes, then the Vickrey payoff is not generally in the core and the proxy auction yields an outcome with higher seller revenues.

Theorem (Ausubel and Milgrom 2002). *If π is a bidder-optimal point in the Core(L,w), then there exists a full-information Nash equilibrium of the proxy auction with associated payoff vector π.*

These equilibria may be obtained using strategies of the form: bid your true value minus a nonnegative constant on every package. We emphasize that this conclusion concerns full-information Nash equilibrium: bidders may need to know π to compute their strategies.

Two important advantages of the proxy auction over the Vickrey auction are that the prices and revenues are monotonic (increasing the set of bidders leads to higher prices) and the payoffs are competitive. To illustrate the comparative weaknesses of the Vickrey auction, suppose there are two identical items and two bidders. Bidder 1 values the pair only at $2.05. Bidder 2 wants a single item only and has a value of $2. The Vickrey auction awards the pair to bidder 1 for a price of $2, which is the opportunity cost incurred by not assigning an item to bidder 2. So far, the outcome is unproblematic.

Let us now add a bidder 3 with the same values as bidder 2. In this case, the Vickrey auction awards the items to bidders 2 and 3. Bidder 2's Vickrey price is the opportunity cost of its good to the other participants, which is $2.05 - 2.00 = $0.05. Bidder 3's price is the same. Total revenues fall from $2.00 to $0.10. Moreover, the new outcome is not in the core, because the coalition of the seller and bidder 1 could both do better by making a private deal, for example by trading the package at a price of $1. By way of contrast, adding a bidder in the proxy auction can never reduce seller revenues.

4. The Clock-Proxy Auction

The clock-proxy auction begins with a clock auction for price discovery and concludes with the proxy auction to promote efficiency.

The clock auction is conducted with the revealed-preference activity rule until there is no excess demand on any item. The market-clearing item prices determine the initial minimum bids for all packages for all bidders. Bidders then submit values to proxy agents, who bid to maximize profits, subject to a relaxed revealed-preference activity rule. The bids from the clock phase are kept live as package bids in the proxy phase. All of a bidder's bids, both clock and proxy, are treated as mutually exclusive. Thus, the auctioneer obtains the provisional winning bids after each round of the proxy phase by including all bids—those submitted in the clock phase as well as those submitted in the proxy phase—in the winner determination problem and by selecting at most one provisional winning bid from every bidder. As usual, the proxy phase ends after a round with no new bids.

4.1. Relaxed Revealed-Preference Activity Rule

To promote price discovery in the clock phase, the proxy agent's allowable bids must be constrained by the bidder's bids in the clock phase. The constraint we propose is a relaxed version of the revealed preference activity rule.

First, we restate revealed preference in terms of packages and the associated minimum bids for the packages. Consider two times s and t ($s < t$). Suppose the bidder bids for the package S at time s and T at time t. Let $P^s(S)$ and $P^s(T)$ be the package price of S and T at time s; let $P^t(S)$ and $P^t(T)$ be the package price of S and T at time t; and let $v(S)$ and $v(T)$ be the value of package S and T. Revealed preference says that the bidder prefers S to T at time s:

$$v(S) - P^s(S) \geq v(T) - P^s(T)$$

and prefers T to S at time t:

$$v(T) - P^t(T) \geq v(S) - P^t(S).$$

Adding these two inequalities yields the revealed preference activity rule for packages:

(RP') $\qquad P^t(S) - P^s(S) \geq P^t(T) - P^s(T).$

Intuitively, the package price of S must have increased more than the package price of T from time s to time t, for otherwise, at time t, S would be more profitable than T.

Notice that the constraint RP' is automatically satisfied at any two times in the proxy phase, because the proxy agent is required to bid to maximize profits. However, an activity rule based on RP' is too strict when comparing a time s in the clock phase with a time t in the proxy phase. Due to the linear pricing in the clock phase, the bidders have an incentive to reduce demands below their true demands. One purpose of the proxy phase is to let the bidders undo any inefficient demand reduction that would otherwise occur in the clock phase and to defect from any collusive split of the items that would otherwise take place. Hence, it is important to let the bidders expand their demands in the proxy phase. The amount of expansion required depends on the competitiveness of the auction.

We propose a *relaxed revealed-preference activity rule*:

(RRP) $\qquad \alpha[P^t(S) - P^s(S)] \geq P^t(T) - P^s(T).$

At every time t in the proxy phase, the proxy agent is permitted to bid on the package T only if RRP is satisfied for every package S bid at time s in the clock phase. The proxy agent bids to maximize profits, subject to satisfying RRP relative to all prior bids.

The auctioneer chooses the parameter $\alpha > 1$ based on the competitiveness of the auction. For highly competitive auctions, little demand reduction is likely to occur in the clock phase and α can be set close to 1. On the other hand, if there is little competition (and high concentration), then a higher α is appropriate.

It is possible to state RRP in terms of a restriction on the value function v reported to the proxy, rather than on the bids. Intuitively, a bidder's reported value for a package is constrained by all of its bids in the clock phase. In particular, if the bidder bid on some package S but not T at some time s, then it may not claim at the proxy phase that a bid on T would have been much more profitable, as formalized by the inequality: $v(T) - P^s(T) \leq \alpha(v(S) - P^s(S))$. Under this version of RRP, a bidder is required to state in the proxy phase a value for each package on which the bidder has already bid in the clock phase. The advantage of this approach is that it allows the proxies to bid accurately according to the bidders' reported values while still imposing consistency across stages.

4.2. Why Include the Clock Phase?

The clock phase provides price discovery that bidders can use to guide their calculations in the complex package auction. At each round, bidders are faced with the simple and familiar problem of expressing demands at specified prices. Moreover, because there is no exposure problem, bidders can bid for synergistic gains without fear. Prices then adjust in response to excess demand. As the bidding continues, bidders get a better understanding of what they may win and where their best opportunities lie.

The case for the clock phase relies on the idea that it is costly for bidders to determine their preferences. The clock phase, by providing tentative price information, helps focus a bidder's decision problem. Rather than consider all possibilities from the outset, the bidder can instead focus on cases that are important given the tentative price and assignment information. Although the idea that bidders can make information processing decisions in auctions is valid even in auctions for a single good (Compte and Jehiel 2000), its importance is magnified when there are many goods for sale, because the bidder's decision problem is then much more complicated. Rather than simply decide whether to buy at a give price, the bidder must decide which goods to buy and how many of each. The number of possibilities grows exponentially with the number of goods. Price discovery can play an extremely valuable role in guiding the bidder through the valuation process.

Price discovery in the clock phase makes bidding in the proxy phase vastly simpler. Without the clock phase, bidders would be forced either to determine values for all possible packages or to make uninformed guesses about which packages were likely to be most attractive. Our experience with dozens of bidders suggests that the second outcome is much more likely; determining the values of exponentially many packages becomes quickly impractical with even a modest number of items for sale. Using the clock phase to make informed guesses about prices, bidders can focus their decision making on the most relevant packages. The bidders see that they do not need to consider the vast majority of options, because the options are excluded by the prices established in the clock phase. The bidders also get a sense of what packages are most promising, and how their demands fit in the aggregate with those of the other bidders.

In competitive auctions where the items are substitutes and competition is strong, we expect the clock phase to do most of the work in establishing prices and assignments— the proxy phase would play a limited role. When competition is weak, demand reduction may lead the clock phase to end prematurely, but this problem is corrected at the proxy stage, which eliminates incentives for demand reduction. If the clock auction gives the bidders a good idea of likely package prices, then expressing a simple approximate valuation to the proxy is made easier. For example, with global economies of scope, a bidder might report to his proxy bidder a value for each item, a fixed cost of operation, and a limit on the number of items acquired. This is just an example, but it serves to highlight that simple valuation functions might serve well once the range of likely package prices is limited.

4.3. Why Include the Proxy Phase?

The main advantage of the proxy phase is that it pushes the outcome toward the core, that is, toward an efficient allocation with competitive payoffs for the bidders and competitive revenues for the seller.

In the proxy phase, there are no incentives for demand reduction. A large bidder can bid for large quantities without the fear that doing so will adversely impact the price the bidder pays.

The proxy phase also mitigates collusion. Any collusive split of the items established in the clock phase can be undone in the proxy phase. The relaxed activity rule means that the bidders can expand demands in the proxy phase. The allocation is still up for grabs in the proxy phase.

The clock-proxy auction has some similarities with the Anglo-Dutch design initially proposed for (but not ultimately used in) the United Kingdom's third-generation mobile wireless auction (Klemperer 2002). Both formats have an ascending auction followed by a sealed-bid last-and-final round. However, the motivation for the last-and-final round is quite different. In the Anglo-Dutch design, the last round has pay-as-bid pricing intended to introduce inefficiency, so as to motivate inefficient bidders to participate in the auction (and perhaps increase auction revenues). In the clock-proxy auction, the last round is more similar to Vickrey pricing and is intended to promote efficiency, rather than prevent it. The relaxed activity rule in the proxy round, however, does encourage the undoing of any tacit collusion in the clock phase, and in this sense is similar to the last-and-final round of the Anglo-Dutch design.

The proxy phase will play a more important role to the extent that competition is limited and complementarities are strong and varied across bidders. Then it is more likely that the clock phase will end prematurely. However, in competitive auctions, the proxy phase may not be needed.

A potential problem with a clock-only auction under our proposed rules arises from a bidder's ability to reduce quantity on products even when the price of a product does not go up. This may appear to create a "free withdrawal" and a potential source of gaming. For example, a bidder might bid up a competitor on a competitor's preferred license to the point where the competitor drops out. Then the strategic bidder reduces quantity on this product. Alternatively, the bidder might bid up the competitor and then drop quantity before the competitor drops out.

Two features mitigate this potential problem. First, the revealed-preference activity rule makes it risky for a bidder to overbid on items that the bidder does not want. Unlike the activity rule based on aggregate quantity, the bidder dropping quantity on a product for which the price has not increased is not given any credit in the RP inequality and hence has no ability to expand demand on another product. Second, the preferred approach would run the winner-determination-problem at the end among *all* prior bids. Hence, the strategic bidder may find that it is obligated to purchase items that it does not want. (Of course, if goods are mostly substitutes, then one simply could prevent quantity reductions for goods that have cleared.)

4.4. Two Examples

We illustrate our answers to "Why include the clock phase?" and "Why include the proxy phase?" with two examples.

In our first example, there are two items and two bidders. Bidder 1 wants just a single item and values it at v_1. Bidder 2 wants up to two items and values each at v_2 (valuing the package of two items at $2v_2$). The private values v_1 and v_2 are drawn independently from the uniform distribution on [0,1]. Each bidder i knows the realization of v_i but only the distribution of $v_j (j \neq i)$. In the clock auction, this is a classic example of demand reduction. For simplicity, assume that the clock price ascends continuously. Bidder 1's weakly dominant strategy is to bid a quantity of 1 at all prices up to v_1 and then to drop to a quantity of 0. Bidder 2 has a choice whether to bid initially for a quantity of two, or to bid for only one unit and cause the price clock to stop at zero. A straightforward calculation shows that bidding for only one unit and obtaining a zero price maximizes bidder 2's expected payoff, establishing that this is the unique equilibrium (Ausubel and Cramton 2002, p. 4).

Thus, conducting only a clock phase is disastrous for the seller; revenues equal zero and the outcome of each bidder winning one unit is inefficient whenever $v_2 > v_1$. However, suppose that the clock phase is followed by a proxy round, using a parameter $\alpha \geq 2$ in the relaxed revealed-preference activity rule. Because the substitutes condition is satisfied in this example, the bidders' dominant strategies in the proxy round are each to bid their true values. Thus, the clock-proxy auction yields the bidder-optimal core outcome, and the seller earns revenues of $\min\{v_1, v_2\}$. Nothing of consequence occurs in the clock phase, and the proxy phase yields the desirable outcome by itself.

In our second example, there are m items and n bidders ($n > m$). Each bidder i values item k at v_{ik}. But bidder i has value for only a single item, and so for example if bidder i received both items k and l, his value would be only $\max\{v_{ik}, v_{il}\}$. The values v_{ik} are random variables with support [0,1]. Each bidder i knows the realization of v_{ik} ($k = 1, \ldots, m$), but only the distribution of $v_{jk} (j \neq i)(k = 1, \ldots, m)$. In the clock auction, because bidders have demand for only a single item, each bidder's dominant strategy is to bid a quantity of one on an item k such that $v_{ik} - p_k = \max_{l=1,\ldots,m}\{v_{il} - p_l\}$ and to bid a quantity of zero on all other items. Therefore, the clock phase concludes at the Vickrey outcome, which is also the predicted outcome of the proxy phase (because the substitutes condition is satisfied). Thus, the clock-proxy auction again yields the bidder-optimal core outcome. This time the clock phase yields the desirable outcome by itself, and nothing further occurs in the proxy phase.

If the bidders find it costly to determine their values, the clock phase may find the outcome without the need for bidders to calculate all their values. For example, suppose $m = 2$ and $n = 3$ and the bidders' estimated value pairs are (2,4), (3,8) and (7,2), but each bidder knows each of its values only to within ±1, without further costly investment. In the clock phase, bidder 1 will be the first to face the need to invest in learning its exact values. If he does so, the auction will end at prices of 2 and 4 without the second and third bidder ever needing to make that investment. Price discovery at the clock phase saves bidders 2 and 3 from the need to determine their full values for the proxy stage.

4.5. Comparison with the Simultaneous Ascending Auction

The simultaneous ascending auction as implemented by the FCC is an important benchmark of comparison, given its common use in auctioning many related items (see Cramton, 2006). The clock auction is a variant of the simultaneous ascending auction in which the auctioneer specifies prices and the bidders name quantities. There are several advantages to the clock implementation.

The clock auction is a simpler process than the simultaneous ascending auction. Bidders are provided the minimal information needed for price discovery—the prices and the excess demand. Bidders are not distracted by other information that is either extraneous or useful as a means to facilitate collusion.

The clock auction also can take better advantage of substitutes, for example, using a single clock for items that are near perfect substitutes. In spectrum auctions, there is a tendency for the spectrum authority to make specific band plans to facilitate the simultaneous ascending auction. For example, anticipating demands for a large, medium, and small license, the authority may specify a band plan with three blocks—30 MHz, 20 MHz, and 10 MHz. Ideally, these decisions would be left to the bidders themselves. In a clock auction, the bidders could bid the number of 2-MHz blocks desired at the clock price. Then the auction would determine the band plan, rather than the auction authority. This approach is more efficient and would likely be more competitive, because all bidders are competing for all the bandwidth in the clock auction. With the preset band plan, some bidders may be uninterested in particular blocks, such as those that are too large for their needs.

Clock auctions are faster than a simultaneous ascending auction. Simultaneous ascending auctions are especially slow near the end, when there is little excess demand. For example, when there are six bidders bidding on five similar licenses, then it typically takes five rounds to obtain a one bid-increment increase on all items. In contrast, in a clock auction, an increment increase takes just a single round. Moreover, intra-round bids allow larger increments, without introducing inefficiencies, because bidders still can express demands along the line segment from the start-of-round prices to the end-of-round prices.

The clock auction limits collusion relative to the simultaneous ascending auction. Signaling how to split up the items is greatly limited. Collusive strategies based on retaliation are not possible, because bidder-specific quantity information is not given. Further, the simultaneous ascending auction can have a tendency to end early when an obvious split is reached, but this cannot happen in the clock auction, because the bidders lack information about the split. Also there are fewer rounds to coordinate a split.

The clock auction, as described here, eliminates the exposure problem. As long as at least one price increases, a bidder can reduce quantity on his other items. The bid is binding only as a full package. Hence, the bidder can safely bid for synergistic gains.

The clock-proxy auction shares all these advantages of the clock auction, and in addition promotes core outcomes. The proxy phase further mitigates collusion and eliminates demand reduction. The cost of the proxy phase is added implementation complexity. Also the absence of linear pricing reduces the transparency of the auction. It is less obvious to a bidder why he lost. Nonetheless, the auctioneer at the conclusion of the auction can disclose sufficient information for the bidders to determine the outcome without revealing any supramarginal values.

4.6. Combinatorial Exchange

Like other package auctions, the clock-proxy auction is designed for settings with a single seller. With multiple sellers and no item prices, there is an additional problem to solve: how to divide the auction revenues. For example, if separate sellers own items A and B, and if all the bidders want to buy items A and B together, with no interest in these separate and separately owned items, the auction itself can provide no information about how to allocate the revenue from the winning bid among the sellers. The revenue-sharing rule has to be determined separately, and there is no simple and completely satisfactory solution to this problem.

The clock-proxy auction can be extended to handle exchanges with one passive seller and many active buyers and sellers. A natural application is the auctioning of encumbered spectrum (Cramton, Kwerel, and Williams 1998; Kwerel and Williams 2002). The spectrum authority would be the passive seller, selling overlay licenses. Incumbents are (potentially) the active sellers, selling their existing rights. In this setting, one can adapt the clock-proxy auction very simply. An incumbent seller's bid would reflect an offer to sell a package. Formally, its bid would specify the goods it offers as negative quantities in the clock phase and would specify negative quantities and prices in the proxy stage. In principle, one could even allow bids in which an incumbent offers to exchange its good for another good plus or minus some compensating payment, where the package is expressed by a vector of positive and negative numbers.

Alternative designs differ in how they divide auction revenues and in what bids sellers are allowed to make. For example, one possibility is to fix the items to be sold at the proxy stage as those that were not acquired by their original owners at the clock stage. Final revenues would then be distributed to sellers in proportion to the prices from the clock stage. Another possibility is to allow the sellers to bid in every stage of the auction, essentially negotiating what is sold and how revenues are to be split through their bidding behavior. A third possibility is to allow sellers to set reserve prices and to use those to divide revenues among the sellers.

These alternative designs split revenues differently, so they create different incentives for incumbents to report exaggerated values. The result will be differences in the likelihood of a successful sale. So far, theory provides little guidance on which choice is best, beyond indicating that the problem can sometimes be a hard one. If there are many sellers whose goods are sufficiently good substitutes, then the problem may not be too severe. This strongly suggests that the most important issue for the FCC in making the package exchange a success is careful attention to the incumbents' rights, to make their goods as substitutable as possible.

4.7. Implementation Issues

We briefly discuss four of the most important implementation issues.

Confidentiality of Values

One practical issue with the proxy phase is confidentiality of values. Bidders may be hesitant to bid true values in the proxy phase, fearing that the auctioneer would somehow manipulate the prices with a "seller shill" to push prices all the way to the bidders' reported values. Steps need to be taken to assure that this cannot happen. A highly transparent auction process helps to assure that the auction rules are followed. Auction software can be tested and certified that it is consistent with the auction rules. At the end of the auction, the auctioneer can report all the bids. The bidders can then confirm that the outcome was consistent with the rules. In addition, there is no reason that the auctioneer needs to be given access to the high values. Only the computer need know.

A further step to protect the privacy of high values is to allow a multi-round implementation of the proxy phase. The critical feature of the proxy phase is that the relative values are locked. If bidders do not want to reveal their final values, that can be handled. In a multi-round version of the proxy phase, bidders must freeze the relative values of the packages they name but can periodically authorize a fixed dollar increase in all of their bids. With this approach, the auction becomes an ascending, pay-as-bid package auction.

Price Increments in the Clock Phase

When auctioning many items, one must take care in defining the price adjustment process. This is especially true when some goods are complements. Intuitively, undersell in the clock phase is minimized by having each product clear at roughly the same time. Otherwise price increases on complementary products can cause quantity drops on products that have already cleared. Thus, the goal should be to come up with a price adjustment process that reflects relative values as well as excess demand. Moreover, the price adjustment process effectively is resolving the threshold problem by specifying who should contribute what as the clock ticks higher. To the extent that prices adjust with relative values the resolution of the threshold problem will be more successful.

One simple approach is to build the relative value information into the initial starting prices. Then use a percentage increase, based on the extent of excess demand. For example, the percentage increment could vary linearly with the excess demand, subject to a lower and upper limit.

Expression of Proxy Values

Even with the benefit of the price discovery in the clock phase, expressing a valuation function in the proxy phase may be difficult. When many items are being sold, the bidder will need a tool to facilitate translating preferences into proxy values. The best tool will depend on the circumstances.

At a minimum, the tool will allow an additive valuation function. The bidder submits a demand curve for each item. The value of a package is then found by integrating the demand curve (adding the marginal values) up to the quantity of the item in the package, and then adding over all items. This additive model ignores all value interdependencies across items; it assumes that the demand for one item is independent of the demand for other items. Although globally (across a wide range of quantities) this might be a bad assumption, locally (across a narrow range of quantities) this might be a reasonable approximation. Hence, provided the clock phase has taken us close to the equilibrium, so the proxy phase is only doing some fine-tuning of the clock outcome, then such a simplistic tool may perform reasonably well. And of course it performs very well when bidders actually have additive values.

A simple extension of the additive model allows the bidder to express perfect substitutes and complements within the additive structure. For example, items A and B may be designated perfect complements in the ratio 1 to 3 (one unit of A is needed for three units of B). Then the bidder expresses a demand curve for A and B (with the one-to-three ratio always maintained). Items C and D may be designated perfect substitutes in the ratio 2 to 1 (two Cs equal one D). Then the bidder expresses a demand curve for C or D (with all quantity converted to C-equivalent). This extension effectively allows the bidder to redefine the items in such a way to make the additive model fit. For example, in a spectrum auction, a bidder for paired spectrum will want to express a demand for paired spectrum. This can be done by designating the upper and lower channels as perfect complements, but then the blocks of paired spectrum as perfect substitutes. A bidder for unpaired spectrum would designate all channels as perfect substitutes, and then express a single demand curve for unpaired spectrum.

Demand curves typically are expressed as step functions, although in some contexts piece-wise linear demand curves are allowed. Bidders should be able to specify whether quantity can be rationed. For example if a bidder drops quantity from 20 to 10 at a price of $5, does this mean the bidder is just as happy getting 14 units as 10 units or 20 units when the price is $5 per unit, or does the bidder only want exactly 10 units at a price of $5, and exactly 20 units at a price of $4.99? Is there a minimum quantity that the bidder must win for the item to have value?

Beyond this, the tool should allow for the inclusion of bidder constraints. Budget constraints are the most common: do not bid more than X. Other constraints may be on quantities: only value A if you win B. This constraint arises in spectrum auctions when a bidder has secondary regions that have value only if the primary regions are won.

The bidders' business plans are a useful guide to determine how best to structure the valuation tool in a particular application. Business plans are an expression of value to investors. Although the details of the business plans are not available to the auctioneer, one can construct a useful valuation tool from understanding the basic structure of these business plans.

Calculating Prices in the Proxy Phase

The proxy phase is a sealed-bid auction. At issue is how best to calculate the final assignment and prices. The final assignment is easy. This is just the value maximizing assignment given the reported values. The harder part is determining the prices for each winning package. The clock phase helps by setting a lower bound on the price of each package. Given these starting prices, one approach would be to run directly the proxy auction with negligible bid increments. With many items and bidders this would require voluminous calculations.

Fortunately, one can accelerate the process of calculating prices using various methods (see Hoffman et al., 2006; Day and Raghavan 2004; Zhong et al. 2003). First, as David Parkes suggested, package prices for all bidders can start at "safe prices," defined as the maximum bid on the package by any losing bidder. Second, prices can increase in discrete jumps to the point where a bidder starts or stops bidding on a particular package. Although these methods have not yet been fully developed, calculating the prices in the proxy phase likely can be done with many items and bidders in an expedient manner.

The precise process for calculating the prices is especially important when some items are complements, because then there will be a set of bidder-optimal points in the core, and the price process will determine which of these points is selected.

5. Conclusion

We propose the clock-proxy auction for auctioning many related items—a simultaneous clock auction followed by a last-and-final proxy round. The basic idea is to use anonymous linear prices as long as possible to maximize price discovery, simplicity, and transparency. The clock phase also greatly facilitates the bidders' valuation analysis for the proxy round, because the analysis can be confined to the relevant part of the price space identified in the clock phase. Finally, unlike the simultaneous ascending auction, the clock auction does not suffer from the exposure problem.

For highly competitive auctions of items that are mostly substitutes, the clock auction without the proxy round will perform well. Indeed a clock auction without a proxy round may be the best approach in this setting, as it offers the greatest simplicity and transparency, while being highly efficient.

With limited competition or items with a complex and varied structure of complements, adding the proxy phase can improve the auction outcome. In particular, a core outcome is achieved. Seller revenues are competitive and the allocation is efficient. The demand reduction incentive present in the clock phase is eliminated. Most importantly, adding the proxy round does no harm: in the simplest settings where the clock auction alone performs well, adding the proxy round should not distort the outcome. The proxy round simply expands the settings in which the auction performs well.

Acknowledgments

This research was inspired by the Federal Communications Commission's efforts to develop a practical combinatorial auction for its spectrum auctions. We are especially grateful to Evan Kwerel for his insights and encouragement.

References

Ausubel, Lawrence M. (2004), "An Efficient Ascending-Bid Auction for Multiple Objects," *American Economic Review*, 94:5, 1452–1475.

Ausubel, Lawrence M. (2006), "An Efficient Dynamic Auction for Heterogeneous Commodities," *American Economic Review*, forthcoming.

Ausubel, Lawrence M. and Peter Cramton (2002), "Demand Reduction and Inefficiency in Multi-Unit Auctions," University of Maryland Working Paper 9607, revised July 2002.

Ausubel, Lawrence M. and Peter Cramton (2004), "Auctioning Many Divisible Goods," *Journal of the European Economic Association*, 2, 480–493, April–May.

Ausubel, Lawrence M., Peter Cramton, and Wynne P. Jones (2002). "System and Method for an Auction of Multiple Types of Items." International Patent Application No. PCT/ US02/16937.

Ausubel, Lawrence M. and Paul Milgrom (2001), "System and Method for a Dynamic Auction with Package Bidding," International Patent Application No. PCT/US01/43838.

Ausubel, Lawrence M. and Paul Milgrom (2002), "Ascending Auctions with Package Bidding," *Frontiers of Theoretical Economics*, 1, 1–45, www.bepress.com/bejte/frontiers/vol1/iss1/art1.

Ausubel, Lawrence M. and Paul Milgrom (2006), "Ascending Proxy Auctions," in Peter Cramton, Yoav Shoham, and Richard Steinberg (eds.), *Combinatorial Auctions*, Chapter 3, MIT Press.

Compte, Olivier and Philippe Jehiel (2000), "On the Virtues of the Ascending Price Auction." Working paper, CERAS-ENPC.

Cramton, Peter (1997), "The FCC Spectrum Auctions: An Early Assessment," *Journal of Economics and Management Strategy*, 6:3, 431–495.

Cramton, Peter (2006), "Simultaneous Ascending Auctions" in Peter Cramton, Yoav Shoham, and Richard Steinberg (eds.), *Combinatorial Auctions*, Chapter 4, MIT Press.

Cramton, Peter, Evan Kwerel, and John Williams (1998), "Efficient Relocation of Spectrum Incumbents," *Journal of Law and Economics*, 41, 647–675.

Day, Robert W. and S. Raghavan (2004), "Generation and Selection of Core Outcomes in Sealed-Bid Combinatorial Auctions," Working Paper, University of Maryland.

Hoffman, Karla, Dinesh Menon, Susara van den Heever, and Thomas Wilson, "Observations and Near-Direct Implementations of the Ascending Proxy Auction" in Peter Cramton, Yoav Shoham, and Richard Steinberg (eds.), *Combinatorial Auctions*, Chapter 17, MIT Press.

Klemperer, Paul (2002), "What Really Matters in Auction Design," *Journal of Economic Perspectives*, 16:1, 169–189.

Kwasnica, Anthony M., John O. Ledyard, Dave Porter, and Christine De Martini (2005), "A New and Improved Design for Multi-Object Iterative Auctions," *Management Science*, forthcoming.

Kwerel, Evan R. and John R. Williams (2002), "A Proposal for the Rapid Transition to Market Allocation of Spectrum," Working Paper, Office of Plans and Policy, FCC.

McAfee, R. Preston and John McMillan (1996), "Analyzing the Airwaves Auction," *Journal of Economic Perspectives*, 10, 159–176.

Milgrom, Paul (2000), "Putting Auctions Theory to Work: The Simultaneous Ascending Auction," *Journal of Political Economy*, 108(2): 245–272.

Milgrom, Paul (2004), *Putting Auction Theory to Work*, Cambridge: Cambridge University Press.

Parkes, David C. and Lyle H. Ungar (2000), "Iterative Combinatorial Auctions: Theory and Practice," *Proceedings of the 17th National Conference on Artificial Intelligence (AAAI-00)*, 74–81.

Porter, David, Stephen Rassenti, Anil Roopnarine, and Vernon Smith (2003), "Combinatorial Auction Design," *Proceedings of the National Academy of Sciences*, 100, 11153–11157.

Zhong, Jie, Gangshu Cai, and Peter R. Wurman (2003), "Computing Price Trajectories in Combinatorial Auctions with Proxy Bidding," Working Paper, North Carolina State University.

CHAPTER 7

Spectrum Auction Design

Peter Cramton[*]

1. Introduction

Fred Kahn recognized the important role of market design in improving how markets work. He believed that prices should be set in an open competitive process, rather than administratively. I had the pleasure of working with Fred on a project to evaluate the pricing rule in California's electricity market. We examined whether the electricity market should use uniform pricing or pay-as-bid pricing (Kahn et al. 2001). In this tribute to Fred Kahn, I also focus on auction design, but in the communications industry.

Spectrum auctions have been used by governments to assign and price spectrum for about 20 years. Over those years, the simultaneous ascending auction, first introduced in the US in 1994, has been the predominant method of auctioning spectrum. The auctions have proved far superior to the prior methods of beauty contests and lotteries (Cramton 1997; Milgrom 2004).

Despite the generally positive experience with the simultaneous ascending auction, several design issues have surfaced. Some were addressed with minor rule changes. For example, bidders' use of trailing digits to signal other bidders and support tacit collusion was eliminated by limiting bids to integer multiples of the minimum increment (Cramton and Schwartz 2002). However, many other design problems remain. In this paper, I identify these problems, and describe a new approach—the combinatorial clock auction—which is based primarily on the clock-proxy auction (Ausubel et al. 2006), which addresses the main limitations of the simultaneous ascending auction.

My focus here is on spectrum auction design, rather than spectrum policy more generally. Certainly, communications regulators face many other critical challenges, such as how best to free up new spectrum for auction (Cramton et al. 1998), or whether an auction is needed at all (FCC 2002). For some allocations, it is better to set aside the

---

[*] I thank my collaborators, Larry Ausubel, Robert Day, and Paul Milgrom for helpful discussions, as well as Nathaniel Higgins, Evan Kwerel, Thayer Morrill, Peter Pitsch, and Andrew Stocking. I thank the staff at Ofcom, especially Graham Louth, Director of Spectrum Markets, whose leadership and intellectual contribution were essential to the successful implementation of the combinatorial clock auction. I am grateful to the National Science Foundation and the Rockefeller Foundation for funding.

spectrum for common property use, as is done with unlicensed spectrum. In particular, for applications that do not create additional scarcity, the commons model is better than the auction model. There are many examples of this: garage door openers, car locks, and other device controllers, but the most important is Wi-Fi. These application require little bandwidth or power, and thus, do not make the spectrum scarce. Scarcity problems are mitigated by operator separation. In contrast, mobile phones require much greater power and bandwidth, creating spectrum scarcity, and hence an auction is needed to assign the scarce resource among the competing carriers.

Spectrum auctions to date have been long-term auctions in which the winner is granted a license for 10 to 25 years, with a strong expectation of renewal following expiration. One might think instead that a spot market for spectrum, much like a spot market for electricity, would be a more flexible and efficient instrument. Someday that will be true. But today's hardware, especially the handset, is not sufficiently flexible to accommodate a real-time spot market. Moreover, carriers must make large specific investments in their networks. These long-term investments are better supported with a long-term license for spectrum, which is a critical input. Over the next 20 years increasingly flexible hardware will be introduced. Eventually it will make sense to organize the spectrum market much like the electricity market. The basic element will be a real-time spot market that establishes the price of bandwidth at a particular time and location. But for now, long-term spectrum auctions are both necessary and desirable.

One of the greatest challenges for the regulator is keeping up with the rapid technological development of wireless communications. Indeed, one of the main reasons for switching from beauty contests, to lotteries, to auctions was that beauty contests and lotteries were too slow. Wireless communications plays an essential role in modern economies, both in developed and developing countries. Slowing the pace of wireless innovation and development has large costs to economic growth. For this reason, regulators must do whatever they can to promote a competitive wireless industry. Allocating sufficient spectrum in a timely manner is paramount.

The combinatorial clock auction described here helps facilitate the spectrum allocation process by enabling the auction to determine how the spectrum is organized, which is called the band plan. Prior methods required that the regulator determine a fixed band plan before the auction began. As a result, before each auction there is a long regulatory process, much like the beauty contests of before, but with the companies' lobbying for particular band plans, rather than for direct spectrum awards. This is the most time-consuming and error-prone element of the spectrum management process. Thus, the new approach promises not only to improve spectrum assignments, but also to improve the band plans within which the assignments fit, and to do so with less delay.

From an auction theory viewpoint, spectrum auctions are both challenging and interesting. The government is auctioning many items that are heterogeneous but similar. Often there are competing technologies as well as companies to provide a wide range of communication services. As a result, the setting has a complex structure of substitutes and complements. This is among the most difficult auction settings that are seen in practice.

The goal for the government should be efficiency, not revenue maximization. The government should focus on ensuring that those who can put the spectrum to its highest use get it. Focusing simply on revenue maximization is short-sighted. Many steps such as technical and service flexibility, and license aggregation and disaggregation, improve efficiency and thereby improve revenues. But short-run revenue maximization by creating monopolies, which would create the highest profits before spectrum fees, and therefore would sustain the largest fees, should be resisted. Indeed, competition, which ultimately will lead to greater innovation and better and cheaper services, will likely generate *greater* government revenues from a long-run perspective. The government can best accomplish this objective with an efficient auction that puts the spectrum to its best use.

The regulator may find it necessary to introduce spectrum caps or other preferences that favor new entrants so as to level the playing field between incumbents and new entrants (Cramton et al. 2011). Incumbents include in their private value the benefit of foreclosing competition, thus driving a wedge between social value and private value. In theory the regulator can correct this externality by favoring the new entrant, but in practice this has proven to be difficult. The FCC's experience with preferences for certain bidders—set-asides, bidding credits, and installment payments—has been disappointing, at least with respect to mobile broadband communication, which is where most of the value lies.

In contrast, a good example of successful intervention was Canada's use of set asides in its 2008 Advanced Wireless Services or AWS auction. As a result, multiple deep-pocketed new entrants came to the auction and bid up the price of not only the set-aside blocks, but also the non-set-aside blocks. The result was a much more competitive auction (with much higher revenues) and the introduction prospectively of some potentially strong new service providers. The approach effectively broke up regional market-splitting by the dominant incumbents. Another successful intervention was the FCC's use of a spectrum cap in early broadband PCS auctions. The cap limited the quantity of spectrum that any one carrier could hold in a geographic area, which addressed the potential market failure of limited competition in the market for wireless services.

Despite these successes in Canada and the US, the FCC's long and sometimes troubled history with bidder preferences is an important case study for other countries that are considering preferences for various parties. Installment payments proved especially problematic, as it led to speculative bidding, bankruptcy, and lengthy delay in the use of the spectrum.

In addition, the regulator must resist the temptation to force more "winners" than the market can efficiently support. Sometimes regulators fragment the spectrum and prohibit aggregation in the auction in an effort to create as many winners as possible. The India 3G spectrum auction may be one example. Aggregation up to a suitable competitive constraint is preferred.

1.1. Three Main Points

There are three main points that I wish to emphasize:

Enhance substitution. First, in terms of the auction design, it is important to enhance the substitution across the items that are being sold. Enhanced substitution is accomplished through both the product design—what is auctioned—and the auction format. Often in the spectrum setting, the product design can be just as important as the auction design.

Encourage price discovery. Second, encouraging price discovery is extremely important. We need a dynamic process, because unlike some situations, in the case of spectrum auctions, there is much uncertainty about what things are worth. The bidders need to do a considerable amount of homework to develop a crude valuation model, and they need the benefit of some collective market insights, which can be revealed in a dynamic auction process, in order to improve their decision-making. The nice thing about a dynamic auction is that through this price process the bidders gradually have their sights focused on the most relevant part of the price space. Focusing bidder decisions on what is relevant is in my mind the biggest source of benefit from the dynamic process. This benefit is generally ignored by economists, because economists assume that the bidders fully understand their valuation models, when in practice bidders almost never have a completely specified valuation model. Yes, they do a lot of homework, but there is still much uncertainty about what spectrum lots are worth, and how they should be valuing the spectrum. The experience of the 3G spectrum auctions in Europe is a good example. The bids were based more on stock prices in a bubble situation, rather than on solid analysis about values.

Induce truthful bidding. The third feature that I wish to emphasize is the importance of inducing truthful bidding. This is accomplished in the auction design through an effective pricing rule and an activity rule. The two rules work together to encourage bidders truthfully to express preferences throughout the entire auction. This truthful expression of preferences is what leads to excellent price discovery and ultimately an efficient auction outcome.

A variety of different pricing rules are used in practice. The two most common are pay-as-bid pricing, where the bidder pays what it bid if it is a winner, and for a homogenous product, uniform pricing, where the bidder pays the market-clearing price. In the particular applications I am discussing here, there generally are not clearing prices, because of strong complementarities and heterogeneous items. As a result, a new kind of pricing rule is needed. The pricing rule that I will describe in detail later is a generalization of Vickrey's second-price rule.

I now give a brief overview of the combinatorial clock auction. The approach may appear complex. Some amount of complexity is required given the complex economic problem. Simpler versions, such as a simultaneous clock auction are possible in settings where all bidders intend to use the same technology. This may well be the case in developing countries that are conducting spectrum auctions for a particular use after the technology battles have been resolved from the experience in developed countries.

1.2. An Overview of the Combinatorial Clock Auction

The combinatorial clock auction is especially useful in situations where the regulator does not know which technology will make the best use of the spectrum. In such cases, the auction itself can determine the ultimate band plan that specifies how the spectrum is organized. Such an auction is said to be technology neutral, since it allows the competing technologies to determine the winning technologies, as well as carriers. A good example is an auction that accommodates both paired and unpaired technologies, such as LTE and WiMAX, respectively. A combinatorial auction is essential in this case, since the two uses require that the spectrum be organized in fundamentally different ways. The combinatorial clock auction is an especially simple, yet powerful, auction that lets competitive bids determine the ultimate band plan.

The combinatorial clock auction has features to address each of my three main points.

First, the product design simplifies the products whenever possible. For example, if bidders primarily care about the quantity of spectrum that they win in a geographic area, the auction should involve generic spectrum (if possible), and the bidders bid for a quantity of spectrum in each area. This simplifies the auction, enhances substitution, and improves competition. The specific assignment of spectrum lots is determined in the last stage of the auction, once the critical decisions have been made (who won how much in each area). This approach also allows a technology neutral auction, which lets the spectrum be organized in different ways for the different technologies. Each bidder indicates the quantity of spectrum and the type of use in its bids. In this case, the first stage of the auction determines not only who won how much in each area, but also the overall quantity of spectrum that is allocated for a particular use in the area.

Second, to encourage price discovery, the auction begins with a "clock" stage (i.e., each auction in the simultaneous auction process has a "clock" that shows the most recent bid price). Prices ascend for each product with excess demand until there is no excess demand for any product. This simple and familiar price discovery process works extremely well when bidders have incentives for truthful bidding. In the important case of substitutes, the clock stage determines an efficient assignment together with supporting competitive equilibrium prices. Moreover, complements are handled with no increase in the complexity of the clock process. Each bid in the clock stage is a package bid, so bidders can bid without fear of winning only some of what they need.

Bidders may find that they are unable to express preferences for all of the desirable packages in the clock stage, so following the clock stage is a supplementary round. Bidders can increase their bids on packages on which they bid in the clock stage and submit new bids on other packages. All of the clock stage bids and the supplementary round bids then are run through an optimizer to determine the value-maximizing assignment of the spectrum. This is the generic assignment.

Third, to induce truthful bidding, the auction uses Vickrey-nearest-core pricing. The efficient assignment is priced to minimize the bidders' total payments subject to competitive constraints (no group of bidders has offered the seller more). In practice, this often implies Vickrey pricing, ensuring truthful bidding. However, because of complements, there may be one or more competitive constraints that cause the payments to be greater than Vickrey payments for some bidders. In this event, the smallest deviations from Vickrey prices are used.

To induce truthful bidding throughout the clock stage, an activity rule based on revealed preference is used. This rule encourages bidders to bid in the straightforward manner of selecting the most profitable package in each round. Deviations from bidding on the most profitable package throughout the clock stage may impose a constraint on subsequent bids, either later in the clock stage or in the supplementary round.

Once the generic assignments are determined and priced, the specific assignment stage is run. Each winner submits top-up bids for each specific assignment that is better than the winner's worst specific assignment. The bids indicate the incremental value for each feasible alternative. Then an optimization program is run to determine the efficient specific assignment. Again the prices for the specific assignments are Vickrey-nearest-core prices. This concludes the auction.

This paper builds on well-developed literatures in auction theory and practice—especially combinatorial auctions and spectrum auctions. Much of the literature on combinatorial auctions is summarized in Cramton et al. (2006). The work of Ausubel et al. (2006), Ausubel and Milgrom (2006a,b), Day and Raghavan (2007), Day and Milgrom (2008), Day and Cramton (2012), Milgrom (2007, 2010), Parkes (2006), and Porter et al. (2003) is especially relevant. On spectrum auctions see Coase (1959) for the original proposal, Ausubel et al. (1997) on synergies, McMillan (1994), Cramton (1995, 1997, 2006), Klemperer (2004), and Milgrom (2004) on the performance of the simultaneous ascending auctions, and Brusco and Lopomo (2002) and Cramton and Schwartz (2002) on collusion. Kagel et al. (2010) experimentally compare the simultaneous ascending auction with a particular ascending combinatorial auction, which differs significantly from the one presented here.

I begin by describing some of the problems of the simultaneous ascending auction. Then I present the combinatorial clock auction, which retains the benefits, while addressing the weaknesses, of the simultaneous ascending auction. I emphasize two essential elements of the combinatorial clock auction: the pricing rule and the activity rule. Along the way, I summarize both experimental and field results with the combinatorial clock auction.

The combinatorial clock auction is of great practical interest. The design has been adopted for major spectrum auctions in many countries over three continents.

2. Simultaneous Ascending Auction

The workhorse for spectrum auctions since 1994 has been the simultaneous ascending auction, which is a simple generalization of the English auction to multiple items in which all items are auctioned simultaneously. Thus, unlike Sotheby's or Christie's auctions in which the items are auctioned in sequence, here all the items are auctioned at the same time.

The process is as follows: Each item or lot has a price that is associated with it. Over a sequence of rounds, bidders are asked to raise the bid on any of the lots that they find attractive, and the auctioneer identifies the provisional winner for each lot at the end of every round. The process continues until nobody is willing to bid any higher. This process was originally proposed by Preston McAfee, Paul Milgrom, and Robert Wilson for the FCC spectrum auctions. Since its introduction in July 1994, the design has undergone numerous enhancements, but the basic design has remained intact in its application worldwide for the vast majority of spectrum auctions.

An important element of the basic design is an activity rule to address the problem of bid sniping: waiting until the last minute to bid seriously (which reduces the amount of information that is generally available to other bidders and that could help them bid efficiently). The rule adopted by the FCC and used in all simultaneous ascending auctions to date is a quantity-based rule: The rule requires a bidder that wants to be a big bidder at the end of the auction must be a big bidder throughout the auction. Each bidder must maintain a level of activity, based on the quantity of spectrum for which the bidder is bidding, in order to continue with that level of eligibility later on. Thus, a bidder cannot play a snake-in-the-grass strategy where the bidder holds back and waits, and then pounces late in the auction, thereby winning without making its true intent known until the last instant.

As mentioned, the simultaneous ascending auction has been used for a long time. The FCC has conducted about 80 simultaneous ascending auctions, since it was introduced in July of 1994. The FCC has gotten good at conducting the auctions, and the design has worked reasonably well. Nonetheless, it is perhaps surprising how quickly inertia set in. The FCC was initially highly innovative in its initial choice of design, but since then the FCC has just made minor incremental improvements in response to obvious and sometimes severe problems with the original simultaneous ascending auction design.

Why has the design held up so well? The simultaneous ascending auction is an effective and simple price discovery process. It allows arbitrage across substitutes. It lets bidders piece together desirable packages of items. And, because of the dynamic process, it reduces the winner's curse by revealing common value information during the auction (Kagel and Levin 1986, Kagel et al. 1996).

But the design does, and has been observed to have, many weaknesses.

- As a result of the pricing rule, there is a strong incentive for large bidders to engage in demand reduction—to reduce the quantity demanded before the bidder's marginal value is reached in order to win at lower prices.
- Especially if there is weak competition, bidders have an incentive to engage in tacit collusion. The bidders employ various signaling strategies, where they attempt to work out deals through the language of the bids. The goal of the strategies is to allocate the items among the bidders at low prices.
- As a result of the activity rule, there are parking strategies. A bidder maintains eligibility by parking its eligibility in particular spots that the bidder is not interested in and then moves to its true interest later.
- The simultaneous ascending auction is typically done without package bids. The bidders are bidding on individual lots, and there is the possibility that a bidder will win some of the lots that it needs for its business plan, but not all. This exposure to winning less than what the bidder needs has adverse consequences on efficiency. Essentially, the bidder has to guess. Either the bidder bids for what it wants, or not. When there are complementarities, this is a tough decision for the bidder to make. The bidder may make the wrong decision and win something it actually does not want or fail to win something it does want.
- The lack of package bids also makes the simultaneous ascending auction vulnerable to hold up, which is basically a speculator stepping in and taking advantage of a bidder (Pagnozzi 2010). The speculator can make it clear to large bidders that it would be expensive to push him out of the way. As a result, the large bidders let the speculator win some desirable lots at low prices, and then the speculator turns around and sells them to the big players after the auction is over. That is the holdup strategy. It is easy to do and effective. Preventing resale would reduce this problem, but resale is desirable in a rapidly changing, dynamic industry.
- There is limited substitution across licenses, which is something I am going to emphasize. The reader might think that it would be easy to arbitrage across the lots, but in fact that is not the case. This is especially true in a large country like the United States, where

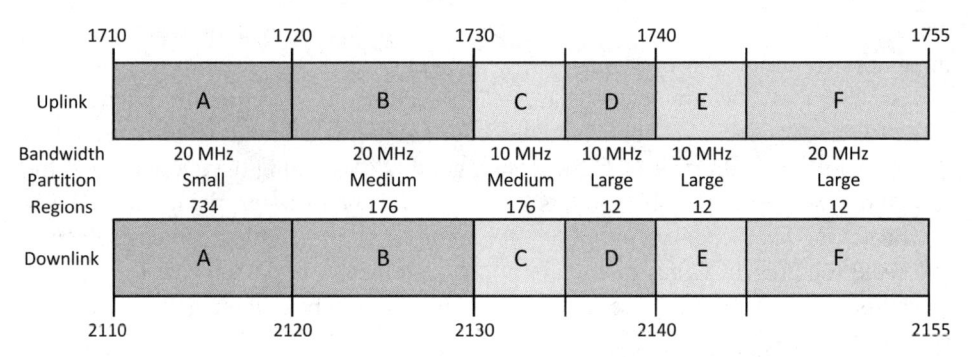

	1710	1720	1730	1740		1755
Uplink	A	B	C	D	E	F
Bandwidth	20 MHz	20 MHz	10 MHz	10 MHz	10 MHz	20 MHz
Partition	Small	Medium	Medium	Large	Large	Large
Regions	734	176	176	12	12	12
Downlink	A	B	C	D	E	F
	2110	2120	2130	2140		2155

Figure 1. The US AWS band plan: something for everyone.

the FCC splits up the frequency bands in different ways, geographically, and the bidders can only bid on individual lots, rather than packages.

As a result of all these factors, the bidding strategies are quite complicated.

2.1. The US AWS and 700 MHz Auctions

The difficulties in arbitraging across substitutes are best illustrated in the two most recent major auctions in the United States: Advanced Wireless Services (AWS) and 700 MHz.

The AWS auction sold 90 MHz of spectrum in 161 rounds in 2006, and raised $14 billion. As in all of its auctions, the FCC began the process by settling on a specific band plan (the product design, as shown in Figure 1), which effectively determined how the available bandwidth in each location was going to be split up into lots. Each lot is a particular frequency band covering a particular geographic area. In the case of the AWS auction, the FCC decided that six frequency blocks of paired spectrum (A-F) were to be auctioned. Three blocks were 20 MHz and three were 10 MHz. Because the US is so large, each frequency block was also partitioned geographically. And because the FCC was attempting to accommodate all types of bidders, the FCC partitioned the blocks in three different ways: for blocks D-F the country was split into 12 large regions; for blocks B and C the country was split into 176 medium-size regions; and for block A the country was split into 734 small regions. Remarkably, the different partitions do not form a hierarchy in the sense that a bidder cannot construct one of the medium-sized lots by aggregating a number of small lots. This inability to aggregate small into medium clearly limits substitution across blocks.

The underlying substitution problem was caused both by the product design—the use of specific blocks that followed three different geographic schemes—and the auction format. Figure 2 illustrates the severe problems that bidders had substituting across blocks in the AWS auction. It shows the price per 10 MHz of spectrum for each of the blocks at the end of critical days in the auction. Recall that there are six blocks, so there are six bars (A through F) at the end of each day. The 20 MHz bars are twice as wide as the 10 MHz bars, so the area of the bar corresponds to revenues at the time indicated. Finally, different shades of gray represent different bidders, so the reader can see who

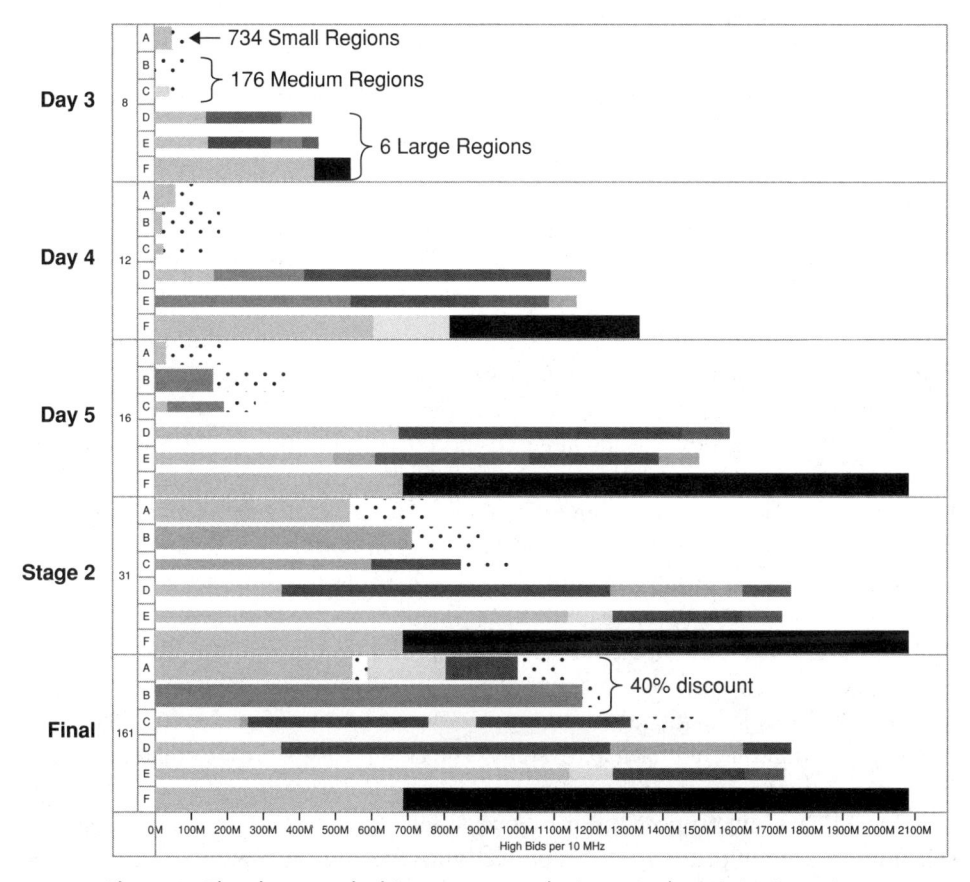

Figure 2. The absence of arbitrage across substitutes in the US AWS auction.

the provisional winners are at the various times in the auction. The two largest bidders are T-Mobile (diagonal stripes) and Verizon (horizontal stripes).

If there was perfect arbitrage across blocks, then the length of the bars would be the same at each time in the auction, which would indicate equal prices across blocks. Over time, the prices would move higher, but the prices would tend to move together across the blocks, as bidders would arbitrage to the cheaper lots per MHz of spectrum.

What happened in the AWS auction is extremely far from that, as is illustrated by the end of day five. At this point, the F block has already reached its final price. The A block is less than one twentieth the price of the F block. If the A block is roughly equivalent to the F block, why wouldn't Verizon, say, switch to the much cheaper A block, instead of placing bids twenty times higher on the F block? The reason has to do with substitution difficulties. When Verizon is bumped off a large F block license, it is easy for Verizon to substitute down to the A block, submitting say the 100 or so bids on the A lots that roughly cover the corresponding F lot. The problem is that once Verizon has shifted down it would be nearly impossible to shift back up to F. The reason is that in subsequent rounds Verizon would only be bumped from some of the corresponding A block lots. Verizon would have to withdraw from many A lots in order to return to F,

Table 1. *Band plan and final prices ($/MHz-pop) for paired spectrum in 700 MHz auction*

Block	A	B	C
Bandwidth	12 MHz	12 MHz	22 MHz
Type	paired	paired	paired
Partition	176	734	12
Price	$1.16	$2.68	$0.76

exposing itself to large withdrawal penalties. In addition on block A, Verizon would be vulnerable to various hold-up strategies, where speculators could pick important holes in a synergistic aggregation of lots.

Since substituting down from large (F, E, D) to small (C, B, A) lots is easier than substituting up, the auction essentially proceeded in a sequential fashion. First, the bidders competed for the large-lot blocks (F, E, D), then they competed for the medium-lot blocks (C and B), and finally the competition fell to the small-lot block (A). This explains the sequential, rather than simultaneous price process across blocks. See Bulow et al. (2009) for more on this auction.

The next major auction in the US was the 700 MHz auction in 2008. The band plan for the paired spectrum is shown above. The FCC did the same thing in this auction. Specific blocks were auctioned, using three different partitions of the US. Again the different partitions did not form a hierarchy. The final prices per MHz-pop (bandwidth times population) range from $0.76 for the C block to $2.68 for the B block, as shown in Table 1. These final prices differ by over a factor of three. We see again that the substitution across blocks is far from perfect. Interestingly, this time it is the small-lot block B that sold for a high price, and the large-lot block C that sold for a low price—which is just the opposite of what happened in the AWS auction.

Although the C-block had an open access provision, which required that the carrier not discriminate against either devices or applications, the terms of open access were sufficiently watered down that I doubt it had much of an impact on the C-block price. In my view, the price difference was because competing bidders thought that competing on the C-block against Verizon (or perhaps AT&T and Verizon) was sufficiently hopeless that it would be better to focus on the A and B blocks. See Cramton et al. (2007) for more on the competitive issues in this auction.

The conclusion from the 20 years of history of spectrum auctions that have used the simultaneous ascending auction is that it works reasonably well in simple situations with a single geographic scheme. However in more complex settings, the approach leads to complex bidding strategies that complicate the auction and may undermine the efficient assignment of spectrum.

3. A Better Way: The Combinatorial Clock Auction

Fortunately, there is a better way. All that is needed is a number of complementary enhancements that ultimately simplify the bidding process, improve its efficiency, and greatly expand its power.

First, much of the game playing, such as tacit collusion and other bid signaling, can be eliminated with a shift to anonymous bids. In a combinatorial clock auction the round-by-round revelation of information is limited to aggregate measures of competition. Limiting round reports to prices and excess demand for each product gives the bidders the information needed to form expectations about likely prices and to resolve common value uncertainty, yet such reports do not allow the signaling strategies that support tacit collusion. Moreover, the streamlined report simplifies bidder decision-making and keeps the bidders focused on what is most relevant: the relationship between prices and aggregate demand.

In most instances, the spectrum lots that cover the same region in adjacent frequencies are nearly perfect substitutes. The bidder primarily cares about the quantity of spectrum in MHz that it has in the region, rather than the exact frequency location. Moreover, to minimize interference problems and maximize data speeds bidders prefer contiguous spectrum within any region. In this setting, it makes sense in the initial stage to auction generic spectrum. The initial stage determines the quantity of contiguous spectrum won in each region. The spectrum is treated as if it were a homogenous good within each region. This is an enormous simplification of what is being sold. The idea is to treat each MHz of spectrum within a geographic region and a particular frequency band as perfect substitutes. The auction first resolves the main question of how much spectrum in each region each winner gets and at what price, before the auction turns to the more subtle and less important question of the exact frequencies.

Of course, there are some auctions where the differences across frequencies are too great to allow this simplified treatment—for example, because of major interference differences by frequency, as the result of incumbents with a right to stay in the particular band. In such cases, the specific spectrum lots can be auctioned from the start; but in most cases, it is desirable to auction generic spectrum first and then determine the specific assignment in a second stage.

The specific assignment stage is simplified, since it only involves winners of the generic stage. The number of specific assignments typically is limited to the number of ways that the winners can be ordered. Thus, if there are m winners there are m! different specific assignments. For example, an auction with four winners in a particular region would have $4! = 4 \times 3 \times 2 = 24$ different possible specific assignments. If we assume separability across regions, each of the four bidders would only need to express preferences among at most 24 different specific assignments. This number is reduced further if we assume that the bidder only cares about its own specific assignment and not the location of the other winners, as is commonly the case. Then for example with four winners of equal size, each winner would only need to express three preferences: the incremental value from the bidder's first, second, and third-best specific assignment compared with its fourth-best.

The use of generic lots, wherever possible, simplifies the auction, enhances substitution, and improves price discovery. Despite these advantages the FCC has chosen in each of its roughly 80 auctions to sell specific lots. This is a common mistake in auction design. Interestingly, even in countries that recognized the advantages of selling generic lots, such as the German 3G auction, the generic lots were auctioned using a method for specific lots; that is, in the German 3G auction, even though the lots were perfect substitutes, the bidders bid on specific lots.

The first innovation is an improved product design, based on generic spectrum in each region, which accommodates multiple types of use.

Once generic lots are adopted the next innovation becomes easier to see: the adoption of simple and powerful techniques that are well-suited to auctioning many divisible goods.

The second innovation is the use of a simultaneous clock auction. This is a simplification of the simultaneous ascending auction. Each product has its own "clock," indicates its current price. Because of generic lots, each product may consist of multiple lots. In each round, the bidder is asked to indicate for each product the quantity of lots desired at the current price. At the end of the round, the auctioneer adds up the individual bids and reports the demand for each product. The price is then increased on any product with excess demand. This process is repeated until there is no excess demand for any product.

The two critical differences between the clock auction and the simultaneous ascending auction are: 1) the bidder only answers demand queries, stating the quantities desired at the announced prices; and 2) there is no need to determine provisionally winning bidders at the end of every round.

The third innovation is more subtle, but extremely powerful. One can interpret the demand vector reported by each bidder in each round as a package bid. The bidder is saying, "At these prices, I want this package of lots." Taking this interpretation seriously yields a combinatorial auction (or package auction) without the need for any optimization. This allows bidders to express complementarities within a simple price discovery process.

Lawrence Ausubel and I have been conducting exactly this sort of package auction since 2001 for electricity and gas products in France, Germany, Belgium, Denmark, Spain, Hungary, and the United States (Ausubel and Cramton 2004). Thus far, we have conducted over 70 high-stakes auctions with this format for assets worth over $10 billion. We also used the approach in a spectrum auction in Trinidad and Tobago in 2005. The approach has been highly successful.

The clock auction may end with some products in excess supply, as a result of complementarities among lots. In addition, since the clock process follows a single price path and only includes a limited number of price points, it is desirable to allow the bidder to specify additional bids in a supplementary round following the clock stage. The purpose is to let the bidder express preferences for additional packages that were missed by the clock process. In addition, the bidder can improve its bids on packages that were already bid on in the clock stage.

Once the clock bids and the supplementary bids are collected, an optimization is run to determine the value-maximizing generic assignment and prices. This two-step process of a clock auction followed by supplementary bids, which I call a combinatorial clock auction, was proposed by Lawrence Ausubel, Paul Milgrom, and me for spectrum auctions at an FCC auction conference in 2003 (Ausubel et al. 2006). We proposed the same approach for spectrum auctions in the UK in 2006, as well as for airport takeoff-and-landing rights in 2003. Meanwhile, Porter et al. (2003) demonstrate in the experimental lab the high efficiency of a closely related approach.

Two critical elements of a successful combinatorial clock auction are the pricing rule and the activity rule. I will discuss both at length. These two important rules work together to ensure that the bids are an accurate expression of bidder preferences throughout the entire auction. The high efficiency of the combinatorial clock auction derives mainly from incentives for nearly truthful bidding. A pricing rule that is based on second pricing encourages truthful bidding; and the activity rule based on revealed preference ensures that these incentives for truthful bidding are felt throughout the clock stage.

4. UK Spectrum Auctions

The need for a technology-neutral auction is commonplace in today's world of rapidly developing communications technologies and applications. Although the regulator can typically identify the viable candidate technologies based on early development, the regulator cannot decide how available spectrum should be split among the technologies without a market test. Examples are numerous, and several will be discussed here.

Ofcom, which is the independent regulator and competition authority for the UK communications industries, was the first to recognize and act on this need for a technology-neutral auction. In spring 2006, Lawrence Ausubel and I proposed to Ofcom a version of the combinatorial clock auction. Since June 2006, I have been working with Ofcom in developing, testing, and implementing the design for a number of its auctions. Two such auctions—the 10-40 GHz auction and the L-band auction—have occurred already. Both went well, and provided a useful field test for the economically much larger 800 MHz and 2.6 GHz auctions. Several countries in addition to the UK have since adopted the design for 4G auctions involving one or many spectrum bands.

Ofcom has three main goals for the auction design: The auction should be technology neutral, which allows alternative viable technologies to compete for the spectrum on an equal basis. The auction should accommodate flexible spectrum usage rights, which permits the user to decide how the spectrum would be used, subject to minimizing interference externalities with neighbors. And the auction should promote an efficient assignment of the spectrum, which puts the spectrum to its best use.

Simplicity and transparency are important secondary objectives. On simplicity, Ofcom recognized that satisfying the main objectives posed serious challenges, which could not be addressed with an auction design that is too simple. Moreover, simplicity has to be assessed in recognition of the complexity of bidder participation. For example, the simultaneous ascending auction has simple rules, but incredibly complicated bidding strategies. In contrast, the combinatorial clock auction has more complex rules, but the rules have been carefully constructed to make participation especially easy. For the most part, the bidder can focus simply on determining its true preferences for packages that it can realistically expect to win. In a combinatorial clock auction it is the auctioneer that needs to do the complex optimization, whereas the bidders can focus on their values for realistic packages.

Revenue maximization was explicitly excluded as an objective. Nonetheless, an efficient auction necessarily will generate substantial revenues. Indeed, my advice to countries is to focus on efficiency. A focus on revenues is short-sighted. In my view, the government is better off finding as much spectrum as possible and then auctioning it so as to put the spectrum to its best use. This approach creates a competitive and innovative market for communications, which has substantial positive spillovers to the rest of the economy. Under this approach, long-term revenues likely will far exceed those that would come from the maximization of short-term auction revenues.

I now explain the details of two essential rules in the combinatorial clock auction: the pricing rule and the activity rule. The rules may appear complex, but the complexity actually simplifies the bidding strategies, which makes it easier for bidders to participate in the auction.

5. The Pricing Rule: Vickrey-Nearest-Core Pricing

Prices are determined at two points in the auction: after the clock stage, including the supplementary bids, to determine the base prices for the winners in the value-maximizing generic assignment; and after the assignment stage to determine the additional payments for specific assignments.

The pricing rule plays a major role in fostering incentives for truthful bidding. Pay-as-bid pricing in a clock auction or a simultaneous ascending auction creates incentives for demand reduction (Ausubel and Cramton 2002). Large bidders shade their bids, in recognition of their impact on price. This bid shading both complicates bidding strategies and also leads to inefficiency.

In contrast, Vickrey pricing provides ideal incentives for truthful bidding. Each winner pays the social opportunity cost of its winnings, and therefore receives 100 percent of the incremental value created by its bids. This aligns the maximization of social value with the maximization of individual value for every bidder. Thus, with private values, it is a dominant strategy to bid truthfully. See Ausubel (2004, 2006) for an analysis in a clock auction.

Unfortunately, as a result of complements, it may be that the Vickrey prices are too low in the sense that one or more bidders would be upset with the assignment and prices paid, claiming that they had offered the seller more. For example, suppose there are two items, A and B, and three bidders. Bidder 1 bids $4 for A, bidder 2 bids $4 for B, and bidder 3 bids $4 for A and B. The Vickrey outcome is for 1 to win A, 2 to win B, and each winner pays $0. Bidder 3 in this case has a legitimate complaint, "Why are you giving the goods to bidder 1 and 2, when I am offering $4 for the pair?" The basic problem is that with complements, the Vickrey outcome may not be in the core. Some coalition of bidders may have offered the seller more than the sum of the Vickrey prices. (The core is defined as a set of payments that support the efficient assignment in the sense that there does not exist an alternative coalition of bidders that has collectively offered the seller more.) This point has been emphasized in Ausubel and Milgrom (2002).

The solution is to increase one or more prices to assure that the prices are in the core. In order to provide the best incentives that are consistent with core pricing, the auctioneer finds the lowest payments that are in the core; that is, such that no alternative coalition of bidders has offered the seller more than the winning coalition is paying.

If we are auctioning a single item, then this is the second-price auction. Suppose the highest bidder bids $100 and the second-highest bidder bids $90. The item is awarded to the highest bidder, who pays the second-highest price of $90—which is the social opportunity cost of awarding the good to the highest bidder. Alternatively, we can think of assigning the item to maximize value, so we assign it to the highest bidder, and then we find the smallest payment that satisfies the core constraints. In this case, the second-highest bidder would be upset if the highest bidder paid less than $90, so $90 is the bidder-optimal core price. When the items are substitutes, then the bidder-optimal core point is unique and identical to the Vickrey prices.

The payment-minimizing core prices, or bidder-optimal core prices, typically are not unique when the Vickrey prices are outside the core. Thus, it will be important to have a method of selecting a unique bidder-optimal core point when there are many such points. One sensible approach that has been adopted in each of the recent Ofcom auctions for both the base prices and the assignment prices is to select the payment minimizing core prices that are closest to the Vickrey prices. This is what I call Vickrey-nearest-core pricing. Since the set of core prices is convex—a polytope formed from the intersection of half-spaces—and the Vickrey prices are always unique, there is a unique vector of core prices that is closest in Euclidean distance to the Vickrey prices. Not only are the prices unique, but since they are bidder-optimal-core prices, they also maximize the incentive for truthful bidding among all prices that satisfy core constraints (Day and Milgrom 2008).

The approach then is to take all of the bids from the clock stage and the supplementary bids, determine the value maximizing assignment, and then determine the payment-minimizing core prices that are closest to the Vickrey prices. It is my experience that bidders are quite happy with this approach: They like the idea of minimizing payments, and they recognize the importance of making sure that the prices are sufficiently high that no coalition of bidders has offered the seller more. Prices are as small as possible subject to the competitive constraints.

Calculating the winning assignments and prices involves solving a sequence of standard optimization problems. The basic problem is the winner determination problem, which is a well-understood set-packing problem. The main winner determination problem is to find the value maximizing assignment. To guarantee uniqueness, there is a sequence of lexicographic objectives, such as: 1) maximize total value; 2) minimize concentration; 3) maximize quantity sold; and 4) randomize. First the auctioneer maximizes total value. Then a constraint that the value equals this maximum value is added, and concentration is minimized. Then another constraint that concentration equals this minimum level is added, and the quantity sold is maximized. Finally, the constraint that the quantity sold equals this maximum quantity is added and an objective based on random values for each bid is maximized. This guarantees uniqueness.

Calculating the prices is a bit more involved. First, the Vickrey prices are determined by solving a sequence of winner determination problems, essentially removing one winner at a time to determine each winner's social opportunity cost of winning its package. Then the bidder-optimal core prices are determined by using a clever constraint generation method that was proposed in Day and Raghavan (2007). Having found the Vickrey prices, another optimization is solved to find the most violated core constraint. If there is none, then the process is finished, since the Vickrey prices are in the core. Otherwise, this most-violated constraint is added, and the optimization is resolved, again finding the most violated core constraint. It is added to the optimization, and again the optimization is resolved. This is continued until there is no violated core constraint, and then the process is finished.

The reason that that Day-Raghavan approach is a highly efficient method of solution is because in practice there are typically only a handful of violated core constraints; thus, the procedure stops after just a few steps. In contrast the number of core constraints grows exponentially with the number of bidders and that makes including all of the core constraints explicitly an inefficient method of solving the problem, both in time and memory.

As mentioned, the tie-breaking rule for prices is important, since typically ties will arise along the southwest face of the core polytope. Finding the prices that are closest to the Vickrey prices involves solving a simple quadratic optimization. This yields a unique set of prices. Uniqueness is important. It means that there is no discretion in identifying the outcome, either in the assignment or the prices.

An example will help illustrate all of these concepts: Suppose that there are five bidders—1, 2, 3, 4, 5—bidding for two lots: A and B. The following bids are submitted:

$b_1\{A\} = 28$
$b_2\{B\} = 20$
$b_3\{AB\} = 32$
$b_4\{A\} = 14$
$b_5\{B\} = 12$

Bidders 1 and 4 are interested in A, bidders 2 and 5 are interested in B, and bidder 3 is interested in the package A and B.

Determining the value maximizing assignment is easy in this example. Bidder 1 gets A and bidder 2 gets B, which generates 48 in total value. No other assignment yields as much. Vickrey prices are also easy to calculate. If we remove bidder 1, then the best assignment gives A to bidder 4 and B to bidder 2, resulting in 34, which is better than the alternative of awarding both A and B to bidder 3, which yields 32. Thus, the social opportunity cost of bidder 1's winning A is $34 - 20 = 14$ (the value lost from bidder 4 in this case). Similarly, if we remove bidder 2, then the efficient assignment is for bidder 1 to get A and bidder 5 to get B, resulting in 40. Then the social opportunity cost of bidder 2's winning B is $40 - 28 = 12$ (the value lost from bidder 5). Hence, the Vickrey outcome is for bidder 1 to pay 14 for A and for bidder 2 to pay 12 for B. Total revenues are $14 + 12 = 26$. Notice that bidder 3 has cause for complaint, since bidder 3 offered 32 for both A and B.

Now consider the core for this example. The core is represented in the payment space of the winning bidders—in this case the payments of bidders 1 and 2. Each bid defines a half-space of the payment space:

- Bidder 1's bid of 28 for A implies 1 cannot pay more than 28 for A.
- Bidder 2's bid of 20 for B implies 2 cannot pay more than 20 for B.
- Bidder 3's bid of 32 for AB implies that the sum of the payments for A and B must be at least 32.
- Bidder 4's bid of 14 for A implies that bidder 1 must pay at least 14 for A.
- Bidder 5's bid of 12 for B implies that bidder 2 must pay at least 12 for B.

The core is the intersection of these half-spaces as shown in Figure 3.

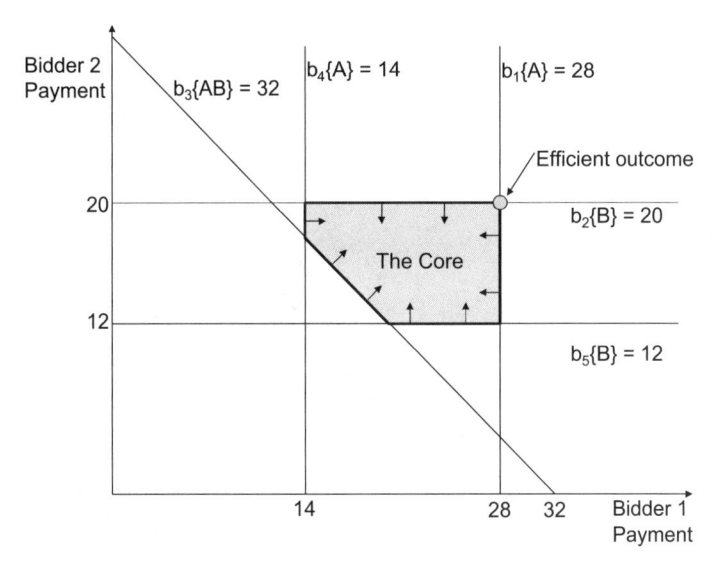

Figure 3. The Core.

This example is quite general. First, in contrast to some economic settings, in an auction the core is always nonempty. The reason is that the core always includes the efficient outcome; all of the constraints are southwest of the efficient point, since the efficient point maximizes total value. Second, the core is always a convex polytope, since it is the intersection of numerous half-spaces. Third, complementarities, like bidder 3's bid for AB, are the source of the constraints that are neither vertical nor horizontal. These are the constraints that can put the Vickrey prices outside the core. Without complementarities, all of the constraints will be vertical and horizontal lines, and there will be a unique extreme point to the southwest: the Vickrey prices.

The graphical representation of the core is also a useful way to see the Vickrey prices. Vickrey is asking how much can each winner unilaterally reduce its bids and still remain a winner. As shown in Figure 4, bidder 1 can reduce its bid to 14 before bidder 1 is displaced by bidder 4 as a winner. Similarly, bidder 2 can reduce its bid to 12 before being displaced by bidder 5. Thus, the Vickrey prices are 14 and 12. The problem is that these payments sum to 26, which violates the core constraint coming from bidder 3's bid of 32 for AB.

Bidder-optimal core prices can also be thought of as maximal reductions in the bids of winners, but rather than reducing the bids of each winner one at a time, we jointly reduce all the winning bids, as shown in Figure 5, until the southwest face of the core is reached. As can be seen, this does not result in a unique core point, since the particular point on the southwest face depends on the rate at which each winner's bids are reduced. The bidder-optimal core points consist of the entire southwest face of the core. If the southwest face is a unique point, then it is the Vickrey prices; if the southwest face is not unique, then the face is a core constraint involving complementarities, and the Vickrey prices lie outside the core.

Nonetheless, there is always a unique bidder-optimal core point that is closest to the Vickrey prices. This is seen in Figure 6, as the bidder-optimal core point that forms a

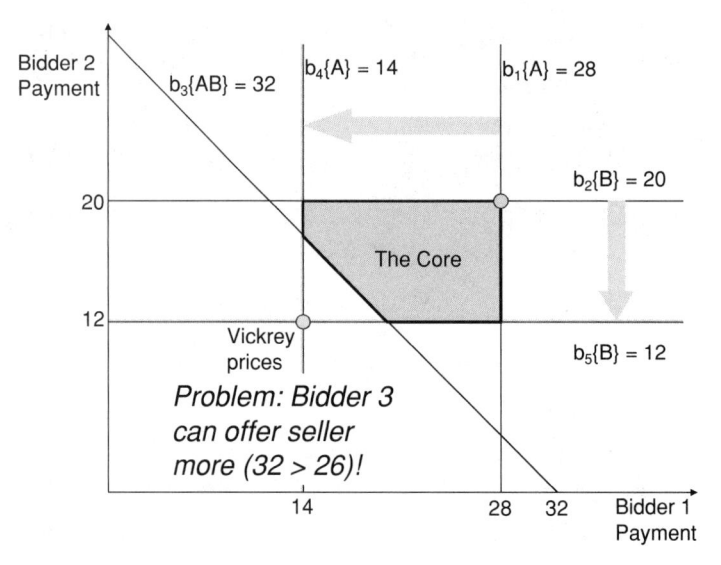

Figure 4. Vickrey prices: how much can each winner's bid be reduced holding others fixed?

90 degree angle with the line that passes through the Vickrey prices. This point minimizes the Euclidean distance from the Vickrey prices.

Vickrey-nearest-core pricing was adopted in each of the UK spectrum auctions and has been adopted in several other auctions. Erdil and Klemperer (2010) argue that marginal incentives for truthful bidding may be improved by using a reference point other than the Vickrey prices for selecting among bidder-optimal core prices. In particular, they recommend a reference point that is independent of the winners' bids. See also Ausubel and Baranov (2010) for additional analysis.

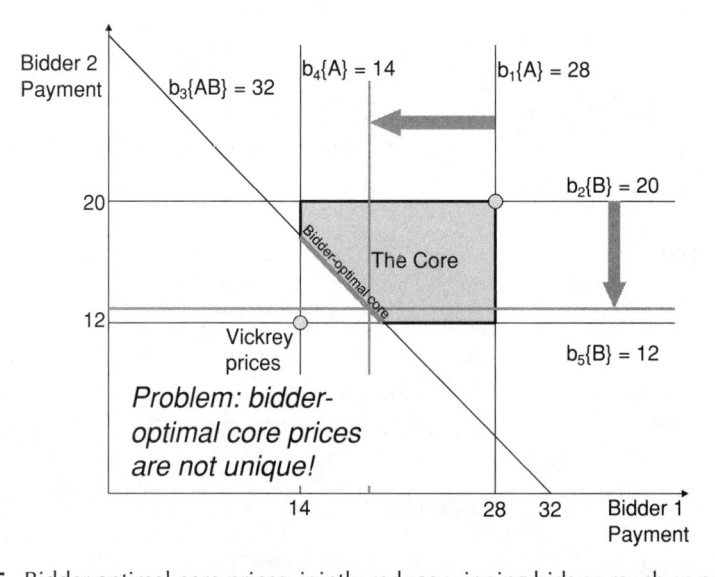

Figure 5. Bidder-optimal core prices: jointly reduce winning bids as much as possible.

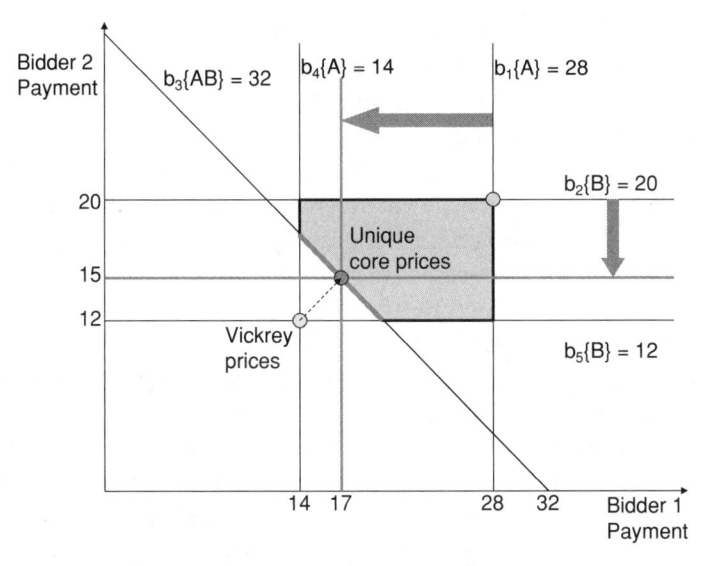

Figure 6. Core point closest to Vickrey prices.

Bidder-optimal core pricing has several advantages. First, it minimizes the bidders' incentive to distort bids in a Pareto sense: There is no other pricing rule that provides strictly better incentives for truthful bidding. Bidder-optimal core pricing implies Vickrey pricing, whenever Vickrey is in the core. For example, when lots are substitutes, Vickrey is in the core, and the bidders have an incentive to bid truthfully. Since the prices are in the core, it avoids the problem of Vickrey prices' being too low as a result of complements.

6. The Activity Rule: Revealed Preference

Good price discovery is essential in realizing the benefits of a dynamic auction. Good price discovery stems from providing incentives for the bidders to make truthful bids throughout the auction process. The pricing rule discussed in the prior section is an essential element, but one also has to be concerned about what is seen on eBay every day: bid sniping—jumping in at the last instance in an auction and thereby holding information back. Absent an activity rule, bidders will have an incentive to hold back to conceal information. The activity rule is intended to promote truthful bidding throughout the auction process.

Nearly all high-stake auctions, such as the FCC spectrum auctions, have an activity rule. The FCC uses a quantity-based rule. This rule has worked reasonably well in the FCC's simultaneous ascending auctions; but in a combinatorial clock auction with Vickrey-nearest-core pricing, we need a more complex rule: one that is based on revealed preference (Ausubel et al. 2006). Such a rule is effective at getting bidders to bid in a straightforward way throughout the clock stage, selecting the most profitable package given the current prices.

The traditional activity rule in both simultaneous ascending auctions and clock auctions has been a quantity-based rule: To be a large winner at the end of the auction, the bidder must be a large bidder throughout the auction. In particular, each lot corresponds to a particular quantity of spectrum, measured in either MHz-pop or in "eligibility points". The bidder starts with an initial eligibility based on the bidder's initial deposit. To maintain this level of eligibility in future rounds, the bidder needs to bid on a sufficiently large quantity of spectrum in the current round, where "sufficiently large" is stated as some percentage, typically between 80% and 100% of the bidder's current eligibility. If the bidder bids on a smaller quantity, the bidder's eligibility is reduced in future rounds. This quantity-based rule has worked reasonably well, although as mentioned, it does create an incentive for parking eligibility on lots that a bidder is not truly interested in, especially if the eligibility points are not a good measure of relative value across lots. (The FCC's MHz-pop measure is especially poor with small lots. Spectrum in New York City is much scarcer than spectrum in Montana. As a result, spectrum values are much higher in New York City on a per MHz-pop basis. Despite this fact, which has been demonstrated in many dozens of spectrum auctions, the FCC still continues to use MHz-pop as the quantity measure in its auctions, which exacerbates parking and other problems that are associated with the activity rule.)

In many clock auctions, an activity requirement of 100% is used, which means that the bidder cannot increase the size of the package, as measured in eligibility points, as prices rise. For the case of a single product, this means that the bidder must bid in a manner that is consistent with a downward-sloping demand curve.

In a combinatorial clock auction, one can use this quantity-based rule in the clock stage, but one also needs to specify how the rule limits bids in the supplementary round. This linkage between the clock bids and the supplementary bids is of critical importance, for otherwise the bidder could snipe: submit all of its bids in the supplementary round.

Ofcom proposed the following, which I call the *eligibility point rule*: During the clock stage the bidder cannot increase the package size. Moreover, whenever the bidder reduces the package size, the bid on all larger packages is capped by the prices at the time of the reduction. For example, if during the clock stage a bidder drops from a package of size 10 to 6 at prices p, then for all packages q of size 7 to 10, the supplementary bid cannot be more than p · q.

The eligibility point rule, which Ofcom used in its first two combinatorial clock auctions, has the advantage of simplicity. For each package there is at most a single linear constraint on the supplementary bid. However, it has a potentially serious problem: The straightforward strategy of bidding on the most profitable package in the clock stage is a poor strategy. A bidder following such a strategy would find that its supplementary bids would be sharply constrained, well below true values. To avoid this problem, the bidder must instead bid in the clock stage to maximize package size, subject to a nonnegative profit constraint. That is, the bidder throughout the clock stage bids on the largest package that is still profitable.

Lawrence Ausubel, Paul Milgrom, and I proposed an alternative activity rule that is based on revealed preference for the combinatorial clock auction (Ausubel et al. 2006). Revealed preference is the underlying motivation for all activity rules. The intent is to require the bidder to bid in a way throughout the auction that is consistent with the bidder's true preferences. Since we do not know the bidder's true preferences, the best we can hope for is for the bidder to bid in a manner that is consistent with its revealed

Table 2. *An example with two bidders and two identical lots*

	Marginal Value		Average Value	
	Bidder A	Bidder B	Bidder A	Bidder B
1 lot	16	8	16	8
2 lots	2	2	9	5

preferences. In the simplest case of a single-product clock auction, this is equivalent to monotonicity in quantity, just like the eligibility point rule, but when we have multiple products the two rules differ in important ways.

For the combinatorial clock auction, the *revealed preference rule* is as follows (see Harsha et al. 2010 for a stronger statement): During the clock stage, a bidder can only shift to packages that have become relatively cheaper; that is, at time $t' > t$, package $q_{t'}$ has become relatively cheaper than q_t:

(P) $$q_{t'} \cdot (p_{t'} - p_t) \leq q_t \cdot (p_{t'} - p_t).$$

Moreover, every supplementary bid $b(q)$ must be less profitable than the revised package bid $b(q_t)$ at t:

(S) $$b(q) \leq b(q_t) + (q - q_t) \cdot p_t.$$

Each clock bid for package q_t, as improved in the supplementary round, imposes a cap on the supplementary bid for package q.

An important advantage of the revealed preference rule is that a bidder that follows the straightforward strategy of bidding on its most profitable package in the clock stage would retain the flexibility to bid its full value on all packages in the supplementary round.

To illustrate the implications of the two activity rules, consider the following example with two bidders and two identical lots (one product) in a setting of substitutes: The bidders' preferences are given in Table 2, which indicates the marginal and average value for 1 lot and 2 lots.

Since the lots are substitutes, both bidders want to bid their true values in the supplementary round. However, consider what happens in the clock stage in response to the two different rules.

With the revealed preference rule, each bidder has an incentive to bid on its most profitable package in each round. Thus, the bidding simply moves up each bidder's marginal value (demand) curve. When the clock price reaches 2, both bidders drop from a package of size 2 to 1, and excess demand drops to zero. The clock stage ends at the competitive equilibrium price of 2 and the efficient assignment. Indeed, there is no need for any supplementary bids in this case. Bidder A can enter supplementary bids of 16 and 18, and bidder B can enter supplementary bids of 8 and 10, but these supplementary bids will not change the outcome in any way. Each bidder wins one lot and pays 2 (the Vickrey price). The supplementary round is unnecessary. The clock stage, by revealing the bidders marginal value information, up to the point of no excess

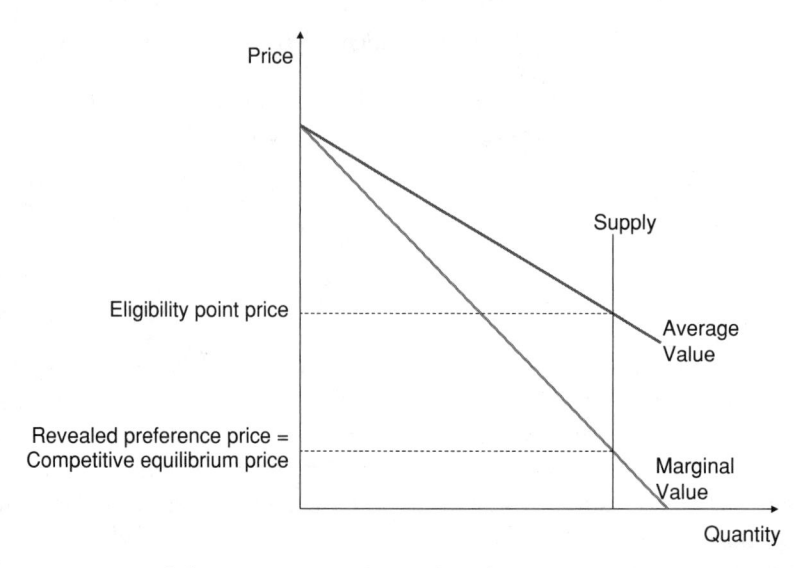

Figure 7. Downward sloping aggregate demand implies average value > marginal value.

demand, has revealed all that is needed to determine and price the efficient assignment. This is a general result with substitutes.

With the eligibility point rule, bidders are forced to distort their bidding away from the straightforward strategy of profit maximization. In order to preserve the ability to bid full values in the supplementary round, the bidders instead bid on the largest package that is still profitable. This entails moving up the average value curve, since when the average value is exceeded a package is no longer profitable. Thus, when the clock price reaches 5, bidder B's average value for 2 is reached, and the bidder drops its demand to 1. Then when the clock price reaches 8, bidder B's average value for 1 is reached and bidder B drops out. At this point there is no excess demand, so the clock stage ends with bidder A demanding 2, bidder B demanding zero, and the clock price at 8. In the supplementary bid round, the bidders again submit their true preferences, and the optimization determines that each bidder should win one lot and should pay 2. The supplementary round was required to determine the efficient assignment and price the goods. Notice that the clock stage did little but mislead the bidders into thinking that bidder A would win all the items at a high price.

The reader might think that I somehow rigged this example to make the eligibility point rule look bad. This is not the case. Whenever lots are substitutes, the same features will be observed. With revealed preference, the clock stage will converge to the competitive equilibrium, revealing the efficient outcome and supporting prices; whereas with the eligibility point rule, the clock stage ends with an assignment that is excessively concentrated and prices that are too high. This result follows from the simple fact that average value exceeds marginal value, whenever aggregate demand is downward sloping, as shown in Figure 7. Having participated in many dozens of major spectrum auctions, I can confirm that this is indeed the typical case.

What is essential for price discovery is the revelation of the marginal value information. This helps bidders make the marginal tradeoffs that are of greatest relevance in

Efficiency and number of bids by simulation

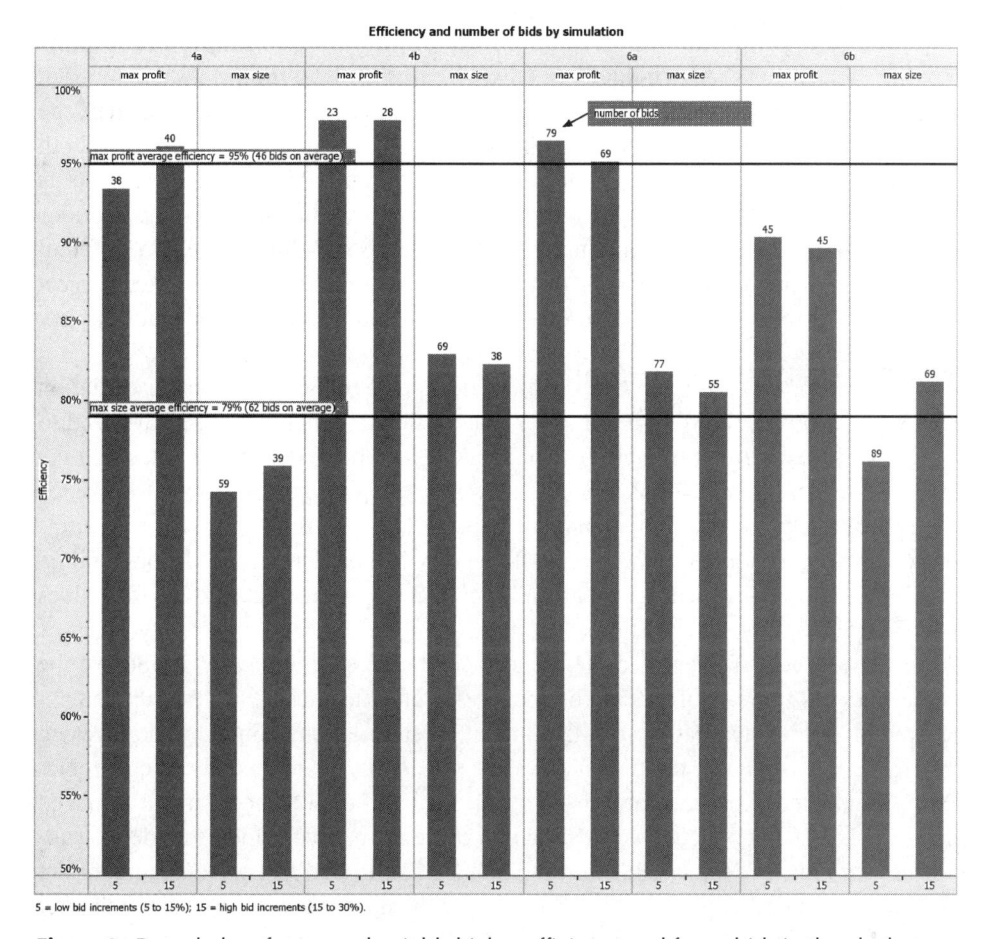

[Bar chart showing efficiency percentages and number of bids across simulations 4a, 4b, 6a, 6b, each with max profit and max size conditions at bid increments 5 and 15. Max profit average efficiency = 95% (46 bids on average); max size average efficiency = 79% (62 bids on average). 5 = low bid increments (5 to 15%); 15 = high bid increments (15 to 30%).]

Figure 8. Revealed preference rule yields higher efficiency and fewer bids in the clock stage.

figuring out what the outcome should be. This is why I believe that the eligibility point rule is a poor choice.

To further test the two activity rules, I conducted numerous simulations that used realistic demand scenarios with significant complementarities from both technological and minimum scale constraints. I assumed that the bidders bid on the most profitable package with revealed preference (max profit) and bid on the largest profitable package with the eligibility point rule (max size). The results are summarized in Figure 8. It is clear that the revealed preference rule achieves substantially higher efficiency in many fewer rounds.

As a final test of the two activity rules, as well as other elements of the auction design, I conducted a series of full-scale tests in the experimental lab. For the tests, the Ofcom auction platform was used and indeed Ofcom staff served as the auctioneer. The subjects in the test were PhD students, who had taken an advanced course in game theory and auction theory, and had prior participation in combinatorial clock auction experiments. I chose such an experienced and expert subject pool, since in the actual spectrum auctions bidders often hire experts and devote substantial time and money to understand the strategic implications of the rules.

Each subject participated in several auctions over a two-week period. In each auction, the subject was given a bidding tool, which calculated the subject's value for each package consistent with the bidder's business plan. The scenarios as represented by the various bidding tools were chosen to be realistic. The valuation models included both substitutes and complements. Complements came from minimum scale constraints as well as technological requirements. A training session was held before the auctions to explain the details of the combinatorial clock auction, including the two different activity rules. All subjects participated in both activity rule treatments. Each subject was paid an amount that was based on her experimental profits. The average subject payment was $420.

The experiments confirmed that the eligibility point rule caused a major deviation from straightforward bidding in the clock stage. Bidders quickly realized the need to bid on the largest profitable package. This undermined price discovery; but, given the private value setting and simple valuation models, the poor performance of the clock stage was largely corrected by the supplementary bids and the optimization that followed. There were some instances of inefficiency when bidders deviated from bidding on the largest profitable package and then found that they were unable to bid full values in the supplementary stage.

In contrast, with the revealed preference rule, bidders almost always followed the straightforward strategy of bidding on the most profitable package. In the supplementary round, bidders typically bid full value and were not constrained by the revealed preference rule. As a result, efficiency was nearly 100%. More recently, Bichler et al. (2011) conducted experimental tests of the combinatorial clock auction that achieved lower levels of efficiency (between 89 and 96 percent), because bidders tended to submit too few bids. For the combinatorial clock auction to perform well, it is important for bidders to submit all relevant bids. The experiments that I conducted did not suffer from "too few bids" because the bidders had a bidding tool that made it easy for them to submit bids on all of the relevant packages. In my experience with real bidders, the bidders have had such tools, and indeed the development of such tools is a big task in the preparations for the auctions.

One issue that was discovered in the lab was the complexity of the revealed preference rule. The few bidders who deviated from bidding on the most profitable package in each round of the clock stage found that they were unable to bid full value in the supplementary round as a result of the revealed preference constraint. These bidders had to make adjustments to bids to satisfy the revealed preference constraints, but it was difficult for them to figure out what changes to make. The challenge for the bidder is to figure out how best to adjust numerous bids in order simultaneously to satisfy many constraints (one per round). Even the brightest PhD students found this to be a daunting task without some computational help.

One solution to the complexity problem is for the auction system to provide the bidder with some help. For example, the bidder could provide the system with its desired bids. The auction system then would indicate a summary of the bids that currently violate revealed preference constraints and suggest an alternative set of bids that satisfies all constraints and is closest (in Euclidean distance) to the desired bids. This is exactly the information that the subjects in the lab were looking for in the few instances of deviations from straightforward bidding. In the lab, the deviations were minor, and the bids would have been easily adjusted with the help of a smart auction system.

In addition to complexity, the revealed preference rule may at times be too strong. Bidders' values may change over the course of the auction—for example, as the result of common value uncertainty, or the bidder may have budget constraints. Thus, there are good reasons to simplify and somewhat weaken the revealed preference rule.

The approach adopted for the 4G auctions in several countries, such as the UK, Canada, and Australia, uses a revealed preference rule that only imposes a subset of the revealed preference constraints. Importantly all bids in the supplementary round must satisfy revealed preference with respect to the final clock round. Ausubel and Cramton (2011) provide further details.

The idea behind the rule is that it may be unnecessary to include all of the revealed preference constraints to get the bidders to adopt straightforward bidding. Since the incentive for bid sniping is not too strong, even the possibility of a revealed-preference constraint may be sufficient to induce the desired behavior. People put coins in parking meters in order to avoid the possibility of a parking ticket. We can hope that a simplified revealed preference rule will have the same effect in the combinatorial clock auction.

Specifically, all supplementary bids $b(q)$ are capped by the revealed preference constraint with respect to the final clock package q_f:

$$(S') \qquad b(q) \leq b(q_f) + (q - q_f) \cdot p_f.$$

One of the desirable features of the rule is that the final package in the clock stage plays an especially important role in limiting bids. Thus, any distortion from profit maximization in the final clock package is especially costly to the bidder. Of course, the bidder never knows which clock round will be the last, so there is always some incentive to bid consistent with profit maximization. Moreover, as excess demand falls, the probability that the current round will be the last tends to increase, strengthening the incentive for straightforward bidding throughout the clock stage.

A second desirable feature of the simplified revealed preference rule is that it makes the final clock assignment and prices much more meaningful, limiting the impact of the supplementary round and motivating aggressive bidding in the clock stage.

Proposition 1. *If the clock stage ends with no excess supply, then the final assignment is the same as the clock assignment. The supplementary round cannot alter the clock assignment.*

Proof. (S') implies that the marginal value of awarding q_f to the bidder rather than q is at least the value of the lots at prices p_f:

$$b(q_f) - b(q) \geq (q_f - q) \cdot p_f.$$

It follows that any change in the final assignment cannot result in a higher total value. □

Proposition 2. *If the clock stage ends with excess supply, then a winner can guarantee that it wins its clock assignment by raising its bid on its clock package by the value of the unsold lots at the final clock prices.*

Proof. (S′) implies that the marginal value of awarding q_f to the bidder rather than q is at least the value of the lots at prices p_f:

$$b(q_f) - b(q) \geq (q_f - q) \cdot p_f.$$

It follows that any change in the final assignment can result in a marginal value of at most $q_u \cdot p_f$, where q_u is the vector of unsold lots in the clock assignment. Thus if a winner increases its bid on q_f by the amount $q_u \cdot p_f$, the final assignment must award the bidder q_f. □

The propositions demonstrate that the clock stage provides excellent price and assignment discovery whenever the final clock assignment has little or no excess supply. Clock winners know how to guarantee their clock assignment. It is not necessary to increase bids to full value. A clock winner only needs to raise its bid on the final clock package by the value of the unsold lots at the final clock prices. Potential clock losers have an incentive to bid until no profitable packages remain, since losing in the clock stage may prevent winning any package.

In the case of substitutes, the clock stage performs perfectly, if we assume a continuous clock. The pricing and activity rules provide incentives for straightforward bidding. The clock stage yields a competitive equilibrium with an efficient assignment and supporting prices. Supplementary bids are not needed to improve the assignment. The final assignment is the same as the clock assignment. The optimization simply reduces prices to reflect opportunity costs.

In the general case, the incentives for straightforward bidding are strong, but not perfect. Complements may push the Vickrey prices outside of the core, creating a threshold problem for some bidders. Nonetheless, if the clock stage ends without excess supply, then the final assignment is the clock assignment. Supplementary bids may affect prices, but not the assignment. If there is excess supply at the end of the clock stage, the winners can guarantee winning at least the clock assignment with a limited raise.

7. Conclusion

The combinatorial clock auction is a large advance over the simultaneous ascending auction. It eliminates the exposure problem; it eliminates most gaming behavior; it enhances substitution; and it encourages competition. The combinatorial clock auction enables a technology-neutral auction. This should be especially valuable in settings where the regulator does not know in advance how the spectrum should be organized. The auction, through the competitive bids, determines how the spectrum is organized, rather than the regulator. In an environment where the regulator has little information about what technology or use is best, letting the auction resolve such matters can greatly expand the realized value of the scarce spectrum resource.

A further advantage of the combinatorial clock auction is that it is readily customized for a variety of settings. Typically, a communications regulator will have a sequence of auctions over many years, as new spectrum gradually is made available. The combinatorial clock auction can be adapted to the unique characteristics of any particular auction. Adopting a consistent and flexible auction platform reduces transaction costs for the government and, more importantly, the bidders.

The auction design also enhances competition. The process is highly transparent and encourages price discovery. There is enhanced substitution both through the product design and the auction format. Bidder participation costs are reduced.

As in any market design problem, an important task for the regulator is to identify and mitigate potential market failures. In this setting and many others, the most important potential failure is market power. This is especially an issue in settings where there already is a highly concentrated communications market. Spectrum is an essential input for any new entrant. The approach here allows the regulator to address this potential market failure, as well as others, with a variety of instruments, such as spectrum caps, set asides, or bidding credits. The instruments must be used with care, or else they may do more harm than good.

One of the greatest harms is delaying the allocation and award of spectrum. Avoiding economic loss from delay should be a main priority of the regulator. Incumbents often will argue that spectrum awards should be put off. Such arguments may simply be a far less costly means of impeding competition than outbidding an entrant in an auction.

Fortunately, the use of a state-of-the-art auction design, such as the combinatorial clock auction and its variants, does not cause delay. These auctions can be designed and implemented, even by developing countries, in short order, provided that the country is using successful techniques that have been adopted elsewhere. The bottleneck is regulatory procedures, not auction design and implementation. Providers of auction services can readily meet deadlines of a few months, if necessary.

The combinatorial clock auction can be applied in many other industries. For example, the approach was proposed and tested for the auctioning of takeoff-and-landing slots at New York City's airports. The approach is well-suited for any setting in which there are many interrelated items, some of which are substitutes and some of which are complements.

More broadly, the approach described here is an example of using auction design to harness the power of markets. The approach leads to improved pricing of a scarce resource and improved decision making—both short term and long term. Innovation is fostered from the better pricing and assignment of the scarce resource.

References

Ausubel, Lawrence M. (2004), "An Efficient Ascending-Bid Auction for Multiple Objects," *American Economic Review*, 94:5, 1452–1475.

Ausubel, Lawrence M. (2006), "An Efficient Dynamic Auction for Heterogeneous Commodities,"*American Economic Review*, 96:3, 602–629.

Ausubel, Lawrence M. and Oleg V. Baranov (2010), "Core-Selecting Auctions with Incomplete Information," Working Paper, University of Maryland.

Ausubel, Lawrence M. and Peter Cramton (2002), "Demand Reduction and Inefficiency in Multi-Unit Auctions," University of Maryland Working Paper 9607, revised July 2002.

Ausubel, Lawrence M. and Peter Cramton (2011), "Activity Rules for the Combinatorial Clock Auction," Working Paper, University of Maryland.

Ausubel, Lawrence M. and Peter Cramton (2004), "Auctioning Many Divisible Goods," *Journal of the European Economic Association*, 2, 480–493, April–May.

Ausubel, Lawrence M., Peter Cramton, R. Preston McAfee, and John McMillan (1997), "Synergies in Wireless Telephony: Evidence from the Broadband PCS Auctions," *Journal of Economics and Management Strategy*, 6:3, 497–527.

Ausubel, Lawrence M., Peter Cramton, and Paul Milgrom (2006), "The Clock-Proxy Auction: A Practical Combinatorial Auction Design," in Peter Cramton, Yoav Shoham, and Richard Steinberg (eds.), *Combinatorial Auctions*, Chapter 5, 115–138, MIT Press.

Ausubel, Lawrence M. and Paul Milgrom (2002), "Ascending Auctions with Package Bidding," *Frontiers of Theoretical Economics*, 1: 1–45, www.bepress.com/bejte/frontiers/vol1/iss1/art1.

Ausubel, Lawrence M. and Paul Milgrom (2006a), "Ascending Proxy Auctions," in Peter Cramton, Yoav Shoham, and Richard Steinberg (eds.), *Combinatorial Auctions*, Chapter 3, 79–98, MIT Press.

Ausubel, Lawrence M. and Paul Milgrom (2006b), "The Lovely but Lonely Vickrey Auction," in Peter Cramton, Yoav Shoham, and Richard Steinberg (eds.), *Combinatorial Auctions*, Chapter 1, 17–40, MIT Press.

Bichler, Martin, Pasha Shabalin, and Jurgen Wolf (2011), "Efficiency, Auctioneer Revenue, and Bidding Behavior in the Combinatorial Clock Auction," Working Paper, TU Munchen.

Brusco, Sandro and Giuseppe Lopomo (2002), "Collusion via Signalling in Simultaneous Ascending Bid Auctions with Heterogeneous Objects, with and without Complementarities," *Review of Economic Studies*, 69, 407–436.

Bulow, Jeremy, Jonathan Levin, and Paul Milgrom (2009), "Winning Play in Spectrum Auctions," Working Paper, Stanford University.

Coase, Ronald H. (1959), "The Federal Communications Commission," *Journal of Law and Economics*, 2, 1–40.

Cramton, Peter (1995), "Money Out of Thin Air: The Nationwide Narrowband PCS Auction," *Journal of Economics and Management Strategy*, 4, 267–343.

Cramton, Peter (1997), "The FCC Spectrum Auctions: An Early Assessment," *Journal of Economics and Management Strategy*, 6:3, 431–495.

Cramton, Peter (2006), "Simultaneous Ascending Auctions," in Peter Cramton, Yoav Shoham, and Richard Steinberg (eds.), *Combinatorial Auctions*, Chapter 4, 99–114, MIT Press.

Cramton, Peter, Evan Kwerel, Gregory Rosston, and Andrzej Skrzypacz (2011), "Using Spectrum Auctions to Enhance Competition in Wireless Services," *Journal of Law and Economics*, 54, 2011.

Cramton, Peter, Evan Kwerel, and John Williams (1998), "Efficient Relocation of Spectrum Incumbents," *Journal of Law and Economics*, 41, 647–675.

Cramton, Peter and Jesse Schwartz (2002), "Collusive Bidding in the FCC Spectrum Auctions," *Contributions to Economic Analysis & Policy*, 1:1, 1–17.

Cramton, Peter, Yoav Shoham, and Richard Steinberg (2006), *Combinatorial Auctions*, Cambridge, MA: MIT Press.

Cramton, Peter, Andrzej Skrzypacz and Robert Wilson (2007), "The 700 MHz Spectrum Auction: An Opportunity to Protect Competition In a Consolidating Industry" submitted to the U.S. Department of Justice, Antitrust Division.

Day, Robert and Peter Cramton (2012), "The Quadratic Core-Selecting Payment Rule for Combinatorial Auctions," *Operations Research*, 60:3, 588–603, May–June.

Day, Robert and Paul Milgrom (2008), "Core-selecting Package Auctions," *International Journal of Game Theory*, 36, 393–407, March.

Day, Robert W. and S. Raghavan (2007), "Fair Payments for Efficient Allocations in Public Sector Combinatorial Auctions," *Management Science*, 53, 1389–1406.

Erdil, Aytek and Paul Klemperer (2010), "A New Payment Rule for Cole-Selecting Package Auctions," *Journal of the European Economic Association*, 8, 537–547.

Federal Communications Commission (2002), "Spectrum Policy Task Force," ET Docket No. 02–135.

Harsha, Pavithra, Cynthia Barnhart, David C. Parkes, and Haoqi Zhang (2010), "Strong Activity Rules for Iterative Combinatorial Auctions," *Computers & Operations Research*, 37:7, 1271–1284.

Kagel, John H. and Dan Levin (1986), "The Winner's Curse and Public Information in Common Value Auctions," *American Economic Review*, 76, 894–920.

Kagel, John H., Dan Levin, and Jean-Francois Richard (1996), "Revenue Effects and Information Processing in English Common Value Auctions," *American Economic Review*, 86, 442–460.

Kagel, John H., Yuanchuan Lien, and Paul Milgrom (2010), "Ascending Prices and Package Bidding: A Theoretical and Experimental Analysis," *American Economic Journal: Microeconomics*, 2:3, 160–185.

Kahn, Afred E., Peter Cramton, Robert H. Porter, and Richard D. Tabors (2001), "Uniform Pricing or Pay-as-Bid Pricing: A Dilemma for California and Beyond," *Electricity Journal*, July, 70–79.

Klemperer, Paul (2004), *Auctions: Theory and Practice*, Princeton University Press.

McMillan, John (1994), "Selling Spectrum Rights," *Journal of Economic Perspectives*, 8, 145–162.

Milgrom, Paul (2004), *Putting Auction Theory to Work*, Cambridge: Cambridge University Press.

Milgrom, Paul (2007), "Package Auctions and Exchanges," *Econometrica*, 75, 935–966.

Milgrom, Paul (2010), "Simplified Mechanisms with Applications to Sponsored Search and Package Auctions," *Games and Economic Behavior*, 70:1, 62–70.

Pagnozzi, Marco (2010), "Are Speculators Unwelcome in Multi-object Auctions?" *American Economic Journal: Microeconomics*, 2:2, 97–131.

Parkes, David C. (2006), "Iterative Combinatorial Auctions," in Peter Cramton, Yoav Shoham, and Richard Steinberg (eds.), *Combinatorial Auctions*, Chapter 2, 41–78, MIT Press.

Porter, David, Stephen Rassenti, Anil Roopnarine, and Vernon Smith (2003), "Combinatorial Auction Design," *Preceedings of the National Academy of Sciences*, 100, 11153–11157.

CHAPTER 8

A Practical Guide to the Combinatorial Clock Auction[*]

Lawrence M. Ausubel and Oleg Baranov

Since its proposal in a 2006 academic paper,[1] the combinatorial clock auction (CCA) has rapidly established itself as one of the leading formats for government auctions of telecommunications spectrum. Its initial implementations were for relatively small auctions and some of these applications may be viewed as experimental. However, in the past few years, usage of the CCA has gained substantial momentum. From 2012 to this writing in 2015, the CCA has been used for more than ten major spectrum auctions worldwide, allocating prime sub-1-GHz spectrum on three continents and raising approximately $20 billion in revenues (see Table 1). Despite the presence of an existing auction format—the simultaneous multiple round auction (SMRA)—which often performs reasonably well, the CCA has the potential of displacing it and becoming the new standard design choice for spectrum auctions.

The CCA design consists of a two-stage bidding process. The first stage, known as the *clock rounds*, is a multiple-round clock auction. In each round, the auctioneer announces prices for all items and bidders respond with quantities demanded at these prices. If aggregate demand exceeds available supply for any items, the auctioneer announces higher prices for these items in the next round. The bidding process continues until prices reach a level at which aggregate demand is less than or equal to supply for every item. The second stage, known as the *supplementary round*, is a sealed-bid auction process in which bidders can improve their bids made in the first stage and submit additional bids as desired for other combinations of items. Throughout the entire auction, all bids are treated as all-or-nothing package bids.

To determine winnings and associated payments, all bids placed during the clock rounds and all bids placed in the supplementary round are entered together into a standard winner determination problem (WDP). Winning packages are determined by finding an allocation that maximizes the total value (as reflected in bids) subject to

---

[*] We are grateful to Maarten Janssen for helpful comments. All errors are our own.
[1] The CCA format was proposed by Ausubel, Cramton and Milgrom (2006), first presented at the FCC's Conference on Combinatorial Bidding in Wye River, Maryland in November 2003. See: http://wireless.fcc.gov/auctions/default.htm?job=conference_summary&y=2003&page=summary.

Table 1. *Combinatorial Clock Auctions to date, as of 2015*

Country and Auction	Year	Revenues
Trinidad and Tobago Spectrum Auction	2005	$25.1 million ($US)
UK 10 – 40 GHz Auction	2008	£1.43 million
UK L-Band Auction	2008	£8.33 million
Netherlands 2.6 GHz Spectrum Auction	2010	€2.63 million
Denmark 2.5 GHz Spectrum Auction	2010	DKK 1.01 billion
Austria 2.6 GHz Spectrum Auction	2010	€39.5 million
Switzerland Spectrum Auction	2012	CHF 996 million
Denmark 800 MHz Spectrum Auction	2012	DKK 739 million
Ireland Multi-Band Spectrum Auction	2012	€482 million
Netherlands Multi-Band Spectrum Auction	2012	€3.80 billion
UK 4G Spectrum Auction	2013	£2.34 billion
Australia Digital Dividend Spectrum Auction	2013	$1.96 billion ($AU)
Austria Multi-Band Spectrum Auction	2013	€2.01 billion
Slovakia 800, 1800 and 2600 MHz Spectrum Auction	2013	€164 million
Canada 700 MHz Spectrum Auction	2014	$5.27 billion ($CA)
Slovenia Multi-Band Spectrum Auction	2014	€149 million
Canada 2500 MHz Spectrum Auction	2015	$755 million ($CA)

feasibility constraints: each item can be sold only once and only one bid from each bidder can be selected as part of the winning allocation. Corresponding payments are found by solving a series of counterfactual WDPs that identify the relevant opportunity costs (second prices) imposed by winners on other participants. In general, a payment rule based on opportunity costs creates incentives for bidders to reveal their true values, facilitating efficient outcomes.

In current practice, it is standard for the CCA also to include a third bidding stage, known as the ***assignment stage***. This stage is added to the CCA in order to significantly simplify bidding in the first two stages. The main idea is to treat several closely-related items as completely identical during the clock and supplementary rounds. For example, the European digital dividend auctions auctioned six distinct licenses in the 800 MHz band. In these auctions, bidders were typically asked to bid for quantities of "generic" 800 MHz spectrum blocks during the first two stages of the auction. In a third stage that takes the winning allocations of the "generic" spectrum as given, bidders had the opportunity to compete for specific frequency assignments within the 800 MHz band. The assignment stage, usually implemented as a sealed-bid auction, determines the mapping from generic spectrum to physical frequencies.[2]

The literature on the CCA is growing rapidly. Ausubel and Baranov (2014) describe the evolution of the CCA, including expected innovations to the design. Cramton (2013) outlines the flaws in the SMRA design and argues how the CCA design solves them. Various pricing mechanisms for combinatorial auctions and their properties, including the pricing rules currently used for CCAs, have been studied by Parkes (2001), Ausubel and Milgrom (2002), Day and Raghavan (2007), Day and Milgrom

---

[2] For more detailed descriptions of the CCA mechanics, see regulator websites such as: www.ic.gc.ca/eic/site/smt-gst.nsf/eng/sf10583.html and www.acma.gov.au/Industry/Spectrum/Digital-Dividend-700MHz-and-25Gz-Auction/Reallocation/combinatorial-clock-auctions-reallocation-acma (last accessed 5 December 2015).

(2008), and Day and Cramton (2012). Possibilities for strategic manipulations of the CCA are explored by Janssen and Karamychev (2013) and Levin and Skrzypacz (2014). Experimental comparisons of the CCA design with other auction designs have been studied by Bichler, Shabalin and Wolf (2013) and Bichler, Goeree, Mayer and Shabalin (2014).

Despite this burgeoning literature, there has been little written to address some of the most pressing practical issues facing regulators. The goal of the current article is to fill some of this void. Section 1 addresses reserve prices. In the consultation processes that frequently precede spectrum auctions, a great deal of attention is dedicated to the *levels* of reserve prices. However, very little thought typically goes into the *implementation* of reserve prices, which in CCAs can be as consequential as the levels. As a result, many CCAs have adopted implementations that unnecessarily distort bidders' incentives. Section 2 discusses endogenous band plans. In current practice, the configuration of the spectrum is largely determined through lobbying by stakeholders in a consultation process preceding the auction. One of the promises of combinatorial auction formats such as the CCA is that they open the possibility of the "market" not only determining the allocation of spectrum but also the underlying band plan. We point out some issues and potential pitfalls in designing a CCA with endogenous band plans. Section 3 examines activity rules in the supplementary round. Activity rules make the early rounds of dynamic auctions informative, while still allowing bidders to switch to bidding on different items in later rounds. In a CCA, they serve the additional role of limiting overbidding and surprise bids in the supplementary round; we provide some guidance to currently used activity rules. Section 4 considers competition policy. When competition policies (or any other side objectives) are integrated into the auction, important properties of the CCA design need to be preserved. We show by example that some recent European CCAs appear to have had incomplete integrations. Section 5 treats briefly the two topics of bidding language and the allocation of the "core burden". While bidding language may appear to be an abstract issue, regulators implementing CCAs have always imposed a maximum number of bids that can be submitted in the supplementary round. Bidding languages provide an opportunity to make the bid limit non-binding. The allocation of the difference between Vickrey-Clark-Groves and core prices may also appear overly academic, but it affects the fundamental fairness between large and small bidders, and it can be improved with a simple fix involving weighting. Section 6 concludes.

1. Reserve Prices

Reserve prices have been employed in the vast majority of spectrum auctions, irrespective of their formats. At least three rationales have been put forward for the use of reserve prices. First, they establish a lower bound for auction revenues. Second, they promote competitive behavior among bidders by reducing possible gains from various strategic manipulations. In public auctions, there is a third rationale: reserve prices internalise the societal opportunity cost of selling assets to private parties now rather than retaining them for the future.

In the consultation processes that frequently precede spectrum auctions, a great deal of attention is dedicated to the *levels* of reserve prices. However, comparatively little thought typically goes into the *implementation* of reserve prices, which in CCAs can be as consequential as the levels.[3] The explanation for the importance of the reserve-price implementation in CCAs is quite straightforward. In previous formats used for spectrum auctions (e.g. the SMRA), there are individual prices paid for each item sold, and so starting at the reserve price for each item has an unambiguous meaning. By contrast, auctions with package bidding (e.g. the CCA) result in prices for packages of items, which do not have linear representations as sums of single-item prices.

In this section, we focus on the two approaches to reserve prices that have been used in CCAs:

Bundle reserve prices: This is the simple requirement that the payment for any bundle of goods must be at least the sum of the reserve prices established for the items contained in the bundle.

Incremental reserve prices: Under this approach, reserve prices are interpreted as minimum incremental costs of acquiring additional items. Incremental reserve prices ensure that an increase in payment for winning extra items is always at least the reserve prices for these items.

We illustrate the difference using a simple example with two items, A and B, each with a reserve price of 10. Suppose that, absent the reserve prices, a bidder would pay 15 for item A, 14 for item B, and 19 for A+B (the combination). Given the reserve prices, a payment of 15 for item A or a payment of 14 for item B is adequate, while a payment of 19 for A+B is too low. Under the bundle reserve prices, the payment for A+B is increased to 20, the sum of the reserve prices for the two items. Under the incremental reserve prices, the payment for A+B is increased to 25, increasing the stand-alone payment of 15 for item A by the reserve price of item B. Following directly from definitions, incremental reserve prices always satisfy the bundle reserve prices, but the reverse implication does not hold.

To implement bundle reserve prices, the regulator might employ an approach known as "bounds only" — the submitted bid for any package must be at least the sum of its component reserve price and, if the opportunity-cost-based price for a bidder ends up being too small, the final payment for any winning package is increased to its component reserve prices. To implement incremental reserve prices, the regulator might use an approach known as "reserve bidders" — fictitious bidders who bid for each individual item at its reserve price are added to the winner determination and price determination processes, thus explicitly applying opportunity costs at the reserve price level to all possible combinations of items.

It might seem that the incremental approach would necessarily generate at least as much revenue as the bundle approach, but Day and Cramton (2012) show that there is no general revenue ranking of the two treatments. The incremental approach would

---

[3] For example, in the run-up to the UK 4G Auction, Ofcom (the UK regulator) devoted 85 paragraphs (¶8.01–¶8.85) of its final major consultation document to the determination of levels of reserve prices, but only two paragraphs (¶7.35–¶7.36) to its change in implementation of reserve prices from the first alternative described below to the second. See: http://stakeholders.ofcom.org.uk/binaries/consultations/award-800mhz/statement/statement.pdf (last accessed 6 Dec. 2015).

indeed generate at least as much revenue as the bundle approach when they both allocate the same set of items to actual bidders. However, the incremental approach may generate lower revenues if it allocates fewer items.

There are two clear benefits of bundle reserve prices for a regulator. First, the bundle approach always results in a weakly larger set of items being sold, and the number of unallocated items is frequently viewed by regulators as a measure of failure. Second, final payments by bidders are less sensitive to the particular choice of reserve prices under the bundle approach. Or putting it differently, reserve prices serve only to protect aggregate revenues — once opportunity-cost prices have become sufficiently high, the exact choice of reserve prices has no further effect on bidders' payments. By contrast, incremental reserve prices might affect bidder payments even with reasonable competition for some of the items. Given that many regulators set reserve prices using nontransparent ad-hoc procedures, a less intrusive implementation is viewed as more desirable.

At the same time, the incremental approach is preferred from the perspective of bidder incentives. Use of opportunity-cost-based pricing in the CCA incentivises bidders to bid their true values.[4] It turns out that these incentives can be damaged by bundle reserve prices as implemented via the "bounds only" approach.

Consider our earlier example with items A and B. Suppose that the values of Bidder 1 are $v_1(A) = 15$, $v_1(B) = 20$, $v_1(AB) = 25$ and that the values for Bidder 2 are $v_2(A) = 15$, $v_2(B) = 19$, $v_2(AB) = 25$. If both bidders bid truthfully, Bidder 2 is awarded item A and her opportunity cost payment is 5 (the value taken from Bidder 1 by Bidder 2's participation).[5] If each item has a reserve price of 10, under the bundle approach, Bidder 2's payment will be increased to 10 to meet the reserve price. Now suppose that Bidder 2 deviates and bids 21 for item B, resulting in an inefficient allocation. Then Bidder 2 will be awarded a more valuable item B while still paying only 10 (the opportunity cost of winning item B). By contrast, the same deviation by Bidder 2 under the incremental approach will result in a payment of 15, rendering the deviation unprofitable. The last observation is general: the incremental approach never disturbs the incentives to bid truthfully and, therefore, is more likely to lead to efficient outcomes.

Intuitively, bundle reserve prices, when binding, generate a "reserve credit" that can be used by bidders to purchase extra items or to exchange current items for more valuable ones. In the previous paragraph, when winning item A, Bidder 2 has a reserve credit of 5, and the incremental cost of getting item B instead of A is also 5. Hence, Bidder 2 can switch from winning item A to winning item B for free by inflating its bid for B.

Generally, overstating values can be risky when opponents' values are private information. In sealed-bid auctions, such uncertainty can mitigate the incentives to overstate values. However, in the CCA, bidders are able to make inferences about opponent's values from the aggregate demand information they get during the clock rounds. For example, if at opening clock prices of (10, 10), Bidder 1 demands item B only,

---

[4] CCAs traditionally use core-selecting payment rules that might differ from the Vickrey payment rule in specific scenarios. When this is the case, bidders sometimes have incentives to understate their values for some bundles.

[5] If Bidder 2 does not participate, Bidder 1 would have won AB instead of B for additional value of $5 (v_1(AB) - v_1(B) = 5)$.

Bidder 2 would infer that Bidder 1's incremental value for A is less than 10, which in turn makes overbidding on item B a relatively safe bet.

The incentives for overbidding can be rather large, as they appear to have been in the 2013 Slovakian 4G spectrum auction. In the 1800 MHz band, three large "B" blocks were set aside for an entrant. In the prime 800 MHz band, the supply of "A" blocks was six, and the three incumbents were subject to spectrum caps of two blocks each. In the less valuable 2.6 GHz band, the supply of "C" blocks was 14, and there was no spectrum cap. The reserve price for an "A" block was €19 million and the reserve price for a "C" block was €1 million.[6]

Observe that, in such a scenario, if only incumbents competed for the "A" blocks, there would be zero opportunity cost associated with winning "A" blocks (since the aggregate demand would never exceed the supply). This implies a €38 million "reserve credit" for each incumbent, that could potentially be used for rampant overbidding on "C" blocks.

Examples of CCAs that adopted a bundle reserve approach were in the UK (two auctions held in 2008), Denmark (2010 and 2012), Switzerland (2012), Ireland (2012) and Slovakia (2013). An incremental reserve price implementation was chosen for CCAs in UK 4G (2013),[7] Australia (2013) and Canada (2014 and 2015).

2. Endogenous Band Plans

Radio spectrum can be used for multiple purposes. In current practice, the configuration and permitted use of the spectrum is largely determined through lobbying by stakeholders in a consultation process preceding the auction. One of the promises of combinatorial auction formats such as the CCA is that they open the possibility of the "market" not only determining the allocation of spectrum but also the underlying band plan. A regulator who wishes to accommodate bidders who plan to use spectrum in mutually-exclusive ways can consider making the band plan endogenous. This approach allows the same physical spectrum to be offered multiple times through mutually-exclusive lots, making the auction attractive for bidders with diverse interests. The CCA design makes it very easy for regulators to embed endogenous band plans into their auctions.

A classic example of an endogenous band plan embedded within a CCA is the initial design for the 2.6 GHz UK auction in 2007.[8] The proposal included 38 blocks (5 MHz each) in the 2.6 GHz band, with endogenous determination of the number of paired lots (spectrum blocks suitable for frequency division duplex or FDD) and the number of unpaired lots (blocks suitable for time division duplex or TDD). Note that FDD lots and TDD lots need to be physically separated from each other, so once the band plan is

---

[6] See the published auction rules at www.teleoff.gov.sk/data/files/35571.pdf and www.teleoff.gov.sk/data/files/33771.pdf (last accessed on 6 December 2015).

[7] Ofcom initially proposed to use the bundle approach for all lots, but later decided against it — see footnote 3. The incremental approach was applied to the A1, A2, C and E lots. The bundle approach was applied to the D1 and D2 lots.

[8] This design was never implemented as the auction was later superseded by the 2013 UK 4G Auction that additionally included the 800 MHz and 1800 MHz bands. The expansion of the supply coupled with a significant time delay warranted a complete redesign of the auction.

determined, the quantities of each are locked in. According to the auction rules, bidders would submit bids specifying both the number of blocks they wanted and whether they wanted them in paired, unpaired, or combined configurations. The auctioneer would then determine the actual split between the two technologies by maximizing the total value of the allocation.

A more exotic example of an endogenous band plan was included in the actual UK 4G Auction held in 2013. There, up to four of the paired lots in the 2.6 GHz band were made available either for low-power shared use (as D1 or D2 lots) or as high-power non-shared use (as C lots). In the low-power use, up to 10 different operators would share the same frequencies. In the high-power use, each C lot would be owned and operated by a single operator. The final decision between low-power and high-power usage was to be determined by comparing the sum of bids from bidders desiring low-power lots with the bids from bidders desiring high-power lots.

While endogenous band plans are conceptually attractive, a regulator should be aware of three types of complications that they may create.

Interaction with Reserve Prices

When a regulator incorporates extra design elements such as endogenous band plans, the interaction of the reserve price policy with the new design elements requires careful reexamination. For example, Ofcom in its UK 4G Auction decided to use the "reserve bidders" approach for all lots except the lots designated for the low-power shared use (D1 and D2). Instead, the low-power lots were subject to the "bounds only" approach. The rationale behind Ofcom's decision appears to be reasonable. On the one hand, Ofcom avoided situations in which the fictitious reserve bidders for D lots would have helped actual bidders bidding on D lots to compete against the C lot bidders. On the other hand, Ofcom's policy introduced a positive bias into its objective of "technological neutrality": fictitious "reserve bidders" for C blocks can displace actual bidders for D lots even if the latter bid above their reserve prices and should be awarded their spectrum.

Free-Rider Problem

All CCAs to date used core-selecting payment rules that are well known to induce certain incentives for free riding amongst bidders. In recent CCAs for spectrum, these incentives were never a major concern.[9] However, the free-rider problem can be quite severe when the auctioneer uses a combination of a core-selecting payment rule and an endogenous band plan. If one set of bidders bids on licenses for use with technology A and another set of bidders bids on the same frequencies for use with technology B, bidders within each group can have strong incentives to limit their bidding and to free ride on other bidders seeking to use the same technology.

---

[9] The typical market structure of the wireless industry, with 3 to 4 major incumbents, combined with common spectrum caps almost always guaranteed that bidders would not benefit from such free-riding strategies. The only potential exceptions to date may have been auctions with regional licenses in countries with strong regional bidders, which naturally may create a free-rider problem.

The free-rider problem in the "FDD vs. TDD" example above appears to be minor, but the "Low-Power vs. High-Power" example looks problematic. Up to 10 bidders for low-power use were effectively invited to compete jointly against one or two bidders for high-power use, inducing a severe free-rider problem among the low-power bidders. All would have strong incentives to bid very conservatively and to let other low-power bidders pick up the tab. Therefore, from the planning stages of this auction, the low-power shared technology was unlikely to prevail. In fact, the bidding data from the auction shows that one bidder, Niche (BT), placed a substantial number of supplementary bids that were suggestive of extreme incentives for free riding.[10]

Strategic Manipulations

Another important issue that arises in the context of endogenous band plans is unintended opportunities for strategic manipulations. In many circumstances, a bidder in a CCA may find it beneficial to manipulate the price trajectory during the clock rounds in order to put itself in a better position for the supplementary round.[11]

The UK 4G Auction was ripe for such strategies. In some circumstances, a bid for a single low-power D2 block — in combination with the price incrementing policy that was used in the auction — would reduce the endogenous supply of high-power C blocks and cause an artificial increase in their clock price.

3. Supplementary-Round Activity Rules

The imposition of activity rules in dynamic auctions has been one of the most important innovations in recent auction design. Generally speaking, activity rules are intended both to speed up the auction process and to curtail "bid sniping" opportunities (bidders concealing their true intentions until the very end of the auction). The traditional notion of "bid sniping" comes from eBay auctions with fixed ending times, where the high-value bidder can sometimes reduce its payment by submitting its winning bid at the very last moment. In a CCA, activity rules serve the additional roles of limiting surprise bids in the supplementary round and the bidder's ability to drive up its opponents' payments by overbidding for packages that can no longer be won.

CCAs to date have had considerable diversity in their activity rules. There are two distinct places in the CCA where the regulator needs to select an activity rule: (1) a clock-round activity rule that limits the set of items on which the bidder can bid in later clock rounds, based on bids in earlier clock rounds; and (2) a supplementary-round activity rule that limits the amounts that the bidder can bid on various packages in the supplementary round. Clock-round activity rules (and issues thereof) are in essence

---

[10] Niche placed 54 (out of 89) bids for packages that included either a D1 or D2 lot, that were just a minimal increment (£1000 or £2000) over its corresponding base bids without D lots. The amounts of the base bids were at least £20 million.

[11] Clock prices affect revealed-preference constraints that in turn set upper bounds for the amounts of supplementary bids (see Section 3 for details).

a special case of general considerations relating to any dynamic auction — the next two paragraphs will provide an overview of the issues. However, supplementary-round activity rules are relatively unique to the CCA. Therefore, the main focus of this section will be to provide the regulator with some guidance on selecting the supplementary-round activity rule.

Historically, spectrum auctions have utilised points-based activity rules: the auctioneer assigns a number of points to each item, before the start of the auction, and requires each bidder to adhere to monotonicity in the total number of points associated with each successive bid. In other words, a bidder must bid for a large bundle in early rounds in order to be allowed to bid for an equally-large bundle in later rounds, limiting possibilities for bid-sniping. Unfortunately, such an approach may also interfere with truthful bidding. For this reason, Ausubel, Cramton and Milgrom (2006) suggested introducing revealed-preference considerations into the clock-round activity rule of the CCA. Early implementations of the CCA tended to ignore their suggestion and to follow the traditional points-based approach in clock rounds. More recently, a hybrid approach, based upon a combination of points and revealed-preference considerations, has been taken in several spectrum auctions.

In Ausubel and Baranov (2014, 2016a), we show that a hybrid approach can actually make matters worse, and we propose instead to base the clock-round activity rule entirely upon the Generalized Axiom of Revealed Preference (GARP) — a well-known rationality concept used in economics.[12] Furthermore, in Ausubel and Baranov (2016b), we show that the GARP-based activity rule can be a foundation for a pricing mechanism that dynamically approximates VCG payoffs and thereby improves bidding incentives in the CCA.

For the supplementary round, the CCA's activity rule is also based on a combination of points and revealed-preference ideas. Current implementations of activity rules rely on the concepts of the Relative Cap, the Intermediate Cap and the Final Cap. All of them use the following concept of revealed-preference constraint:

Revealed-Preference Constraint: The revealed preference constraint for package x with respect to the clock round t is $b(x) \leq b(x_t) + p_t(x - x_t)$, where $b(x)$ is the bid amount for package x, x_t is the package demanded in Round t, $b(x_t)$ is the final bid amount for package x_t, and p_t is the vector of clock prices for Round t.[13]

Intuitively, the revealed-preference constraint states that it is unreasonable for the bidder to claim a high value for package x relative to package x_t, given that the package x_t was revealed to be preferred to package x in Round t. The Relative Cap, Intermediate Cap and Final Cap rules are defined in terms of revealed-preference constraints with respect to certain clock rounds as follows:

---

[12] GARP imposes revealed-preference constraints against the bidder's demands in all clock rounds. To put it differently, the GARP activity rule requires the bidder to exhibit rational behavior in all of its demand choices. One significant advantage of the GARP activity rule to regulators is that it completely eliminates the need for points — and therefore it completely eliminates the need to assign points. Furthermore, the GARP activity rule does not require a monotonic price trajectory.

[13] More precisely, this is the constraint imposed by the Weak Axiom of Revealed Preference (WARP) for a bidder with a quasilinear payoff function.

Relative Cap: A bid for the package x should satisfy the revealed-preference constraint with respect to the last clock round in which the bidder's eligibility, as measured by points, was at least the total points associated with package x.

Intermediate Cap: A bid for the package x should satisfy the revealed-preference constraints with respect to all eligibility-reducing rounds starting from the last clock round in which the bidder's eligibility, as measured by points, was at least the total points associated with package x.

Final Cap: A bid for the package x should satisfy the revealed-preference constraint with respect to the final clock round.

The Final Cap is the most natural constraint of revealed preference theory coming out of the clock rounds. Yet many regulators decide to rely only on the Relative Cap, and view the Final Cap as giving excessive allocation stability between the clock stage and the supplementary round. In the extreme case, it may be impossible for the supplementary bids to change the allocation of the final clock round; consequently, opportunity cost pricing based on these bids may become unreliable due to poor incentives. Recently, both the ACMA (the Australian regulator) and Ofcom (the UK regulator) used this rationale to decide against the Final Cap.

However, a simple omission of the Final Cap can be costly. While the absence of the Final Cap does mitigate the incentive issue, it does so at the expense of leaving substantial bid sniping opportunities and thus risking elimination of many advantages of a dynamic auction. The power of the Final Cap comes from its use of revealed-preference constraints generated *for all bidders at the same time* (more specifically, *using the same price vector*). In contrast, both the Relative Cap and Intermediate Cap use bidder-specific eligibility-reducing rounds to generate revealed-preference constraints.

To illustrate the importance of the Final Cap, we compare the extent of bid sniping opportunities available to bidders in the supplementary round under several activity rules. For the UK 4G Auction (see Table 2), we have calculated "theoretical"[14] bid amounts needed to protect final clock packages for all major bidders[15] under the Relative Cap (the actual activity rule used in the auction), the Intermediate Cap (considered, but never used), the Relative Cap + Final Cap (used in Ireland), and the Intermediate Cap + Final Cap (used in Canada). The exposure numbers were generally high in the UK 4G Auction, mostly due to the large value of unsold lots at the end of the clock stage. For this reason, we report two sets of numbers. "Exposure" provides the theoretical amounts required to protect the final clock package using the actual amounts of unsold lots in the auction, while "Net Exposure" provides the theoretical amounts when the effect of unsold lots on exposure is removed (as if the auctioneer had placed bids for

---

[14] These "theoretical" exposures are calculated as maximum incremental values that the bidder's opponents could place on the lots in the bidder's final clock package and unsold lots on top of values expressed for their final clock packages. We calculated them based on the full history of clock bids; while the disaggregated bids were unavailable to bidders during the auction, the calculation could be approximated using aggregate demand, which was disclosed. Also, since the low-power D1 and D2 lots did not enter into the final allocation, we excluded any effects associated with D1 and D2 lots from this calculation.

[15] They are calculated for all bidders who had non-empty final clock packages in the UK 4G Auction with the exception of Three. Under the competition policy employed in this auction, Three was guaranteed to win one of the designated minimum spectrum portfolios and therefore never needed to protect its bids for any of these packages (and was made aware of this fact during the auction).

Table 2. *Exposure Calculation for the UK 4G Auction (2013)*

Bidder/Final Clock Package	Type	Relative Cap (used in UK)	Intermediate Cap	Relative Cap + Final Cap (used in Ireland)	Intermediate Cap + Final Cap (used in Canada)
Vodafone (2–A1, 3–C)	Exposure	245%	188%	170%	170%
	Net Exposure	176%	118%	100%	100%
Telefonica[16] (1–A2)	Exposure	290%	216%	192%	192%
	Net Exposure	198%	124%	100%	100%
EE (9–E)	Exposure	715%	456%	456%	456%
	Net Exposure	359%	100%	100%	100%
Niche(BT) (2–C)	Exposure	1103%	637%	524%	524%
	Net Exposure	679%	212%	100%	100%

all unallocated lots at the final clock prices). All exposure numbers are reported as percentages of the bidders' final clock prices. As can be seen from the table, the Final Cap consistently reduces the amount that any bidder needs to bid to guarantee its winnings, due to reduced bid sniping opportunities. Furthermore, the reduction is rather significant. Even without undersell (Net Exposure), bidders might need to increase their bids by three-quarters or more in order to protect their final clock packages under the Relative Cap. As an extreme example, Niche would have needed to increase its bid by almost a factor of seven in order to assure that it won its final clock package. But the need to substantially increase bids after the clock rounds undermines the purpose of conducting the clock rounds at all. Our calculations demonstrate that the Relative and Intermediate Caps by themselves provide rather weak protection against bid sniping in the realistic setting of the actual UK 4G Auction data.

4. Competition Policy

One of the important tasks of a spectrum regulator is to incorporate competition policy into spectrum auctions. To be clear, in discussing "competition policy," we are generally not referring to introducing competition into the spectrum auctions themselves, but to injecting competition into the *downstream market* for mobile voice and data services. Spectrum is a scarce input into the provision of mobile services, and concentration in spectrum holdings would generally lead to concentration in the downstream market. To the extent that the spectrum auction facilitates new entry and reduces concentration in spectrum holdings, it can help to improve the competitive performance of the downstream market. As we will now see, the CCA is quite robust to the introduction of standard competition measures, and readily allows more flexible measures ("virtual set-asides") than are typically used today. However, the CCA becomes more sensitive when adding novel instruments — and it is also easy for the well-intentioned regulator to introduce inadvertent design flaws into the process.

---

[16] In the actual UK 4G Auction, Telefonica was bidding for a (1–A2, 1–D2) package in rounds 41-52. For simplicity, we are omitting D1 and D2 lots from these calculations and assuming that Telefonica was bidding for (1–A2) lot in rounds 41-52. See also footnote 14.

Historically — and generally in conjunction with use of the SMRA format — the two most frequent instruments for competition policy in spectrum auctions have been the *set-aside* and the *spectrum cap*:

Set-aside: One or more spectrum blocks are reserved for a particular class of bidders. The most common form that this has taken is that a specific block of spectrum is set aside for "entrants"; while "incumbents" are excluded from bidding on the set-aside block.[17]

Spectrum cap: A limit is placed on the quantity of spectrum within a given group of bands that a bidder is permitted to acquire or hold. Spectrum caps may be applied to winnings within a given auction or they may apply cumulatively to specified existing holdings as well as to acquisitions within the given auction. In countries such as the US, India, Canada and Australia, where spectrum licenses are regional, spectrum caps have been applied to each and every geographic region.[18]

However, due to the nature of older auction formats (e.g. the SMRA), there have been two frequent limitations on these policy instruments. First, since the bids have been for specific (as opposed to generic) licences, set-asides have needed to be for specific licences. This required regulators to make unnecessarily intrusive decisions (e.g. which exact frequencies should be reserved for entrants). Second, since the determination of high bids has historically been done at the individual license level, spectrum caps were applied individually to each incumbent (as opposed to establishing an aggregate cap on the overall acquisitions or holdings of all incumbents). This potentially had the effect of impeding competition in the auction among incumbent bidders and thereby depressing auction revenues. Both of these preconditions change under the CCA.

Virtual Set-asides/Aggregate Spectrum Caps

In new auction formats such as the CCA, it is possible to go beyond the two aforementioned limitations and to provide for a *virtual set-aside* (which may equivalently be viewed as an *aggregate spectrum cap*):

---

[17] Set-asides have been used in auctions in many countries, including the US and the UK. In the US Broadband PCS spectrum auctions which began in 1994, the FCC set aside two of the original six blocks of broadband PCS spectrum for "designated entities" in an effort to promote small business ownership of spectrum. The implementation of set-asides did not work out well, as the set-aside was bundled with other policy instruments, such as instalment payments, that were intended to favour small businesses. The result of this exercise was that one of the largest winners of set-aside spectrum entered into bankruptcy, and there was a roughly ten-year period when most of the set-aside spectrum went unused. By contrast, the UK had an apparently successful experience with set-asides. In the UK 3G Auction of 2000, Ofcom set aside one of five 3G licenses for a new entrant. This resulted not only in actual entry and hence competition in the downstream market for wireless services, but in increasing competition in the auction and probably increasing auction revenues.

[18] Spectrum caps have been prevalent in spectrum auctions worldwide. At the time of the US Broadband PCS spectrum auctions, the FCC established a 45-MHz cap for commercial mobile radio spectrum, including both existing cellular licenses and the new PCS licenses being auctioned. By contrast, in many of the European 3G auctions in 2000, bidders were limited to winning at most a specified number of spectrum blocks, irrespective of their existing spectrum holdings. In recent 4G auctions, bidders have often been subject to an overall constraint on existing and new holdings for spectrum generally, together with a tighter constraint on the prime sub-1-GHz spectrum. For example, in the 2013 UK 4G spectrum auction, there was both a 105-MHz overall cap on existing and new holdings (which proved to be binding for incumbent Everything Everywhere) and a 27.5-MHz cap on sub-1-GHz spectrum (which proved to be binding for incumbents Vodafone and Telefonica).

Virtual set-aside/aggregate spectrum cap: If a quantity of N generic spectrum blocks is offered, then a particular class of bidders (e.g. "incumbents") is limited, in aggregate, to winning a quantity of $N-K$ of these generic blocks. Such a policy instrument can be called an *aggregate spectrum cap*. Equivalently, this can be viewed as reserving a quantity of K generic spectrum blocks for another class of bidders (e.g. "entrants"). Such a policy instrument can be called a *virtual set-aside*.

One key advantage of this policy instrument (as compared to the implicit spectrum reservation of a standard spectrum cap) is that, for any quantity reserved for entrants, there can now be greater competition among incumbents. For example, suppose that there are two incumbent mobile operators and six blocks of spectrum available. The regulator can implicitly reserve two blocks for entrants by limiting each incumbent to an in-auction spectrum cap of two blocks. However, unless entrants demand greater than two blocks, this instrument would eliminate all competition for the incumbents. Instead, the regulator can establish an aggregate cap of four blocks on the incumbents. Each one is free to bid for and win up to four blocks. With a virtual set-aside, there are still two blocks reserved for entrants, but competition between the two incumbents is permitted to occur.

A virtual set-aside in the CCA has several advantages over the standard set-aside in the SMRA. First, it is more flexible, and allows the market to determine which specific frequency blocks are won by entrants. Second, with the CCA's price determination mechanism rather than implicit uniform pricing, a single entrant may be capable of winning the entire set-aside, as opposed to being vulnerable to pressure from a second entrant who may force it to surrender some of the set-aside. (And the regulator may, for competition reasons, prefer that a single entrant win the entire set-aside, as a single substantial entrant is likely to pose a greater competitive threat to incumbents than a diffuse competitive fringe.) Third, in some situations, a standard set-aside may be viewed as an excessively strong policy and of questionable legality. While the aggregate spectrum cap is isomorphic to a virtual set-aside, the aggregate spectrum cap (as a limit on incumbents, rather than a reservation for entrants) may be considered more acceptable and less subject to legal challenge.

Spectrum Floor

The UK 4G Auction of 2013 went a step further beyond a virtual set-aside and established a "spectrum floor". Ofcom (the regulator) decided that either a 2×5 MHz block of 800 MHz spectrum or a 2×20 MHz block of 2.6 GHz spectrum would be reserved for an entrant. This went further than a virtual set-aside in two respects. First, it allowed the market to determine not only which frequency blocks would constitute the set-aside, but whether they would belong to the more valuable 800 MHz band or the less valuable 2.6 GHz band. Second, it institutionalized that the entire set-aside would be required to be won by a single entrant.

Spectrum floor: Two or more alternative sets of spectrum ("minimum spectrum portfolios") are reserved for a particular class of bidders. The choice of which minimum spectrum portfolio is awarded, and to which bidder, is decided endogenously by the solution to a modified winner determination problem: maximize the value of accepted bids, subject to feasibility and to awarding one of the minimum spectrum portfolios to an eligible bidder.

With a CCA, two issues are likely to arise when spectrum is reserved for entrants. The first issue can be referred to as "infeasible bids". Given that certain spectrum blocks are reserved for entrants, certain other combinations of blocks might never be part of a feasible winning bid. In this event, placing bids for infeasible combinations of blocks might become a stalling tactic or a way to drive up prices of certain spectrum bands; therefore, it is essential that the auction rules prevent the submission of infeasible bids. The second issue relates to the supplementary bids. Under the rules generally used in CCAs today, the bidder is permitted to submit a supplementary bid that raises its final clock package (the package that it bid for in the final round of the clock stage) by any arbitrary amount. However, there is one exception: no supplementary bids are permitted to be submitted for the null set, even if it is the bidder's final clock package (i.e. if the bidder has dropped out of the clock stage by the final clock round). When spectrum is reserved for entrants, similar logic would suggest that the bidder should not be allowed to place a supplementary bid on the reserved package.[19]

5. Other Considerations

Compact Bidding Languages

The standard bidding language for CCAs treats each of a bidder's bids as an all-or-nothing package bid, with the restriction that only one bid per bidder can win. This bidding language is fully expressive, in that it allows bidders to communicate all possible valuations. However, it comes at a steep price. The bidding language is non-compact in that it may require a very large number of bids to express simple (e.g. additive) preferences.[20]

While the bidding language may appear to be an abstract issue, regulators implementing CCAs have always needed to impose a limit on the number of bids that each bidder can submit in the supplementary round.[21] Bidding languages provide an opportunity to make the bid limit non-binding.

One approach for auctions with large numbers of items offered, used successfully in Canada's 2500 MHz Auction (with 318 licences grouped into 106 categories), is to allow bidders to place non-mutually-exclusive "OR bids" as increments to their final

---

[19] More generally, a clean solution to this problem might be to require the entire collection of highest bids to be consistent with a GARP-based activity rule (see the discussion in Section 3).

[20] See Nisan (2006) for a general overview of bidding languages in combinatorial auctions.

[21] The reason for this limit is to assure that the winner determination problem (which is NP-complete) can be solved in reasonable time. The UK 4G Auction set a limit of 4000 bids, while Canada's 700 MHz Auction set a limit of 500 bids. The limit was probably non-binding in the UK (with only 28 licences grouped into 4 categories, excluding low-power D licences), but may have been binding in Canada (with 98 licences grouped into 56 categories).

Table 3. *"Core Burden" in Canadian 700 MHz Auction (in mil $CA)*

Bidder	Base Vickrey Payment	Nearest-Vickrey Core Payment Split	Nearest-Vickrey Core Payment Payment	Weighted Nearest-Vickrey Core Payment Split	Weighted Nearest-Vickrey Core Payment Payment
SaskTel	$2.755 mil	$5.188 mil	$7.943 mil	$4.802 mil	$7.557 mil
MTS	$3.198 mil	$5.188 mil	$8.386 mil	$5.574 mil	$8.772 mil

clock packages. The list of must-address issues for the regulator then includes the interaction of OR bids with the CCA activity rules and the treatment of OR bids by the winner-determination problem.

Payment Rules: Allocation of the "core burden"

Core-selecting payment rules, in which winners are charged the opportunity costs of their winnings subject to the necessary "core" adjustments, have always been at the heart of the CCA design. In general, there are many payment rules that will satisfy this principle and it is important for the regulator to make an appropriate choice of core adjustment. While the core adjustment may appear to be an overly academic topic for regulators, it affects the fundamental fairness of the treatment of large versus small bidders.

To date, all implementations have used some variant of a "Nearest-Vickrey" payment rule that selects a unique set of payments from the minimum revenue frontier by minimizing the Euclidian distance to the Vickrey-Clark-Groves payments. Both Australia (2013) and Canada (2014 and 2015) made a simple improvement by employing a weighted version of the Nearest-Vickrey rule that minimizes a *weighted* Euclidian distance to the VCG payments. The strongest rationale for including weights is that, in auctions with regional licenses, the winning bidders might receive disparate winnings. Day and Cramton (2012) had motivated the Nearest-Vickrey rule, in part, by the fairness of bidders sharing the "core burden" equally. However, with disparate winnings, equal sharing might be unfair to those bidders with significantly smaller winnings.

The effects of the weighted Nearest-Vickrey rule can be illustrated using the actual data of Canada's 700 MHz Auction (see Table 3). In this auction, the VCG payments of two bidders, SaskTel and MTS, needed to be adjusted by a total of $10.376 mil ($CA) to satisfy the "blocking constraint" of their rivals. Under the Nearest-Vickrey rule, both bidders would have borne 50% of the core burden ($5.188 mil each). Instead, the weighted Nearest-Vickrey rule required MTS to bear approximately 54% of the burden, shifting the "core burden" towards MTS, which won greater value (as measured by the reserve price of the licences won).

6. Conclusion

The CCA has quickly become one of the standard techniques in the toolbox of spectrum auction designers. Most academic effort is currently devoted to examining the major theoretical issues surrounding this auction format, while many of the questions important for practical applications are overlooked. Yet these small details can prove decisive for the overall success of the auction. Frequently, practitioners need to customize the basic CCA design in order to accommodate their specific objectives and unique environments. Fortunately, the CCA is a highly versatile design that allows integrating complex secondary objectives. Unfortunately, naïve integrations can introduce unintended consequences, damaging the auction's performance as a whole.

Recent CCAs have demonstrated that the details of implementing features even as standard as a set-aside can be far from trivial. And, when a regulator seeks to superimpose more complex side objectives on top of the standard CCA design, the nuances of the implementation can become critical. While the goal of accommodating additional features of the regulatory environment within a customised CCA is laudable, some of the custom design changes that have been implemented in various CCAs appear to contradict basic principles. For example, bidders in one auction were able to directly affect their own payments through a newly introduced option, contrary to the principles of opportunity cost pricing. Changes made to the winner determination process have created new gaming opportunities and modified activity rules have allowed some bidders to switch back and force between packages with very different degrees of competitiveness.

In this article, we have attempted to summarise and analyse some of the most common choices that a regulator needs to make when implementing the CCA. Regulators are encouraged to innovate on the basic CCA design to advance their novel objectives, but they are cautioned to pay heed of the basic principles underlying the CCA and thereby to avoid introducing novel design flaws.

References

Ausubel, Lawrence M., and Oleg V. Baranov. 2014. "Market Design and the Evolution of the Combinatorial Clock Auction." American Economic Review: Papers & Proceedings, 104(5): 446-451.

Ausubel, Lawrence M., and Oleg V. Baranov. 2016a. "Revealed Preference and Activity Rules in Auctions." Working paper.

Ausubel, Lawrence M., and Oleg V. Baranov. 2016b. "Vickrey-Based Pricing in Iterative First-Price Auctions." Working paper.

Ausubel, Lawrence M., Peter Cramton and Paul Milgrom. 2006. "The Clock-Proxy Auction: A Practical Combinatorial Auction Design." In Combinatorial Auctions, eds. Peter Cramton, Yoav Shoham, and Richard Steinberg, 115–138. Cambridge: MIT Press.

Ausubel, Lawrence M., and Paul Milgrom. 2002. "Ascending Auctions with Package Bidding." Frontiers of Theoretical Economics 1(1): Article 1.

Bichler, Martin, Jacob Goeree, Stefan Mayer and Pasha Shabalin. 2014 "Spectrum auction design: Simple auctions for complex sales." Telecommunications Policy 38(7), 613–622.

Bichler, Martin, Pasha Shabalin and Jürgen Wolf. 2013. "Do core-selecting combinatorial clock auctions always lead to high efficiency? An experimental analysis of spectrum auction designs." Experimental Economics 16(4): 511–545.

Cramton, Peter. 2013. "Spectrum Auction Design." Review of Industrial Organization 42(2): 161–190.

Day, Robert W., and Peter Cramton. 2012. "Quadratic Core-Selecting Payment Rules for Combinatorial Auctions." Operations Research 60(3): 588–603.

Day, Robert, and Paul Milgrom. 2008. "Core-Selecting Package Auctions." International Journal of Game Theory 36: 393–407.

Day, Robert W., and S. Raghavan. 2007. "Fair Payments for Efficient Allocations in Public Sector Combinatorial Auctions." Management Science 53: 1389–1406.

Janssen, Maarten and Vladimir Karamychev. 2013. "Spiteful Bidding and Gaming in Combinatorial Clock Auctions." Working paper.

Levin, Jonathan and Andrzej Skrzypacz. 2014. "Are Dynamic Vickrey Auctions Practical?: Properties of the Combinatorial Clock Auction." Working paper.

Nisan, Noam. 2006. "Bidding Languages for Combinatorial Auctions." In Combinatorial Auctions, eds. Peter Cramton, Yoav Shoham, and Richard Steinberg, 215–231. Cambridge: MIT Press.

Parkes, David C. 2001. "Iterative Combinatorial Auctions: Achieving Economic and Computational Efficiency." Univ. of Pennsylvania doctoral dissertation.

CHAPTER 9

Market Design and the Evolution of the Combinatorial Clock Auction[†]

Lawrence M. Ausubel and Oleg Baranov[*]

The combinatorial clock auction (CCA) is an important recent innovation in market design. It has progressed rapidly from a 2003 academic paper to real-world adoption. In the past few years, it has been used for more major spectrum auctions worldwide than any other auction format. As such, the CCA is the first format that has the potential to eclipse the simultaneous multiple-round auction (SMRA) as the standard for spectrum auctions.[1]

The defining characteristic of the CCA is a two-stage bidding process. The first stage is a dynamic clock auction: the auctioneer announces prices for the items in the auction; and bidders respond with quantities desired at the announced prices. Bidding in this stage progresses in multiple rounds as prices increase until aggregate demand is less than or equal to supply for every item. In the second stage, bidders have the opportunity to submit a multiplicity of supplementary bids, both to improve upon their bids from the clock rounds and to express values for other packages.

Following the second stage, the bids from both the clock rounds and the supplementary round are entered into winner determination and pricing problems. The winner determination problem treats these bids as package bids, and determines the value-maximizing allocation of the items among the bidders. The pricing problem is based on second-price principles.

In most applications, there is also a third stage of bidding. Generally, several items in the auction are treated as identical during the first two stages. For example, in the European digital dividend auctions, there have generally been six distinct licenses offered

---

[*] We are grateful to Peter Cramton, Bob Day, Jon Levin, Paul Milgrom, Bruna Santarossa, Andy Skrzypacz, and Hal Varian for helpful comments. All errors are our own.

[†] Go to http://dx.doi.org/10.1257/aer.104.5.446 to visit the article page for additional materials and author disclosure statement(s).

[1] The CCA format was proposed by Ausubel, Crarmton, and Milgrom (2006), first presented at the FCC's Wye River Conference in October 2003. The first practical implementations were the Trinidad and Tobago Spectrum Auction, in 2005, and the UK's 10–40 GHz and L-Band Auctions, in 2008. In recent years, it has been utilized for digital dividend auctions in Austria, Australia, Canada, Denmark, Ireland, the Netherlands, Slovakia, Switzerland, and the United Kingdom. Another account of the CCA is provided by Cramton (2013).

in the 800 MHz band. In the auction's first two stages, bidders simply indicate quantities of "generic" 800 MHz spectrum that they wish to purchase. The third stage takes bidders' winnings of generic spectrum as given, and bidders express values for specific 800 MHz licenses. Thus, it determines the assignment from generic spectrum to physical frequencies.

The CCA addresses many concerns that had been raised in prior spectrum auctions. One clear weakness of the SMRA and other older auction formats has been that only single-item bids were permitted. A bidder who would achieve synergies from acquiring a New York license and a Washington, DC license is exposed to significant risk from bidding above the stand-alone value of either license; the bidder could be outbid on one license, while remaining obliged to purchase the other license. This is often referred to as the "exposure problem." The CCA eliminates the exposure problem by explicitly incorporating package bids. Another weakness of many older auction formats has been that the pricing rule creates incentives for strategic demand reduction and, consequently, has tended to produce inefficient outcomes. By moving in the direction of Vickrey pricing, the CCA mitigates incentives for demand reduction.

At the same time, the CCA's performance may be impeded by several limitations. First, with a large number of product categories (e.g., in spectrum auctions in countries with regional licenses), it may be difficult or impossible for bidders to express their values for all relevant combinations of items in the supplementary round. Second, the activity rules imposed on bidders may at once be too stringent, preventing straightforward bidding, and too weak, permitting manipulative and exploitative pricing of opponents. Third, unlike most other auction formats, the current CCA is "iterative second-price" rather than "iterative first-price": bidders generally pay less than the amounts of their winning bids. This creates a fundamental tension between the clock rounds and the supplementary round.

Since its initial proposal in 2003, the CCA has been in almost continual evolution. In this paper, we review a few of the most important changes that have already occurred and we propose three additional enhancements.

I. Early Evolution of the CCA

This section reviews a few of the most important evolutionary changes to the CCA.

A. Opportunity Cost Pricing

Since the CCA's inception, opportunity-cost pricing has been one of its main principles. Opportunity cost is formalized by the Vickrey-Clark-Groves (VCG) mechanism. However, to avoid "uncompetitive" pricing, the initial CCA proposal, as well as all implementations to date, have employed variants on the VCG mechanism that generate core payoffs (relative to the submitted bids).[2]

---

[2] When goods are not substitutes, VCG prices may lie outside the core. To avoid this, the initial CCA proposal used the "ascending proxy auction" (Ausubel and Milgrom 2002; Parkes 2001).

Several recent academic papers have emphasized the desirable incentive properties of the bidder-optimal frontier of the core (Day and Raghavan 2007; Day and Milgrom 2008). Mechanisms selecting outcomes from this frontier are referred to as "core-selecting mechanisms." Due to the multiplicity of bidder-optimal core outcomes, Day and Cramton (2012) suggested selecting the bidder-optimal core point that minimizes the Euclidean distance from VCG payoffs. This is known as the "nearest-Vickrey" mechanism.

This choice of pricing rule has not been entirely uncontroversial. Several papers have contrasted the performance of nearest-Vickrey with other core-selecting mechanisms in a stylized sealed-bid environment and have argued that other mechanisms perform better.[3] However, these results are not necessarily applicable to the CCA, an auction format that includes a dynamic clock stage.

In practice, nearest-Vickrey pricing or a weighted version of this mechanism has been used in all CCA implementations to date.[4]

B. Reserve Prices

Reserve prices have been employed in most spectrum auctions, irrespective of format. The initial CCA proposal was not explicit about the reserve price implementation. Day and Cramton (2012) observed that a relevant design choice is whether the reserve prices are applied at the item level or at the package level. If applied at the item level, it is *as if* the auctioneer includes a collection of single-item bids, each bid from a distinct fictitious bidder, at the reserve price ("reserve bidders"). If applied at the package level, the auctioneer merely requires that the payment for each winning package bid must be at least the price of the items in the package evaluated at the reserve prices ("bounds only").

Day and Cramton (2012) favored the "bounds only" over the "reserve bidders" approach, for at least two reasons. First, holding reserve prices and actual bids fixed, the reserve bidders approach is more likely to result in items being withheld by the seller. Second, outcomes determined by the bounds only approach are less sensitive to choices of reserve prices. Largely on this basis, early CCAs in the United Kingdom and elsewhere adopted a bounds only approach.

However, under the bounds only approach, bidders might be able to acquire marginal items at very low incremental costs (sometimes zero). This violates one of the general rationales for a reserve price in a public auction: the reserve price should reflect the societal opportunity cost of selling the item today rather than saving it for later. Moreover, bidders in some scenarios may find it optimal to bid above their values for some items, knowing that they have unspent "reserve capacity" which will absorb the cost. This is not possible under a reserve bidders approach.

Consequently, several of the most recent CCAs, including those in Australia, Canada, and the United Kingdom, have adopted the reserve bidders approach.

---

[3] See Erdil and Klemperer (2010) and Ausubel and Baranov (2013).
[4] Roughly speaking, the weighted version allocates the core burden among winners based upon the relative size of their winnings.

C. Assignment Stage

In a traditional spectrum auction, if six nearly identical frequency blocks were offered, the seller would ask bidders to submit bids on the specific licenses A to F. A more economical approach is for bidders to indicate the quantity of blocks they would like to purchase during the main part of the auction. Only after the winning quantities have been determined do bidders need to submit bids for specific frequencies. This approach both speeds the progress of the auction and reduces the complexity of bidding.

The main insights behind introducing the CCA's assignment stage were: (i) it is preferable to replace any administrative decisions about the assignment of specific licenses with a bidding process; but (ii) bidding options should be limited only to assignments that would be considered as outcomes of administrative decisions. All other things being equal, a contiguous assignment is considered to create greater value than a checkerboard assignment. Consequently, the assignment stage is limited to assignments where all winners receive contiguous spectrum within a region.

The first CCA with an assignment stage was the 2005 Trinidad and Tobago Spectrum Auction. Essentially all CCAs to date have adopted this approach.

II. Future Evolution of the CCA

This section proposes and discusses three evolutionary enhancements for future CCAs.

A. Bidding Language

One of the critical elements of the CCA design is the explicit use of package bids. The initial CCA proposal used a bidding language under which all bids are treated as mutually exclusive ("XOR bids"). The XOR bidding language is understood to be fully expressive, but non-compact.[5]

Until recently, the compactness issue was not a practical concern in the CCA context, since most CCAs were used to allocate relatively small numbers of items. This has changed with the CCA's adoption in countries with regional licenses, such as Australia and Canada. With sufficiently many items, XOR bids might prevent bidders from expressing their values for the relevant packages.

The issue of compactness of bidding language has been studied extensively in the context of combinatorial sealed-bid auctions.[6] One of the main prescriptions is the use of non-mutually-exclusive bids ("OR bids"). More generally, permitting flexible combinations of OR and XOR bids provides multiple ways to improve compactness. In the CCA context, the natural course of evolution is to incorporate a flexible bidding language directly into the supplementary round.

The CCA format presents a somewhat novel environment for designing an effective bidding language. Frequently, a CCA bidder views the supplementary round as

---

[5] A bidding language is said to be fully expressive if it can be used to communicate any valuation profile. While XOR bids are fully expressive, an astronomical number of XOR bids can be required to express even very simple preferences.

[6] See Nisan (2006) for a review of the literature on bidding language in combinatorial auctions.

an opportunity to express its marginal values for incremental items relative to its final clock package. Furthermore, conditional on winning its final clock package, the bidder may view the incremental items as being locally additive in value. Using this insight, one natural way to introduce OR bidding flexibility into the CCA is to allow a bidder to specify various OR bids as increments on top of its bid for the final clock package.

In conjunction with the compact bidding language, it is important to provide bidders with various controls that they may exercise over their submitted OR bids. For example, a bidder can be allowed to specify a total size limit or a total budget limit for the collection of OR bids that will be considered.

The integration of OR bids into the CCA is not completely straightforward. One (perhaps the greatest) challenge is to design an appropriate activity rule for OR bids. Typical CCA activity rules are formulated in terms of whole packages, and are not trivially extended to handle OR components.[7] This issue creates a need for strong and robust activity rules, which can be consistently applied both to XOR and OR bids, such as the activity rule proposed in the next subsection.

B. Revealed-Preference Activity Rules

The most fundamental innovation of the SMRA, when it was introduced for spectrum auctions 20 years ago, was the imposition of activity rules on bidders. Activity rules are intended to prevent "bid sniping": the phenomenon widely observed in auctions such as eBay where bidders conceal their true intentions until the very end of the auction. Bid sniping effectively converts a dynamic auction into a sealed-bid auction and thereby works at cross-purposes to a dynamic auction.

Standard implementations of activity rules in SMRAs are based upon "points." The auctioneer assigns a number of points to each item in the auction, most commonly based upon the population of a license region or some other measure of value.[8] The activity rule is then a variant on the requirement that the bidder's total points bid must be non-increasing as the auction progresses. Thus, it requires bidders to submit serious bids in early rounds of the auction in order to retain the right to submit bids for an equivalent quantity (as measured by points) in later rounds.

Point-based activity rules are too weak, in that they create a number of opportunities for strategic bidding. Most notoriously, points give rise to "parking": placing bids on items that one is not interested in buying, for the purpose of stockpiling points for future use.

Less appreciated is the fact that point-based activity rules are, in other respects, too strong. For any choice of points, there exist valuations and price histories such that the point-based activity rule prevents the bidder from bidding straightforwardly according to its values.

---

[7] Given the potentially vast number of values communicated through OR bids, a sensible approach would be to move validation of the activity rule from the bid entry process to the winner determination process. Bids for individual OR components would be entered subject to the usual activity rules, but bids for combinations of OR bids would be capped automatically by the solver (rather than being validated and possibly rejected at the time of bid entry).

[8] For example, a New York license might be assigned 100 points, whereas a Washington, DC license might be assigned 25 points.

Motivated by these considerations, the original CCA proposal suggested incorporating the Weak Axiom of Revealed Preference (WARP) into the activity rule. Nonetheless, the initial implementations of the CCA, as well as recent applications in the United Kingdom and Slovakia, used point-based activity rules.[9]

In a current working paper (Ausubel and Baranov 2014), we propose introducing activity rules based upon the Generalized Axiom of Revealed Preference (GARP), while completely eliminating any role for points. Use of GARP had apparently been overlooked by prior researchers, on account that WARP already appeared to be quite strict and to risk leaving bidders in "dead-ends" (i.e., situations where the only legal next bid is to drop out of the auction). But it turns out that imposing the stricter activity rule may actually be doing bidders a favor by preventing them from getting into trouble in the first place. A WARP dead-end is possible after any bidding history of nonzero bids only if the bidding history itself contains a GARP violation.

The *GARP-based activity rule* is the requirement that, after a price and bid history $(p_1, x_1), \ldots, (p_{t-1}, x_{t-1})$, the bidder is permitted to bid x_t in round t if and only if the history $(p_1, x_1), \ldots, (p_t, x_t)$ is consistent with GARP.

Ausubel and Baranov (2014) prove that the GARP-based activity rule always permits truthful bidding and guarantees that the bidder will never reach a "dead-end."

Most auction research assumes quasilinear bidder values. The above results hold irrespective of whether bidders' values are restricted to be quasilinear. However, a pure GARP-based activity rule (without quasilinearity) appears too weak for practical purposes; in practice, one would probably require consistency with both GARP and quasilinearity.

Such activity rules also are computationally practical. Algorithms are known for validating against GARP (with or without quasilinearity) that are polynomial in the size of the data.

Ausubel and Baranov (2014) also raise the possibility of relaxations or refinements of the quasilinear GARP-based activity rule. Since budget constraints could cause inconsistent bidding, a reasonable relaxation may be to admit the possibility of budget constraints, but otherwise to impose full quasilinear GARP. At the same time, the Strong Axiom of Revealed Preference (SARP) may work as a refinement that deters some strategic manipulations.

C. Iterative Pricing

Most dynamic auction formats, both in the literature and in real-world applications, could be characterized as "iterative first-price" auctions. Consider, for example, the English auction for a single item. While it is often modeled as a sealed-bid second-price auction, bidders submit bids which, if they turn out to win, specify the actual amounts that will be paid. While effectively a second-price auction, it is *literally* an iterative first-price auction.

---

[9] Several other recent CCAs (Ireland, Australia, and Canada) have utilized hybrid activity rules based upon both points and WARP.

The same statement also applies to the SMRA and most other dynamic formats. However, the CCA is different. Bidders' clock round bids and supplementary bids are entered into winner determination and pricing problems, and the winners' payments may be substantially lower than the nominal amounts of their bids. The CCA as it has evolved today is an iterative second-price auction.

In the CCA, there exists a tension between strict activity rules and second pricing. With strict activity rules and no undersell, bidders are guaranteed to win their packages from the final clock round. In such scenarios, bidders lack incentives to place supplementary bids for winner determination purposes. At the same time, there are minimal consequences to a bidder who inflates its expressed opportunity cost for pricing purposes. Thus, the resulting prices may be either too low or too high.

In Ausubel and Baranov (2014), we propose to evolve the CCA further by transforming it in the direction of an iterative first-price auction. We proceed by asking the question: Given opponents' bids and given the activity rule, what is the maximum amount that a bidder could ever have to pay to win a package if this were the final clock round of the auction? We refer to the answer as the "exposure calculation." Generally, the highest possible payment is less than the nominal bid.

The nominal bid amount for a package could be discounted, based upon the bidder's exposure calculation. Furthermore, one could then utilize the discounted bid amounts for pay-as-bid pricing. In principle, these changes would convert the CCA from an iterative second-price to an iterative first-price auction.

There are some clear advantages in evolving to an iterative first-price approach. With second pricing, supplementary bids serve both an allocation and a pricing role, giving rise to the internal tension within the CCA. With first pricing, attention is focused on the allocation role. The change would also tend to reduce the importance of bidders' budget constraints.

At the same time, there may be other potential approaches besides first pricing to resolve the tensions within the CCA. As noted above, the current CCA design assures a very high degree of stability and predictability in going from the clock rounds to the supplementary round. The stability may be seen as so great that, in some scenarios, second pricing becomes problematic.

Alternatively, one could try to reduce any excessive stability in the current design. This stability has two sources: the strict activity rules, which limit bidders' latitude in placing supplementary bids; and the absolute (rather than relative) interpretation of bids from the clock rounds. When clock bids are taken at face value and supplementary bids are constrained, the winning bids will tend to come from the clock rounds. However, to the extent that clock bids are discounted, there is greater scope for the supplementary bids to change the outcome.

Finally, one should not lose sight that there are advantages to ensuring stability between the clock rounds and supplementary round. When the rules provide stability, bidders have the greatest incentive for truthful bidding in the clock rounds, and the dynamic auction process is the most informative. By contrast, when the outcome can change substantially in the supplementary round, bid sniping becomes effective, the clock rounds lose meaning, and the auction effectively becomes "sealed bid."

III. Conclusion

In this paper, we have proposed three enhancements for future CCAs. Introducing an OR bidding language in the supplementary round would be the most incremental change. In January 2014, the Canadian Government announced that it will adopt this enhancement in its upcoming 2,500 MHz auction. The need is quite clear, as with 318 licenses, grouped into 106 categories, this will be the largest CCA in scale to date.

Incorporating a GARP-based activity rule would also be a modest evolutionary step, as revealed-preference considerations have already been used. Transforming the CCA, from an iterative second-price to an iterative first-price auction, would be the largest step.

Current activity rules may lead to exposure calculations that equal or exceed the final clock prices, while GARP-based activity rules yield exposures more in line with opportunity-cost pricing. Thus, the new activity rule is not only a compelling evolutionary change on its own, but it also facilitates the evolution of the CCA to an iterative first-price auction.

References

Ausubel, Lawrence M., and Oleg V. Baranov. 2013. "Core-Selecting Auctions with Incomplete Information." Unpublished.

Ausubel, Lawrence M., and Oleg V. Baranov. 2014. "The Combinatorial Clock Auction, Revealed Preference and Iterative Pricing." Unpublished.

Ausubel, Lawrence M., Peter Cramton, and Paul Milgrom. 2006. "The Clock-Proxy Auction: A Practical Combinatorial Auction Design." In *Combinatorial Auctions*, edited by Peter Cramton, Yoav Shoham, and Richard Steinberg, 115–38. Cambridge, MA: MIT Press.

Ausubel, Lawrence M., and Paul R. Milgrom. 2002. "Ascending Auctions with Package Bidding." *Frontiers of Theoretical Economics* 1 (1).

Cramton, Peter. 2013. "Spectrum Auction Design." *Review of Industrial Organization* 42(2): 161–90.

Day, Robert W., and Peter Cramton. 2012. "Quadratic Core-Selecting Payment Rules for Combinatorial Auctions." *Operations Research* 60 (3): 588–603.

Day, Robert, and Paul Milgrom. 2008. "Core-Selecting Package Auctions." *International Journal of Game Theory* 36 (3–4): 393–407.

Day, Robert W., and S. Raghavan. 2007. "Fair Payments for Efficient Allocations in Public Sector Combinatorial Auctions." *Management Science* 53 (9): 1389–1406.

Erdil, Aytek, and Paul Klemperer. 2010. "A New Payment Rule for Core-Selecting Package Auctions." *Journal of the European Economic Association* 8 (2–3): 537–47.

Nisan, Noam. 2006. "Bidding Languages for Combinatorial Auctions." In *Combinatorial Auctions*, edited by Peter Cramton, Yoav Shoharn, and Richard Steinberg, 215–31. Cambridge, MA: MIT Press.

Parkes, David C. 2001. "Iterative Combinatorial Auctions: Achieving Economic and Computational Efficiency." PhD diss., University of Pennsylvania.

CHAPTER 10

Quadratic Core-Selecting Payment Rules for Combinatorial Auctions

Robert Day and Peter Cramton

1. Introduction

Combinatorial auctions represent one of the most prominent areas of research in the intersection of Operations Research (OR) and Economics. First proposed for practical governmental applications by Rassenti et al. (1982), a combinatorial auction (CA) is an auction for many items in which bidders submit bids on *combinations* of items, or packages. CAs also are referred to as "package auctions" or auctions with "package bidding." In a general CA, a bidder may submit bids on any arbitrary collection of packages. The "winner-determination problem" identifies the value maximizing assignment given the package bids. This problem is as complex as the Weighted Set-Packing problem, and hence NP-hard (see Rothkopf et al. 1998).

Thus, in the many real-world applications of CAs, the computational techniques of OR facilitate more efficient economic outcomes in environments too complex for classical (i.e., non-computational) economic theory. Conversely, the game-theoretic framework surrounding CAs provides a host of new computational challenges and optimization problems for OR.

One critical element of any CA is the pricing rule, which determines what each winner pays for the package won. In this paper, we present a new class of optimization-based pricing rules for combinatorial auctions in general, demonstrate some of their unique features, and elaborate upon some properties of the larger class of core-selecting mechanisms. We also describe the use of this algorithm for recent and upcoming spectrum-license auctions in the United Kingdom, for upcoming spectrum auctions in several European countries (e.g., the Netherlands, Denmark, Portugal, and Austria), and for use in the United States for the Federal Aviation Administration's (FAA) proposed allocation of landing rights to control congestion at airports.[1] Further, we provide the relevant economic interpretation and theoretical basis for our algorithm's various features.

---

[1] In December 2008, one month prior to the actual auction of landing rights for the three New York City airports, a Federal court stayed the auction. The plans for auctioning slots are now uncertain.

2. Background

The use of auctions for allocating spectrum-license-rights to telecommunications companies gained prominence in 1994 when the Federal Communications Commission (FCC) began to use a Simultaneous Ascending Auction (SAA) to sell spectrum licenses in the United States. The initial design, which is still used today with only slight modifications, avoided the idea of a "combinatorial" or "package" auction, in which bidders bid on packages of licenses because of the inherent computational difficulty.

A main difficulty with the SAA and other auctions that only allow bids on individual lots is the *exposure problem*. A bidder finds it risky to bid on a collection of lots, because of the risk of receiving an incomplete package of complements. For example, a bidder may need both A and B. If the bidder is allowed to bid only on individual lots as in the SAA, the bidder risks winning only one of the required lots. A CA avoids this problem by letting the bidder bid on the package {A, B} with no risk of winning just A or just B. For a thorough discussion of the strengths and drawbacks of the SAA and its implementation by the FCC, see Cramton (2006). Also, for a general introduction to CAs, see the edited volume by Cramton, Shoham, and Steinberg (2006).

To maintain many of the strengths of the SAA while mitigating its primary weaknesses, several authors have proposed hybrid auction formats, that combine the simple price-discovery process of a "price-clock" with the efficiency and exposure-problem-elimination of a CA. (See Porter et al. 2003, Ausubel et al. 2006, and Cramton 2009.) Here we present the latest development in this line of research, a combinatorial-clock auction *with quadratic-core-pricing*. For the remainder of the paper, we focus on this pricing rule as adopted for several upcoming governmental auctions.

In §2.1–2.3 we introduce notation and a general CA model, and motivate the use of core-pricing as the most usable generalization of the second-price concept from single-item auctions. In §3.1 we describe the core of the CA game formally, and elaborate upon some of the alternative representations of the core. In §3.2 we describe the particular core-selection rule that we proposed for the U. K. spectrum license market, describing some of its properties in §3.3 and §3.4. The implementation of seller reserve values is described in §4. In §5 we briefly describe our experience implementing these pricing rules for real-life auctions, while §6 presents results from computational experiments. Conclusions follow in §7.

Additionally, supplementary material is provided in an appendix, with appendix A.1 demonstrating how semi-sincere strategies eliminate a form of envy, A.2 and A.4 providing technical details on computational implementations, and A.3 providing the proof of a results in the main text. The final appendix, A.5, outlines some practical considerations regarding real-life implementations.

2.1. The Environment: Heterogeneous Goods and Bidders

We consider an environment in which a government intends to sell many interrelated heterogeneous items. The heterogeneity of spectrum licenses arises from varying geographical coverage, as well as technological considerations, such as interference with adjacent frequency bands, etc. The primary goal of the government agency is assumed to be efficiency: the items should be sold to those who value them the most. Stated differently, the government's objective is the maximization of social welfare.

In the case of spectrum-licenses, bidders may have complex preferences over the items being auctioned, with some bidders considering certain items to be substitutes, while others treat the same items as complements. Differing technologies may give rise to such heterogeneity among the bidders' preferences. One bidder may treat any two items as substitutes because her communication technology is neutral to the spectrum on which it is transmitted, while another bidder might require a pair of adequately separated licenses as uplink and downlink frequencies for two-way communication. The latter bidder thus treats certain pairs as complements.

With a variety of new communication technologies emerging, it is important that the auction design be technology neutral. If, for example, bidders were homogeneous in their desire for "paired" licenses with a certain optimal spacing between uplink and downlink frequencies, it would be appropriate for the auction design to specify that licenses be sold as bundled pairs. If instead some bidders desire a single contiguous strip of unpaired licenses while others desire pairs, this pre-bundling of licenses into pairs would be inappropriate. This is the case for the U. K.'s 2.6 GHz auction, in which Ofcom determined that bidders could bid on contiguous blocks of either paired or unpaired spectrum licenses, or some combination of both; the strength of the bids themselves would determine the quantity of spectrum of each type. In general, this flexibility of package bidding, the cornerstone of CAs, provides an opportunity for OR tools to improve economic outcomes; the problem of determining the optimal set of bids to accept is generally complex and closely related to the NP-hard set-packing problem. (See Rothkopf et al. 1998, and deVries and Vohra 2003.)

Package bidding alone can often represent a daunting challenge to both the bidders and the bid-taker. In the U. K.'s 2.6 GHz spectrum auction, for example, there are 39 unique licenses offered for sale, and thus 2^{39} packages for each bidder to consider placing a bid on. In practice, the auctioneer cannot accept this full set of package bids from each bidder, so instead limits the number of package bids it will accept (in the U. K. auctions the number of bids is usually capped in the thousands). The bidders thus face the difficulty of deciding which are the "best" packages to bid on, in addition to the problem of deciding their value for any single package.

For the remainder of the paper we therefore address the computation of prices following the final sealed-bid round in a two-stage hybrid design known as the "clock-proxy auction," as proposed by Ausubel et al. (2006). In this design a final sealed-bid CA is preceded by a preliminary "clock stage," used as a preference elicitation tool, allowing the bidders to learn about market competition and discover valuable information about which packages seem most profitable to bid on as competitive prices are revealed. We propose the use of quadratic programming in conjunction with constraint generation to determine the best set of final prices in the final sealed-bid auction of such a design, which takes all clock-stage bids and any other "supplementary" package bids made by the bidders as exclusive package offers.

2.2. Winner Determination

Here, we consider bidders that have participated in a clock auction (or, to be more precise, the clock-phase of a two-phase hybrid auction) and have submitted any supplementary package bids, and consider the auctioneer's problem of determining the final set of package bids to accept and the payments to collect from each bidder. Though some mechanisms considered in the mechanism design literature consider the possibility of outcomes that are not efficient with respect to submitted bids (for example, Myerson 1981, or Goldberg and Hartline 2003), for the governmental allocation of public resources, we consider efficiency to be essential and indispensable in order to achieve the government's goals and to promote the perceived fairness of the auction outcome.

Let $M = \{1, 2 \ldots, m\}$ represent the set of m items being auctioned and $N = \{1, 2, \ldots n\}$ represent the set of n bidders. Each bidder has submitted a collection of bundle bids, with $b_j(S)$ representing bidder j's monetary bid on any bundle $S \subseteq M$. The efficient winner determination problem over the set of bidders N is defined by the following integer program, which maximizes the value of accepted bids without selling the same item to more than one bidder:

$$wd(N) = \max \sum_{j \in N} \sum_{S \subseteq M} b_j(S) \cdot x_j(S) \tag{WD}$$

subject to $\displaystyle\sum_{S \supseteq \{i\}} \sum_{j \in N} x_j(S) \leq 1, \quad \forall i \in M$ (1)

$$\sum_{S \subseteq M} x_j(S) \leq 1, \quad \forall j \in N \tag{2}$$

$$x_j(S) \in \{0, 1\}, \quad \forall (S, j) \text{ such that a bid } b_j(S) \text{ was submitted} \tag{3}$$

Additionally, we note that this formulation implies a specific "XOR" bidding language, in which, in accordance with constraint set (2), no two bids made by the same bidder may be accepted by the auctioneer. Though a host of alternative bidding languages have been described in the literature (see Nisan 2006), we maintain this XOR formulation because it is general enough to describe any other bidding language (albeit exhaustively). Further, it is this bidding language that has been used in practice in the U. K. spectrum auctions, because in general, the clock phase narrows the number of bids which will need to be bid upon substantially, keeping this formulation from growing too large, and because the implication of each bid is most easily understood by the bidders in this setting, allowing little room for confusion regarding the implication of any bid made. Each bid is an exclusive offer that cannot be recombined with any other bids of the same bidder.

2.3. Payment determination: Second price rules and Core-selection

We now consider the algorithm for the determination of payments in the final sealed-bid auction. First, we motivate core pricing as the appropriate generalization of the "second price" rule.

A fundamental development of early auction theory is the equivalence (under the assumption of private values) of the outcomes in the well-known English auction (in which an item is offered at increasing prices until only one bidder continues to indicate willingness to purchase) and the second-price sealed-bid auction for a single item. Krishna 2002 provides an overview of auction theory. In the *second-price sealed-bid auction*, bidders submit a sealed-bid for the single item being auctioned, with the highest bid winning the item, and the winner paying the amount of the *second* highest bid. The second-price sealed-bid auction (for a single item) is well-known to satisfy each of the following properties:

1. Individual Rationality: each bidder expects a non-negative payoff for participating. In the case of auctions, this simply means that non-winners do not pay, and that each winner pays an amount less than or equal to her bid.
2. Efficiency: the highest valued bid wins. In the combinatorial case, this will be interpreted as: the winning bidders form an optimal solution to (WD).
3. Dominant Strategy Incentive Compatibility: misreporting one's value for the item(s) *never* gives an advantage.
4. The "Core" property: no coalition (subset of all players) can form a mutually beneficial renegotiation among themselves. In the case of an auction, this simply means that the seller would not prefer to ignore the outcome dictated by the auction and renegotiate with a subset of the bidders.

It is also well-known that the Vickrey auction, also known as the Vickrey-Clarke-Groves or VCG mechanism, is the unique mechanism in the combinatorial setting that satisfies properties 1, 2 and 3 from this list. The VCG outcome implements the efficient solution described by (WD), and each winning bidder j receives a discount from her winning bid amount, equal to $wd(N) - wd(N\setminus\{j\})$, which induces her to bid honestly. Unfortunately, it is easily shown that property 4 does not hold for the VCG mechanism. The reader may easily verify that in a two item auction for items A, and B, with bids by three bidders $b_1(A) = 2, b_2(B) = 2, b_3(A, B) = 2$, the VCG payments are both zero for winning bidders 1 and 2, despite a competing bid of 2 on the items they win. Thus this simple example (from Ausubel and Milgrom 2002) illustrates that the core property is not upheld by the VCG auction, since both the seller and bidder 3 would prefer to renegotiate for both items at any price in the open interval (0,2).

Given the beauty of the VCG mechanism in its ability to elicit truthful revelations of preferences from the bidders, it is not surprising that it has received a great amount of attention in the literature. However, several authors have noted that the VCG auction is not practical for actual implementation. (The reader may refer to Rothkopf 2007, Ausubel and Milgrom 2006, or Rothkopf et al. 1990, for example.) We instead contribute to the growing literature that "core-selecting mechanisms" or "auctions with core pricing" provide the most usable generalization of the second-price sealed-bid auction paradigm to the combinatorial setting. In this category of CAs, we eschew the approach of the VCG mechanism (which treats properties 1–3 as constraints and ignores property 4) and instead treat properties 1, 2, and 4 as constraints while minimizing (with respect to some metric) the deviation from property 3. We do not, however, completely disregard the VCG outcome, but instead use it as a baseline for incentive compatibility; the closer we get to the VCG payments, the incentives to distort one's bids become less and less. Further, when the VCG mechanism does happen to satisfy property 4, our mechanism will also produce the VCG outcome.

A *core-selecting mechanism* is one that satisfies property 4 when bids are treated as true values for the corresponding bundles. (This distinction about true values is necessary, since property 3 does not hold in general, and so we cannot claim that bids are equal to true values.) Before delving into the technical description of how we propose to compute core prices in the aforementioned governmental auction applications, we now briefly summarize some of the properties of core selecting mechanisms which motivate our claim that bidder-optimal core prices provide a useful notion of second prices for the combinatorial setting.

To begin, for any core-selecting mechanism:

- An allocation must be efficient with respect to reported preferences (see Milgrom 2004).
- No bidder can ever earn more than her payoff under the VCG auction by disaggregating and using false-name or shill-bidders (see Day and Milgrom 2008). Note that this is not true of the VCG auction.
- Determining a core outcome is NP-hard whenever the winner-determination problem is NP-hard (see Day and Raghavan 2007).
- For any profile of opponents' bids, each bidder has a best reply that is a semi-sincere strategy, i.e., given true utility $u_j(S)$ for each item set S, each bidder j has a best strategy of the form $b_j(S) = \max(0, u_j(S) - \alpha_j)$ with the same $\alpha_j \geq 0$ for each bundle $S \subseteq M$ (see Day and Milgrom 2008).

This last point says that there exists a semi-sincere strategy among any bidder's set of optimal strategies (so-called as a bidder is truthful about the relative values of bundles receiving positive bids), which is elsewhere referred to alternatively as either a truncation strategy (by analogy to truncation strategies in matching markets) or a profit-target strategy (as a bidder j targets an amount of profit α_j and cannot receive less than this amount of profit when among the winning bidders). In appendix A.1, we provide provide further motivation for the use of semi-sincere strategies in core selecting auctions, in their ability to eliminate ex post envy. Similar connections between envy reduction and core-selection are provided in the concurrent work of Othman and Sandholm (2010).

Next, we consider *bidder-optimal* core mechanisms, which are optimal, or efficient, in the Pareto sense. That is, if the auction determines an efficient allocation and prescribes payment vector p, then there is no alternative payment vector p' also in the core, such that $p' \leq p$. As is typical in Pareto-optimality, this can be read as: no bidder can be made better off without another being made worse off. If the core-selecting mechanism is also bidder-optimal, we have:

- The incentives to unilaterally misreport are not dominated by any other core-selecting mechanism (see Day and Milgrom 2008).
- If the buyer-submodularity condition holds, then the Vickrey outcome is the unique bidder-optimal core point. Thus any bidder-optimal core-selecting mechanism is equivalent to the Vickrey mechanism whenever bids and valuations satisfy buyer-submodularity, in which case the auction is dominant-strategy incentive-compatible. The same result holds if the more restrictive gross-substitutes condition is satisfied for bids and valuations (see Ausubel and Milgrom 2006). Further, if it is common knowledge that the buyer-submodularity condition holds (for valuations) then truth-telling by all bidders is a Nash equilibrium in any core-selecting auction, even if bidders are free to use shill bidders. (This follows from Theorem 1 of Day and Milgrom, 2008, which implies that any player's payoff is no more than her induced Vickrey payoff in any core-selecting auction, even if using shills.)

- Any bidder-optimal core payoff vector induces a semi-sincere strategy that is a full-information Nash equilibrium (see Day and Milgrom 2008, or Day and Raghavan 2007).
- Any Nash equilibrium in which winners use semi-sincere strategies and losers bid truthfully achieves a bidder-optimal core point with respect to the true valuations of the bidders. Thus at any such full-information equilibrium an outcome must be efficient with respect to true preferences, not just relative to reported preferences/bids (see Day and Milgrom 2008, or Day and Raghavan 2007).

These last two points elucidate what might be seen as a "strategic correction" property of bidder-optimal core-selecting auctions. Bernheim and Whinston (1986) show a similar theorem that the bidder-optimal core points are precisely the full-information Nash equilibria in a first-price (i.e., pay-as-bid) format. Thus in a core-selecting auction, if all bidders bid truthfully, the auction makes them pay an amount equal to what they "should have bid" in a first-price format, effectively correcting their strategies. By analogy, the second price sealed-bid auction corrects the winner's bid to what she should have bid to just tie the bid of the next highest bidder, if she had know how much that was. The main difference in the combinatorial setting is that, there are *many* bidder-optimal outcomes, and thus many equilibrium strategies, so the auction additionally helps the bidders by *selecting* an equilibrium to coordinate to. In this paper, we explore various attractive criteria for selecting such an equilibrium outcome.

This point regarding strategic correction is worth emphasizing for its relevance in response to regulators worried about the adoption of a core-selecting auction, who may ask (as they did at the FAA when adopting a core-selecting rule), "Aren't the core-prices overly complicated? Wouldn't the bidders prefer the simplicity of a pay-as-bid bid rule?" The results on the Nash equilibria in these auctions allows us to answer that in fact the opposite is true. In a first-price auction, it is difficult to determine the correct bidding strategy which will result in payments as small as possible fixing the other bidders' bids, and every dollar bid above this optimal amount is a dollar wasted. However, in a core-selecting auction, the auctioneer effectively corrects the bids for you, so that at the conclusion, having seen the payments of the other bidders, each bidder would agree to bid exactly her payment amount, given that all other bidders bid exactly their payment. Further, this is exactly what they would have liked to have bid as a group in a pay-as-bid auction. This point will be further illustrated numerically with Figure 1 and Example 1 provided in the following section.

3. Selecting a Core Outcome

As noted in §2.3, the bidder-optimal points in the core represent satisfactory outcomes from the auction. Bidders are satisfied that they are just paying enough to beat out competitors, and that no one can be made better off without another being made worse off. The seller is satisfied to receive competitive revenues determined by competition, and for which no readily apparent better alternative is available. We now define the core formally, noting a few interesting alternative formulations and their uses, and demonstrate some new techniques for selection among core outcomes.

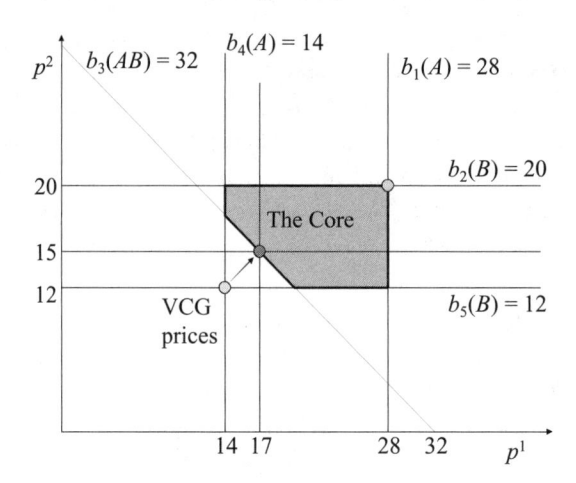

Figure 1. The core point closest to VCG payments.

In addition to notation already introduced, let payment vector $p \in R^n_+$ represent the non-negative vector of payments for each bidder, and let $\pi_j = b_j(S_j) - p_j$ represent the observable surplus or profit experienced by bidder j when the auction awards bidder j set S_j. Bidders are said in this case to have quasi-linear utility (in that their profit is linear in payment). Also, one may note that we are dealing only with observable surplus, not true net utility $u_j(S_j) - p_j$, since without a guarantee of incentive compatibility, the auctioneer will have no knowledge of these amounts. Also, we may write $\pi_0 = \sum_{j \in N} p_j$ for the profit of the seller. (For now, we assume that prices are normalized by reserve amounts, so that we need not subtract the value of each item from the seller's profit. Stated differently, the seller has no value for keeping the items herself.)

An *outcome* is represented by a feasible solution to problem (WD), which we will specify by the set of awarded (possibly empty) bundles $\{S_j\}$ for each bidder and a payment vector p, thus inducing a profit vector π. An outcome is said to be *blocked* by coalition $C \subseteq N$ if there is some alternative outcome with awarded bundles $\{\bar{S}_j\}$ and payments $\bar{p}$, such that $\bar{\pi}_j = b_j(\bar{S}_j) - \bar{p}_j \geq \pi_j$ for all $j \in C$, and for which $\bar{\pi}_0 = \sum_{j \in C} \bar{p}_j > \pi_0$. An outcome that is not blocked in this context is said to be in the *core with respect to the submitted bids b*. For this paper, we may simply say that the outcome is in *the* core, since we do not consider the underlying utility functions. Also, since in other economical settings the core is not always guaranteed to exist, it is worth noting that in this setting the pay-what-you-bid point is always in the core, and thus the core is always non-empty.

It may be helpful at this point to consider an example. Let $m = 2$ items, A and B, $n = 5$ bidders, and let bids be as follows (each bidder submits only one bid):

Example 1:
$$b_1(A) = 28 \quad b_2(B) = 20$$
$$b_3(AB) = 32 \quad b_4(A) = 14 \quad b_5(B) = 12$$

It is easy to determine that the unique winners in the efficient allocation are bidders 1 and 2, and that the VCG payments are $p_1^{VCG} = 14$, and $p_2^{VCG} = 12$. The core itself can be graphed in payment space as in Figure 1.

Here we note that, due to the simplicity of the example, the constraints defining the core are simply the bids of the losing bidders (this is not always the case). In particular, since bidder 4 would always object (block) if bidder 1 paid less than 14 for item A, we have the constraint, $p_1 \geq 14$. Similarly, bidder 5 dictates $p_2 \geq 12$. Bidder 3 would object if bidders 1 and 2 together did not beat his bid on the items they have won, thus $p_1 + p_2 \geq 32$. Upper bounds on payments are given by the bids themselves, consistent with our assumption of individual rationality. Next, one will note from the picture that we are in a situation for which the VCG outcome is not in the core; bidder 3 alone forms a blocking coalition.

Using the technique of Day and Raghavan (2007), one can guarantee bidder-optimality by minimizing total payments over the core, and so for Example 1 we could determine any payment vector on the line segment connecting the point (14,18) to (20,12), any of which represents a bidder-optimal core point. This simple example also clearly illustrates the statements from §2.3 regarding Nash equilibria in a first price auction. Fixing the bid b_2 of bidder 2 anywhere in the range [12,18], bidder 1's optimal strategy in the first price auction is to bid $32 - b_2$, and conversely, fixing b_1 within the range [14,20], bidder 2's best bid is $32 - b_1$. In practice, however, each bidder will not know the bid of the other, making coordination to any one of these points difficult if not impossible. Thus a risk-averse bidder would typically have to bid more than the optimal amount (i.e., where he would bid if he knew what the other were bidding). In the first price auction, this problem, caused by lack of information, costs the bidder in a one-to-one fashion. From bidder 1's perspective, every dollar bid over $32 - b_2$ is a dollar wasted. In any core-selecting auction, on the other hand, the bidders need only bid somewhere within the core (including bidding truthfully) and the auction will charge them a total of 32.

But still, there is a lack of precision because the Day and Raghavan (2007) algorithm does not specify which of these bidder-optimal points should be chosen. We are motivated, however, by the observation of Parkes et al. (2001), that the difference between a final payment and the VCG payment represents a measure of "residual incentive to misreport," and so should be minimized. As one method to achieve a simple to compute minimization of the groups' incentive to deviate from truth-telling, we propose the following refinement of the Day and Raghavan (2007) procedure: over all total-payment minimizing core points, select the one that minimizes the sum of square deviations from the VCG payment point. Of course, minimization of this amount is equivalent to minimizing the positive square root of this amount, so one may rightly describe this selection as the core point with minimum Euclidean distance from VCG. This rule can be referred to as a VCG-nearest or Vickrey-nearest rule.

For Example 1, this results in the unique payment outcome (17,15). Interestingly, this outcome is unchanged (for Example 1) as long as bidder 1 bids at least 20 and bidder 2 bids at least 18, fixing the bids of the losing bidders. The values 20 and 18 are the minimum amounts that could be bid, respectively, without the bid of 32 emerging in the VCG computations. If bidders 1 and 2 bid more than 32 in total, and less than 20 and 18 respectively, than the VCG point can in fact move, causing a slightly different outcome. For example, if the bids for 1 and 2 changed to 19 and 16, the VCG point shifts to (16,13), and final payments become (17.5, 14.5). Thus, when the VCG point moves, the relative payments can also change slightly, but the payoff to the seller remains unchanged.

One may also note that bidders 1 and 2 in Example 1 each pay an equal amount (3 units) above their VCG payments in order to match the blocking bid made by bidder 3. This is indeed a general phenomenon (based on the Karush-Kuhn-Tucker optimality conditions) and one which we describe in detail in §3.4. First, however, we describe a few distinct presentations of the core and the interesting implications/economic interpretations of each formulation.

3.1. Core Formulations

First, working straight from the definition, the coalitional core constraints are most commonly modeled (in the Economics literature) as in Day and Milgrom (2008):

$$\sum_{j \in C \cup 0} \pi_j \geq wd(C) \quad \forall C \subseteq N \quad (4)$$

emphasizing that final payoffs (on the left) must exceed the value that each coalition C can generate if they alone deal with the seller (on the right). Yet from the point of view of computation, this formulation hides the (discrete) selection of a bundle for each bidder, and is thus not guaranteed to be convex in π-space, making it a difficult formulation for use in a direct computational implementation. In practice, we take a divide-and-conquer approach, first solving the winner-determination problem and then computing core payments once a particular set of winning bundles $\{S_j\}$ has been determined. Substituting in these bundles, canceling payments that are duplicated in the π_0 term, and recognizing that $wd(N) = \sum_{j \in N} b_j(S_j)$ yields an alternative formulation:

$$\sum_{j \in W} p_j \geq wd(C) - \sum_{j \in C}(b_j(S_j) - p_j) \quad \forall C \subseteq N \quad (5)$$

where W represents the set of bidders who win non-empty bundles. Here the right-hand-side reflects what coalition C is willing to offer to the seller at payment vector p; they will offer as much as can be obtained from them as a group, $wd(C)$, less the profit they are already making at payment vector p, which is $\sum_{j \in C}(b_j(S_j) - p_j)$.

As shown by Day and Raghavan (2007) this formulation is convenient from an algorithmic point-of-view, when we treat computations modularly (i.e., with a blackbox mindset). If we already have code (a blackbox) that solves winner-determination problems, and we are considering whether a particular payment vector p is in the core, we can simply reduce each bid by the surplus at p and re-run the winner determination. This will find the coalition making the highest offer to the seller, and if this is more than the current total payments, then a violated core constraint has been identified. (In fact this finds the most violated constraint.) This complexity equivalence between separation and winner determination is helpful to demonstrate that finding core outcomes is indeed of equivalent complexity as winner determination (see Day and Raghavan 2007). Also, this formulation is noteworthy because it is in this form that the "core" was defined legally within the regulations for the U. K. spectrum auctions. Rather than defining the core in terms of possible renegotiations, this separation formulation gives a clearly-defined, mechanically-checkable stopping criterion, or provides a certificate that a payment vector is not in the core.

Finally, from a math programming standpoint, it is most helpful to segregate decision variables and constants on their respective sides of the inequality. This yields the following formulation, which follows from the previous formulation by simply canceling payment terms appearing on both sides:

$$\sum_{j \in W \setminus C} p_j \geq wd(C) - \sum_{j \in C} b_j(S_j) \qquad \forall C \subseteq N \qquad (6)$$

We use this formulation for our actual computations of core prices, which we find by quadratic optimization over the core. This formulation, too, has its own interesting economic interpretation, lending further credence to core-mechanisms as selecting "fair" payments. Given the efficient allocation $\{S_j\}$, the right-hand-side of (6) finds what coalition C would pay to get everything, $wd(C)$, minus what they would pay for what they actually get, and is thus equal to the most coalition C would be willing to pay to take away what the complementary set of bidders is getting. Thus each core constraint says that any set of bidders pays at least as much as their opponents would pay to take their stuff away from them, a competitively pleasing, and arguably fair proposition indeed.

3.2. Quadratic Rules for Payment Determination

Letting $\beta_C = wd(C) - \sum_{j \in C} b_j(S_j)$, and denoting the vector of all such β_C values as β, formulation (6) can be written more compactly as:

$$pA \geq \beta$$

where each column a_C is the characteristic vector of the complementary set of winners. (That is, the jth entry in a_C equals 0 if bidder j is in set C and equals 1 if bidder j is not in C. Since non-winners never pay, the dimension of each a_C is $|W| \times 1$, rather than $n \times 1$.) The core-selection region is defined by these constraints, as well as the individual rationality constraints: $p \leq b$, where each component b_j in the vector b is given by $b_j = b_j(S_j)$.

We now present a class of algorithms for core-selection based on quadratic programming. Motivated by the concurrent work of Erdil and Klemperer (2009), these rules may be referred to as reference rules, in which payments are determined by minimizing the Euclidean distance to a reference vector of prices. A p^0-*reference rule* finds final payments p^* that minimize the sum of squared deviations from payment reference point p^0, which may be either constant or dynamically determined, but is constant with respect to the following optimization:

$$\min (p - p^0)(p - p^0)^T \qquad (7)$$

$$pA \geq \beta \qquad (8)$$

$$p \leq b \qquad (9)$$

Also, Day and Raghavan (2007) provided some motivation that payment minimization over the core may deter certain types of group deviation, and that a Threshold rule (as described by Parkes et al. 2001) without explicit payment minimization may not result in payment minimization. Similarly, a reference rule as just described may not minimize total payments over the core, unless this payment minimization is enforced explicitly. (Example 2 on page 221 provides an example of this phenomenon.) We therefore also describe *MRC-reference rules* in which the feasible set of payments is limited to those core points which minimize total revenue, referred to as the Minimum Revenue Core or MRC by Erdil and Klemperer (2009). To employ such a rule, we first find minimal core payments by solving the LP:

$$\mu = \min p1 \tag{LP}$$

$$pA \geq \beta \tag{10}$$

$$p \leq b \tag{11}$$

Then determine final payments p^* as the optimal solution to the following QP:

$$\min(p - p^0)(p - p^0)^T \tag{QP}$$

$$pA \geq \beta \tag{12}$$

$$p \leq b \tag{13}$$

$$p1 = \mu \tag{14}$$

This last MRC-reference rule with $p^0 = p^{VCG}$ is the auction format adopted by Ofcom for spectrum license auctions in the U.K.

In practice, evaluating each β_C requires the solution of a winner-determination problem, so with $2^n - 1$ non-empty coalitions to consider, it is advantageous to employ a core-constraint generation procedure as in Day and Raghavan (2007), which we will henceforth abbreviate CCG. Starting at the payment vector p^0, reduce each bid by the current surplus, i.e., for all $S \subseteq M$, let $b_j(S) = \hat{b}_j(S) - b_j(S_j) + p_j^0$, where $\hat{b}$ represents the fixed, submitted bid. Then solve (WD) with these new bids, finding the first violated coalition C_1, the set of bidders winning nonempty bundles in this altered version of (WD). We then let our first approximation of matrix A be simply $A_1 = a_{C_1}$, and let $\beta^1 = \beta_{C_1}$. Next we solve formulation (LP) with A_1 and β^1 replacing A and β, yielding minimum payment solution μ_1, and then solve formulation (QP) with A_1, β^1, and μ_1 replacing A, β, and μ, labeling the solution to (QP) as p^1. The algorithm continues in this fashion, finding a new violated constraint $pa_{C_t} \geq \beta_{C_t}$ at p^{t-1}, and concatenating the corresponding column to A_{t-1} and new entry to β^{t-1} forming A_t and β^t, as long as this solution to the surplus-reduced (WD) exceeds $p^{t-1}1$. If the solution to the surplus-reduced (WD) does not exceed $p^{t-1}1$, then we may set $p^* = p^{t-1}$ and terminate with a solution to (QP), representing final payments in the auction. Further discussion on the efficacy of this CCG approach is given in appendix A.2.

3.3. Robustness and Constant Reference Rules

It has been observed by Ott (2009) and Lamy (2009) that revenues may in some cases decrease when bids increase in any bidder-optimal core-selecting auction, contradicting an erroneous proposition put forth in Day and Milgrom (2008). Here we provide a result regarding the relative insensitivity of the auction outcome (including total revenues of the seller) under a special set of assumptions, helping to motivate the use of certain reference rules, and provide further insight into the types of strategies which lead to profitable deviation from truth-telling.

Theorem 1. *Consider a reference rule (or an MRC-reference rule) in which the reference vector p^0 is independent of winners' bids. Suppose that bidders restrict to semi-sincere strategies, and that for a fixed set of bids, the individual rationality constraints are never binding, i.e., suppose that $p^* < b$. Then, the auction outcome does not change for any uniform increase in a semi-sincere strategy made by a winning bidder.*

Proof. See appendix A.3. □

Though the exploration of appropriate reference vectors which satisfy the condition of independence from winners' bids remains open, it is easy to see that a constant reference vector selected in advance by the auctioneer would be independent, satisfying that hypothesis of the theorem. Interestingly, this theorem states that if a bidder focuses on semi-sincere strategies (perhaps to relieve envy possibilities via Lemma 7.2) then if she bids enough that she will have paid more than she had bid, she may as well have bid honestly.

Also, this theorem can also be stated "locally" for a single winning bidder, that if p^0 does not change with a uniform increase in semi-sincere strategy by bidder $j \in W$, and if $p_j^* < b_j$, then the overall auction outcome does not change with any uniform increase in semi-sincere bidding strategy by bidder j. Indeed, this local result applies under a VCG-nearest rule when it is the case that $p_j^* < b_j$ and that bidder j participates in the efficient solution *even when any other single bidder is removed*. In that case, each VCG payment for a bidder $\bar{j} \neq j$ remains unchanged following the semi-sincere increase via cancellation of the increase, and thus p^0 is unaffected by the increase. The reader may verify that this is the case for Example 1 from page 202, that any bid increase by a winning bidder leaves the outcome unchanged.

But of course, the hypothesis that $p^* < b$ is indeed a strong assumption, and considering a violation of such an assumption elucidates situations in which there is scope for profitable deviation from truth-telling. For example, let us revisit Example 1 from page 202, but now considering a situation in which rather than a dynamic VCG-nearest rule, the auctioneer had arbitrarily selected the point (14,12) as p^0, resulting in a confirmation of the independence assumption. (The following explanation would change little if we used another constant vector as p^0.) We see that the outcome of the auction is unchanged as long as bidders 1 and 2 bid any amount greater than or equal to (17,15) respectively on their bundles of interest. (Notice that any bids less than or equal to value on the bundle of interest constitute semi-sincere strategies when bidders are single-minded.) The scope for profitable bid-shading only occurs when one of the bidders bids below her final payment under truth-telling, and only when the other winning bidder bids at least as much in the other direction, strictly above her own final payment. For example, if bidder 1 knows that bidder 2 will bid 19, she can bid any amount down to 14, which (assuming preference in a tie-breaking rule) results in no change of allocation. (We see that in addition to being sure that bidder 2 bids enough to make the sum of their bids exceed 32, bidder 1 would also like to be sure to beat the competing bid of bidder 4.) If this new bid amount for bidder 1 is less than 17, her payment if honest, she will pay exactly as bid, violating the assumption of Theorem 1, and bidder 2 will be forced to pick up the difference $32 - b_1$. Of course, without knowledge of bidder 2's bid, and bidder 4's bid, this shading below 17, which is profitable, also carries the risk of missing the efficient allocation, resulting in zero payoff.

This example shows the limitations of Theorem 1, but since a bidder j will often not have enough knowledge to safely shade to a point where $p_j^* = b_j(S_j)$, the theorem is likely to be relevant in many situations. Roughly speaking, if we consider pay-as-bid outcomes to be unlikely, then this theorem states that decreases in α_j to reduce risk will likely not be costly. Also, this alteration of Example 1 seems to suggest that the most attractive combinatorial auction would be one in which the auctioneer knew the VCG point based on true valuations, and used this true VCG point as the reference price vector p^0. But it is hard to imagine a situation in which the auctioneer would have enough foreknowledge to predict the true VCG point accurately, yet still feel the need to conduct an auction.

Still, this motivates the goal of the auctioneer conducting a reference rule auction, to attempt to select an independent reference point that approximates the true VCG point as well as possible, in an effort to maintain the desired notion of "near-truthfulness." Prior to the influence of Erdil and Klemperer (2009), we did consider constant-p^0-reference rules, but rejected this idea in our consultation to Ofcom for U. K. spectrum auctions, due to the distortions caused in the final payments, favoring large bidders, and because this approach makes the final distribution of payments highly dependent on the assumptions and actions of the auctioneer. Let us elaborate on these points, again by example.

It is easy to see with a two-winner auction that a constant reference rule, such as $p^0 = \vec{0}$, favors larger bidders. If we were to apply such a rule to the data for Example 1, for example, bidders 1 and 2 both pay 16, an equalization of payments despite higher marginal competition on item A from bidder 4. The following more extreme example demonstrates how this problem can get worse as the situation becomes more lopsided.

$$\text{Example 3:} \quad \begin{matrix} b_1(A) = 100 & b_2(B) = 20 \\ b_3(AB) = 60 & b_4(A) = 50 \end{matrix}$$

In this example, which is quite similar to Example 1, the VCG payments for efficient winners 1 and 2 of (50,0) are not in the core; the two must raise their combined payments to 60 in order to keep bidder 3 from blocking. If the reference rule $p^0 = 0$ is used, the result is that bidder 2 will pay the entire burden of this total payment increase; final payments become (50,10), while a VCG-nearest rule (MRC or not) results in a sharing of this burden, with payments (55,5). This problem can be mitigated by using a more sophisticated constant value (for example, selecting the reference vector with each term set equal to the sum of the reserves values for each item in the respective winning bundles) but the problem persists that bundles with high value relative to the auctioneer's expectations pay less of the burden of overcoming a competing coalitional offer. Since the underlying assumption of an auction is often that the seller has poor a priori knowledge of value relative to that of the bidders, we are motivated to select an outcome based on good (bid-based) information over poor (seller prior) information.

Further, a one-for-one change on the part of the seller often results in a corresponding one-for-one change in payments under a constant reference rule, making the outcomes highly sensitive to the pre-auction actions of the seller. For example, considering Example 1 with a seller predefined constant reference rule $p^0 = (14,12)$, we end at the payment vector $p^* = (17,15)$. But if the seller had instead selected the reference vector $p^0 = (15,11)$, a one-for-one change, the resulting payment vector is $p^* = (18,14)$, a corresponding one-for-one change. This sensitivity puts a great deal of pressure on the auctioneer in the selection of the constant reference point, and opens the possibility of post-auction lawsuits if the criteria for the reference point selection cannot be adequately justified. A zero-reserve (or a bound-only reserve as will be discussed in §4) VCG-nearest rule, however, does not suffer from this sensitivity to auctioneer selection and may therefore be seen as a safer design choice on the part of the seller.

3.4. The Karush-Kuhn-Tucker Optimality Conditions

Employing a typical tool from the nonlinear programming toolkit (see for example Bazaraa, Sherali, and Shetty 1979) we derive the Karush-Kuhn-Tucker (KKT) conditions for the optimality of problem (QP)[2]. These conditions are necessary and sufficient, since the constraint defining functions are linear (hence quasiconvex) and the objective is convex, as long as the reference point $p^0 \leq p$ for all p in the core[3]. Letting $\tilde{A}$ be the submatrix of A consisting of the columns that are tight at p, then the KKT conditions indicate that p is an optimal solution to (QP) if and only if there exist a vector $z \geq 0$, a vector $w \geq 0$, and a scalar $v \geq 0$, such that:

$$p = p^0 + z\tilde{A}^T - v\vec{1} - wI_p \qquad \text{(KKT)}$$

where the matrix I_p contains a row of the identity matrix e_j for each bidder j who pays-as-bid at p.

Thus the final payment vector p^* can be decomposed as follows for each bidder:

$$p_j^* = p_j^0 + \sum_{a_C \in \tilde{A} | j \notin C} z_C - v - w_j$$

that is, each bidder j pays her VCG value, plus a penalty for any marginally unblocking (i.e., tight) coalition C that j does not belong to (and this penalty is equal for all bidders not in C) minus an offset term v that is equal across all bidders and serves to guarantee payment minimization, and minus a personal offset term w_j to guarantee individual rationality for a pay-as-bid bidder j. The equity of these z terms across bidders contributes to the "fairness" of this payment rule; payments are based on equal contributions to overcome a competitive challenge from other bidders, except where individual rationality constraints cap the contributions of a bidder, in which case the personalized offset w_j takes affect. If a non-MRC-reference rule is employed, then the v terms disappear from this decomposition, as the relevant constraint disappears from the derivation of these KKT conditions. If additionally individual rationality constraints are not binding, the simplest payment decomposition emerges; in that case winners only pay equal

---

[2] Of course, we multiply the objective by $\frac{1}{2}$ prior to taking a gradient, as is typical in quadratic optimization.
[3] Note that this is always true for the VCG-nearest rule, while other selections of p^0 would require additional arguments to justisfy p^0 as a lower bound on payments.

penalties for a coalition C they do not belong to, z_C, with no universal adjustment v, and no personalized adjustment w_j.

If one were to consider using the same feasible region but a different strictly-convex objective function, the KKT derivation changes only in the objective gradient terms. So strictly speaking, with a new objective function $f(p - p^0)$, rather than a linear decomposition of $p - p^0$, we get a linear decomposition $\nabla f(p - p^0) = z\tilde{A}^T - wI_p - v\mathbf{1}$. So if we instead minimized $\sum_{j \in W}(p_j - p_j^0)^4$, we would have for each bidder a linear decomposition of $(p_j^* - p_j^0)^3$, seeming to only add confusion to the breakdown of payments, and further motivating the quadratic objective as the most simple convex objective function to interpret.

It is also worth noting that the optimal solution p^* to (QP) is unique since we are minimizing an L_2-distance to a fixed point over the convex set of payment minimizing core points (and if there were multiple optima, the triangle inequality would verify that a convex combination of these "optima" had a lower objective value, a contradiction). But the vector (v, w, z) on the other hand, which decomposes these payments may not be unique. We elaborate upon this phenomenon in appendix A.4. Also, the KKT conditions allow us to identify a quick solution technique for solving (QP) at intermediate stages of the constraint generation algorithm, which we also elaborate upon in appendix A.4.

4. Seller Reserves

Here we consider the subtleties of applying the quadratic core-selection approach when the seller has a non-zero reserve value for some or all of the items being sold. For example, in many single-item auction environments, reserve prices from the seller are easily modeled by simply having the seller submit a "dummy" bid equal to the reserve amount on the item; if the seller wins, then the item is kept. Here we note that this approach may be misapplied in the context of a core-selecting auction if the proper care is not taken.

Let us consider a seller who has an additive reserve-value r_i for each item i in the auction, or collectively, the seller has a reserve vector r, which will be treated additively. That is, the seller's net payoff is given by $\pi_0 = \sum_{j \in W} p_j - r \cdot x$, where x is the characteristic vector of the items sold. Letting $r_S = \sum_{i \in S} r_i$ for any $S \subseteq M$, the social welfare maximizing objective of the winner determination problem then becomes $wd(N) = \max \sum_{j \in N} \sum_{S \subseteq M} [b_j(S) \cdot x_j(S)] - r \cdot x$ which is in turn equivalent to $\max \sum_{j \in N} \sum_{S \subseteq M} [b_j(S) - r_S] \cdot x_j(S)$. Thus as is standard when considering VCG mechanisms with reserves (see, for example, Ausubel and Cramton 2003), the auction outcome can be computed by first reducing each package bid by the total reserve amount for the package, then proceeding as if the seller had zero reserves, and finally adding the bundle reserve back in to determine final payments.

Consider[4] the following simple two-bidder, two item example:

Example[4] 4: $b_1(A) = 40, b_2(AB)$

$= 40$, and the seller has a reserve value of 10 for each item.

---

[4] We thank Larry Ausubel for helping to devise this simple example.

Efficiency demands that item A be sold to bidder 1, while B remains unsold. With the reduction of bids approach, we may begin by reducing each bid by the total reserve amount for the package, resulting in reduced bids, $b_1^r(A) = 30$ and $b_2^r(AB) = 20$. We then compute a VCG payment of 20 for bidder 1, and applying CCG, we note that bidder 2 does not block this VCG outcome. These reserve amounts must be added back into any final payment to produce the actual final payment for bidder 1, $p_1 = 20 + 10 = 30$. (Notice that if $p_1 = 20$, the seller and bidder 2 would prefer an exchange of both items for any price in the open interval (30,40). Also note that if the seller instead had a zero reserve value, either bid could be accepted as an efficient solution, and in either case the winner would pay 40.)

This procedure may seem to contain a redundancy, given that we first reduce each package bid by the reserve amount for the package, and then add this package reserve amount back into any final payment. Given that seller reserves are adequately modelled using "dummy bids" in the context of single-item auctions, for example, one might be tempted to try the following approach: leave bids in their "unreduced" form, treating the seller as if his reserve value was zero, and insert a "dummy bidder" or "reserve bidder" to represent the interests of the seller, bidding the reserve amount for each item. In this example, we would add bids $b_3(A) = 10$ and $b_4(B) = 10$. This approach does necessarily lead to the correct determination of the efficient solution; the bids made by 1 and 4 win, with a winning reserve bidder indicating that item B stays with the seller. But a naïve application of the CCG algorithm proceeds as follows: following determination of the efficient solution, VCG payments are computed for bidder 1 and (dummy) bidder 4 as $p_1 = 30$, $p_4 = 0$. But this set of payments is blocked by bidder 2, who would be willing to pay up to 40 to take both items away from the winners. Minimizing the distance to the VCG point (for example) after applying the relevant core constraint payments are adjusted to $p_1 = 35$, $p_4 = 5$, which is in the core with respect to these four bids, given that the dummy bidders are treated just as any other bidder.

But this treatment of the dummy bidders just as any other bidder is at the heart of the problem, and as we can see this misapplication causes bidder 1 to pay more. Though the determination of the efficient solution proceeds correctly when the seller's reserve amount is replaced with a seller-dummy in the objective of the WDP, the surplus reduction step in the generation of core constraints proceeds incorrectly; it fails to consider that the seller loses the value of item B if it were to be reallocated to bidder 2, thus overstating the seller's willingness to form a blocking coalition with bidder 2.

But the use of seller-dummy bidders is intuitively appealing; the seller wants to leave open the possibility of buying back some of its own property if competition is too low, and wishes the competition for its own property to be reflected in both the determination of winners and payments. An easy fix is available that maintains this intuitively appealing use of dummy bidders to reflect reserves, however, and it is this variation which was used in the rules published as part of the December 2008 FAA slot-auction bidder seminar. Those rules included the following (paraphrased) treatment of seller reserves:

- The seller will specify a reserve amount for each item, r_i, stipulating that any bid $b_j(S)$ must not be less than r_S, and that any payment made by any bidder for package S must not be less than r_S.

- The seller will introduce into the auction, for every item i, a reserve bidder bidding the amount r_i for the package $\{i\}$.

The main discrepancy here is that any payment (including the "payments" made by reserve bidders) must be greater than or equal to the package reserve amount. This forces the "payment" within the algorithm for any reserve bidder to be exactly equal to the reserve amount r_i, which in turn forces the seller to be fully compensated by a potential blocking coalition if it involves a reserve bidder. For example, applied to the previously considered scenario, we would still have $p_1 = 30$, but with a VCG payment for reserve bidder 4 under the reserve amount 10, we are forced to set the initial payment $p_4 = 10$. The bid by bidder 2 is no longer blocking, as she cannot overcome the additional 10 units which must be compensated to the seller to obtain item B, which appear as payments from the fictional bidder 4. Though logically equivalent, the two rules above were deemed more appealing than the following possibility based on the "reduced bid," which may be thought of as "moving the origin" according to the reserve vector, and then "moving it back" at the end of the auction:

- The seller will specify a reserve amount for each item, r_i, stipulating that each bid $b_j(S)$ be reduced by the amount r_S prior to the winner determination and CCG implementation.
- Each bidder j winning package S_j will pay the amount p_j determined using reduced bids, plus the base reserve amount r_S.

Relative to the equivalent former pair, this latter pair of rules seems a bit more confusing for participating bidders, as it introduces a second reduction of bids that is different from the "surplus-based" reduction of bids that occurs in the CCG computation of a core outcome. It may also be unclear at first glance that the "movement of the origin" in bid-space prior to running the algorithm, followed by an equivalent "move-back" in payment-space at the end of the auction is a non-trivial operation.

Next, we note a different reserve-setting procedure, appropriate for a seller who has zero value for keeping any item, but wishes to set reserve payments in order to ensure adequate compensation is received when there is a lack of competition on a particular item or bundle. (For example, a spectrum authority may not have any value for holding a spectrum license unsold, but also would like to charge a nominal fee.) This procedure is implemented by simply setting a lower bound r_S on any bundle S, without the insertion of dummy bidders. In this case we see that if we compared a situation where an item i went unsold and bidder j is awarded set S_j with bid amount $b_j(S_j) = a$, to an alternative solution in which bidder j is awarded set $S_j \cup \{i\}$ with a bid of $b_j(S_j \cup \{i\}) = a$, there would be no change in social welfare, consistent with a seller who literally has zero value for keeping item i.

Next we note via example, a peculiar finding regarding the use of dummy "reserve bidders" (as previously outlined) and the more simple use of "bounds-only" reserves on any bundle payment (as in the previous paragraph). Consider the following:

Example 5:

	Bids		Reserve Bidders	Bounds Only
	$b_1(AB) = 100$	$b_2(CD) = 100$	$p_1 = 55$	$p_1 = 45$
	$b_3(BC) = 90$	seller reserve for each item, $r_i = 10$	$p_2 = 55$	$p_2 = 45$

If the seller employs reserve bidders (and bounds as outlined above) or equivalently runs the auction after reducing bids by reserve amounts and adds the reserve amounts back into final payments, the result is higher payments than if the seller simply set a lower bound on any bundle payment of $10|S|$. This result seems intuitive; if the seller is aggressively bidding for items, this may drive up the prices on those items. But the result is reversed in the following example:

Example 6:

	Bids		Reserve Bidders	Bounds Only
	$b_1(A) = 100$	$b_2(B) = 100$	$p_1 = 35$	$p_1 = 45$
	$b_3(ABCD) = 90$	seller reserve for each item, $r_i = 10$	$p_2 = 35$	$p_2 = 45$

In both Examples 5 and 6, the reserve bounds are always loose for the actual winners. For Example 5, we see that the dummy bidders participate in coalition formation, but are not part of the efficient solution; losing bidder 3 forces the winning bidders to pay at least 90 on her own, but at least 110 with the help of reserve bidders who are willing to buy back the items A and D at 10 each.

In Example 6 we see the reverse phenomenon; the active reserve bidders participate in the efficient solution, but not in coalition formation for price setting. Without reserve bidders, the two winners need to raise 90 units of revenue to ensure that bidder 3 is not blocking. When the reserve bidders are present, however, the winning bidders "get help" from the reserve bidders, who contribute 20 to buy back the items C and D, and thus 20 less is collected from the actual winning bidders. Note that depending on the actual utility function of the seller, both outcomes are logically consistent. If the seller actually perceives a loss of 10 units of value for giving away an item, then for Example 6, the seller perceives 90 units of utility from her revenue combined with her value for keeping items C and D. If the seller instead had no value for the items she keeps, then she gets 90 units of utility from revenue only.

Thus these two approaches to reserve setting are highly dependent on the utility of the seller for keeping items, and the revenue implications of choosing one method over the other are not always clear. So for a telecommunications authority like Ofcom, we recommended "bounds only" reserve setting, as they did favor allocating as much spectrum as possible. But for the FAA we recommended "reserve bidders" as the potential reduction of delay from unallocated slots had value according to the FAA's objectives, who planned to (potentially) retire slots that were not sold at auction.

We note finally that the reserve bidder technique, since equivalent to a reduction of each bundle bid by the bundle reserve amount, does indeed represent a "shift of the origin" in payment space, and thus like a constant-p^0-reference results in payments that are highly sensitive to changes in the reserve structure set by the auctioneer before the auction. For example, a one-for-one shift of the reserve amounts so that $r_A = 11$ and $r_B = 9$ in Examples 5 and 6, results in the identical shift in payments by the bidders when using a reserve-bidder scheme. Final payments become (56,54) and (36,34) in Examples 5 and 6, respectively. But with a bounds-only technique payments are relatively insensitive to changes in the seller-specified reserve amount. In general, unless a bidder is paying an amount exactly equal to the seller specified bundle reserve amount, the choices made by the auctioneer regarding bundle reservation value are inconsequential under a bounds-only method, while the previous argument showed that this is not true of a reserve-bidder approach.

5. Applications

The practical applicability of the techniques proposed here is limited mainly by our computational ability to solve larger and more complex winner determination problems. For assurance of a timely auction in real-life, we would usually like to guarantee worst-case run times for any winner determination problem within a few minutes or hours. In practice, computational run time was not an issue at all during our testing for Ofcom, in which we reviewed several hundred test cases in anticipation of the U. K.'s three spectrum auctions. In all testing, we implemented the algorithm described in §3.2 using CPLEX 11.1, and test cases were run in parallel by consulting company dotEcon and by associates at the Smith Institute for Industrial Mathematics and System Engineering. Run times were consistently under 20 minutes for even the worst cases, with the median cases taking a few seconds or less to solve. For the two real-world auctions, run times for winner and payment determination were around 1 second.

A few other practical points regarding the applications of the techniques proposed here are outlined in appendix A.5. One minor point of caution did arise in our testing which we will mention here, however. In the assignment stage of the 10–40 GHz or of the 2.6 GHz auctions, Ofcom wanted to assure that any unsold spectrum was kept as one contiguous block, so that it could be readily used or resold at a later date. This condition was solved easily enough with an IP formulation by having an appropriately sized space-holding bid for unsold blocks, and having the IP determine a partition of the spectra within a category, since the quantities of lots had already been determined in the principal stage. But care must be taken with a partitioning formulation when computing either the VCG payments or the core payments.

Specifically, if we tried to find a VCG payment by *removing* a bidder, the partitioning IP became infeasible. The simple solution is to not remove a bidder, but instead lower all of his bids to zero for his VCG computation. Similarly, when attempting to separate a violated core constraint, we reduce each bid by the bidder's current surplus. When solving the ensuing partitioning formulation IP, we noticed that it was necessary to replace any negative bid with zero, or else certain blocking coalitions would be ignored. Further, the algorithm as proposed in §3.2 said that a constraint was generated with a 1 for every bidder not receiving items in the separation IP solution, but *every* bidder *must* receive items in the assignment stage partitioning formulation. Again, the simple fix is to place a 1 into the constraint for any bidder who is forced to take a zero-valued bundle (under the surplus adjusted bids). These are fairly straightforward modifications of the core-selection algorithm, but we include these facts for completeness, that a slightly different procedure must be taken in winner-determination problems for which each bidder *must* receive items.

6. Computational Experiments

In this section we describe the results of a set of computational experiments performed using data generated by the Combinatorial Auction Test Suite (CATS) as introduced by Leyton-Brown et al. (2000). The CATS software simulates bidding behavior in a number of realistic economic environments, for example, when bidders are interested only in bundles of contiguous geographic regions in a spectrum license auction, or in bundles that form a path in a shipping-lane auction, etc. We used the same instances used by Day and Raghavan (2007) (which are available at http://users.business.uconn.edu/bday/CATS-CCG.zip) restricting to the auctions for 16, 32, and 64 items. Among these instances with three different sizes for the number of auction items, we allowed the CATS number-of-bids parameter to vary among the values {10, 25, 50, 100, 250, 1000}, and replicated each of these parameter values 50 times, for a total of 1050 randomly generated auction instances.

All instances were run using CPLEX 11.1 on a Windows Vista, AMD Turion 64 2 GHz processor with 2GB RAM. Relative to the earlier computations performed by Day and Raghavan (2007) on these instances, all worst-case and average run times (with one exception) actually decreased; the computational gains from an upgrade to CPLEX 11.1 from CPLEX 9.0 more than outweighed the increased burden of solving quadratic programs to select among MRC-points, and the increased burden of the randomized tie-breaking rule put in place for the FAA slot auction, which itself constituted an additional winner-determination run. The only exception was in the worst-case run time for 64 items and 1000 bids, which increased from 3703 to 7583 seconds, but this entire increase could be accounted for by the additional winner-determination needed to implement the randomized tie-breaking rule (6472 seconds was spent in the winner-determination phase, which included both runs). When this single instance is removed, all worst-case run times are an improvement over the Day and Raghavan (2007) results. All 16 item auctions concluded in under 2.5 seconds, all 32 item auctions in under 30 seconds, and all 64 item auctions in under 2.5 hours (under 49 minutes when the single worst case was removed). The average performance for the largest 1000 bid cases were 0.96, 5.6, and 611 seconds, for 16, 32, and 64 items, respectively, indicating that this class of algorithms does indeed perform in a comfortable time-scale for these auction sizes.

Also as in Day and Raghavan (2007), we found that a large minority (roughly 42%) of these CATS instances result in VCG outcomes when a VCG-nearest rule is applied. Considering all instances in this study, the VCG outcome delivers an average of 42% surplus to the bidders, while the VCG-nearest MRC rule delivers 33% surplus, leaving about 9% of value as a potential benefit from unilateral misrepresentation of preferences, since the VCG payment gives the maximum amount of benefit available from a unilateral deviation. Restricting only to instances in which the VCG outcome differs from an MRC outcome (about 58% of instances), the average bidder surplus becomes 33% for VCG, 18% for the VCG-nearest MRC rule, leaving a maximum of about 15% of value available through unilateral misrepresentation. Thus the majority of the possible benefits of bid shading are removed by this quadratic core-selecting rule, relative to a pay-as-bid rule. These results are summarized in Table 1. Also,

Table 1. *Average bidder surplus as a % of value*

Method	All instances	MRC ≠ VCG instances
VCG	42%	33%
MRC	33%	18%

though 15% of value may seem substantial, one should remember that this measures the maximum possible gain from deviation, assuming that the bidder knows to shade by this amount and not more, and that opponents do not shade their bids as well.

6.1. VCG-Nearest vs. Zero-Nearest

With Example 3 on page 208, we showed that the use of a zero-nearest reference rule (i.e., when $p^0 = \vec{0}$) can result in a high-valued winner shouldering little if any of the monetary burden of overcoming a blocking coalition. Here we show that this phenomenon is not peculiar to a carefully constructed example, but instead that it occurs frequently when using a random data-set, in this case the CATS data. Towards this end, we duplicated the runs described above using a zero-nearest MRC reference rule, rather than a VCG-nearest MRC rule. Then, we looked at the difference between the final MRC payment and the VCG payment for each bidder. For any instance in which this amount was positive, and for which there were at least two winners, we isolated the highest-valued and lowest-valued winning bidder, and measured the percentage of increase from the VCG total that was paid for by each of these two bidders. That is, we computed:

$$\frac{p_{\bar{j}}^* - p_{\bar{j}}^{VCG}}{\sum_{j \in N}(p_j^* - p_j^{VCG})} \tag{15}$$

where $\bar{j}$ was the index of the highest-valued winning bidder, and the lowest-valued winning bidder, respectively. Using this measure, we confirmed that the intuition shown by Example 3 did indeed persist. (In that example, this statistic took the values 0 and 1, for the highest- and lowest-valued winners, respectively, under the zero-nearest rule, and (0.5, 0.5) under the VCG-nearest rule.) For the VCG-nearest computations, this statistic (15) averaged roughly 20% for the highest-valued winner, while the zero-nearest computations resulted in a value of about 6% for the highest-valued winner; the use of a zero-nearest rule results in high-valued winners shouldering less of the burden of overcoming blocking coalitions when the VCG outcome is not in the core. Similarly, we found that the lowest-valued winner paid about 7% of the burden under a VCG-nearest rule, while they paid 12% under the zero-nearest implementation. Further, the extreme behavior of Example 3, in which the higher-valued winner paid none of the burden of overcoming a blocking coalition (beyond the VCG payment) was also observed in the CATS data. This phenomenon, in which the statistic (15) equalled zero for the highest-valued winning bidder, occurred in only 8% of the relevant instances under a VCG-nearest rule, but over 32% of the instances using a zero-nearest rule showed this extreme lopsided-ness. Figure 2 visually indicates the overall disproportionate burden placed on the lowest valued winner relative to the highest valued winner

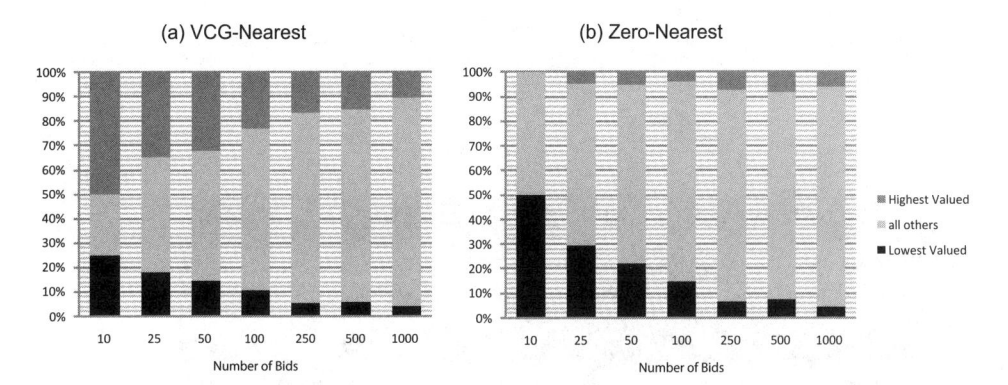

Figure 2. When total payments are more than the total VCG payments, these graphs show the proportion of the difference paid (on average) by the winner with the highest valued bundle, the lowest valued bundle and all other winners, under (a) the VCG-Nearest rule, and (b) the Zero-Nearest rule, as a function of the number of bids parameter.

under a zero-nearest rule. Clearly from viewing these graphs, this disparity between the two approaches is most pronounced when the number of winners is small. (Using the CATS data, the presence of more bids make it more likely to have more winners.)

6.2. Reserve Bidders vs. Bounds-Only Reserves

To observe the effect of the two seller-reserve formats discussed here, we ran each CATS auction instance using a VCG-nearest MRC reference rule, once using the bounds-only approach, and once using the reserve-bidders approach. Though Examples 5 and 6 demonstrated that drastically different outcomes could occur in some specially constructed examples, and that neither was a universally better approach, we wanted to demonstrate that the discrepancies between the two approaches persisted in the robust environment provided by the CATS data. Since the CATS data did not generate a seller-reserve value for each item, we arbitrarily asserted a reserve value of $r_i = 20$ monetary units for half of the items in each auction instance, and reserve value $r_i = 40$ for the other half of the items. After this we pruned out any bids that did not meet the implied (additive) bundle reserve value, and removed any auction instances for which less than two bids remained, leaving us with 706 or roughly 67% of the original auction instances for further investigation.

As expected, a bounds-only approach tended to cause an increase in the number of units sold, but only a modest increase; for this data less than 66% of items were sold if reserve bidders were used and just above 70% were sold under the bounds-only approach. If considering a seller who has no value for keeping an item, and thus whose utility is specified exactly by the amount of revenue generated, we found that about 56% of the instances experienced an increase in revenue under the bounds-only approach, though the overall performance showed a small (1%) average *reduction* in revenue with the bounds-only approach. In the extreme cases, the largest positive effect of switching to bounds-only was a 78% increase in revenue, while the largest negative effect was a nearly 40% reduction in revenue with bounds-only. But mostly, the revenue effects were noticeable but modest, averaging a 6% absolute deviation

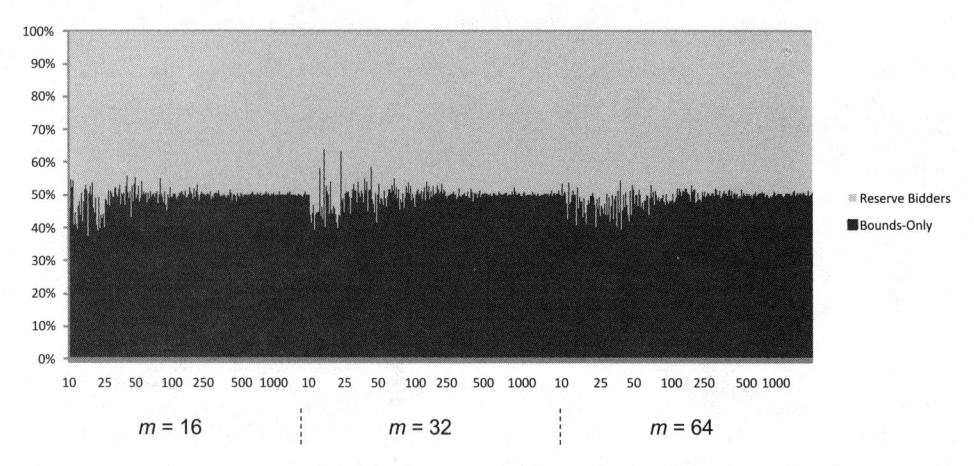

Figure 3. Revenue comparison of the Reserve Bidder and Bounds-Only approaches over all relevant instances.

between the two reserve formats. These results indicate that in specific circumstances, the effect of reserve-format selection may be substantial in either direction, but that these large discrepancies are not typical (79% of the instance has an absolute revenue change of 10% or less from switching approaches). Also as may be expected, the revenue effect of the reserve-format selection had less of an impact on auctions with a larger number of bids. When auctions with fewer than 250 bids were removed from our analysis, 99.6% had a revenue change of less than 10% when switching approaches, and here we saw a 1% average *increase* in revenue over all instances when switching to bounds-only. Also, when these auctions with a small number of bids were removed, the extreme cases were much less extreme; revenue decreased by no more 7%, and increased by no more than 11% when switching to bounds-only.

The full set of total revenue comparisons is given in Figure 3, where each vertical bar represents one of the 706 instances, grouped by the number of items in the auction (16, 32, or 64) and ordered within a group by the number of bids as indicated on the horizontal axis. Values on the vertical axis indicate total auction revenue as a percentage of the revenue sum of the two methods for that instance; thus 50% indicates equal revenue across the two methods, 66.6̄% indicates that one method had twice the revenue of the other, etc. This picture indicates that the direct revenue comparison is indeed unclear, but seems visually to slightly favor the reserve-bidder approach if just considering revenue, consistent with the overall 1% revenue disadvantage of bounds-only indicated above. Also, we see that regardless of the number of items being sold, the discrepancy between the two approaches becomes less pronounced with more bids in the auction, indicated by the convergence to the 50% line as we move from left to right in each item-quantity group.

When we consider a seller who *does* value keeping unsold items, the seller's net utility from the auction is then the value of the revenue received minus the value of the items sold. In this case our results more clearly favor a reserve-bidder format for this type of seller, though again, not in all instances. In about 7% of these instances (including auctions with both many and few bids) a seller would relinquish 100% of her utility if choosing a bounds-only approach over a reserve-bidder approach. Though a single extreme instance did show a nearly 105% increase in seller utility under bounds-only, 71% of all instances showed a reduction of seller utility when switching to bounds-only. Further, of the roughly 29% showing an increase in seller utility, 95% showed only a modest increase of less than 15%. On average, switching to a bounds-only approach resulted in a 15% loss of seller utility. This average effect was much less for auctions with less than 250 bids, only a 1.5% reduction in average seller-utility when switching to bounds-only. Further, the scope for large increases in seller utility in extreme cases nearly disappeared when looking only at these many-bid instances; the biggest increase when switching to bounds-only then became about 9%.

7. Conclusions

We presented a general algorithm for selecting among core outcomes for use in any combinatorial auction, and described the many beneficial properties of the approach. We motivated the idea that quadratic core-pricing is simple to understand (via Example 1), but that it is also general enough and extensible enough to handle the full complexity of any combinatorial auction problem. For instances in which the winner-determination problem can be solved in a reasonable time, then "fair" payments can also be computed in a reasonable time. This development represents an important milestone in Operations Research, in which a computational, algorithmic development opens the door to efficient solutions for a wide class of economic resource-allocation problems.

The prices we generate represent a natural generalization of the second-price paradigm from single-item auctions. Among core-selecting mechanisms, the pricing rule minimizes the incentives for bidders as a whole to misreport their true values for packages. Using standard nonlinear programming tools, we demonstrated how the underlying mathematics induces an equitable decomposition of payments, so that different bidders each contribute an equal amount for any payment goal that they must achieve as a group. We also showed the computational simplicity of the specific quadratic programming problem we encounter, and demonstrated several different interpretations and formulations of the core, each one providing a different perspective to help motivate the core property as a natural requirement in combinatorial auctions.

Our computational experiments lent further support for the practical viability of these computational techniques, using a standard benchmark from the CA literature. These experiments also helped to demonstrate some of the seeming distortions that arise when a zero-nearest version of the algorithm is implemented, in which the proportion of the burden needed to overcome a blocking coalition is unevenly spread across winners. Since any constant reference rule is equivalent to a zero-nearest reference rule with the corresponding change of coordinates, this motivates our slight-preference for VCG-nearest rules in practice. As the concurrent work of Erdil and Klemperer (2009) begins to hint though, there may be alternative dynamic selection rules, which are, like the VCG-nearest rule, not skewed by pre-auction parameter settings made by the auctioneer, but this remains the subject of future study. Also, our computations indicated that the selection of a reserve scheme may be less-trivial than it first appears, and this as well remains a interesting avenue of future research.

As the benefits of this class of quadratic payment determination algorithms become more well-known, we expect further applications to emerge. Based on preliminary presentations of this research and the early successes of the auctions held in the U. K., the FAA adopted this pricing rule for the auction of landing-slot rights in the three New York City airports. In order to minimize disruption to the status quo, the plan was to auction only a small portion of time-slots at the three airports. As a result, the associated winner-determination problems solve easily using standard, off-the-shelf-software such as CPLEX. The greatest challenge for the landing-slot application, however, has proven to be the politics of auctioning, not any difficulties in computing winners or payments.

Appendix to "Quadratic Core-Selecting Payment Rules for Combinatorial Auctions," by Day and Cramton

A.1. Semi-sincere Strategies Eliminate Incremental Envy

To further motivate why a bidder would choose to implement a semi-sincere strategy (since there may exist other best-response strategies) we note the following:

Definition 7.1. Given an auction outcome, a bidder is *incrementally envy-free* if she would not prefer to additionally buy any collection of her opponents' awarded bundles at the respective prices they pay.

Lemma 7.2. *If a winning bidder employs a semi-sincere strategy in any core-selecting auction, then she is incrementally envy-free.*

Proof. Without loss of generality, suppose that each bidder j wins bundle S_j, and that bidder 1 incrementally-envies the bundles of bidders 2, 3, ...k. That is, bidder 1 who is awarded the bundle S_1 and pays p_1, would prefer to receive the bundle $S_1 \cup S_2 \cup \cdots S_k$ and pay $p_1 + p_2 + \cdots p_k$. Letting $v_j(S)$ represent bidder j's value for a bundle S, this means that:

$$v_1(S_1 \cup S_2 \cup \cdots S_k) - \sum_{j=1}^{k} p_j > v_1(S_1) - p_1 \qquad (16)$$

$$b_1(S_1 \cup S_2 \cup \cdots S_k) - b_1(S_1) > \sum_{j=2}^{k} p_j \qquad (17)$$

with (17) following from (16) by the definition of the semi-sincere strategy. But employing a constraint of the form (6) as defined on page 205, where W represents the set of all bidders receiving non-empty bundles, we have:

$$\sum_{j=2}^{k} p_j \geq wd(N\setminus\{2,3,\ldots k\}) - \sum_{j \in W\setminus\{2,3,\ldots k\}} b_j(S_j) \qquad (18)$$

$$b_1(S_1 \cup S_2 \cup \cdots S_k) - b_1(S_1) > wd(N\setminus\{2,3,\ldots k\}) - \sum_{j \in W\setminus\{2,3,\ldots k\}} b_j(S_j) \qquad (19)$$

$$b_1(S_1 \cup S_2 \cup \cdots S_k) + \sum_{j \in W\setminus\{1,2,3,\ldots k\}} b_j(S_j) > wd(N\setminus\{2,3,\ldots k\}) \qquad (20)$$

where (18) is simply the core constraint of the form (6); (19) follows transitively from (17), and (20) following after canceling terms. But (20) is a contradiction; a collection of disjoint bundle bids made by a subset of the players in $N\setminus\{2, 3, \ldots k\}$ is greater than $wd(N\setminus\{2, 3, \ldots k\})$, which is defined as maximal over such combinations of bids. □

This lemma motivates not only that a bidder can find a utility-maximizing strategy among her possible semi-sincere strategies, but also that she will not envy any collection of her opponents' bundles, a property not captured in her quasi-linear utility function. We also note that the existence of a best reply in semi-sincere strategies does not imply incremental-envy-freeness, as the VCG auction always has a best reply in semi-sincere strategies (with $\alpha_j = 0$ for all bidders) but may not satisfy this property when outcomes are not in the core.

A.2. Efficacy of Core Constraint Generation

Next we provide a few details that are of theoretical interest, concerning the efficacy of this CCG approach. One natural question to ask is whether the constraints generated ever end up to be redundant, or loose constraints in the final solution. For example, is it ever the case that at iteration t we find violated constraint $pa_{C_t} \geq \beta_{C_t}$, but later find, at termination, that $p^*a_{C_t} > \beta_{C_t}$? In our initial experiments, under a mild additional condition of minimality of a_{C_t} at each iteration, the answer seemed to be no, but the subsequent work of Lubin et al. (2015) revealed that generated constraints could in fact become loose. Still the minimality condition may be of some interest, given that it helps avoid redunancy in the constraint generation process.

To illustrate this minimality condition consider the following example, which is a slight modification of the main example from Day and Raghavan (2007), under a VCG-nearest MRC-reference rule. Here, $m = 3$, the item-set is $\{A, B, C\}$, and $n = 9$ (single-minded) bidders have the following bids:

$$\text{Example 2:} \quad \begin{array}{lll} b_1(A) = 20 & b_2(B) = 20 & b_3(C) = 20 \\ b_4(AB) = 28 & b_5(AC) = 26 & b_6(BC) = 23 \\ b_7(A) = 10 & b_8(B) = 10 & b_9(C) = 10 \end{array}$$

The unique efficient allocation is to accept the single bids of bidders 1, 2, and 3, with $p^0 = (10, 10, 10)$. Reducing each bid by current surplus, we find two candidates for C_1; both $\{4, 7\}$ and $\{3, 4\}$ would be willing to offer the seller 38 at the current prices, but the first selection results in $a_{C_1} = (1, 1, 1)^T$, $\beta_{C_1} = 38$, while the other results in $a_{C_1} = (1, 1, 0)^T$, $\beta_{C_1} = 28$. If we follow the algorithm through to termination using either selection, we do always arrive at the (unique) solution $p^* = (15.5, 12.5, 10.5)$. But if we generate the first of these possible constraints $p_1 + p_2 + p_3 \geq 38$, we see that in the final solution this constraint becomes loose: $p_1^* + p_2^* + p_3^* = 38.5 > 38$. The latter selection for C_1, on the other hand, has $(1, 1, 0)^T < (1, 1, 1)^T$, thus satisfying minimality among maximally violated coalitions at p^0, and indeed remains tight at termination; $p_1^* + p_2^* = 28$. Also, one may note for this example that the use of a VCG-nearest reference rule without explicit payment minimization results in payments $(14.\bar{6}, 13.\bar{3}, 11.\bar{3})$, which has a greater sum $39.\bar{3} > 38.5$, verifying that the *explicit* minimization of payments is necessary, if desired, and does not follow in general from the minimization of distance to a reference point, even though Example 1 and other small examples seem to suggest the reverse.

A.3. Proof of Theorem 1

Proof. Suppose for example (WOLOG) that winning bidder 1 uniformly increases her semi-sincere bids, i.e., if she originally submitted $b_1(S) = \max(0, u_1(S) - \alpha_1)$ for all bundles S, now she submits $\bar{b}_1(S) = \max(0, u_j(S) - \alpha_j + \delta)$. First, we note that such an increase could not change the efficient[5] solution, as the efficiency of any allocation involving bidder 1 increases by the same amount, and those not involving her do not change.

Now consider the various cases for each coalitional constraint in the core-selection program, $\sum_{j \in W \setminus C} p_j \geq wd(C) - \sum_{j \in C} b_j(S_j)$. If Bidder 1 does not belong to coalition C, clearly this constraint remains unchanged. Further, if bidder 1 does belong to C, and bidder 1 would receive items in the winner-determination problem solved over C, then the terms $wd(C)$ and $b_1(S_1)$ would both increase by δ, resulting in a cancellation, and the constraint again remains unchanged.

Finally, if bidder 1 belongs to C but does not receive items under $wd(C)$, the C-indexed constraint will be relaxed by δ when bidder 1 switches from bids b to $\bar{b}$, but this relaxation is inconsequential since the constraint must already be loose at optimality, and the relaxation of loose constraints will not affect the optimization. To see that the C-indexed constraint is indeed loose at p^* in this case, note that since bidder 1 receives no goods in the solution to $wd(C)$, we have $wd(C) = wd(C \setminus \{1\})$. By the feasibility of p^* we have $\sum_{j \in W \setminus (C \setminus \{1\})} p_j^* \geq wd(C \setminus \{1\}) - \sum_{j \in (C \setminus \{1\})} b_j(S_j)$, and then subtracting $p_1^* < b_1(S_1)$ which holds by assumption, we recover $\sum_{j \in W \setminus C} p_j^* > wd(C) - \sum_{j \in C} b_j(S_j)$, as desired. Since individual rationality constraints were also loose at optimality, we have shown that all constraints defining the core either remain the same or are the relaxation of already loose constraints at optimality. Together with the assumption that the reference vector p^0 has not changed, this results in no change to p^* following the increase in semi-sincere strategy. □

A.4. Nonuniqueness of a Payment Decomposition and Solving (QP) Iteratively as Constraints are Generated

If the constraint qualification condition arises that the columns of $\tilde{A}$ together with the 1-vector are linearly independent, then the (v, w, z) combination (guaranteed to exist by KKT) is unique. But even in a simple example, such as Example 2, this constraint qualification may fail and the payment decomposition will not be unique.

For Example 2, we find that:

$$\tilde{A} = \begin{bmatrix} 1 & 1 & 0 \\ 1 & 0 & 1 \\ 0 & 1 & 1 \end{bmatrix}$$

---

[5] This assumes the use of a tie-breaking rule that is robust against such an increase. In the U. K. spectrum auctions, for example, each item in each bundle bid received a randomly generated number, and the sum became the objective function in a secondary WD optimization, constrained to be efficient.

with $p^0 = (10, 10, 10)$, $p^* = (15.5, 12.5, 10.5)$, and an infinite number of (v, z) satisfying (KKT), including, for example, $(v^1, z^1) = (-4.5, 6, 4, 1)$ and $(v^2, z^2) = (-3.5, 5.5, 3.5, 0.5)$. In this situation, one can linearly minimize the magnitude of v, in order to maximize the information in the z variables, in this case yielding $(v^3, z^3) = (-2.5, 5, 3, 0)$. In our experience, however, the existence of a unique decomposition seems to be the rule rather than the exception.

In order to solve problem (QP) iteratively as constraints are generated, let $\bar{A}_t = (1, A_t)$, and after separating a new constraint, specified by a_{C_t} and β_t, suppose we have solved the corresponding instance of (LP), determining μ_t. Previous payment vector p^{t-1} satisfies all known core constraints, except the new C_t constraint, and may also not have high enough total payments to satisfy $p1 = \mu_t$. Given a decomposition (v^{t-1}, z^{t-1}) of p^{t-1}, we may find the next decomposition (v^t, z^t) by finding a feasible solution to the following linear system:

$$(dv, dz)\bar{A}_t^T \bar{A}_t = (\delta_0, 0, \ldots 0, \delta_t)$$

$$dz \geq -z^{t-1}$$

and setting $(v^t, z^t) = (v^{t-1}, z^{t-1}, 0) + (dv, dz)$, where $\delta_0 = \mu_t - \mu_{t-1}$, and $\delta_t = \beta_t - p^{t-1}a_{C_t}$. This result holds because of the interesting combinatorial form of $\bar{A}_t^T \bar{A}_t$, whose entries can be characterized as follows. First, let $R_i = W \setminus C_i$ represent the set that is responsible for overcoming a blocking offer made by C_i, and whose characteristic vector is given by a_{C_i}. Then defining $R_0 = W$, and indexing the rows and columns of the symmetric matrix $\bar{A}_t^T \bar{A}_t$ starting at zero, it is easily shown that the row r, column c entry of $\bar{A}_t^T \bar{A}_t$ is given by $|R_r \cap R_c|$. Then if we assess the impact of changing (v, z) values by (dv, dz) we see that the changes add up appropriately so that the C_tth entry of $(dv, dz)\bar{A}_t^T \bar{A}_t$ gives the change in the total payments made by the group R_t, and thus the solution to this linear system provides an updated decomposition vector satisfying the KKT conditions. In practice, the constraint generation procedure typically only generates linearly independent core constraints, resulting in an invertible $\bar{A}_t^T \bar{A}_t$, in which case a unique solution to the linear system always exists. If it did happen that a linearly-dependent, loose constraint did arise (a solution will still exist if a lin. dep. constraint can remain tight), the addition of a single vector can cause at most one such dominated constraint, and we could, for example, remove constraints one at time until a solution exists, removing said dominated constraint from all remaining A_{t+i} matrices.

A.5. Practical Design Considerations for the U. K. Spectrum Auctions

In the 10–40 GHz auction, there were 27 lots auctioned, divided into 7 categories. Ofcom took an approach of first auctioning generic lots in a "principal stage," where each bid simply indicates quantities for each of the 7 categories. Each bidder then wins some quantity of generic (i.e., unspecified) licenses from each category, and we determined "base prices" for each of these bundles using the quadratic core-selection rule described here. This principal stage was then followed by an "assignment stage" in which a bidder who, for example, won 4 lots in Category A, could bid for *which* contiguous strip of 4 lots she actually wanted. Each category then had its own sealed-bid assignment stage auction with "top-up" bids, and the winners again paid the corresponding quadratic core prices, which were then listed as "additional prices" on top of their base price from the principal stage. As one would expect, these additional prices were small in comparison (sometimes zero) as there is little difference among the various specific assignments within a category.

A similar design is proposed for the upcoming 2.6 GHz auctions (in both the U. K. and the Netherlands) but rather than fixed categories, the bidders specify instead the quantities of "paired" lots (separated by a fixed bandwidth) and how many singleton or "unpaired" lots they are interested in. Further, they can specify different preferences for groups of unpaired licenses that are split among two zones separated by a block of paired licenses. Again, a principal auction is used to determine who gets how many of each type of license, and is followed by an assignment stage to determine exactly which licenses are awarded to which bidders. Though the winner-determination problem for this auction is not as straight forward as formulation (WD), as long as we have some algorithm for solving winner determination problems the quadratic core-payment mechanism can be implemented. Indeed, even with different "blackboxes" for winner determination (for example, we used CPLEX's branching-based IP algorithm while dotEcon employed a dynamic programming technique) the uniqueness of the core selection strategy we present here guarantees that any technique will eventually produce the same outcome, making the validity of the process easy to verify.

The design choice to separate the auction into a principal generic stage and a specific assignment stage is helpful to guarantee computational quickness in the winner-determination problems we solve over the course of the algorithm, but is probably more valuable to reduce complexity for the bidders by not forcing them bid separately on nearly indistinguishable items. In the U. K.'s L-Band auction, this was not an issue, as the items being offered were 17 unique lots. Here the computational challenge was a separation requirement between high power and lower power usages, so that a bidder was required to reveal the intended usage of a package of lots within her bid, and the auction had to be sure that lots with dissimilar usages had at least two unused "guard bands" between them. Once a suitably tight IP formulation was determined, however, we were quickly able to solve winner determination problems for this 17 item auction with as many as 50,000 bids in just a few minutes, even in the worst case.

References

Ausubel, L., and P. Cramton. 2003. Vickrey Auctions with Reserve Pricing. *Economic Theory*, 23: 493–505.

Ausubel, L., P. Cramton, and P. Milgrom. 2006. The clock-proxy auction: A practical combinatorial auction design. In P. Cramton, Y. Shoham, and R. Steinberg, editors, *Combinatorial Auctions*, chapter 5, 115–138. MIT Press.

Ausubel, L., and P. Milgrom. 2002. Ascending auctions with package bidding. *Frontiers of Theoretical Economics*, 1(1): 1–42.

Ausubel, L., and P. Milgrom. 2006. The lovely but lonely Vickrey Auction. In P. Cramton, Y. Shoham, and R. Steinberg, editors, *Combinatorial Auctions*, chapter 1, 17–40. MIT Press.

Bazaraa, M., H. Sherali, and C. Shetty. 1979. *Nonlinear Programming: Theory and Algorithms*. Wiley and Sons.

Bernheim, B., and M. Whinston. 1986. Menu Auctions, Resource Allocation and Economic Influence. *Quarterly Journal of Economics*, 101: 1–31.

Bichler, M., P. Shabalin and A. Pikovsky. 2008. Computational analysis of iterative combinatorial auctions with linear prices. *Information Systems Research.* Online publication 2008.

Cramton, P. 2009. Spectrum Auction Design. Working Paper, University of Maryland.

Cramton, P. 2006. Simultaneous Ascending Auctions. In P. Cramton, Y. Shoham, and R. Steinberg, editors, *Combinatorial Auctions*, chapter 4, 99–114. MIT Press.

Cramton, P., Y. Shoham, and R. Steinberg, editors. 2006. *Combinatorial Auctions*, MIT Press.

Day, R., and P. Milgrom. 2008. Core-Selecting Package Auctions. *International Journal of Game Theory*, 36(3): 393–407.

Day, R. and S. Raghavan. 2007. Fair payments for efficient allocations in public sector combinatorial auctions. *Management Science.* 53(9): 1389–1406.

deVries, S. and R. Vohra. 2003. Combinatorial Auctions: A Survey. *INFORMS J. of Computing*, 15(3): 284–309.

Erdil, A. and P. Klemperer. 2010. A New Payment Rule for Core-Selecting Package Auctions. *Journal of the European Economic Association*, to appear.

Goldberg, A. and J. Hartline. 2003. Envy-Free Auctions for Digital Goods. *ACM Conference on Electronic Commerce (EC '03)*, Association for Computing Machinery, Inc., San Diego, CA, June 2003.

Krishna, V. 2002. *Auction Theory*. Academic Press.

Lamy, L. 2009. Core-selecting package auctions: a comment on revenue-monotonicity, Working Paper, Paris-Jourdan Sciences Économiques (PSE).

Leyton-Brown, K., M. Pearson, Y. Shoham. 2000. Towards a universal test suite for combinatorial auction algorithms. EC '00: Proc. 2nd ACM Conf. Electronic Commerce. ACM Press, New York, 66–76.

Lubin, B., B. Bünz, and S. Seuken. 2015. New Core-Selecting Payment Rules with Better Fairness and Incentive Properties, Working Paper, Boston University.

Milgrom, P. 2004. *Putting Auction Theory to Work*. Cambridge University Press.

Myerson, Roger B. 1981. Optimal Auction Design. Mathematics of Operations Research, 6: 58–73.

Nisan, N. 2006. Bidding Languages in Combinatorial Auctions. In P. Cramton, Y. Shoham, and R. Steinberg, editors, *Combinatorial Auctions*, chapter 9, 215–232. MIT Press.

Othman, A. and Sandholm, T. 2010. Envy Quotes and the Iterated Core-Selecting Combinatorial Auction. *Proceedings of the National Conference on Artificial Intelligence* (AAAI).

Ott, M. 2009. Second-Price Proxy Auctions in Bidder-Seller Networks. Doctoral Dissertation, Karlsruhe Institute of Technology, Karlsruhe, Germany.

Parkes, D., J. Kalagnanam and M. Eso. 2001. Achieving Budget-Balance with Vickrey-Based Payment Schemes in Exchanges. In *Proc. 17th Int. Joint Conf. on Artificial Intelligence* (IJCAI'01) 1161–1168.

Porter, D., S. Rassenti, A. Roopnarine, and V. Smith. 2003. Combinatorial Auction Design. *Proceedings of the National Academy of Sciences*, 100: 11153–11157.

Rassenti, S., V. Smith, and R. Bulfin. 1982. A Combinatorial Auction Mechanism for Airport Time Slot Allocation. *The Bell Journal of Economics*, 13(2): 402–417.

Rothkopf, M. 2007. Thirteen reasons why the Vickrey-Clarke-Groves mechanism is not practical. *Operations Research.* 55(2): 191–197.

Rothkopf, M., A. Pekeč, and R. Harstad. 1998. Computationally manageable combinatorial auctions. *Management Science,* 44(1): 131–1147.

Rothkopf, M., T. Teisberg, and E. Kahn. 1990. Why are Vickrey auctions rare? *Journal of Political Economy*, 98: 94–109.

CHAPTER 11
Core-Selecting Package Auctions[*]

Robert Day and Paul R. Milgrom

1. Introduction

Recent years have seen several new and important applications of matching procedures in practical applications, including school assignments in New York and Boston and new designs for life-saving organ exchanges. The mechanisms that have been adopted, and sometimes even the runner-up mechanisms, are stable matching mechanisms. Recall that *stable matches* are matches with the property that no individual can do better by staying unmatched and no pair can both do better by matching to one another. Since pairs are the only significant coalitions in this theory, stable matches are a kind of core allocation. *Stable matching mechanisms* are direct mechanisms that select a stable match with respect to the reported preferences; the definition does not require that the mechanism be incentive-compatible.

Evidence suggesting that stable matching mechanisms remain in use long after unstable mechanisms have been abandoned is found both in empirical studies (Roth and Xing 1994) and in laboratory experiments (Kagel and Roth 2000). If stable mechanisms actually lead to stable matches, then these mechanisms have the important practical advantage that no couple that would prefer to renege after the mechanism is run in favor of some alternative pairing, because no such agreement can be better for both members of the couple than the outcome of a stable matching mechanism. Even for a stable mechanism, with enough uncertainty, there might be pairs that could increase their expected payoffs by matching in advance, but the resulting unstable match would be vulnerable to defections by parties who might find a better alternative.

---

[*] This paper evolved from Milgrom (2006), which reported a portion of Milgrom's Clarendon lectures for 2005. The authors subsequently discovered that versions of what is here Theorem 3 appeared both in that paper and one produced independently by Day and Raghavan (2006). We have collaborated on this revision; in particular, nearly all of the material on shill bidding is new.

 Milgrom received financial support for this research from National Science Foundation under grant ITR-0427770. We thank Roger Myerson for suggesting the connection to Howard Raiffa's observations about bargaining, Yeon-Koo Che for comments on an earlier draft, and Manuj Garg for proofreading.

A similar analysis applies to core-selecting auction mechanisms.[1] An individually rational outcome is in the core of an auction game if and only if there is no group of bidders who would strictly prefer an alternative deal that is also strictly better for seller. Consequently, an auction mechanism that delivers core allocations has the advantages that there is no individual or group that would want to renege *after* the auction is run in favor of some allocation that is feasible for it and any non-core agreement made before the auction is vulnerable to defections, as the seller attracts better offers afterwards.

For both matching and auction mechanisms, the preceding arguments have full force only if the procedures actually result in stable or core allocations, which in turn depends on the participants' strategies. Casual evidence suggests that participant behavior in real mechanisms varies widely from naïve to sophisticated, and the most sophisticated participants do not merely make reports in the mechanism. Instead, they also make decisions about whether to make pre-emptive offers before the auction, to enter the auction as a single bidder or as several, to stay out of the auction and try to deal with the winners afterwards, to buy with the intent to resell some or all of what is acquired, to renege on deals, or even to attempt to persuade the seller to alter the timing or rules of the mechanism itself. All of these elements can be important in some auction settings.

Despite the great variety of important constraints real auction settings, it is customary in mechanism design theory to impose incentive constraints first, investigating other aspects of performance only later. It is, of course, equally valid to begin with other constraints and such an approach can be useful. To the extent that optimization is only an approximation to the correct behavioral theory for bidders, it is interesting to investigate how closely incentive constraints can be approximated when other constraints are imposed first. For example, while it is known that there exists no strategy-proof two-sided stable matching mechanism[2] and that the only strategy-proof efficient auction mechanism, which is the (generalized) Vickrey auction,[3] suffers from several severe practical drawbacks, some of which are described below,[4] there has so far been little work on the nature of the trade-off. One can usefully ask: by how much do the incentives for truthful reporting fail when other design objectives are imposed as constraints?

The modern literature does include some attempts to account for multiple performance criteria even when incentives are less than perfect. Consider, for example, the basic two-sided matching problem, commonly called the *marriage problem*, in which men have preferences over women and women have preferences over men. The literature often treats stability of the outcome as the primary concern while still evaluating the incentive properties of the mechanism. In the marriage problem, there always exists a unique *man-optimal* match and a unique *woman-optimal match*.[5] The man-optimal

---

[1] The core is always non-empty in auction problems. Indeed, for any profile of reports, the allocation that assigns the items efficiently and charges each bidder the full amount of its bids selects a core allocation. This selection describes the "menu auction" analyzed by Bernheim and Whinston (1986). Other core-selecting auctions are described in Ausubel and Milgrom (2002) and Day and Raghavan (2006).

[2] Roth (1982).

[3] A result of Green and Laffont (1979), as extended by Holmstrom (1979), shows that for any path-connected set of valuations (for environments with quasi-linear preferences), the only strategy-proof direct auction mechanism that selects total-value-maximizing choices is the Vickrey mechanism.

[4] For a more thorough treatment, see Ausubel and Milgrom (2005).

[5] As Gale and Shapley first showed, there is a stable match that is Pareto preferred by all men to any other stable match, which they called the "man optimal" match.

mechanism, which is the direct mechanism that always selects the man-optimal match, is strategy-proof for men but not for women[6] and the reverse is true for the woman-optimal mechanism. Properties such as these are typically reported as advantages of the mechanism,[7] although these incentives fall short of full strategy-proofness. Even when strategy-proofness fails, finding profitable deviations may be so hard that many participants find it most attractive just to report truthfully. A claim of this sort has been made for the pre-1998 algorithm used by National Resident Matching Program, which was not strategy-proof for doctors, but for which few doctors could gain at all by misreporting and for which tactical misreporting was fraught with risks (Roth and Peranson 1999).[8]

The analysis of multiple criteria is particularly important for the design of *package auctions* (also called "combinatorial auctions"), which are auctions for multiple items in which bidders can bid directly for non-trivial subsets (packages) of the items being sold, rather than being restricted to submit bids on each item individually. In these auctions, several criteria besides incentive compatibility merit the attention of a practical mechanism designer. Revenues are an obvious one. Auctions are commonly run by an expert auctioneer on behalf of the actual seller and any failure to select a core allocation with respect to reported values implies that there is a group of bidders who have offered to pay more in total than the winning bidders, yet whose offer has been rejected. Imagine trying to explain such an outcome to the actual seller or, in a government sponsored auction, to a skeptical public![9] Monotonicity of revenues with respect to participation is another important property of auction mechanisms, because its failure could allow a seller to increase sales revenues by disqualifying bidders after the bids are received.[10] Another important desideratum is that a bidder should not profit by entering and playing as multiple bidders, rather than as a single one.[11]

We illustrate these three desiderata and how they fail in the Vickrey auction with an example of two identical items are for sale. The first bidder wants both items and will

---

[6] Hatfield and Milgrom (2005) identify the conditions under which strategy-proofness extends to cover the college admissions problem, in which one type of participant (colleges) can accept multiple applicants, but the other kind (students) can each be paired to only one college. Their analysis also covers problems in which wages and other contract terms are endogenous.

[7] For example, see Abdulkadiroglu et al. (2005).

[8] There is quite a long tradition in economics of examining approximate incentives in markets, particularly when the number of participants is large. An early formal analysis is by Roberts and Postlewaite (1976).

[9] McMillan (1994) describes how heads rolled when second-price auctions were used to sell spectrum rights in New Zealand and the highest bid was sometimes orders of magnitude larger than the second highest bid.

[10] It is quite common in auctions for final evaluations of bidder qualifications to be made after bids are received to ensure the winning bidder's ability to close the deal. Sellers may carefully study financing and regulatory constraints before accepting a bid. In the US radio spectrum auctions, bidders typically submit a "short form application" before each auction and, after the bidding but before the results are finalized, the winning bidders make an additional cash deposit and submit a "long form" that is checked in detail to ensure their qualifications to buy.

[11] Yokoo et al. (2004) were the first to emphasize the importance of "false name bidding" and how it could arise in the anonymous environment of Internet auctions. The problem they identified, however, is broader than just anonymous Internet auctions. For example, in the US radio spectrum auctions, several of the largest corporate bidders (including AT&T, Cingular, T-Mobile, Sprint, and Leap Wireless) have at times had contracts with or financial interests in multiple bidding entities in the same auction, enabling strategies that would not be possible for a single, unified bidder.

pay up to ten for the pair; it has zero value for acquiring a single item. The second and third bidders each have values of 10 for either one or two items, so their marginal values of a second item are zero. The Vickrey auction outcome assigns the items to the second and third bidders for prices of zero. Given that any of the three bidders would pay ten for the pair of items, a zero price is surely too low: that is the low revenue problem. Generally, the low revenue problem for the Vickrey auction is that its payments to the seller may be less than those at any core allocation.[12] Notice, too, that the seller could increase its sales revenue in this example by disqualifying bidder 3, thereby raising the total Vickrey price to 10. This illustrates the disqualification problem created by revenue non-monotonicity. Finally, suppose that the second and third bidders are both controlled by the same player whose actual values are 10 for one item or 20 for two. If the bidder were to participate as a single entity, it would win the two items and pay a price of ten. By bidding as two entities, each of which demands a single item for a price of 10, the player reduces its total Vickrey price from ten to zero: that is the shill bidding problem. These vulnerabilities are so severe that practical mechanism designers must investigate when and whether relaxing the incentive compatibility objective can alleviate the problems.

We have discussed matching and package auction mechanisms together not only because they are two of the currently mostly active areas of practical mechanism design but also because there are some remarkable parallels between their equilibrium theories. One parallel connects the cases where the workers in the match are substitutes for hospital and when the goods in the auction are substitutes for the bidders. In these cases, the mechanism that selects the doctor-optimal match is *ex post* incentive-compatible for doctors and a mechanism, the ascending proxy auction of Ausubel and Milgrom (2002), which selects a bidder-optimal allocation (a core allocation that is Pareto optimal for bidders), is *ex post* incentive-compatible for bidders.[13]

A second important connection is the following one: for every stable match x and every stable matching mechanism, there exists an equilibrium in which each player adopts a certain *truncation strategy*, according to which it truthfully reports its ranking of all the outcomes at which it is not matched, but reports that it would prefer to be unmatched rather than to be assigned an outcome worse than x. What is remarkable about this theorem is that *one single profile of truncation strategies is a Nash equilibrium for every stable matching mechanism*. We will find that a similar property is true for core-selecting auctions, but with one difference. In matching mechanisms, it is usual to treat all the players are strategic, whereas in auctions it is not uncommon to treat the seller differently, with only a subset of the players—the *bidders*—treating as making decisions strategically. We are agnostic about whether to include the seller as a bidder or even whether to include all the buyers as strategic players. Regardless of how the set of strategic players is specified, we find that for every allocation on the Pareto-frontier of the core for the players who report strategically, there is a single

---

[12] In this example, the core outcomes are the outcomes in which two and three are the winning bidders, each pays a price between zero and ten, and the total payments are at least ten. The seller's revenue in a core-selecting auction is thus at least 10.

[13] This is also related to results on wage auctions in labor markets as studied by Kelso and Crawford (1982) and Hatfield and Milgrom (2005), although these models do not employ package bidding.

profile of truncation strategies that is an equilibrium profile for *every* core-selecting auction.[14]

The preceding results hinge on another similarity between package auctions and matching mechanisms. In any stable matching mechanism or core-selecting auction and given any reports by the other players, a player's best reply achieves its maximum core payoff or best stable match given its actual preferences and the reported preferences of others. For auctions, there is an additional interesting connection: the maximum core payoff is exactly the Vickrey auction payoff.

Next are the inter-related results about incentives for *groups* of participants. Given a core-selecting auction, the incentives for misreporting are minimal for individuals in a particular group S if and only if the mechanism selects an S-best core allocation. If there is a unique S-best allocation, then truthful reporting by members of coalition S is an *ex post* equilibrium. This is related to the famous result from matching theory (for which there always exists a unique man-optimal match and a unique woman-optimal match) that it is an *ex post* equilibrium for men to report truthfully in the man-optimal mechanism and for women to report truthfully in the woman-optimal mechanism.

Another result is that any auction that minimizes the seller's revenue among core allocations results in seller revenue being a non-decreasing function of the bids. As argued above, revenue-monotonicity of this sort is important because, without it, a seller might have an incentive to disqualify bids or bidders to increases its revenues and a bidder might have an incentive to sponsor a shill, whose bids reduce prices.

The remainder of this paper is organized as follows. Section 2 formulates the package auction problem. Section 3 characterizes core-selecting mechanisms in terms of revenues that are never less than Vickrey revenues, even when bidders can use shills. Section 4 introduces definitions and notation and introduces the theorems about best replies and full information equilibrium. Section 5 states and proves the theorem about the core-selecting auctions with the smallest incentives to misreport. Section 6 shows that the revenue-minimizing core-selecting auction is revenue-monotonic. Various corresponding results for the marriage problem are developed in Sect. 7, while Sect. 8 concludes.

2. Formulation

We denote the seller as player 0, the bidders as players $j = 1, \ldots, J$, and the set of all players by N. Each bidder j has quasi-linear utility and a finite set of possible packages X_j. Its value associated with any feasible package $x_j \in X_j$ is $u_j(x_j) \geq 0$. For convenience, we formulate our discussion mainly in terms of bidding applications, but the same mathematics accommodates much more, including some social choice problems. In the central case of package bidding for predetermined items, x_j consists of a package of items that the bidder may buy. For procurement auctions, x_j could also usefully incorporate information about delivery dates, warranties, and various other product

---

[14] These truncation strategies also coincide with what Bernheim and Whinston (1986) call "truthful strategies" in their analysis of a "menu auction", which is a kind of package auction.

attributes or contract terms. Among the possible packages for each bidder is the null package, $\emptyset \in X_j$ and we normalize so that $u_j(\emptyset) = 0$.

For concreteness, we focus on the case where the auctioneer is a seller who has a feasible set $X_0 \subseteq X_1 \times \cdots \times X_J$ with $(\emptyset, \ldots, \emptyset) \in X_0$—so the no sale package is feasible for the seller—and a valuation function $u_0 : X_0 \to \mathbb{R}$ normalized so that $u_0(\emptyset, \ldots, \emptyset) = 0$. For example, if the seller must produce the goods to be sold, then u_0 may be the auctioneer-seller's variable cost function.

For any coalition S, a goods assignment $\hat{x}$ is *feasible* for coalition S, written $\hat{x} \in F(S)$, if (1) $\hat{x} \in X_0$ and (2) for all j, if $j \notin S$ or $0 \notin S$, then $\hat{x}_j = \emptyset$. That is, a bidder can have a non-null assignment when coalition S forms only if that bidder and the seller are both in the coalition.

The *coalition value function* or *characteristic function* is defined by

$$w_u(S) = \max_{x \in F(S)} \sum_{j \in S} u_j(x_j) \qquad (1)$$

In a *direct auction mechanism* (f, P), each bidder j reports a valuation function $\hat{u}_j$ and the profile of reports is $\hat{u} = \{\hat{u}_j\}_{j=1}^J$. The outcome of the mechanism, $(f(\hat{u}), (P_j(\hat{u}))) \in (X_0, \mathbb{R}_+^J)$, specifies the choice of $x = f(\hat{u}) \in X_0$ and the payments $p_j = P_j(\hat{u}) \in \mathbb{R}_+$ made to the seller by each bidder j. The associated payoffs are given by $\pi_0 = u_0(x) + \sum_{j \neq 0} p_j$ for the seller and $\pi_j = u_j(x) - p_j$ for each bidder j. The payoff profile is individually rational if $\pi \geq 0$.

A *cooperative game* (with transferable utility) is a pair (N, w) consisting of a set of players and a characteristic function. A payoff profile π is feasible if $\sum_{j \in N} \pi_j \leq w(N)$, and in that case it is associated with a feasible allocation. An *imputation* is a feasible, non-negative payoff profile. An imputation is in the *core* if it is efficient and unblocked:

$$\text{Core}(N, w) = \left\{ \pi \geq 0 \bigg| \sum_{j \in N} \pi_j = w(N) \text{ and } (\forall S \subseteq N) \sum_{j \in S} \pi_j \geq w(S) \right\} \qquad (2)$$

A direct auction mechanism (f, P) is *core-selecting* if for every report profile $\hat{u}$, $\pi_{\hat{u}} \in \text{Core}(N, w_{\hat{u}})$. Since the outcome of a core-selecting mechanism must be efficient with respect to the reported preferences, we have the following:

Lemma 1. *For every core-selecting mechanism (f, P) and every report profile $\hat{u}$,*

$$f(\hat{u}) \in \arg\max_{x \in X_0} \sum_{j \in N} \hat{u}_j(x_j) \qquad (3)$$

The payoff of bidder j in a Vickrey auction is the bidder's marginal contribution to the coalition of the whole. In cooperative game notation, if the bidders' value profile is u, then bidder j's payoff is $\bar{\pi}_j = w_u(N) - w_u(N - j)$.[15]

---

[15] A detailed derivation can be found in Milgrom (2004).

3. Revenues and Shills: Necessity of Core-Selecting Auctions

We have argued that the revenues from the Vickrey outcome are often too low to be acceptable to auctioneers. In order to avoid biasing the discussion too much, in this section we treat the Vickrey revenues as a just-acceptable lower bound and ask: what class of auctions have the properties that, for any set of reported values, they select the total-value maximizing outcome and lead always to bidder payoffs no higher than the Vickrey payoffs, even when bidders may be using shills? Our answer will be: exactly the class of core-selecting auctions.

In standard fashion, we call any mechanism with the first property, namely, that the auction selects the total-value-maximizing outcome, "efficient".

Theorem 1. *An efficient direct auction mechanism has the property that no bidder can ever earn more than its Vickrey payoff by disaggregating and bidding with shills if and only if it is a core-selecting auction mechanism.*

Proof. Fix a set of players (seller and bidders) N, let w be the coalitional value function implied by their reported values, and let π be the players' vector of reported payoffs. Efficiency means $\sum_{j \in N} \pi_j = w(N)$. Let $S \subseteq N$ be a coalition that excludes the seller. These bidders could be shills. Our condition requires that they earn no more than if they were to submit their merged valuation in a Vickrey auction, in which case the merged entity would acquire the same items and enjoy a total payoff equal to its marginal contribution to the coalition of the whole: $w(N) - w(N - S)$. Our restriction is therefore $\sum_{j \in S} \pi_j \leq w(N) - w(N - S)$. In view of efficiency, this holds if and only if $\sum_{j \in N-S} \pi_j \geq w(N - S)$. Since S was an arbitrary coalition of bidders, we have that for every coalition $T = N - S$ that includes the seller, $\sum_{j \in T} \pi_j \geq w(T)$. Since coalitions without the seller have value zero and can therefore never block, we have shown that there is no blocking coalition. Together with efficiency, this implies that $\pi \in Core(N, w)$. □

4. Truncation Reports and Equilibrium

In the marriage problem, a *truncation report* refers to a reported ranking by person j that preserves the person's true ranking of possible partners, but which may falsely report that some partners are unacceptable. For an auction setting with transferable utility, a truncation report is similarly defined to correctly rank all pairs consisting of a non-null goods assignment and a payment but which may falsely report that some of these are unacceptable. When valuations are quasi-linear, a reported valuation is a truncation report exactly when all reported values of non-null goods assignments are reduced by the same non-negative constant. We record that observation as a lemma.

Lemma 2. *A report $\hat{u}_j$ is a truncation report if and only if there exists some $\alpha \geq 0$ such that for all $x_j \in X_j$, $\hat{u}_j(x_j) = u_j(x_j) - \alpha$.*

Proof. Suppose that $\hat{u}_j$ is a truncation report. Let x_j and x'_j be two non-null packages and suppose that the reported value of x_j is $\hat{u}_j(x_j) = u_j(x_j) - \alpha$. Then, $(x_j, u_j(x_j) - \alpha)$

is reportedly indifferent to $(\emptyset, 0)$. Using the true preferences, $(x_j, u_j(x_j) - \alpha)$ is actually indifferent to $(x'_j, u_j(x'_j) - \alpha)$ and so must be reportedly indifferent as well: $\hat{u}_j(x_j) - u_j(x_j) - \alpha = \hat{u}_j(x'_j) - u_j(x'_j) - \alpha$. It follows that $u_j(x'_j) - \hat{u}_j(x'_j) = u_j(x_j) - \hat{u}_j(x_j) = \alpha$.

Conversely, suppose that there exists some $\alpha \geq 0$ such that for all $x_j \in X_j$, $\hat{u}_j(x_j) \equiv u_j(x_j) - \alpha$. Then for any two non-null packages, the reported ranking of (x_j, p) is higher than that of (x'_j, p') if and only if $\hat{u}(x_j) - p \geq \hat{u}(x'_j) - p'$ which holds if and only $u(x_j) - p \geq u(x'_j) - p'$. $\square$

We refer to the truncation report in which the reported value of all non-null outcomes is $\hat{u}_j(x_j) = u_j(x_j) - \alpha_j$ as the "α_j truncation of u_j".

In full information auction analyses since that of Bertrand (1883), auction mechanisms have often been incompletely described by the payment rule and the rule that the unique highest bid, when that exists, determines the winner. Ties often occur at Nash equilibrium, however, and the way ties are broken is traditionally chosen in a way that depends on bidders' values and not just on their bids. For example, in a first-price auction with two bidders, both bidders make the same equilibrium bid, which is equal to the lower bidder's value. The analysis assumes that the bidder with the higher value is favored, that is, chosen to be the winner in the event of a tie. If the high value bidder were not favored, then it would have no best reply. As Simon and Zame (1990) have explained, although breaking ties using value information prevents this from being a feasible mechanism, the practice of using this tie-breaking rule for analytical purposes is an innocent one, because, for any $\varepsilon > 0$, the selected outcome lies within ε of the equilibrium outcome of any related auction game in which the allowed bids are restricted to lie on a sufficiently fine discrete grid.[16]

In view of lemma 1, for almost all reports, assignments of goods differ among core-selecting auctions only when there is a tie; otherwise, the auction is described entirely by its payment rule. We henceforth denote the payment rule of an auction by $P(\hat{u}, x)$, to make explicit the idea that the payment may depend on the goods assignment in case of ties. For example, a first-price auction with only one good for sale is any mechanism which specifies that the winner is a bidder who has made the highest bid and the price is equal to that bid. The mechanism can have any tie-breaking rule to be used so long as (3) is satisfied. In traditional parlance, the payment rule P defines an *auction*, which comprises a set of mechanisms.

Definition. $\hat{u}$ is an equilibrium of the auction P if there is some core selecting mechanism (f, P) such that $\hat{u}$ is a Nash equilibrium of the mechanism.

For any auction, consider a tie-breaking rule in which bidder j is *favored*. This means that in the event that there are multiple goods assignments that maximize total reported value, if there is one at which bidder j is a winner, then the rule selects such a one. When a bidder is favored, that bidder always has some best reply.

Theorem 2. *Suppose that (f, P) is a core-selecting direct auction mechanism and bidder j is favored. Let $\hat{u}_{-j}$ be any profile of reports of bidders other than j. Denote j's actual value by u_j and let $\bar{\pi}_j = w_{\hat{u}_{-j}, u_j}(N) - w_{\hat{u}_{-j}, u_j}(N - j)$ be j's corresponding*

---

[16] See also Reny (1999).

Vickrey payoff. Then, the $\bar{\pi}_j$ truncation of u_j is among bidder j's best replies in the mechanism and earns a payoff for j of $\bar{\pi}_j$. Moreover, this remains a best reply even in the expanded strategy space in which bidder j is free to use shills.

Proof. Suppose j reports the $\bar{\pi}_j$ truncation of u_j. Since the mechanism is core-selecting, it selects individually rational allocations with respect to reported values. Therefore, if bidder j is a winner, its payoff is at least zero with respect to the reported values and hence at least $\bar{\pi}_j$ with respect to its true values.

Suppose that some report $\hat{u}_j$ results in an allocation $\hat{x}$ and a payoff for j strictly exceeding $\bar{\pi}_j$. Then, the total payoff to the other bidders is less than $w_{\hat{u}_{-j}, u_j}(N) - \bar{\pi}_j \le w_{\hat{u}_{-j}, u_j}(N - j)$, so $N - j$ is a blocking coalition for $\hat{x}$, contradicting the core-selection property. This argument applies also when bidder j uses shills. Hence, there is no report yielding a profit higher than $\bar{\pi}_j$, even on the extended strategy space that incorporates shills.

Since reporting the $\bar{\pi}_j$ truncation of u_j results in a zero payoff for j if it loses and non-negative payoff otherwise, it is always a best reply when $\bar{\pi}_j = 0$.

Next, we show that the truncation report always wins for j, therefore yielding a profit of at least $\bar{\pi}_j$ so that it is a best reply. Regardless of j's reported valuation, the total reported payoff to any coalition excluding j is at most $w_{\hat{u}_{-j}, \hat{u}_j}(N - j) = \max_{x=(\emptyset, x_{-j}) \in X_0} \sum_{i \in N-j} \hat{u}_i(x)$. If j reports the $\bar{\pi}_j$ truncation of u_j, then the maximum value is at least $\max_{x \in X_0}(\sum_{i \in N-j} \hat{u}_i(x) + u_j(x)) - \bar{\pi}_j = w_{\hat{u}_{-j}, u_j}(N) - \bar{\pi}_j$, which is equal to the previous sum by the definition of $\bar{\pi}_j$. Applying lemma 1 and the hypothesis that j is favored establishes that j is a winner. □

Definition. An imputation π is *bidder optimal* if $\pi \in Core(N, u)$ and there is no $\hat{\pi} \in Core(N, u)$ such that for every bidder j, $\pi_j \le \hat{\pi}_j$ with strict inequality for at least one bidder (by extension, a feasible allocation is *bidder optimal* if the corresponding imputation is so).

Next is one of the main theorems, which establishes a kind of equilibrium equivalence among the various core-selecting auctions. We emphasize, however, that the strategies require each bidder j to know the equilibrium payoff π_j, so what is being described is a full information equilibrium but not an equilibrium in the model where each bidder's own valuation is private information.

Theorem 3. *For every valuation profile u and corresponding bidder optimal imputation π, the profile of π_j truncations u_j is a full information equilibrium profile of every core selecting auction. The equilibrium goods assignment x^* maximizes the true total value $\sum_{i \in N} u_i(x_i)$, and the equilibrium payoff vector is π (including π_0 for the seller).*[17]

Proof. For any given core-selecting auction, we study the equilibrium of the corresponding mechanism that, whenever possible, breaks ties in (3) in favor of the goods assignment that maximizes the total value according to valuations u. If there are many such goods assignments, any particular one can be fixed for the argument that follows.

---

[17] Versions of this result were derived and reported independently by Day and Raghavan (2006) and Milgrom (2006). The latter paper has been folded into this one.

First, we show that no goods assignment leads to a reported total value exceeding π_0. Indeed, let S be the smallest coalition for which the maximum total reported value exceeds π_0. By construction, the bidders in S must all be winners at the maximizing assignment, so $\pi_0 < \max_{x \in X_0, x_{-S} = \emptyset} u_0(x_0) + \sum_{i \in S-0}(u_i(x_i) - \pi_i) \leq w_u(S) - \sum_{i \in S-0} \pi_i$. This contradicts $\pi \in Core(N, w_u)$, so the winning assignment has a reported value of at most $\pi_0 : w_{\hat{u}}(N) \leq \pi_0$. If j instead reports truthfully, it can increase the value of any goods allocation by at most π_j, so $w_{u_j, \hat{u}_{-j}}(N) \leq \pi_0 + \pi_j$.

Next, we show that for any bidder j, there is some coalition excluding j for which the maximum reported value is at least π_0. Since π is bidder optimal, for any $\varepsilon > 0$, $(\pi_0 - \varepsilon, \pi_j + \varepsilon, \pi_{-j}) \notin Core(N, w_u)$. So, there exists some coalition S_ε to block it: $\sum_{i \in S_\varepsilon} \pi_i - \varepsilon < w_u(S_\varepsilon)$. By inspection, this coalition includes the seller but not bidder j. Since this is true for every ε and there are only finitely many coalitions, there is some S such that $\sum_{i \in S} \pi_i \leq w_u(S)$. The reverse inequality is also implied because $\pi \in Core(N, w_u)$, so $\sum_{i \in S} \pi_i = w_u(S)$.

For the specified reports, $w_{\hat{u}}(S) = \max_{x \in X_0} \sum_{i \in S} \hat{u}_i(x_i) \geq \max_{x \in X_0} u_0(x_0) + \sum_{i \in S-0} (u_i(x_i) - \pi_i) \geq w_u(S) - \sum_{i \in S-0} \pi_i = \pi_0$. Since the coalition value cannot decrease as the coalition expands, $w_{\hat{u}}(N - j) \geq \pi_0$. By definition of the coalition value functions, $w_{\hat{u}}(N - j) = w_{u_j, \hat{u}_{-j}}(N - j)$.

Using Theorem 2, j's maximum payoff if it responds optimally and is favored is $w_{u_j, \hat{u}_{-j}}(N) - w_{u_j, \hat{u}_{-j}}(N - j) \leq (\pi_0 + \pi_j) - \pi_0 = \pi_j$. So, to prove that the specified report profile is an equilibrium, it suffices to show that each player j earns π_j when these reports are made.

The reported value of the true efficient goods assignment is at least $\max_{x \in X_0} u_0(x_0) + \sum_{i \in N-0}(u_i(x_i) - \pi_i) = w(N) - \sum_{i \in N-0} \pi_i = \pi_0$. So, with the specified tie-breaking rule, if the bidders make the specified truncation reports, the selected goods assignment will maximize the true total value.

Since the auction is core-selecting, each bidder j must have a reported profit of at least zero and hence a true profit of at least π_j, but we have already seen that these are also upper bounds on the payoff. Therefore, the reports form an equilibrium; each bidder j's equilibrium payoff is precisely π_j, and that the seller's equilibrium payoff is $w_{\hat{u}}(N) - \sum_{i \in N-0} \pi_i = \pi_0$. □

5. Minimizing Incentives to Misreport

Despite the similarities among the core-selecting mechanisms emphasized in the previous section, there are important differences among the mechanisms in terms of incentives to report valuations truthfully. For example, when there is only a single good for sale, both the first-price and second-price auctions are core selecting mechanisms, but only the latter is strategy-proof.

To evaluate various bidders' incentives to deviate from truthful reporting, we introduce the following definition.

Definition. The *incentive profile* for a core-selecting auction P at u is $\varepsilon^P = \{\varepsilon_j^P(u)\}_{j \in N-0}$ where $\varepsilon_j^P(u) \equiv \sup_{\hat{a}_j} u_j(f_j(u_{-j}, \hat{u}_j)) - P(u_{-j}, \hat{u}_j, f_j(u_{-j}, \hat{u}_j))$ is j's maximum gain from deviating from truthful reporting when j is favored.

Our idea is to minimize these incentives to deviate from truthful reporting, subject to selecting a core allocation. Since the incentives are represented by a vector, we use a Pareto-like criterion.

Definitions. A core-selecting auction P provides *suboptimal incentives* at u if there is some core selecting auction $\hat{P}$ such that for every bidder j, $\varepsilon_j^{\hat{P}}(u) \leq \varepsilon_j^{P}(u)$ with strict inequality for some bidder. A core selecting *auction provides optimal incentives* if there is no u at which it provides suboptimal incentives.

Theorem 4. *A core-selecting auction provides optimal incentives if and only if for every u it chooses a bidder optimal allocation.*

Proof. Let P be a core-selecting auction, u a value profile, and π the corresponding auction payoff vector. From Theorem 2, the maximum payoff to j upon a deviation is $\bar{\pi}_j$, so the maximum gain to deviation is $\bar{\pi}_j - \pi_j$. So, the auction is suboptimal exactly when there is another core-selecting auction with higher payoffs for all bidders, contradicting the assumption that π is bidder optimal. $\square$

Recall that when the Vickrey outcome is a core allocation, it is the unique bidder optimal allocation. So, Theorem 4 implies that any core selecting auction that provides optimal incentives selects the Vickrey outcome with respect to the reported preferences whenever that outcome is in the core for those reports. Moreover, because truthful reporting then provides the bidders with their Vickrey payoffs, Theorem 2 implies the following.

Corollary. *When the Vickrey outcome is a core allocation, then truthful reporting is an ex post equilibrium for any mechanism that always selects bidder optimal core allocations.*

We note in passing that any incentive profile that can be achieved by any mechanism is replicated by the corresponding direct mechanism. There is a "revelation principle" for approximate incentives, so one cannot do better than the results reported in Theorem 4 by looking over a larger class of mechanisms, including ones that are not direct.

6. Monotonicity of Revenues

The core allocations with respect to the reports that minimize the seller's payoff are all bidder optimal allocations, so a mechanism that selects those satisfies the conditions of Theorem 4. That mechanism has another advantage as well: its revenues are nondecreasing in the bids.

Theorem 5. *The seller's minimum payoff in the core with bidder values $\hat{u}$ is nondecreasing in $\hat{u}$.*

Proof. The seller's minimum payoff is

$$\min_{\pi \geq 0} w_{\hat{u}}(N) - \sum_{i \in N-0} \pi_i \text{ subject to } \sum_{i \in S} \pi_i \geq w_{\hat{u}}(S) \text{ for all } S \subseteq N \quad (4)$$

The objective is an expression for π_0; it incorporates the equation $w_{\hat{u}}(N) = \sum_{i \in N} \pi_i$ which therefore can be omitted from the constraint set. The objective is increasing in $w_{\hat{u}}(N)$ and the constraint set shrinks as $w_{\hat{u}}(S)$ increases for any coalition $S \neq N$. Hence, the minimum value is non-decreasing in the vector $(w_{\hat{u}}(S))_{S \subseteq N}$. It is obvious that the coalitional values $w_{\hat{u}}(S)$ are non-decreasing in the reported values $\hat{u}$, so the result follows. □

The theorems established above do not extend to auctions with bidder budget constraints.[18]

7. Connections to the Marriage Problem

Even though Theorems 2–5 in this paper are proved using transferable utility and do not extend to the case of budget-constrained bidders, they do all have analogs in the non-transferable utility marriage problem.

Consider Theorem 2. Roth and Peranson (1999) have shown for a particular algorithm in the marriage problem that any fully informed player can guarantee its best stable match by a suitable truncation report. That report states that all mates less preferred than its best achievable mate are unacceptable. The proof in the original paper makes it clear that their result extends to any stable matching mechanism, that is, any mechanism that always selects a stable match.

Here, in correspondence to stable matching mechanisms, we study core-selecting auctions. For the auction problem, Ausubel and Milgrom (2002) showed that the best payoff for any bidder at any core allocation is its Vickrey payoff. So, the Vickrey payoff corresponds to the best mate assigned at any stable match. Thus, the auction and matching procedures are connected not just by the use of truncation strategies as best replies but by the point of the truncation, which is at the player's best core or stable outcome.

Theorem 3 concerns Nash equilibrium. Again, the known results of matching theory are similar. Suppose the participants in the match in some set S^c play non-strategically, like the seller in the auction model, while the participants in the complementary set S,

---

[18] John Hegeman has produced an example showing that the theorem does not extend to the case of binding budget constraints. The following table shows values and budgets for three bidders and three items. Only bidder 2 has a binding budget constraint.

Bidder	A	B	C	AB	BC	AC	ABC	Budget
1	10	10	0	17	10	10	17	17
2	0	0	0	14	19	19	19	14
3	0	0	9	0	9	9	9	9

If bidders 2 and 3 report their values and budgets truthfully, then bidder 1's best truncation depends on how the core point is chosen in this example without transferable utility. If the bidder 1 reduces its reported values by 3 or more, then there is a core allocation in which 1 loses, while 2 buys AB for 14 and 3 buys C for 9. However, if bidder 1 reduces its reported value by less than 6, then there is also a core allocation at which it is a winner, buying A at price 4 while 2 buys BC at price 14.

whom we shall call bidders, play Nash equilibrium. Then, for bidder-optimal stable match,[19] the profile at which each player in S reports that inferior matches are unacceptable is a full-information Nash equilibrium profile of *every* stable matching mechanism and it leads to that S-optimal stable match. This result is usually stated using only men or women as the set S, but extending to other sets of bidders using the notion of bidder optimality is entirely straightforward.

For Theorem 4, suppose again that some players are non-strategic and that only the players in S report strategically. Then, if the stable matching mechanism selects an S-optimal stable match, then there is no other stable matching mechanism that weakly improves the incentives of all players to report truthfully, with strict improvement for some. Again, this is usually stated only for the case where S is the set of men or the set of women, and the extension does require introducing the notion of a bidder optimal match.

Finally, the last result states that increasing bids or, by extension, introducing new bidders increases the seller's revenue if the seller pessimal allocation is selected. The matching analog is that adding men improves the utility of each woman if the woman-pessimal, man-optimal match is selected—a result that is reported by Roth and Sotomayor (1990).

8. Conclusion

We motivated our study of core-selecting auctions by comparing them to stable matching mechanisms, which have been in long use in practice. Both in collected case studies and in the Kagel-Roth laboratory experiments, participants stopped using unstable matching mechanisms, preferring to make the best match they could by individual negotiations, even when congestion made that process highly imperfect. In contrast, participants continued to participate in stable matching mechanisms for much longer, and many such mechanisms continue in use today. They have the practical advantage of being able to find stable allocations for the reported preferences, which would be a difficult task for them in the limited time typically available both in the experiments and in real markets.

These observations, however, compare only some particular stable and unstable matching mechanisms, so even the generalization to all matching mechanisms remains untested. If one can imagine experiments with other matching mechanisms, one can do the same with auctions. When is it likely that parties will reach agreement before the auction and when will they simply bid?

Despite the theoretical similarities, we need to confront the reality that stable matching mechanisms are now often used in practice while core-selecting auctions remain rare. It is possible that this is about to change: the computations required by core-selecting auctions are, in general, much harder than those for matching and computational tractability for problems of interesting scale has only recently been achieved.

Yet, there are other reasons to doubt the practicability of core-selecting auctions. In an environment with N items for sale, the number of non-empty packages for which

---

[19] This is defined analogously to the bidder optimal allocation.

a bidder is called to report values is $2^N - 1$. That is unrealistically large for most applications if N is even a small two-digit number. For the general case, Segal (2003) has shown that communications cannot be much reduced without severely limiting the efficiency of the result.

Although communication complexity is an important practical issue, it need not definitively rule out core-selecting package auctions. In many real-world settings, there is substantial information about the kinds of packages that make sense and those can be incorporated to allow bidders to express a good approximation of their values with comparative ease. For the case where goods are substitutes, some progress on compact expressions of values has already been made.[20] An auctioneer may know that a collection of airport landing rights between 2:00 and 2:15 are valued similarly to ones between 2:15 and 2:30, or that complementarities in electrical generating result from costs saved by operating continuously in time, minimizing time lost when the plant is ramped up or down. Practical designs that take advantage of this knowledge can still be core-selecting mechanisms, where the reported values can be compactly expressed if they satisfy predetermined constraints.

If the problem of communication complexity can be solved, then core-selecting auctions appear to provide a practical alternative to the Vickrey design. The class includes the pay-as-bid "menu auction" design studied by Bernheim and Whinston (1986), the ascending proxy auction studied by Ausubel and Milgrom (2002) and Parkes and Ungar (2000), and any of the mechanisms resulting from the core computations in Day and Raghavan (2006). Within this class, the auctions that select bidder-optimal allocations conserve as far as possible the advantages of the Vickrey design—matching the Vickrey auction's *ex post* equilibrium property when there is a single good, or goods are substitutes, or most generally when the Vickrey outcome happens to lie in the core—and avoiding the low revenue and monotonicity problems of the Vickrey mechanism.

From the perspective of pure theory, the most interesting part of this analysis is that all of the main results about core-selecting auctions have analogues in the theory of stable matching mechanisms that David Gale helped to pioneer. The deep mathematical reasons for this similarity remain to be fully explored.

References

Abdulkadiroglu A, Pathak P, Roth A, Sonmez T (2005) The Boston Public School Match. In: AEA papers and proceedings, pp 368–371

Ausubel L, Milgrom P (2002) Ascending auctions with package bidding. Front Theor Econ 1(1), Article 1

Ausubel L, Milgrom P (2005) The lovely but lonely Vickrey auction. In: Cramton P, Shoham Y, Steinberg R (eds) Combinatorial auctions. MIT Press, Cambridge

Bernheim BD, Whinston M (1986) Menu auctions, resource allocation and economic influence. Q J Econ 101:1–31

Bertrand J (1883) Théorie Mathématique de la Richesse Sociale. J des Savants 69:499–508

---

[20] Hatfield and Milgrom (2005) introduced the *endowed assignment valuations* for this purpose.

Day RW, Raghavan S (2006) Fair payments for efficient allocations in public sector combinatorial auctions. Manag Sci (Forthcoming)

Green J, Laffont J-J (1979) Incentives in public decision making. North Holland, Amsterdam

Hatfield J, Milgrom P (2005) Matching with contracts. Am Econ Rev 95(4):913–935

Holmstrom B (1979) Groves schemes on restricted domains. Econometrica 47:1137–1144

Kagel J, Roth A (2000) The dynamics of reorganization in matching markets: a laboratory experiment motivated by a natural experiment. Q J Econ:201–235

Kelso A, Crawford V (1982) Job matching, coalition formation, and gross substitutes. Econometrica 50:1483–1504

McMillan J (1994) Selling spectrum rights. J Econ Perspect 8:145–162

Milgrom P (2004) Putting auction theory to work. Cambridge University Press, Cambridge

Milgrom P (2006) Incentives in core-selecting auctions. Stanford University

Parkes D, Ungar L (2000) Iterative combinatorial auctions: theory and practice. In: Proceedings of the 17th national conference on artificial intelligence, pp 74–81

Reny P (1999) On the existence of pure and mixed strategy nash equilibria in discontinuous games. Econometrica 67(5):1029–1056

Roberts J, Postlewaite A (1976) The incentives for price-taking behavior in large exchange economies. Econometrica 44(1):115–129

Roth AE (1982) The economics of matching: stability and incentives. Math Oper Res 7:617–628

Roth AE, Peranson E (1999) The redesign of the matching market for American physicians: some engineering aspects of economic design. Am Econ Rev 89:748–780

Roth AE, Sotomayor M (1990) Two-sided matching: a study in game-theoretic modeling and analysis. Cambridge University Press, Cambridge

Roth AE, Xing X (1994) Jumping the gun: imperfections and institutions related to the timing of market transactions. Am Econ Rev 84:992–1044

Segal I (2003) The communication requirements of combinatorial auctions. In: Cramton P, Shoham Y, Steinberg R (eds) Combinatorial Auctions. Princeton University Press, Princeton

Simon LK, Zame WR (1990) Discontinuous games and endogenous sharing rules. Econometrica 58:861–872

Yokoo M, Sakurai Y, Matsubara S (2004) The effect of false-name bids in combinatorial auctions: new fraud in internet auctions. Games Econ Behav 46(1):174–188

CHAPTER 12

A New Payment Rule for Core-Selecting Package Auctions*

Aytek Erdil and Paul Klemperer

Although the Combinatorial Clock Auction has been used in several spectrum auctions, the theoretical justification for the usual rule used to compute bidders' payments is weak if "minimum revenue core" prices are not unique in the final (core-selecting package auction) stage. So this paper proposes an alternative way to compute bidders' payments in that case. Specifically, we propose a new, easy-to-implement, class of payment rules: "Reference Rules". In our simple model, Reference Rules give bidders lower marginal incentives to deviate from "truthful bidding" than the usual ("Vickrey-nearest") payment rule, and are as robust as the usual rules to large deviations. (By contrast, small, almost-riskless, profitable deviations from "truthful bidding" are often easy for bidders to find under the usual rules.) Other considerations, including fairness and comprehensibility, also seem to support the use of Reference Rules. So although we take no position on the general merits, or otherwise, of Combinatorial Clock Auctions, we believe that using Reference Rules could improve their design.

1. Introduction

Day and Milgrom (2008) recently proposed a novel multi-object auction form – the "core-selecting package auction" – that seems sufficiently attractive, in particular in its handling of complementarities between objects, that it has already been adopted by regulators in several countries. The United States planned to use it for auctioning airport takeoff and landing slots, and the United Kingdom and other European countries have

---

* This chapter was originally written in 2009, and published in the Journal of the European Economic Association, 2010, 8(2–3), 537–547. It is reproduced here with the kind permission of the European Economic Association and the MIT Press. A new introductory paragraph and the appendix of our earlier working paper (Erdil and Klemperer, 2009) have been added to the original paper, and minor additional revisions have been made. We are grateful to Jeremy Bulow, Kalyan Chatterjee, Myeonghwan Cho, Peter Cramton, Bob Day, Gerhard Dijkstra, Jacob Goeree, Rocco Macchiavello, Daniel Marszalec, Meg Meyer, Paul Milgrom, Marco Pagnozzi, David Parkes, an anonymous referee, the editor and many other friends and colleagues for helpful discussions and advice.

used it for auctioning radio spectrum.[1] However Day and Milgrom's original work did not completely specify the auction's payment rules. This paper helps fill that gap.

A core-selecting auction takes sealed bids, identifies the "efficient" allocation (i.e., the allocation that would be value-maximising if all bids were actual values), and chooses associated payments so that the final (non-negative) payoffs are in the core (i.e., no set of bidders can join with the seller to form a "blocking coalition"). That is, a core-selecting auction allocates goods in the same way as a Vickrey-Clarke-Groves (henceforth, Vickrey) auction but substitutes core payments for Vickrey payments.

Economic theorists have, of course, traditionally favoured Vickrey pricing because bidders' dominant strategies are then to bid their actual values for packages in simple private-value environments.[2] However, Vickrey pricing can lead to very low revenues, extreme incentives to merge and demerge, collusive possibilities that are hard to guard against (in particular, successful collusion requires only two bidders to participate), and the possibility that the auctioneer and/or bidders can gain from using "shills" (see, e.g., Ausubel and Milgrom (2006)). Substituting core pricing for Vickrey pricing mitigates these problems.

Of course, substituting core pricing for Vickrey pricing also creates incentives for bidders to deviate from bidding their values. Day and Milgrom therefore suggested using *minimum-revenue core* (MRC) pricing to "minimise" bidders' incentives to deviate from "truthful bidding" (when each bidder knows all other bidders' actual values and assumes the others are bidding their true values); specifically, Day and Raghavan (2007) showed that among all core-payment vectors, the MRC selections minimise the sum-across-bidders of each bidder's maximum possible gain from unilaterally deviating from bidding her actual value.

However, unless the Vickrey point is in the core, the MRC is typically not a single point, and in this case Day and Milgrom provide no guidance about how to choose the payment vector. In realistic environments, bidders' payments can be extremely sensitive to this choice.[3] The planned auctions to date of which we are aware follow Day and Cramton's (2008) suggestion of selecting the MRC point that is closest in Euclidean distance to the Vickrey payments. But the justification for this "Vickrey-nearest" rule is unclear, since *all* the points in the MRC minimise the sum of bidders' incentives to deviate from truth-telling.[4]

One can also question the significance of a bidder's *maximum* possible gain from a deviation. Achieving this would require the bidder to deviate (arbitrarily close) to the point at which her allocation would change, so arbitrarily small changes by other

---

[1] The radio spectrum auctions mostly have a clock-proxy first stage based on Ausubel et al. (2006). We do not address issues about this stage in this paper. The planned airport-slot auction was canceled.

[2] This result assumes no budget constraints, no worries about revealing information to third parties, no externalities between bidders (unless bidders can make bids that depend on other bidders' allocations), etc.

[3] Experiments simulating real contexts have led to outcomes in which total bidders' payments (i.e., total MRC revenues) are of the order of ten times Vickrey payments which means different MRC-selecting payment rules might reallocate as much as 90% of total payments among bidders. Cramton (2009, p. 25) reports experiments in which subjects typically bid truthfully in a Vickrey-nearest MRC-selecting package auction.

[4] Alternatively, if the Euclidean distance from the Vickrey payments is a better aggregate measure of incentives to deviate from truth-telling, as Parkes et al (2001) suggested, then why restrict attention to the MRC? (The MRC may not include the core point that minimises the Euclidean distance, see Day and Raghavan, 2007. Note 19 gives a simple example.)

bidders could cause her a net loss, and even a perfectly-informed bidder would be taking a risk.

By contrast, a bidder that reduces her winning bid for a package by a *small* amount below her actual value obtains an almost-riskless profit in any core-selecting auction in which payments are increasing in winning bids. Thus, for example, for the Vickrey-nearest rule an ϵ reduction gives the bidder a first-order gain by reducing her payment for her current allocation by (the order of) ϵ with probability close to 1 whenever the Vickrey point is outside the core (the payment is unaffected when the Vickrey point is in the core); and generates only a second-order loss, since with probability (of order) ϵ she would change the allocation, but that could happen only when her actual value is at most ϵ more than what she needs to pay to be a winner. Furthermore, in a complex environment, bidders are unlikely to understand the full space of alternatives, but may have a clearer view of where and how to gain from smaller deviations. So *marginal* incentives for "truth telling" probably matter most for a mechanism's robustness.

This paper, therefore, argues that payment rules for core-selecting auctions should minimise *marginal* incentives to deviate from "truth-telling."

Specifically, we propose a class of "Reference Rules" in which bidders' payments are, roughly speaking, determined independently of their own bids as far as possible. More precisely, a reference rule selects the MRC payments that are closest to a reference point (a vector of reference payments) chosen by the auction designer, based on any information available to her except the winners' bids (in particular, the reference point can be based upon the losers' bids).

We show that in the environments we consider, there always exists a Reference Rule that dominates the Vickrey-nearest rule in the sense that the Reference Rule has a lower sum-across-bidders of marginal deviation incentives for all possible valuation vectors (and has a strictly lower sum for some valuation vectors).[5] A regulator who has any prior information about the distributions of bidders' valuations can, of course, do even better (in expected terms) by choosing an appropriate Reference Rule. Alternatively, a regulator can pursue an objective such as "fairness," by using a Reference Rule that implements "minimally discriminatory" prices that are as close to equal as possible for identical packages.

Of course, the standard economists' approach to choosing the best payment rule would be to ask which rule is most socially-efficient when bidders play (Bayesian-Nash) equilibrium in response to the rule. Erdil and Klemperer (2009) show the "Vickrey-nearest" rule is the ex-ante welfare-maximising MRC-selecting rule in probably the simplest example possible, with uniform priors, and when – crucially – the values of the runner-up bids are commonly known in advance.[6] But this result does

---

[5] Our use of a *sum-across-bidders* criterion follows most of the literature. Reference Rules also perform well on the *probability-that-bidders-have-zero-marginal-incentive* criterion. (This criterion might suit a regulator keen to minimise the probability of bidder-regret, and is in the spirit of Lubin and Parkes (2009); Parkes et al.'s (2001) "Small rule" likewise maximises the number of bidders who have zero incentive to deviate.) The Vickrey-nearest rule sometimes performs well on a *minimax-marginal-incentives* criterion, but not always (and this criterion in any case seems inconsistent with selecting from the MRC, rather than from the "bidder-optimal core" – see Erdil and Klemperer, 2009).

[6] See also the Appendix.

not extend beyond this unrealistically simple environment,[7] so it provides only limited support for the Vickrey-nearest rule, or any other rule. Finding the "optimal" rule in any more realistic environment seems an intractable problem.[8] Furthermore, no real-world regulator could use any rule that depends on detailed information about priors about distributions of bidders' values, and relying on Bayesian-Nash equilibrium responses in these *extremely* complex many-object environments is in any case questionable.[9]

The logic that proponents of core-selecting auctions espouse is *not*, therefore, a simple equilibrium one. They note that bidders use rules of thumb and approximations. If regulators, supported by theory, practice, and experiments, can demonstrate that "truthful" bidding is approximately optimal, then bidders who face huge uncertainties and are usually risk-averse may bid at least roughly this way. Thus Cramton, Day, Milgrom and Raghavan's advocacy of MRC selection was based on the objective of minimising (maximum-possible) incentives to deviate from truthful reporting, rather than on any equilibrium analysis. Our approach in this paper is in a similar spirit, but focusing on marginal – rather than maximal-possible – incentives to deviate from "truth-telling."[10]

Section 2 describes the environment, and defines Reference Rules. Section 3 formalises the notion of a bidder's *marginal* incentive to deviate from "truth-telling." Section 4 shows that extremely simple Reference Rules perform well on our marginal incentives criterion in an elementary example. Section 5 shows the intuition for this example extends to more general environments, and Section 6 concludes.

2. The Environment

Consider an auction of two indivisible objects, X and Y. Each "type-X" bidder makes (only) a single bid for X; each "type-Y" bidder makes a single bid for Y; and each "type-Z" bidder makes a single bid for the package of both goods. There are arbitrarily many bidders of each type. Their valuations for the object (or package) they bid for are drawn from intervals $[\underline{x}, \overline{x}]$, $[\underline{y}, \overline{y}]$, and $[\underline{z}, \overline{z}]$, respectively, with their joint distribution being nonzero everywhere.[11]

We write x_i, y_i, and z_i for the i-th highest among the bids made by type-X, -Y, and -Z bidders, respectively. A **payment rule** P specifies winners' payments as a function of all bids, such that only winners pay and no winner pays more than her bid. We write

---

[7] It is particularly implausible that the values of the runner-up bids are commonly known ex ante; in this case it is unclear whether an auction would be needed, or at least whether the runner-up bids would ever actually be made.

[8] For a sense of the difficulties, see Ledyard (2007).

[9] The analysis would anyway be incomplete in ignoring incentives to merge, demerge, bid using multiple entities, collude, etc. It is also unclear why we would want to restrict an equilibrium analysis to rules that select from the MRC, or even from the core. See the Appendix for more discussion of the equilibrium approach.

[10] Similarly, the designers of frequently repeated Internet-advertising auctions are interested in mechanisms (such as the generalised second price auction, Edelman, Ostrovksy, and Schwarz, 2007, Varian, 2007) which yield "locally-envy-free" equilibria, so behaviour is stable in the sense that no bidder can gain from making a small change to his previous (equilibrium) behaviour if all other bidders stick to their previous behaviour.

[11] Erdil and Klemperer (2009) discuss some extensions to this model.

P_j for the payment made by a type-j winner, and V_j for her Vickrey payment. For simplicity, we restrict ourselves to rules in which the payment vector is differentiable in all the bids.[12] A **monotonic** payment rule is one in which every bidder's payment is weakly increasing in her bid.

An outcome, i.e., an allocation and an associated vector of payments, is in the **core** if no subset of bidders can jointly offer an alternative allocation and a vector of payments which makes the seller and all bidders in the subset weakly better off, and makes at least one of them strictly better off. Slightly abusing terminology, we refer to the payment vectors from core outcomes as core vectors. The **Minimum Revenue Core (MRC)** consists of the core vectors which minimise the sum of payments among all core vectors.

We consider **MRC-Selecting Auctions**, so:

(1) the objects are assigned to maximise total value, that is, the highest type-Z bidder wins both objects if and only if $z_1 > x_1 + y_1$; and the highest type-X and type-Y bidders each win the object they bid for otherwise,[13]
(2) the payments are such that the outcome is in the MRC.

If the Vickrey vector is in the core, it is the unique MRC vector (it minimises all winners' core payments simultaneously). If, however, the Vickrey point is not in the core, there is in general no core vector which minimises everyone's payment simultaneously.

In an MRC-selecting auction, therefore, if a type-Z bidder wins, she pays $P_Z = V_Z = \max\{z_2, x_1 + y_1\}$, since the Vickrey vector is then always in the core. If, however, the highest type-X and type-Y bidders are the winners, $V_X = \max\{x_2, z_1 - y_1\}$ and $V_Y = \max\{y_2, z_1 - x_1\}$, but the sum of their actual payments must be at least z_1 to be in the core. So if $z_1 \leq V_X + V_Y$, they pay $P_X = V_X$ and $P_Y = V_Y$. However, if $z_1 > V_X + V_Y$, the MRC consists of all payment vectors (P_X, P_Y) such that $P_X + P_Y = z_1$, $x_2 \leq P_X \leq x_1$, and $y_2 \leq P_Y \leq y_1$ – Figure 1(a) illustrates the Vickrey point, the core, and the MRC, for this case, when there is a single bidder of each type.

The **Vickrey-nearest rule** picks the point in the MRC that is closest, in Euclidean distance, to the Vickrey vector, so $P_X = V_X + (z_1 - V_X - V_Y)/2$ and $P_Y = V_Y + (z_1 - V_X - V_Y)/2$. Figure 1(a) illustrates this rule, when there is a single bidder of each type.

A **Reference Rule** with **Reference Payments** r picks the closest point of the MRC to r, in which r is independent of the winners' bids, but can depend on any other information (e.g., losers' bids), or criteria (e.g., equal prices for identical objects).

For example, a possible reference vector is (r_X, r_Y, r_Z) with $r_Z = \max\{z_2, x_1 + y_1\}$; $r_X = x_2$ and $r_Y = y_2$ if $z_1 \leq x_2 + y_2$; but if $z_1 > x_2 + y_2$ then $r_Y = z_1 - r_X$, and r_X is any function of $\{x_2, x_3, \ldots, y_2, y_3, \ldots, z_1, z_2, \ldots\}$ such that $x_2 \leq r_X \leq z_1 - y_2$. With this rule, when the highest type-X and type-Y bidders win, $P_X = r_X$ and $P_Y = r_Y$ if both $x_1 \geq r_X$ and $y_1 \geq r_Y$, but $P_X = x_1$ if $x_1 < r_X$, and $P_Y = y_1$ if $y_1 < r_Y$. Figure 1(b) illustrates this reference rule when there is a single bidder of each type.

---

[12] This ensures the concept of marginal incentive is well-defined. We also allow the regulator to use randomisations over payment rules, in which case we deal with expected payments.
[13] For simplicity, we ignore ties.

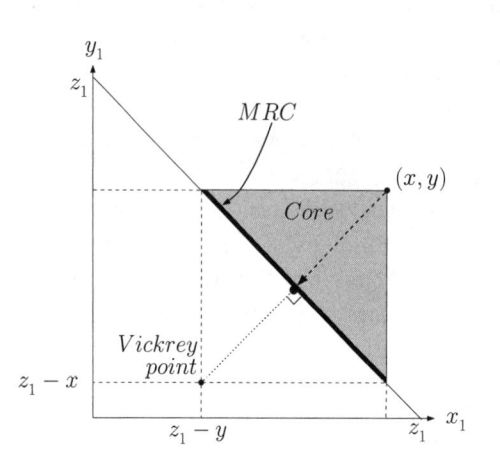

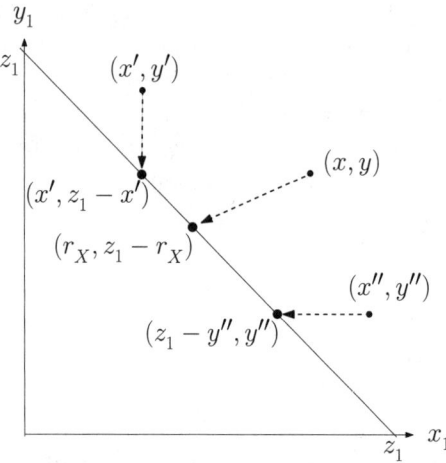

(a) Vickrey point, core, Minimum Revenue Core (MRC), and Vickrey-nearest pricing for a winning bid vector (x, y)

(b) Reference Rule pricing with Reference Payments $(r_x, z_1 - r_x)$ for three different winning bid vectors $(x, y), (x', y'), (x'', y'')$

Figure 1. Mapping of winning bid vectors to MRC points for the Vickrey-nearest rule, and the $(r_X, z_1 - r_X)$-reference rule when there is a single bidder of each type. (Note that the Vickrey-nearest payments are sensitive to any perturbations of (x, y), but the reference rule payments are robust to small perturbations of (x, y).)

3. Incentives for Small Deviations from Truth-Telling

We focus on "incentives for small deviations." That is, we assume all bidders bid their actual values, and we then ask what profit increase a winner could have earned if she had made a small change in her bid, holding all other bidders' bids constant.

Let an ϵ-**deviation** for bidder j with valuation v_j be a bid of v'_j such that $|v_j - v'_j| \leq \epsilon$. Her **marginal incentive to deviate under payment rule** P **at valuation vector** v is then[14]

$$\Delta_j^P(v) = \lim_{\epsilon \to 0} \left(\frac{\text{Bidder } j\text{'s maximum profit increase from an } \epsilon\text{-deviation}}{\epsilon} \right).$$

Because a type-Z winner always pays her Vickrey payment in a MRC-selecting rule, it is immediate that she has zero incentive to deviate from truthful bidding. We therefore focus on the marginal incentives of type-X and type-Y winners only. For the example of the Vickrey-nearest rule, illustrated in Figure 1(a), $\Delta_X = \partial(V_X + (z_1 - V_X - V_Y)/2)/\partial x_1 = 1/2$. And similarly $\Delta_Y = 1/2$.

We say a payment rule P **dominates** another rule P' if the marginal incentive to deviate under P is weakly lower than under P' for every vector of valuations, and for each bidder.

---

[14] We omit the dependence on P and v when it leads to no confusion. Losing bidders' marginal incentives are, of course, all zero.

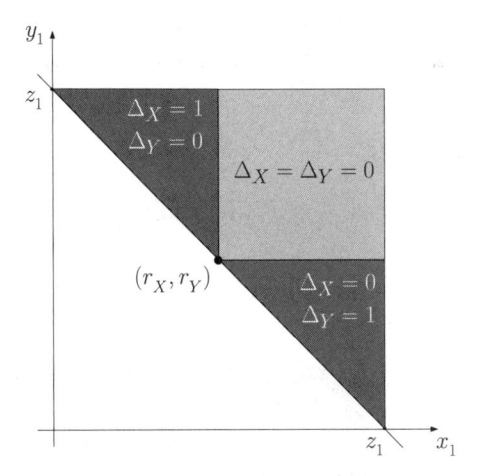

Figure 2. Marginal deviation incentives, Δ_X and Δ_Y, of type-X and type-Y winners, for the reference rule with reference payments (r_X, r_Y), in the Example.

It is not hard to see that a monotonic MRC-selecting rule is undominated,[15] so a Pareto-like criterion based on marginal incentives does not refine the set of admissible rules. Previous authors' criterion of minimising the *sum-across-bidders* of *maximum* gains from unilateral deviations also restricts the admissible set of rules no further in this environment. But looking at the *sum-across-bidders* of *marginal* incentives to deviate does have "bite:"

We say a payment rule P **sum-dominates** another rule P', if the sum-across-bidders of marginal incentives to deviate is weakly less under P than under P' for every vector of valuations.

To illustrate that a reference rule can perform well, and much better than the Vickrey-nearest rule, on this criterion, we now examine a simple special case.

4. Example

If there is a single bidder of each type, z_1 is commonly known, and $x_1, y_1 \in [0, z_1]$, then any reference rule with reference payments (r_X, r_Y) such that $r_X + r_Y = z_1$ sum-dominates the Vickrey-nearest rule.

Figure 2 shows the marginal deviation incentives, Δ_X and Δ_Y, of type-X and type-Y winners, for each valuation vector (x_1, y_1), for the reference rule with reference vector (r_X, r_Y). In the lightly shaded area, the payments under the reference rule are completely insensitive to small deviations, so the sum of marginal incentives is zero. In the darker shaded area, the sum of marginal incentives is one for this reference rule, but for the Vickrey-nearest rule, by contrast, the sum of marginal incentives is one *everywhere*, as shown in the previous section.

---

[15] If the bid vector is in the MRC, it is the unique MRC point, so all MRC-selecting rules choose it. If P' dominates P then along the path on which x_1 is continuously reduced to V_X, holding all other bids constant, so that the bid vector is in the MRC, the total decrease in the type-X winner's payment under P' cannot be higher than that under P. So for any given vector of valuations $P'_X \leq P_X$. Likewise, $P'_Y \leq P_Y$ everywhere. So, since total payments are the same in all MRC-selecting rules, P and P' are identical everywhere.

Thus, *any* reference rule for which $r_X + r_Y = z_1$ improves on the Vickrey-nearest rule in this example. A simple rule of this kind that may also seem "fair" – it selects the MRC payments that are closest to equal – is $r_X = r_Y = z_1/2$.[16]

The intuition of this example extends to our general environment.

5. More General Environments

Proposition. *The Vickrey-nearest rule is sum-dominated by a reference rule (with strict domination if $\bar{z} \geq \min\{\bar{x} + \underline{y}, \underline{x} + \bar{y}\}$).*

Proof. It suffices to specify the reference payments when the winning bids are x_1 and y_1. (1) If $z_1 \geq \bar{x} + y_2$ and $z_1 \geq \bar{y} + x_2$, the Vickrey-nearest rule is strictly sum-dominated by any reference rule for which the winners' reference payments are $(r_X, r_Y) = (q, z_1 - q)$, where q is any function of the losing bids such that $q < \bar{x}$ and $z_1 - q < \bar{y}$. (2) If $z_1 \geq \bar{x} + y_2$ and $z_1 < \bar{y} + x_2$, the rule with reference vector $(x_2, z_1 - x_2)$,[17] strictly sum-dominates the Vickrey-nearest rule. (3) Likewise, when $z_1 \geq \bar{y} + x_2$ and $z_1 < \bar{x} + y_2$, the rule with reference vector $(z_1 - y_2, y_2)$ strictly sum-dominates the Vickrey-nearest rule. (4) Finally, if $z_1 < \bar{x} + y_2$ and $z_1 < \bar{y} + x_2$, then the rule that randomises equally between the reference vectors $(x_2, z_1 - x_2)$ and $(z_1 - y_2, y_2)$, is equivalent, in expectation, to the Vickrey-nearest rule. □

The intuition for the result is that conditional on any x_2 and y_2, the variables x_1 and y_1 are distributed on $[x_2, \bar{x}]$ and $[y_2, \bar{y}]$. Equivalently, changing variables to $x = x_1 - x_2, y = y_1 - y_2, x^* = \bar{x} - x_2$, and $y^* = \bar{y} - y_2$, we have that the "winning bids" x and y are distributed on $[0, x^*]$ and $[0, y^*]$ – which essentially returns us to the Example of the previous section.[18]

Note the proof shows that the selected reference rule *strictly* sum-dominates the Vickrey-nearest rule, unless cases (1), (2), and (3) all have zero probability.

Figure 3(a) shows the marginal deviation incentives for each valuation vector (x_1, y_1) under the Vickrey-nearest rule. The last part of the proof above uses the fact that these are always identical to the marginal deviation incentives for a reference rule that randomises equally between the reference vectors $(x_2, z_1 - x_2)$ and $(z_1 - y_2, y_2)$. So even a regulator with no information about the distribution of bidders' valuations can pick a reference rule that replicates the Vickrey-nearest rule, and a designer with any distributional information can, of course, do better still by choosing the better-performing of the two reference vectors rather than randomising between them.

Choosing a reference rule with a reference vector that is an appropriately chosen convex combination of the two reference vectors is likely to yield a further improvement: Figure 3(b) shows the marginal deviation incentives under such a reference

---

[16] In this example, Parkes and Ungar's (2000) "iBundle auction" and Ausubel and Milgrom's (2002) equivalent "ascending proxy auction" are equivalent to this rule, but these authors' auctions are not always MRC-selecting.

[17] Note that this rule makes the payment more sensitive to the bidder whose valuation has larger support, so likely higher information rents, consistent with results we would expect from an equilibrium analysis.

[18] Case 1 of the proof corresponds to the trivial extension of the Example to $x^*, y^* \leq z$, in which $z = z_1 - x_2 - y_2$ (the Example assumed $x^* = y^* = z$); the reference rule we use corresponds to the Example's with reference payments $(q, z - q)$ with $q < x^*$ and $z - q < y^*$. The remaining cases correspond to the Example with reference vectors $(0, z)$ (case 2, corresponding to $x^* \leq z, y^* > z$); $(z, 0)$ (case 3, corresponding to $x^* > z, y^* \leq z$); and equal randomisation between $(0, z)$ and $(z, 0)$ (case 4, corresponding to $x^* > z, y^* > z$).

(a) Vickrey-nearest Rule

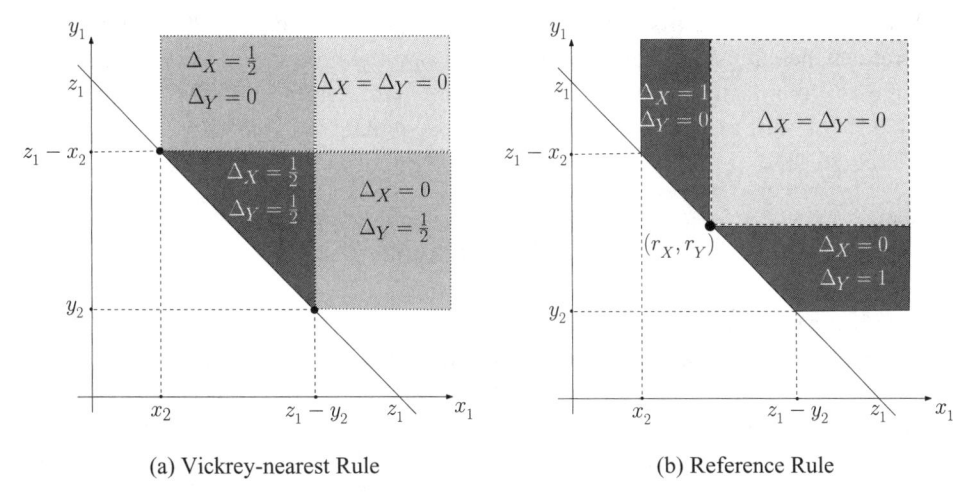

(b) Reference Rule

Figure 3. Marginal deviation incentives, Δ_X and Δ_Y, of type-X and type-Y winners, for the Vickrey-nearest rule, and (r_X, r_Y)-reference rule with $r_Y = z_1 - r_X$.

rule. Although there is now no sum-dominance relationship between the Vickrey-nearest rule in Figure 3(a) and the reference rule in Figure 3(b), the latter clearly does better if there is a high probability of (x_1, y_1) lying close to the 45° line through (r_X, r_Y).

These figures thus illustrate how the reference rule can be chosen to achieve robustness over the auction designer's preferred domain, using the information available to her, including what can be learnt from losing bids.

Extending the analysis beyond our two-objects/three-bidder-types environment raises a number of additional issues.

One important issue is that it is not clear whether a core selecting auction should select from the MRC or from the "bidder-optimal core" – the set of core-payment vectors such that no bidder's payoff can be strictly increased without reducing any other's payoff. Although this coincides with the MRC in our environment, it is often a strict superset of it in more general environments.[19]

Day and Milgrom's original program of selecting among *all* core-selecting payment rules on the Pareto-like criterion of being undominated in terms of maximal incentives refined the set of admissible payment rules only to the bidder-optimal core-selecting rules. Similarly, if we select among core-selecting rules using Pareto-like criteria based on marginal incentives, we refine to rules selecting from the bidder-optimal core. Minimising the sum-across-bidders of marginal incentives then further refines the set of admissible payment rules to a set including reference rules that choose particular bidder-optimal (but not necessarily MRC) points when possible.[20]

---

[19] For example, if each of three winning bidders has value 2 for any one of three objects A, B, and C, and a fourth bidder has value 3 for either of the packages $\{A, B\}$, or $\{B, C\}$ (but has no interest in just B, or in the package $\{A, C\}$), the unique MRC point has payments 1, 2, 1 for A, B, C, respectively, whereas the bidder-optimal core consists of vectors $(p, 3 - p, p)$ with $1 \leq p \leq 2$. The closest core point to the Vickrey payments is $(4/3, 5/3, 4/3)$.

[20] See Erdil and Klemperer (2009). Likewise, Parkes and Ungar's (2000) and Ausubel and Milgrom's (2002) core-selecting auction is bidder-optimal core- (but not necessarily MRC-) selecting.

However, the sum-across-bidders of maximum-possible incentives to deviate criterion refines the admissible set from the bidder optimal core to the MRC when these differ. So if minimising *both* maximum-possible *and* marginal incentives is desired, we should continue to use our sum-across-bidders of marginal incentives criterion after restricting to MRC points.[21]

6. Conclusion

Our approach of looking for a rule in which each winner's payment is as far as possible independent of her bid seems in the original spirit of Vickrey pricing. Bidders' *marginal* incentives for unilateral deviations may be more relevant to whether they bid close to "truthfully" than the maximum-possible incentives to deviate that previous authors focused on; and analysing bidders' marginal incentives suggests Reference Rules are natural payment rules.

Other practical considerations may argue the same way. If the packages being priced comprise identical, or close-to-identical, objects, a Reference Rule can set "minimally discriminatory" prices by specifying reference payments proportional to the number of slots acquired in an airport-slot auction, or proportional to the bandwidth acquired in a radio spectrum auction.[22] This approach of assigning prices that are as close to equal as possible for equal packages may both seem fair, and also be easy to explain to observers. But our analysis above is clearly only preliminary; much more research remains to be done.[23]

A. Appendix

A Bayesian-Nash equilibrium analysis of Section 4's Example for the special case in which it is common knowledge that x_1 and y_1 are independently, identically and uniformly distributed yields:

---

[21] By contrast, as discussed in the Introduction, using a Vickrey-nearest rule after restricting to the MRC is harder to justify. Furthermore, with more than two objects, Vickrey-nearest pricing can recreate some of the strange incentives for merger, demerger, and collusion, which core pricing was intended to mitigate. In a simple three-unit example, bidders i, j and k each have value 6 for a single unit (but no interest in more than one). A "package" bidder has value 12 for the package of all three units (but no use for fewer than three). With truthful bidding, and Vickrey-nearest pricing, i, j and k each pay 4 if they all bid independently, but i and j together pay 9 if they bid as a single entity (and independently of k) – so merger hurts (and *de*merger pays) here.

[22] Or if the core-selecting auction is part of a dynamic package auction, the reference payments might be set proportional to the "activity points."

[23] Nor does our research yet support the more general use of core-selecting auctions. No core-pricing rule fully eliminates the problems that plague the Vickrey auction. For example if two bidders i and j have values for objects A, B, and the package $\{A, B\}$, of 2,1,3 and 1,2,3, respectively, then allocating A to i, and B to j is efficient. The Vickrey payment vector is (1,1). If i increases her bid for A to 3, the same allocation is chosen with Vickrey payments (1,0). If j also increases her bid for B to 3, the allocation remains unchanged with Vickrey payments (0,0). Even though all the Vickrey payment vectors are in the core, revenues are non-monotonic in bids (which opens the way for shilling), and collusively raising bids can lower revenues.

While complementarities create problems for the simultaneous ascending auction that is now commonly used for allocating, e.g., radio spectrum (see, e.g., Klemperer, 2004), concerns about this auction's costliness and susceptibility to market power are addressed for "substitutes" environments by Klemperer's (2008, 2010) and Milgrom's (2009) recently-proposed auction designs.

If bidders choose the bids corresponding to the welfare-maximising Nash equilibrium given the pricing rule, then the welfare-maximising MRC-selecting mechanism is Vickrey-nearest pricing.

To understand this result, note that choosing MRC payments for type-X and type-Y winners is equivalent to the "public goods" problem of deciding how they should share the cost of raising $z_1 - (V_X + V_Y)$, the amount they are required to pay on top of their Vickrey payments. Since a type-Z bidder pays her Vickrey payment, so has a dominant strategy of bidding her commonly-known valuation, z_1, it follows that any MRC payment rule P is equivalent (in payoffs for the type-X and type-Y bidders) to a direct-revelation trading rule, where the type-X bidder corresponds to a seller with value $s = z_1 - x_1$, the type-Y bidder corresponds to a buyer with value $b = y_1$, and when trade occurs it does so at price $p(b, s) = P_Y(z_1 - s, b)$. Myerson and Satterthwaite (1983) show that the linear increasing equilibrium of the 1/2-double auction identified by Chatterjee and Samuelson (1983) maximises the expected gains from trade subject to the incentive compatibility and interim individual rationality constraints. So the optimal MRC payment rule would pick $p(b, s) = (b + s)/2 = (y_1 + z_1 - x_1)/2$, which is identical to the type-Y bidder's payment under Vickrey-nearest pricing.

Note however that (i) the Vickrey-nearest rule supports a continuum of other Nash equilibria in this example, and many of these yield *very* much lower welfare.[24] At least, if we depart from this very special case, there is no obvious way of choosing among equilibria.[25] (ii) Even the welfare-maximising equilibrium yields not much more of the available surplus than is achieved by the simplest "reference rule."[26] (iii) Even small generalisations seem to render the example intractable.[27]

References

Ausubel, Lawrence, Peter Cramton, and Paul Milgrom (2006). "The clock-proxy auction: a practical combinatorial auction design." In *Combinatorial Auctions*, edited by Peter Cramton, Yoav Shoham, and Richard Steinberg, MIT Press.

Ausubel, Lawrence, and Paul Milgrom (2002). "Ascending Auctions with Package Bidding." *Frontiers of Theoretical Economics*, 1:1, Article 1.

Ausubel, Lawrence, and Paul Milgrom (2006). "The lovely but lonely Vickrey auction." In *Combinatorial Auctions*, edited by Peter Cramton, Yoav Shoham, and Richard Steinberg, MIT Press.

Chatterjee, Kalyan, and William Samuelson (1983). "Bargaining under Incomplete Information." *Operations Research*, Vol. 31, No. 5, 835–851.

Cramton, Peter (2009). "Spectrum Auction Design." U Maryland Working Paper.

---

[24] The multiplicity corresponds to the multiplicity of equilibria of the Chatterjee-Samuelson mechanism. The fact that many equilibria yield much lower *revenue* (as well as lower welfare) than the unique equilibrium of the Vickrey auction casts doubt on the merits of this core-selecting auction.

[25] In our special case it seems reasonable to focus on the unique linear increasing equilibrium.

[26] The latter uses reference payments $r_X = r_Y = z_1/2$, $r_Z = x_1 + y_1$, and in its simplest equilibrium both bidders bid $z_1/2$ if $v \geq z_1/2$, and zero otherwise, capturing 75% of the total available surplus, compared with the 84% captured by the best equilibrium under the Vickrey-nearest rule. See Erdil and Klemperer (2009) for further discussion.

[27] In particular, any relaxation of the assumption of common knowledge of z_1 seems too hard. For a sense of the difficulties, see Ledyard's (2007) impressive analysis of our environment, though without imposing core (or MRC)-pricing constraints.

Day, Robert W., and Peter Cramton (2008). "The Quadratic Core-Selecting Payment Rule for Combinatorial Auctions." U Maryland Working Paper.

Day, Robert W., and Paul Milgrom (2008). "Core-selecting package auctions." *International Journal of Game Theory*, 36:3–4, 393–407.

Day, Robert W., and S. Raghavan (2007). "Fair Payments for Efficient Allocations in Public Sector Combinatorial Auctions." *Management Science*, 53:9, 1389–1406.

Edelman, Benjamin, Michael Ostrovsky, and Michael Schwarz (2007). "Internet Advertising and the Generalized Second-Price Auction: Selling Billions of Dollars Worth of Keywords." *American Economic Review*, 97:1, 242–59.

Erdil, Aytek, and Paul Klemperer (2009). "Alternative Payment Rules in Core-Selecting Package Auctions." Unpublished notes, Oxford University.

Klemperer, Paul (2004). *Auctions: Theory and Practice. (The Toulouse Lectures in Economics)*, Princeton University Press.

Klemperer, Paul (2008). "A New Auction for Substitutes: Central-Bank Liquidity Auctions, 'Toxic Asset' Auctions, and Variable Product-Mix Auctions." mimeo, Oxford University.

Klemperer, Paul (2010). "The Product-Mix Auction: A New Auction Design for Differentiated Goods." *Journal of the European Economic Association,* 8:2–3, 526–536.

Ledyard, John O. (2007) "Optimal combinatoric auctions with single-minded bidders." *EC'07: Proc. 8th ACM conference on Electronic commerce*, pp. 237–42.

Lubin, Benjamin, and David C. Parkes (2009). "Quantifying the Strategyproofness of Mechanisms via Metrics on Payoff Distributions." *Proc. 17th National Conference on Artificial Intelligence (AAAI-00)*, pp. 74–81.

Milgrom, Paul (2009). "Assignment Messages and Exchanges." *American Economic Journal: Microeconomics*, 1:2, 95–113.

Myerson, Roger B., and Mark A. Satterthwaite (1983). "Efficient mechanisms for bilateral trading." *Journal of Economic Theory*, 29, 265–281.

Parkes, David C., Jayant Kalagnanam, and Marta Eso (2001). "Achieving Budget-Balance with Vickrey-Based Payment Schemes in Exchanges." *Proc. 17th Int'l Joint Conf. Artificial Intelligence (IJCAI 01)*, pp. 1161–1168.

Parkes, David C., and Lyle H. Ungar (2000). "Iterative Combinatorial Auctions: Theory and Practice." *Proc. 17th National Conference on Artificial Intelligence (AAAI-00)*, pp. 74–81.

Varian, Hal (2007). "Position auctions: Theory and Practice." *International Journal of Industrial Organization*, 25, 1163–78.

CHAPTER 13

On the Impossibility of Core-Selecting Auctions

Jacob K. Goeree and Yuanchuan Lien[*]

1. Introduction

Practical auction design is often complicated by institutional details and legal or political constraints. For example, using bidder-specific bidding credits or reserve prices may be considered discriminatory and unlawful in some countries making it impossible to implement an optimal auction design. More generally, the use of sizeable reserve prices may cause political stress due to the fear that it slows down technological progress when licenses remain unsold. While constraints of this nature are common and important in practice, mechanism design theory has typically treated them as secondary to incentive constraints.

Recent work by Day and Raghavan (2007), Day and Milgrom (2008), and Day and Cramton (2008) breaks with this tradition and asks how close incentive constraints can be approximated if other (institutional or political) constraints are put first. In particular, these authors have proposed an alternative payment rule to fix some drawbacks of the well-known Vickrey-Clarke-Groves mechanism, or "Vickrey auction" for short. When goods are substitutes the Vickrey auction produces an outcome, i.e. an allocation and payoffs for the seller and bidders, that is in the core. However, when goods are complements, the Vickrey outcome, while efficient, is not necessarily in the core and seller revenue can be very low as a result. Furthermore, non-core outcomes are "unfair" in that there are bidders willing to pay more than the winners' payments, which makes the auction vulnerable to defections as the seller can attract better offers afterwards. The low revenue, perceived unfairness, and instability of Vickrey outcomes can create legal and political problems, which the alternative payment rule seeks to avoid.

The types of auctions proposed by Day et al. employ a payment rule that insures that outcomes are in the core with respect to *reported* values, i.e. the final allocation maximizes the total reported value and no coalition of bidders can block the outcome with respect to bidders' reports. Unless reported values are always truthful, however, it does *not* imply that core outcomes are produced with respect to bidders' *true* preferences. To

---

[*] We would like to thank Larry Ausubel, Peter Cramton, Paul Milgrom, John Wooders, two referees and especially the Editor, Faruk Gul, for their suggestions. Goeree gratefully acknowledges financial support from the European Research Council (ERC Advanced Investigator Grant, ESEI-249433).

distinguish these two cases we use core to mean "core with respect to true values" and core* to mean "core with respect to reported values," and refer to the auctions proposed by Day et al. as core*-selecting auctions.

Ideally, a core*-selecting auction produces outcomes that are in the core so that they are stable. The well-known Green-Laffont (1977) and Holmstrom (1979) theorem states that the Vickrey auction is the only efficient auction with a dominant-strategy equilibrium. Thus, the Vickrey auction is the only candidate for core implementation in dominant strategies. Under the weaker notion of Bayesian Nash equilibrium, the Vickrey auction is no longer the unique efficient mechanism (e.g. in the single-unit, symmetric case all standard auction formats are efficient in equilibrium). But we show that *any* core-selecting auction must be equivalent to the Vickrey auction in the sense that, for every possible valuation profile, the seller's revenue and bidders' profits are ex post identical. In other words, imposing dominant strategies comes at no cost for core implementation.

Our approach builds on previous work by Krishna and Perry (1998), Williams (1999), and Krishna and Maenner (2001) who generalize Myerson's "payoff equivalence" result and establish equivalence between the expected revenue of the Vickrey auction and the *expected* revenue of any efficient mechanism. Moreover, Ausubel and Milgrom (2002) show that the core constraints imply that the *ex post* revenue of any efficient mechanism can be no less than that of the Vickrey auction. Combining these results enables us to show that, in equilibrium, any efficient core-selecting auction yields the same ex post profits for the seller and bidders as the Vickrey auction. In other words, if the Vickrey outcome is in the core then it is the unique Bayesian Nash (and unique dominant-strategy) implementable outcome and otherwise no core outcome can be implemented.

For the case of substitutes it is well known that the Vickrey outcome is in the core. Our results imply that other mechanisms that result in different outcomes, including core*-selecting auctions, yield outcomes outside the core. When goods are not substitutes and the Vickrey outcome is not in the core, our results imply that *no* auction produces equilibrium outcomes that are in the core. In addition, as a competitive equilibrium with linear prices is a core outcome that may differ from the Vickrey outcome, our results imply that the competitive equilibrium outcome cannot always be implemented.

We also illustrate that core*-selecting auctions may perform worse than the Vickrey auction. We consider the BCV mechanism[1] proposed by Day and Cramton (2008), which has the following properties: (i) the outcome is in the core*, (ii) bidders' profits are maximized, and (iii) payments are as close as possible to the original Vickrey payments. For complete-information environments it has been shown that the BCV auction yields the Vickrey outcome when it is in the core and results in higher seller revenues when it is not. Moreover, with complete information, the BCV auction minimizes the maximal gain from deviating from truthful bidding (e.g. Day and Milgrom, 2008).[2]

These positive results, however, rely crucially on the assumption that bidders' values and, hence, their bids are commonly known. In most practical applications, bidders'

---

[1] BCV stands for bidder-optimal, core*-selecting, and Vickrey-nearest.

[2] It is important to point out that bidders' maximal incentives for (possibly large) deviations are minimized. Erdil and Klemperer (2010) show that bidders always have marginal incentives to deviate from truthful bidding, and they propose the introduction of reference prices to reduce bidders' incentives for such marginal deviations.

values constitute proprietary information. We therefore consider a simple incomplete-information environment where two local bidders interested in single items compete with a global bidder interested in the package. Bidders' values are privately known and uniformly distributed. We show that the BCV auction results in revenues on average lower than Vickrey revenues and outcomes that are inefficient and on average *further from the core than Vickrey outcomes*.[3] The reason for these negative results is that bidders no longer have a dominant strategy to bid truthfully.

To summarize, recent literature on "practical auction design" relaxes incentive constraints and focuses on the stability and fairness of outcomes. In particular, core*-selecting auctions yield stable and fair outcomes with respect to the bids, which are treated as exogenous parameters without making any assumptions about bidders' behavior, i.e. how their bids relate to their private information. In contrast, a core-selecting auction produces stable and fair outcomes with respect to bidders' true values assuming Bayesian Nash equilibrium behavior. Our results show that if a core-selecting auction exists it is identical to the Vickrey auction. Hence, no core-selecting auction exists if the Vickrey auction is not core selecting, in which case, the Vickrey auction may outperform the recently proposed core*-selecting auctions.

This paper is organized as follows. In the next section we introduce the bidding environment and prove that core-selecting auctions are generally not possible. In Section 3 we explain the construction of BCV prices and analyze the Bayesian Nash equilibrium of the BCV auction and evaluate its performance in terms of revenue and efficiency. Section 4 concludes.

2. Model and Main Result

The model follows Ausubel and Milgrom (2002) and Krishna and Maenner (2001). The set of auction participants, L, consists of the seller, labeled $l = 0$, and the buyers, labeled $l = 1, \ldots, |L| - 1$. Buyer l's type is K-dimensional with $t_l \in [0, 1]^K$ and buyers' types are independently distributed according to a probability measure with full support. Let $t = (t_l \in [0, 1]^K; l \in L \setminus 0)$ denote the profile of types. There are N indivisible items for sale, with Ω being the set of items. Let 2^Ω denote the powerset of Ω, i.e. the set of all subsets of Ω. An allocation is a set $x = \{x_l \mid x_l \in 2^\Omega; l \in L \setminus 0\}$, where x_l denotes the package or bundle assigned to buyer l. An allocation is feasible if for any $\omega \in \Omega$ and $l, l' \in L \setminus 0$ with $l \neq l'$, $\omega \in x_l$ implies $\omega \notin x_{l'}$. The set of all feasible allocations is denoted by X. Buyer l values allocation $x \in X$ at $v_l(x, t) \in \mathbb{R}$, while the seller has no value for the items: $v_0(\cdot, \cdot) = 0$. A buyer with type 0 has zero value, $v_l(x, 0) = 0$, as does a buyer who receives no items, $v_l(\{\emptyset, x_{-l}\}, t) = 0$. As in Krishna and Maenner (2001, Hypothesis I) we assume that buyer l's valuation function $v_l(x, \{t_l, t_{-l}\})$ is convex in t_l.[4] Finally, utilities are quasi-linear, i.e. when the allocation is x and buyer l pays a price $h \in \mathbb{R}_+$ her utility is $u_l(h, x, t) = v_l(x, t) - h$.

---

[3] Our results do not rule out that the BCV auction may outperform the Vickrey auction in other environments. For example, Ausubel and Baranov (2010) show that when local bidders' values are correlated the BCV auction performs better than the Vickrey auction if the degree of correlation is sufficiently high.

[4] Together with convexity of the type space this assumption guarantees payoff equivalence, see the proof of Proposition 1 in Krishna and Maenner (2001). For alternative sufficient conditions for payoff equivalence, see Hypothesis II in Krishna and Maenner (2001) or Williams (1999) who shows that equivalence requires that the interim expected valuation $V_l(t_l^* | t_l) \equiv E_{t_{-l}}[v_l(x(t_l^*, t_{-l}), t_l)]$ of each agent is continuously differentiable in (t_l^*, t_l) at points that satisfy $t_l^* = t_l$. Note that $v_l(x, \cdot)$ depends only on buyer l's own type in Williams' paper.

An (indirect) auction mechanism consists of a feasible allocation rule $\hat{x}_l(b)$ and a payment rule $\hat{h}_l(b)$ based on bids, or reported values, $b = (b_l(x_l) \in \mathbb{R}_+; x_l \in 2^\Omega, l \in L \setminus 0)$ with $b_0(\cdot) = 0$.[5] The coalitional value for $S \subset L$ with respect to reported values is given by $B(S) = \max_{x \in X} \Sigma_{l \in S} b_l(x_l)$ if $0 \in S$ and $B(S) = 0$ otherwise. With this definition of the coalitional value, the core* can be defined as

$$\text{core}^*(L, B) = \left\{ \pi \mid \sum_{l \in L} \pi_l = B(L), (\forall S \subset L) \, B(S) \leq \sum_{l \in S} \pi_l \right\}.$$

Definition 1. A core*-selecting auction $\{\hat{x}_l(b), \hat{h}_l(b)\}$ is such that $\hat{\pi}_l(b) \equiv b_l(\hat{x}_l(b)) - \hat{h}_l(b)$ for $l = 1, \ldots, |L| - 1$ and $\hat{\pi}_0(b) \equiv \Sigma_{l \in L \setminus 0} \hat{h}_l(b)$ satisfy $(\hat{\pi}_l(b); l \in L) \in \text{core}^*(L, B)$.

This definition takes bids as exogenous parameters. To define core-selecting auctions that yield allocations in the core with respect to bidders' true valuations, an assumption is needed about how bidders' private information translates into bids. We assume equilibrium behavior. Let $b^*(t)$ denote the Bayesian Nash equilibrium bids of an auction $\{\hat{x}_l(b), \hat{h}_l(b)\}$ and let $x_l(t) = \hat{x}_l(b^*(t))$, $\pi_l(t) = v_l(\hat{x}(b^*(t)), t) - \hat{h}_l(b^*(t))$ for $l = 1, \ldots, |L| - 1$ and $\pi_0(t) = \Sigma_{l \in L \setminus 0} \hat{h}_l(b^*(t))$ be the Bayesian-Nash equilibrium allocations and payoffs respectively. We refer to $\{x_l(t), \pi_l(t)\}$ as the (equilibrium) *outcome* of the auction mechanism.

The set of possible allocations is compact so $\arg\max_{x \in X} \Sigma_{l \in S} v_l(x, t)$ is non-empty for all t. Given the type profile, t, the coalitional value is given by $w(S, t) = \max_{x \in X} \Sigma_{l \in S} v_l(x, t)$ if $0 \in S$ and $w(S, t) = 0$ otherwise. With this definition of the coalitional value, the core can be defined as

$$\text{core}(L, w, t) = \left\{ \pi \mid \sum_{l \in L} \pi_l = w(L, t), (\forall S \subset L) \, w(S, t) \leq \sum_{l \in S} \pi_l \right\}.$$

Definition 2. An auction $\{\hat{x}_l(b), \hat{h}_l(b)\}$ is core-selecting if the outcome $\{x_l(t), \pi_l(t)\}$ satisfies $(\pi_l(t); l \in L) \in \text{core}(L, w, t)$ for all t.

A direct corollary of this definition is that a core-selecting auction is efficient, losing bidders pay nothing, and winning bidders' payoffs are non-negative.

Recall that two auctions, $\{\hat{x}_l(b), \hat{h}_l(b)\}$ and $\{\hat{x}'_l(b), \hat{h}'_l(b)\}$, are *interim* payoff equivalent if, for all $t_l \in [0, 1]^K$ and $l \in L$, the associated equilibrium payoffs satisfy $E_{t_{-l}}[\pi_l(t)] = E_{t_{-l}}[\pi'_l(t)]$. The stronger notion of *ex post* payoff equivalence requires that, for all t, $\pi_l(t) = \pi'_l(t)$.

Proposition 1. *Any core-selecting auction is ex post payoff equivalent to the Vickrey auction.*[6] *Hence, if the Vickrey auction is not core-selecting, no core-selecting auction exists.*

---

[5] Note that in core*-selecting auctions, a bidder's reported values only depend on her own allocation and not that of others, i.e. reported values do not reflect allocative externalities (even if actual values do). If a bidder does not report a value for a certain package then this package is assumed to have zero value.

[6] Recall that the Vickrey auction is a direct auction mechanism $\{x_l^V(t), h_l^V(t)\}$ that truthfully implements an efficient allocation $x(t)$ in dominant strategies. Specifically, $x_l^V(\cdot) = x_l(\cdot)$ and $h_l^V(t) = v_l(x(t), t) - w(L, t) + w(L \setminus l, t)$.

Proof. Bidders' types are revealed ex post and Ausubel and Milgrom (2002, Theorem 5) show that bidder l's Vickrey payoff $\pi_l^V(t) = w(L, t) - w(L \setminus l, t)$ is her highest payoff in the core. Thus, the payoff from any core-selecting auction satisfies $\pi_l(t) \leq \pi_l^V(t)$ for all l and t. The model assumptions detailed at the start of the section match those in Krishna and Maenner (2001), which implies that we can apply their interim payoff equivalence result (see their Proposition 1): the expected payoff function of a Bayesian incentive compatible mechanism is, up to an additive constant, determined by the allocation rule. The assumption that a buyer with the lowest type 0 has zero value implies that the additive constant is 0.[7] In addition, any core-selecting auction employs the same efficient allocation rule.[8] Interim payoff equivalence thus implies that bidder l's interim expected payoff in any Bayesian incentive compatible, efficient mechanism is the same as in the Vickrey auction: $E_{t_{-l}}[\pi_l(t)] = E_{t_{-l}}[\pi_l^V(t)]$. Therefore, $\pi_l(t) = \pi_l^V(t)$ for almost all types. □

Remark 1. Only a small subset of core constraints is required for the above result, namely the ones that involve L, $L \setminus l$, and l for all $l \in L$.

Remark 2. Our results differ from the Green and Laffont (1977) and Holmstrom (1979) theorem, which establishes conditions under which the Vickrey auction is the unique efficient auction with a dominant-strategy equilibrium. We relax dominant strategy equilibrium to Bayesian Nash equilibrium and replace the condition that losing bidders pay nothing with the requirement that the outcome must be in the core – we show that the only possibility is still the Vickrey auction.[9]

Remark 3. Auction papers that employ incomplete-information environments like the one studied here typically focus on only a single core constraint: ex post efficiency. The rationale is that the auction "should put the items in the hands of those that value them the most," i.e. no further gains from trade are possible and no aftermarket is needed. The main idea behind imposing additional ex post core constrains is that sales prices should be such that the seller cannot benefit from forming a coalition with a set of bidders different from the winning bidders. Recall that, by the revelation principle, the seller is able to infer bidders' values ex post and will want to sell to a different set of bidders if the outcome is not in the core. To summarize, the notion of ex post core is useful in incomplete information environments because it guarantees that the seller does not renege on the auction outcome and no aftermarket is needed.

When does the Vickrey auction result in a core outcome? Ausubel and Milgrom (2002, Theorem 12) show that the sufficient and necessary condition is that goods are substitutes (see also Gul and Stacchetti, 1999).[10] Specifically, following Ausubel and

---

[7] When bidder l's type is 0 the Vickrey payoff, $\pi_l^V(t) = w(L, t) - w(L \setminus l, t)$, is zero since no value results from assigning items to bidder l. Since the Vickrey payoff is the highest payoff in the core, $\pi_l(t) \leq \pi_l^V(t)$, this implies that the core payoff of a bidder with type 0 is necessarily zero.

[8] When there are multiple efficient allocations, e.g. in case of a tie, we assume that the Vickrey and core-selecting auction select the same allocation.

[9] We also assume independence of types, which is obviously not required with ex post incentive compatibility.

[10] Here we assume that buyers' valuations do not depend on packages received by others. Recall that given t, goods are substitutes to a buyer if whenever under certain prices the buyer's demand includes an item, he never drops the item if the prices for other items increase. Specifically, given a vector of item prices $p \in \mathbb{R}_+^{|\Omega|}$, the

Milgrom (2002), let V be a collection of valuation profiles, where a valuation profile $\{v_l(\cdot)\}_{l \in L}$ is a set containing the seller and bidders' valuation functions. Denote by V_{Sub} the set contains all valuation profiles such that goods are substitutes for all bidders and types. Similarly, denote by V_{Add} the set of all valuation profiles containing solely additive valuations.[11]

Corollary 1. *For $V \supseteq V_{Add}$, a core-selecting auction exists if and only if $V \subseteq V_{Sub}$.*

This result follows from Theorem 12 in Ausubel and Milgrom (2002), which shows that for $V \supseteq V_{Add}$ the Vickrey auction is core-selecting if and only if $V \subseteq V_{Sub}$.[12] This fact, together with Proposition 1, yields Corollary 1.

Our results also have implications for the existence of mechanisms that lead to Walrasian, or competitive equilibrium (CE), outcomes. The definition below follows the notation in Definition 2.

Definition 3. An auction $\{\hat{x}_l(b), \hat{h}_l(b)\}$ is CE-selecting if there exist item prices $\{\hat{p}_j(b)\}_{j \in \Omega}$ such that $\hat{h}_l(b) = \sum_{j \in \hat{x}_l(b)} \hat{p}_j(b)$ and the outcome $\{x_l(t), \pi_l(t)\}$ satisfies $x_l(t) \in \arg\max_{y \subseteq \Omega} \{v_l(y, t) - \sum_{j \in y} p_j(t)\}$ and $\pi_l(t) = v_l(x_l(t), t) - \sum_{j \in x_l(t)} p_j(t)$ for all $l \in L \setminus 0$ and all t where $p_j(t) \equiv \hat{p}_j(b^*(t))$.

Since a CE-selecting auction is necessarily core-selecting, Proposition 1 implies:

Corollary 2. *Any CE-selecting auction is ex post payoff equivalent to the Vickrey auction. Hence, if the Vickrey auction is not CE-selecting, no CE-selecting auction exists.*

To illustrate, suppose there are $2M$ identical items (with $M \geq 2$) and two bidders with valuation functions $v_i(m, t) = v(m)t_i$ for $i = 1, 2$, where the t_i are independently distributed and $v(m)$ is a strictly concave function for $0 \leq m \leq 2M$. Consider type profiles for which the types are "close," e.g. $t_1 = t(1 + \epsilon)$ and $t_2 = t(1 - \epsilon)$ with ϵ small. An efficient outcome dictates that both bidders get M units and the resulting Vickrey payments are approximately $(v(2M) - v(M))t$ for both bidders. In other words, the Vickrey per-unit price is

$$p^V = \frac{v(2M) - v(M)}{M} t.$$

A lower bound for the competitive equilibrium price, p, follows from the requirement that at price p neither bidder desires an additional unit: $v(M)t - Mp \geq v(M+1)t - (M+1)p$, or

$$p \geq \frac{v(M+1) - v(M)}{1} t,$$

which exceeds the Vickrey per-unit price by strict concavity of $v(\cdot)$. Thus, in any CE-selecting auction, if it exists, the competitive equilibrium prices are higher than the

---

buyer's demand $x_l^*(p, t)$ solves $\max_{x_l \subseteq \Omega} (v_l(x_l, t) - \sum_{k \in x_l} p_k)$. Goods are substitutes to buyer l if for any t and p, item $m \in x_l^*(p, t)$ implies $m \in x_l^*(p', t)$ for any p' with $p'_m = p_m$ and $p'_j \geq p_j$ for $j \neq m$.

[11] With additive valuations, the value of a package is the sum of the values for individual items in that package.

[12] If $V \not\subseteq V_{Sub}$, then goods are not substitutes for some bidder in some valuation profile in V. Given such a bidder's valuation function, Ausubel and Milgrom (2002, Theorem 12) show how to construct additive valuations (or choose valuations from V_{Add}) for three additional bidders such that under the valuation profile consisting of these four bidders, the Vickrey auction is not core-selecting.

Vickrey prices for a positive measure of types, resulting in lower bidder payoffs than those in the Vickrey auction. Therefore, Corollary 2 implies that a CE-selecting auction does not exist.

This result has important implications for equilibrium behavior in commonly used auction formats. For example, it implies that straightforward bidding cannot be an equilibrium of the simultaneous ascending auction for all possible valuation profiles, since straightforward bidding results in competitive equilibrium prices (e.g. Milgrom, 2004).

3. BCV Auction: An Example

Consider an environment with two local bidders and a single global bidder who compete for two items labeled **A** and **B**. Local bidder 1 (2) is interested only in acquiring item **A** (**B**), for which she has value $v_1 \in \mathbb{R}$ ($v_2 \in \mathbb{R}$). The global bidder 3 is interested only in the package **AB** consisting of both items, for which she has value $V \in \mathbb{R}$. For simplicity, we assume bidders can only bid on packages they are interested in. Let b_i be local bidder i's bid (or reported value) for the item she is interested in and B be the global bidder's bid for package **AB**.

In the Vickrey auction, bidders simply bid their values for the object they are interested in: $b_i(v_i) = v_i$ and $B(V) = V$. This yields a fully efficient outcome, i.e. the global bidder wins all items iff $B > b_1 + b_2$, or equivalently, iff $V > v_1 + v_2$. While the Vickrey auction generates full efficiency, it is well known that it may result in low revenues when the outcome is not in the core (e.g. Ausubel and Milgrom, 2006).

The BCV auction $\{\hat{x}_l(b), \hat{h}_l(b)\}$ selects bidder-optimal points[13] in the core* and minimizes the distance, $(\sum_{l \subset L \setminus 0}(\hat{h}_l(b) - \hat{h}_l^V(b))^2)^{1/2}$, between the Vickrey payment $\hat{h}_l^V(b)$ and the BCV auction payment $\hat{h}_l(b)$, where $b \equiv \{b_1, b_2, B\}$.[14]

Definition 3. For the simple environment studied here the BCV auction $\{\hat{x}_l(b), \hat{h}_l(b)\}$ is characterized by payments $\hat{h}_3(b) = (b_1 + b_2)\mathbf{1}_{B > b_1 + b_2}$ for the global bidder and

$$\hat{h}_i(b) = \left(\max(0, B - b_{-i}) + \frac{1}{2}(B - \max(0, B - b_i) - \max(0, B - b_{-i}))\right)\mathbf{1}_{b_1 + b_2 \geq B}$$

for local bidders $i \in \{1, 2\}$. The allocation rule $\hat{x}_l(b)$ is the same as in the Vickrey auction: the global bidder wins the package **AB** if $B > b_1 + b_2$, otherwise each local bidder wins the item she is interested in.

When the global bidder wins, the outcome is always in the core* with respect to submitted bids so there are no adjustments to the global bidder's payment. As a result, the global bidder's strategy is unaffected, i.e. truthful bidding remains optimal. For the local bidders truthful bidding is no longer optimal, however, since their own bids affect their payments.

---

[13] The outcome $(\hat{\pi}_l; l \in L)$ in the core* is bidder-optimal in the core* if no other outcome $(\hat{\pi}'_l; l \in L)$ in the core* exists such that $\hat{\pi}'_l \geq \hat{\pi}_l$ for all $l \in L \setminus 0$ and $\hat{\pi}'_l > \hat{\pi}_l$ for some l.
[14] See Day and Cramton (2012) for more details.

The next result shows the degree to which local bidders "shade" their bids in response to the change in payment rule. We assume that local bidders' values are uniformly distributed on [0, 1] and the global bidder's value for the package is uniformly distributed on [0, 2].

Result 1: A Bayesian Nash equilibrium of the BCV auction is given by $B(V) = V$ and

$$b(v) = \max(0, v - \alpha) \tag{3.1}$$

where $\alpha = \frac{1}{2}E(b(v)) = 3 - 2\sqrt{2}$.

Result 1 (see the Appendix for a proof) shows that the introduction of BCV prices creates incentives for local bidders to "free ride," i.e. each local bidder wants to win but prefers other local bidders to bid high. The consequences for the auction's performance are easy to calculate. A direct computation shows that the relative average efficiency of the core*-selecting auction is: $E_{BCV}/E_V = 98\%$, relative average revenue is: $R_{BCV}/R_V = 91\%$, and the relative average Euclidean distance to the core is: $d_{BCV}/d_V = 126\%$.[15] We can thus conclude:

Result 2: Compared to the Vickrey auction, the BCV auction has lower expected efficiency and revenue and it produces outcomes that are on average further away from the core.

The main insight of Result 1, i.e. that truthful bidding is not an equilibrium in the BCV auction, holds under more general conditions.[16]

4. Conclusion

The BCV auction has some remarkable properties in complete-information environments where bidders' values and, hence, their bids are commonly known (e.g. Day and Milgrom, 2008). In particular, when bids are completely predictable, the introduction of BCV prices ensures outcomes that are "fair" and seller revenues that are not embarrassingly low.

This paper considers the performance of the BCV auction for the realistic case when bidders' values are privately known and, hence, their bids are not perfectly predictable. If in such incomplete-information environments truthful bidding would be optimal then the BCV auction would reliably outperform the Vickrey auction. However, our analysis shows that truthful bidding is *not* an equilibrium. We show that the BCV auction may result in lower expected revenue and efficiency as well as outcomes that are on average *further* from the core than Vickrey outcomes.

---

[15] In this example, the average distance from the core is $\sqrt{E_t[\sum_{S \subset L} \max\{0, w(S,t) - \sum_{l \in S} \pi_l(t)\}^2]}$, where $\pi_l(t)$ is agent l's profit from the mechanism in consideration, specifically, $\hat{\pi}_l(t)$ or $\pi_l^V(t)$.

[16] For general distributions of the local bidders' values, $F(v)$, the optimal bid function is given by $b(v) = \max(0, v - \alpha)$ where now $\alpha = \frac{1}{2}\int_\alpha^1 (v - \alpha)dF(v)$. For some distributions, the resulting performance measures of the BCV auction are *worse* than those of Result 2. Likewise, the assumption that the global bidder's value is uniformly distributed can be relaxed. The resulting bidding function for the local bidders is no longer linear but free riding still occurs.

We study whether core-allocations can be achieved by any mechanism in an environment where goods can be substitutes and/or complements. We prove that the ex post surplus, seller's revenue, and bidders' profits in any core-selecting auction are identical to their Vickrey counterparts, i.e. any core-selecting auction is equivalent to the Vickrey auction. A fortiori, if the Vickrey outcome is outside the core, no core-selecting auction exists.

Our impossibility result is akin to Myerson and Satterthwaite's (1983) finding that efficient bilateral trade is not generally possible. In both cases the intuition is that incentive compatibility requires that market participants have information rents (reflecting their private information). In the two-sided setting studied by Myerson and Satterthwaite, traders' expected information rents may exceed the surplus generated by trade. In the one-sided setting studied here, bidders' information rents imply an upper-bound on the seller's revenue – an upper-bound that generally conflicts with some of the core constraints.

Core allocations are possible when all goods are substitutes. The interest in core*-selecting auctions, however, derives from environments in which the substitutes assumption is relaxed. In this case, competitive equilibrium does not necessarily exist and "...the conception of auctions as mechanisms to identify market clearing prices is fundamentally misguided" (Milgrom, 2004, p. 296). The core, which always exist in these one-sided applications, seems the natural and relevant solution concept since "...competitive equilibrium outcomes are always core outcomes, so an outcome outside the core can be labeled uncompetitive" (Milgrom, 2004, p. 303). Our results, however, demonstrate that with incomplete information, core assignments are not generally possible unless all goods are substitutes.

A. Appendix: Proof

Proof of Result 1. As under the VCG mechanism, the global bidder's payment under BCV does not depend on his bid. So, the dominant strategy is to bid $B(V) = V$. Suppose local bidder 1 with value v_1 bids $b_1 \geq 0$. Consider a deviation to $b_1 + \varepsilon$ with $\varepsilon > 0$. The expected gain of this deviation takes place when it turns the local bidder from a losing bidder to a winner, i.e. when the global bidder's value lies between $b_1 + b_2(v_2)$ and $b_1 + \varepsilon + b_2(v_2)$, where $b_2(\cdot)$ is bidder 2's bidding function. Since the global bidder's bid and the sum of the local bidders' bids are equal (up to order of ε) in this case, bidder 1's BCV payment is simply $B - b_2(v_2) = b_1$, i.e. her own bid, and the profit is $v_1 - b_1$. (As a result, if $v_1 - b_1 < 0$, a reduction of bid decreases the chance of winning and thus reduces the loss. A reduction of bid also reduces the expected payment as shown in the following. So, $b_1 > v_1$ cannot be optimal.) Since the distribution of the global bidder's value is uniform on $[0, 2]$ and we have $b_i \leq v_i \leq 1$, the probability that B lies between $b_1 + b_2(v_2)$ and $b_1 + \varepsilon + b_2(v_2)$ is equal to $\frac{1}{2}\varepsilon$. Hence, the expected gain from deviation is

$$\frac{1}{2}\varepsilon(v_1 - b_1).$$

To determine the expected cost of deviation, note that the only term in local bidder 1's payment affected by an increase in bidder 1's bid is the $\frac{1}{2}\max(0, B - b_1)$ term. Hence, when bidder 1 raises her bid by ε her payment goes up by $\frac{1}{2}\varepsilon$ if and only if the global bidder's value is greater than b_1 and less than $b_1 + b_2(v_2)$ (since otherwise the local bidders do not win). So the expected cost of local bidder 1's deviation is simply

$$\frac{1}{2}\varepsilon \int_0^1 \int_{b_1}^{b_1+b_2(v_2)} \frac{1}{2} dV dv_2 = \frac{1}{4}\varepsilon E(b_2(v_2)).$$

Thus, for the bid $b_1 \geq 0$ to be optimal, the gain from deviation to $b_1 + \varepsilon$ with $\varepsilon > 0$ must not be greater than the loss. We have

$$v_1 - b_1 \leq \frac{1}{2} E(b_2(v_2)).$$

Similarly, for $b_1 > 0$, a deviation to $b_1 + \varepsilon$ with $\varepsilon < 0$ leads to the same result except with opposite signs of changes,

$$v_1 - b_1 \geq \frac{1}{2} E(b_2(v_2)).$$

Combining the above two conditions, we have when $v_1 > \frac{1}{2} E(b_2(v_2))$, the optimal bid $b_1^* = v_1 - \frac{1}{2} E(b_2(v_2))$. When $v_1 \leq \frac{1}{2} E(b_2(v_2))$, b_1^* cannot be positive, otherwise the second inequality is violated and deviation to $b_1 + \varepsilon$ is profitable. Thus, $b_1^* = 0$ for $v_1 \leq \frac{1}{2} E(b_2(v_2))$. Note that the above conditions also guarantee the global optimality of b_1^* because at any $b_1 \neq b_1^*$, locally moving towards b_1^* yields higher profit. We conclude that local bidder i's unique optimal response is $b_i(v_i) = \max(0, v_i - \frac{1}{2} E(b_{-i}(v_{-i})))$. Let $\alpha_1 = \frac{1}{2} E(b_2(v_2))$. We have $E(b_1(v_1)) = (1 - \alpha_1)^2/2 \equiv 2\alpha_2$. Similarly, $(1 - \alpha_2)^2/2 = 2\alpha_1$. Therefore, the equilibrium with $\alpha_1 = \alpha_2 = 3 - 2\sqrt{2}$ can be derived. □

References

Ausubel, Lawrence M. and Oleg V. Baranov (2010) "Core-Selecting Auctions with Incomplete Information," Working Paper.

Ausubel, Lawrence M. and Paul Milgrom (2002) "Ascending Auctions with Package Bidding," *Frontiers of Theoretical Economics*, 1(1), Article 1.

Ausubel, Lawrence M. and Paul Milgrom (2006) "The Lovely But Lonely Vickrey Auction," in Peter Cramton, Yoav Shoham, and Richard Steinberg, eds., *Combinatorial Auctions*, MIT Press.

Day, Robert W. and Peter Cramton (2012) "The Quadratic Core-Selecting Payment Rule for Combinatorial Auctions," *Operations Research*, 60:3, 588–603, 2012.

Day, Robert W. and Paul Milgrom (2008) "Core-Selecting Package Auctions," *International Journal of Game Theory 36*, 3–4, 393–407.

Day, Robert W. and Subramanian Raghavan (2007) "Fair Payments for Efficient Allocations in Public Sector Combinatorial Auctions," *Management Science*, 53(9), 1389–1406.

Erdil, Aytek and Paul Klemperer (2010) "A New Payment Rule for Core-selecting Package Auctions," *Journal of the European Economic Association*, 8:2–3, April-May, 537–547.

Green, Jerry and Jean-Jacques Laffont (1977) "Characterization of Satisfactory Mechanisms for the Revelation of Preferences for Public Goods," *Econometrica*, 45, 427–438.

Gul, Frank and Ennio Stacchetti (1999) "Walrasian Equilibrium with Gross Substitutes," *Journal of Economic Theory*, 87, 95–124.

Holmstrom, Begnt (1979) "Groves Schemes on Restricted Domains," *Econometrica*, 47, 1137–1144.

Krishna, Vijay and Eliot Maenner (2001) "Convex potentials with an application to mechanism design," *Econometrica*, 69(4), 1113–1119.

Krishna, Vijay and Motty Perry (1998) "Efficient Mechanism Design," Working Paper, Pennsylvania State University.

Milgrom, Paul (2000) "Putting Auction Theory to Work: the Simultaneous Ascending Auction," *Journal of Political Economy*, 108(2), 245–272.

Milgrom, Paul (2004) *Putting Auction Theory to Work*, Cambridge University Press.

Myerson, Robert B. and Mark A. Satterthwaite (1983) "Efficient Mechanisms for Bilateral Trading," *Journal of Economic Theory*, 29, 265–281.

Williams, Steven R. (1999) "A Characterization of Efficient, Bayesian Incentive Compatible Mechanisms," *Economic Theory*, 14, 155–180.

CHAPTER 14

Ascending Combinatorial Auctions with Risk Averse Bidders

Kemal Guler, Martin Bichler, and Ioannis Petrakis

1. Introduction

The need to buy or sell multiple objects arises in areas such as industrial procurement, logistics, or when governments allocate spectrum licenses or other assets. It is a fundamental topic and the question how multiple indivisible objects should be allocated via an auction has enjoyed renewed interest in recent years (Airiau and Sen 2003; Cramton et al. 2006; Day and Raghavan 2007; Xia et al. 2004). One of the key goals in this research literature is to develop mechanisms that achieve high (allocative) efficiency with a *strong* game-theoretical solution concept such as a dominant strategy or an ex-post Nash strategy, such that bidders have no incentive to misrepresent their valuations. In other words, the strategic complexity for bidders is low as they do not need prior information about other bidders' valuations. Allocative efficiency measures whether the auctioned objects finally end up with the bidders who value them the most, thus, representing a measure of social welfare.

Combinatorial auctions are the most general types of multi-object market mechanisms, as they allow selling or buying a set of heterogeneous items to or from multiple bidders (Cramton et al. 2006). Bidders can specify package (or bundle) bids, i.e., prices are defined for the subsets of items that are auctioned. The price is only valid for the entire bundle, and the bid is indivisible. For example, in a combinatorial auction a bidder might want to buy a bundle, consisting of item x and item y, for a bundle price of $100, which might be more than the sum of the item prices for x and y, when sold individually. We will say that bidder valuations for both items are *complementary* in this case.

Many publications have focused on the computational complexity of the allocation problem in combinatorial auctions (Lehmann et al. 2006; Rothkopf et al. 1998). Computational complexity is often manageable in real-world applications with a low number of items and bidders. In terms of strategic complexity, the standard solution in mechanism design is the Vickrey–Clarke–Groves (VCG) mechanism, which achieves efficiency in dominant strategies, i.e., bidders cannot increase their payoff by deviating from a truthful revelation of their valuations (Clarke 1971; Groves 1973; Vickrey 1961).

The Vickrey–Clarke–Groves mechanism is actually the unique direct revelation mechanism with a dominant strategy equilibrium (Green and Laffont, 1977), but it is rarely used in practice. One of the problems of the VCG mechanism is that the outcomes might not be in the *core*. This means that a set of losing bidders was willing to pay more than what the winners had to pay. Non-core outcomes lead to very low revenue for the auctioneer and possibilities for shill bidding (Ausubel and Milgrom 2006b). Also, in many markets bidders are simply reluctant to reveal their true valuations to an auctioneer in a single-round sealed-bid auction, and they prefer an ascending (multi-round) auction format which is more transparent and conveys information about the competition in the market.

Much recent research has therefore focused on ascending combinatorial auctions, i.e., generalizations of the single-item English auction where bidders can outbid each other iteratively (Drexl et al. 2009; Schneider et al. 2010; Xia et al. 2004). Such auctions are selecting a core-outcome wrt. the bids, because a coalition of losing bidders can always increase their bids to become winning as long as bidders still have a positive payoff.

Unfortunately, no ascending combinatorial auction format allows for a strong solution concept for general types of bidder valuations (Bikhchandani and Ostroy 2002). In other words, with sufficient prior information about other bidders' valuations, there could always be valuations where a bidder can profit from not bidding truthfully up to his valuation. Let's consider a simple market with two identical items and three bidders, which we refer to as the *threshold model*. One "global" bidder is only interested in the bundle of two items for a value of up to $100, while each of the two "local" bidders wants only one of the items and has a value of $80. Since the local bidders together are stronger than the global bidder (i.e., the sum of their one-item valuations is higher than the global bidder's two-item valuation), they could try to free-ride on each other by pretending to have lower valuations. For example, one of the local bidders could pretend to have a value of $21 only and hope that the second local bidder outbids the global bidder together with his bid of $21. Sano (2011) has analyzed a threshold model with two items and a simple button auction. He showed that with uncertainty about other local bidders' valuations, ascending combinatorial auctions can even lead to inefficient solutions, where a local bidder drops out too early such that the global bidder wins although this is not the efficient solution. The model requires a few simplifying assumptions to yield a closed-form perfect Bayesian equilibrium strategy, but it nicely captures the strategic challenges that bidders face.

In the threshold model with strong local bidders also the outcome of the VCG auction is not in the *core* and the winners pay less than what the losers could have paid (Goeree and Lien 2013). The total of the VCG payments in the above example is $40, while the global bidder would have bid up to $100. Only with strong restrictions on the bidder valuations such as if goods are substitutes, an ascending combinatorial auction can achieve a strong game-theoretical solution concept such as an ex-post equilibrium (Ausubel and Milgrom 2006a), where straightforward bidding maximizes bidders' utility independent of the valuations of other bidders such that there are no incentives for manipulation.

This can be considered a negative result, because combinatorial auctions are typically used when bidders do have complementary valuations, which casts doubts on the efficiency of the various combinatorial auction formats that have been developed in the recent years. Note that this result is not restricted to linear (item-level) prices or non-linear anonymous prices (Drexl et al. 2009; Schneider et al. 2010; Xia et al. 2004), but it is relevant for the entire literature on ascending multi-object auctions.

The result actually applies to any type of core-selecting combinatorial auction. For example, a two-stage core-selecting combinatorial clock auction (CCA) has been used in several countries world-wide recently to sell spectrum (Cramton 2013). In the CCA, bidder-optimal core-selecting payments are computed after an ascending and a sealed-bid auction phase. Cramton and Day (2012) developed a quadratic core-selecting payment rule, which minimizes the Euclidean distance from the Vickrey payments and which is nowadays in use in spectrum auctions around the world.

Overall, this suggests that ascending combinatorial auctions in the field might be prone to manipulation and inefficiency. For example, there are spectrum auctions in countries with many local operators bidding only on regional spectrum licenses and a few national bidders, who are interested in larger packages. These environments can be seen as straightforward extensions of the three-bidder threshold model and similar manipulation might well matter and lead to inefficiencies.

In contrast to the negative result by Sano (2011), lab experiments have consistently showed high efficiency of ascending combinatorial auctions when subjects had complementary valuations. Many of these experiments used simple threshold models similar to the one described above (Banks et al. 2003; Goeree and Holt 2010; Kwasnica et al. 2005; Scheffel et al. 2011). In this paper, we want to analyze if risk-aversion can explain high efficiency in ascending combinatorial auctions. As we will see, the impact of risk aversion on efficiency in this environment is not as obvious as in single-object auctions. Modeling risk aversion is a non-trivial extension of the results in Sano (2011), but one that is central for understanding bidder behavior in such auctions in the field.

1.1. Risk Aversion

In high-stakes spectrum auctions it is unlikely that bidders are risk-neutral as the outcome of such an auction can impact the economic fate of a telecom substantially. So, smaller payoffs have a higher utility as long as bidders win, which can be modeled with a concave utility function. *Risk aversion* is an important phenomenon to be considered in auctions due to the uncertainties faced by bidders in auctions in general. It is a fundamental concept in expected utility theory (Arrow 1965; Pratt 1964) and widely used to explain overbidding behavior in first-price sealed-bid auctions (Chen and Plott 1998; Cox et al. 1988; Kirchkamp and Reiss 2011). Risk aversion leads to different revenue rankings of single-item auctions and is likely to have an ample effect on equilibrium strategies in core-selecting combinatorial auctions. In single-item first price sealed-bid auctions, Riley and Samuelson (1981) show that risk aversion leads to uniformly higher bids and thus higher revenue. Bidders increase their bids since they want insurance against the possibility of losing. Risk aversion was also analyzed in the context of all-pay single-item auctions (Fibich et al. 2006), and single-item auctions with infinitely many bidders (Fibich and Gavious 2010).

In core-selecting auctions, the experiments in Goeree and Lien (2009) show that local bidders do not drop out at a price of null in experiments, as theory would suggest, but they continue to bid further. One might expect that risk aversion serves as a natural explanation for this overbidding phenomenon also in core-selecting auctions. However, the equilibrium bidding strategies in single-item auctions *cannot easily be extended* to combinatorial auctions. While risk aversion does not affect the equilibrium strategy under the single-item English auction, it might well have an impact on the condition for a non-bidding equilibrium in an *ascending* core-selecting combinatorial auction. The higher one local bidder bids, the higher is the probability of the other local bidders to drop out. Due to the *non-increasing equilibrium bidding strategies*, it is unclear if a risk-averse bidder should actually overbid in equilibrium and if risk-aversion can recover the efficiency losses and the low revenue observed by Goeree and Lien (2013) for sealed-bid and Sano (2011) for ascending core-selecting combinatorial auctions.

1.2. Contributions

While a large part of the literature on combinatorial auctions focuses on computational questions, we complement this literature with a game-theoretical analysis. The contributions of this paper are the following: We consider the environments that have been analyzed in the Bayesian literature on sealed-bid and ascending core-selecting combinatorial auctions by Sano (2011) but analyze the *impact of risk aversion* on equilibrium strategies, revenue, and efficiency. Understanding risk aversion is important for practical applications and the impact is not obvious as indicated above.

First, we characterize the necessary and sufficient conditions for a *perfect Bayesian equilibrium* of the ascending core-selecting auction mechanism to have the small bidders to drop at the reserve price. We do so for general environments with risk-averse bidders as well as arbitrary asymmetries across bidders with respect to initial wealth, value distributions, and risk attitudes. Our first result is a generalization of the condition for a non-bidding equilibrium in Sano (2011), which allows for arbitrary concave utility functions, reserve prices, and differences in initial wealth. This flexibility in all three parameters allows for the analysis of realistic scenarios.

Second, we provide comparative statics and show that risk aversion and bidder asymmetries affect the equilibrium outcomes in ways that can be systematically analyzed. The impact of risk aversion on equilibrium bidding strategies is not obvious, since bidding higher can allow the other local bidder to drop out. Unlike supermodular games or games fulfilling the single-crossing property (Athey 2001), such as the first price single-item sealed-bid auction, *the impact of risk aversion is not obvious in the ascending auction game*. The game does not fulfill these properties, as we will show, and bidders cannot simply buy insurance against the possibility of losing by increasing their bids. Increasing bids may lead to a lower probability of winning. Theorem 3 is central and it indicates that even in this case, risk-aversion reduces the scope of the non-bidding equilibrium in the sense that dropping at the reserve price ceases to be an equilibrium as the bidders become more risk averse. We also analyze different wealth levels, asymmetries across local bidders, and stochastic dominance orderings of the distributions of valuations and their impact on non-bidding. Similar to Goeree and Lien (2009) and Sano (2011) we do not analyze the continuation game, when the condition for a non-bidding equilibrium does not hold. So far, there are no Bayesian models of ascending combinatorial auctions in general to our knowledge, and it is likely that no closed-form solutions of Bayesian equilibria exist. The analysis of the continuation game is a significant challenge beyond the focus of this paper. However, our results already shed light on the impact of risk aversion on efficiency.

The remainder of the paper is structured as follows. Section 2 contains a formulation of the model with descriptions of the environment and the auction procedure, as well as some preliminary lemmas on the model elements. In Section 3, we characterize the necessary and sufficient conditions for an ascending core-selecting auction to have a perfect Bayesian equilibrium in which small bidders stop bidding at the reserve price. Section 4 discusses parametric cases with specific distribution and utility functions. Finally, Section 5 provides conclusions. Throughout the paper, proofs that are technical in nature are placed in the "Appendix".

2. The Model

We consider a stylized multiple-unit auction environment with two objects and three bidders. Two objects are to be sold to one or two of three potential buyers through a core-selecting auction. The two objects are indistinguishable from the bidders' valuation perspective. As in Goeree and Lien (2009) and Sano (2011) the environment has an a priori asymmetry in one of the bidders, referred to as the 'large bidder', who demands two units of the object and has value zero for a single unit. Each of the other bidders, referred to as 'small bidders', demand a single unit and has zero valuation for the second unit.

We consider an independent private values (IPV) environment where each buyer has a private value for the object(s) which is unknown to the others. In addition to the asymmetry of small and large bidders with respect to the number of units demanded, we allow a full range of ex ante asymmetries across bidders with respect to the distribution of valuations, utility functions and initial wealth levels.

We first describe the elements of the bidding environment followed by the details of the auction mechanism.

2.1. The Environment: Information, Valuations, and Utility Functions

The auction environment with three bidders, $N = \{1, 2, 3\}$, is represented by the collection $e = \{(F_1, F_2, G), (u_1, u_2, u_3), (\omega_1, \omega_2, \omega_3)\}$, where the three tuples denote the valuation distributions, the utility functions, and the wealth levels, respectively, of the three bidders. We provide the details of the two parts of an environment next.

2.1.1. Information and Valuations

Each buyer has a private value for the object(s) which is unknown to the others. We denote the valuations by V and index it by the bidders. Bidder i's valuation is a random variable V_i with distribution function $F_i(\cdot)$ for $i = 1, 2$ or $G(\cdot)$ for $i = 3$ which has a strictly positive and continuously differentiable density function $f_i(\cdot)$ or $g(\cdot)$ on its support $[\underline{v}_i, \bar{v}_i]$. We set $\underline{v}_1 = \underline{v}_2 = \underline{v}_3 = 0$, $\bar{v}_1 = \bar{v}_2 = 1$, $\bar{v}_3 = 2$. Bidders $i = 1, 2$ are only interested in a single unit, while bidder $i = 3$ is only interested in the package of two units. Each bidder is interested in a single package only, and which package bidders are interested in is assumed to be public knowledge. All already referenced papers modeling core-selecting auctions as a Bayesian game are using the same environment, which can be considered the simplest scenario where the Vickrey auction is not in the core. In our paper, we want to understand what risk aversion does to core-selecting auctions in this environment, where we still have closed form solutions for risk-neutral bidders.

2.1.2. Utility Functions and Risk Aversion

Bidders are expected utility maximizers. Bidder i has von-Neumann–Morgenstern utility function $u_i : \mathbb{R} \to \mathbb{R}$. If a bidder with value v and current wealth ω wins and pays a price b, his utility is $u_i(v - b + \omega)$; his utility is $u_i(\omega)$ if he loses. We assume u_i is twice continuously differentiable, with $u'_i(x) > 0$ and $u''_i(x) < 0 \forall x$. Therefore, u_i is concave and i risk-averse. Since von-Neumann–Morgenstern utility functions are unique up to affine transformations, i.e., $u(x)$ and $a + cu(x)$ represent the same underlying preferences for any choice of real numbers a and $c > 0$, we normalize the utility functions so that $u(\underline{\omega}) = 0$ and $u(\bar{\omega}) = 1$ for some wealth levels $\underline{\omega}$ and $\bar{\omega}$ with $0 \leq \underline{\omega} < \bar{\omega}$.

We use the measure $A(x) = \frac{-u''(x)}{u'(x)} \in \mathbb{R}^+$ introduced by Pratt in his seminal work (Pratt 1964), to compare the risk aversion of different bidders. Bidder i is more risk-averse than bidder j iff $A_i(x) > A_j(x) \forall x$. For the main theorems and corollaries we do not assume a parametric form of the utility function. We will, however, discuss certain results also for CARA, DARA, CRRA utility functions, which we briefly introduce in a form also considering initial wealth of bidders. Also these utility functions are used for the analysis of parametric cases.

A bidder exhibits constant relative (to his wealth) risk aversion if his utility function is of the form[1]:

$$u(x; \underline{\omega}, \bar{\omega}, \rho) = \frac{x^{1-\rho} - \underline{\omega}^{1-\rho}}{\bar{\omega}^{1-\rho} - \underline{\omega}^{1-\rho}} \text{ where } 0 < \rho \neq 1, 0 \leq \underline{\omega} < \bar{\omega}.$$

The corresponding Arrow–Pratt measure is $A(x) = \frac{-u''(x)}{u'(x)} = \frac{\rho x^{-\rho-1}}{x^{-\rho}} = \frac{\rho}{x}$.

For the case with $\rho = 1$, taking the limit $\rho \to 1$ we obtain the logarithmic utility function[2] $u(x; \underline{\omega}, \bar{\omega}) = \ln(x/\bar{\omega})/\ln(\underline{\omega}/\bar{\omega})$. CRRA is a case of the decreasing absolute risk aversion family (DARA), the Arrow–Pratt measure of which has the form $A(x) = \frac{1}{ax+b}, a > 0$.

---

[1] This representation follows from using the base utility function $\frac{x^{1-\rho}}{1-\rho}$ and selecting the parameters a and c in the representation $u(x) = c\left(\frac{x^{1-\rho}-a}{1-\rho}\right)$ so that $u(\underline{\omega}) = 0$ and $u(\bar{\omega}) = 1$. Specifically, $u(\underline{\omega}) = 0 \Rightarrow a = \underline{\omega}^{1-\rho}$, and $u(\bar{\omega}) = c\left(\frac{\bar{\omega}^{1-\rho}-\underline{\omega}^{1-\rho}}{1-\rho}\right) = 1 \Rightarrow c = \frac{1-\rho}{\bar{\omega}^{1-\rho}-\underline{\omega}^{1-\rho}}$.

[2] This is obtained by using positive affine transformations of the base utility function ln(x), i.e., u(x) = c ln(x) + a with the normalization $u(\underline{\omega}) = 0$ and $u(\bar{\omega}) = 1$.

A bidder exhibits constant absolute (independent to his wealth) risk aversion if his utility function is of the form[3]:

$$u(x; \underline{\omega}, \bar{\omega}, \lambda) = \frac{e^{-\lambda \underline{\omega}} - e^{-\lambda x}}{e^{-\lambda \underline{\omega}} - e^{-\lambda \bar{\omega}}} \text{ where } \lambda > 0, 0 \leq \underline{\omega} < \bar{\omega}. \text{ Hence } A(x) = \lambda.$$

2.2. The Ascending Core-Selecting Auction

We study a multi-unit clock auction with package bidding as in Goeree and Lien (2009) and Sano (2011), and follow the notation and exact rules of these papers. The economic environment is identical to Sano (2011), and we intentionally kept the notation and model description, because this allows for easier comparison of the results.

Again, before the auction starts, bidders decide whether they participate or not and make this decision public. In this ascending clock auction, a single clock visible to all bidders is used to indicate the per-unit price p of the items. The clock starts at an initial price $p = r$ and increases continuously, such that the reserve price for the package is $2r$.

Each bidder responds with the demand at the current prices. Bidders are restricted by the activity rule: they can never increase their demands. Demand is known, and we assume that small bidders can demand either one unit or zero and large bidders can demand either the package of two units or zero. For the specific environments where each bidder can demand 0 or k units, bidder messages can take only two values and a bidder can indicate whether she is 'in' or 'out' (equivalently, 'continue' or 'stop') by pressing or releasing a button. In the case we consider, a bidder pressing the button at a price p indicates his decision to stop bidding, that is, that his demand is zero at that price. Let p_i be the price at which bidder i drops from the auction.

The auctioneer raises the price until the allocation is determined on the basis of these reported values. As in Goeree and Lien (2009) and Sano (2011) we use the following tie-breaking rule: if two or more local bidders drop out at the same time, one is selected to drop out randomly, i.e., he has to leave the auction without winning, while the others are allowed to continue. The auction rules of this button auction and the known demand of each bidder describe a simple market, which allows for a formal analysis and closed form equilibrium strategies. Still these assumptions capture the strategic challenges that bidders face in the threshold model. Let us briefly revisit the possible outcomes of this auction as introduced by Sano (2011):

- **Case 1** Bidder 3 first stops at p_3 such that bidders 1 and 2 each win and they both pay p_3.
- **Case 2** Bidder 1 first stops at p_1 and bidder 3 stops next at $p_3 > p_1$. Note that the efficient allocation has not yet been determined. The price continues to increase. If bidder 2 is active until $2p_3 - p_1$, then bidders 1 and 2 each win one unit, since the total value for small bidders is $p_1 + p_2 > 2p_3$. Bidder 1 pays p_1, and bidder 2 pays $2p_3 - p_1$.

---

[3] This representation is based on positive affine transformations of the base utility function $-e^{-\lambda x}$. Selecting the parameters a and c in the representation $u(x) = u(x) = c(a - e^{-\lambda x})$ so that $u(\underline{\omega}) = 0$ and $u(\bar{\omega}) = 1$ yields the functional form we use. Specifically, $u(\underline{\omega}) = 0 \Rightarrow a = e^{-\lambda \underline{\omega}}$, and $u = (\bar{\omega}) = c(e^{-\lambda \underline{\omega}} - e^{-\lambda \bar{\omega}}) = 1 \Rightarrow c = \frac{1}{e^{-\lambda \underline{\omega}} - e^{-\lambda \bar{\omega}}}$.

- **Case 3** Bidder 1 first stops at p_1 and bidder 3 stops next at $p_3 > p_1$. If bidder 2 stops at $p_2 < 2p_3 - p_1$, bidder 3 wins both units. Since the total value for small bidders is $p_1 + p_2 < 2p_3$, bidder 3 pays the amount $p_1 + p_2$ by the bidder-optimal core discounting.
- **Case 4** Bidders 1 and 2 stop first and second at p_1 and p_2 respectively. Then, bidder 3 wins both units with price $p_1 + p_2$ by bidder-optimal core discounting.

Bidder-optimal core discounting is adopted in order to promote truthful bidding. This discounting affects the bidders' incentives. However, as we will see later, a considerable part of the analysis is, in fact, independent of the discounting. Indeed, in the model with two local and one global bidder, the combinatorial clock auction by Porter et al. (2003), the auction iBundle by Parkes and Ungar (2000), the Ascending Proxy Auction by Ausubel and Milgrom (2002), and the two-stage CCA (Cramton 2013) would all lead to the same result.

Sano (2011) shows that the large bidder has a weakly dominant strategy of truthful bidding independent of the history of other bidders' bids. He also shows that a small bidder has a weakly dominant strategy of truthful bidding after the other small bidder stops bidding. All bidders in his analysis are assumed to be risk neutral.

An interesting implication of the analysis in Sano (2011) is that the payment rule does not matter for the ascending mechanism we study. A winning small bidder 2 in case 2 would always drop out at the price p_2 where $p_1 + p_2 > 2p_3$. Bidder 2 would not have an incentive to increase his bid in the supplementary sealed-bid phase as it is used in the combinatorial clock auction (Bichler et al. 2013), and the VCG payment would be in the core with respect to the bids submitted. The same is true for the winners in all other cases, who would always pay what they bid.

3. Non-bidding Equilibrium in Ascending Auctions

In this section we derive the non-bidding equilibrium condition for general types of risk-averse utility functions and extend the analysis of risk-neutral bidders of Goeree and Lien (2009) and Sano (2011). Subsequently we describe the non-bidding equilibrium with respect to asymmetry in the small bidders' utility functions, wealth levels, and value distributions. Our initial analysis concerns an economy with two small and one large bidder but we will generalize the results to many bidders.

Throughout the paper, we use the term '*non-bidding equilibrium*' to refer to a perfect Bayesian equilibrium in which the equilibrium bid $\beta_i(v_i, \emptyset) = \min\{v_i, r\}$ for each $i = 1, 2$ *and all* $v_i \in V = [0, 1]$. The second argument refers to the bid history and $\beta_i(v_i, \emptyset)$ describe the dropping price of i under empty history, i.e., assuming no other bidder has dropped. Second, we will use the terms symmetry and asymmetry to refer to ex ante symmetry and asymmetry of small bidders. Thus, '*symmetric bidders*' is used instead of more cumbersome 'ex ante symmetric small bidders' to refer to an environment where small bidders are identical ex ante with respect to valuation distributions, risk attitudes and wealth levels, i.e., when $F_1 = F_2 = F$, $u_1 = u_2 = u$, and $\omega_1 = \omega_2 = \omega$.

3.1. The Threshold Model

The following Theorem 1 is a generalization of Theorem 1 by Sano (2011) for bidders with arbitrarily risk-averse utility functions. This is an important but non-trivial extension as can be seen in the "Appendix".

Theorem 1. *In environments with ex ante symmetric small bidders the necessary and sufficient condition for existence of a non-bidding equilibrium is*

$$\int_r^1 G(t+r)\left(\frac{f(t)}{1-F(r)} - L(u; 1, t, r, \omega)\right) dt \geq 0. \tag{1}$$

For a utility function $u(z)$ the functional $L(u; x, y, z) = \frac{u'(x+z)}{u(y+z)-u(z)}$, for $0 \leq x \leq y$, plays a central role in the statement and derivation of our main result. Some properties of this functional, such as its relation to Arrow–Pratt measure of absolute risk aversion, may be of independent interest. We collect some observations on this functional before we state our results.

$$L(u; v, t, r, \omega) = \frac{u'(v-t+\omega)}{u(v-r+\omega) - u(\omega)}$$

For a utility function in CRRA family this functional takes the form

$$L(u; v; t, r, \omega) = \begin{cases} \frac{(1-\rho)(v-t+\omega)^{-\rho}}{(v-r+\omega)^{1-\rho} - \omega^{1-\rho}} = \frac{(1-\rho)(\frac{v-t}{\omega}+1)^{-\rho}}{\omega\{(\frac{v-r}{\omega}+1)^{1-\rho} - 1\}} & if \rho \neq 1 \\ \frac{\ln(\omega/\tilde{\omega})}{(v-t+\omega)\ln((v-r+\omega)/\omega)} & if \rho = 1 \end{cases}$$

For the CARA family of utility functions, it becomes

$$L(u; v, t, r, \omega) = \frac{\lambda e^{-\lambda(v-t+\omega)}}{e^{-\lambda \omega} - e^{-\lambda(v-r+\omega)}} = \frac{\lambda e^{-\lambda(v-t)}}{1 - e^{-\lambda(v-r)}}$$

When small bidders are ex ante symmetric and risk-neutral, the necessary and sufficient condition for non-bidding equilibrium reduces to the condition in Theorem 1 of Sano[4]:

$$\int_r^1 G(t+r)\left(\frac{f(t)}{1-F(r)} - \frac{1}{1-r}\right) dt \geq 0 \quad (2) \tag{2}$$

If we suppose zero reserve prices and risk neutrality, Eq. (2) reduces to $\int_0^1 g(t)(F(t) - t)dt \leq 0$, i.e., F first-order stochastically dominates the uniform distribution. In environments with risk-aversion this is not a sufficient condition: Suppose F, G are uniform distributions, $r = 0$ and u is a CRRA with $\rho = 1, \omega = 1$. The non-bidding equilibrium condition is violated [the integral in (1) evaluates to -0.029 whereas under risk-neutrality to 0]. This fact impacts positively on both efficiency and revenue of the package clock auction since the bidder who successfully stops at r, now reveals his valuation to a greater extent.

---

[4] The condition in Sano is $\int_r^1 (G(t+r) - G(2r))\left(\frac{f(t)}{1-F(r)} - \frac{1}{1-r}\right) dt \geq 0$ but the term $-G(2r)$ is redundant since $\int_r^1 G(2r)\left(\frac{f(t)}{1-F(r)} - \frac{1}{1-r}\right) dt = G(2r)(1-1) = 0$.

So far, we have assumed complete symmetry among the small bidders. In the following, we deviate from this assumption and discuss the impact of asymmetry with respect to the small bidders' utility function, wealth levels, and value distributions.

Corollary 1. *In environments with asymmetric small bidders the necessary and sufficient condition for existence of a non-bidding equilibrium is*

$$\int_r^1 G(t+r)\left(\frac{f_2(t)}{1-F_2(r)} - L(u_1; 1, t, r, \omega_1)\right) dt \geq 0 \qquad (3a)$$

$$\int_r^1 G(t+r)\left(\frac{f_1(t)}{1-F_1(r)} - L(u_2; 1, t, r, \omega_2)\right) dt \geq 0 \qquad (3b)$$

3.2. Many Bidders Case

In this section, we generalize the case with three bidders and analyze cases with m small and l large bidders. As in Theorem 3 in Sano (2011), large bidders retain the dominant strategy property. Also, the free-rider problem does not arise when there are many small bidders, and there is competition among small bidders. However, this generalization to risk-averse bidders yields substantial differences in the proof provided in the "Appendix".

Theorem 2. *In environments with $m \geq 2$ ex ante symmetric small bidders and $l \geq 1$ ex ante symmetric large bidders, if*

$$\int_s^1 (G(t+r) - G(2s))^l \left(\frac{f(t)}{1-F(s)} - L(u; 1, t, s, \omega)\right) dt \geq 0 \qquad (4)$$

for each $l = 1, \ldots, n$ and all $s \in [r, 1]$, then the following strategies constitute a perfect Bayesian equilibrium:

- *Each large bidder follows a truthful strategy.*
- *If more than two small bidders continue bidding, then each small bidder follows a truthful strategy.*
- *If only two small bidders continue bidding (along with a large bidder), then each small bidder stops immediately.*
- *If a bidder is the only active small bidder, then he follows a truthful strategy.*

3.3. Comparative Statics and Discussion of the Ascending Core-Selecting Auction

We will now discuss the impact that different levels of risk aversion, different wealth levels, and stochastic orderings of valuation distributions have on the result.

3.3.1. Different Levels of Risk Aversion

We compare two environments $e, \tilde{e}$ which differ only in the risk-aversion of small bidders, with $\tilde{e}$ being the less risk-averse one. The risk-aversion of the large bidder is dispensable in our analysis since he always has the dominant strategy of truthful bidding.

Our main question is whether the increase of the risk-aversion of small bidders makes their non-bidding strategy less attractive. The ascending auction game does not fulfill the single crossing property (Athey 2001) and a bid increase does not lead to an increase of the probability of winning, as we will show with the next example. Therefore, the impact of risk-aversion cannot simply be explained as in the first-price single-item auction.

Suppose bidder 1 is a local bidder with a low valuation while bidder 2 is a local bidder with a high valuation. If bidder 1 increases his bid from b to $b + \varepsilon$, i.e., stays longer in the auction, then the probability that bidder 2 becomes the first bidder to stop increases, since his bid may be in the interval $[b, b + \varepsilon]$. However, bidder 1 has a higher probability of winning if he is the first to stop and in this way forces bidder 2 with the high valuation to truthfully compete with the large bidder. If bidder 2 stops first, the probability that bidder 1 wins over the global bidder is lower, but there is a trade-off with bidder 2's payoff, if he is not allowed to drop out. Therefore, increasing the bid does not necessarily lead to an increase in the probability of winning. Also the single crossing property, which demands $\frac{\partial^2 \pi(b,v)}{\partial b \partial v} > 0$ where $\pi(b, v)$ is the expected payoff of a bidder with valuation v and bid b, is violated since the derivative can be negative.

Despite these facts, Theorem 3 answers affirmatively our main question. There are situations where risk-averse bidders continue bidding whereas risk neutral (or less risk-averse) bidders do not. As a corollary, risk aversion has a positive impact on the efficiency and revenue of the auction with non-bidding at all being the worst case scenario.

Theorem 3 now analyzes the impact of risk aversion assuming arbitrary concave utility functions. It describes the main result of this paper. To compare risk-aversion the Arrow–Pratt measure is employed. Following Pratt (1964) bidder i is considered as more risk-averse than j iff the Arrow–Pratt measure of his utility function is greater or equal than j's for any wealth level.

Theorem 3. *If the environment $e = \{F_1, F_2, G, u_1, u_2, u_3, \omega_1, \omega_2, \omega_3\}$ admits a non-bidding equilibrium then so does the environment $\tilde{e} = \{F_1, F_2, G, \tilde{u}_1, \tilde{u}_2, u_3, \omega_1, \omega_2, \omega_3\}$ where $\tilde{u}_i(x)$ is such that $\tilde{A}_i(x) < A_i(x)$, for $\forall\, x$ and $i = 1, 2$, i.e., where small bidders are less risk-averse.*

In what follows, we provide some intuition about the result of Theorem 3. We examine how the utilities change for the two strategies "drop at the reserve price r" and "continue". In the risk-neutral case the profit of bidder 1 when dropping at r is $(v_1 - r)$ multiplied by the probability of winning when dropping at r. It is important to observe that this probability is independent of his valuation v_1 since bidder 1 bids the amount r and whether he wins or not, depends solely on the other two bidders' bids, who behave truthfully, thus on their valuations. Additionally, this probability of winning is the same in both environments e, $\tilde{e}$. On the other hand, if bidder 1 decides to continue, his profit is $(v_1 + r - v_3)$ times the probability of winning which is equal to the probability of the large bidder v_3 being in the interval $[2r, v_1 + r]$, since bidders 1 and 3 bid truthfully after bidder 2 has dropped at r. Also this probability is invariant in environments e, $\tilde{e}$. Since $v_3 \in [2r, v_1 + r]$ when bidder 1 wins, his profit of continuing given that he wins is in the interval $[0, v_1 - r]$ and is smaller than when dropping at r. The non-bidding equilibrium condition holds when the profits of dropping at r are greater or equal than

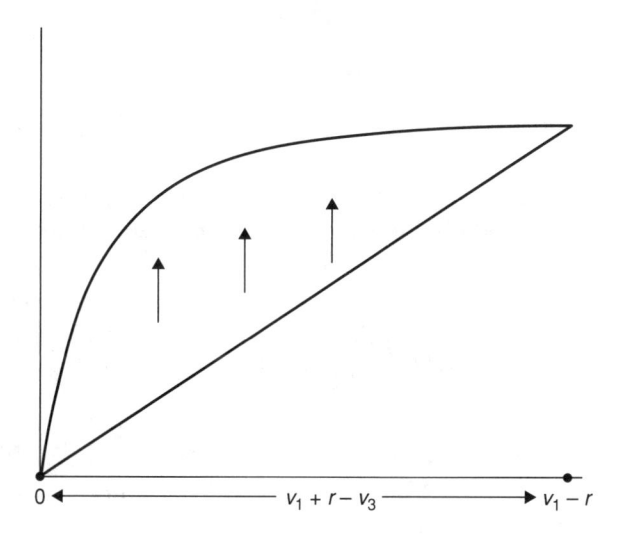

Figure 1. Utility of wealth gained when winning by dropping at r($v_1 - r$) and continuing ($v_1 + r - v_3$) for a risk-neutral and a concave utility function.

of continuing:

$$(v_1 - r)Prob(win\backslash drop) \geq (v_1 + r - v_3)Prob(win\backslash cont).$$

We argued that changing the utility function from risk neutral to a strictly concave doesn't affect $Prob(win|drop)$, $Prob(win|cont)$ but only the utility of the quantities $(v_1 - r)$, and $(v_1 + r - v_3)$. Figure 1 shows how these change. For expositional purposes the two curves intersect at $(v_1 - r)$.

The shape of the utility functions provides a rationale for our main result in this section. The utility when winning by dropping at r is the same in both environments e, ẽ. On the contrary, the utility when winning by continuing at r is higher in the risk-averse environment e for any monetary amount received. Therefore, the strategy of continuing becomes more attractive for risk-averse bidders and non-bidding ceases to be an equilibrium.

Corollary 2. *Theorem 3 holds if $\tilde{u}_1(x) = a + bx$, $\tilde{u}_2(x) = a' + b'x (a, b, a', b, \in \mathbb{R})$, and $u(x)$ is strictly concave for at least one $i = 1, 2$ (otherwise $u_i(x) = \tilde{u}_i(x))$, i.e., in environment ẽ both local bidders are risk-neutral and in e at least one of them is risk averse.*

Corollary 3. *Theorem 3 holds if $u_1, u_2, \tilde{u}_1, \tilde{u}_2$ belong to the family of CARA utility functions and $\lambda_1 \leq \tilde{\lambda}_1, \lambda_2 \leq \tilde{\lambda}_2$.*

Corollary 4. *Theorem 3 holds if $u_1, u_2, \tilde{u}_1, \tilde{u}_2$ belong to the family of CRRA utility functions and $\rho_1 \leq \tilde{\rho}_1, \rho_2 \leq \tilde{\rho}_2$.*

Corollary 5. *The efficiency and revenue in environment e are greater than or equal to the efficiency and revenue in environment ẽ.*

Wealth levels and stochastic orderings of distributions can have ample effect on the results and are discussed in the following.

3.3.2. Wealth Levels and Non-bidding Equilibrium

We now examine the case where the initial wealth of one bidder increases by a positive amount δ. CARA utility functions are independent of wealth. If bidders exhibit a CRRA or DARA utility function, then their risk aversion decreases whereas if they exhibit an IARA utility function, then risk aversion increases with higher wealth.

We will show that the left hand side of the non-bidding equilibrium (1) decreases if we add a positive amount of wealth δ as a corollary of Theorem 3. The proof will follow the one of Theorem 3. However, we cannot apply the results of Theorem 3, because the utility function now remains the same and we cannot leverage $\tilde{A}_i(x) < A_i(x) \forall x$. Instead, we depart from the condition $A_i(\omega + \delta) < A_i(\omega)$, $\delta > 0$ (this is the case for CRRA, DARA).

Corollary 6. *Suppose u_1 is a CRRA or DARA utility function. If the environment $e = \{F_1, F_2, F_3, u_1, u_2, u_3, \tilde{\omega}_1, \omega_2, \omega_3\}$ admits a non-bidding equilibrium then so does the environment $\tilde{e} = \{F_1, F_2, F_3, u_1, u_2, u_3, \omega_1, \omega_2, \omega_3\}$ where $\tilde{\omega}_1 > \omega_1$.*

3.3.3. Stochastic Dominance Orderings and Non-bidding Equilibrium

The conditions of first order stochastic dominance (FSD) and second order stochastic dominance (SSD) are:

$$\tilde{H} \succcurlyeq_{FSD} H \iff \tilde{H}(x) \leq H(x) \quad \forall x \text{ (with strict inequality for some x)}$$

$$\tilde{H} \succcurlyeq_{SSD} H \iff \int_0^x (H(t) - \tilde{H}(t))dt \geq 0 \quad \forall x \text{ (with strict inequality for some x)}$$

FSD implies SSD but not vice versa, hence FSD is a stronger condition. Variable x first order stochastically dominates y iff the probability that x is higher than an amount z is higher than the probability that y is higher than this amount, for any z. Hence x is stochastically larger than y (it has a higher expected value). SSD mirrors the riskiness of two variables. If x second order stochastically dominates y, then x is less risky. Additionally, if the distributions of the random variables satisfy the single crossing property, then x is also stochastically larger than y.

We examine the impact of changing the distributions of the bidders' valuations to new ones that stochastically dominate the former ones. We show that changing the distribution of a valuation of a small bidder to a distribution which stochastically dominates the former leads to the non-bidding equilibrium being more probable. The reason is that the incentives to free ride increase in accordance with the probability that one small bidder outbids alone the large bidder.

Theorem 4. *Suppose $r = 0$. If the environment $e = \{F_1, F_2, G, u_1, u_2, u_3, \omega_1, \omega_2, \omega_3\}$ admits a non-bidding equilibrium then so does the environment $\tilde{e} = \{\tilde{F}_1, F_2, G, u_1, u_2, u_3, \omega_1, \omega_2, \omega_3\}$ where $\tilde{F}_1 \succcurlyeq_{FSD} F_1$.*

Theorem 5. *Suppose additionally g is non-increasing. Then the weaker condition $\tilde{F}_1 \succcurlyeq_{SSD} F_1$ is sufficient for Theorem 4 to hold.*

Theorems 4 and 5 hold for any reserve price r if we impose first order stochastic dominance to the left-truncated versions of the cumulative distributions: $\tilde{F}_1(\cdot|r) \succcurlyeq_{FSD}$

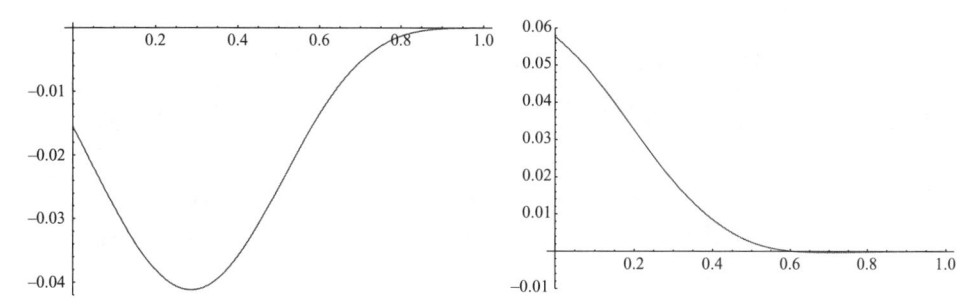

Figure 2. Condition (2) for different values of r with $F \sim N$ (**0.5,0.25**) on the *left* and $F \sim N(\mathbf{0.8,0.4})$ on the *right* with risk-neutral bidders.

$F_1(\cdot|r)$. Note that this condition is not implied by $\tilde{F}_1 \succcurlyeq_{FSD} F_1$ since the stochastic dominance is not necessarily preserved after truncating.

4. Analysis of Parametric Cases

In what follows, we will analyze parametric cases of the sealed-bid and ascending core-selecting auctions.

4.1. The Non-bidding Equilibrium Condition in the Ascending Auction

We will illustrate selected parametric cases of the ascending auction assuming different types of distribution functions. Letting bidders be *risk-neutral* and all v_i be drawn from a uniform distribution, (2) evaluates to 0 for all $r \in [0, 1]$ leading to a non-bidding equilibrium. If F is a truncated Gaussian distribution $F \sim N(0.5, 0.25)$ on the interval $[0\ldots1]$ and $G \sim N(1, 0.5)$, then condition (2) is negative for $r \in [0\ldots1]$, and risk-neutral bidders would continue bidding. Now if we increase the valuations of the small bidders and $F \sim N(0.8, 0.4)$ then condition (2) is positive for most values of r and the small bidders would try to drop (Fig. 2). The reason for free riding is that the probability for the competing small bidder to outbid the large bidder increases.

It is now interesting to understand, how *risk aversion* impacts this condition. If v_i are uniformly distributed, the value of (1) is negative for both CARA and CRRA utility functions, and as opposed to the risk neutral case, the bidders would continue to bid (Fig. 3).

Figure 4 shows condition (2) for different values of r when F is a truncated Normal distribution $F \sim N(0.5, 0.25)$ or $F \sim N(0.8, 0.4)$ on the interval $[0\ldots1]$, $G \sim N(1, 0.5)$ and we have a CARA utility function with $\lambda = 0.9$. In Fig. 5 we change the utility function to a CRRA with $\rho = 0.9$, $\omega = 1$.

Last but not least, we vary the parameters λ and ρ and keep $r = 0$, $F \sim N(0.8, 0.4)$, $G \sim N(1, 0.5)$ (Fig. 6). Condition (1) holds and leads to a non-bidding equilibrium for all values of ρ and for $\lambda < 1.4$. Higher values of λ lead to such high risk aversion that the non-bidding equilibrium is eliminated.

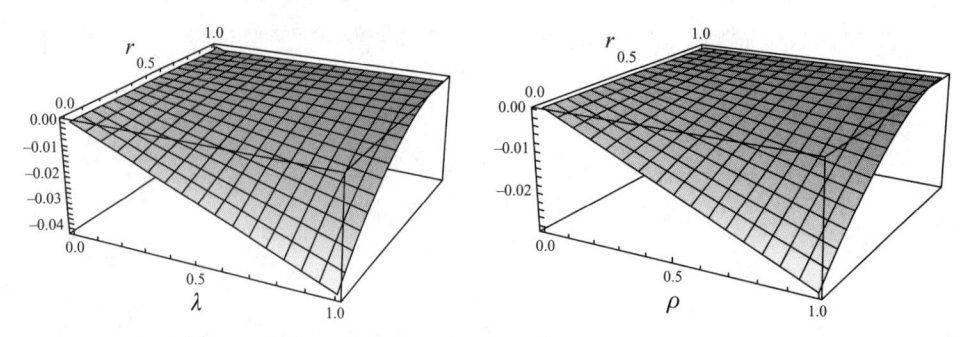

Figure 3. Condition (1) for different values of r, $F \sim U(0,1)$, $G \sim U(0,2)$. On the *left* for a CARA with $\lambda = 0 \ldots 1$ and on the *right* CRRA utility function with $\rho = 0 \ldots 1$, $\omega = 1$.

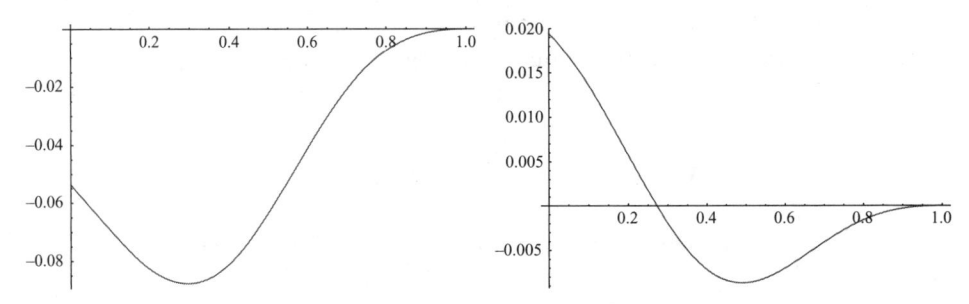

Figure 4. Condition (2) for different values of r with $F \sim N(0.5, 0.25)$ on the *left* and $F \sim N(0.8, 0.4)$ on the *right* with CARA utility functions.

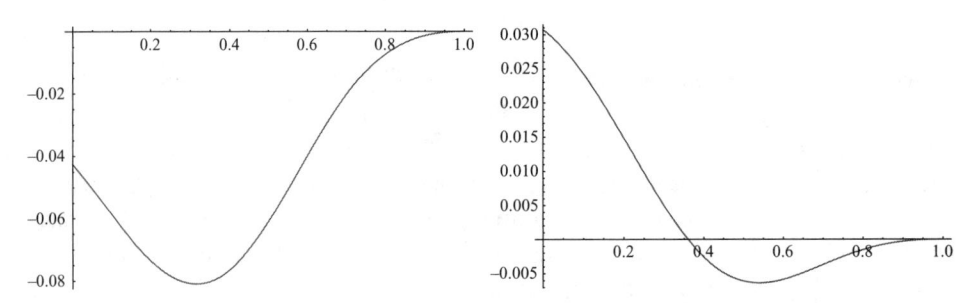

Figure 5. Condition (1) for different values of r with $F \sim N(0.5, 0.25)$ on the *left* and $F \sim N(0.8, 0.4)$ on the *right* with CRRA utility functions and $\rho = 0.9$, $\omega = 1$.)

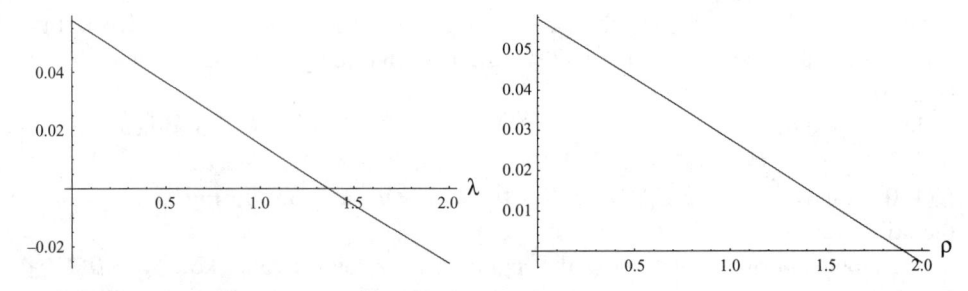

Figure 6. Condition (1) with $F \sim N(0.8, 0.4)$ for $r = 0$ and different parameters of a CARA or CRRA utility function.

5. Conclusions

Ascending combinatorial auctions have led to a fruitful stream of research on pricing rules in such auctions. The recent game-theoretical literature in this area casts doubts that such auctions are efficient in the field. The analysis of a simplified button auction in a threshold model rule shows that there can even be inefficient non-bidding equilibria (Goeree and Lien 2009; Sano 2011). Overall, the threshold model illustrates that there is potential for profitable manipulation and truthful bidding might not be in the interest of bidders. This model can easily be extended to real-world markets where local bidders compete against bidders interested in a global or national coverage as it is regularly the case in spectrum auctions. The game-theoretical results are in sharp contrast to the high allocative efficiency that combinatorial auctions achieved in lab experiments.

Previous game-theoretical work assumes risk-neutral bidders. Risk aversion is an important phenomenon in high-stakes auctions and it is important to understand its impact on equilibrium strategies, revenue and efficiency of these auctions. We stick to the environments and auction mechanisms, which have been analyzed so far, but take risk aversion into account. The free-rider problem among the two local bidders is such that the impact of risk aversion is not obvious, however. In this paper, we show that risk aversion, reserve prices, and bidder asymmetries affect the equilibrium outcomes in ways that can be systematically analyzed. Risk aversion reduces the scope of the non-bidding equilibrium in ascending core-selecting auctions.

Interesting questions remain open. The nature of equilibria when the non-bidding condition fails is perhaps the most important gap to be filled. Also, environments with more items or multi-minded bidders require further analysis.

Acknowledgments

The authors gratefully acknowledge funding from the German National Science Foundation (DFG BI-1057-7). Kemal Guler's work was undertaken during a visit to Bilkent University supported under a fellowship Grant from the TUBITAK BIDEP 2236 Co-Circulation Program Project Number 114C020. He gratefully acknowledges the financial support of TUBITAK and the hospitality of Bilkent University.

Appendix

Theorem 1. *In environments with ex ante symmetric small bidders the necessary and sufficient condition for existence of a non-bidding equilibrium is*

$$\int_r^1 G(t+r)\left(\frac{f(t)}{1-F(r)} - L(u; 1, t, r, \omega)\right) dt \geq 0 \tag{1}$$

Proof of Theorem 1. Suppose bidder 2 selects to drop at r. For bidder 1, if he also selects to drop at r, his expected payoff is determined by the randomization that determines who gets to continue. If bidder 1 drops at r, and bidder 2 is selected to continue bidding in the tie-breaking lottery, bidder 1's expected utility is

$$\pi(drop; v, r) = u(v - r + \omega)Prob\{1 \text{ wins with bid } r|((r, v), \beta_{-1})\}$$
$$+ u(\omega)Prob\{1 \text{ loses with bid } r|((r, v), \beta_{-1})\}$$
$$= u(\omega) + (u(v - r + \omega) - u(\omega))Prob\{1 \text{ wins with bid } r|((r, v), \beta_{-1})\}$$
$$= u(\omega) + \frac{u(v - r + \omega) - u(\omega)}{1 - G(2r)} \left\{ \int_r^1 G(t + r) \frac{f_2(t)}{1 - F_2(r)} dt - G(2r) \right\} \quad (5)$$

where we used the following derivations to evaluate the probability term,

$$Prob\{1 \text{ wins with bid } r|((r, v), \beta_{-1})\} = Prob\{v_2 + r > v_3|\bar{v}_2 \geq c_2 \geq r, v_3 \geq 2r\}$$

$$= \frac{Prob\{v_2 + r > v_3, \bar{v}_2 \geq v_2 \geq r, v_3 \geq 2r\}}{Prob\{\bar{v}_2 \geq v_2 \geq r, v_3 \geq 2r\}} = \frac{Prob\{v_2 + r > v_3, \bar{v}_2 \geq v_2 \geq r, v_3 \geq 2r\}}{Prob\{\bar{v}_2 \geq v_2 \geq r\}Prob\{v_3 \geq 2r\}}$$

$$= \frac{Prob\{v_2 + r > v_3 \geq 2r, \bar{v}_2 \geq v_2 \geq r\}}{(1 - F_2(r))(1 - G(2r))} = \frac{\int_r^{\bar{v}_2} \int_{2r}^{v_2 + r} dG(v_3) dF_2(v_2)}{(1 - F_2(r))(1 - G(2r))}$$

$$= \frac{\int_r^{\bar{v}_2} (G(v_2 + r) - G(2r)) dF_2(v_2)}{(1 - F_2(r))(1 - G(2r))} = \frac{\int_r^{\bar{v}_2} G(t + r) f_2(t) dt - G(2r)(1 - F_2(r))}{(1 - F_2(r))(1 - G(2r))}$$

$$= \frac{\int_r^{\bar{v}_2} G(t + r) \frac{f_2(t)}{1 - F_2(r)} dt - G(2r)}{1 - G(2r)}$$

If bidder 1 does not drop at r, or if he drops but he is selected to continue in the tie-breaking lottery, his expected utility is

$$\pi(continue; v, r) = u(\omega) + \frac{u(v - r + \omega) - u(\omega)}{1 - G(2r)}$$
$$\times \left\{ \int_r^v \frac{u'(v - y + \omega)}{u(v - r + \omega) - u(\omega)} G(y + r) dy - G(2r) \right\} \quad (6)$$

where we used the fact that 1 wins in the continuation game against bidder 3 in the event that $\{v + r > w_3\}$ and pays $w_3 - r$ when he wins.

$$\pi(continue; v, r) = \frac{\int_{2r}^{v+r} u(v + r - s + \omega) g(s) ds + \int_{v+r}^{\bar{\omega}_3} u(\omega) g(s) ds}{1 - G(2r)}$$

$$= u(\omega) + \frac{\int_{2r}^{v+r} (u(v + r - s + \omega) - u(\omega)) g(s) ds}{1 - G(2r)}$$

Integration by parts, setting $y := s - r$ and rearranging terms gives (6).

If the ties are broken via a lottery that selects bidder 1 with probability q and bidder 2 with probability $1 - q$, expected utility of dropping at r for bidder 1 is

$$EU(drop; v, r) = q\pi(drop; v, r) + (1 - q)\pi(continue; v, r)$$
$$= \pi(continue; v, r) + q(\pi(drop; v, r) - \pi(continue; v, r))$$

Therefore, the expected utility difference between the two actions for bidder 1 is

$$\Delta EU(v, r) := EU(drop; v, r) - EU(continue; v, r)$$

$$= q(\pi(drop; v, r) - \pi(continue; v, r))$$

$$= q\left(\frac{u(v - r + \omega) - u(\omega)}{1 - G(2r)}\left\{\int_r^1 G(t + r)\frac{f_2(t)}{1 - F_2(r)}dt - G(2r)\right\}\right.$$

$$- \frac{u(v - r + \omega) - u(\omega)}{1 - G(2r)}$$

$$\left. \times \left\{\int_r^v \frac{u'(v - y + \omega)}{u(v - r + \omega) - u(\omega)}G(y + r)dy - G(2r)\right\}\right)$$

$$= q\frac{u(v - r + \omega) - u(\omega)}{1 - G(2r)}\left\{\int_r^1 G(t + r)\frac{f_2(t)}{1 - F_2(r)}dt\right.$$

$$\left. - \int_r^v \frac{u'(v - y + \omega)}{u(v - r + \omega) - u(\omega)}G(y + r)dy\right\}$$

$\square$

Remark. Note that the sign of the expected utility difference is independent of the value of q.

$$\text{sign } \Delta EU(v, r) = \text{sign}\left\{\int_r^1 G(t + r)\frac{f_2(t)}{1 - F_2(r)}dt - \int_r^v L(v, t, r, \omega)G(t + r)dt\right\}$$

Thus the condition that bidder 1's expected utility from dropping at r is at least as high as his expected utility from continuing becomes

$$\Delta EU(v, r) > 0 \, \forall_v \iff \int_r^1 G(t + r)\frac{f_2(t)}{1 - F_2(r)}dt$$

$$> \int_r^v L(v, t, r, \omega)G(t + r)dt \forall_v$$

$$\iff \int_r^1 G(t + r)\frac{f_2(t)}{1 - F_2(r)}dt$$

$$> \max_v \int_r^v L(v, t, r, \omega)G(t + r)dt$$

We show in Lemma 1 that $T(v) := \int_r^v L(v, t, r, \omega)G(t + r)dt$ is monotone increasing in v and thus it is maximized when $v = 1$ with maximum value that is equal to $\int_r^1 L(1, t, r, \omega)G(t + r)dt$.

Therefore,

$$\Delta EU(v, r) > 0 \, \forall_v \iff \int_r^1 G(t + r)\frac{f_2(t)}{1 - F_2(r)}dt > \int_r^1 L(1, t, r, \omega)G(t + r)dt$$

$$\iff \int_r^1 G(t + r)\left(\frac{f_2(t)}{1 - F_2(r)} - L(1, t, r, \omega)\right)dt > 0$$

Lemma 1. $T(v) := \int_r^v L(v, t, r, \omega) G(t + r) dt$ is monotone increasing in v.

Proof Lemma 1. First we prove some claims. □

Claim 1.

(a) $\dfrac{\partial L(v, t, r, \omega)}{\partial v} = -\dfrac{\partial L(v, t, r, \omega)}{\partial t} - L(v, t, r, \omega) L(v, r, r, \omega)$

$= \dfrac{u''(v - r + \omega)}{u(v - r + \omega) - u(\omega)} - L(v, t, r, \omega) L(v, r, r, \omega)$

(b) $L(v, r, r, \omega) - L(v, v, r, \omega) = \dfrac{u'(v - r + \omega) - u'(\omega)}{u(v - r + \omega) - u(\omega)}$

Proof of Claim 1. Directly from $L(v, t, r, \omega)$ definition. □

Claim 2.

$\int_r^v \dfrac{u''(v - t + \omega)}{u(v - r + \omega) - u(\omega)} G(t + r) dt > G(v + r) \dfrac{u'(v - r + \omega) - u'(\omega)}{u(v - r + \omega) - u(\omega)}$

Proof of Claim 2.

$G(t + r) < G(v + r) \iff \dfrac{u''(v - t + \omega)}{u(v - r + \omega) - u(\omega)} G(t + r)$

$> \dfrac{u''(v - t + \omega)}{u(v - r + \omega) - u(\omega)} G(v + r)$

$\iff \int_r^v \dfrac{u''(v - t + \omega)}{u(v - r + \omega) - u(\omega)} G(t + r) dt$

$> G(v + r) \int_r^v \dfrac{u''(v - t + \omega)}{u(v - r + \omega) - u(\omega)} dt$

$\iff \int_r^v \dfrac{u''(v - t + \omega)}{u(v - r + \omega) - u(\omega)} G(t + r) dt > G(v + r) \dfrac{u'(v - r + \omega) - u'(\omega)}{u(v - r + \omega) - u(\omega)}$

□

Claim 3. $T(v) < G(v + r)$

Proof of Claim 3. Using the fact that $G(x)$ is increasing, we get

$T(v) = \int_r^v L(v, t, r, \omega) G(t + r) dt = \dfrac{\int_r^v u'(v - t + \omega) G(t + r) dt}{u(v - r + \omega) - u(\omega)}$

$< \dfrac{G(v + r) \int_r^v u'(v - t + \omega) dt}{u(v - r + \omega) - u(\omega)} = G(v + r)$

Now we prove Lemma 1 by showing $T'(v) > 0$

$$T'(v) = L(v, v, r, \omega)G(v+r) + \int_r^v \frac{\partial L(v, t, r, \omega)}{\partial v} G(t+r) dt$$

$$= L(v, v, r, \omega)G(v+r) - \int_r^v \frac{-u''(v-t+\omega)}{u(v-r+\omega) - u(\omega)} G(t+r) dt +$$

$$- L(v, r, r, \omega) \int_r^v L(v, t, r, \omega) G(t+r) dt \quad \text{(Claim 1a)}$$

$$= L(v, v, r, \omega)G(v+r) + \int_r^v \frac{u''(v-t+\omega)}{u(v-r+\omega) - u(\omega)} G(t+r) dt$$

$$- L(v, r, r, \omega)T(v) \quad (T(v) \text{ definition})$$

$$> L(v, v, r, \omega)G(v+r) + G(v+r)\frac{u'(v-r+\omega) - u'(\omega)}{u(v-r+\omega) - u(\omega)}$$

$$- L(v, v, r, \omega)T(v) \quad \text{(Claim 2)}$$

$$= L(v, v, r, \omega)G(v+r) + G(v+r)(L(v, r, r, \omega) - L(v, v, r, \omega))$$

$$- L(v, r, r, \omega)T(v) \quad \text{(Claim 1b)}$$

$$= L(v, r, r, \omega)(G(v+r) - T(v)) > 0 \quad \text{(Claim 3 and } u(x) \text{ concave)}$$

$\square$

Theorem 2. *In environments with m ex ante symmetric small bidders and l ex ante symmetric large bidders, if*

$$\int_s^1 (G(t+r) - G(2s))^l \left(\frac{f(t)}{1 - F(s)} - L(u; 1, t, s, \omega) \right) dt \geq 0 \quad (4)$$

for each $l = 1, \ldots, n$ and all $s \in [r, 1]$, then the following strategies constitute a perfect Bayesian equilibrium:

1. *Each large bidder follows a truthful strategy.*
2. *If more than two small bidders continue bidding, then each small bidder follows a truthful strategy.*
3. *If only two small bidders continue bidding (along with a large bidder), then each small bidder stops immediately.*
4. *If a bidder is the only active small bidder, then he follows a truthful strategy.*

Proof of Theorem 2. Each large bidder follows obviously a truthful strategy. If more than two small bidders continue bidding, each of them follows a truthful strategy (otherwise who stops loses immediately). Hence, it only needs to be shown that if two small bidders continue bidding (along with a large bidder), then each small bidder stops immediately.

Let s be the current price level.

$H_1(\cdot|\cdot)$ denotes the conditional CDF of $w^{(1)}$ among 1 valuations and $h_1(\cdot|\cdot)$ the corresponding pdf

$$\text{e.g., } H_l\left(v_2 + s | w^{(l)} \geq 2s\right) = \frac{(G(v_2 + s) - G(2s))^l}{(1 - G(2s))^l} \text{ and } h_l\left(v_2 + s | w^{(l)} \geq 2s\right)$$

$$= \frac{g(v_2 + s)(G(v_2 + s) - G(2s))^{l-1}}{(1 - G(2s))^l}$$

If bidder 1 stops at s and if it is accepted, his ex interim payoff at s is:

$$\pi(drop; v, s) = u(v - s + \omega) Prob\{1 \text{ wins with bid } s | ((s, v), \beta_{-1})\}$$
$$+ u(\omega) Prob\{1 \text{ loses with bid } s | ((s, v), \beta_{-1})\}$$
$$= u(\omega) + (u(v - s + \omega) - u(\omega))$$
$$\times Prob\{1 \text{ wins with bid } s | ((s, v), \beta_{-1})\}$$
$$= u(\omega) + \frac{u(v - s + \omega) - u(\omega)}{(1 - F_2(s))(1 - G(2s))^l}$$
$$\times \left\{\int_s^1 (G(v_2 + s) - G(2s))^l f_2(v_2) dv_2\right\} \quad (7)$$

where we used the following derivations to evaluate the probability term

$$Prob\{1 \text{ wins with bid } s | ((s, v), \beta_{-1})\}$$
$$= Prob\left\{v_2 + s > w^{(1)} | \bar{v}_2 \geq v_2 \geq s, w^{(l)} \geq 2s\right\}$$
$$= \frac{\int_s^{\bar{v}_2} H_l(v_2 + s | w^{(l)} \geq 2s) f_2(v_2) dv_2}{1 - F_2(s)}$$
$$= \frac{\int_s^{\bar{v}_2} (G(v_2 + s) - G(2s))^l f_2(v_2) dv_2}{(1 - F_2(s))(1 - G(2s))^l}$$

If bidder 1 does not stop at s, he bids up truthfully since bidder 2 stops. Bidder 1 wins in the continuation game against bidder 3 in the event that $\{v + s > w^{(1)}\}$ and pays $w^{(1)} - s$ when he wins.

$$\pi(continue; v, s) = \int_{2s}^{v+s} u\left(v + s - w^{(1)} + \omega\right) h_l\left(w^{(1)} | w^{(l)} \geq 2s\right) dw^{(1)}$$
$$+ u(\omega) \left(1 - \int_{2s}^{v+s} h_l\left(w^{(1)} | w^{(l)} \geq 2s\right) dw^{(1)}\right)$$
$$= u(\omega) + \int_{2s}^{v+s} \left(u\left(v + s - w^{(1)} + \omega\right) - u(\omega)\right)$$
$$\times h_l\left(w^{(1)} | w^{(l)} \geq 2s\right) dw^{(1)}$$
$$= u(\omega) + \int_{2s}^{v+s} (u(v + s - t + \omega) - u(\omega)) h_l(t \backslash t \geq 2s) dt$$

$$= u(\omega) + (u(v - s + \omega) - u(\omega))(-H_l(2s|2s \geq 2s))$$
$$+ \int_{2s}^{v+s} u'(v + s - t + \omega)H_l(t\backslash t \geq 2s)dt$$
$$= u(\omega) + \int_{2s}^{v+s} u'(v + s - t + \omega)H_l(t\backslash t \geq 2s)$$
$$\times dt \text{ (since } H_l(2s\backslash 2s \geq 2s) = 0) \tag{8}$$

The difference (8)–(7) is:

$$\Delta EU(v, s) := \pi(continue; v, s) - \pi(drop; v, s)$$
$$= \frac{u(v - s + \omega) - u(\omega)}{(1 - F_2(s))(1 - G(2s))^l} \left\{ \int_s^1 (G(v_2 + s) - G(2s))^l f_2(v_2)dv_2 \right\}$$
$$- \int_{2s}^{v+s} u'(v + s - t + \omega) \frac{(G(t) - G(2s))^l}{(1 - G(2s))^l} dt$$

$$\text{sign } \Delta EU(v, r) = \text{sign} \left\{ \frac{u(v - s + \omega) - u(\omega)}{(1 - F_2(s))} \left(\int_s^1 (G(v_2 + s) - G(2s))^l f_2(v_2)dv_2 \right) \right.$$
$$\left. - \int_s^v u'(v - y + \omega)(G(y + s) - G(2s))^l dy \right\}$$
$$\times \text{ (change variable } y = t - s)$$
$$= \text{sign} \left\{ (u(v - s + \omega) - u(\omega)) \left(\int_s^1 (G(v_2 + s) - G(2s))^l \right. \right.$$
$$\left. \left. \times \frac{f_2(v_2)}{(1 - F_2(s))} dv_2 - \int_s^v \frac{u'(v - y + \omega)}{u(v - s + \omega) - u(\omega)} \right. \right.$$
$$\left. \left. \times (G(y + s) - G(2s))^l dy \right) \right\}$$
$$= \text{sign} \left\{ \int_s^1 (G(t + s) - G(2s))^l \frac{f_2(t)}{(1 - F_2(s))} dt \right.$$
$$\left. - \int_s^v L(v, t, s, \omega)(G(t + s) - G(2s))^l dt \right\}$$

We show below (Lemma 2) that $Q(v) := \int_s^v L(v, t, s, \omega)(G(t + s) - G(2s))^l dt$ is monotone increasing in v and thus it is maximized at $v = 1$.

Therefore we show $\Delta EU(v, s) > 0 \forall_v$

$$\iff \int_s^1 (G(t + s) - G(2s))^l \frac{f_2(t)}{(1 - F_2(s))} dt$$
$$> \int_s^1 L(1, t, s, \omega)(G(t + s) - G(2s))^l dt$$
$$\iff \int_s^1 (G(t + s) - G(2s))^l \left(\frac{f_2(t)}{(1 - F_2(s))} - L(1, t, s, \omega) \right) dt > 0$$

$\square$

Lemma 2. $Q(v) := \int_s^v L(v, t, s, \omega)(G(t+s) - G(2s))^l dt$ is monotone increasing in v.

Proof Lemma 2. First some claims:

Claim 4
$$\int_r^v \frac{u''(v-t+\omega)}{u(v-s+\omega) - u(\omega)}(G(t+s) - G(2s))^l dt$$
$$> (G(v+s) - G(2s))^l \frac{u'(v-s+\omega) - u'(\omega)}{u(v-s+\omega) - u(\omega)}$$
□

Proof Claim 4. Similar to Claim 2 □

Claim 5 $(G(v+s) - G(2s))^l > Q(v)$

Proof Claim 5. Similar to Claim 3 □

Now we prove the lemma by showing $Q'(v) > 0$

$$Q'(v) = L(v, v, s, \omega)(G(v+s) - G(2s))^l$$
$$+ \int_s^v \frac{\partial L(v, t, s, \omega)}{\partial v}(G(t+s) - G(2s))^l dt$$
$$= L(v, v, s, \omega)(G(v+s) - G(2s))^l$$
$$- \int_r^v \frac{\partial L(v, t, s, \omega)}{\partial t}(G(t+s) - G(2s))^l dt$$
$$- L(v, s, s, \omega) \int_r^v L(v, t, s, \omega)(G(t+s) - G(2s))^l dt \quad (\text{claim 1a})$$
$$= L(v, v, s, \omega)(G(v+s) - G(2s))^l$$
$$+ \int_r^v \frac{u''(v-t+\omega)}{u(v-s+\omega) - u(\omega)}(G(t+s) - G(2s))^l dt - L(v, s, s, \omega)Q(v)$$
$$> L(v, v, s, \omega)(G(v+s) - G(2s))^l$$
$$+ (G(v+s) - G(2s))^l \frac{u'(v-s+\omega) - u'(\omega)}{u(v-s+\omega) - u(\omega)}$$
$$- L(v, s, s, \omega)Q(v) \quad (\text{claim 4})$$
$$= L(v, v, s, \omega)(G(v+s) - G(2s))^l$$
$$+ (G(v+s) - G(2s))^l(L(v, s, s, \omega) - L(v, v, s, \omega))$$
$$- L(v, s, s, \omega)Q(v) \quad (\text{claim 1b})$$
$$= L(v, s, s, \omega)\left((G(v+s) - G(2s))^l - Q(v)\right) > 0$$
$$(\text{claim 5 and } u(x) \text{ concave})$$

Theorem 3. *If the environment $e = \{F_1, F_2, G, u_1, u_2, u_3, \omega_1, \omega_2, \omega_3\}$ admits a non-bidding equilibrium then so does the environment $\tilde{e} = \{F_1, F_2, G, \tilde{u}_1, \tilde{u}_2, u_3, \omega_1, \omega_2, \omega_3\}$ where $\tilde{u}_i(x)$ is such that $\tilde{A}_i(x) < A_i(x)$, for $\forall x$ and $i = 1, 2$, i.e., where small*

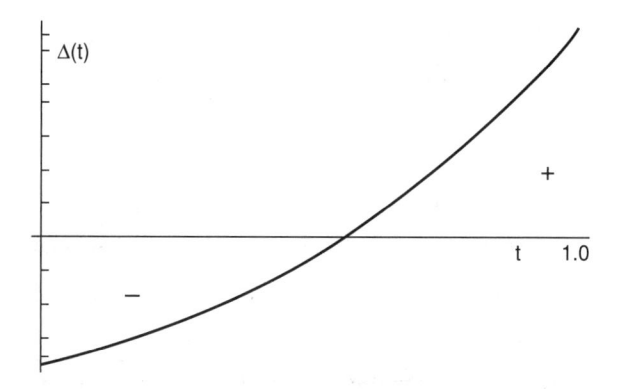

Figure 7. Shape of $\Delta(t)$.

bidders are less risk-averse. If the environment $\tilde{e}$ does not admit non-bidding in a perfect Bayesian equilibrium, neither does the environment e.

Proof of Theorem 3. We will show that the left hand side of the non-bidding equilibrium condition (1) decreases as $A(x)$ increases. First, we describe the left hand side as a function of the utility function

$$NB(u) := \int_r^1 G(t+r)\left(\frac{f(t)}{1-F(r)} - L(u; 1, t, r, \omega)\right) dt$$

Let $\tilde{u}(x), u(x)$ be two utility functions and $\tilde{A}(x), A(x)$ the corresponding Arrow–Pratt measures such that $\tilde{A}(x) < A(x)$ for $\forall x$. We need to show

$$NB(\tilde{u}) > NB(u) \iff \int_r^1 G(t+r)(L(u; 1, t, r, \omega) - L(\tilde{u}; 1, t, r, \omega))dt > 0 \quad (9)$$

Define $\Delta(t) := L(u; 1, t, r, \omega) - L(\tilde{u}; 1, t, r, \omega)$
$\int_r^1 L(u; 1, t, r, \omega)dt = 1 \forall u, t, r, \omega$, therefore

$$\int_r^1 \Delta(t)dt = 0 \quad (10)$$

We will show that $\Delta(t)$ has always the form in Fig. 7. Precisely we will show $\Delta(1) > 0$ and $\Delta(t) = 0$ has exactly one root. Due to (10) the positive and the negative areas are equal. Since $G(x)$ is increasing, the positive area is multiplied by greater values than the negative area, hence inequality (9) is true.

To show $\Delta(1) > 0$ we make use of the following lemma, proven in (Pratt, 1964):

$$A_2(x) > A_1(x) \forall x \iff \frac{u_2(y) - u_2(x)}{u_2'(w)}$$
$$< \frac{u_1(y) - u_1(x)}{u_2'(w)} \text{for } \forall(w, x, y) : \omega \leq x \leq y$$

Let $w = x = \omega$, $y = \omega + 1 - r$, then we get:

$$A(x) > \tilde{A}(x) \forall x \iff \frac{u(\omega + 1 - r) - u(\omega)}{u'(\omega)} < \frac{\tilde{u}(\omega + 1 - r) - \tilde{u}(\omega)}{\tilde{u}'(\omega)}$$

$$\iff \frac{u'(\omega)}{u(\omega + 1 - r) - u(\omega)} > \frac{\tilde{u}'(\omega)}{\tilde{u}(\omega + 1 - r) - \tilde{u}(\omega)}$$

$$\iff L(u; 1, t, r, \omega) > L(\tilde{u}; 1, t, r, \omega) \iff \Delta(1) > 0$$

We proceed to show $\Delta(t) = 0$ has exactly one root:

$$\frac{\partial L(v, t, r, \omega)}{\partial t} = \frac{-u''(v - t + \omega)}{u(v - r + \omega) - u(\omega)} = \frac{-u''(v - t + \omega)}{u'(v - t + \omega)} \cdot \frac{u'(v - t + \omega)}{u(v - r + \omega) - u(\omega)}$$

$$= A(v - t + \omega) L(v, t, r, \omega)$$

$$\frac{\partial \Delta(t)}{\partial t} = A(1 - t + \omega) L(u; 1, t, r, \omega) - \tilde{A}(1 - t + \omega) L(\tilde{u}; 1, t, r, \omega)$$

Suppose there are two values of t, t_1^* and t_2^* with $t_2^* > t_1^*$ such as $\Delta(t_1^*) = \Delta(t_2^*) = 0$.

Since $L(u; 1, t_1^*, r, \omega) = L(\tilde{u}; 1, t_1^*, r, \omega)$ and $A(1 - t_1^* + \omega) > \tilde{A}(1 - t_1^* + \omega)$, it follows $\frac{\partial \Delta(t_1^*)}{\partial t} > 0$. Thus:

$$\forall_t \in (t_1^*, t_2^*) \Delta(t) > 0 \tag{11}$$

Using the mean value theorem of differential calculus, we get that $\exists t^c \in (t_1^*, t_2^*)$ with

$$\frac{\partial \Delta(t^c)}{\partial_t} = \frac{\Delta(t_2^*) - \Delta(t_1^*)}{t_2^* - t_1^*} = 0$$

$$\iff A(1 - t^c + \omega) L(u; 1, t^c, r, \omega) - \tilde{A}(1 - t^c + \omega) L(\tilde{u}; 1, t^c, r, \omega) = 0$$

$$\iff L(u; 1, t^c, r, \omega) < L(\tilde{u}; 1, t^c, r, \omega) \quad (\text{since } A(1 - t^c + \omega)$$
$$> \tilde{A}(1 - t^c + \omega))$$

$$\iff \Delta(t^c) < 0 \text{ which is a } \textbf{contradiction} \text{ to (11)}.$$

Thus there cannot exist two (or more) values of t, t_1^* and t_2^* with $t_2^* > t_1^*$ such as $\Delta(t_1^*) = \Delta(t_2^*) = 0$. Due to (10) and (11) there is at least one root. Thus there is exactly one root. The last step of the proof is to formally show that the positive area is multiplied by greater values than the negative area, which implies that the integral in (9) is positive.

Define t* as $\Delta(t^*) = 0$. Then,

$$-\int_r^{t^*} G(t + r) \Delta(t) dt < -G(t^* + r) \int_r^{t^*} \Delta(t) dt < \int_{t^*}^1 G(t + r) \Delta(t) dt$$

(since for any t in $[r, t^*] G(t + r) < G(t^* + r)$ and G increasing)

$$\iff \int_r^{t^*} G(t + r) \Delta(t) dt + \int_{t^*}^1 G(t + r) \Delta(t) dt > 0$$

$$\iff \int_r^1 G(t + r) \Delta(t) dt > 0 \iff \int_r^1 G(t + r)(L(u; 1, t, r, \omega)$$
$$- L(\tilde{u}; 1, t, r, \omega)) dt > 0$$

$\square$

Corollary 6. *Suppose u_1 is a CRRA or DARA utility function. If the environment $e = \{F_1, F_2, F_3, u_1, u_2, u_3, \omega_1, \omega_2, \omega_3\}$ admits a non-bidding equilibrium then so does the environment $\tilde{e} = \{F_1, F_2, F_3, u_1, u_2, u_3, \tilde{\omega}_1, \omega_2, \omega_3\}$ where $\tilde{\omega}_1 > \omega_1$.*

Proof of Corollary 6. First, we express the left hand side as a function of ω as

$$NB(\omega) := \int_r^1 G(t+r) \left(\frac{f(t)}{1-F(r)} - L(u; 1, t, r, \omega) \right) dt$$

We need to show $NB(\omega + \delta) > NB(\omega) \iff \int_r^1 G(t+r)(L(u; 1, t, r, \omega) - L(u; 1, t, r, \omega + \delta))dt > 0$ □

Define $\Delta(t, \delta) := L(u; 1, t, r, \omega) - L(u; 1, t, r, \omega + \delta)$
We first show $\Delta(1, \delta) > 0 \forall \delta > 0$:
Integrating $A(\omega + \delta) < A(\omega)$ from x to y we get

$$\int_x^y \frac{u''(\omega + \delta)}{u'(\omega + \delta)} d\omega > \int_x^y \frac{u''(\omega)}{u'(\omega)} d\omega \log\left(\frac{u'(y+\delta)}{u'(x+\delta)}\right) > \log\left(\frac{u'(y)}{u'(x)}\right)$$

$$\iff \frac{u'(y+\delta)}{u'(x+\delta)} > \frac{u'(y)}{u'(x)} \text{ for } x < y, \delta > 0$$

$$q(y) := \frac{u(y) - u(x)}{u'(x)} - \frac{u(y+\delta) - u(x+\delta)}{u'(x+\delta)}$$

$$q(y) = \frac{u'(y)}{u'(x)} - \frac{u'(y+\delta)}{u'(x+\delta)} < 0 \text{ due to the last inequality}$$

Apply the mean value theorem on $[x, y]$:

$$\frac{q(y) - q(x)}{y - x} < 0 \iff q(y) < 0 \text{ since } q(x) = 0, y - x > 0$$

Now replace in $q(y) < 0$ y with $\omega + 1 - r$ and x with ω:

$$\frac{u(\omega + 1 - r) - u(\omega)}{u'(\omega)} - \frac{u(\omega + 1 - r + \delta) - u(\omega + \delta)}{u'(\omega + \delta)} < 0$$

$$\iff L(1, 1, r, \omega + \delta) < L(1, 1, r, \omega) \iff \Delta(1, \delta) > 0$$

We proceed to show $\Delta(t) = 0$ has exactly one root:

$$\frac{\partial L(v, t, r, \omega)}{\partial t} = \frac{-u''(v - t + \omega)}{u(v - r + \omega) - u(\omega)} = \frac{-u''(v - t + \omega)}{u'(v - t + \omega)} \frac{u'(v - t + \omega)}{u(v - r + \omega) - u(\omega)}$$

$$= A(v - t + \omega) L(v, t, r, \omega)$$

$$\frac{\partial \Delta(t)}{\partial t} = A(1 - t + \omega) L(u; 1, t, r, \omega) - \tilde{A}(1 - t + \omega + \delta) L(u; 1, t, r, \omega + \delta)$$

Suppose there are two values of t, t_2^* and t_1^* with $t_2^* > t_1^*$ such as $\Delta(t_1^*) = \Delta(t_2^*) = 0$.
Since $L(u; 1, t_1^*, r, \omega) = L(u; 1, t_1^*, r, \omega + \delta)$ and $A(1 - t_1^* + \omega) > A(1 - t_1^* + \omega + \delta)$, it follows $\frac{\partial \Delta(t_1^*)}{\partial t} > 0$. Thus:

$$\forall_t \in (t_1^*, t_2^*) \Delta(t) > 0 \tag{12}$$

Using the mean value theorem of differential calculus, we get that $\exists t^c \in (t_1^*, t_2^*)$ with

$$\frac{\partial \Delta(t^c)}{\partial t} = \frac{\Delta(t_2^*) - \Delta(t_1^*)}{t_2^* - t_1^*} = 0$$

$$\iff A(1 - t^c + \omega) L(u; 1, t^c, r, \omega)$$
$$- A(1 - t^c + \omega + \delta) L(u; 1, t^c, r, \omega + \delta) = 0$$

$$\iff L(u; 1, t^c, r, \omega) < L(u; 1, t^c, r, \omega + \delta) \quad \text{(since } A(1 - t^c + \omega)$$
$$> A(1 - t^c + \omega + \delta))$$

$$\iff \Delta(t^c) < 0 \quad \text{which is a } \textbf{contradiction} \text{ to (12)}$$

Theorem 4. *Suppose $r = 0$. If the environment $e = \{F_1, F_2, G, u_1, u_2, u_3, \omega_1, \omega_2, \omega_3\}$ admits a non-bidding equilibrium then so does the environment $\tilde{e} = \{\tilde{F}_1, F_2, G, u_1, u_2, u_3, \omega_1, \omega_2, \omega_3\}$ where $\tilde{F}_1 \succcurlyeq_{FSD} F_1$.*

Proof of Theorem 4.

$$NB(\tilde{F}_1) > NB(F_1) \iff \int_0^1 G(t) \left(\tilde{f}_1(t) - f_1(t)\right) dt \geq 0$$

$$\iff G(1) \left(\tilde{F}_1(1) - F_1(1)\right) - \int_0^1 g(t) \left(\tilde{F}_1(t) - F_1(t)\right) dt \geq 0$$

$$\iff \int_0^1 g(t) \left(\tilde{F}_1(t) - F_1(t)\right) dt \leq 0 \tag{13}$$

The last inequality holds since $g(t) > 0$ and $\tilde{F}_1 \succcurlyeq_{FSD} F_1$. □

Theorem 5. *Suppose additionally g is non-increasing. Then the weaker condition $\tilde{F}_1 \succcurlyeq_{SSD} F_1$ is sufficient for Theorem 4 to hold.*

Proof of Theorem 5. If $g(t)$ is non-increasing, then only the weaker second-order stochastic dominance is required:

$$\Delta(t) := F_1(t) - \tilde{F}_1(t).$$

Since $\Delta(0 + \varepsilon) \geq 0$ (for a small ε—due to SSD), if Δ has up to one roots, (13) holds immediately. Suppose now Δ has $n + 1$ roots $t_0 < \ldots < t_n$. $\tilde{F}_1 \succcurlyeq_{SSD} F_1 \Rightarrow \int_{t_0}^{t_i} \Delta(t) dt \geq 0 \forall i$.

$\Delta(t) \geq 0 \forall_t \in [t_{i-1}, t_i]$, if i odd and $\Delta(t) \leq 0 \forall_t \in [t_i, t_{i+1}]$ if i even.

Let $I_i := \int_{t_{i-1}}^{t_i} g(t) \Delta(t) dt$. Since g non-increasing, $I_{2k-1} \geq g(t_{2k-1}) \int_{t_{2k-2}}^{t_{2k-1}} \Delta(t) dt \geq 0$ and

$$I_{2k} \geq g(t_{2k-1}) \int_{t_{2k-1}}^{t_{2k}} \Delta(t) dt \quad \forall k \in \mathbb{N}$$

It can be observed that the sign of $\Delta(t)$ is positive in $[t_{i-1}, t_i]$, if i is odd, else negative. The inequalities concerning the integrals I_i hold since g is non-increasing (see Fig. 8)

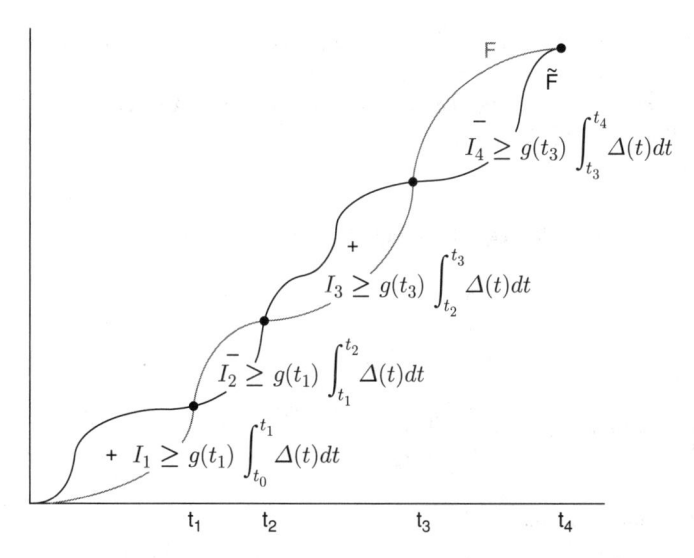

Figure 8. Second order stochastic dominance $\tilde{F} \succcurlyeq_{SSD}$ but not $\tilde{F} \succcurlyeq_{FSD} F$.

Now let $S_n := \sum_{i=1}^{n} I_i$. We will show by induction that $S_n \geq 0 \forall n$ [this immediately implies (13)].

$$S_1 \geq 0 \text{ since } S_1 = I_1 \geq g(t_1) \int_{t_0}^{t_1} \Delta(t)dt \geq 0$$

$$S_2 = I_1 + I_2 \geq g(t_1) \int_{t_0}^{t_2} \Delta(t)dt \geq 0 \text{ due to } \tilde{F}_1 \succcurlyeq_{FSD} F_1.$$

We assume $S_{2k} \geq g(t_{2k-1}) \int_{t_0}^{t_{2k}} \Delta(t)dt \geq 0$ and show $S_{2k+2} \geq g(t_{2k+1}) \int_{t_0}^{t_{2k+2}} \Delta(t)dt \geq 0$

$$S_{2k+2} = S_{2k} + I_{2k+1} + I_{2k+2} \geq S_{2k} + g(t_{2k+1})$$

$$\times \int_{t_{2k}}^{t_{2k+2}} \Delta(t)dt \geq g(t_{2k-1}) \int_{t_0}^{t_{2k}} \Delta(t)dt$$

$$+ g(t_{2k+1}) \int_{t_{2k}}^{t_{2k+2}} \Delta(t)dt \geq g(t_{2k+1})$$

$$\times \int_{t_0}^{t_{2k}} \Delta(t)dt + g(t_{2k+1}) \int_{t_{2k}}^{t_{2k+1}} \Delta(t)dt$$

$$g(t_{2k+1}) \int_{t_0}^{t_{2k+2}} \Delta(t)dt \geq 0$$

In addition $S_{2k+1} = S_{2k} + I_{2k+1} \geq 0$ □

References

Airiau S, Sen S (2003) Strategic bidding for multiple units in simultaneous and sequential auctions. Group Decisi Negot 12(5):397–413

Arrow KJ (1965) *Aspects of the theory of risk-bearing*. Yrjö Jahnssonin Säätiö

Athey S (2001) Single crossing properties and the existence of pure strategy equilibria in games of incomplete information. Econometrica 69(4):861–889

Ausubel LM, Milgrom P (2006a) Ascending proxy auctions. Comb Auctions 79–98

Ausubel LM, Milgrom P (2006b) The lovely but lonely Vickrey auction. Comb Auctions 17–40

Ausubel LM, Milgrom PR (2002) Ascending auctions with package bidding. BE J Theor Econ 1(1):1

Banks J, Olson M, Porter D, Rassenti S, Smith V (2003) Theory, experiment and the federal communications commission spectrum auctions. J Econ Behav Organ 51(3):303–350

Bichler M, Shabalin P, Wolf J (2013) Do core-selecting combinatorial clock auctions always lead to high efficiency? An experimental analysis of spectrum auction designs. Exp Econ 16(4):511–545

Bikhchandani S, Ostroy JM (2002) The package assignment model. J Econ Theory 107(2):377–406

Chen K-Y, Plott CR (1998) Nonlinear behavior in sealed bid first price auctions. Games Econ Behav 25(1):34–78

Clarke EH (1971) Multipart pricing of public goods. Public Choice 11(1):17–33

Cox JC, Smith VL, Walker JM (1988) Theory and individual behavior of first-price auctions. J Risk Uncertain 1(1):61–99

Cramton P (2013) Spectrum auction design. Rev Ind Organ 42(2):161–190

Cramton P, Day R (2012) The quadratic core-selecting payment rule for combinatorial auctions. Oper Res 60(3):588–603

Cramton P, Shoham Y, Steinberg R (2006) Combinatorial auctions

Day RW, Raghavan S (2007) Fair payments for efficient allocations in public sector combinatorial auctions. Manag Sci 53(9):1389–1406

Drexl A, Jømsten K, Knof D (2009) Non-linear anonymous pricing combinatorial auctions. Eur J Oper Res 199(1):296–302

Fibich G, Gavious A (2010) Large auctions with risk-averse bidders. Int J Game Theory 39(3):359–390

Fibich G, Gavious A, Sela A (2006) All-pay auctions with risk-averse players. Int J Game Theory 34(4):583–599

Goeree J, Lien Y (2009) An equilibrium analysis of the simultaneous ascending auction

Goeree JK, Holt CA (2010) Hierarchical package bidding: a paper & pencil combinatorial auction. Games Econ Behav 70(1):146–169

Goeree JK, Lien Y (2013) On the impossibility of core-selecting auctions. Theor Econ

Green J, Laffont JJ (1977) Characterization of satisfactory mechanisms for the revelation of preferences for public goods. Econ J Econ Soc 427–438

Groves T (1973) Incentives in teams. Econ J Econ Soc 617–631

Kirchkamp O, Reiss JP (2011) Out-of equilibrium bids in auctions: wrong expectations or wrong bids. Econ J 121(557):1361–1397

Kwasnica AM, Ledyard JO, Porter D, DeMartini C (2005) A new and improved design for multiobject iterative auctions. Manag Sci 51(3):419–434

Lehmann D, Müller R, Sandholm T (2006) The winner determination problem. Comb Auctions 297–317

Parkes DC, Ungar LH (2000) Iterative combinatorial auctions: theory and practice. MIT Press, Cambridge

Porter D, Rassenti S, Roopnarine A, Smith V (2003) Combinatorial auction design. Proc Natl Acad Sci USA100(19):11153

Pratt JW (1964) Risk aversion in the small and in the large. Econ J Econ Soc 122–136

Riley JG, Samuelson WF (1981) Optimal auctions. Am Econ Rev 71(3):381–392

Rothkopf MH, Pekeč A, Harstad RM (1998) Computationally manageable combinational auctions. Manag Sci 44(8):1131–1147

Sano R (2011) Non-bidding equilibrium in an ascending core-selecting auction. Games Econ Behav

Scheffel T, Pikovsky A, Bichler M, Guler K (2011) An experimental comparison of linear and nonlinear price combinatorial auctions. Inf Syst Res 22(2):346–368

Schneider S, Shabalin P, Bichler M (2010) On the robustness of non-linear personalized price combinatorial auctions. Eur J Oper Res 206(l):248–259

Vickrey W (1961) Counterspeculation, auctions, and competitive sealed tenders. J Finance 16(l):8–37

Xia M, Koehler GJ, Whinston AB (2004) Pricing combinatorial auctions. Eur J Oper Res 154(1):251–270

CHAPTER 15

Properties of the Combinatorial Clock Auction

Jonathan Levin and Andrzej Skrzypacz[*]

In this paper we study some properties of a new auction design, the combinatorial clock auction (or CCA). The CCA was proposed by Ausubel, Cramton and Milgrom (2006). It is essentially a dynamic Vickrey auction. The Vickrey auction is central to economic theory as the unique auction that provides truthful incentives while achieving an efficient allocation. Yet it is often viewed as impractical for real-world applications because it requires bidders to submit bids for many possible packages of items.[1] Economists think of dynamic auctions as having an advantage in this regard because bidders can discover gradually how their demands fit together — what Paul Milgrom has called the "package discovery" problem.

The CCA combines an initial clock phase, during which prices rise and bidders state their demands in response to the current prices, with a final round in which bidders submit sealed package bids. The seller uses the final bids to compute the highest value allocation and the corresponding Vickrey payments.[2] Ideally, bidders demand their most desired package at every stated price in the clock phase, allowing for information revelation. Then in the final round, they bid their true preferences, leading to an efficient allocation with truthful Vickrey prices. The question we address is whether this is the likely equilibrium outcome of the CCA; that is, whether the desirable incentive properties of the Vickrey auction are retained.

---

[*] We thank Chiara Farronato, Maarten Janssen, Paul Klemperer, Erik Madsen, Paul Milgrom, David Salant, Ernesto Wandeler and Bob Wilson for useful comments and discussions. The authors have advised bidders and the U.S. Federal Communications Commission on radio spectrum auctions. No party had the right to review this paper prior to circulation.

[1] Ausubel and Milgrom (2006) offer additional reasons why sealed-bid Vickrey auctions have not caught on in practice, including vulnerability to collusion even by losing bidders, incentives for shill bidding, and the potential for low revenue.

[2] The original CCA of Ausubel, Cramton and Milgrom (2006) and the rules used in practice actually call for a slight modification of Vickrey pricing, where the Vickrey prices are adjusted upwards if the outcome is outside the core. This "core-adjustment" will not be relevant in our model, but in general it means the truth-telling properties of the Vickrey auction may not apply. For papers on core-adjustment see for example Day and Raghavan (2007), Day and Milgrom (2008), Edril and Klemperer (2009), Ausubel and Baranov (2010), Beck and Ott (2013) and Goeree and Lien (2016).

The practical motivation for our study is the recent and widespread adoption of CCA bidding to sell radio spectrum licenses.[3] Spectrum auctions have provided the motivation for some important recent innovations in auction design, starting with the simultaneous ascending auction pioneered by the FCC in the early 1990s and subsequently adopted in many other countries (Klemperer, 2004; Milgrom, 2004). The FCC design allows for gradual information revelation, but it does not easily accommodate package bidding, and it creates incentives for demand reduction because winners pay their bids (Cramton, 2013). In principle, the CCA addresses both of these issues.

As it turns out, the CCA does have an equilibrium in which bidding is truthful and the outcome is efficient. But it is a rather tenuous equilibrium, and we identify two distinct reasons to doubt it. The first arises from the fact that CCA bidders are asked to submit their demands twice: during the clock phase and then in the final round. These demands are linked by activity rules, which are essential in a dynamic auction to make early bids meaningful and prevent bidders from holding back like eBay snipers (Ausubel, Cramton and Milgrom, 2006). The CCA activity rules have the feature that the clock phase bids can pin down exactly the allocation of items (Ausubel and Cramton, 2011; Ausubel and Baranov, 2013). Then the final bids determine the payments. But with Vickrey pricing, a bidder cannot affect her own payment unless she changes the allocation. So each bidder may be completely indifferent across her permissible final bids, despite the choice affecting the prices paid by rivals and hence incentives in the clock phase. To support the truthful and efficient equilibrium, bidders must maximally raise their final bids. But because bidders may adopt different strategic postures in their final bids, there are also a wide range of other, inefficient, ex post equilibria.

The second issue we investigate involves predatory bidding. With a Vickrey pricing rule, bidders have a strict incentive to bid truthfully only if their bid has a positive probability of winning. Yet in a CCA, bidders may have the opportunity in the clock phase to exaggerate their demand with essentially no risk of winning. By doing so, they can relax the activity rule constraints on their final bids and raise rival payments. A bidder who anticipates this type of predation has an incentive to reduce demand to avoid paying predatory prices, again leading to inefficient outcomes. While the "two demands" problem is rather specific to the CCA rules, the potential for predation seems likely to arise in any dynamic implementation of the Vickrey auction. In the CCA case, the upshot is that bidders must behave "just right" to support the truthful and efficient equilibrium, raising their final bids to the limit allowed by the activity constraints, but taking no action in the clock phase to purposefully relax these constraints.

We develop these points in a series of simple models. We start in Section I by describing the CCA rules and providing an example of how they work. We then focus on a standard allocation problem where bidders have linear downward sloping demand curves. In Section II, we show how different strategic postures in the final round each give rise to a range of ex post and typically inefficient equilibria. The multiplicity arises even though we focus on (linear) proxy strategies in which bidders do not condition on rival

---

[3] Countries that have used CCAs to sell radio spectrum licenses include Australia, Austria, Canada, Denmark, Ireland, the Netherlands, Slovenia, Slovakia, Switzerland, and the UK. For more background on spectrum auctions, see Cramton (2013) or Loertscher, Marx and Wilkening (2015).

bidding behavior. Section III then considers the possibility that bidders may prefer to drive up rival prices if they can do so without reducing their own payoff. We characterize an ex post equilibrium in which one bidder is predatory in the clock phase while the other restricts attention to (linear) proxy strategies and reduces his demand to keep the price down. The Appendix also considers a version of the model in which both bidders are able to relax the final bid activity constraints, and equilibrium outcomes involve demand reduction in the clock phase and are again inefficient.

The models suggest that CCA rules permit a wide range of plausible behaviors and outcomes. In Section IV, we provide some evidence from recent auctions for radio spectrum. We find that even large and sophisticated bidders have adopted widely varying strategic postures in their CCA final bids. And in at least one high-stakes auction, bidders appear to have taken actions in the clock phase that served to relax final bid constraints and raise rival prices.

Our paper relates to an extensive literature on multi-item auctions (e.g. Milgrom, 2004, 2007). One of its central messages is that multi-item auction design involves hard trade-offs. The standard auctions that have been considered - simultaneous ascending and clock auctions, pay-as-bid combinatorial auctions - have well-documented limitations. In this sense, it is hardly surprising that the CCA also appears to have some drawbacks from a strategic perspective. In fact, one point we emphasize in the conclusion is that our analysis highlights a serious challenge for any attempt to implement a dynamic Vickrey auction, namely that it may be difficult to provide incentives for bidders to submit truthful bids for packages that they are very unlikely to win, despite these bids potentially being important for setting rival prices.

More narrowly, the novelty of the CCA design means there are not many directly related papers. The closest are Janssen and Karamychev (2013) and Janssen and Kasberger (2015). The first of these papers analyzes bidding incentives in a CCA with discrete quantities and multiple products. It shows that when bidders have preferences for raising rivals' costs (modeled similarly to what we do in Section III), bidders have an incentive to submit large final round bids and bid aggressively in the clock phase. It also considers the implications of bidder budget constraints. The second paper extends our analysis by characterizing equilibria in non-linear proxy strategies when bidders prefer to raise rivals' costs. A nice insight is that in some cases, it is possible to construct equilibria where bidders engage in predation but the outcome is nonetheless efficient. However, under some conditions this is not possible and all symmetric proxy strategy equilibria are inefficient. Finally, Salant (2014) provides a broader review of practical auction design that covers the CCA as well as competing formats.[4]

I. The Combinatorial Clock Auction

The combinatorial clock auction can be used with multiple bidders and multiple products in different quantities. For concreteness, we will assume there are two bidders, $i = 1, 2$, and a single product. The product is perfectly divisible and there is unit supply.

---

[4] Additionally, Knapek and Wambach (2012) discuss why bidding in a CCA may be strategically complicated; Bichler et.al. (2013) present experimental results on the CCA.

The auction consists of an initial clock phase and a final bid round. In the clock phase, the seller gradually increases the price p. In response, the bidders announce demands x_1, x_2. A bidder may reduce her demand, but not increase it, as the auction proceeds. The price increases until there is no excess demand. If at this point $\sum_i x_i = 1$, we say there is market clearing. Alternatively if $\sum_i x_i < 1$, there is excess supply. Our analysis will focus on the case where the starting price is low and bidders reduce their demands continuously, so there is never excess supply.

After the clock phase, the bidders submit final (sealed) bids $S_1(x_1)$ and $S_2(x_2)$ that express valuations for all possible quantities (i.e. each bid S_i is a function). The seller computes the allocation and winner payments based on these final bids. She selects the allocation that maximizes $\sum_i S_i(x_i)$ subject to the feasibility constraint that $\sum_i x_i \leq 1$. She then computes the Vickrey payment for each bidder. Bidder i's Vickrey payment is $\max_{x \in [0,1]} S_j(x) - S_j(x_j^*)$, where x_j^* is j's winning quantity.[5] If S_j is increasing, i pays $S_j(1) - S_j(x_j^*)$.

The final bids are constrained by activity rules that tie them to bids in the clock phase. The activity rules we describe correspond to those used in recent CCA auctions.[6] There are three parts. First, bids in the clock phase are binding. If at price p, bidder i demanded x, then bidder i's final bid for x units must be at least px, i.e. $S_i(x) \geq px$ for any p that was quoted in the clock phase. Second, final bids must satisfy revealed preference with respect to the last clock bids. If the clock phase ends at a price p^*, with i demanding x_i^*, then for any $x \neq x^*$, $S_i(x) - p^*x \leq S_i(x^*) - p^*x^*$. Third, for quantities $x \geq x^*$, i's final bids must satisfy an additional local form of revealed preference. If p was the highest price at which i demanded x or more units, then i cannot express an incremental value greater than p for obtaining slightly more than x, i.e. $\lim_{\varepsilon \to 0^+} \frac{S_i(x+\varepsilon) - S_i(x)}{\varepsilon} \leq p$.[7]

We illustrate these activity rules with an example. For this example, we assume that during the clock phase bidder 2 behaves as if she has a value for x_2 units equal to $V_2(x_2) = 2x_2 - \frac{1}{2}x_2^2$, and a diminishing marginal value $v_2(x_2) = 2 - x_2$.[8] So when the price is p, she demands $x_2(p) = 2 - p$ units irrespective of bidder 1's behavior. The quantity $1 - x_2(p)$ is the residual supply available to bidder 1 at price p. The clock price starts at $p = 1$. It rises so long as bidder 1 demands $x_1(p) > 1 - x_2(p)$, and stops as soon as $x_1(p) \leq 1 - x_2(p)$. Suppose the clock phase ends at a price p^* with bidder 2 demanding $x_2^* = 2 - p^*$, and bidder 1 demanding $x_1^* = 1 - x_2^*$.

What are the permissible final bids for bidder 2? Despite the activity rules, she retains considerable flexibility. Figure 1 illustrates the possibilities. The lower curve shows

---

[5] We do not consider reserve prices, but they can be incorporated by adding a "bidder 0" that bids $S_0(x)$ where $S_0(1) - S_0(x)$ is the required revenue to sell $1 - x$ units.

[6] It is exactly the rule used in Canada, and equivalent to the rules used in Switzerland, Ireland, Netherlands, and UK if those auctions had been run for only a single category of licenses. The exact rules used in these auctions vary somewhat in how they handle multiple license categories.

[7] An obvious candidate for an activity rule would be to require global revealed preference. That is, if i demanded x at price p, then for any $z \neq x$, final bids must satisfy $S_i(x) - px \geq S_i(z) - pz$. In our model, this rule would *impose* what we later will call consistent bidding. However, as noted by Ausubel and Cramton (2011), such an approach seems unworkable with multiple categories because it can lead to "dead ends". In response to this, Ausubel and Baranov (2013) have suggested a global approach based on GARP. We discuss their proposal in the conclusion.

[8] For this example it is not important whether V_2 is bidder 2's actual valuation. In our equilibrium analysis below, bidders generally will not bid truthfully in the clock phase.

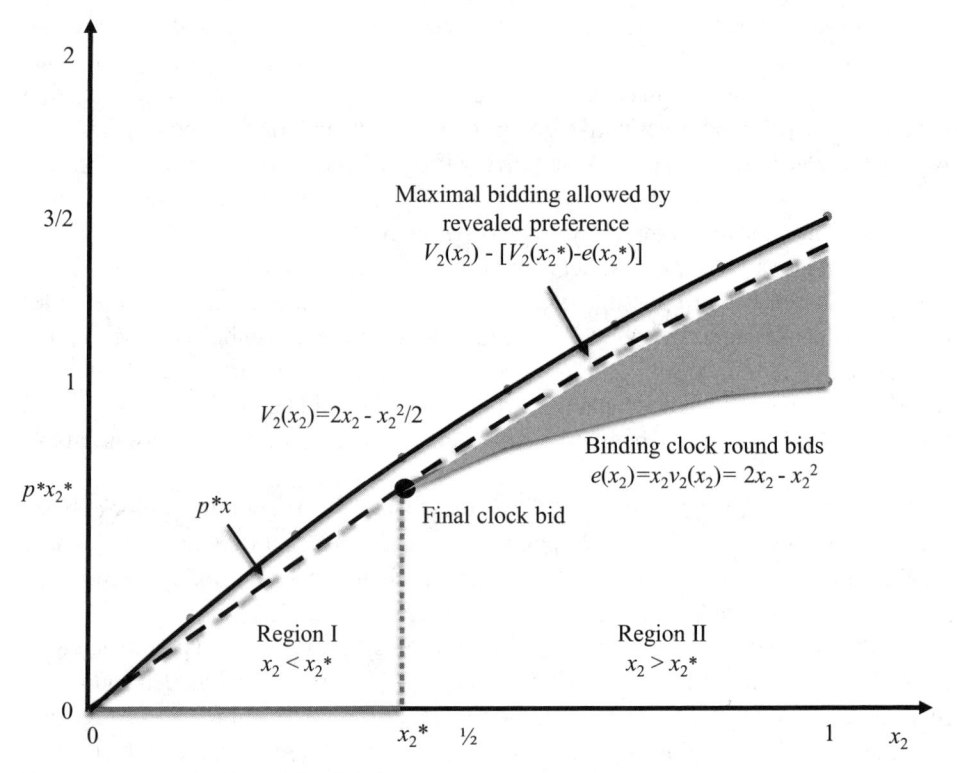

Figure 1. Activity Rule and Final Bid Options.
Note: Figure shows the flexibility that bidder 2 has in choosing final bids in the final bid round. The lower curve shows bidder 2's clock phase bids, which place a lower bound on her final bids. The dashed curve is the upper bound for final bids (assuming bidder 2 leaves her last clock bid unchanged). The top curve is the valuation that guides bidder 2's clock round bidding. The dashed upper bound is parallel to this curve for quantities above x_2^*.

the bids that bidder 2 submitted during the clock phase. As the price rose, bidder 2 reduced her demand continuously from 1 to x_2^*. For quantities below x_2^*, bidder 2 has not recorded any bid during the clock phase, so her bid is zero. For quantities above x_2^*, bidder 2 demanded x_2 units when the price was $p = v_2(x_2)$. When she did this, she submitted a bid of $x_2 v_2(x_2)$ for x_2 units. Of course this is less than the value function $V_2(x_2)$ that guided bidder 2's clock bidding because the auctioneer records the revenue from bidder 2's bid at price $p = v_2(x_2)$, and does not include her consumer surplus.

The activity rules state that bidder 2's final bid S_2 must lie everywhere above the bids she recorded during the clock phase. Also, above x_2^* the final bid function cannot rise more steeply than V_2. It must satisfy the local revealed preference restriction: $S_2'(x_2) \leq v_2(x_2)$. So at the upper extreme, bidder 2 can raise her bids to the dashed curve, which rises at a slope $v_2(x_2)$ from her final clock bid.[9] The shaded area in Figure 1 shows that

---

[9] Our discussion presumes that bidder 2 does not increase her clock bid for x_2^*, i.e. sets $S_2(x_2^*) = x_2^* v_2(x_2^*)$. We will assume this throughout for simplicity. If bidder 2 raises her bid for x_2^* by Δ she is permitted to translate all her bids up by this same amount, so there is an analogue of quiet and consistent bidding for any choice of Δ. Of course, if bidder 2 raises her bid for x_2^*, but does not raise her other bids, then bidder 1 could end up paying very little, and in fact pays zero if S_2 achieves its maximum at x_2^*. This creates the possibility that bidder 2's final bids might be chosen in a way that drives down bidder 1's payment relative to quiet bidding, but we will not focus on it.

space in which the final bid function must lie, for quantities above x_2^*. Below x_2^*, bidder 2 may submit any non-negative bid S_2 so long as it lies below the dotted line with slope p^* that runs straight from the origin to bidder 2's final clock bid. These bids, however, are not important for pricing.[10]

Two types of final bids will be important later. We say that bidder 2 is *consistent* if she raises her final bid above x_2^* to its maximum amount. When she does this, she expresses marginal values that correspond exactly to her (inverse) demand in the clock phase. That is, her demands are consistent in the two parts of the auction. In contrast, we say that bidder 2 is *quiet* if she does not raise her final bids at all, so that $S_2(x_2) = xv_2(x_2)$. Under quiet bidding, S_2 corresponds to the revenue generated by bidder 2's clock phase demand, and the slope of S_2 is the marginal revenue associated with assigning more units to bidder 2, rather than the marginal values implicit in bidder 2's clock demand. In addition to consistent and quiet bidding, there are intermediate cases as well, since in the range $x_2 \geq x_2^*$ (Region II in the figure), bidder 2 can select any $S_2(x_2)$ that lies in the shaded area and rises less steeply than $V_2(x_2)$.

Now we turn to an important implication of the activity rules that holds whether bidders are consistent, quiet or something intermediate. Suppose the clock phase ends with market clearing. Then the final clock demands will be value-maximizing for *any* permissible final bids. To see why, let x_1^*, x_2^* be the final clock demands. Consider an alternative assignment where each bidder i receives $x_i = x_i^* + \varepsilon_i$, with $\sum_i \varepsilon_i \leq 0$ required for feasibility. From the second activity rule requirement,

$$\sum_i S_i(x_i) \leq \sum_i S_i(x_i^*) - p^* \cdot (x_i^* - x_i) = \sum_i S_i(x_i^*) + p^* \cdot \sum_i \varepsilon_i \leq \sum_i S_i(x_i^*). \tag{1}$$

This has the following consequence noted by Ausubel and Cramton (2011). If the clock phase ends with market clearing and ties are resolved in favor of the clock phase allocation, bidder i's quantity and payment do not depend on her final bids. The payment part follows from Vickrey pricing: fixing i's quantity, her payment depends only on the bids of others. Therefore if a bidder is maximizing her own individual profit she will be *completely indifferent* across all permissible final bids. However, these bids are very important for prices paid by the other bidder. To see this, assume for simplicity that S_2 is increasing. Then bidder 1 must pay $S_2(1) - S_2(x_2^*)$ — bidder 2's bid for all units minus her bid for the units she wins. This amount is higher if bidder 2 is consistent than if she is quiet.

II. Bidding in the CCA: Discounts and Demand Expansion

In this section and the following ones, the environment is the same. There is a single divisible unit to be allocated. Each bidder $i = 1, 2$ has marginal value for an xth unit given by $u_i(x) = a_i - b_i x$, where $a_i \geq b_i > 0$. Bidder i's total value for x units is $U_i(x) = \int_0^x u_i(z)dz = a_i x - \frac{1}{2}b_i x^2$. Throughout, we use lower case to denote marginal

---

[10] That is because the activity rules prevent bidder 2 from claiming to value $x_2 < x_2^*$ more than she values x_2^*. So the maximum of S_2 will be achieved at a quantity $x_2 \geq x_2^*$.

values and upper case to denote total values. We assume the a_i's are private information and take values between $[\underline{a}_i, \overline{a}_i]$, while the b_i's are common knowledge. To avoid messy corner solutions, we assume $\overline{a}_i - b_i < \underline{a}_j$, so that efficient allocation is interior and satisfies $u_1(x_1) = u_2(1-x_1)$.

A *proxy strategy* for bidder i consists of two functions: marginal values $v_i(x)$ for the clock phase (assumed to be decreasing and continuous), and marginal values $s_i(x)$ for the final bid.[11] In the clock phase, bidder i demands x when $p = v_i(x)$ (or $x = 1$ if $p < v_i(1)$). This expresses a bid $xv_i(x)$ for x units. If the clock phase ends at a price p^*, with i demanding x_i^*, then in the final round, she bids $S_i(x) = x_i^* p^* + \int_{x_i^*}^{x} s_i(z)dz$ for quantities $x \geq x_i^*$. To satisfy the activity rules, the proxy values must satisfy, for all x: (i) $S_i(x) \geq xv_i(x)$, and (ii) $s_i(x) \leq v_i(x)$. The latter condition ensures revealed preference with respect to the final clock bid for $x \geq x_i^*$: it implies that $s_i(x) \leq p^* = v_i(x_i^*)$. For quantities $x < x_i^*$, bidder i can set $S_i(x) = 0$ — any bids that satisfy the activity rules in this region cannot affect allocation or pricing.

We restrict bidders to use proxy strategies and focus on equilibria in *linear* proxy strategies, in which each bidder bids a linear demand curve in the clock phase, and adopts a mixture of quiet and consistent behavior in the final round. Formally, bidder i specifies a linear demand for the clock phase

$$v_i(x) = A_i - B_i x, \tag{2}$$

with $A_i \geq B_i > 0$, and associated marginal revenue curve

$$m_i(x) = \frac{d}{dx} x v_i(x) = A_i - 2B_i x. \tag{3}$$

Bidder i also specifies a linear demand for the final bid round:

$$s_i(x) = (1 - \gamma_i) v_i(x) + \gamma_i m_i(x) \tag{4}$$
$$= A_i - (1 + \gamma_i) B_i x$$

The parameter $\gamma_i \in [0, 1]$ captures the extent to which bidder i is consistent ($\gamma = 0$) versus quiet ($\gamma = 1$). Higher values of γ_i mean lower marginal prices for bidder j, because i is expressing less value for any particular unit. It is easy to see that for any value of γ_i, $s_i(x)$ satisfies the activity rules.

A linear proxy strategy requires the choice of three parameters: A_i, B_i and γ_i. Provided the clock price starts sufficiently low, the clock phase will end with market clearing. So for any A_j, B_j, γ_j, bidder i's payoff will depend only on her choice of A_i, B_i, and will be independent of her choice of γ_i. Therefore to characterize equilibria, we fix γ_1 and γ_2 as parameters, and solve for equilibrium choices of A_1, A_2, B_1, B_2. We focus on equilibria that are ex post, in that for any a_i, the strategy adopted by bidder i is a best response to j's strategy for every value of $a_j \in [\underline{a}_j, \overline{a}_j]$.

---

[11] We call these proxy strategies because bidders express preferences (i.e. marginal values) that get transformed into demands in the auction. This is the same approach and terminology used by Ausubel and Milgrom (2002) in their analysis of ascending auctions with package bidding. We focus on proxy strategies in order to emphasize issues that are specific to the CCA. The CCA also admits a rich set of contingent or bootstrapped equilibria, in which bidder j makes a certain demand in equilibrium because she believes that if she doesn't, bidder i will punish her as the auction proceeds. However, these types of dynamics are familiar from other dynamic auctions and are not specific to the CCA.

A. Proxy Best Responses

We start by deriving best responses for bidder 1. We show that if bidder 2 uses a linear proxy strategy, where A_2 varies with her private information a_2, but B_2 and γ_2 do not, then bidder 1 always has an ex post best response that involves using a linear proxy strategy in which A_1 is the only parameter to vary with a_1 (so that iterating on the best-response correspondence keeps us within this class of strategies).

Suppose bidder 2 bids according to $v_2(x) = A_2 - B_2 x$. If bidder 1 bids according to $v_1(x)$, and the clock allocation is interior, then bidder 1 obtains quantity x_1 such that

$$v_1(x_1) = v_2(1 - x_1). \tag{5}$$

Then after the final bid round, bidder 1 will pay $S_2(1) - S_2(1 - x_1)$. His final payoff will be:

$$U_1(x_1) - [S_2(1) - S_2(1 - x_1)]. \tag{6}$$

A necessary condition for bidder 1's strategy to be ex post optimal is that knowing v_2, he does not prefer to purchase a slightly larger or smaller quantity than x. The marginal benefit of additional quantity is $u_1(x)$, and the marginal price is $s_2(1 - x)$. So a necessary condition for ex post optimality is that:

$$u_1(x) = s_2(1 - x). \tag{7}$$

This condition is also sufficient for ex post optimality if it holds for all v_2, x that satisfy (5).[12]

Substituting (5) into (7), and using the fact that bidder 2 is playing the linear proxy strategy $s_2(1 - x) = v_2(1 - x) - \gamma_2 B_2(1 - x)$, we obtain a best response for bidder 1:

$$v_1(x) = u_1(x) + \gamma_2 B_2(1 - x). \tag{8}$$

Therefore bidder 1 can follow the linear proxy strategy $v_1(x) = A_1 - B_1 x$, with

$$A_1 = a_1 + \gamma_2 B_2 \quad \text{and} \quad B_1 = b_1 + \gamma_2 B_2, \tag{9}$$

and this is a best response for every a_2.

Note that in general the best response of bidder 1 deviates from truth-telling and the optimal deviation depends on the behavior of bidder 2 in the final bid round. The closer is bidder 2 to a quiet strategy in the final round, the more aggressive is the best response of bidder 1. To gain intuition, recall that the clock phase determines allocations and the final bid determines prices. Bidder 1's price for his final unit equals bidder 2's final marginal bid for this unit, $s_2(x_2^*)$. Under consistent bidding $s_2(x_2) = v_2(x_2)$. Under quiet bidding $s_2(x_2) = \frac{d}{dx_2} x_2 v_2(x_2)$. Consistent bidding means that bidder 1 must

---

[12] Note that our derivation assumes for notational convenience that $S_2(x)$ is maximized at $x = 1$, but does depend on this. To see that condition (7) is sufficient, note that given v_2, bidder 1's global best response problem is first to choose x_1 that maximizes $U_1(x) - \{\max_y S_2(y) - S_2(x)\}$, which is a concave problem, and then choose some decreasing v_1 such that $v_1(x_1) = v_2(1 - x_1)$. By this reasoning, the strategies we characterize remain best responses even if we remove the restriction to continuously-decreasing proxy strategies. For example, even if bidders could drop demand discontinuously to end the clock phase with excess supply, they would not find it profitable.

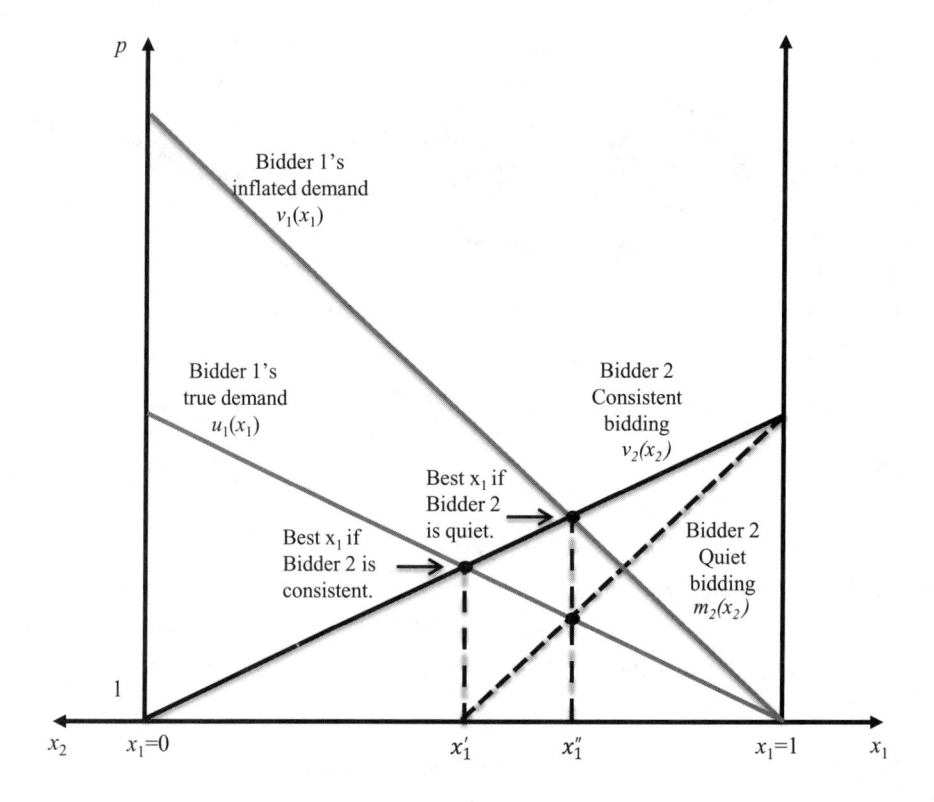

Figure 2. Best Responses to Quiet and Consistent Bidding.
Note: Figure shows bidder 1's best responses to bidder 2 using a consistent or quiet strategy. If bidder 2 is consistent, bidder 1 optimally intersects his clock demand with bidder 2's clock demand. If bidder 2 is quiet, bidder 1's marginal price is lower than the clock price. The best response is to purchase the quantity at which bidder 1's marginal value equals bidder 2's marginal revenue, which can be done by inflating demand in the clock phase, as shown in the figure.

pay the clock price at which bidder 2 gave up the final unit. Quiet bidding means that bidder 1 must pay only the (smaller) marginal revenue reduction.

Figure 2 shows the best response problem for bidder 1, and how it depends on bidder 2's behavior. The x-axis shows the allocation: as we move to the right, we increase x_1 and decrease x_2. The y-axis is dollars. The solid green line plots bidder 2's clock demand $v_2(x_2)$. If bidder 2 is consistent in the final round, this is also her final bid, and the marginal prices faced by bidder 1, so bidder 1's best response is to purchase the quantity x_1', at which $u_1(x_1') = v_2(1 - x_1')$. He can do this by bidding truthfully. The dotted green line shows the marginal revenue $m_2(x_2)$ associated with v_2. If bidder 2 is quiet in the final round, this line represents bidder 2's final bids. Bidder 1's best response is to purchase the quantity x_1'' at which $u_1(x_1'') = m_2(1 - x_1'')$. To do this, he needs to inflate his clock round bid, so that $v_1(x_1'') = v_2(1 - x_1'')$, as shown in the picture.

The figure shows the optimization problem for bidder 1 for a single value of a_2. However, bidder 1 wants the clock phase to end with his marginal benefit for additional quantity $u_1(x_1)$ just equal to the marginal price $s_2(x_2)$ for each realization of a_2. To make this happen, bidder 1 needs to have $v_1(x_1) = v_2(x_2)$ at the relevant x_1, x_2. Solving the

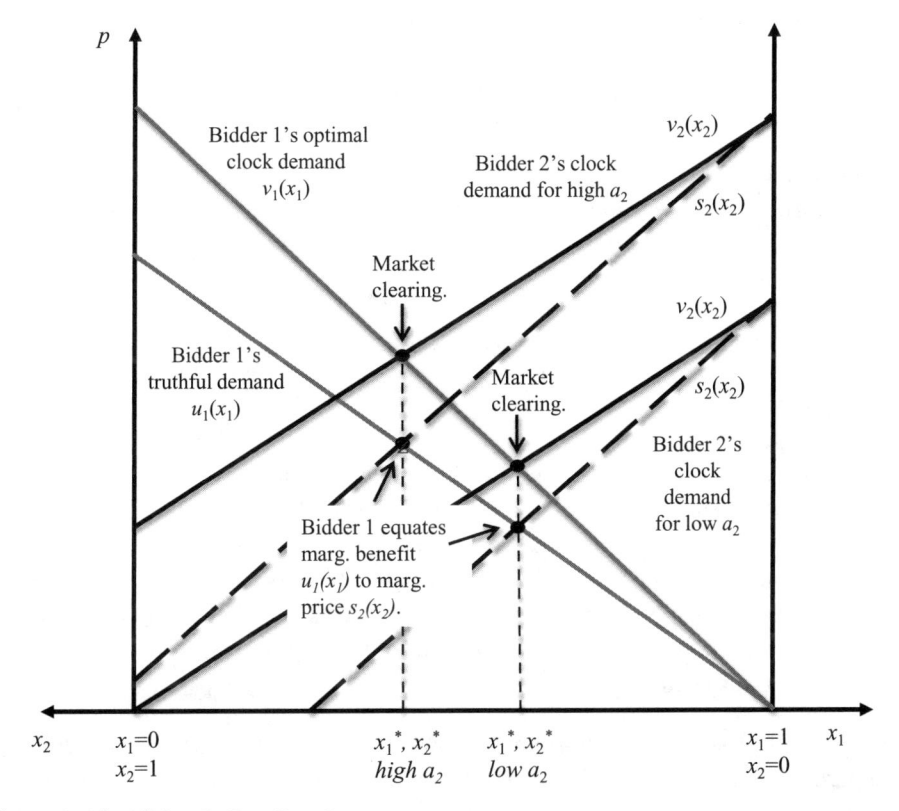

Figure 3. Identifying Ex Post Best Responses.
Note: Figure shows the derivation of bidder 1's ex post best response. Bidder 1's optimal clock demand is chosen so that for any realization of bidder 2's demand, the clock phase ends (market clearing) at the allocation where bidder 1's true marginal valuation just equals his marginal price, which he correctly anticipates will be set by bidder 2's final bid. Bidder 1's clock demand inflates his true demand.

optimization problem for each a_2 traces out bidder 1's best-response demand curve $v_1(x_1)$. This is illustrated in Figure 3.

The best-response derivation highlights a key strategic issue in the CCA. On the one hand, if a bidder cares only about his own payoff, he is completely indifferent across permissible final bids (any γ_i is a best response). On the other hand, the way the indifference is resolved is very important for determining rival incentives in the clock phase. A bidder will want to bid truthfully in the clock phase if his rival uses a consistent final bid strategy, but overstate his clock demand if he anticipates a quiet final bid strategy.

B. Proxy Equilibria

We now solve for an ex post equilibrium in linear proxy strategies. To do this, we combine the best response conditions (9) for bidders 1 and 2. Then, so long as γ_1 and γ_2 are not both equal to one,

$$A_1 = a_1 + \frac{\gamma_2(\gamma_1 b_1 + b_2)}{1 - \gamma_1 \gamma_2} \equiv a_1 + \lambda_1, \qquad (10)$$

and

$$B_1 = b_1 + \frac{\gamma_2(\gamma_1 b_1 + b_2)}{1 - \gamma_1\gamma_2} \equiv b_1 + \lambda_1. \tag{11}$$

Proposition 1. *Fix any* $\gamma_1, \gamma_2 \in [0, 1]$ *with* $\gamma_1\gamma_2 < 1$. *The CCA has an ex post equilibrium in linear proxy strategies, in which bidder i bids according to* $v_i(x)$ *in the clock phase and* $s_i(x)$ *in the final round, with*

$$v_i(x) = u_i(x) + \lambda_i(1 - x),$$
$$s_i(x) = v_i(x) - \lambda_j x,$$

and $\lambda_i = \frac{\gamma_j(\gamma_i b_i + b_j)}{1-\gamma_i\gamma_j}$.

Remark 1. The above proposition describes equilibria for $\gamma_1\gamma_2 < 1$. What if both bidders are completely quiet? Then the best responses (9) imply $B_1 = b_1 + B_2$, and $B_2 = b_2 + B_1$. The system "explodes" as $\gamma_1 = \gamma_2 \to 1$ and there is no equilibrium. This non-existence is a consequence of assuming two bidders and is familiar from other models with linear demands (e.g., Kyle 1989 or Vives 2011). For example, adding to the model a small non-strategic third bidder would allow us to construct linear proxy equilibria even for $\gamma_1 = \gamma_2 = 1$.

Figure 4 illustrates the equilibrium. It shows the equilibrium bids $v_1(x_1)$ and $v_1(x_2)$ for a particular realization of a_1, a_2, along with the final bids $s_1(x_1)$ and $s_1(x_2)$. As the clock price rises, the bidders reduce demand along their proxy demand curves $v_1(x_1)$ and $v_2(x_2)$ until reaching market clearing at p^*. At the end of the clock phase, the final allocation is determined to be x_1^*, x_2^*. Then the final bids are made. Bidder 1 pays $S_2(1) - S_2(x_2^*) = \int_{x_2^*}^{1} s_2(x_2)dx_2$, which is the shaded area in the figure. The price bidder 1 pays for his last unit is $s_2(x_2^*)$.[13]

C. Properties of Equilibria

Bidding Behavior. The equilibrium involves bidders engaging in demand expansion during the clock phase because they perceive that their true marginal prices will be discounted from the clock price if their opponent is less than consistent, $\gamma_2 > 0$. The residual supply available to bidder 1 will be x when the clock price is $p = v_2(1 - x)$. Yet bidder 1's true marginal price for buying the xth unit is only

---

[13] We have described equilibria in the game in which the players choose proxy strategies. A natural question is whether the same behavior could be supported as equilibria if players choose their clock demands strategically in response to a continuously increasing price. This is relatively easy to establish if the bidders only observe the price and not their opponent's current demand. In the latter case, formalizing the game in continuous time is cumbersome. However, we can sketch how off-path equilibrium beliefs might be chosen to support the behavior above. If j initially bids according to the proposed strategy of some type $a_j \in [\underline{a}, \overline{a}]$, but then deviates, one can use "renewal beliefs" where i assumes that from demand x_j at price p, j will bid according to the linear strategy of the type a_j that would have bid x_j at p (these are called renewal beliefs because they are applied regardless of j's prior behavior up to p). To make this work for all deviations, one can expand the interval of possible bidder j types by adding zero probability types so that for any x_j chosen at p, there is some a_j that would have chosen x_j under the linear strategy described above.

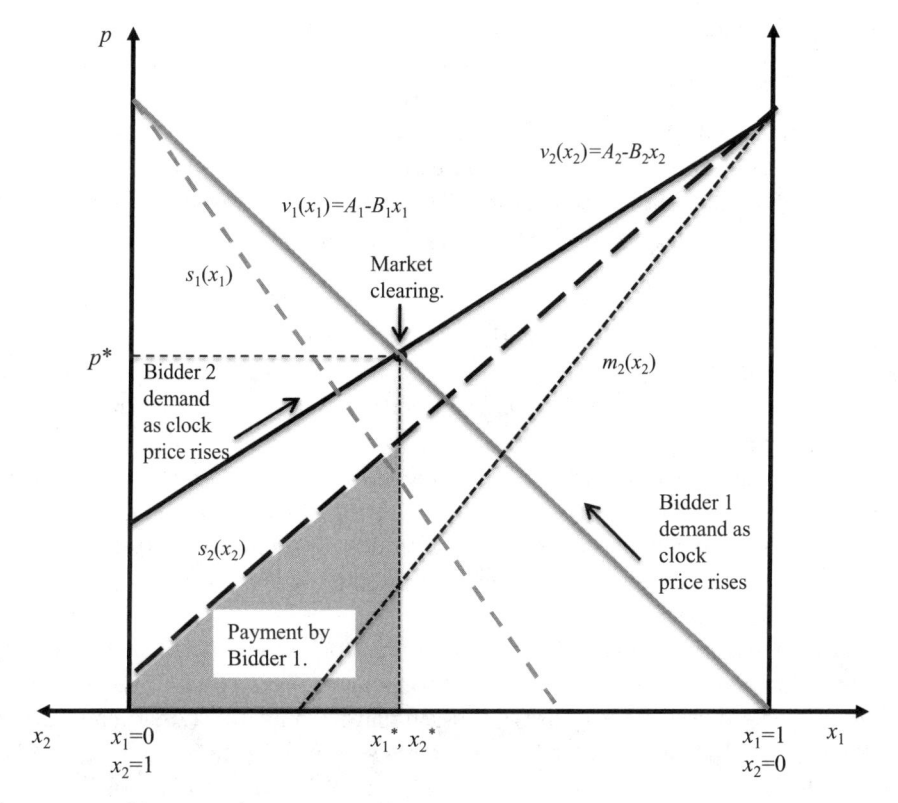

Figure 4. Equilibrium in the CCA Auction.
Note: Figure shows equilibrium behavior for a single realization of values. The solid lines represent the equilibrium clock demand curves of the bidders; the dashed lines the final round demand curves. Bidder 2's clock round demand is intermediate between her clock demand and marginal revenue; she bids partway between quiet and consistent.

$s_2(1 - x) = p - \lambda_1(1 - x)$. The discount is greatest if bidder 1 is buying a small quantity, and smaller for large quantities. So unless bidder 2 is consistent, the equilibrium response is to engage in demand expansion.

Allocation and Revenue. The equilibrium allocation is generally not efficient and the revenue differs from what would occur in a Vickrey auction with truthful bidding (unless both bidders are truthful and consistent, in which case allocation is efficient, and each bidder pays its truthful Vickrey price).

We can obtain a fairly sharp characterization for symmetric equilibria in which $\gamma_1 = \gamma_2 = \gamma$ and $b_1 = b_2 = b$. In this case, bidder i's clock phase strategy is $v_i(x) = u_i(x) + \frac{\gamma}{1-\gamma} b(1 - x)$. Provided that $\gamma > 0$, the equilibrium allocation will be distorted toward $1/2$ because a low-value bidder inflates her clock phase demand more than a high-value bidder. In particular, the efficient allocation (where $u_1(x_1) = u_2(1 - x_1)$) occurs at $x_1^e = \frac{1}{2} + \frac{a_1 - a_2}{2b}$. The equilibrium outcome x_1^* solves $v_1(x_1) = v_2(1 - x_1)$, which means

$$x_1^* = \frac{1}{2} + \frac{a_1 - a_2}{2b}(1 - \gamma) = \frac{1}{2}\gamma + (1 - \gamma)x_1^e. \qquad (12)$$

It is also possible to show for the case of symmetric equilibria that the CCA revenue is decreasing in γ. Since the outcome with $\gamma = 0$ corresponds to a Vickrey auction with truthful bidding, this means that every symmetric equilibrium that involves any degree of quiet final round bidding ($\gamma > 0$) generates lower expected revenue than a truthful Vickrey auction. The derivation of this result requires some calculations that we go through in the Appendix. The Appendix also shows an example demonstrating that the revenue ranking is ambiguous for asymmetric equilibria.

III. Predatory Behavior

The previous section emphasized that each bidder is indifferent between alternative final bids, yet her choice matters for how her opponent should behave in the clock phase. The result is that even if we restrict attention to relatively non-strategic proxy strategies, there are many equilibria involving varying amounts of demand expansion in the clock phase, to compensate for the price reductions offered in the final bid round. The equilibrium is generally not efficient, unless both bidders raise their bids fully in the final round.

In practice, bidders may not be truly indifferent. A bidder might benefit from raising rival costs, or from looking good relative to opponents. This possibility is discussed by Morgan, Steiglitz and Reis (2003), Janssen and Karamychev (2013) and Janssen and Kasberger (2015), among others. Such a bidder will want to be consistent in her final bid. However, she may also have an incentive to relax her activity constraints by exaggerating demand in the clock phase. Figure 5 illustrates the potential for this type of predatory bidding. In this figure, we start from the equilibrium in which both bidders are truthful and consistent. The solid area shows bidder 1's payment. If instead bidder 2 overstates her clock demand, by demanding $x_2 = 1$ until p^*, she does not change the allocation or her own payment. However, by bidding consistently with this inflated clock demand in the final round, she forces bidder 1 to pay the clock price p^* for all of his units. As we now show, if bidder 1 anticipates this predatory behavior, he will want to engage in demand reduction, leading to an inefficient outcome where the predatory bidder obtains an advantage.

A. Proxy Bidding and Predatory Best Responses

We now consider a version of the model in which bidder 2 first maximizes her own profit, and second attempts to make her rival pay more. We assume that bidder 1 bids some proxy strategy $v_1(x)$ in the clock phase, and then bids consistently in the final round, $s_1(x) = v_1(x)$.[14] We allow bidder 2 to use any bidding strategy so long as she does not create excess supply. To model this, we allow bidder 2 to drop her demand discretely, say from x_2' to x_2, at a given price p, but assume that if she does this, she

---

[14] The equilibrium we derive below is consistent with bidder 1 also having a lexicographic preference for raising bidder 2's payment, so long as either bidder 1 restricts attention to linear proxy strategies, or restricts attention to a continuous proxy strategy and there is sufficiently rich support on the possible equilibrium allocations. See Janssen and Kasberger (2015) for a follow-up to this paper which offers a more complete analysis of the case when both bidders have lexicographic preferences.

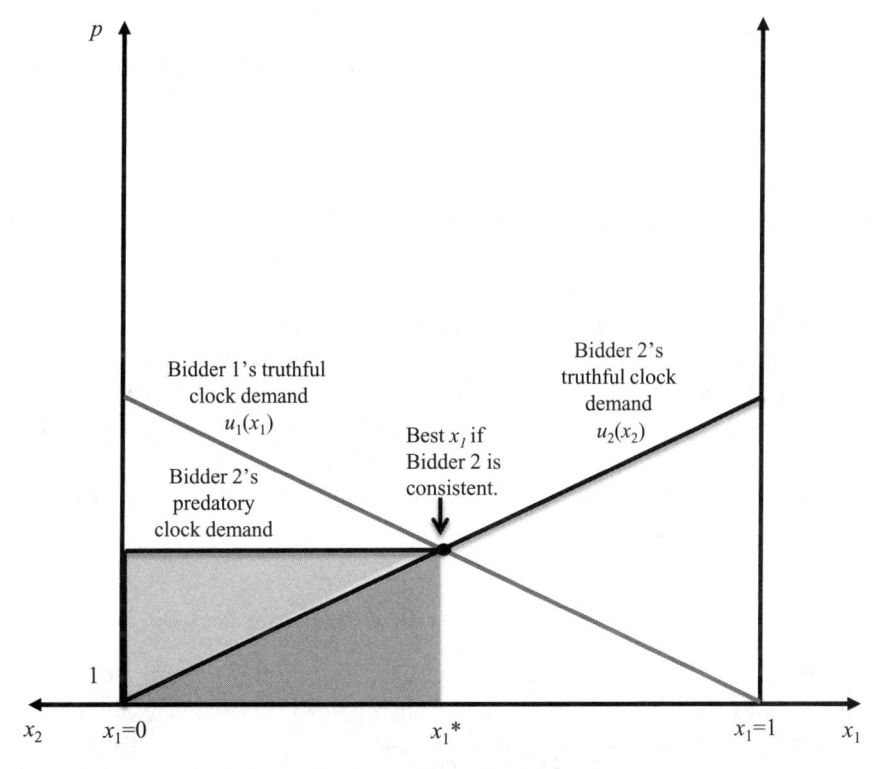

Figure 5. Predatory Clock Phase Bidding to Raise Rival Prices.
Note: Figure shows truthful clock demands with solid shaded area representing bidder 1's payment. If bidder 2 maintains a higher (predatory) clock demand and submits final bids consistent with this higher demand, the allocation is unchanged but bidder 1 pays the entire shared rectangle.

offers to buy any intermediate quantity at that price. The last assumption ensures there will be market clearing.[15]

Bidder 2's most effective strategy is to keep her demand at 1 until the price p and residual supply $1 - x_1$ reach a level at which

$$u_2(1 - x_1) = p = v_1(x_1), \qquad (13)$$

and then reduce her demand to $1 - x_1$ ending the auction. She can then submit her maximal final bid of $S_2(z) = pz$ for $z \geq 1 - x_1$, and make bidder 1 pay the clock price for all x_1 units.

Why is this optimal for bidder 2? First, consider bidder 2's problem of maximizing her own profit. To buy x_2 units and obtain value $U_2(x_2)$ she must pay $S_1(1) - S_1(1 - x_2) = V_1(1) - V_1(1 - x_2)$.[16] So she wants to choose a quantity x_2 that maximizes:

$$U_2(x_2) - [V_1(1) - V_1(1 - x_2)]. \qquad (14)$$

---

[15] In particular, if bidder 1 is demanding x_1 at p and $1 - x_1$ is strictly between x_2' and x_2, then when bidder 2 drops her demand from x_2' to x_2, bidder 2 will be assigned $1 - x_1$.

[16] This formula assumes that $v_1(x) > 0$ for all x. If $v_1(x)$ is negative for large x, then bidder 2 pays $\max_x V_1(x) - V_1(1 - x_2)$. Either way, bidder 2's best response is to select x_2 such that $u_2(x_2) = v_1(1 - x_2)$.

Therefore, her ex post optimal quantity is the unique solution to $u_2(x_2) = v_1(1-x_2)$. Moreover, conditional on buying x_2 units and ending the auction with market clearing, the most she can possibly make bidder 1 pay for his $x_1 = 1 - x_2$ units is $x_1 v_1(x_1)$, which she achieves.

Remark 2. If we allowed bidder 2 to create excess supply at the end of the clock phase, she could increase bidder 1's payment even more. For example, suppose bidder 1 follows a proxy strategy $v_1(x) = 1 - x$. Then, bidder 2 with valuation $u_2(x) = 1 - x$ could demand $x_2 = 1$ until the price reaches (almost) 1 and then drop demand to $x_2 = 1/2$. She would then submit a final round bid with $S_2(1) = 1$ and $S_2(1/2) = 5/8$. In this way the final allocation would be $(1/2, 1/2)$ as in the best response we described above, but bidder 1 would end up paying his full value: $S_2(1) - S_2(1/2) = 3/8 = V_1(1/2)$. Such extreme predatory behavior is even more difficult to execute and even more risky for bidder 2 than what we describe. Moreover, analyzing equilibria in this case is difficult, so we maintain the assumption that bidder 2 is not allowed to create excess supply in the clock phase.

B. Ex Post Equilibrium with a Predatory Bidder

How does bidder 2's predatory behavior affect the auction? Because bidder 1 pays the full market clearing clock prices, rather than the Vickrey payment, he optimally responds by reducing demand.

For bidder 1 to purchase x, he must pay $xu_2(1-x)$. So bidder 1's ex post best response solves

$$\max_x U_1(x) - xu_2(1-x). \tag{15}$$

The unique optimal x satisfies

$$u_1(x) = u_2(1-x) - xu_2'(1-x) = u_2(1-x) + b_2 x. \tag{16}$$

This implies demand reduction: $-xu_2'(1-x) > 0$, so in equilibrium bidder 1 gets a quantity that is smaller than efficient (efficiency is where $u_1(x) = u_2(1-x)$).

To implement the optimal x, bidder 1 needs the clock phase to end when $v_1(x) = u_2(1-x)$. Therefore the following linear proxy strategy is ex post optimal against any u_2 parameterized by a_2:

$$v_1(x) = u_1(x) - b_2 x = a_1 - (b_1 + b_2)x. \tag{17}$$

Proposition 2. *Suppose bidder 2 has a lexicographic preference for making its rival pay more. Then there is an ex post equilibrium in which bidder 1 uses the proxy strategy $v_1(x) = u_1(x) - b_2 x$ in the clock phase, and bidder 2 maintains demand of 1 until dropping demand immediately to $1 - x$ when $p = v_1(x) = u_2(1-x)$ and then both bidders are consistent in the final round.*

We emphasize that the cause of the inefficiency in the above equilibrium is distinct from what we identified in Proposition 1. In the previous section, inefficient ex post equilibria arise because bidders, in a situation of indifference, understate their final demand relative to their clock bidding. Here both bidders are consistent in their final round bidding. The inefficiency arises because bidder 2 is able to inflate her inframarginal clock demand and manipulate bidder 1's payment with no direct consequence for her own allocation or payment.

C. Properties of the Equilibrium with Predatory Bidder

Allocation. In the equilibrium we described the allocation is skewed inefficiently in favor of bidder 2. In particular, the equilibrium allocation is:

$$x_1^* = \frac{a_1 - a_2 + b_2}{b_1 + 2b_2} = x_1^e \frac{b_1 + b_2}{b_1 + 2b_2} < x_1^e, \tag{18}$$

where x_1^e is the efficient allocation. Here the intuition is straightforward: given the demand submitted by the bidder 1, the predatory bidder 2 wants to choose the same quantity as if she was bidding truthfully, whereas bidder 1 engages in defensive demand reduction.

Revenue. There are two opposing effects. First, bidder 1 reduces her demand relative to truthful bidding in the clock phase. Second, bidder 2 forces bidder 1 to pay the final clock price for all her units. With some algebra, one can show that revenue may be above or below the truthful Vickrey revenue. For example, if $b_1 = b_2$ and $a_1 = a_2$, the demand reduction effect dominates and revenue is lower than in the truthful equilibrium. If $b_1 = b_2$ but $a_1 \gg a_2$ (so bidder 1 gets almost all units), the predatory effect dominates and revenues are higher than in the truthful equilibrium.

Mutual Predation. The model we have described has a single predatory bidder. What happens if both bidders are predatory and attempt to push rival payments up toward the final clock prices? In the Appendix, we develop a version of the model in which each bidder has the ability to relax its final bid constraint, and increase its rival's cost. We show that this leads bidders in the clock phase to engage in demand reduction, which again creates inefficiency in the allocation. The modeling approach we adopt in the Appendix is motivated by features of CCA sales with multiple categories, where in practice bidders have a fair amount of flexibility to relax the final round revealed preference constraints. We will discuss an example of this below.

IV. Evidence on CCA Bidding Behavior

Given the ambiguous nature of incentives in the CCA, a natural question is whether data from past auctions can tell us more about bidder behavior. Either bidding data or summary reports are publicly available for several CCA sales of radio spectrum licenses. This evidence suggests a striking degree of heterogeneity across bidders and across auctions. Some bidders have submitted minimal final round bids, as in the quiet strategy described above. Some have submitted final bids that express clear valuation increments and resemble the consistent strategy described above. Others seem to have followed strategies designed to make rivals pay prices that are close to the linear prices at the end of the clock phase, as in our predatory bidding example.[17]

A. Early UK Auctions (2008)

The United Kingdom held two early CCA sales in 2008: for spectrum licenses in the 10-40 GHz range and then for L Band licenses. The auctions had combined revenue of around £10 million. We have information on these sales from reports released by the UK government (Cramton, 2008a,b; Jewitt and Li, 2008).[18]

In the 10-40 GHz auction, there were ten bidders competing for 27 available licenses. There were 2.2 bids on average across licenses in the initial round of the clock phase, and it took 17 rounds to reach market clearing. However, there was relatively little activity in the final bid round, despite all ten bidders winning licenses. Only two bidders submitted final round bids on large numbers of packages. In Cramton's (2008a) description of the auction, the others "simply increased their clock bids, and added a handful of supplementary [i.e. final] bids on packages closely related to their bids in the latter part of the clock stage." Moreover, as Jewitt and Li (2008) explain, "all but one of the bidders made their highest supplementary bid either on their final clock package, or on a subset of it." This is an extreme form of quiet bidding. Most bidders expressed zero (or in fact negative) value for incremental spectrum beyond what they actually won!

In the L Band auction, there were eight bidders and 17 licenses for sale. Bidders could demand arbitrary packages of these licenses. Again, there was a fair amount of competition in the clock phase. The average demand for the licenses in the first round was 3.8 and the market cleared after 32 rounds. Again, however, there was little activity in the final bid round. Six of the bidders submitted final bids on just zero, one or two packages (Cramton, 2008b). Only two bidders submitted significant numbers of final bids. So again, the behavior of most bidders could be described as quiet, and there seems to have been some of the same behavior flagged by Jewitt and Li above. For example, Cramton (2008b) writes: "It is difficult to understand why WorldSpace [which entered only two new final bids] did not enter a more complete set of supplementary bids. Based on its bidding in the clock stage, it would appear to value nearly any set of three small lots at its upper limit of 2,614."

B. UK 4G Auction (2013)

The United Kingdom's subsequent auction for 800 and 2600 MHz spectrum involved much more valuable licenses, with the auction generating over £2 billion in revenue.

---

[17] We discuss evidence from five past auctions below. There is also some recent data available from the Canadian 700 MHz auction conducted in early 2014. Our analysis of that data further supports the claim that there can be a great deal of heterogeneity in bidder behavior. Of the three most active bidders, two (Bell Canada and Telus) submitted final bids for a large number of different license packages (close to 500) at essentially the maximum amount allowed by the activity rules, whereas the third (Rogers), which ended up paying much more for the licenses it won, submitted only a single final round bid, with which it increased its bid for its winning package.

[18] The UK government also published a note explaining how the bids in the 10-40 GHz auction determined the winner payments: http://stakeholders.ofcom.org.uk/binaries/spectrum/spectrum-awards/completed-awards/10-28-32-40-ghz-awards/baseprices.pdf.

Table 1. *Bidding in the UK 800/2600 Auction*

Bidder	Packages Bid Clock	Packages Bid Final	MHz Won 800	MHz Won 2600	Payment
EE	6	48	10	70	£589M
Niche (BT)	7	89	–	20	£186M
H3G	7	12	10	–	£225M
MLL	8	8	–	–	–
HKT	8	8	–	–	–
Telefonica	7	6	20	–	£550M
Vodafone	11	94	20	65	£791M

There were four 10 MHz licenses and a 20 MHz license available in the 800 MHz band. There were also multiple licenses available at 2600 MHz.

Table 1 shows the number of distinct packages bid on by each bidder in the 52 clock rounds, and subsequently in the final round.[19] If a bidder bid for the same package in multiple rounds, we count it just once in the first column; if a bidder bid for a package in the clock phase and raised the bid in the final round, it counts in both columns.

Two of the bidders, MLL and HKT, dropped their demands to zero during the clock phase. Two other bidders, Telefonica and H3G, were active throughout the clock phase but submitted just a few final round bids. In contrast, EE, Niche and Vodafone bid for large numbers of packages in the final round.

Figure 6 shows the full set of package bids submitted by Vodafone. The bars represent the amount of spectrum demanded in different bands, and the line above shows the amount of each bid. Vodafone submitted bids for essentially all combinations of licenses that involved 20 MHz of low-frequency spectrum (the most it was allowed to bid for) and the most desirable high-frequency spectrum. The bids are highly systematic. Vodafone expressed a value for the 20 MHz low-frequency block nearly equivalent to its value for two 10 MHz blocks, and appears to have expressed clear incremental values for the high-frequency blocks.

The incremental values expressed in these final bids are consistent with Vodafone's demand reductions in the clock phase. For instance, in clock round 37, when the price was £87.6 million for each 10 MHz license, Vodafone reduced demand from 4 to 3 C band licenses. Later in its final round bidding, it expressed an incremental value of £87.6 million for a fourth C band license. Toward the end of the clock phase, Vodafone was bidding for 20 MHz at 800 MHz and 30 MHz in band C, and reduced its demand from 7 to 5 to 4 to 3 and then to 0 licenses in band E. The prices at which it made the reductions are consistent with the final bids shown in Figure 6. In this sense, Vodafone appears to have bid in a way that approximates fairly closely the consistent behavior described above.

Telefonica, which bid for similar amounts of spectrum during the clock phase, behaved very differently in its final round bidding. Figure 7 shows its complete set of

---

[19] The numbers in this section come from our own analysis of the bidding data, which is available at: http://stakeholders.ofcom.org.uk/spectrum/spectrum-awards/awards-archive/completed-awards/800mhz-2.6ghz/auction-data/.

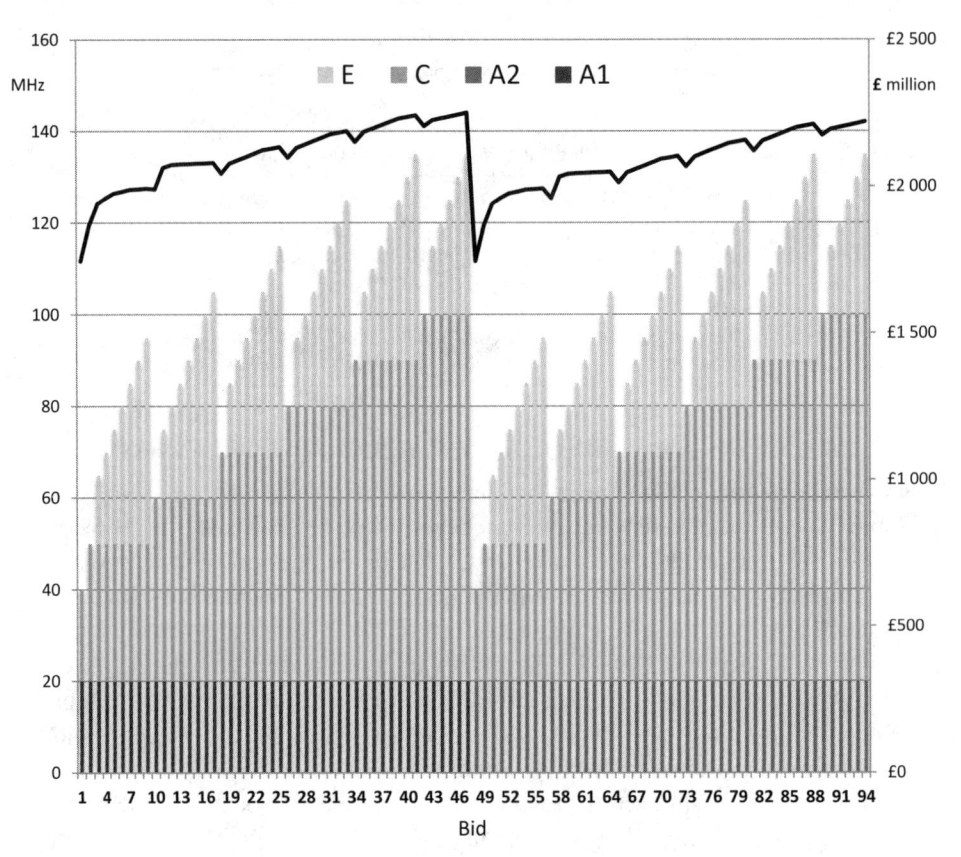

Figure 6. Final Bids by Vodafone in the UK 800/2600 MHz Auction.
Note: Figure shows all of Vodafone's final bids in the UK 800/2600 auction. The solid bars show the composition of each bid in terms of the MHz demanded in each of the four color-coded bands (E, C, A1 and A2). The solid line above shows the value of the bid in GBP (£). Vodafone's bids place consistent value on spectrum increments corresponding to clock behavior.

package bids. Telefonica bid for 7 different packages in the clock rounds. In the final round, it added four new packages (bids 1–4), raised its bid for two packages from the clock phase (bids 5–6), and left five clock phase bids unchanged (bids 7–11). The bids it left unchanged are dominated. They could not have been winning bids or mattered for rival prices as each is for a larger package than bid 4, and offers less money.

Telefonica's five meaningful bids were quite similar in terms of spectrum demanded and amount offered. Bid 2 was Telefonica's winning bid. So Telefonica expressed very little value for packages larger than what it won — the incremental values expressed in its final bids are much lower than the prices at which it reduced demand in the clock rounds. In this sense, its bidding behavior was much closer to the quiet strategy described above than to consistent bidding.

C. Austrian 4G Auction (2013)

The Austrian 4G auction involved the sale of high-value spectrum licenses in the 800, 900 and 1800 MHz bands. The only bidders were the three major wireless companies

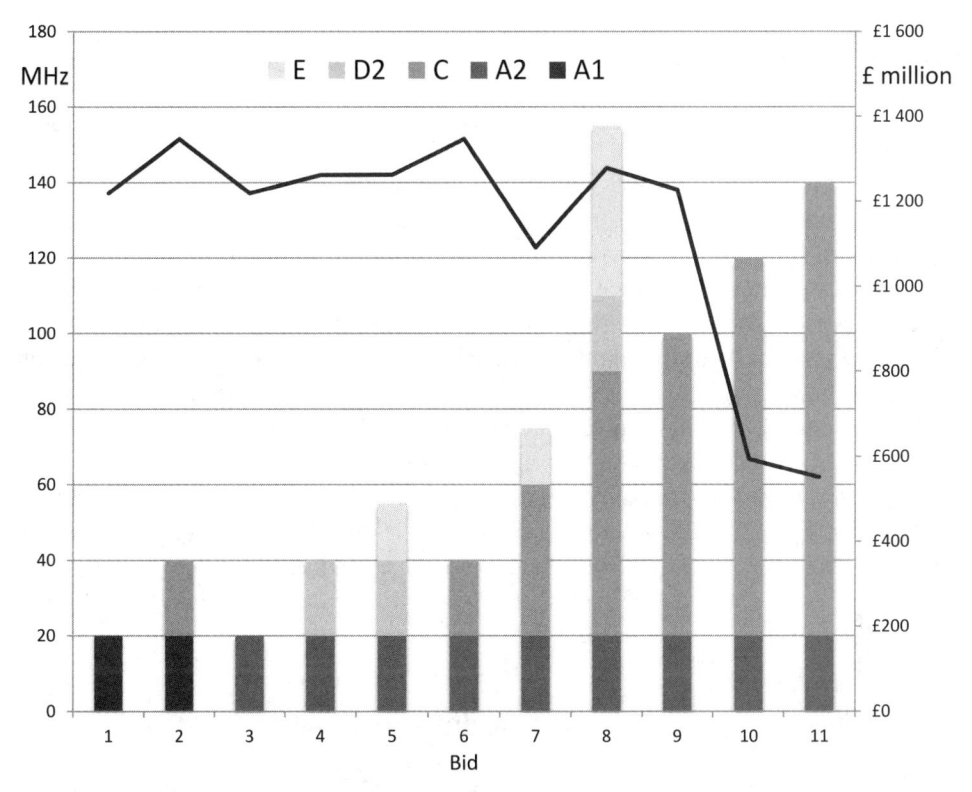

Figure 7. Final Bids by Telefonica in the UK 800/2600 MHz Auction.
Note: Figure shows all of Telefonica's final bids in the UK 800/2600 auction. The solid bars show the composition of each bid in terms of the MHz demanded in each of the four color-coded bands (E, C, A1 and A2). The solid line above shows the value of the bid in GBP (£). Telefonica submitted very few serious bids in the sealed bid round, with much smaller incremental valuations than it revealed during the clock phase, closer to a quiet strategy.

in Austria. Each was limited to bidding on no more than 50% of the available licenses. This still allowed any two bidders to submit a combined bid for all licenses in the auction, i.e. it did not imply that winners automatically received some spectrum at reserve prices.

The auction yielded revenue of just over €2 billion, which far exceeded forecasts.[20] A report released by the regulatory authority after the auction cited aggressive final round bids as a key factor in the high prices paid by the winners:[21] The report reads:

> During [the final bid] stage every bidder was allowed to submit as many as 3,000 supplementary bids. (...) The three bidders actually submitted a total of more than 4,000 supplementary bids. More than 65% of these supplementary bids were submitted for the largest permissible combinations of frequency blocks, with a share of some 50% of available frequencies. In addition, the bidders

---

[20] Press coverage after the auction quoted one industry CEO as saying that the high prices were "a bitter pill to swallow," and another as claiming that the outcome was "a disaster for the industry as a whole." (www.fiercewireless.com/europe/story/austrian-operators-file-complaints-over-spectrum-auction-800-mhz-900-mhz-an/2013-11-27).

[21] The report, titled "Result of the 2013 multiband auction driven by consistently offensive bidding strategy on the part of all three contenders" is available at www.rtr.at/en/pr/PI28102013TK, along with a presentation containing the numbers quoted below.

utilised almost to the full the price limits that had applied to these large packages during the sealed-bid [i.e. final] stage. (...) These supplementary bids submitted on large frequency packages had a significant effect on the prices offered by the other bidders. At the same time, such bids generally only have a marginal likelihood of winning out in the end. If these bids for very large numbers of frequencies had been ignored when determining the winners and prices, the revenue from the auction would have settled at a level of about EUR 1 billion.

A remarkable feature of the Austrian auction is that the final revenue ended up quite close to the total license prices at the end of the clock phase, which were €2.07 billion. Had the bidders submitted no final round bids (i.e. been quiet), the winners would have paid €765 million. Instead they paid €2.01 billion. If bidding in both stages of the auction was truthful, average license prices under the Vickrey formula only would be as high as prices at the end of the clock phase if bidders were willing to pay for all their incremental spectrum at the same rate as for a marginal license. It seems very likely therefore that in this auction bidders took steps to relax their final bid activity constraints. The Appendix describes a version of the model with this feature, in which due to mutually aggressive behavior both bidders can end up paying nearly full clock prices for every unit.

V. Conclusion

Our analysis highlights two properties of the combinatorial clock auction. First, the activity rules used to encourage truthful bidding mean that a bidder's final round bids may have no effect at all on her own payoff. Yet if bidders do not increase their final bids to levels consistent with their expressed demand in the clock phase (and they have no strict incentive to do so), this leads to price discounts and incentives for demand expansion in the initial clock phase. The result is a wide range of ex post equilibria, with no guarantee of an efficient allocation or truthful Vickrey prices.

Second, the auction provides bidders with the opportunity to raise rival prices with little or no risk to their own payoff by relaxing the constraints on their final bids. We have illustrated how this can lead not just to higher payments, but to distorted incentives in the clock phase and inefficient allocations. In Section III, a single predatory bidder maintains high demand during the clock phase before dropping demand to clear the market, leading to an equilibrium in which the second bidder reduces demand to avoid high payments. Janssen and Kasberger (2015) expand this analysis by showing that in our model, if both bidders have lexicographic preferences to raise rival costs, an efficient equilibrium in proxy strategies may not exist at all.

A loose way to summarize these points is that in order to support a truthful equilibrium as we expect in a Vickrey auction, the CCA relies on bidders behaving "just right": raising their final round bids maximally so that the revealed preference activity constraints bind, but not taking actions in the clock phase to purposely relax these constraints. Our examples show how, if bidders are not sufficiently aggressive, or are overly aggressive, incentives for demand expansion and/or reduction appear and outcomes need not be efficient, even if behavior is completely understood and bidders play minimally strategic ex post equilibria.

Our analysis makes several simplifying assumptions. We mostly restrict attention to proxy strategies in which bidders do not condition their bidding on rival behavior. With contingent strategies, our model admits many more equilibria. These include highly collusive equilibria in which bidders split the market and drive each other's prices to zero, using the threat of aggressive final bids to punish deviations. However, these types of equilibria also arise in traditional clock auctions and are not special to the CCA.[22] Our assumptions also implied that the clock phase ends with market clearing. If bidders can drop demand discontinuously, the clock phase may end with some units unallocated. This can be useful for package bidders who want to avoid exposure problems — a potentially important benefit of the CCA that is not captured in our model[23] — but also creates new strategic possibilities.[24]

Another important point is that while our analysis shows some limitations of the CCA, other multi-item auction designs have their own drawbacks. Ausubel et al. (2014) have shown in great generality that uniform price auctions create incentives for demand reduction and inefficiency. Ausubel and Milgrom (2006) have catalogued problems with the sealed-bid Vickrey auction, such as incentives for collusion, and the fact that Vickrey outcomes may lie outside the core. Our analysis points to a further problem with the Vickrey auction that we view as equally serious. If bidders understand that the allocation will almost certainly lie in a particular range (the relevant situation in most radio spectrum auctions), their incentives to bid truthfully outside of this range may be very weak despite these bids potentially being crucial for pricing. Of course, in a static Vickrey auction, bidding truthfully is still weakly dominant.[25] However in a dynamic implementation of Vickrey pricing, the dominant strategy property is lost. In our model of Section III, bidder 1's strict best response to bidder 2's predation is to engage in demand reduction, leading to inefficiency.

From a practical standpoint, an auction designer choosing between a CCA and a uniform price auction (e.g. a simultaneous multi-round or clock auction) faces a set of trade-offs. The need for flexible package bidding favors a CCA design. So does the potential for highly inefficient demand reduction. In the other direction, the CCA is arguably a more complicated design and can create situations where there is considerable ambiguity about the prices a bidder faces at any point in the auction. Dealing with this potentially requires a high level of bidder sophistication. As we have seen in the paper, there is also the possibility for widely varying prices within an auction depending on the strategic postures adopted by bidders.[26]

---

[22] In fact, Riedel and Wolfstetter (2006) show that the simultaneous multiple round auction with complete information has an essentially unique subgame perfect equilibrium in which the bidders immediately demand their efficient allocation and the auction ends.

[23] See Bulow, Levin and Milgrom (2009) and Cramton (2013) for discussions of the exposure problem faced by package bidders in traditional clock auctions.

[24] If there are unallocated units in the final round of the CCA, bidders will have an incentive to bid truthfully for these units, but not necessarily for units that they cannot possibly win. The ability to create excess supply also creates new opportunities for predation, as a predatory bidder can potentially drop its demand to zero and subsequently raise its rival's payment as illustrated in Remark 2.

[25] The Vickrey auction also has many equilibria in weakly dominated strategies (see, e.g. Blume et al., 2009).

[26] There are several examples in past spectrum CCAs of bidders paying disparate prices for similar amounts of spectrum. In the Canadian 700 MHz auction in 2014, Telus paid roughly twice the amount of Bell Canada

A final question is whether different CCA activity rules could resolve some of the issues we have flagged, and lead to a successful dynamic Vickrey implementation. A recent and very interesting paper by Ausubel and Baranov (2013) suggests one possibility, which is to require that clock phase bids satisfy GARP, and then to use an algorithm based on the Afriat inequalities to fill in the final bids. In our setting, this would amount to requiring global rather than local revealed preference. This would resolve the quiet bidding problems illustrated in Section II, but not the predatory bidding problems illustrated in Section III.[27] Nevertheless, this proposal and others seem to merit further investigation.

VI. References

Ausubel, Lawrence M. 2004. "An Efficient Ascending-Bid Auction for Multiple Objects." *American Economic Review* 94(5): 1452–1475.

Ausubel, Lawrence M. and Oleg Baranov. 2010. "Core-Selecting Auctions with Incomplete Information." University of Maryland Working Paper.

Ausubel, Lawrence M. and Oleg Baranov. 2013. "An Enhanced Combinatorial Clock Auction." University of Maryland Working Paper.

Ausubel, Lawrence M. and Peter Cramton. 2011. "Activity Rules for the Combinatorial Clock Auction." University of Maryland Working Paper.

Ausubel, Lawrence M., Peter Cramton and Paul Milgrom. 2006. "The Clock-Proxy Auction: A Practical Combinatorial Auction Design." In *Combinatorial Auctions*, edited by Peter Cramton, Yoav Shoham and Richard Steinberg, 115–138, MIT Press.

Ausubel, Lawrence M., Peter Cramton, Marek Pycia, Marzena Rostek and Marek Weretka. 2014. "Demand Reduction and Inefficiency in Multi-Unit Auctions." *Review of Economic Studies* 81(4): 1366–1400.

Ausubel, Lawrence M. and Paul Milgrom. 2002. "Ascending Auctions with Package Bidding." *Frontiers of Theoretical Economics* 1(1): Article 1.

Ausubel, Lawrence M. and Paul Milgrom. 2006. "The Lovely but Lonely Vickrey Auction." In *Combinatorial Auctions*, edited by Peter Cramton, Yoav Shoham and Richard Steinberg, 17–40, MIT Press.

Beck, Marissa and Marion Ott. 2013. "Incentives for Overbidding in Minimum-Revenue Core-Selecting Auctions." Stanford University Working Paper.

Bichler, Martin, Pasha Shabalin, and Jürgen Wolf. 2013. "Do Core-Selecting Combinatorial Clock Auctions Always Lead to High Efficiency? An Experimental Analysis of Spectrum Auction Designs." *Experimental Economics* 16(4): 511–545.

($1.14 billion CAN versus $0.57 billion CAN) for roughly similar amounts of spectrum. In the 2012 Switzerland auction, Sunrise paid 33% more than Swisscom (482 million CHF versus 360 million CHF) for a smaller package of spectrum. Of course with Vickrey pricing such anomalies can occur even with truthful bidding. However, given the particular details of the auctions (for instance Bell Canada and Telus run very comparable business operations), it seems likely that differences in strategic posture, such as those described in the paper, were also a factor.

[27] A similar point about incentives for aggressive bidding also would apply in the Ausubel (2004) clinching auction, which in our setting would be an alternative dynamic VCG implementation. Ausubel, Cramton and Milgrom (2006) also proposed a relaxed version of revealed preference, which forces bidders to increase their final clock rounds bids in order to "guarantee" their package from the last clock round. However, this rule also doesn't appear to provide strong incentives to submit "correct" bids for losing packages.

Bulow, Jeremy, Jonathan Levin and Paul Milgrom. 2009. "Winning Play in Spectrum Auctions," Stanford University Working Paper.

Cramton, Peter. 2008a. "A Review of the 10–40 GHz Auction." Office of Communications, United Kingdom.

Cramton, Peter. 2008b. "A Review of the L Band Auction." Office of Communications, United Kingdom.

Cramton, Peter. 2013. "Spectrum Auction Design." *Review of Industrial Organization* 42(2): 161–190.

Day, Robert and Paul Milgrom. 2008. "Core-Selecting Package Auctions," *International Journal of Game Theory* 36: 393–407.

Day, Robert. and S. Raghavan. 2008. "Fair Payments for Efficient Allocations in Public Sector Combinatorial Auctions," *Management Science* 53: 1389–1406.

Erdil, Aytek and Paul Klemperer. 2009. "A New Payment Rule for Core-Selecting Auctions." *Journal of the European Economic Association* 8: 537–547.

Goeree, Jacob and Yuanchuan Lien. 2016. "On the Impossibility of Core-Selecting Auctions." *Theoretical Economics* 11: 41–52.

Janssen, Maarten and Vladimir Karamychev. 2013. "Gaming in Combinatorial Clock Auctions," Tinbergen Institute Discussion Paper 13–027/VII.

Janssen, Maarten and Bernhard Kasberger. 2015. "On the Clock of the Combinatorial Clock Auction." University of Vienna Working Paper.

Jewitt, Ian and Zhiyun Li. 2008. "Report on the 2008 UK 10–40 GHz Spectrum Auction." Office of Communications, United Kingdom.

Klemperer, Paul. 2004. *Auctions: Theory and Practice*. Princeton University Press.

Knapek, Stephan and Achim Wambach. 2012. "Strategic Complexities in the Combinatorial Clock Auction." CESifo Working Paper 3983.

Kyle, Albert S. 1989. "Informed Speculation With Imperfect Competition." *Review of Economic Studies* 56(3): 317–355.

Loertscher, Simon, Leslie Marx and Tom Wilkening. 2015. "A Long Way Coming: Designing Centralized Markets with Privately Informed Buyers and Sellers." *Journal of Economic Literature* 53(4): 857–897.

Milgrom, Paul. 2004. *Putting Auction Theory to Work*. Cambridge University Press.

Milgrom, Paul. 2007. "Package Auctions and Exchanges." *Econometrica* 75: 935–965.

Morgan, John, Ken Steiglitz and George Reis. 2003. "The Spite Motive and Equilibrium Behavior in Auctions." *Contributions to Economic Analysis & Policy* 2(1): Article 5.

Riedel, Frank and Elmar Wolfstetter. 2006. "Immediate Demand Reduction in Simultaneous Ascending-Bid Auctions: A Uniqueness Result." *Economic Theory* 29(3): 721–726.

Salant, David. 2014. *A Primer on Auction Design, Management, and Strategy*. MIT Press.

Vives, Xavier. 2011. "Strategic Supply Function Competition With Private Information." *Econometrica*, 79: 1919–1966.

CHAPTER 16

Budget Constraints in Combinatorial Clock Auctions[*]

Maarten Janssen, Vladimir A. Karamychev, and
Bernhard Kasberger

1. Introduction

Combinatorial Clock Auctions (CCAs) are multi-object auctions where bidders make package bids in a clock phase followed by a supplementary round. CCAs have been recently used around the world to allocate spectrum frequencies for mobile telecommunication purposes. CCAs were introduced by Ausubel et al. (2006) and are the subject of quite a few recent investigations (see, e.g., Ausubel and Baranov, 2014; Bichler et al., 2013; Janssen and Karamychev, 2016; Knapek and Wambach, 2012; Levin and Skrzypacz, 2016; and papers in this volume).

One of the issues that is under-explored in the literature on CCAs is the impact of budget constraints on the bidding behavior of participating bidders and, consequently, on the efficiency properties of the auction. It is difficult to obtain direct evidence of the fact that budget constraints play an important role in real-life CCAs. It is also difficult to believe, however, that bidders in recent spectrum auctions have not been financially constrained. The amount of money typically paid is in the billions of euros, and even though a firm may think it will earn that money back in the years after the auction, it is likely to have to borrow the money in one way or another. Also, casual empiricism suggests that share prices of companies participating in a long-lasting auction decline during and after the auction (*cf.* the share prices of KPN in the Netherlands in 2012 and A1 in Austria in 2013).

It can be useful to distinguish between hard and soft budget constraints. On one hand, a hard budget constraint implies that bidders cannot pay more than a certain exogenously determined amount of money. On the other hand, senior management can set a soft budget constraint and inform the company's bidding team that it is not allowed to spend more than a certain amount of money. That is, under soft budget constraints,

---

[*] Parts of this paper were first included in another paper, entitled "Spiteful Bidding and Gaming in Combinatorial Clock Auctions". Over the years, we have benefited from discussions with Martin Bichler, Jacob Goeree, Paul Klemperer, Jon Levin, Emiel Maasland, Tuomas Sandholm, David Salant, Andrzej Skrzypacz, and Achim Wambach. The first two authors have advised bidders in recent auctions in Europe. The research of Janssen and Kasberger was supported by funds of the Oesterreichische Nationalbank (Oesterreichische Nationalbank, Anniversary Fund, project number: 15994).

senior management may set aside a certain amount of money to be invested in acquiring spectrum. A soft budget may be updated during the auction when it turns out that a certain desired package cannot be obtained with the agreed budget. If soft budget constraints do affect bidding behavior, it is an interesting question to ask how these constraints can be optimally chosen or whether it is optimal not to have these constraints at all.[1] This paper considers, however, hard budget constraints only.

This paper points at different effects of hard budget constraints in a CCA under two alternative sets of assumptions concerning bidders' preferences. First, we consider "standard" preferences, where bidders only care about the spectrum they win and the price they have to pay for that spectrum. Second, we consider bidders having alternative preferences: in addition to their own surplus, they also care about the price other bidders pay for the spectrum they win. Under these alternative preferences, bidders have a spite motive and, *ceteris paribus*, prefer outcomes where rivals pay more for their winning allocation. Janssen and Karamychev (2016) argue that comments by the Austrian regulator RTR can be interpreted as saying that in the supplementary round bidders have made many bids on very large packages the bidders knew they were unlikely to win, and that these bids were effective in raising the prices other bidders had to pay. Janssen and Karamychev (2016) provide two arguments why real-world bidders in spectrum auctions are likely to engage in spiteful bidding, in addition to the evidence quoted from the Austrian regulator.[2]

We model the spite motive in a lexicographic way, i.e., a bidder always prefers outcomes with a larger intrinsic surplus (the value of the winning package minus the payment); rival payments only come into consideration to distinguish between outcomes with identical intrinsic surplus. This lexicographic preference ordering is an elegant way to select among the many equilibria of the supplementary round resulting in the same spectrum allocation, but different payments.[3]

Paradoxically (maybe), having a hard budget constraint does not mean that, in a CCA, a bidder should avoid bidding above budget. In particular, a bidder may insert bids in the supplementary round that are not winning ones (and therefore do not affect the allocation). Bidders with a spite motive make these bids only to raise rivals' costs. Thus, we distinguish three ways in which a bidder can satisfy his budget constraint, depending on how much risk he accepts that he has to pay more than his budget: (i) a conservative, (ii) a neutral, or (iii) a risky way. A *conservative* bidder never bids above budget. A *neutral* bidder only makes bids that are such that whatever feasible bids the other bidders make from that moment onwards in the clock or the supplementary phase, he will never pay more than his budget. As we will explain below, given the auction rules that apply, a bidder can calculate that certain bids cannot be winning. In a multi-bidder auction with many units being auctioned, this may require complicated combinatorial calculations. The difference between a conservative and a neutral bidder is that the first may not trust his own calculations (or the algorithm that his advisors

---

[1] See Burkett (2015) for endogenous budget constraints and Burkett (2016) for the principal's optimal choice of a budget in the single-unit case.

[2] Milgrom (2004) argues that bidders dislike different prices for similar packages. Raising other bidders' costs might come from these motives as well.

[3] If bidders have lexicographic preferences for raising rivals' costs, then the set of equilibria should coincide with the set of equilibria in which no bidder can raise other bidders' costs without decreasing his expected surplus.

are using). A *risky* bidder goes one step further than the first two types of bidders. A risky bidder has certain expectations of rivals' (future) bidding behavior, and given these expectations does not have to pay more than his budget, and these expectations are correct in equilibrium. As there may be multiple equilibria in a CCA, this type of bidder may have to pay more than budget, if his expectations turn out to be incorrect.

We obtain the following results for standard preferences. First, when considering the supplementary phase as a standard VCG (Vickrey–Clarke–Groves) mechanism, we show that under a budget constraint, the VCG mechanism no longer has a weakly dominant strategy. Instead, we characterize a range of bids on different packages that remain undominated. Bidders face the following trade-off: bidding full budget on all packages with a value larger than budget increases the chances of winning at least one package, but it may not be the most profitable package to win (given the bid strategy of others). This implies that even if bidders' behavior in the clock phase is such that the constraints on supplementary round bids are not binding, optimal bidding in the supplementary round may be nontrivial and dependent on bidders' expectations of rival bids.

We next analyze some aspects of clock phase bidding by means of two examples. A first example shows that the information during earlier clock rounds may be such that a bidder knows he can safely bid above budget without running the risk of winning that package and having to pay above budget. Bidding above budget may be beneficial as it relaxes the constraints on supplementary round bidding, allowing the bidder to bid true marginal values in that round. Bulow et al. (2009) define a bidder's exposure as the maximal amount a bidder has to pay if all bids become winning. Due to the pay-as-bid pricing rule, a bidder's exposure in the SMRA is simply equal to the sum of his bids. In the CCA, however, a bidder's exposure is the VCG price of the currently demanded package. The example shows that the exposure can be sufficiently different from the bid. In the example, the (positive) role of the clock phase is to provide bidders with information of their exposure, and thus, of the possibility of making bids without the risk of winning them. This role of the clock phase has so far been neglected. A second example shows that in CCAs where multiple bands are allocated the clock phase may actually last longer (depending on bidding behavior and how bidders react to the budget constraint) if bidders are budget-constrained. Moreover, in the clock phase bidders may face similar considerations as the ones we discussed above for the VCG mechanism indicating bidding in the clock phase under a budget constraint is strategically complex.

Finally, we show that in a CCA the spite motive interacts in a complicated way with budget constraints. In the context of an example, conservative bidders (those bidders without bids above budget) may have to pay more for identical packages than their risk-taking competitors pay. Ironically, conservative bidding is associated with the risk of having to pay more than competitors! In another example, budget constraints lead to multiple equilibria with a Hawk–Dove type flavor: aggressive, very risky, bidders perform well against neutral, or risky, but less aggressive bidders, but their bidding leads to payments above budget if all bidders are aggressive.

Cramton (1995) and Salant (1997), among others, highlight the importance of budget constraints in spectrum auctions. However, most academic papers on multi-unit auctions ignore budget constraints despite their practical importance. Che and Gale

(1998) and Benoit and Krishna (2001) are early papers discussing single- and multi-unit auctions with budget-constrained bidders respectively. If bidders are budget-constrained, the single-unit second-price auction has a weakly dominant strategy (e.g. Krishna, 2010). On the contrary, the multi-unit version of the VCG auction no longer has an equilibrium in weakly dominant strategies (see Ausubel and Milgrom, 2006, for an example where a bidder's optimal bid depends on the bid of a competitor). This already indicates the problems that budget constraints impose on auction designers and bidders.

For the Simultaneous Multi-Round Auction (SMRA), Brusco and Lopomo (2008) show that private budget constraints may lead to strategic demand reduction and therefore to potentially inefficient outcomes. In a subsequent paper, Brusco and Lopomo (2009) analyze a simple model (two bidders, two units) of the SMRA with complementaries and known budget constraints. Without budget constraints there exists an efficient non-collusive equilibrium, but with budget constraints, the exposure problem might arise. In equilibrium, the bundle can be assigned to the bidder with lower budget and lower valuation for the bundle. A positive use of bidders' budget constraints is exemplified in Bulow et al. (2009). The authors describe a way to forecast relatively early in the auction the final revenue based on budget constraints in an SMRA. Moreover, they present a real-world example in which this information was successfully used in a high-stake spectrum auction. Ausubel and Milgrom (2002) introduce an ascending pay-as-bid auction. The pay-as-bid payment rule facilitates bidding under a budget constraint, since there is no uncertainty about the final price if a bid becomes winning.

Ausubel (2004) puts forward a dynamic version of the VCG mechanism and illustrates that bidding under a budget constraint might be easier and more efficient in the dynamic version than under the sealed-bid VCG mechanism. In an example much like our Example 3, he shows that efficiency can be hard to obtain in the sealed-bid version, but relatively easy in his dynamic "clinching" auction. In the clinching auction, bidders learn their VCG price during the auction. If at least aggregate demand is revealed in every round, bidders know at which prices they clinched some goods. Therefore, they know the price they have to pay for their current clinches and can calculate the difference between budget and the price for current clinches. If this difference is above the current price, bidders can keep demanding truthfully. Ausubel's paper restricts its attention to bidders with decreasing marginal values.[4] We focus on CCAs with possible complementaries across units where the exposure problem might arise. The CCA is another dynamic version of the sealed-bid VCG auction. Unlike Ausubel's (2004) auction, the CCA is a package auction that solves the exposure problem. In the CCA bidders do not directly learn parts of the allocation and final prices during the clock phase, but they can compute upper and lower bounds on final VCG prices. This information on bidders' exposure is provided through the activity rule and can be used to forecast a range of possible final prices. We show that if the forecast indicates that the final price cannot be above budget, neutral bidders (in the sense

---

[4] Dobzinski et al. (2012) show that in a setting very much like in Ausubel (2004) and with publicly known budget constraints, an "adaptive" version of Ausubel's auction is the unique mechanism that is simultaneously pareto-optimal and incentive-compatible. However, if the budgets are private information, there is no incentive-compatible and pareto-optimal auction.

distinguished above) may find it optimal to bid above budget. Kroemer et al. (2017), Gretschko et al. (2017), and Fookes and McKenzie (2017) look at other aspects of bidding under budget constraints in a CCA.

The rest of the paper is organized as follows. Section 2 determines the set of strategies that are not dominated in a VCG mechanism where bidders are budget-constrained and have standard preferences. Section 3 discusses the two examples illustrating the different optimal behaviors in the clock phase under a budget constraint. Section 4 discusses the complexities of combining spiteful bidding with a budget constraint. Section 5 concludes with a discussion. Proofs are in the Appendix.

2. Budget-Constrained Bidders in VCG

A well-known result for second-price auctions is that bidders have a weakly dominant strategy to bid their value. For one-unit auctions, this result has an analog when bidders are budget-constrained: bidding the minimum of the value of the object and the budget is a weakly dominant strategy (see, e.g., Krishna, 2010, Proposition 4.2). This section shows that this result does not generalize to VCG auctions. Accordingly, bidding under a budget constraint is a non-trivial exercise in a multi-unit second-price auction.

To show which strategies are weakly dominated, and which cannot be eliminated as weakly dominated strategies, we use the following notation. Let there be K different types of objects to auction with n_k objects of type $k = 1, \ldots, K$. We use x to denote generic packages, and use Greek letter superscripts to refer to specific packages, e.g., x^α. The set of all feasible packages is denoted by X, and the aggregate supply is denoted by $\bar{x} = (n_1, \ldots, n_K) \in X$. There are n bidders, and the intrinsic valuation of bidder i for any package x^α is denoted by $v_i^\alpha = v_i(x^\alpha)$. The set of all valuations of bidder i is denoted by $V_i = \{(x^\alpha, v_i^\alpha) : x^\alpha \in X\}$. Let $\Psi_i \subseteq X$ be a subset of packages that bidder i bids on in a VCG mechanism. Accordingly, let $\Phi_i = \{(x^\alpha, b_i^\alpha) : x^\alpha \in \Psi_i\}$ be the set of bidder i's bids in the VCG mechanism, where $b_i^\alpha = b_i(x^\alpha)$ is the monetary amount b_i^α that bidder i bids on package x^α. A feasible auction allocation is denoted by $A = (x_1^A, \ldots, x_n^A)$. Bidder i has a budget ω_i, which is assumed to be a hard budget restriction. When no confusion is possible we drop subscript i.

In the following proposition, we state which strategies (set of bids) are weakly dominated in the VCG mechanism, and which set of bids are not under the assumption that all bids are potentially pivotal (Milgrom, 2004, p. 50).

Proposition 1. *Let all bids be potentially be pivotal and let $x^{\max}$ be the most valuable package of bidder i, i.e., $v_i(x_i^{\max}) \geq v_i(x)$ for all $x \in X$, and $v_i^{\max} = v_i(x^{\max})$ be the corresponding value. Then, a collection of VCG bids Φ_i is weakly dominated if, for some package x^α:*

1. $b_i^\alpha > \min(v_i^\alpha, \omega_i)$, or
2. $b_i^\alpha < \max\{\min(v_i^{\max}, \omega_i) + (v_i^\alpha - v_i^{\max}), 0\}$.

The set of undominated bids consists of:

1. $b_i^{\max} = \min(v_i^{\max}, \omega_i)$ on the most valuable package $x_i^{\max}$, and
2. $b_i^\alpha \in [\max\{\min(v_i^{\max}, \omega_i) + (v_i^\alpha - v_i^{\max}), 0\}, \min(v_i^\alpha, \omega_i)]$ on all other packages x^α.

The proposition can be relatively easily understood. Under a hard budget constraint it is never optimal to bid above value or above budget. In an optimal strategy a bidder bids the full budget on his most valuable package, and the bid difference between this bid and the bids on all other packages will *not* be larger than the difference in valuations. For these other packages, a bidder faces the trade-off between winning at least one package (in which case they will bid full budget on less valuable packages as well), or winning the most profitable package (in which case they will bid full budget minus the value difference on less valuable packages as well).

The following example, which also will be used in the next section on strategic bidding under a budget constraint in the clock phase, illustrates Proposition 1.

Example 1. Undominated strategies in the VCG auction with a budget constraint
There are three bidders competing for one band in which four units are for sale. The set of feasible packages is, for simplicity, $X = \{1, 2, 3, 4\}$, and bidders' realized values are:

Package x^α	(1)	(2)	(3)	(4)
Values of bidder 1: v_1^α	5.9	12	12	12
Values of bidder 2: v_2^α	5	9.5	10	10
Values of bidder 3: v_3^α	5	8	8	9

for one, two, three and four units respectively. Bidder 1 has a budget of $\omega_1 = 9$. Bidders do not know each other's valuation (private information scenario). In particular, bidder 1 knows that values of his rivals are either as stated above, or as summarized below:

Package x^α	(1)	(2)	(3)	(4)
Possible values of bidder 2: $\widehat{v}_2^\alpha$	3.5	10	10	10
Possible values of bidder 3: $\widehat{v}_3^\alpha$	4	9.5	10	10

We refer to the first set of valuations as "actual" and to the second set as "alternative." The winner determination problem is to find a feasible allocation $A = (x_1, x_2, x_3)$ that maximizes the function $b(A) = b_1(x_1) + b_2(x_2) + b_3(x_3)$ such that $x_1 + x_2 + x_3 \leq 4$.

Table 1 shows the values of the sum of all the bids for the two possible sets of valuations for different bids of the first bidder and truthful bidding of players 2 and 3, respectively. The bold entry indicates the highest sum of bids for the given bidding behaviors of the three bidders. If all bidders bid truthfully in the VCG auction, then the final auction allocation is (2, 1, 1), and bidder 1's VCG price is

$$p_1^{VCG} = 17.5 - 10 = 7.5.$$

In the world of alternative preferences, bidder 1 still wins 2 units in the efficient (and final) allocation, but now he has to pay 9.5, which is above the budget of 9. Bidder 1 does not have a weakly dominant strategy. With actual preferences, he wants to win 2 units because the VCG prices is below budget. Therefore, it is better to submit a bid that makes it more likely that he wins 2 units. Preserving the true increase in utility

Table 1. *The impact of the bidding of a budget-constrained bidder on the final allocation in the VCG auction*

	Actual values of the other bidders					
Bids of Bidder 1	$b(2, 2, 0)$	$b(2, 1, 1)$	$b(1, 2, 1)$	$b(1, 1, 2)$	$b(2, 0, 2)$	$b(0, 2, 2)$
$b_1 = (5.9, 12, 12, 12)$	21.5	**22**	20.4	18.9	20	17.5
$b_1 = (5.9, 9, 9, 9)$	18.5	19	**20.4**	18.9	17	17.5
$b_1 = (2.9, 9, 9, 9)$	18.5	**19**	17.4	15.9	17	17.5
	Alternative values of the other bidders					
Bids of Bidder 1	$b(2, 2, 0)$	$b(2, 1, 1)$	$b(1, 2, 1)$	$b(1, 1, 2)$	$b(2, 0, 2)$	$b(0, 2, 2)$
$b_1 = (5.9, 12, 12, 12)$	**22**	19.5	19.9	18.9	21.5	19.5
$b_1 = (5.9, 9, 9, 9)$	19	16.5	**19.9**	18.9	18.5	19.5
$b_1 = (2.9, 9, 9, 9)$	19	16.5	16.9	15.9	18.5	**19.5**

Notes: The rows present different bids of bidder 1. The columns indicate the respective sum of bids for an allocation if bidders 2 and 3 bid truthfully. The bold number indicates the maximal sum for a given bid of bidder 1 and truthful bids of the other two bidders.

from 1 to 2 units makes it more likely to win 2 units. If he bids

$$\Phi_1 = \{(1, 2.9), (2, 9), (3, 9), (4, 9)\}$$

the marginal increase from 1 unit to 2 units is higher, therefore it is more likely to win 2 units. On the other hand, he risks winning nothing. This can be seen in the lower panel of Table 1. Under actual preferences, he would win two units, but under the alternative preferences he would not win anything. A bid that makes it more likely that he wins 1 unit is

$$\Phi_1 = \{(1, 5.9), (2, 9), (3, 9), (4, 9)\}.$$

If he submits this bid, the marginal bid on 1 is relatively large, therefore it is more likely that he will win 1 in the end. For both sets of preferences of rival bidders, he wins one unit, which is his optimal share as the VCG price for two units is above bidder 1's budget.

Thus, the example confirms that bidding under uncertainty and a budget constraint is a non-trivial task and the optimal bid depends on the beliefs about the other bidder's valuation and play.

Another important aspect of budget constraints in a VCG (and the CCA) is that they may affect the prices competitors pay, even if the budget itself is not binding in the sense that a bidder may not need to pay his full budget. In addition, the bidder that is most budget-constrained (often the smallest bidder), may be the one that in the end pays the most!

Example 2. Non-binding budget constraints may benefit competitors In this second example, there are three bidders competing for two bands with supply $\bar{x} = (6, 3)$ and the set of feasible packages is $X = \{(1, 1), (2, 1), (1, 2), (3, 1), (3, 2)\}$. Let all bidders have the following (symmetric) values:

$$V_i = \{((1, 1), 30.5), ((2, 1), 45), ((1, 2), 39.5), ((3, 1), 49), ((3, 2), 52)\}.$$

If bidders bid truthfully and bid their values on all packages, $\Phi_i = V_i$, then it is easy to see that the auction allocation is $A = ((2, 1), (2, 1), (2, 1))$ with an auction price of

$$p_i^{VCG}(2, 1) = (b(3, 2) + b(3, 1)) - (b(2, 1) + b(2, 1)) = 11.$$

Each bidder gets a surplus of 34.

Suppose now that bidder 1 has a budget of $\omega_1 = 45$ and that he uses this budget so as to maximize his chances of winning the most profitable package without paying more than his budget. He will do so by bidding

$$\Phi_1 = \{((1, 1), 30.5), ((2, 1), 45), ((1, 2), 39.5), ((3, 1), 45), ((3, 2), 45)\}.$$

Note that the budget is much higher than what he has to pay if all bidders bid their value. If the other two bidders continue to bid their values, $\Phi_2 = V_2$ and $\Phi_3 = V_3$, then the winning allocation is unaffected and so is the price the budget-constrained bidder has to pay. However, the two other bidders gain from the budget-constraint of their competitor and now only pay 7, instead of 11, since

$$p_2^{VCG}(2, 1) = (b_3(3, 2) + b_1(3, 1)) - (b_3(2, 1) + b_1(2, 1)) = 7.$$

Thus, the only budget-constrained bidder in this example is the one that pays the most even though all bidders acquire identical packages.

The next section adds a clock phase to these two examples to show that sometimes the clock phase may provide bidders with information that allow them to infer that the VCG price is below budget so that they may actually bid above budget (which may restore efficiency).

3. Budget-Constraints in the clock of the CCA

This section concentrates on some aspects of strategic bidding under a budget constraint that explicitly involve the clock phase of the CCA. First, we show that the outcome of the CCA can be efficient, whereas the outcome of the VCG auction is not necessarily efficient. This may be the case when bidders are willing to bid above budget as long as they can compute that their exposure is not more than their budget (this is what we have called neutral bidding). The VCG mechanism does not provide bidders with information about which bids of the other bidders are feasible, and therefore it may be the case that bidders have to pay more than their budget if they make some bids above budget. The dynamic nature of the clock phase of the CCA, paired with the activity rule that links the clock phase to the supplementary phase, allows bidders to compute upper bounds on the final VCG price if they themselves choose to make certain bids. This is sometimes called a bidder's exposure. If the exposure is below budget, bidders may safely bid above budget without running the risk of having to pay more than their budget.

Second, if during the clock phase bidders infer that their exposure is above their budget constraint, similar considerations to the ones we identified in Section 2 for the VCG mechanism apply during the clock. A bidder may reduce demand to 0 in an attempt maximize his chances to get his most preferred outcome, but may also bid on the most

Table 2. *Observed bids in the clock phase*

p	$D_1(p)$	$D_2(p)$	$D_3(p)$
1	2	2	2
2	2	2	2
3	2	2	2
4	2	2	1

profitable package for which it still can ensure it will not pay above budget to maximize the chance of getting at least one package. Moreover, we show that in CCAs where multiple bands are allocated the clock may actually last longer (depending on bidding behavior and how bidders react to the budget constraint) if budget-constrained bidders choose the latter option.

In the examples we use the following notation: the clock round prices are denoted by p^t, the clock round demand by D_i^t, and the final auction price bidder i pays for package x^α by $p_i^{VCG}(x^\alpha)$.

3.1. An Efficiency Restoring Role of the Clock Phase

The activity rules of a CCA translate bidders' clock demand to constraints on the admissible bidding function in the supplementary phase. Bidders can use the information that is revealed in the clock phase to compute upper (and lower) bounds on the other bidders' supplementary bidding function. This information allows them to forecast the maximal VCG price, i.e., their exposure. Thus, depending on the development of the clock phase and the monetary value of the budget, bidders may learn that even if they bid above budget in the clock phase, and subsequently in the supplementary phase, they never have to pay above budget. The applicability of this observation crucially depends on the activity rules and the information disclosure policy.[5] The more information on the other bidder's clock demand is revealed, the better budget-constrained bidders are able to forecast future final prices. The next example shows how this may work.

Example 3. The possible efficiency restoring role of the clock phase We reconsider Example 1, but now introduce a clock phase. For simplicity, we assume that bidders learn the individual demands after each clock round. Table 2 summarizes the demands at the given price. The clock starts at a price of 1. Due to excess demand the price increases up to 4. At the price $p = 4$, bidder 1 observes that bidder 3 reduced demand to one. Since there is still excess demand, the price will be increased to 5 in the next clock round. Under truthful bidding, bidder 1 would demand two units at this price. In this case he has to bid at least 10, which is above budget, for two units in the supplementary phase. He wants to bid above budget only if he knows that the final VCG price is below budget. Bidder 1 can infer from the observed history that in all admissible continuations of the clock, the final VCG price is below budget.

If bidder 1 demands two units at a price of 5, then the clock can end at $p = 5$ with five possible final clock allocations that are consistent with the observed history. First,

---

[5] The activity rules used in this section are the final and the relative cap.

Table 3. *Transformation of clock demands into constraints on the supplementary bidding function*

Possible clock demand histories						
$D_i(1)$	$D_i(2)$	$D_i(3)$	$D_i(4)$	$D_i(5)$	Constraints on $b_i(1)$	Constraints on $b_i(2)$
2	2	2	2	2	$b_i(1) \leq b_i(2) - 5$	$10 \leq b_i(2)$
2	2	2	2	1	$5 \leq b_i(1)$	$8 \leq b_i(2) \leq b_i(1) + 5$
2	2	2	2	0	$b_i(1) \leq 5$	$8 \leq b_i(2) \leq 10$
2	2	2	1	1	$5 \leq b_i(1)$	$6 \leq b_i(2) \leq b_i(1) + 4$
2	2	2	1	0	$4 \leq b_i(1) \leq 5$	$6 \leq b_i(2) \leq b_i(1) + 4$

the clock can end with market clearing, in which case the final clock round demands are either (2, 2, 0) or (2, 1, 1). Second, it can end with excess supply, in which case the final clock round demands are either (2, 1, 0), or (2, 0, 1), or (2, 0, 0). The clock continues only if (2, 2, 1) is demanded in round 5.

Table 3 summarizes the constraints on the supplementary bidding function for all possible clock demands that are consistent with the observed demands in Table 2 and the clock ending at $p = 5$. The constraint on $b_i(3)$ and $b_i(4)$ are the same for all bidders. In the supplementary phase, it must be true that $b_i(3) \leq b_i(2) + 1$ and $b_i(4) \leq b_i(2) + 2$.

Bidder 1 can now compute the possible final VCG prices if the clock ends at $p = 5$ by him demanding two units. Similarly, bidder 1 can compute possible final VCG prices if he demands two units at $p = 5$ and the clock continues and he bids such that he respects his budget constraints at $p > 5$. In this case it must be that the demand equals (2, 2, 1) at a price of 5. The detailed calculations in the Appendix show that, in either case, bidder 1 will not have to pay more than his budget. Intuitively, if the final allocation is (2, 1, 1), then the other bidders can make him pay at most 9, since bidders 2 and 3 cannot raise their bids on two units more than 5 and 4 respectively.

If the clock finishes at $p = 5$, by bidding for two units at the final clock price bidder 1 is able to bid his true marginal values on all packages in the supplementary round. Thus, using the VCG pricing rule he will always acquire the bundle with the highest intrinsic value, and he does not want to deviate from the truthful bidding strategy.

The example depends on the activity rule and the information policy of the auction. The analysis was performed under the assumption that bidders learn the other bidders' past demand after every clock round. In the current example, revelation of aggregate demand would suffice, however, to obtain the same result. In more complicated auctions with more bidders and multiple bands, the more information is revealed about demand the better bidders are able to compute VCG prices (and their exposure) accurately. If no information about demand is revealed (like in the beginning of the clock phase in the 2013 Austrian auction), bidders cannot compute their exposure.

The auction designer faces a trade-off in the choice of informational policy. Revealing more information about demand can foster collusion or spiteful behavior (as in Janssen and Karamychev, 2016, and Levin and Skrzypacz, 2016), but it can also enable efficient outcomes when bidders face budget constraints.

3.2. Budget Constraints May Prolong the Clock

If bidders cannot infer that their exposure is smaller than their budget, budget-constrained bidders have to adjust their clock phase bidding. The next example of a CCA with multiple bands demonstrates that bidders have to make strategically difficult decisions during the clock phase. By significantly reducing demand or by dropping out of the clock phase altogether, a bidder may maximize his chances to get his most preferred outcome. On the other hand, such a decision also increases the chance of not winning anything and, from this perspective, it may be better to simply bid on the most profitable package for which the calculated exposure is below budget. With multiple bands, bidders may adjust their bidding behavior in such a way that the clock may actually last longer if budget-constrained bidders choose to bid on the most profitable package for which the calculated exposure is below budget.

Example 4. Budget constraints may extend the clock if a bidder's exposure is above budget We reconsider Example 2 with a clock phase. Suppose that the eligibility points for the first band equal 1 and that they equal 2 for the second band. We assume that the CCA begins with reserve prices $p^1 = (10, 1)$, and price increments in both bands are equal to one. The following table represents the clock phase development. If bidders bid truthfully in the clock phase their behavior is given by the next table. It is clear that round $t = 5$ is the final clock round.

Round, t	Prices, p^t	D_1^t	D_2^t	D_3^t
1	(10, 1)	(1, 2)	(1, 2)	(1, 2)
2	(10, 2)	(1, 2)	(1, 2)	(1, 2)
3	(10, 3)	(1, 2)	(1, 2)	(1, 2)
4	(10, 4)	(1, 2)	(1, 2)	(1, 2)
5	(10, 5)	(2, 1)	(2, 1)	(2, 1)

If bidders also bid truthfully in the supplementary round, they bid their values on all packages, $\Phi_i = V_i$, so that, like in Example 2, all bidders pay 11 for the package (2, 1), and their surplus equals 34.

Suppose now that bidder 1 has a budget of $\omega_1 = 23$. The exposure of bidding $D_i^5 = (2, 1)$ in round $t = 5$ is larger than the available budget. If the auction ended at that round, the others could maximally raise their bid $b(3, 2)$ to $b(2, 1) + 15$ and their bid $b(3, 1)$ to $b(2, 1) + 10$ so that together the other bidders can raise the price bidder 1 has to pay for obtaining package (2, 1) to 25. As a result of the budget constraint, bidding $D_i^5 = (2, 1)$ in round $t = 5$ is not feasible for him. Among the (still feasible) packages, (1, 1) and (1, 2), package (1, 2) is the most profitable one at prices $p^5 = (10, 5)$ so that one may assume bidder 1 bids $D_i^5 = (1, 2)$. If this happens, the price for band 2 keeps increasing until round $t = 7$, when bidder 1 switches to the package (1, 1) and the clock stops consequently. The following table represents the clock phase development for rounds $t = 5, \ldots, 7$ if bidder 1 has a budget constraint of $\omega_1 = 23$.

Round t	Prices p^t	D_1^t	D_2^t	D_3^t
5	(10, 5)	(1, 2)	(2, 1)	(2, 1)
6	(10, 6)	(1, 2)	(2, 1)	(2, 1)
7	(10, 7)	(1, 1)	(2, 1)	(2, 1)

The relative cap rule imposes the following restrictions on the supplementary round bids of bidder $i = 1$:

$$b_1(2, 1) \leq b_1(1, 1) + 10, \; b_1(1, 2) \leq b_1(1, 1) + 7,$$
$$b_1(3, 1) \leq b_1(1, 1) + 20, \; b_1(3, 2) \leq b_1(1, 2) + 20.$$

If bidders 2 and 3 bid truthfully in the supplementary round, and bidder 1 bids according to his budget:

$$b_1(1, 1) \in [17, 23], \text{ and } b_1(2, 1) = b_1(1, 2) = b_1(3, 1) = b_1(3, 2) = 23,$$

bidder 1 wins (1, 1) at price

$$p_1^{VCG}(1, 1) = (b_2(3, 2) + b_3(3, 1)) - (b_2(2, 1) + b_3(3, 1)) = 52 - 45 = 7$$

and obtains a surplus of 23.5. In this case, bidders $i = 2, 3$ win (2, 1) and (3, 1) at prices

$$p_2^{VCG}(2, 1) = (b_3(3, 2) + b_1(3, 1)) - (b_3(3, 1) + b_1(1, 1)) = 26 - b_1(1, 1),$$
$$p_3^{VCG}(3, 1) = (b_2(3, 2) + b_1(3, 1)) - (b_2(2, 1) + b_1(1, 1)) = 30 - b_1(1, 1),$$

with surplus $(19 + b_1(1, 1))$ from both packages.

Alternatively, having observed that bidders 2 and 3 have switched to (2, 1) in round $t = 5$, bidder $i = 1$ can drop to (0, 0) in round $t = 6$. The following table represents the clock phase development for rounds $t = 5, 6$.

Round t	Prices p^t	D_1^t	D_2^t	D_3^t
5	(10, 5)	(1, 2)	(2, 1)	(2, 1)
6	(10, 6)	(0, 0)	(2, 1)	(2, 1)

The relative cap rule imposes the following restrictions on the supplementary round bids of bidder 1:

$$b_1(2, 1) \leq 26, \; b_1(1, 2) \leq 22, \; b_1(3, 1) \leq 36, \; b_1(3, 2) \leq b_1(1, 2) + 20.$$

If bidders $i = 2, 3$ bid truthfully in the supplementary round, and bidder 1 bids $b_1(2, 1) = 23$, he wins (2, 1) at price

$$p_1^{VCG}(2, 1) = (b_i(3, 2) + b_j(3, 1)) - (b_i(2, 1) + b_j(2, 1)) = 11,$$

as in the case of no budget. Therefore, bidder 1 has a chance to win (2, 1) by dropping out of the clock phase altogether. He pays a price smaller than his budget and obtains a surplus that is larger than the surplus that he obtains by bidding (1, 2) in round 6 and (1, 1) in round 7.

Example 4 shows that budget-constrained bidders may face strategically difficult decisions in the clock phase of the CCA. Depending on the bidding behavior of their competitors, it may be optimal to drop out of the clock phase or to continue bidding for packages whose exposure is within the budget limit. Depending on how a bidder resolves this dilemma, we have illustrated by means of an example that the clock phase may, paradoxically, be extended if bidders are budget-constrained.

4. Budget-Constrained Bidders Under the Spite Motive

In this section, we extend the analysis to include bidders with a preference to raise rivals' costs. As we explained in Section 1, there are good reasons why in spectrum auctions bidders are not indifferent across auction outcomes that differ in terms of what competitors pay for their spectrum. As bidders' payments in a CCA depend on competitors' bids, this implies that each bidder may want to investigate to what extent they are able to raise rivals' costs. We will say that bidders bid to raise rivals' costs if the bid difference between two packages is larger than the value difference between the same two packages. In general, bidders do not want to bid in such a way that their bids on packages that were intended to raise rivals' costs end up winning and that the marginal payment they have to make for winning that package is larger than the value difference. Under a budget constraint, there is an alternative concern, namely that bidders do not want to pay more than their budget.

In the Section 1, we argued that there are different ways to satisfy the budget constraint. First, bidders may bid in such a way that independent of the behavior of their competitors, they will never have to pay more than their budget. Second, there is an equilibrium interpretation that allows bidders to bid in such a way that given correct expectations of their competitors' behavior, bidders bid in such a way that they do not have to pay above budget. In this section, we will exemplify these notions by considering an example where all bidders have a budget constraint.

The motive to raise rivals' costs is modeled in the following way. By the end of the supplementary round, any bidder has bid on a set of packages. A bidder either wins one of these packages and pays the opportunity cost of winning this package imposed on others, or he does not win. A bidder's intrinsic pay-off of bidding equals the standard surplus (value − payment). If, for a fixed strategy profile of other bidders, the intrinsic pay-off of two strategies is identical, bidders prefer the strategy that raises the sum of rivals' payments most. Thus, we can say a strategy σ dominates another strategy σ' if:

1. for any bids of the other bidders, σ never results in a lower intrinsic pay-off, or in the same intrinsic pay-off and a lower sum of rivals' payments than σ'; and
2. for some of the others' bids, either σ results in a larger intrinsic pay-off, or it results in the same intrinsic pay-off but raises the sum of rivals' payments, as compared to σ'.

In previous sections we have seen that under some conditions bidders can place bids above budgets on certain packages and win those packages at prices that are below the budgets. Without making such bids, bidders could not have won those packages. Under the spite motive, bidders have yet another reason to bid above budgets, namely to raise prices that other bidders pay. In other words, bidders may find it optimal to place bids above budgets on certain packages with no intention of winning them.

Example 5. Raising auction prices by bidding above budget We reconsider Example 4. Let all three bidders have budget $\omega_i = 23$. Under the assumption that bidders bid according to values satisfying the budget constraint, the clock phase will develop as in the next table:

Round, t	Prices, p^t	D_1^t	D_2^t	D_3^t
1–6	$(10, t)$	$(1, 2)$	$(1, 2)$	$(1, 2)$
7	$(10, 7)$	$(1, 1)$	$(1, 1)$	$(1, 1)$

Relative bid caps and feasibility restrictions on the supplementary round bids are as follows:

$$17 \leq b_1(1, 1)$$
$$b_1(2, 1) \leq b_1(1, 1) + 10$$
$$22 \leq b_1(1, 2) \leq b_1(1, 1) + 7$$
$$b_1(3, 1) \leq b_1(1, 1) + 20$$
$$b_1(3, 2) \leq b_1(1, 2) + 20.$$

The maximal safe bid $\widehat{b}$ for a package is the maximal bid such that the package never becomes winning. The table

Package x^α	$(1, 1)$	$(2, 1)$	$(1, 2)$	$(3, 1)$	$(3, 2)$
Maximal safe bids $\widehat{b}_1^\alpha$	23	23	30	23	40

specifies the maximal safe bids of bidder 1. Any feasible bid on $(1, 2)$ is never winning. Indeed, $b_1(1, 2)$ can only be winning if

$$b_1(1, 2) \geq b_1(1, 1) + b_i(1, 1) \geq b_1(1, 1) + 17,$$

but the feasibility constraint on $b_1(1, 2)$ says that it should not be larger than $b_1(1, 1) + 7$. Thus, bidding above budget on certain packages may be without risk of actually winning the package.

On other packages, it is not difficult to see that by exceeding the maximal safe bid, bidder 1 runs a risk of winning the package and paying above budget. For example, if bidder 1 increases $b_1(3, 1)$ by $x > 0$ above 23 the auction allocation may be $A = ((3, 1), (0, 0), (3, 2))$, and the price bidder 1 pays equals

$$p_1^{VCG}(3, 1) = (b_2(3, 1) + b_3(3, 2)) - b_3(3, 2) = 23 + \frac{1}{2}x > \omega_i,$$

if other bidders bid as below:

Package x^α	(1, 1)	(2, 1)	(1, 2)	(3, 1)	(3, 2)
Bids of bidder 1, b_1^α	23	23	30	$23 + x$	40
Bids of bidder 2, b_2^α	$17 + \frac{1}{2}x$		22	$23 + \frac{1}{2}x$	
Bids of bidder 3, b_3^α	17		24	36	43

Similarly, bidding $b_1(3, 2) = 40 + x$ brings a risk of winning (3, 2) if others bid as in the next table:

Package x^α	(1, 1)	(2, 1)	(1, 2)	(3, 1)	(3, 2)
Bids of bidder 1, b_1^α	23	23	30	23	$40 + x$
Bids of bidder 2, b_2^α	17		22	29	$40 + \frac{1}{2}x$
Bids of bidder 3, b_3^α	17		22	29	$40 + \frac{1}{2}x$

In this case, the price bidder 1 would pay equals

$$p_1^{VCG}(3, 2) = (b_2(3, 1) + b_3(3, 2)) - b_2(3, 1) = 40 + \frac{1}{2}x > \omega_i.$$

Suppose all three bidders bid the maximal safe bids (to be more precise, let them bid $b_i(2, 1) = 23$, and let bids $b_i(1, 1)$ and $b_i(3, 1)$ be marginally lower than 23). Then, the allocation is $A = ((2, 1), (2, 1), (2, 1))$ with price $p_1^{VCG}(2, 1) = 17$. The high bid on package (3, 2) does not run the risk of being a winning package, but is actually very effective in raising rivals' costs as it is used to determine the prices other bidders pay. The neutral bidder we identified in Section 1 may make such a bid and does not risk winning it. If all bidders are conservative and do not make bids above 23, then all will win the package (2, 1) at price $p_1^{VCG}(2, 1) = 0$. One neutral bidder could raise the prices of the others to 17 without raising his own price.

A problem with the assumed bidding behavior is that it is not an equilibrium. In particular, one of the bidders, say bidder 1, may increase his bids $b_1(3, 1)$ and $b_1(3, 2)$ beyond maximal safe bid in order to raise his rivals' costs. Our final example shows how equilibrium bidding in the supplementary phase may involve all bidders making risky bids, but one of them being more aggressive than the others. The equilibrium is asymmetric and has a Hawk–Dove flavor where the more aggressive "Hawk" bidder pays the lowest prices. The equilibrium presented satisfies the budget constraint "in equilibrium," but all bidders run the risk of having to pay more than their budgets if they fail to coordinate on this specific equilibrium.

Example 6. Hawk–Dove equilibria of the supplementary phase We continue the previous example. Let $x \in [0, 4]$, and bidders bid in the supplementary round

$$\Phi_1 = \{((1, 1), 17), ((2, 1), 23), ((1, 2), 24), ((3, 1), 29), ((3, 2), 40 + x)\},$$

$$\Phi_i = \{((1, 1), 17), ((2, 1), 23), ((1, 2), 24), ((3, 1), 29 - x), ((3, 2), 40)\},$$

where $i = 2, 3$. Here again, bids on packages $(3, 1)$ and $(3, 2)$ are marginally lower than the reported amounts and, therefore, not winning. Alternatively, there is a tie-breaking rule in place such that if multiple allocations maximize the total sum of bids, the allocation with the maximal number of winning bidders is chosen. In both cases, the winning allocation is $A = ((2, 1), (2, 1), (2, 1))$ with prices

$$p_1^{VCG}(2, 1) = (b_2(3, 1) + b_3(3, 2)) - (b_2(2, 1) + b_3(2, 1)) = 23 - x,$$
$$p_i^{VCG}(2, 1) = (b_1(3, 1) + b_j(3, 2)) - (b_1(2, 1) + b_j(2, 1)) = 23.$$

Bidders get surpluses of $22 + x$, 22, and 22 correspondingly.

For any $x \in [0, 4]$, this bidding behavior constitutes an equilibrium of the supplementary round. By increasing his bid on $(1, 1)$, bidder i wins $(1, 1)$ at price 17 with surplus 13.5. By increasing his bid on $(3, 1)$, bidder $i = 2, 3$ wins $(3, 1)$ at price $29 - x$ with surplus $20 + x$, but the price is well above budget. Finally, by increasing his bid on $(3, 2)$, bidder $i = 2, 3$ wins $(3, 2)$ at price 40 with surplus 12. Similar arguments establish that bidder 1 cannot benefit from deviating. Thus, no player can deviate profitably.

Note that there is a continuum of these asymmetric equilibria. In each of these equilibira, bidder 1, the Hawk, bids more aggressively than bidders $i = 2, 3$, the Doves. Being aggressive pays off, as the more aggressive bidder pays less for identical spectrum. Thus, there is not only a coordination problem as to which value of x to stick to, but also who of the three bidders plays the role of the more aggressive bidder. If bidders do not coordinate and at least two of them play the aggressive strategy of bidder 1, then the winning allocation is that two bidders win $(3, 1)$ and $(3, 2)$, respectively, while each has to pay an amount that is much larger than their budget.

The examples illustrates the different ways to bid in a CCA under a budget constraint. First, conservative bidders who make all their bids below their budgets may pay more than others when they do not express a higher bid for larger packages. Second, a neutral bidder calculates the maximal bids he can make on different packages in order to be certain not to win them and bids accordingly to raise rivals' costs. In many instances, the bidding strategies of neutral bidders do not form an equilibrium, however, and risky bidders may want to increase their bids further when these higher bids will not be winning, given (rational) expectations of what others will bid. The CCA encourages bidders to be aggressive in case others are restrained as aggressive bidders may pay considerably less for identical packages.

5. Conclusion

Mainly by analyzing illustrative examples, this paper has considered the implications of bidders in a combinatorial clock auction to be budget constrained. As the last phase of the CCA is a kind of Vickrey–Clark–Groves (VCG) mechanism, we first have considered the implications of a budget constraint in the standard VCG mechanism. We have shown the range of bids that are and are not weakly dominated. In general, a bidder faces a trade-off between bidding budget on many possible packages to acquire at least one of them or to differentiate the bids on different packages by their value difference to acquire the most profitable one. Trying to get the most profitable package runs the risk of not winning any package as all bids are too low to be selected by the winner determination algorithm. We have also shown that by bidding budget on many possible packages, a budget-constrained bidder may pay more for an identical package than unconstrained competitors.

We then considered the clock phase of the CCA and argued that it may be beneficial to provide bidders with detailed information about the bidding behavior of their competitors. With this information, bidders are able to calculate their exposure at the beginning of each clock round. Knowing their exposure is below budget, this information allows them to bid above budget. In an example, we show that this may improve the efficiency of the auction. Compared to the VCG mechanism, this also provides a positive, unexplored role for the clock phase. We also show that if a bidder's exposure is above budget, then bidders face the same trade-off in the clock phase of the CCA as in the VCG mechanism (or in the supplementary round of the CCA).

Finally, we show that if, *ceteris paribus*, bidders have an incentive to raise rivals' costs (what is also called a *spite motive*), they always want to increase their bids on non-winning packages to the maximal extent possible. We have identified that in a CCA this does not imply that bidders do not bid above budget. Budget-constrained bidders have two alternative ways they can satisfy their budget constraint. First, bidders are in the position to calculate for each package the maximal bid they can make such that it cannot be a winning bid *no matter what their competitors are bidding*. Spiteful bidders want to make these bids in order to raise rivals' costs. Second, bidders may make bids such that in equilibrium no bidder pays an amount that is larger than their budget. In this case, bidders have specific and correct expectations concerning the bidding behavior of their competitors and given these expectations their bids are optimal such that their final payment stays within their budget. Depending on the specific environment, the second interpretation of bidding under a budget constraint allows bidders to be (much) more aggressive than the first one and may lead to asymmetric equilibria of a Hawk–Dove nature, where the aggressive Hawk is much better off than the more peaceful Doves (given that they coordinate in the prescribed way). If coordination does not happen and multiple bidders bid according to the Hawkish strategy profile, they win large packages and have to pay prices above budget.

To sum up, the clock can be beneficial to the bidders because they can use the demand of the other bidders to compute upper and lower bounds of the final VCG price for a package. This information can be used eventually to place bids above budget (but below value) in order to win the most desired package. This can restore efficiency. However, the information released in the clock can also be used to raise rivals' costs.

6. Appendix

Proof of Proposition 1 We omit the subscript i. Any bid above the value, $b^\alpha > v^\alpha$, is dominated by $b^\alpha = v^\alpha$. Next, any bid above budget, $b^\alpha > \omega$, is dominated by $b^\alpha = \omega$. This proves part (1) of the proposition. In order to prove part (2), we define $z^\alpha = \min(v^{\max}, \omega) + (v^\alpha - v^{\max}) - b^\alpha$, and $Z = \{x^\alpha : \forall x^\beta \in \Psi : z^\alpha \geq z^\beta\}$. In other word, Z

is a subset of packages that generate the largest surplus had the bidder win them at VCG prices equal to their bids b^α. Consider an alternative set of bids $\widetilde{\Phi} = \{\widetilde{b}^\alpha : x^\alpha \in \Psi\}$ where $\widetilde{b}^\alpha = b^\alpha$ if $x^\alpha \notin Z$ and $\widetilde{b}^\alpha = b^\alpha + \varepsilon$ if $x^\alpha \in Z$. In other words, we raise bids on all packages from Z by a small amount $\varepsilon > 0$. It is easy to verify that $\widetilde{\Phi}$ dominates Φ.

Detailed calculations for Example 3 Suppose first that the clock phase ends at $p = 5$ and bidder 1 has demanded two units. In the determination of the final allocation and the VCG prices, no bidder's bid on 4 can ever play a role. To see this, note that

$$b_i(4) \leq b_i(2) + 2 < b_i(2) + 6 \leq b_i(2) + b_j(2),$$

that is, the bid on 4 is always smaller than the bid on 2 by the same bidder and the bid on 2 by another bidder. If the clock ends with demands $(2, 2, 0)$, then this is the final allocation by the final cap rule. Since the bid on 4 plays no role in the determination of VCG prices, there are in principle three possible ways to construct the VCG price for bidder 1. However, only the cases

$$p_1^{VCG} = b_2(3) + b_3(1) - b_2(2),$$
$$p_1^{VCG} = b_2(2) + b_3(2) - b_2(2)$$

are relevant, since

$$b_2(2) + b_3(2) \geq b_2(2) - 5 + b_3(2) + 1 \geq b_2(1) + b_3(3).$$

In the first case, the VCG price is at most $p_1^{VCG} \leq 6$. This constraint comes from the fact that the bid on 3 can be at most 1 larger than the bid on 2 and from the last line in Table 3. In the second case, $p_1^{VCG} = b_3(2) \leq 9$, which is equal to the budget.

If the clock ends with $(2, 1, 1)$, then the final clock allocation is the final allocation. The possible VCG prices are

$$p_1^{VCG} = b_2(3) + b_3(1) - b_2(1) - b_3(1) \leq b_2(2) + 1 - b_2(1) \leq 6,$$
$$p_1^{VCG} = b_2(2) + b_3(2) - b_2(1) - b_3(1) \leq b_2(1) + 5 - b_2(1) + b_3(1) + 4 - b_3(1) = 9.$$

If the clock ends with excess supply and if bidder 1 bids $b_1 = (5.9, 10, 10, 10)$, then in any final allocation in which bidder 1 gets two units, he does not pay more than budget. If the final allocation is $(2, 2, 0)$ or $(2, 1, 1)$, the same considerations as above apply. If the final allocation is $(2, 0, 2)$, then

$$p_1^{VCG} = b_2(2) + b_3(2) - b_3(2) = b_2(2) \leq 10.$$

But if $b_2(2) \geq \omega_1$, then $(2, 0, 2)$ is no longer implemented by the auctioneer since

$$b_1(1) + b_2(2) + b_3(1) \geq 5.9 + 9 + b_3(1) \geq 10 + b_3(1) + 4 \geq b_1(2) + b_2(2)$$

is true. Moreover, bidder 1 never gets three units since

$$b_1(3) + b_j(1) \leq b_1(2) + b_j(1) \leq b_1(2) + b_j(1) + b_{5-j}(1),$$

for $j = 2, 3$. Even if $b_2(2) = 0$, it is true that $b(1, 2, 1) > b(3, 0, 1)$.

Consider then the case where bidder 1 demands two units at $p = 5$ and the clock continues, so that the demand at $p = 5$ must be $(2, 2, 1)$. If the clock ends at some higher price $p > 5$, bidder 1 also never has to pay more than budget. If the clock continues, bidder 1 can drop demand to 0 at $p = 6$ and submit the supplementary bid $b_1 = (5.9, 10, 10, 10)$. Since he never has to pay more than his bid, by doing so the only possibility in which he has to pay more than budget is if he wins two units. But in no final allocation in which he wins two units, does he have to pay more than budget. First, note that $(2, 2, 0)$ is never the final allocation since

$$b_1(1) + b_2(2) + b_3(1) \geq 5.9 + b_2(2) + 5 > b_1(2) + b_2(2) = 10 + b_2(2).$$

Second, also $(2, 0, 2)$ is never winning since

$$b_1(1) + b_2(2) + b_3(1) \geq 5.9 + 10 + b_3(1) \geq 10 + 4 + b_3(1) \geq b_1(2) + b_3(2).$$

Third, $(2, 1, 1)$ is winning if

$$b_1(1) + b_2(2) + b_3(1) \leq b_1(2) + b_2(1) + b_3(1),$$

$$b_2(2) \leq 4.1 + b_2(1).$$

In this case, the VCG price is either

$$p_1^{VCG} = b_2(2) + b_3(2) - b_2(1) - b_3(1) \leq 4 + 4.1 < 9,$$

or

$$p_1^{VCG} = b_2(3) + b_3(1) - b_2(1) - b_3(1) \leq 5.1 < 9,$$

which is less than budget in both cases.

As a result, bidder 1 can safely demand two units at a price of 5 in the clock phase. He knows that the final VCG price is never above budget.

References

Ausubel, L. M. (2004). An efficient ascending-bid auction for multiple objects. *American Economic Review*, 94(5):1452–1475.

Ausubel, L. M. and Baranov, O. V. (2014). Market design and the evolution of the combinatorial clock auction. *American Economic Review: Papers & Proceedings*, 104(3):446–451.

Ausubel, L. M., Cramton, P., and Milgrom, P. (2006). The clock-proxy auction: A practical combinatorial auction design. In Cramton, P., Shoham, Y., and Steinberg, R., editors, *Combinatorial Auctions*, chapter 5. Cambridge, MA: MIT Press.

Ausubel, L. M. and Milgrom, P. (2002). Ascending auctions with package bidding. *Frontiers of Theoretical Economics*, 1(1):1534–1563.

Ausubel, L. M. and Milgrom, P. (2006). The lovely but lonely Vickrey auction. In Cramton, P., Shoham, Y., and Steinberg, R., editors, *Combinatorial Auctions*. Cambridge, MA: MIT Press.

Benoit, J.-P. and Krishna, V. (2001). Multiple-object auctions with budget constrained bidders. *The Review of Economic Studies*, 68(1):155–179.

Bichler, M., Shabalin, P., and Wolf, J. (2013). Do core-selecting combinatorial clock auctions always lead to high efficiency? An experimental analysis of spectrum auction designs. *Experimental Economics*, 16(4):511–545.

Brusco, S. and Lopomo, G. (2008). Budget constraints and demand reduction in simultaneous ascending-bid auctions. *The Journal of Industrial Economics*, 56(1):113–142.

Brusco, S. and Lopomo, G. (2009). Simultaneous ascending auctions with complementarities and known budget constraints. *Economic Theory*, 38(1):105–125.

Bulow, J., Levin, J., and Milgrom, P. (2009). Winning play in spectrum auctions. Working paper.

Burkett, J. (2015). Endogenous budget constraints in auctions. *Journal of Economic Theory*, 158:1–20.

Burkett, J. (2016). Optimally constraining a bidder using a simple budget. *Theoretical Economics*, 11(1):133–155.

Che, Y.-K. and Gale, I. (1998). Standard auctions with financially constrained bidders. *The Review of Economic Studies*, 65(1):1–21.

Cramton, P. (1995). Money out of thin air: the nationwide narrowband pcs auction. *Journal of Economics and Management Strategy*, 4(2):267–343.

Dobzinski, S., Lavi, R., and Nisan, N. (2012). Multi-unit auctions with budget limits. *Games and Economic Behavior*, 74:486–503.

Fookes, N. and McKenzie, S. (2017). Impact of budget constraints on the efficiency of multi-lot spectrum auctions. In Bichler, M. and Goerer, J. K., editors, *Handbook of Spectrum Auction Design*. Cambridge: Cambridge University Press.

Gretschko, V., Knapek, S., and Wambach, A. (2017). Bidding complexities in the combinatorial clock auction. In Bichler, M. and Goerer, J. K., editors, *Handbook of Spectrum Auction Design*. Cambridge: Cambridge University Press.

Janssen, M. C. W. and Karamychev, V. A. (2016). Gaming in combinatorial clock auctions. *Games and Economic Behaviour*, 100:186–207.

Knapek, S. and Wambach, A. (2012). Strategic complexities in the combinatorial clock auction no. 3983. CESIFO Working Paper.

Krishna, V. (2010). *Auction Theory*, 2nd edn. Harlow, Essex: Academic Press.

Kroemer, C., Bichler, M., and Goetzendorff, A. (2017). (Un)expected bidder behavior in spectrum auctions: about inconsistent bidding and its impact on efficiency in the combinatorial clock auction. *Group Decision and Negotiation*, 25(1):31–63.

Levin, J. and Skrzypacz, A. (2016). Properties of the combinatorial clock auction. *American Economic Review*, 106(9):2528–2551.

Milgrom, P. (2004). *Putting Auction Theory to Work*. Cambridge: Cambridge University Press.

Salant, D. (1997). Up in the air: GTEs experience in the MTA auction for personal communication services licenses. *Journal of Economics and Management Strategy*, 6(3):549–572.

CHAPTER 17

(Un)expected Bidder Behavior in Spectrum Auctions

About Inconsistent Bidding and its Impact on Efficiency in the Combinatorial Clock Auction

Christian Kroemer, Martin Bichler, and Andor Goetzendorff

1. Introduction

The design of auction protocols and systems has received considerable academic attention in the recent years and found application in industrial procurement, logistics, and in public tenders (Airiau and Sen, 2003; Bellantuono et al., 2013). Spectrum auction design is one of the most challenging and visible applications. It is often seen as a pivotal example for the design of multi-object markets and successful auction designs are likely role-models for other markets in areas such as procurement and logistics.

Efficiency, revenue, and strategic simplicity for bidders are typical design goals that a regulator has in mind. In theory, the Vickrey-Clarke-Groves (VCG) auction is the only strategy-proof and efficient auction but for practical reasons, it has rarely been used so far (Rothkopf, 2007). Several other auction formats have been designed and used for selling spectrum. The most prominent example is the Simultaneous Multi-Round Auction (SMRA) which has been used since the mid-90s to sell spectrum licenses world-wide. The more recent Combinatorial Clock Auction (CCA) is a two-phase auction format with an initial ascending clock auction and a sealed-bid supplementary bid phase afterward. It has lately been used to sell spectrum in countries such as Australia, Austria, Canada, Denmark, Ireland, the Netherlands, Slovenia, and the UK.

The CCA draws on a number of elegant ideas inspired by economic theory. A revealed preference activity rule should provide incentives for bidders to *bid straightforward* or *consistent*, i.e., to bid truthfully on one of the payoff-maximizing packages in each round of the clock phase. If bidders fail to maximize utility and bid on a package with a less than optimal payoff, we will also refer to this as *inconsistent bidding behavior*, i.e., bids which are not consistent with the assumption of utility maximization. A second-price rule should set incentives to bid all valuations truthfully in the second sealed-bid phase. It can be shown that if bidders respond to these incentives in both phases of the CCA, then the outcome is efficient and in the core (Ausubel et al., 2006). However, bidders might not have incentives to bid truthful in both phases, and this can lead to inefficiencies.

1.1. Reasons for Inefficiency in the CCA

The CCA is used in high-stakes auctions and much recent research tries to better understand when it is efficient in theory and in the lab. For the former, Goeree and Lien (2013) highlight possibilities for profitable manipulation and deviations from truthful bidding in core-selecting auctions in a market with several local and one global bidder. They show that the Bayesian Nash equilibrium outcome in this market can be further from the core than that of the VCG auction in a sealed-bid auction, and that in their model truthful bidding is never an equilibrium in a core-selecting auction. Sano (2012) analyzes the same market situation and shows that in ascending auctions a core-selecting payment rule can lead to an inefficient perfect Bayesian equilibrium where local bidders drop out at the start. Janssen and Karamychev (2013) and later Levin and Skrzypacz (2014) provide a complete information analysis of the CCA rules considering the activity rules of the CCA and show that there are multiple equilibria with no guarantee for efficiency. The equilibria depend on assumptions about bidders' incentives to drive up prices of competitors, which is risk-free in the CCA as was shown in Bichler et al. (2013a) (see Section 2.4).

Lab experiments yielded low revenue and low efficiency for the CCA in a market with a larger number of licenses (Bichler et al., 2013a). Interestingly, also the CCA conducted in the UK in 2013 achieved a revenue below the expectations, leading to an investigation by the UK National Audit Office (Arthur, 2013), whereas some other CCAs such as the one in Austria in 2013 achieved high revenue. It turns out that one reason for low efficiency and revenue in the experiments was that bidders submitted only a small subset of the thousands or millions of packages they could bid on. This can have strategic but also very practical reasons. In larger combinatorial auctions such as the Canadian 700 MHz auction in 2013 with 98 licenses, national bidders could potentially bid up to 18^{14} packages. It will only be possible to submit bids on a small subset of all possible packages for any bidder. All other packages are treated by the winner determination in the CCA as if bidders had no valuation for these combinations, which is unlikely.

In contrast, the SMRA uses an "OR" bidding language, where bidders can have multiple winning bids. During the winner determination, bids on different items provide an estimate for the value that a bidder has for every possible combination of bids on individual items. Also in the British auction in 2013 only a low number of package bids was submitted. Problems due to the exponential growth in the number of packages can sometimes be addressed by a compact bid language, as was discussed in Bichler et al. (2014). The rules for the CCA in Canada in 2015 try to address this problem by allowing for restricted OR bids in the supplementary stage. Of course, the bid language does not solve the strategic reasons for bidders to bid on many or only a few packages in the supplementary stage. We will discuss some of these reasons in Section 2.4.

1.2. Contribution of this Paper

In this paper, we show that apart from missing bids in the supplementary phase, also inconsistent bidding in the clock phase can be a source of inefficiency. We show that bidders in the lab and in the field (Canada and UK) do not bid straightforward in the clock phase. There are actually several reasons for inconsistent bidding behavior. For example, bidders might have budget constraints (Shapiro et al., 2013) or values might be interdependent, which can lead to inconsistent bidding as bidders revise their valuations when they learn about other bidders' valuations during the auction. Even if bidders have independent and private values without budget constraints, there can be incentives to reduce or inflate demand in the clock phase in order to drive up payments of competitors (Bichler et al., 2013a; Janssen and Karamychev, 2013; Levin and Skrzypacz, 2014).

However, the revealed preference activity rule prohibits bidders from bidding truthfully up to their valuation in the supplementary phase, if they do not bid straightforward in the clock phase, and this can lead auctioneers to select an inefficient allocation. We provide evidence from the lab and from the field showing that the resulting inefficiencies can be significant, while being much less obvious at the same time. Measuring this inefficiency due to restrictions on the supplementary bid prices is straightforward in the lab, where the values of bidders are available. But also the analysis of the field data from the British LTE auction in 2013 and from the Canadian 700 MHz auction in 2014 suggests that inconsistent bidding was an issue. We introduce metrics based on Afriat's Efficiency Index, which allow measuring the level of inconsistency. Numerical simulations based on data from the lab and from the UK indicate that the impact of inconsistent bidding on efficiency can be substantial. In the lab we found an overall efficiency loss of around 5%, which can be attributed to inconsistent bidding in the clock phase. In the data from the field, where we don't know the bidders' true valuations, we also found a surprising large number of supplementary stage bids at the bid price limit imposed by the clock phase. This can be seen as an indication that these bids were also below the true valuation, although one can assume that bidders in these countries tried to bid up to their true valuation.

A strong activity rule, which forces bidders to be consistent across auction rounds, appears to be an intuitive solution to fix the problems discussed in this paper. However, in the conclusions we will outline issues which arise when a regulator tries to force bidders to bid straightforward.

1.3. Outline

The remainder of this paper is structured as follows: After briefly introducing the rules of the CCA in Section 2, we will discuss Afriat's Efficiency Index to analyze whether bidders in the CCA are bidding straightforward in Section 3. We will use this metric to analyze bidders in the lab in Section 4 and bidders from the British and the Canadian auction in Section 5. Finally, we will use computer simulations to analyze the impact of these deviations on the auction's final outcome in Section 6. Section 7 discusses stronger activity rules to force consistent bidding in the clock phase and potential problems arising from such rules.

2. The Combinatorial Clock Auction

Used for the first time in 1994, the SMRA has been the de facto standard auction format for spectrum sales for almost 20 years (Milgrom, 2000). A number of well-known strategic problems have led to substantial research on alternative auction formats. In particular, the exposure problem turned out to be central. Bidders are often interested in specific combinations or packages of licenses. Their value for these packages can be much higher than the sum of the individual license values in this package. As the SMRA allows only bidding on single items, a bidder risks winning only part of his package, having to pay more than what the subpackage is worth to him. Combinatorial auctions address this problem by allowing bidders to submit bids on packages rather than on single items. In 2008, the British regulator Ofcom decided on the two-stage Combinatorial Clock Auction (CCA) (Ausubel et al., 2006), a format which has been used in many countries world-wide in the past years.

First, we briefly describe the overall auction process that was the same in the recent auctions. Then we discuss the activity rules, and draw on the latest version used in Canada in 2014.[1]

2.1. The CCA Auction Process

In the *clock phase*, the auctioneer announces ask prices for all licenses at the beginning of each round. In every round bidders communicate their demand for each item at the current prices. At the end of a round, the auctioneer determines a set of over-demanded licenses for which the bidders' demand exceeds the supply. The price for all over-demanded lots is increased by a bid increment for the next round. This clock phase continues until there are no over-demanded lots left.

The *supplementary stage* is designed to eliminate incentives for demand reduction and other inefficiencies in the combinatorial clock auction due to the limited number of bids that bidders can submit in the first phase. In this sealed-bid stage bidders are able to increase bids from the clock phase or submit bids on bundles they have not bid on so far. Bidders can submit as many bids as they want, but the bid price is restricted subject to the CCA activity rule (see next subsection). Finally, all bids from both phases of the auction are considered in the winner determination and the computation of payments for the winners. The winner determination is an NP-hard combinatorial optimization problem (Lehmann et al., 2006). For the computation of payments, a Vickrey-nearest bidder-optimal core-pricing rule is used (Day and Cramton, 2012).

With certain assumptions on the bidders' valuations it is possible to determine the efficient allocation and the VCG payments, even if bidders do not bid up to their true valuation in the supplementary stage. For example, if bidders have independent and decreasing marginal valuations for homogeneous items and all bidders bid straightforward then it is possible to determine Vickrey payments even if bidders would not increase their bids after the clock phase. Under these assumptions bidders have strong incentives to bid truthful as the clock auction is ex post incentive compatible. However, combinatorial auctions are typically used when bidders have complementary valuations and this is when the clock auction loses its favorable properties. Without substitutes valuations, an efficient outcome can not be guaranteed in a clock auction, not even with

---

[1] The auction rules of the Canadian 700 MHz auction in 2014 can be found at www.ic.gc.ca/eic/site/smt-gst.nsf/eng/sf10583.html. The auction rules of the British auction in 2013 can be found at http://stakeholders.ofcom.org.uk/spectrum/spectrum-awards/awards-archive/completed-awards/800mhz-2.6ghz/.

fully straightforward bidding by all participants. Actually, simple examples show that the clock phase can have very low efficiency, if all bidders bid straightforward (see Section 7). Even if valuations were gross substitutes no ascending auction can always impute Vickrey prices (Gul and Stacchetti, 1999), i.e., payments for which bidders have no incentives to shade their true valuations.[2]

Apart from the observation that bidders in spectrum auctions often have complementary valuations, a number of other reasons can cause differences between the true VCG payments and the payments computed in the CCA. For example, in larger auctions with many licenses bidders might be unable to submit supplementary bids on all possible packages. However, such missing bids can have an impact on the payments of others. There are also differences to the VCG payments, if bidders bid higher or lower than their valuation for strategic reasons, and there can be multiple non-truthful equilibria in this auction (Levin and Skrzypacz, 2014).

2.2. Activity Rules in the CCA

The CCA combines two auctions in the clock and in the supplementary phase. This requires additional rules setting incentives to bid truthfully in both phases. Without activity rules, bidders might not bid actively in the clock phase, but wait for the other bidders to reveal their preferences, and only bid in the supplementary phase. Originally, the clock phase of the CCA employed a simple monotonicity rule which does not allow to increase the size of the package in later rounds as prices increase. It has been shown that with substitutes preferences straightforward bidding is impossible with such an activity rule (Bichler et al., 2011, 2013a). Later versions use a hybrid activity rule using a monotonicity rule and a revealed preference rule (Ausubel et al., 2006). Revealed preference rules allow bidders to bid straightforward in the clock phase. If they do, then bidders are able to bid on all possible packages up to their true valuation in the supplementary stage (Bichler et al., 2013a). In the following we describe the latest version of the activity rules as they have been used in the Canadian 700 MHz auction in 2014. These rules have also been used in our simulations in Section 6.

First, an eligibility points rule is used in the clock phase to enforce activity in the primary bid rounds. The number of bidder's eligibility points is non-increasing between rounds, such that bidders cannot bid on more licenses when the prices rise. A bidder may place a bid on any package that is within his current eligibility. Second, in any round, the bidder is also permitted to bid on a package that exceeds his current eligibility provided that the package satisfies revealed preference with respect to each prior eligibility-reducing round. Bidding on a larger package does not increase the bidder's eligibility in subsequent rounds.

The revealed preference rule works as follows: A package in clock round t satisfies revealed preference with respect to an earlier clock round s for a given bidder if the bidder's package x_t has become relatively less expensive than the package bid on in

---

[2] Ausubel (2006) showed that there is an ascending auction with multiple price trajectories and item-level prices, which is efficient and yields the VCG allocation and payments. The auction runs one ascending auction with all bidders, and one with each bidder excluded in turn. However, this auction format is quite different from the clock auctions used in the field so far.

clock round s, x_s, as clock prices have progressed from the clock prices in clock round s to the clock prices in clock round t. x_s and x_t are vectors where each component describes the number of licenses demanded in the respective category, i.e., region or spectrum band. The revealed preference constraint is:

$$\sum_{i=1}^{m}(x_{t,i} * (p_{t,i} - p_{s,i})) \leq \sum_{i=1}^{m}(x_{s,i} * (p_{t,i} - p_{s,i}))$$

where:

- i indexes the licenses;
- m is the number of licenses;
- $x_{t,i}$ is the quantity of the ith license bid in clock round t;
- $x_{s,i}$ is the quantity of the ith license bid in clock round s;
- $p_{t,i}$ is the clock price of the ith license bid in clock round t; and
- $p_{s,i}$ is the clock price of the ith license bid in clock round s.

A bidder's package, x_t, of clock round t is consistent with revealed preference in the clock rounds if it satisfies the revealed preference constraint with respect to all eligibility-reducing rounds prior to clock round t for the given bidder.

2.3. Activity Rules in the Supplementary Phase

Under the activity rule for the supplementary round, there is no limit on the supplementary bid amount for the final clock package. All supplementary bids on packages other than the final clock package must satisfy revealed preference with respect to the final clock round regardless of whether the supplementary bid package is smaller or larger, in terms of eligibility points, than the bidder's eligibility in the final clock round. This is referred to as the *final cap rule*.

In addition, supplementary bids for packages that exceed the bidder's eligibility in the final clock round must satisfy revealed preference with respect to the last clock round in which the bidder was eligible to bid on the package and every subsequent clock round in which the bidder reduced eligibility. This is also called the *relative cap rule*.

Let x denote the package on which the bidder wishes to place a supplementary bid. Let x_s denote the package on which the bidder bid in clock round s and let b_s denote the bidder's highest monetary amount bid in the auction on package x_s, whether the highest amount was placed in a clock round or the supplementary round.

A supplementary bid b on package x satisfies revealed preference with respect to a clock round s, if b is less than or equal to the highest monetary amount bid on the package bid in clock round s, that is, b_s plus the price difference in the respective packages, x and x_s, using the clock prices of clock round s. Algebraically, the revealed preference limit is the condition that:

$$b \leq b_s + \sum_{i=1}^{m}(p_{s,i} * (x_i - x_{s,i}))$$

where:

- x_i is the quantity of the ith license in package x;
- b is the maximum monetary amount of the supplementary bid on package x; and
- b_s is the highest monetary amount bid on package x either in a clock round or in the supplementary round.

In addition, for supplementary bid package x, let $t(x)$ denote the last clock round in which the bidder's eligibility was at least the number of eligibility points associated with package x.

A given bidder's collection of supplementary bids is consistent with the revealed preference limit if the supplementary bid for package x, with a monetary amount b for the given bidder satisfies the following condition: for any package x, the monetary amount b must satisfy the revealed preference constraint, as specified above with respect to the final clock round and with respect to every eligibility-reducing round equal to $t(x)$ or later.

Note that, in the application of the formula above, the package x_s may itself be subject to a revealed preference constraint with respect to another package. Thus, the rule may have the effect of creating a chain of constraints on the monetary amount of a supplementary bid for a package x relative to the monetary amounts of other clock bids or supplementary bids.

2.4. Incentives for Strategic Manipulation and the CCA's Prisoner's Dilemma

These activity rules have strategic implications, which have been analyzed in a number of papers. Possibilities for spiteful bidding have been shown in Bichler et al. (2011) and later in Bichler et al. (2013a), who demonstrate that standing bidders after the clock phase can determine bid prices in the supplementary round (aka. safe supplementary bids) such that their standing bid from the clock phase becomes winning with certainty. Consequently, the allocation cannot change anymore after the clock phase providing little incentives for bidding truthful in the second phase assuming independent and private values.

In reality, bidders might often care about the prices others have to pay and consequently their payoff, i.e., bidders might be spiteful. Since the allocation cannot change anymore, the CCA provides possibilities for supplementary bids which drive up the competitors' payments, but at no risk of losing the standing bid from the clock phase (Bichler et al., 2013a). Also, they cannot pay more for this bid than what they have bid. In recent spectrum auction implementations, the regulator decided not to reveal excess supply in the last round, in order to make spiteful bidding risky. It depends on the market specifics, if this risk is high enough to eliminate spiteful bidding.

Another issue in both the VCG auction and the CCA is that they violate the law of one price. This means, two bidders might win identical allocations at different prices. We introduce a brief example following Bichler et al. (2013a) to illustrate this point: Suppose there are two bidders and two homogeneous units of one item. Bidder 1 and bidder 2 both have preferences for only one unit and a standing bid of $5 on one unit after the clock phase. If both bidders only bid on one unit, they both pay zero. Now, according to the CCA activity rules, the allocation cannot change any more. Suppose, bidder 2 also bids $9 for two units in the CCA, although he does not have such a valuation for two units. As a consequence, bidder 2 would still pay zero, while bidder 1 would pay $4. However, outcomes where bidders get the same allocation at very different prices are typically perceived as problematic (see Section 5.3), no matter if they are due to spiteful bids or truthful bidding.

Violations of the law of one price and possibilities for riskless spiteful bidding introduce a situation much like in a prisoner's dilemma: If a bidder does not want to pay more for his allocation relative to competitors, he can bid high on losing package bids to drive up payments of competitors after the clock phase. If all bidders follow this strategy, then the payments will be at their bid prices. Often there is excess supply after the clock phase, and the standing clock bids need to be increased by the price of the unsold licenses in the final clock round to win with certainty (Bichler et al., 2013a). This, of course, can also drive up their own payments to the level of this safe supplementary bid.

Janssen and Karamychev (2013) show in a complete information analysis that bidders with an incentive to raise rivals' costs can submit large final round bids and aggressive bids in the clock phase. Levin and Skrzypacz (2014) recently provided an elegant complete information model characterizing the ex post equilibria and resulting inefficiencies that can arise in the CCA. First, they show that the CCA can have many ex post equilibria if bidders have independent private values. If several bidders try to raise each others payments spitefully, then they show that there are again multiple equilibria featuring demand reduction in the clock phase with no guarantee of efficiency. Knapek and Wambach (2012) discuss strategic complexities partly related to an earlier version of the CCA activity rule.

3. Revealed Preference Theory and Straightforward Bidding in Auctions

As outlined in the introduction, straightforward bidding is a central assumption for the two-stage CCA to be efficient (Ausubel et al., 2006). Note that the revealed preference activity rules in the CCA are such that bidders can be limited in the amount they bid in the supplementary round if they do not bid straightforward in the clock phase (Bichler et al., 2011, 2013a). This can also lead to inefficiency, as we will show. Ausubel and Baranov (2014) draw on the theory of revealed preference as a rationale for the activity rules used in the latest version of the CCA in the Canadian 700 MHz auction and for future versions. They show that the current version is based on the Weak Axiom of Revealed Preference (WARP), while future versions should be based on the General Axiom of Revealed Preference (GARP) and eliminate eligibility-point-based activity rules. In what follows, we will revisit important concepts of revealed preference theory and then discuss how they relate to straightforward bidding in an auction. We will also introduce a version of Afriat's Efficiency Index, which allows us to measure straightforward bidding in empirical bid data.

The concept of revealed preferences was originally introduced by Samuelson in order to describe rational behavior of an observed individual without knowing the underlying utility function. He described the simple observation that "if an individual selects batch one over batch two, he does not at the same time select two over one" (Samuelson, 1938). The term "select over" relates to a concept which is nowadays known as "revealed preferred to" and can be defined as follows:

Definition 1. Given some vectors of prices and chosen bundles (p_t, x_t) for $t = 1, \ldots, T$, x_t is directly revealed preferred to a bundle x ($x_t R_D x$) if $p_t x_t \geq p_t x$. Furthermore, x_t is strictly directly revealed preferred to x ($x_t P_D x$) if $p_t x_t > p_t x$. The relations R and P are the transitive closures of R_D and P_D, respectively.

Intuitively, a selected bundle x_1 is directly revealed preferred to bundle x_2 if given x_1 and x_2, both at price p, x_1 is chosen. This definition implies some sort of budget (or income) for each observation. Consider a world with only two bundles x_1 and x_2, x_1 being the more expensive one. If an individual chooses to consume x_1 nevertheless, we know that she prefers it over x_2 such that $x_1 R_D x_2$. This implies that as a rational utility maximizer, she will never strictly prefer x_2 when x_1 is affordable at the same time. More formally, this is known as the Weak Axiom of Revealed Preference (WARP).[3] If she chooses x_2, though, we do not know if that decision is due to an actual preference or a budget constraint below the price of x_1. Hence, there is also no way to predict which choice will be made in another observation where she might have a higher income or face different prices as we have learned nothing about the relation R_D.

In a setting with more than two bundles, WARP is not enough to determine if a consumer is a rational utility maximizer. A set of choices $\{x_1 R_D x_2, x_2 R_D x_3, x_3 R_D x_1\}$ is not violating WARP but is possibly irrational. In order to detect this inconsistency, we need to consider the transitive closure R which also includes $x_1 R_D x_3$, possibly contradicting $x_3 R_D x_1$. Therefore, in a world with more than two bundles the consumption data of a rational utility maximizer needs to satisfy the Strong Axiom of Revealed Preference (SARP)[4] or, if indifference between distinct bundles is valid, the Generalized Axiom of Revealed Preference (GARP).[5] Varian (2006) provides an extensive discussion of WARP, SARP, and GARP.

Applying these axioms to the clock phase of the CCA is straightforward: In each clock round (observation), there is a single known price vector for which each bidder submits a single demand vector. Hence, we can easily build the revealed preference relation R_D and its transitive closure R for every bidder. For the supplementary round S, we know the bid prices $p^S x$ even without an explicit price vector p^S, as bidders bid on bundles instead of single items. As only at most one of the bidder's bids will win, for any pair of supplementary bids $\{x_1^S, x_2^S\}$, the bidder reveals her preference for the higher bid. This allows us to infer $x_1^S R_D x_2^S$ if the bid on x_1^S is higher or equal to the bid x_2^S, or vice versa.

Table 1 provides a simple example of CCA bidding data for an auction with 3 clock rounds and a supplementary phase. In each round of the clock phase, the considered bidder reveals her preference of the chosen bundle over all other affordable bundles. In the supplementary phase, bundles with higher bids are preferred over those with lower bids. The given data is consistent with a set of valuations such as (75, 60, 55)

---

[3] If $x_t R_D x_s$ then it must not be the case that $x_s P_D x_t$ for WARP to be satisfied.
[4] If $x_t R x_s$ then it must not be the case that $x_s R x_t$ for SARP to be satisfied.
[5] If $x_t R x_s$ then it must not be the case that $x_s P x_t$ for GARP to be satisfied.

Table 1. Applying the revealed preference theory to a bidder's CCA bids

	Prices p_t		Bundle Prices $p_t x$			
Round t	A	B	$x_1 = (2, 2)$	$x_2 = (2, 1)$	$x_3 = (1, 1)$	Revealed Preference
1	10	10	40*	30	20	$x_1 P_D x_2, x_1 P_D x_3$
2	20	20	(80)	60*	40	$x_2 P_D x_3$
3	30	20	(100)	(80)	50*	
S			75		55	$x_1 P_D x_3$
	Valuations		100	100	100	

Chosen bundles x_t are marked with *, bundles in brackets are assumed non-affordable as they are more expensive than the chosen one.

Table 2. Example for non-straightforward bidding behavior with GARP

	Prices p_t		Bundle Prices $p_t x$		
Round t	A	B	$x_1 = (1, 0)$	$x_2 = (0, 1)$	Revealed Preference
1	10	50	10*	(50)	
2	30	80	30	80*	$p_2 x_2 > p_2 x_1 \Rightarrow x_2 P_D x_1$

This example is no violation of GARP. It can be explained by an increase in income from $t = 1$ to $t = 2$.

for the three bundles. However, it is not consistent with the assumed actual valuations (100, 100, 100) that would require to always choose the cheapest of the three packages. When using the actual valuations to infer revealed preferences as well[6], the resulting relation violates GARP in this case, but it cannot be detected without knowing the true valuations.

Afriat's Theorem says that a finite set of data is consistent with utility maximization (i.e., straightforward bidding) if and only if it satisfies GARP (Afriat, 1967). However, GARP allows for changes in income or budget across different observations (see Table 2) as traditional revealed preference theory is based on the assumption of an idealized individual who "confronted with a given set of prices and with a given income [...] will always choose the same set of goods" (Samuelson, 1938).

The auction literature typically assumes that bidders have quasi-linear utility functions such that they maximize their payoff given the prices. Quasi-linear utility functions imply that there are no binding budget constraints or "infinite income." Ausubel and Baranov (2014) argue that a GARP-based activity rule would require GARP and quasi-linearity. Also, the efficiency results for the CCA in Ausubel and Milgrom (2002) and Ausubel et al. (2006) only hold if bidders are quasi-linear and they bid straightforward. Unfortunately, Table 2 shows that the traditional definition of GARP allows for changes in income and therefore allows substantial deviations from straightforward bidding if we assume quasi-linear utility functions.

---

[6] $x_i R_D x_j$ for any pair (i, j) as all valuations are equal.

Therefore, we aim for a stronger definition of revealed preference with non-binding budgets, as they are assumed in theory. With this assumption, the different bids in an auction also reveal how much one bundle is preferred to another one:

Definition 2. Given some vectors of prices and chosen bundles (p_t, x_t) for $t = 1, \ldots, T$ and a constant income, we say x_t is revealed preferred to a bundle x by amount c (written $x_t R_c x$) if $p_t x_t \geq p_t x + c$.

Intuitively, $x_t R_c x$ can be interpreted as "x_t is chosen over x if it costs no more than the price of x plus c". We will refer to this definition of revealed preference as *GARP with quasi-linear utility (GARPQU)*. Note that c will be negative in all cases where x is more expensive than x_t, which would be ignored in the traditional definition of revealed preferences (see Definition 1). The result of applying this definition to a set of bid data will be a family of relations R_c instead of a single revealed preference relation R. R_c has several properties:

- $x_1 R_c x_2$ implies $x_1 R x_2$ if $c \geq 0$ (definition)
- $x_1 R_c x_2$ implies $x_1 P x_2$ if $c > 0$ (definition)
- $x R_c x$ for all $c \leq 0$ (reflexivity)
- $x_1 R_{c_1} x_2$ and $x_2 R_{c_2} x_3$ imply $x_1 R_{c_1 + c_2} x_3$ (transitivity)
- $x_1 R_{c_1} x_2$ implies $x_1 R_{c_2} x_2$ if $c_1 > c_2$ (derived from transitivity and reflexivity of $R_{c_1 - c_2}$)

These properties are sufficient to derive a contradiction $x R_c x$ with $c > 0$ ("$u(x) > u(x)$") for any non-straightforward bidding behavior that can be detected without knowing the actual utility function u. For example, it is easy to see that the choices in Table 2 do not describe straightforward bidding because they are not consistent under the above properties of R_c: $(x_1 R_{-40} x_2 \wedge x_2 R_{50} x_1 \Rightarrow x_1 R_{10} x_1)$.

Note that the result of such an analysis of a series of bids is always binary: either a set of data satisfies GARPQU or it does not. In revealed preference theory, measures such as Afriat's Efficiency Index (AI) were developed to describe how well a set of consumer choices conforms to utility maximization. The AI is a goodness of fit metric that spans the range [0; 1] with 1 indicating perfect compliance with a tested axiom (Afriat, 1973). It requires a variable e in all revealed preference inequations (see Definition 1 and Definition 2):

- $x_t R_D x$: $p_t x_t \geq p_t x$ becomes $e \cdot (p_t x_t) \geq p_t x$
- $x_t P_D x$: $p_t x_t > p_t x$ becomes $e \cdot (p_t x_t) > p_t x$
- $x_t R_c x$: $p_t x_t \geq p_t x + c$ becomes $e \cdot (p_t x_t) \geq p_t x + c$

Applying the axioms with $e < 1$ leads to a relaxed version that is easier to satisfy. For instance, assume $e = 0.9$: If bundle x_t was chosen for a price of \$100 the pair (x_t, x) will only be included if $p_t x \leq \$90$. The AI is equal to the maximum value of e which satisfies the tested axiom. We will use a graph-based algorithm based on Smeulders et al. (2012) for computing the AI. There are related metrics such as the Varian Index (VI) which follows the same principle as the AI but uses a vector instead of a single constant value e (Varian, 1990). Unfortunately, the computation of VI is NP-hard (Smeulders et al., 2012).

4. Evidence from the Lab

In a lab experiment we can not only observe the bids, but also know the induced valuations of bidders. In what follows, we will analyze straightforward bidding in the lab and draw on the data from experiments conducted by Bichler et al. (2013a). We will focus on 16 auctions with 4 bidders in a multi-band value model with 24 blocks in 4 different bands. This means, bidders could submit up to 2400 package bids. This experimental setup is comparable to multi-band auctions with national licenses as they were conducted in Austria, Ireland, the UK, and Switzerland, although the number of bands differed from country to country.

4.1. Missing Bids

The auctions in the lab suffered from the missing bids problem with only 8.3 supplementary bids per bidder on average. Bichler et al. (2013a) argue that this has contributed to the low efficiency of only 89.3% observed in the CCA, which was substantially lower than that of the auctions with SMRA, which achieved an average efficiency of 98.5%. In comparison with the standing clock bids, the allocation changed after the supplementary phase in 14 auctions by 34.9% of all licenses on average. Significant changes in the allocation could also be observed in the British auction after the supplementary stage.

4.2. Inconsistent Bidding

Figure 1 shows the AI based on GARPQU for all 64 bidders participating in a CCA in the lab experiments. The left-hand box plot describes bids from the clock phase only, the middle box plot the bids submitted in both phases, and the right box plot the clock bids and all true valuations for all packages of a bidder. A median AI of 1.000 for clock bids shows that there is no evidence for significant deviations from straightforward bidding in the bids during the clock rounds. When including data from the supplementary

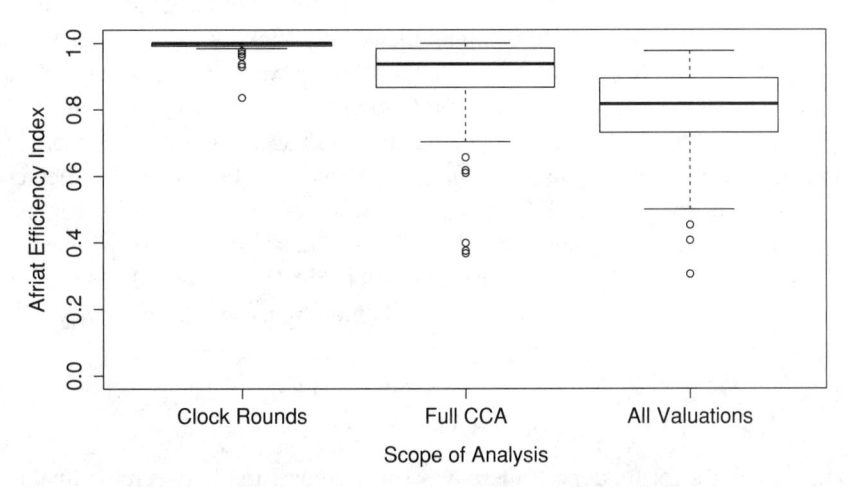

Figure 1. Boxplot for AI of 16 · 4 Bidders from Lab auctions.

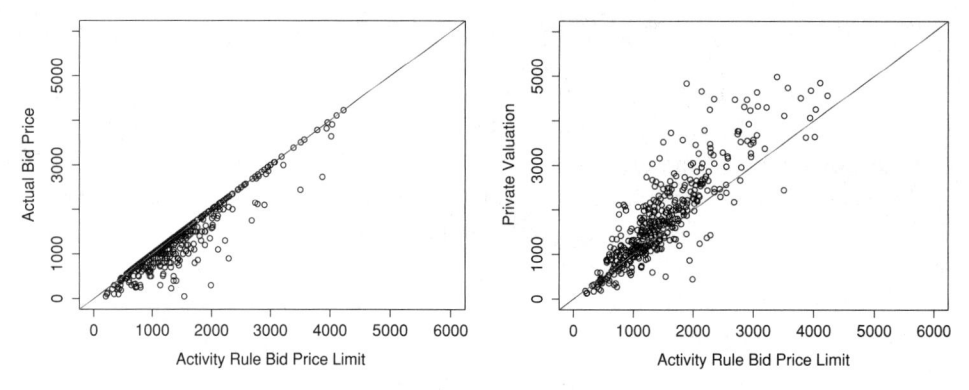

Figure 2. Scatterplots for supplementary bids of 4 bidders in 16 auctions in the lab (left) in comparison to their induced private valuations (right).

round, however, the median AI drops to 0.938, indicating deviations from straightforward bidding in the supplementary phase. The AI with truthful supplementary bids for all possible bundles, which is described in the third boxplot (All valuations) drops to 0.816 and suggests that bidders did indeed not bid straightforward with respect to their true valuations in the clock phase.

Deviations from straightforward bidding such as those indicated by boxplot 3 can limit the possible bid amount in the supplementary phase substantially. For the lab data we can see how high bidders have bid in the supplementary phase relative to their bid price limit.

The left scatter plot in Figure 2 shows that bidders often bid close to the bid price limit (correlation 0.9448). The right scatter plot illustrates the private valuations with respect to the bid price limit imposed by the activity rule and their behavior in the clock phase. For 57.2% of all submitted supplementary bids, the bid price limit was lower than their valuation for the corresponding bundle and hence it did not allow bidders to bid their valuation truthfully in the supplementary phase.

Figure 2 deserves further explanation. As described in Section 2, if bidders had independent and decreasing marginal valuations, then they would not need to bid up to their true valuation in the second phase and even if bidders did not bid at all after the clock phase, the auctioneer could compute the correct Vickrey payments. The valuations of bidders in the lab were complements and there was often excess supply after the clock phase. Given the uncertainty that bidders faced in the lab, their most likely strategy was to bid truthful on their supplementary packages if possible. Bidders in the lab knew in which order they had to submit supplementary bids such that they could maximize the bids for supplementary packages. However, there were significant differences between the final payments of bidders in the clock stage and the payments one would get if bidders submitted all their valuations truthfully in a sealed-bid auction. In Section 6 we analyze the impact that inconsistent bidding has on the efficiency of these auctions.

4.3. Clock Prices

It would be helpful for bidders, if there was some connection between the final clock prices and the core payments, because this could give bidders a useful hint on how high they need to bid in the second phase. However, the final prices from the clock phase can differ substantially from the payments. We compared the core payments of all winning bids with the corresponding linear bundle prices in the final clock round and found that the average payment was only 59.1% of the last clock price. The standard deviation of this ratio in the lab was 22.6%. For the British LTE auction in 2013 this average payment was at 56.5% of the final clock prices. Also in simulations with straightforward bidders who bid truthful in the supplementary round the clock prices do not necessarily provide an indication for payments or winning supplementary bids.

5. Evidence from the Field

The British regulator Ofcom was the first to publish the bid data on a CCA in 2008 and 2013 (Ofcom, 2013a). We will primarily focus on the 2013 multi-band spectrum auction as it is closest to auctions in other countries and similar to the environment analyzed in the lab (Bichler et al., 2013a). Then we will discuss the Canadian 700 MHz auction in 2014, where bid data was revealed as well, before we summarize public information about CCA applications in some other countries. Although all these auctions used a CCA, there are important differences in the caps used, in the licenses and the band plan, and in details of the auction rules, which requires caution in the comparison of the results. Of course, we cannot know the true valuations of bidders in this auction, however, we highlight some patterns which are similar to what we found in the lab data. In particular, bidders only bid on a small subset of all possible packages and there was a very high number of supplementary package bids at the bid price limit and not below, which can be seen as an indication of bidders over-constraining themselves in the supplementary phase due to inconsistent clock bids.

5.1. The British LTE Auction in 2013

In the British auction in 2013, 28 licenses in the 800 MHz and 2.6 GHz bands were sold, and the bid data was released to the public. There were 4 A1 blocks of paired spectrum in 800 MHz and another A2 block with a coverage obligation. In addition, there were 14 blocks of paired spectrum in the 2.6 GHz band, and another 9 blocks of unpaired spectrum in the 2.6 GHz band. The unpaired spectrum was considered less valuable than paired spectrum bands. There were seven bidders, Vodafone, Telefonica, Everything Everywhere, Hutchinson, Niche, HKT, and MLL. A spectrum cap was put on the 800 MHz band for Vodafone and Telefonica, who are considered large bidders. The detailed rules can be found at (Ofcom, 2013b).

The bid data reveals the main interests of these seven bidders. Vodafone and Telefonica bid on 800 MHz and both 2.6 GHz bands. They consistently bid on two 2×5 MHz blocks in 800 MHz spectrum throughout the clock phase and both won two blocks. Everything Everywhere and Hutchinson also bid on the valuable 800 MHz spectrum, but ceased to bid on 800 MHz in the clock phase. Niche, MLL, and HKT can be considered smaller players. MLL and HKT only bid on the unpaired spectrum in 2.6 GHz and they did not win anything. Niche bid on both 2.6 GHz bands and also won blocks in both bands. More details on the auction can be found in Appendix A, where we describe the valuations of bidders for our numerical experiments.

5.1.1. Missing Bids

Let us now provide some statistics to shed light on the missing bids problem in the British auction, which might be one of the reasons for the low revenue encountered (Arthur, 2013; Smith, 2013), before we discuss straightforward bidding. With all the caps considered, larger bidders such as Vodafone and Telefonica could bid on 750 packages in this auction. However, after 52 clock rounds in which the seven bidders selected 7.7 distinct bundles on average, they submitted only 39.6 supplementary bids per bidder on average (277 bids in total). Bidders always submitted higher bids on the packages submitted in the clock phase, but bid on average on 31.9 new bundles only in the supplementary phase. Telefonica submitted no more than 11 supplementary bids, while Vodafone submitted 94 of 750 supplementary bids mostly covering combinations of licenses with 20 MHz in low frequency bands. Everything Everywhere submitted 84 supplementary bids, and Hutchinson only 17 bids.

Note that the winner determination treats a missing bid as if the valuation of a bidder for this package was zero in a CCA. It is questionable if bidders had no value for all the other packages or a value below the reservation prices. In this case, the missing bids problem appears to have been an issue.

The total revenue from the bidder-optimal core prices of £2.23bn is equivalent to the Vickrey payments in this auction, which is also due to the low number of supplementary bids which led to a lower number of core constraints when computing the bidder-optimal core payments (Day and Cramton, 2012). Consequently, the discounts were very high. The sum of the bids in the revenue maximizing allocation amounts to £5.25bn.

It is interesting to note that with only the bids from the clock phase and without the supplementary phase the auction had a revenue of £1.92bn, which is only 13.9% less than the final result including the bids of the supplementary phase. The supplementary phase did change the allocation considerably, however, which might have come as a surprise to some bidders. 19.3% of all winning licenses from the clock phase (weighted by their eligibility points) were re-allocated after the supplementary round.

5.1.2. Inconsistent Bidding

Next, we analyze straightforward bidding in the British auction using Afriat's index as we have discussed it in Section 3. Table 3 shows the AI per bidder for the clock phase only and for all bids including the supplementary phase. Although the median AI is high (0.995) for bids in the clock phase only, it decreases to 0.811 when we also consider supplementary bids. Note that this value is lower than what we have found in the lab auctions even though it is an upper bound for the "true AI". If the true valuations of each bidder are taken into account the AI can be considerably lower as we have seen in Section 4 and in the example in Table 1.

The reason for low auctioneer revenue after the supplementary phase might, however, also have been due to limits on the bid prices imposed by the activity rule.

Table 3. *Afriat's Index (AI) based on bids of the clock phase and based on all bids including the supplementary stage in the British LTE auction*

Bidder	AI in clock phase	AI of all bids
EE	0.811	0.811
H3G	0.988	0.845
HKT	1	0.626
MLL	1	0.619
Niche	0.995	0.814
Telefonica	1	0.467
Vodafone	0.946	0.946

Figure 3 compares the supplementary bid prices and the corresponding bid price limit imposed by the activity rule and shows that the bids of many bidders are very close to this limit. The bid data is highly correlated with the bid price limit imposed by the activity rule (Pearson correlation coefficient of 0.9824) and yields a median ratio of the bid price to the bid price limit of 92.3% (mean: 80.5%). Interestingly, this ratio was particularly high for the big bidders Vodafone and Telefonica with a median of 98.1% and 96.5% respectively. For supplementary bids of the remaining five bidders, the median ratio was only 83.0%, which might be due to the fact that these were

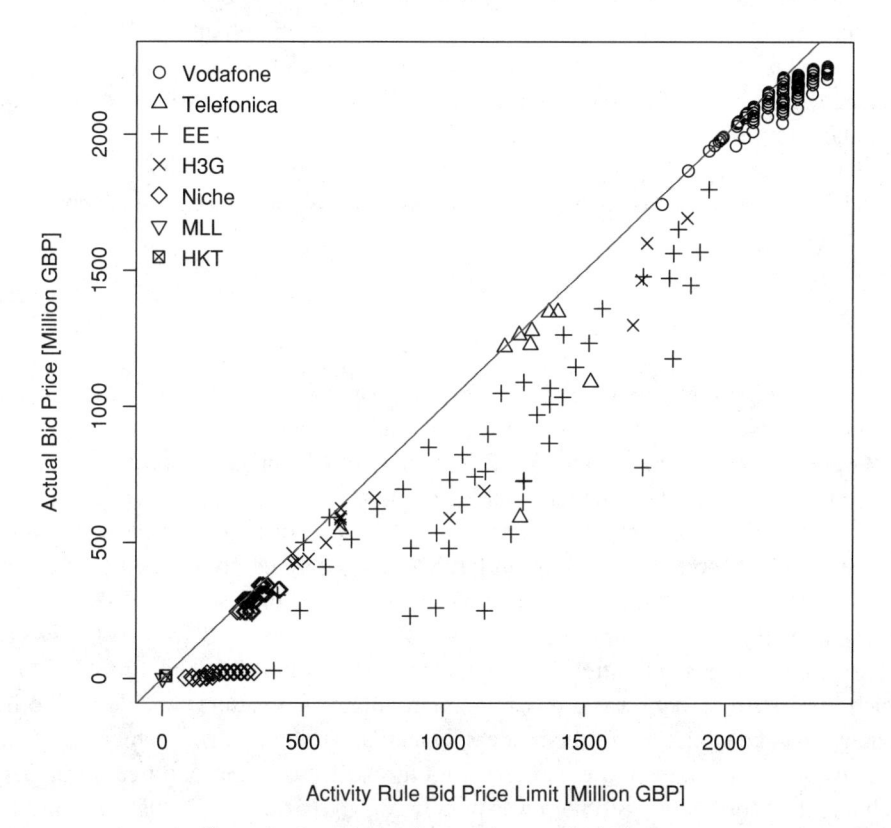

Figure 3. Scatterplot for supplementary bids (UK).

financially weaker bidders. In this auction spiteful bidding (bidders submitting high losing package bids to drive up payments of competitors) did not seem to be an issue such that the more likely explanation is that bidders could not bid up to their valuations.

5.2. The Canadian 700 MHz Auction in 2014

The Canadian 700 MHz auction in 2014 comprised 5 paired spectrum licenses (A, B, C, C1, and C2), and two unpaired licenses (D, E) in 14 service areas. B and C as well as C1 and C2 were treated as generic licenses. Although the licenses are all in the 700 MHz band, they are technically not similar enough to sell all of them as generic licenses of one type.

The total revenue of $5.27bn from the bidder-optimal core prices was 32.4% less than $7.14bn, the sum of provisionally winning bids after the final clock round. The sum of the bids in the revenue maximizing allocation was $9.13bn. Again, the clock prices provided little guidance for what might constitute a winning bid in the supplementary phase.

The auction was dominated by three national carriers Bell, Rogers, and Telus. Rogers was the strongest bidder and contributed 62.45% to the overall revenue, while Telus paid 21.69% and Bell 10.73%. Rogers did not bid on C1/C2 and aimed for licenses in A, and B/C throughout the auction, while Bell and Telus also bid on C1/C2 in certain service areas. The smaller bidders mainly bid on remaining C1/C2 blocks. Bell and Telus had to coordinate and find an allocation such that they both got sufficient coverage in the lower 700 MHz band (A, B and C blocks), which explains much of the bid data. There was a disparity in how much bidders had to pay for different packages, which can be explained by different valuations that bidders placed on packages and the payment rule. Still, due to the high competition and revenue the auction is considered successful.

5.2.1. Missing Bids

Overall, the high competition among the three national telecoms Bell, Rogers, and Telus and the clever spectrum caps for them explains much of the result. All eight bidders were restricted to at most 2 paired frequency blocks in each service area. Large national wireless service providers such as Rogers, Bell, and Telus were further limited in that they could only bid on one paired license in each service area among licenses B, C, C1 and C2. This cap on large wireless service providers did not, however, include block A. Still, the national bidders could bid on $2*3*3 = 18$ packages per region including the empty package, which leads to $18^{14} \approx 3.75*10^{17}$ packages in all regions. Rogers submitted 12 supplementary bids, Bell 543 and Telus 547 bids, which suggests that there was a missing bids problem as it is questionable if all other packages had no valuation for the bidders. Note that only one license remained unsold after the auction. Rogers bid consistently on the A licenses and one license in B/C, such that the coordination problem was largely solved by Bell and Telus, who split the regional service areas on B/C and C1/C2.

Table 4. *Afriat's Index (AI) based on bids from the clock phase and based on all bids including the supplementary stage in the Canadian 700 MHz auction*

Bidder	AI in clock phase	AI of all bids
Bell	0.871	0.159
Bragg	0.722	0.151
Feenix	0.873	0.730
MTS	0.893	0.450
Novus	0.949	0.379
Rogers	0.977	0.454
SaskTel	0.499	0.384
TbayTel	0.857	0.716
Telus	0.930	0.235
Videotron	0.728	0.493

5.2.2. Inconsistent Bidding

It is interesting to understand straightforward bidding in the Canadian 700 MHz auction as well. Although the regulator disclosed all the bid data, the clock prices were not made public. We used a linear program which helped us reconstruct clock prices from the bid data. There are some assumptions in this linear program and we cannot compute the price trajectories and the resulting AI with certainty, such that the numbers in Table 4 are only estimates. However, the order of magnitude in the AI was similar for different price trajectories that we could derive. The numbers suggest that bidders deviated substantially from straightforward bidding. One explanation is that bidders such as Bell and Telus actively tried to coordinate and agree on non-overlapping packages of licenses. It is also interesting to note that some small local bidders bid on competitive service areas in A and B/C outside the service area in which they operate. One conjecture is that this was done in an attempt to park eligibility rights and keep clock prices low in their own service area. As the regulator did not disclose the excess supply after the clock phase and due to the uncertainty in this large scenario, it is not unreasonable to believe that bidders tried to bid up to their true valuation in the supplementary stage. Actually, in Canada the supplementary bid on the final clock package was substantially higher than the final clock round bid for many bidders. Figure 4 shows that, again, a very large proportion of the other supplementary bids are exactly at their bid price limit indicating that they might have been truncated due to restrictions imposed by the activity rule. The low AIs for the different bidders provide further evidence.

5.3. Observations from other Countries

Apart from Canada and the UK, bids were not made public in other countries. As mentioned earlier, the UK also released data for two earlier CCAs in 2008, the L-band auction with 17 licenses, and the 10–40 GHz auction with 27 licenses. In the L-band auction bidders submitted between 0 and 15 bids in the supplementary phase also indicating missing bids from at least some of the bidders. In this auction with much less valuable spectrum than in 2013, one bidder won all 17 lots with a bid of £20m. The bidder only had to pay £8.334m, which was the revenue of the best coalition of bidders

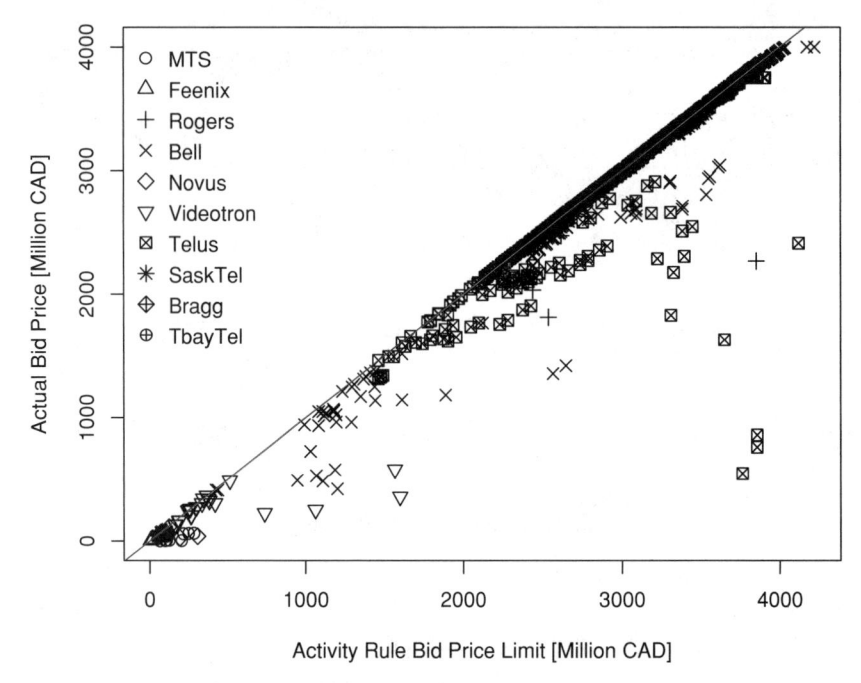

Figure 4. Scatterplot for supplementary bids (Canada).

without the winner (Cramton, 2008). In the 10–40 GHz auction all but one bidder made their highest supplementary bid either on the final clock package, or on a subset thereof (Jewitt and Li, 2008).

The Swiss auction in 2012 was remarkable, because one bidder payed almost 482 million Swiss Francs, while another one payed around 360 million Swiss Francs for almost the same allocation. This can happen in a Vickrey auction as well as in a CCA when one bidder contributes more to the overall revenue with his bids than another bidder (see Section 2.4).

The Austrian Auction in 2013 on the 800 MHz, 900 MHz, and 1800 MHz bands is another interesting case. Bidders could potentially submit up to 12,810 package bids. The regulator reported that the three bidders actually submitted 4000 supplementary bids in total. The regulator also disclosed that most of these bids were submitted on very large packages (RTR, 2013). This large number of supplementary bids can be seen as one reason for the high prices paid in Austria. The attempt to drive up prices of other bidders and avoid having to pay more for an allocation than ones competitors, as it happened in Switzerland, can serve as one explanation for this bidding behavior. However, it leads to the prisoner's dilemma discussed in Section 2.4.

6. Estimating the Impact of Missing Bids and Inconsistent Bidding

We performed computer simulations of the CCA for the lab value model as well as for the British 4G auction. For the latter, we estimated valuations for the seven bidders from the bid data with base valuations, intra-band and inter-band synergies. The estimated valuations are described in Appendix A. We did not perform this analysis for the large

Canadian auction with 98 licenses, because this would require many more assumptions due to the regional structure and the many licenses involved. Data from 16 auctions in the lab and 10 sets of synthetic valuations for the British scenario were used. All significance tests reported in this section are using a Wilcoxon signed-rank sum test.

Efficiency and revenue of an auction are typically used as primary metrics. Throughout the rest of this paper, we will use the terms *allocative efficiency*:

$$E = \frac{\text{actual surplus}}{\text{optimal surplus}} \times 100\%$$

and auctioneer's *revenue share*:

$$R = \frac{\text{auctioneer's revenue}}{\text{optimal surplus}} \times 100\%$$

The revenue share shows how the resulting total surplus is distributed between the auctioneer and the bidders. Optimal surplus describes the resulting revenue of the winner-determination problem if all valuations of all bidders were available, while actual surplus considers the true valuations for those packages of bidders selected by the auction. In contrast, auctioneer's revenue describes the cumulative payments of the bids selected by the auction, not their underlying valuations.

In the following subsections we analyze the impact of missing bids and inconsistent bidding in the clock phase. The main results are summarized in Table 5. A baseline for this analysis are the simulations with truthful bidders, i.e., bidders who bid straightforward in each clock round and submit truthful supplementary bids on all bundles. As expected, all simulations where bidders submitted all package bids truthfully were 100% efficient in contrast to the efficiency of 89.3% we measured in the lab.

Table 5. *Mean efficiency and revenue for the British and the lab value models with the CCA and bidders bidding truthful on a subset of the packages in the supplementary round up to their true valuation or a restricting bid price limit, resp.*

supplementary bids in simulations		British VM (straightforward)		Lab VM (straightforward)		Lab VM (actual clock bids)	
		Efficiency	Revenue	Efficiency	Revenue	Efficiency	Revenue
	none	94.7%	28.1%	95.3%	40.5%	86.3%	37.2%
	1	94.8%	29.3%	95.7%	63.8%	89.2%	54.0%
	2	94.8%	29.5%	96.0%	64.5%	90.4%	54.5%
	3	95.2%	30.2%	96.2%	64.3%	91.7%	54.0%
	5	95.5%	31.2%	96.4%	64.7%	92.8%	53.3%
	10	96.1%	33.3%	97.5%	64.8%	92.5%	54.4%
	20	96.9%	36.5%	98.1%	65.5%	93.7%	56.1%
	50	99.2%	42.8%	99.1%	68.5%	95.1%	60.0%
	100	100%	61.3%	99.6%	71.8%	94.3%	63.4%
	200	100%	61.4%	99.8%	73.6%	94.5%	65.6%
	500	100%	61.4%	99.8%	75.0%	94.3%	66.6%
	all	100%	62.9%	100%	75.7%	95.0%	71.4%
						89.3%	41.0%
Human subjects		39.6 suppl. bids/bidder Estimated values				8.3 suppl. bids/bidder Induced values	

6.1. Impact of Missing Bids in the Supplementary Phase

We first evaluate the impact of the missing bids problem. In this set of simulations, the simulated bidders bid straightforward in the clock phase (see Figure 5 and the first two column-pairs in Table 5), such that they could bid up to their valuation in the supplementary phase. As human bidders only submit a small subset of possible supplementary bids, there are just a few core constraints leading to lower prices and hence lower auctioneer revenue. In order to better understand the impact of this effect, we restricted our bidders in the number N of additional packages they can bid on in the second phase.

More precisely, bidders always started the supplementary phase with truthful bids on all clock bundles in reverse order of submission which allows them to maximize the amount they can bid on other packages without violating the activity rule. Then they submitted additional truthful bids on bundles chosen after a heuristic which we observed in the British auction. First, bidders do not demand more units of a certain band than they did in the clock phase, and second, the bidders Telefonica and Vodafone do not submit any bids without two A blocks. Out of this pre-selection, bidders selected up to N of their $2N$ strongest bids. We define the strength of a bid as the valuation divided by the bundle size in terms of the corresponding eligibility points. We have also tested different bundle selection heuristics, but the differences in efficiency were minor. The artificial bidders were bidding truthful as far as they could, such that only limitations in the number of bids submitted and constraints from inconsistent clock bids matter.

For the simulations with the lab value model, there is a significant difference in revenue between no supplementary bids at all (first line in Table 5) and the submission of one new bid. This is due to the fact that in the treatment without additional supplementary bids, we just evaluated the bids submitted in the clock phase. In the treatment with one addition bid, clock bids were updated to their true valuation. The number of bids affects auctioneer revenues in both value models. Even for 50 additional bids, the revenue share is still significantly lower than with supplementary bids on all bundles (p-value $= .0000$), which is due to missing bids.

With 50 additional bids, the efficiency was beyond 99% in both value models. While we found an average efficiency of 89.3% for the CCA in the lab with 8.3 supplementary bids on average, simulated auctions with no bids in the second phase at all yielded an efficiency of 95.3%, which is significantly higher than in the lab (p-value $= .0052$). A substantial part of this difference can be attributed to inconsistent bidding in the clock phase which we will discuss next.

6.2. The Impact of Inconsistent Bidding in the Clock Phase

As we have discussed in Section 4, bidders in the lab and in the field did not bid straightforward in the clock phase and were therefore limited by the activity rule in the supplementary phase. Now, we want to understand how much efficiency loss can be attributed to these limitations in the simulations. In the British value model we only have the bid data of a single instance, which is why we only report on the lab value model.

For all 16 instances we replicated the bids of human bidders in the clock phase. In the supplementary phase the agents tried to bid their true valuations on additional bundles

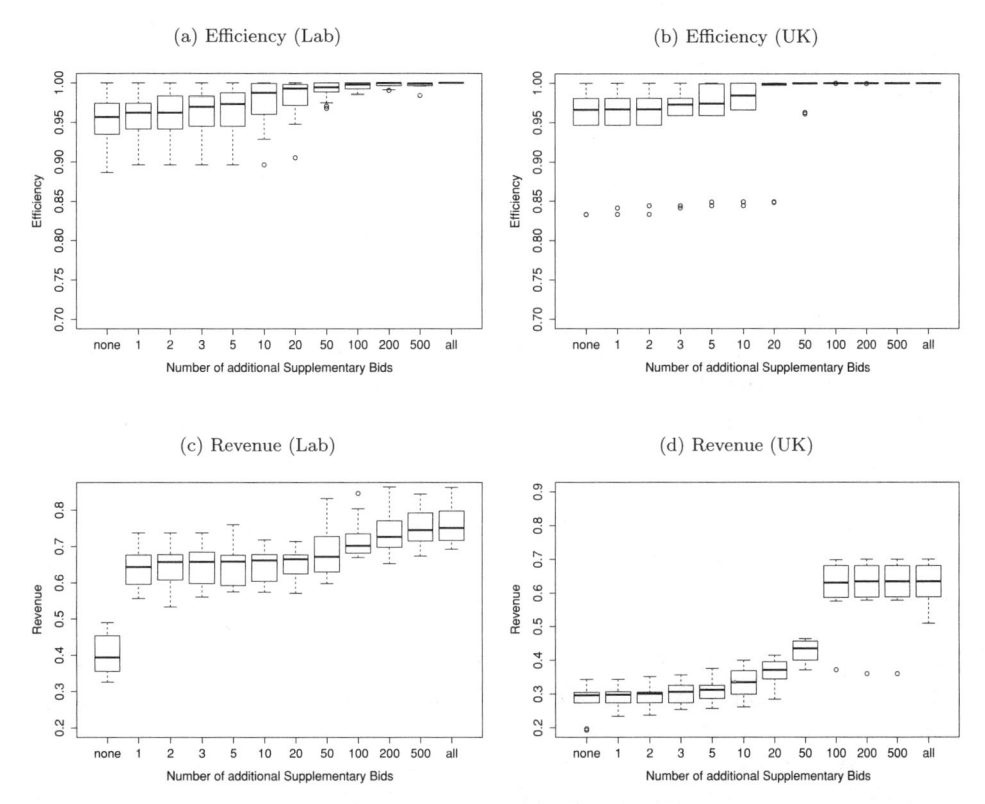

Figure 5. Mean efficiency and revenue for the British and the lab value models with the CCA.

like in the previous subsection. If this was impossible due to the revealed preference activity rule, they chose the highest possible bid price instead.

Figure 6 and the third column pair of Table 5 summarizes the results. Even for supplementary bids on all possible bundles, the efficiency was only 95.0% on average. This is not significantly different (p-value $> .95$) to the mean efficiency of 95.3% that we measured with straightforward bidders but without any supplementary bids. These

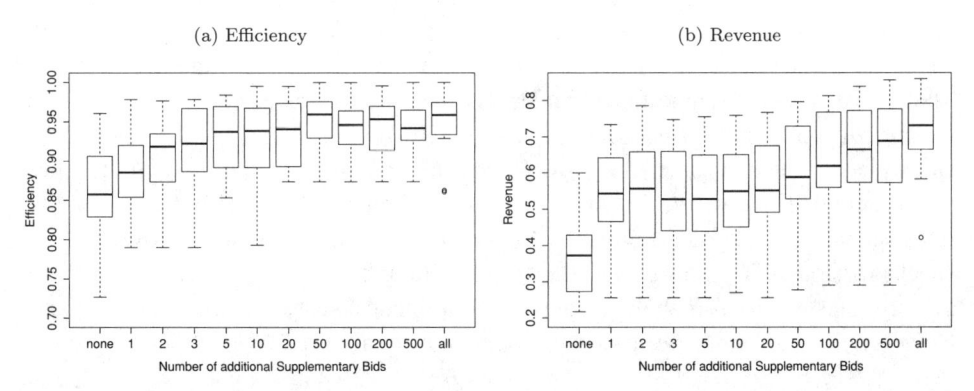

Figure 6. Efficiency and revenue for lab instances with different numbers of (as possible) truthful supplementary bids after actual clock round bids.

findings provide evidence that non-straightforward bidding in the first phase reduces efficiency of the final outcome. The auctioneer revenue share was significantly lower as well. For some simulations the average differences in revenue share were more than 10%, which was only due to inconsistent bidding in the clock phase.

7. Can Strong Activity Rules Serve as a Remedy?

Our analysis of bids in a CCA in the lab and in the recent British and Canadian spectrum auctions indicates that bidders do not bid straightforward in the clock phase of the CCA. This inconsistent bidding with respect to their true valuations can lead to inefficiencies, because the deviations from straightforward bidding in the clock phase restricts bidders from bidding up to their true valuations in the supplementary phase. The difference in efficiency and revenue in simulations with bidders bidding on their bid price limit induced by the activity rule and bidders bidding truthful is substantial, even if we assume the same number of supplementary bids being submitted by the bidders. If bidders do not bid up to their true valuations in the supplementary stage, this can have an impact on payments and the allocation of bidders as simulations show. Both, the missing bids problem and restrictions due to inconsistent bidding can lead to payments, which are quite different from the VCG or core payments if bidders submitted their valuations truthfully.

Efficiency, simplicity, transparency, and robustness against manipulation are often considered design goals for spectrum auctions. No auction format is perfect and there are always trade-offs that an auctioneer needs to make. For example, a Vickrey auction exhibits dominant strategies, but the payments of bidders are not anonymous and it can happen that two bidders with similar allocations pay vastly different prices, which can cause envy. In a similar way, non-core outcomes can be considered unfair, however, core-selecting auctions cannot have dominant strategies for general valuations. For regulators it is important to understand the properties of different auction formats and make an informed choice. Giving up anonymous prices and the transparency of a simple ascending auction format should only be done if the resulting auction achieves higher efficiency and has stronger incentives for bidders to bid truthful.

The CCA has developed over the recent years and a number of suggestions have been picked up to improve the design. For example, new versions of the CCA will allow for a restricted set of OR bids to address the missing bids issue in large auctions. There have also been suggestions to address problems such as dead ends arising from the current activity rule (Ausubel and Baranov, 2014) via stronger activity rules in the clock phase, which enforce straightforward bidding. While the current activity rules can be derived from the Weak Axiom of Revealed Preference (WARP), future activity rules should be based on the General Axiom of Revealed Preference (GARP), which checks for consistency throughout the entire bidding history of a bidder. Such strong activity rules would also avoid problems due to inconsistent bids in the clock phase. However, there are a number issues that need to be considered.

– First, straightforward bidding with a larger number of licenses is challenging for human bidders and possibly requires automated bidding agents for larger auctions with dozens or hundreds of licenses, let alone that there are reasons for bidders not to bid straightforward, such as budget constraints mentioned in the introduction or interdependencies in the valuations of bidders. One might be able to address budget constraints during the auction such that automated agents could be a remedy. However, they would effectively turn the clock phase into a sealed-bid auction, which is then followed by another supplementary sealed-bid stage in the current CCA design. These two sealed-bid auctions deserve some discussion.

– Second, Bichler et al. (2013b) show that the efficiency of a clock auction with certain types of bidder valuations and straightforward bidding can be close to zero.[7] Not only that the standing bids after the final clock round do not provide an indication for the efficient allocation, also the clock prices do not provide helpful information about the final payments, as can be seen in data from the field and the lab. At least, it is not obvious how bidders should use these price signals from the clock phase.

Both points raise the question, what value the clock phase really adds to the auction. One argument in favor of an ascending or dynamic multi-object auction is that bidders do not need to provide all their valuations on exponentially many packages in one step. Levin and Skrzypacz (2014) writes that "economists think of dynamic auctions as having an advantage in this regard because bidders can discover gradually how their demands fit together." Although the single-stage combinatorial clock auction was shown to be highly efficient in lab experiments apparently helping bidders to find efficiency-relevant bundles, bidders in the lab did not bid straightforward (Scheffel et al., 2012). Overall, using GARP with a traditional clock auction exhibits some challenges.

Many regulators have adopted an ascending auction over sealed-bid alternatives for efficiency reasons. Evan Kwerel, senior economist at the FCC, explained the decision of the US Federal Communications Commission (FCC) to adopt an ascending auction format for selling spectrum licenses by saying: "In the end, the FCC chose an ascending bid mechanism, largely because we believed that providing bidders with more information would likely increase efficiency and, as shown by Paul Milgrom and Robert J. Weber, mitigate the winner's curse" (Milgrom, 2004). The argument draws on the *linkage principle*, which implies that ascending auctions generally lead to higher expected prices than sealed-bid auctions with interdependent bidder valuations (Milgrom and Weber, 1982). In contrast to bidders with independent values, bidders with interdependent values might not always bid consistent as their valuations can change and GARP can be too strong to allow for these changes.

Transparency is also an important argument for ascending auctions as a bidding team needs to set expectations throughout the auction and inform stakeholders. Bidders in SMRA see the final allocation and prices develop throughout the auction, which

---

[7] Let's introduce a simple example to better illustrate how straightforward bidding can lead to inefficiency in the clock auction: Consider a market with two items {A, B} and three bidders. Bidder 1 has a value of $10 for A, bidder 2 has a value of $4 for B and $10 for {A, B}, and bidder 3 only has a value of $10 for the package {A, B}. If all bidders bid straightforward starting with prices of zero and unit increments, then bidder 2 will never reveal his valuation for A, leading to 71% efficiency. Bidders 2 and 3 would actually drop out at a price of $5 for both items in the clock stage, which is when bidder 1 still bids on item A. It is easy to extend the example and achieve very low revenue.

typically takes several weeks. However, this type of transparency is much reduced in the CCA. How much bidders finally have to pay depends on the bids submitted in the supplementary stage and is a result of a quadratic optimization problem which is almost impossible to predict given the many possible packages bidders can bid on and the missing bids problem. If they are unable to submit a safe supplementary bid, then the allocation can change substantially after the clock phase, as it has happened in the British LTE auction. This makes the outcome of the CCA hard to predict during the auction.

One advantage that an ascending auction still has over a sealed-bid auction is the fact that winners do not need to reveal their valuation for the winning package to the regulator. Regulators need to decide whether this feature outweighs the added complexity stemming from a two-stage CCA. Ascending combinatorial auctions can certainly be of help for bidders in coordinating with other bidders and finding a feasible allocation. However, if an activity rule enforces straightforward bidding, the possibilities for such coordination will be much reduced.

Designing efficient multi-item auctions is difficult when a regulator needs to consider conflicting design goals such as incentive-compatibility, simplicity, efficiency, and the law-of-one-price. The bid language, the payment rule, and the decision to use a sealed-bid or an ascending format are design choices, which all have substantial impact on efficiency and revenue of an auction. A simple bid language can have a significant positive impact on the efficiency of an auction as was shown in lab experiments (Bichler et al., 2014), and it is not unreasonable to assume similar effects in the field. The pros and cons of different activity rules considering realistic assumptions about bidder preferences in a spectrum auction are still a fruitful area for future research.

A. Details on the Value Model and the Simulations based on the British 4G Auction

The value model used in our simulations in Section 6 is based on the British 4G auction in 2013 in which the 800 MHz as well as the 2.6 GHz band were sold (Ofcom, 2013b). We will provide a brief description of the British auction and how we derived the value model for each bidder in our simulations, mirroring the main characteristics of this market. The valuations can be made available upon request.

Table 6. *Overview of auctioned lots in the UK 4G Auction (Ofcom (2012))*

Lot	Amount	Description	EPs	Start Price
A(i)	4	2 × 5 MHz paired spectrum in the 800 MHz band	2250	$225mn
A(ii)	1	2 × 10 MHz paired spectrum in the 800 MHz band with coverage obligation	4500	£250mn
C	10/12/14	2 × 5 MHz paired spectrum in the 2.6 GHz band	150	£15mn
D(i)	≤10	2 × 10 MHz paired spectrum in the 2.6 GHz band (shared low power)	30	£3mn
D(ii)	≤10	2 × 20 MHz paired spectrum in the 2.6 GHz band (shared low power)	60	£6mn
E	9	5 MHz unpaired spectrum in the 2.6 GHz band	1	£0.1mn

Table 7. *Overview of auctioned lots for the simplified UK 4G Auction scenario*

Lot	Amount	Description	EPs	Start Price
A	6	2 × 5 MHz paired spectrum in the 800 MHz band	2250	£225mn
B	14	2 × 5 MHz paired spectrum in the 2.6 GHz band	150	£15mn
C	9	5 MHz unpaired spectrum in the 2.6 GHz band	1	£0.1mn

A.1. Licenses up for Sale

Table 6 illustrates the lots used in the auction. We simplified this band plan to allow for an easier analysis. The 800 MHz spectrum was split into two generic lots A(i) and A(ii) where A(ii) has twice as much bandwidth and eligibility points. Furthermore, the winner of A(ii) is obliged to use his spectrum to build a nationwide network. For simplicity, we neglected these legal details in our experiments and combined A(i) and A(ii) into one generic lot A with 6 licenses and the specifications of A(i).

The paired 2.6 GHz spectrum was split into three generic lots C, D(i), and D(ii) with amounts dependent on the bids submitted in the auction. D(i) and D(ii) are shared low power lots of different bandwidth whose winners will jointly use the same frequencies. The British auction rules allowed three different outcomes: First, up to 10 units of D(ii) are sold along with 10 units of C; second, up to 10 units of D(i) are sold along with 12 units of C, or third, the entire paired 2.6 GHz spectrum is sold in 14 units of C. Whichever allocation maximizes revenues wins. Based on the fact that 14 units of C were sold in the British auction and almost no bids containing shared low power lots were submitted, we discarded D(i) and D(ii) in our numerical experiments, as they did not seem to be important for this market.

For the unpaired 2.6 GHz spectrum (band E), only one lot with 9 units was used. However, the number of licenses that can actually be used is lower and dependent on the number of winners since one reserved block per winning package is required as a protection ratio between any two different users. In our simulations, we ignored this limitation and assumed 9 fully useable blocks as C is the least important band in the auction. The resulting list of bands used in our simulations can be found in Table 7.

In addition, a number of spectrum caps were imposed, some of which were also based on existing spectrum holdings in the British auction. For simplicity, we only used one simple spectrum cap that limits the amount of 800 MHz spectrum assigned to a single bidder to no more than 4 blocks or 2 × 20 MHz which is common in similar auctions.[8]

A.2. Bidders in the British 4G Auction

Seven bidders participated in the auction and five of them won at least one license. As expected, the most valuable target - a pair of A blocks for building a nationwide network with maximum reach - was won by the two big incumbents Vodafone and

---

[8] e.g. German 800 MHz/1.8 GHz/2.0 GHz/2.6 GHz auction in 2010 (Bundesnetzagentur (2010)), Danish 800 MHz auction in 2008 (Danish Business Authority (2012)).

Table 8. *Final allocation of the British 4G Auction in the simplified band plan*

	A (800 MHz paired)	B (2.6 GHz paired)	C (2.6 GHz unpaired)	
Vodafone	2	4	5	Primary Bidder
Telefonica	2	–	–	
Everything Everywhere	1	7	–	Secondary Bidder
Hutchison 3G	1	–	–	
Niche Spectrum Ventures	–	3	4	2.6 GHz Bidder
MLL Telecom	–	–	–	Small Bidder
HKT Company	–	–	–	

Telefonica. Table 8 shows the results of the British auction using the simplified lots introduced in the previous section.

Since all auction data has been made publicly available, the segmentation of participants into four generic bidder types is based not only on the results, but also on actual bidding behavior throughout the auction. As illustrated in Figure 7, there are four bidders who competed in all three bands: the primary bidders Vodafone and Telefonica as well as the secondary bidders Everything Everywhere and Hutchison 3G. The reason for separating them into two groups is the obvious strength of the primary bidders with regard to lot A. For these lots, the primary bidders' bids were much higher than the final clock prices in the supplementary round, compared to both secondary bidders.[9] The 2.6 GHz Bidder Niche Spectrum Ventures was focused on the 2.6 GHz lots only while the small bidders MLL Telecom and HKT Company only competed for the licenses in the C band.

A.3. A Value Model for the Simulations

Based on the public bid data, we derived a value model, i.e., valuations for each bidder, which allowed us to run simulations and estimate the impact of different bidding strategies in Section 6. First, we defined individual base valuations for each bidder indicating how much he is willing to pay for a single license in a band. Second, intra-band synergies were defined for any package with more than one license within the same band up to a certain limit (e.g., 2 blocks of A, 4 blocks of B or C). More licenses of the same band exhibit decreasing marginal value beyond these limits. For the expensive A blocks we even assumed that no bidder is interested in winning more than two licenses. Third, inter-band synergies were defined increasing the value of a bundle comprising licenses from bands A and B. The mean base valuations were defined based on the final clock prices and the supplementary bids and can be found in Table 9. The primary bidders had a much higher valuation in the A band compared to other bidders, while we assumed similar valuations for the B and C bands. Even though the true valuations of bidders are unknown, this allowed for a reasonable sensitivity analysis in Section 6.

---

[9] Since EE is the largest mobile service provider in the UK (Ofcom, 2011), it might be surprising to describe them as secondary bidders. However, the classification was solely made based on the bids in this particular auction.

17 (UN)EXPECTED BIDDER BEHAVIOR IN SPECTRUM AUCTIONS 365

Table 9. Mean base valuations for the national licenses scenario

	A (800 MHz paired)	B (2.6 GHz paired)	C (2.6 GHz unpaired)
Primary Bidder	£300mn	£70mn	£15mn
Secondary Bidder	£200mn	£70mn	£15mn
2.6 GHz Bidder	–	£70mn	£15mn
Small Bidder	–	–	£8mn

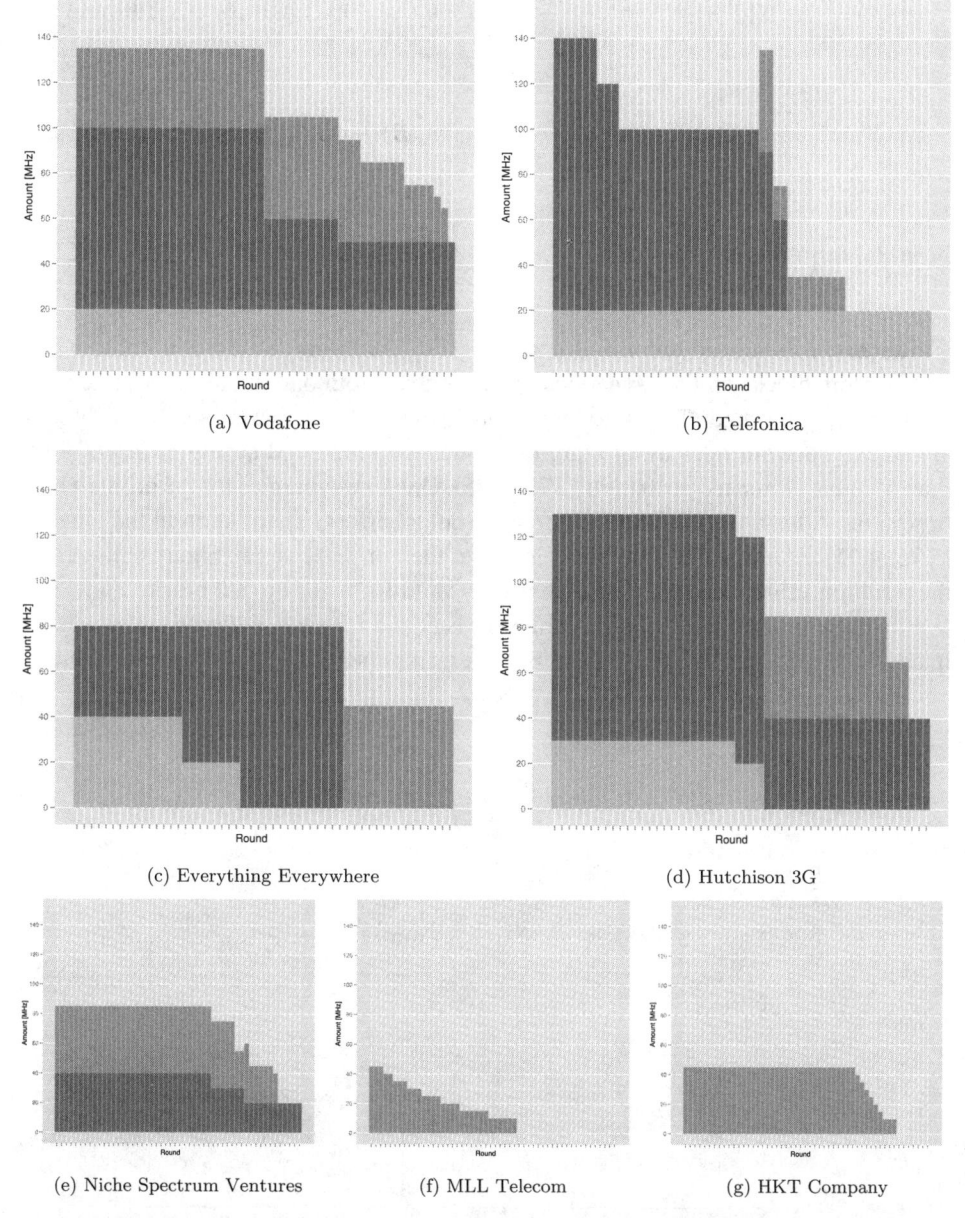

Figure 7. Visualization of bidding behavior throughout the clock rounds of the British 4G auction with a simplified band plan (A = light grey, B = dark grey, C = medium grey).

Based on the mean base valuations v in Table 9 the valuations for each simulation were determined based on two parameters, the relative strength of a bidder s_i and a random influence r_i. Both values are drawn from a uniform distribution in the interval [∗0.75, ∗1.25] and multiplied with the mean base valuations v for each band. For example, consider a primary bidder with relative strength 0.8 and random influence for blocks A and B of 1.1 and 1.0, resp. His valuations are $v_i(A) = 0.8 \cdot 1.1 \cdot £300mn = £264mn$ and $v_i(B) = 0.8 \cdot 1.0 \cdot £70mn = £56mn$. Then we determined the valuation $v_i(nX)$ for different bundles with n licenses within a band X.

$$v_i(nA) = \left(\min\{2, n\} + \min\left\{\frac{1}{2}, \frac{n-1}{n}\right\} \cdot syn_i(A) + \max\{0, ln(n-1)\} \right) \cdot v_i(A) \quad (1)$$

$$v_i(nB) = \left(\min\{4, n\} + \min\left\{\frac{3}{4}, \frac{n-1}{n}\right\} \cdot syn_i(B) + \max\{0, ln(n-3)\} \right) \cdot v_i(B) \quad (2)$$

$$v_i(nC) = \left(\min\{4, n\} + \min\left\{\frac{3}{4}, \frac{n-1}{n}\right\} \cdot syn_i(C) + \max\{0, ln(n-3)\} \right) \cdot v_i(C) \quad (3)$$

The first and second summand correspond to the linear increase in value when adding more blocks and the synergies on top of that. Both only increase in value until a threshold is reached. The third summand is only relevant when additional blocks are added, but with decreasing marginal value. All intra-band synergies $syn_i(X)$ are drawn from a uniform distribution [1.75; 2.25]. Only the synergies for the primary bidders in the A band were higher and drawn from a uniform distribution in the interval [3.75; 4.25] assuming that two blocks in A was their primary target. Figure 8 illustrates the resulting valuation function which is only valid for A for up to 2 blocks.

Finally, a uniformly distributed parameter is drawn for each bidder to determine his inter-band synergies for bands A and B. Synergies across these bands can be assumed

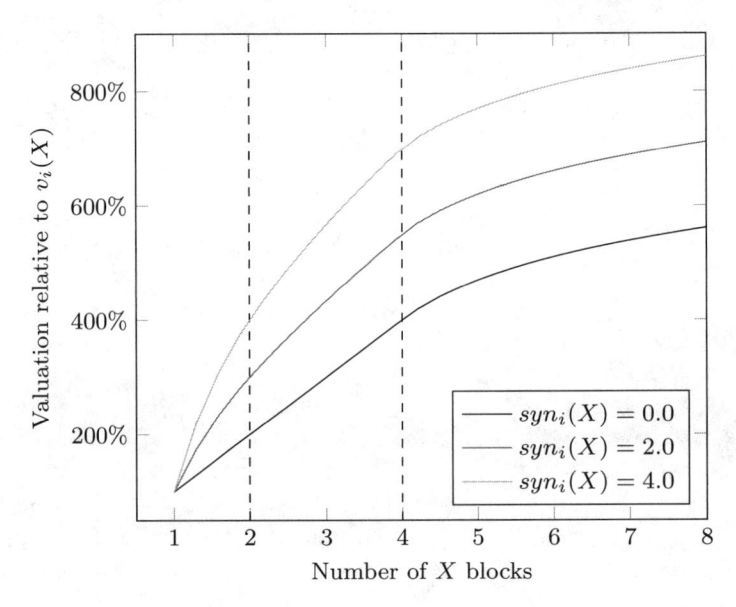

Figure 8. Plot of intra-band valuations.

to be much lower than intra-band synergies, and we use a uniform distribution in the interval [0.0; 0.2]. The valuation for a bundle containing licenses from bands A and B is now computed as the sum of the valuations for intra-band bundles multiplied with $(1 + syn_i)$. For example, a bidder's valuation for a bundle $ABBC$ comprised of one block of A, two blocks of B, and one license in band C is $v_i(ABBC) = (v_i(A) + v_i(2B) + v_i(C)) \cdot (1 + syn_i)$. Based on the random variables above, we generated 10 different instances of the value model.

The correlation between the supplementary bids in the British 4G auction and the valuations generated with the above model is 0.957, indicating that the generated valuations are a reasonable basis for our simulation study.

References

Afriat SN (1967) The construction of utility functions from expenditure data. International Economic Review 8(1):67–77

Afriat SN (1973) On a System of Inequalities in Demand Analysis: An Extension of the Classical Method. International Economic Review 14(2):460–472

Airiau S, Sen S (2003) Strategic bidding for multiple units in simultaneous and sequential auctions. Group Decision and Negotiation 12(5):397–413, DOI 10.1023/B:GRUP.0000003741.29640.ac, URL http://dx.doi.org/10.1023/B:GRUP.0000003741.29640.ac

Arthur C (2013) 4G auction to be investigated by audit office after poor return. URL www.theguardian.com/technology/2013/apr/14/4g-auction-national-audit-office

Ausubel L, Baranov O (2014) Market design and the evolution of the combinatorial clock auction. American Economic Review: Papers & Proceedings 104(5):446–451

Ausubel LM (2006) An efficient dynamic auction for heterogeneous commodities. The American economic review pp 602–629

Ausubel LM, Milgrom PR (2002) Ascending auctions with package bidding. Frontiers of Theoretical Economics 1(1):1–42, URL www.ausubel.com/auction-papers/ascending-proxy-auctions.pdf

Ausubel LM, Cramton P, Milgrom P (2006) The clock-proxy auction: A practical combinatorial auction design. In: Cramton P, Shoham Y, Steinberg R (eds) Combinatorial Auctions, MIT Press, chap 5, pp 115–138

Bellantuono N, Ettorre D, Kersten G, Pontrandolfo P (2013) Multi-attribute auction and negotiation for e-procurement of logistics. Group Decision and Negotiation pp 1–21, DOI 10.1007/s10726-013-9353-7, URL http://dx.doi.org/10.1007/s10726-013-9353-7

Bichler M, Shabalin P, Wolf J (2011) Efficiency, auctioneer revenue, and bidding behavior in the combinatorial clock auction. In: Second Conference on Auctions, Market Mechanisms and Their Applications (AMMA), New York, NY, USA

Bichler M, Shabalin P, Wolf J (2013a) Do Core-Selecting Combinatorial Clock Auctions always lead to high Efficiency? An Experimental Analysis of Spectrum Auction Designs. Experimental Economics 16:511–545

Bichler M, Shabalin P, Ziegler G (2013b) Efficiency with linear prices? A theoretical and experimental analysis of the combinatorial clock auction. INFORMS Information Systems Research pp 394–417

Bichler M, Goeree J, Mayer S, Shabalin P (2014) Spectrum auction design: Simple auctions for complex sales. Telecommunications Policy 38:613–622

Bundesnetzagentur (2010) Frequency Award 2010. URL www.bundesnetzagentur.de/cln_1931/EN/Areas/Telecommunications/Companies/FrequencyManagement/ElectronicCommunications Services/FrequencyAward2010_Basepage.html?nn=324044

Cramton P (2008) A Review of the L-band Auction. Tech. Rep. August, URL http://works.bepress.com/cramton/11/

Danish Business Authority (2012) Information Memorandum - 800 MHz Auction. URL http://erhvervsstyrelsen.dk/file/251159/information-memorandum-800mhz-auction.pdf

Day RW, Cramton P (2012) Quadratic core-selecting payment rules for combinatorial auctions. Operations Research 60(3):588–603

Goeree JK, Lien Y (2013) On the impossibility of core-selecting auctions. Theoretical Economics

Gul F, Stacchetti E (1999) Walrasian equilibrium with gross substitutes. Journal of Economic Theory 87(1):95–124

Janssen M, Karamychev V (2013) Gaming in combinatorial clock auctions. Games and Economic Behaviour 100:186–207

Jewitt I, Li Z (2008) Report on the 2008 uk 10-40 ghz spectrum auction. Tech. rep., URL http://stakeholders.ofcom.org.uk/binaries/spectrum/spectrum-awards/completed-awards/jewitt.pdf

Knapek S, Wambach A (2012) Strategic Complexities in the Combinatorial Clock Auction. CESifo Working Paper Series 3983, CESifo Group Munich, URL http://ideas.repec.org/p/ces/ceswps/_3983.html

Lehmann D, Müller R, Sandholm T (2006) The Winner Determination Problem. In: Cramton P, Shoham Y, Steinberg R (eds) Combinatorial Auctions, MIT Press, chap 12, URL www.cs.cmu.edu/~sandholm/winner-determination-final.pdf

Levin J, Skrzypacz A (2014) Are dynamic Vickrey auctions practical?: Properties of the combinatorial clock auction. Stanford University Working Paper September

Milgrom P (2000) Putting Auction Theory to Work: The Simultaneous Ascending Auction. DOI 10.1086/262118

Milgrom P (2004) Putting Auction Theory to Work. Cambridge University Press

Milgrom PR, Weber RJ (1982) A theory of auctions and competitive bidding. Econometrica 50(5):1089–1122

Ofcom (2011) Everything Everywhere becomes the UKs largest network in terms of revenue. URL http://stakeholders.ofcom.org.uk/market-data-research/market-data/communications-market-reports/cmr11/telecoms-networks/5.48

Ofcom (2012) The Wireless Telegraphy (Licence Award) Regulations 2012. URL www.legislation.gov.uk/uksi/2012/2817/contents/made

Ofcom (2013a) 800 MHz & 2.6 GHz Auction Data. URL http://stakeholders.ofcom.org.uk/spectrum/spectrum-awards/awards-archive/completed-awards/800mhz-2.6ghz/auction-data/

Ofcom (2013b) 800 MHz & 2.6 GHz Combined Award. URL http://stakeholders.ofcom.org.uk/spectrum/spectrum-awards/awards-archive/completed-awards/800mhz-2.6ghz/

Rothkopf MH (2007) Thirteen reasons why the Vickrey Clarke Groves process is not practical. Operations Research 55(2):191–197

RTR (2013) Multiband Auction 800/900/1800 MHz. Rundfunk und Telekom Regulierungs-GmbH, URL www.rtr.at/en/tk/multibandauktion

Samuelson PA (1938) A note on the pure theory of consumer's behaviour. Economica 5(17):61–71, URL www.jstor.org/stable/10.2307/2548836

Sano R (2012) Non-bidding equilibrium in an ascending core-selecting auction. Games and Economic Behavior 74:637–650

Scheffel T, Ziegler A, Bichler M (2012) On the impact of package selection in combinatorial auctions: An experimental study in the context of spectrum auction design. Experimental Economics 15:667–692

Shapiro R, Holtz-Eakin D, Bazelon C (2013) The economic implications of restricting spectrum purchases in the incentive auctions. Georgetown University Washington Working Paper

Smeulders B, Spieksma FCR, Cherchye L, De Rock B (2012) Goodness of fit measures for revealed preference tests: Complexity results and algorithms. aghedupl V(212):1–16, URL http://home.agh.edu.pl/~faliszew/COMSOC-2012/proceedings/paper_11.pdf

Smith C (2013) Did Ofcom's 4G auction rules cost the UK an extra 3 billion? URL www.techradar.com/news/phone-and-communications/did-ofcom-s-4g-auction-rules-cost-the-uk-an-extra-3-billion-1138302

Varian HR (1990) Goodness-of-fit in optimizing models. Journal of Econometrics 46:125–140, URL www.sciencedirect.com/science/article/pii/030440769090051T

Varian HR (2006) Revealed Preference. In: Szenberg M, Ramrattan L, Gottesman AA (eds) Samuelsonian economics and the twenty-first century, January 2005, Oxford University Press, Oxford, chap 6, pp 99–115

PART III
Alternative Auction Designs

CHAPTER 18

A Combinatorial Auction Mechanism for Airport Time Slot Allocation*

Stephen J. Rassenti, Vernon L. Smith, and Robert L. Bulfin

1. The Problem of Allocating Airport Slots

In 1968 the FAA adopted a high density rule for the allocation of scarce landing and take-off slots at four major airports (La Guardia, Washington National, Kennedy International, and O'Hare International). This rule establishes slot quotas for the control of airspace congestion at these airports.

Airport runway slots, regulated by these quotas, have a distinguishing feature which any proposed allocation procedure must accommodate: an airline's demand for a take-off slot at a flight originating airport is not independent of its demand for a landing slot at the flight destination airport. Indeed, a given flight may take off and land in a sequence of several connected demand interdependent legs. For economic efficiency it is desirable to develop an airport slot allocation procedure that allocates individual slots to those airline flights for which the demand (willingness to pay) is greatest.

Grether, Isaac, and Plott (hereafter, GIP) (1979, 1981) have proposed a practical market procedure for achieving this goal. Their procedure is based upon the growing body of experimental evidence on the performance of (1) the competitive (uniform-price) sealed-bid auction and (2) the oral double auction such as is used on the organized stock and commodity exchanges. Under their proposal an independent primary market for slots at each airport would be organized as a sealed-bid competitive auction at timely intervals. Since the primary market allocation does not make provision for slot demand interdependence, a computerized form of the oral double auction (with block transaction capabilities) is proposed as an "after market" to allow airlines to purchase freely and sell primary market slots to each other. This continuous after market exchange would provide the institutional means by which individual airlines would acquire those slot packages which support their individual flight schedules. Thus, an airline that acquired slots at Washington National which did not flight-match the slots acquired at O'Hare could either buy additional O'Hare slots or sell its excess

---

* We are grateful to the National Science Foundation for research support under a grant to the University of Arizona (V.L. Smith, principal investigator).

Washington slots in the after market. Although GIP's proposed after market permits airlines to exchange slots freely and thereby acquire the appropriate slot packages, it suffers from two disadvantages.

(1) Individual airlines may experience capital losses and gains in the process of trading airport slots in the after market. Thus, an airline with an excess of A slots and a deficiency of B slots may discover in the after market that the going price of B slots is unprofitably high (for that particular airline), while excess A slots can be sold only at a loss.
(2) It costs resources to trade in the after market. Hence, to the extent that slots are not allocated to the appropriate packages in the primary market, the cost of participating in the combined primary-after market mechanism is increased.

Ideally, the primary market would allocate slots in the appropriate packages initially with the after market performing only two functions (i) marginal corrections in primary market misallocations, and (ii) slot allocation adjustments due to new information not available at the time of the primary auction. Thus, a sudden grounding of all DC-10 aircraft would leave Continental Airlines with a surplus of O'Hare runway slots, which could be sold in the after market to airlines not affected by the DC-10 grounding.

In this article, we address the problem of designing a "combinatorial" sealed-bid auction to serve as the primary market for allocating airport slots in flight-compatible packages for which individual airlines would submit package bids. The objective is to allocate slots to an individual airline *only* in the form of those combinations and subject to those contingencies that have been prespecified by the airline.

2. The Auction Optimization Mechanism

To increase the overall efficiency of the slot allocation mechanism suggested by GIP (1979), and to decrease its reliance on an after market, we have developed an optimization model with the following features for use in a computer-assisted primary sealed-bid auction market: (a) direct maximization of system surplus in the criterion function; (b) airport coordination through consideration of resource demands in logically packaged sets; (c) scheduling flexibility through contingency bids on the part of airlines.

Consider the following integer programming problem:

$$(P) \begin{cases} \text{Maximize} & \sum_j c_j x_j \\ \text{Subject to:} & \sum_j a_{ij} x_j \leq b_i \ \forall \, i, \\ & \sum_j d_{kj} x_j \leq e_k \ \forall \, k, \\ & x_j \in \{0, 1\}; \end{cases}$$

where

$i = 1, \ldots, m$ subscripts a resource (some slot at some airport);
$j = 1, \ldots, n$ subscripts a package (set of slots) valuable to some airline;
$k = 1, \ldots, l$ subscripts some logical constraint imposed on a set of packages by some

airline;

$$a_{ij} = \begin{cases} 1 & \text{if package } j \text{ includes slot } i, \\ 0 & \text{otherwise;} \end{cases}$$

$$d_{kj} = \begin{cases} 1 & \text{if package } j \text{ is in logical constraint } k, \\ 0 & \text{otherwise;} \end{cases}$$

e_k = some integer ≥ 1,
c_j = the bid for package j by some airline.

The contingency bids expressed in the set of logical constraints have one of two format types: "Accept no more than p of the following q packages." or "Accept package V only if package W is accepted." The first type is identical in format to any of the resource constraints. For example, suppose an airline bids c_a on package a, c_b on package b, and specifies either a or b, but not both. The added constraint is then written $x_a + x_b \leq 1$. The second type can be converted to this format through a simple variable transformation. For example, suppose an airline bids as in the previous example, but specifies b only if a. By creating the package ab with $c_{ab} = c_a + c_b$, package b can be eliminated from consideration. The added constraint becomes $x_a + x_{ab} \leq 1$.

In a manner analogous to the parallel independent slot auctions suggested by GIP (1979), sealed bids (c_j) for packages (j) and the contingency constraints (k) specified by each airline are used to parameterize the model and determine an "optimal" primary allocation. The problem which results is recognized as a variant of the set packing problem with general right-hand sides. It can be solved, as was done for the experiments reported in Section 3, with a specialized algorithm developed by Rassenti (1981). A problem of the enormous dimensions dictated by even a four-city application (perhaps 15,000 constraints and 100,000 variables) will present a significant challenge for the finest configuration of hardware and software available. Fortunately, a practicable solution within 1 or 2% of the linear optimum, and very often the optimum itself in the discrete solution set, is almost assuredly achievable in a reasonable amount of time.

Given the solvability of P and its potential for ensuring an efficient primary allocation, there remain several questions: how to induce bidding airlines to reveal their true values; how to price allocated slots; and how to divide income among the participating airports. We suggest a resolution of these concerns with the following procedure: (1) Determine a complete set of marginal (shadow) prices, one for each slot offered. (2) Charge any airline whose package j was accepted in the solution to P a price for j equal to the sum of the marginal prices for the slots in that package. This provides the uniform price feature that has demonstrated good demand revelation behavior in single commodity experiments (GIP, 1979). (3) Return to any airport whose slot i was included in some accepted package j an amount equal to the marginal price for i. Such a scheme will guarantee that the price paid for an accepted package is less than (or rarely equal to) the amount bid for that package.

If problem P were a linear program, the determination of the suggested set of shadow prices would be a trivial and well-solved matter. Discrete programming problems, however, present special difficulties with respect to shadow pricing. Consider, for example,

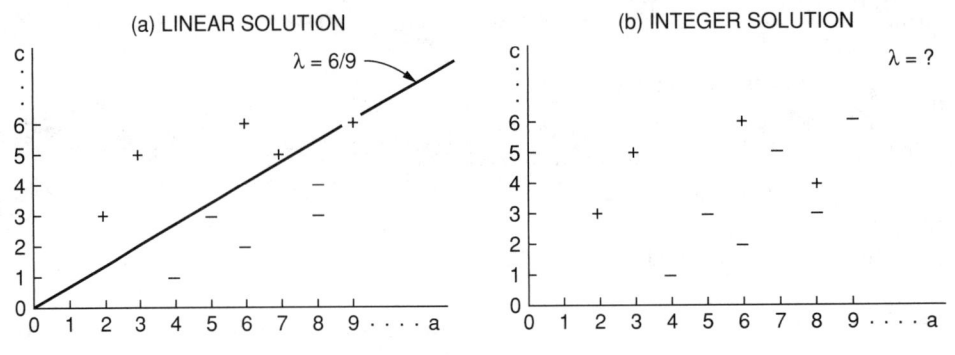

Figure 1.

(a) LINEAR SOLUTION, $\lambda = 6/9$

(b) INTEGER SOLUTION, $\lambda = ?$

a discrete project selection problem with a single resource constraint:

$$(K) \begin{cases} \text{Maximize} & 5X_1 + 3X_2 + 6X_3 + 5X_4 + 6X_5 + 3X_6 \\ & + 4X_7 + 3X_8 + 2X_9 + X_{10} = Z; \\ \text{Subject to:} & 3X_1 + 2X_2 + 6X_3 + 7X_4 + 9X_5 + 5X_6 \\ & + 8X_7 + 8X_8 + 6X_9 + 4X_{10} \leq 24; \\ & X_j \in \{0, 1\} \; \forall \, j = 1, \ldots, 10. \end{cases}$$

If the choice space for X_j is relaxed to its linear programming equivalent, $0 \leq X_j \leq 1$, then the solution is trivially given by $(Z, X_1, X_2, \ldots, X_{10}) = (23, 1, 1, 1, 1, .66, 0, 0, 0, 0, 0)$. The critical return rate, $\lambda = 6/9$ for project 5, is the optimal Lagrangian multiplier or shadow price for the resource. But the discrete problem has the optimal solution $(21, 1, 1, 1, 0, 0, 1, 1, 0, 0, 0)$, and obviously no critical ratio exists which separates projects that are chosen from those that are not. Figure 1 illustrates these solutions.

In the traditional sense, Lagrangian multipliers for an integer program may not exist;[1] that is, no set of prices will support the optimal division of packages into accepted and rejected categories. Therefore, it is possible that a package bid that is greater than its shadow resource cost will be rejected, while another package bid that is also greater than its shadow cost is accepted. In the experiments reported in Section 3, we provided subject bidders with a guideline explanation of these cases. This allows subjects to select strategic or best reply (Cournot) responses if they wish.

With this problem in mind, the following two pseudo-dual programs to P were developed to define bid rejection prices (problem D_R) and acceptance prices (problem D_A) that will serve as bidding guidelines for individual agents.

$$(D_R) \begin{cases} \text{Minimize} & \sum_R y_r \\ \text{Subject to:} & \sum_j w_i a_{ij} \leq c_j \; \forall \, j \in A, \\ & y_r \geq c_r - \sum_i w_i a_{ir} \; \forall \, r \in R, \\ & y_r \geq 0, \quad w_i \geq 0; \end{cases}$$

---

[1] Wolsey (1981) presents a "state of the art" discussion of price functions in integer programming.

where

the optimal solution to P is $\{x_j^*\}$;
the set of accepted packages is $A = \{j|x_j^* = 1\}$;
the set of rejected packages is $R = \{r|x_r^* = 0\}$;
the set of lower bound slot prices (prices charged) to be determined is $\{w_i^*\}$;
the amount by which a rejected bid exceeds the market price (if at all) is y_r.

$$(D_A) \begin{cases} \text{Minimize} & \sum_A y_j \\ \text{Subject to:} & \sum_i v_i a_{ir} \geq c_r \ \forall \, r \in R, \\ & y_j \geq \sum_i v_i a_{ij} - c_j \ \forall \, j \in A, \\ & y_j \geq 0, \quad v_i \geq 0; \end{cases}$$

where

the set of upper bound slot prices to be determined is $\{v_i^*\}$;
noindent the amount by which an accepted bid is below the upper bound slot prices (if at all) is y_j.

Problem D_A is the complement of D_R with respect to the accept-reject dichotomy. If unambiguous separating prices exist, the solutions to D_A and D_R coincide. In Figure 2 the analogous pseudo-dual problems for the project selection problem K above are schematically solved for the obvious upper and lower bound return ratios.

The following categorization of bids can now be made. (i) If a bid was greater than the sum of its component values in the set $\{v_i^*\}$, it was definitely accepted. (ii) If a bid was less than the sum of its component prices in the set $\{w_i^*\}$, it was definitely rejected. (iii) All bids in between were in a region where acceptance or rejection might be considered independent of relative marginal value and determined by the integer constraints on efficient resource utilization. The bids in category (iii) correspond to the core of the integer programming problem P. They comprise a small percentage of all bids and are known to decrease in relative number as problem size increases. Figure 3 gives the regions analogous to (i), (ii), and (iii) for the project selection example.

What can be said about the theoretical incentive properties of this proposed computer assisted sealed-bid auction? Certainly it is *not* generally incentive compatible;

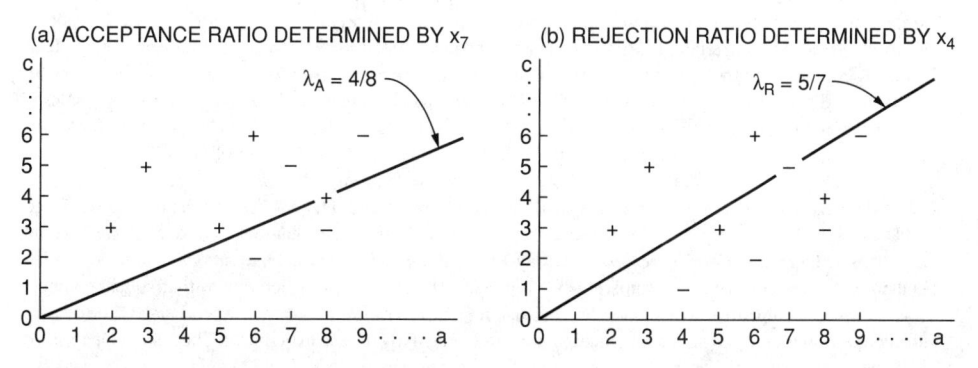

Figure 2.

Categorization for problem K

[Figure 3: Graph showing lines $\lambda_R = 5/7$, $x_j = 1$ (i); $x_j = ?$ (ii); $\lambda_A = 4/8$, $x_j = 0$ (iii)]

Figure 3.

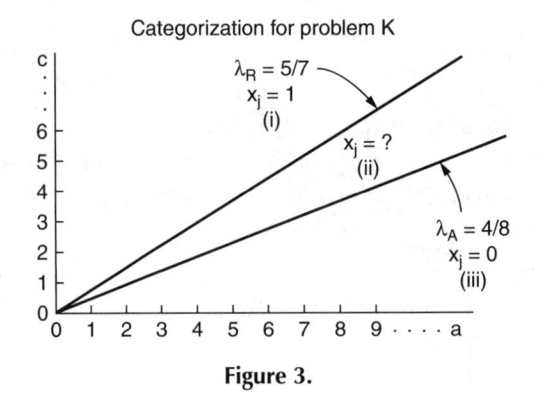

that is, if any bidder desires to acquire multiple units of any given package or multiple units of the same slot, then the door is open to the possibility of strategically underbidding the true value of certain packages (Vickrey 1961). However, strategic behavior is fraught with risks for the individual, even in simple multiple unit auctions for a single commodity, because individuals do not know the bids and the true valuations of their competitors.[2] We conjecture that this is why GIP do not observe significant underrevelation of demand in laboratory experiments with one commodity auctions.

Since the combinatorial auction we suggest for the airport slot problem is far more complex than any of the single commodity auctions that have been studied, we would expect to observe at least as much demand-revealing behavior in our auction as in the others. Since this is both an open and a behavioral question, we devised a laboratory experimental design to compare our combinatorial auction procedure with the procedure proposed in GIP (1979).

3. Experimental Results

Eight experiments were conducted using students with economics and engineering backgrounds. Each experiment consisted of a sequence of market periods in which

---

[2] A referee suggests that "if this mechanism were implemented, it would be used over a long period of time with substantial sums at stake and one would expect some investment in learning about the use of strategic maneuvers." Some comment on this view is important because it represents a widely shared belief among economists. We think it is at least as likely—the evidence seems to suggest that it is more likely—that the airlines would compete away any rents which they now capture from airport slot resources. In fact, a very reasonable hypothesis, given the immense uncertainty as to what airport slot combinations are actually worth (these are well defined in our experiments), might be that airline bids would actually exceed those levels that would sustain long-run profitability. If the recent, and continuing, vigorous price competition among airlines in passenger ticket sales is any indication, then this last hypothesis is quite likely to be supported. Braniff Airlines has just filed for Chapter 11 bankruptcy in an environment in which "it has been widely asserted that part of Braniff's financial ills stemmed from steep fare cuts to raise ridership—at the expense of profit" (*Wall Street Journal*, May 17, 1982, p. 4). In less than a week after Braniff's collapse, Midway Air announced entry and "set the fare between Chicago and Dallas-Fort Worth at a cut rate of $89 one way, despite industry hopes that the collapse of Braniff would end extensive fare wars" (*Wall Street Journal*, May 17, 1982, p. 1). These field observations of noncooperative behavior are consistent with the results of several hundred experiments in which two to four sellers compete away all except competitive rents in a variety of distinct pricing institutions. In the experiments reported here subjects were quite active in attempting manipulative strategies early in each experiment. These strategies tended to be abandoned over time, as indicated by the tendency for resource prices (Table 3) to increase in successive periods.

objective economic conditions remained constant in successive periods for the six participants. The first period was always considered a learning period (no payoff), and the number of periods completed was time dependent (3-hour limit). Individual subjects were paid the difference between the assigned redemption value of packages and the prices paid for packages in the market (Smith 1976). Subjects' earnings varied between $8 and $60, depending on individual endowed valuations and subject and group bidding behavior.

A 2 × 2 × 2 experimental design was employed. The three factors were: (1) GIP (control) versus our RSB (treatment) primary auction; (2) subjects "experienced" versus "inexperienced;" (3) combinatorial "complexity" (easy versus difficult) of resource utilization.

The GIP mechanism employed copied that suggested and used by GIP (1979). The first two RSB experiments used marginal package pricing and the "easy" combinatorial design, while the second two used the marginal item pricing scheme described in Section 2 and the "difficult" combinatorial design.[3] The term "experience" indicates previous participation in either a control or treatment version of an experiment, but even "inexperienced" subjects generally had some experience with other less complicated decisionmaking experiments such as a single commodity auction. Combinatorial "complexity" refers to the degree of difficulty a subject would encounter in attempting to shuffle slots from one package use to another for the purpose of making a redemption claim or a purchase or sale in the after market. It has two components: the amount of package repetition among various agents and the amount of item repetition within any agent's packages.

Table 1 gives the observed efficiency after primary and secondary markets for each cell of the design. The trial (zero) period results are not listed. Efficiency is defined as total subject payoff (realized system surplus) divided by total theoretical payoff (system surplus computed by assuming full demand revelation). The data support several important hypotheses. The overall efficiency of the RSB mechanism is generally greater than that of the GIP mechanism. It is achieved without the uniformly strong dependence on the secondary market displayed by the GIP mechanism. The ability of subjects to include contingency bids in appropriate situations, though not included in the experiments conducted, should serve to accentuate this effect. Market experience seems a significant factor in determining the efficiency of either mechanism. Learning is in evidence during the multiperiod course of each experiment. The RSB mechanism, however, seems to require less learning and displays quicker achievement of high efficiency. This fact suggests that the RSB mechanism will adapt more efficiently to changing economic conditions—an important criterion in judging the performance of any mechanism.

The sample size is too small, with one observation per cell, to test for the significance of each "treatment." But by aggregation across all treatments except the bidding mechanism, we can report a nonparametric sign test of the null hypothesis that the

---

[3] Appendix A contains the instructions and forms used in the experiments in which the primary market priced items marginally. The instructions and forms used in our GIP experiments and in the two RSB experiments that priced packages marginally can be obtained by writing Smith at the University of Arizona. Appendix B contains the contrasting package valuation designs used in the "easy" and "difficult" combinatorial treatment.

Table 1. Efficiency, by Period and Treatment Condition, in the Primary and After Market

GIP

	Period	Primary Market, P	After Market, A	Period	Primary Market, P	After Market, A
Experienced Subjects	1	0.904	0.974	1	0.609	0.730
	2	0.871	0.920	2	0.695	0.864
	3	0.871	0.953	3	0.709	0.851
	4	0.907	0.983	4	0.752	0.903
				5	0.804	0.919
				6	0.795	0.969
Inexperienced Subjects	1	0.853	0.861	1	0.721	0.917
	2	0.778	0.942	2	0.726	0.831
	3	0.650	0.865	3	0.602	0.798
	4	0.685	0.911	4	0.408	0.829
	5	0.763	0.907	5	0.463	0.923
				6	0.465	0.902
		Easy			Difficult	

RSB

	Period	Primary Market, P	After Market, A	Period	Primary Market, P	After Market, A
Experienced Subjects	1	0.832	0.923	1	0.986	0.986
	2	0.898	0.944	2	0.978	0.987
	3	0.935	0.973	3	0.985	0.985
	4	0.971	0.971	4	0.985	0.985
	5	0.986	0.986	5	0.986	0.986
				6	0.991	0.993
Inexperienced Subjects	1	0.884	0.923	1	0.951	0.965
	2	0.918	0.951	2	0.860	0.940
	3	0.936	0.977	3	0.976	0.979
	4	0.967	0.977	4	0.931	0.931
	5	0.869	0.870	5	0.984	0.984
		Easy			Difficult	

fourth period difference in efficiency between RSB and GIP is equally likely to be positive or negative, as against the research hypothesis that there is a positive difference. From Table 1, for the primary market, all four paired differences are positive, and the null hypothesis is rejected at $p = .0625$. In the after market, three of the four paired differences are positive, and the null hypothesis can be rejected only at $p = .25$. This is consistent with our prior expectation that the principal advantage of RSB over GIP is to improve primary market allocation sufficiently to make after market exchange unnecessary. Thus, comparing after market efficiency with primary market efficiency in period 5 across all RSB experiments, we observe no difference.

The more detailed breakdown of surplus presented in Table 2 adds further support to these observations. Uniformly more agents were in debt after the GIP primary market. On several occasions, an agent who needed to participate in multilateral trades for gains in the GIP secondary market was caught short of completion (e.g., GIP-inexperienced-easy period 1). This is a difficulty to be reckoned with in any independent slot marketing scheme.

Speculation is defined as the purchase of a package by an individual for whom the package has no redemption value. Such a purchase can be profitable only through resale at a higher price in the after market. Because of the extremely high efficiency realized in the primary RSB allocation, speculative behavior is very risky (e.g., RSB-inexperienced-easy period 5 and difficult period 4). Primary prices and allocations are too near optimal to yield much speculative profit, and combinatorial problems are encountered with after market multilateral trade. Table 2 also reinforces the notion that experience is less important in the RSB mechanism. Under the more incentive-compatible conditions of RSB, there was 95% demand revelation for nonspeculative bids in the final period.

The difficulty with determining item values by independent auctions is emphasized by Table 3, which traces prices by period. The absolute deviation from theoretical values is much larger for the GIP mechanism. In fact, the task of estimating item value from package redemption value proved too much for inexperienced GIP subjects under difficult conditions where a market collapse occurred. Unless there is demand overrevelation (bidding in excess of value), it is impossible for this situation to occur in the RSB mechanism, where the optimization routine makes an "intelligent" pricing decision.

Finally, combinatorial complexity seems to lower GIP mechanism efficiency by a significant amount, while the performance of the RSB mechanism appears not to deteriorate and perhaps even to improve. This is to be expected if we are correct in our conjecture that the decision costs potentially associated with this factor are borne by the computer in the RSB mechanism.

4. Alternatives for Implementing an Airport Slot Auction

To our knowledge, this study constitutes the first attempt to design a "smart" computer-assisted exchange institution. In all the computer-assisted markets known to us in the field, as well as those studied in laboratory experiments, the computer passively records bids and contracts and routinely enforces the trading rules of the institution. The RSB mechanism has potential application to any market in which commodities are composed of combinations of elemental items (or characteristics). The distinguishing

Table 2. Breakdown of Surplus by Market, Period, and Treatment Condition (no. agents in debt/negative agents' surplus/positive agents' surplus)

	GIP		RSB		
Period	Primary	After	Primary	After	
1	1/ −.17/36.11	0/ 0.00/45.20	1/ −1.18/33.66	0/ .00/44.57	
2	1/ −4.15/33.64	1/ −.63/36.70	0/ .00/40.69	0/ .00/46.86	
3	4/ −5.11/10.70	1/ −1.77/18.24	1/ −1.83/43.64	0/ .00/46.79	Experienced
4	3/ −352/15.55	0/ .00/21.32	0/ .00/41.75	0/ .00/41.75	Subjects
5			0/ .00/43.04	0/ .00/43.04	
6					Easy
1	0/ .00/18.95	1/ −3.71/23.66	1/ −.39/27.18	0/ .00/31.92	
2	3/ −10.54/ 6.99	2/ −2.99/21.19	0/ .00/32.02	0/ .00/36.39	
3	3/ −30.00/ 3.06	3/ −9.54/11.18	0/ .00/30.12	0/ .00/35.66	Inexperienced
4	4/ −34.30/ 2.86	3/ −5.84/ 4.31	0/ .00/31.83	0/ .00/33.22	Subjects
5	4/ −15.90/ 3.57	2/ −6.36/13.13	1/ −5.17/25.13	1/ −5.17/25.29	
6					
1	5/ −11.18/ 1.86	3/ −4.63/10.36	0/ .00/34.52	0/ .00/34.52	
2	4/ −7.65/ 7.54	1/ −1.19/32.44	0/ .00/28.46	0/ .00/29.58	
3	4/ −8.84/11.84	1/ −.13/20.80	0/ .00/21.56	0/ .00/21.56	Experienced
4	2/ −12.30/14.43	1/ −2.70/23.52	0/ .00/19.53	0/ .00/19.53	Subjects
5	2/ −2.75/ 9.38	2/ −.41/21.28	0/ .00/18.94	0/ .00/18.94	
6	3/ −10.27/ 9.71	0/ .00/21.02	0/ .00/15.04	0/ .00/15.21	
					Difficult
1	3/ −31.20/ 1.67	3/ −7.39/ 2.04	1/ −.71/26.72	0/ .00/27.78	
2	4/ −37.68/ .00	4/ −27.07/ 2.32	3/ −6.29/ 7.28	1/ −1.93/12.79	
3	5/ −60.51/ .00	5/ −36.69/ .34	0/ .00/11.30	0/ .00/11.59	Inexperienced
4	5/ −87.40/ .00	5/ −35.27/ .00	1/ −6.51/13.74	1/ −6.51/13.74	Subjects
5	4/ −7.05/ .00	3/ −23.13/ 3.09	0/ .00/15.21	0/ .00/1521	
6	3/ −18.82/19.26	0/ .00/54.65			

Table 3. Market Prices by Period under Difficult Conditions (using marginal item pricing in RSB mechanism)

			GIP							RSB				
		Item Prices							Item Prices					
Period	A	B	C	D	E	F		A	B	C	D	E	F	
1	2.30	2.50	2.00	2.10	1.00	2.00		2.27	2.56	2.11	2.39	1.06	1.73	Experienced
2	2.31	2.75	2.00	2.26	1.00	2.00		3.07	2.58	2.24	2.67	.94	1.93	Subjects
3	2.50	2.76	2.00	2.01	1.00	2.00		3.10	3.03	2.78	2.63	.95	1.88	
4	3.00	3.00	2.00	2.26	.50	2.26		3.35	2.90	2.95	2.70	1.15	1.60	
5	3.00	3.05	2.00	2.50	.25	2.50		2.56	3.19	2.98	2.74	1.17	1.81	
6	3.00	3.25	2.50	2.76	.50	2.00		2.88	3.13	3.23	2.57	1.63	1.97	
Theoretical Prices	2.66	3.16	3.09	2.66	2.51	2.49		2.66	3.16	3.09	2.66	2.51	2.49	
1	3.00	2.75	2.50	3.00	2.75	3.00		1.96	3.64	2.44	2.29	.71	2.08	Inexperienced
2	3.20	3.20	3.00	3.10	2.90	3.00		2.25	3.80	2.50	3.32	1.25	2.13	Subjects
3	3.50	4.00	3.30	3.50	2.00	2.90		3.24	3.93	2.78	2.97	1.29	1.47	
4	4.00	5.00	4.00	3.30	1.50	2.50		2.89	3.78	2.73	3.03	1.59	1.58	
5	4.51	5.00	4.51	3.20	1.00	1.00		2.76	3.49	2.80	2.95	1.86	1.39	
6	4.51	.01	3.50	.01	.01	.01		2.78	3.47	2.77	2.79	2.02	1.97	
Theoretical Prices	2.66	3.16	3.09	2.66	2.51	2.49		2.66	3.16	3.09	2.66	2.51	2.49	

feature of our combinatorial auction is that it allows *consumers* to define the commodity by means of the bids tendered for alternative packages of elemental items. It eliminates the necessity for producers to anticipate, perhaps at substantial risk and cost, the commodity packages valued most highly in the market. Provided that bids are demand revealing, and that income effects can be ignored, the mechanism guarantees Pareto optimality in the commodity packages that will be "produced" and in the allocation of the elemental resources. The experimental results suggest that: (a) the procedures of the mechanism are operational, i.e., motivated individuals can execute the required task with a minimum of instruction and training; (b) the extent of demand underrevelation by participants is not large, i.e., allocative efficiencies of 98–99% of the possible surplus seem to be achievable over time with experienced bidders. This occurred despite repeated early attempts by inexperienced subjects to manipulate the mechanism and to engage in speculative purchases.

The problem of allocating airport time slots requires improved methods (GIP, 1981), and the problem has grown from bad to worse in the aftermath of the recent strike attempt by the air traffic controllers. We think the RSB mechanism, or some variant that might be developed from it, has potential for ultimate application to the time slot problem. But as we view it, before such an application can or should be attempted, at least two further developments are necessary. First, at least two additional series of experiments need to be completed. Another series of laboratory experiments should be designed, using larger numbers of participants, resources, and possible package combinations. The subjects in these new experiments should be the appropriate operating personnel of a group of cooperating airlines. Depending on the results of such experiments, the next step might be to design a limited scale field experiment with only a few airports and airlines.

Second, there should be extensive discussion and debate within the government, academic, and airline communities concerning alternative means of implementing the combinatorial auction. There is a wide range of choice here. Our discussion, as well as the reported experiments, were based on the assumption that airline bids would be denominated in U.S. currency and that the revenue would be allocated to the airports. There are, however, other alternatives; we offer just a few to stimulate discussion. (1) If it is believed that airport revenue should not be based on the imputed rents from scarce time slots, then bids in the combinatorial auction could be denominated in "slot currency" or vouchers issued in fixed quantities to each airline. These vouchers could be freely bought and sold among the airlines but would only be redeemable in time slots. (2) Alternatively, each airline could continue to be given some "historical" allocation of slots, with the RSB mechanism modified to become a two-sided sealed bid-offer combinatorial auction. In such an auction each airline would submit package bids for slots to be purchased, and package offers of slots to be sold. Under this form of implementation, the rent imputed to airport slots would of course be retained as "revenue" by the airlines. (3) An important question not addressed in either the GIP or RSB procedures is the pricing of airline seats, which directly affects the willingness-to-pay for airport slots. We would suggest that the idea of a computerized continuous double (bid-offer) auction of seats be considered along with the combinatorial auctioning of slots. All the major airlines are computerized down to the boarding gate, so that the computerized trading of seats may be technically feasible, and could provide a more flexible means of increasing airline revenue while lowering passenger cost through improved load factors. (4) Finally, we should note that we think there may be an inherent contradiction in the attempt to allow free (deregulated) entry by the airlines, but not permit free entry by airports. Ultimately, a pricing system for airport slots, which returns revenue to the airports, could allow not only for package bids from the airlines, but slot price offers from the airports, with each airport subject to competition from new regional, suburban, and national airports.

Appendix A

Instructions

RSB Instructions. This is an experiment in the economics of market decisionmaking. Various research organizations have provided funds for the conduct of this research. The instructions are simple, and if you follow them carefully and make good decisions, you might earn a considerable amount of money, which will be paid to you in cash after the experiment. In this experiment we are going to conduct two kinds of markets to distribute six distinct items among you in a sequence of periods or market days. The six distinct items are represented by the letters: A, B, C, D, E, F. At the end of the experiment we shall redeem (that is, buy) certain packages of items you have acquired during each period. The amounts to be paid to you as an individual can be determined from your payoff sheet included with the instructions. The payoff tables may differ among individuals. This means that the patterns of payments differ and the monetary amounts may not be comparable. The first market is the primary market and is of the sealed-bid type. In this market each of you may bid to buy items offered in fixed quantities. The second market will be a secondary market of the oral-bid-offer type. In this market you may buy or sell items obtained in the primary market to one another if you wish. Alternatively, you may simply keep what you have for the experimenters to redeem. In all sales, whether to the experimenters or to other participants, you may keep any profits you earn. For each sale you make, your profits are computed as follows: your earnings = sale price − purchase price.

Redemption Values. In your folder there is a sheet labelled "Redemption Values." This sheet indicates the amount the experimenter will pay you for given packages of items at the end of the period. Suppose for example you ended the period with 2A, 2C, and 2F items, and your redemption values were as follows:

package	\multicolumn{6}{c}{items included}	value
	A	B	C	D	E	F	
1	1	1	1				1.20
2	1		1				.40
3	1		1			1	1.20
4	1					1	.72
5			1			1	.70
6	1					1	.60

Since you may claim each item in only one package, you may legitimately claim for the set [AC, CF, AF] which will redeem .40 + .70 + .72 = $1.82. But the set [ACF, CF] with one leftover item A is a better claim since it redeems 1.20 + .70 = $1.90. In this case your period profit may have been increased by previously selling off the leftover item *A* in the secondary market.

Primary Market. Each period there will be a limited number of units of each kind of item available. As a buyer you purchase packages of one or more units by submitting bids which may be accepted or rejected. You will decide each period how many bids to submit for which packages in what amounts. Suppose you wish to bid for packages *AF* at .72, *AF* at .48, and *AC* at .37. Then your bid sheet for the primary market should look like this:

package	\multicolumn{6}{c}{items included}	value
	A	B	C	D	E	F	
1	1					1	.72
2	1					1	.48
3	1		1				.37

Bids are accepted or rejected each period as follows. The bid sheets are collected from all buyers. All bids are fed into a computer program which selects the set of bids which are most valuable without violating any constraint on the number of units of each item available. The program also gives two sets, low and high, of item unit values. Each accepted bid represents the purchase of one package at a total price equal to the sum of its low item values. Your purchase price will always be less than or equal to your bid price, since any bid less than the sum of its low item values was definitely rejected. Any bid greater than the sum of its high item values was definitely accepted. Consider the above set of bids. Suppose the low and high item unit values for *A*, *C*, and *F* were given as (.25, .10, .32) and (.25, .16, .34). Then package 1 was definitely accepted since .72 > .25 + .34. The market price for the package *AF* was .25 + .32 = .57. Package 2 was definitely rejected since .48 < .57. Package 3 might have been rejected or accepted since .25 + .10 < .37 < .25 + .16. At the close of the primary market bid sheets will be returned to each buyer indicating which of his bids were accepted. The low and high sets of item values will be posted.

Secondary Market. The secondary market provides an opportunity to buy additional units or sell units from the inventory acquired in the primary market. This is an oral auction. You may announce a bid (offer) to buy (sell) any package of one or more items for a specified amount. This bid (offer) will be placed on the board until it is accepted by some other participant or you cancel it. You are free to make as many bids and offers as you wish. Many may remain unaccepted but you are free to keep trying. *Note*: You may not sell what you do not have in inventory. Each purchase or sale in which you participate should be recorded on a separate line in the sequence of occurrence on your secondary market balance sheet. From your final inventory at the end of the secondary market, you specify a set of item packages that you want to redeem for cash.

Profits. Period profit is calculated as: profit = redemption revenue + sales revenue from secondary market − purchase costs from both markets. After each period has ended, make the appropriate entries on your payoff sheet. The experimenters will pay you all you have earned during all periods at the conclusion of the experiment.

18 A COMBINATORIAL AUCTION MECHANISM

Primary Market   Agent #1   Period #1
Agent's Bids for Packages

Package	Item Included						Bid	Accepted Yes or No	Market Price
	A	B	C	D	E	F			
1									
2									
⋮									
8									
# Units								Total Cost	
	Totals for Auction Accepted Bids Only								www

Secondary Market   Agent #1   Period #0

Note: Before the secondary market begins, copy the # of units of each item brought in the primary market into the spaces labelled inventory for transaction 0.

	Package Dealt								Inventory					
Transaction	Items Included						Sold For	Bought At	# Units Each Item					
	A	B	C	D	E	F	$	$	A	B	C	D	E	F
0	✗	✗	✗	✗	✗	✗								
⋮														
6														
Total Sales YYY														
Total Costs XXX														
Final Inventory														

Redemption Values   Agent #1   Period #0

Note: Before claiming redemptions, copy your final inventory from the secondary market into the following table. Make sure all packages you intend to redeem are covered by this inventory. Remember that any unit of a given item can only be used in one package.

Item	A	B	C	D	E	F
# Units						

			Items Included					
Package	A	B	C	D	E	F	Value	Claimed Yes or No
1	1	1					6.27	
2			1	1			5.77	
3	1				1		5.06	
4	1		1			1	8.25	
5		1	1		1		8.34	
Total Value of Redemptions =								ZZZ

Payoff Sheet   Agent #1

Your profit from each period is calculated as follows:
Period Profit = Redemption Value (ZZZ) + Sales in Secondary Market (YYY)
               − Costs in Secondary Market (XXX) − Costs in Primary Market (WWW)
After each period concludes, make the proper entries in the following table:

Period	Red. Value ZZZ	Sales Sec. YYY	Costs Sec. XXX	Costs Pri. WWW	Profit
0		+	−	−	=
1		+	−	−	=
2		+	−	−	=
⋮		+	−	−	=
6		+	−	−	=
Total Profit Over All Periods				PPP	

I acknowledge receipt of the above amount (PPP) from the experimenters:

Appendix B

Agent Value Information

Easy Resource Utilization Design.

Agent	Package	Value	Item A	Item B	Item C	Item D	Item E	Item F
1	1	5.98	1	1				
1	2	9.46	1	1	1			
1	3	5.17		1	1			
2	4	6.32	1			1		
2	5	6.63		1	1			
2	6	9.51	1	1	1			
3	7	8.77	1	1				1
3	8	5.95	1		1			
3	9	5.15			1	1		
3	10	8.85	1	1	1			
4	11	5.46	1		1			
4	12	9.83	1		1			1

Agent	Package	Value	Item A	Item B	Item C	Item D	Item E	Item F
4	13	5.69	1	1				
4	14	6.03		1	1			
5	15	6.42	1	1				
5	16	4.50	1				1	
5	17	4.98		1	1			
5	18	9.13	1		1		1	
5	19	4.76	1		1			
6	20	5.76	1		1			
6	21	8.02		1	1	1		
6	22	4.39		1	1			
6	23	9.45	1		1			1
6	24	6.17	1		1			
6	25	5.20	1	1				
# Units Demanded			18	15	18	2	2	3
# Units Available			13	11	15	1	2	3

Difficult Resource Utilization Design.

Agent	Package	Value	Item A	Item B	Item C	Item D	Item E	Item F
1	1	6.27	1	1				
1	2	5.77			1	1		
1	3	5.06	1				1	
1	4	8.25	1		1			1
1	5	8.34		1	1		1	
2	6	5.31	1	1				
2	7	5.56			1	1		
2	8	5.76	1		1			
2	9	6.44		1			1	
2	10	5.84			1		1	
2	11	8.86	1				1	1
3	12	5.17	1	1				
3	13	5.76			1	1		
3	14	8.87	1		1		1	
3	15	9.40		1	1			1
4	16	5.98	1	1				
4	17	6.27			1	1		
4	18	5.78	1					1
4	19	5.78		1		1		
4	20	5.56				1		1
4	21	8.61		1			1	1
5	22	5.60	1	1				
5	23	5.82			1	1		
5	24	5.65		1				1
5	25	8.34		1		1	1	
5	26	7.82	1			1		1
6	27	5.07	1	1				
6	28	5.65			1	1		
6	29	8.33		1		1		1
6	30	9.59	1			1	1	
# Units Demanded			14	14	12	12	9	9
# Units Available			7	7	7	7	7	7

References

Grether, D., Isaac, M., and Plott, C. "Alternative Methods of Allocating Airport Slots: Performance and Evaluation." CAB Report. Pasadena, Calif.: Polynomics Research Laboratories, Inc., 1979.

——,——, and ——. "The Allocation of Landing Rights by Unanimity among Competitors." *American Economic Review*, Vol. 71 (May 1981), pp. 166–171.

Rassenti, S. "0–1 Decision Problems with Multiple Resource Constraints: Algorithms and Applications." Unpublished Ph.D. thesis, University of Arizona, 1981.

Smith, V. "Experimental Economics: Induced Value Theory." *American Economic Review*, Vol. 66 (May 1976).

Vickrey, W. "Counterspeculation, Auctions, and Competitive Sealed Tenders." *Journal of Finance* (March 1961).

"Midway Air Sets Chicago-Dallas Ticket at $89." *Wall Street Journal* (May 17, 1982), pp. 1, 4.

Wolsey, L. "Integer Programming Duality: Price Functions and Sensitivity Analysis." *Mathematical Programming*, Vol. 20 (1981), pp. 173–195.

CHAPTER 19

A New and Improved Design for Multi-Object Iterative Auctions

Anthony M. Kwasnica, John O. Ledyard, David P. Porter, and Christine DeMartini

1. Introduction

Theory, experiment and practice suggest that, when bidder valuations for multiple objects are super-additive, combinatorial auctions are needed to increase efficiency, seller revenue, and bidder willingness to participate (Bykowsky et al. 2000, Rassenti et al. 1982, Ledyard et al. 2002). A combinatorial auction is an auction in which bidders are allowed to express bids in terms of packages of objects. The now famous FCC spectrum auctions are a good example of the relevance of these issues. In 41 auction events from 1994 to 2003, the FCC used what is known as a Simultaneous Multiple Round (SMR) auction to allocate spectrum and raise over $40 billion in revenue. This auction format does not allow package bidding. The FCC auctions also divide the spectrum by geographic location. It is reasonable to expect that some bidders might receive extra benefits by obtaining larger, more contiguous portions of the spectrum. A firm might enjoy cost savings if they could purchase two adjacent locations. However, without package bidding, a bidder cannot express that preference, potentially lowering the efficiency and revenue of the auction. If the bidder attempts to acquire both licenses through bidding on the licenses individually, they might be forced to expose themselves to potential losses. The high number of bidder defaults on payments might, in part, be evidence of losses caused by the lack of package bidding.[1] In response to these difficulties, the FCC plans to allow package bidding in future auctions (Federal Communications Commission 2002, Dunford et al. 2001). In particular, the FCC in its auction #31 for the upper 700 MHz band, affords bidders the ability to submit bids for packages of licenses. The particular design presented in this paper was developed prior to the FCC package auction design. Indeed one of the major features of the FCC design was clearly influenced by the pricing rules we developed herein. Specifically, the FCC will use a "current price estimate" in auction #31 that will provide a price for each license and these prices will be used to determine the minimal

---

[1] An extreme example can be found in the PCS C Block auctions where there were $874 million in defaults. See "Airwave Auctions Falter as a Source of Funds for U.S." in *The New York Times*, April 3, 1997.

acceptable bids in the next round of bidding. We discuss this in more detail below in Section 3.3.

While the potential utility of combinatorial auctions is considerable, combinatorial auctions have not yet reached their full potential in practice.[2] The successful implementation of a combinatorial auction requires one to overcome a number of hurdles. Two widely recognized and discussed issues are:

1. The computational complexity of the winner determination problem, and
2. The complexity of the bidding environment for the bidder.

Computational complexity comes from the fact that determining a set of winning bids—those that maximize the sum of the bid prices subject to feasibility constraints—is NP-complete. Rothkopf et al. (1998) have shown that computational issues can be reduced via limitations on acceptable bids and other strategies. Others have found promising algorithms (Sandholm et al. 2001), and Andersson et al. (2000) have shown that CPLEX software is fairly effective in solving the winner determination problem in combinatorial auction simulations. For problems of reasonable size, the computational complexity of the winner determination problem is simply not the limiting factor.[3]

The computational complexity of the bidders' problems is more of an issue in practice. Bidders must determine their valuations for all subsets of items they are interested in (up to a maximum of 2^K values if they are interested in K items). Then they must formulate an optimal bidding strategy given those valuations. If the bidders make incomplete or incorrect calculations, the efficiency or revenue an auction will generate can be significantly reduced.

Several approaches have been taken to reduce these difficulties. Some have proposed using the Vickrey sealed-bid auction. Under a Vickrey auction, bidders have a dominant strategy to truthfully report their values to the auctioneer. While this would eliminate strategic complexity, it would not reduce the complexity associated with valuation determination.[4] Some have suggested a pay-what-you-bid, sealed-bid, one-shot auction (Rassenti et al. 1982), but this brings back the strategic complexity without reducing the valuation computation complexity. Others have suggested using progressive auctions, similar to an English auction, to reduce the cognitive burden on

---

[2] There are an increasing number of exceptions including Sears Logistics Services (Ledyard et al. 2002), the Automated Credit Exchange (Ishikida et al. 2001), the course registration auction at the Chicago Business School (Graves et al. 1993), and the Mars IBM procurement auction (Davenport et al. 2003).

[3] We can also report some additional data from Net Exchange (nex.com), based on simulations with a 200 MHz PC. They created test runs where the number of bids were four time the number of items, where 1/3 are multi-item bids, where 5% of the bids involve more than 3 items, and where OR groups were allowed. For problems based on hundreds of items (including runs in the 500–800 range), using DASH Express, the optimum was always found very quickly. For 1000 items, 50% of the problems computed to the optimum in less than 30 seconds. 90% computed to the optimum in less than 30 minutes. 95% computed to within 2% of the best upper bound (the relaxed linear programming solution) in less than 30 minutes. And, in one other observation, Ledyard et al. (2002), the winner determination problems for the Sears logistics auctions for 850 items always solved in less than 30 minutes using now totally out-dated 1992 technology. With modern technology and algorithms, fairly large problems are almost always easily solved in a timely fashion.

[4] There are other problems with Vickrey auctions, many of which were originally noted, discussed and explained by Groves & Ledyard (1977a) and Groves & Ledyard (1977b) in the context of public goods. Banks et al. (1989) also discuss the problems and investigate an iterative version of Vickrey. It was the unsatisfactory performance of that mechanism that led to the development of AUSM.

the bidders in both valuation and strategic computation (Banks et al. 1989, Parkes 1999). Two candidate auction designs in this area are the continuous package bid auction called the Adaptive User Selection Mechanism (AUSM) first proposed by Banks et al. (1989) and the Simultaneous Multi-Round (SMR) auction used by the FCC (Milgrom 2000). The idea in each of these is that bidders need only compute valuations when necessary, and that bidders have time to focus on and compute strategies.

In this paper, we take the best features of these auctions, add a new element, and create a new design we call Resource Allocation Design (RAD) that produces in our experiments higher efficiencies, higher revenues, and a shorter duration than the original designs. We take the issues of computational and cognitive complexity seriously and formulate a combinatorial auction mechanism that attempts to ease those burdens. The features we borrow are (1) package bidding from AUSM, and (2) an iterative format, eligibility, and minimum bid increments from SMR. The feature we add is (3) a method to provide *prices* that will guide bidders to desirable outcomes. We use an iterative auction that gives bidders, at each iteration, a vector of prices—one for each object—that new bids must beat in order to be accepted. Bidders need not consider separate prices for each subset of objects—a subset price will simply be the sum of the prices for each item in the subset—so the information the bidders need to process is, in some sense, small. (See Nisan & Segal (2003) for a precise analysis of the size of the messages.) Bidders need not bother computing valuations for items whose prices are obviously higher than the valuations would be, thus reducing that dimension of computational complexity.

We use the test bed approach of experimental economics to establish the performance improvements of RAD over SMR and AUSM. There is no theory to use to compare auctions, especially if one wants to take account of the limitations of bidders' cognitive skills. We have found that computer simulations also fail to capture many of the important details of human cognition. One cannot use the data from auctions that occur in practice because it is not possible to know the fundamentals—the true values of the items for each bidder.[5] So, if progress is to be made we must adopt the approach of an engineer's wind tunnel and turn to the laboratory for data. The use of the laboratory as a test bed for complex auctions in complex environments began with Ferejohn et al. (1979), Smith (1979), Grether et al. (1981) and Rassenti et al. (1982). This methodology has proven to be fairly successful in providing guidance for the design of a variety of implemented auctions (Plott 1997, Ishikida et al. 2001, Ledyard et al. 1997). Building on knowledge from theoretical and practical experience, one can create test bed environments in the laboratory which exhibit as much complexity or simplicity as one wishes. In these environments, one can test any auction. With laboratory control, one can calculate performance measures unknowable in the field. One can precisely answer questions such as: did the highest value bidders win the items, was there a bidder who wanted a particular configuration and did not get it, and were there bidders who, because of the auction design, bid more for an item than it was truly worth to them?

We evaluate the RAD design in the lab using both a complex and a simple test bed. We are able to compare the performance of RAD to a version of the SMR auction used

---

[5] One cannot econometrically estimate them from the data unless one knows what the strategic behavior of the agents was and that behavior is invertible.

by the FCC. Since we used the same test bed as in previous experiments, we are also able to compare the performance of RAD, when possible, with that of the Adaptive User Selection Mechanism (AUSM) combinatorial auction proposed by Banks et al. (1989), which is widely regarded as one of the first combinatorial auction mechanisms.

In Section 2, we describe the background of our search for a high performance multi-object auction design. In Section 3, we formally describe the SMR and AUSM designs and the RAD auction. In Sections 4 and 5, we describe the test bed and the performance measures we use to evaluate the design. In Section 6, our findings are offered. Finally, in Section 7, we provide our conclusions and work that remains to be done.

2. The Context

As most theorists realize, it is relatively simple to describe a demand-revealing, efficient auction. A natural extension of the famous Vickrey sealed-bid auction will award the objects to the highest valuing bidders and eliminate any incentive for them to misrepresent their preferences. If we accept that the winner determination problem is not an issue, then the Vickrey auction appears to eliminate the strategic complexity facing the bidder. However, the bidder still faces the complexity of calculating and expressing these valuations.[6] If K items are being auctioned, each agent's bid would need to be 2^K numbers—potentially creating a very large, very complex communication problem. Further, if there is any affiliation in the values of bidders then sealed bid auctions of this sort are thought to be less efficient than auctions that allow bidders to learn as they bid (Milgrom & Weber 1982).[7] Even when only one object is for sale, bidders in experimental sessions often do not understand the demand revealing incentives of the Vickrey auction (Kagel et al. 1987).

Progressive auctions, such as the English auction, usually perform quite well in the laboratory (Coppinger et al. 1980). There are two types of progressive auctions one might consider: iterative and continuous auctions. A continuous auction is similar to the classic English auction where bids can be submitted at any time. Iterative auctions proceed in a series of rounds, which last a specified period of time. During a round, bidders have the opportunity to place bids before the auctioneer considers any of the bids placed in the round. Once a bid is submitted, in the case of a continuous auction, or at the end of a round, in the case of an iterative auction, the auctioneer processes the bid(s) and identifies *provisionally winning* or standing bids. These are the bids that will win if no new bids are forthcoming. In all cases, the auctioneer then provides information back to the bidders. The process repeats until a *stopping rule* is satisfied. At that time the provisionally winning bids become winning bids.

To understand the possibilities and choices facing the designer of multi-object auctions, we begin by recalling the key features of two vastly different designs: the Simultaneous Multiple Round (SMR) design (Milgrom 2000) and the Adaptive User

---

[6] In fact, Sandholm (2000) has shown that, if valuation computation is costly, the positive strategic implications of the Vickrey auction may not hold. Also, see Larson & Sandholm (2001b) and Larson & Sandholm (2001a) for a further discussion of the strategic complexity of auctions.

[7] Dasgupta & Maskin (2000) show that the Vickrey auction could, in theory, be extended to this setting by allowing for bids to be functions that allow each bidder to state what his value would have been were the other bidders' information revealed.

Selection Mechanism (AUSM) (Banks et al. 1989). The SMR design allows only single-item bids, is iterative, and has an eligibility based stopping rule (i.e., a *use-it-or-lose-it* feature) driven by a minimum increment requirement for new bids. The SMR design was used extensively by the FCC to run early bandwidth auctions. On the other hand, AUSM allows package bids, is continuous, and is stopped at the discretion of the auctioneer. An iterative version of AUSM was used by Sears Logistics Services to procure trucking services (Ledyard et al. 2002). Three aspects of the design are the same for each: Winning bidders pay what they bid, provisionally winning bids are determined by maximizing potential revenue subject to feasibility, and provisionally winning bids remain as a standing commitment until replaced by another provisional winner.

Both the SMR and AUSM auctions represent a compromise, the result of a sequence of design choices. Each choice often leads to one side of a seeming unavoidable trade-off. Therefore each auction process has its potential weaknesses. In this paper we focus on potential failures in performance in the areas of efficiency, revenue, bidder losses, complexity, and the time to complete an auction.

In most discussions of the design of multi-object auctions, the primary goals have either explicitly or implicitly been high efficiency and/or high revenue. The goals of maximizing efficiency and maximizing revenue are not antithetical. In fact, the amount of revenue collected is generally limited by the efficiency of the auction. In single-item auctions, contingent upon sale, maximal revenue usually occurs by maximizing efficiency and then extracting as much of the surplus as possible (Myerson 1981). This approach does not always work in multi-object auctions.[8] In environments without income effects, such as quasi-linear preferences, what trade-off there is can be most easily seen in the following identity:

Efficiency × Maximal Possible Surplus ≡ Seller's Revenue + Bidders' Profits.

High efficiency and low revenue can occur if and only if bidder profits are high, which might occur under collusion. And high revenue and low efficiency can occur if and only if bidders incur losses.

Both the SMR and AUSM auction processes have a difficult time consistently generating 100% efficiency across a variety of environments (Ledyard et al. 1997, Kwasnica et al. 1998). The SMR mechanism, because it only allows single-item bids, faces the exposure problem. The exposure problem occurs in situations where bidders' values are super-additive. In order to win a package, which the bidder values more than the sum of the individual items in the package, the bidder might need to bid above her value on the individual items. If the bidder does not end up winning the package, this can expose the bidder to losses. Bidders who are aware of this problem might stop bidding in order to avoid the risk of losses causing low efficiencies and seller revenue. To combat the exposure problem, the FCC allowed provisional winners to withdraw with a penalty. Porter (1999) analyzes the effect of this rule and finds that although efficiencies are higher, so too are bidder losses. The AUSM mechanism, because it allows package bids, does not suffer from the exposure problem but faces the threshold problem. The threshold problem occurs when a number of bidders for small packages must

---

[8] In spite of Williams (1999) who identifies the optimal, efficient auction to be a Vickrey-Groves mechanism, we do not know that the optimal, revenue maximizing auction is always efficient. In fact Armstrong (2000) suggests it may not be so.

coordinate their efforts to unseat a bidder for a big package. In this situation each bidder has the incentive to allow the other bidders to be the ones who increase their bid in order to displace the big bidder. In principle, all bidders may fail to raise their bids allowing the big package to win even if it should not have. The threshold problem may cause low efficiencies as collections of small bidders may not be able to coordinate their bids to dislodge a large, inefficient bidder. To combat the threshold problem, AUSM is often used with a standby queue—a public bulletin board on which potentially combinable bids can be displayed.[9] The use of a queue, however, shifts the computational burden to the bidders; they must now consider the bids in the queue when making a new bid.

While it is easy to measure efficiency, seller revenue and bidder losses, it is harder to measure the complexity of a mechanism or the costs of the length of time to complete the auction.[10] Nevertheless, we can make a few observations about the performance of the SMR and AUSM designs. Because the SMR auction proceeds in measured steps and because bidders seem to have a relatively simple information processing problem at each step, most consider it a simple mechanism.[11] But, because of this slow but steady approach, SMR auctions can take a very long time to complete. The FCC's Broadband PCS D,E, and F Block auctions lasted 276 rounds spanning 85 days. AUSM proceeds in a seemingly disorganized manner with bids allowed in any order, stopping when no new bids are forthcoming or the auctioneer deems the auction to be at an end. Because of this, AUSM finishes quickly. But many feel that this places a difficult information processing burden on bidders that, together with the standby queue, makes AUSM a very complex mechanism.

So each mechanism has both desirable and undesirable performance characteristics. The obvious question then is: can we do better than both? In particular can we take the successful design aspects of each, perhaps augment them a bit, and create a hybrid that dominates both? Based on the research reported in this paper, we suggest that the answer is yes.

3. The Auctions

Rather than providing a fully general framework, in this paper we will focus on the particular designs we evaluate. Let $I = \{1, \ldots, N\}$ represent the set of bidders, $K = \{1, \ldots, K\}$ represent the set of objects to be sold and $t = 1, 2, 3, \ldots$ represent the iterations or *rounds*. In general, a bid can be a very abstract entity involving complex contingent logic.[12] In this paper, we restrict our attention to very simple bids. A bid is a pair $b = (p, x)$ where p is positive real number representing the bid price and $x \in \{0, 1\}^K$ represents the items desired.[13] A bid here signifies, "I am willing to pay up

---

[9] It is shown in Banks et al. (1989) that the queue increases both efficiency and revenue in continuous auctions.
[10] In the field, it is difficult, if not impossible, to measure any of these variables. In the lab, since we know the induced valuations of bidders, we can directly measure efficiency, revenue and losses.
[11] Formulating an optimal strategy to win packages of items when bidding is restricted to single-item bids is actually quite difficult.
[12] See, for example, Ishikida et al. (2001), Rassenti et al. (1982), and Grether et al. (1981). Recent work by computer scientists emphasizes expressiveness (Nisan 2000). We discuss some of this in Section 3.3.
[13] This structure can easily be generalized to cases where there are multiple-copies of items available. We treat each $k = 1, \ldots, K$ as a single indivisible object.

to p for the collection of objects for which $x_k = 1$ if and only if I get all of them." In the auctions we analyze, winners will actually pay what they bid.

Begin by assuming we are in round t and all N bidders have submitted their bids. The set of bids placed by bidder i in round t is B_t^i, and $B_t = \cup_{i \in I} B_t^i$ is the set of all submitted bids. An arbitrary element of B_t is expressed as $b_j = (p^j, x^j)$.

All the auction designs use a straight-forward allocation rule: provisionally award the items to the collection of bids that would yield the highest revenue. We solve the following allocation problem:

$$\max \sum_{j \in B_t} p^j \delta^j \quad (1)$$

subject to

$$\delta^j \in \{0, 1\} \text{ for all } j \in B_t$$

and

$$\sum_{j \in B_t} x_k^j \delta^j \leq 1 \text{ for all } k = 1, \ldots, K.$$

If there is only single-item bidding, this simply selects the highest bidder for each item. With package bidding, the combinatorial optimization problem is equivalent to a set-packing problem on a hypergraph. We acknowledge that it is well known that if the number of objects and bids is large, then one cannot necessarily guarantee that an optimal solution will be found in a reasonable amount of time. But it should also be noted though that computation is increasingly less of an issue in combinatorial auctions (see footnote 3). Of course, computation was never an issue in any of the results reported in this paper due to the relatively small scale of the test cases examined (10 objects and 5 bidders).

Let δ_t^* be a solution to this problem. If $\delta_t^{*j} = 1$ we say that bid j is *provisionally winning* in round t. Let $W_t = \{(p^j, x^j) \in B_t \mid \delta_t^{*j} = 1\}$ be the collection of provisionally winning bids. Then i's winning bids are the set $W_t^i \equiv B_t^i \cap W_t$. There is no restriction placed on the number of winning bids for each particular bidder; each bid placed by an individual is considered independently from the other bids placed. An obvious initial condition is to have $W_0 = \emptyset$.

If the auction stops at this round, for each $j \in W_t^i$, bidder i will receive the items for which $x_k = 1$ and will pay p^j to the auctioneer. If the auction does not stop, then all provisional winners are automatically resubmitted in round $t + 1$, so $W_t^i \subseteq B_{t+1}^i$ for all i.

3.1. The SMR Auction

The basic SMR auction design requires only a few new rules in addition to those from above. First, only single-item bids are allowed. That means for all i, and t

$$\sum_{k \in K} x_k = 1 \text{ for all } (p, x) \in B_t^i. \quad (2)$$

Second, SMR uses eligibility to force active and meaningful bidding. Introduced by Paul Milgrom as the truly unique part of the SMR design, eligibility rules are designed to encourage active bidding while not allowing the auction to stop too quickly. A soft close makes an efficient allocation more likely.

Eligibility limits the number of items a bidder can bid on in a round as a function of the bidder's past bidding behavior. Specifically, a bidder's eligibility is the number of distinct objects he is allowed to bid on in a round.[14] Let A^i_{t-1} be the number of distinct items i bid on in round $t-1$,[15]

$$A^i_{t-1} = |\{k | x^j_k = 1 \text{ for some } (p, x) \text{ in } B^j_{t-1}\}|. \tag{3}$$

Initially bidders are allowed to bid on all items, $A^i_0 = K$. In round t, a collection of bids B^i_t for i satisfies eligibility if and only if

$$A^i_t \leq A^i_{t-1}. \tag{4}$$

That is, a collection of bids is eligible if and only if the new bids plus last round's winning bids are placed on no more than A^i_{t-1} objects.[16] Eligibility can easily be checked incrementally as each new bid is offered. Since eligibility limits the items a bidder bids on by the number of items they bid on in the previous round, eligibility encourages early bidding.

The stopping rule is obvious once eligibility is imposed.

$$\text{Stop at the end of } t \text{ if } \sum_{i \in I} A^i_t \leq K. \tag{5}$$

If eligibility satisfies this constraint, no bidder will be able to bid on anything other than the items they are currently provisionally winning. Therefore, ownership will not change in any subsequent period.[17]

While eligibility encourages bidders to place bids, a bidder could repeatedly submit a small bid for the package of all items in order to maintain her eligibility. Then, $A^i_t = K$ and the auction never ends. To drive the auction to finish we also need to force new bids to be serious. So a *minimum increment* rule is imposed. It is based on a vector of single-item prices λ^t, which are known at the start of round t. In the SMR design, the price vector λ^{t+1} is simply the high price from t. That is,

$$\lambda^{t+1}_k = p_k \text{ if } (p, x) \in W_t \text{ and } x_k = 1. \tag{6}$$

We let $\lambda^1_k = 0$ for all k but one could allow λ^1 to be any reserve prices. Let $N^i_t = B^i_t \setminus W^i_{t-1}$ be the set of new bids. Then we require that

$$p \geq \sum_{k \in K} x_k (\lambda^t_k + M) \text{ for all } (p, x) \in N^i_t. \tag{7}$$

where M is a minimum bid increment chosen by the auctioneer.[18]

---

[14] In the FCC spectrum auctions, a *weighted* measure of eligibility was used. Objects were weighted by their MHz Pops. Let w_k be the weight assigned to k. Let $\alpha^i_{t-1} = \{k | i \text{ has an active bid on } k \text{ in round } t-1\}$. A bidder's eligibility in t is then $A^i_{t-1} = \sum_{k \in \alpha^i_{t-1}} w_k$.

[15] We use $|A|$ to indicate the cardinality of the set A.

[16] In the early rounds of the FCC spectrum auctions bidders were allowed to bid on more than A^i_{t-1} items by multiplying A^i_{t-1} by $r > 1$.

[17] As a referee correctly pointed out, this stopping rule can allow the auction to stop early if all bidders do not bidding on some object(s).

[18] The minimum increment could be altered over time, but we forego that degree of freedom in this paper.

So at the start of each round, each bidder $i \in I$ knows the objects for sale K, the prices on each object λ^t, her winning bids from the previous round W_{t-1}^i, and her eligibility A_{t-1}^i. Each bidder then chooses new bids N_t^i satisfying Equations (4) and (7). By the resubmittal rule, $B_t^i = W_{t-1}^i \cup N_t^i$. Using the revenue maximizing allocation rule described by Equation (1) the auctioneer computes W_t. The auctioneer computes A_t^i for all i. Using (5) the auction is then stopped or continues to round $t+1$.

The rules given by Equations (1)–(7) describe what we have called the SMR design.

3.2. The AUSM Design

Since the AUSM design is a continuous auction, it is somewhat more difficult to formally define the AUSM rules using the notation developed earlier. Banks et al. (1989) provide a more complete definition. One can think of a continuous auction as an iterative auction where only a single bid is placed in each round, or $|\bigcup_{i \in I} N_t^i| = 1$. As in the SMR design, in the basic AUSM design, this new bid is considered along with the previous provisionally winning bids $B_t^i = W_{t-1}^i \cup N_t^i$ in solving the allocation problem given by Equation (1). This rule makes it very difficult for a new bid to win since it must independently raise the surplus of the allocation problem. The AUSM with a standby queue avoids this problem by requiring that all bids placed in previous rounds are considered in each iteration: $B_t^i = B_{t-1}^i \cup N_t^i$.[19]

The basic AUSM design does not place any restrictions on the types of bids placed. AUSM does not use an eligibility calculation, prices λ^t, or a minimum increment requirement to drive bidding. While the actual stopping rule may be at the final discretion of the auctioneer, a typical rule will take the form of a decision to stop the auction if no new bids have been submitted in a certain time period.

3.3. The RAD Design

This design represents a serious attempt to make package bidding work in the context of a multi-object, iterative auction. It shares a number of similarities with the auction designs discussed previously. Like the SMR design, the RAD design is iterative, has an eligibility based stopping rule, forces a minimum bid increment, and computes prices for each item for sale. Like the AUSM design, the RAD design allows package bidding. The key difference in design from the SMR approach is that package bids are allowed and a new pricing rule is introduced. Allowing package bids is accomplished by simply eliminating equation (2) as a restriction on new bids.

Some more recent auction designs allow bidders to submit "exclusive or" (XOR) bids that allow a bidder to identify a subset of her bids and require that at most one of the bids in that subset be accepted.[20] Parkes (1999) and Ausubel & Milgrom (2002) are recent examples that allow this more expressive bidding language. In our structure, we allow any number of a bidder's bids to be accepted. Although we were aware of the

---

[19] In practice, the auctioneer posted B_{t-1}^i and required the new bidder to declare the combination of previously placed, but not provisionally winning, bids that beat the current winning bids when combined with her new bid. Thus the computational burden was shifted to each bidder.

[20] The value of XOR bids was actually recognized in even the most early combinatorial auction design (Rassenti et al. 1982). See Sandholm (2002a) and Sandholm (2002b) for a formal introduction to XOR bidding languages.

utility and power of XOR bids at the time we were running the experiments reported in this paper, we believed that the results would be more informative and persuasive if we modified the SMR in as few ways as possible.

Pricing is a bit more subtle. Game theory provides one suggestion—construct Vickrey prices for the auction. Vickrey prices are personalized prices that have been shown to eliminate all strategic incentives for bidders. Assuming bidders have correctly formulated their valuations, bidders should be willing to submit a full, honest report of those values to an auctioneer who has committed to Vickrey prices. So, in theory, if bidders are faced with Vickrey prices they should realize that their only strategy is to simply bid their value for each particular combination. In fact if Vickrey prices are used, then the auction can be run in one round (effectively a sealed-bid auction). But this approach has a number of drawbacks. First, it is not clear that bidders will interpret these bids correctly. In laboratory experiments with Vickrey auctions for only one object, bidders systematically deviate from the ideal strategy of simply stating one's value. Second, in an auction of any size bidders may not be able to submit all of their potentially desired packages in one round. It would require messages of 2^K numbers. Finally, if one views bidders as 'learning' about their valuations and profitable bids in the course of the auction, Vickrey prices provide little information to the bidder about potentially profitable combinations of new bids. One could, of course, try to create an iterative Vickrey type auction. This was done in Banks et al. (1989) with little success. The interested reader should consult the discussion there.

A second potential approach is closely related to the economic theory of competitive equilibrium. A set of prices—one for each object—is said to constitute a competitive equilibrium if, given these prices, the supply of objects equals demand (i.e., excess demand is zero). This approach was explored a bit by Bykowsky et al. (2000) where it was shown that if one simply prices the items for auction, competitive equilibrium prices may not exist because of the non-convexities caused by complementarities. Bikhchandani & Ostroy (2002) took this further and looked at personalized (individual specific) prices on packages. They were able to provide several possibility results. But from the point of view of auction design, both of these approaches leave something to be desired. First, from Bickchandani and Ostroy, we learn that we may need 2^K prices and so would be back in the communicatively difficult world of Vickrey. Second, competitive equilibrium is just that—an equilibrium theory. It is silent on the dynamics of price discovery—unless one wants to adopt the Walrasian tâtonnement, a process that does not work well in the laboratory because it requires re-contracting which opens up all sorts of possibilities for non-constructive manipulation. It is our belief that the advantage of well-designed iterative auctions over one-shot auctions is that they allow orderly discovery of alternatives and prices because important feedback information is provided to bidders between rounds.

So we turn to a third and, ultimately, more productive approach. We restrict ourselves to only pricing items so as to keep communication complexity to a minimum. We then look for a pricing rule that will convey information to bidders about opportunities in the next bidding round. Three properties seem important for this: (a) all accepted bids should, if they were to pay these prices, pay something less than or equal to what they bid, (b) all losing bids should, if they were to pay these prices, pay something more than what they bid—indicating they needed to bid higher in order to win, and (c) new bids that are willing to pay more than the price of their bundle at those prices should have a good opportunity to win—that is the prices ought to "guide" new bids to collections and values that can increase revenues.[21]

To insure (a), to keep computation simple, and to retain the "pay what you bid" nature of SMR, we chose to require that prices λ_k^t satisfy $\sum_{k \in K} \lambda_k^t x_k^j = p^j$ for all winners. Insuring (b) is a bit more difficult. Let $L_t = B_t \setminus W_t$ be the losing bids at t. To have (b) we would need a set of prices, λ^t, such that $p^j = \sum_{k \in K} \lambda_k^t x_k^j$ for all $j \in W_t$ and $p^j \leq \sum_{k \in K} \lambda_k^t x_k^j$ for all $j \in L^t$. If prices satisfy these equations, then the winning bidders would be paying their bid and losing bidders would see that the prices were greater than their bid. Unfortunately, once package bidding is allowed and Equation (1) is used to decide winners, it can no longer be guaranteed that such prices exist. So we must turn to an approximation of the ideal. There are many ways to do this. We choose one we believe, and the experimental evidence supports these beliefs, also provides good inter-round signals about opportunities, and lack of opportunities, for successful bids. To compute RAD prices λ^{t+1}, we begin by solving the following problem:[22]

$$\min_{\lambda^t, Z, g} Z \qquad (8)$$

subject to

$$\sum_{k \in K} \lambda_k^t x_k^j = p^j \quad \text{for all} \quad (p^j, x^j) \in W_t$$

$$\sum_{k \in K} \lambda_k^t x_k^j + g^j \geq p^j \quad \text{for all} \quad (p_i^j, x^j) \in L_t$$

$$0 \leq g^j \leq Z \quad \text{for all} \quad (p_i^j, x^j) \in L_t$$

$$\lambda^t \geq 0.$$

Problem (8) selects a set of prices that ensures that revenue collected from the prices exactly equals the dollar amount for each winning bid; for losing bids it attempts to set prices that keep the package out with as little distortion as possible. The variable g^j is the amount for each losing package which ensures that the bid is not affordable.[23] We then want to find the smallest such value across all losing packages so that the distortion of the information is minimized. Let g^* and Z^* be a solution to (8). At the prices λ^t there may be some losing bids for which $\sum_{k \in K} \lambda_k^t x_k^j \leq p^j$, falsely signaling a possible winner. Such is the nature of package bidding. On the positive side, such bids can be resubmitted if $p^j - (\sum_{k \in K} \lambda_k^t x_t^j)$ is large enough. Further, Equation (8) is

---

[21] A related approach is to design an agent that efficiently queries bidders about their valuations (Conen & Sandholm 2001). However, in our experimental setting, the issue of value elicitation was not an issue. One still needs to design an auction that helps bidder know what to do with their valuations.

[22] One must recognize the pioneering work of Rassenti et al. (1982). They proposed a sealed-bid combinatorial auction (RSB auction) to solve an airport slot allocation problem and introduced the use of prices computed from a relaxed problem. They also provided experimental evidence as to the capability of their mechanism. The computation we use is a bit different from theirs, primarily for one reason. Since theirs is not an iterative process, the RSB prices are simply a way to collect dollars from the bidders and are not used as an information device. We need our prices to generate good signals during iteration.

[23] This is a little bit like 2-part pricing, a well-known solution to pricing with non-convexities, but since the losers never pay, the second part—the g—are never really collected.

designed to minimize the maximum violation of the inequalities for losing bundles. In fact, if ideal prices exist, they will be the solution and $g^{j*} = 0$ for all $b^j \in L_t$.

At this point it may still be possible to further lower some of the g^j that, in the first solution, satisfy $0 \leq g^j \leq Z^*$. So, to further the computation of λ^t, a sequence of iterations of Equation (8) is performed. We lexicographically lower as many g^j as possible. So at this point we have satisfied the desired property (a) and have done about as well as we can on property (b). What about (c) which asks that prices provide good signals about new bids with good opportunities to win? If the solution to Equation (8) after lexicographic minimization is unique, there is no more we can do. But in many cases, the solution will not be unique and we have an opportunity to improve. We know that the prices indicate, for all of the packages that were submitted in the previous round, what one would have to bid to have any chance of inclusion in the next round, assuming all other bids are resubmitted. So the only way to improve on this is to signal where a new package might be successful. New packages will be successful if they can be combined with losers from the last round to bump a winner from the last round out of the solution. So we will finish the price computation in a way that provides relevant information. For each winning bundle we lexicographically maximize the minimum price in the bundle subject to the constraints of Equation (8) at the g^* we solved for earlier. The formalities are provided in the Appendix.[24]

Why this works may seem mysterious, so we turn to three examples that illustrate what is happening here. The following examples help explain the ability of the RAD pricing rule to convey such information.[25]

Example 1. Let there be two objects labeled $\{A, B\}$ and three bidders labeled 1,2, and 3. Suppose that the following is true:

- Bidder 1 is the high bidder on the package $\{A, B\}$ with a bid of 10.
- Bidder 2 bid 8 for $\{A\}$.
- Bidder 3 has not bid but is willing to pay 4 for $\{B\}$.

In this situation, bidder 1 holds the provisionally winning bid, but bidder 2 and 3 could combine to outbid the current standing bid. Any prices such that $\lambda_A + \lambda_B = 10$ and $\lambda_A \geq 8$ will satisfy Equation (8). However, if we choose $\lambda_A = 10$ and $\lambda_B = 0$, then bidder 3 may bid 1 for $\{B\}$ in the next round only to find out that they lose. If $\lambda_A = 8$ and $\lambda_B = 2$, bidder 3 will know that they have to bid at least 2 in order to become provisionally winning. If bidder 3 bids 3 on $\{B\}$ and bidder 2 resubmits his bid, then the new provisional winners will be bidder 2 with object $\{A\}$ and bidder 3 with object $\{B\}$. The prices that would be generated by RAD would be 8 for A and 2 for B.

Example 2. Let there be two objects labeled $\{A, B\}$ and three bidders labeled 1,2, and 3. Suppose that the following is true:

- Bidder 1 is the high bidder on the package $\{A, B\}$ with a bid of 10.
- Bidder 2 bid 4 for $\{A\}$.
- Bidder 3 has not bid but is willing to pay 6 for $\{B\}$.

---

[24] A side benefit of this procedure is that we end up with a unique set of prices. This is important for "respectability." It is important for bidder confidence that we get the same answer if we rerun the algorithms.

[25] The following examples assume the minimum increment is zero. It is possible that a large minimum increment might upset some of the usefulness of these prices.

If we select $\lambda_A = 4$, then it must be that $\lambda_B = 6$. Given this information, bidder 3 will assume that it is not profitable for them to bid. In a sense, it puts all the burden of ousting the current standing bid on bidder 3. This could exacerbate the threshold problem. The more natural and fair decision is to 'split the difference' by setting $\lambda_A = 5$ and $\lambda_B = 5$.

The appropriate prices identified in Examples 1 and 2 are obtained, when ideal prices exist, by minimizing the maximum of λ_A, λ_B subject to the prices satisfying Equation 8.

Example 3. Let there be three objects labeled $\{A, B, C\}$ and four bidders labeled 1, 2, 3, and 4. Suppose that the following is true:

- Bidder 1 is the high bidder on the package $\{A, B, C\}$ with a bid of 30.
- Bidder 2 bid 25 for $\{A, B\}$.
- Bidder 3 bid 25 for $\{B, C\}$.
- Bidder 4 bid 22 for $\{A, C\}$.
- Bidder 3 is willing to pay 15 for $\{C\}$ but has not bid.

Bidder 1 is the provisional winner. The prices we want, if they exist, satisfy $\lambda_A + \lambda_B + \lambda_C = 30$, $\lambda_A + \lambda_B \geq 25$, $\lambda_B + \lambda_C \geq 25$, and $\lambda_A + \lambda_C \geq 22$. Since the last three inequalities imply that $\lambda_A + \lambda_B + \lambda_C \geq 36$, no such prices can exist. We try to get as close as possible. We choose $\lambda_A, \lambda_B, \lambda_C$ and g^1, g^2, g^3 such that $\lambda_A + \lambda_B + g^1 \geq 25$, $\lambda_B + \lambda_C + g^2 \geq 25$, and $\lambda_A + \lambda_C + g^3 \geq 22$, and we want g^1, g^2, and g^3 to be small. We could minimize $g^1 + g^2 + g^3$, or we could minimize the maximum of g^1, g^2, g^3. We could pick $\lambda_A = \lambda_B = \lambda_C = 10$ yielding $g^1 = g^2 = 5$ and $g^3 = 2$. We could also pick $\lambda_A = \lambda_C = 11$ and $\lambda_B = 8$ yielding $g^1 = g^2 = 6$ and $g^3 = 0$. In the second case, relative to the first, $g^1 + g^2 + g^3$ is the same but the maximum is more. In the second case, a bidder for $\{B\}$ knows exactly how much they must bid to become provisionally winning. However, the prices overvalue what someone must bid on either $\{A\}$ or $\{C\}$ to become provisionally winning. The prices in the first case overstate the value of the bid required for all single-items to become winning, but the difference is less for $\{A\}$ or $\{C\}$ as compared to the second case. The RAD pricing rule picks the first case. In either case, bidder 3 can bid for $\{C\}$ and, assuming bidder 2 resubmits her bid, become a provisionally winning bid.

If, on the other hand, bidder 3's value for $\{C\}$ was only 10.5, then bidder 3 would only find it profitable to bid when the price is $\lambda_C = 10$. If bidder 3's value was 7, then he would not be willing to bid in either case despite the fact that a bid of 7 could unseat the bidder 1's current high bid.

These prices help ease the two practical design issues discussed earlier. First, the computation of prices occurs by completing a series of nearly instantaneous linear programs. Therefore, the auctioneer needs to conduct only one NP-complete computation, the winner determination itself. Second, the prices present information on the level of bidding for all objects in a manner that is simple and natural for the bidders. Instead of looking at prices on all subsets, the bidders are presented with a price for each object. There is one possible complaint one might register about our pricing rule; individual prices will not necessarily be increasing over time. This is because over time the opportunities for new packages to combine with old rejected packages to displace provisional winners will change. This is an unavoidable feature of environments with complementarities when only prices on items are used. It is important to remember however that the sum over all prices is always increasing and our experimental tests of RAD reported below, indicate that subjects had no problem with this feature.

There is no reason, in theory, to expect these prices to work well or badly. However, we demonstrate, through the use of human subject experiments, that RAD can perform quite well across a number of reasonable performance measures. As the reader will see in the data below, this combination of pricing and stopping rule work very well together to eliminate strategic problems caused by the threshold problem. Changing the SMR design to allow package bidding with the particular pricing rule we designed generates a significant increase in performance in environments with multiple objects with complementarities.[26] There is also no degradation of performance in the goods with no complementarities.

4. The Experimental Design

The environment used as a test bed for all auctions in this paper was created by combining features of the *spatial fitting* environment originally utilized by Ledyard et al. (1997) and an additive environment. Since the two value environments are combined into one environmental test bed we can see if there are spill-over problems from items with complementarities to those without complementarities. Specifically, the five participants were allowed to bid on ten heterogeneous items labeled $A, B, C, D, E, F, G, H, I$, and J. Bidder values for the first six items were highly super-additive. Five separate draws of valuations were determined in the following manner:

- The single-item packages, (A, B, C, D, E, F), had integer values drawn independently from a uniform distribution with support $[0, 10]$.
- The two-item packages, $(\{A, B\}, \{A, C\}, \ldots \{E, F\})$, took integer values drawn independently from a uniform distribution with support $[20, 40]$.
- The three-item packages, $(\{A, B, C\}, \ldots \{D, E, F\})$, had integer values determined independently by draws from a uniform distribution with support $[140, 180]$.
- The value for the six-item package, $\{A, B, C, D, E, F\}$, was drawn uniformly from $[140, 180]$.

For each period, a total of 25 unique packages and valuations were generated by the previous steps. Each bidder was randomly given five of the packages. In order to obtain the value for a particular package, the bidder had to obtain all objects in the package, and the bidder could not include the same object in multiple packages. The bidder's value for a disjoint combination of packages was given by the summation of the package values. All other packages had zero value to the bidder. In general, a combination of two three-item packages formed the largest total value. However, the optimal package configuration is typically overlapped by other competing packages. Therefore, these valuations were meant to be a difficult test of any allocation mechanism. An indicator of that difficulty is that, in period 3 and 5 *competitive* equilibrium prices did

---

[26] Cybernomics (2000) reports on experiments in which a package bidding extension for the SMR was tested that did not use prices but provided eligibility benefits for certain bids. In a different class of environments, they find efficiency results similar to ours, but the amount of time required to complete an auction was generally longer. They were aware of our results prior to their work but chose not to adopt the pricing feature of our design.

Table 1. *Values in a Spatial Fitting Example*

Bidder 1	Packages:	{F}	{C, D}	{B, C, F}	{B, D, E}	{A, B, E}
	Values:	9	22	128	130	120
Bidder 2	Packages:	{B}	{D, F}	{A, E}	{A, F}	{A, B, D}
	Values:	8	28	24	27	130
Bidder 3	Packages:	{C}	{A}	{D}	{B, D}	{A, B, F}
	Values:	2	3	8	20	119
Bidder 4	Packages:	{E}	{A, B, C}	{A, D, F}	{B, D, F}	{A, E, F}
	Values:	10	117	112	128	125
Bidder 5	Packages:	{C, F}	{D, E}	{C, E, F}	{B, E, F}	{A, B, C, D, E, F}
	Values:	29	25	117	125	142

not exist. In Table 1, we provide a sample set of spatial fitting valuations (Period 2). In this example, the efficient package combination is {A, B, D} for bidder 2 and {C, E, F} for bidder 5.

The valuations for the remaining four objects (G, H, I, J) were determined in an additive manner. Each bidder had a valuation for each individual object between 40 and 180. If a bidder obtained more than one of these items, they received the sum of their valuations. Therefore, competitive equilibrium prices lie between the highest and the second highest valuation for each of the objects. These items were added to the spatial fitting environment for two reasons. First, since we suspected that under some auction designs bidders would be making net losses on the first six objects, these objects would serve as a convenient tool to ensure that bidders' overall payoffs for the auction were not negative.[27] Second, performance in these markets could provide a quick check of any auction's proficiency in the easiest of environments.

All sessions were conducted using members of the Caltech community, primarily undergraduates. Five subjects participated in each experimental session. In each session, the number of auctions (or periods) completed varied. No session lasted longer than three hours. Subjects received new redemption value sheets at the beginning of each new auction.

Bidder values were kept private. At the end of each auction, subjects calculated their profits and converted the token values into dollars. Subjects were paid privately at the end of the experimental session. In addition to participating in a practice auction, all subjects had prior experience with the general auction format; they had all participated in training sessions that utilized simplified auction rules and environments.

A total of 25 RAD and 17 SMR auctions were completed in 15 experimental sessions. The AUSM data come from 12 auctions completed in previous experiments reported by Ledyard et al. (1997).[28]

5. Performance Measures

When choosing an auction design, a variety of criteria and measures may be used. In general there will be trade-offs between these measures; different auctions will perform

---

[27] In reality, in many of the SMR auction sessions even these four additive objects were not enough.
[28] Due to the second-hand nature of the AUSM data, we could not compare AUSM to RAD and SMR in all cases. Likewise, we were unable to distinguish the exact period observations for the AUSM data as reported in Table 2.

better depending on which measure one focuses. For example, high efficiency may sometimes come at the cost of seller revenue and the time to complete the auction.

5.1. Efficiency

Efficiency is the most obvious choice of a performance measure. It was, in fact, the original policy goal of the FCC PCS auction design. In any environment, each bidder has a set of valuations that can be indicated as a (payoff) function $V^i : \{0, 1\}^K \to \mathbb{R}$ where $V^i(y)$ is bidder $i's$ redemption value, the amount the experimenter will pay that bidder, if they hold the combination of objects indicated by y at the end of the auction. The maximal possible total valuation is:

$$V^* = \max \sum_{i=1}^{I} V^i(y^i)$$

subject to

$$\sum_{i=1}^{N} y_k^i \leq 1 \text{ for all, } k = 1, ., K$$

$$y^i \in \{0, 1\}^K.$$

If $\{\hat{y}^i\}_{i=1}^{I}$ is the final allocation chosen in an auction, the *efficiency* of that auction is

$$E = \frac{\sum_i V^i(\hat{y}^i)}{V^*}.$$

It is true (see Ledyard et al. (1997)) that the absolute level of efficiency can be deceptive since one can increase the percentage by simply adding a constant amount to each V^i function. This leaves the efficient allocation unchanged but increases E when $E < 1$. However, we will only use efficiency to compare performance across auctions in the same environment. So this is not a problem for us.

5.2. Seller's Revenue

If the auction designer happens to also be the seller, he may be interested in maximizing revenue: the sum of the final bids. Since revenue can vary significantly across environments, we used the percentage of the maximum possible surplus (V^*) which is actually captured by the seller as our measure of seller revenue. Revenue as percentage of maximum possible surplus is given by

$$R = \frac{\sum_i \sum_k \lambda_k^* \hat{y}^i}{V^*}$$

where λ^* is the vector of final prices. As with efficiency, it is not the absolute value we care about but relative performance across auctions.

5.3. Bidder Profit

Bidder profit is another possible performance measure. With the presence of significant complementarities, some auction mechanisms can cause some bidders to lose money (Bykowsky et al. 2000). A high probability of losses can lead to a variety of performance failures. Bidders may be unwilling to participate in auctions they know they are likely to lose money in. They may not bid aggressively and thereby cause efficiency losses. Losses may also lead a bidder to default on payment contracts, which in turn undermines the credibility of the auction. Increasing the surplus to the bidders can, however, conflict with a goal of high revenue for the seller. All other things being equal (including efficiency of the auction), any increase in bidder profits must come at the expense of seller revenue. Therefore, while it may not be clear why a designer would want to maximize bidder profitability, there does seem to be a compelling reason to avoid bidder losses. In all of the experimental sessions we report on in this paper, a bidder's profit on the bid i is given by

$$P^i = V^i(\hat{y}^i_t) - \lambda^* \cdot \hat{y}^i$$

where λ^* is the vector of final prices.

5.4. Net Revenue

Because of the possibility of bidder losses, we also measure what the auctioneer might expect to actually collect at the end of an auction. It is likely that bidders who made losses would default on their payments after the auction is over. Assume that any bidder that would experience losses by completing the deal does default on at least the portion of their bid that is not profitable.[29] What the auctioneer would actually collect under these circumstances is given by net revenue as a percentage of maximum possible revenue,

$$NR = \frac{(\sum_i \sum_k \lambda^*_k \hat{y}^i) + \sum_i L^i}{V^*}$$

where

$$L^i = \begin{cases} P^i & \text{if } P^i < 0 \\ 0 & \text{otherwise.} \end{cases}$$

In other words, revenue is only generated from the portion of sales that are profitable for the bidder as well.

5.5. Auction Duration

When analyzing iterative auctions, the duration of the auction becomes a relevant concern. In this paper we measure auction duration by the number of iterations (rounds)

---

[29] Our measure of net revenue is designed to be conservative in favor of revenue generation in auctions with losses. Therefore, we only subtract the actual losses from the revenue amount. It might be reasonable to assume that a bidder would default on their entire payment if they ended the auction at a loss. This would obviously greatly reduce the net revenue calculation.

before the auction is completed. Increased iterations can reduce seller profitability because each iteration typically has some fixed administrative cost as well as the possible opportunity costs of foregone rental revenue on the objects. Obviously, one could hold an auction in one iteration as a sealed bid auction, but that generally leads to lower efficiency and revenue. There is a possible trade-off between auction duration and efficiency. Increased iterations may allow high value bidders to *find* the right package thus increasing efficiency.

Since the spatial fitting and additive environments were run simultaneously, the number of iterations until the entire auction closed is not necessarily an accurate performance measure of auction duration for either environment. In order to determine the auction duration for the additive markets, we identified the round that these four markets would have closed if there were no spatial fitting markets. For example, an auction may have lasted 12 iterations, but the last new bid on any of the additive valued items occurred in the sixth iteration. Then, the auction for the additive environment would be said to have ended in iteration seven since, assuming bidding would have been identical, the auction for just the additively valued objects would have ended after no new bids were placed in the seventh round. While it is possible that the addition of the spatial fitting environment may have altered bidding behavior on the additive items, and vice versa, this measure seems to be a reasonable proxy for the speed of the auction in the additive environment. The symmetric measure was used for the spatial fitting environment.

6. Results

In this section, we compare the performance of the RAD, SMR and AUSM mechanisms. The bottom line is that in complex environments, RAD yields higher efficiencies, higher net revenues, and lower bidder losses than does SMR, and RAD does it in many fewer iterations. Also, we find that AUSM lies somewhere in between RAD and SMR in performance in complex environments. There is no performance difference in simple additive environments.

6.1. Results From the Spatial Fitting Test Bed

We begin by considering efficiency. The average auction efficiency across periods under the SMR design was 67%. The average efficiency for the RAD design was 90%. The continuous AUSM obtained an average efficiency of 94%, which is not significantly different from the results for RAD. Table 2 gives the results of Wilcoxon-Mann-Whitney Rank-Sum pair-wise comparisons of these three institutions.[30] RAD significantly improves the efficiency of the allocation over SMR in all periods.

---

[30] The Wilcoxon-Mann-Whitney Rank-Sum test is a powerful nonparametric substitute to the standard *t*-test when data has at least ordinal measurement (Siegel & Castellan 1988). When examining data generated from human subjects, it is typical to assume that the data does not meet the assumptions required for a *t*-test. A high test statistic, z, indicates that the second institution is stochastically larger (in terms of the performance measure) than the first. The reported p's are the p-values associated with the null hypothesis that the first institution is greater than or equal to the second institution.

Table 2. *Spatial Fitting Wilcoxon-Mann-Whitney Rank-Sum Test Results*

Performance Measure	Institutions Compared		
	SMR v. AUSM	SMR v. RAD	AUSM v. RAD
Efficiency	$z = 3.29$	$z = 3.55$	$z = .332$
	$p = .000$	$p = .000$	$p = .371$
Bidder Profits	$z = 3.05$	$z = 2.83$	$z = .584$
	$p = .002$	$p = .006$	$p = .280$
Net Revenue	$z = 1.28$	$z = 2.23$	$z = 1.23$
	$p = .100$	$p = .013$	$p = .109$

Conclusion 1. RAD yields efficiency at least as high as AUSM and significantly higher than SMR.

These results appear to provide compelling evidence that package bidding is an essential part of an auction if complementarities exist and one desires allocative efficiency. As further evidence of this, 20 out of 25 (80%) auctions under RAD and 10 out of 12 (83%) auctions under AUSM led to full efficiency. This is true in only 4 out 17 (24%) auctions using the SMR design. The fact that both auctions that allow package bidding yield dramatically higher efficiencies than the SMR design suggests an obvious conclusion.

Conclusion 2. Package bidding significantly increases efficiency.

When package bidding was not allowed (SMR) bidders, as a whole, averaged losses of $7.73 in each period for the markets with complementarities. In RAD where package bidding was permitted, bidders earned positive profits on average. ($z = 2.83$, $p = .006$). Total bidder profit averaged $4.23 in RAD and $5.68 in AUSM. Table 2 gives the results of Wilcoxon-Mann-Whitney Rank-Sum pair-wise comparisons of these three institutions. On an individual level, 30 out of 85 (35%) bidders lost money under the SMR auction. Under RAD, only 4 out of 125 (3.2%) bidders ended an auction with losses. Under AUSM, only 1 out 60 (1.7%) bidders ended an auction with losses.

Conclusion 3. Package bidding significantly increases average bidder profits and reduces individual losses.

While the number of bidders with losses decreased when package bidding was allowed, it is surprising that any bidders made losses. Without package bidding losses are to be expected. In order to win a package, bidders must put themselves at risk of only obtaining part of the package. However, when package bidding is allowed, bidders have no incentive to bid for packages above their values. After closely examining the data from experiments where bidders were allowed to bid on packages, we have some conjectures as to why losses occurred. First, eligibility management encourages bidders to bid on as many items as possible in order to keep option open. It is possible that bidders thought that an easy and relatively risk free method to keep their eligibility high was to place small bids on single-item packages even if that bid were higher than its true value. They may have thought that it was very likely that someone would value the object above their small bid and therefore they would not lose money from this bid. However, at times, these small bids were sufficiently large to be winners. In a few experiments, we observed behavior consistent with this rationale. The strategic implications of eligibility management remains to be seriously studied. However, it is clear that it leads to bids that are inconsistent with short-run value maximization.[31]

Second, if a bidder makes a mistake in bidding in early iterations, it may be difficult to escape from it. For whatever reasons, bidders occasionally placed bids that were inconsistent with their valuations. A simple example of this occurs if a bidder had a value of 100 for the package $\{A, B, C\}$, and a value of no more than 25 for any 2 item subset. If that bidder intended to place a bid of 50 on $\{A, B, C\}$ but, through negligence, missed indicating C, they would have a bid of 50 on $\{A, B\}$ yielding a loss of 25 if no one ever bids higher.[32] If those bids were sufficiently high, no other bidder would be able to *rescue* them by out bidding them.[33] One interesting but little studied aspect of practical auction design is the prevention of "typos": unintentional errors in data entry. The hard part is separating "typos" from strategic moves later claimed to be mistakes. We do not pursue this here.

Seller net revenue as a percentage of maximum possible revenue was 74% and 69% under RAD and AUSM respectively. The SMR auction, on the other hand, averaged a net revenue of only 61%. Both RAD and AUSM yield significantly higher net revenue than does SMR. In Table 2, we provide Wilcoxon-Mann-Whitney Rank-Sum pairwise comparisons of the net revenue of the three auction types. The package bid auctions generally are expected to collect all of the revenue generated by the auction. But the revenue for SMR is reduced from the apparent amount of 96% to the realistic expectation of 61% due to the substantial bidder losses. Since there are few instances of bidder losses under package bidding, actual revenue under RAD and AUSM is close to net revenue at 79% and 71% respectively.

Conclusion 4. Package bidding significantly increases seller net revenues—those revenues the seller can expect to collect.

RAD yields somewhat higher net and absolute revenue than AUSM, which implies that the RAD design is able to extract more of the surplus from the bidders. Using a Wilcoxon-Mann-Whitney Rank-Sum test, we find that revenue as a percentages of maximum surplus is significantly greater under RAD ($z = 1.62$, $p = .055$). We conjecture this is because the second highest bidders, the ones whose bids drive the winners to increase their bids, are more easily able to find and express their willingness to pay in the iterative mode than in a continuous mode.

Auction duration, measured as the number of iterations before completion of the auction, was significantly shorter under RAD. Using the same price increment rule in both SMR and RAD, the SMR auctions averaged 16.2 iterations as compared to 3.32 for RAD. A Wilcoxon-Mann-Whitney Rank-Sum Test indicates that the auction duration is significantly shorter under the RAD design than in the SMR auction ($z = 4.98$, $p = .000$). In fact, it often took longer to complete the additive markets than the

---

[31] The use of XOR bids is one obvious solution to this problem.

[32] This actually happened to one of the authors during early software tests.

[33] Under the iterative design, if a bidder realized his mistake before the completion of that round, he could delete the bid. It is easy to imagine that a similar errors could be made in a continuous auction without any hope for correction.

spatial fitting items (see Conclusion 9). Since the AUSM mechanism was a continuous auction, it is obviously not possible to directly compare the speed of these two formats.

Conclusion 5. Auction duration is shortest under RAD.

6.2. Results from the Additive Test Bed

In this section, we report on the results for the four objects that had additive valuations for all bidders. In general, the efficient allocation would require only single-item bids among the additive objects. So package bids would occur only if bidders were attempting a sophisticated strategy[34] to capture a larger share of the objects. But this rarely happened. In only 3 of 25 RAD auctions do the final winning bids contain packages of additive objects. Further, in these 3 auctions the final allocations involve package bids across additive and other objects.[35]

Conclusion 6. Package bids rarely occur among the winning bids in the additive environment.

Although bidders are clearly willing to bid on packages in the additive environment, they are rarely able to use this ability to their advantage as is evidenced by the extremely high levels of efficiency achieved in the additive environment. A 100% efficient auction indicates that all the possible gains from trade (surplus) have been captured by either the bidders or the seller. As expected, all auctions did quite well in terms of efficiency in this environment. In most of the auctions, the four objects were allocated to the highest valuing bidders: 20 of 25 (80%) for RAD and 15 of 17 (88%) auctions for the SMR auction. The AUSM design lead to full efficiency in the additive environment in 10 out of 12 (83%) auctions. There were no significant differences in the level of efficiency achieved by any of the mechanisms. Therefore, package bidding auctions, specifically RAD, do not seem to degrade auction performance in simple settings.

Conclusion 7. In the additive environment, under RAD, SMR, and AUSM, efficiency is very near 100%. There are no discernible differences between the auctions.

In the additive environment, there was very little difference in the revenue collected by the seller under SMR and RAD. The SMR and RAD mechanisms averaged 69.96% and 71.96% of the maximum possible revenue respectively. A rank-sum test also shows no significant difference between the observed revenues.

Conclusion 8. In the additive environment, the auction institutions yield similar seller revenue.

RAD yielded lengths that were significantly shorter than the SMR. The average auction duration under the SMR auction was 11.7 iterations but only 6.1 for RAD.[36] A

---

[34] Such a strategy might be to create an artificial *threshold* that would yield a possible problem for others allowing the bidder to get the items even if it were not an efficient allocation.

[35] We used the final prices to estimate the portion of the bid occurring in the additive environment.

[36] These results are, of course, confounded by the fact that the length the additive part of the auction is the round after which no new bid is made on an additive object. This is not necessarily independent of the existence of the spatial part of the auction.

Wilcoxon-Mann-Whitney Rank-Sum Test indicates that the auction duration is significantly shorter under the RAD design than in the SMR auction ($z = 4.67$, $p = .000$). As before, there is no direct comparison with the speed of AUSM.

Conclusion 9. In the additive environment, auction duration is shortest under the RAD design.

Since package bidding has no advantage in the additive environment, and assuming all bids in the additive markets are on the individual items the prices, we were surprised by this result. A potential explanation for this difference is that RAD allows bidders to quickly learn about the outcomes in the spatial fitting portion of the market. They can then turn their attention to the additive markets.

7. Conclusions and Open Issues

7.1. Conclusions

The experimental test results point to two clear conclusions:

1. The option to bid for packages clearly improves performance in difficult environments, and does not degrade performance in simple environments.
2. RAD dominates SMR in efficiency, net revenue auction duration, and protecting bidders from losses.

The general principle that package bidding is an important option for multi-object auctions in environments with significant complementarities is reaffirmed by the evidence. Auctions that only allow bidding on single-items almost always exhibit lower levels of allocative efficiency and higher bidder losses. When auctions are run in an iterative mode, single-item only bidding can also lead to much longer auctions. The only redeeming feature of these auctions seems to be their revenue generating capabilities. Unfortunately, much of that revenue comes from losses to bidders as opposed to increased surplus extraction. This may be acceptable in the short run if it can be collected. However, if the design is used repeatedly, bidders will learn to avoid these losses, perhaps by avoiding the auction altogether, and efficiency and revenue will ultimately suffer.

But we have gone further here than simply establishing that package bidding is sensible. We have provided a new auction design, RAD, which clearly outperforms others. Relative to the SMR design, RAD produces higher efficiency, greater net revenue, greater bidder profits and a much quicker time to completion. It even produces similar efficiencies to and higher revenues than the continuous AUSM with a standby queue. Since RAD uses a pricing rule instead of a queue to mitigate the threshold problem, it is no more complex from a bidder's point of view than the SMR auction and significantly simpler than the continuous AUSM. Finally, there is no evidence of degradation in performance when RAD is used in simple, additive environments.

Why do we think RAD worked so well? We believe it is the decentralizing influence of the prices. Under the SMR mechanism, prices were only calculated from single-item bids. Therefore, if bidders were not bidding above their valuations, in this environment, it is guaranteed that the single-item prices would be much lower than the actual winning bids for the packages. If we consider the sample parameters given in Table 1, the maximum prices for bidders unwilling to expose themselves to potential losses are: 3, 8, 2, 8, 16, and 9 for the first six items.[37] The competitive prices, which do exist in this case, are 38, 49, 30, 43, 38, and 49. If we examine the data for this parameter set (period 2), we find that this difference between stand alone and competitive prices is prevalent experimentally. The RAD prices, however, are close to the competitive prices. In general we would expect final prices to be somewhat lower than the competitive prices due to the bid increment requirement, which made the true price higher than that reported here. Once the mandatory bid increment is considered, the RAD prices are not significantly different from the competitive prices for five of the six objects. In the RAD mechanism, prices are calculated using all bids. Therefore, in general, they will more closely represent the level of competition for an item. Since the prices are typically calculated in order to indicate the level of competition below the winning packages, they can indicate to bidders markets where bidding is thin. Thus, prices aid in finding an appropriate fit.

7.2. Open Issues

It would seem that the RAD design would be a natural candidate for use as a multi-object iterative auction in its current form. However, in spite of the excellent performance in our tests, there are at least two problem areas that might be considered for redesign. The first, and simplest to fix, is a result of the eligibility rule. If bidders have budgets for items, they may find themselves bidding for and even winning items which have little value to them simply to preserve eligibility. While this problem has generally been recognized even when there is no package bidding, we know of no papers that purport to provide a solution. There is, nevertheless, a straight-forward solution: the use of exclusive "XOR" bids in an iterative auction. Bidders would be allowed to place a bid saying "I bid $1000 for A,B and C OR I bid $800 for D and E". The appropriate constraints would be added to the allocation problem (1) and the rest of the mechanism would be left as is. This could be done to the SMR rules as well as to RAD and others. XOR bids may appear to increase a bidder's problem complexity a bit, but such bids do eliminate the anxiety and confusion raised by the need to find "safe places" to preserve eligibility.

A second problem with the RAD design is that, although the pricing rule seems to guide and coordinate small bidders to solve the threshold problems, it can also orphan some bidders at early stages even through they belong in the efficient allocation. An example will illustrate.

Example 4. Suppose in round five there are four bids submitted as follows:

- Bid #1 for $\{A, B, C\}$ at 99.
- Bid #2 for $\{A, B\}$ at 75.

---

[37] These are simply the maximal single-item values. The only way prices under the SMR design could be higher is if someone bid on a single-item above their value.

- Bid #3 for $\{A, C\}$ at 75.
- Bid #4 for $\{B, C\}$ at 75.

Under the RAD design bid #1 wins and the prices[38] are $(\lambda_A \lambda_B \lambda_C) = (33, 33, 33)$. Now suppose there is a bidder who is willing to pay 30 for $\{A\}$. Had they bid 28 for $\{A\}$ in round five they would have been a winning bid along with bid #4. But now they can't bid since $\lambda_A > 30$. This may lower efficiency. There are several features of RAD that work against such *orphaning*. First, if this bidder had bid in round five, they would not have been orphaned. So aggressive participation helps. Second, suppose in round six, bid #3 is not resubmitted, but 1,2, and 4 are. Then 1 still wins and $\lambda = (24, 51, 24)$.[39] If it gets to this stage our bidder for A can reenter the fray if they still have the eligibility to do so. Of course, if the auction stops in round 5, which it will if there are no additional new bids, it will end at an inefficient allocation.

The crucial point to remember is that, in spite of these problems, RAD attains high efficiency and outperforms SMR. Clearly, the problems facing bidders in SMR, like the exposure problem, are more severe than those problems that face them in RAD.

Acknowledgments

The authors thank the associate editor and two anonymous referees for careful, constructive reviews. The authors thank Elena Katok, Evan Kwerel, Paul Milgrom, Charles Plott, and Rakesh Vohra for their helpful comments. This is a significantly revised version of DeMartini et al. (1999).

Appendix

In Section 3.3, we indicated that the RAD auction pricing algorithm (8) might not yield a unique price vector. We use the following lexicographic routines to eliminate that ambiguity.[40]

Let g^*, λ^*, and Z^* solve (8). If $Z^* = 0$ then go to problem (10) below. If $Z^* > 0$, let $J^* = \{j \in L_t \mid Z^* = g*^j\}$. If $J^* = L_t$ then go to (10) below. Otherwise,

$$\min_{\lambda^t, Z, g} Z \qquad (9)$$

---

[38] Notice that these are not separating prices, which is what causes a problem.

[39] These are separating prices. This also shows that prices are not necessarily monotonically increasing (since $24 < 33$). The sum of prices is however always increasing.

[40] An alternative approach would minimize $\sum_j (g^j)^2$ in (8), which would avoid iteration. We chose to stick with linear programs for computational simplicity and a desire to minimize the number of bids missed rather than the total size of the miss.

subject to

$$\sum_{k \in K} \lambda_k^t X_k^j = p^j \text{ for all } (p^j, x^j) \in W_t$$

$$\sum_{k \in K} \lambda_k^t X_k^j + g^{*j} = p^j \text{ for all } (p_i^j, x^j) \in J^*$$

$$\sum_{k \in K} \lambda_k^t X_k^j + g^j = p^j \text{ for all } (p_i^j, x^j) \in L_t \setminus J^*$$

$$0 \le g^j \le Z \text{ for all } (p_i^j, x^j) \in L_t \setminus J^*$$

$$\lambda^t \ge 0.$$

Let $\hat{Z}, \hat{g}, \hat{\lambda}$, be the solution to (9). If $\hat{Z} = 0$ go to 10 below. Otherwise, let $\hat{J} = \{j \mid \hat{Z} = \hat{g}^j\}$. If $J^* \cup \hat{J} = L_t$ then go to problem (10) below. Otherwise, let $J^* = J^* \cup \hat{J}$ and go to (9) again.

When the iteration on (9) is complete we will have prices which approximate our "ideal" but not always obtainable prices. They may still not be unique. So, we go through a sequence of iterations which eliminate non-uniqueness and which create prices to guide bidders to solve the threshold problem. Let $\hat{g}$ be the solution from the last iteration of (9). Let $\hat{K} = K$.

$$\max_{Y, \lambda} Y \quad (10)$$

subject to

$$\sum_{k \in K} \lambda_k^t x_k^j = p^j \text{ for all } (p^j, x^j) \in W_t$$

$$\sum_{k \in K} \lambda_k^t x_k^j + \hat{g}^j = p^j \text{ for all } (p^j, x^j) \in L_t$$

$$\lambda_k^t \ge Y \text{ for all } k \in \hat{K}. \quad (11)$$

Let Y^*, λ^* solve (10). Let $K^* = \{k \in K \mid \lambda_k^t = Y^*\}$. Let $\tilde{K} = \hat{K} \setminus K^*$. If $\tilde{K} \ne \emptyset$, return to (10) and solve it replacing (11) with

$$\lambda_k^t \ge Y \text{ for all } k \in \tilde{K}$$

$$\lambda_k^t = \lambda_k^{*t} \text{ for all } k \in K \setminus \tilde{K}.$$

When $\tilde{K} = \emptyset$, we are done and the prices $\lambda^* = \lambda^{t+1}$. These are unique, approximate the ideal prices, and provide signals about thresholds.

References

Andersson, A., M. Tenhunen, F. Ygge. 2000. Integer programming for combinatorial auction winner determination. *Proceeding of the Fourth International Conference on Multiagent Systems.*

Armstrong, M. 2000. Optimal multi-object auctions. *Review of Economic Studies* **67**(3), 455–81.

Ausubel, L., P. Milgrom. 2002. Ascending auctions with package bidding. *Frontiers of Theoretical Economics* **1**(1).

Banks, J. S., J. O. Ledyard, D. P. Porter. 1989. Allocating uncertain and unresponsive resources: an experimental approach. *Rand Journal of Economics* **20**(1), 1–25.

Bikhchandani, S., J. M. Ostroy. 2002. The package assignment model. *Journal of Economic Theory* **107**(2), 377–406.

Bykowsky, M. M., R. J. Cull, J. O. Ledyard. 2000. Mutually destructive bidding: The FCC auction problem. *Journal of Regulatory Economics* **17**(3), 205–28.

Conen, W., T. Sandholm. 2001. Preference elicitation in combinatorial auctions: Extended abstract'.

Coppinger, V. M., V. L. Smith, J. A. Titus. 1980. Incentives and behavior in english, dutch and sealed-bid auctions. *Economic Inquiry* **43**, 1–22.

Cybernomics. 2000. An experimental comparison of the simultaneous multiple round auctions and the CRA combinatorial auction, Technical report. Report to the Federal Communications Commission.

Dasgupta, P., E. Maskin. 2000. Efficient auctions. *Quarterly Journal of Economics* **115**(2), 341–388.

Davenport, A. J., C. An, E. Ng, G. Hohner, G. Reid, H. Soo, J. R. Kalagnanam, Rich, J. 2003. Combinatorial and quantity-discount procurement auctions benefit mars, incorporated and its suppliers. *Interfaces* **33**(1), 23–35.

DeMartini, C., A. Kwasnica, J. Ledyard, D. Porter. 1999. A new and improved design for multi-object iterative auctions, Social Science Working Paper 1054, California Institute of Technology, Pasadena, CA.

Dunford, M., M. Durbin, K. Hoffman, D. Menon, R. Sultana, T. Wilson. 2001. Issues in scaling up the 700 MHz auction design. http://wireless.fcc.gov/auctions/conferences/combin2001/presentations.html. Presentation.

Federal Communications Commission 2002. Auction of licenses in the 747–762 and 777–792 MHz bands scheduled for june 19, 2002: Further modification of package bidding procedures and other producerdures for auction no. 31, Technical Report AUC-02-31-B, Federal Communications Commission.

Ferejohn, J., R. Forsythe, R. Noll. 1979. An experimental analysis of decision making procedures for discrete public goods: A case study of a problem in institutional design, *in* V. L. Smith, ed., Research in Experimental Economics. Vol. 1, JAI Press, pp. 1–58.

Graves, R., L. Schrage, J. Sankaran. 1993. An auction method for course registration. *Interfaces* **23**(5), 81–92.

Grether, D., M. Isaac, C. Plott. 1981. The allocation of landing rights by unanimity among competitors. *American Economic Review* **71**, 166–171.

Groves, T., J. Ledyard. 1977*a*. Some limitations of demand-revealing processes. *Public Choice* **29**(2), 107–124. Special Supplement to Spring 1977 Issue.

Groves, T., J. Ledyard. 1977*b*. Some limitations of demand-revealing processes – reply. *Public Choice* **29**(2), 139–143. Special Supplement to Spring 1977 Issue.

Ishikida, T., J. Ledyard, M. Olson, D. Porter. 2001. The design of a pollution trading system for souther california's RECLAIM emissions trading program, *in* R. M. Isaac, ed., Research in Experimental Economics. Greenwich, CT.

Kagel, J. H., R. M. Harstad, D. Levin. 1987. Information impact and allocation rules in auctions with affiliated private values: A laboratory study. *Econometrica* **55**, 1275–1304.

Kwasnica, A. M., J. O. Ledyard, D. Porter, J. Scott. 1998. The design of multi-object multi-round auctions, Social Science Working Paper 1045, California Institute of Technology, Pasadena, CA.

Larson, K., T. Sandholm. 2001*a*. Computationally limited agents in auctions, *in* AGENTS-01 Workshop of Agents for B2B. pp. 27–34.

Larson, K., T. Sandholm. 2001*b*. Costly valuation computation in auctions. *TARK: Theoretical Aspects of Reasoning about Knowledge* **8**.

Ledyard, J. O., M. Olson, D. Porter, J. A. Swanson, D. P. Torma. 2002. The first use of a combined value auction for transportation services. *Interfaces* **32**(5), 4–12.

Ledyard, J. O., D. Porter, A. Rangel. 1997. Experiments testing multiobject allocation mechanisms. *Journal of Economics and Management Strategy* **6**, 639–675.

Milgrom, P. R. 2000. Putting auction theory to work: The simultaneous ascending auction. *Journal of Political Economy* **108**(2), 245–272.

Milgrom, P. R., R. J. Weber. 1982. A theory of auctions and competitive bidding. *Econometrica* **50**(5), 1089–1122.

Myerson, R. B. 1981. Optimal auction design. *Mathematics of Operations Research* **6**, 58–73.

Nisan, N. 2000. Bidding and allocation in combinatorial auctions, *in* ACM Conference on Electronic Commerce. pp. 1–12.

Nisan, N., I. Segal. 2003. The communication requirements of efficient allocations and supporting lindahl prices. manuscript.

Parkes, D. C. 1999. iBundle: An efficient ascending price bundle auction, *in* Proceeding ACM Conference on Electronic Commerce. pp. 148–57.

Plott, C. R. 1997. Laboratory experimental testbeds: Applications to the PCS auction. *Journal of Economics & Management Strategy* **6**, 605 – 638.

Porter, D. P. 1999. The effect of bid withdrawl in a multi-object auction. *Review of Economic Design* **4**(1), 73–97.

Rassenti, S. J., V. L. Smith, R. L. Bulfin. 1982. A combinatorial auction mechanism for airport time slot allocation. *Bell Journal of Economics* **13**, 402–417.

Rothkopf, M. H., A. Pekec, R. M. Harstad. 1998. Computationally manageable combinatorial auctions. *Management Science* **44**(8), 1131–1147.

Sandholm, T. 2000. Issues in computational Vickrey auctions. *International Journal of Electronic Commerce* **4**(3), 107–129.

Sandholm, T. 2002*a*. Algorithm for optimal winner determination in combinatorial auctions. *Artificial Intelligence* **135**(1–2), 1–54.

Sandholm, T. 2002*b*. eMediator : a next generation electronic commerce server. *Computational Intelligence* **18**(4), 656–676. Special Issue on Agent Technology for Electronic Commerce.

Sandholm, T., S. Suri, A. Gilpin, D. Levine. 2001. Cabob: A fast optimal algorithm for combinatorial auctions, *in* International Joint Conference on Artificial Intelligence. pp. 1102–8.

Siegel, S., N. J. Castellan. 1988. *Nonparametric Statistic for the Behavioral Sciences*, McGraw-Hill.

Smith, V. L. 1979. Incentive compatible experimental processes for the provision of public goods, *in* V. L. Smith, ed., Research in Experimental Economics. Vol. 1, JAI Press, pp. 59–168.

Williams, S. 1999. A characterization of efficient, bayesian incentive compatible mechanisms. *Economic Theory* **14**(1), 155–80.

CHAPTER 20

Hierarchical Package Bidding: A Paper & Pencil Combinatorial Auction

Jacob K. Goeree and Charles A. Holt[*]

1. Introduction

Auctions with multiple items are typically conducted in an environment in which bidders' values depend on acquiring combinations, e.g. networks of broadcast licenses or timber rights for adjacent tracts of land. Concerns for economic efficiency and revenue enhancement have led the Federal Communications Commission (FCC) to run auctions simultaneously for large numbers of licenses in a series of bidding rounds, with provisional winners being announced after each round. Under the simultaneous multi-round auction format (SMR), the highest bid on each license becomes the provisional price that must be topped in a subsequent round. This approach has been copied in other countries with considerable success, but experimental evidence indicates that efficiency and revenue may be reduced when bidders hesitate to incorporate synergy values into their bids for fear that they will end up winning only part of a desired package (see, for instance, the references in Brunner et al., 2010).

The "exposure problem" is a major concern in what is arguably the auction of a lifetime, i.e. the upcoming FCC 700 MHz spectrum auction.[1] This spectrum has better propagation and penetration properties than any spectrum sold before and is extremely valuable for wireless applications (the FCC has set *minimum* prices at over 10 billion). More importantly, the wireless industry is concentrated and the 700 MHz auction provides the last opportunity for a new firm to enter the market. For an entrant to be

---

[*] We are grateful to Leslie Marx, Paul Milgrom, Rudolf Müller, and Martha Stancill, two anonymous referees and the co-editor for helpful suggestions. We thank Christoph Brunner, Maggie McConnell, Lindsay Osco, and Kevin Watts for research assistance, and Dash Optimization for the free academic use of their Xpress-MP software. We acknowledge partial financial support from the Federal Communications Commission (FCC contract 05000012), the Alfred P. Sloan Foundation, the Gordon and Betty Moore Foundation, the Gates Grubstake Fund, the National Science Foundation (SBR 0551014), and the Dutch National Science Foundation (NWO-VICI 453.03.606).

[1] The 700 MHz spectrum has been coined the "FCC's crown jewels" by FCC commissioner Adelstein and is more generally referred to as "beach front property." According to former FCC chief of staff Blair Levin, "...the 700 MHz auction will be the biggest spectrum auction ever held."

successful in the wireless market, however, it has to acquire a nationwide footprint, which is virtually impossible with the current SMR format because of exposure risk.

Some pre-packaging of licenses into larger groups may help solve the exposure problem. In FCC Auction 65, for example, bids were proportionally higher for large blocks of bandwidth for air-to-ground communications; a block with 3 times as much bandwidth as a smaller block sold for about 4.5 times as much. In Auction 66 that closed in the fall of 2006, a 20 MHz band divided into over 700 local areas sold for 2.27 billion dollars, but the same amounts of bandwidth sold for 2.45 billion when divided into 176 regions, and for 4.17 billion when divided into 12 regions. Of course, pre-packaging generally disadvantages small bidders who might be part of an optimal allocation due to incumbency or efficient local operations. Instead of dividing licenses into large or small groups, there may be considerable gains in allowing competition that determines the packages, e.g. between bidders on individual licenses, regional groups, and a single national license. A combinatorial auction solves this problem by allowing bids on packages of various sizes and letting the bidding competition determine the market structure.

There are two steps to be done after each round of a combinatorial auction: (i) the determination of provisional winners or "assignment" part, (ii) and the information provision or "pricing" part. The assignment part, which involves finding the non-overlapping bids that maximize seller revenue, is easy to explain but NP-hard (non-deterministic polynomial-time hard) to do. The number of possible allocations grows exponentially, and with many objects for sale there is no guarantee the best allocation is found in a reasonable amount of time – this is commonly referred to as the computational complexity problem. The problem is manageable in the sense that the computer can run for a fixed amount of time and the best solution at that point can be chosen. Then this solution can be used as a lower bound from where next round's solution can be found ("branch-and-bound"). However, some experts claim they can "wreck any combinatorial auction with enough licenses" by abusing computational issues. In any case, bidders will not be able to reproduce the outcome of a round to understand why their bids did not win, unless they solve an NP-hard problem quickly. Approximations must be particularly worrisome for public officials who anticipate that losing bidders may later complain about the assignments and prices in a particular round.

Rothkopf, Pekeč, and Harstad (1998) propose a type of "hierarchical" pre-packaging to avoid computational issues in combinatorial auctions. In this approach, there are several hierarchy levels of varying package sizes. For instance, if there are only three levels then the lowest could contain individual licenses, the middle level could contain non-overlapping regional packages, and the highest level could contain the national package. With this tree structure, the revenue maximization problem is recursive and can be solved in a linear manner, since revenue-maximizing "winners" at one level can be compared with those at the next level up in the hierarchy.

This paper proposes and tests a simple pricing formula for hierarchical combinatorial auctions. If a bid on an individual license is provisionally winning, then that bid would become the price for the license, as is the case for SMR. The idea underlying the pricing is that prices for individual licenses would have to be scaled up by lump-sum "taxes" to share the burden of unseating a provisionally winning package bid. For example, if a bid on a regional package is provisionally winning, prices for individual licenses would be increased so that bidders on individual licenses in that region would know how high they have to bid to unseat the provisional regional winner. In this sense, prices help these bidders solve a coordination or "threshold problem," since each would prefer that someone else bear the cost of unseating the package bid.

The key feature of the proposed mechanism is that the resulting license prices are composed of intuitive and easy to compute components that match the transparency of the recursive revenue-maximizing allocation rule. These simple hierarchical allocation and pricing rules are the basis for the "hierarchical package bidding" (HPB) format that the FCC will use for the upcoming 700 MHz auction. In the Procedures Public Notice (October 5, 2007) the FCC motivates their choice as follows "... *we will use HPB in part because the mechanism for calculating prices is significantly simpler than other package bidding formats* ..."[2]

Even though a simple tree structure with hierarchical package bidding provides straightforward price indicators for how high bids on individual licenses and packages must be to "get into the action," there is a concern that the definitions of non-overlapping packages at each level may not match the interests of particular bidders. For example, suppose that a bidder has super-additive values for multiple licenses that are not spanned by a particular regional package definition. In this case, package bidding does not fully protect from exposure risk, and the "second-best" nature of the constraints added by hierarchical package bidding may result in lower rather than higher revenues and efficiencies. This possibility motivated the laboratory experiments to be discussed below. For purposes of comparison, we also consider the SMR format currently used by the FCC, and a fully flexible auction format that does not restrict package bids that can be submitted. This fully flexible format is a version of the FCC's Modified Package Bidding (MPB) that uses the Resource Allocation Design (RAD) pricing mechanism proposed by Kwasnica et al. (2005).[3] RAD prices are essentially approximations to shadow prices determined by the dual of a constrained revenue-maximization problem, as explained in more detail below.

In the next section we provide a general description of the assignment and pricing rules for auctions with hierarchically pre-defined packages. Section 3 discusses the experiment design and procedures. Section 4 provides aggregate statistics on efficiency, revenue, and bidder profits for the three auction formats. Section 5 contains a discussion of the exposure problem we observe in the SMR format and the threshold problem that occurs with flexible package bidding. In spectrum auctions, it is often the case that licenses that cover different blocks of spectrum within the same geographic region are viewed as substitutes while licenses that cover different regions are viewed as complements. In Section 6, we present results for two additional experiments based on related designs with a mix of substitutes and complements. Section 7 concludes. A summary data table can be found in the Appendix.

---

[2] See http://fjallfoss.fcc.gov/edocs_public/attachmatch/DA-07-4171A1.pdf. The specific pricing and assignment rules are described in Appendix H. An earlier FCC Public Notice that invited feedback about our mechanism mentions the importance of the experiments and the relative success of HPB, see http://fjallfoss.fcc.gov/edocs_public/attachmatch/DA-07-3415A1.pdf. We would like to stress that the HPB format resulted from independent research. We designed the auction prior to proposing it to the FCC who responded by asking us to test it in the lab.

[3] The MPB experiments reported below employ the RAD design described by Kwasnica et al. (2005).

2. Simple Combinatorial Assignment and Pricing

The hierarchical structures considered here can be formed by repeatedly breaking up larger packages into smaller ones. For example, the top level could consist of a single nationwide package, which is divided into smaller regional packages at the second level, which in turn contain many individual licenses belonging to the first level. More generally, let there be $H \geq 1$ hierarchy levels, labeled by $h = 1, \ldots, H$, where level h contains I_h packages. The lowest level contains the smallest units, e.g. individual licenses, and higher levels consist of bigger packages. Level-h packages are denoted $P_{i_h}^h$ for $i_h = 1, \ldots, I_h$ and cover $\alpha_{i_h}^h$ bidding units (e.g. population times bandwidth). The number of packages within a hierarchy falls as we go up in the hierarchy tree, i.e. $I_{h'} \leq I_h$ for $h' \geq h$. Packages within a hierarchy are non-overlapping, i.e. for level-h packages $P_{i_h}^h \neq P_{j_h}^h$ we have $P_{i_h}^h \cap P_{j_h}^h = \emptyset$. Furthermore, a package from a lower hierarchy level is contained in exactly 1 package from each of the higher levels, i.e. for each $P_{i_h}^h$ and for each $h' > h$ there is a unique level-h' package $P_{j_{h'}}^{h'}$ such that $P_{j_{h'}}^{h'} \supset P_{i_h}^h$. The number of bidding units covered by a package equals the number of bidding units covered by the level-1 items it contains

$$\sum_{P_{i_1}^1 \subset P_{i_h}^h} \alpha_{i_1}^1 = \alpha_{i_h}^h. \qquad (1)$$

In particular, each hierarchy level covers the entire nation: $\sum_{i_h=1}^{I_h} \alpha_{i_h}^h = \alpha$ for each level $h = 1, \ldots, H$, with α the total number of bidding units in the nation.

The combinatorial auction with hierarchically structured packages is "simple" in the sense that the assignment problem can be solved in a linear manner. First, for each hierarchy level h, we find the highest bids on the packages within the hierarchy. Let these best bids be denoted by $b^{\max}(P_{i_h}^h)$. To find the optimal assignment and revenue we follow a recursive algorithm, defining revenues $R(P_{i_h}^h)$ for packages of all levels.

1. Set $h = 1$ and define revenues for packages in this level to be $R(P_{i_1}^1) = b^{\max}(P_{i_1}^1)$ for $i_1 = 1, \ldots, I_1$. Furthermore, label these high bids "provisionally winning."
2. If $h < H$, increase h by 1 and do step 3, otherwise quit.
3. If $b^{\max}(P_{i_h}^h) > \sum_{P_{i_{h-1}}^{h-1} \subset P_{i_h}^h} R(P_{i_{h-1}}^{h-1})$, where the sum is over all level-$(h-1)$ packages contained in $P_{i_h}^h$, then $R(P_{i_h}^h) = b^{\max}(P_{i_h}^h)$ and $b^{\max}(P_{i_h}^h)$ is labeled provisionally winning and bids from all lower levels on packages that overlap with $P_{i_h}^h$ are unmarked. Otherwise, $R(P_{i_h}^h) = \sum_{P_{i_{h-1}}^{h-1} \subset P_{i_h}^h} R(P_{i_{h-1}}^{h-1})$. Return to step 2.

By construction, the maximum revenue is the sum of revenues for items in the top level, $R = \sum_{i_H=1}^{I_H} R(P_{i_H}^H)$, and the provisionally winning bids are those that are still marked after the algorithm finishes. Note that the total number of comparisons is linear in the number of predefined packages.

Prices are assigned to the smallest possible objects, i.e. objects in the lowest hierarchy level. The main idea is to match the recursive approach of the assignment part and add a "tax" to level-1 prices (proportional to the number of bidding units covered) if the revenue of a lower level falls short of that of the next level up. This tax is such that for every hierarchy level, h, the best revenue $R(P_{i_h}^h)$ associated with a level-h package

$P_{i_h}^h$ can be obtained by summing the prices of the level-1 packages contained in $P_{i_h}^h$. A fortiori, summing the prices of all level-1 packages yields the best possible revenue. Let $p(P_{i_1}^1)$ denote the price of level-1 package $P_{i_1}^1$.

1. Set $h = 1$ and define prices for packages in this level to be $p(P_{i_1}^1) = b^{\max}(P_{i_1}^1)$ for $i_1 = 1, \ldots, I_1$.
2. If $h < H$, increase h by 1 and do step 3, otherwise quit.
3. For each level-h package $P_{i_h}^h$, add $\tau^h(P_{i_h}^h) = \frac{\alpha_{i_1}^1}{\alpha_{i_h}^h}(R(P_{i_h}^h) - \sum_{P_{i_{h-1}}^{h-1} \subset P_{i_h}^h} R(P_{i_{h-1}}^{h-1})) \geq 0$ to the price $p(P_{i_1}^1)$ of each level-1 package $P_{i_1}^1$ contained in $P_{i_h}^h$. Return to step 2.

The recursive algorithms used to determine prices and allocations are similar and could be combined to run simultaneously after each round of the auction. Furthermore, they are trivial from a computational viewpoint.

The level-1 prices that result from this recursive approach can be neatly summarized as:

$$p(P_{i_1}^1) = b^{\max}(P_{i_1}^1) + \sum_{P_{i_h}^h \supset P_{i_1}^1} \frac{\alpha_{i_1}^1}{\alpha_{i_h}^h}\left(R(P_{i_h}^h) - \sum_{P_{i_{h-1}}^{h-1} \subset P_{i_h}^h} R(P_{i_{h-1}}^{h-1})\right), \tag{2}$$

where the (outer) sum is over all higher-level packages that contain $P_{i_1}^1$. There is exactly one such package at each higher level, so the sum contains a single term for each $h > 1$, which is the level-h tax $\tau^h(P_{i_1}^1)$, defined above. The right-side of (2) is therefore equal to a "base price" augmented with non-negative taxes: $p(P_{i_1}^1) = b^{\max}(P_{i_1}^1) + \sum_{h>1} \tau^h(P_{i_1}^1)$. This interpretation was used in the instructions phase of the experiments reported below.

Proposition. *The prices defined in (2) have the following properties:*

(i) *Prices reduce to standard SMR prices in the absence of package bids. Taxes, if any, are proportional to the number of bidding units covered.*
(ii) *Prices signal how high bids must be to unseat current winners at any hierarchy level.*
(iii) *Prices are "market clearing," i.e. the sum of individual license prices within a package exceeds (equals) the amount of a losing (winning) bid for the package.*

Proof. Without package bids $R(P_{i_h}^h) = \sum_{P_{i_1}^1 \subset P_{i_h}^h} b^{\max}(P_{i_1}^1)$ for all packages $P_{i_h}^h$ so (2) reduces to $p(P_{i_1}^1) = b^{\max}(P_{i_1}^1)$. To show (ii), define the level-h revenue $R(h) = \sum_{i_h=1}^{I_h} R(P_{i_h}^h)$. Using (1), it is readily verified that the linear prices in (2) sum up to level-H revenues:

$$\sum_{i_1=1}^{I_1} p(P_{i_1}^1) = R(1) + \sum_{h>1}(R(h) - R(h-1)) = R(H),$$

which is the maximum revenue by construction. Hence, prices of the level-1 items sum to the winning bids, irrespective of the levels in which the winning bids occur. Property (iii) follows by construction and reflects the fact that the assignment problem has an integer solution, so dual prices exist. ☐

The proportional tax property in (i) ensures that small licenses do not get overpriced.[4] The second property shows how the prices help smaller bidders avoid the "threshold" problem that potentially occurs when they have to coordinate their bids in response to an aggressive package bid. The third property does not generally hold with flexible (non-hierarchical) package bidding.[5]

3. Experimental Design

3.1. Auction Formats

Our main design involves 7 bidders and 18 licenses, with three alternative auction formats. All experiments were conducted using *j*Auctions, which enables bidders to create "custom" packages in order to see the value complementarities associated with winning combinations of licenses.[6] Under flexible package bidding (MPB), bidders could bid on these self-created packages. Under the baseline simultaneous multi-round auction (SMR), these custom packages would be shown but could not be placed into the bidding basket. Finally, under hierarchical package bidding (HPB) bidders are permitted to submit bids for pre-defined packages but not for custom packages.

Without package bidding (SMR), the highest bids submitted for each license in a round become the provisionally winning bids. With package bidding (either MPB or HPB), the provisionally winning bids for licenses or packages are those that maximize seller revenue. At the end of each round, bidders receive information on all provisionally winning bids (for licenses and packages) and the corresponding ID numbers. Bidders also see the prices for all licenses, the sum of their own values for the licenses and packages that they are provisionally winning, and the sum of prices that would be paid for those licenses and packages if the auction had ended. Bidding continues from round to round until no new bids are submitted (or withdrawn under SMR), at which time the provisionally winning bids become the final bids that determine allocations and prices paid. Under SMR, bidders have limited opportunities to withdraw provisionally winning bids: in at most two rounds of the auction, bidders can withdraw as many provisionally winning bids as they wish.[7] Withdrawals are subject to penalties that compensate the seller for lower prices obtained: if the license is sold, the penalty

---

[4] Consider a two-level hierarchy with 1 national package divided into three licenses, labeled A, B, and C, which cover different bidding units: A and C are small ($\alpha_A = \alpha_C = 1$) while B is large ($\alpha_B = 10$). Suppose that values for A and C are somewhere in the [5,20] range while the value for B is somewhere in the [50,100] range. Furthermore, suppose there are three small bidders and one large bidder: bidder 1 wants A, bidder 2 wants B, and bidder 3 wants C, while bidder 4 values only the package ABC. If opening bids of the small bidders are 4 on licenses A–C while bidder 4 places a bid of 60 on ABC then without correcting for bidding units, prices are $p_A = p_B = p_C = 20$. This would cause bidders 1 and 3 to drop out since their values are below 20. In other words, small licenses get over-priced, thereby eliminating small bidders from the auction. In contrast, the pricing formula (2) yields prices of $p_A = p_C = 4 + \frac{1}{12}(60 - 12) = 8$ and $p_B = 4 + \frac{10}{12}(60 - 12) = 44$, giving small bidders a chance to compete.

[5] Consider, for example, the following four bids: $b_{ABC} = 30$ and $b_{AB} = b_{AC} = b_{BC} = 24$. No market clearing prices exist in this case.

[6] *j*Auctions is a JAVA-based suite of auction programs developed by Jacob Goeree.

[7] Since package bidding protects bidders from the exposure problem, withdrawals were not permitted under MPB and HPB.

is equal to the difference between the withdrawn bid and the final sales price (if this difference is positive, otherwise the penalty is 0). If the license goes unsold, the penalty is 25% of the withdrawn bid.

The bids in each round are used to construct prices that place lower bounds on the bids that can be submitted in the subsequent round. Under SMR, prices are simply equal to the highest bids for the licenses. Under MPB, the prices for licenses are set so that the losing bids on licenses (or packages) are less than or equal to the corresponding license prices (or sum of prices for individual licenses in that package) and the winning bids are equal to the corresponding prices. Although the intuition behind these "Walrasian" constraints is clear, the presence of complementarities in license values may preclude the existence of such dual prices. In this case, prices are approximated in a manner that minimizes some measure of the extent to which the constraints are not satisfied.[8] As shown in the previous section, HPB prices are Walrasian prices (not approximations). These prices are computed in a straightforward recursive manner by scaling up high bids at each level so that they sum to the high bid at the next level of the hierarchy.[9]

New bids at the start of a round must exceed provisionally winning bids under SMR by at least one bid increment (3 points in the experiment), whereas new bids under the two package bidding formats must exceed the price of a license or sum of prices for licenses in a package by at least one bid increment for each license in the package. For example, the next bid on a package of 2 licenses with prices 5 and 10 would have to be $21(= 5 + 10 + 3 + 3)$, but higher bids of 24, 27, etc. would also be allowed.[10]

In the FCC auctions, financial pre-qualifications determine initial bidding activities for each participant, where activity is measured in terms of population and bandwidth ("MHz-pop" units). Each license in the experiment has one activity unit to avoid the extra complexity. Activity has a "use it or lose it" feature, except that a provisionally winning bid does not need to be resubmitted to maintain activity. Therefore, activity *in this round* is defined in terms of the number of different licenses for which a bidder

---

[8] In addition, there may be multiple solutions to the constraints, and a second constrained optimization problem is run to resolve the indeterminacy, e.g. maximizing the minimum price, or minimizing the sum of squared deviations from previous round prices (Kwasnica et al., 2005).

[9] Under SMR, new and current provisionally winning bids are considered to determine provisional winners for the next round. Under MPB and HPB all bids received are considered (to prevent cycles). Of course, retaining old bids implies they may become winning at a later stage, and the FCC is considering allowing bidders to "drop" non-provisionally winning bids in a single round. To maintain comparability with SMR where regional bidders could win at most 4 licenses due to activity constraints we imposed purchase limits of 4 for regional bidders, which were used in the revenue maximization routines in MPB and TPB. These limits are absent when eligibility is determined by the number of licenses bidders register for prior to the start of the auction, in which case revenue maximization follows from the simple recursive structure described above. In the presence of purchase constraints, the computed HPB prices that determine minimum bids for the next round may differ from current winning bids – the revenues reported in this paper are based on bids submitted, not on computed prices.

[10] Under MPB, after a round in which news bids were submitted but revenue did not increase, the bid increment needs to be raised to avoid cycling. Consider the following scenario where the bid increment is assumed to be 3: suppose there is a winning bid of 30 on the ABC package and three other bids of 29 on AB, AC, and BC. The computed prices for A, B, and C are 10 each. In the next round the minimum acceptable bids are $26(= 10 + 10 + 3 + 3)$ for each package of two licenses. So the losing bidders could resubmit their bids of 29, and minimum acceptable bids would again be 26 for each license etc. This cycling behavior can occur whenever a package bid is winning and losing bidders are bidding on sub-packages. The solution is to increase the bid increment (from 3 to 6 to 9, etc.).

20 HIERARCHICAL PACKAGE BIDDING

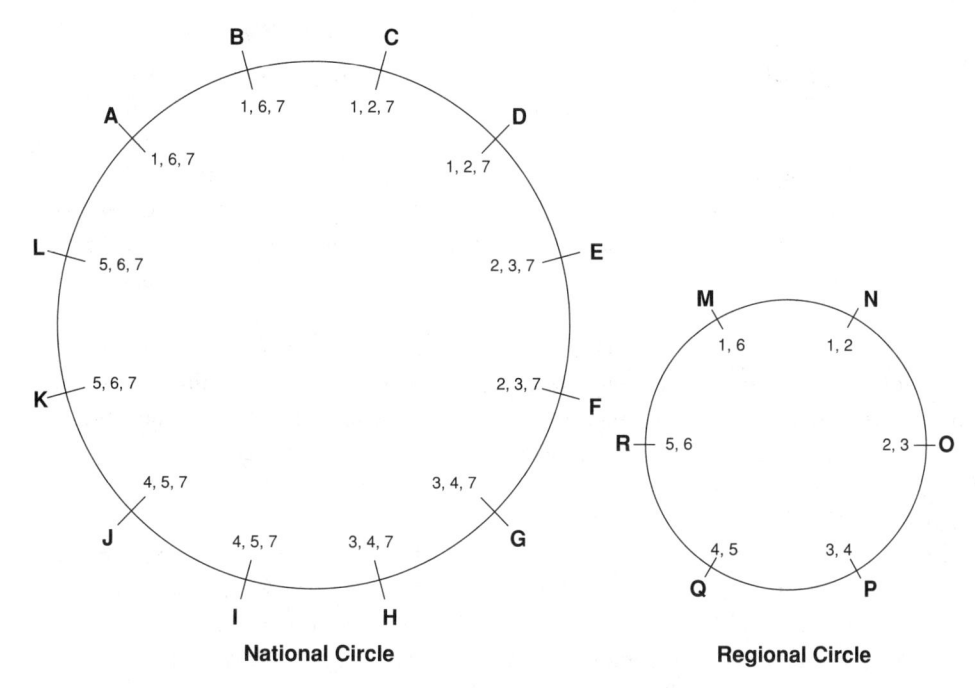

Figure 1. Regional bidders 1–6 are interested in 4 licenses from the national circle and 2 licenses from the regional circle. National bidder 7 is interested in all 12 licenses from the national circle.

is the provisional winner or for which a bid is submitted *in the previous round*, either individually or as part of a package. Under MPB and HPB, a bid from a previous round which was not then a winning bid might become a winning bid later as a result of others' bidding behavior. However, this does not raise a bidder's activity. In other words, bidders' activities can only decrease over the course of the auction. If a person's bidding activity is below the permitted level, then the permitted activity falls to that lower level in the subsequent round.

3.2. Sessions and Bidder Interests

Each auction involved 7 participants: six "regional" bidders (labeled 1 through 6) and one "national" bidder (labeled 7). A graphical representation of bidders' interests is shown in Figure 1. The large circle (licenses A through L) on the left contains licenses of interest to the national and regional bidders. For example, license A is of interest to regional bidders 1 and 6 and to the national bidder 7. Note that each regional bidder has an interest in 4 adjacent licenses, with partial overlap in these interests. The smaller circle on the right (licenses M through R) contains licenses only of interest to regional bidders, e.g. license P is of interest to regional bidders 3 and 4. Activity and purchase limits were such that regional bidders can acquire at most four licenses, and the national bidder can acquire up to twelve licenses on the larger circle. One useful feature of the smaller circle was to reduce earnings inequities in cases where the national bidder managed to win a national license or a large share of licenses, which was expected to happen more frequently with package bidding.

The license values for each bidder were randomly determined, and values for combinations were determined by scaling up the values of the individual licenses in the combination. For the national bidder, the baseline draw distributions are uniform on the range [0, 10] for licenses A-D and I-L and uniform on the range [0, 20] for licenses E-H. For regional bidders, the baseline draw distributions are uniform on the range [0, 20] for licenses A-D and I-L and uniform on the range [0, 40] for licenses E-H. Finally, for licenses M-R the baseline draw distributions are uniform on the range [0, 20]. These value distributions (not the actual draws) were common knowledge among the bidders. Note that the E-H region of the national circle is, on average, worth more to the national and regional bidders. This asymmetry allows us to measure the impact of "pre-packaging" on regional bidders of different strengths. In particular, the regional bidder with an interest in the high-value licenses E-H would often be a strong competitor with the national bidder, and a threshold problem could arise if the other regional bidders drop out of the bidding without coordinating a strong response to an aggressive national bid.

In all formats, bidders could bid on individual licenses. Recall that bidders cannot bid on packages under SMR, and they have full flexibility in bidding on packages under MPB, subject to activity constraints. Under HPB, the admissible packages have a hierarchical structure consisting of a single national package ABCDEFGHIJKL in our HPB$_2$ design, where the subscript indicates the number of levels in the hierarchy. We also implemented a three-level hierarchy, with an additional middle level consisting of three non-overlapping regional packages: ABCD, EFGH, and IJKL in our HPB$_3$-odd design and ABKL, CDEF, and GHIJ in our HPB$_3$-even design, see Table 1 (licenses E, F, G, and H are shown in bold to indicate their higher values). In other words, in HPB$_3$-odd the odd numbered regional bidders can bid on their preferred packages but not the even numbered bidders, while in HPB$_3$-even the reverse is true. In HPB$_2$, regional bidders could only submit bids for individual licenses, while the national bidder could bid on individual licenses and/or on the national package.[11] Each of these 5 treatments

---

[11] The hierarchy structures employed in the experiment simplified the explanation of the HPB pricing rule. Consider, for example, the three-level hierarchy used in HPB$_3$-odd and HPB$_3$-even. Let the 12 items on the national circle be indexed by $r = 1, \ldots, 3$ to denote the region and by $i = 1, \ldots, 4$ to denote a license within the region. Let $b_{ir}^{\max}$ denote the best bid for license i in region r, $b_r^{\max}$ the best package bid for region r, and $b^{\max}$ the best package bid for the nation-wide license. Define the revenue for region r as $\text{Rev}_r = \max(b_r^{\max}, \sum_{i=1}^{4} b_{ir}^{\max})$ and the national revenue as $\text{Rev} = \max(b^{\max}, \sum_{r=1}^{3} \text{Rev}_r)$. The price p_{ir} of license i in region r is simply the maximum bid for the license plus possibly a "regional tax" (if the sum of individual bids falls short of the regional profit) plus possibly a "national tax" (if the sum of regional profits falls short of the national profit):

$$p_{ir} = b_{ir}^{\max} + \frac{\alpha_{ir}}{\alpha_r}\left(\text{Rev}_r - \sum_{i'=1}^{4} b_{i'r}^{\max}\right) + \frac{\alpha_{ir}}{\alpha}\left(\text{Rev} - \sum_{r'=1}^{3} \text{Rev}_{r'}\right),$$

with $\alpha_r = \sum_{i=1}^{4} \alpha_{ir}$ the number of bidding units in region r and $\alpha = \sum_{r=1}^{3} \alpha_r$ the number of bidding units nationwide. (For simplicity, each license covers $\alpha_{ir} = 1$ bidding unit in the experiment.) As an example, suppose the high bids are 3 on each of the 12 items, the regional bids are 24, and the national bid is 84. In this case, the national bid is winning and prices for each of the items are the high bids plus a regional and a national tax: $3 + \frac{1}{4}(24 - 4 \times 3) + \frac{1}{12}(84 - 3 \times 24) = 7$, which determines bids on each item needed to unseat the winning national bid. Finally, for the two-level hierarchy used in HPB$_2$ the above pricing rule further simplifies to $p_i = b_i^{\max} + \frac{\alpha_i}{\alpha}(\text{Rev} - \sum_{i'=1}^{12} b_{i'}^{\max})$, where $i = 1, \ldots, 12$ now labels the twelve licenses on the national circle. In other words, the price p_i of license i is now simply the maximum bid for the license plus possibly a "national tax."

Table 1. *Complementarities Design*

	# Groups	# Bidders (Activity) Regional	# Bidders (Activity) National	Licenses Regional	Licenses National	Available Packages	License Synergy
SMR	5	6(4)	1(12)	A-L, M-R	A-L	None	20%
HPB$_2$	5	6(4)	1(12)	A-L, M-R	A-L	ABCDEFGHIJKL	20%
HPB$_3$-odd	5	6(4)	1(12)	A-L, M-R	A-L	ABCD, EFGH, IJKL, ABCDEFGHIJKL	20%
HPB$_3$-even	5	6(4)	1(12)	A-L, M-R	A-L	CDEF, GHIJ, ABKL, ABCDEFGHIJKL	20%
MPB	5	6(4)	1(12)	A-L, M-R	A-L	All	20%

was replicated 5 times as indicated in Table 1, for a total of 25 sessions (using different groups of 7 participants in each session). For each replication or "wave," we generated new sets of random values for all the bidders; we used the same sets of values across auction formats to reduce performance differences due to the random draws.

Value complementarities were chosen so that the optimal allocation would involve a single national package on the large circle for some sets of value draws and a combination of regional package awards and individual license awards in others. For both national and regional bidders, each license acquired goes up in value by 20% (with two licenses), by 40% (with three licenses), by 60% (with four licenses), etc., and by 220% if the national bidder wins all twelve licenses A-L. In waves 1–4, complementarities occur among all licenses while in wave 5 they occur only among licenses from the national circle. For example, if bidder 1 wins the combination ABM, the value synergies apply to licenses A and B and M in waves 1–4 and only to A and B in wave 5. The national bidder can acquire up to twelve licenses and has value complementarities for all licenses in all five waves.

Participants were recruited from the Caltech student population for sessions that lasted about an hour and a half. Including the experiments reported in Section 6, we conducted 58 sessions using 340 subjects.[12] Each session began with an instruction period and 3 practice auctions, followed by 6 auctions used to determine their earnings. The number of auctions was announced in advance. Bidder roles were reassigned randomly after each auction in order to attenuate earnings differences across national and regional bidders and to help bidders understand the strategic considerations faced by both types of bidders. Earnings were calculated by converting points to dollars at a rate of 2 points per dollar. Average earnings were about $40 per person, including a $5 show-up fee, and payments were made in cash immediately after the final auction.

---

[12] The results for 20 of these sessions were submitted to the FCC in a consulting report that is cited in their decision to adopt HPB for the package bidding segment of the upcoming 700 MHz auction (FCC Public Notice, August 2007).

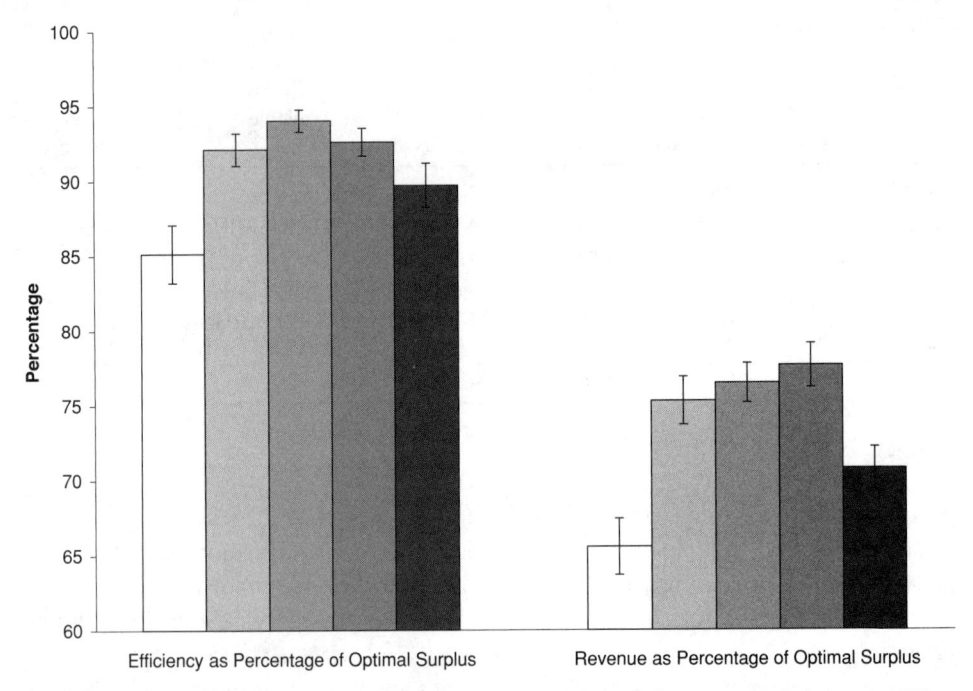

[Figure 2 bar chart]

Figure 2. Efficiencies and Revenues by Auction Format. SMR (white), HPB$_2$ (light gray), HPB$_3$-odd (medium gray), HPB$_3$-even (dark gray), MPB (black).

4. Results: Aggregate Data

In this section we report the main indicators for auction performance: efficiency, revenues, and bidders' profits.

4.1. Efficiency

Market efficiency is defined as the value of the allocation obtained in the auction (the actual surplus, S_{actual}) divided by the value of the best possible allocation (the maximum possible surplus, $S_{optimal}$):[13]

$$efficiency = \frac{S_{actual}}{S_{optimal}} \times 100\%. \tag{3}$$

The summary Table A1 in the Appendix lists average efficiencies across sessions and auction formats.

Package bidding is designed to help bidders avoid the "exposure problem" of bidding high for licenses with high complementarities. As expected, switching from SMR to the package auction formats raises efficiency: from 85.1% to 89.7% for MPB, to 92.1% for HPB$_2$, and to 92.6% and 94.0% for HPB$_3$-even and HPB$_3$-odd respectively. The left part of Figure 2 shows average efficiencies across formats: the white bar corresponds to SMR, the light-gray bar to HPB$_2$, the medium-gray bar to HPB$_3$-odd, the

---

[13] The total number of possible allocations with this setup is 27,433,982.

dark-gray bar to HPB$_3$-even, and the black bar to MPB. The standard deviations are indicated by the bracketed intervals at the top of each bar. Notice that efficiencies are higher and less variable under the HPB formats as compared to SMR and MPB.

The performance differences suggested by Figure 2 are supported by a statistical analysis. As illustrated by Figure 2, the HPB environments yield very similar results (in terms of efficiency and revenue) and, hence, we will pool the observations from the HPB environments. Below $\sim$ indicates a pair-wise ordering that is not significant, $\succ^*$ indicates significance at the 10% level, $\succ^{**}$ indicates significance at the 5% level, and $\succ^{***}$ indicates significance at the 1% level.

Result 1: Efficiencies are ranked HPB $\succ^{***}$ SMR and HPB $\succ^*$ MPB.

Support. See the Appendix for an overview of session averages across auction formats. There are five averages for SMR and MPB, corresponding to the five value waves, and fifteen averages for HPB after pooling the HPB environments. The non-parametric test employed is a Wilcoxon matched-pairs signed-rank test. The difference in ranks between MPB and SMR is 9 (five observations), which is not significant. The difference in ranks between HPB and MPB is 90 (fifteen observations), which is significant ($p = 0.09$). Finally, the difference in ranks between HPB and SMR is 120 (fifteen observations), which is significant at the one-percent level ($p < 0.001$). □

One cause of efficiency reductions with SMR is the incidence of unsold licenses, which happens at a rate of 2.1 licenses (out of 18) when averaged over all sessions. Likewise, on average 1.0 license is unsold under MPB while there are virtually no unsold licenses with HPB, see Figure 3.

Result 2: The numbers of unsold licenses are ranked SMR $\sim$ MPB $\succ^{***}$ HPB.

Support. The higher rate at which licenses are awarded under HPB is clear from the right-most set of bars in Figure 3. The difference between HPB and MPB in terms of license sales rates is significant with a Wilcoxon matched-pairs signed-rank test ($p < 0.001$ with fifteen observations) while the difference between MPB and SMR is (borderline) insignificant ($p = 0.13$ with five observations). □

4.2. Revenues

Revenues are also normalized by the value of the best possible allocation:

$$revenue = \frac{R_{actual}}{S_{optimal}} \times 100\%. \qquad (4)$$

The introduction of package bidding enhances revenues as shown in the right part of Figure 2. Switching from SMR to the package auction formats raises revenue from 65.6% to 70.8% for MPB, to 75.3% for HPB$_2$, and to 76.5% and 77.9% for HPB$_3$-odd and HPB$_3$-even respectively. As before, these comparisons can be evaluated with a Wilcoxon test based on the session averages reported in the Appendix.

Result 3: Revenues are ranked HPB $\succ^{***}$ MPB $\succ^{**}$ SMR.

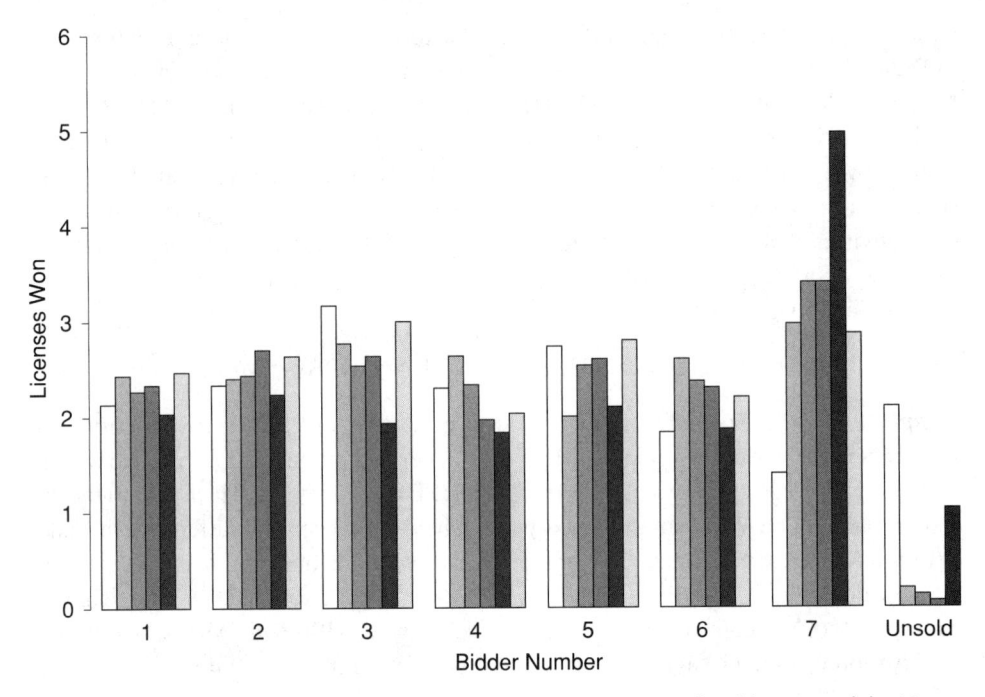

Figure 3. Licenses Acquired by Regional Bidders (1–6), the National Bidder (7), and the Number of Unsold Licenses. Optimal Awards of Licenses Shown in Yellow. SMR (white), HPB$_2$ (light gray), HPB$_3$-odd (medium gray), HPB$_3$-even (dark gray), MPB (black).

Support. The difference in ranks between MPB and SMR is 15 (five observations), which is significant ($p = 0.04$). The difference in ranks between HPB and MPB is 120 (fifteen observations), which is significant at the one-percent level ($p < 0.001$). □

The brackets on each of the bars on the right side of Figure 2 indicate the standard deviations of the revenues across formats. Note that SMR and MPB result in lower and more variable revenues as compared to HPB. Also, it is apparent from Figure 2 that the average efficiencies and revenues with only two tiers are slightly lower than for the three-tier hierarchies. Even though this difference is not significant, it suggests that in some situations it is important to have enough hierarchy levels to allow small bidders to compete effectively when they are part of the efficient allocation.

4.3. Bidders' Profits

Consistent with the definitions of revenue and efficiency, bidders' profits are normalized by the value of the best possible allocation:

$$profits = \frac{\sum_i \pi^i_{actual}}{S_{optimal}} \times 100\%. \tag{5}$$

This profit is the difference between actual surplus and seller's revenue, except for SMR where possible penalties from withdrawing winning bids are recorded separately (see the Appendix). Rather than simply reporting the profits for the bidders as a group it is

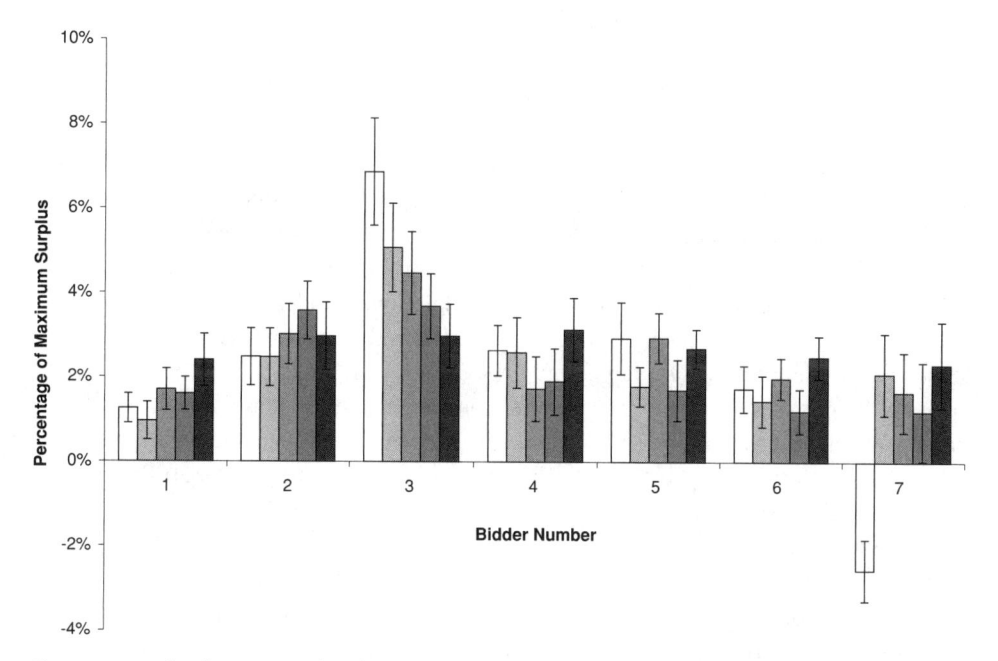

Figure 4. Profits for Regional Bidders (1–6) and the National Bidder (7). SMR (white), HPB$_2$ (light gray), HPB$_3$-odd (medium gray), HPB$_3$-even (dark gray), MPB (black).

useful to show them by bidder type since this highlights the impact that package bidding and/or pre-packaging has on different types of bidders. Figure 4 displays bidders' profits by treatment and bidder number, using the same color-coding as in Figure 2. Again, the standard deviations are indicated by the brackets at the top of each bar.

The ability to bid for combinations allows national bidders to bid high on large packages and avoid the exposure problem, resulting in positive profits for the national bidder (7) in the MPB and HPB auctions. In contrast, the national bidder loses money (in all waves, see the Appendix) when the SMR format is used. These losses are not surprising given prior results on SMR experiments; they result from the value complementarities that create an exposure problem for the national bidder. In the experiment, the effects of negative earnings on individual behavior were mitigated by the fact that bidder roles were randomly assigned in each auction. The differences in profits for the national bidder are corroborated by non-parametric tests. In contrast, the differences in profits for regional bidders (slightly higher under SMR than MPB and HPB) are not significant.

Result 4: The national bidder's profit is ranked MPB $\sim$ HPB $\succ$*** SMR. The regional bidders' profits (as a group) are ranked SMR $\sim$ MPB $\sim$ HPB.

Support. For the national bidder's profit the difference in ranks between HPB and SMR is 110 (fifteen observations), which is significant ($p = 0.004$). The difference in ranks between MPB and HPB is 62 (fifteen observations), which is not significant. For the regional bidders' profits the difference in ranks between SMR and MPB is 88 (fifteen observations), which is (borderline) insignificant ($p = 0.11$). The difference in ranks between SMR and MPB is 7 (five observations), which is not significant. □

5. Discussion

5.1. The Exposure Problem in SMR

One reason for the efficiency and revenue advantages conferred by package bidding vis-à-vis SMR is apparent from looking at the outcomes of the first four auctions of wave 5 in which the optimal allocation frequently involved awarding large-circle licenses to the national bidder: 12, 12, 12, and 11 licenses in auctions 1–4 respectively.[14] Under SMR, the numbers of licenses actually obtained by the national bidder in these four auctions were 7, 2, 5, and 1, which shows that the national bidder was unable to overcome the exposure problem and obtain large networks even when it was optimal to do so. The national bidder was much more successful for the package bidding auctions with the same value draws; the numbers of licenses obtained by the national bidder were 12, 12, 12, and 12 licenses under MPB and HPB$_3$-odd, 12, 12, 12, and 11 under HPB$_2$, and 12, 7, 12, 12 under HPB$_3$-even.[15]

The effects of exposure are apparent in the first SMR auction of wave 1. The optimal allocation involved awarding all 12 licenses on the large circle to the national bidder whose values for the licenses are listed in Table 2 together with the final prices. For all licenses, the final price exceeded the national bidder's stand-alone value for the license, and the total cost (249) was much higher than the sum of the individual values (102). Suppose the national bidder had conserved all activity up to the final round and was willing to top the current prices of Table 2, which would involve a minimum expenditure of 285 (= 249 + 36) for a national package worth 326 (= 3.2 × 102). If the other bidders (who all have sufficient activity to respond) came back with increases of one additional bid increment, the cost of acquiring the national package would exceed its value. In this case, the national bidder could not profitably win any of the licenses. An even worse outcome would result when only some of the regional bidders respond and the national bidder would win a less valuable subset. For example, suppose the national bidder would win only the high-value EFGH combination, which, including synergies, is worth 102 = (19 + 17 + 10 + 18) × 1.6, but would require an expenditure of at least 177 to top the prices listed in Table 2.

The national bidder has to evaluate the risk of being "exposed" during the course of the auction, not just at the end, since decisions to maintain activity are made on a round-by-round basis. In the experiment, the national bidder gradually lost activity and was the provisional winner for fewer licenses as the auction proceeded. At the start of round 8, for example, the national bidder was winning licenses A, C, E, and G with a combined value of 55 (including synergies) and prices of 9, 3, 24, and 9 respectively for

---

[14] Recall that in wave 5, small bidders enjoy synergies only for licenses acquired from the national circle.

[15] Performance differences are also apparent in the awards of blocks of licenses to regional bidders. There were 13 cases where the optimal allocation provided at least 3 of the 4 high-value licenses (E-H) to bidder 3, the only bidder who had an interest in all of these and could bid on them as a package in HPB$_3$-odd. Package bidding generally does better in these cases, with overall average efficiency of 85% for SMR, as compared with 88% for MPB and 93% for HPB. In both package bidding formats, the EFGH package was sometimes awarded to bidder 3 when it should not have been, but the efficiency consequences were small and certainly smaller than the consequences of not awarding one or more of these high-value licenses at all as happened several times with SMR (a total of six cases out of 13, with three of the four high-value licenses unsold in auction 5 of wave 3).

Table 2. *National Bidders' Values and Final Prices for Licenses A–L*
SMR Auction 1 in Wave 1

License	A	B	C	D	E	F	G	H	I	J	K	L	Total
Value	9	5	2	5	19	17	10	18	8	3	3	3	102
Final Price	12	18	3	9	42	60	15	48	12	12	12	6	249

a total cost of 45. At the start of round 9, the national bidder was only winning license C because regional bidders raised their bids to 12, 27, and 12 on licenses A, E, and G. The total cost of acquiring the ACEG combination became 63 and exceeded the value of 55. At this point, the national bidder (rationally) decided to let the activity drop and to withdraw the provisionally winning bid of 3 on license C since it was worth only 2 to this bidder.

In this auction, license C remained unsold even though its price fell back to the minimum opening bid of 3. The reason is that the two regional bidders (1 and 2) with an interest in C had low values for it (2 and 4), and had no additional activity above what they were provisionally winning or actively competing for. More generally, the price following a withdrawal falls back to the second-highest bid for the license, which tends to be high since most withdrawals occur late in the auction. Bidders who are constrained in their activity, as is typically the case late in the auction, are often unwilling or unable to pick up a high-priced unsold unit.

To summarize, unsold licenses in the SMR auctions are a consequence of the interaction between exposure risk and the limited bid-withdrawal option intended to alleviate this problem. For example, Brunner et al. (2010) report *no* unsold licenses in their low-complementarities treatment without exposure risk, and there are no unsold licenses in the no-synergy experiment discussed in Section 6.2. Banks et al. (2003) report SMR experiments without withdrawal options, in which case there are no unsold licenses but bidders incur large losses. These observations also suggest that all licenses are more likely to be sold when bidders' value draws are such that the exposure problem is less severe. In the experiment, there were no unsold licenses in 7 of the SMR auctions. But even for these auctions, the average revenue for SMR was 10% lower compared to HPB (69.6% versus 76.5%) while their efficiencies were virtually the same (93.6% versus 93.4%). Finally, the option to withdraw provisionally winning bids does not protect the national bidder from losses, see Figure 4.

The reduced performance of SMR in our design with value complementarities corroborates previous results regarding the adverse effects of exposure risk on efficiency and revenue (e.g. Bykowsky, Cull, and Ledyard, 2000). In addition, our findings suggest that the SMR procedure is disadvantageous for efficient providers for whom it is essential to acquire a national license (e.g. a major new entrant).

5.2. The Threshold Problem in MPB

The awarding of national licenses also provides a perspective on why HPB yields higher revenues and efficiencies than the more flexible package bidding format, MPB. There were several MPB auctions in which the national bidder won many licenses on the

Table 3. *Round-By-Round Outcomes for MPB Auction 2, Wave 1*

	National Circle			Regional Circle	
Round	National	Regional	Unsold	Regional	Unsold
1	12	0	0	2	4
2	12	0	0	2	4
3	0	7	5	5	1
4	0	7	5	5	1
5	12	0	0	2	4
6	12	0	0	3	3
7	12	0	0	3	3
8	12	0	0	3	3
9	12	0	0	4	2
10	12	0	0	4	2
11	12	0	0	4	2
12	12	0	0	4	2
13	12	0	0	4	2
14	12	0	0	6	0
15	12	0	0	6	0
16	12	0	0	6	0
17	12	0	0	6	0
18	12	0	0	6	0
19	12	0	0	6	0
20	12	0	0	6	0
21	0	10	2	5	1
22	12	0	0	6	0
23	12	0	0	6	0
Optimal	0	12	0	6	0

large circle when it was not optimal to do so, whereas there are relatively few such cases under HPB. To see how the national bidder was sometimes able to obtain all licenses under fully flexible package bidding (MPB) even when it was not optimal to do so, consider the round-by-round results of auction 2 in wave 1, shown in Table 3. The optimal allocation involved only a single license for the national bidder, but there were only three rounds in which the national bidder was not the provisional winner on all 12 licenses. In each of these three rounds, the regional bidders were not able to coordinate a very strong response in the sense that their provisionally winning bids left numerous provisionally unsold licenses (5 out of 12 licenses in rounds 3 and 4, and 2 out of 12 licenses in round 21). With fully flexible bidding, the regional bidders were bidding on "home-made" overlapping packages that did not "fit" in the sense that the revenue maximizing allocation left unsold licenses, which made it easier for the national bidder to regain provisional winner status in the subsequent rounds. In contrast, when HPB was used with the same draws, the regional bidders were able to effectively block the national bidder, and the resulting efficiency was 10 percentage points higher: 83% in MPB versus 93% in HPB (efficiency was 94% in HPB_2, 95% in HPB_3-odd, and 89% in HPB_3-even).

Motivated by this example, we focused on rounds in which the national bidder wins nothing and counted the number of licenses provisionally won by the regional bidders

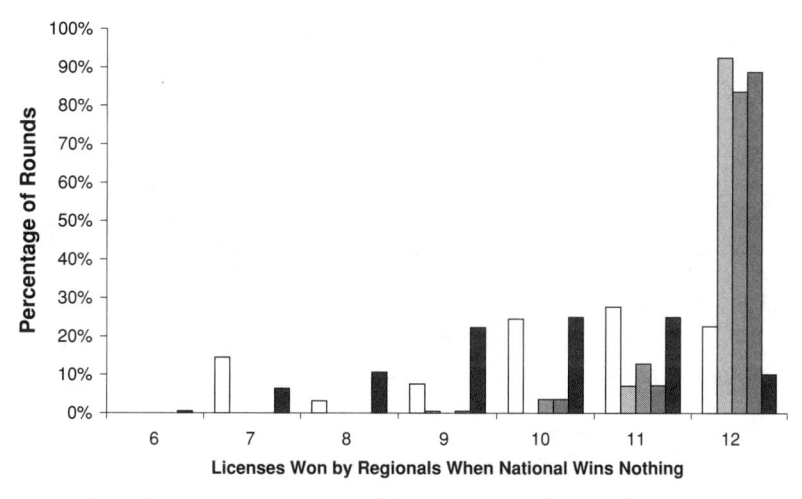

Figure 5. Number of Licenses Regional Bidders Provisionally Win from the Large National Circle (Licenses A-L) in Rounds Where the National Bidder Wins Nothing. SMR (white), HPB$_2$ (light gray), HPB$_3$-odd (medium gray), HPB$_3$-even (dark gray), MPB (black).

from the large circle (licenses A-L). The results are shown in Figure 5. Under MPB, the regional bidders are able to coordinate their bids such that they provisionally win all 12 licenses only 10% of the time when the national bidder is not winning any licenses. More than 65% of the time they provisionally win 10 licenses or less (out of 12), resulting in prices for the 12 licenses that can easily be topped by the national bidder. In contrast, under HPB, the regional bidders are able to coordinate and provisionally win 11 or 12 licenses more than 95% of the time. Our experiments nicely demonstrate that the threshold problem we observe with flexible package bidding is not merely a "free-rider" problem but also a coordination problem – indeed, the latter is more pronounced in our data. To our knowledge, our paper is the first to clearly demonstrate the existence of a threshold problem for smaller bidders in auctions with flexible package bidding.

We are not claiming that HPB will yield better performance in terms of efficiency and revenue in all environments. For instance, if the hierarchical pre-packaging completely mismatches bidders' preferences, the resulting exposure problem that all bidders face would likely reduce bids and revenues. Alternatively, if there is no bidder with an interest in the national package, mis-coordination would be more easily resolved by regional bidders who would be provisional winners in all rounds. The design of our experiment is based on the belief that the FCC will be able to craft economically relevant packages for at least some of the bidders and that there would be one or more bidders (e.g. *de novo* entrants) interested in a national package. Furthermore, as the results from treatment HPB$_3$-even demonstrate, the performance of HPB is robust to some degree of "mis-packaging." Even an extremely simple hierarchy structure such as HPB$_2$ leads to improved performance relative to the other formats. The experiments discussed in the next section provide a further robustness check in environments with richer sets of value synergies.

Table 4. *Synergy Factors for Licenses Won within a Band*

		\multicolumn{6}{c}{Licenses Won within the Band}
		1	2	3	4	5	6
	0	1.00	1.43	1.80	2.13	2.40	2.63
	1	0.95	1.35	1.70	2.00	2.25	2.45
Licenses Won	2	0.90	1.28	1.60	1.88	2.10	2.28
in the other	3	0.85	1.20	1.50	1.75	1.95	2.10
Band	4	0.80	1.13	1.40	1.63	1.80	1.93
	5	0.75	1.05	1.30	1.50	1.65	1.75
	6	0.70	0.98	1.20	1.38	1.50	1.58

6. Two Follow-Up Experiments

6.1. Complements and Substitutes

The SMR was ideally designed to deal with substitutes since bidders can switch back and forth between equally desired licenses in response to changes in prices. Substitutes are likely to play some role when multiple blocks of identically sized bandwidth are sold in the same regions, as is often the case in spectrum auctions.[16] We also tested an alternative design with 2 separate bands of 6 licenses, with positive synergies for licenses within a band and negative synergies between bands. The "upper band" consists of licenses A through F and the "lower band" consists of licenses G through L. Table 4 lists the synergy factors that can be used to determine the value of any combination of licenses. The synergy factor in the table is multiplied by the sum of values for the licenses in a combination. For example, if a bidder wins three licenses within the upper band and no licenses in the lower band then the column labeled 3 and the row labeled 0 apply. In this case, the total value for the package equals the sum of the individual values plus an additional 80%. Likewise, if the bidder wins three licenses in the upper band and two in the lower band, then the value of the combination is the sum of the values in the upper band times 1.6 plus the sum of the values in the lower band times 1.2.

Note that the numbers in Table 4 are increasing within each row to reflect complements, while substitution effects are captured by the decreasing numbers within each column. To see how bidders' valuations are super-additive for some combinations, consider a national bidder who acquires licenses A and B from the upper band. For simplicity, assume the stand-alone values for the licenses are $v(A) = 15$ and $v(B) = 15$, which is the average over the range [10, 20] used in the experiment. It follows from

---

[16] As a matter of theory, it is possible to design simultaneous ascending auctions that converge monotonically to Walrasian package prices in the pure complements case discussed above. This result assumes "straightforward bidding," i.e. allocating bidding activity to licenses and packages with the highest profit margins based on current prices. In practice, bidders typically do not bid straightforwardly (Brunner et al., 2010). More importantly, the auction would require non-linear pricing, i.e. the price for a package is not necessarily equal to the sum of the prices for the licenses it contains. As a result, the auctioneer and bidders would have to keep track of many, many prices (e.g. 262,143 in the experiment reported above).

the the top row in Table 4 that the combination is worth $v(AB) = 43$ (calculated as $= 1.43 \times 30$), which exceeds the sum of the individual values: $v(AB) > v(A) + v(B)$. Similarly, the value of acquiring licenses G and H from the bottom band is 43 on average. Next consider the value for acquiring the combination ABGH, which is readily calculated as $v(ABGH) = 77 (= 1.28 * 60)$ and is less than the sum of the values of the two combinations from each line: $v(ABGH) < v(AB) + v(GH)$. The subadditive nature of bidders' valuations across bands is more dramatic for larger combinations of licenses. For example, the value of winning a combination of 6 licenses from a single band is 236 when each of the 6 licenses is worth 15 individually. Winning an additional 6 licenses from the other band raises the total value only to 284, i.e. $v(ABCDEFGHIJKL) = v(ABCDEF) + v(GHIJKL) - 188$.

Subjects in the experiment did not have to multiply synergy factors times sums of values. Rather, the *j*Auctions bidder interface showed the total value (including positive and/or negative synergies) for all the pre-defined packages and for any "custom" package of interest. In our HPB treatment, bids could be submitted for individual licenses and for pre-defined packages (AB, CD, EF and ABCDEF for the upper band and GH, IJ, KL and GHIJKL for the lower band). Under SMR, no bids could be placed for the pre-defined packages although bidders could see their values. Under all three auction formats, bidders could view the values of any desired combination of licenses but only under MPB were bidders allowed to submit bids for such custom packages.

As before, bidders were assigned roles randomly at the start of each auction, with IDs 1 through 3 for regional bidders and IDs 4 and 5 for national bidders. Regional bidders are interested in four licenses in each band with values drawn independently from a uniform distribution on [10, 30].[17] National bidders have values for all 6 licenses in each band; these draws are from a uniform distribution on [10, 20]. These ranges of values were selected to ensure that the optimal allocations involved awards of a national license in some but not all auctions. We conducted 6 waves of SMR, MPB and HPB sessions with 6 auctions in each session, using new random draws for each wave. In all auctions, regional bidders started with an activity limit of 8 and national bidders with an activity limit of 12. Other procedural elements (cash conversion rate, bid increment, common knowledge of value distributions, subject pool, number of practice periods) were unchanged from the experiment reported above. One procedural change that we implemented, however, was that bidders were unable to see the IDs and associated bids made by others ("anonymous bidding").

Efficiencies and revenues for this mixed-synergy design are shown in the middle part of Table 5 (the results for the complementarities treatment are shown in the top part for comparison, with pooled results for the different HPB treatments, and the bottom part shows the results of an experiment discussed in the next subsection). In this mixed-synergy design, revenue and efficiency increases of roughly 10% were observed when switching from SMR to HPB, which are about the same magnitude as those observed in the complementarities design. Again, the improved performance of HPB

---

[17] Bidder 1 values A, B, C, and F in the "upper" A–F band and licenses G, H, I, and L in the "lower" G–L band. Bidder 2 has values for B, C, D, E and H, I, J, K and bidder 3 has values for A, D, E, F and G, J, K, L.

Table 5. *Summary Statistics for the Three Synergy Designs*

	Efficiency	Revenue	Regional's Profit	National's Profit	Rounds	Unsold
Complementarities Design						
SMR	85.1%	65.6%	3.0%	−2.6%	17.8	2.1
HPB	93.0%	76.5%	2.4%	1.6%	12.2	0.1
MPB	89.7%	70.8%	2.8%	2.3%	15.1	1.0
Mixed-Synergy Design						
SMR	85.4%	78.0%	1.7%	1.1%	15.4	1.5
HPB	93.2%	85.7%	0.7%	2.6%	12.8	0.0
MPB	94.5%	84.6%	1.1%	3.3%	15.3	0.1
No-Synergy Design						
SMR	99.4%	77.1%	7.1%	0.4%	5.8	0.0
HPB	99.2%	77.3%	7.0%	0.5%	6.0	0.0
MPB	98.8%	78.3%	6.5%	0.4%	5.8	0.0

is accomplished in fewer rounds on average, as can be seen from the "Rounds" column in Table 5.

Result 5: With mixed synergies, efficiencies are ranked HPB $\sim$ MPB $\succ$*** SMR and revenues are ranked HPB $\succ$** SMR and MPB $\sim$ SMR. The national bidder's profit is ranked HPB $\sim$ MPB $\succ$*** SMR while a regional's profit is ranked HPB $\sim$ MPB $\sim$ SMR. The numbers of unsold licenses are ranked SMR $\succ$*** HPB $\sim$ MPB.

Support. See Table A1 in the Appendix for an overview of session averages across auction formats. There are six averages for each auction format, corresponding to the six value waves, which are evaluated using a Wilcoxon matched-pairs sign-rank test. Comparing efficiency levels, the difference in ranks between HPB and SMR and between MPB and SMR is 21 ($p = 0.01$). Comparing revenue levels, the difference in ranks between HPB and SMR is 19 ($p = 0.04$) and between MPB and SMR is 15 ($p = 0.17$). For the national bidder's profit, the difference in ranks between HPB and SMR and between MPB and SMR is 21 ($p = 0.01$). For a regional bidder's profit, the difference in ranks is 13 ($p = 0.30$) between SMR and HPB and 11 ($p = 0.46$) between SMR and MPB. Finally, comparing numbers of unsold licenses the difference in ranks between SMR and HPB and between SMR and MPB is 21 ($p = 0.01$). □

6.2. No-Synergy Design

A final experiment we ran is based on a *no-synergy* design with payoffs that are strictly linear or additive, i.e. $v(P_1 \cup P_2) = v(P_1) + v(P_2)$ for any two packages P_1 and P_2. This design can be implemented by replacing all the synergy factors in Table 4 by 1. We conducted 5 waves of SMR, MPB and HPB sessions using the exact same setup as in the mixed-synergies design (e.g. two bands, two national bidders and 3 regional bidders, no information about others' IDs and bids, etc.). Giving bidders the opportunity to bid on packages has no adverse effects for efficiency and revenue in the no-synergies design, as the bottom part of Table 5 shows.

Result 6: Without synergies, efficiencies are near perfect for all three auction formats and are not significantly different across formats, nor are the revenues, bidders' profits, the number of unsold licenses, and the auction's duration.

Support. See Table A1 for an overview of session averages across formats. There are five averages for each format, corresponding to the five value waves. Evaluating differences with a Wilcoxon test reveals that none are significant at the 10% level. □

7. Conclusion

In this paper we investigate how the constrained structure of hierarchical package bidding (HPB) compares with alternatives that involve flexible package bidding (MPB) or no package bidding (SMR). The design entails an environment with value complementarities, which are likely significant for *de novo* entrants wishing to establish a national footprint. In addition, the design features a "spike" of adjacent high-value licenses, which introduces an asymmetry that exaggerates the threshold problem. Since different blocks of spectrum within the same region may be considered substitutes, we conducted two additional experiments: one with a stylized model that incorporates both complements and substitutes and the other one without any synergy effects. Within this general modeling framework, we used sets of randomly generated values that induce a wide range of possible market structures.

With complementarities, the results of the laboratory auctions reveal a clear advantage for HPB even though the pre-made packages allowed under HPB did not match the preferred packages for half of the regional bidders (while those bidders were part of the optimal assignment). HPB yields significantly higher auction revenues and efficiencies than SMR and MPB for all pairwise comparisons. Comparable improvements in revenues and efficiencies of HPB over SMR were observed in the mixed-synergy design, while the relative performance of HPB was not significantly reduced in the no-synergy design. These performance differences were not due to increased fine-tuning over a large number of bidding rounds, since the HPB auctions actually tended to have fewer rounds.

The lower efficiencies and revenues observed for SMR in the presence of value complementarities could have been anticipated from prior work. The option of withdrawing bids in SMR, which was introduced as a partial remedy of the exposure problem, generated a higher incidence of unsold licenses compared to the other auction formats. Previous experiments that did not allow bid withdrawals resulted in fewer or no unsold licenses, but there the effects of exposure risk are indicated by the low efficiencies and the negative earnings observed. And in our experiments, even in those cases where no license went unsold, SMR yields less revenue than HPB.

What came as a surprise was the relative ranking of HPB and the more flexible MPB.[18] One factor that contributed to this difference is that the home-made packages

---

[18] In an interesting recent contribution, Milgrom (2007) shows how limiting bidders' strategy spaces in combinatorial auctions, as in our HPB design, can eliminate certain undesirable (low-efficiency, low-revenue) outcomes.

constructed under the flexible MPB format tended to overlap, causing a "fitting problem" that made it difficult for strong regional bidders to unseat a national package bid. Indeed, the number of licenses awarded to the national bidder was much higher than the optimal number under MPB, but not under HPB and SMR. More importantly, in rounds when the national bidder won nothing, regional bidders were unable to coordinate their bids under MPB while their coordination problems were virtually non-existent with pre-defined hierarchically-structured packages. Pre-packaging has the obvious disadvantage that the chosen packages may not be optimal, but in a non-overlapping hierarchical structure they are chosen to "fit," which enables bidders to coordinate their bids and avoid threshold problems with positive effects for efficiencies and revenues.

Combining results from the three synergy designs provides a clear picture. The SMR auction, which has been used by the FCC for more than a decade (and copied by similar regulatory agencies around the world), performs well in the linear environment it was designed for. But so does the HPB auction, whose simple assignment and pricing rules reduce to those of the SMR auction when the absence of synergies results in license-by-license competition. Moreover, when bidders are interested in aggregating combinations of licenses to build a regional or national network, the HPB auction significantly outperforms SMR. It enables efficient *de novo* entrants to establish a national footprint, a level of aggregation that is virtually impossible under the commonly used SMR format due to exposure risk.

The adverse effects of exposure risk are not just of academic concern – they played an important role in the design of the upcoming 700 MHz auction. The current wireless industry is highly concentrated and the 700 MHz auction provides the last opportunity for a new firm to enter the market. For an entrant to be successful in the wireless market, however, it has to acquire a nationwide footprint. The solution to this important market design problem is to allow for package bidding. But package auctions can be complex and can result in coordination or "threshold" problems for smaller bidders (as our MPB data indicate). While package auctions have been discussed for more than a decade, the FCC never adopted any of the existing formats due to concerns about complexity.

The hierarchical package bidding format proposed here is a 'paper & pencil' package auction: trivial to implement with assignment and pricing rules that are transparent and easily verifiable by bidders as the auction proceeds. The predefined packages are chosen to fit, thereby eliminating coordination problems. Of course, one has to be careful in generalizing the relative performance of HPB to other environments. But the experiments show that its improved performance is robust to some degree of package misspecification and occurs even for a simple two-layer hierarchy with a single nationwide license. HPB is a prime example of "economic engineering," i.e. the combination of applied mechanism design with wind-tunnel laboratory testing. It offers a simple and transparent solution to a complex market design problem, and puts economic research right at the heart of the FCC's most important auction to date.

Appendix A. Summary Data Table

Table A1. *Average Performance Measures by Session and Treatment*

	Complementarities Design					Mixed-Synergy Design			No-Synergy Design		
	SMR	HPB$_2$	HPB$_3$-odd	HPB$_3$-even	MPB	SMR	HPB	MPB	SMR	HPB	MPB
Efficiency											
Wave1	88.5%	90.0%	92.4%	91.5%	88.7%	89.1%	92.8%	96.8%	99.6%	99.3%	99.9%
Wave 2	91.3%	92.0%	94.6%	92.4%	86.6%	89.0%	92.2%	94.2%	99.4%	99.4%	97.2%
Wave 3	80.2%	88.3%	94.6%	92.0%	89.0%	73.9%	91.8%	96.0%	99.4%	99.9%	99.6%
Wave 4	81.9%	94.5%	92.4%	92.3%	86.1%	87.5%	97.9%	90.6%	98.9%	99.5%	99.7%
Wave5	83.8%	95.8%	96.0%	94.8%	98.2%	85.2%	89.9%	95.1%	99.5%	97.7%	97.4%
Wave 6						87.7%	94.4%	94.0%			
Average	**85.1%**	**92.1%**	**94.0%**	**92.6%**	**89.7%**	**85.4%**	**93.2%**	**94.5%**	**99.4%**	**99.2%**	**98.8%**
Revenue											
Wave1	67.3%	79.0%	71.9%	81.5%	70.3%	81.2%	83.2%	88.7%	77.5%	76.9%	79.7%
Wave 2	64.2%	73.2%	76.8%	77.2%	65.3%	77.1%	84.9%	80.1%	74.4%	76.5%	74.1%
Wave 3	65.5%	74.0%	77.0%	73.5%	71.9%	65.1%	86.1%	86.6%	76.9%	75.1%	76.2%
Wave 4	62.1%	72.6%	78.7%	78.7%	71.0%	79.6%	90.1%	88.1%	76.8%	75.8%	77.5%
Wave5	68.6%	77.5%	77.9%	77.3%	75.5%	81.9%	86.9%	84.4%	80.1%	82.2%	84.0%
Wave 6						83.3%	82.7%	79.7%			
Average	**65.6%**	**75.3%**	**76.5%**	**77.7%**	**70.8%**	**78.0%**	**85.7%**	**84.6%**	**77.1%**	**77.3%**	**78.3%**
Penalties	**4.3%**	**0.0%**	**0.0%**	**0.0%**	**0.0%**	**3.2%**	**0.0%**	**0.0%**	**0.0%**	**0.0%**	**0.0%**
Profit Nationals											
Wave1	−1.3%	−0.2%	0.8%	−2.6%	1.7%	2.3%	4.2%	2.5%	0.8%	0.7%	0.6%
Wave 2	−0.8%	0.0%	0.0%	0.3%	0.1%	2.9%	4.1%	4.0%	0.3%	0.3%	0.7%
Wave3	−4.5%	0.1%	−0.7%	−0.4%	1.2%	0.9%	2.8%	2.1%	0.2%	0.4%	0.2%
Wave 4	−3.9%	0.8%	−0.3%	2.0%	0.3%	1.2%	1.8%	2.1%	0.3%	0.5%	0.3%
Wave5	−2.3%	9.9%	8.4%	6.5%	8.3%	−1.9%	−0.5%	3.4%	0.3%	0.4%	0.4%
Wave 6						1.4%	3.4%	5.4%			
Average	**−2.6%**	**2.1%**	**1.6%**	**1.1%**	**2.3%**	**1.1%**	**2.6%**	**3.3%**	**0.4%**	**0.5%**	**0.4%**
Profit Regionals											
Wave1	3.0%	1.8%	3.3%	2.1%	2.8%	1.1%	0.4%	1.1%	6.8%	7.0%	6.3%
Wave 2	4.4%	3.0%	3.0%	2.4%	3.6%	2.1%	−0.3%	2.1%	8.1%	7.4%	7.2%
Wave 3	2.0%	2.3%	3.0%	3.1%	2.6%	2.4%	0.0%	1.8%	7.3%	8.0%	7.7%
Wave 4	2.8%	3.6%	2.3%	1.9%	2.5%	1.8%	1.4%	−0.6%	7.2%	7.5%	7.2%
Wave5	2.6%	1.5%	1.6%	1.8%	2.4%	2.4%	1.3%	1.3%	6.3%	4.9%	4.2%
Wave 6						0.5%	1.6%	1.1%			
Average	**3.0%**	**2.4%**	**2.6%**	**2.2%**	**2.8%**	**1.7%**	**0.7%**	**1.1%**	**7.1%**	**7.0%**	**6.5%**
Unsold Licenses											
Wave1	2.2	0.3	0.2	0.3	1.2	1.2	0.0	0.0	0.0	0.0	0.0
Wave 2	1.5	0.0	0.2	0.0	1.8	1.2	0.0	0.0	0.0	0.0	0.0
Wave3	2.7	0.3	0.3	0.0	1.2	2.8	0.0	0.0	0.0	0.0	0.0
Wave 4	2.7	0.3	0.0	0.0	0.8	1.5	0.0	0.3	0.0	0.0	0.0
Wave5	1.5	0.0	0.0	0.0	0.2	1.5	0.0	0.0	0.0	0.0	0.0
Wave 6						1.0	0.0	0.0			
Average	**2.1**	**0.2**	**0.1**	**0.1**	**1.0**	**1.5**	**0.0**	**0.1**	**0.0**	**0.0**	**0.0**
# Rounds											
Wave1	20.0	11.5	8.8	12.0	20.2	11.8	13.5	15.8	4.8	6.0	4.5
Wave 2	18.0	13.2	16.7	13.8	14.0	17.8	13.8	14.3	6.5	5.8	6.2
Wave 3	17.0	13.0	12.7	12.0	15.3	20.0	11.3	15.2	7.5	6.2	6.7
Wave 4	19.5	8.5	14.5	14.8	16.3	11.7	12.3	11.0	4.0	6.3	4.7
Wave5	14.5	8.0	13.3	10.0	9.7	14.3	15.3	19.8	6.3	5.8	6.7
Wave 6						16.8	10.7	15.5			
Average	**17.8**	**10.8**	**13.2**	**12.5**	**15.1**	**15.4**	**12.8**	**15.3**	**5.8**	**6.0**	**5.8**

References

Banks, J. S., J. O. Ledyard, and D. P. Porter (1989) "Allocating Uncertain and Unresponsive Resources: An Experimental Approach," *Rand Journal of Economics*, 20(1), 1–25.

Brunner, C., J. K. Goeree, C. A. Holt, and J. O. Ledyard (2010) "An Experimental Test of Combinatorial FCC Spectrum Auctions," *American Economic Journal: Micro-Economics*, 2(1), February, 39–57.

Bykowsky, M. M., R. J. Cull, and J. O. Ledyard (2000) "Mutually Destructive Bidding: The FCC Auction Design Problem," *Journal of Regulatory Economics*, 17(3), May, 205–228.

FCC Public Notice (August 17, 2007) "Auction of 700 MHz Band Licenses Scheduled for January 16, 2007: Comment Sought on Competitive Bidding Procedures for Auction 73," AU Docket No. 07-157. See http://fjallfoss.fcc.gov/edocs_public/attachmatch/DA-07-3415A1.pdf.

FCC Procedures Public Notice (October 5, 2007) "Auction of 700 MHz Band Licenses Scheduled For January 24, 2008," AU Docket No. 07-157. See http://fjallfoss.fcc.gov/edocs_public/attachmatch/DA-07-4171A1.pdf.

Kwasnica, A. M., J. O. Ledyard, D. P. Porter, and C. DeMartini (2005) "A New and Improved Design for Multi-Object Iterative Auctions," *Management Science*, 51(3), March, 419–434.

Milgrom, P. (2007) "Simplified Mechanisms with Applications to Sponsored Search and Package Auctions," Working Paper, Stanford University.

Rassenti, S. J., V. L. Smith, and R. L. Bulfin (1982) "A Combinatorial Auction Mechanism for Airport Slot Allocation," *Bell Journal of Economics*, 13, 402–417.

Rothkopf, M., A. Pekeč, and R. Harstad (1998) "Computationally Manageable Combinatorial Auctions," *Management Science*, 44, 1131–1147.

CHAPTER 21

Assignment Messages and Exchanges[†]

Paul R. Milgrom[*]

In abstract mechanism theory, the designer is often presumed able to create a direct mechanism in which each participant reports its "type," revealing the participant's preferences along with anything else the participant may know. In practice, these details can be too numerous to report. For example, in Federal Communciations Commission (FCC) Auction No. 66 with 1,132 licenses for sale, a type includes a vector of values for every subset of licenses. Reporting that vector would have entailed reporting 2^{1132} numbers.

One approach to mitigating the length-of-report problem is to simplify reporting by limiting the message space. The National Resident Matching Program uses this approach. It limits hospitals' reports to a number of positions and a rank order list of candidates. If a hospital has 10 openings and interviews 50 candidates, it reports the number 10 and a list of 50—a manageably short message. In contrast, because the number of classes of 10 or fewer doctors from among 50 is about 1.3×10^{10}, a general type report, including a rank order list of all those classes, would be impracticably long.

This paper introduces and analyzes a new message space—the space of *assignment messages*—designed for use in auctions, exchanges, and other applications where goods are substitutes. Assignment messages describe preferences indirectly as the value of a linear program for which the set of constraints is describable as a structured collection of trees or hierarchies. We show that if the constraints have this form, then the goods are substitutes, regardless of the various parameters. Conversely, if the constraints describing substitution among different goods do not respect the tree

---

[*] The assignment auction and exchange is an invention of the author for which a patent is pending. Support for research into the exchange's theoretical properties was provided by National Science Foundation grant SES-0648293. I thank Marissa Beck for outstanding research assistance and Joshua Thurston-Milgrom for editorial help. Eduardo Perez and Clayton Featherstone made helpful comments on an earlier draft of this paper, and two referees also contributed. Any opinions expressed here are those of the author alone. An earlier version of this paper was entitled "Assignment Auctions."

[†] To comment on this article in the online discussion forum, or to view additional materials, visit the articles page at: www.aeaweb.org/articles.php?doi=10.1257/mic.1.2.95.

443

structure, then there exist parameters such that goods are not substitutes. In that sense, the constraint structure employed by assignment messages is the most general one consistent with substitutable preferences in linear programming.

An *assignment exchange* is a simplified direct Walrasian mechanism in which participants are restricted to report their preferences using assignment messages. The properties of assignment exchanges are discussed below.

Among the parameters reported by a bidder in an assignment message are ones that specify local rates of technical substitution among goods.[1] *Integer assignment messages* restrict those rates to be zero or one, and restrict any bounds on groups of quantities to be integers. If all traders' preferences can be described in this way, then there is an efficient allocation that is an integer vector.[2] Consequently, the *integer assignment exchange*, which is the assignment exchange restricted to integer assignment messages, transacts in integer quantities.

The assignment exchange shares important aspects of its price and payoff structure with its namesake, the assignment mechanism of Lloyd S. Shapley and Martin Shubik (1971).[3] The integer assignment exchange has the further property that all equilibrium quantities are integers and extends the Shapley-Shubik mechanism in three important ways. First, participants in an integer assignment exchange may buy or sell multiple types of goods simultaneously, instead of just one type. Second, they may trade any integer number of units of each type of good, instead of just one unit. And third, they may buy some goods and sell others, instead of being restricted to just one role as a buyer or a seller.

The integer allocation property can be important for a variety of applications, including those in which commodities are shipped most efficiently by the truckload or container. Even when goods are perfectly divisible, contracts are often denominated and traded in whole numbers of units, so the ability to respect integer constraints may be useful even in those applications.

The restriction of local rates of technical substitution to zero or one is a strong one, but it is surprisingly often a reasonable approximation for practical applications. For example, an electric utility delivering retail power to its customers might acquire wholesale power from generators at three different locations, A, B, and C, but may be limited in its ability to utilize power from each source by its source-specific transmission capacities. When additional transmission capacity is available at source A, one unit of power from A can substitute for one unit from any other source. When capacity is not available, an additional unit of power at A is unusable. It replaces zero units of power from other sources. Similarly, a cereal maker may be able to substitute bushels of grain delivered today for bushels delivered tomorrow up to a limit imposed by its grain-storage

---

[1] Strictly speaking, because the model is one of preferences rather than production, rates of "technical substitution" are not defined. However, assignment messages report constraints resembling production constraints as well as parameters to determine the slopes of those constraints, so it is convenient and intuitive to describe the slopes of constraints using the language of producer theory.

[2] When bundles necessarily consist of integer quantities and goods are substitutes, a version of the limited one-for-one substitution property is implied. See Faruk Gul and Ennio Stacchetti (1999) and Milgrom and Bruno Strulovici (2009).

[3] In both mechanisms, goods are substitutes, and the set of market-clearing goods prices is a nonempty, closed, convex sublattice. Consequently, there is a seller-best, buyer-worse equilibrium price vector and a seller-worst, buyer-best equilibrium price vector.

capacity, or it may substitute one unit of a particular type or grade of grain for one of another type within limits specified by the product-formulation requirements. A similar substitution pattern is sometimes found among sellers, as when a manufacturer can deliver several versions of the same processed good in a total amount that is limited by the overall capacity of its factory.[4] This pattern of *limited one-for-one substitution* can be a useful approximation whenever lots differ in attributes such as time and location of availability, grade, degree of processing, delivery and contract terms, or some combination of these.

General assignment messages extend the integer assignment messages by allowing participants to specify local rates of technical substitution besides zero and one. For example, in markets for electric power, if the transmission losses in shipping power from A are higher than from B, then one unit of power from A replaces less than one unit from B—the rate of technical substitution is positive but less than one. Using integer assignment messages, a bidder can account for such transmission losses only approximately, by treating the power from different sources as having different money values, but general assignment messages allow an exact representation.

An important attribute of assignment messages is that they allow not only bids to buy or sell one of several different goods, but also "swap" bids. For example, in a securities market, a swap could specify that an offer to buy shares of stock is executed only if an offer to sell certain call options on that stock is also executed. Such a linkage can be especially valuable in markets with limited liquidity because it eliminates execution risk.[5]

The ability to report swap bids makes the integer assignment exchange applicable to some resource allocation problems involving *complementary* goods for which package exchange mechanisms might have been thought to be necessary.[6] This is, perhaps, surprising given that assignment messages can directly only express substitutable preferences. Figure 1 displays an example.

Points A, B, and C, in Figure 1, represent physical locations (in southeast Wyoming) where wind farms produce electrical power carried by new long-range transmission lines. Point D represents a node (in northwest Colorado) where the power is injected into the existing transmission grid. For a producer located at A, transmission capacity along lines AC and CD are Leontief complements; the producer is constrained by the minimum of the capacity acquired on AC or CD. Similarly, producers at B regard BC and CD as Leontief complements. The power producers located at A, B, and C compete to acquire capacity on the CD link. Let us assume that there are one or more separate capacity suppliers for each link and that the costs for any suppliers that can supply more than one link are additively separable across links.

---

[4] The National Resident Matching Program, with its fixed number of slots at each hospital, imposes one-for-one substitution but excludes resident wages from the process. An assignment auction could be suitable for that application, provided that wages are made endogenous. Vincent P. Crawford (2008) proposes a simultaneous ascending auction mechanism for the same application.

[5] Some traders call this "leg risk" because the danger is that one "leg" of a transaction is executed while the other is not.

[6] See Milgrom (2007) for an introduction to the economic package allocation problem; Noam Nisan (2006) for an analysis of some message spaces that might be used in package auctions; and Peter Cramton, Yoav Shoham, and Richard Steinberg (2006) for a collection of related articles.

```
        A                    B
         \                  /
          \                /
           \              /
            \            /
             \          /
              \        /
               \      /
                \    /
                 \  /
                  \/
                  /\C
                  |
                  |
                  |
                  |
                  |
                  D
```

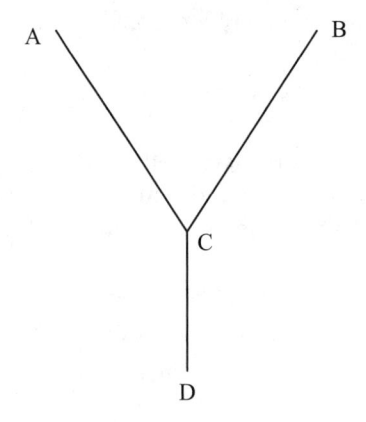

Figure 1. A Y-Shaped Electrical Transmission Grid.

Despite the technical complementarities among successive links, preferences of both buyers of transmission links and suppliers of capacity can be expressed using integer assignment messages. The key lies in the way *lots* are defined. Suppose the exchange is organized to trade three kinds of lots. Each lot is a package of links sufficient to transmit a unit of energy from one of the points A, B, or C to point D (AD, BD, or CD, respectively). With lots defined in that way, each energy producer/capacity buyer can bid on the lot connecting its location to the root at D, so these participants can express their preferences accurately. A supplier who wishes to offer capacity on one of the single links AC or BC can do that using a *swap* bid that links offers to buy and sell. For example, an offer to sell capacity on AC at a price of at least X is represented as a swap that links an offer to sell capacity on the AD lot with a bid to buy equal capacity on the CD lot at a price difference of at least X. Thus, with the specified lots, both buyers and sellers can express preferences accurately. The theorems about assignment exchanges apply. Despite complementarities and indivisible lots, which often preclude the existence of supporting prices, this is a special case in which the existence of market-clearing prices is guaranteed.[7]

Restricting the messages available to participants in a mechanism can affect incentives and performance. In a general simplification, some message profiles may be equilibria of the simplified mechanism even though they were not equilibria of the original, extended mechanism. A *tight* simplification is one with the property that, for every profile of participant preferences in some specified set and every $\varepsilon \geq 0$, all of the full-information, pure ε-Nash equilibria of the simplified mechanism are also ε-Nash equilibria of the original mechanism (see Milgrom (forthcoming)). Assignment exchanges are tight simplifications of general Walrasian exchange mechanisms for any preference that can be represented by a continuous, real-valued utility function whose arguments are the bidder's assigned quantity vector and the price vector. Thus, even though participants may have preferences that are not well described by assignment messages, the

---

[7] A similar construction can be used in any acyclic network by identifying one node in each component of the graph as a root, and expressing all lots in terms of flows from a node to a root. Demand need not be located only at the roots for this construction to work, but the demanded packages of links must lie in sequence on one side of the root.

restriction to assignment messages never introduces any pure ε-Nash equilibrium that was not already present in the full Walrasian mechanism.

The remainder of this paper is organized as follows. Section I introduces the assignment message space and reports three theorems about it. The first is that the assignment messages express only substitutable preferences. The second is that when all preferences are expressed by assignment messages, the set of market-clearing prices is a nonempty, closed, convex sublattice. The third is that if all participants' preferences are expressed with *integer assignment messages*, then there is an efficient allocation using only integer quantities of all goods. Section II provides a partial converse to two of these theorems. Assignment messages require that the constraints connecting different goods form a "tree." If that constraint is relaxed at all, then the conclusions of the first two theorems of Section I are no longer valid. Section III discusses tightness. Its main conclusion is that the assignment exchanges, as well as many further simplifications of these exchanges, are tight simplifications of a Walrasian mechanism. Section IV discusses the connections between the assignment exchange and two familiar mechanisms: the single-product double auction and the Vickrey auction. Section V discusses some of the most likely applications.

I. Assignment Messages

Consider a resource allocation problem with goods indexed by $k = 1, \ldots, K$ and participants are indexed by $n = 1, \ldots, N$. If participants' preferences are quasi-linear, then the utility for a trade is expressed as the value $V_n(q_n)$ of the bundle $q_n \in \Re^K$ acquired plus any net cash transfer. The set of demanded bundles at price vector p is arg $\max_{q_n} V_n(q_n) - p \cdot q_n$, where q_n may include both positive and negative components. A direct mechanism must specify a message space for describing V_n. Assignment messages model demand as originating from multiple sources, describing each q_n as the sum of scalars x_j for $j \in J(n)$, where j is the serial number of a bid and $J(n)$ is the set of serial numbers for bids submitted by bidder n.

Formally, an assignment message consists of a collection of bids and constraints.[8] Each bid by bidder n consists of a 5-tuple $(k_j, v_j, \rho_j, l_j, u_j)$ where k_j identifies the type of product, v_j identifies the "value" of the bid, $\rho_j > 0$ identifies the "effectiveness," and the remaining two terms are lower and upper bounds on quantity: $l_j \leq 0 \leq u_j$. The role of the effectiveness coefficient, which is to allow general local rates of technical substitution, will be formalized shortly.

In addition to the bids, participant n's assignment message expresses quantity constraints of two kinds. First are the *single-product bid group constraints* for each good k:

$$l_{kS} \leq \sum_{j \in S} x_j \leq u_{kS} \quad \text{for } S \in \mathfrak{F}_{nk}, \tag{1}$$

---

[8] A related precursor to this message space is the space of *endowed assignment messages*, introduced by John William Hatfield and Milgrom (2005).

where $\Im_{nk}$ includes all singletons $S = \{j\}$ for which $k_j = k$ and may include other subsets of $R_{nk} = \{j \in J(n) | k_j = k\}$. For the singletons, $l_{k_j\{j\}} \equiv l_j$ and $u_{k_j\{j\}} \equiv u_j$. Second are the *multi-product bid group constraints* indexed by the set $\Im_{n0}$. These are of the form

$$l_{0S} \leq \sum_{j \in S} \rho_j x_j \leq u_{0S} \quad \text{for } S \in \Im_{n0}. \tag{2}$$

Unlike the sets used in the single-product group constraints, the sets $S \in \Im_{n0}$ may include bids on multiple products. Also, unlike the sums in (1), those in (2) are weighted by the effectiveness coefficients, to parameterize the rates of technical substitution among the different products. Note that these constraints can apply to bids to buy ($l_{kS} = 0$), bids to sell ($u_{kS} = 0$), bids to buy or sell ($l_{kS} < 0, u_{kS} > 0$), and swaps between multiple products ($l_{0S} = u_{0S} = 0$).

To simplify notation, we suppress the bidder index n while we are analyzing the reports and preferences of a single bidder. The index will reappear later when we analyze allocations for multiple participants. Using the bids and constraints, bidder n's message is interpreted to report a value for any feasible bundle of products $q = (q_1, \ldots, q_K)$ as follows:

$$V(q) = \max_x \sum_{j \in J} v_j x_j \quad \text{subject to}$$

$$l_{kS} \leq \sum_{j \in S} x_j \leq u_{kS} \quad \text{for } S \in \Im_k, \quad k = 1, \ldots, K$$

$$l_{0S} \leq \sum_{j \in S} \rho_j x_j \leq u_{0S} \quad \text{for } S \in \Im_0$$

$$\sum_{j \in R_k} x_j = q_k \quad \text{for } k = 1, \ldots, K. \tag{3}$$

Because the vector $(q, x) \equiv 0$ satisfies all the constraints in (3), the zero bundle $q = 0$ is feasible. By a theorem of linear programming, the set of vectors q for which the problem is feasible is a closed, bounded, convex set $Q \subseteq \Re^K$, and V is a continuous, concave function on that set.

The next step is to put more structure on the single- and multi-product bid constraints to complete the definition of assignment messages. To describe this structure, we need to define three more concepts: trees, constraint forests, and extended predecessor functions.

As we described above, assignment messages allow two kinds of constraints. There is a set of constraints that describes substitution among products. These are required to form a tree. In addition, for each product k, there may be a set of constraints limiting the quantities assigned to each bid. These, too, must form a tree. Together, these trees form a constraint forest. To describe the relevant trees in a compact notation, we define an extended predecessor function that not only maps sets into their predecessors in the tree, but also maps bids into the smallest set in the tree that contains that bid. These concepts, and others essential to the theorems of this section, are defined below.

Definitions:

1. The *demand correspondence* for V is $D(p) \equiv \arg\max_{q \in Q} V(q) - p \cdot q$.
2. The *indirect profit function* for V is $\pi(p) \equiv \max_{q \in Q} V(q) - p \cdot q$.
3. The valuation V is *substitutable* if for all prices $p, p' \in \Re^K_+$ and all $k = 1, \ldots, K$, if $D(p) = \{x\}$ and $D(p_{-k}, p'_k) = \{x'\}$ are singletons, and $p'_k > p_k$, then $x'_{-k} \geq x_{-k}$.
4. A collection of sets $\Im$ is a *tree* if (1) for any two nondisjoint sets $S, S' \in \Im$, either $S \subset S'$ or $S' \subset S$ and (2) $\Im$ contains a largest set—the union of all its elements. That largest set is the *root* of $\Im$.
5. Given a tree of sets $\Im$, its *extended predecessor function* (P) maps each element of $\Im$, excluding the root R, into its unique predecessor (the smallest set in $\Im$ which contains it), and maps each $j \in R$ into the smallest set S satisfying $j \in S \in \Im$. Below, P_k denotes the extended predecessor function for tree $\Im_k$.
6. A *constraint forest* is a collection of trees and associated bounds $\{\Im_0, \ldots, \Im_K, \{(l_{kS}, u_{kS}) | S \in \Im_k, k = 0, \ldots, K\}\}$ with all $l_{kS} \leq 0 \leq u_{kS}$. The trees satisfy:
    (a) The root of $\Im_0$ is $R_0 = J$ and, for $k = 1, \ldots, K$, the root of $\Im_k$ is $R_k = \{j \in J | k_j = k\}$.
    (b) For $k = 1, \ldots, K$, the terminal nodes of tree $\Im_k$ are the singleton sets $\{j\}$ with $j \in J$ and $k_j = k$.
    (c) All bounds except the root bounds are finite, $0 \geq l_{kS} > -\infty$ and $0 \leq u_{kS} < +\infty$, but the bounds on the roots may be infinite, $0 \geq l_{kR_k} \geq -\infty$ and $0 \leq u_{kR_k} \leq +\infty$.
    (d) For any singleton set $\{j\} \in \Im_{k_j}$, $l_{k_j\{j\}} = l_j$ and $u_{k_j\{j\}} = u_j$.
7. An *assignment message* consists of a collection of bids $(k_j, v_j, \rho_j, l_j, u_j)$ and a constraint forest $\{\Im_0, \ldots, \Im_K, \{(l_{kS}, u_{kS}) | S \in \Im_k, k = 0, \ldots, K\}\}$.
8. An *integer assignment message* is an assignment message with each $\rho_j = 1$ and with all bounds l_{kS} and u_{kS} integers.
9. An *assignment exchange* is a mechanism mapping profiles of assignment messages for each bidder n to an outcome pair $(q_1^*, \ldots, q_N^*, p^*)$, where $q^* \in \arg\max_{\{q | q_n \in Q_n\}} \sum_{n=1}^{n} V_n(q_n)$ subject to $\sum_{n=1}^{N} q_{nk} = 0$ for $k = 1, \ldots, K$ and p^* is a supporting price vector. That is, for $n = 1, \ldots, N$, $q_n^* \in \arg\max_{q \in Q_n} (V_n(q) - p^* \cdot q)$ (equivalently, $p^* \in \arg\min_p \pi_n(p) + p \cdot q_n^*$).
10. An *integer assignment exchange* is an assignment exchange in which the messages are restricted to be integer assignment messages.

The integer assignment messages extend the set of messages allowed by the Shapley-Shubik mechanism. In the Shapley-Shubik mechanism, each participant occupies just one role, as a buyer or a seller. Each seller message includes just one bid ($|J(n)| = 1$), and each buyer message includes just one bid for each product. If participant n is a seller, then the constraints on its one bid are $l_{n1} = -1$ and $u_{n1} = 0$. If participant n is a buyer, then its constraint bounds for each bid are $l_{nj} = 0$ and $u_{nj} = 1$, and its one multi-product group constraint has bounds $l_{n0R_{n0}} = 0$ and $u_{n0R_{n0}} = 1$. The integer assignment message space extends this Shapley-Shubik message space by allowing more bids, more constraints, and general integer bounds.

The three main results of this section can now be stated. Proofs follow.

Theorem 1. *If participant n reports an assignment message, then its valuation $V: q \to \Re$, as given by (3), is continuous, concave, and substitutable, and its indirect profit function is submodular.*

Theorem 2. *If every participant n reports a continuous, concave substitutable valuation on a convex, compact set Q_n, then the set of market-clearing prices for the report profile is $\arg\min_p \sum_{n=1}^{N} \pi_n(p)$. This set is a nonempty, closed, convex sublattice.*

Theorem 3. *If every participant reports an integer assignment message, then there is an integer vector $q^* \in \arg\max_{\{q|q_n \in Q_n\}} \sum_{n=1}^{N} V_n(q_n)$ subject to $\sum_{n=1}^{N} q_{nk} = 0$ for all k.*

The proof of Theorem 1 makes use of two lemmas, which are of independent interest.

Lemma 1. *Suppose that the valuation function V is such that the corresponding indirect profit function π is well defined. Then V is substitutable if and only if its indirect profit function π is submodular.*[9]

Lemma 2. *Suppose $\pi(p) = \min_z g(z)$ subject to $(z, p) \in S$, where g is submodular, S is a sublattice in the product order, and p is a parameter. Then, π is submodular.*

Proof of Lemma 1. Since π is convex on $\Re^K$, it is locally Lipschitz and differentiable almost everywhere. By Hotelling's lemma, the demand set is a singleton $D(p) = \{x(p)\}$ at exactly those points of differentiability, and $\pi_k(p) \equiv \partial \pi(p)/\partial p_k = -x_k(p)$. Substitutability is equivalent to the condition that for $k = 1, \ldots, K$, $x_k(p)$ is nondecreasing in $p_{k'}$ for $k' \neq k$. Submodularity is equivalent to the condition that, on the same domain, $\pi_k(p)$ is nonincreasing in $p_{k'}$ for $k' \neq k$.

Proof of Lemma 2. Let p and p' be two price vectors, and let z and z' be corresponding optimal solutions, so that $\pi(p) = g(z)$, $\pi(p') = g(z')$, and $(z, p), (z', p') \in S$. Since S is a sublattice, $(z \wedge z', p \wedge p'), (z \vee z', p \vee p') \in S$. By the definition of π, $\pi(p \wedge p') \leq g(z \wedge z')$, and $\pi(p \vee p') \leq g(z \vee z')$. Since g is submodular, $g(z \wedge z') + g(z \vee z') \leq g(z) + g(z')$. Hence, $\pi(p \wedge p') + \pi(p \vee p') \leq \pi(p) + \pi(p')$.

Proof of Theorem 1. We will use the dual program corresponding to (3) to show that the indirect profit function π satisfies the assumptions of Lemma 2.

In program (3), let λ_{kS}^u denote the dual price of the upper-bound (k, S)-constraint, λ_{kS}^l the dual price of the corresponding lower-bound constraint, and μ_k the dual price of the product k constraint. Since only one of λ_{kS}^u and λ_{kS}^l can be nonzero, both can be inferred from $\lambda_{kS} = \lambda_{kS}^u - \lambda_{kS}^l$. Using the duality theorem of linear programming (e.g., see David Gale (1960)) in the third inequality below, the indirect profit function

---

[9] Earlier versions of this result, as in Lawrence M. Ausubel and Milgrom (2002) or Milgrom and Strulovici (2009), impose additional restrictions, such as discreteness of the goods, which are appropriate for those contexts. This version drops the unnecessary additional assumptions.

corresponding to V is

$$\pi(p) = \max V(q) - p \cdot q$$

$$= \max_{q,x} \sum_{j \in J} v_j x_j - \sum_{k=1}^{k} p_k q_k \qquad \text{subject to}$$

$$l_{kS} \leq \sum_{j \in S} x_j \leq u_{kS} \qquad \text{for } S \in \mathfrak{I}_k, k = 1, \ldots, K$$

$$l_{0S} \leq \sum_{j \in S} \rho_j x_j \leq u_{0S} \qquad \text{for } S \in \mathfrak{I}_0$$

$$\sum_{j \in R_k} x_j - q_k = 0 \qquad \text{for } k = 1, \ldots, K$$

$$= \min_{\lambda, \mu} \sum_{k=0}^{K} \sum_{S \in \mathfrak{I}_k} (u_{kS} \lambda_{kS}^u - l_{kS} \lambda_{kS}^l) \qquad \text{subject to}$$

$$\sum_{\{S \in \mathfrak{I}_k | j \in S\}} \lambda_{kS} + \mu_k + \rho_j \left(\sum_{\{S \in \mathfrak{I}_0 | j \in S\}} \lambda_{0S} \right) \geq v_j \quad \text{for all } j \in J, k = k_j$$

$$\mu_k \leq p_k \qquad \text{for } k = 1, \ldots, K. \qquad (4)$$

A change of variables reveals the lattice structure in (4). For $k = 0$ and $S \in \mathfrak{I}_0$, define $\Lambda_{0S} = -\sum_{\{\hat{S} \in \mathfrak{I}_0 | S \subseteq \hat{S}\}} \lambda_{0\hat{S}}$ and $f_{0S}(\lambda) = u_{0S} \max(-\lambda, 0) + l_{0S} \min(-\lambda, 0)$. For $k = 1, \ldots, K$ and $S \in \mathfrak{I}_k$, define $\Lambda_{kS} = \mu_{kS} + \sum_{\{\hat{S} \in \mathfrak{I}_k | S \subseteq \hat{S}\}} \lambda_{k\hat{S}}$ and $f_{kS}(\lambda) = u_{kS} \max(\lambda, 0) + l_{kS} \min(\lambda, 0)$. For all $k = 0, \ldots, K$ and $S \in \mathfrak{I}_k$, f_{kS} is nonnegative and convex. Substituting into (4), we obtain

$$\pi(p) = \min_{\Lambda, \mu} \sum_{k=0}^{K} \sum_{S \in \mathfrak{I}_k - \{R_k\}} f_{kS}(\Lambda_{kS} - \Lambda_{kP_k(S)}) + \sum_{k=1}^{K} f_{kR_k}(\Lambda_{kR_k} - \mu_k)$$

$$+ f_{0R_0}(\Lambda_{0R_0})$$

subject to

$$\Lambda_{k_j\{j\}} - \rho_j \Lambda_{0P_0(j)} \geq v_j \quad \text{for all } j \in J$$

$$\mu_k \leq p_k \quad \text{for } k = 1, \ldots, K. \qquad (5)$$

Notice that the dual constraints and objective simplify to this form because of the tree structure we have imposed. For $k_j = k$, the sets in tree $\mathfrak{I}_k$ that include j are exactly $\{j\}, P_k(\{j\}), P_k(P_k(\{j\})), \ldots, R_k$, and similarly for tree zero.

Because each f_{kS} is convex, each term of the objective in (5) is submodular in (Λ, μ, p) using the product order. The objective is a sum of submodular functions and therefore is itself submodular. A set $\{y | l \leq a \cdot y \leq u\}$ is a sublattice in the product order if and only if any two nonzero elements of the a-vector have opposite signs.[10] So, each constraint in problem (5) defines a sublattice on the set of possible (Λ, μ, p)-vectors,

---

[10] This property of the *rows* of the *dual* constraint matrix, that no two nonzero entries have the same sign, is in remarkable correspondence with the condition required in the proof of Theorem 3 that no two nonzero entries in the *columns* of the constraint matrix of the *primal* problem have the same sign. The dual constraint matrix is obtained from the primal constraint matrix essentially by transposition, so the two conditions coincide. That is why the structure of assignment messages can be useful for proving the substitutes conclusion of Theorem 1 and the integer allocation conclusion of Theorem 3.

and the intersection of sublattices is a sublattice. Hence, by Lemma 2, $\pi(p)$ is submodular. And therefore, by Lemma 1, V is substitutable.

Proof of Theorem 2. Since the corresponding primal problem can be represented as a continuous, concave maximization on a compact set, the maximum exists and coincides with the minimum of the dual. Since the valuations are concave, the set of market-clearing prices is the set of solutions to the dual problem: $\arg\min_p \sum_{n=1}^N \pi_n(p)$. Since each π_n is continuous and convex, the set of minimizers of the dual problem is closed and convex. Since each π_n is submodular, by a theorem of Donald M. Topkis (1978), the set of minimizers of the dual problem is a sublattice.

Proof of Theorem 3. To find $q^* \in \arg\max_{\{q|q_n \in Q_n\}} \sum_{n=1}^N V_n(q_n)$ subject to $\sum_{n=1}^N q_{nk} = 0$, we substitute from (3) and introduce variables x_{nkS} as the sums of their successors in the tree (the elements of the set $P_{nk}^{-1}(S)$), so that the optimization is converted into one in which every inequality constraint involves just one variable. Because the bid j, and not just the set $S = \{j\}$, can be a successor to sets in the constraint trees under the extended predecessor function, define $x_{nkj} \equiv x_j$ for all $n = 1, \ldots, N$ and $k = 1, \ldots, K$. The tree structure allows us to show something stronger than claimed by the theorem, namely, that there is an integer optimal solution x^* to the resulting problem:

$$\max_q \sum_n V_n(q_n) \quad \text{subject to } \sum_n q_{nk} = 0 \quad \text{for } k = 1, \ldots, K$$

$$= \max_x \sum_n \sum_{j \in J(n)} v_j x_j \quad \text{subject to}$$

$$l_{nkS} \leq \sum_{j \in S} x_j \leq u_{nkS} \quad \text{for } S \in \Im_{nk}, k = 1, \ldots, K, n = 1, \ldots, N$$

$$l_{n0S} \leq \sum_{j \in S} x_j \leq u_{n0S} \quad \text{for } S \in \Im_{n0}, n = 1, \ldots, N$$

$$\sum_n \sum_{j \in R_{nk}} x_j = 0 \quad \text{for } k = 1, \ldots, K$$

$$= \max_x \sum_n \sum_{j \in J(n)} v_j x_j \quad \text{subject to}$$

$$-x_{nkS} + \sum_{S' \in P_{nk}^{-1}(S)} x_{nkS'} = 0 \quad \text{for } S \in \Im_{nk}, k = 1, \ldots, K, n = 1, \ldots, N$$

$$x_{n0S} - \sum_{S' \in P_{n0}^{-1}(S)} x_{n0S'} = 0 \quad \text{for } S \in \Im_{n0}, n = 1, \ldots, N$$

$$l_{nkS} \leq x_{nkS} \leq u_{nkS} \quad \text{for } S \in \Im_{nk}, k = 0, \ldots, K, n = 1, \ldots, N$$

$$\sum_n x_{nkR_{nk}} = 0 \quad \text{for } k = 1, \ldots, K. \tag{6}$$

The sign restrictions $l_{nkS} \leq 0$ and $u_{nkS} \geq 0$ ensure that $x \equiv 0$ satisfies the constraints of the problem, so the problem is feasible. The bounds on each variable imply that the constraint simplex is bounded. For a feasible, bounded linear program, there is always an optimal solution at a vertex of the constraint simplex.[11] Hence, to prove the theorem, it is sufficient to show that every vertex of the simplex defined by the constraints in (6) is an integer vector.

---

[11] See, for example, Gale (1960).

Each vertex of the constraint simplex is determined by a set of binding upper and lower bound constraints of the form $x_{nkS} = u_{nkS}$ or $x_{nkS} = l_{nkS}$ and the equation $Ax = 0$, which describes the equality constraints in (6). Fix any vertex and denote the right-hand sides of the binding upper and lower bound constraints by $\bar{u}$ and $\bar{l}$, which, by hypothesis, are integer vectors. Write the vector x in the form $(\hat{x}, \bar{x}_l, \bar{x}_u)$, where the binding inequality constraints are $\bar{x}_l = \bar{l}, \bar{x}_u = \bar{u}$, which we write as $\bar{x} = (\bar{u}, \bar{l}) \equiv \bar{b}$. Let $\bar{A}$ and $\hat{A}$ be the matrices consisting of the columns of A corresponding to $\bar{x}$ and $\hat{x}$, respectively. Then the equation $Ax = 0$ can be written as $0 = Ax = \hat{A}\hat{x} + \bar{A}\bar{x} = \hat{A}\hat{x} + \bar{A}\bar{b}$. Taking $b \equiv -\bar{A}\bar{b}$, the equality constraints can be written as $\hat{A}\hat{x} = b$. Observe that b is an integer vector, because $\bar{A}$ is an integer matrix and $\bar{b}$ is an integer vector.

It is therefore sufficient to show that for every nonsingular submatrix $\hat{A}$ of A and every integer vector b, there is an integer solution $\hat{x}$ to $\hat{A}\hat{x} = b$. For this, it suffices to show that A is *totally unimodular*.[12] According to a theorem attributed to Alan J. Hoffman (see I. Heller and C. B. Tomkins (1956)), a matrix is totally unimodular if two conditions are satisfied: all the entries of A are elements of the set $\{0, +1, -1\}$, and any two nonzero entries in the same column have opposite signs. We finish by verifying these Hoffman conditions.

Examine the columns of A as represented in (6) which correspond to the variables x_{nkS}. For $k = 0$ and $S = R_{n0}$, the root of tree $\Im_{n0}$ for some participant n, x_{n0S} appears in only one equality constraint in (6), and so has the single entry $+1$ in its column. For $k = 1, \ldots, K$, each of the variables $x_{nkR_{nk}}$ appears twice (once in its defining equation and, again, in the market-clearing constraint for k), and its two coefficients, ± 1, have opposite signs. For $k = 1, \ldots, K$, and all sets $S \in \Im_{nk} - \{R_{nk}\}$, x_{nkS} appears twice: once with coefficient -1 in the equation defining x_{nkS} and once with coefficient $+1$ in the equation defining $x_{nkP_{nk}(S)}$. For $k = 0$ and $S \in \Im_{n0} - \{R_{n0}\}$, x_{n0S} appears twice: once with coefficient $+1$ in its defining equation and once with coefficient -1 in the equation defining $x_{n0P_{n0}(S)}$. Last are the x_j variables. Recall that by our extended definition of predecessor, $j \in P_{nk}^{-1}(S)$ for exactly two sets, one in $\Im_{nk_j}$ with coefficient $+1$ and one in $\Im_{n0}$, with coefficient -1. Hence, the Hoffman conditions are satisfied.

II. Partial Converse to Theorems 1 and 2

The structure of assignment messages allows bidders to report values and effectiveness coefficients without limitations but restricts the form of constraints to be a constraint forest. This section shows that if one weakens the restriction that $\Im_{n0}$ is a tree, then the conclusions of Theorems 1 and 2 fail.

The problem can be illustrated with an example of a buyer for whom the lower bounds l_j and l_{kS} are all zero. Suppose that there are three goods and that this buyer has three bids, $j = 1, 2, 3$, each with $v_j = 2.9$, $k_j = j$, and $u_j = 2$. Suppose that the multiproduct group constraints in the problem are $x_1 + x_2 \leq 3$ and $x_2 + x_3 \leq 3$, violating the tree structure. Then, for the price vector $(0, 1, 2)$, the corresponding demand is $(2, 1, 2)$ and for the price vector $(3, 1, 2)$, the corresponding demand is $(0, 2, 1)$; raising

---

[12] See the *Wikipedia* entry (http://en.wikipedia.org/unimodular_matrix) on "unimodular matrix" for an accessible treatment of the relevant mathematics.

the price of good 1 reduces the demand for good 3, violating the substitutes condition. Moreover, if the available quantities are one unit of good 2 and two units each of goods 1 and 3, then the market clears for price vectors (0, 1, 2) or (2, 1, 0) but not for the join, which is (2, 1, 2), so the set of market-clearing prices in this example is not a sublattice.

More generally, given *any* set of constraints $\mathfrak{I}_{n0}$ that fails to have the tree structure, we can find a similar counter example as follows. Since the constraints do not form a tree, there are two sets, $S, S' \in \mathfrak{I}_{n0}$, such that each of the three disjoint sets $S - S'$, $S \cap S'$, and $S' - S$ are nonempty. Let goods 1, 2, and 3 denote elements of these three sets and specify that the values of any other goods are zero. Let the bounds constraining these goods be given as in the preceding paragraph and let the bounds on all other constraints be very large, so that those constraints do not bind. This specification reproduces the example of the preceding paragraph starting from any $\mathfrak{I}_{n0}$ that is not a tree. That proves the following theorem.

Theorem 4. *If the set $\mathfrak{I}_{n0}$ is not a tree, then there exist bids and integer bounds for each $S \in \mathfrak{I}_{n0}$ and supplies for the other participants, such that the valuation V_n is not a substitutes valuation, the indirect profit function π_n is not submodular, and the set of market-clearing prices is not a sublattice.*

III. Tightness

A *direct mechanism* is a triple (N, M, ω), where N is the set of participants, M is the product space of types ("message profiles"), and $\omega: M \to \Omega$, where Ω is the set of possible outcomes. The mechanism $(N, \hat{M}, \omega)$ is a simplification of the mechanism (N, M, ω) provided $\hat{M} \subseteq M$. For tightness analysis, it is assumed that $\Omega \subseteq \times_{n \in N} \Omega_n$ where each Ω_n is a topological space, and that each player n's payoff is represented by a continuous function $u_n : \Omega_n \to \mathfrak{R}$.

A simplified direct mechanism has the *outcome closure property* if, for every player n, strategy profile $\hat{m}_{-n} \in \hat{M}_{-n}$, strategy $m_n \in M_n$, and every open set $O \subset \Omega_n$ such that $\omega_n(m_n, \hat{m}_{-n}) \in O$, there is a strategy $\hat{m}_n \in \hat{M}_n$, such that $\omega_n(\hat{m}) \in O$. This means that when other participants are limited to using simplified messages, limiting n to do the same has little or no effect on the set of outcomes that n can produce. The mechanism $(N, \hat{M}, \omega)$ is a tight simplification of (N, M, ω) if for all utility profiles $u = (u_n)_{n \in N}$ and every $\varepsilon \geq 0$, every pure-strategy profile that is an ε-Nash equilibrium of the simplified mechanism is also an ε-Nash equilibrium of the original, extended mechanism. The Simplification Theorem of Milgrom (forthcoming) asserts that if $(N, \hat{M}, \omega)$ has the outcome closure property with respect to (N, M, ω), then the simplification is tight.

For this application, we take $\omega_n = (q_n, p)$. This specification permits each participant to care about his own goods assignment and the prices, but not about the goods assigned to others. In standard equilibrium theory, preferences for a participant n depend only on $(q_n, p \cdot q_n)$, his goods assignment, and payment. By including the price vector in a more general way, the tightness analysis allows that a participant may prefer that its competitor's product commands a low price or that its partner's product commands a high price. It also allows a participant to have any preference for which the preferred sets are all closed and convex, but participants are not limited to such preferences and certainly not just to the preferences that are describable using assignment messages.

The next theorem applies not just to the full assignment exchange, but also to mechanisms that limit the messages participants can use to a subset of the assignment messages. To describe the permissible limitations on messages, let us say that an assignment message m_n is *minimally constrained* if its only finite constraint bounds (l_{kS}, u_{kS}) correspond to the singleton sets $S = \{j\}$. An *elementary assignment message* m_n for participant n is an assignment message that is minimally constrained and includes, at most, two bids for any product k: $|\{j \in J(n) : k_j = k\}| \leq 2$ for $k = 1, \ldots, K$. A *full Walrasian exchange* is any mechanism that accepts messages describing, for each participant, closed convex preferences over net trades and a feasible consumption set with the null trade in its interior; and maps any message profile into a corresponding competitive equilibrium outcome, whenever one exists.

Theorem 5. *Any simplified Walrasian exchange in which each bidder n's message space contains only assignment messages, and contains all elementary assignment messages, satisfies the outcome closure property with respect to any full Walrasian exchange and (hence) is a tight simplification.*

Theorem 5 is proved by showing that for any price vector and goods assignment that can be obtained by some general message, a buyer can acquire nearly the same bundle and bring about nearly the same prices with an elementary assignment message that bids for the equilibrium quantities at slightly higher than the equilibrium prices and that bids for additional quantities at slightly lower prices. For the full proof, details are added to ensure that this construction applies not only to buyers but also to sellers and to participants who bid to buy some items and to sell others. This establishes the outcome closure property. The tightness conclusion then follows from the Simplification Theorem.

Proof: Let $\hat{M}_n$ be bidder n's simplified message space, and let M_n be the message space used by a full Walrasian mechanism, as described above. Fix a participant n and messages $\hat{m}_{-n} \in \hat{M}_{-n}$ and $m_n \in M_n$. Let $(p, q) \equiv \omega(\hat{m}_{-n}, m_n)$. We now construct the elementary message described informally in the preceding paragraph.

Let $\sigma_{nk} = sign(q_{nk}) \in \{-1, 0, 1\}$ and fix $\varepsilon > 0$. Since n's message space includes all elementary assignment messages, it includes the message $\hat{m}_n$ with bids $j = 1, \ldots, 2K$ as follows. For $j = 1, \ldots, 2K$, let $k_j = \lceil j/2 \rceil$ (the smallest integer weakly exceeding $j/2$) and set $v_{2k-1} = p_k + \sigma_{nk}\varepsilon$, $v_{2k} = p_k - \sigma_{nk}\varepsilon$, $u_{2k-1} = u_{2k} = \max(0, q_{nk})$ and $l_{2k-1} = l_{2k} = \min(0, q_{nk})$. The message $\hat{m}_n$ specifies no other finite bounds. Let $(\hat{p}, \hat{q})$ be the competitive equilibrium outcome selected by the full Walrasian mechanism when the message profile is $\hat{m}$.

Since (p, q) is a competitive equilibrium for the report profile $(\hat{m}_{-n}, m_n)$, $q_n \in$ arg max$_{y_n}[\max_{\{y_{-n}|\sum_{l \neq n} y_l = -y_n\}}(V_n(y_n|m_n) + \sum_{l \neq n} V_l(y_l|\hat{m}_l))]$. And since n demands q_n at prices p, (p, q) is also a competitive equilibrium for report profile $\hat{m}$. From that and the fact that $\varepsilon > 0$, q_n *uniquely* solves arg max$_{y_n}[\max_{\{y_{-n}|\sum_{l \neq n} y_l = -y_n\}}$ $(V_n(y_n|\hat{m}_n) + \sum_{l \neq n} V_l(y_l|\hat{m}_l))]$. Hence, even though there may be multiple competitive equilibria for the message profile $\hat{m}$, all assign the bundle q_n to participant n: $\hat{q}_n = q_n$. Moreover, since every market-clearing price vector supports this choice by n, the price vector $\hat{p}$ must satisfy $p_k - \varepsilon \leq \hat{p}_k \leq p_k + \varepsilon$ for every product k. Since ε can be arbitrarily small, the outcome closure property is proved. Tightness then follows from the Simplification Theorem cited above.

IV. Connections to Two Familiar Mechanisms

In case $K = 1$, each participant's assignment message describes a step supply or demand function. The assignment exchange is then a familiar double auction, in which the allocation is determined by intersecting single-product supply and demand curves. When the market-clearing prices or quantities are not unique, any selection rule is consistent with the assignment exchange.[13] In general, the assignment exchange extends the single-product double auction by allowing multiple products and a rich set of substitution possibilities among them.

The integer assignment exchange is connected to the Vickrey auction. In a Vickrey auction, if a participant n acquires a single unit of a single good k, its payment is the opportunity cost of that good, which is equal to the incremental value of one additional unit of good k to the coalition of all *other* participants. In the linear program for the integer assignment exchange, the lowest market-clearing price p_k for good k is its lowest dual price—the amount by which the optimal value would increase if an additional unit of good k were made available to the coalition of *all* players. If participant n has demand for just one unit in total and acquires a unit of good k, then the additional unit for the coalition of all participants is actually assigned to someone besides n, so p_k is the increased optimal value of that unit to the other participants—n's Vickrey price.

Theorem 6. *Suppose that some participant n bids to acquire, at most, one unit in an integer assignment exchange, and that the exchange selects the price vector p that is the minimum market-clearing price vector. Then, if n acquires a unit of good k, the price p_k is equal to n's Vickrey payment.*

A symmetric statement can be made about participants who *sell* one unit and exchanges that select the *maximum* market-clearing price vector.

V. Likely Applications

The most immediate opportunity for application of the assignment exchange technology is to auction off two or more substitute products for which the length-of-report problem is important. Paul D. Klemperer (2008) has independently proposed a simple version of the assignment auction design. For this section, an *auction*

---

[13] In one-sided cases (with just bids to buy and a fixed supply, or bids to sell and a fixed demand), the kinds of problems found in share auctions (Robert B. Wilson (1979)) can present themselves. Typical solutions to these problems, such as those proposed in David McAdams (2002) and Ilan Kremer and Kjell G. Nyborg (2004), can be adapted to the assignment exchange.

is simply an exchange with one seller and many buyers or one buyer and many sellers.[14]

The previous best-practice mechanisms for dealing with the length-of-report problem were sequential mechanisms—the simultaneous ascending and descending clock auctions (Ausubel (2001)). In simultaneous clock mechanisms, bidders are asked to report supplies or demands at each of a sequence of announced prices, and the reported information is used to find approximate market-clearing prices and allocations. Because demands are announced for only a finite number of price vectors, the information reported is much less than that of a full direct mechanism.

Simultaneous ascending or descending multi-product auctions of various kinds have been used for several high-value applications, most commonly ones involving radio spectrum, electricity, or natural gas (Milgrom (2004)), but also for real estate transactions and certain agricultural commodities markets.[15] When the goods for sale are substitutes and participants bid myopically, various versions of the simultaneous ascending or descending auctions have been found not only to economize on communications but also to identify allocations that are efficient or stable or to find minimum or maximum market-clearing prices (Alexander S. Kelso, Jr. and Crawford (1982); Gul and Stacchetti (2000); Milgrom (2000); Ausubel (2004), Milgrom and Strulovici (2009)). This property makes these auctions directly comparable to assignment auctions.

Because simultaneous ascending and descending auctions economize on communications and enable bidders to substitute in response to changing prices, they have important advantages over independent auctions of different goods. But they also have properties that make them unsuitable for many applications. Four of these disadvantageous properties are high participant costs, long times-to-completion, imprecise computations, and difficulties of scheduling. Any multi-round, real-time process adds the cost of real-time bidding to the costs of preparing for the auction. In current practice, dynamic auctions for gas and electricity take several hours to reach completion, while spectrum auctions take days, weeks, or even months. Such long times-to-completion cripple these mechanisms for the most time-sensitive markets, such as hour-ahead power markets, where only minutes are available to complete an exchange. In practice, ascending and descending auctions fail to identify exact market-clearing prices because they change the direction of price increments only a small number of times, using discrete price increments.[16] Finally, in export markets, where potential buyers may reside in a dozen or more different time zones, scheduling a convenient time for several hours of real-time bidding may be impossible. These four problems are avoided by direct mechanisms, including simplified direct mechanisms like the assignment auction.

---

[14] Assignment auctions have several variations, mirroring the variations common in other sealed-bid auctions. For example, the auctioneer (whether buyer or seller) may move first, possibly announcing target quantities or reserve prices, or a supply or demand curve, or perhaps announcing rates of substitution among products. Or, there may be multiple stages, for example, a qualifying stage with just some bidders invited to the second stage.

[15] In a simultaneous ascending auction, prices can be called by the auctioneer (these are the so-called "clock auctions") or by individual bidders.

[16] Ausubel and Cramton (2004) show how a clock auction with a richer message space ("intra-round bidding") can avoid some of the disadvantages of discrete price increments.

The two main practical limitations of assignment exchanges arise because the message space may be too narrow to express bidders' actual preferences and because, as a static mechanism, the auction provides no opportunity for bidders to learn from competing bids. The latter can be significant when there is uncertainty about a common factor that raises or lowers all values together, or when a bidder's preferred trades depend on the trades made by other bidders.

Even the integer assignment messages, with their limited one-for-one substitution, allow ample expressiveness for some applications. Suppose, for example, that an electricity buyer can purchase power from any of three sources, $k \in \{1, 2, 3\}$, subject to transmission costs (t_1, t_2, t_3) and transmission capacity limits (U_1, U_2, U_3). If the buyer needs to buy P units of power and the value per unit is α, then bids $j = 1, 2, 3$ with $k_j = j$, $v_j = \alpha - t_j$, $u_j = U_j$, $l_j = 0$, and one constraint for $S = \{1, 2, 3\}$ with $u_{0S} = P$ and $l_{0S} = 0$ accurately express the bidder's demand. If there are also significant transmission losses from some source j, a general assignment message accommodates those by allowing the bidder to set $\rho_j < 1$.

In a double-auction with multiple buyers and sellers of electric power, other kinds of assignment messages can be valuable. For even if a buyer has already filled all of its power needs for some time period, it may be willing to sell up to β units of power at source 1 and buy the same quantity at 2 or 3, provided the price differential is favorable. This swap can be encoded with three bids and the constraints: $0 \geq x_1 \geq -\beta$, $\beta \geq x_2, x_3 \geq 0$, $x_1 + x_2 + x_3 = 0$.

Swap bids have the potential to add liquidity to an exchange hindered by lack of volume. Investigating this fully is beyond the scope of this paper. It requires a theory of why owners do not constantly participate in, and provide liquidity to, markets. Nevertheless, it is clear that in a market with modest liquidity, swaps encourage participation by limiting the risk that one part of an intended transaction might be executed without the other parts. With separate markets, a swapper with a budget limit might have to sell one commodity before buying the other in order to raise funds to transact, leaving the swapper exposed to the risk of not finding a seller for the other part of the planned transaction. By eliminating such risks, swaps make participation safer, increasing liquidity.

The power of simple assignment messages in the preceding example is important because simplicity is often a design goal. One might simplify the general assignment exchange by limiting the number of bids, constraints, or levels in the constraint trees. Theorems 1, 2, 3, and 5 have been constructed to apply even to exchanges that incorporate such additional simplifications.

One common limitation imposed by auctioneers is a credit limit on buyers. Buyers might also want to express a budget limit. The assignment message space does not allow this to be done directly, but it does allow surrogates, such as a limit on the maximum total bid from a bidder, or on the maximum quantities that can be demanded.

Maximum quantity limits on some bidder or set of bidders can also be useful for a government auctioneer when bidder market power is a concern, or when there is a goal of promoting entry. Sometimes, this goal is best implemented by careful product definitions. For example, if the auctioneer wants to limit bidders 1 and 2 to purchase no more than half of the available units of good 1, it can accomplish that by splitting good 1 into types 1A and IB and restricting bidders 1 and 2 from bidding on type IB. This procedure is similar to the set-asides used by the FCC to restrict purchases by incumbents in some radio spectrum auctions.[17]

Whether the assignment messages are sufficiently encompassing is likely to vary by application. Certainly, scale economies and complements among lots are sometimes important and cannot generally be solved by redefining lots. For example, in electricity, generating plants typically have large fixed costs that require all or nothing decisions about whether to use their power capacity. While such limits are not directly expressible using assignment messages, it is often possible to use the assignment exchange as part of a solution. One ad hoc procedure is to operate the exchange in two or more rounds to allow preliminary price discovery to guide bids at the final round. This does not entirely eliminate the fixed-cost problem, but it may sometimes mitigate it sufficiently. Staged dynamics of this sort may also be helpful when there are important common value elements or when bidders can invest in information gathering during the process, as in Olivier Compte and Philippe Jehiel (2000) or Leonardo Rezende (2005).

Three key properties of assignment and integer assignment messages—that they are simple to use, express only substitutable preferences, and that integer assignment messages lead to efficient integer solutions—make them potentially valuable for use with other mechanisms in addition to the Walrasian exchange. For example, two principal disadvantages of "standard" Vickrey auctions—the length-of-report problem and "low" seller revenues (less than in any core allocation)—hinge on the requirement to report a separate value for each possible package and the availability of messages that report nonsubstitutable values, respectively.[18] A simplified Vickrey auction in which bidders are limited to reporting assignment messages escapes both of these disadvantages. There may also be applications to matching problems, without cash transfers, such as the problems of assigning students to courses or flight attendants to routes, where integer allocations are essential and the substitutes structure may be a reasonable approximation.[19]

Simplification represents a promising approach to applied mechanism design, and assignment messages show high potential for use in simplified mechanisms for trading substitutable goods. Exchanges that utilize assignment messages are tight, easy for bidders to use, quick to run, precise in determining both equilibrium prices and goods assignments, and adaptable to settings that require integer solutions. The assignment exchange design is *robust*, in the sense that its key properties remain intact even when the assignment message space is further restricted in any way that does not eliminate any *elementary* assignment messages. It is also *maximal* in the sense that no extension of the constraint tree architecture is possible without destroying the key substitutes property of the message space. Taken together, these attributes make the assignment exchange an attractive candidate for the many practical applications.

---

[17] The FCC combined this with restrictions on post-auction transfers to limit gaming of the system.
[18] Milgrom (2004), sections 2.5 and 8.1 and Ausubel and Milgrom (2006).
[19] Eric Budish et al. (2008) have begun to study this problem.

References

Ausubel, Lawrence M. 2001. "System and Method for an Efficient Dynamic Multi-Unit Auction." US Patent 7,165,046, filed July 5, 2001, and issued January 16, 2007.

Ausubel, Lawrence M. 2004. "An Efficient Ascending-Bid Auction for Multiple Objects." *American Economic Review*, 94(5): 1452–75.

Ausubel, Lawrence M., and Peter Cramton. 2004. "Auctioning Many Divisible Goods." *Journal of the European Economic Association*, 2(2–3): 480–93.

Ausubel, Lawrence M., and Paul R. Milgrom. 2002. "Ascending Auctions with Package Bidding." *Frontiers of Theoretical Economics*, 1(1): Article 1.

Ausubel, Lawrence M., and Paul R. Milgrom. 2006. "The Lovely but Lonely Vickrey Auction." In *Combinatorial Auctions*, ed. Peter Cramton, Yoav Shoham, and Richard Steinberg, 17–40. Cambridge, MA: MIT Press.

Budish, Eric, Yeon-Koo Che, Fuhito Kojima, and Paul R. Milgrom. 2008 "Implementing Random Assignments: A Generalization of the Birkhoff-Von Neumann Theorem." Unpublished.

Compte, Olivier, and Philippe Jehiel. 2000. "On the Virtues of the Ascending Price Auction: New Insights in the Private Value Setting." www.enpc.fr/ceras/compte/ascend.pdf.

Cramton, Peter, Yoav Shoham, and Richard Steinberg. 2006. *Combinatorial Auctions*. Cambridge, MA: MIT Press.

Crawford, Vincent P. 2008. "The Flexible-Salary Match: A Proposal to Increase the Salary Flexibility of the National Resident Matching Program." *Journal of Economic Behavior & Organization*, 66(2): 149–60.

Gale, David. 1960. *The Theory of Linear Economic Models*. Chicago, IL: University of Chicago Press.

Gul, Faruk, and Ennio Stacchetti. 1999. "Walrasian Equilibrium with Gross Substitutes." *Journal of Economic Theory*, 87(1): 95–124.

Gul, Faruk, and Ennio Stacchetti. 2000. "The English Auction with Differentiated Commodities." *Journal of Economic Theory*, 92(1): 66–95.

Hatfield, John William, and Paul R. Milgrom. 2005. "Matching with Contracts." *American Economic Review*, 95(4): 913–35.

Heller, I., and C. B. Tompkins. 1956. "An Extension of a Theorem of Dantzig' s." In *Linear Inequalities and Related Systems*, ed. Harold William Kuhn and Albert William Tucker, 247–54. Princeton, NJ: Princeton University Press.

Kelso, Alexander S., Jr., and Vincent P. Crawford. 1982. "Job Matching, Coalition Formation, and Gross Substitutes." *Econometrica*, 50(6): 1483–1504.

Klemperer, Paul D. 2008. "A New Auction for Substitutes: Central Bank Liquidity Auctions, the U.S. TARP, and Variable Product-Mix Auctions." www.nuff.ox.ac.uk/users/klemperer/substsauc_NonConfidentialVersion.pdf.

Kremer, Ilan, and Kjell G. Nyborg. 2004. "Divisible-Good Auctions: The Role of Allocation Rules." *RAND Journal of Economics*, 35(1): 147–59.

McAdams, David. 2002. "Modifying the Uniform-Price Auction to Eliminate 'Collusive-Seeming Equilibria.'" www.nyu.edu/sed2002/pdfs/at2-2-txt.pdf.

Milgrom, Paul. Forthcoming. "Simplified Mechanisms with an Application to Sponsored-Search Auctions." *Games and Economic Behavior.*

Milgrom, Paul. 2000. "Putting Auction Theory to Work: The Simultaneous Ascending Auction." *Journal of Political Economy*, 108(2): 245–72.

Milgrom, Paul. 2004. *Putting Auction Theory to Work*. Cambridge, MA: Cambridge University Press.

Milgrom, Paul. 2007. "Package Auctions and Exchanges." *Econometrica*, 75(4):935–65.

Milgrom, Paul, and Bruno Strulovici. 2009. "Substitute Goods, Auctions, and Equilibrium." *Journal of Economic Theory*, 144(1): 212–47.

Nisan, Noam. 2006. "Bidding Languages for Combinatorial Auctions." In *Combinatorial Auctions*, ed. Peter Cramton, Yoav Shoham, and Richard Steinberg, 215–31. Cambridge, MA: MIT Press.

Rezende, Leonardo. 2005. "Mid-Auction Information Acquisition." https://netfiles.uiuc.edu/lrezende/www/noise.pdf.

Shapley, Lloyd S., and Martin Shubik. 1971. "The Assignment Game I: The Core." *International Journal of Game Theory*, 1(1): 111–30.

Topkis, Donald M. 1978. "Minimizing a Submodular Function on a Lattice." *Operations Research*, 26(2): 305–21.

Wilson, Robert B. 1979. "Auctions of Shares." *Quarterly Journal of Economics*, 93(4): 675–89.

CHAPTER 22

The Product-Mix Auction: A New Auction Design for Differentiated Goods[1]

Paul Klemperer

The "Product-Mix Auction" is a single-round auction that can be used whenever an auctioneer wants to buy or sell multiple differentiated products. So potential applications include spectrum sales, and selling close-substitutes "types" of energy, as well as applications in finance such as the one discussed below.

The Product-Mix Auction effectively takes multiple single-round auctions that are each for (many units of) a single variety of the good, and combines all these auctions into one auction. Bidders express their relative preferences *between* varieties, as well as for alternative quantities of specific varieties, in simultaneous "sealed bids". All the information from all the bids is then used, in conjunction with the auctioneer's own preferences, to set *all* the prices *and* the total quantities of all the goods allocated.

The Product-Mix Auction is like the "simultaneous multiple-round auction" (SMRA) in that both auctions find competitive equilibrium allocations consistent with the preferences that participants express in their bids. However, the Product-Mix Auction has several advantages over the SMRA. It is (obviously) much faster than an

---

[1] This chapter was originally written in 2008, and published in the Journal of the European Economic Association, 2010, 8(2–3), 526–536. It is reproduced here with the kind permission of the European Economic Association and the MIT Press. A new first section has been added, and minor additional revisions made, to the original paper.

I invented this auction for the Bank of England when it consulted me at the beginning of the financial crisis in 2007, and I also advised the Bank on the updated version it introduced in 2014. The Bank now uses it regularly to auction loans of funds secured against different varieties of collateral. It runs the auction more often at times of potential stress–the Governor of the Bank (Mervyn King) wrote that "[it] is a marvellous application of theoretical economics to a practical problem of vital importance to financial markets," and remarks by a Deputy Governor were published by a major British newspaper here: www.guardian.co.uk/science/video/2013/jul/12/geometry-banking-crisis-video. I have been a *pro bono* adviser to the Bank of England since autumn 2007, and I have also given *pro bono* advice to the US Treasury, other central banks, government agencies, etc., about these issues. I thank the relevant officials for help, but the views here are my own and do not represent those of any organisation. I am very grateful to Elizabeth Baldwin, Jeremy Bulow and Daniel Marszalec for their help in advising the Bank of England. I also particularly benefited from discussions with Marco Pagnozzi, and thank Olivier Armantier, Eric Budish, Vince Crawford, Aytek Erdil, Meg Meyer, Moritz Meyer-ter-Vehn, Tim O'Connor, Rakesh Vohra, the editor and anonymous referees of the JEEA, and many other friends and colleagues for helpful advice.

SMRA's iterative process (for example, the UK's 3G mobile-phone SMRA auction took 150 rounds over seven weeks), and it is therefore also simpler to use and less vulnerable to collusion. Moreover, unlike a standard SMRA, it allows the auctioneer as well as the bidders, to specify how the relative quantities of the different varieties to be sold should depend on their relative prices.

I invented the auction at the beginning of the financial crisis, in response to the 2007 Northern Rock bank run–Britain's first bank run since the 1800's. The Bank of England has used it regularly and successfully since, to improve its allocation of funds to banks, building societies, etc. In the Bank's auction, the different "types of goods" are loans of funds secured against different types of collateral, and the "prices" are interest rates, that is, the interest rate a borrower pays depends on the quality of the collateral it offers.[2]

One indication of the auction's success is that the then-Governor of the Bank of England (Mervyn King) wrote, after the Bank had been using it regularly for over eighteen months and auctioned £80 billion worth of repos using it, that "the Bank of England's use of Klemperer auctions in our liquidity insurance operations is a marvellous application of theoretical economics to a practical problem of vital importance to financial markets"; he made a similar statement to the *Economist* a year later; and an Executive Director of the Bank described the auction as "A world first in central banking... potentially a major step forward in practical policies to support financial stability".[3]

The Bank of England is enthusiastic about the way not only the prices (i.e., interest rates) but also–by contrast with the SMRA–the *quantities* of the different goods (loans against different collaterals) are determined within the auction, as a function both of bids on all the collaterals, and of its own preferences (expressed through a set of "supply functions"). Specifically, the Bank doesn't want to accept very much low-quality collateral unless the markets are stressed, so the quantities of poorer collaterals that its auction accepts are increasing functions of the differences between the bids offering them and the bids offering better collaterals.

So the Product-Mix Auction is better than many central banks' approaches of treating all collateral as identical, or pre-determining the interest-rate spreads between different collaterals. Fixed price differences take no account of the auctioneer's preferences about the relative amounts of each variety it would like to sell. Nor do fixed price differences make any use of the market's information about the "correct" relative prices for different collaterals—a central bank often has very imperfect information, and even when it has some information, it might not want to reveal its views to the market. Finally, the auction gives the central bank useful insight into the market's views about relative prices.

The Product-Mix Auction is also superior to, e.g., the U.S. Federal Reserve's 2008–9 Term Securities Lending Facility (TSLF) that held separate auctions for loans

---

[2] High-quality collateral would include UK sovereign debt; poorer collateral would include mortgage-backed securities.

[3] A Deputy Governor's remarks were published by a major British newspaper (the *Guardian*) here: www.guardian.co.uk/science/video/2013/jul/12/geometry-banking-crisis-video.

See Bank of England (2010, 2011), Fisher (2011), Milnes (2010), Fisher, Frost, and Weeken (2011), Frost, Govier, and Horn (2015), the *Economist* (2012), and the Bank of England's website.

against different types of collateral in different weeks. Running separate auctions means neither the quantity of funds allocated, nor the interest rate, on any collateral can depend on the bids on *other* collaterals. Moreover, by handling multiple goods simultaneously, the Product-Mix Auction gives bidders less market power and less ability to manipulate outcomes than if the goods were auctioned separately. So outcomes better approximate the perfectly competitive ideal.

The Product-Mix Auction approach also helps bidders. For example, a commercial bank might be interested in borrowing £500 million pounds, and be willing to pay up to 3% interest if it could use poor collateral, but prefer to use good collateral if that reduced its interest rate by 1% or more. Such a bidder can make a bid expressing these preferences—and the auction will then always do what is best for the bidder given the final auction prices. So, by contrast with the TSLF, such a bidder never has to guess which auction to bid in.

Crucially, the auction is simple to understand. The bidding rules, and the general principles of the auction, are easily comprehensible. Moreover, in the two-variety case the diagrammatic solution method is simple enough that the top officials of the Bank were quickly able to understand it, and so became comfortable that the approach was robust. The simplicity of the auction was critical both to getting the auction accepted for practical use, and to it working efficiently (i.e., bidders finding it easy to participate, and also choosing their bids correctly).

The fact that the diagrammatic solution method creates and intersects (relative) demand and supply curves to find the competitive equilibrium, and that this then specifies the auction's solution, also makes it appealing to practitioners who have basic economics training. Indeed, the visualisations that the Bank of England uses during each auction emphasises the familiar economic logic of efficient competitive equilibrium. In the Bank's initial two-variety implementation the (relative) demand curve created from the bids was shown superimposed on the Bank's supply curve (as illustrated in figure 2 below). That visualisation therefore looked exactly like the standard supply-and-demand diagram that demonstrates competitive equilibrium in an elementary textbook (although we will see that the Bank could immediately read off *both* final quantities, as well as the final price difference between the two goods, off this diagram).

More than two varieties requires more visualisations. (For the second-generation implementation, the Bank set up four simple charts to guide them during the auction.) However, experience of implementing and using the simple two-variety case was crucial in making the Bank's officials comfortable about updating the auction to the many-variety case.[4]

Another recent modification to the Bank's auction means that the *total* quantity allocated (summed over all goods) now responds in a more sophisticated way to the information in the bidding. And although the auction was designed in response to the crisis,

---

[4] With more than two varieties a purely graphical solution method would be both inefficient and hard to understand, so the Bank of England uses a simple linear programme to solve the many-variety case, plus a set of visualisations to describe the solution. (See Frost, Govier, and Horn (2015) and—especially–Baldwin and Klemperer (in prep.) for more details.)

the Bank wanted a solution that would be used in normal times too (in part, so that the use of a specific auction design would convey no information). So the auction is now used regularly; indeed the current Governer, Mark Carney, has announced plans for its greater use.[5]

The remainder of this paper is a tiny revision of my 2008 paper. After the introduction (section 1), section 2 describes a two-good implementation that corresponds closely to the Bank of England's initial implementation.[6] Section 3 describes easy extensions.[7] Section 4 describes further extensions and the relationship to the SMRA, and section 5 concludes.

1. Introduction

How should goods that both seller(s) and buyers view as imperfect substitutes be sold, especially when multi-round auctions are impractical? This was the Bank of England's problem in autumn 2007 as the credit crunch began.[8] The Bank urgently wanted to supply liquidity to banks, and was therefore willing to accept a wider-than-usual range of collateral, but it wanted a correspondingly higher interest rate against any weaker collateral it took. A similar problem was the U.S. Treasury's autumn 2008 Troubled Asset Recovery Program (TARP) plan to spend up to $700 billion buying "toxic assets" from among 25,000 closely-related but distinct sub-prime mortgage-backed securities.

Because financial markets move fast, in both cases it was highly desirable that any auction take place at a single instant. In a multi-stage auction, bidders who had entered the highest bids early on might change their minds about wanting to be winners before the auction closed,[9] and the financial markets might themselves be influenced by the evolution of the auction, which magnifies the difficulties of bidding and invites manipulation.[10]

---

[5] The auctions were initially monthly for up to £5 billion of 3 month repos (or £2.5 billion for 6 month repos). The auctions are now monthly for up to a potentially much larger total quantity which itself depends in a sophisticated way on the bidding, of 6 month repos. (More frequent auctions have been held at times of potential stress; they were held weekly after the 2016 "Brexit" vote.) Actual quantities have varied enormously depending on market conditions and other Central Bank policies. As an illustration, in 2010–11, the spreads between high and low quality collateral varied between 9 and 30 bps for 3 month repos; between 16 and 53 bps for 6 month repos.

[6] I do *not* give full details of the Bank's objectives and constraints here, and not all the issues I discuss are relevant to it. See Bank of England (2010, 2011), Fisher (2011), Fisher, Frost, and Weeken (2011), and the Bank of England's website.

[7] In 2014 we extended the Bank of England's auction to more goods (and used a different solution method that is simpler for the many-good case) and to variable total quantity (section 3.3); future auctions may use others of the enhancements described in section 3. See Frost, Govier, and Horn (2015), the Bank of England's website, and—especially–Baldwin and Klemperer (in prep.).

[8] The crisis began in early August 2007, and a bank run led to Northern Rock's collapse in mid-September. Immediately afterwards, the Bank of England first ran four very-unsuccessful auctions to supply additional liquidity to banks and then consulted me. I got valuable assistance from Jeremy Bulow and Daniel Marszalec.

[9] Some evidence is that most bids in standard Treasury auctions are made in the last few minutes, and a large fraction in the last few seconds. For a multi-round auction to have any merit, untopped bids cannot be withdrawn without incurring penalties.

[10] The Bank of England insisted on a single stage auction. Ausubel and Cramton (2008) argued a multi-stage auction was feasible for the U.S. Treasury.

An equivalent problem is that of a firm choosing its "product mix": it can supply multiple varieties of a product (at different costs), but with a total capacity constraint, to customers with different preferences between those product varieties, and where transaction costs or other time pressures make multiple-round auctions infeasible.[11] The different varieties of a product could include different points of delivery, different warranties, or different restrictive covenants on use.

This paper outlines a solution to all these problems – the Product-Mix Auction, which I invented for the Bank of England, and also subsequently proposed to the U.S. Treasury.[12]

My design is straightforward in concept – each bidder can make one or more bids, and *each* bid contains a *set* of mutually exclusive offers. Each offer specifies a price (or, in the Bank of England's auction, an interest-rate) for a quantity of a specific "variety". The auctioneer looks at all the bids and then selects a price for each "variety". From each bid offered by each bidder, the auctioneer accepts (only) the offer that gives the bidder the greatest surplus at the selected prices, or no offer if all the offers would give the bidder negative surplus. All accepted offers for a variety pay the same (uniform) price for that variety.

The idea is that the menu of mutually-exclusive sets of offers allows each bidder to approximate a demand function, so bidders can, in effect, decide how much of each variety to buy *after* seeing the prices chosen. Meanwhile the auctioneer can, if it wishes, look at demand *before* choosing the prices; allowing it to choose the prices ex-post creates no problem here, because it allocates each bidder precisely what that bidder would have chosen for itself given those prices.[13] (In practice, the Bank of England precommits to a supply function that will uniquely determine the prices as a function of the bids.[14])

Importantly, offers for each variety provide a competitive discipline on the offers for the other varieties, because they are all being auctioned simultaneously.

---

[11] That is, the Bank of England can be thought of as a "firm" whose "product" is loans; the different "varieties" of loans correspond to the different collaterals they are made against, and their total supply may be constrained. The Bank's "customers" are its counterparties, and the "prices" they bid are interest rates.

[12] After I had proposed my solution to the Bank of England, I learned that Paul Milgrom was independently pursuing related ideas. He and I therefore made a joint proposal to the U.S. Treasury, together with Jeremy Bulow and Jon Levin, in September-October 2008. Other consultants, too, proposed a static (sealed-bid) design, although of a simpler form, and the Treasury planned to run a first set of simple sealed-bid auctions, each for a related group of assets, and then enhance the design using some of the Bulow-Klemperer-Levin-Milgrom ideas in later auctions. However, it then suddenly abandoned its plans to buy subprime assets (in November 2008). Note also, however, that Larry Ausubel and Peter Cramton (who played an important role in demonstrating the value of using auctions for TARP, see e.g., Ausubel et al. (2008)) had proposed running dynamic auctions, and the possibility of doing this at a later stage was also still being explored.

Milgrom (2009) shows how to represent a wide range of bidders' preferences such that goods are substitutes, and shows a linear-programming approach yields integer allocations when demands and constraints are integer, but my proposal seems more straightforward and transparent in a context such as the Bank of England's.

[13] That is, it chooses prices like a Walrasian auctioneer who is equating bidders' demand with the bid-taker's supply in a decentralized process (in which the privately-held information needed to determine the allocation is directly revealed by the choices of those who hold it).

The result assumes the conditions for "truthful" bidding are satisfied – see below.

[14] In the many-variety version of the auction in use since 2014, the Bank of England precommits to a *set of* supply functions.

Compare this with the "standard" approach of running a separate auction for each different "variety". In this case, outcomes are erratic and inefficient, because the auctioneer has to choose how much of each variety to offer before learning bidders' preferences, and bidders have to guess how much to bid for in each auction without knowing what the price-differences between varieties will turn out to be; the wrong bidders may win, and those who do win may be inefficiently allocated across varieties. Furthermore, each individual auction is much more sensitive to market power, to manipulation, and to informational asymmetries, than if all offers compete directly with each other in a single auction. The auctioneer's revenues are correspondingly generally lower.[15] All these problems also reduce the auctions' value as a source of information. They may also reduce participation, which can create "second-round" feedback effects furthering magnifying the problems.[16]

Another common approach is to set fixed price supplements for "superior" varieties, and then auction all units as if they are otherwise homogenous. This can sometimes work well, but such an auction cannot take any account of the auctioneer's preferences about the proportions of different varieties transacted.[17] Furthermore, the auctioneer suffers from adverse selection.[18]

Section 2 shows how the Product-Mix Auction implements my alternative approach in a way that is simple, robust, and easy to understand, so that bidders are happy to participate. Section 3 discusses extensions. In particular, it is easy to include multiple buyers and multiple sellers, and "swappers" who may be on either, or both, sides of the market. Section 4 compares the Product-Mix Auction with the simultaneous multiple-round auction (SMRA). The Product-Mix Auction yields the same results as a "proxy" SMRA whose implementation is modified to permit the auctioneer to specify how the quantities traded will depend on the final prices. The Product-Mix Auction's static (i.e., single round of bids) design means it is also simpler, cheaper, and less susceptible to collusion and other abuses of market power, than is a (standard dynamic) SMRA. Section 5 concludes.

2. A Simple Two-Variety Example

The application this auction was originally designed for provides a simple illustration. A single seller, the Bank of England (henceforth "the Bank") auctioned just two

---

[15] Thus, for example, if the U.S. Treasury had simply predetermined the amount of each type of security to purchase, ignoring the information about demand for the large number of closely-related securities, competition would have been inadequate. There were perhaps 300 likely sellers, but the largest 10 held of the order of two-thirds of the total volume, and ownership of many individual securities was far more highly concentrated.

[16] The feedback effects by which low participation reduces liquidity, which further reduces participation and liquidity, etc., are much more important when there are multiple agents on both sides of the market – see Klemperer (2008).

[17] Moreover, as noted above, a Central Bank might not want to signal its view of appropriate price-differentials for different collaterals to the market in advance of the auction.

[18] If, for example, the U. S. Treasury had simply developed a "reference price" for each asset, the bidders would have sold it large quantities of the assets whose reference prices were set too high – and mistakes would have been inevitable, since the government had so much less information than the sellers.

"goods", namely a loan of funds secured against strong collateral, and a loan of funds secured against weak collateral. For simplicity I refer to the two goods as "strong" and "weak".[19] In this context, a per-unit price is an interest rate. The rules of the auction are as follows:

1. Each bidder can make any number of bids. *Each bid* specifies a *single* quantity and an offer of a per-unit price for *each* variety. The offers in each bid are mutually exclusive.
2. The auctioneer looks at all the bids and chooses a minimum "cut-off" price for each variety – I will describe later in this section how it uses the construction illustrated in Figures 1a, 1b, and 2 to determine these minimum prices uniquely, for any given set of bids, and given its own preferences.
3. The auctioneer accepts all offers that exceed the minimum price for the corresponding variety, *except* that it accepts at most one offer from each bid. If both price-offers in any bid exceed the minimum price for the corresponding variety, the auctioneer accepts the offer that maximizes the bidder's surplus, as measured by the offer's distance above the minimum price.[20]
4. All accepted offers pay the minimum price for the corresponding variety – that is, there is "uniform pricing" for each variety.[21]

Thus, for example, one bidder might make three separate bids: a bid for £375 million at {5.95% for (funds secured against) weak OR 5.7% for (funds secured against) strong}; a bid for an additional £500 million at {5.75% for weak OR 5.5% for strong}; and a bid for a further £300 million at {5.7% for weak OR 0% for strong}. Note that since offers at a price of zero are never selected, the last bid is equivalent to a traditional bid on only a single collateral.[22]

An example of the universe of all the bids submitted by all the bidders is illustrated in Figure 1a. The prices (i.e., interest rates) for weak and strong are plotted vertically and horizontally, respectively; each dot in the chart represents an "either/or" bid. The number by each dot is the quantity of the bid (in £millions). The three bids made by the bidder described above are the enlarged dots highlighted in bold.

---

[19] We assume (as did the Bank) that there is no adverse selection problem regarding collateral. For the case in which bidders have private information regarding the value of the collateral they offer, see Manelli and Vincent (1995).

[20] See notes 23 and 26 for how to break ties, and ration offers that equal the minimum price.

[21] Klemperer (2008) discusses alternative rules. In particular, the product-mix auction could be run with discriminatory ("pay your bid") pricing. However, discriminatory pricing leads to less efficient outcomes than uniform pricing, even if bidders bid optimally. Furthermore, because it is harder for new bidders to understand how to bid, discriminatory pricing may discourage entry of needy banks in times of crisis. Moreover, the information that the auction provides the central bank is better with uniform pricing. The reason is that with uniform pricing, if the number of bidders is not too small, a bidder that knows its own valuation should choose bids that (approximately) represent its true preferences (see the last paragraph of this section, and note 28). The collateral a bidder is allocated on any bid is then always the one that is best for it given the prices the auction sets. With discriminatory pricing, optimal bidding is not close to "truthful", and would generally not always result in the bidder being allocated the collateral that is best for it if both price-offers in a bid exceed the minimum price for the corresponding variety. So the product-mix auction's "either/or" bids exacerbate the difficulties that discriminatory pricing can cause, relative to uniform pricing.

[22] A bidder can, of course, restrict each of its bids to a single variety. Note also that a bidder who wants to guarantee winning a fixed total quantity can do so by making a bid at an arbitrarily large price for its preferred variety, and at an appropriate discount from this price for the other variety.

22 THE PRODUCT-MIX AUCTION 469

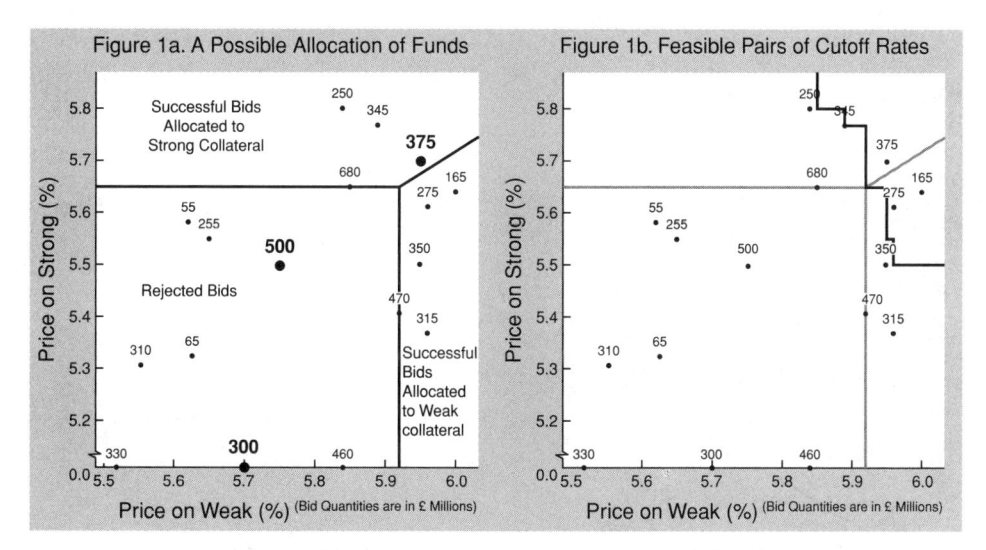

Figure 1. An example of bids in the Bank of England's auction.

The cut-off prices and the winning bids are determined by the Bank's objectives. If, for example, the Bank wants to lend £2.5 billion, and there are a total of £5.5 billion in bids, then it must choose £3 billion in bids to reject.

Any possible set of rejected bids must lie in a rectangle with a vertex at the origin. Figure 1a shows one possible rectangle of rejected bids, bounded by the vertical line at 5.92% and the horizontal line at 5.65%. If the Bank were to reject this rectangle of bids, then all the accepted bids – those outside the rectangle – would pay the cut-off prices given by the boundaries: 5.92% for weak, and 5.65% for strong.

Bids to the north-east of the rectangle (i.e, those which could be accepted for either variety) are allocated to the variety for which the price is further below the offer. So bids that are both north of the rectangle, and north-west of the diagonal 45° line drawn up from the upper-right corner of the rectangle, receive strong, and the other accepted bids receive weak.

Of course, there are many possible rectangles that contain the correct volume of bids to reject. On any 45° line on the plane, there is generally exactly one point that is the upper-right corner of such a rectangle.[23] It is easy to see that the set of all these points forms the stepped downward–sloping line shown in Figure 1b.[24] This stepped line is

---

[23] Moving north-east along any 45° line represents increasing all prices while maintaining a constant difference between them. Because the marginal bid(s) is usually rationed, there is usually a single critical point that rejects the correct volume of bids. But if exactly £3 billion of bids can be rejected by rejecting entire bids, there will be an interval of points between the last rejected and the first accepted bid. As a tie-breaking rule, I choose the most south-westerly of these points.

[24] The initial vertical segment starts at the highest price for weak such that enough can be accepted on weak when none is accepted on strong (this price is the weak price of the bid for 680), and continues down as far as the highest price bid for strong (the strong price of the bid for 250). At this point some strong replaces some weak in the accepted set, and there is then a horizontal segment until we reach the next price bid for weak (the weak price of the bid for 345) where more strong replaces weak in the accepted set and another vertical segment begins, etc.

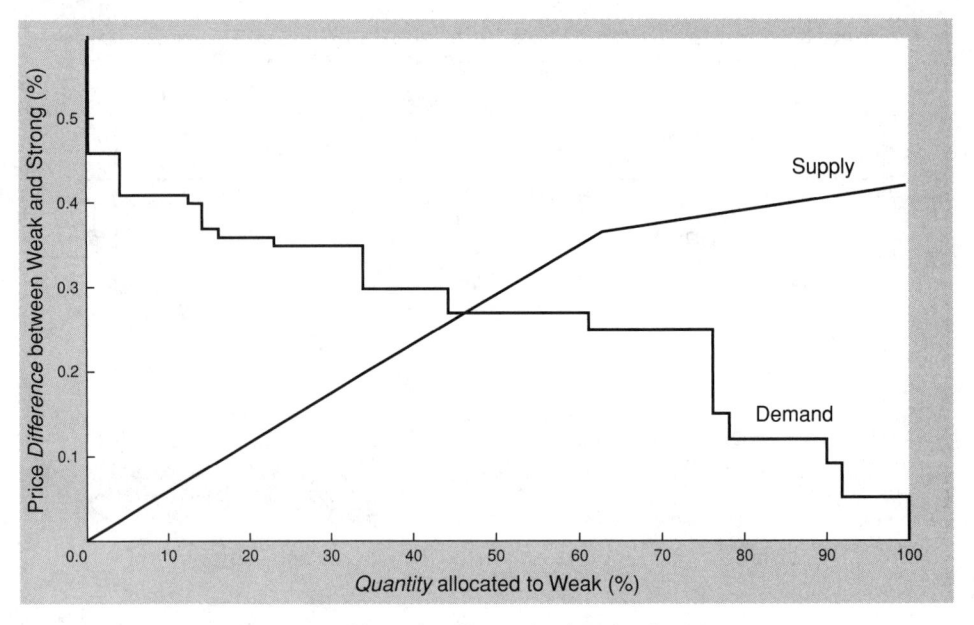

Figure 2. Equilibrium in the Bank of England's auction.

therefore the set of feasible pairs of cut-off prices that accept exactly the correct volume of bids.

Every point on Figure 1b's stepped line (i.e., every possible price pair) implies both a price-difference and (by summing the accepted bids below the corresponding 45° line) a proportion of sales that are weak. As the price-difference is increased, the proportion of weak sales decreases. Using this information we can construct the downward-sloping "demand curve" in Figure 2.

If it wished, the auctioneer (the Bank) could give itself discretion to choose any point on the "demand curve" (equivalently, any feasible rectangle in Figures 1a, 1b) after seeing the bids. In fact, the Bank prefers to precommit to a rule that will determine its choice. That is, the Bank chooses a "supply curve" or "supply schedule" such as the upward-sloping line in Figure 2 so the proportion allocated to weak increases with the price-difference.[25]

The point of intersection between the Bank's supply curve and the "demand curve" constructed from the bids determines the price differential and the percentage of weak sold in the auction. With the supply curve illustrated, the price difference is

---

[25] The proposal for the U.S. TARP to employ a "reference price" for each asset corresponds to choosing the multi-dimensional equivalent of a horizontal supply curve; buying a predetermined quantity of each asset corresponds to using a vertical supply curve. As I noted above, both these approaches are flawed. Choosing an upward-sloping supply curve maintains the advantage of the reference-price approach, while limiting the costs of mispricing. (The optimal choice of supply-curve slope involves issues akin to those discussed in Poole (1970), Weitzman (1974), Klemperer and Meyer (1986), etc.; maintaining the reserve power to alter the supply curve after seeing the bids protects against collusion, etc., see Klemperer and Meyer (1989), Kremer and Nyborg (2004), Back and Zender (2001), McAdams (2007), etc.)

0.27% and the proportion of weak is 45% – corresponding to the outcome shown in Figure 1a.[26]

This procedure ensures that bidders whose bids reflect their true preferences[27] receive precisely the quantities that they would have chosen for themselves if they had known the auction prices in advance. So unless a bidder thinks its own bids will affect the auction prices, its best strategy is to bid "truthfully"; if bidders all do this, and the Bank's supply curve also reflects its true preferences, the auction outcome is the competitive equilibrium.[28]

3. Easy Extensions

3.1. Multiple Buyers and Multiple Sellers

It is easy to include additional potential sellers (i.e., additional lenders of funds, in our example). Simply add their maximum supply to the total that the auctioneer sells, but allow them to participate in the auction as usual. If a potential seller wins nothing in the auction, the auctioneer has sold the seller's supply for it. If a potential seller wins its total supply back, there is no change in its position.

3.2. "Swappers" Who Might Want to be on Either Side of the Market

Exactly the same approach permits a trader to be on either, or both, sides of the market. If, for example, letting the auctioneer offer its current holdings of strong, a bidder in the auction wins the same amount of weak, it has simply swapped goods (paying the difference between the market-clearing prices).

3.3. Variable Total Quantity

Making the total quantity sold (as well as the proportions allocated to the different varieties) depend upon the prices is easy. (The Bank of England's 2014 implementation does this.)

The auctioneer might, for example, precommit to the total quantity being a particular increasing function of the price of strong. Using the procedure of Section 2 to solve

---

[26] By determining the proportion of weak, Figure 2 also determines what fractions of any bids on the rectangle's borders are filled, and the allocation between goods of any bids on the 45° line.

[27] This does not require pure "private value" preferences, but does not allow bidders to change their bids in response to observing others' bids.

   We can extend our mechanism to allow bidders with "common values" to update their bids: the auctioneer takes bids as above, and reports the "interim" auction prices that would result if its supply were scaled up by some pre-determined multiple (e.g., 1.25). It then allows bidders to revise the prices of any bid that would win at the interim prices, except that the price on the variety that the bid would win cannot be reduced below that variety's interim price. Multiple such stages can be used, and/or more information can be reported at each stage, before final prices and allocations are determined – we offered such an option to the U.S. Treasury, though it was not our main recommendation.

[28] Because on the order of 40 commercial banks, building societies, etc., bid in the Bank of England's auctions, it is unlikely that any one of them can much affect the prices. I assume the Bank's supply curve is upward sloping so, given our tie-breaking rule (see note 23), if there are multiple competitive equilibria the outcome is the unique one that is lowest in both prices.

for the strong price corresponding to every possible total quantity yields a weakly-decreasing function, and the unique intersection of the two functions then determines the equilibrium.

3.4. Other Easy Extensions

There can, of course, be more than two goods, with a cut-off price for each, and a bid rejected only if *all* its offers are below the corresponding cut-off prices.

Bidders can also be allowed to ask for different amounts of the different goods in a bid.

Relatedly, Iceland recently planned a Product-Mix Auction in which each of a bidder's bids could specify cut-off prices for three alternative goods, and the amount of money to be spent by the bid. So the amount of a good bought by a winning bid would be the amount to be spent divided by the chosen good's price; the type of good (if any) bought would depend on all goods' prices, as usual.[29]

Bidders can be permitted to specify that a total quantity constraint applies across a group of bids.

Several other extensions are also easy.

Importantly, bidders can express more complex preferences by using several bids in combination: for example, a bidder might be interested in £100 million weak at up to 7%, and £80 million strong at up to 5%. However, even if prices are high, the bidder wants an absolute minimum of £40 million. This can be implemented by making all of the following four bids, if negative bids are permitted:

1. £40 million of {weak at maximum permitted bid OR strong at maximum permitted bid *less* 2%}.
2. £100 million of weak at 7%.
3. £80 million of strong at 5%.
4. *minus* £40 million of {weak at 7% OR strong at 5%}.

The point is that the fourth (negative) bid kicks in exactly when one of the second and third bids is accepted, and then exactly cancels the first bid for £40 million "at any price" (since 2% = 7% − 5%).[30]

---

[29] The government announced in June 2015 that it would hold a Product-Mix Auction to permit blocked "offshore" Icelandic funds to be spent on one or more of three alternative financial instruments. A further announcement was planned for April 2016. But on April 5, 2016 the Prime Minister resigned after the "Panama papers" revealed his wife held over 500 million "offshore" Icelandic crowns (around £3 million). There is no reason to believe he or his wife would have benefited from the auction design (nor, to my knowledge, was this suggested), but the Product-Mix Auction was not held.

[30] A bidder can perfectly represent any preferences across all allocations by using an appropriate pattern of positive and negative bids if the goods are imperfect substitutes such that the bidder's marginal value of a good is reduced at least as much by getting an additional unit of that good as by getting an additional unit of the other good (i.e., if $V(w,s)$ is the bidder's total value of £w of weak plus £s of strong, then $\partial^2 V/\partial w^2 \leq \partial^2 V/\partial w \partial s \leq 0$ and $\partial^2 V/\partial s^2 \leq \partial^2 V/\partial w \partial s \leq 0$). More general preferences than this require more complex representations–but the important point, of course, is that preferences can typically be well-approximated by simple sets of bids.

The geometric techniques used in the analysis of the product-mix auction also yield new results in the multidimensional analysis of demand: see Baldwin and Klemperer (2012, 2015).

4. Further Extensions, and the Relationship to the Simultaneous Multiple Round Auction

My auction is equivalent to a static (sealed-bid) implementation of a simplified version of a "two-sided" simultaneous multiple round auction (SMRA). (By "two-sided" I mean that sellers as well as buyers can make offers – see below.)

Begin by considering the special case in which the auctioneer has predetermined the quantity of each variety it wishes to offer, and the bids in my auction represent bidders' true preferences. Then the outcome will be exactly the same as the limit as bid increments tend to zero of a standard SMRA if each bidder bids at every step to maximize its profits at the current prices given those preferences,[31] since both mechanisms simply select the competitive-equilibrium price vector.[32]

The general case in which the auctioneer offers a general supply curve relating the proportions of the different varieties sold to the price differences is not much harder. We now think of the auctioneer as acting *both* as the bid-taker selling the maximum possible quantity of both varieties, *and* as an additional buyer bidding to buy units back to achieve a point on its supply curve. That is, in our example in which the Bank auctions £2.5 billion, we consider an SMRA which supplies £2.5 billion weak *and* £2.5 billion strong, and we think of the Bank as an additional bidder who has an inelastic total demand for £2.5 billion and who bids in exactly the same way as any other bidder.[33,34]

So my procedure is equivalent to a "proxy SMRA", that is, a procedure in which bidders submit their preferences, and the auctioneer (and other potential sellers) submit their supply curves, and a computer then calculates the equilibrium that the (two-sided) SMRA would yield.[35] However, my procedure restricts the preferences that the auction participants can express. Although I can permit more general forms of bidding

---

[31] In a SMRA the bidders take turns to make bids in many ascending auctions that are run simultaneously (e.g., 55% of 2.5 billion = 1.375 billion auctions for a single £1 of strong, and 45% of 2.5 billion = 1.125 billion auctions for a single £1 of weak). When it is a bidder's turn, it can make any new bids it wishes that beat any existing winning bid by at least the bidding increment (it cannot top up or withdraw any of its own existing bids). This continues until no one wants to submit any new bids. (For more detail, including "activity rules" etc., see, e.g., Milgrom (2000), Binmore and Klemperer (2002), and Klemperer (2004).)

[32] An exception is that an SMRA may not do this when bidders' preferences are such that they would ask for different amounts of the different goods in a single bid in my procedure. All the other types of bids discussed above reflect preferences such that all individual units of all goods are substitutes for all bidders (so bidding as described above in an SMRA is rational behaviour if the number of bidders is large). I assume the auctioneer also has such preferences (i.e., the Bank's supply curve is upward sloping), so if there are multiple competitive equilibria, there is a unique one in which all prices are lowest and both mechanisms select it (see note 28 and Crawford and Knoer (1981), Kelso and Crawford (1982), Gul and Stacchetti (1999), and Milgrom (2000)).

[33] That is, whenever it is the Bank's turn to bid, it makes the minimum bids to both restore its quantity of winning bids to £2.5 billion and win the quantity of each variety that puts it back on its supply curve, given the current price-difference. (It can always do this to within one bid increment, since the weak-minus-strong price difference can only be more (less) than when it last bid if its weak (strong) bids have all been topped, so it can increase the quantity of strong (weak) it repurchases relative to its previous bids, as it will wish to do in this case.)

[34] If there are other sellers (or "swappers") add their potential sales (or "swaps") to those offered in the SMRA, and think of these participants as bidding for positive amounts like any other bidders.

[35] Although Section 2's description may have obscured this, our procedure is symmetric between buyers and sellers. (It is not quite symmetric if the auctioneer doesn't precommit to its supply curve, but if bidders behave competitively their bids are unaffected by this.)

than those discussed above (see Klemperer (2008)),[36] some constraints are desirable. For example, I am cautious about allowing bids that express preferences under which varieties are complements. (The Bank of England's 2014 implementation permits the auctioneer to express some complementary preferences.)[37]

Importantly, exercising market power is much harder in my procedure than in a standard SMRA, precisely because my procedure does not allow bidders to express preferences that depend on others' bids. In particular, coordinated demand reduction (whether or not supported by explicit collusion) and predatory behaviour may be almost impossible. In a standard dynamic SMRA, by contrast, bidders can learn from the bidding when such strategies are likely to be profitable, and how they can be implemented – in an SMRA, bidders can make bids that signal threats and offers to other bidders, and can easily punish those who fail to cooperate with them.[38,39]

Finally, the parallel with standard sealed-bid auctions makes my mechanism more familiar and natural than the SMRA to counterparties. In contexts like the Bank of England's, my procedure is much simpler to understand.

5. Conclusion

The Product-Mix Auction is a simple-to-use, sealed-bid, auction that allows bidders to bid on multiple differentiated assets simultaneously, and bid-takers to choose supply functions across assets. It can be used in environments in which a simultaneous multiple-round auction (SMRA) is infeasible because of transaction costs, or the time required to run it. The design also seems more familiar and natural than the SMRA to bidders in many applications, and makes it harder for bidders to collude or exercise market power in other ways.

Relative to running separate auctions for separate goods, the Product-Mix Auction yields better "matching" between suppliers and demanders, reduced market power, greater volume and liquidity, and therefore also improved efficiency, revenue, and quality of information. Its applications therefore extend well beyond the financial contexts for which I developed it.

---

[36] I could in principle allow any preferences subject to computational issues; these issues are not very challenging in the Bank of England's problem.

[37] The difficulty with complements is the standard one that there might be multiple unrankable competitive equilibria, or competitive equilibrium might not exist (see note 32), and an SMRA can yield different outcomes depending upon the order in which bidders take turns to bid. In independent work, Milgrom (2009) explores how to restrict bidders to expressing "substitutes preferences". Crawford (2008)'s static mechanism for entry-level labor markets (e.g., the matching of new doctors to residency positions at hospitals) adddresses related issues in a more restrictive environment. See also Budish (2004).

[38] In a standard SMRA, a bidder can follow "collusive" strategies such as "I will bid for (only) half the lots if my competitor does also, but I will bid for more lots if my competitor does", see, e.g., Klemperer (2002, 2004), but in our procedure the bidder has no way to respond to others' bids. Of course, a bidder who represents a significant fraction of total demand will bid less than its true demand in *any* procedure, including mine, which charges it constant per-unit prices. But it is much easier for a bidder to (ab)use its market power in this way in an SRMA.

[39] A multi-round procedure (either an SMRA, or an extension of our procedure – see note 27) may be desirable if bidders' valuations have important "common-value" components, but may discourage entry of bidders who feel less able than their rivals to use the information learned between rounds.

References

Ausubel, L., and Cramton, P. (2008). 'A Troubled Asset Reverse Auction'. Mimeo: University of Maryland.

Ausubel, L., Cramton, P., Filiz-Ozbay, E., Higgins N., Ozbay, E. and Stocking, A. (2008). 'Common-Value Auctions with Liquidity Needs: An Experimental Test of a Troubled Assets Reverse Auction'. Working Paper: University of Maryland.

Back, K., and Zender, J. (2001). 'Auctions of Divisible Goods With Endogenous Supply'. *Economics Letters*, 73: 29–34.

Baldwin, E., and Klemperer, P. (2012). 'New Geometric Techniques to Analyse Demand'. Mimeo: Oxford University.

Baldwin, E., and Klemperer, P. (2015). 'Understanding Preferences: "Demand Types", and the Existence of Equilibrium with Indivisibilities'. Working Paper: Oxford University.

Baldwin, E., and Klemperer, P. (in prep.). 'The multi-dimensional product-mix auction'.

Bank of England (2010). 'The Bank's new indexed long-term repo operations'. *Bank of England Quarterly Bulletin*, 50/2: 90–91.

Bank of England (2011). 'The Bank's indexed long-term repo operations'. *Bank of England Quarterly Bulletin*, 51/2: 93.

Binmore, K. and Klemperer, P. (2002). 'The Biggest Auction Ever: the Sale of the British 3G Telecom Licenses'. *Economic Journal*, 112: C74–C96.

Budish, E. (2004). 'Internet Auctions for Close Substitutes'. M.Phil Thesis: University of Oxford.

Crawford, V. P. (2008). 'The Flexible-Salary Match: A Proposal to Increase the Salary Flexibility of the National Resident Matching Program'. *Journal of Economic Behavior & Organization*, 66/2: 149–60.

Crawford, V. P., and Knoer, E. M. (1981). 'Job Matching with Heterogeneous Firms and Workers'. *Econometrica*, 49: 437–50.

The *Economist*. "A golden age of micro". The *Economist*, Free Exchange, www.economist.com/blogs/freeexchange/2012/10/microeconomics. October 19, 2012.

Fisher, P. (2011). 'Recent developments in the sterling monetary framework'. www.bankofengland.co.uk/publications/speeches/2011/speech487.pdf

Fisher, P., Frost, T., and Weeken, O. (2011). 'Pricing central bank liquidity through product-mix auctions–the Bank of England's indexed long-term repo operations'. Working Paper: Bank of England.

Frost, T., Govier N., and Horn T. (2015). 'Innovations in the Bank's provision of liquidity insurance via indexed long-term repo (ILTR) operations'. *Bank of England Quarterly Bulletin*, 55/2: 181–188.

Gul, F. and Stacchetti, E. (1999). 'Walrasian Equilibrium with Gross Substitutes'. *Journal of Economic Theory*, 87: 95–124.

Kelso, A. S. Jr., and Crawford, V. P. (1982). 'Job Matching, Coalition Formation, and Gross Substitutes'. *Econometrica*, 50: 1483–1504.

Klemperer, P. (1999). 'Auction Theory'. *Journal of Economic Surveys*, 13/2: 227–86. [Also reprinted in *The Current State of Economic Science*, (1999). S. Dahiya, ed. 711–66.]

Klemperer, P. (2002). 'What Really Matters in Auction Design'. *Journal of Economic Perspectives*, 16: 169–89.

Klemperer, P. (2004). *Auctions: Theory and Practice*, Princeton: Princeton University Press.

Klemperer, P. (2008). 'A New Auction for Substitutes: Central Bank Liquidity Auctions, the U.S. TARP, and Variable Product-Mix Auctions'. Mimeo: Oxford University.

Klemperer, P. and Meyer, M, (1986). 'Price Competition vs. Quantity Competition: The Role of Uncertainty'. *Rand Journal of Economics*, 17: 618–38.

Klemperer, P. and Meyer, M. (1989). 'Supply Function Equilibria in Oligopoly under Uncertainty'. *Econometrica*, 57: 1243–77.

Kremer, I. and Nyborg, K. (2004). 'Underpricing and Market Power in Uniform Price Auctions'. *Review of Financial Studies*, 17: 849–77.

Krishna, V. (2002). *Auction Theory*. Academic Press: New York.

McAdams, D. (2007). 'Uniform-Price Auctions with Adjustable Supply'. *Economics Letters*, 95: 48–53.

Manelli, A. M. and Vincent, D. (1995). 'Optimal Procurement Mechanisms'. *Econometrica*, 63: 591–620.

Menezes, F. M. and Monteiro, P. K. (2005). *An Introduction to Auction Theory*. Oxford University Press: Oxford.

Milgrom, P. R. (2000). 'Putting Auction Theory to Work: The Simultaneous Ascending Auction'. *Journal of Political Economy*, 108: 245–72.

Milgrom, P. R. (2004). *Putting Auction Theory to Work*. Cambridge University Press: Cambridge.

Milgrom, P. R. (2009). 'Assignment Messages and Exchanges.' *American Economic Journal: Microeconomics*, 1: 95–113.

Milnes, A. (2010). "Creating Confidence in Cash". *Blueprint*, October.

Poole, W. (1970). 'Optimal Choice of Monetary Policy Instruments in a Simple Stochastic Macro Model'. *Quarterly Journal of Economics*, 84: 197–216.

Weitzman, M. (1974). 'Prices vs. Quantities'. *Review of Economic Studies*, 41: 477–91.

CHAPTER 23

The Continuous Combinatorial Auction Architecture

Charles R. Plott, Hsing Yang Lee, and Travis Maron

1. Introduction and Background

The history of the continuous combinatorial auction marks the evolution of the mechanism. The concept of a combinatorial auction is due to Rassenti, Smith, and Bulfin (1982) who were motivated by the use of simultaneous ascending price auctions to allocate landing rights (Grether, Isaac, and Plott (1979 and subsequently published in 1989)). The ideas were generalized by Banks, Ledyard, and Porter (1989) to include the concept of a "standby queue" which serves a function similar to non-winning bids in the current system.

The first example of a continuous combinatorial auction is found at Brewer and Plott (1996). They demonstrate that representations in terms of binary confects of packages afford both the flexibility for widespread application and the computational speed required to support the auction. This early mechanism depended heavily on the existence of a fixed set of packages on which bids could be placed. The computer could quickly compute non-intersecting packages that maximized the value of the sale and permitted the auction to proceed as a type of continuous, simultaneous, ascending price auction. The packages played the role of items on which bids were placed. The non-intersecting packages that produced the most revenue from the auction were declared the leading bids at each instant of time. That first mechanism was followed by slight generalization to a procurement problem in which the buyer organized sellers to minimize procurement cost and sellers could offer endogenous packages. The organization was a simultaneous, decreasing price auction.

In the 1990s, the FCC was considering the adoption of a combinatorial auction as a replacement for the simultaneous, rounds-based, ascending price auction that the FCC had used to auction parts of the electromagnetic spectrum. The initial research focused on a hybrid process that consisted of rounds followed by a continuous phase. (See the report of Charles R. Plott, FCC Why River Conference, May 5–7, 2000). Experiments with the hybrid revealed that, that most of the adjustment and efficiently came from the continuous phase. That discovery led to the study of combinatorial auctions that operated only in continuous time. Of course, the architectures of the continuous combinatorial auctions and the traditional auctions based on rounds have some similarities but they have many differences that required exploration.

A difference between the continuous auction and rounds auctions is the role and form of activity and eligibility requirements. The continuous auction uses neither, and the rounds based auctions use both. In the rounds auction, "eligibility" is a limitation on the packages on which bids can be placed and activity requirements dictate a reduction in eligibility if bidding activity is not adequate. It is a type of "use it or lose it" condition. By contrast, the continuous auction is based on special ending clocks that play an incentive role similar to activity and eligibility requirements but are much different in substance and performance.

2. Items and Bids

<u>Items.</u> n items are for sale at the auction indexed $Y = \{1, 2, \ldots, n\}$; Let $S \in \{0, 1\}^n$ be a combination, set, or package of items.

<u>Individuals.</u> m individuals participate in bids $M = \{1, 2, \ldots, m\}$. Each bidder has an Identification Number known only to the bidder. It is possible to give a bidder several ID numbers if the bidder wants them.

<u>Bids.</u> A bid is a price and a package of items of the form $b_j^i(S)$ where i is the index of the individual submitting the bid, j is the bid number as recorded in the system and $S \subseteq Y$ is a package of items.

Let $b_q^i(S_q)$ be bid number q, where i is the bidder and $S_q \in \{0, 1\}^n$. That is, the qth bid was placed by i, for a dollar amount b_q^i for a package of items S_q.

<u>Bid Properties.</u> (i) Bids are submitted under "all or none" conditions. Either the entire package is accepted as a provisional winner or none of it is accepted. (ii) Multiple bids can be tendered. (iii) All bids remain in the system and can be selected as provisional winners unless cancelled. (iv) Provisional winning bids cannot be cancelled.

3. Rules and Procedures

<u>Provisional Winners.</u> After each bid is submitted, the system publishes the set of provisionally winning bids. The provisional winners are bids in the set of bids that would maximize the value of the sale if the auction concluded at that moment subject to the condition that no item is contained in more than one provisionally winning bid.

B = all bids submitted and not cancelled.

$x_q \in \{0, 1\}$ indicates whether or not the qth bid was accepted as a provisional winner of the auction.

W = Provisionally winning bids. Provisional winners are the subset of all bids, B, that maximize the value of the sale subject to the fact that no item is contained in more than one provisional winning bid.

That is, provisional winning bids are q: x_q = argmax R.
$R = \text{Max} \sum x_q b_q^i$ Subject to $\sum x_q S_q \leq (1, 1, 1, \ldots, 1)$
$x_q : q \in B$

Winning Bids. The bids that are not provisional winners remain in the system and play an important role. Notice that the computation of the provisional winners includes an examination of all bids in the system.

Thus, a new bid can be partnered with a non-provisionally winning bid such that an existing package is broken and the new bid and the partnering bid become provisional winners. The implication is that non-winning bids exist in the system as potential partners or as the pieces of a complex coalition that can be assembled to replace large package bids as provisional winners. By placing a non-winning bid, the bidder is revealing a willingness to pay for a package, which, theoretically, could be a maximum willingness to pay in the absence of mistakes or conspiracies.

Notice that this is a complex calculation that could require an examination of all families of subsets of bids as candidates for provisionally winning. Obviously, this can be computationally challenging. Given the sizes of existing auctions and tests, it has not presented a problem.

Increment Requirements. The increment requirement represents a major departure for standard auctions. When bids are on a single unit, increment requirements state that new bids must be some fixed increment above the price of the currently winning bids on the item alone. Thus, the bidding is progressively upward. The increment requirement of the continuous combinatorial auction when the bid is a package of items is much different. In order to be submitted, a bid need not be high enough to become a provisional winner. Furthermore, the increment requirement is not based on the sum of implicit prices computed for individual items.

The function of the increment rule as is the case with all increment rules, is to encourage bids to move the system to an equilibrium and to do so at a fast pace. Without special rules regarding increments, the system could be filled with bids that are dominated by existing bids. Let $v(S)$ be the maximum value for which the set S could be sold given the bids in the system. This value is determined by computing the winning bids from all bids submitted given that the sale of items was restricted to the set S. Let k be the (constant) increment required for bidding on a single item. For a bid $b(S)$ to meet the increment, it must meet the condition $b(S) \geq v(S) + k|S|$.

Stopping Rules and Warning Lights. Two clocks are used: a new bid clock and a new provisional winner clock. The new bid clock resets with each new bid and starts a countdown. Typically, the reset time is from three to five minutes. The time can be shortened as the auction progresses. The new winner clock resets each time a bid is placed that determines a new pattern of provisional winners. Typically, the reset time is between ten and fifteen minutes. Like the new bid clock, the time can be shortened as the auction proceeds.

The auction ends if either clock reaches zero. Basically, the new bid clock forces a flow of bids, similar to offers in a negotiation and does so under the threat of the auction ending. The new winner clock forces concessions of sufficient magnitude to advance the value of the sale. In essence, the new bid clock says "You must make an offer. You must make an offer." The new winner clock says "You must make an offer sufficiently to get a deal done or the auction ends anyway." Thus, the ultimatum feature of game theory is operational in both clocks.

If either clock gets within one minute of zero, red "railroad lights" begin to blink on the screen. In essence, the system is constantly pressing for revenue gains by using the threat of ending the auction. While bidders who are not provisional winners do face a dominant strategy of bidding as the clock counts down, bid need not be large and there is no advantage to waiting until the last moment to bid. Last moment bids just give competitors more time.

Special Bids, Robots and Either/Or. Bidders are able to place an either/or bid. If one bid becomes a provisional winner, the other cannot. This feature allows expression of indifference across sets.

Robots are available for bidding on single items. The bidder can instruct the robot to bid no more than a stated amount that the bidder can change at any time. The robot will place a bid at the minimum increment any time the bid in question is no longer a provisional winner. Because bidding on sets requires fashioning bids, the robots are not available for bidding on sets.

4. Information and Query

Bidders use the information and query functions to fashion bids. In particular, the bidders are given information needed about an entire package as opposed to some measure of implicit prices of the items in the package. In contrast to standard rounds based auctions, the system does not compute a measure of individual prices, the sum of which will indicate whether or not the package will be a provisional winner. Of course, since the system does not compute item prices, substantial information must be made available to bidders in some other form. Some of this can best be understood by a study of the interfaces presented in Section 5 but a key list is included here.

As will become clear, the computations are complex, basically NP complete. However, unlike many auction architectures, the system need not compute prices or temporary equilibria based on preferences over all items, submitted by all bidders at the same time. Instead, the system responds to a single addition to the bids that exist in the system. Conventions exist to respond to computations that are taking too long for fast progression of the auction.

Provisional Winners. A table is published that contains the provisional winner of each item, whether the provisionally winning bid is a package bid or a single, the highest bid placed on the item as a single, and the highlights of the bids of the bidder who is doing the bidding.

The provisionally winning table is updated with each new bid. New provisionally winning bids are accompanied by a small red dot that disappears in a few seconds. All items in a new winning package appear with the red dot.

New non-winning bids are shown as a small black dot on the provisionally winning table. The dots also disappear after a few seconds. These black dots signal the possibility that a bidder wants part of a package but cannot bid enough to become a provisional winner and seeks partners to bid on the rest of the package from which the bidder wants a portion. The black dots carry information that serves to coordinate bidding by coalitions of small bidders who want to break up a large package bid.

All Bids. A page of all bids in the system is published. It includes the bid number, the bidder ID, the items in the package, the time of submission, and the amount of the bid.

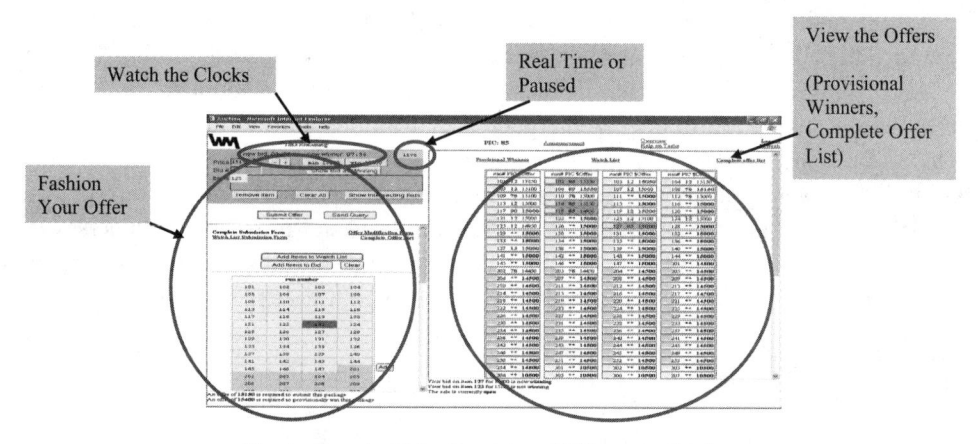

23 THE CONTINUOUS COMBINATORIAL AUCTION ARCHITECTURE

Illustration 1. Bids, Provisional Winners, Clocks.

Query. The query system and related functionality serve to replace the role of prices in the fashioning of bids. Important queries can be exercised at the time the bidder formulates a bid but before submission. When potential package is selected for a potential bid, the bidder is immediately shown both the minimum amount that can be bid as dictated by the increment requirements and the minimum amount it would take for the item to become a provisional winner. These two operations serve as tools to help bidders explore how to fashion bids in relation to the bids in the system. By adding or removing items from a package, the query can be used to determine the marginal cost

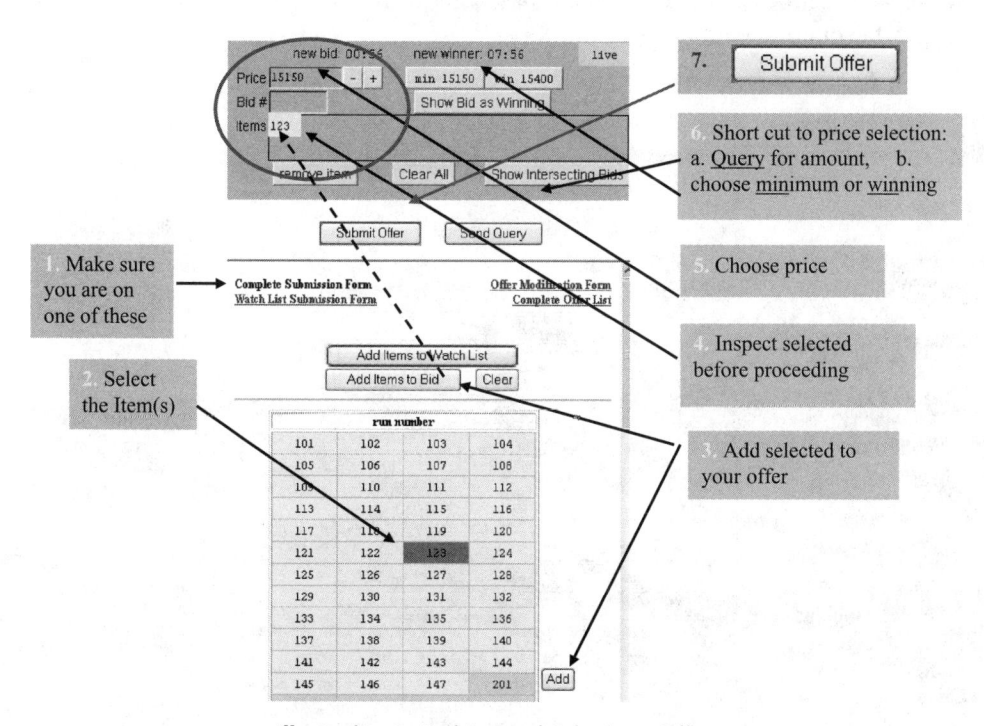

Illustration 2. Fashion and Submit an Offer.

of adding items to a package. By removing a single item from a package, the bidding required to become a provisional winner can be significantly reduced.

Show as Winning. After a bid is fashioned but before it is submitted, the bidder can choose this option to display the pattern of winning and non-winning bids that will be the consequence of the submission. It will show all new provisional winners, all bids that were provisional winners and remain as such, all bids that were provisional winners and now are not, and all bids that were not provisional winners and would be provisional winners if the bid was submitted. This allows bidders to search more efficiently for partners and avoid adding items to a package that would be too costly.

5. Interfaces

The interfaces presented here reflect what we have learned about what bidders want to know, aided by strategic considerations from game theory. When observing individual behavior in experiments, we follow the principle that the individual is an optimizer subject to perception of conditions and options available. Behavior that is not consistent with the incentives that we know exist are viewed as mistakes or misperceptions that the properly designed interfaces should prevent.

The interfaces produced in the illustrations below reflect the experience gained from experiments. The best way to explain interfaces seems to be to simply show them. The next seven pages are screen shots and explanations of the major functions and how they relate to bidder decisions. The illustrations begin with the home screen and provide a map to the other screens.

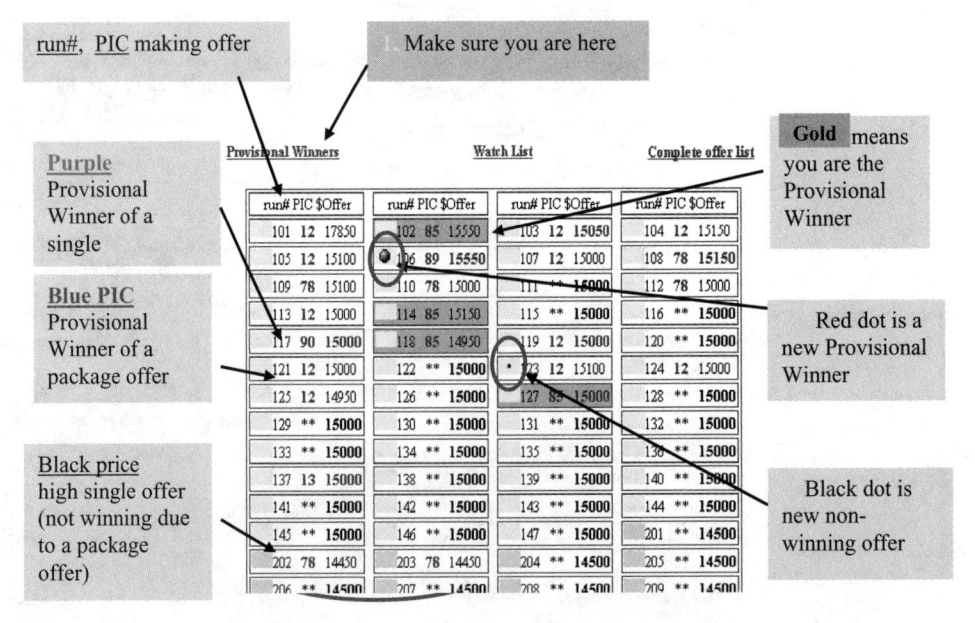

Illustration 3. View the Offers (Provisional Winners).

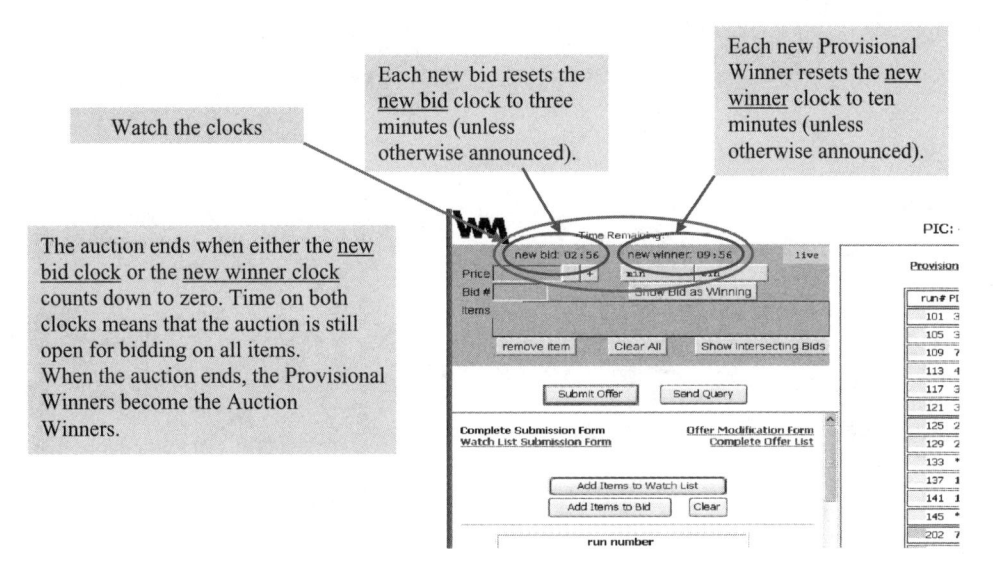

Illustration 4. Ending the Auction.

6. Performance

Performance of the continuous combinatorial auction is addressed in three sections. The first section describes the experimental parameters. The section assumes an understanding of preference inducement, the nature of subject training and instruction. Much of what we know and can measure is derived from experiments and experimental testbeds. The second section is a sketch of the parameters in experiments and the third is a highlight of some experimental results. The third section reports the major properties of two field applications.

Experimental testbed methodology reflects an attempt to learn about mechanisms and environments that have never existed before in naturally occurring environments. Data from the field does not exist and appropriate data might even be impossible to get.

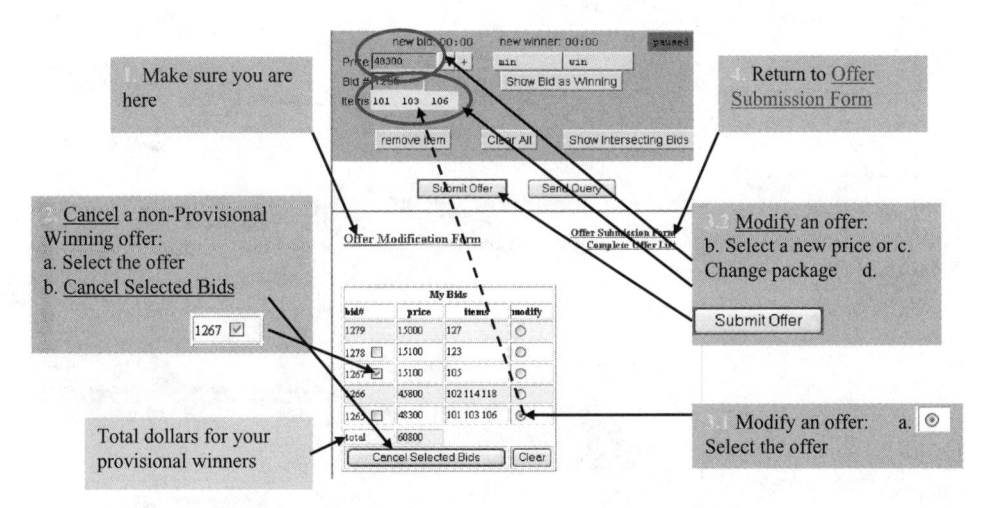

Illustration 5. Offer Management (Offer Modification).

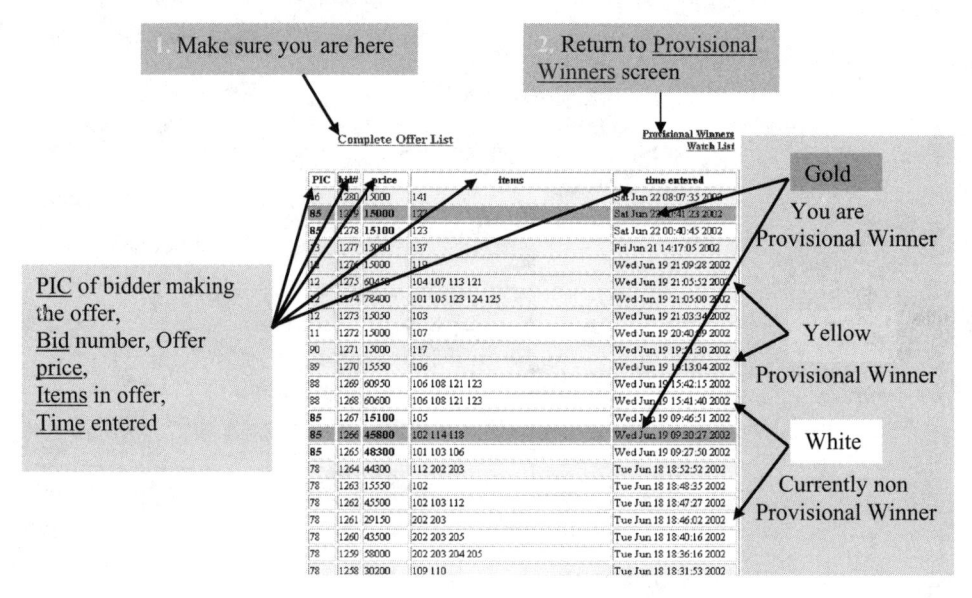

Illustration 6. View the Offers (Complete Offer List).

Furthermore, the method operates in a world in which theory is suggestive but limited. The fact that the theory is incomplete suggests that "theory testing" is not necessarily a testbed objective because the answer to the question of whether the theory is true or not is already known. Certainly, it is not true because it is incomplete and therefore vulnerable to a variety of sources of rejection.

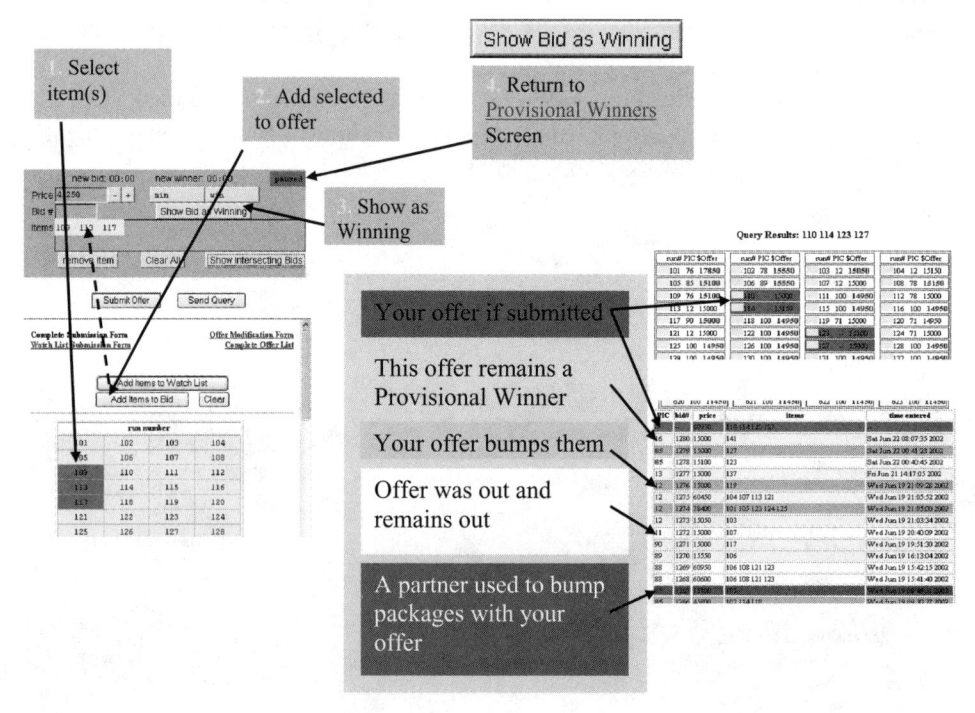

Illustration 7. Strategy Tools.

Three questions are posed in a testbed. (1) Does the mechanism do what it is supposed to do? The question asks for a demonstration of proof of principle. (2) Does the mechanism do it does for understandable reasons? The question asks about a test of design consistency. The question asks if the result reflects the theory that was used in the design or are they simply random? Clearly, this is a basic question because it asks about the possibility that the design will scale. (3) Will the mechanism work in the proposed field environment? Of course, this third question is a key. It asks about the robustness of the theory when applied to possibly unknown conditions. It calls for tests under a variety of environments that could challenge the performance. The test environments might look nothing at all like "the real world" because the real world in which one might imagine the mechanism being deployed might not have conditions that theory suggests are stressful. On the other hand, testing in environments that might closely resemble the application environment might prove valuable in uncovering interactions with institutions and aspects of the environment that might not be anticipated by theory. Institutional facts and environmental features can interact in surprising ways and have negative effects on performance. Examples of both types of environments are reported in the second section.

Experimental Parameters

Standard experimental economics techniques are used to induce incentives. Of course, explaining preferences with synergies is a bit of a trick. Special techniques were used for that task.[1]

Three classes of parameters existed for stress tests of the mechanism. The optimal allocations for the first sets are shown in Figures 1 and 3. Twenty items are to be allocated to five bidders. A representative indifference curve is drawn in the figure for each of the five bidders. Each of the bidders has an incentive to buy all twenty items should the prices be sufficiently low. The arrows are rough indicators of the gradient directions. The indifference curves cannot illustrate the synergies among the items, but the complementarities exist except when the items are "far" away from the maximum. For items near the maximum (about nine items), the purchase of any pair of items produces a value more than the sum of the values of the two items when the items are evaluated independently.

Efficiency is the measure used to assess performance. Assume that each individual bids the actual value of all packages. Under such assumptions, the revenue that would be produced by the auction is a measure of potential "social benefits" and since the revenue is maximized it serves as a measure of the maximum possible benefits. In an experimental auction, the values are induced and are thus known to the experimenter. Thus, at the conclusion of the auction the allocations are known and the value of items

---

[1] We will call our method of inducing preferences over sets "The basis method for synergies inducement". A subject is given an array of subsets with a value attached to each that operates as a basis from which the value of any subset can be computed. The array is determined by the specific preference over sets, the pattern of preference synergies that the experimenter wants to induce. The array is a type of basis for computing the value of any subset. To find the value of an arbitrary set S start with the first element of the array and continue along the array to the very first subset of S, call it S_1. Record the value of S_1. Continue along the array to the first subset of $S \setminus S_1$. Call it S_2. Record the value of S_2. Continue along the array to the first proper subset of $S \setminus S_1 \setminus S_2$ and call it S_3. Record the value of S_3. Continue the process until all units of S have been included in a subset. Add the recorded values to get the value of S.

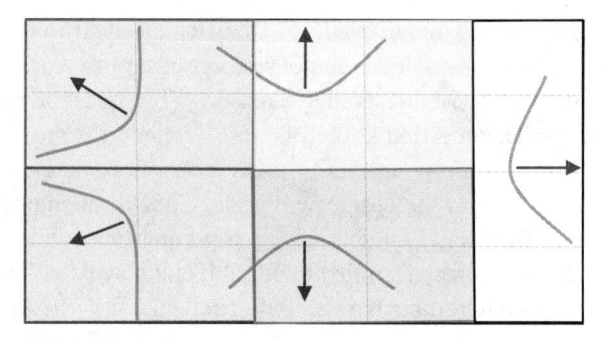

Figure 1. Relatively easy parameters.

to the bidder to whom they are allocated can be computed and summed. Call it the "total value received" by the bidders, independent of the prices paid.

auction efficiency = total value received/maximum possible surplus.

The efficiency is a type of cost/benefit measure only in the case of the auction in which the cost to the seller plays no role, there is no social cost, only benefits of the demand side. Typically, prices are an issue of income distribution as opposed to efficiency in allocation so the net benefits, value minus cost to bidders, are not part of the measurement.

The patterns of efficient allocations for the relatively easy parameters are Figure 1. As can be seen, four participants should acquire four adjacent items, resulting in four square patterns of allocations. The fifth bidder should acquire all items in the column to the right.

In the relatively hard parameters of Figure 2, four bidders have exactly the same preferences as existed in Figure 1. Three bidders are added and the resulting optimal pattern is illustrated in Figure 2. One of the new bidders should win the four units in the center. The two other new bidders should each acquire a unit at the extreme of the fifth column.

Experimental Tests

While many experiments were conducted, most lead to changes in the instructions and the interfaces and functionality of the mechanism. Nine small scale experiments and reported in Table 1. Five experiment in the table are based on the easy parameters

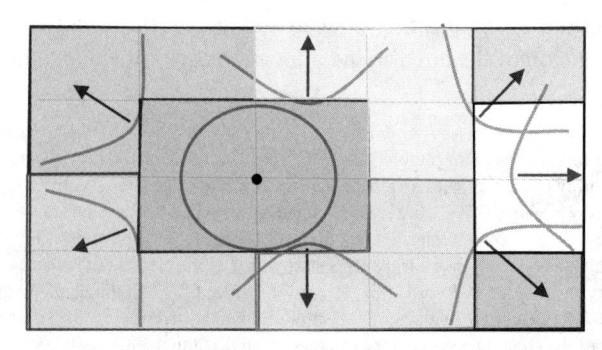

Figure 2. Relatively hard parameters.

Table 1. *Data*

DATA	Number of bidders	Efficient allocation value	Actual allocation value	Efficiency	note: experimental parameters
060411A	5	14296	14296	100%	easy
06041B	5	14296	14296	100%	easy
060517A	5	14926	14926	100%	easy
060519A	5	14296	14296	100%	easy
060519B	5	14296	14296	100%	easy
060511	8	10000	10000	100%	hard
060524	8	10000	10000	100%	hard
060525A	8	10000	8550	85.5%	hard–A person bought almost all – lost money
060525B	8	10000	6900	69%	hard – A key person bought only 1 and lost money on it. The two poor formers are included as examples that demonstrate the nature of the insights produced by testbed experiments.

and four use the hard parameters. In addition to the experiments in the table, the results of one reasonably large scale experiment are reported.

The five easy case parameter experiments were conducted near the end of the testing phase, when both software and instruction procedures had become stabilized. As can be seen, the experiments were regularly producing efficiencies near 100 %. Four hard parameter experiments are reported. The experiments on 060511 and the two on 060524 were with experienced subjects and resulted in an efficiency of 100%. The two experiments on 060525 were not conducted with appropriately trained subjects and

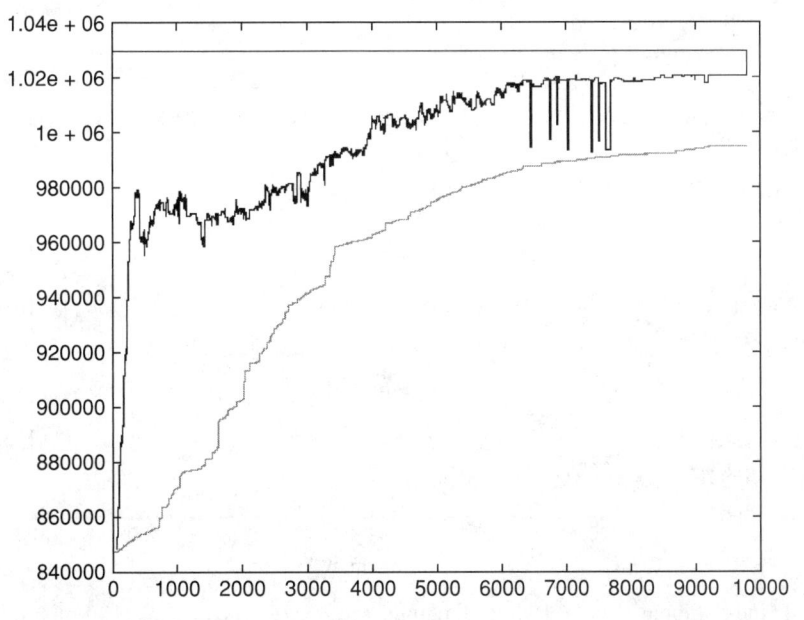

Figure 3. Experiment 060526 50 items 30 bidders: Revenue and Efficiency.

are included as examples of what can be revealed by the testbed. The sources of the obvious subject misunderstandings were addressed following these experiments and were incorporated in modified training procedures as the project moved toward the applications.

As can be viewed in the figure, the revenue starts at a low level and rapidly increases, following an almost concave path and finally asymptotes at the level where the auction ends naturally as dictated by the clocks. The efficiency level converges to a level of approximately 90%. Revenues clearly approach an asymptote but as is the case with combinatorial auctions, there is some ambiguity about the appropriate equilibrium concept so the predicted revenue is not known.

Field Applications

The results of two field applications are reported in this section. Of course, the details of parameters are unknown so efficiencies and maximum possible revenues are unknown. However, the time series are instructive and hold the impression of similarity to experiments conducted under laboratory conditions.

The results of an auction for 100 metric ton pallets of natural rubber are contained in Figure 4. Four internet bidders located around the world competed for 22 pallets of natural rubber located in a warehouse in Vietnam. The auction was conducted by the United Nations International Natural Rubber Organization, which had accumulated the pallets as part of a price stabilization program and was prepared to release the natural rubber back to private companies. Buyer identities were not public information.

The pallets were from different plantations. Natural rubber from a given plantation is a homogeneous product but rubber from different plantations has different and well known qualities. Starting bids were tendered by bidders as sealed bids. These bidders were accustomed to bidding in sealed bid environments and the initial bids are similar to other sealed bids that the administrator for the INRO auction had observed and the

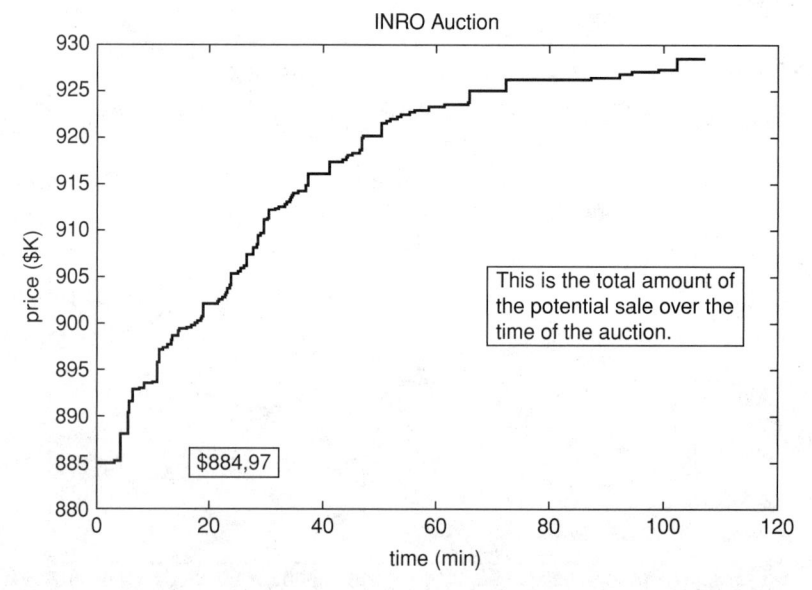

Figure 4. International Natural Rubber Organization 4 bidders 22 items (100 metric tons of natural).

bids were approximately market prices that exist in public markets. The initial bids on the rubber from different plantations reflected the difference of quality among the different plantations. Bids tended to be the same for rubber from the same plantation but bids differed for rubber from different plantations. Scale preferences were also evident. Bids for packages were frequently tendered. One bidder wanted all of the rubber in the warehouse and at the auction opening placed the high bid on all items for sale. This bidder ended the auction with ten pallets while the three other winners ended with 6, 5 and 1 pallets respectively.

Package bidding followed quickly after the initial bids. Some bidders expressed values for rubber from a limited set of plantations and others seemed to be interested in a mix with some sensitivity to price and quantity. The black dots appeared throughout the auction, signaling a bid on part of an existing larger package bid. The auction took about two hours and the "railroad lights" tended to appear signaling a threat to end the auction in the absence of bidding.

Total revenue in the INRO auction follows an approximately concave movement over time. If the starting revenue of $884,975 is assumed to be the revenue that would have been produced by a sealed bid when compared with the $927,000 auction revenue, the combinatorial auction produced about 5.5% more revenue. The flat places in the time series reveal instances of no bids (and thus the warning clock flashes) which increase in frequency as the auction progressed. Such patterns exist in experimental data.

The second field application examined is an auction for aquaculture sites located in Port Phillip Bay near Melbourne, Australia. The sites are appropriate for the growing of bivalve shellfish. The state of Victoria decided to auction eighteen sites. A total often bidders participated and bid for 18 sites. Seven bidders were winners producing $575,000 in revenue.

The sites were scattered across six locations. Bidders were interested in scale since they must meet regular demand for deliveries. They are also interested in a portfolio of sites reflecting a diversity of location due to currents, winds, possible diseases, and location relative to home base and delivery points. Thus, multiple synergies existed and package bid were used frequently[2].

The revenue and timing from the aquaculture auction are displayed in Figures 5a, 5b and 5c. Figures 5a demonstrates the typical concavity of revenue when displayed in clock time over the approximate 2 hours of the auction. The delay in the middle of the figure reflects an equipment problem that delayed the auction for about five minutes[3]. Bids are entered rapidly at first (shown in Figure 5b) and then slow down as the auction advances and become very slow at the end (shown in Figure 5b). This pattern is very reminiscent of the behavior of continuous auctions.

---

[2] The total number of bids was 300 of which 129 were bids for packages of items and 171 were bids on single items. The auction lasted about 7000 seconds which means that a bid or ask arriving every 5 seconds or so. There were 1032 query, about 3.5 query per bid, so something was happening about every 4 seconds not counting cancellations and other activities.

[3] Someone unplugged to power to the server at the remote location where the auction was held. The event illustrates the need for review and testing of every mode in which an online auction might fail. The fact that the recovery was complete and fast, with no apparent disruption to the auction suggests the existence of considerable background research not covered in this brief paper.

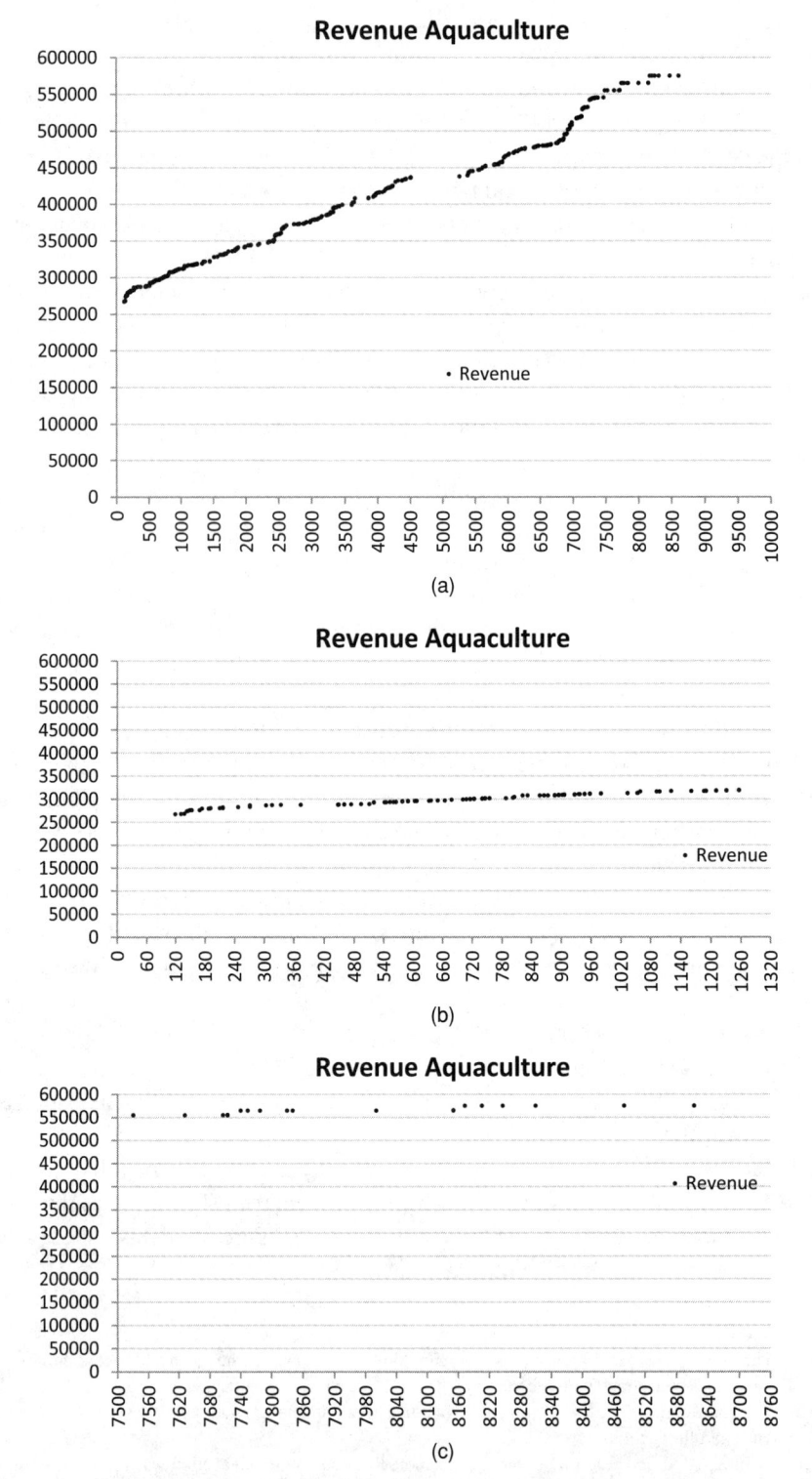

Figure 5. (a) Aquaculture Revenue and Timing of all Bids. (b) Timing of Bids (first 20 minutes). (c) Aquaculture Timing of Bids (last 20 minutes).

7. Summary and Observations

Several features of the mechanism are worth emphasis. The absence of a concept of a price per item is a departure from tradition. Replacing the measures contained in prices are queries and displays that can respond to human pattern recognition and crafted information needs. Obviously, the economic content of a concept of prices is working in the background but the operation of the mechanism is not based on their use. The use of clocks is important. They carry key public information and create the proper level of incentives for coordination. A bidder need only meet an increment requirement to keep a negotiation alive before facing an all or none choice of implementing a "contribution" to the public good of breaking up a large bid or collections of bids and becoming a provisional winner. The dots provide feedback by calling attention to actions of others and the possible intentions that underlay the actions of others play. This type of information that contributes to coalition formation plays a key role.

Computational problems can clearly pose problems as the size of the auction grows but the ability to solve big problems depends on the structure of the problem and the computing technology. Continuous combinatorial auctions much larger than reported here have been conducted. The computations times we encountered were all measured in fractions of seconds even when hundreds of bids exist in the set of all bids. Of course, ways exist to reduce the computational problem at the expense of limitations on permissible bids. Such restrictions have been tested but not used.

The testbed methods have some departures from what an untutored theorist might expect. The methods are designed to address problems where the theory is not complete and might be no more than suggestive. Tests of such theories do not make much sense when research is confronted by a scale of limited budget, limited time, and an unbounded infinity of variables. Yet, the role of theory plays a fundamental role. Theory, regardless of how incomplete it might be, is the tool that takes the analysis from the limited observations under controlled conditions to the substantially unknown conditions of the field. The theory must be robust and the testbeds help establish that.

References

Banks, J. S., J. O. Ledyard, and D. P. Porter, (1989). "Allocating uncertain and unresponsive resources: An experimental approach". *Rand Journal of Economics*, 20(1):1–25.

Brewer, Paul J. and Charles R. Plott, (1996). "A Binary Conflict Ascending Price (BICAP) Mechanism for the Decentralized Allocation of the Right to Use Railroad Tracks," *International Journal of Industrial Organization*, 14:857–886.

Brewer, Paul J. and Charles R. Plott, (2002). "A Decentralized, Smart Market Solution to a Class of Back-haul Transport Problems: Concept and Experimental Testbeds", *Interfaces*, 32(5):13–36.

Grether, David, R., Mark Isaac, and Charles R. Plott, "Alternative Methods of Allocating Airport Slots: Performance and Evaluation," Paper prepared for the Civil Aeronautics Board. Pasadena: Polinomics Research Laboratories, Inc., 1979.

Grether, David M., R. Mark Isaac, and Charles R. Plott, *The Allocation of Scarce Resources: Experimental Economics and the Problem of Allocating Airport Slots*, Westview Press Underground Classics in Economics series, 1989.

Plott, Charles R., "A Combinatorial Auction Designed for the Federal Communications Commission", presented at the Combinatorial Auction Conference 5/18/2000. FCC, Why River Conference.

Plott, Charles R., Nilanjan Roy, and Baojia Tong, "Marshall and Walrus, Disequilibrium Trades and the Dynamics of Equilibration in the Continuous Double Auction Market". *Journal of Economic Behavior and Organization (forthcoming.)*

Rassenti, S. J., V. L. Smith, and R. L. Bulfin, (1982). "A combinatorial auction mechanism for airport time slot allocation", *Bell Journal of Economics* 13(2):402–417.

CHAPTER 24

Coalition-based Pricing in Ascending Combinatorial Auctions

Martin Bichler, Zhen Hao, and Gediminas Adomavicius

1. Introduction

Due to the advances in modern computing and communication capabilities which allow to adopt advanced auction mechanisms in increasingly broader and larger-scale online settings, the theory on how multiple indivisible objects should be allocated via an auction has enjoyed renewed interest in recent years (Krishna 2002, Cramton et al. 2006, Bichler et al. 2010). One of the key goals in this research literature is to develop mechanisms that achieve high (allocative) efficiency. *Allocative efficiency* measures whether auctioned objects end up with the bidders who have the highest valuations for them, representing a measure of social welfare.

In this paper we aim to design highly efficient ascending combinatorial auctions. Among the problems that make this goal hard to achieve, the *coordination* of bidders is a key problem that has been largely underexplored in prior work. Specifically, the currently losing bidders have a task of identifying individually profitable and collectively complementary item bundles to bid on from an exponentially-sized set of all possible bundles (and determining appropriate bid prices for them), which together stand a chance of becoming a winning bid set in the next round. Identifying such bundle bids is possible either via a large number of auction rounds or by means of a highly non-trivial coordination among coalitions of losing bidders, which is difficult without appropriate price feedback. We address this challenge by proposing a *coalitional pricing rule*, which is able to draw on the data that the auctioneer collects about bidders' preferences throughout the auction and, as a result, helps currently losing bidders to coordinate. The proposed combinatorial auction mechanism exhibits substantially improved convergence and increased efficiency in lab experiments.

1.1. The Need for Ascending Combinatorial Auctions

Combinatorial auctions are among the most general types of multi-object market mechanisms, as they allow selling (or buying) a set of heterogeneous items to (or from) multiple bidders. Bidders can specify package (or bundle) bids, i.e., prices are defined individually for subsets of items that are auctioned (Cramton et al. 2006). The price is only valid for the entire bundle, and the bid is indivisible. For example, in a combinatorial auction a bidder might be willing to buy a bundle, consisting of items A and B, for a bundle price of €100, which might be more than the sum of the item prices for A (€30) and B (€50) that the bidder is willing to pay, if items are bought individually. The ability to submit bundle bids allows the bidders to express their economic preferences precisely, which is valuable in settings where bidders may have superadditive (or subadditive) valuations, i.e., when a bidder's valuation of the entire bundle is higher (or lower) than the sum of individual item valuations, as in the above example. We will refer to a bidding language as a set of allowable bid types (e.g., bundle bids or bids on individual items only) in an auction. If bidders can win multiple bids, this is referred to as an OR bidding language. If they can only win a single bid at most, then it is an XOR bidding language.

In simultaneous multi-object auctions where only individual item bids are allowed, bidders incur the risk that they may end up winning only a subset of items from their desired bundle, and that they may end up paying too much for this subset. This is called the *exposure problem* (Rothkopf et al. 1998). While bidding on bundles in combinatorial auctions solves this problem, the design of these auctions leads to several types of complexity. One type is computational complexity when determining an optimal allocation. Other types are strategic complexity for bidders, and communication complexity. Strategic complexity describes the difficulty for bidders to find an optimal bidding strategy, while communication complexity describes the number of messages (i.e., price announcements and bid submissions) which need to be exchanged between the auctioneer and the bidders in order to determine the efficient allocation. It has been shown that the communication complexity to find the efficient solution in combinatorial auctions is exponential with respect to the number of items in the worst case (Nisan and Segal 2006).

Computational complexity is manageable in real-world applications with a low number of items, bidders, and submitted bids; e.g., a winner determination problem with 20–30 items and 10 bidders can typically be solved in seconds. In terms of strategic complexity, one possible solution is to use the Vickrey-Clarke-Groves (VCG) mechanism, which achieves efficiency in dominant strategies, i.e., bidders cannot increase their payoff by deviating from a truthful revelation of their valuations. Unfortunately, VCG is rarely used due to a number of practical problems (Ausubel and Milgrom 2006b). In particular, in many markets bidders are simply reluctant to reveal their true valuations to an auctioneer in a single-round sealed-bid auction, and they prefer an ascending (multi-round) auction format which is more transparent and conveys information about the competition in the market. In a recent paper, Levin and Skrzypacz (2016) write that dynamic auctions have an advantage in multi-item settings, because bidders can gradually find out how their demands fit together. This property is important but not necessarily given in all multi-item auction designs.

1.2. Inefficiency in Ascending Combinatorial Auctions

As a result, much recent research has focused on ascending multi-object auctions, i.e., generalizations of the single-item English auction where bidders can outbid each other iteratively. However, it was recently shown that no ascending multi-object auction format can be incentive-compatible for general types of bidder valuations when modeled as a Bayesian game (Sano 2012, Goeree and Lien 2014). In other words, with sufficient prior information about other bidders and allowing any type of valuations, it is always possible that a bidder may profit from not bidding truthfully up to his valuation. Let us consider a simple example with two identical items and three bidders. One "global" bidder is only interested in the bundle of two items, while each of the two "local" bidders wants only one of the items. If the local bidders together are stronger than the global bidder (i.e., the sum of their one-item valuations is higher than the global bidder's two-item valuation), then they could always try to free-ride on each other. Suppose that the global bidder has a valuation of $10 for the two-item bundle, and each local bidder has a valuation of $8 for his item of interest. One local bidder could drop out at a price of $2.5, such that the other local bidder is forced to bid up to $7.5 to become winning. Sano (2012) has shown that, without complete information but having prior distributional information about bidder valuations in a Bayesian game, a local bidder might drop out too early resulting in the auction being inefficient in equilibrium. In any case, this and similar types of manipulations are only possible with sufficient prior information about other bidders' valuations.

In many real-world markets, the information set available to bidders is quite different from markets modeled under complete information or as Bayesian games with single-minded bidders. Bidders are interested in multiple packages, and it is unknown to a bidder which packages are of interest to his competitors. Also, the *common prior* assumption in Bayesian games, which has long been a concern in game theory (Wilson 1987), is particularly troublesome in combinatorial auctions with exponentially many possible packages a bidder can bid on. Bidders would need to have the same prior distributions for all possible packages, which is unrealistic in all but very small combinatorial auctions. In addition, in many auctions in procurement applications or on the Internet, the number of competitors is unknown, and there can always be a new bidder throughout the auction. Bid shading is less of a concern in such environments.

Even if bidders do not shade their bids due to a lack of prior information about others' valuations, this does not automatically lead to efficient outcomes because of the communication complexity of combinatorial auctions. If bidders do not bid on all bundles of positive value to them, but only a small subset thereof, then the auction may not end with an efficient outcome. This restricted bundle selection has been experimentally shown to be the biggest barrier to efficiency across auction formats (Scheffel et al. 2012).

Due to the exponential growth of possible bundles, even in combinatorial auctions with only 20 items bidders would not be able to reveal over a million possible bundle valuations, i.e., to submit all possible bundle bids. Recent combinatorial auctions used for spectrum sales had 100 licenses simultaneously on sale. It is clearly impossible to enumerate all the exponentially many bundles for a bidder. Finding promising bundles, i.e., bundles that stand a chance of becoming winning when combined with the bids of other bidders, becomes the central strategic problem for bidders in such auctions, which has largely been ignored in the game-theoretical literature on combinatorial auctions. This *coordination problem* requires different theoretical models.

1.3. Contributions and Outline

The main contribution of this work is to propose an auction format that leverages the bidding information that the auctioneer collects throughout an ascending auction about losing, but high-revenue coalitions. We select such high-revenue coalitions and propose ask prices to the members of each coalition such that together they can outbid the current winning coalition. This new type of pricing rule is called "coalitional winning level (CWL)." The auctioneer can provide such prices based on the bids that he collected in past rounds, and he can distribute the additional amount needed to make the losing coalition winning in a fair manner using the cost-sharing rule based on the Shapley value (Dehez 2007). In particular, we show that such a cost-sharing rule not only satisfies fairness axioms, but also results in cost sharing among the bidders in a coalition that is *in the core*, i.e., it does not create incentives to deviate for any subset of a bidder coalition.

In this work, we are focusing on markets where bidders typically do not have reliable prior information about other bidders' valuations or the number of their competitors. The coordination problem introduced earlier becomes the central strategic challenge for bidders in such markets. Bidders have an exponential number of packages to choose from, but want to coordinate on competitive equilibrium in a low number of rounds.

We focus on markets without reliable prior distributional information, so that we can restrict our attention to the bidder's decision problem in a single bidding round. The auctioneer in this auction model acts like a third party providing a public signal in a correlated equilibrium. This helps bidders to coordinate on one out of multiple equilibria. Also, in an online combinatorial auction in the field, without prior distributional valuation information and with hundreds of packages to bid on, the ask prices (CWLs) of the auctioneer provide a recommendation how complementary bids of losing bidders can become winning in the next round.

Our experimental results show that the the CWL auction has significantly higher efficiency, and at the same time communication with the auctioneer is substantially reduced compared to ascending auction designs from prior literature. In our experiments we do not see free-riding behavior, i.e., the bidders indeed take advantage of ask prices to coordinate. This rapid convergence of the auction increases the practical applicability of the mechanism to a broad set of application settings.

The paper is structured as follows. In Section 2, we discuss related literature. In Section 3, we introduce the auction format and describe various properties. Section 4 presents the experimental design, while Section 5 summarizes the results of the numerical simulations. These simulations provide an indicator for the outcome of such auctions with truthful bidders who bid on their payoff-maximizing bundles in each round. In Section 6 we summarize and discuss the findings of our lab experiments, before concluding the article with Section 7.

2. Related Literature

Let us briefly survey the relevant literature in this section. As mentioned earlier, the well-known Vickrey-Clarke-Groves (VCG) mechanism achieves efficiency in dominant strategies. Its central limitation is that the auction outcome might not be *in the core* (Ausubel and Milgrom 2006b), i.e., the winning coalition of bidders might have to pay less than what a losing coalition of bidders was willing to pay. This is possible due to the Vickrey discount which the winning bidders are given (Goeree and Lien 2014). We provide a simple example with two items (A and B) and three bidders (1, 2, and 3) to make this apparent. Suppose bidder 1 only wants A for which he has bid his value of \$7, bidder 2 only wants B for which he has bid his value of \$8, and bidder 3 only wants the package AB with a value of \$10. The auctioneer declares bidder 1 and 2 to be winners and the maximum of total valuations of the sale to be \$15. In a VCG mechanism the winners get a discount, which incentivizes truthful bidding. As a result, the Vickrey payment for bidder 1 is \$10 − \$8 = \$2 and that of bidder 2 is \$10 − \$7 = \$3. Consequently, the auction revenue is \$5, although bidder 3 was willing to pay \$10. In many applications, such as high-stakes government auctions, such an outcome might be difficult to justify. Therefore, such high-stakes auctions are typically conducted as open-cry ascending auctions, rather than sealed-bid events. For these reasons, during the recent years there has been an increasing interest in core-selecting auctions (Day and Milgrom 2008), i.e., auctions where there cannot be a losing coalition of bidders that together could have outbid the winners based on their submitted bids.

iBundle (Parkes and Ungar 2000), the ascending proxy auction (APA) (Ausubel and Milgrom 2006a), and dVSV (de Vries et al. 2007) are examples of ascending core-selecting auction formats which provide allocatively efficient solutions when bidders follow a *straightforward* bidding strategy, i.e., when they truthfully bid on their payoff-maximizing bundle(s) in each round until the prices stop because a bidder becomes winning. If bidder valuations are *buyer submodular*, then this strategy is even an ex post Nash equilibrium, which is a strong solution concept where bidders do not need to reason about other bidders' valuations. Buyer submodularity requires that, if a bidder is added to a smaller coalition, then he adds more to the overall revenue than if added to a larger coalition with more bidders (Parkes 2006).

Note that super-additive valuations violate buyer submodularity. In the above example with three bidders, the third bidder has super-additive valuations since both items are complements for him (i.e., he does not want each item individually, only their combination), thus violating buyer submodularity. Also, it is easy to see that it is not an ex post Nash equilibrium strategy to bid truthfully until the price clock stops or the valuation is reached in this example. Bidder 1 might drop out before his price for one unit stops increasing for his one item and he might free-ride on bidder 2, who then needs to outbid bidder 3 in an ascending auction. In such cases, only a Bayesian Nash equilibrium is possible, which requires prior distributional information about other bidders' valuations. In this market, when bidders only have prior distributions about the valuations, the Bayes-Nash equilibrium strategy can even lead to non-bidding of the local bidders, and consequently, to inefficient outcomes (Sano 2012, Guler et al. 2013). We will refer to the family of efficient ascending multi-object auctions, which allow for an ex post Nash equilibrium at least for buyer submodularity, as *bidder-optimal ascending core-selecting* (BACS) auctions.

Adomavicius and Gupta (2005) introduce deadness (DLs) and winning levels (WLs) both as pricing rules and information feedback to bidders in combinatorial auctions and evaluate them in the lab (Adomavicius et al. 2013). Deadness levels are the lowest prices above which a bid can still potentially become winning in any future auction state (depending on the arrival of complementary bids from other bidders), winning levels are prices above which a bid would immediately become winning. Petrakis et al. (2013) showed that ascending combinatorial auctions with deadness levels as ask prices (the *DL* auction) belong to the above family of BACS auctions and share the same ex post Nash equilibrium strategy as BACS auctions.

BACS auctions can be thought of as algorithms designed to provide an exact solution to a hard computational problem. However, they typically lead to a huge number of auction rounds (Schneider et al. 2010), and lab experiments provide evidence that human bidders substantially deviate from straightforward bidding (Scheffel et al. 2011) so that efficiency is no longer guaranteed. More recent experimental research shows that restricted bundle selection due to the exponential growth of bundles is the main reason for inefficiency in combinatorial auctions (Scheffel et al. 2012, Bichler et al. 2013), while bid shading is much less of an issue. Rather than shading their bids optimally, bidders in such auctions are primarily concerned with finding the right bundle which, together with the bids of other bidders, will end up in a winning coalition. We will refer to this problem as the "coordination problem" and to markets with little or no distributional information about bidder valuations as "online" markets. We use the term *online markets* related to the concept of online algorithms or online mechanisms from the literature in computer science (Parkes 2007).

Our proposed combinatorial auction mechanism is different from the ones mentioned above. In particular, the auctioneer targets losing coalitions by proposing *coalitional winning levels* (*CWLs*) as ask prices to the members of these coalitions, which would allow them to jointly outbid the currently winning coalition. The semantics of *CWLs* is intuitive for bidders and provides guidance in what is arguably *the central problem* that bidders face in each round of a combinatorial auction with many items: the selection of promising bundles, which stand a chance of becoming winning together with the bids of other bidders.

3. The Auctions

In what follows, we will briefly describe the *DL* auction as a representative of BACS (i.e., bidder-optimal ascending core-selecting) auctions. We will then introduce the proposed *CWL* auction and discuss some theoretical underpinnings of such an auction. Before we do this, we provide an example of different pricing rules as they have been discussed in the literature to better illustrate the different approaches.

3.1. An Introductory Example with Different Pricing Rules

The following example extends the one used by Petrakis et al. (2013) to illustrate *DLs*. We compare *CWLs* with *iBundle*, *DLs*, *WLs*, and an auction format with linear ask prices, RAD (Kwasnica et al. 2005).

The top two rows of Table 1 describe six bids from different bidders (i.e., bidders 1 to 6), submitted on subsets of three items (A, B, and C). In this example we will assume

Table 1. *Example with six bids and different ask prices*

bundles	AB	BC	AC	B	C
bids	$22^*_1, 16_2$	24_3	20_4	7_5	8^*_6
DL	$22_1, 16_2$	24_3	20_4	7_5	8_6
WL	$22_1, 22_2$	30_3	23_4	10_5	8_6
iBundle	$22_1, 17_2$	25_3	21_4	8_5	8_6
RAD	22	24	14	16	8
CWL	$22_1, 22_2$	30_3	**21.5**$_4$	**8.5**$_5$	8_6

that, at this point in the auction, bidders are only interested in those bundles for which they have submitted bids so far.

The bottom five rows of Table 1 describe bundle prices in different auction formats at this stage in the auction (i.e., after 6 bids have been submitted). Subscripts indicate bidders, i.e., 22_1 indicates a bid of €22 from bidder 1. Ask prices have subscripts only if they differ among bidders in this example. Asterisks denote the provisional winning bids. In this example, we assume an XOR bid language, where each bidder can win at most one bundle. For such languages, it is known that *DL*s and *WL*s for a given bundle may have different values for different bidders, i.e., their computation needs to be personalized (Petrakis et al. 2013). Losing bidders need to bid higher than these values by a minimum bid increment. As mentioned earlier, the *WL* for a given bundle describes the lowest bid price above which a submitted bid would instantly become winning, i.e., without needing any new complementary bids from other bidders. However, it is clear from the example that bidders 4 and 5 could possibly become winning even at lower prices than their current *WL*s, if they coordinate and form a coalition, indicated by prices in bold type. Ask prices in iBundle (Parkes and Ungar 2000) are in line with *DL*s, but they add a bid increment (€1) for losing bids. Linear programming-based heuristics for computing linear prices (i.e., where a bundle price is simply a sum of individual item prices) such as *RAD* (Kwasnica et al. 2005) are an alternative. Unfortunately, *RAD* prices can be lower than a losing bid (see the *RAD* ask price on *AC* for bidder 4) or unnecessarily much higher than the sufficient winning bid (see the *RAD* ask price on *B* for bidder 5) (Bichler et al. 2009).

In an online market, bidders typically start out bidding on their highest valued packages in order to find out if this package can become winning together with the bids of others. After the winner determination, the auctioneer can evaluate which losing bidders would, in combination, achieve high revenue and have a potential to outbid the current winning coalition. *CWL*s can be seen as a way to derive personalized and non-linear ask prices in-between *DL*s and *WL*s designed to quickly find a *competitive equilibrium*, i.e., a state where there is no coalition of bidders that can outbid the currently winning coalition of bidders at these ask prices, as defined below.

Definition 1 (Competitive Equilibrium, CE (Parkes 2006)). Prices α and allocation X^* are in competitive equilibrium if allocation X^* maximizes the payoff of every bidder and the auctioneer revenue given prices α. The allocation X^* is said to be *supported* by prices α in CE.

It has been shown that competitive equilibrium and the core, mentioned in Section 2, refer to the same concept in multi-object auctions (Bikhchandani and Ostroy 2002). *CWL*s are a way to find prices such that, at the end of an ascending auction, there is no coalition of losing bids who could make themselves better off. *CWL*s leverage the information that is available about losing coalitions during the auction and provide tailor-made prices to bidders in these coalitions, i.e., proposals on how they can jointly outbid the currently winning coalition. The coalition of bidders 4 and 5 in our example would only need to increase their bids by a combined €3 plus increment in order to become winning. Both bidders would become winning, if bidder 4 bids above €21.5 and bidder 5 bids above €8.5, for example. The proposed *CWL* feedback is designed to help coordinating bidders who form a high-revenue coalition, and it is particularly useful if bidders are interested in many packages. In what follows, we will describe *DL*s and *WL*s in a more formal way before we introduce *CWL*s and their properties.

3.2. The *DL* Auction

We will first introduce the necessary notation and then describe the *DL* auction as an example of a BACS auction. There is set $\mathcal{K}$ of m indivisible items indexed with k, which are auctioned among a set $\mathcal{I}$ of n bidders. Let $i, j \in \mathcal{I}$ denote the bidders and $v_i : 2^{\mathcal{K}} \to \mathbb{R}$ denote a value function of bidder i, which assigns a real value to every subset $S \subseteq \mathcal{K}$ of items. The bundle that is assigned to bidder i in allocation X is denoted as $X_i \subseteq \mathcal{K}$. We denote $X = (X_1, \ldots, X_n)$ as an allocation of the m items among bidders, with $X_i \cap X_j = \emptyset$ for every $i \neq j$, with $i, j \in \mathcal{I}$. A coalition is defined as a set of bidders whose bids constitute a feasible allocation. A winning coalition is the coalition of bidders whose bids constitute the revenue maximizing allocation, and a losing coalition is any coalition except the winning coalition. In other words, a bidder can be a member of the winning coalition and, at the same time, be a member of multiple losing coalitions, based on the bids he submitted. We denote a losing coalition L, and the set of all losing coalitions $\mathcal{L}$. Let Γ denote the set of all possible allocations, then $X^L \in \Gamma$ denotes an allocation of items among a losing coalition $L \in \mathcal{L}$.

The social welfare of an allocation $X = (X_1, \ldots, X_n)$ is $\sum_{i \in \mathcal{I}} v_i(X_i)$, and an efficient allocation X^* maximizes social welfare among all allocations X, i.e. $X^* \in \arg\max_X \sum_{i \in \mathcal{I}} v_i(X_i)$. The revenue maximizing allocation, $\overline{X}$, is such that $\overline{X} \in \arg\max_X \sum_{i \in \mathcal{I}} b_i(X_i)$, where b_i is the bid price of bidder i for the bundle assigned to him in allocation X.

We focus on ascending combinatorial auctions (CAs), which consist of different rounds and where an ask price $\alpha_i(S)$ is available for each bundle S and each bidder i in each round. A round defines a certain time period during which the auctioneer collects new bids from bidders and at the end of which a new allocation and new ask prices for the next round are computed. The *DL* auction uses the XOR bidding language, i.e., at most one bundle bid from a given bidder could be winning at any given time. Let B^t denote the bids submitted in round $t \in \mathbb{N}$, and $\overline{X}^t$ denote the revenue maximizing allocation at the end of round t, based on the set of all bids B submitted in the auction so far. Note that, in theory, a round could close after each newly submitted bid, such that t may also refer to a single bid. In our experiments each bidder can submit multiple

Algorithm 1: The *DL* auction algorithm

Result: $\overline{X}$ and bid prices $b_i(\overline{X}_i)$

1 Initialization
2 **for** *i=1* **to** *n* **do**
3   **foreach** *S* **do** $\alpha_i(S) \leftarrow \epsilon$
4     $X_i \leftarrow \emptyset$
5 **end**
6 *termination* $\leftarrow$ false
7 $t \leftarrow 0$
8 $B \leftarrow \emptyset$
9 **while** *(¬ termination)* **do**
10     $t \leftarrow t+1$
11   Bidders submit bids B^t where each $b_i(S) \in B^t$ satisfies $b_i(S) \geq \alpha_i(S)$
12   **if** $(B^t = \emptyset)$ **then** *termination* $\leftarrow$ true
13   **else**
14         $B \leftarrow B \cup B^t$
15   Compute $\overline{X}^t \in \arg\max_X \sum_{i \in \mathcal{I}} b_i(X_i)$
16   **foreach** $b_i(S) \in B_l^t$ **do** $\alpha_i(S) \leftarrow DL^t(i,S) + \epsilon$
17   **end**
18 **end**

bids in a round. B_l^t denotes the set of all losing bundle bids, even losing bids from a winning bidder (because a winning bidder can only win at most one bundle, but may well submit multiple bids in each round), after round t. ϵ describes a minimum bid increment per round. We next define deadness levels as ask prices:

Definition 2. The *deadness level, DL*, of a bundle S for bidder i at round t, $DL^t(i, S)$, is the minimal price that bidder i has to overbid to maintain a chance to win S at some future round $t' > t$.

So *DL*s are the highest prices at or below which a bid cannot become winning in any future auction state. Therefore, they constitute a lower bound for acceptable new bids. The *DL* auction uses only *DL* ask prices (i.e., DL + ϵ), and belongs to the family of BACS auctions, as was shown in Petrakis et al. (2013). Algorithm 1 outlines the *DL* auction.

As shown in Algorithm 1, the *DL* auction is conducted in a round-based ascending format. The auction begins with ask prices of 0. At the start of each subsequent round, each bidder is given the following information: ask prices on all bundles he has bid on so far in the auction (but can request ask prices of any other bundles on demand, if desired) and information whether he is currently winning any bundle he has previously bid on. This is the only information made available to each bidder, and he does not know anything else about other bidders, including the bundles they have bid on so far, or if they are winning some bundles. A bidder can then submit as many bids as he likes in this round. However, the bids need to be higher than or equal to the respective ask prices. A bid higher than the ask price is called a jump bid. When all bidders finish submitting new bids, the round closes and a new allocation with the current winning coalition as well as new ask prices are computed. The auction terminates when no new bid is submitted in a round, and winning bidders pay what they bid.

We have also implemented two additional auction rules: an activity rule to incentivize bidders to stay active from the start, and the possibility for bidders to submit a so-called "*last-and-final bid*," which helps to avoid small efficiency losses due to bid increments. Both are described below and complete our description of the *DL* auction.

In general, ascending multi-object auctions can enforce different types of activity rules. In our experiments, we used an activity rule in line with earlier experiments in the combinatorial auction literature (Scheffel et al. 2012). If a currently losing bidder in round t does not submit any new bids in round $t + 1$, then he is not allowed to bid in any future rounds $t' > t$. All his previous active bids will still be considered relevant for the auction, but he may not submit any new bid for the auction. This activity rule does not apply to currently winning bidders, as their inactivity does not necessarily imply that they are not interested in the auction any more. We have used this soft activity rule in both the *DL* and the proposed *CWL* auction.

In addition, it is possible to have a situation, where the new ask price $\alpha_i(S)$ is too high for a bidder because the bid increment was too big. In this case, the *DL* auction allows for a *last-and-final bid* between these bounds (i.e., between the bidder's last bid on the package and the new ask price $\alpha_i(S)$) (Parkes 2006). Suppose bidder i has submitted a bid of €18 on bundle AB in round t. In the next round $t + 1$, he sees that he has not won AB, and the new ask price for AB, based on the *DL* plus an increment, is €22. Assuming the bidder's true valuation for AB is €20, then he can now submit a last-and-final bid of €20 − ϵ, where ϵ is a profit margin he wants to achieve. However, after this round he would not be allowed to bid on AB anymore.

We described the key details of the *DL* auction, and will now turn to one of the main properties satisfied by the *DL* auction, as it is related to the *straightforward* bidding strategy.

Definition 3. A *straightforward* bidder i only bids ask prices on his demand set $D_i = \{S \subseteq \mathcal{K} : v_i(S) - \alpha_i(S) \geq v_i(S') - \alpha_i(S'), \forall S' \subseteq \mathcal{K}\}$ in each round, i.e., on those bundles which maximize his payoff, based on given ask prices.

Importantly, in the *DL* auction format, straightforward bidding is an ex post equilibrium if bidders' valuations are submodular (Parkes 2006, Petrakis et al. 2013). This is because with such valuations, the auctions end up in VCG prices for the winners, and bidders do not have an incentive to shade their bids (Parkes 2006). Unfortunately, straightforward bidding also leads to a large number of auction rounds as *all* losing package valuations get elicited from all bidders via minimum bid increments in each round. While this process allows to prove efficiency of the allocation, the number of bids which need to be submitted by bidders is beyond what human bidders can be expected to do, except in auctions with only very few items. Schneider et al. (2010) have shown using numerical simulations that, with straightforward bidding, even small auctions with only 9 items can easily lead to 150 and more auction rounds. With, say, 10 minutes per round, this would lead to 25 hours, which would be unacceptable in most applications.

3.3. Coalitional Winning Levels

In addition to deadness levels (*DL*s), Adomavicius and Gupta (2005) also defined winning levels (*WL*s) as a form of information feedback to bidders. $WL + \epsilon$ is the minimum bid price for a bidder on a bundle, such that this bundle bid becomes winning at round $t + 1$, if no other bid was submitted. As indicated in the introduction, *WL*s can be prohibitively high for small bidders in larger auctions with many items, since *WL*s reflect an amount a bidder needs to bid to become winning unilaterally, i.e., without the help of any new bids of other bidders.

The coalitional winning level (*CWL*) extends the concept of a *WL* ask price from an individual bidder to a group of losing bidders. It is an ask price that would make a losing coalition winning, if accepted by all members of the coalition. This is valuable feedback for overcoming coordination problems inherent to all combinatorial auctions, as illustrated in the following example.

Example 1. Consider four small bidders, each one bidding €10 on a different single item, and a large bidder, bidding €100 on the bundle containing all four of these items. The valuation of each small bidder for their respective single item is €50. By definition, the *WL* ask price faced by each small bidder is €70 + ϵ for the desired item, indicating the scenario where each small bidder competes with the large bidder individually. The *WL*-based ask price is higher than the small bidders' valuations; as a result, the small bidders would not bid anymore, and the efficient allocation is not achieved. For comparison, the *CWL* for the losing coalition is €100 in total. So if, for example, each small bidder in the coalition receives an individual *CWL* ask price of €25 + ϵ and bids on it, the coalition would outbid the large bidder. Finally, the *DL* of each bidder would be €10 only, as each bidder could become winning at €10 if the other small bidders outbid the large bidder.

Example 1 illustrates that the spread between *DL*s (here €10) and *WL*s (here €70) can be very large. In examples with many items and bidders being interested in many bundles, *CWL*s can give bidders useful information about bundles for which complementary bids exist. In addition, they can help bidders focus on a few (rather than all) of their bundles with positive valuations (i.e., where the bidder valuation for a bundle is higher than its current ask price).

Let L denote a coalition of losing bidders, where bidder i desires the bundle S_i, with $S_i \cap S_j = \emptyset$ for all $i, j \in L$. Denote the collection of the desired bundles as $S^L = \bigcup_{i \in L} S_i$.

Definition 4. The *coalitional winning level*, *CWL*, of coalition L for the desired bundles S^L at a particular round t is the minimal price that the coalition must bid in aggregate to win these bundles at auction state $t + 1$: $CWL^t(L, S^L) = \min \sum_{i \in L} b_i(S_i)$, such that $S_i \in \overline{X}^{t+1} \forall i \in L$, assuming all new bids of this coalition come in the next round, i.e., $t + 1$.

In this definition we assume that only the losing bidders $i \in L$ submit bids in round $t + 1$, so that $\overline{X}^{t+1}$ describes the revenue maximizing allocation in round $t + 1$. The *CWL* value for any losing coalition L can be computed as follows:

$$CWL^t(L, S^L) = CAP^t(\mathcal{K}) - CAP^t(\mathcal{K}, S^L). \tag{1}$$

$CAP^t(\mathcal{K})$ denotes the optimal value of the winner determination problem (CAP), and $CAP^t(\mathcal{K}, S^L)$ the optimal value of CAP in which each bidder $i \in L$ wins his desired bundle $S_i \in S^L$ for free. There is substantial literature on the computational hardness of CAP (Leyton-Brown et al. 2006), but instances of up to 20–30 items and 10 bidders can typically be solved in seconds. The computation of CWLs in Equation 1 can easily be derived from the proof for $WL^t(i, S) = CAP^t(\mathcal{K}) - CAP^t(\mathcal{K}, S_i)$ in Petrakis et al. (2013), where the desired bundles S^L of all bidders $i \in L$ are treated as if they were one single bundle S_i of one single losing bidder i. Similar computations have also been described by Adomavicius and Gupta (2005).

3.4. Computing Individual CWLs

Once $CWL^t(L, S^L)$ is computed for a losing coalition L, we still face the question of how to distribute this price among members of L. Let us denote $CWL_i^t(L, S_i)$ as the amount that transforms the $CWL^t(L, S^L)$ to individual ask prices for every member i of losing coalition L. Here, S_i describes the package assigned to bidder $i \in L$. There are different ways how bidders in a coalition can share the additional amount $\Delta^t = CWL^t(L, S^L) - \sum_{i \in L} b_i(S_i)$ that is needed to outbid the current winning coalition. One could think of many *cost sharing functions* $\Delta_i^t = g_i(\Delta^t)$ to distribute Δ^t among the bidders $i \in L$ such that $\sum_{i \in L} \Delta_i^t = \Delta^t$. For example,

- $\Delta_i^t = \Delta^t \times \frac{|S_i|}{|S^L|}$ based on the bundle size $|S_i|$ of a bundle S_i within a coalition;
- $\Delta_i^t = \Delta^t \times \frac{b_i(S)}{\sum_{i \in L} b_i(S)}$ based on the level of the bid prices $b_i(S)$ within a coalition;
- $\Delta_i^t = \Delta^t \times \frac{1}{|L|}$ based on the number of members in a coalition (a.k.a. uniform distribution).

These heuristic cost sharing functions are simple because their calculations require only a few simple arithmetic operations. Aside from computational simplicity, a *fair* division of Δ^t among the bidders $i \in L$ would be a natural design goal. The Shapley value is arguably the most well-known solution concept for coalitional games (Dehez 2007), and it is considered fair, as it satisfies a number of fairness axioms including symmetry and additivity (Shoham and Leyton-Brown 2009).

Let's briefly review the Shapley value. Let L be a coalition of $|L|$ bidders, and $M \subseteq L$ be some sub-coalition. Let $w(M)$ denote the coalitional value of M that needs to be distributed among its members. Coalitional value can also be the cost that a coalition has to bear. The Shapley value θ_i provides a unique distribution (among the players) of the value generated by the coalition of all bidders $i \in L$ and is defined as:

$$\theta_i = \sum_{M \subseteq L \setminus \{i\}} \frac{|M|!(|L| - |M| - 1)!}{|L|!} (w(M \cup \{i\}) - w(M)) \qquad (2)$$

Overall, the Shapley value has a number of general properties, which are desirable. For example, it distributes the total value of a coalition. Bidders with the same contribution to the coalitional value get the same Shapley value. Bidders who do not contribute to the coalitional value get a zero Shapley value. However, if designed appropriately as a convex game, there are two properties of our coalitional game that make the Shapley value particularly desirable.

In *super-additive* games with $w(L \cup L') \geq w(L) + w(L')$ and $L \cap L' = \emptyset$, where L and L' are two losing coalitions, the Shapley value guarantees each participant a payoff of at least the amount that he could achieve by not forming a coalition. An important subclass of super-additive games are *convex* games. A game is convex if $w(L \cup L') \geq w(L) + w(L') - w(L \cap L')$. For every convex game, the core is nonempty, and the Shapley value is also in the core for convex games. This means that, based on the Shapley value, there cannot be a losing sub-coalition of $M \subset L$ that can make themselves better off as compared to a situation where all members of the coalition L accepted the Shapley value. In other words, sub-coalitions do not have an incentive to deviate, and the coalition can be considered stable as neither individuals nor groups of bidders in L have an incentive to deviate.

Note that sharing a given $\Delta^t = CWL^t(L, S^L) - \sum_{i \in L} b_i(S_i)$ in a round among a losing coalition of bidders is neither a convex nor a super-additive game. However, instead of distributing Δ^t, one can distribute the overall savings that the coalition experiences compared to the sum of the winning levels of each bidder: $\Psi^t = \sum_{i \in L} WL^t(i, S) - CWL^t(L, S^L) = \sum_{i \in L} WL^t(i, S) - \sum_{i \in L} CWL_i^t(L, S_i)$. We require that $WL^t(i, S) \geq CWL_i^t(L, S_i)$ for all $i \in L$ for sharing functions, where this is not always satisfied. We can now use the Shapley value to derive Ψ_i^t from Ψ^t, and this game is super-additive and convex as the following results show.

Lemma 1. *The game of distributing $\Psi^t(L) = \sum_{i \in L} WL^t(i, S) - CWL^t(L, S^L)$ to individual bidders $i \in L$ is super-additive.*

Proof. Let $K = CAP^t(\mathcal{K})$ be the revenue achieved by the winning coalition at auction state t, i.e., the threshold for any losing coalition at state t. The winning level of a single-minded bidder $i \in L$ is $WL^t(i, S) = K - \sum_{j \in L \setminus i} b_j$, where b_j are the bids. The coalitional winning level (CWL) of any losing sub-coalition $L' \subseteq L$ is $CWL^t(L', S^{L'}) = K - \sum_{j \in L \setminus L'} b_j$.

Super-additivity is defined as $\Psi(L) \geq \Psi(L') + \Psi(L'') \; \forall L', L'' \subseteq L, L' \cap L'' = \emptyset$. We dropped the superscript t for brevity.

Now we have

$$\Psi(L') = \sum_{i \in L'} WL^t(i, S) - CWL^t(L', S^{L'}) \tag{3}$$

$$= \sum_{i \in L'} (K - \sum_{j \in L \setminus i} b_j) - (K - \sum_{j \in L \setminus L'} b_j) \tag{4}$$

$$= K(|L'| - 1) - \sum_{i \in L'} \sum_{j \in L \setminus i} b_j + \sum_{j \in L \setminus L'} b_j \tag{5}$$

$$= K(|L'| - 1) - (|L'| \sum_{j \in L} b_j - \sum_{j \in L'} b_j) + \sum_{j \in L \setminus L'} b_j \tag{6}$$

$$= K(|L'| - 1) - |L'| \sum_{j \in L} b_j + \sum_{j \in L} b_j \tag{7}$$

$$= (|L'| - 1)(K - \sum_{j \in L} b_j) \tag{8}$$

Note that $(K - \sum_{j \in L} b_j) \geq 0$ as L is a losing coalition, and it is the same for $\Psi(L)$, $\Psi(L'')$, and $\Psi(L')$. It remains to show that $|L| \geq |L'| + |L''| - 1$ which is true for all subcoalitions $L', L'' \subseteq L$, because L' and L'' are disjoint. □

Lemma 2. *The game of distributing* $\Psi^t(L) = \sum_{i \in L} WL^t(i, S) - CWL^t(L, S^L)$ *to individual bidders* $i \in L$ *is convex.*

Proof. Convexity is defined as $\Psi(L) \geq \Psi(L') + \Psi(L'') - \Psi(L' \cap L'')$, $\forall L', L'' \subseteq L$. If $L' \cap L'' = \emptyset$, then $\Psi(L' \cap L'') = 0$ and convexity reduces to super-additivity. Else assume $|L' \cap L''| > 0$. Based on Lemma 1 it remains to show that $(|L| - 1) \geq (|L'| - 1) + (|L''| - 1) - (|L' \cap L''| - 1)$, which is equivalent to $|L| \geq |L'| + |L''| - |L' \cap L''| = |L' \cup L''|$ which is true for all $L', L'' \subseteq L$. □

Let's now define the Shapley value (SV) based computation of individual *CWLs*.

Definition 5. A Shapley value based $CWL^t_{i,SV}(L, S_i)$ is defined as $WL^t(i, S) - \Psi^t_{i,SV}$, where

$$\Psi^t_{i,SV} = \sum_{M \subseteq L \setminus \{i\}} \frac{|M|!(|L| - |M| - 1)!}{|L|!} (\Psi^t(M \cup \{i\}) - \Psi^t(M)) \tag{9}$$

In this definition, $\Psi^t(M)$ is the total coalitional value or savings of coalition M. This leads to the following proposition.

Proposition 1. *Consider only the members* $i \in L$ *of a losing coalition, who need to derive individual CWLs from* $CWL^t(L, S^L)$. *No sub-coalition* $M \subset L$ *can make itself better off as compared to when all members of the coalition accepted* $CWL^t_{i,SV}(L, S_i)$ *given that* $v_i(S_i) \geq CWL^t_{i,SV}(L, S_i)$ *for all* $i \in \mathcal{I}$.

Proof. Lemma 2 shows that the computation of $CWL^t_{i,SV}(L, S^L)$ constitutes a convex coalitional game. Every convex game has a nonempty core, and in every convex game the Shapley value is in the core (Shoham and Leyton-Brown 2009, p. 394). □

Explaining the Shapley value and its properties to subjects in the lab takes a substantial amount of time. Since it is important in an economic experiment that subjects fully understand the mechanism, in our experiments, we have therefore decided to use a cost-sharing rule based on a uniform distribution for simplicity.

3.5. The *CWL* Auction

Based on the definition of *CWLs*, we will now describe the *CWL* auction. The auction process is identical to Algorithm 1 and the *DL* auction, including the availability of last-and-final bids and the activity rule. However, the ask prices are different. Instead of the computation of the *DLs* on line 16 in Algorithm 1, *CWLs* are computed for some of the highest revenue coalitions which are currently losing.

In case some members of a losing coalition do not accept the *CWL*, then this coalition would not become winning in the very next round, but the members (other than the ones who submitted a last-and-final bid on a relevant package) can always update their bids in a new round. All new bids from the previous round are taken into account at the end of the round, and they can be the foundation for new coalitions to be built.

Therefore, this process implicitly supports the collaborative search for a competitive equilibrium.

There are some degrees of freedom in how the auctioneer selects losing coalitions. The auctioneer could select any number between one or all losing coalitions in each round from the list of those losing coalitions with high revenue. How many losing coalitions are selected in each round depends on the size of the auction. In auctions with a few items and bidders one could select all coalitions, but this will not scale to large auctions as there are too many bidders and packages. Another implementation choice for the auctioneer is whether bidders are required to respond to a *CWL* immediately (i.e., in the next round) or not. For example, in order to proactively discourage free-riding behavior, the auctioneer may choose to select one losing coalition in each round, and require a response from each member of this coalition. This strict rule was not used in our experiments.

If more coalitions are provided with a *CWL*, we find that this can further reduce the number of auction rounds. In our experiments, where we are restricted to smaller auctions, we computed a *CWL* for every losing $b_i(S) \in B_l^t$. This specific implementation leads to the fact that sometimes a given bid can be part of multiple coalitions. In order to determine minimal core prices and avoid coalitions having to pay too much, conservatively we selected the minimum of these ask prices across different coalitions: $\alpha_i(S_i) \leftarrow \min_{L \in \mathcal{L}} CWL_i^t(L, S_i) + \epsilon$.

It is interesting to point out that finding the minimum $\min_{L \in \mathcal{L}} CWL_i^t(L, S_i)$ can be directly computed as part of the computation of winning levels $WL^t(i, S) = CAP^t(\mathcal{K}) - CAP^t(\mathcal{K}, S_i)$. The result of $CAP_i^t(\mathcal{K}, S_i)$ returns the highest-revenue coalition with bidder i winning S_i. This provides all information necessary to compute the $\min_{L \in \mathcal{L}} CWL_i^t(L, S_i)$ for any of the cost sharing functions to compute individual *CWL*s described in the previous subsection.

Selecting the lowest possible CWL_i^t for each bidder comes at a cost. In some cases, a coalition might not win even if all members agree to the *CWL* ask prices. In other words, we intentionally avoid having bidders paying more than what would have been necessary to win at the potential expense of additional auction rounds. In contrast, if the highest *CWL* for a bundle across all coalitions, i.e., $\max_{L \in \mathcal{L}} CWL_i^t(L, S_i^L)$, was selected, then it could happen that members of the coalition could pay more than what is necessary to become winning. In the simulations, the additional number of auction rounds caused by our proposed conservative pricing rule was very low, which is why we use it in our experiments.

3.6. Bidding Strategy and Bundle Selection

A bidding strategy in an auction involves two decisions in each round: which packages to bid on and how high a bidder should bid on the selected packages. Straightforward bidding (see Definition 3) is one such strategy, where bidders always bid on the package maximizing (absolute) payoff at the ask prices. We have discussed that, in a *DL* auction, straightforward bidding is an ex post equilibrium at least for some types of valuations. Bidding straightforwardly would take hundreds of auction rounds (Schneider et al. 2010), and bidders with a positive cost for participating in a round are unlikely to bid straightforwardly. Moreover, straightforward bidding was not reported in earlier experiments (Scheffel et al. 2012, Adomavicius et al. 2013); in contrast, the authors describe jump bidding and different forms of bundle selection.

In a *CWL* auction, an auctioneer aids the coordination of bidders using adequate information feedback. Although this coordination avoids unnecessary auction rounds, a losing coalition needs to outbid the winning coalition such that the process still leads to a competitive equilibrium. First, let us provide an illustrative example on how the information feedback in a *CWL* auction can help reduce the number of auction rounds.

Example 2. Consider the sale of 18 pieces of land ($A - R$) on a shore line. One developer (bidder 1) needs three adjacent pieces of land for a small hotel, while the other developer (bidder 2) plans for a large resort and needs 15 adjacent pieces. Both compete against bidder 3, who is interested in all 18 pieces of land. Let's assume that, in the first round, bidder 1 submits XOR bids of €3 on bundles $A - C, D - F, G - I, J - L, M - O$ and $P - R$ for which he has the same preference, while bidder 2 bids on $A - O$ for €9, and bidder 3 on $A - R$ for €20. The *CWL* for bidder 1's bid on $P - R$ will be €7, while the *CWL* for bidder 2's bid will be €13 (using uniform distribution sharing rule for simplicity). In contrast, the *CWL* for all other bundle bids for bidder 1 will be €20. Therefore, the *CWL* information can serve as a signal that, right now, bidder 1 can focus on bundle $P - R$. In contrast, with only *DL*s available, bidder 1 could not see the difference of $P - R$ to his other five bundles of interest and might bid on other bundles (i.e., all DLs would be €3), which cannot become winning given the valuations. He could be trying to bid on these other packages over multiple rounds and increase the ask prices, but the allocation would not change. Furthermore, if *WL* information is available, the *WL* for $P - R$ on the other hand would be €11, which indicates the entire cost that is needed to outbid bidder 3 without taking into account possible coalitions.

In the absence of reliable prior distributional information about the valuations of competitors in an online market, a rational bidder will not submit a last-and-final bid below the *CWL*, because he does not have sufficient information to decide whether the bid is just high enough such that he becomes winning in expectation. As outlined in the introduction, straightforward bidders are also able to coordinate in a DL auction, and a competitive equilibrium arises. However, such a competitive equilibrium comes at the cost of a huge number of auction rounds. In DL auctions with straightforward bidders, all package values of losing bidders need to be revealed, and this might be an unrealistic assumption given the exponential number of packages bidders can bid on. In contrast, CWL auctions can lead to high efficiency in markets with many more objects, because bidders are able to coordinate more effectively.

In the next section, we provide experimental results which demonstrate that bidders accept *CWL*s and, as a consequence, the number of auction rounds is reduced substantially, while efficiency is significantly higher than in a *DL* auction in the lab.

4. Experimental Design

Several ascending combinatorial auction formats have been analyzed in the past (see discussion in Section 2). An experimental comparison with all of these formats would be beyond the scope of a single paper. Because BACS auctions satisfy a strong solution concept (at least for a restricted set of bidder preferences) and, in particular, because the *DL* auction is a BACS auction that has been the focus of much recent research in IS, the *DL* auction represents a natural candidate to compare against. In addition, the proposed *CWL* auction is an extension of the literature on *DL* auctions (Adomavicius and Gupta 2005, Petrakis et al. 2013, Adomavicius et al. 2013).

In what follows, we will introduce three different bidder value models for which we compare the *DL* and *CWL* auction formats. Another set of value models and the respective simulations are described in Appendix A. They yield the same results and, due to space constraints, in the main paper we only included those value models for which we conducted both computational simulation and lab experiments with human participants.

4.1. Value Models

We use three different value models (VMs) in our experiments. These are the Threshold (**Thr**) VM, the **Mix** VM, and the Symmetry (**Sym**) VM. The Sym VM is based on an earlier work by Adomavicius et al. (2013) in their experimental studies on *DL* and *WL*. We added the **Thr** VM and the **Mix** VM in order to understand if the results carry over to other environments. The **Thr** VM models a threshold problem with a single global bidder and several local bidders only interested in small packages. Such environments have received much attention, as they could lead to free-riding behavior as outlined in the introduction.

In addition to the value models described in this section, in Appendix A we provide numerical simulations with three additional value models, two with 18 and one with 9 items, which resemble the ones used in lab experiments by Scheffel et al. (2011) and Goeree and Holt (2010). These value models are modeled after spectrum auction markets with regional licenses and real-estate markets. There is no significant difference in efficiency the simulations, while the efficiency of the *CWL* auctions is significantly higher than that of the *DL* auctions in the lab experiments. In the lab and in simulations the number of bids and the number of auction rounds in *CWL* auctions are substantially reduced.

4.1.1. The Threshold Value Model

In this value model, we consider a market with a single global bidder and two local bidders facing a threshold problem. They compete for 6 items labelled A to F. The global bidder is defined to be single-minded and has interest only in the bundle containing all 6 items. Each of the local bidders is interested in various bundles of smaller sizes, but the experimental subjects did not know the specific bundles that were of interest to other bidders. Bidders also did not have distributional information about the other bidders' valuations before the auction. They only knew that two local bidders were competing against a global bidder. The bidder valuations for the individual bundles are drawn from uniform distributions based on pre-specified intervals, which was not known by the bidders. Table 2 represents the basic bidder preferences and value distributions for all bundles. This model is designed in such a way that the local bidders could potentially overcome the threshold posed by the global bidder if they could coordinate and

Table 2. *Preference structure of the Threshold VM. Global bidder is interested in 1 bundle only, while local bidder 1 and local bidder 2 are both interested in 5 different bundles each*

Items	A	B	C	D	E	F
Global Bidder	[60,65]					
Local Bidder 1	[50,55]					
			[50,55]			
			[40,50]			
				[40,50]		
					[40,50]	
Local Bidder 2	[20,30]					
		[20,30]				
			[20,30]			
				[20,30]		
						[20,30]

form a coalition containing either the bundles *ABCD* from bidder 1 and *EF* from bidder 2, or the bundles *CDEF* from bidder 1 and *AB* from bidder 2. Other combinations of package bids from bidder 1 and bidder 2 did not stand a chance of winning if the global bidder bids up to his true valuation.

4.1.2. The Mix Value Model

Similarly to the Threshold model, the Mix value model was designed to analyze the ability of two local bidders to enter into a successful winning coalition that can outbid the global bidder. Compared to the Threshold value model, the two local bidders have more bundles of interest, making a successful coalition more challenging.

In particular, we again have 6 items labelled *A* to *F*. The global bidder is only interested in the bundle containing all 6 items. The local bidders are interested in all 6 items and all bundles of size up to 4 items. For each local bidder, there exists a "preferred item" (which is chosen randomly) that has a higher value for that bidder than other items. We introduce local complementarities by implementing an additional (i.e., bonus) value of 10% for each adjacent item in a bundle.

The values for the individual bundles are randomly determined from different intervals, depending on the baseline draws for single-item valuations. In particular, for each bidder, we first determined the "preferred item", for which the baseline draw is uniform from the range [90,110]. The adjacent items to this preferred item then had valuations drawn uniformly in the range [40,60]. Their neighboring items next had valuations drawn uniformly in the range [20,30], while the last remaining item, i.e., the item with the greatest distance from the preferred item, had valuation drawn uniformly from [7,17]. Once these single-item valuations were drawn for each bidder, the bundle valuations were computed, and a list of all valuations was provided to the subjects privately. Figure 1 shows an example of baseline draws for a local bidder in this setting. In this example, the bidder has a valuation of $(90 + 45 + 20) * 1.2 = 155 * 1.2 = 186$ for the bundle *BCD*, as this bundle contains 2 adjacent neighbors and, therefore, gets

```
        55
        A
  30         90
   F    B
   
   E    C
  10         45
        D
        20
```

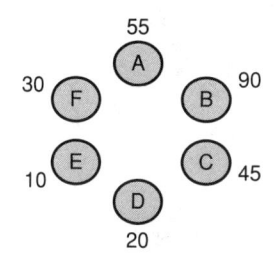

Figure 1. An example for the item valuations for a local bidder in the Mix VM.

20% bonus. The number of bidders and the fact that there are two local and a global bidder were common knowledge among bidders. However, bidders did not know the preferred item for other local bidders.

4.1.3. *The Symmetry Value Model*

This was one of the value models (Setup 1) used by Adomavicius et al. (2013) to test and evaluate the impact of deadness and winning levels as price feedback in the lab. It allows us to compare our result to theirs as it does not include random draws. In the Sym VM, bidders have equal strength in their valuations and there is no threshold problem as in Thr VM and Mix VM.

There are again 6 items labelled A to F, which need to be auctioned among three bidders. A distinct item, designated the "preferred item", is identified for each bidder participating in the auction. This item has the highest value (100 monetary units) for the bidder, with the value of each remaining item decreasing by 50% the further the item is from the preferred item. There are complementarities among items by creating super-additive valuations for bundles with adjoining items in them. This is accomplished by adding 10% to the additive valuation of the items for each adjoining item in the bundle, as in the Mix VM. Adomavicius et al. (2013) motivated this setting with the real-world scenario of real-estate properties around a lake, where local complementarities arise. To preserve the symmetry between bidders, we picked the items A, C and E as preferred items for each of the three bidders respectively, as in Adomavicius et al. (2013).

The identity of the preferred item is private information to each bidder. Bidders were not told how many other participants were in their specific auction, as in Adomavicius et al. (2013). Furthermore, while the rules for generating the valuations of the items were common knowledge, and each bidder in an auction knew the distribution of his own values, participants had no explicit knowledge of the valuations of other bidders as they did not know the respective preferred items of other bidders. Figure 2 shows an example for the private valuations for all bidders. In this example, bidder 1 has a valuation of $(100 + 50) * 1.1 + 25 = 190$ for the bundle ABE, as the bundle contains only 1 adjacent neighbor, and therefore gets 10% bonus.

4.2. Treatments

We have used a fully factorial design with the two treatment variables, the auction format and the value model. Six sessions were conducted for each auction format in

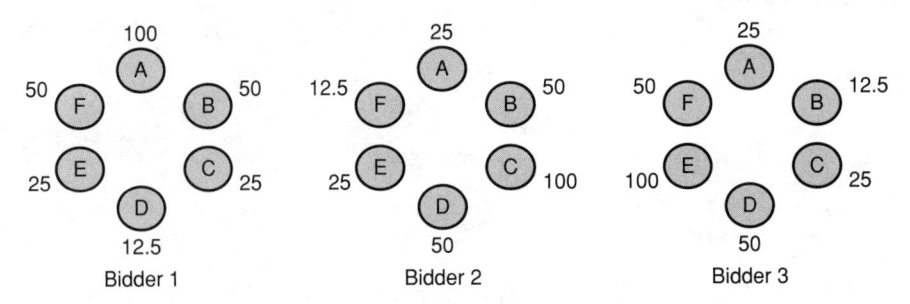

Figure 2. An example for the item valuations for all bidders in the Symmetry VM, where preferred item for bidder 1 is A, for bidder 2 is C, and for bidder 3 is E.

the Threshold, Symmetry, and Mix value model. Every session only used one auction format, and every session consisted of two waves with different value draws. During a session, the two waves were run in parallel with two groups of students. Each wave consisted of three auctions, one for each of the value models (i.e., Thr, Sym, and Mix). These three auctions were conducted sequentially, and in each wave we used a different sequential order of these three auctions to level out learning aspects. We used the same 12 waves for the first 6 sessions testing the *DL* auctions and for the last 6 sessions testing the *CWL* auction to allow for better comparison. Table 3 provides an overview of the sessions and waves in the experiments. In this table we consider the three VMs (Thr, Sym and Mix) together, as the respective auctions are always conducted together in a wave. Overall, there were 12 auctions per treatment combination (72 auctions in 6 treatments) and 72 total participants, each of which having participated in 3 auctions.

The *DL* and *CWL* auction formats have been implemented as described in the previous section. In all experiments we used the same user interface and round-based auction process. We displayed *WL*s in the *DL* auction as additional information feedback (i.e., in addition to the ask prices based on *DL*s) in order to allow for better comparison to Adomavicius et al. (2013). We used a round-based auction process and did not determine the winners after every single submitted bid. Also, we used an XOR bid language throughout, because this bid language is able to express any valuations, i.e., it is fully expressive compared to an OR bid language (Nisan 2006). The bid increment was 15 Francs per item in the Mix and Symmetry VMs, and 3 Francs per item in the Threshold VM. We used Francs as a name for our experimental currency. A bid increment of 3 Francs per item means a bundle increment for a bundle containing three items is

Table 3. Treatments, sessions, and participants for the different test combinations

Treatment	AF	VM	# Sessions	Auctions per Session	# Auctions	# Participants per Auction	# Participants
1	DL	Thr					
2	DL	Sym	6	6	36	3	36
3	DL	Mix					
4	CWL	Thr					
5	CWL	Sym	6	6	36	3	36
6	CWL	Mix					
					72		72

9 Francs. We used a per-item bid increment rather than a bundle bid increment because a per-item bid increment takes bundle sizes into account when raising prices. Therefore, a per-item bid increment can help bidders more effectively to focus on smaller and, thus, more coalition-prone bundles during an auction. Also, as valuations in the Threshold VM are generally smaller than those in the other two VMs, we scaled down the increment appropriately.

4.3. Procedures for Human Subject Experiments

All experiments were conducted from November 2012 to July 2013 with students at a major European university. Each session started with a presentation in which we explained the auction format to be used, the pricing rules, and the different value models in detail. This presentation was also provided to students as a hand-out. Then subjects participated in one training auction to get to know the auction environment, the software, and the user interface. Afterwards, we repeated the main rules, and subjects were asked questions in a quiz to make sure they understood all rules and were familiar with the auction procedure.

The number of auctions was announced in advance. The first auction round was not time-restricted and only ended when every bidder announced they were ready to enter into the second round. In all subsequent rounds, bidders had at most 5 minutes to place their bids. This was perceived to be sufficient by the participants. Bidder roles (e.g., global vs. local) were randomly assigned for each auction in order to alleviate earning differences across different bidders. Earnings were calculated by converting the experimental payoff amounts to EUR by 3:1, i.e., bidders were paid on their economic performance in the auctions. The resulting earnings were between the minimum of €5 (i.e., show-up fee) and the maximum of €50 per subject across all waves. Average earnings were €29, and the average duration of a wave was 1 hour and 19 minutes without the introductory part and without the training auction. This includes the time for breaks between auctions.

5. Simulation Results

In this section we present the results of numerical simulations based on two different bidding strategies: (i) straightforward bidders (s), as described earlier in the paper, and (ii) heuristic bidders (h). Heuristic bidders bid on 5 bundles in each round, where these bundles are randomly chosen from among their 10 best (i.e., payoff-maximizing) bundles at that time. Heuristic bidders model bidder behavior that is based on observations from lab experiments reported in prior literature (Scheffel et al. 2012). It models a "trembling hand", where bidders want to bid on their best bundles by payoff, but they make small mistakes. We have chosen 10 bundles, because in Scheffel et al. (2012) it was shown that bidders typically focus on a small set of bundles, independent of the number of possible bundles in an auction. We have analyzed variations of these bidders, e.g., bidders who bid on their best three packages, but the differences were minor. For our analysis, these artificial bidders serve as a baseline. In the lab experiments in the next section, we analyze whether the differences between *DL* and *CWL* auctions

Table 4. *Average efficiency achieved for all simulated bidder behavior and lab results*

Efficiency	AF	Strategy	Threshold VM	Mix VM	Symmetry VM
Simulation Results	DL	s	100.0%	100%	100.0%
	DL	h	100.0%	99.1%	100.0%
	CWL	s	100.0%	99.8%	100.0%
	CWL	h	100.0%	98.9%	100.0%
Lab Results	DL		96.3%	97.2%	98.3%
	CWL		100.0%	98.1%	100.0%

carry over to the lab, where bidders are heterogeneous and follow different strategies on how they select packages or how they use jump bids.

Throughout, we use **allocative efficiency** E as a primary aggregate measure for comparing different auction mechanisms.[1] In addition, we measure **auctioneer's revenue share** R, which shows how the resulting total surplus is distributed between the auctioneer and the bidders.[2]

Optimal surplus describes the resulting revenue of the winner-determination problem if all valuations of all bidders were available, while actual surplus considers the true valuations for those packages of bidders selected by the auction. In contrast, auctioneer's revenue used in the revenue distribution describes the sum of the bids selected by the auction, not their underlying valuations.

Table 4 shows the efficiency results, averaged over all auction instances used in the lab experiments. In particular, we simulated the two different agent bidding strategies described above for each auction instance which was later used in the lab. The average of these simulated values is then compared to results of the lab experiments using the three value models.

Overall, efficiency was very high in both the simulations and the lab experiments. Only the average efficiency in the Mix VM simulation was lower, as bidders had up to 56 bundles of interest, which makes the coordination problem harder. The Symmetry and the Threshold VM had 100% efficiency throughout, with the single exception of the *DL* auction in which the efficiency was significantly worse in the lab for the Threshold and the Mix VMs, as compared to the simulations. This might be due to the bundle selection and the jump bidding of bidders in the lab, which will be described below.

Table 4 provides an initial comparison of efficiency results between all simulated agent behaviors and real bidder behaviors in the lab. Although there are differences between the simulation and the lab, these differences are minor at an aggregate level. As we will show, at an individual level, bidder behavior with respect to bundle selection and jump bidding exhibit some differences compared to the simulations. Table 5 shows a comparison of all aggregate simulation results including the auctioneer revenue share, the number of rounds, and the number of bids submitted.

The differences in efficiency between *DL* and *CWL* auctions across all simulations were insignificant (t-test, $\alpha = 0.05$). Differences in revenue were mostly significant

---

[1] We measure efficiency as $E = \frac{\text{actual surplus}}{\text{optimal surplus}} \times 100\%$.
[2] We measure measure auction revenue share as $R = \frac{\text{auctioneer's revenue}}{\text{optimal surplus}} \times 100\%$.

Table 5. *Average measures of auction performance: aggregate and for each of the three value models Threshold, Mix and Symmetry*

VM	AF	Strategy	Efficiency	Revenue Share	No. of Rounds	No. of Bids
all	DL	s	100%	97.3%	100%	100%
all	CWL	s	99.9%	98.1%	37.0%	45.3%
all	DL	h	99.6%	96.4%	28.3%	93.6%
all	CWL	h	99.5%	97.8%	16.2%	44.3%
Threshold	DL	s	100%	99.1%	100%	100%
Threshold	CWL	s	100%	99.7%	40.9%	48.4%
Threshold	DL	h	100%	98.5%	32.0%	98.1%
Threshold	CWL	h	100%	99.1%	23.8%	58.5%
Mix	DL	s	100%	97.4%	100%	100%
Mix	CWL	s	99.8%	98.0%	31.8%	42.9%
Mix	DL	h	99.1%	96.5%	22.6%	90.1%
Mix	CWL	h	98.9%	97.8%	8.3%	30.6%
Symmetry	DL	s	100%	74.3%	100%	100%
Symmetry	CWL	s	100%	79.6%	50.8%	35.7%
Symmetry	DL	h	100%	71.3%	51.7%	82.2%
Symmetry	CWL	h	100%	82.6%	19.2%	38.9%

but small. Most importantly, however, (although there are only small differences in efficiency and revenue between the *DL* and *CWL* auctions) the communication between the auctioneer and the bidders in the *CWL* auction is substantially reduced. The average number of rounds is significantly smaller in the *CWL* auction, and so is the number of bids submitted throughout the auction. We normalized the numbers so that the *DL* auction with straightforward bidders describes 100% of the rounds and number of bids. In the next section, using these results as a conjecture, we investigate whether the same pattern of auction outcomes emerges in the lab experiments. Appendix A presents the results of a number of additional numerical simulations on larger value models, which are in line with the results presented in this section.

6. Experimental Results

Next we will present our results of the lab experiments. First, we will look at the results at an aggregate level, concentrating on the comparison of the two auction formats *DL* and *CWL*, using metrics such as allocative efficiency and auctioneer's revenue share as well as the number of bids submitted and rounds required by the bidders. For the pairwise comparisons of aggregate metrics we use the Wilcoxon rank sum test. In the second phase, we will analyze the bidders' bundle selection and jump bidding behavior during the auctions.

Result 1: (High efficiency of the proposed CWL auctions) The allocative efficiency of the *CWL* auction is significantly higher than that of the *DL* auction.

It is interesting to see that the efficiency of the *CWL* auction tends to be even higher than the already high efficiency of the *DL* auction. We have fitted a regression model with efficiency as the dependent variable and control for auction format, session, and

Table 6. *Average aggregate measures of auction performance in all eight combinations of auction formats and value models. Superscript A refers to results from Adomavicius et al. (2013)*

AF	VM	E	R	Avg. no. of all bids	Avg. bundles	Avg. bid improvements	Avg. time (min.)	Avg. no. of rounds
DL	Thr	96.3%	80.8%	30.0	4.9	8.5	14.7	8.3
CWL	Thr	100.0%	80.8%	19.2	4.9	3.1	10.6	5.4
DL	Mix	97.2%	89.8%	114.0	29.3	25.7	37.0	15.3
CWL	Mix	98.1%	88.4%	83.8	29.3	10.9	25.3	9.4
DL	Sym	98.3%	78.4%	166.6	31.4	22.6	45.8	13.6
CWL	Sym	100.0%	63.8%	116.6	29.7	8.8	28.7	10.5
DL[A]	Sym	93.5%	65.3%	n.a.	n.a.	n.a.	n.a.	n.a.

value model. The coefficient for the *DL* auction format is actually significant (*p*-value 0.003) and negative, while the different value models did not have a significant influence on efficiency.

While we did not find signs of fatigue among the subjects in their responses after the auction or in the bid data, an influence of fatigue on the results can always be an issue. Fatigue could help explain lower efficiency in the *DL* auction, but the long auction durations in the *DL* auction are actually also a concern for applications in the field.

The auctioneer's revenue share of the *CWL* auction equals that of the *DL* auction in the Thr VM, but is lower than the *DL* auction for the Mix and Sym VM. In the Sym VM bidders did not have to outbid a large bidder and the preferred items were disjoint, such that they could coordinate faster at lower prices. The aggregate results are presented in Table 6. Figure 3 provides box plots of the efficiency and revenue share.

The auctioneer's revenue share is higher in the numerical simulations compared to those in the lab. This is due to the fact that, in the simulation, straightforward bidders reveal all valuations of all losing bundle bids truthfully, often using last-and-final bids. Also, in the simulation, heuristic bidders in simulation reveal their preferences to a large degree. In the lab, subjects typically want to get a payoff and sometimes drop out of the bidding process for a bundle before they reveal their true valuation.

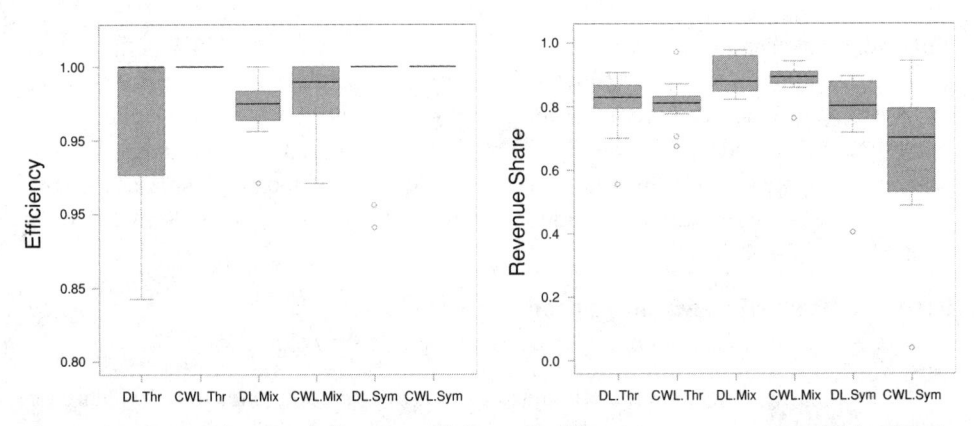

Figure 3. Left: Distribution of allocative efficiency in different treatment groups; Right: Distribution of revenue share in different treatment groups.

Note that the revenue share in the Sym VM is lower than in the other two value models. In the Thr and Mix VMs the smaller bidders need to outbid the global bidder, who drives up prices, which leads to a higher revenue share. In contrast, in the Sym VM, it can happen that bidders coordinate very early leading to low prices and consequently a low revenue for the auctioneer. This is also a reason for the higher variance in revenue in the Sym VM.

We have included the efficiency and revenue results from Setup 1 (Sym VM) in the experiments by Adomavicius et al. (2013). Note that even though we used the same value model, the experiments are not fully comparable because Adomavicius et al. (2013) used a continuous auction and a bid increment of 1 monetary unit, i.e., not a round-based format with a bid increment of 15 monetary units and last-and-final bids. They also used an OR bidding language instead of an XOR bid language, but this should not matter in the experiments as the valuations in the Sym VM are super-additive. We conjecture that the round-based auction process has advantages for convergence, because bidders do not get updates in the prices and allocations continuously. Instead, they receive new prices after each round. As this information does not change until the round is over, this might lead to a more structured way of decision making. Adomavicius et al. (2013) analyzed an online environment where bidders join throughout the auction and such a round-based mechanism might not be possible.

Result 2: (Fast convergence of the proposed CWL auctions) In the *CWL* auction, the number of bids submitted, the time to finish the auction, and the number of rounds was substantially lower than in the *DL* auction.

As in the numerical simulations, significantly less communication was required to achieve high efficiency. We provide the average number of all bids per auction in Table 6. In addition, we provide the average number of bundles that the small bidders bid on (Avg. Bundles), and their average number of bundle bids improving a starting bundle bid throughout the auction (Avg. Bid Improvements). We exclude the big bidder, as he is only interested in one bundle and, thus, only bids on and improves this one bundle. Interestingly, the number of Avg. Bundles is identical or similar in all value models, meaning that in both auction formats small bidders bid on a similar number of new bundles, mostly in the first rounds of the auctions. In contrast, the number of Avg. Bid Improvements was much lower in the *CWL* auction than in the *DL* auction across all three value models, suggesting that the *CWL* format enabled small bidders to improve their bundle bids in a much more structured and focused way. Finally, we report the average auction duration time in minutes and the average number of bidding rounds required – these numbers are significantly higher in the *DL* auction. It is particularly interesting to see that the savings in the average time increase with the complexity of the value models; i.e., the *CWL* format saves 4.1 minutes on average in the Thr VM, 11.7 minutes in the more complex Mix VM, and 17.1 minutes in the Sym VM.

In the simulations, the number of bids in the *CWL* auctions was mostly less than half of those submitted in the *DL* auction. The number of bids was also reduced substantially in the *CWL* auctions in the lab, and *CWL* auctions took 64–74% of the bids in the *DL* auction. Differences from the simulation can be explained by the use of jump bids in

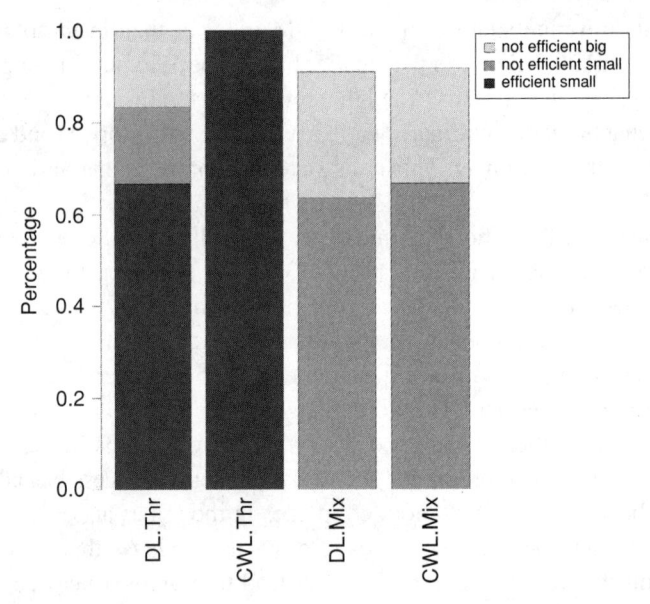

Figure 4. Auctions where the small bidders won in a 100% efficient or an inefficient allocation, or they lost to the large bidder, as percentages of the number of auctions where the small bidders should have won in an efficient allocation.

the *DL* auction and the deviations from straightforward bidding – these characteristics are representative of human bidding behavior. In both the numerical simulations and the lab experiments we used the same bid increments, and last-and-final bids were available to mitigate the potential efficiency losses due to bid increments.

Finally, we look at the threshold problem posed by the big bidder in the value models Thr and Mix.

Result 3: In the Thr VM, small bidders always find the efficient allocation in a *CWL* auction in those auctions where a coalition of small bidders is efficient. In contrast, in the *DL* auction, 40% of these auctions are not efficient, i.e., in 40% of the *DL* auctions where a coalition of small bidders should have won in the efficient allocation, the big bidder won instead. In the Mix VM, there was no significant difference among the auction formats.

In Figure 4 we show the percentage of auctions where an efficient allocation favored a coalition of small bidders. Of those auctions we show the percentage where a coalition of small bidders won with the 100% efficient allocation or with an allocation that was not 100% efficient, and where the threshold bidder ("big" bidder) won instead. There was a large number of possible allocations with the small bidders winning in the Mix VM compared to the Thr VM. This is the reason why the auctions never resulted in the 100% optimal solution in the Mix VM, but only in the solutions that were close to optimal.

We will also report on bundle selection and jump bidding, because both describe bidder strategies and both can influence the efficiency and revenue of an auction. Bundle

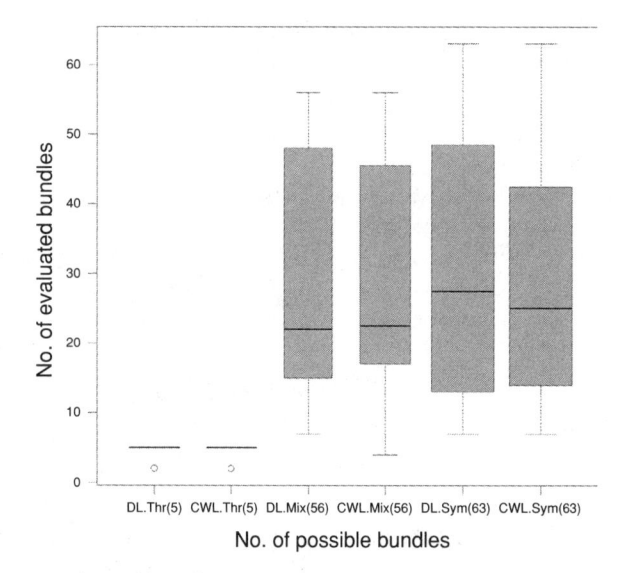

Figure 5. Average number of bundles evaluated by small bidders relative to all possible numbers of bundles for them, indicated by numbers in parentheses.

selection in the lab varied a lot across bidders and was different from straightforward or heuristic bidder strategies used in the simulations.

Result 4: In the Mix and Sym VMs with more than 50 or 60 bundles of interest resp., bidders submitted bids on 20 to 25 packages in the auctions (median). The number of packages varied significantly across bidders. A high payoff and a high valuation of a package relative to other bundles have a positive impact on the likelihood of a bidder selecting relevant bundles in both auction formats.

Figure 5 shows the number of bundles on which bidders submitted bids in all value models. While in the Thr VM there are only five bundles of interest for the small bidders, and they typically bid on all of them, the number of possible bundles of interest for small bidders in the Mix VM is 56, and in the Sym VM it is 63 for all bidders, and bidders typically bid only on a subset of possible bundles.

We also analyzed significant covariates for bundle selection using a logistic regression. For this, we generated a table of all bundles with a positive payoff (i.e., relevant bundles) that a bidder could have bid on in each round. The dependent variable describes whether a given relevant bundle has been selected by the bidder or not (i.e., whether the bidder submitted a bid on a given bundle with positive payoff). The model includes covariates such as the auction format, rank of a bundle by valuation, and rank of a bundle by payoff in a round. We used bidder ID dummy variables to control for fixed effects. We also control for the round of the auction, the number of the auction within an experimental session, and the value model. All covariates were significant. The probability of a bundle being selected decreases with a lower rank by payoff or a lower rank by valuation (Table 7). In the Thr VM bidders have a higher likelihood to bid on a bundle, while in the Sym VM bidders have a lower likelihood, compared to the Mix VM. This is induced by the comparatively low number of bundles of interest (i.e., 5)

Table 7. *Logistic regression of the bidder's likelihood to bid on a bundle*

	Estimate	Std. Error	z value	Pr(>|z|)
(Intercept)	−0.011	0.151	−0.075	0.940
DL auction	−0.747	0.197	−3.786	0.000
Rank by value	−0.018	0.001	−17.344	0.000
Rank by payoff	−0.010	0.001	−7.396	0.000
Auction round	−0.219	0.005	−40.596	0.000
Auction no. in session	−0.073	0.027	−2.704	0.007
Sym VM	−0.231	0.045	−5.152	0.000
Thr VM	1.055	0.077	13.664	0.000
Bidder ID	...	...	...	...
Null deviance:	35533	on 68386	deg. of freedom	
Residual deviance:	29451	on 68303	deg. of freedom	
AIC:	29619			

in the Thr VM and vice versa in the Sym VM. Most importantly, there are also significant bidder-specific idiosyncrasies, as the dummy variables for Bidder ID revealed. Overall, payoff influenced the bid selection, but the analysis indicates that pure straightforward bidding cannot fully explain the bidding behavior, and bidder idiosyncrasies matter. This is in line with earlier experimental work on bidder idiosyncrasies in the bundle selection in BACS auctions and the various factors influencing this selection (Scheffel et al. 2012).

Result 5: Bidders in the *CWL* auction explore more bundles in the initial rounds of the auctions, but they submit fewer bids in later rounds.

Table 6 includes statistics on the average number of bundles that small bidders bid on (Avg. bundles) and the number of bundle bids improving a starting bundle bid throughout the auction (Avg. bid improvements). Figure 6 shows the average submissions of new bids and improved bids for all three value models over time. This figure shows a consistent progress for bid submissions both for the new bundles and the improved bids on the previous bundles. In all value models, new bids are mostly submitted during the first 10–20% of the auction process, with bidders generally submitting new bids for more rounds in the *CWL* auction compared to the *DL* auction. Much of the coordination takes place in these initial rounds. In contrast, bidders in the *CWL* auction need to submit fewer bids on bundles which they have already bid on (i.e., improved bids "Imp") and bidding is more focused.

Bidders were allowed to submit bids that are higher than the ask price in both auctions. Such jump bids represent a possibility to signal high-valued packages, which can become part of a coalition of winning bidders. We wanted to understand how bidders use jump bidding (because high jump bids could impact revenue and also efficiency) and whether we can find differences between the two auction formats.

Result 6: Bidders submitted most jump bids in the first half of the auction rounds. There was no significant difference in the proportion of jumps in both auction formats across the value models.

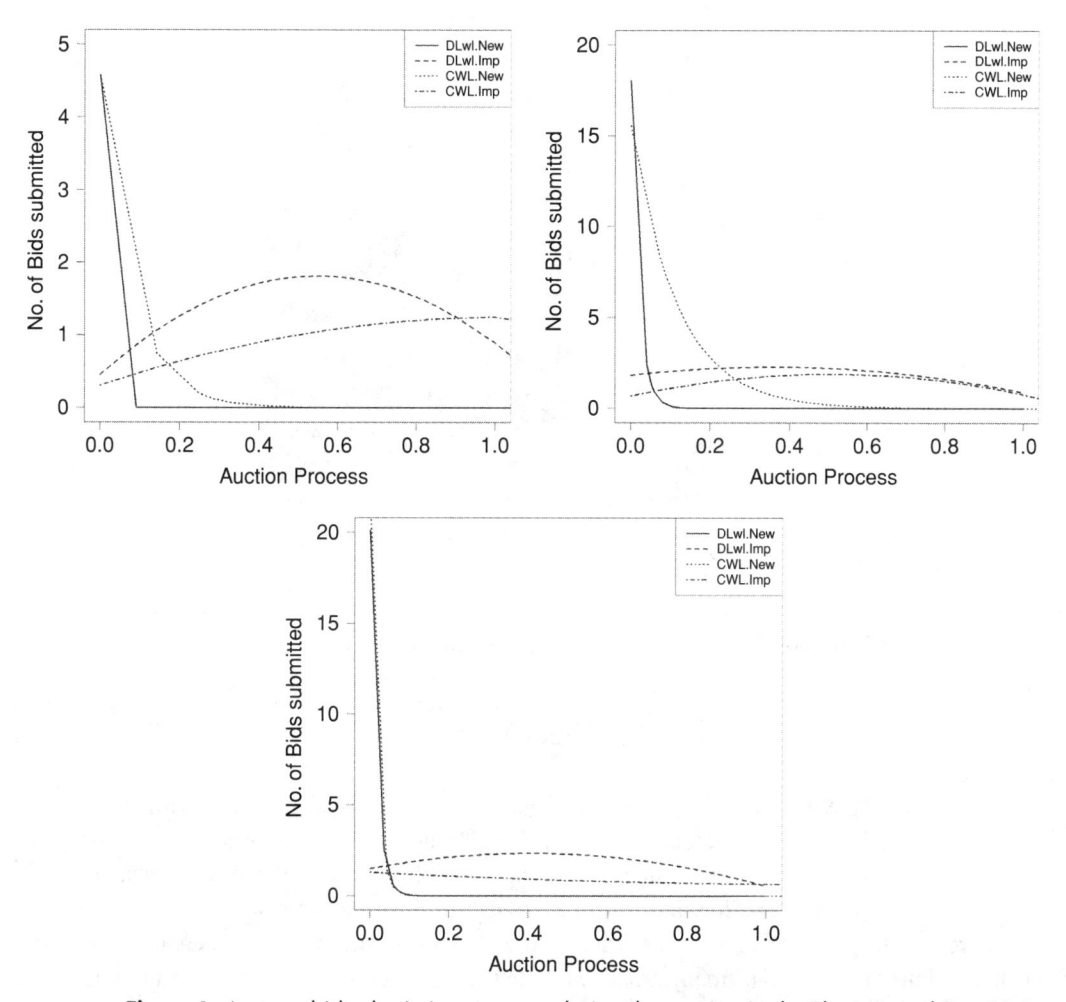

Figure 6. Average bid submission progress during the auction in the Thr, Mix and Sym VM.

The bar chart in Figure 7 provides an overview regarding what proportion of jump bids was submitted in the first, second, third, and fourth quarter of the auctions on average. It shows that in both auction formats the majority of jump bids is submitted in the first quarter of the auction. We conjecture that bidders try to signal packages of high value in the early rounds and find bidders interested in complementary packages this way. They submit fewer jump bids towards the end, where bidders only focus on a few bundles that they are trying to win.

Overall, there was heterogeneity in the number of package bids a bidder selected in each round and in the number of jump bids. The high efficiency in the auctions can be seen as an indicator for robustness against these differences in bidding behavior.

We also ran some lab experiments with instances of a large value model with 18 items and found qualitatively similar results. Efficiency was higher than 99% for the *DL* and *CWL* auctions, but the number of bids and rounds in the latter were substantially reduced.

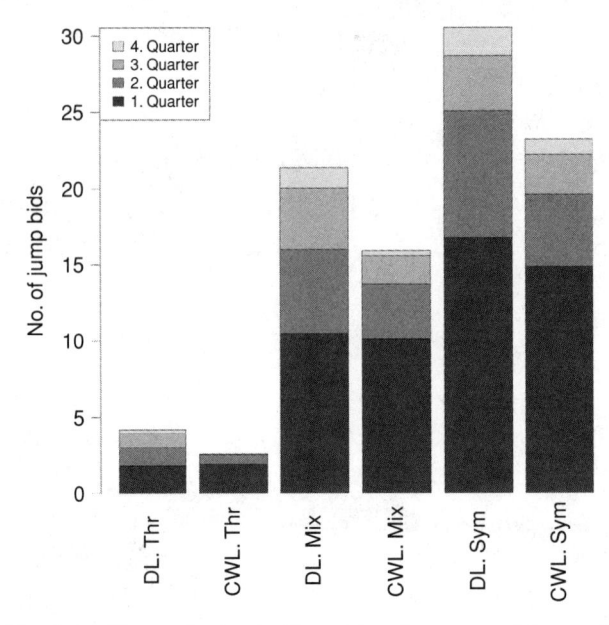

Figure 7. Number of jumps in the the first to fourth quarter of the auction rounds.

7. Conclusions

The design of efficient multi-object auctions is a fundamental problem with many applications in e-sourcing and other domains. Ascending combinatorial auctions do not require bidders to reveal valuations on exponentially many packages as in sealed-bid auctions, but rather allow bidders to discover winning packages iteratively, which is preferred to sealed-bid auctions (such as the Vickrey-Clarke-Groves mechanism) in many online markets. The main strategic challenge in ascending combinatorial auctions is coordination: how can bidders find the right packages from a large set, which together with bids of other bidders could become a winning coalition? This type of coordination among losing bidders has largely been ignored, but it is arguably a pivotal problem for bidders in larger online combinatorial auctions.

Our approach leverages information about losing coalitions, which can be collected by the auctioneer throughout the auction. Coalitional winning levels are provided in each auction round to help losing bidders coordinate implicitly and outbid a coalition of standing winners. They provide an implicit proposal on how much to bid in order to become winning jointly. The results of numerical simulations and corresponding lab experiments with realistic value models indicate substantial savings in the number of auction rounds and bids, and even higher efficiency in the lab. This type of information feedback helps bidders coordinate with much less communication, which makes combinatorial auctions a viable mechanism in many more practical applications.

We would like to note that the proposed auction mechanism has been designed with online markets in mind, i.e., markets with little or no publicly available distributional information about other bidders valuations. The empirical performance of this auction format in numerous experiments with real bidders provides evidence that the proposed coalition-based pricing mechanism indeed facilitates the intended performance improvements, e.g., substantially accelerated convergence with high efficiency. However, it has been well-documented that strategic bidder behavior does occur in numerous settings (free-riding, jump bids, sniping, etc.), especially where some information about valuations of other bidders is available or can be learned through repeated interactions. For example, if bidders had precise information about the preferences of their competitors, then bidders often have possibilities to manipulate strategically. However, in a combinatorial auction with exponentially many packages and many bidders, this would presume the availability of a lot of information. In summary, it is important for an auctioneer to understand the information available to bidders in a market before deciding on a specific auction format.

Appendix A. Simulation Studies with Three Additional Complex Value Models

In this appendix, we provide the results of numerical simulations with three additional value models which were different to those in the main part of the paper. Two value models have 18 items, one has 9 items. The two global-synergy value models (with 18 and 9 items respectively) are based on an earlier work by Goeree and Holt (2010) and use global complementarities (GSVM). The last value model uses local complementarities between 18 items (LSVM), modeled after a real-estate auction. We simulated 64 instances of the GSVM and 160 of the LSVM.

A.1. The 18-Item Global-Synergy Value Model (18GSVM)

The global-SVM is based on the experiments in Goeree and Holt (2010) and involves seven bidders and 18 items. Figure 8 represents the bidders' preferences. The six regional bidders (labeled 1 through 6) are each interested in four adjacent items of the national circle (consisting of items A through L) and two items of the regional circle (consisting of items M through R) while for the national bidder (labeled 7) the twelve items of the national circle are relevant. This information was common knowledge,

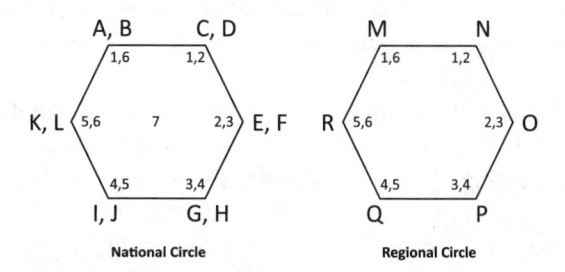

Figure 8. Competition structure of the Global-SVM. Regional bidders 1–6 are interested in four items from the national circle and two items from the regional circle. National bidder 7 is interested in all twelve items from the national circle (Goeree and Holt 2010).

but it was not known which bidders were interested in a particular item. For example, experimental subjects did not know that, besides bidder 7, also 3 and 4 were interested in item H.

The values for the individual items are randomly determined. For the national bidder the baseline draw distributions are uniform from the range [0, 10] for items A-D and I-L, and uniform from the range [0, 20] for items E-H. For regional bidders, the baseline draw distributions are uniform from the range [0, 20] for items A-D, I-L, and M-R, and uniform from the range [0, 40] for items E-H. These value distributions (not the actual draws) were common knowledge among the experimental subjects. For comparison, we use the same draws as in Goeree and Holt (2010), which the authors have kindly provided. For both bidder types the value of items in a package increases by 20% (with two items), 40% (with three items), 60% (with four items), etc. and by 220% for the package containing twelve items. For the computation of the complementarities the identity of the items does not matter. For example, a bidder's valuation of a package of items increases by the same percentage independent of the adjacency of the items; this means, this is a value model in which global complementarities apply.

Activity and purchase limits are such that regional bidders can bid on and acquire at most four items in a single round, while the national bidder is able to acquire all his twelve items of interest. The number of $2^{18} - 1 = 262,143$ possible packages for the national bidder did not allow for a pure straightforward strategy in each round of the simulation. Therefore, the bidders were restricted with respect to the packages they could select. From the national circle they only included packages containing both items on a vertex. This means for example that a bidder only bids on packages including, say, both *A* and *B*, but not packages with *A* alone, as both *A* and *B* are on the same vertex of the national circle. As a result, even straightforward bidders in the *DL* auction could not achieve full efficiency. In order to be able to simulate a straightforward bidder, we designed a version of this value model with 9 items, wich allows us to analyze all possible packages in each round in the 9-item global-synergy value model described next.

A.2. The 9-Item Global-Synergy Value Model (9GSVM)

Figure 9 represents the bidders' preferences. The three regional bidders (labeled 1 through 3) are each interested in four adjacent items of the national circle (consisting of items A through F) and one item of the regional circle (consisting of items G through I) while for the national bidder (labeled 4) all six items of the national circle are relevant.

The values for the individual items are randomly determined. For the national bidder the baseline draw distributions are uniform from the range [0, 15] for items A,B,E,F and uniform from the range [0, 30] for items C,D. For regional bidders, the baseline draw distributions are uniform from the range [0, 20] for items A,B,E,F and G,I,H and uniform from the range [0, 40] for items C,D. For both bidder types the value of items in a package increases by 20% with each additional item. For the calculation of the complementarities the identity of the items does not matter, i.e., this is a value model in which global complementarities apply. Activity and purchase limits are such that

A, B
/1, 3\
  4
/2, 3   1, 2\
E, F   C, D
National Circle

G
/1\
/3   2\
I   H
Regional Circle

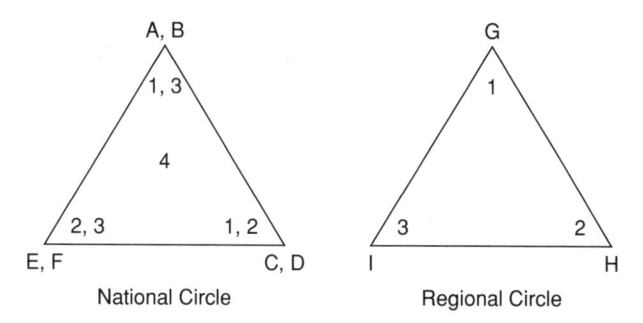

Figure 9. Competition structure of the Global-SVM. Regional bidders 1–3 are interested in four items from the national circle and one item from the regional circle. National bidder 4 is interested in all six items from the national circle.

regional bidders can bid on and acquire at most three items in a single round, while the national bidder is able to acquire all his six items of interest.

A.3. The 18-Item Local-Synergy Value Model (18LSVM)

The local-synergy value model consists of 18 items arranged quadratically, and considers the scenario in which complementarities are gained from spatial proximity. In this value model items are placed on a quadratic map. The arrangement of items matters for the calculation of the complementarities, which only arise if the items are neighboring.

This model also contains two different bidder types: one national bidder, interested in all bundles consisting of at least 7 items, and three regional bidders. Each regional bidder is interested in a randomly determined preferred item and all horizontal and vertical adjacent items. This means that a regional bidder is interested in three to five items. Examples are shown in Table 8, in which the preferred item of a regional bidder is Q or K, and all gray shaded items in the proximity of the preferred item have a positive valuation. For each bidder i we draw the valuation $v_i(k)$ for each item k in the proximity of the preferred item from a uniform distribution. Item valuations for the national bidder are from the range [3, 9] and for regional bidders from the range [3, 20].

We assume that bidders experience only low complementarities on small packages, but complementarities increase heavily with the number of adjacent items. We further assume that adding items to already large packages do not increase complementarities anymore. The rationale for these assumptions is the lack of economies of scale with

Table 8. *Local-SVM with the preferred items Q and K of two regional bidders. All their positive valued items are shaded*

A	B	C	D	E	F
G	H	I	J	K	L
M	N	O	P	Q*	R

A	B	C	D	E	F
G	H	I	J	K*	L
M	N	O	P	Q	R

Table 9. *Average measures of auction performance*

VM	AF	Strategy	Efficiency	Revenue	Rounds	Number of Bids
9GSVM	DL	s_{DL}	100%	77.5%	100%	100%
9GSVM	CWL	s_{CWL}	100%	68.2%	81.0%	14.7%
9GSVM	DL	h_{DL}	100%	61.2%	178.8%	77.2%
9GSVM	CWL	h_{CWL}	100%	64.7%	33.3%	18.6%
18GSVM	DL	s_{DL}	97.87%	84.28%	100%	100%
18GSVM	CWL	s_{CWL}	95.66%	76.71%	40.33%	39.02%
18GSVM	DL	h_{DL}	98.00%	72.16%	36.37%	78.22%
18GSVM	CWL	h_{CWL}	96.95%	76.64%	14.72%	33.99%
18LSVM	DL	s_{DL}	100%	83.9%	100%	100%
18LSVM	CWL	s_{CWL}	100%	83.5%	51.5%	25.7%
18LSVM	DL	h_{DL}	100%	82.7%	108.3%	100.6%
18LSVM	CWL	h_{CWL}	100%	84.7%	35.6%	36.5%

small packages and a saturation of this effect with larger packages. Therefore, complementarities are modeled based on a logistic function, which assigns a higher value to larger packages than to smaller ones. Valuations used in the simulations are available on request.

A.4. Efficiency and Revenue

As in the main text of our paper, we analyze straightforward and heuristic bidding stragegies: two types of straightforward bidders (s_{DL} and s_{CWL}) and two types of heuristic bidders, who randomly select 5 from their best 10 bundles (h_{DL} and h_{CWL}). Table 9 summarizes the results of the simulations. The first observation is that with last-and-final bids all auctions with the value models 9GSVM and 18LSVM were fully efficient. This is not surprising for s_{DL}, where there are proofs for full efficiency. It is, however, interesting to see that even those auctions where the *CWL*s were used as ask prices yielded full efficiency. Differences in revenue were significant (t-test, $\alpha = 0.05$) but small.

As indicated earlier, in the 18GSVM value model the number of packages explored in each round was reduced to keep the package selection tractable, which led to an efficiency loss in all auctions. However, the differences in efficiency and revenue are hard to interpret due to the resulting randomness in the package selection. Still, the number of auction rounds and the number of bids were reduced substantially with *CWL* auctions as compared to *DL* auctions.

Acknowledgments

The financial support from the Deutsche Forschungsgemeinschaft (DFG) (BI 1057/1-4) is gratefully acknowledged.

References

Adomavicius, G., S. Curley, A. Gupta, P. Sanyal. 2013. Impact of information feedback in continuous combinatorial auctions: An experimental study of economic performance. *MIS Quarterly* **37** 55–76.

Adomavicius, G., A. Gupta. 2005. Toward comprehensive real-time bidder support in iterative combinatorial auctions. *Information Systems Research (ISR)* **16** 169–185.

Ausubel, L., P. Milgrom. 2006a. Ascending proxy auctions. P. Cramton, Y. Shoham, R. Steinberg, eds., *Combinatorial Auctions*, chap. 3. MIT Press, Cambridge, MA, 79–98.

Ausubel, L., P. Milgrom. 2006b. The lovely but lonely Vickrey auction. Peter Cramton, Yoav. Shoham, Richard Steinberg, eds., *Combinatorial Auctions*, chap. 1. MIT Press, Cambridge, MA, 17–40.

Bichler, M., A. Gupta, W. Ketter. 2010. Designing smart markets. *Information Systems Research* **21**(4) 688–699.

Bichler, M., P. Shabalin, A. Pikovsky. 2009. A computational analysis of linear-price iterative combinatorial auctions. *Information Systems Research* **20**(1) 33–59.

Bichler, M., P. Shabalin, J. Wolf. 2013. Do core-selecting combinatorial clock auctions always lead to high efficiency? An experimental analysis of spectrum auction designs. *Experimental Economics* 1–35.

Bikhchandani, S., J. M. Ostroy. 2002. The package assignment model. *Journal of Economic Theory* **107**(2) 377–406.

Cramton, Peter C, Yoav Shoham, Richard Steinberg, et al. 2006. *Combinatorial Auctions*, vol. 475. MIT Press Cambridge.

Day, R., P. Milgrom. 2008. Core-selecting package auctions. *International Journal of Game Theory* **36** 393–407.

de Vries, Sven, James Schummer, Rakesh V Vohra. 2007. On ascending Vickrey auctions for heterogeneous objects. *Journal of Economic Theory* **132**(1) 95–118.

Dehez, P. 2007. Fair division of fixed costs defines the shapley value. Tech. rep., University of Louvain, CORE. URL http://dx.doi.org/10.2139/ssrn.970910.

Goeree, J., C. Holt. 2010. Hierarchical package bidding: A paper & pencil combinatorial auction. *Games and Economic Behavior* **70**(1) 146–169. doi:10.1016/j.geb.2008.02.013.

Goeree, J., Y. Lien. 2014. On the impossibility of core-selecting auctions. *Theoretical Economics*.

Guler, K., J. Petrakis, M. Bichler. 2013. Can risk aversion mitigate inefficiencies in core-selecting combinatorial auctions. Working Paper 2263667, SSRN. URL http://papers.ssrn.com/sol3/papers.cfm?abstract_id=2263667.

Krishna, V., ed. 2002. *Auction Theory*. Elsevier Science, San Diego, CA, USA.

Kwasnica, T., J. O. Ledyard, D. Porter, C. DeMartini. 2005. A new and improved design for multi-objective iterative auctions. *Management Science* **51**(3) 419–434.

Levin, J., A. Skrzypacz. 2016. Properties of the combinatorial clock auction. *American Economic Review* **106**(9) 2528–2551.

Leyton-Brown, K., E. Nudelman, Y. Shoham. 2006. Empirical hardness models for combinatorial auctions. P. Cramton, Y. Shoham, R. Steinberg, eds., *Combinatorial Auctions*. MIT Press, Cambridge, MA.

Nisan, N. 2006. Bidding languages. P. Cramton, Y. Shoham, R. Steinberg, eds., *Combinatorial Auctions*. MIT Press, Cambridge, MA.

Nisan, N., I. Segal. 2006. The communcation requirements of efficient allocations and supporting prices. *Journal of Economic Theory* **129** 192–224.

Parkes, D. 2006. Iterative combinatorial auctions. P. Cramton, Y. Shoham, R. Steinberg, eds., *Combinatorial Auctions*. MIT Press, Cambridge, MA.

Parkes, D. 2007. *Online Mechanisms*. 411–439.

Parkes, David C, Lyle H Ungar. 2000. Iterative combinatorial auctions: Theory and practice. *AAAI/IAAI*. 74–81.

Petrakis, I., G. Ziegler, M. Bichler. 2013. Ascending combinatorial auctions with allocation constraints: On game-theoretical and computational properties of generic pricing rules. *Information Systems Research (ISR)* **24** 768–786.

Rothkopf, M. H., A. Pekec, R. M. Harstad. 1998. Computationally manageable combinatorial auctions. *Management Science* **44** 1131–1147.

Sano, R. 2012. Non-bidding equilibrium in an ascending core-selecting auction. *Games and Economic Behavior* **74** 637–650.

Scheffel, T., A. Pikovsky, M. Bichler, K. Guler. 2011. An experimental comparison of linear and non-linear price combinatorial auctions. *Information Systems Research* **22** 346–368.

Scheffel, T., A. Ziegler, M. Bichler. 2012. On the impact of package selection in combinatorial auctions: An experimental study in the context of spectrum auction design. *Experimental Economics* **15** 667–692.

Schneider, S., P. Shabalin, M. Bichler. 2010. On the robustness of non-linear personalized price combinatorial auctions. *European Journal on Operational Research* **206** 248–259.

Shoham, Y., K. Leyton-Brown. 2009. *Multiagent systems: Algorithmic, Game-Theoretic, and Logical Foundations*. Cambridge University Press.

Wilson, R. 1987. Game-theoretic analyses of trading processes. *Advances in Economic Theory: Fifth Word Congress*. Cambridge University Press, 33–70.

PART IV
Experimental Comparisons of Auction Designs

CHAPTER 25

Experiments Testing Multiobject Allocation Mechanisms

John O. Ledyard, David P. Porter, and Antonio Rangel

1. Introduction

During the discussion and evaluation of proposals for the design of the Federal Communications Commission (FCC) mechanism to sell the spectrum, over 130 auctions were run under controlled conditions at Caltech for the National Telecommunications and Information Administration (NTIA), the FCC, and others.[1] While these data were used in those debates, we do not intend to relive that process here. Instead, in this paper, we reexamine these data and try to extract some useful information for those who may, in the future, be involved in the difficult task of creating mechanisms to auction multiple items.

The two major design questions we can say something about are (1) should the items be auctioned off sequentially or simultaneously? and (2) should package bidding be allowed? Our main conclusion is that, over a very wide range of environments, package bidding mechanisms (weakly) dominate simultaneous mechanisms, which in turn (weakly) dominate sequential mechanisms. This conclusion is based on three observations derived from a close look at the data.

First, in environments with multiple items to be allocated, if those items are homogeneous and substitutes, then little coordination between buyers is needed and the only role of the mechanism is to sort bidders with high values from bidders with low values. Both the sequential and simultaneous mechanisms seem to work very well at finding efficient allocations in these "easy" environments.

Second, in environments with multiple items to be allocated, if those items are heterogeneous, then some coordination among bidders is necessary to achieve high-value allocations even if there are only low synergy values. Simultaneous auctions provide a first step at this coordination that sequential auctions might have difficulty in providing.

---

[1] Some of the trials and the data generated are described in a report to the FCC. See Ledyard et al. (1994). For a discussion of the role of experimentation in the FCC design process, see Plott (1996). We would like to thank Robin Hanson for his design of the spatial environments.

Third, in environments with heterogeneous goods exhibiting complementarities, significant coordination is required for an auction or allocation mechanism to perform well with respect to efficiency or revenue. Sequential auctions perform poorly. Simultaneity is clearly necessary but not sufficient to attain high efficiencies. The simultaneous, one-price-per-item auction tends to produce outcomes that are either high in efficiency, revenue, *and* losses or low in efficiency, revenue, and losses. Package bidding seems to help a lot in systematically attaining high efficiency, high revenue, and no losses.

In the rest of this paper we explain and detail the collection of experiments. In Section 2, we provide information on the environments covered, the mechanisms tested, and the performance measures used. In Section 3, we describe the data and our findings. In Section 4, we address some of the questions and issues left unanswered by these data.

Because we expect that many readers may not have a background in the methodology of applied mechanism design, we have included a brief introduction in an appendix. Policy makers and theorists interested in applying the results in this paper should read Section A.1 carefully.

2. What Did We Do?

During the actual FCC design process, a wide range of questions were continuously thrown at the experimentalists who were trying to provide insights and data, as fast as possible, about situations for which theory had virtually nothing to say. Whenever the experimentalists found a problem with a current manifestation of the proposed designs, new proposed solutions were immediately put forward. No careful theoretical analysis or experimental design was followed, nor could one be, given the urgency of the situation. Nevertheless, we think that the experiments that were done can be organized in a reasonably coherent fashion and that, while they do not cover the entire territory one may wish they had, some fairly straightforward conclusions can be drawn for future designs.

2.1. Questions and Method

There were two major design questions with respect to the auction rules about which the data we have reveal some information:

1. Should the items be auctioned off sequentially or simultaneously?[2]
2. Should package bidding be allowed?

There were other issues that achieved some relative importance at various times during the design process but for which there still is neither any convincing theory nor

---

[2] A hybrid design was also considered, which involved comparing the results of a simultaneous sealed bid of all items and a sequential open outcry auction of each item. See Plott (1997) for a description of the process and data on its comparative performance. Milgrom (1995) also has a description of this proposal. We do not cover that design here.

enough experimental evidence on which to base a judgment. Should there be a withdrawal rule or not, and if so, in what form?[3] What should be the appropriate stopping rule? Should activity rules be required, and if so, what should they be? How many waivers should be allowed? While there are some data that might provide light, we feel that more experiments and theory are needed before anything conclusive can be said, and so we will not address these secondary questions in this paper.

Experimental methods in economics provide a type of "wind tunnel" within which to test mechanism designs. These tests can be a valuable source of scientific information that one can use to determine the likely performance of new mechanisms in new environments. The process is simple and very similar to the testing of airfoils in wind tunnels or the testing of hull shapes in towing tanks. One first simulates the environment, in our case by inducing the constellation of participants' valuations and the information they each have about these valuations. Then a mechanism is provided and allowed to operate within the testbed environment. Performance is measured. With enough variation in the environments and enough variation in the mechanisms, one can begin to reach some conclusions about details in design that affect performance. Hunches and arguments loosely based on inappropriate theory can be replaced by facts.

Testbed experiments can be a valuable source of data about the performance of newly designed mechanisms for which there are no examples in operation. As an illustration, see Ledyard et al. (1994b) for the research that led to the Cassini trading mechanism—a bulletin-board trading system now in use as a project management device in the design and construction of the Cassini spacecraft for a mission to Saturn.[4] It may well have been the very first active mechanism for trading worldwide over the Internet. For other illustrations, see Plott (1994). As with any evidence, including theory, testbed data must be weighed carefully. But if used intelligently they can eliminate bad designs, provide comparative performance data, and actually help a decisionmaker come to good conclusions during the design process.

2.2. The Economic Environments We Used[5]

All of the *economic* environments reported in this paper are derived from the following generic setup. A set of n objects, labeled $x_1, \ldots, x_n$, are to be allocated to m agents. Agent i's profit function is $V_i(x_{1i}, \ldots, x_{ni})$ where $x_{ki} = 1$ if and only if agent i is awarded item k. Thus, an agent knows what they will be paid if they successfully acquire any particular subset of the items. In some cases below, agents will be assumed to have common knowledge about aspects of others' values. In other cases agents will know

---

[3] There were data and theory on one proposed rule, to allow withdrawal at any time for free. These suggested that such a rule would destabilize the auction and produce low efficiencies in the allocation and low revenue. [See, for example, Banks et al. (1989) and Milgrom (1995).] The rule was eventually eliminated from further consideration. Porter (1996) provides an experimental analysis of the withdrawal rule currently used in the FCC auction and finds that there is a positive relationship between individual losses and allocative efficiency when the rule is imposed.

[4] Another mechanism that is successfully running in practice and that was developed with the aid of laboratory testbeds is the ACE market. ACE is now operating in Los Angeles at least four times per year as a call market for trading emissions credits—a very complex process. Those interested in the details can go to the WorldWideWeb page at www.ace.mkt.com, which includes a link to download the client software.

[5] More details can be found in the Appendix, Section A.3.

nothing *a priori* except their own valuations. In creating this generic setup, we realized we were abstracting from correlated and asymmetric information. We did so, not because we thought such information was unimportant, but because we wanted to concentrate the limited resources we had on the performance of the proposed auctions when problems such as the winner's curse were absent. In the presence of those problems some fundamental performance properties might have been either obscured or exacerbated. This abstraction is, of course, a defect of the existing research that can and should be corrected in future work.

In this class of environments the most efficient allocations solve the problem

$$\max_x \sum_i V_i(x_{1i}, \ldots, x_{ni})$$

subject to

$$x_{ji} = 0 \text{ or } 1,$$

$$\sum_j x_{ji} = 1 \quad \text{for each } i.$$

For purposes of reporting the test results, we split the environments we used into three somewhat arbitrary classes. The first, which we refer to as *easy*, involve constellations of values for which no reasonable mechanism should have any problems achieving efficient allocations. If a proposed mechanism had failed to perform well in these situations, one would have been fairly sure that it would also fail to perform in more complicated environments. One can think of these as minimal competency tests. These environments have at least two features that make the allocation problem easy for a mechanism: there are no significant coordination issues that require the mechanism to fit complementary demands together, and there is a competitive equilibrium (CE) price vector with one price per item that is sufficient to support the 100%-efficient allocation.

2.2.1. Easy Environments

The first easy class of environments used is one in which values are additive, i.e., the value function is of the form $V_i(x_{1i}, \ldots, x_{ni}) = \sum_j V_i(x_{ji})$. In the experiments, there were six items for sale to six demanders. Each demander knew that his value and those of other participants were to be drawn uniformly, with replacement, from a fixed list of ten value sheets.[6]

A second easy class of environments used had items that were homogenous and had decreasing marginal values, i.e., $V_i(x_{1i}, \ldots, x_{ni}) = V_i(\sum_j x_{ji})$ with $V_i'(\sum_j x_{ji}) < 0$. In the experiments, there were eight participants and ten units to be allocated. Each participant had decreasing demands for up to four units. In addition, subjects knew the number of units being auctioned, the number of participants, and who bid on what items; however, they did not know the distribution over which values were drawn.

The final easy class of environments we used was one in which items were homogenous but had increasing marginal values (superadditive values), i.e. $V_i'(\sum_j x_{ji}) > 0$. We

---

[6] More details can be found in the Appendix, Section A.3.

also constructed the value functions so that a competitive equilibrium price existed. In the experiment, there were eight participants and ten units to be allocated. Each participant had increasing demands for up to as many as four units. Subjects knew the number of units being auctioned, the number of participants, and who bid on what items. However, subjects did not know the distribution over which values were drawn or even if a single price would clear the market.

2.2.2. Moderate Environments

The second sets of values we consider are cases in which the degree of difficulty for mechanisms is raised a bit. We consider this a move towards the actual possibilities. We introduce heterogeneity into the environment. We believe, and the data support, that heterogeneity can significantly increase the difficulty any auction design has producing efficient allocations. By effectively increasing the dimension of the commodity space from one to n, prices must now not only separate high-value bidders from low-value bidders; prices must also coordinate the demands of bidders across commodities. Prices in one market affect the demands in another, and general equilibrium phenomena become important. Theory and data and intuition from environments with homogenous objects are not sufficient background for analyzing environments with heterogeneity.

The first moderate environment we used was similar to the easy superadditive case described above, except that values were selected so that there did not exist a single market clearing price for the items. So unless the mechanism can produce nonlinear pricing, either the outcome must result in losses to at least one bidder, or some participant must forgo the pursuit of potentially profitable opportunities. As in the easy superadditive case, subjects knew the number of units being auctioned, the number of participants, and who bid on what items. However, subjects did not know the distribution over which values were drawn or if a single price could clear the market.

The second moderate environment we used was the assignment problem that incorporates heterogeneity but allows buyers *to redeem one and only one item*. The mechanism is presented with a coordination problem that has features of what might happen if bidders had budgets. The individual payoff in an assignment environment is given by

$$V_i(x_{1i}, \ldots, x_{ni}) = \sum_j^n V_i(x_{ji}) \cdot \theta_{ji},$$

$$\theta_{ji} \in \{0, 1\} \quad \text{and} \quad \sum_j^n \theta_{ji} = 1.$$

The only problem facing the mechanism is the coordination of the demanders. In the experiments, there were six items for sale to six demanders. Each demander knew that his value and those of other participants were to be drawn uniformly, with replacement, from a fixed list of ten value sheets (see Appendix, Sec. A.3, for more details).

The final moderate environment we used was one in which individuals value the heterogeneous items offered more in groups than singly. That is, for some items and same agents, preferences may have the property that $V(\{a, b\}) > V(\{a\}) + V(\{b\})$. This

Table I. *Simple Fitting Environment*

| Packages | Values |||
	Subject 1	Subject 2	Subject 3
a	2	2	4
b	2	4	2
c	4	2	2
ab	23	24	27
bc	24	27	24
ac	27	23	23
abc	42	32	32

structure of preference is often characterized as possessing "complements" or "synergies." The theory that guided the creation of these testbed environments can be found in Bykowsky et al. (1995). These experiments cover a variety of cases in which the mechanism must coordinate the bidders and guide them to best fit together. In some of the cases those trying to assemble packages can be exposed to losses if they try to build a particular package but are eventually outbid for a piece of it.

A constellation of values with this potential risk can be seen in the environment provided in Table I. There, three subjects are competing for three heterogeneous items (*a*, *b*, and *c*) with the values listed. Subject 1 has the highest value for each of the items. However, subjects 2 and 3 have high values for packages that overlap at item *b*. It is easy to see that there are no competitive equilibrium prices for this case. We also included some constellations of values with synergies and for which a simple competitive equilibrium existed.[7]

Subjects knew the number of units being auctioned and the number of participants and who bid on what items. However, subjects did not know the distribution over which package values were drawn.

2.2.3. Hard Environments

Finally we turn to three classes of environments that were intentionally constructed to test the limits and robustness of each of the mechanisms. In these environments value synergies over specific packages of items (spatial demands) are predominant. This forces bidders to coordinate their bids to find the highest-value fit among packages of items.

The first hard environment we used had three items and three demanders. Each demander had values for single items and also a synergy value for all three items. The values of bidders were determined as follows:

1. The integer values for the single items were drawn independently from a triangular distribution with support [0,98].

---

[7] At the time we thought that the performance of some of the proposed auctions would differ according to whether a competitive equilibrium existed or not. As it turns out, that conjecture was shown to be wrong by the data, and so we have reported the data from both situations together.

Table II. *Values in a Spatial Fitting Example*

Bidder 1	Packages:	f	cd	bcf	bde	abe
	Values:	9	22	128	130	120
Bidder 2	Packages:	b	df	ae	af	abd[a]
	Values:	8	28	24	27	130
Bidder 3	Packages:	c	a	d	bd	abf
	Values:	2	3	8	20	119
Bidder 4	Packages:	e	abc	adf	bdf	aef
	Values:	10	117	112	128	125
Bidder 5	Packages:	cf	de	cef[a]	bef	abcdef
	Values:	29	25	117	125	142

[a] Optimal fit.

2. The value for the three-item package was determined by adding a number randomly selected from the interval [0,149] to the highest value for item a, b, or c drawn in step 1.

The efficient outcome can have either the entire set of items going to one demander or one item going to each demander. This class of environments was designed to contrast the interests of a user who wants a major package with the interests of some single-item bidders. We created this environment after hearing many suggestions in the policy analysis of the FCC auction design that package bidding would bias the results in favor of those wanting many items. We thought it important to study that unsubstantiated claim and, in doing so, to give that argument its best chance. Subjects in this experiment knew the distribution from which the values were drawn.

The second hard environment we used had five bidders and six items (a, b, c, d, e, and f) to be allocated. Table II shows the nature of the problem faced by an allocation mechanism. Bidders 2 and 5 have three-item packages that exactly fit together and have the highest possible value of any combination of packages. The problem is that every one has high-value, three-item packages, all of which overlap. The task of the mechanism is to guide the owners of the components of the optimal allocation to find each other. This environment was created to give package bidding its best chance. The distribution over which the values were drawn was given as common information to subjects in this experiment.

The third hard environment used was designed to investigate the boundary case described in Banks et al. (1989) in which multidimensional demands of lumpy sizes must fit into a box with fixed dimensions (a *network* problem). This environment involves values in which bidders receive payoffs only for packages that are highly interrelated and must fit with the packages of other similar demanders. In particular, there were two resources with fixed supply of 20 units each. Each subject had nine two-dimensional packages of the fixed resources they could select from, for which they would receive value.

2.3. Allocation Mechanisms Tested

Into these multifaceted environments we threw three mechanisms: a sequential ascending-bid auction, a simultaneous ascending-bid batch auction, and a continuous package bidding auction. The mechanism designs were taken from the early debates in the winter of 1993 over which mechanism the FCC could or should use to allocate PCS spectrum. The debate dealt primarily with the allocative efficiency of the mechanisms and their revenue-generating properties. We describe below each of the mechanisms as we implemented them in the various testbeds.

2.3.1. The Sequential Ascending-Bid Auction

As its name implies, the sequential auction mechanism allocates one unit at a time in some sequential order. The method used in the laboratory testbed was to auction them off in random order. All participants knew, before the bidding began, the order in which the units would be auctioned off. Each unit was allocated using an ascending-oral-bid auction. Of all the mechanisms we tested, this was clearly the easiest to implement.

2.3.2. The Simultaneous Ascending-Bid Batch Auction

This mechanism operates across of a series of rounds. Once each round, individuals submit a sealed bid on each of as many items as they wish. After a round closes, the highest bid submitted for each item, the *standing bid*, is identified and displayed along with all other bids submitted. An allocation is made when bidding stops.

2.3.2.1. Activity and Update Rules

In order to be able to submit a bid in the round, a participant must have been *active* in the previous round. To be active a participant must have submitted an acceptable bid in the previous round or have had the standing bid two rounds back.[8] In order for a bid to be *acceptable* in a round, it had to be at least 10% higher than the standing bid for the item.

A *second-stage activity rule* was imposed if the auction did not close before round 8.[9] This rule restricted the number of items for which a participant could bid in a round. The restriction was that a participant could bid for at most a total number of items equal to (1) the number of acceptable bids placed in the previous round for items for which the participant did not have the standing bid for the item, plus (2) the number of items for which the participant had the standing bid two rounds back but no longer had the standing bid. In addition, the participant could always bid on those items for which he currently had the standing bid.

These activity rules are exactly those that were employed in the FCC's auction for nationwide narrowband PCS licenses.

---

[8] In some experiments participants were provided with two waivers that they could use to stay active. The original purpose of the waivers was to ensure that a bidder with logistical problems in entering a bid was not penalized. In our experiments logistical problems were not an issue. Still, over half of the allotted waivers were used by participants, presumably for strategic reasons.

[9] In the spatial environment the second-stage activity rule was not used.

2.3.2.2. Withdrawal Rule

A withdrawal rule allowed participants to delete any of their standing bids before a round began. After such a withdrawal, the price of that item was dropped to zero and that bid became the standing bid of the experimenter. An individual who with drew his bid paid a *penalty* equal to the maximum of the difference between the amount of the bid he withdrew and the highest bid submitted after his withdrawal, and zero.[10]

2.3.2.3. Stopping, Pricing, and Allocation Rule

The simultaneous ascending-bid batch auction stopped if no acceptable bids were submitted in a round or if the process reached some round after round 13. In the latter case, the actual round the process was to be stopped at was announced two rounds ahead of time.[11] When the process stopped, the items were awarded to the participants with the standing bids in that round. Withdrawal penalties were also calculated at this time.

2.3.3. The Package Bidding Mechanism

This is the AUSM mechanism described in Banks et al. (1989).[12] It is similar to the continuous ascending-bid auction[13] but with two special features. First, participants are allowed, *but not required*, to submit bids for packages of items as well as for individual items.[14] That is, they can say "I am willing to pay $100 for the package $\{a, b, c\}$ and not have to identify a separate bid for each item. With such a bid, they are requesting to be allocated a and charged $100 if and only if they are also allocated b and c. This bid is accepted if and only if $100 is more than the sum of the standing bids for the packages that contain a, b, and c. So, for example, if there is a standing bid of $35 for a, a standing bid of $50 for b, and a standing bid of $5 for c, then the bid of $100 for $\{a, b, c\}$ wins. If, however, a bid of $75 for $\{b, c\}$ is made before the bid of $100 for $\{a, b, c\}$ is submitted, then the standing bids are the bid for a and the bid for $\{b, c\}$. The $100 bid for $\{a, b, c\}$ is then no longer large enough to become a standing bid. It would need to be greater than $110. Because it can sometimes take several small package

---

[10] In the spatial environments, a slightly different withdrawal rule was used. In those experiments, one could withdraw all of one's bids when the auction stopped. The withdrawn items were then offered in a random but sequential order to the next highest bidder on that item. This was an early idea put forward by some for the FCC auction design.

[11] In the spatial environments a hard stop rule was not used. Instead, the mechanism was allowed to run its course. That is, the auction stopped and all markets closed simultaneously if and only if no new bids were entered in a round. This was the stopping rule eventually chosen by the FCC.

[12] Other references with details about AUSM include Ledyard et al. (1996) and Bykowsky et al. (1995).

[13] AUSM can and has been run as a batch process. This requires that an optimization routine be run after each round, but with modern computers and software and with economic incentives driving the structure of the bids, there have been no computational problems in practice.

[14] There seems to be a widely held misperception that AUSM and related package bidding mechanisms require that each bidder submit a bid for every possible package or 2^n bids. This is wrong. Just as in the simultaneous ascending-bid auction, bidders need only bid on those items they truly want and think they have a chance of winning. In fact, if package bidding is allowed, in equilibrium fewer bids are needed to support an efficient allocation than in the simultaneous ascending-bid auction. Data from testbeds and from real-world use suggest that package bidding generates no more serious bids per person than any other mechanism and, indeed, may actually generate fewer.

Table III. *Experimental Design*

Mechanism	Environment	Number of Experimental Sessions	Subjects	Comments
Sequential	Additive	1[a]	Inexperienced	
	Decreasing	1[a]	Inexperienced	Conducted at end
	Assignment	2[b]	Inexperienced	of simultaneous
	Spatial fitting	2[b]	Experienced	sessions
Simultaneous batch	Additive	1[c]	Experienced	
	Decreasing	4[c]	Experienced	
	Superadditive CE[d]	4[c]	Experienced	Fixed ending
	Superadditive without CE	5[c]	Experienced	round used
	Assignment	2[c]	Experienced	
	Fitting	3[c]	Experienced	No second stage and no fixed ending round
	Spatial	4[c]	Experienced	As above;
	Spatial fitting	3[c]	Experienced	withdrawal at end only
	Network	1[a]	Experienced	No withdrawal
AUSM	Fitting	2[a]	Experienced	
	Spatial	5[b]	Experienced	Standby queue
	Spatial fitting	4[b]	Experienced	used
	Network	2[a]	Experienced	

[a] David Porter designed and conducted these experiments.
[b] Robin Hanson and David Porter designed and conducted these experiments.
[c] Antonio Rangel, David Porter, and John Ledyard designed and/or conducted these experiments.
[d] Competitive equilibrium.

bids to displace a large package bid, a second special feature of the continuous AUSM mechanism[15] is a bulletin board on which bidders can post small bids that are not large enough to displace a current winner but that might be part of a collection of bids that would be large enough. This standby queue of bids is always available for bidders to combine with to displace a large package bid.

2.4. Summary of Procedures

All of the experiments were conducted at the California Institute of Technology using the student population as the subject pool. All of the mechanisms, other than the sequential auction, were computerized. For the AUSM and simultaneous batch auctions, experienced subjects were used who had three hours of training in the rules of the mechanisms and software. In Table III are listed the relevant information for each of the experimental sessions.

---

[15] When AUSM is run in the batch format, this feature is not necessary, because every bid is processed simultaneously. Even when it is available, it is rarely used. See Kwasnica et al. (1997) for a comparison of a batch AUSM mechanism with a continuous AUSM mechanism.

3. What Did We Find?

We present the results in three parts. First, we provide a brief summary of the measured performance, both efficiency and revenue, of the mechanisms in the various testbed environments. Second, we look more closely at some results from the hard testbeds, the ones designed to stress the limits of the mechanisms. Finally, we present some observations based on our reading of the totality of the evidence. We also answer the two design questions raised in Section 2.

In organizing the data, we use three standard performance measures: efficiency, revenue, and bidders' surplus.[16] Since we are working in environments in which value is measured in terms of profit, we measure efficiency in the usual way as the aggregate value achieved by the mechanism as a percentage of the maximum possible. It has been correctly pointed out by some that the absolute value of this measure is not particularly illuminating.[17] However, comparing values across mechanisms in the same environments can be informative. For example, if I were to tell you that mechanism M1 produced observed efficiencies between 90% and 96% and that mechanism M2 produced efficiencies between 86% and 89%, for the same structure of payments, you would be justified in concluding that M1 outperformed M2 on that class of environments with respect to attaining efficient allocations. We provide such comparative data below.

Our second measure of performance, revenue, is simply the number of dollars collected from the participants in payment for the items. Again the absolute numbers do not necessarily provide any guidance for the designer. The usual solution is to use revenue as a percentage of the predicted market equilibrium prices. Unfortunately, in some of the environments we report on below, no such prices exist. So, in our analysis, we will use revenue as a percentage of the maximum value attainable. Again, it is the relative values of revenue collected across mechanisms in the same environments that can be informative.

Finally, because it is a measure of the user's gains from participation and because it can serve as an explanation for why some potential participants argued for particular designs, we include data on bidders' surplus. In this case, this is simply the difference between the value attained and the revenue paid. Bidders' surplus as a percentage of the maximum value attainable is thus simply the difference between the efficiency and revenue percentages.

3.1. Basic Performance—Efficiency and Revenue

3.1.1. Easy Environments

In Figures 1 and 2 are plotted the efficiency and revenue percentages achieved by the mechanisms tested in environments with additive values, with decreasing returns, and

---

[16] We of course also look at other dimensions of interest. Of particular interest in the debate about package bidding is the extent to which a mechanism biases the outcomes in favor of bidders who only want a small number of items as opposed to those who get high value from large packages.

[17] In the Appendix, Section A.2, we expand and explain this problem.

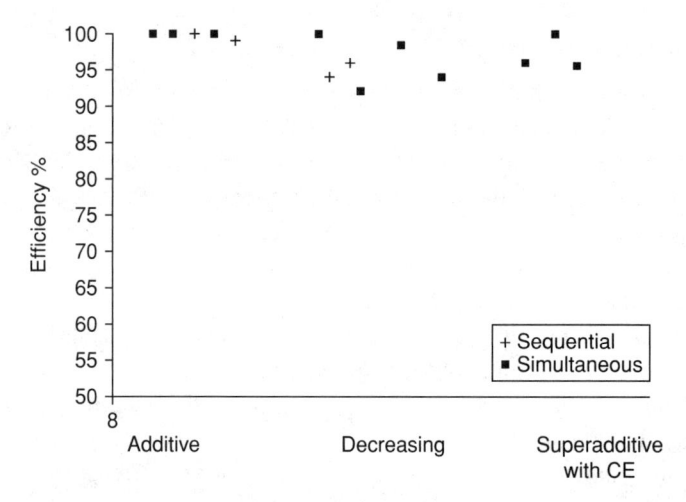

Figure 1. Mechanism Efficiency—Easy Environments.

with increasing returns with competitive equilibria. These data confirm our prior intuition that all of the mechanisms would do well in such unchallenging situations. The variation observed in the performance of the batch process, relative to what would be efficient, should be expected in light of the requirement that bids increase by at least 10% in each round.

3.1.2. Moderate Environments

In Figure 3 and 4 are the data from the tests in which the mechanisms were exposed to moderately more difficult parametric conditions. These were one with homogeneous goods with increasing marginal values and no competitive equilibrium, the assignment problem, and a simple fitting problem.

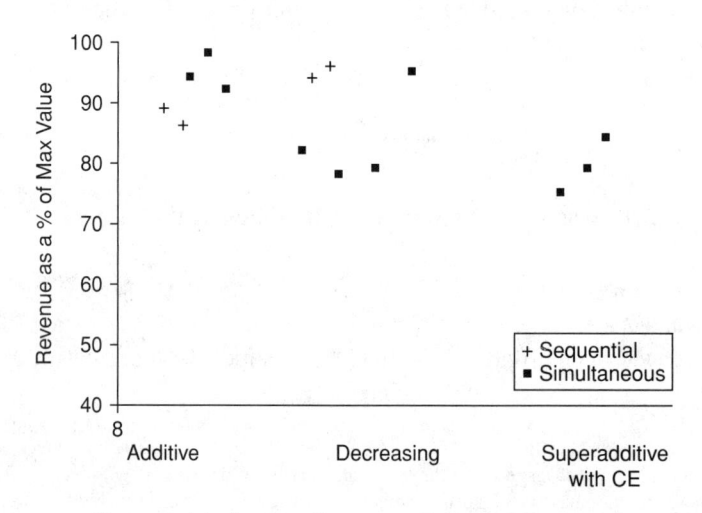

Figure 2. Mechanism Revenue—Easy Environments.

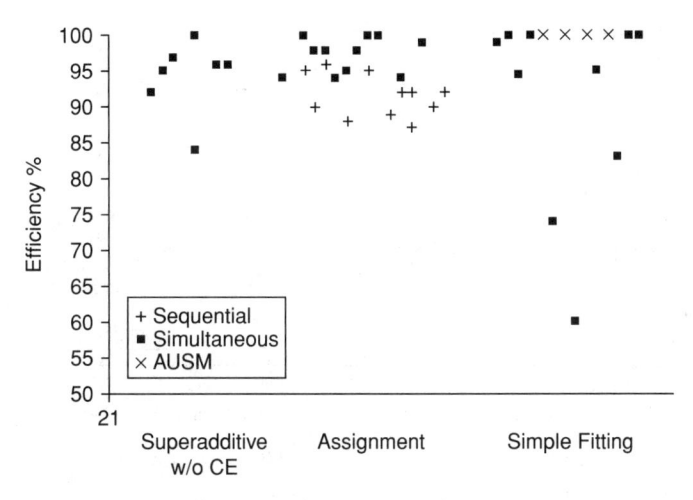

Figure 3. Mechanism Efficiency—Moderate Environments.

The mechanisms now begin to separate themselves. The simultaneous mechanisms do a better job of finding efficient allocations and produce more revenue than the sequential mechanism. However, relative to the easy environments, both the simultaneous and sequential auctions now exhibit some losses in efficiency. The packaging mechanism seems to work very well in the simple fitting environment. One finding of some interest is that, in the simple fitting case, the revenue produced by the simultaneous auction can actually exceed the maximum value of the final allocation. This means that one or more of the bidders has sustained significant losses—has paid more for the items won than they are actually worth to that bidder. This is not a winners'-curse phenomenon: remember that there is no correlated information. It is the result of using an inappropriate mechanism in an environment with complementarities among heterogeneous items.[18]

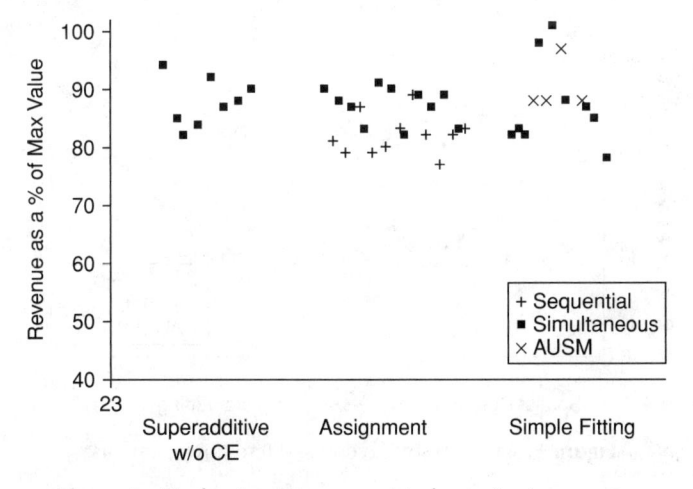

Figure 4. Mechanism Revenue—Moderate Environments.

---

[18] See Bykowsky et al. (1995) for how this might happen.

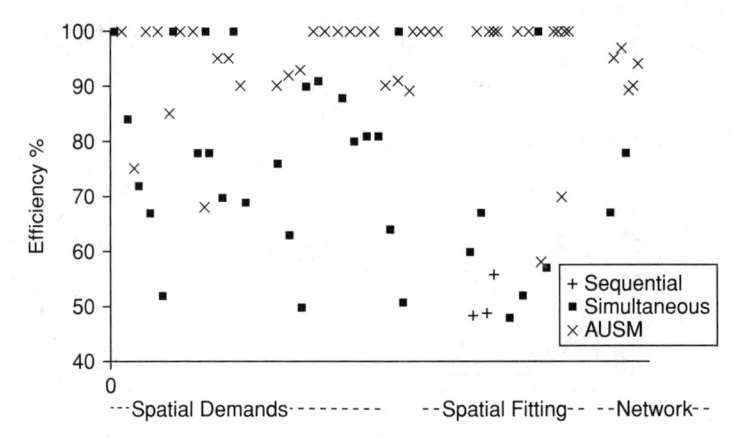

Figure 5. Mechanism Efficiency—Hard Environments.

3.1.3. Hard Environments

In Figure 5 and 6 are the data from the testbeds that used the most difficult coordination environments. These data are perhaps the most revealing about the relative performance capabilities of each mechanism. Stress tests often highlight strengths and weaknesses missed under more normal conditions.

The results seem very clear from these figures. First, without package bidding, there are major losses in allocative efficiency. Thus, it appears that simultaneous auction processes are necessary but not sufficient to coordinate demanders. A simultaneous auction does eliminate single-item efficiencies. But in complex environments with nonconvexities arising from heterogeneous spatial returns to scale, one price per item is simply not enough information to guide bidders to efficient allocations.[19] Opportunities for economies of scale, scope, and fit are easily missed. Allowing bids for packages leads to improvements in efficiencies and revenue, and losses are controlled. Efficiencies are

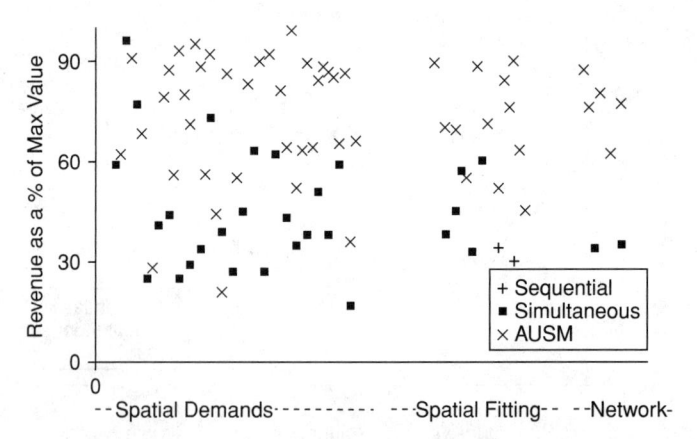

Figure 6. Mechanism Revenue—Hard Environments.

---

[19] The theory behind this observation can be found in Calsamiglia (1977), Jordan (1987), and Mount and Reiter (1996).

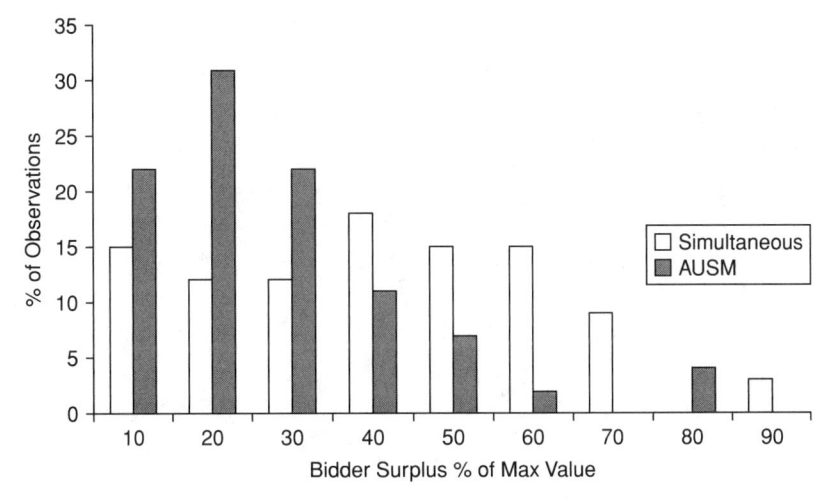

Figure 7. Distribution of Bidder Surplus.

improved because bidders can find and bid on those packages with significant complementarities without bearing the risk of losing part of that package. Revenue is improved for the same reason.[20] However, the gain in efficiency and revenue from allowing package bidding appears to come at some expense to bidders' surplus. In Figure 7 are displayed the distribution of bidders' surplus (net profit)[21] as a percent of the maximum value.

3.2. A Slightly Deeper Look

There is a three-way tradeoff in the design of mechanisms between efficiency, revenue, and bidders' surplus. In Figures 8 and 9 we exhibit this tradeoff separately for two classes of the hard environments: the spatial and the spatial fitting. In the figures, points to the northeast represent higher efficiencies, since revenue percentage plus surplus percentage equals efficiency percentage. The distribution of points along the lines of equal efficiency represents the distribution of the surplus between the seller (revenue) and the buyers (bidders' surplus). We chose these tests because they appear to generate the starkest differentiation in performance between the simultaneous and the packaging mechanisms. Such differentiation allows some insight into the particular strengths of each mechanism.

3.2.1. Giving the Simultaneous Auction Its Best Chance

The spatial testbed is probably the ideal example of a situation that gives the simultaneous ascending-bid auction its best shot at outperforming the package-bidding auction.

---

[20] In ascending-bid auctions, winning prices are driven by the values of the second best allocations, since winners must bid enough to ration the losers out. With packaging, the second-best allocations can be more easily found and bid, and, with synergies, these are worth more than the sum of the values of the single-item allocations. Thus package bidding generally yields higher revenue without losses.

[21] Net profit is simply $V_i(x_{1i}, \ldots, x_{ni}) - \sum_j b_{ji}$. Here, b_{ji} is the amount paid by i for item j.

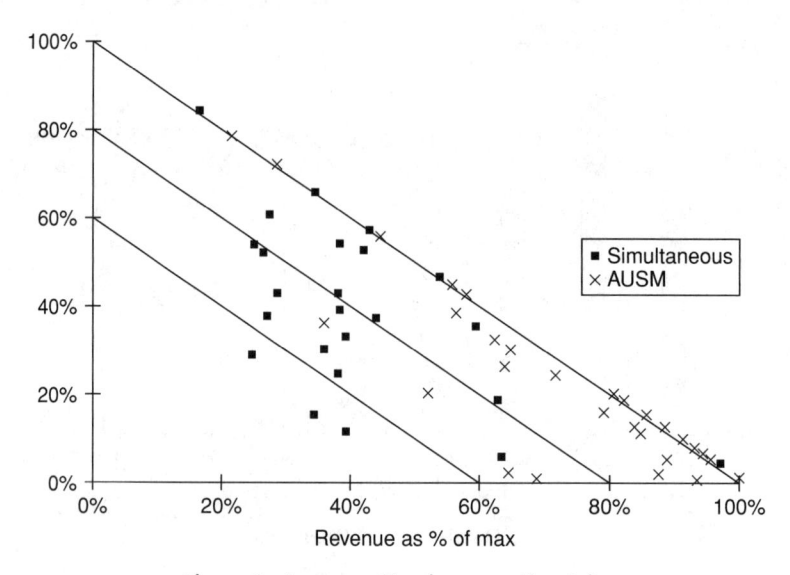

Figure 8. Ausm vs. Simultaneous (Spatial).

It places the interests of single-item demanders in direct conflict with those of demanders who wish the entire collection of items. To make that conflict even starker, the highest-valued demander of the collection is also one of the single-item demanders. Thus, in situations in which the single-item demanders should win (i.e., when that is the efficient outcome), not only must they outbid the demander with the highest value

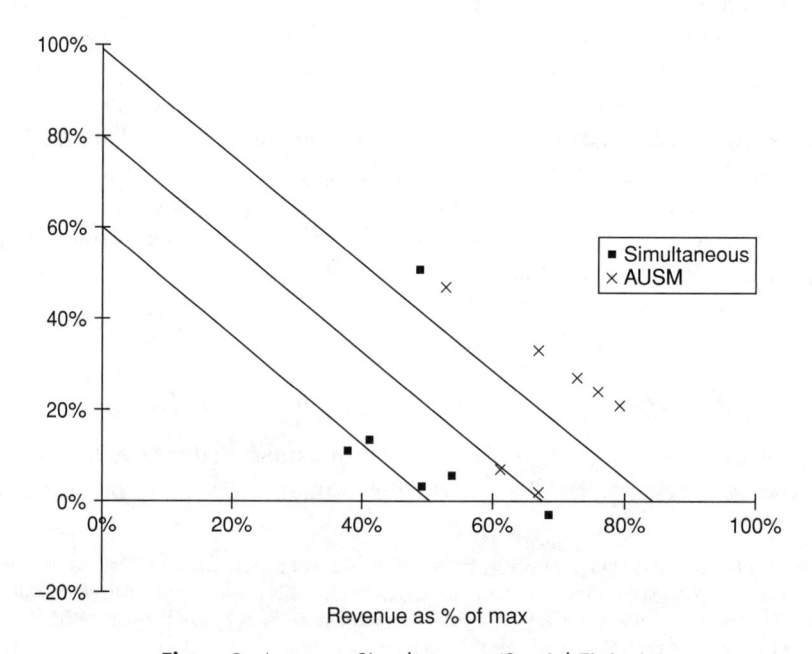

Figure 9. Ausm vs. Simultaneous (Spatial Fitting).

for the whole collection, but also the demander of the whole must be part of the effort to outbid himself.[22]

In these tests, AUSM tends to generate outcomes that are, relative to those of the simultaneous mechanism, high in both efficiency and revenue but low in surplus. The simultaneous ascending-bid auction does better in generating bidders' surplus, but does so in many tests at a serious loss in efficiency. The numbers are straightforward. In 90% of its tests, AUSM yielded more than 80% efficiency. In only 33% of its tests did the simultaneous mechanism yield more than 80% efficiency. On a relative basis, using efficiency as the appropriate measure, the data seem to reject the charges that AUSM has a threshold problem. A closer examination, however, reveals some evidence to the contrary. In 22 of the 35 tests of AUSM in this environment, the 100% efficient outcome was for the single-item demanders to win. AUSM produced that outcome only 45% of the time. In the other 13 tests the 100%-efficient outcome was for the demander of the whole to win. AUSM produced that outcome 100% of the time. It appears that in this extreme test for the existence of a threshold problem, there are signs that AUSM has one but that its effect on efficiency and revenue is low. The numbers for the simultaneous mechanism are almost the exact opposite of those for AUSM. In 17 tests of the simultaneous mechanism in this environment, the 100%-efficient allocation was for the single-item demanders to win. The simultaneous mechanism produced that outcome only 75% of the time.[23] In the other 7 tests, the demander of the whole should win to produce 100% efficiency. The simultaneous mechanism produced that outcome only once, or 14% of the time.[24] Clearly the data from this extreme testbed highlight the fact that each of the two mechanisms possesses an unmistakable bias. AUSM seems to be biased slightly in favor of large-package demanders, while the simultaneous mechanism seems to be biased seriously in favor of single item demanders. Nevertheless, if one is interested in generating highly efficient allocations, then AUSM clearly dominates in these tests.

Looking at the other data from the spatial tests, we note that in 90% of the tests AUSM produced revenue in excess of 50% of the maximum possible. In only 20% of its tests did the simultaneous mechanism yield more than 50% of the maximum possible revenue. With respect to bidders' surplus, the simultaneous mechanism yielded over 40% of the maximum possible surplus in over 50% of its tests, and over 30% of the maximum surplus in over 70% of its tests. AUSM, on the other hand, yielded over

---

[22] This should certainly lead to the exposure of a threshold problem for AUSM, if one exists. It has been the speculation of some that package bidding creates a situation in which those bidders who value large packages have an advantage over bidders who only value single items. The speculation is that when it is efficient for single-item demanders to win, AUSM will let the large package bidders win instead, leading to low efficiencies. This is called the threshold problem, because the single-item bidders have to coordinate to jointly overcome the threshold provided by the large package bid. The counter bias, that the simultaneous mechanism may let single-item bidders win in tests when it is efficient for large packages to win, is generally not mentioned in these discussions.

[23] This seemingly high failure rate of 25%, in situations for which the mechanism seems particularly suited (compare that with AUSM's 0% in its "good" environments), occurs because of the internal conflict faced by the bidder who is the high-value large-package demander. That bidder must choose at some point during the auction to go for one unit instead of all three. If the bidder waits too long to withdraw from his pursuit of three, then a misallocation of the single items can occur and that bidder can actually face losses. That seems to have happened quite often in these tests.

[24] Compare that with AUSM's 45% in its "bad" environments.

40% of the maximum surplus in only 15% of its tests, and over 30% of the maximum in only 20%. A straightforward policy observation follows. If the surplus attained is all that is important to the potential participants in an auction, and if efficiency is allowed to take a back seat to self-interest, then bidders should be expected to argue for the simultaneous auction, while the seller should be expected to argue for the inclusion of package bidding.

3.2.2. Giving AUSM Its Best chance

The spatial fitting test-bed is probably the ideal example of a situation that gives the package-bidding auction its best shot at outperforming the simultaneous ascending-bid auction. It highlights situations in which the efficient allocations involve no single-item buyers and no buyers who want the entire collection. Rather, the efficient allocation is usually one in which two buyers each buy three of the six items. Also, and as importantly, the second-best allocation (remember, that's the one that drives prices) also involves two buyers buying three items each but usually in a different configuration than that of the first-best allocation. So coordination is the key to success in these testbeds.

Turning to the data, one can see that the conflict between bidder and seller is no longer as sharp as it was in the spatial tests and, in a certain sense, can be said to be not there at all. Here, with rare exceptions, package bidding leads simultaneously to higher efficiency higher revenue and higher bidders' surplus. The efficiency increase that occurs by including package bidding apparently creates enough surplus to allow both sides of the market to be better off. The numbers are again straightforward. In only two tests did AUSM fail to achieve 100% efficiency, while in only one of its tests did the simultaneous mechanism exceed 70% efficiency. So much coordination is needed to find both the best allocation and the next-best allocations that the single-price-per-item structure of the simultaneous auction doesn't have a chance. In the spatial fitting tests, AUSM yielded bidders' surplus less than 20% only twice. The simultaneous mechanism, on the other hand, managed to yield higher than 20% surplus only once. In fact, because bidders are exposed as they try to acquire packages in the simultaneous auction, many bidders actually lose money. In one test, the losses were so high that there was a *negative* bidders' surplus and only 70% efficiency: that is, bidders paid out more in revenue to the seller than they would make as owners of the items they bid on and won. These losses did not, however, yield the highest revenue to the seller. In most cases that was accomplished by allowing packaging.[25] In 90% of its tests, AUSM exceeded 50% of the maximum possible revenue. In only 20% of its tests did the simultaneous mechanism exceed 50% of the maximum possible revenue.

3.3. Summary of Our Observations and Their Support

In Section 1, we identified two major design choices that the data might help provide answers for. These were (1) sequential or simultaneous, and (2) package bidding or

---

[25] As a side note, the data also support the claim that allowing package bidding does not give the package of the whole any particular advantage.

not. We have exhibited a lot of data from various testbed combinations of mechanisms and environments, many of which were created to influence those choices. Because the experiments were designed "on the fly," the data may not be as definitive as one might wish. On the other hand, we believe there is still enough evidence to allow us to make some observations that will stand up to further examination.

Observation 1: If the items are homogeneous, then the answer to (1) is that it doesn't matter. The answer to (2) in these environments is unknown but probably unimportant.

The support for this observation comes from the data in Section 3.1.1. In environments with multiple items to be allocated, if those items are homogeneous and substitutes, then little coordination between buyers is needed and the only role of the mechanism is to sort bidders with high values from bidders with low values. Both the sequential and simultaneous mechanisms seem to work very well at finding efficient allocations in these "easy" environments.[26]

Observation 2: If the items are heterogeneous, then the answer to (1) is that simultaneous is better. If the extent of complementarity between items is small, the answer to (2) is that it probably doesn't matter.

The support for this observation comes from the data in Section 3.1.2. In environments with multiple items to be allocated, if those items are heterogeneous, then some coordination among bidders is necessary to achieve high-value allocations even if there are only low synergy values. Simultaneous auctions provide a first step toward this coordination in a way that sequential auctions are unable to.[27] Packaging doesn't seem to either help or hurt relative to the simultaneous mechanism.

Observation 3: If there are significant complementarities, then the answer to (2) is that package bidding is significantly better.

The support comes from the data in Section 3.1.3. In environments with heterogeneous goods exhibiting complementarities, significant coordination is required for an auction or allocation mechanism to perform well with respect to efficiency or revenue. Sequential auctions perform poorly. Simultaneity is clearly necessary but not sufficient to attain high efficiencies. The simultaneous one-price-per-item auction seems to produce outcomes that are either high in efficiency, revenue, *and* losses or low in efficiency, revenue, and losses. Package bidding seems to help a lot in attaining high efficiency, high revenue, and no losses.

---

[26] We do not have any data on the performance of AUSM, the package-bid mechanism, in these simple environments with homogeneity, but we find it plausible that packaging could actually hurt here by guiding bidders to try to attain allocations that are inefficient. Of course, it is also plausible that AUSM, like the other mechanisms, will also perform well in these easy environments.

[27] See Milgrom (1995, pp. 14, 15) for a concise discussion of why this might be so.

4. What Next?

In this section we try to point to some of the future research that we think is vital to the creation of better designs of complex auctions. There is a major gap between theory, scientific evidence, and practice in the design of these mechanisms. Until there are some serious breakthroughs in the theory of heterogeneous, multiunit auctions, it is also likely that experimental evidence will have to suffice.

4.1. Stopping, Activity and Withdrawal Rules

The simplest and most needed research is the testing of various straightforward variations in existing designs. Among these variations are withdrawal rules and stopping rules, including the various activity rules that have been proposed or used. Nothing systematic has yet been done to provide the research needed to answer questions that came up in the design of the FCC auction. For example, there are no serious theoretical discussions about whether withdrawal rules do any good at all and, if so, what are the better rules.[28] Discussions are naive. From an individual's standpoint, the possibility of withdrawal allows a bidder to be more aggressive,[29] to try risky fitting strategies at lower risk, and (maybe) to avoid losses incurred "by mistake." From a strategic point of view (i.e., when the reactions of the other players are also considered), some of these benefits may disappear. Losses occur for sure only when prices are high and the end of the auction is near, in exactly those cases in which no one is left to bail the loser out. However, it is still possible that the apparent reduction in risk will increase efficiency and revenue at the cost of increased losses. A second strategic effect is less benign. The lowering of the risk of loss could lead an opponent to try to drive the price of an item up to force you to give it up and, more importantly, because of that to release another item at a loss. This type of strategy can lower both efficiency and revenue. What will really happen remains 10 be carefully studied.

With respect to stopping rules, we also have virtually no systematic data or theory that can provide guidance about what stopping rule should be used in which situation.[30] Stopping and its corollary, the encouragement of active bidding, remain very much an art both in the lab and in practice. The FCC chose to allow bidding to continue until no new bids are entered, over another serious proposal: that bidding stop item by item when no new bids are entered for that item. Their choice in turn required that some form of activity rule be designed that would force participation, since otherwise all bidders would have an incentive to wait for others to go first.[31] On the other hand, in markets for emissions permits for the Los Angeles basin we have, seemingly successfully, used a much different stopping rule.[32] Those auctions close at the end of a round if the aggregate value of the standing bids does not increase by more than 5% over the

---

[28] A first step in the direction of providing some scientific evidence about the effect of withdrawal rules can be found in Porter (1996).
[29] It lowers the *expected* cost of not acquiring a piece of a package in a simultaneous auction.
[30] A first step in this direction can be found in Kwasnica et al. (1997).
[31] Banks et al. (1989) has some relevant observations on this phenomenon.
[32] More details can be found at the World Wide Web address vwvw.ace-mkt.com.

previous round.[33] This yields a much faster closure to the process and requires no activity rules other than the very natural and simple one that all high bids are binding and must be improved on to be displaced. What is not known with any certainty is whether this faster stopping creates more or less revenue or more or less efficiency than, say, the FCC rules, and whether such a finding would depend in any systematic way on the environment. With the extreme importance of these issues, it is very surprising that there is virtually no theoretical or experimental research.

4.2. Complexity

4.2.1. Computational and Strategic Complexity

Another level of open questions in the design of multiple-unit, heterogeneous goods auctions involves issues of complexity in both mechanism and environment. This is especially important when one begins to anticipate the impact of scaling up the experimental tests to something closer to the actual application. One of the issues one must face in comparing mechanisms, as in the question of whether to allow package bidding or whether to use a continuous or batch process, is how to judge the computational and strategic complexity of each approach. These concepts lie behind some of the discussions during the FCC design process but have never been satisfactorily defined and measured.[34] For example, batch processing gives all bidders time to think through their next response and so seemingly simplifies their problems; on the other hand, because of the sealed-bid nature of each round of batch processing, bidders have to anticipate their competitors' responses and never know for sure what a provisionally winning incremental bid is. As another example, package bidding provides bidders with a strategically easier way to coordinate their own bids and minimizes their exposure to losses; on the other hand, if used in batch mode, package bidding requires the auctioneer to use an optimization algorithm and confronts bidders with some coordination complexity when collections of small bids are needed to produce efficiency. There are solutions to many of these problems, but a complete study of the tradeoffs, including developing methods to measure the effects of the strategic complexity on bidders, is long overdue.

4.2.2. Environmental Complexity

Proposed mechanisms must be studied in more complex environments. Two variations that are obvious to consider include environments in which bidders have budget constraints and environments in which there are correlated or affiliated values across multiple, heterogeneous items. The first of these is important to study because it appears there were a number of bidding teams in the FCC auctions who were given a budget by their senior management and told to do as well as they could within that constraint.[35] One might expect this to be a common situation in large, complex auctions. It is our

---

[33] There is also a provision for a maximum number of rounds. In the LA emissions markets that we designed, this number is currently 5.

[34] Some initial research has begun in this direction. For a leading example, see Rothkopf et al. (1995).

[35] There are many reasons why such a constraint might exist, but a leading candidate would be principal-agent problems.

conjecture, and that of several others, that in an environment with budgets, package bidding is going to yield better performance than the simultaneous auctions. To the contrary, a few others have opined that with a good withdrawal rule the simultaneous auction will do as well as a packaging auction. The correct answer awaits further study.

Auctions for single items have been studied in great detail, both in theory and in experiment, in environments with correlated values. Evidence of the winner's curse has been found, and it has been shown that ascending-bid auctions allow better information aggregation than sealed-bid auctions. This was one of the compelling reasons behind the decision by the FCC to use an ascending-bid auction instead of a sealed-bid auction. One might hope that having multiple heterogeneous items would not change these results a great deal. But since sequential, simultaneous, and package-bidding auctions all provide different information to bidders during the auction, it is possible that systematic differences in performance could appear in environments with correlated values that would reverse the findings above. Our conjecture is that this will not happen, but this needs to be studied both theoretically and experimentally before we can be sure.

4.3. Designs?

Finally, some purely speculative thoughts on what the future will bring in the design of auctions to price and allocate a large number of multiple heterogeneous items at one time. Under the rubric of moderate fixes, we think there are two that are the easiest and most productive. One would be the development of really good stopping rules. These would drive bidding activity without using complex eligibility requirements and activity rules, they would cause convergence to equilibrium reasonably rapidly, and they would not impose a lot of strategic complexity on the bidders. The second development would be user-friendly package bidding. This would reduce the seeming complexity facing the bidder while allowing the significant improvements in revenue, efficiency, and bidder surplus that such bidding creates.[36]

In the more speculative realm of really new approaches, we suggest one. In Banks et al. (1989) we considered a number of mechanisms and chose AUSM on the basis of the evidence there. It has proven to be a flexible, successful mechanism in many applications. But there was another mechanism we considered, the iterative Vickrey mechanism. In that design, we tried to capture some of the demand-revealing aspects of Vickrey's original mechanisms (see Vickrey, 1961), while introducing some of the cognitively easier aspects of simple iterative bidding found in standard English auctions.[37] We were looking for demand revelation because we believed that if there were little strategic loss from bidding one's true values, then the strategic complexity of simultaneously bidding for multiple items would be significantly reduced and good

---

[36] We continue to believe that the unsupported claims by some that package bidding is "too complex" are exaggerated and unfounded. Our experience in the applications of large auctions with truckers (an 800-item auction for Sears) and with environmental engineers (the ACE pollution permit market) suggests the contrary.

[37] Some of these ideas can be found in Rassenti et al. (1982), but they only consider a sealed-bid mechanism, which doesn't appear to do the desired job. Some of these ideas can be found in Ausubel (1996), but the mechanism there seems extremely complex and difficult to implement even in a very simple testbed.

performance would be more likely. We also believed that with the appropriate iterative procedure, bidders would not need to submit bids for all packages in each round (a possibility that would destroy the implementability of the mechanism). Because we were unable to provide appropriate commitment rules,[38] the mechanism did not perform as well as we had hoped. However, we believe there is still an iterative Vickrey design to be found that will minimize strategic complexity, allow package bidding, and provide excellent efficiency and revenue performance. If so its performance could easily surpass that of all of the mechanisms studied in this paper.

Appendix

A.1. Some Background Methodology on Applied Mechanism Design

The FCC auction designers' problem was to create a mechanism to allocate and price a number of heterogeneous items. The goal, at least as initially stated by the FCC, was to allocate those items to the highest-value users.[39] The basic problem, common to most mechanism design efforts, was that the information needed to solve this problem (the values of the items) was best known, if at all, by the various potential users and not by the FCC. Further, none of these potential users had any incentive to precisely reveal their information to the FCC. There is a standard solution to this problem that has been developed over a number of years of basic research in economics and other disciplines: *If one can predict the performance of various mechanisms over a range of possible user values for the items, then one does not need to know the details of the specific values to achieve one's goals; one need only select the appropriate mechanism to achieve the desired outcome.*

The idea is simple. A mechanism, such as a particular auction format, works as follows. Participants bring their own information and valuations to the auction. The auction is then held, and the participants use their information to determine how they interact with each other through the mechanism. The interaction between individual behavior and the auction rules produces an allocation of the items and payments for those items. Operating the same auction on a different constellation of values will generally produce a different allocation and payment distribution. Operating a different auction format on the same constellation of values across individuals will generally also produce a different allocation and payment distribution. We refer to the relationship a mechanism creates between the particular constellation of values and the allocation and payments as the *performance of the mechanism*.

Consider the simple diagram in Figure 10. E is to be thought of as a set of possible constellations of values with a single point in E representing the true valuations. We call elements of E environments. X is to be thought of as the set of possible allocations of items and payments for those items that might result. We call elements of X

---

[38] These would be rules that require bids to be somewhat binding so as to prevent cheap-talk uses of the bidding process, which in turn could prevent the auction from converging.
[39] See, e.g., Milgrom (1995, pp. 13–14).

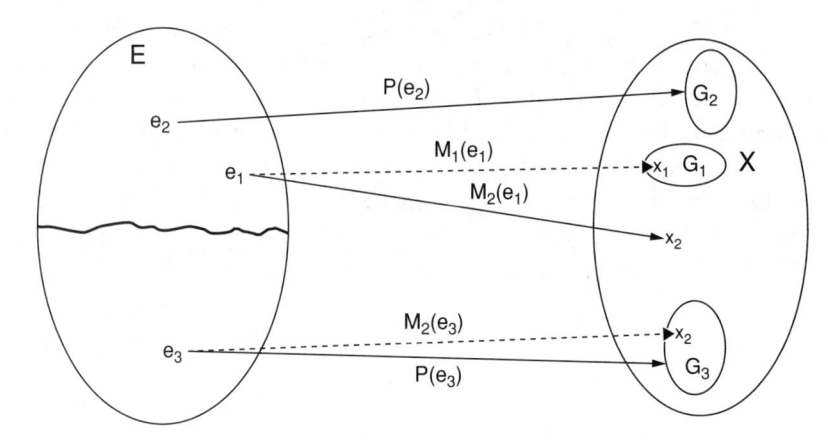

Figure 10.

outcomes. A mechanism's *performance* is then a mapping from E to X. So, for example, mechanism $M1$ produces outcome $x1$ if the environment is $e1$, while mechanism $M2$ produces outcome $x2$ in that same environment. The policy issue in mechanism design is to determine the standard of performance that is desired for the mechanism to be chosen. For example, should the mechanism try to produce an outcome that maximizes the aggregate value of the allocation? Such a standard will usually be another mapping from some part of E to X and will look like P in Figure 10. So the performance standard P asks the outcome to be something in the set $G2$ if the true environment is $e2$ and something in the set $G3$ if the environment is $e3$. If the scientific evidence can establish the performance of various mechanisms and if the performance of at least one of those mechanisms is consistent with the performance standard of the decisionmaker over the part of E within which the decisionmaker thinks the true valuations lie, then the design problem has been solved. In Figure 10, if the decisionmaker thinks that the true e is somewhere in the upper half of E, if that decisionmaker has the performance standard P in mind, and if $M1$ produces something in $P(e)$ for all of the e in the top half, then *even though no one knows the true constellation of values*, the mechanism design problem is solved by using $M1$.[40]

The *science* of applied mechanism design, then, is focused on providing the best evidence possible on the performance of mechanisms in a variety of environments. The *policy* of applied mechanism design is focused on using those findings to pick the most appropriate mechanism for the situation. In this paper we concentrate on the science only.

---

[40] Decisionmakers involved in policy should be forewarned. Even with the best scientific evidence about the performance of mechanisms, arguments about the possible location of the true state of the world can derail good intentions. If the future participants are asked to provide advice during the design phase and if these participants know something about the true e, they may have an incentive to provide arguments intended to improve their final allocation. So in Figure 10, a potential participant may know that e is truly in the top half of E. If that participant likes $x2$ better than $x1$ and if that participant knows the policymaker's performance standard is P, then that participant may argue strongly that "e must be in the bottom half." If that argument is successful, then the decisionmaker will select mechanism $M2$, since $M2(e)$ is in $P(e)$ for e in the lower half of E. But then when the mechanism $M2$ is run in conjunction with the true world, say $e1$, the outcome $x2$ occurs. The participant is better off; the policymaker may not be.

A.2. Possible Problems with Efficiency as a Performance Measure

It has been correctly pointed out by some that the absolute value of this measure is not particularly illuminating. For example, even if you knew that a mechanism produces 95% efficiency on average over a class of environments, there would still be no basis for you to know whether this is good or bad. An example easily illustrates this. Suppose there are two items to allocate, A and B. Further suppose bidder 1 is to be paid $6 if she gets A, $10 if she gets B, and $20 if she gets both. Suppose bidder 2 is to be paid $4 if he gets A, $15 if he gets B, and $20 if he gets both. The optimal allocation is that 1 gets A and 2 gets B, for a total profit (before payments) of $21. If, instead, the actual outcome were that 1 gets B and 2 gets A, the profit (before payments) would be $14, for an efficiency of 66%. Now suppose I wanted to make this look a little better. I could simply add $300 to each possible payoff. This would not change the incentives to each agent (it is only a lump-sum payment), but it would yield a significantly better-looking efficiency measure, $(310 + 304)/(315 + 306) = 98.87\%$. So the absolute number is of little value. However, comparing values across mechanisms in the same environments can be informative. For example, if I were to tell you that mechanism $M1$ produced observed efficiencies between 90% and 96% and that mechanism $M2$ produced efficiencies between 86% and 89%, for the same structure of payments, you would be justified in concluding that $M1$ outperformed $M2$ on that class of environments with respect to attaining efficient allocations.

A.3. Details on the Environments

A.3.1. Easy Environments

A.3.1.1. Additive Values

The first easy environment examined is one in which values are additive, i.e., the value function is of the form $V_i(x_{1i}, \ldots, x_{ni}) = \sum_j V_i(x_{ji})$ In the experiments, there were six items for sale to six demanders. Each demander knew that his value and those of other participants were to be drawn uniformly, with replacement, from a fixed list of ten value sheets, shown in Table IV. In this environment, a competitive equilibrium always exists for each item, and its competitive-equilibrium price is equal to the second-highest value for that item. For example, if a participant drew sheet 4, his profit before any payment for item c would be 900 and his gross profit for item a would be only 100. If he obtained both items a and c he would be paid a gross profit of 1000. The subjects also knew the distribution of the possible value draws. That is, they knew Table IV and the process by which values were assigned to each subject.

A.3.1.2. Decreasing Values

Items in this environment are *homogenous*, i.e., $V_i(x_1, x_2, \ldots, x_n) = V_i(\sum_j x_j)$ with $V_i'(\sum_j x_j) \leq 0$, so that we have a downward-sloping demand with a competitive equilibrium (CE). One of the demand conditions we used is given in Figure 11. In this environment there were six participants and ten units to be allocated. Each participant had decreasing demands for up to four units.

Table IV. *Value-Sheet Space*

Item Sheet	a	b	c	d	e	f
1	900	450	400	350	300	250
2	400	600	800	600	400	200
3	800	600	400	200	400	600
4	100	100	900	400	300	200
5	400	800	400	200	0	200
6	900	600	300	200	100	0
7	300	300	300	300	300	900
8	750	250	250	750	400	400
9	400	200	400	600	800	600
10	850	350	350	650	150	150

A.3.1.3. Superadditive Values with Competitive Equilibrium

The items are homogenous but we have $V_i'(\sum_j x_j) \geq 0$. In addition, there is a price that clears the market. An example is given in Figure 12. Thus, while there exists a risk of assembling the package of items, a single (CE) price exists. In the figure, each step represents a participant's marginal return function for two units. Thus, there were eight participants in the experiments.

A.3.2. Moderate Environments

A.3.2.1. Superadditive Values without Competitive Equilibrium

An example is given in Figure 13. There are eight participants. In the figure, each step represents a participant's marginal return function for the first three units. The marginal return for more than three units is zero. Notice that in this environment there is no single price equilibrium. As soon as bids go over 100, losses must occur. When price is above

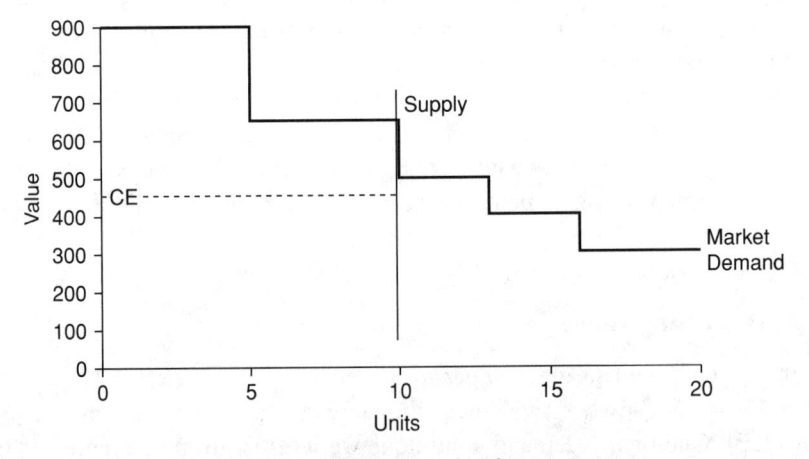

Figure 11. Decreasing-Marginal-Value Environment.

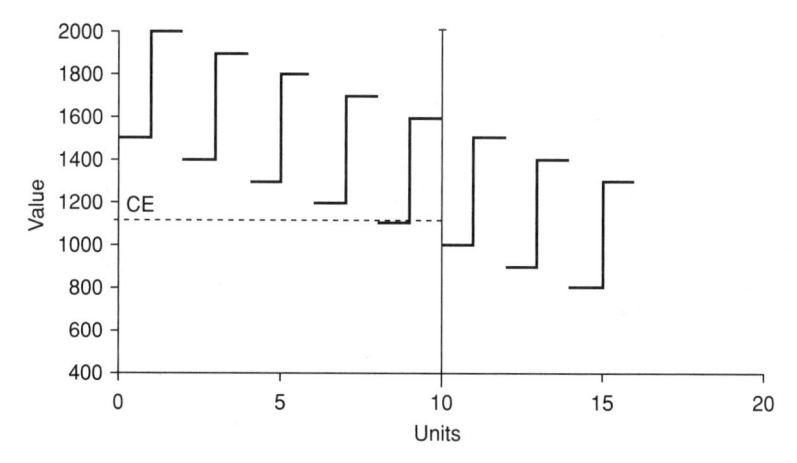

Figure 12. Superadditive Environment with CE.

100, bidder 5, for example, might then withdraw his bid and accept any price above 70. This result occurs because the mechanism does not allow nonlinear pricing. Thus, either the outcome must result in losses, or at least one bidder must forgo the pursuit of potentially profitable opportunities.

In the experiments, subjects were provided with the following common information about their environment: the number of units being auctioned, the number of participants, and who had the standing bid on which items. The subjects did not know the distribution over which the values were drawn or even if a single price could clear the market.

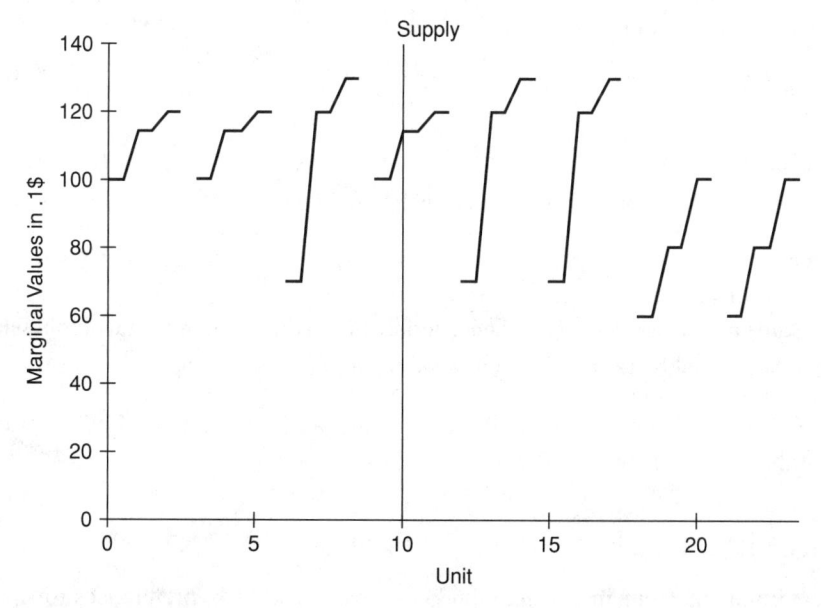

Figure 13. Superadditive Environment with no CE.

Figure 14.

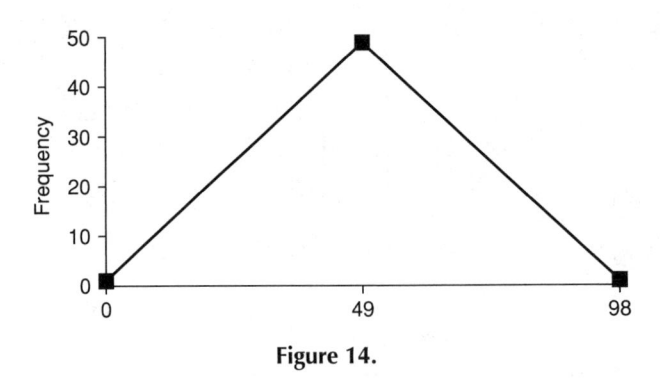

A.3.2.2. The Assignment Problem

The environment here is exactly that described in Olson and Porter (1994). In these experiments we used the same parameters as in the additive-value environment (see Table II for values) with the added restriction that demanders can use one and only one item to make a profit. In addition, we used the same common knowledge structure where each demander knew that his value and those of other participants were to be drawn uniformly, with replacement, from the fixed list of ten value sheets in Table IV.

A.3.3. Hard Environments

These environments are to provide boundary cases that test the robustness of mechanisms that do not allow individuals to package demands. In general they are extensions of the case described above with more constraints and priors over values provided to participants. We detail the three specific environments we used below.

A.3.3.1. Spatial Demands

Values for three items called a, b, and c along with a value for the full package abc were drawn from a common knowledge distribution as follows:

1. The integer values for the single items are drawn independently from a triangular distribution with support [0,98] (Fig. 14).
2. The value for the package abc is then determined by adding a number randomly selected from the interval [0,149] to the highest value for a, b, or c in step 1.

This parameter set can generate values in which a competitive equilibrium price for each of the items a, b, and c exists or not.

A.3.3.2. Spatial Fitting

This environment is one in which individuals have nonadditive preferences for specific packages of items and must find how they fit together. There are five participants and six heterogeneous items to allocate. The structure of the problem is as follows:

1. The single-item packages called a, b, c, d, e, and f have their integer values drawn independently from the uniform distribution with support [0,10].
2. The two-item packages $\{a, b\}$, $\{a, c\}$, ..., $\{e, f\}$ have their integer values drawn independently from the uniform distribution with support [20,40].
3. The three-item packages $\{a, b, c\}$, ..., $\{d, e, f\}$ have their integer values drawn independently from the uniform distribution with support [110,140].
4. A single value is given for the six-item package $\{a, b, c, d, e, f\}$ drawn from [140,180].

A total of 25 unique packages from the total possible generated from steps 1–4 above were given to participants. The main point to note is that two three-item packages clearly form the largest total value. However, this optimal package configuration is likely to be overlapped by many other competing packages. The task of the mechanism is to guide the owners of the optimal packages to find each other.

References

Ausubel, L.M., 1996, "An Efficient Ascending-Bid Auction for Dissimilar Objects," Preliminary paper, University of Maryland, January.

Banks, J., J.O. Ledyard, and D. Porter, 1989, "Allocating Uncertain and Unresponsive Resources: An Experimental Approach," *RAND Journal of Economics*, 20, 1–23.

Bykowsky, M.M., R.J. Cull, and J.O. Ledyard, 1995, "Mutually Destructive Bidding: The FCC Auction Design Problem," Caltech Social Science Working Paper 916, January.

Calsamiglia, X., "Decentralized Resource Allocation and Increasing Returns," *Journal of Economic Theory*, 14(2), 263–283, April.

Jordan, J.S., 1987, "The Informational Requirements of Local Stability in Decentralized Allocation Mechanisms," in T. Groves, R. Radner, and S. Reiter, eds, *Information, Incentives, and Economic Mechanisms*, Minneapolis: University of Minnesota Press.

Kwasnica, A.M., J.O. Ledyard, D. Porter, and J. Scott, 1997, "The Design of Multi-round Multi-object Auctions," Preliminary Paper, delivered to the Public Choice Society, March 21.

Ledyard, J.O., C.R. Plott, and D. Porter, 1994a, "A Report to the FCC and to Tradewinds International Inc.: Experimental Tests of Auction Software, Supporting Systems and Organization," Final Report for Subcontract Number 005 (Prime Contract FCC-94-12), September.

——, D. Porter, and A. Rangel, 1994b, "Using Computerized Exchange Systems to Solve an Allocation Problem in Project Management," *Journal of Organizational Computing*, 4(3), 271–296.

——, C. Noussair, and D. Porter, 1996, "The Allocation of a Shared Resource Within an Organization," *Economic Design*, 2, 163–192.

Milgrom, P., 1995, "Auctioning the Radio Spectrum," in *Auction Theory for Privatization*, Cambridge University Press, Chapter 1.

Mount, K.R., and S. Reiter, 1996, "A Lower Bound on Computational Complexity Given by Revelation Mechanisms," *Economic Theory*, 7(2), 237–266.

Olson, M. and D. Porter, 1994. "An Experimental Examination into the Design of Decentralized Methods to Solve the Assignment Problem with and without Money," *Economic Theory*, 4, 11–40.

Plott, C.R., 1994, "Market Architectures, Institutional Landscapes and Testbed Experiments," *Economic Theory*, 4, No. 1.

——, 1997, "Laboratory Experimental Testbeds: Application to the PCS Auction, "*Journal of Economics & Management Strategy*, this issue.

Porter, D., 1996, "The Effect of Bid Withdrawal in a Multi-Object Auction," Working Paper 982, California Institute of Technology.

Rassenti, S.J., V.L. Smith, and R.L. Bulfin, 1982, "A Combinatorial Auction Mechanism for Airport Time Slot Allocation," *Bell Journal of Economics* 13, 402–417.

Rothkopf, M.H., A. Pekec, and R.M. Harstad, 1995, "Computationally Manageable Combinatorial Auctions," Preliminary Paper, April.

Vickrey, W., 1961, "Counterspeculation, Auctions and Competitive Sealed Tenders." *Journal of Finance*, Vol. 16 (May), 8–37.

CHAPTER 26

Laboratory Experimental Testbeds: Application to the PCS Auction

Charles R. Plott

1. Introduction

The use of laboratory experimental methods in economics has been growing rapidly. With each application, new insights are gained into how the methodology can be used to supplement the more traditional forms of research. Such was the case with the development of the Federal Communications Commission (FCC) policy for the auction of licenses for personal communication systems (PCS). At several different stages the laboratory experimental methods of economics were used. The application differed at each of these stages, representing the different types of relationships that can exist among theory, observation, and policy. This paper is a brief account of the applications.

The use of laboratory experimental data began only after many major decisions had been made by the FCC. The big questions whether or not there should be an auction, what was to be auctioned, and when the auctions were to take place had all been answered. The government had decided that an auction mechanism should be used in place of the classical, administered methods of granting broadcast licenses. The structure of the licenses had been determined, and the time frame and sequences of auctions had been determined.

Wisely, the government separated the decision about the rules that were going to govern the auction from the other decisions. An independent rulemaking process was used to decide what the rules of the auction should be. The FCC's rulemaking process took place in fall of 1993. Decisions were made in the winter and spring of 1994. The first auction of nine (nationwide narrowband) PCS licenses took place in July 1994, and it was followed by the auction of thirty regional narrowband PCS licenses in the fall of 1994. In addition, the FCC conducted an auction of licenses for interactive video data services (IVDS) in July 1994.

By the fall of 1993 the business world was fully aware of the rulemaking process and had engaged many groups of consultants to help them position themselves. Businesses understood that the rules and form of the auction could influence who acquired what and how much was paid. The rules of the auction could be used to provide advantages to themselves or to their competitors. Thus, a mixture of self-interest and fear motivated many different and competing architectures for the auctions as different businesses promoted different rules. The position of the FCC was that the efficient allocation of the licenses was to be the primary criterion for deciding among the competing options. The criterion was not to maximize revenue, and it was not to simply mimic historically accepted methods of conducting auctions. This attitude of the FCC colored the whole rulemaking process, shaped the debates, and generally influenced the character of the rules that were proposed. The efficiency criterion and the openness to new types of auctions also opened the door for experimental methods. As will be outlined in the pages that follow, the experimental methodology provides a noncontroversial, inexpensive, and fast method for getting data on how various types of auctions might perform. While experiments could not remove all controversy, they could at least remove part of it.

The first experiments were conducted in the fall of 1993. In January 1994, a conference was held at Caltech in which much experimental data were reviewed. A group of Caltech experimentalists were hired in the spring of 1994 to help test rules and to help with the actual implementation of the auctions. That relationship lasted through the fall of 1994 and the regional narrowband auctions of thirty licenses. After that the FCC had its own software, procedures, auction team, and experience. The report that follows describes what was done and what was learned during that year of auction decisions and development.

2. Laboratory Methods

Perhaps, before discussing the FCC auctions, it would be instructive to provide a word or two on the nature of experimental economics as applied to the study of auctions. The basic idea is to use substantial financial incentives to create simple and well-controlled auction processes. The methods are used to study a wide variety of decision processes, so it is possible to compare behavior of many different people in many different contexts. The people engaging in the auctions make money that is theirs to keep. The characteristics of the people, the nature of the incentives, the rules of the auctions, what people are told, etc., are all carefully considered and may differ among experiments, depending upon the purpose and what one wants to know. The experimental procedures, like the ones employed in the experiments reported here, are exactly the same as ones that have been subjected to thousands of studies. The results of the experiments are compared internally against theory and other types of experiments, so confidence is built that the results are not due to some sort of special or isolated feature.

The general idea of a laboratory experiment is to study the operation of rules, such as auction rules, in very simple cases. The simplicity assures that the nature of any problems detected can be identified and studied. The variables studied reflect human behavior in the use of the rules, the relationship of behavior to the technology used to implement the rules of the auction, and the reliability of the technology itself. An experimental *testbed* is a simple working prototype of a process that is going to be employed in a complex environment. The creation of the prototype and the study of its operation provides a joining of theory, observation, and the practical aspects of implementation.

Two questions are posed. The first is: *does the auction work*, in the sense that it produces outcomes and efficiencies that are generally acceptable? The second question is *does the auction operate according to the theory that led to its creation?* If a mechanism does not work acceptably in a simple case created in a laboratory, then there may be no reason to think that it will work in the complex cases found in a field application. Such failures are viewed as a failure of *proof of concept*. However, even if a mechanism passes a proof of concept, it might have done so for accidental reasons. Unless the performance is reasonably consistent with theory, unless the mechanism works for understandable reasons, then again, there may be no reason to think that it would work in complex field applications. The second question reflects a requirement that a mechanism meet a test of *design consistency*.

3. An Overview and the Context of Applications

The most intense use of experimental methods in the FCC applications occurred at three different stages of the policymaking process. At each of these stages natural research partnerships could be identified. Initially, the experimental research was focused on broad aspects of the rules that might be put in place. This was the first stage of *testbedding*, as the properties of substantially different types of rules were examined. The second stage of testbedding evolved as the rules began to take a more definite form. The study shifted to detailed features of particular rules and was then expanded to include assessments of the operational form of specific rules as they were implemented in the software. Rules, stated as policy, can be very different when they are put in operational form as procedures and software. Simple laboratory environments provide an inexpensive method of discovering practical problems with the rules, as they are found in the real software setting, that could prove to be very expensive if they surfaced during the operation of a multibillion-dollar auction. Thus, at this stage, the experimental methods were, in a sense, part of debugging. The final stage of application occurred during the actual operation of the auction. Theory, modified by experience in use, is very useful when attempting to make decisions in rapidly changing circumstances. The observations from experiments were used as a source of judgment about events that were taking place during the auctions and the possible implications of changing features of the auction ("improvements") as the auction was taking place. The paper addresses decisions that were made during the auctions and how they were influenced by experiments.

The final rules used by the FCC had several key elements. First, the rules were implemented electronically with decentralized bidders. Second, a separate market was opened for each license. The FCC had reason to think that complementarities existed among licenses. In some cases, the value of a license depended upon the other licenses held, so bidders wanted to commit to buy simultaneously. All markets were open simultaneously. Third, the markets proceeded in "rounds" of bidding. Within a round, bids were submitted for all licenses. Fourth, activity rules were in place. These rules were used to force bidders to bid rather than waiting to see how others bid before submitting their own bids. The activity rules imposed requirements on the number of items on which a bidder could bid. If the bidder failed to bid (unless the bidder was the high bidder), then the right to bid in future rounds could be reduced. The auctions had different stages in which different activity rules were imposed, and these rules became more stringent in later stages when the FCC might be attempting to get the markets to close. Finally, the auction had a withdrawal option. Suppose a bidder valued packages or groups of licenses and that the value fell sharply if any member of the package was missing. A bidder who held part of a package and felt that the other parts would be too expensive could withdraw from the licenses held, letting the price fall to the lowest past bid. If the final price was less than the bid of the withdrawing bidder, then the bidder paid the difference. In addition, the auction had waiver features that allowed bidders to skip rounds a limited number of times without the penalty imposed by activity rules and thereby allow time to assess complicated bidding strategies or seek additional financing.

4. Research on Rules: The First Stage

The first stage focused on the rules of the auction and the consequent behavior that might be expected under various conditions. Different rules can induce different patterns of outcomes in terms of efficiency and distribution, depending upon the underlying economic conditions. Much of the theory that existed at the time of the design of the auction was incomplete and untested. No single theory existed about which there was a consensus. The first stage of experimental work was thus closely connected to the development of theory and a sensitivity to the differences of opinions that existed among theorists.

Research during this first stage was difficult. The rules were not determined. Up to and even during the actual auction, the rules were constantly and rapidly evolving. By practical necessity, and because of the need for information, the experiments typically addressed a feature of the rules, or features of classes of rules, as opposed to fully testing some well-defined set of rules and procedures. Furthermore, because of time constraints, decisions had to be made on very small numbers of observations. The environment in which the actual auctions would take place was similarly uncertain. It was assumed that the items auctioned involved complementarities, and that a large number of participants would be bidding. But, the full implications of these assumptions were never fully explored. Similarly, many relevant environments were not studied at all. For example, while it is well established experimentally that uncertain common values of the items auctioned can result in a winner's curse, the special problems that might surface for FCC rules in common-value environments were not pursued. Time and resource constraints prevented the study of many interesting and important problems.

Testimony during the FCC decision-making process provided a focus for early experiments. Three major issues that surfaced in the testimony were chosen for experimental examination. The first issue was whether the auction would be one of the forms commonly implemented by professional auction houses, such as oral auctions, or would be something completely new. In part, this first issue seemed to turn on whether it was technologically feasible to do something new and completely different from the time-tested methods of auctioning things. The second issue was focused more narrowly on the definitions and characteristics of particular classes of rules: whether the licenses should be sold sequentially or simultaneously. The third issue was similarly focused on the details of the rules, and the type of behavior that might be observed under different rules. As will be discussed, this third issue was closely associated with the economic environments that might exist for the FCC auctions. The expected behavior can depend dramatically upon the environmental features present. The research focused on rules that might be able to operate in troublesome environments.

4.1. Something Old or Something New

The most commonly used auction rules, as implemented by professional auction houses, are sealed bids and oral, ascending-bid auctions in which items are sold sequentially. Experimental research suggested that sealed-bid auctions would not function as desired. Almost all experimental work suggests that some sort of iteration is necessary for processes to have the efficiency and price-discovery properties suggested by pure theory. Equilibration (and thus disequilibrium) seems to be a fact of life. While no experimental work was conducted to explore this particular issue, many experimental sealed-bid auctions have been conducted, and that literature was used for reference.

One early issue was whether the classical oral auction should be used, as opposed to a more technologically oriented process. Some voices in the FCC were skeptical of the advisability of using new technologies that had no track record in the field. The question was whether or not people could operate in the type of technological environment characteristic of new types of processes that were being suggested by theory. The fear was that the behavioral/cognitive demands required by the processes would render them infeasible.

These early issues were brought into focus at a meeting held at the California Institute of Technology in January of 1994. Experimentalists addressed this issue directly by demonstrating the operation of decentralized electronic auction processes. Computerized auctions have been operating for years in laboratories, where they have been used in experiments. Laboratory experiments have demonstrated conclusively that people are generally capable of dealing with the "technologically intensive" processes that are applied in modern electronic and computerized auctions. They have also demonstrated that the necessary software and hardware exist in operational form and can be made reliable. Thus, at the Caltech conference, the operation of new processes, based on electronic technology, was demonstrated.

In addition, conference presentations were made by the Pacific Stock Exchange and other parties familiar with the operations of electronic and computerized market processes. Thus, while the issue of the tried and true against something new continued to be raised in some in the debates, experimental data existed that could be used as an answer to those concerned about the issue and, in some respects, clearly demonstrated that the problem of information and cognitive limitations of people would not be an insurmountable obstacle to the implementation of new types of auctions. It also became clear that technology was not an obstacle. The consensus developed that new rules could be used, and the discussion moved to consider the forms that they might take.

4.2. Simultaneous Auctions vs. Sequential Japanese Auctions with a Package Bid

Very early on the discussions became narrowed to two different auction architectures. These two competing architectures were the focus of many of the early experiments.

The experiments were designed to inquire about the properties of simultaneous auctions in comparison with sequential auctions that are accompanied by bids on predetermined packages. In the latter architecture, a specific set of items would be offered for sale, either as a package or individually. Sealed bids would be tendered for the package. The winning bid for the package would be announced. After the announcement, the markets for individual items would be opened and the items would then be individually auctioned. Whether the sale was made by package of items to the winning sealed bid, or by individual items to the winners of the individual-item markets, was to be determined by which would generate the most money. The details of the institutions studied are as follows:

4.2.1. Simultaneous, Continuous, Ascending-Price Auctions for All Items (With and Without Release-to-Market Provision)

Within this set of rules, each license would be identified in a separate market. All markets would be open simultaneously. Bids could be tendered at any time the bidder desired. An accepted bid must be higher than the existing bid. All markets would close at the same time, when no market had received a bid for some predetermined period of time. That is, if no bids were tendered in any market for a set period, then all markets would close simultaneously; but if a bid occurred in any market, then all markets would remain open for at least the predetermined period.

The method of ending the auction is very important and figures heavily in the rules finally adopted by the FCC. The FCC auctions were not continuous, but instead proceeded in rounds. In a continuous market, the continuous threat of the end serves to force bidders into action. If there is no action, then the markets close, and the faster the bidding, the sooner the auction will be over. The introduction of rounds gives bidders an incentive to wait. Thus, the FCC adopted activity rules and rules governing eligibility that are not part of the earliest experiments.

The release-to-market provision (withdrawal) gave bidders an opportunity to withdraw from units on which they had the high bid. If a unit is released to the market by a bidder, then the bid price is dropped and the bidding can then start from the lower level. The bidder who withdrew from the item would pay the FCC the difference between the bidder's bid and the final bid at which the item sold. The idea is that a bidder who failed to get a package would be able to sell the partial package already acquired back to the market. Reselling during the auction, as opposed to after the auction closed, might be advantageous because during the auction the demand might be expected to be strong due to the assembled buyers at the auction.

4.2.2. Sequential, Continuous Auctions with a Sealed Bid for Packages of Predesignated Collections of Items

Under this proposal, the sealed bids would be opened prior to the opening of the markets in which individual items would be auctioned. Two different possibilities existed to govern the sequence chosen for the individual item auctions: (1) Japanese auctions would be conducted for each item, with items sold in random order; (2) Japanese auctions would be conducted for each item, with individual items auctioned in the order from highest expected value to the lowest.

Four institutional features need to be emphasized. First, the auctions for individual items are continuous. There are no rounds or stages in the bidding, so the termination rule is that the auction remains open until only one person is left. Second, the Japanese auction is an ascending-price auction that differs from an ordinary one only in the way that bids are tendered. The price goes up at a pace determined by the auctioneer. All bidders are considered to be "in," that is, agreeing to purchase at the stated price, unless the bidder has explicitly chosen to "drop out." A bidder who has "dropped out" is no longer an active bidder on the item and has no standing to buy it, regardless of the final price. The price is determined as soon as only one bidder remains "in." That is, the auction stops when the next to the last bidder "drops out." The person valuing the item most gets it at the value of the bidder with the second-highest bid value.

The third dimension of the rule is the sequence. The items are sold one at a time in order. In one case, the order is randomly determined. In another case, the items are sold in the order beginning with the item with the highest expected value. In the experiment the values of items are randomly drawn with publicly known distributions conditional on the item. Thus, the item for which the expected value is the highest is sold first. In the field, there is often common agreement about the items that are likely to bring the highest prices when offered. That feature of common agreement is captured by the experimental procedure.

The fourth dimension of the rule is a sealed bid for a package that is opened before the auction. That is, the results of the sealed-bid auction are to be made public before bidding on the individual items begins. The collection that constitutes the package is designated prior to the auction. If the items, when auctioned individually, do not command prices that total more than the items would bring if sold at the winning sealed bid, then they are sold as a package to the bidder tendering the highest sealed bid. If the items command a sum of prices from the individual auctions that is greater than the sealed bid, then they are sold individually.

4.3. Rules and Performance

The overriding question posed for research was related to the efficiency of the allocations fostered by the auction rules. Closely related questions concerned the mechanisms through which the rules operated. What were the sources of any observed inefficiencies? In particular, could packages be efficiently assembled from independent markets, or could bidders for independent components compete successfully against a bid for a package? Who was advantaged or disadvantaged in different architectures? What was the revenue-generating potential of the different rules?

Experiments were conducted with seven and with nine items for auction. Each agent had a private value induced for each of the items offered for auction. In some cases, agents had a superadditive value for a collection of all items. *Superadditive* means that the value of the collection of all items was greater for the agent than the sum of the values of the items when considered individually. In other cases, agents had superadditive values for a specific collection of three of the items.

The key feature of parametric configurations was whether the superadditive value for the collection was greater than the sum of the highest (first) values of the items considered independently, or was less. If the superadditive value of the collection is greater than the sum of the first values, then the efficient allocation is that the items should be sold as a package to the agent with the superadditive value. If the sum of the first values is greater than the superadditive value, then the efficient allocation is that items be sold individually. If the auction fails to deliver the items to the hands of the agents valuing them the most, then the efficiency of the auction suffers. An auction that operates at 100% efficiency has managed to deliver the items exactly to the agents who value them the most.

In order to make data comparable across institutional treatments, the same environmental parameters were conducted in the same sequence of periods for the institutions compared. Thus, subjects in different institutional treatments were exposed to the same sequence of values. Of course, the subjects differed across institutional treatments. If the experiment involved sequential auctions, with items auctioned from highest (expected) value to lowest value, then the agents were informed of the probability distribution from which agent's values were drawn.

Pressures of time and money substantially limited the amounts of experimental data that could be collected. Had the FCC developed a research and funding strategy to facilitate a confident and fully scientific approach, the data would be much richer and decisions would have been made on much more reliable evidence, but that was not the case. The research environment was much more akin to a management situation in which judgments were to be made and having some data is better than having no data. The strategy was to select certain key aspects of the parameter/theory space and collect such data as one could. The experiments were chosen to highlight and explore key points of interaction among competing theories. Exactly how these selections were made and the nature of the statistical arguments that might be made involve detail that cannot be reported here. Because of the limitations on data, the key results are labeled as "observations" and the data are "illustrative," as opposed to the frequently used terminology of "result" and "support."

In spite of the limitations on the number and variety of data, the patterns that do exist in the data are unambiguous. The first observation summarizes the overall pattern by using efficiency comparisons. The auction system efficiency suffers in the presence of a sealed package bid and sequential Japanese auction procedures as compared to the simultaneous auction alternative.

Observation 1: The overall efficiency of the simultaneous auctions (with a release provision) is higher in all experiments than the Japanese auctions with a sealed bid.

Data illustration. Figure 1 contains a good example of the data. Compared here are two experiments under the same conditions in each period, for a sequence of periods. The individual parameters changed each period, and shown at the bottom of the figure is a notation that indicates if the efficient allocation has items allocated to different individuals (I) or has a collection of items sold to one individual (C). Notice that in twelve out of fifteen periods the efficiency of the simultaneous auctions is at least as high as the Japanese counterpart, and in seven periods the efficiency of the sequential

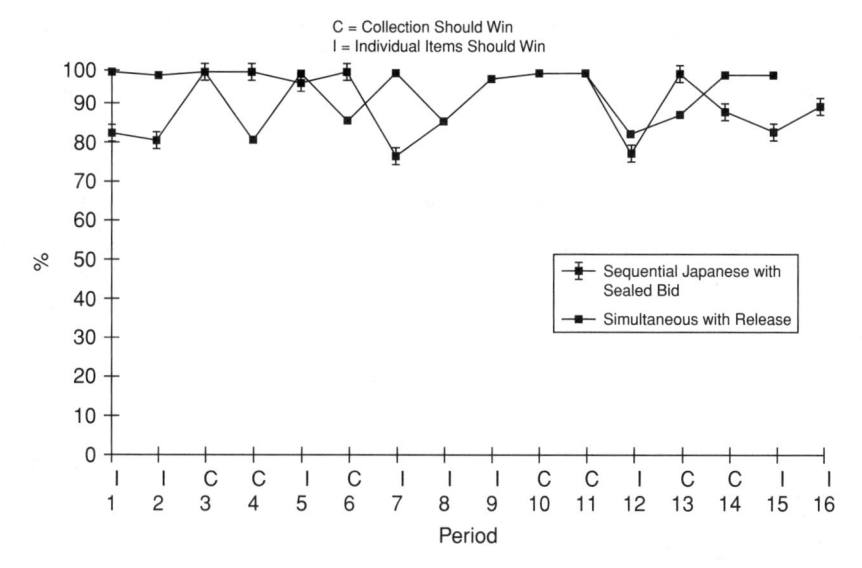

Figure 1. Efficiency by period: the simultaneous process has an efficiency edge.

auctions are strictly better. In only three of the fifteen periods is the efficiency of the Japanese auction higher than that of the simultaneous auction with release.

Figure 1 contains hints for additional observations. Notice that the instances in which the efficiency of the Japanese auction exceeds the efficiency of the simultaneous auctions are almost always those in which the sealed-bid package is supposed to win according to the efficient allocation. This suggests that the sealed-bid option creates an advantage for the package. The next result makes that property clear. □

Observation 2: The existence of the package bid option creates an advantage for the agent who wants the collection defined by the package.

Data illustration. Tables I–III contain relevant data. The data are divided into two cases. The first case is one in which the value of the collection to some single agent is greater than the sum of the highest agent values of the items considered individually (and thus should be sold as package to a single agent). The second case is where the maxima of agent values, when considering the items individually, sum to more than the value of the collection to any single agent. In this second case, the items should be sold individually to different agents. Under all conditions studied, when a package bid exists as part of the rules, the collection is sold as a package to a single agent when it should be sold that way, but in about half of the cases in which the items should be sold individually, they are nevertheless sold to the single individual who wants the collection as a package. By comparison, under the other auction rules, in which no package bid is tendered, the collection is never sold as a package when it should not be. The relative advantage of a package bidding process to the agent wanting the package is clear. □

Comparative experiments between the simultaneous auctions and sequential Japanese auctions with a package sealed bid yield the properties summarized by the next two observations.

Table I. Seven-Item Experiments with Two Collections of Three Items with Superadditive Values[a]

		Japanese Auction		Simultaneous Auction with Withdrawal Provision
		Sealed Bid for Package, Random Order	No Sealed Bids, Random Order	
Value of the Package is greater than the sum of the first values	Collection was successfully assembled	2 successes/ 2 periods	2 successes/ 2 periods	5 successes/ 6 periods
	Collection assembled was profitable	2 of 2 assembled collections were profitable	1 of 2 assembled collections were profitable	5 of 5 assembled collections were profitable
Value of the package is less than the sum of the first values	Collection was successfully assembled	5 successes/ 10 periods	0 successes/ 10 periods	0 successes/ 6 periods
	Collection assembled was profitable	5 of 5 assembled collections were profitable		

[a] (Number of times event occurred)/(number of times event was possible).

Table II. Nine-Item Experiments with Superadditive Values for All Nine Items[a]

		Japanese Auction with Sealed Bid for Package, Random Order	Simultaneous Auction	
			No Withdrawal Provision	Withdrawal Provision
Value of the package is greater than the sum of the first values	Collection was successfully assembled	2 successes/ 3 periods	4 successes/ 9 periods	3 successes/ 4 periods
	Collection assembled was profitable	2 of 2 assembled collections were profitable	2 of 2 assembled collections were profitable	3 of 3 assembled collections were profitable
Value of the package is less than the sum of the first values	Collection was successfully assembled	5 successes/ 15 periods	0 successes/ 8 periods	0 successes/ 3 periods
	Collections assembled was profitable	5 of 5 assembled collections were profitable		

[a] (Number of times event occurred)/(number of times event was possible).

Table III. *Nine-Item Experiments with Superadditive Values for all Nine Items*[a]

		Japanese Auction with Sealed Bid for package, Ordered by Expected Values	Simultaneous Auction with Withdrawal Provision
Value of the package is greater than the sum of the first values	Collection was successfully assembled	7 successes/ 7 periods	4 successes/ 6 periods[b]
	Collection assembled was profitable	7 of 7 assembled collections were profitable	4 of 4 assembled collections were profitable
Values of the package is less than the sum of the first values	Collection was successfully assembled	6 successes/ 9 periods	0 successes/ 8 periods
	Collection assembled was profitable	6 of 6 assembled collections were profitable	

[a] (Number of times event occurred)/(number of times event was possible).
[b] Two never tried to get the collection.

Observation 3: A bidder wanting the package is advantaged by the Japanese auction with a sealed bid for a package, as compared to the same bidder operating under the simultaneous auction rules:

(i) The sealed-bid process always produces an assembled package when one should be assembled, while the simultaneous auctions sometimes fail to produce a successfully purchased collection.
(ii) The sealed-bid process frequently produces an assembled package when it should not produce one, while the simultaneous auction never produces a package when it should not.

Data illustration. Tables I–III contain relevant data. In Table I the results of experiments with three item packages are shown. First, consider the sequential Japanese auction with and without a package bid. When it was possible to submit a bid for a package, the package always won when it should have won (2 of 2 possibilities) but it also won half of the time when it should not have won (5 of 10 possibilities). The package was always profitable. Consider now a comparison with the simultaneous auction. In the three-item collection experiments, the collection was almost always assembled when it should have been (5 of 6 possibilities) and was never assembled when it should not have been (0 of 6 possibilities). In Table II, the nine-item collection was successfully assembled when it should have been in most of the instances under both the Japanese auctions and the simultaneous auctions with the release provision. However, in the cases in which the collection should not have been assembled, it was nevertheless successfully assembled in one-third of the possibilities (5 of 15 possibilities) under the Japanese auction with the sealed bid, but it was never successfully assembled when it should not have been—when the auction was operating under the simultaneous auction rules. □

Observation 4: The existence of a sealed bid harms the profits on the items that come late in a sequence of auctions.

Data illustration. Under the sequential auction rules, those that win the early auctions have an incentive to bid up the prices of the items that come later in the sequence. These bidders do not want to win the items, but they do want the prices to be high, so the total of the collection will be above the sealed bid. For example, in a paired experiment of identical parameters, the price of the final item auctioned under the sequential Japanese rules was higher than the same item sold under the simultaneous auction rules (with no package bid) in eight of fifteen periods, while the reverse was true in only three of the fifteen periods. In the remaining four cases, the prices were essentially the same. □

The next observation is implicit in the discussion of the observations stated above. It is simply stated without a review of the data.

Observation 5: An order of individual item auctions, from the highest-to the lowest-valued items, creates an advantage for the sealed bid.

The final observation explores the sources of inefficiency in the simultaneous auctions that were studied. The data show that the inefficiencies were not so much from a failed attempt at putting together a package, as from a failure to make any attempt at all. Inefficiency was not so much due to the difficulty of coordination, as it was to the perceived risk in the attempt.

Observation 6: Inefficiencies in the simultaneous auctions are due primarily to a failure of the agent with the highest value for the collection to attempt to buy the collection. The release rule reduces the perceived risk of attempting to buy the collection and thereby improves efficiency.

Data illustration. Package assembly, under the parameters studied (a competitive equilibrium exists), is almost always successful when attempted, and successful packages have always been profitable. In the three-item cases the collections were purchased in five of the six instances in which it was efficient to do so, and in the nine-item case it occurred in three of the four times. The most dramatic departure from success was in the seven-item experiments, in which only four attempts were made in seven instances. The power of the release provision is shown by the nine-item experiments in which no release provision existed. Without the release provision, in only four of the nine instances in which the collection should have been sold to a single individual did the individual attempt the purchase, and then only two of these attempts were profitable. □

The exact behavior of the auctions can be sensitive to very subtle details of how the auction process operates. For example, as is noted in Observations 4 and 5, there is a tendency to drive up the prices of competitors, especially when it may help increase the sum of individual values of a package to exceed the sealed bid on the package. This strategy is risky in that a bidder may end up a winner of unwanted items. In the Japanese auctions, bidders seem to become emboldened when they have information about the number of other bidders that are "in." A bidder seeing several other bidders "in" is willing to stay in "a little longer"—contributing to a type of *bubble* that drives the price up further than it would have been if information about the number of other bidders had not been present.[1] Even if the information is not officially available as part of the organized auction, the procedures may be such that it can be inferred. For example, if all bidders are in the same room, and if exit from the auction is accompanied by a click of a key, or a blink of a screen, or any number of other subtle sources of information, such bubbles might exist even when efforts are made to prevent them. The discovery of such phenomena underscores the need to study the operational details of auctions.

Summarizing all observations leads to the following conjecture about the implication of the rules when implemented in environments such as those in the experiments:

1. The simultaneous auctions with release are more efficient than are the Japanese auctions with a sealed bid for the package.
2. The existence of the sealed bid for a package creates an advantage for the agent wanting the collection, and it creates a disadvantage for those wanting items late in the auction sequence (perhaps the smallest agents).
3. Inefficiencies in the simultaneous auctions are primarily due to the fact that agents who would have a collection in the efficient allocation never attempt to assemble the package because of risk aversion. When such agents do attempt to get the collection, they can succeed.

4.4. Lurking Problems and Alternative Rules

Soon after the Caltech conference, the FCC began to focus on simultaneous auctions with withdrawal provisions as the appropriate set of rules. As the rules began to take form throughout the rulemaking process, the research began to focus on related issues. A primary concern of researchers, but not necessarily the FCC, was the sensitivity of the behavioral characteristics of the auction process to the environment in which it might be operating. While much information had been produced about the general properties of the simultaneous ascending-price auction with a release provision in comparison with other rules involving sequences of auctions, questions remained about how this set of rules might perform in special environments that might reasonably be expected to be present in the circumstances in which the FCC auction would be operating. How were the rules going to perform under the economic environmental circumstances that were thought to exist? Do potential problems exist (in light of nonconvexities, superadditive values, and uncertain common values)? If problems exist, are they generic in the sense that they would almost certainly be encountered, and does the interdependence that they foster promote other types of behavior, such as collusion?

It is well known that nonconvexities and superadditive (complementary) values can destroy the existence of the equilibrium in the competitive model and can also cause instabilities. However, very little is known about what might happen in actual markets with these properties, and during the early stages of rulemaking, nothing was known about the behavior of the particular rules ultimately adopted by the FCC. Figure 2 can

---

[1] There seems to be no theoretical foundation for this phenomenon, since expectations of the actions of others could cause the same behavior. Nevertheless, in experiments with the information removed such bubbles were less pronounced if they existed at all. Of course, the numbers of experiments are very small here, so the data can be relied on little more than developing an intuition about what a more complete study would show.

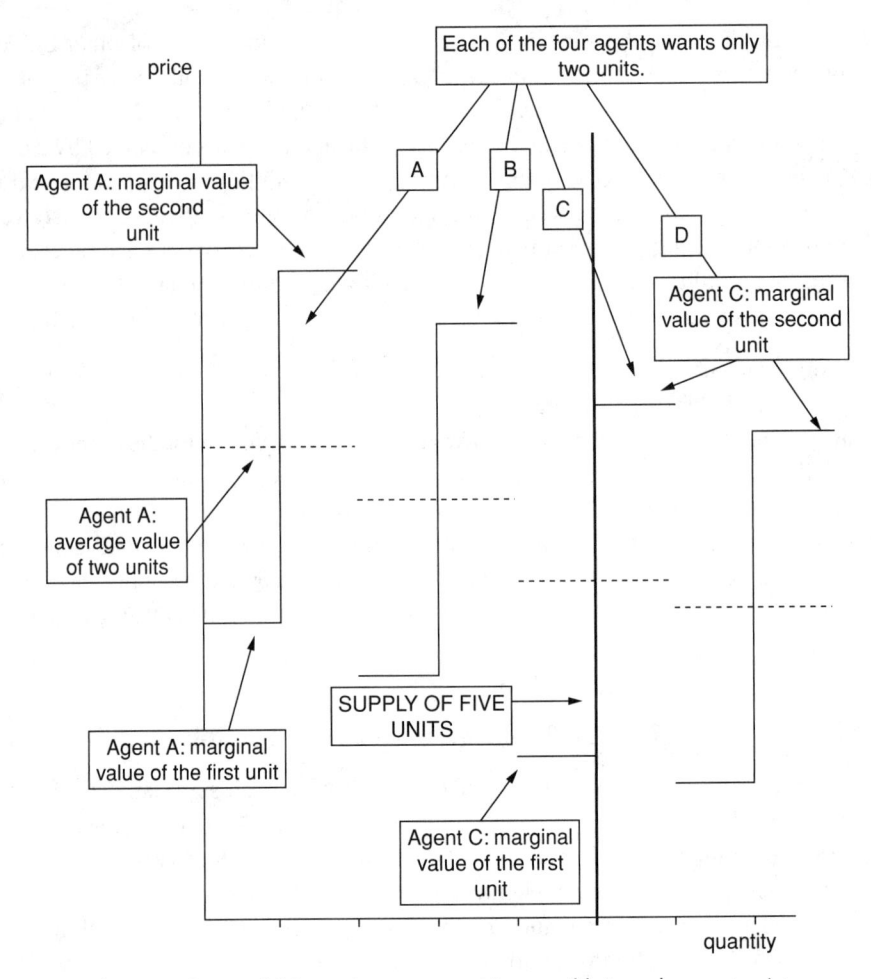

Figure 2. Superadditive values: competitive equilibrium does not exist.

be used to demonstrate the nature of some of the lurking problems in a very simple example, which can be applied to both the competitive model and the FCC auction rules.

Suppose the world consisted of four agents with strong complementary tastes for only two units. The essence of complementarity is that values of sets of items are greater than the sum of the items when considered independently. In this case, pairs of licenses have strong complementarities. The complementarity can be seen by the fact that the value of an item depends on whether it is the only item held or is held in the presence of another item. One could say that a special synergy exists for two units, but agents place zero marginal value on the third unit. In the figure the marginal values of each of the four agents are displayed in the order of the average value of two units, starting with the agent with the highest average value, agent A, and continuing to the agent with the lowest average value, agent D. Only five units are offered for sale.

The example and the similar examples that follow, can be used to make two points: (1) the auction can lead to losses by participants; (2) there may be instabilities or "cycles" that delay termination or even prevent the auction from stopping.

In order to develop an intuition about how such phenomena might occur, notice first that no competitive equilibrium exists. There is no price that equates supply with demand. At any price below the average value for agent C (the marginal buyer), demand exceeds supply, and at any price at or above the average value for C, supply exceeds demand. The fact that licenses are lumpy creates a nonconvexity in the environment that destroys the existence of that type of equilibrium.

Now, notice that the complementarities create an instability at the margin that has a marginal propensity to push prices up. This property can lead to a loss. Suppose agent C acquires one unit in a simultaneous ascending-bid auction; then, if the agent follows a local, marginal adjustment, (s)he is willing to bid prices up to the marginal value of the second unit. Since all units are identical, there must only be one price in the market, and at the price of the second unit for agent C, all agents, including C, lose money. The price is above the average value of all agents. Thus, as one can see, in this environment these rules have a potential, theoretically, for leading bidders into circumstances in which they can suffer a loss even though all "local" decisions are profit-improving to the deciding agent.

If the rules contain a release clause, then theory gives no guide to how the process might stop. If prices ascend to above the average value of any of the agents, then (s)he may want to release the unit to the market and take a certain loss, rather than test the competition into larger regions of loss. When the price is sufficiently low, it could attract a new buyer (such as D) and start the spiral upward again. Thus, from a naive, theoretical point of view, the auction could experience a series of withdrawals over long periods of time and never attain a natural closing within an acceptable time frame. Prices would just cycle.

Experiments demonstrated that the theoretical possibility of losses is also a real possibility. Figure 3 contains the data from an experiment with parameters of the form discussed above. The rules were very similar to those ultimately used by the FCC, except that the auctions were continuous as opposed to involving rounds, and there were no activity rules because of the nature of the stopping rule. A separate market was open for each item, and all markets were simultaneously open. The rule for each market was an ascending-price auction in which any bidder could submit a bid on any item at any time as long as the bid was greater than the previous bid. All markets closed simultaneously when no market showed activity (bids) for some fixed amount of time (e.g., a minute). This stopping rule seemed to eliminate the need for activity rules. Individuals could withdraw or release an item to the market. When a withdrawal occurred, the price was reduced to zero and all bidding started over. If the final price was below the bid that had been made by the withdrawing bidder, the bidder paid the difference.

In Figure 3 the horizontal lines represent some of the important parameters of the experiment. The top horizontal line is the counterpart of the marginal value of the second unit for agent C. If the price is bid to the level of the top horizontal line, then losses will certainly occur. The two middle lines are the counterparts of the average value for C and D, respectively. The data represent the bids on all items and the time of submission, measured as the number of seconds that elapsed from the opening of the auction. As can be seen, the time series of bids begins low and continues along what appears to be an exponential path, with some interesting waves, until the auction ends. As is readily observable, the prices of items tend to equate and finally settle near the average

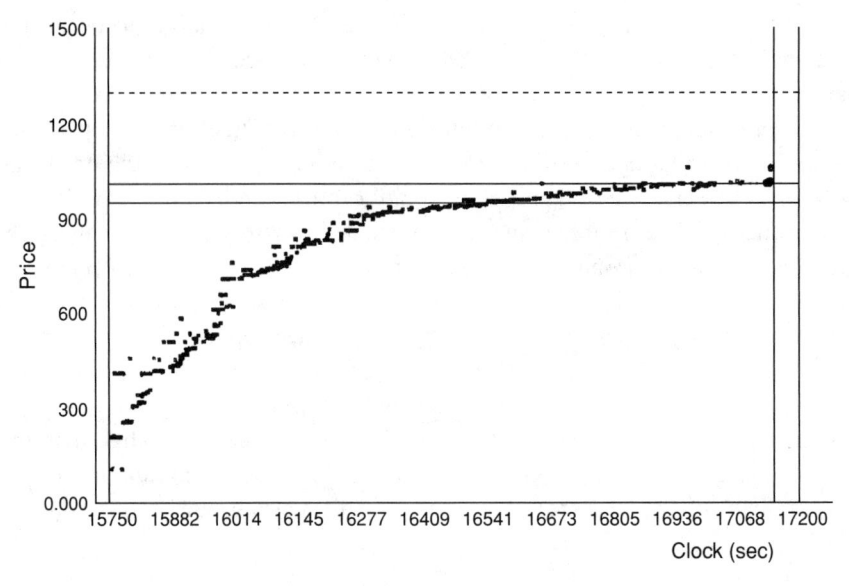

Figure 3.

value of the marginal agent. Of course, since the marginal agent had a unit at this price, (s)he lost money.

Experiments also demonstrated that the theoretical possibility of cycles is also real. Figure 4 contains data from two additional experiments. Again, the data follow a roughly exponential path toward the average value of the marginal agent. Again, jumps or waves are present. However, in panel A of Figure 4 an item is released. According to the rules of this auction, any released item begins at a price of zero, so a sequence of

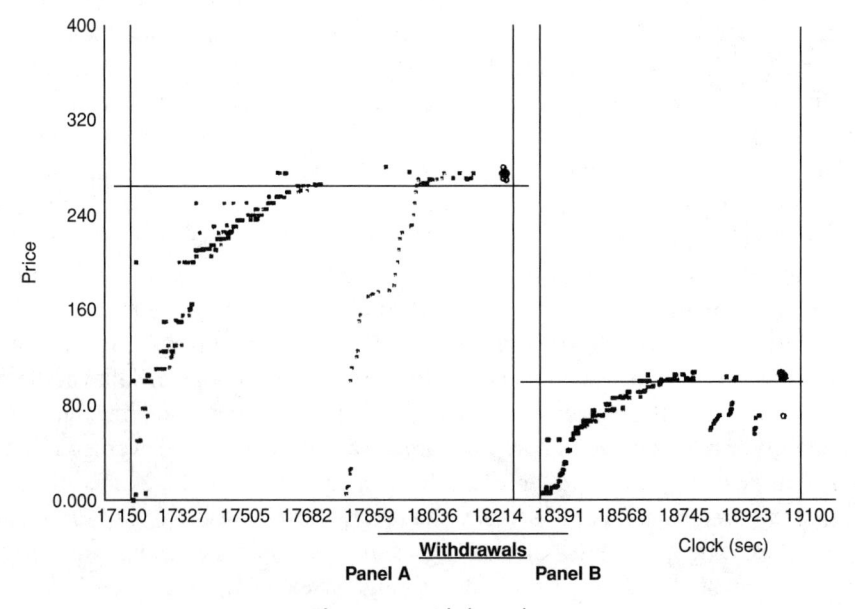

Figure 4. Withdrawals.

bids that bring the price of the released items up to the levels of the others is observed. Since the new price of the item is above the average value of the marginal person, the new holder lost money. Panel B shows that releases can occur more than once during an auction. As can be seen in that experiment, the item was released two times, leading to a cycle of length three.

The existence of strong complementarities creates a special kind of competition that does not facilitate efficiency. Complex synergies foster complex *fitting* problems from which substantial gains are possible. This process of coordination that is necessary for fitting might be accompanied by other activities considered undesirable. If a competitor detects that a rival or two have managed to coordinate their actions so they fit, the firm can damage them both by getting into the bidding action. This is a type of *destructive competition* in which one agent can make it difficult for a competitor to obtain a package, even when the package represents an efficient allocation. An agent might try to damage others that (s)he views as competitors by driving up the prices, or by acquiring key elements of a package that is of special value to a competitor or to the fit of a group of competitors. Evidence of the ability and the willingness of agents to engage in bidding activities intended only to influence the allocations of other agents was contained in Result 4. Further evidence was exhibited in by the willingness of competitors to drive up the price to rivals in the Japanese auction. No systematic evidence exists on the nature of this type of competition. Indeed, we do not understand the role it might play in the allocation process. But, what we do have suggests that it cannot be simply dismissed as implausible.

If selective competition is thought to be a problem, one response might be to shield the identities of the agents. Firms would then be unable to collude and would be unable to identify particular rivals that they would be willing to damage. However, such a response has its difficulties. If a firm is attempting to put together some sort of package, then other firms might be advised to explore packages that fit. Successful fitting would be difficult to achieve by simple random bidding: it requires an understanding of the firm's intentions. Indeed, it might be important to call attention to a coordination problem or to force coordination concessions by tendering bids that could be interpreted as destructive. However, to understand the other firm's strategy, it is necessary to be able to identify that firm through bids. Multiple identification numbers have been suggested in this context. A firm could use some identification numbers to signal intentions where needed and other identification numbers to hide. These complex strategic possibilities have not been explored. We only know that the potential exists and that there might be institutional "fixes."

The final environmental problem stems from the fact that the PCS licenses are thought to have a common but uncertain value. It has been well established experimentally that in such environments a *winner's curse* phenomenon can exist [for a review of the literature, see Kagel and Roth (1995)]. Each bidder has private information about the common value. If this information is distributed with the true value as a mean, then the highest privately estimated value is greater than the true value. If propensity to bid is positively related to privately estimated value, then the high bidder will be the agent whose private value is the highest above the true value. Unless this property of auctions is recognized, the agent will bid more for the item than its value and as a result suffer a loss from the auction. How this might work out when there is a sequence of bids and complementarities is simply unknown. No experiments have been conducted that provide an assessment of what the dimensions of the problem might be.

In summary, the FCC auction is exploring domains of economic environments about which very little is known: nonconvexities, complementarities, and asymmetric information. The rules of the auction were designed specifically to cope with parts of these environmental conditions. The simultaneous nature of the auctions and the withdrawal feature were specifically implemented to facilitate efficient allocations in the presence of such features. Nevertheless, problems are lurking. It is easy to find parameter values at which agents suffer losses, the system cycles, and the results are not even close to 100% efficient. Does withdrawal help in the very difficult cases, or does it simply lure people into traps in which they can lose money?

Modifications of the simultaneous auctions are still a subject of research. One issue is whether or not the process should be continuous, as opposed to the stages that are now used. Continuous auctions have many advantages. In particular, with continuous auctions there is no obvious need for activity rules and the related complex stopping rules. The stages seem be a response to businesses expressing a need to have time to make decisions, garner the capital for big purchases, and assemble the information needed for bidding. The stages also seem to reflect some doubt about the ability of technology to facilitate a continuous auction.

A lack of confidence in technology, as well as a lack of theory, seemed to dampen enthusiasm for the implementation of a "smart market" that would be capable of dealing with complex bids for packages of licenses. Many experiments have explored the use of package bidding in the context of electronic markets.[2] Such markets have demonstrated a capability of solving very hard coordination problems. How they might be made to manage common-value problems or destructive competition remains to be determined.

5. The Development and Implementation of Auction Technology: The Second Stage

The second stage of applications involved testing the implementation of the specific rules selected to govern the auction. The problems that were addressed stemmed from three sources.

First, the language of lawyers and those writing policy is not precise from the point of view of game theorists, who attempt to model the behavior of the system. Terms that make sense from the point of view of the law can be very imprecise and, depending upon interpretation, could have dramatic effects on the structure of the auction and consequent behavior.

Second, complex systems of rules involve subtle interactions and ambiguities. Rules must be internally consistent, and they must be complete in the sense that an outcome is produced by the process under all circumstances. The complex ways in which the rules interact, and the presence of ambiguities, do not become evident until one tries to actually implement the rules in an operational environment. Thus, as part of the research

---

[2] There are different styles of such markets. For examples see Banks et al. (1989), Brewer and Plott (1996), Plott and Porter (1997), Rassenti et al. (1994), and Rassenti et al. (1982).

it was necessary to evaluate the consistency and completeness of the rules themselves, as opposed to any assessment of behavior within the context of the rules. A laboratory experiment requires the translation of policy concepts to operational concepts, so the process of experimentation actually produces a working auction prototype.

The third source of problems stemmed from the software and hardware. It was necessary to determine if the software and procedures of the auction successfully implemented the rules, as stated in policy. Even if the rules were complete and consistent in one implementation, they might not be in another. Software implementation and auction procedures must be explored from the point of view of game theory and the strategic opportunities that a real environment fosters.

A group of economists from Caltech[3] was contracted to test the software and advise on rules and their implementation. Cantor-Fitzgerald was contracted to develop the software for the first narrowband auction in July, 1994. After the July auction, another contractor was selected to develop software for subsequent auctions. Evidently, the FCC wanted to own the software and this was not consistent with the interests of Cantor-Fitzgerald, which used modifications of software that the company uses for market making in the bond industry. The Caltech team was associated with the development and implementation of both technologies.

There are several problems that exist in the many steps between policy conception and operational implementation in the field. First, the exact "rules" were always in a state of evolution. There is a learning that takes place as the rules are implemented. The interactions among rules are subtle. A conflict exists between fairness and proper price discovery, and as this conflict is discovered, there is a tendency for the latter to give way to the former under the pressure of politics. For example, if the auction is taking too long and must be stopped, how should that be accomplished? Many reasonable answers to such questions advanced, such as a single best and final offer, closing specific markets in which no new bids have been tendered, requiring bidders to bid only on items on which they previously bid, etc.

It was perceived that a person might have special reasons not to want to bid, so waivers were invented, and then automatic waivers were required of the software. The concepts of withdrawals, eligibility, increments, and announcement of stage changes all involve reasonable-sounding concepts when considered alone, but there remain questions about how they might interact with each other, with other policies, and with the realities of software performance. Can one waive and bid at the same time? What happens if you withdraw at the end of the auction: should the auction remain open so the withdrawal can be cleared? How shall a withdrawal be priced? How is eligibility of everyone influenced by withdrawals? Should it go up so anyone can buy the item released to the market? How is eligibility influenced by increments: should eligibility be lost if increments are reduced because of lack of bids? As these interactions become discovered, there is a tendency to change the policy.

In the first FCC auctions, there was a tendency for policy changes to take place without a full recognition of what they might mean for software development and the time needed for that development and testing, as well as the likelihood that small changes in policy would create a need for further changes that would be discovered only as the

---

[3] They were John Ledyard, Charles Plott, and Dave Porter.

implementation advanced. More importantly, the technical complexity and subtleties of game implementation were not fully recognized and reflected in the procedures for communicating with software developers. Moreover, the importance and the technical complexity of the link between policy and rigorous institutional design was not fully appreciated, and that led to problems that were potentially very severe.

When the Caltech team first tested the Cantor-Fitzgerald software, they discovered that Cantor had not been properly informed about the time line within an auction. The contractor was unaware of the existence of rounds. It was also unaware of the time line within a round (that there was a bidding period followed by a computation period that was followed by a withdrawal period, etc., within each round). The policy language was not sufficiently precise to identify exact actions with time. For example, Cantor had been led to believe that phase 3 was a "stay in your own lane" policy in the sense that bidders could only bid on the exact item on which they had been bidding previously. In fact, phase 3 required that the number/size of licenses could not be expanded, except possibly by very little. A week before the first auction, the Cantor programming team met with the Caltech team and the FCC. From that meeting, a complete reprogramming of the auction software was undertaken according to an architecture that previously had not been communicated to Cantor.

Time pressure before the first auction gave very little opportunity to test the Cantor software before the July auction took place. Because of these difficulties, the FCC felt the need to provide a nonelectronic backup system for the July auction. It hired the Caltech team to develop one. This backup was made possible by the fact that all bidders were at the same location, a hotel meeting room. The Caltech experimentalists had practice and experience in implementing such auctions in laboratory environments. To the Caltech group it was simply a bigger experiment than the ones that they had been conducting all along. When implemented during the July auction, the backup system ran in parallel with the Cantor electronic system, which operated successfully. The backup system was just about as fast as the Cantor system, and also operated without flaw.

After the July auction, the FCC contracted with new software developers. The Caltech group agreed to test the new software to make sure the rules were appropriately implemented. However, testing was made very difficult by FCC policies. The FCC adopted a policy of not letting the Caltech team have access to the final software, study (or see) the code, or even talk directly to a software developer. Thus, the auction process was a "black box" from the point of view of the Caltech testers.

The strategy for dealing with the problem of testing was to implement a three-part system. First, the overall strategy adopted by the experimentalists was to use the software as it would be used in an experiment. Preferences were induced by application of standard experimental economics techniques. Subjects then used the software in a series of actual experimental auctions that lasted several days. During these experiments the subjects were at Caltech, but the computers and the FCC auctioneers were in Washington, DC. This methodology facilitated learning derived from user experiences who were engaging the equipment as it would be engaged in practice.

The second part of the strategy was to get problems identified and documented as they were revealed. The heart of the second part of the strategy involved payments (sizable bounties of one hundred dollars or more) to subjects able to find errors in any facet of the auction system. Student subjects from Caltech were trained at the beginning of the summer. They completely understood the details of rules and subtle variations of the rules. The same subjects were used over and over for anything that dealt with software tests. The subjects were paid for keeping notebooks and diaries, so a clear record was maintained about the time and the state of the system when errors were (asserted to be) found. This second procedure allowed us to utilize the special knowledge and skills of this trained subject pool. The subjects explored the rules, the auction setup procedures, and even the user-friendliness of the software. These *user bounty* procedures are commonly used by experimentalists when developing software for laboratory use. From the point of view of the users, the experiments revealed many practical software problems that could have caused serious difficulties if discovered in the field during the FCC auctions.[4]

A third system of checks was a system of "parallel checking." Since the FCC auction software and procedures were used to conduct experiments, the data from the experimental auction were available. These data were fed into a parallel auction software system for computation and comparisons. The parallel system took the raw bid data and from them recomputed all numbers computed by the FCC auction computers. These computations were made during the experiment and afterwards. The parallel system operated from a program that we developed ourselves and for which we were virtually certain that the proper rules and computations had been implemented. All computations made by the FCC programs were rechecked. This method of checking proved valuable in several instances when we were able to reverse-engineer the FCC system to identify the source of programming errors.[5]

6. The Auctions and a Retrospective on Performance: The Third Stage

The final stage of the research was to provide advice during the operation of the actual FCC auction. Regardless of the amount of preparation and testing, things happen. Behavior might not be as anticipated due to environmental surprises. In spite of testing, rules can be incomplete and policy must sometimes be made on the spot. Decisions must be made on the spot from experience and judgment. During the first auctions, the experimentalists were the only ones that had studied the actual operations of auctions

---

[4] Some examples of the types of tests are: What happens if you stay logged on after the initial withdrawal; what happens if you log in from multiple locations at the same time; what happens if you enter 0000 rather than 0; what happens if you are theoretically inactive but nevertheless log on after various events; what happens if you log in at the last second of a session or have a power failure; what happens if you follow local software installation exactly to the letter of the instructions? The complaints about friendliness were enough to create enemies: my screen scrolls too fast, too much, too slow; response is too fast, too slow; etc. The test experiments put substantial pressure on the whole FCC auction organization to do rounds quickly, which was important, since the speed of the rounds is a variable that might be used to speed the termination of the overall auction.

[5] Important rounding errors were discovered. A miscomputation of eligibility, a type of double counting after a withdrawal, was discovered. Difficulties with eligibility computations after waivers and between phases were checked with this method. The subjects were attempting many unexpected combinations of actions in their attempt to find errors themselves, and this variability in behavior provided an excellent opportunity to check the internal operations of the "black box" that we were given to reverse-engineer. Many bugs were found and corrected, but one can never be positive about software reliability.

like the one implemented by the FCC. The hope was that the insights that resulted from observing laboratory experiments would be helpful in the field application.

The active participation of the experimentalists during an auction occurred only in the first auction in July 1994. It was here that the procedures for interacting with the bidders and the rules for the realtime operation of the auctions began to take form. An increment committee was formed by the FCC. The job of this committee was to provide policy advice about the minimum allowable bids, the speed of rounds, announcements, the implementation of stages, etc. Plott was a member of the committee, in addition to participating in the backup process.

The experiences gained from laboratory experiments informed the management of the first auction in two ways. Firstly, the experiments had produced many examples of the interaction of procedures, rules, and events that could cause problems in the auction itself, or invite litigation afterwards. Secondly, the patterns of activity in the FCC could be interpreted in the light of the behavior of experiments to gain insights about what was taking place and what might be expected in future rounds.

Laboratory experiments had demonstrated a propensity for agents to misunderstand subtle aspects of the rules. If this happened in the FCC auction, it might be the foundation for a court case. In order to prevent this possibility, during each round the auctioneer made announcements to bidders about critical aspects of the rules, especially those regarding withdrawal and the role of eligibility. Near the end of the auction these general announcements were clearly unnecessary, but those in charge of the auction kept a close eye on agents who might be exhibiting confusion. Help was made available for clarifications of the rules.

The speed of the auction was a general concern. The increment committee was formed to force bids upward and thus speed the process to termination by determining for each license the minimum acceptable bid increment. Very early on, the committee chose to demonstrate a willingness to use no fixed rule. This established the right of the committee to make such judgments, creating some uncertainty on the part of bidders that might be useful for managing the auction, and it provided some flexibility in controls. Some in the FCC thought that a combination of increment rules and stages was sufficient control to speed up the auction and bring it to an efficient termination. A tension existed between the idea that more time to make considered bids, coupled with higher increments on acceptable bids, would speed the auction, and on the other hand, the idea that the termination of the auction should be governed primarily by the number of rounds. Thus, one theory would have the auction take more time between rounds, and the other theory would suggest that the rounds be more frequent. Experimental evidence suggested that frequent rounds could be relied upon to generate an efficient and rapid termination. Many of the early experiments that were allowed to terminate naturally involved continuous-time processes without stages. Examination of these data suggested that the FCC auction could go through as many as a hundred rounds. The more rapid the rounds, the sooner would be the termination. Experiments had also produced evidence of the capacity of large increments to be *demand-killing:* A bidder failing to bid because of a large increment could lose eligibility. If the increment was subsequently reduced, the bidder might not have the eligibility to allow the purchase that (s)he would have otherwise made at the lower price.

Fear of such an event placed the experimentalists in the camp of those against the use of large increment requirements on bids. The possibility of demand-killing policies was very slight as long as the process operated in stage 1 or 2. However, if stage 3 were implemented, demand killing would be a real possibility. A disagreement existed between those who felt that stage 3 would operate like "brakes" and speed the process to termination, and those who felt that more frequent rounds were a safer way to bring the process to termination.

The first auction was held in a Washington, DC, hotel that the FCC had rented for only a limited amount of time. If the auction failed to terminate within the time frame of the rental, then the whole auction would need to be moved to another location. Because such a move would involve the transfer of equipment and electronic configurations, it involved risks to the smooth functioning of the auction. As the final date approached, support grew for taking the first auction from stage 2 to stage 3 in an attempt to bring it to a close. Plott argued against this change: (1) in stage 3 the possibility of demand-killing increments was the greatest; (2) a possibility of withdrawal existed (recall Fig. 4, panels A and B, above) and if it occurred, demand killing might result; (3) the software for stage 3 had not been tested (in retrospect there was a bug); (4) the time path of the bids suggested that the process was converging to an orderly termination and that there was only a need to speed the rounds. After consultation with the bidders, the rounds were speeded and stage 3 was never implemented.

As the auction proceeded, there were continued attempts to make judgments about the state of the system and where it might go. What were the patterns that were reminiscent of laboratory auctions? Was there evidence of scale backs by bidders because of budgets, and were marginal buyers evident? Are there similarities between the laboratory market and the FCC auction? If there are similarities, what can we conclude about the FCC auction? The FCC auctions certainly produced a lot of money, but how would one know if they worked to allocate the licenses efficiently? What would the experiments tell us to look for, and under what circumstances should we look? Unfortunately, research has not yet produced good answers. Here we only examine one of the major features of the data.

Experiments exhibit equilibration to predictable magnitudes. Figure 5 reports data from an experiment that was used to test the FCC software prior to the October auction. The parameters chosen for the experiment were similar to those that might exist in the auction, and the general conditions of the experiment were among those that were thought to possibly exist for the actual auction. Thirty licenses were offered in the experiment, exactly as in the October auction.

The demands and supplies are shown in Figure 6. The demand curve for any particular license was derived from the assumption that all other markets were at the competitive equilibrium price and that the agents allocated a fixed budget among licenses to maximize the induced preference. The values of the demand prices are shown above the curve. The units are on the horizontal axis, and the vertical lines are the supplies. A different market demand is shown for each of ten types of licenses. Within a license type the licenses are homogeneous. The competitive equilibrium prices are shown as the prices that are at the intersection of the market demand and supply.

The revenues from the experiment are shown in Figure 5. The revenue predicted by the competitive equilibrium model is the horizontal line. As can be seen, the revenue

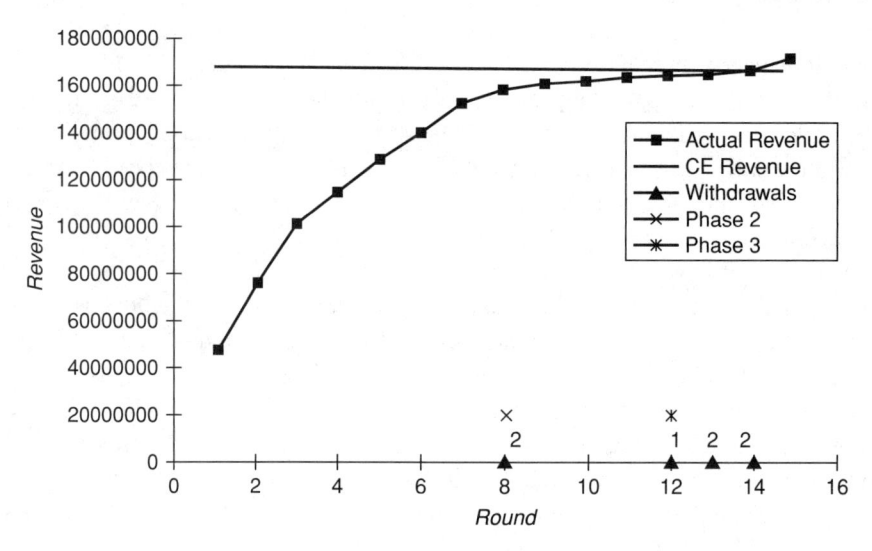

Figure 5. Test, September 20–26, 1994.

moves upward in a somewhat wavy fashion along what appears to be an exponential path with a jump at the end that is probably caused by an FCC intervention in the experiment.[6] This experiment, like many others, converges to near the competitive equilibrium.

Figure 7 shows the time path of the revenues generated by the October FCC auction for 30 licenses. As can be seen, there the path has the same qualities of an exponential path and converges. It also has the bumps and waves along the path that are evident in the experimental data. The bottom curve of Figure 7 helps us see the underlying nature of the revenue-generating process. It is a plot of the value of excess bids and shows the adjustments in bidding as relative prices increase. The structure of these changes suggests that the agents operated with budgets, and when prices got so high that the budget would have been exceeded, they scaled back to cheaper items. The FCC auctions seem to have a property of equilibration, and if the principles of their operation are the same as those in the experiments, then the convergence will be near the competitive equilibrium—should it exist.

In Figure 6 the allocations in the experiment are shown as the black squares under the demand curves. The demand curve above the square is the marginal value of the unit to the buyer, and the number next to the square is the price paid for the unit. As can be seen, the prices in the experiment for equivalent items are similar. That is also a property of the FCC auction data. From the experimental allocations, we can also determine that the experimental auctions were relatively efficient. Notice that the units under the demand curve tend to be the ones acquired, and when the units are the external margins and beyond, the ones that they replace are simply the units on the internal

---

[6] The FCC phoned during the test experiment and informed us that the test must stop because they needed to move the equipment for a demonstration. The experiment was terminated shortly after that. A second test conducted in October suffered the same fate when the FCC called and informed us that "the next round would be the last." The person in charge had evidently concluded (incorrectly) that the FCC software was working properly. The parallel computation procedure found an error in the FCC software later.

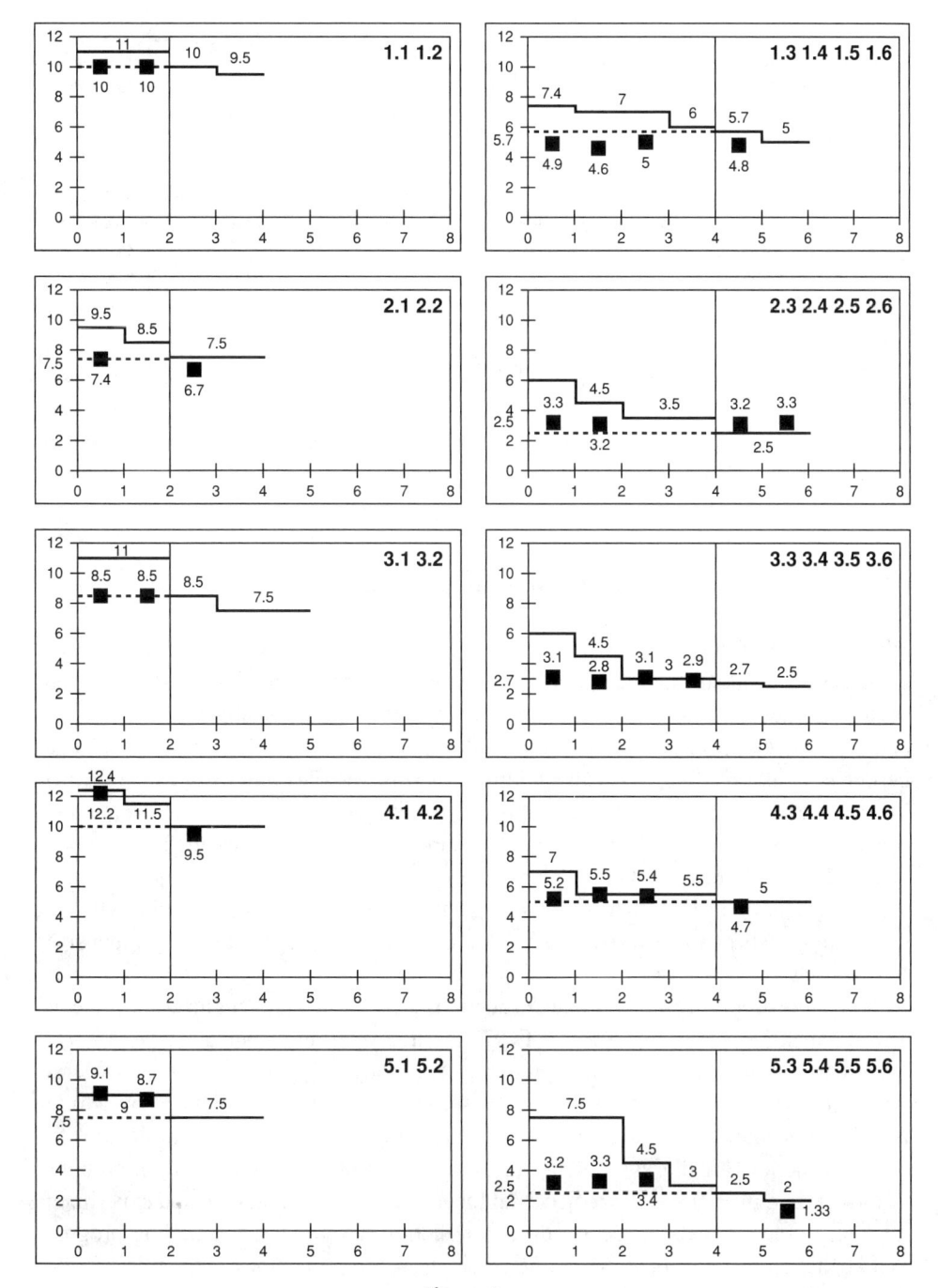

Figure 6.

margin, so the efficiency loss is not very large. Thus, if the FCC auctions are operating by the same principles, we can conclude that they are fairly efficient.

Thus, the FCC auctions and the laboratory experimental auctions have several qualitative features in common. If indeed the same principles were operating in the FCC

Figure 7.

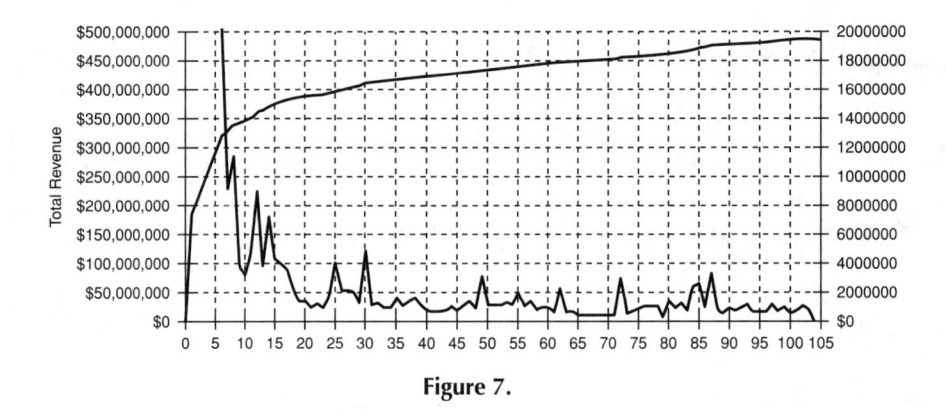

environment, then the FCC auction converged to near the competitive equilibrium and exhibited high efficiency.

7. Concluding Remarks

While the use of laboratory experimental methodology is still in its infancy, it seems clear that the value of the techniques was decisively demonstrated in the development of the FCC auctions. The overall success of the auctions must be attributed to others— economic theorists, applied economists, FCC lawyers, and the FCC staff. However, at certain critical junctures, experimental methods supplied data and insights that helped identify and solve problems that could have caused serious damage to the overall auction effort if they had gone undetected. The laboratory methods provided an inexpensive and timely source of data and experience that supplemented the major efforts in the policymaking process. The laboratory methods uncovered problems of a type that could not have been discovered by any other method, except (possibly very expensive) field testing.

It is interesting to compare the FCC auctions with large-scale engineering projects. The rulemaking procedures that the FCC inherits by virtue of being a governmental agency would never be used for engineering decisions. Imagine building a spacecraft with detailed engineering decisions made through the processes dictated by administrative procedures. Yet, in many respects, decisions regarding the detail of rules for complex auctions are like engineering decisions. The problems are certainly as complex as those found in engineering projects, and there exist solid theoretical and experimental foundations for making decisions about auction design. The astounding progress that has taken place in basic scientific research in economics has made this application possible. It would seem as though some alternative process should be created for institutional design problems that permit designs to reflect scientific considerations as opposed to political considerations. Nevertheless, in spite of what would seem to be the cumbersome and antiquated procedures dictated by administrative processes, the FCC produced a system that has so far operated effectively. Hopefully, the problems that are known to be lurking and are known to be important within the types of rules the FCC has adopted will not arise in future applications. Objective analysis can be applied, and laboratory testing can be used. Possibly, in the future, the policymaking process can systematically incorporate these scientific advances into decisions.

References

Banks, J.S., J.O. Ledyard, and D.P. Porter, 1989, "Allocating Uncertain and Unresponsive Resources: An Experimental Approach," *Rand Journal of Economics* 20(1), 1–25.

Brewer, P.J. and C.R. Plott, 1996, "A Binary Conflict Ascending Price (BICAP) Mechanism for the Decentralized Allocation of the Right to Use Railroad Tracks," *International Journal of Industrial Organization*, 14, 857–886.

Kagel, J.H. and A.E. Roth, eds., 1995, *The Handbook of Experimental Economics*, Princeton University Press.

Plott, C.R. and D.P. Porter, 1996, "Market Architectures and Institutional Testbedding: An Experiment with Space Station Pricing Policies," *Journal of Economic Behavior and Organization*, 31, 237–272.

Rassenti, S.J., V.L. Smith, and R.L. Bulfin, 1982, "A Combinational Auction Mechanism for Airport Time Slot Allocation," *Bell Journal of Economics*, 13(2), 402–417.

Rassenti, S.J., S.S. Reynolds, and V.L. Smith, 1994, "Cotenancy and Competition in an Experimental Auction Market for Natural Gas Pipeline Networks," *Economic Theory*, 4, 41–65.

CHAPTER 27

An Experimental Test of Flexible Combinatorial Spectrum Auction Formats[*]

Christoph Brunner, Jacob K. Goeree, Charles A. Holt, and John O. Ledyard

Simultaneous auctions for multiple items are often used when the values of the items are interrelated. An example of such a situation is the sale of spectrum rights by the Federal Communications Commission (FCC). If a telecommunications company is already operating in a certain area, the cost of operating in adjacent areas tends to be lower. In addition, consumers may value larger networks that reduce the cost and inconvenience of "roaming." As a consequence, the value of a collection of spectrum licenses for adjacent areas can be higher than the sum of the values for separate licenses.[1] Value complementarities arise naturally in many other contexts, e.g. aircraft takeoff and landing slots, pollution emissions allowances for consecutive years, and coordinated advertising time slots. This paper reports a series of laboratory experiments to evaluate alternative methods of running multi-unit auctions, in both high and low-complementarities environments.

Various auction formats have been suggested for selling multiple items with interrelated values. The most widely discussed format is the simultaneous multiple round (SMR) auction, first used by the FCC in 1994. In the SMR auction, bidders are only allowed to bid on single licenses in a series of "rounds," and the auction stops when no new bids are submitted on any license. To win a valuable package of licenses in this type of auction, bidders with value complementarities may have to bid more for some licenses than they are worth individually, which may result in losses when only a subset

---

[*] We acknowledge partial financial support from the Federal Communications Commission (FCC contract 05000012), the Alfred P. Sloan Foundation, the Bankard Fund, and the Dutch National Science Foundation (VICI 453.03.606). The conclusions and recommendations of this paper are those of the authors alone and should not be attributed to the funding organizations. We would like to thank Martha Stancill, Bill Sharkey, and Mark Bykowski for useful discussions. We are grateful to Raj Advani, Charlie Hornberger, Jou McDonnell, Anton Shuster, Walter Yuan, *Arzyx*, and *Eastek Design* for programming support, to Lauren Feiler, Joel Grus, Guido Maretto, Alan Miller, Laura Panattoni, Brian Rogers, and David Young for helping run the experiments, and to *Dash Optimization* for the free academic use of their Xpress-MP software.

[1] There can also be important synergies in the spectrum frequency dimension, where adjacent bands may improve capacity and reduce interference. For instance, in the FCC auction for air-to-ground communications frequencies in May 2006, a package of three bandwidth units sold for about 4.5 times as much as a single unit, and similar synergies were implied by unsuccessful bids.

is won. Avoidance of this "exposure problem" may lead to conservative bidding, lower revenue, and inefficient allocations.[2]

The obvious solution to the exposure problem is to allow bidding for packages of items. In such combinatorial auctions, bidders can make sure they either win the entire package or nothing at all. As a result, bids can reflect value complementarities, which should raise efficiency and seller revenue. Combinatorial bidding, however, may introduce new problems. Consider a situation in which a large bidder submits a package bid for several licenses. If other bidders are interested in buying different subsets of licenses contained in the package, they might find it hard to coordinate their actions, even if the sum of their values is higher than the value of the package to the large bidder (the threshold problem). Thus, there is no clear presumption that package bidding will improve auction performance. The FCC has increasingly relied on laboratory experiments to evaluate the performance of alternative spectrum auctions (see also Jacob K. Goeree and Charles A. Holt, 2010). The next section summarizes the main features of the auction formats to be considered.

I. Alternative Auction Formats

The various combinatorial auctions to be considered are best understood in terms of how they differ from the incumbent standard, the FCC's simultaneous multi-round auction procedure. Therefore, we will begin by explaining how the SMR auction was implemented in the experiments. Each auction consists of multiple rounds in which bidders have a fixed amount of time to submit their bids. Once the round ends, the highest bid on each license is announced as a provisional winner. Once no more bids are submitted, the auction stops and the provisionally winning bids become the final winning bids.

There are two constraints on bidding. The first constraint is the FCC "activity limit" that determines the maximum number of different licenses for which a bidder can submit bids. Each bidder is assigned a pre-specified activity limit at the beginning of the auction. A bidder's activity limit falls if the number of submitted bids (plus the number of provisionally winning bids in the previous round) is less than the bidder's activity limit in the previous round. Activity is transferable, so a bidder with a limit of 3 could bid on licenses A, B, and C in one round and on licenses E, F, and G in the next round, for example. The second restriction is that each bid must exceed the previous high bid for that license by a specified bid increment. This requirement is a minimum, and new bids can exceed the "provisionally winning" bid by up to eight bid increments. The only exception to the increment rule is that the provisionally winning bidder is not required to raise that bid. Bidders can observe others' previous bids and can see which of those were provisionally winning.[3] The effect of activity limits and bid increments

---

[2] In the recent AWS auction (FCC auction 66), for example, the total cost of acquiring 20 MHz of nationwide coverage was $2.268 billion for all 734 individual licenses in the "A-block" while the total cost was $4.174 billion for the 12 larger regions in the "F-block". Presumably there was a larger exposure problem in the A-block because it consisted a larger number of small licenses.

[3] In most FCC auctions to date, bidders' identities are revealed during the auction. More recently, the FCC has contemplated revealing bid amounts but not bidder identities (anonymous or "blind" bidding).

is to force bids upward, although there are limited opportunities for withdrawing bids.[4] The auction stops after a round in which no new bids are submitted and no withdrawals occur.

This multi-round procedure can be adapted to allow for bids on both individual licenses and packages, and this approach has been shown to improve auction performance in some cases.[5] With package bidding, the relevant price of a license is not necessarily the highest bid on that license; indeed there may not even be a (non-package) bid on a particular license. One approach is to calculate the revenue-maximizing allocation of licenses after each round, and to use "shadow" prices that represent marginal valuations in terms of maximized revenue. Then the price of a package is the sum of the prices for individual items, and new bids in the subsequent round then have to improve on these prices by some minimum increment that depends on the size of the package. As with SMR, bidders are given the option of selecting one of a series of pre-specified higher increments. This approach, known as RAD (Resource Allocation Design) pricing, is due to Anthony M. Kwasnica, Porter, Ledyard, and Chrstine DeMartini (2005). One advantage of the RAD approach is that prices may convey information about how high a bidder must go to "get into the action" on a particular license or package.[6] The revenue maximization at the close of each round uses all bids for all completed rounds. This maximization routine results in provisionally winning bids (on licenses or packages) and associated RAD prices. As in the SMR auction, a specified bid increment is added to the RAD price to determine the minimum acceptable bid for the license in the next round.[7] The minimum acceptable bid for a package is simply the sum of minimum acceptable bids for the licenses it contains. Bidders were allowed to submit multiple bids on licenses and/or packages. The treatment of activity limits is analogous to SMR, with activity being calculated as the number of different licenses being bid for or being provisionally won in the previous round (separately or as part of a package). The auction stops when no new bids are submitted, and the "provisional winning bids" for that round become the final winning bids (withdrawals are not needed with this format).

The FCC developed a variant of RAD pricing, called SMRPB. Of the four formats considered, SMRPB is the only auction procedure that employs an "XOR" bidding rule, which means that each bidder can have at most one winning bid. For example, the XOR rule means that a bidder who is interested in both licenses A and/or B must bid on A, B, and the package AB, since a bid on AB alone would preclude winning either license separately while bids on A and B only would preclude winning the package.

---

[4] As a partial remedy to the exposure problem, the FCC allows bidders to withdraw their provisionally winning bids in at most two rounds, at a penalty that equals the difference between their withdrawn bids and the subsequent sale price if that is lower. David P. Porter (1999) reports laboratory data showing that the introduction of this withdrawal rule increases the efficiency of the final allocation as well as the seller's revenue.

[5] Stephen J. Rassenti, Vernon L. Smith, and Robert L. Bulfin (1982) first used experiments to compare the performance of sealed-bid auctions with and without package bidding. John O. Ledyard, Porter, and Antonio Rangle (1997) provide data comparing several iterative processes. The combinatorial auction produces higher efficiencies in both designs.

[6] Ideally, the license prices should represent the revenue value of relaxing the constraint that there is only one of each license. The discreteness in license definitions may, however, result in nonexistence of dual prices, and Kwasnica et al. (2005) propose a method of computing approximate prices.

[7] In order to prevent cycles, the bid increment is raised after a round in which revenue does not increase.

Since XOR bidding typically calls for making bids on lots of combinations, the activity rule used with the FCC's SMRPB auction is based on the size of the largest package bid, so a bidder with activity 3 could bid on both ABD and ABC, for example, but not on ABCD. Another difference with RAD concerns the pricing rule: in the SMRPB version, prices adjust slower in response to excess demand because they are "anchored" with respect to prices in the previous round.[8]

An alternative approach to the pricing problem is to have prices rise automatically and incrementally in response to excess demand, via a "clock" mechanism (Porter, Rassenti, Anil Roopnarine, and Smith, 2003). In each round of the combinatorial clock (CC) auction, the price of a combination is the sum of the prices for each component, and bidders can indicate demands for one or more individual items or combinations of items. If more than one bidder is bidding for an item in the current round, either separately or as part of a package, the clock price for that item rises by the bid increment. Otherwise, the price remains the same. There are no provisional winners, but other aspects of this auction are analogous, e.g., activity is defined in terms of the number of different licenses for which a bidder indicates a demand.[9] The auction typically stops when there is no longer any excess demand for any item.[10] One possible advantage of an incremental clock auction is that it prevents aggressive "jump bids," which have been observed by Kevin McCabe, Rassenti, and Smith (1988) in the laboratory and by R. Preston McAfee and John McMillan (1996) in an FCC auction.[11] The clock-driven price increments may also alleviate the threshold problem of coordinating small bidders' responses to large package bids. In addition it is possible to add a final round of sealed-bids to the clock phase. This final or shootout phase could be structured as a first-price (pay-as-bid) auction or a second-price auction (proxy bidding, see Lawrence M. Ausubel, Peter Cramton, and Paul Milgrom, 2005).

---

[8] See Appendix D in the Goeree and Holt (2005) experiment design report for more details.

[9] Note that Porter et al. (2003) did not use activity limits in their combinatorial clock auctions.

[10] When there is no more excess demand for any of the licenses but some are in excess supply, the revenue maximizing allocation is calculated using *all bids in the current and previous rounds*. If this process results in a failure to sell to the remaining bidder for an item, the clock is restarted to let that bidder have another chance to obtain the item. This restart procedure can be illustrated with a simple three-license example, which is taken from the instructions to subjects. Suppose bidder 1 only wants license A and is willing to bid up to 40 for A, bidder 2 only wants license B and is willing to bid up to 40 for B, and bidder 3 only wants license C and is willing to bid up to 80 for C. Finally, bidder 4 only wants package ABC and is willing to bid up to 150 for ABC. Initially there is excess demand for all licenses, which causes prices to rise. Bidders 1 and 2 drop out when prices rise to 45, 45, 45, but since there is still competition for license C its price continues to rise. Bidder 4 is willing to keep bidding on ABC as long as the price of C does not exceed 60. So when the price of C rises to 65, bidder 4 drops out. At prices of 45, 45, 65 no one is bidding for licenses A and B. At this point, bidder 3 is the only one bidding on C but the computer finds it better to assign ABC to bidder 4 (for a total of $45 + 45 + 60 = 150$) than to assign A to bidder 1, B to bidder 2, and C to bidder 3 (for a total of $40 + 40 + 65 = 145$). To allow bidder 3 (who has a value of 80 for C) to get back into the action on license C, the computer will raise the price of C further to 70, 75, ... until (i) either bidder 3 drops out or (ii) the revenue from assigning A to bidder 1, B to bidder 2, and C to bidder 3 exceeds that of assigning ABC to bidder 4. In this manner the price of a license can rise even though only one bidder is still bidding for it. Also, bidders may be assigned a license or package even though they were no longer bidding in the final round (as for bidders 1 and 2 in the example).

[11] In the recent AWS auction, for example, one of the bidders made the maximum allowed jump bid for the hotly contested Northeast and West regional licenses, effectively doubling the prices to about $1.5 billion. The main competitors for these licenses ceased bidding immediately afterwards.

Results of laboratory experiments suggest that these and other forms of package bidding may enhance performance measures, especially in environments with high complementarities.[12,13] In the Porter et al. (2003) experiment, for example, the combinatorial clock auction attained 100 percent efficiency in 23 sessions and 99 percent efficiency in two other sessions. Previous experiments have mainly focused on specific auction formats. This paper provides a systematic and parallel consideration of SMR and its most widely discussed alternatives, including the one developed by the FCC.[14]

II. Experimental Design

Our design involves groups of eight bidders and 12 licenses, a size that was selected to provide enough added complexity, while still permitting us to obtain sufficient independent observations for a broad range of auction formats and value structures. Bidders' values for the licenses were randomly determined for each auction, which resulted in a rich variety of market structures. There are two types of bidders in this design: small "regional" bidders (labeled 1 through 6) and large "national" bidders (labeled 7 and 8). A graphical representation of bidders' interests is shown in Figure 1. Each diamond represents a different region, and the licenses along the center line (A, D, E, H, I, and L) are the ones of interest to the two national bidders. In the diamond shaped region on the far left, for example, the regional bidders, 1, 2, 5 and 6, are interested in licenses B and C, and in addition, each is interested in one of the licenses (A or D) that are targets for the two national bidders. Similarly, in the middle region, small bidders 1, 2, 3 and 4, are interested in licenses F and G, and each one is interested in one of the licenses (E and H) that are also of interest to the national bidders. The far-right diamond shaped region has a similar structure. Notice that each regional bidder has interests in two adjacent

---

[12] Jeffrey S. Banks, Ledyard, and Porter (1989) proposed a different type of combinatorial auction, called Adaptive User Selection Mechanism (AUSM). In this auction, bidders can submit bids for individual licenses and packages in continuous time. A new bid becomes provisionally winning if revenue can be increased by an allocation that includes the new bid. Kwasnica, Ledyard, Porter, and DeMartini (2005) compare RAD and AUSM to SMR in a laboratory setting. Efficiencies observed with RAD and AUSM are similar and higher than those for SMR, but revenue is higher in SMR since many bidders lose money due to the exposure problem. (If we assume that bidders default on bids on which they make losses and thus set the prices of such bids to zero, revenues are in fact higher under AUSM and RAD than under SMR.)

[13] Charles River and Associates also developed a combinatorial auction, called Combinatorial Multi-Round Auction (CMA). In this auction, only bids that are sufficiently high allow bidders to maintain their activity. A bid is sufficiently high when it is at least 5 percent higher than the currently highest combination of bids that spans the same licenses. Banks, Mark Olson, Porter, Rassenti, and Smith (2003) ran an experiment to compare the CMA and SMR auction formats. They find that the CMA leads to more efficient allocations but less revenue since many bidders incur losses in their SMR auction experiments due to the exposure problem. Porter et al. (2003) also compare CMA to SMR auctions and also find that CMA tends to lead to more efficient allocations.

[14] A common feature of the combinatorial formats discussed in this paper is that they permit "flexible" package bidding, i.e. bidders can construct arbitrary "customized" packages. An alternative approach is to restrict bidding to pre-specified packages as was done in the FCC air-to-ground auction in 2006. Michael H. Rothkopf, Aleksandar Pekeč, and Ronal M. Harstad (1998) have suggested hierarchically structured sets of pre-defined packages to reduce the complexity of the (revenue-based) assignment problem. Goeree and Holt (2010) propose a simple pricing mechanism for hierarchically structured packages and test the resulting auction in the lab.

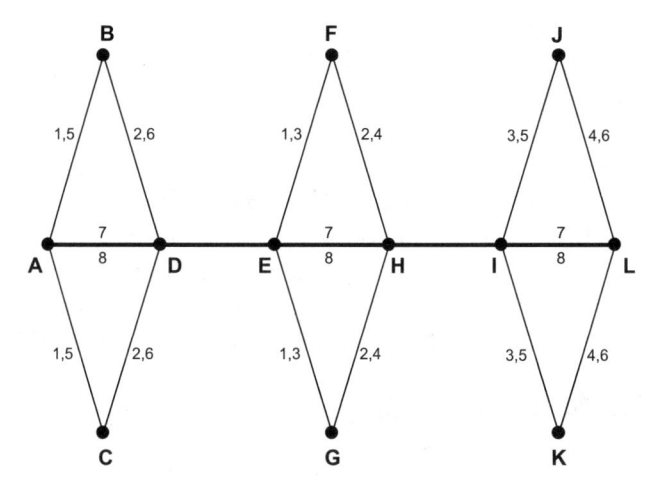

Figure 1. Eight-Bidder Design with Three Regions. Regional bidders (1–6) are interested in one side of one of two diamond-shaped regions. National bidders (7–8) are interested in the middle line connecting all three diamond-shaped regions.

regions, e.g. the left and center diamonds for bidders 1 and 2. Subjects' ID numbers stayed the same throughout the experiment, and, hence, so did their roles as regional or national bidders.

Regional bidders can acquire at most three licenses, and complementarities occur only when licenses in the same region are acquired. For example, if bidder 1 wins the combination ABE, then the value synergies would only apply to A and B, which are in the same region in Figure 1. Since value synergies do not apply across regions, a group of licenses in one region is a substitute for a group of licenses from another region, which creates an interesting "fitting problem." For example, under the SMR procedure, bidder 1 with an activity of 3 could either bid on licenses A, B, and C in the left region or E, F, and G in the middle region to capture the regional synergies. Likewise, under the RAD and CC procedures, bidder 1 could either bid to obtain synergies for the ABC package or the EFG package. The "XOR" rule used in SMRPB facilitates the regional bidders' "choice of region" problem because it allows them to bid on packages from *both* regions knowing that at most one bid can be winning. An additional advantage of the "XOR" rule is that bidders always know the maximum financial liability they face, i.e. the highest dollar amount of any of their bids.

National bidders can acquire up to six licenses and they have value complementarities for all six licenses in some treatments and for only four licenses in other treatments. The larger number of licenses subject to complementarities creates a larger exposure problem for the national bidders. The total number of possible allocations with this setup is 13,080,488.

Auction Formats: The four auction formats are described in detail in Appendices A–C. They include three combinatorial formats (SMRPB, RAD, and CC) and one non-combinatorial format (SMR). The main modification of the basic SMR procedures described above is that bid withdrawals were permitted in at most two rounds of an auction. For example, a bidder who withdraws any number of bids in rounds 8 and 10 would not be able to make any withdrawals in subsequent rounds. If a withdrawn bid caused the final sale price to go down, the bidder had to pay the difference. If a license with a withdrawn bid went unsold, however, then the bidder was only responsible for 25 percent of the withdrawn bid, which represents a penalty intended to mimic the effect of having to pay the difference between a withdrawn bid and a lower sale price in a subsequent auction. A key feature of the withdrawal provisions is that the seller (FCC) becomes the provisionally winning bidder at the second highest bid (minus a bid increment), so that the person who originally made the second highest bid would be able to re-enter at that level if the bidder has activity and interest to do so. This provision can benefit a bidder whose interests have changed, perhaps to a different region.

For each auction format, the experiments cover four different treatments: high/low overlap in national bidders' interests (HO versus LO) and high/low complementarities (HC versus LC). For example, treatment HOHC has high overlap and high complementarities. We next describe the treatment variations in more detail.

Complementarities: Payoffs in the experiment were expressed in terms of "points," where each point was worth $0.40 to subjects. (The bid increment was 5 points in all auctions.) The baseline draw distributions are uniform on the range [5, 45] for each license of interest to national bidders, and on the range [5, 75] for each license of interest to regional bidders. Synergies between licenses are modeled in a linear manner: when a bidder acquires K licenses the value of each goes up by a factor $1 + \alpha(K - 1)$. In the high-complementarities treatment, the synergy factor (α) for national bidders was 0.2. Thus each license acquired by a national bidder goes up in value by 20 percent (with two licenses), by 40 percent (with three licenses), by 60 percent (with four licenses), by 80 percent (with five licenses) and by 100 percent (with all six licenses). With low complementarities, these numbers are 1 percent, 2 percent, 3 percent, 4 percent and 5 percent, corresponding to $\alpha = 0.01$. With high complementarities (HC), each license acquired in the same region by a regional bidder goes up in value by 12.5 percent (with two licenses in the same region), and by 25 percent (with three licenses in the same region), so $\alpha = 0.125$. With low complementarities, these numbers are 1 percent and 2 percent for regional bidders. These minimal complementarities in the LC treatment allowed us to maintain parallelism in instructions and procedures. Participants were informed about the synergies that applied to regional and national bidders and about the distributions of possible values (but not about others' value draws).

Overlap: With high overlap (HO), each national bidder, 7 and 8, has value draws from the same distribution for all six licenses on the base of Figure 1, and the complementarities apply equally to all six licenses. In this sense, each national bidder is equally strong across the line. With low overlap (LO), national bidder 7 only receives complementarities for the four licenses on the left side of the base (A, D, E, and H). Conversely, national bidder 8 receives complementarities for the four licenses on the right side (E, H, I, and L). Thus with high complementarities and low overlap, each national bidder has a natural focus of interest that only partially overlaps with the other national bidder's area. One issue of interest is whether this type of partial separation may yield tacit collusion and less aggressive bidding in the center.

Treatment Structure: The two-by-two treatment design yields four treatments for each of the four auction formats, for a total of 16 treatments. We used the same value draws across auction formats so that differences cannot be attributed to specific sequences of value draws. Each session consisted of one or two practice auctions and a series of six auctions for cash payments. The treatment and auction type was unchanged for all auctions in a session, but the randomly generated value draws changed from one auction to the next. In addition, we used new sequences of random draws for each of three "waves" of 16 sessions that spanned all treatments. To summarize, there were 18 (3 waves times 6 auctions) independent sets of value draws that were used in all four auction formats.

Subjects and Sessions: Before conducting the sessions that form waves 1–3, we trained over 128 Caltech subjects in 16 sessions of eight people. These inexperienced sessions ("wave 0") involved both SMR and combinatorial auctions and were conducted to familiarize subjects with the auction software and bidding environment.[15] For these inexperienced sessions, we promised to pay each person a $60 bonus (in addition to other earnings) if they returned three more times.[16] This decision to use experienced bidders was based on the complexity of the auction formats and on earlier pilot experiments. For the subsequent data analysis, only the data from waves 1–3 but not from wave 0 is used. In waves 1–3, earnings averaged $50 per person per session, including $10 show-up fees and bonuses, for sessions that lasted from one and a half to two hours, depending on the number of auctions.[17] In total, there were 16 training sessions and 48 sessions (3×16) with experienced subjects, each involving a group of eight subjects.

III. Results

One way to measure market efficiency is to divide the sum of all bidder values for licenses they won, the actual surplus (S_{actual}), by the maximum possible surplus ($S_{optimal}$). It is well known that this simple efficiency measure may be difficult to interpret. For example, adding a constant to all value amounts will tend to raise this efficiency ratio, since efficiency losses are affected by differences in valuations, not absolute levels. A more natural measure of efficiency is calculated on the basis of the difference between the actual surplus and the surplus resulting from a random allocation (S_{random}), this being normalized by the maximum such difference.

$$efficiency = \frac{S_{actual} - S_{random}}{S_{optimal} - S_{random}} * 100\%$$

---

[15] The experiments were run using *jAuctions*, which has been developed at Caltech by Goeree. The *jAuctions* software consists of a flexible suite of Java-based auction programs designed to handle a wide range of auction formats and bidding environments, including combinatorial auctions with bid-driven or clock-driven prices, private and common valuations, etc. Instructions, which are available on request, were structured around relevant screen shots of the *jAuctions* program.

[16] As a consequence, most subjects participated in more than one auction format.

[17] In some cases, subjects ended the session with negative earnings, and these subjects were only paid the show-up fee. An alternative would have been to rotate bidder roles during the session, which would have avoided negative final earnings. This is procedure followed in Goeree and Holt (2010).

The value of a random allocation can be computed by taking the average of the surplus over all possible allocations, of which there are 13,080,488 in total for the design in Figure 1.[18] This definition of efficiency measures how much the auction raises surplus relative to a random allocation mechanism. In the analysis that follows, we will use these normalized efficiency measures.

Similarly, revenues will be measured as the difference between actual auction revenue and the revenue from a random allocation in which bidders pay their full values for all licenses and packages they receive ($R_{random} = S_{random}$). This difference is then divided by the difference between the maximum possible revenue ($R_{optimal} = S_{optimal}$) and the revenue from a random allocation. Note that the optimal revenue benchmark is the revenue obtained if bids equal full value on all licenses and packages leaving zero profits for the bidders, i.e. full rent extraction.

$$revenue = \frac{R_{actual} - R_{random}}{R_{optimal} - R_{random}} * 100\%$$

Since $R_{random} = S_{random}$ and $R_{optimal} = S_{optimal}$, the denominators of the normalized efficiency and revenue measures are equal, and the normalized sum of bidders' profits is simply equal to the difference between efficiency and revenue:

$$profit = \frac{S_{actual} - R_{actual}}{S_{optimal} - R_{random}} * 100\% = \frac{\sum_i \pi^i_{actual}}{S_{optimal} - S_{random}} * 100\%$$

All efficiency, revenue, and profit measures reported below are normalized in this manner for the specific value sequences used in each auction for each of the three waves of sessions with experienced bidders.[19]

Efficiency: Package bidding is designed to help bidders avoid the "exposure problem" of bidding high for licenses with high complementarities. As expected, switching from SMR to a combinatorial format raises efficiency in the high complementarities treatments. The differences between SMR and the combinatorial formats occur for both of the high complementarities treatments, as can be seen from the left side of Figure 2. In this and subsequent figures the color-coding is as follows: from light to dark the bars correspond to SMR, CC, SMRPB, and RAD respectively.

In contrast, the switch to combinatorial auctions reduces efficiency when complementarities are minimal (our "low complementarities" treatment). The efficiency levels are now 97 percent for SMR and 89 percent, 92 percent and 96 percent for SMRPB, CC, and RAD respectively. Again, this difference shows up in both LC treatments shown on the right side of Figure 2. Result 1 summarizes our findings, where we use the following notation: $\sim$ implies a pair wise difference is not significant, $\succ^*$ indicates significance at

---

[18] Without the restriction that regional bidders can acquire at most three licenses, the total number of allocations would simply be $4^{12} = 16,777,216$.

[19] There were occasional glitches in the data recording, i.e. when a bidder's computer would temporarily be offline. A detailed analysis of all the bid books reveals that less than 1 percent of all bids were lost. Since a bidder's activity was determined by the bids she submitted (not according to the bids recorded by the server) this had no adverse effects for the bidder's activity. Unless the round in which this occurred was the final round, the effect of lost bids was negligible.

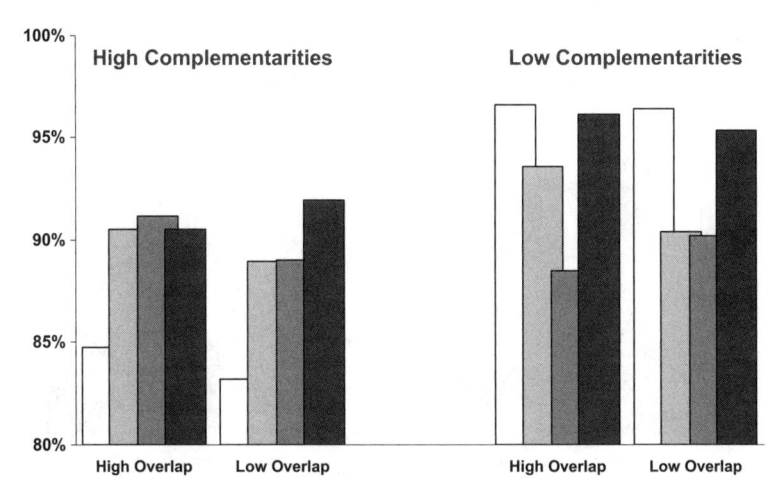

Figure 2. Efficiency by Auction Format. The bars from light to dark (left to right) correspond to SMR, CC, SMRPB and RAD respectively.

the 10 percent level, $\succ^{**}$ indicates significance at the 5 percent level, and $\succ^{***}$ indicates significance at the 1 percent level.

Result 1: With high complementarities, efficiency levels are highest for the three combinatorial formats and are ranked

$$RAD \sim SMRPB \sim CC \succ^{**} SMR$$

With low complementarities, efficiency levels are ranked

$$RAD \sim SMR \succ^{*} CC \sim SMRPB$$

Pooling the low and high complementarities treatments, efficiency levels are ranked

$$RAD \succ^{*} CC \sim SMRPB \sim SMR$$

Support: Session averages are grouped by wave and auction format in the online Appendix D. For example, consider the efficiencies for the HC treatments (pooling high and low overlap) shown in the eight columns on the left side of online Appendix D (top three rows). It is important to compare the auction formats for the same wave, since the valuation draws change from one wave to another. All six of the paired comparisons for the HC treatments show higher efficiencies for any of the package bidding auctions compared to SMR. This effect is significant using a Wilcoxon matched-pairs signed-rank test ($p = 0.03$). These results are generally reversed with low complementarities, where all paired comparisons between SMR and CC and SMRPB go in the opposite direction (higher efficiency for SMR): this effect is significant ($p = 0.03$). The only combinatorial auction that is not statistically different from SMR in terms of efficiency is RAD ($p = 0.41$). When pooling the data from the low and high complementarities treatments, RAD is more efficient than SMRPB ($p = 0.02$), CC ($p = 0.09$), and SMR ($p = 0.09$). There are no significant differences between SMR, SMRPB, and CC.

In addition to being statistically significant, the differences in observed efficiencies are also economically significant. With high complementarities (combining the low

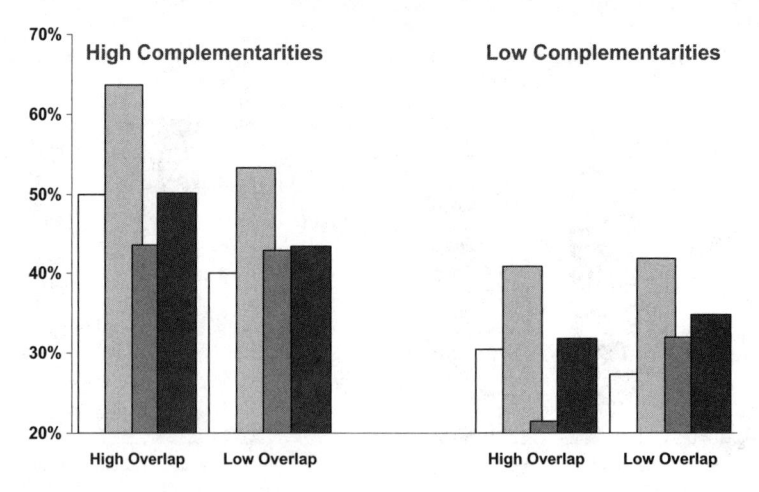

Figure 3. Revenue by Auction Format. The bars from light to dark (left to right) correspond to SMR, CC, SMRPB and RAD respectively.

and high overlap treatments and data from all three waves), the average efficiency in SMR is 84 percent while it is 90 percent, 90 percent and 91 percent in SMRPB, CC, and RAD respectively.

One reason why SMR leads to low efficiencies with high complementarities is the incidence of unsold licenses, which happens with rates of 4 percent and 7 percent in the high and low overlap treatments respectively, see the online Appendix D. Unsold licenses typically result from withdrawals late in the auction when a bidder realizes that it will not be possible to obtain the value synergies associated with multiple licenses. After a withdrawal, recall that the seller becomes the provisional winner at the second highest bid, and the person who made that bid previously may have lost activity or interest in that license, which causes it to go unsold.[20] Withdrawals are not permitted in the combinatorial auctions, where the exposure problem is addressed directly by allowing package bids, so these auctions do not result in unsold licenses. The difference between SMR and any of the combinatorial formats in terms of license sales rates is significant with a Wilcoxon matched-pairs signed-ranks test ($p = 0.05$).

Revenues: Figure 3 shows the revenues by auction format and treatment averaged across sessions (session averages for each parameter/experience wave can be found in rows 4–6 of the table in online Appendix D). What is obvious from Figure 3 is that the combinatorial clock auction extracts more rents for the seller in all treatments, even when it is less efficient than other formats.

---

[20] Since the seller's value for a license is assumed to be 0, an unsold license was given a value of 0 in the efficiency calculations. This calculation provides a lower bound for the efficiency since unsold licenses are typically sold in later auctions (there is, however, an efficiency loss associated with delays in spectrum use). Alternatively, a rough estimate of the upper bound for the efficiency would be to scale the actual efficiency by $1 + x$ where x is the proportion of unsold licenses. By scaling up the efficiencies for the SMR auction in the high-complementarities treatments using an "x factor" of 0.05 (proportion of unsold licenses averaged across treatments) raises average efficiency from 84 percent to 88 percent. This scaled up efficiency is only slightly lower than those for the combinatorial formats, indicating that a large part of the efficiency loss in the SMR auction is due to unsold licenses. Note that these calculations ignore "selection effects," i.e. low-value licenses are more likely to go unsold.

Result 2: Revenues are highest for the combinatorial clock auction and are ranked

$$CC \succ^{***} RAD \sim SMRPB \sim SMR$$

Support: There are three rows in the Revenue section of online Appendix D, one for each wave of parameter values. In each row, there are 4 paired comparisons between CC and a particular alternative format, so overall there are 12 paired comparisons. The CC provides higher revenue in all 12 pair-wise comparisons with each of the alternatives, except for RAD where CC yields higher revenues in 11 of 12 cases. These comparisons are significant using a Wilcoxon matched-pairs signed-rank test ($p = 0.001$). Basically, CC is higher than the others with both low and high complementarities ($p = 0.03$), except for the comparison with RAD in the low-complementarities (LC) treatment ($p = 0.06$). Revenue under RAD is border-line significantly higher than SMRPB when we pool all data ($p = 0.109$). Revenue under RAD is not significantly different from SMR, and SMR and SMRPB raise the same revenues.

These revenue differences are also significant economically: averaging over all treatments and all waves, the revenue from the combinatorial clock format is 50 percent as compared to 37 percent, 40 percent, and 35 percent for the SMR, RAD, and SMRPB auctions respectively.

With high overlap, national bidders own more licenses and, hence, can create bigger packages with higher associated values especially when complementarities are high. The result is that revenues are higher for the High Overlap and High Complementarities bars on the left side of Figure 3. Moreover, national bidders earn more in the high overlap and high complementarities treatments, while the regional bidders do worse (see the table in online Appendix D). These revenue and profits results with our prior expectation that there could be more tacit collusion in the low overlap treatments where there is less head-to-head competition between the national bidders.

Profits: Figure 4 shows bidders' profits by auction format and by treatment. The ability to bid for combinations allows bidders to bid high on packages and avoid the exposure problem, an effect that is mainly relevant with high complementarities. But if all bidders bid higher, the effect on bidder profits is unclear.

Result 3: Bidders' profits are lowest in the combinatorial clock auction and are ranked

$$RAD \sim SMRPB \sim SMR \succ^{***} CC$$

Support: Normalized profits are calculated as the differences between entries in the efficiency and revenue rows of online Appendix D. With three waves and four treatments, there are 12 paired profit comparisons between CC and a particular alternative format, and the CC provides lower profits in all 12 pair-wise comparisons with SMR, and for 11 of the 12 comparisons with RAD and SMRPB. These comparisons are significant using a Wilcoxon matched-pairs signed-rank test ($p = 0.001$). Averaged over treatments, profits for CC are 40 percent while the profits for the other formats are all in a narrow range from 53 percent to 55 percent.

The exposure problem can be alleviated to some extent by the (limited) bid withdrawal provisions built into the SMR bidding rules under consideration. In this manner a bidder may compete aggressively for a package and then decide to withdraw, paying a penalty

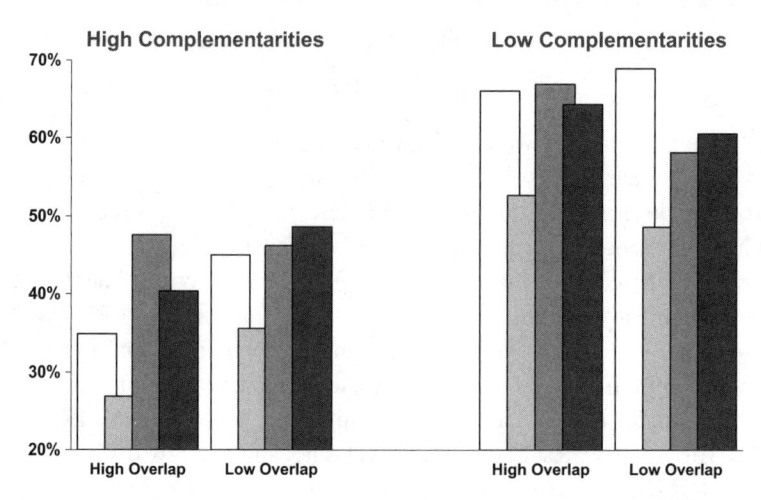

Figure 4. Bidders' Profits by Auction Format. The bars from light to dark (left to right) correspond to SMR, CC, SMRPB and RAD respectively.

equal to the difference between the withdrawn bid and the final sale price if it is higher. Withdrawals are more frequent (and the associated penalties higher) with high complementarities, as would be expected. While the possibility of bid withdrawals helps bidders deal with the exposure problem to some extent, some losses did occur when bidders decided not withdraw or had to pay a penalty after a withdrawal.[21]

Summary: Pooling data across treatments and sessions, the revenue and efficiency results by auction format are given in Table 1. In terms of seller revenues and bidder profits[22], the combinatorial clock auction is best for the seller and worst for the bidders, but these results are not caused by bidder losses, which are not present in the CC sessions (see the losses rows for Nationals and Regionals in online Appendix D). In a comparison with the other formats, the SMRPB auction with XOR bidding is the worst from the seller's point of view (lowest revenue and efficiency), and it is the best from the bidders' point of view (sum of profits for regionals and nationals).

The bottom row of Table 1 provides a perspective on the levels of the realized bidders' profits. The percentages in this row are calculated as ratios of actual bidders' profits (national profits plus regional profits) to profits that would result under collusion, i.e. when all bidders drop out at zero price levels resulting in a random allocation with the corresponding surplus, S_{random}, being divided among the bidders. Note that realized profits are far from collusive levels, especially for the combinatorial clock format.

One reason why the SMRPB auction performs the worst in terms of efficiency is that, in the presence of minimal complementarities, the requirement that bidders can only have one bid accepted (the XOR rule) may reduce efficiency, since bidders have to bid on many combinations of licenses to find all possible efficiency gains. (Even

---

[21] National bidders' penalties averaged 2 percent in the HC treatments and were negligible in the LC treatments. Regional bidders' penalties averaged 1 percent in the HC treatments and were negligible in the LC treatments.

[22] In Table 1, the "Profit Regionals" row lists the total profit (as a percentage) for the *group* of 6 regional bidders, while the "Profit Nationals" row lists the combined profits (as a percentage) for the 2 national bidders.

Table 1. *Summary Statistics by Auction Format*

	SMR	CC	RAD	SMRPB
Efficiency	90.2 percent	90.8 percent	93.4 percent	89.7 percent
Revenue	37.1 percent	50.2 percent	40.2 percent	35.1 percent
Profit Regionals	52.0 percent	38.5 percent	50.0 percent	51.3 percent
Profit Nationals	1.5 percent	2.3 percent	3.5 percent	3.5 percent
Profits/Collusive Profits	34.2 percent	26.0 percent	34.1 percent	34.9 percent

though our design, in which regional bidders face a "choice of region" problem, favors the XOR rule.) RAD, in contrast, more or less reduces to SMR with low complementarities while it enables bidders to extract the extra efficiency gains when complementarities are high. Another consideration may be that the inertia in the SMRPB price adjustment algorithm could exaggerate the threshold problem, since attempts to unseat large package bids may have delayed effects due to inertia. The next section explores the treatment differences in greater detail.

IV. Individual Bidding Behavior

This section provides an analysis of bidding patterns to explain the main qualitative features of the aggregate data. In particular, with high complementarities, efficiency is significantly lower in SMR than in the three combinatorial auction formats. This suggests that bidders are competing conservatively for larger packages when package bids are not allowed, which could lead to an inefficient allocation.

In order to quantify the effect of "exposure risk" on bidder behavior in SMR, we consider a conditional logit model in which bidders choose among all combinations of licenses that are still feasible given their current activity limits. The conditional logit model includes four variables to explain the choice of a specific bidding basket, see Table 2. Since some sources of noise are individual-specific, we estimate robust standard errors allowing for correlation among observations generated from the same subject.

The "profit" variable is the difference between the value of the basket (the value of the combination of licenses that the bidder is either bidding on or provisionally winning) and the minimum required bid. The value calculations include possible synergies. The coefficient of this "Profit" variable shown in the top row of Table 2 is highly significant; as expected, bidders are more likely to bid on a basket when it yields a higher profit.[23]

The second row of the table shows the effect of the binary variable "PW," which assumes a value of 1 if the bidder is already provisionally winning at least one of the licenses in the basket. The highly negative coefficient indicates that subjects are not likely to raise their bids on licenses they are already provisionally winning, which is

---

[23] These estimations are based on all bids, including those that could result in negative profits. Such bids are more prevalent in the high-complementarities treatments (9.2 percent) than in the low-complementarities treatments (6.3 percent).

Table 2. *Bidding behavior in SMR*

	Conditional (fixed-effects) logistic regression			
N	184,884			
Wald chi2(4)	1642		Log pseudolikelihood	−6853
Prob > chi2	0		Pseudo R2	0.69
Bid	Coef.	Robust Std. Err.	Z	P > |z|
Profit	0.09	0.008	10.9	0
PW	−5.22	0.412	−12.6	0
Inertia	1.44	0.081	17.7	0
Exposure	−0.13	0.035	−3.7	0
Exposure * HC	0.08	0.036	2.2	0.03

again intuitive. The third variable, "Inertia" is a dummy variable that is 1 if the set of licenses that the bidder was provisionally winning or bidding for in the last round is the same as the set of licenses the bidder is bidding for or provisionally winning in the current round.

Finally, "Exposure" is measured as the largest possible loss that a bidder might sustain when bidding on a certain combination of licenses.[24] We include an interaction term "Exposure * HC" to allow for the possibility that exposure has less of an effect with high complementarities. Note that exposure is significant and negative in both treatments[25], suggesting that bidders are less likely to bid on baskets that entail the risk of winning licenses at prices above private values. The sign of the coefficient is robust across a variety of different specifications.

To illustrate the importance of exposure, let us consider a simple example for the case of high complementarities. Suppose the national bidder is interested in winning either the national package ADEHIL or nothing at all (and is not the provisional winner on any license). License values are 25 on average so the national package is worth 300 on average. Consider a situation where license prices are 42 each so the minimum required bid would be 47 for each license, totaling 282 for the package. In this case, profit would be 18 but exposure would be 36, i.e. when the national bidder ends up winning only 3 out of the 6 possible licenses. Hence, the national bidder prefers stop bidding for the national package when its price is 252 (6 times the license price of 42) even though the value of the package is 300.

The second qualitative feature of the data is that efficiency is higher in RAD than in SMRPB for the minimal complementarities treatment. While activity in RAD is maintained by bidding or provisionally winning a sufficiently large number of different licenses, bidders in SMRPB have to bid on sufficiently large packages to maintain activity. As a result, bidders in RAD can simply bid on single licenses when there are no complementarities. In SMRPB, however, they typically bid on some profitable

---

[24] Profit always refers to the difference between value and the minimum required bid for *all* licenses that the bidder is bidding on or provisionally winning. As noted earlier, we only consider baskets that yield a positive profit. However, when the bidder ends up winning only *some* of the licenses that he is currently provisionally winning or bidding for, he might sustain a loss.

[25] Running separate regressions for the low and high complementarities treatments yields exposure coefficients of −0.05 (0.007) with high complementarities and −0.11 (0.0038) with low complementarities.

Table 3. *Average Bid Characteristics with Low Complementarities (standard deviation)*

Auction	Bidding Activity	Number of bids	Size of bids
SMRPB	2.91 (0.09)	1.62 (0.08)	2.44 (0.13)
SMR	2.60 (0.07)	2.60 (0.07)	1.00 (0.00)
RAD	2.69 (0.09)	2.11 (0.28)	1.60 (0.24)
CC	2.66 (0.07)	3.15 (0.52)	1.69 (0.13)

large package in order to maintain activity and are not also bidding on the subsets of that large package. Therefore, the number of possible allocations in SMRPB tends to be far lower than in RAD. If a bidder has high values for licenses A, B and C, for example, that bidder will typically bid on all three licenses separately in RAD. In SMRPB, the same bidder typically bids on package ABC only.

To evaluate why efficiency is lower in SMRPB when complementarities are low, we compare bidding behavior in terms of numbers and sizes of bids. The bidding activity column in Table 3 indicates that with low complementarities, subjects are bidding for or provisionally winning about three different licenses on average in all four formats. However, in SMRPB, bidders do so by bidding on fewer packages of a larger average size than in the other formats. The average number of bids submitted is lower in SMRPB than in any other auction format in every single one of the six sessions with low complementarities. Using a Wilcoxon matched-pairs signed rank test, this difference is therefore significant ($p = 0.03$, $n = 6$). Similarly, the average size of the bids under SMRPB is higher in all six sessions. This bid size effect is significant in comparisons between SMRPB and any of the other auction formats ($p = 0.03$, $n = 6$). The consequence of having fewer bids of larger size is to create a fitting problem under SMRPB. This problem is not present for the other formats where activity can be maintained by submitting many smaller bids.

The seller's revenue is higher in the combinatorial clock auction than in any other auction format in all treatments of our experiment. In SMR, efficiency and thus also revenue is negatively affected by the exposure problem when complementarities are high. Moreover, the option to withdraw bids leads to a higher fraction of unsold licenses in SMR, which further reduces the seller's revenue.

In RAD and SMRPB, the threshold problem can potentially reduce the seller's revenue, since large bidders may be able to win packages at low prices when small bidders are unable to coordinate their actions. In order to test for the effects of the threshold problem in SMRPB and RAD, we look at whether small bidders bid up to their values in periods in which they end up winning nothing. Since the threshold problem only pertains to small bids, we only look at bids on individual licenses and packages containing two licenses.

Recall that the combinatorial clock auction solves the threshold problem by forcing bidders to increase bids together on licenses for which there is excess demand. On average, small losing bids are closer to bidders' values in the combinatorial clock auction than in either RAD or SMRPB ($p < 0.001$ for both comparisons using a Wilcoxon matched-pairs signed rank test with 12 observations). The differences between RAD and SMRPB are not significant ($p = 0.23$, $n = 12$), see also Table 4.

Table 4. *Bidding up to value*

Auction	Mean	Standard Deviation
CC	99.2 percent	9.9 percent
SMRPB	87.1 percent	5.3 percent
RAD	85.8 percent	3.0 percent

Table 5. *Size of jumpbid (bid − minimum required bid)*

Auction	Treatment	Mean	Standard Deviation
CC	all	0	0
RAD	HC	3.1	2.2
RAD	LC	1.3	0.8
SMRPB	HC	4.1	2.1
SMRPB	LC	2.6	1.2

If the threshold problem is indeed the reason why small bidders fail to bid up to their values, one would expect large bidders in SMRPB and RAD to submit aggressive "jump bids," i.e. to bid more than the minimum required bid in the early rounds of the auction. Taking the average across all bids submitted during the first five rounds, the difference between the bid price and the minimum required bid is higher in SMRPB than in RAD both with high and low complementarities, see Table 5. However, these differences are not significant (pooling data from low and high complementarities yields $p = 0.15$ using a Wilcoxon matched-pairs signed rank test with 12 observations). Since the combinatorial clock auction does not allow jump bids, both these differences are higher for RAD and SMRPB than for CC (see Table 5).

V. Conclusions

The simultaneous multi-round auction is considered to be a remarkably successful application of game theory, with careful attention to the details of implementation by policy makers. This auction format is currently used around the world, and government officials in other agencies now routinely consult the FCC on auction design matters. Concerns about the effects of value complementarities have convinced many people that new procedures need to be developed and tested. In particular, the FCC developed a package bidding variant of the SMR auction, known as SMRPB. This paper compares the performance of these two alternatives and that of two other package-bidding formats: the combinatorial clock (Porter et al., 2003) and the RAD auction (Kwasnica et al., 2005).

The experiments were conducted with a common *jAuctions* bidder interface and parallel sets of value draws, for an array of structural and auction format treatments. The combinatorial auction procedures used (RAD, SMRPB, and CC) all result in higher efficiency than the currently used SMR procedure when value complementarities are present. It is important to emphasize that value complementarities are not just a theoretical possibility; a package of three bandwidth segments sold for about five times as much as a single segment in a recent FCC auction that offered a very limited menu of pre-specified package bidding options. Complementarities are almost surely significant for other potential applications of package bidding such as emissions permits for successive years.

However, of the three combinatorial auction types, SMRPB performed worst in terms of revenue and efficiency. One distinguishing feature of SMRPB is the XOR rule, which allows each bidder to have at most a single winning bid. A bidder who is interested in obtaining one or more licenses in a certain region thus has to bid on all possible combinations of those licenses. In the experiment, however, bidders submit only a few bids per round in which case the additional constraint of at-most-one-winning bid per bidder becomes detrimental for efficiency and revenue. The poor performance of SMRPB reported here was a main factor in the FCC's decision not to implement it.

The FCC subsequently decided to implement package bidding for one of the five blocks in the recently conducted 700 MHz auction. Unlike the fully flexible package bidding formats considered in this paper, the FCC opted to use a simple format with a single 50-state package and two additional packages (Atlantic and Pacific). This setup is a simple version of the Hierarchical Package Bidding mechanism proposed by Goeree and Holt (2010). Under this mechanism, predefined packages are structured in a hierarchical manner and after each round of bidding, prices for all licenses and packages are determined such that they signal the bid amounts required to unseat the current provisional winners.

Without extensive knowledge of bidders' valuations, there will be some efficiency loss associated with using predefined packages. However, the simplicity of the hierarchical package structure (e.g. individual licenses, non-overlapping regional packages, and a single nationwide package) avoids fitting problems that can arise with fully flexible package bidding. For example, if a nationwide package bid is winning then the non-overlapping nature of the regional packages together with the simple pricing feedback allows regional bidders to avoid the threshold problem. This approach was tested using laboratory experiments based on two-layer and three-layer hierarchies and the results were cited by the FCC as a factor in their decision to use Hierarchical Package Bidding in the recent 700 MHz auction.

References

Ausubel, Lawarence M., Peter Cramton, and Paul Milgrom. 2005. "The Clock-Proxy Auction: A Practical Combinatorial Design," in *Combinatorial Auctions*, Eds. P. Cramton, R. Steinberg and Y. Shoham, MIT Press.

Banks, Jeffrey S., John O. Ledyard, and David P. Porter. 1989. "Allocating Uncertain and Unresponsive Resources: An Experimental Approach," *Rand Journal of Economics*, 20(1): 1–25.

Banks, Jeffrey S., Mark Olson, David P. Porter, Stephen J. Rassenti, and Vernon L. Smith. 2003. "Theory, Experiment and the Federal Communications Commission Spectrum Auctions," *Journal of Economic Behavior and Organization*, 51: 303–350.

Goeree, Jacob K., and Charles A. Holt. 2005. "Comparing the FCC's Combinatorial and Non-Combinatorial Simultaneous Multi-Round Auctions: Experimental Design Report,"

Report prepared for the Wireless Communications Bureau of the Federal Communications Commission.

Goeree, Jacob K., Charles A. Holt, and John O. Ledyard. 2006. "An Experimental Comparison of the FCC's Combinatorial and Non-Combinatorial Simultaneous Multiple Round Auctions," Report prepared for the Wireless Communications Bureau of the Federal Communications Commission.

Goeree, Jacob, K., and Charles A. Holt. 2010. "Hierarchical Package Bidding: A 'Paper & Pencil' Combinatorial Auction," *Games and Economic Behavior*, 70(1): 146–169.

Kwasnica, Anthony M., John O. Ledyard, David P. Porter, and Christine DeMartini. 2005. "A New and Improved Design for Multi-Object Iterative Auctions," *Management Science*, 51(3): 419–434.

Ledyard, John O., David P. Porter, and Antonio Rangel. 1997. "Experiments Testing Multiobject Allocation Mechanisms," *Journal of Economics and Management Strategy*, 6(3): 639–675.

McAfee, Preston R., and John McMillan. 1996. "Analyzing the Airwaves Auction," *Journal of Economic Perspectives*, 10(1): 159–175.

McCabe, Kevin, Stephen J. Rassenti, and Vernon L. Smith. 1989. "Designing 'Smart' Computer Assisted Markets," *European Journal of Political Economy*, 5: 259–283.

Porter, David P. 1999. "The Effect of Bid Withdrawal in a Multi-Object Auction," *Review of Economic Design*, 4(1): 73–97.

Porter, David P., Stephen J. Rassenti, Anil Roopnarine, and Vernon L. Smith. 2003. "Combinatorial Auction Design," *Proceedings of the National Academy of Sciences*, 100(19): 11153–11157.

Rassenti, Stephen J., Vernon L. Smith, and Robert L. Bulfin. 1982. "A Combinatorial Auction Mechanism for Airport Time Slot Allocation," *Bell Journal of Economics*, 13: 402–417.

Rothkopf, Michael H., Aleksandar Pekeč, and Ronal M. Harstad. 1998. "Computationally Manageable Combinational Auctions," *Management Science*, 44: 1131–1147.

CHAPTER 28

On the Impact of Package Selection in Combinatorial Auctions: An Experimental Study in the Context of Spectrum Auction Design

Tobias Scheffel, Georg Ziegler, and Martin Bichler

1. Introduction

Designing *combinatorial auction* (CA) markets is a formidable task: Many theoretical results are negative in the sense that it seems quite unlikely to design fully efficient and practically applicable CAs with a strong game-theoretical solution concept (Cramton et al., 2006a). Experimental research has shown that *iterative combinatorial auction* (ICA) formats with linear prices achieve very high levels of efficiency (Porter et al., 2003; Kwasnica et al., 2005; Brunner et al., 2009; Scheffel et al., 2010). Most of these experiments are based on value models with only a few items of interest. While it is important to understand bidder behavior in small CAs, we need to know whether the promising results carry over to larger auctions, since applications of CAs can easily have more than 10 items. We ran laboratory experiments with those CA formats that have been used or analyzed for the sale of spectrum licenses, the most prominent and most thoroughly investigated application domain for CAs. However, the analysis is not restricted to spectrum sales and the results are also relevant to the design of auctions in other domains, such as procurement and transportation. We do not only observe which package bids bidders submit, but also which packages they evaluate. Interestingly, we find that auction design rules matter, but the cognitive barriers of bidders, i.e. the number of packages they evaluate, are the biggest barriers to full efficiency. These cognitive barriers have not been an issue in the experimental literature so far.

1.1. Spectrum auctions

There has been a long and ongoing discussion on appropriate auction mechanisms for the sale of spectrum rights in the USA (Porter and Smith, 2006). Since 1994, more than 70 spectrum auctions were run using the simultaneous multiround auction (*SMR*), an auction format which is based on work by Paul Milgrom, Robert Wilson, and Preston McAfee (Milgrom, 2000). While in the SMR auction several items are sold in a single auction, package bidding is not allowed. This leads to a number of strategic problems for bidders (Cramton, 2013), such as limited substitution of spectrum due to the activity rules employed and the *exposure problem*. The latter refers to the phenomenon that a bidder is exposed to the possibility that he may end up winning a collection of licenses that he does not want at the prices he has bid, because the complementary licenses have become too expensive. CAs allow for bids on packages of items, avoiding exposure problems. The design of such auctions, however, led to a number of fundamental design problems, and many contributions during the past few years (Cramton et al., 2006b).

Since 2005 countries such as Trinidad and Tobago, the UK, the Netherlands, Denmark, and the USA have adopted CAs for selling spectrum rights (Cramton, 2013). The USA used an auction format called Hierarchical Package Bidding (***HPB***) (Goeree and Holt, 2010), in which only restricted combinatorial bidding is allowed and bidders can only bid on hierarchically structured pre-defined packages. Other countries used a version of the Combinatorial Clock (***CC***) auction (Porter et al., 2003) or the Clock-Proxy auction (Ausubel et al., 2006), which extends the CC auction by a sealed-bid phase.[1] The pricing in the CC auction is rather simple, the price of over-demanded items raises by the specified increment and bidders can only place package bids at the current prices. A third type of CA formats, which has been analyzed for spectrum auctions in the US, uses pseudo-dual linear prices, i.e., they use an approximation of the dual variables of the winner determination problem. Different versions have been discussed in the literature (Brunner et al. (2009); Bichler et al. (2009)), which are all based on the Resource Allocation Design by Kwasnica et al. (2005) and earlier by Rassenti et al. (1982). We refer to these auction formats as pseudo-dual price (***PDP***) auctions in the following.

All auction formats which have been used for selling spectrum use linear (i.e., item-level) ask prices. This is an interesting observation, as there are negative theoretical results by Kelso and Crawford (1982), and Bikhchandani and Ostroy (2002) which show that only non-linear and personalized competitive equilibrium prices can always exist for general valuations. Disadvantages of non-linear price CAs have been studied by Scheffel et al. (2010), who show that these auction formats are impractical, due to the enormous communication complexity when selling more than only a few items. The linearity of ask prices, however, also seems to be an important requirement in other domains such as electricity markets, which allow for package bids (Meeus et al., 2009). There has so far been no game-theoretical equilibrium analysis on any of the linear-price auction formats discussed above.

In a Public Notice, the US Federal Communications Commission (FCC) justified their choice of HPB by stating that "the mechanism for calculating CPEs [current price

---

[1] A specific version of a two-stage clock auction with a supplementary bids phase has been implemented by Ofcom in the UK and other European countries. The clock phase allows for a single package bid per bidder in each round. Then, the supplementary bids phase allows for multiple package bids in a sealed bid phase. After this second stage the winners are determined, and bidder Pareto-optimal core prices are calculated (Day and Raghavan, 2007). This two-stage clock auction design in the UK is considerably different from single-stage formats discussed in this paper. There is also an active discussion on core-selecting auctions (Lamy, 2009; Goeree and Lien, 2016). We focus on the single-stage combinatorial clock auction, as it has been described by Porter et al. (2003), and do not discuss more complex multi- stage and core-selecting auction formats.

estimates] is significantly simpler than other package bidding pricing mechanisms that adequately address coordination issues" (FCC, 2007, §221). While this argument is true when comparing HPB against PDP, it does not necessarily hold when comparing HPB to the CC auction. Apart from the simplicity of the respective auction formats, the main arguments of the FCC were based on a set of lab experiments conducted by Brunner et al. (2009) and, more specifically, Goeree and Holt (2010). In their main value model, specific auction rules like purchase and activity limits apply and the result in a different value model that does not offer such an obvious hierarchical pre-packaging might be different.

Given the importance of spectrum auctions for the telecommunications sector and the significant costs that companies have to bear for licenses, these auction formats demand more empirical results.

1.2. Contributions

Based on the influence of the experiments by Goeree and Holt (2010) on the FCCs decision, we intentionally decided to partly replicate their experiments but extended them with the CC auction. Since in their value model simple simulations indicate that even without package bids high levels of efficiency can be obtained, we extend our experiments further with a new value model that shows different characteristics and captures the local synergies of licenses, which can be well motivated from observations in the field (Ausubel et al. (1997); Moreton and Spiller (1998)).

From these extensions we derive new insights on bidder behavior in three important ICA formats, which have been used or proposed for selling spectrum licenses: CC, HPB, and PDP. The way ask prices are determined in these auction formats varies considerably, and there has been a long discussion on the pros and cons of each approach (FCC, 2007). While we find statistically significant differences in efficiency and revenue among the auction formats, these differences are small. We replicate some important findings of Goeree and Holt (2010) in their value model with global synergies, namely that HPB achieves higher efficiency and auctioneer revenue than PDP. Interestingly, HPB achieved similar results to PDP even in a second value model with local synergies, which is less suitable to a hierarchical pre-packaging. The fully combinatorial CC auction, however, realizes the highest median efficiency and revenue in both value models.

The reasons for inefficiencies in linear-price ICAs are not easy to explain, but our analysis reveals that the bidders' restricted package selection is the largest barrier to efficiency, while the differences in auction rules had much less impact. Bidders mostly pre-selected packages in the first round and then focused on this small subset of packages throughout the auction, even though they had enough time in each round and adequate monetary incentives to analyze more packages. They hardly evaluated new packages based on the price feedback throughout the auction.

Participants of spectrum auctions typically spend substantial time and money to prepare for the auction and one might challenge that results from the lab can be transferred to the field. However, even if they were able to evaluate more combinations, with sufficiently many items of interest in the auction the number of possible packages increases exponentially and the main phenomenon can probably be observed in all but small auctions.[2] It might be even more pronounced in other domains such as procurement or transportation, where bidders cannot invest the same amount of time to prepare for an auction. The result suggests that the positive experimental results on efficiency in small CAs do not necessarily carry over to larger instances. Issues such as pre-packaging and bidder decision support might play a much bigger role in the design of large CAs and deserve more attention in research and in practical auction design.

HPB has been used for selling spectrum by the US FCC and was therefore a main treatment in our experiments. Due to the limitation of package bids in HPB, bidders frequently took an exposure risk but still had more difficulties in coordinating and outbidding the big bidder in a threshold problem. In the fully combinatorial auction formats CC and PDP, small bidders were able to outbid the national bidders more frequently.

Apart from these main results, we report on a number of observations relevant to auctioneers and policy makers. For example, we find that opening bids on individual items have a significant positive effect on efficiency, while activity limits have a negative impact.

The paper is structured as follows: Section 2 describes the related literature and gives an overview of experimental studies on CAs so far. Section 3 describes the experimental environment, and the CA formats analyzed in this paper are introduced. In Section 4 we present our results both on an aggregate level and on the level of individual bidder behavior. A summary and conclusions follow in Section 5.

2. Related Literature

A number of experimental studies deals with CAs and their comparison to simultaneous or sequential auctions. Their main finding is that in the presence of complementarities between the items, package bidding in CAs leads to higher efficiencies (cf. Banks et al. (1989), Ledyard et al. (1997) and Banks et al. (2003)).

Brunner et al. (2009) compare the SMR auction to different kinds of CAs in values models with high and low complementarities. They find the CC auction to achieve the highest revenues among the CAs, which is consistent with the results of our experiments. In the high complementarity case, PDP and CC auctions significantly outperform SMR in terms of efficiency, while the opposite is the case with low complementarities. Their analysis of bidder behavior shows that bidders are more likely to bid on packages with high profit. We, however, also explain which factors determine the bidders' package selection, which helps understand inefficiencies in CAs.

Goeree and Holt (2010) introduce the HPB auction, in which bidders are allowed to place package bids only on predefined packages. They suggest defining predefined packages in a hierarchical manner, which allows solving the winner determination problem in linear time. They also suggest a pricing mechanism which is based on the hierarchical pre-packaging. In an experimental comparison of HPB, SMR, and PDP, they find HPB to achieve higher efficiency and revenue than SMR and PDP. They also

---

[2] For example, in the US Advanced Wireless Service auction in 2006, the FCC sold 1122 licenses, which leads to $2^{1122} - 1$ possible packages.

analyze the threshold problem in PDP and HPB and find that smaller bidders can more effectively solve the threshold problem with HPB due to the pre-packaging. While they focus on the comparison between simultaneous auctions, restricted and fully combinatorial auctions in a value model with global synergies, we investigate only linear-price CAs and extend the analysis also to a value model with local synergies. We further analyze the bidder behavior in CC, in addition to HPB and PDP and provide insights into the reasons for inefficiencies in those CAs. In contrast to Goeree and Holt (2010), we find small bidders in CC and PDP to have less problems to coordinate in a threshold problem.

Finally, Chen and Takeuchi (2009) compare the Vickrey auction against an ascending version of the Vickrey auction in a small value model consisting of 16 packages. They find the ascending auction to generate higher bidder profit and efficiency, while the sealed bid auction generates a higher revenue. A large part of their study deals with the comparison of human bidders versus artificial bidding agents. In this context, they study the bidder behavior with different levels of information about the artificial agents strategies and find that bidders learn the bidding strategy from the agents which use straightforward bidding. In their analysis of the bidder behavior, they also find that it is more likely that bidders bid on packages with higher profits, which is consistent with the findings of Brunner et al. (2009). Whereas their analysis shows the bidder behavior in a Vickrey and an ascending version of the Vickrey auction, we investigate the bidder behavior in linear-price auctions.

3. Experimental Design

In the following, we provide an overview of the auction formats analyzed, the valuations of bidders, and the behavioral assumptions. Then we discuss the treatments used in our experiments, and the experimental procedures.

3.1. Market Mechanisms

The main difference between the CC, HPB, and the PDP auction is the calculation of the ask prices. All auction formats are multi-round auctions, meaning that the auction runs over several rounds, in which bidders are able to place new bids according to the current prices. At the end of each round, bidders are provided with information about their own provisionally winning and losing bids (referred to as standing or active bids in the CC auction), as well as with ask prices on items and packages for the new round. The bidders use an OR bidding language in our reported experiments. This means, they can submit and win more then one bid, with the restriction that each item can be included at most once in the winning allocation. This allows for comparability with the results in Goeree and Holt (2010).

Bidders in spectrum auctions are usually forced to be active and place bids early in the auction through an activity rule. We used a monotonicity rule that is based on eligibility points in all treatments. This means, the bidders were not able to increase the number of distinct items they are bidding on from round to round. Similar rules are used in spectrum auctions in the field (Cramton, 2009).

3.1.1. The Combinatorial Clock Auction

A core issue of running CAs is the computational complexity of the winner determination. Several iterative CA designs solve the winner determination problem after each round. In contrast, Porter et al. (2003) propose the CC auction, which alleviates some of the computational requirements; In the best case the auctioneer is not required to solve the winner determination problem at all. The price calculation in the CC auction is simple. Prices for all items are initially zero, and in each round of an auction, bidders bid on the packages they desire at the current prices. Jump bidding is not allowed. The auction moves to the next round as long as at least one item is demanded by two or more bidders. In this case the respective item prices are increased by a fixed increment. In a simple scenario, in which supply equals demand, the items are allocated to the corresponding bidders of the last round, and the auction terminates. If at some point there is excess supply for at least one item and no other item is over-demanded, the auctioneer solves the winner determination problem by considering all bids submitted so far. If all bidders who are active in the last round are included in the allocation, the auction terminates. Otherwise, the prices on those items not allocated to an active bidder are further increased and the auction continues. The feedback in the CC auction is restricted to the new price vector; additionally, our experimental software shows bidders if on any of their packages the price remains the same, which means that the respective bidder is the only one left, who demands these items at the current prices.

3.1.2. The Pseudo-Dual Price Auction

The PDP auction uses a restricted dual formulation of the winner determination problem to derive ask prices after each auction round. Ask prices for individual items are determined via linear programming such that they sum up to the provisionally winning (package) bids and are lower than the losing bids after each round. Unfortunately, such item-level prices do not always exist in the presence of complementarities. In these cases, prices are set, such that violations of the constraints mentioned above are minimized. In the next round the losing bidders need to bid more than the sum of ask prices for a desired package plus a fixed minimum increment for each item in the package. All bids remain active throughout the auction, and bidders are allowed to submit jump bids as in a single-item English auction. Details of the ask price calculation can be found in (Bichler et al., 2009).

To avoid cycling the minimum increment has to be raised after a round in which new bids were submitted but auctioneer revenue did not increase. In our case the minimum increment is raised from 3 to 6 to 9 etc. until the auctioneer revenue increases at which point it is decreased to its initial value. It is possible that prices on an item also decrease during the course of an auction. If the bidders have placed a bid on the desired package in earlier rounds, they have to overbid their bid-price as well by the minimum increment, not only the current ask price. As soon as no bidder submits a bid in a round any more, the auction terminates and the winning bids become the final bids that determine the allocation and prices to be paid. The round feedback in the PDP auction includes information, whether a bid is provisionally winning, and the new individual-item prices as well as the minimum increment.

3.1.3. The Hierarchical Package Bidding Auction

Goeree and Holt (2010) introduce HPB, an auction format which imposes a hierarchical structure of allowed package bids. This hierarchy reduces the winner determination problem to a computationally simple problem, which can be solved in linear time (Rothkopf et al., 1998). Goeree and Holt (2010) propose a simple and transparent pricing mechanism, which calculates item-level prices. They use a algorithm to determine new ask prices, which starts with the highest bid on every individual item as a lower bound, adding a surcharge if the next level package in the hierarchy received a bid, which is higher than the sum of the individual-item bids contained in the package. The difference is distributed uniformly across the respective item prices or bids on the lower level. This procedure is recursively applied to each level of the hierarchy and ends by evaluating the package(s) at the root level. In each round the bidders have to bid more than the sum of ask prices for a desired package plus a fixed minimum increment for each item in the package. Also, all bids remain active and can become winning throughout the auction. The auction terminates when bidding ends and provisionally winning bids become final bids that determine the allocation and prices to be paid. Similar to PDP, the feedback after each round in the HPB auction includes information, whether a bid is provisionally winning as well as the new individual-item prices.

3.2. Economic Environment

Ausubel et al. (1997, Page 499) define "*local synergies* as those gains in value that specifically arise from obtaining two or more geographically neighboring licenses [...], and] *global synergies* as those gains in value that accrue from obtaining increased numbers of licenses". Both Moreton and Spiller (1998) and Ausubel et al. (1997) find evidence for global and local synergies (complementarities) in spectrum auctions. While Goeree and Holt (2010) only analyze global synergies, we believe that due to the regional awarding of licenses in countries like the USA it is important to understand whether the results transfer to such value models with local synergies. We study these two different types of complementarities in our experiments. One of our value models uses global complementarities, so we refer to it here as global-synergy value model (*global-SVM*). Another value model uses local complementarities between items (*local-SVM*). In contrast to the global-SVM, the topology of items matters in the local-SVM.

3.2.1. The Global-Synergy Value Model

The global-SVM is a replication of the first experiment in Goeree and Holt (2010), which was very influential on the FCC's decision for using HPB, and involves seven bidders and 18 items. Figure 1 represents the bidders' preferences. The six regional bidders (labeled 1 through 6) are each interested in four adjacent items of the national

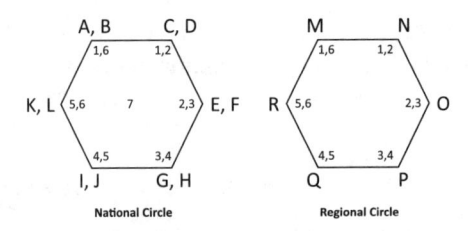

```
        A, B    C, D              M       N
       ╱1,6   1,2╲              ╱1,6   1,2╲
  K, L⟨5,6    7   2,3⟩E, F   R⟨5,6        2,3⟩O
       ╲4,5   3,4╱              ╲4,5   3,4╱
        I, J    G, H              Q       P
       National Circle         Regional Circle
```

Figure 1. Competition structure of the Global-SVM. Regional bidders 1–6 are interested in four items from the national circle and two items from the regional circle. National bidder 7 is interested in all twelve items from the national circle (Goeree and Holt, 2010).

circle (consisting of items A through L) and two items of the regional circle (consisting of items M through R) while for the national bidder (labeled 7) the twelve items of the national circle are relevant. This information was common knowledge, but it was not known which bidders were interested in a particular item. For example experimental subjects did not know that besides bidder 7 also 3 and 4 were interested in license H.

The values for the individual items are randomly determined. For the national bidder the baseline draw distributions are uniform on the range [0, 10] for items A-D and I-L and uniform on the range [0, 20] for items E-H. For regional bidders the baseline draw distributions are uniform on the range [0, 20] for items A-D, I-L and M-R and uniform on the range [0, 40] for items E-H. These value distributions (not the actual draws) were common knowledge among the experimental subjects. For comparison, we use the same draws as in Goeree and Holt (2010), which the authors have kindly provided. For both bidder types the value of items in a package increases by 20% (with two items), 40% (with three items), 60% (with four items), etc. and by 220% for the package containing twelve items. For the calculation of the complementarities the identity of the items does not matter, e.g. a bidders' valuation of a package of items increases by the same percentage independent of the adjacency of the items; i.e, this is a value model in which global complementarities apply.

Activity and purchase limits are such that regional bidders can bid on and acquire at most four items in a single round, while the national bidder is able to acquire all his twelve items of interest. The allocation constraints such as the tight activity and purchase limits lead to additional difficulties, as we see in Section 4.4. As a consequence, we ran another set of experiments with another value model.

3.2.2. *The Local-Synergy Value Model*

The local-SVM consists of 18 items arranged rectangularly and considers the scenario in which complementarities are gained from spatial proximity. In this value model items are placed on a rectangular map. The arrangement of items matters for the calculation of the complementarities, which only arise if the items are neighboring.

This model also contains two different bidder types: one national bidder, interested in all items, and five regional bidders. Each regional bidder is interested in a randomly determined preferred item, all horizontal and vertical adjacent items, and the items adjacent to those. This means that a regional bidder is interested in six to eleven items with local proximity to their preferred item. Examples are shown in Table 1, in which the preferred item of a regional bidder is Q or K, and all gray shaded items in the

Table 1. *Local-SVM with the preferred items Q and K of two regional bidder. All their positive valued items are shaded*

A	B	C	D	E	F
G	H	I	J	K	L
M	N	O	P	Q*	R

A	B	C	D	E	F
G	H	I	J	K*	L
M	N	O	P	Q	R

proximity of the preferred item have a positive valuation. For each bidder i we draw the valuation $v_i(k)$ for each item k in the proximity of the preferred item from a uniform distribution. Item valuations for the national bidder are in the range of [3, 9] and for regional bidders in the range of [3, 20]. These value distributions (not the actual draws), the procedure to determine the preferred item and the number of bidders participating in the auction were common knowledge among the experimental subjects.

We assume that bidders experience only low complementarities on small packages, but complementarities increase heavily with a certain amount of adjacent items. We further assume that adding items to already large packages do not increase the complementarities anymore. The explanation for these assumptions is the lack of economies of scale with small packages and a saturation of this effect with larger packages. Therefore, complementarities arise based on a logistic function, which assigns a higher value to larger packages than to smaller ones:

$$v_i(S) = \sum_{C_h \in P} \left(\left(1 + \frac{a}{100 \left(1 + e^{b-|C_h|} \right)} \right) * \sum_{k \in C_h} v_i(k) \right)$$

P is a partition of S containing maximally connected subpackages C_h, i.e. any item $k \in C_h \subseteq S$ has no vertical or horizontal adjacent item $l \in S \setminus C_h$. For our experiments we choose $a = 320$ and $b = 10$ for the national bidder and $a = 160$ and $b = 4$ for all regional bidders.

3.3. Behavioral Assumptions and Differences of the Value Models

Bidding in CAs is complex due to the auction rules (pricing and activity rules) and the large number of potential bids, which is exponential in the number of items. If bidders face this complicate situation, we conjecture that they rather select packages at the beginning of the auction and then concentrate on their bidding strategy.

Conjecture 1. Bidders select a limited number of packages at the beginning of the auction, which they focus on throughout the auction due to cognitive limits (Miller (1956)).

However, at the beginning of the auction, bidders have only the information about their own valuations, i.e. their baseline valuations and the rules to calculate their complementarities. They do not have any information about the market prices and have only little information about the preferences of their competitors.

Conjecture 2. Bidders select packages according to their baseline valuations at the start of the auction.

Table 2. *Bidding space concerning value model, auction format and number of baseline draws*

	Global-SVM			Local-SVM		
	# items of interest	# packages		# items of interest	# packages	
		CC/PDP	HPB		CC/PDP	HPB
National bidder	12	4,095	16	18	262,143	28
Regional bidder	6	56	7	6–11	63–2,047	7–15

Since complementarities among items in the global-SVM are the same for all bidders and all packages, eliminating the possibility to bid on packages has only a small effect on efficiency. Figure 2 describes the efficiency of numerical experiments with sealed-bid auctions and truth-revealing bidders, who only bid on individual items. The numerical experiments are based on all draws which are used in our experiments. Unlike in the global-SVM, truthful bidding on individual items in the local-SVM only results in an average efficiency of around 60%. This is due to the fact that complementarities occur only on adjacent items and the function is neither linear in the number of items nor the same for both types of bidders. In this case the availability of package bids is more important to achieve high levels of efficiency.

Conjecture 3. Efficiency in the global-SVM is higher as bidders know their market power through their baseline valuations. It is sufficient to stick (only) to the high valued items (packages) to achieve high levels of efficiency.

Table 2 shows the number of possible bids a specific type of bidder could submit in the different value models and auction formats. A large number of possible bids makes it more difficult in the CC and PDP auctions to select the "right" bids and coordinate with other bidders. This is particularly true, since no game-theoretical solution concept

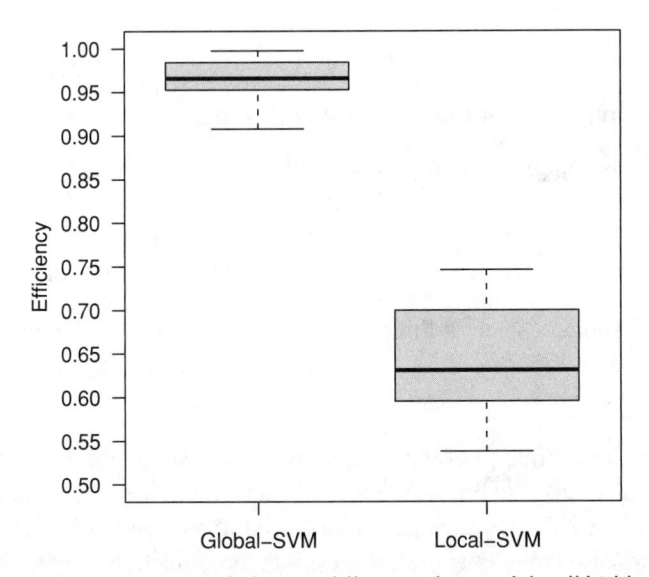

Figure 2. Numerical experiments with the two different value models; all bidders are reporting their true valuations on individual items.

is known for any of the auction formats. The local-SVM provides even more alternatives for each bidder to choose from in each round. Miller (1956) show that the amount of information which people can process and remember is often limited.

Conjecture 4. Bidders evaluate and bid on the same amount of packages regardless of the number of possible package bids due to cognitive limits.

A crucial factor to achieve high levels of efficiency in HPB is that the hierarchy meets the bidders' preferences; if this is the case, the auction is likely to achieve efficient outcomes, and reduces the strategic complexity for bidders as there are less exposure problems. In other words, it is easier for bidders to coordinate in a threshold problem if fewer overlapping packages are available. Otherwise HPB reduces to a set of parallel auctions as bidders cannot use the predefined packages. To be able to compare the results in HPB over the different value models, we use a similar pre-packaging in both value models. Bidders should be able to bid on all individual items and we tried that they are able to bid their preferences.

Since the competition structure is deterministic in the global-SVM it is simple to create a good hierarchy that fits the bidders' preferences. We take the structure from Goeree and Holt (2010) which yield higher efficiencies in their results, so the odd-numbered bidders were able to bid on their desired package of four items in the national circle directly. Hence, the hierarchical structure of possible package bids in HPB is the following: Bidders may bid on the package of items A-L on the top level, A-D, E-H and I-L on the middle level, and on all individual items on the bottom level.

Due to the random competition structure in the local-SVM, it is harder for the auctioneer to design a good hierarchy of allowed packages in HPB. With random competition structure we mean that we cannot tell a priori which bidders are interested in which items due to the randomly determined preferred item. We assume the auctioneer to have only a basic understanding of bidder valuations in the local-SVM and designed a hierarchy that allows bidders much flexibility with a four-level hierarchy:

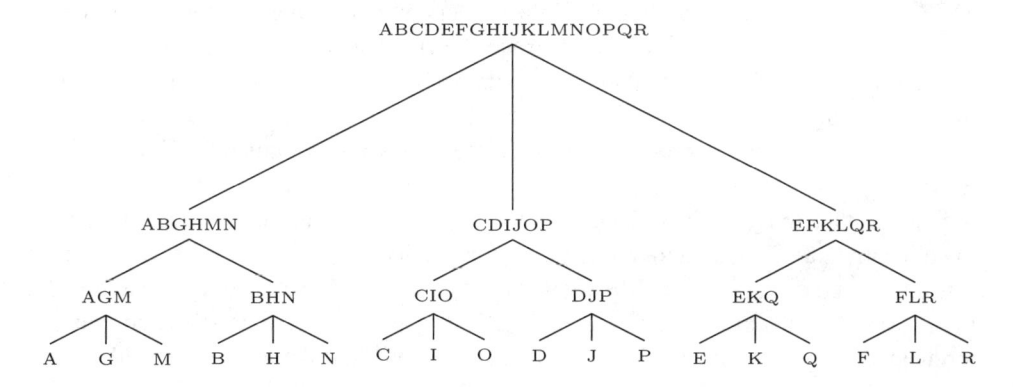

Since the exposure problem is a well known phenomenon in the SMR auction, we expect this problem to occur in the HPB auctions as bidders are not able to express their preferences, due to the restriction of package bids.

Conjecture 5. In HPB bidders have to deal with the exposure problem and take the risk of paying more for a package than it is worth to them.

Table 3. *Treatment structure*

Treatment No.	Auction format	Bidding language	Opening bids	Value model
1	CC	OR	No	global-SVM
2	HPB	OR	No	global-SVM
3	PDP	OR/XOR	No	global-SVM
4	CC	OR	Yes	local-SVM
5	HPB	OR	No	local-SVM
6	PDP	OR	Yes	local-SVM

There are always two regional bidders and the national bidder competing for an item in the national circle, and two regional bidders competing on the items in the regional circle. In real applications, there are often hot spots with a lot of competition, while other areas exhibit very low competition. The threshold problem becomes more crucial in the local-SVM due to the stochastic competition structure. Goeree and Holt (2010) find that the threshold problem is more a coordination problem than a "free rider" problem and that HPB helps regional bidders in coordinating their bids to outbid the national bidder in the global-SVM. While this might be true when bidders are able to reveal their preferences through the predefined packages in HPB, we believe that as soon as they cannot reveal them through package bids, the exposure risk regional bidders have to take compensate the coordination advantage in HPB.

Conjecture 6. The threshold problem is harder to overcome in the local-SVM due to the stochastic competition structure.

3.4. Treatment Structure

Table 3 provides an overview of all six treatment combinations. For both the global-SVM and the local-SVM, we provided the bidders with a simple and easy to use tool to determine the value of a particular package. The user interface and the basic auction process was similar across all treatments[3]. The bid increment was three francs per item in our experiments.

A phenomenon that can occur in CAs is that items remain unsold, although bidders have positive valuations for these items. Since bidders are not restricted to bid on individual items only, their package bids might overlap so that the revenue maximizing allocation does not contain all items, which results in efficiency losses. Due to the restriction on package bids in HPB we find bidders are more likely to bid on individual items, which results often in complete sales. This is not the case in the fully combinatorial auctions, which often terminate with unsold items. In order to avoid unsold items in our comparison of the CAs and to guide bidders in solving their coordination problem, we introduce *opening (singleton) bids*; this means, in the first round bidders have to place individual-item bids on those items they want to be able to bid on in future rounds. Such a rule is regularly used in procurement, in which auctions are typically

---
[3] All experiments were conducted using the *MarketDesigner* software, which was developed at the TU München. It is a web-based software that was already applied for other lab experiments (Scheffel et al. (2010)). The software will be provided for replication of the experiments.

preceded by an initial request for quotes, i.e., a tender on individual items. To keep the experimental setup as close as possible to the experiments by Goeree and Holt (2010), we introduce opening bids only in the local-SVM. In treatment 3 we deviated from the OR bidding language for the regional bidders. Instead they used XOR bidding, i.e. each bidder can win at most one of his bids. This decision was based on the fact that regional bidders have only a moderate number (56) of bundles to choose from in each round and with XOR bidding they can directly express their purchase limits.

In summary, we conducted 24 sessions with 156 participants. For each treatment combination we had four sessions. Each session consisted of two practice auctions and a series of six auctions in the global-SVM and four auctions with the local-SVM. In a session a single treatment combination was tested, but the value draws changed from one auction to the next in the same session. We used four different sets of value draws for each treatment combination, but the same set of draws for other treatments.

3.5. Procedures

All experiments were conducted from November 2008 to July 2009 with undergraduate students of the TU München. Each session started by giving subjects printed instructions which were read aloud. The subjects participated in two training auctions to ensure that they were familiar with the market design and user interface. Subjects took part in a short exam and were encouraged to ask questions to ascertain that they understood the economic environment as well as the market rules. The following four to six auctions were used to determine the earnings of each bidder. The number of auctions was announced in advance. Every auction started with period of five minutes in which bidders received their new valuations simultaneously and submitted their first round bids. In all subsequent rounds, bidders had at most three minutes for placing bids. Bidder roles were reassigned randomly after each auction in order to attenuate earning differences across national and regional bidders and to help bidders understand the strategic considerations faced by both bidder types. Earnings were calculated by converting the achieved gains to EUR by 2:1 in the auctions with the global-SVM and 1:1 in the auctions with the local-SVM. The earnings were distributed between the €10 show-up fee and a maximum of €80 per subject; average earnings were €43.55. The average duration of a session was 4 hours and 17 minutes.

4. Results

First we present our results on an aggregate level. In order to better understand and explain these results we analyze the bidders' package selection in a second step, and continue with the analysis of the threshold problem in our experiments. Finally, we provide more detailed insights about the individual bidder behavior.

4.1. Aggregated Measures

We use ***allocative efficiency*** (or simply ***efficiency***) as a primary aggregate measure. Given an allocation X and price set $\mathcal{P}_{pay}$, let $\pi_i(X, \mathcal{P}_{pay})$ denote the payoff of the bidder i for the allocation X and $\pi_{all}(X, \mathcal{P}_{pay}) := \sum_{i \in \mathcal{I}} \pi_i(X, \mathcal{P}_{pay})$ denote the ***total payoff***

Table 4. *Average aggregate measures of auction performance in the global-SVM. The results of Goeree and Holt (2010) are labeled with a superscript GH*

Auction Format	$E(X)$	$R(X)$	Payoff Nationals	Payoff Regionals	Number of Bids	Unsold Items	Rounds
CC	89.6%	74.6%	0.2%	2.5%	224.3	1.0	15.6
HPB	91.2%	70.5%	0.7%	3.3%	82.1	0.1	13.3
HPB^{GH}	94.0%	76.5%	1.6%	2.6%	n.a.	0.1	13.2
PDP	85.3%	64.8%	0.5%	3.3%	70.6	2.5	11.2
PDP^{GH}	89.7%	70.8%	2.3%	2.8%	n.a.	1.0	15.1
SMR^{GH}	85.1%	64.6%	−2.6%	3.0%	n.a.	2.1	17.8

of all bidders for an allocation at the prices $\mathcal{P}_{pay}$. Further, let $\Pi(X, \mathcal{P}_{pay})$ denote the *auctioneer's revenue*. We measure efficiency as the ratio of the total valuation of the resulting allocation X to the total valuation of an efficient allocation X^* (Kwasnica et al., 2005):

$$E(X) := \frac{\Pi(X, \mathcal{P}_{pay}) + \pi_{all}(X, \mathcal{P}_{pay})}{\Pi(X^*, \mathcal{P}_{pay}) + \pi_{all}(X^*, \mathcal{P}_{pay})} \in [0, 1]$$

In addition, we measure **revenue distribution**, which shows how the overall economic gain is distributed between the auctioneer and bidders. Given the resulting allocation X and prices $\mathcal{P}_{pay}$, the **auctioneer's revenue share** is measured as the ratio of the auctioneer's revenue to the total sum of valuations of an efficient allocation X^*:

$$R(X) := \frac{\Pi(X, \mathcal{P}_{pay})}{\Pi(X^*, \mathcal{P}_{pay}) + \pi_{all}(X^*, \mathcal{P}_{pay})} \in [0, E(X)] \subset [0, 1]$$

We also include the payoff or revenue share of different bidders in our analysis, which is measured by the proportion of the efficient solution that a single bidder gains. For the pairwise comparisons of selected metrics we use the *nonparametric Wilcoxon rank sum test* (Hollander and Wolfe, 1973). $\sim$ indicates an insignificant order, $\succ^*$ indicates significance at the 10% level, $\succ^{**}$ indicates significance at the 5% level, and $\succ^{***}$ indicates significance at the 1% level.

4.1.1. Aggregate Performance Metrics of the Global-SVM

On an aggregated level, our analysis covers mainly the average numbers of efficiency, auctioneer's revenue, and the payoff of regional and national bidders. We further provide the average number of bids, the number of unsold items, and the number of auction rounds. In Table 4, we provide the results of these measures in the global-SVM. We show the distribution of the efficiency and the auctioneer's revenue in the box plots in Figure 3.

Result 1 *(Efficiency and revenue distribution in the global-SVM).* The CC auction and HPB achieve significantly higher efficiency than the PDP auction (HPB $\succ^{***}$ PDP, CC $\succ^{**}$ PDP, CC $\sim$ HPB). CC leads to the highest auctioneer's revenue share; auctioneer's revenues are ranked CC $\succ^*$ HPB $\succ^{**}$ PDP. The differences in the bidders' payoff is mainly insignificant; only the payoff of the regional bidders is significantly lower in CC compared to PDP (PDP $\succ^{**}$ CC).

(a) Efficiency

(b) Auctioneer's revenue

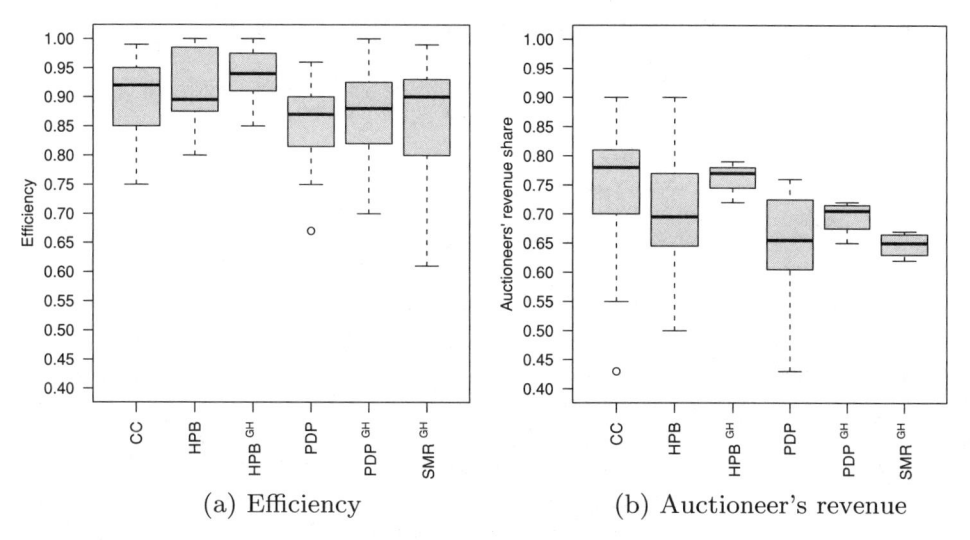

Figure 3. Efficiency and auctioneer's revenue in the global-SVM. The results of Goeree and Holt (2010) are labeled with a superscript GH.

Support. We performed Wilcoxon rank sum tests, and find significant differences in efficiencies between the PDP auction and CC ($p = 0.049$), and PDP and HPB ($p = 0.006$). We do not find a significant difference between CC and HPB ($p = 0.452$). In Figure 3, the box plot shows the highest median auctioneer's revenue share for the CC auction. A Wilcoxon rank sum test confirms that the CC auction leads in our experiments to the highest auctioneer's revenue. HPB results in a significantly lower revenue ($p = 0.078$). PDP is outperformed by both CC ($p = 0.001$) and HPB ($p = 0.047$). We find only small differences in the bidder payoffs (cf. Table 4); only the PDP auction has significantly higher regional bidders' payoffs compared to the CC auction ($p = 0.039$). □

The number of auction rounds is often used as a metric for speed of convergence. If an auction format requires too many rounds to close, it is impracticable in the field and also in the laboratory. While we find significant differences in the number of auction rounds (CC $\succ^{**}$ PDP, CC $\succ^{*}$ HPB), the average is less than 20 rounds in all sessions and never caused a problem. Only the auction duration in some of the CC auctions kept us from conducting all auctions in two of the CC sessions (on one occasion we conducted only four auctions, and in another session only five auctions).

4.1.2. Comparison to the Experiments by Goeree and Holt (2010)

Since our global-SVM is a replication of the first experiment in Goeree and Holt (2010), we continue the analysis with the comparison to their results. The hierarchical prepackaging in our experiments is the same as in the HPB_{odd} treatment in Goeree and Holt (2010), which resulted in the highest efficiency in their experiments.

Result 2 *(Global-SVM: Comparison to Goeree and Holt (2010) experiment one).* We find significant differences between our experiments and those of Goeree and Holt (2010). Efficiencies are ranked $HPB^{GH} \succ^{***}$ HPB, but we replicated their main finding

Table 5. *Average aggregate measures of auction performance in the local-SVM*

Auction Format	$E(X)$	$R(X)$	Payoff Nationals	Payoff Regionals	Number of Bids	Unsold Items	Rounds
CC	92.6%	85.2%	0.5%	1.3%	472.2	0.0	12.7
HPB	86.7%	80.8%	3.2%	0.5%	91.1	0.0	6.5
PDP	89.9%	82.4%	1.1%	1.3%	138.7	0.0	9.0

that a pre-packaged auction format (HPB) achieves higher efficiencies than a PDP auction (HPB $\succ^{***}$ PDP, HPB^{GH} $\succ^{*}$ PDP^{GH}). The fully combinatorial CC auction achieves similar outcomes in efficiency (CC $\sim$ HPB) and achieves higher auctioneer revenue than HPB (CC $\succ^{*}$ HPB).

Support. The average numbers of both experiments are included in Table 4 and the box plots in Figure 3 show the distribution of the efficiencies and auctioneer's revenue. The Wilcoxon rank sum tests confirm the visual observation of the box plot, that HPB^{GH} achieves a significantly higher efficiency compared to HPB ($p = 0.004$) and auctioneer's revenue ($p = 0.006$). We find the CC auction to achieve similar efficiency as HPB ($p = 0.452$), but the auctioneer's revenue is higher in CC ($p = 0.078$). □

It is noteworthy that the main finding in Goeree and Holt (2010), the difference between HPB and PDP, is replicated. Although bidders in the PDP treatment have additional complexity due to the bidding language, we find no significant differences in efficiency and auctioneer's revenue between PDP and PDP^{GH}, however, the number of unsold items is significantly higher in PDP compared to PDP^{GH}. This difference can be explained by the use of the XOR bidding language, which allows only one winning bid per bidder. XOR bidding increases the chance for unsold items due to the reduction of possible allocations compared to OR bidding.

The significant differences in the results could be due to a number of smaller differences in the laboratory procedures and software. In our experiments, bidders were unable to see the other bidders' IDs and bids. We have also used a maximum round duration of three minutes, while in GH bidders had only 40 seconds to submit their bids. There are also differences in the user interface that might cause the differences in the observed results.

4.1.3. Aggregate Performance Metrics of the Local-SVM

In addition to the experiments with the global-SVM, we performed a series of experiments with the local-SVM to see if the results transfer to a different value model, which can be well motivated from observations in the field (Ausubel et al. (1997); Moreton and Spiller (1998)).

In Table 5 we provide the average results of the auctions in the local-SVM and Figure 4 shows the distribution of efficiency and the auctioneer's revenue share in the local-SVM.

Result 3 *(Efficiencies and revenues in the local-SVM).* CC leads to significantly higher efficiency (CC $\succ^{**}$ HPB) and auctioneer's revenue (CC $\succ^{*}$ HPB) than HPB . There was

(a) Efficiency   (b) Auctioneer's revenue

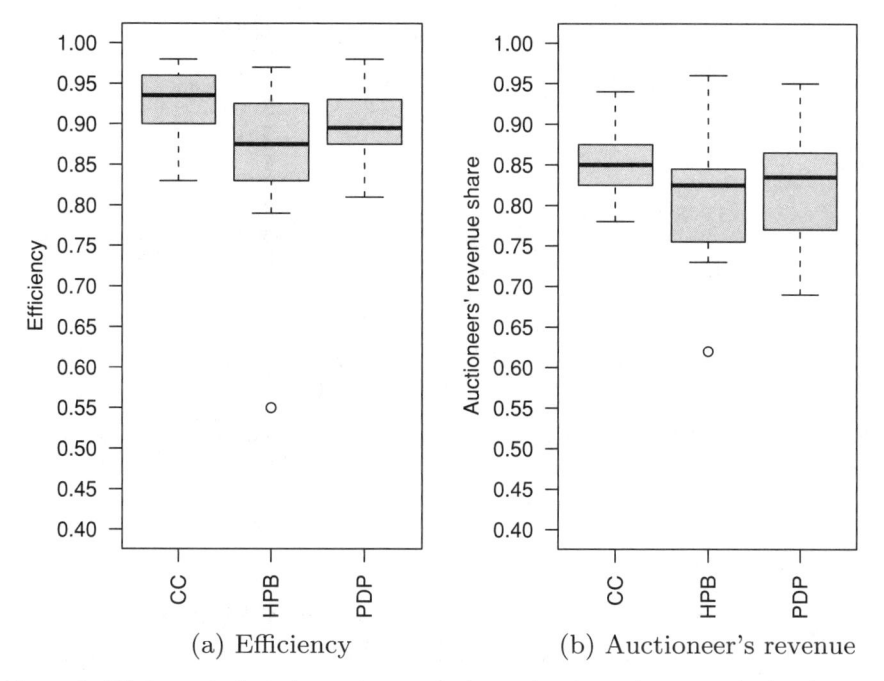

Figure 4. Efficiency and auctioneer's revenue for each auction format in the local-SVM.

no significant difference between HPB and PDP in efficiency and auctioneer's revenue in the local-SVM.

Support. A Wilcoxon rank sum test on the efficiencies and auctioneer's revenue reveals that the CC auction outperforms the HPB auction in efficiency ($p = 0.026$) and auctioneer's revenue ($p = 0.054$). Further comparisons are insignificant. □

4.1.4. Comparisons of the Auction Formats in the Different Value Models

Although the comparison of efficiency and auctioneer's revenue share is difficult across two different value models, we observe that the three auction formats achieved similar levels of efficiency in the global-SVM and the local-SVM. This result confirms Conjecture 3 only partly. While in HPB, which is closer to our SMR simulation, on which the conjecture is based, the statement is true, it does not hold for the fully CAs CC and PDP. The comparison of the experiments in the two value models reveals that there are differences in the ranking of auction formats in terms of efficiency and auctioneer's revenue, but those differences are small.

Result 4 *(Comparison of auction formats in different value models).* The results of the value model with global synergies do not fully carry over to a value model with local synergies. The ranking of efficiencies in the global-SVM (HPB $\succ^{***}$ PDP) and revenue (HPB $\succ^{***}$ PDP) cannot be replicated in the local-SVM, in which differences in efficiencies (HPB $\sim$ PDP) and revenues (HPB $\sim$ PDP) are insignificant. The CC auction achieves the highest median efficiency and auctioneer's revenue in both value

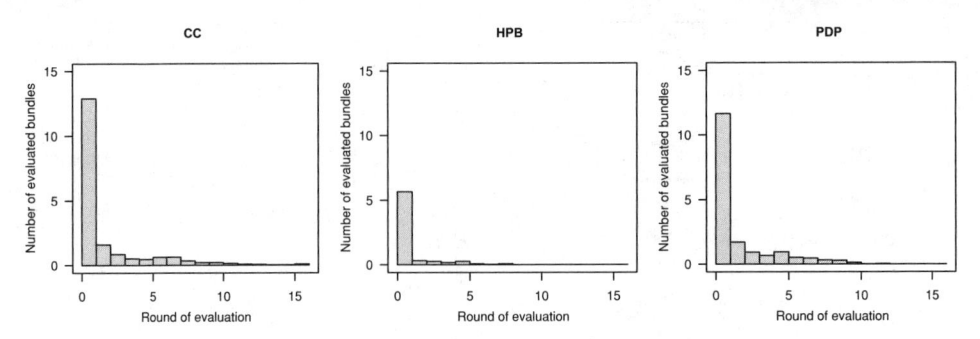

Figure 5. Number of packages evaluated by a bidder in the auctions on the local-SVM.

models. Overall, the differences in these aggregate performance metrics between the different auction formats are small.

Support. Results 1 and 3 show the different rankings of the auction formats in the different value models. □

The ranking of HPB and PDP changes from the global-SVM to the local-SVM. There are several factors causing the different rankings. First, we find HPB to collect more bids in the global-SVM, in contrast to the local-SVM in which PDP collects more bids. Second, the number of unsold items is high in the PDP auction in the global-SVM. This number dropped from ⌀ 2.46 in the global-SVM to ⌀ 0.0 in the local-SVM (cf. Table 5). Opening bids in the local-SVM assures that all items are sold in the PDP as well. Third, the hierarchical structure of the predefined packages in HPB does not fit the bidders' preferences as well in the local-SVM as in the global-SVM. Finally, the tight activity limits of the regional bidders in the global-SVM have a considerable negative impact on PDP but not on HPB. We further elaborate on this in Section 4.4.

In both value models, the CC auctions achieve highest median efficiency and auctioneer's revenue share. One reason for this is the number of bids submitted. In both value models, bidders submit more bids in the CC auction than in any other auction format. We conjecture that the absence of temporary winners during the auction causes this phenomenon. Bidders are not aware of their current position in the auction. Since bidders in CC do not learn about the current allocation, they actively submit bids in each round. In HPB and PDP, if bidders provisionally win all the items they want in a round, there is no reason to submit additional bids and the bidders reveal less information about their preferences to the auctioneer. Another explanation is that jump bidding is not possible in the CC auction, which assures a high number of demand queries.

4.2. Package Selection and Package Evaluation

Efficiency is very high in the experiments but still there is a gap to full efficiency. There might be several reasons for the efficiency losses, as specific auction rules, the price calculation or the bidding strategies. Since we observe coordination problems among the regional bidders, we go on with a deeper analysis of the bidder behavior to find the reason for the efficiency losses. We identify the bidders' selection of packages to be a crucial factor.

Table 6. *Efficiency in the laboratory experiments (LAB) compared to different simulations on packages selected by the bidders (EVALALL, EVALEARLY)*

Local-SVM	CC	HPB	PDP
LAB	92.4%	86.7%	90.0%
EVALALL	93.8%	90.8%	93.2%
EVALEARLY	91.6%	90.1%	90.3%

Both value models are challenging for bidders due to the large number of possible packages, especially in the local-SVM, in which bidders have to select from up to 262,143 different packages with positive valuation. Our experimental software provides a feature to calculate the value of a particular package, which gives us the opportunity to observe package evaluation. Since the calculation of package valuations in the global-SVM can easily be done without the tool, we concentrate on the local-SVM, where we can observe all evaluations. We conducted simulations with artificial bidding agents, who truthfully reveal their valuations on all individual items and packages that the bidders have evaluated in the actual lab auction (***EVALALL*** treatment). In another simulation treatment, ***EVALEARLY***, we consider the individual items and only those packages bidders have evaluated in the first and the second round of the auction.

Result 5 *(Preselection of packages).* Bidders select and evaluate most packages in the very first round, which is consistent with Conjecture 1. New price information throughout the auction does not lead to the evaluation of many more packages later on. Further, package selection after the second round does not have a significant impact on the auction efficiency.

Support. We measure the number of evaluated packages in each round; the result is displayed in Figure 5. Bidders select over 80% of the packages in the first three rounds in CC and PDP and about 91% in the HPB auctions. To understand the impact of the late evaluated packages, we compare the simulations on the package selection, namely the treatments EVALALL and EVALEARLY. The results of EVALALL and EVALEARLY are not significantly different, which means that the late evaluated packages do not increase the possible efficiency (the average results are included in Table 6). □

It is interesting to understand whether the efficiency loss is due to the package selection in a CA or whether it is mostly a result of the bidding strategies and auction rules. We compare the outcome in our laboratory experiments (***LAB***) to the results of the EVALALL treatment. While Table 6 contains all average values of our laboratory experiments and different numerical simulations, Figure 6 also contains the distribution of the values for the treatments LAB and EVALALL.

Result 6 *(Efficiency loss due to package selection).* The selection of packages (and not auction rules or bidding strategies) accounts for most of the efficiency losses in the local-SVM. Only in the PDP auction the bidders in the lab achieve a worse solution than in the simulations with a direct revelation mechanism and restricted package selection (PDP-EVALALL $\succ^{**}$ PDP-LAB).

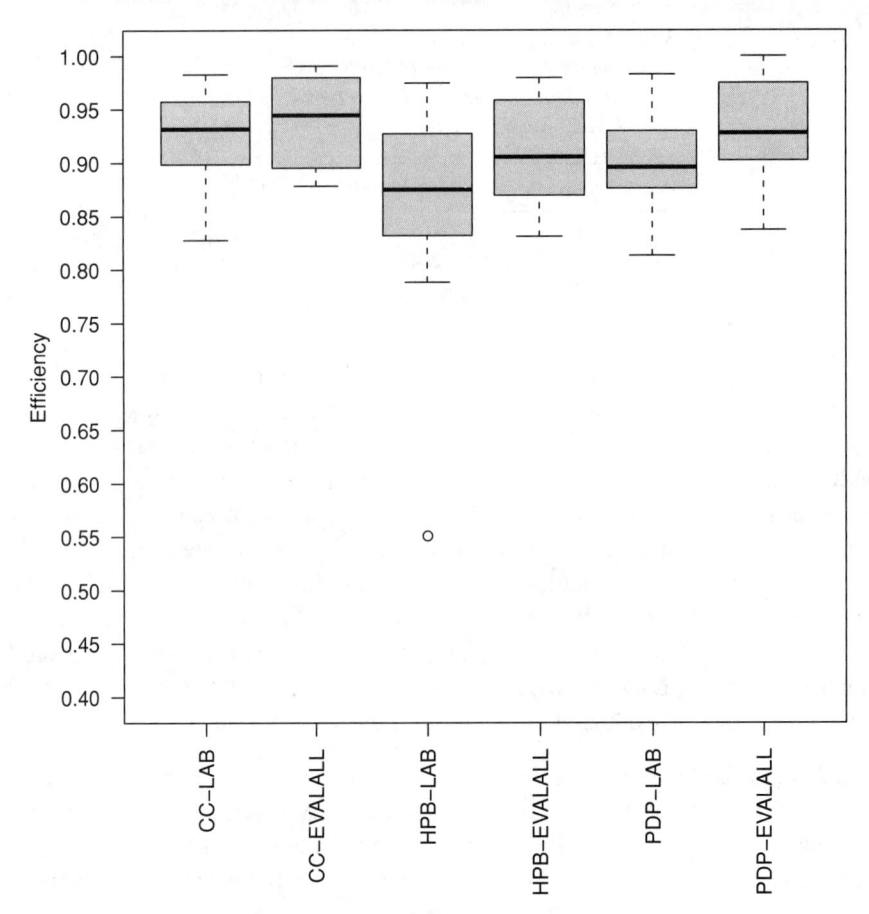

Figure 6. Comparison of the efficiency in different auction formats in the laboratory (LAB) and simulation (EVALALL). Using a sealed bid auction in our simulations, we calculate the best allocation if bidders bid their true value on all packages they have evaluated during the actual lab auction.

Support. The results in Figure 6 and Table 6 show that much of the efficiency loss in the local-SVM is due to package selection of the bidders. In CC and HPB the results in our laboratory experiments (LAB) are not statistically different from the *EVALALL* treatments. This means that, in these two auction formats, the auction ends up in an allocation which is similar to the simulation EVALALL. Only in PDP do we find the simulation (EVALLALL) to achieve significantly higher efficiency than the LAB result ($p = 0.028$). □

Result 6 implies that package selection accounts to a large extent for the efficiency loss in CAs. Therefore, we performed an analysis of the package selection in the local-SVM by using a logistic regression (Hastie and Pregibon, 1992), which describes the bidders' package selection. We only take the evaluations from regional bidders in CC and PDP auctions into account, since these auction formats are fully CA formats and are not biased by limitations of the auction mechanisms in terms of package selection.

To denote the bidders' selection of a package, we list all packages the bidder has a positive valuation for and assign a binary variable, which indicates 1 if the bidder has

Table 7. *Logistic regression of the regional bidders' likelihood to evaluate a package. The dependent variable is the decision whether or not a bidder has evaluated a particular package*

	Estimate	Std. Error	z value	Pr(>|z|)
(Intercept)	−0.0482	0.1466	−0.33	0.7421
MEANBASEVAL	4.6504	0.1856	25.06	0.0000
CONN	−2.4844	0.0561	−44.32	0.0000
ITEMS7	−1.5378	0.1083	−14.20	0.0000
ITEMS8	−2.1002	0.0843	−24.92	0.0000
ITEMS9	−2.3324	0.0812	−28.72	0.0000
ITEMS10	−3.3378	0.0909	−36.71	0.0000
ITEMS11	−3.8165	0.0897	−42.57	0.0000
N	94992	$-2LogL_0$		25259
Freq. of 1:	2801	$-2LogL_{full}$		17547
Freq. of 0:	92191	AIC		17563

evaluated the value of the respective package and zero otherwise. We identify several independent variables which have an impact on the bidders' package selection. The variable ***MEANBASEVAL*** holds the package's mean baseline valuation divided by the bidder's highest baseline valuation. This means that, this variable shows whether the package contains items with high baseline valuations, relative to the bidders highest baseline valuation.

Another important dimension in the bidders' decision for evaluating a package is the number of not connected subpackages within the selected package. Since complementarities only arise for maximally connected subpackages, we count the number of them within the selected package in the variable ***CONN***, i.e. if all items are connected through neighboring items, this variable is 1, otherwise if for example the package can be divided in two sub-packages in which none of the items are horizontally or vertically neighboring, this variable is 2 for the respective package.

Consequently we investigate whether the number of packages a bidder is interested in has an effect on the number of evaluated packages. We model this attribute as a factor (***ITEMS6–ITEMS11***).

Result 7 *(Package selection by baseline valuations).* Bidders select packages according to the relative hight of the baseline valuations. Additionally, they take into account that the whole package has to be maximally connected.

Support. Table 7 shows the results of the logistic regression among several potential factors which can explain the bidders' package selection. We find only the relative hight of the baseline valuations (MEANBASEVAL) to have a significant positive impact on the bidders' package selection, i.e. the higher the relative baseline valuations of the items are, the higher is the likelihood that the bidders evaluate the value of the package. The coefficient for the CONN variable is significantly negative, meaning it is more likely that bidders select packages with a low number of maximally connected sub-packages. □

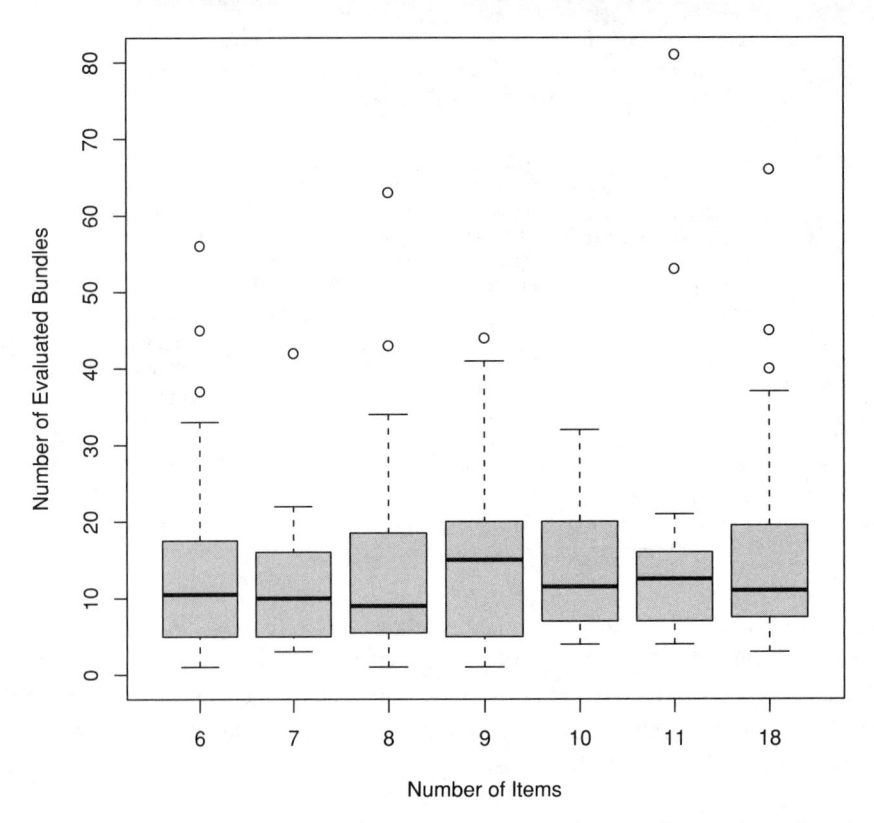

Figure 7. Impact of number of items bidders are interested in on the number of packages evaluated by bidders in the local-SVM.

As Conjecture 2 states, bidders follow the two important dimensions in selecting the packages. They know the baseline valuations of the individual items and they know that the complementarities are higher if the package consists only of one maximally connected subpackage. A further significant factor is the number of items the bidder is interested in, as the number of packages increases exponentially in the number of items, bidders do not evaluate the same ratio of packages as the number of packages increases, according to the coefficients of the variables ITEMS7–ITEMS11 in the logistic regression.

Therefore, we analyze the number of packages which bidders have evaluated in the auction, and compare those to the number of items they have a positive valuation for. Note that the regional bidders are interested in six to eleven items, while the national bidder is interested in 18 items. Figure 7 shows that the number of evaluated packages is independent of the number of items a bidder is interested in. We conjecture that this has to do with cognitive limits in the number of items people can concentrate on as have been analyzed in psychology for years (Miller, 1956).

Result 8 *(Package selection: Impact of number of possible packages).* Independent of the number of possible packages, bidders only evaluate approximately 14 packages including two or more items. The average number of packages evaluated does not increase with the number of items of interest, even though the number of possible packages increases exponentially with the number of items.

Support. In Figure 7 we see bidders select on average the same number of packages independent from the number of items they are interested in. For example, the national bidder has $2^{18} - 1 = 262,143$ packages with a positive valuation; however, bidders typically evaluate only around 14 packages including two or more items. □

The observations about the bidders' package selection deserves a more in-depth discussion. First, one could argue that for the specific case of high-stakes spectrum auctions, bidders will spend the time to evaluate a much larger number of packages than in a lab experiment with students. In an auction with 20 items of interest bidders have more than one million possible packages. Teams in spectrum auctions have usually one or two hours per round and typically multiple people are involved in the decision. We do not expect that bidders are able to evaluate a few hundreds or even thousands of packages and bid on those.

Second, some auctioneers do not only reveal prices, but detailed information about other winning bids, and sometimes even all bids and the bidder identities. When a bidder assumes that another bidder is strong and there is a low probability to overbid this bidder on a particular package, this might in fact provide additional incentives to evaluate one or the other relevant packages during the auction. However, as the number of packages grows exponentially in the number of items, we consider it unlikely that even with different information feedback rules the observed phenomenon of the bidders' package selection vanishes. The development of additional bidder decision support and adequate information feedback is certainly important for future research.

4.3. Threshold, Coordination and Exposure Problem

We analyze the coordination problem by considering two different aspects. First, when some small bidders have to coordinate their bids to outbid a bidder interested in a package covering many items, they are confronted with the threshold problem (Rothkopf et al., 1998). In our value models, we have several regional bidders, interested only in a small set of items, competing against a national bidder. Second, since in our experiments every item is sold in the efficient solution, we can also evaluate the coordination problem in measuring the number of unsold items. We know that some bidders have a positive valuation for unsold items, which means ex-post they regret not having bid on these items at least a price of ϵ.

Figure 8 shows the percentage of auctions in which the national bidder managed to win a huge share of items, even though based on their valuations the regional bidders were able to outbid him. In the global-SVM we had 24 auctions and in the local-SVM 16 auctions for each auction format. In HPB, regional bidders obviously have more difficulties coordinating among themselves in order to outbid the national bidder in a threshold problem. This is due to the fact that the package bids are restricted in HPB. The national bidder has an advantage, as there the package of all items is at the top hierarchy level, which fits his preferences. This is not the case for regional bidders. While in HPB pre-packaging guides bidders to solve their coordination problem by limiting the number of overlapping bids, bidders with a preference structure not matching the pre-packaging possibly face an exposure risk. In our data we find, contrary to Goeree and Holt (2010), evidence that the bidders' exposure risk dominates the coordination advantage in HPB in both value models.

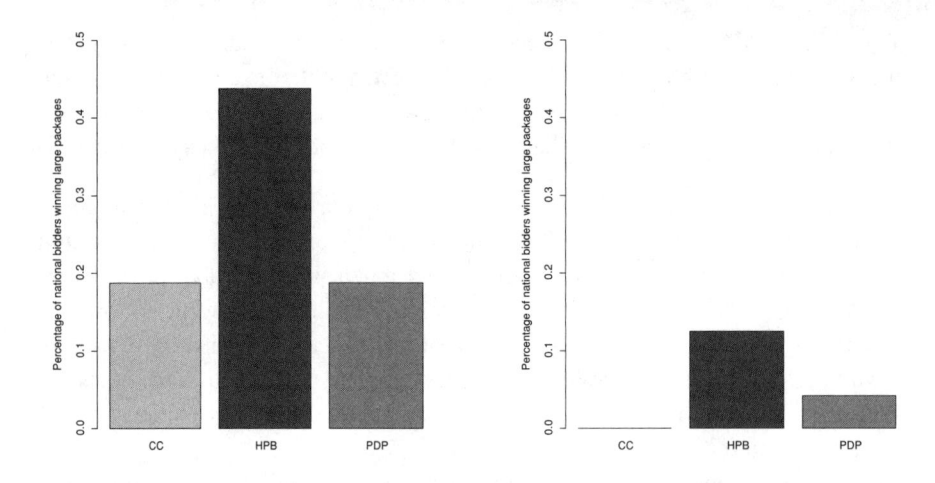

(a) Auctions in which the national bidder wins more than 16 items in the local-SVM

(b) Auctions in which the national bidder wins all 12 items of the national circle in the global-SVM

Figure 8. Threshold problem.

Result 9 *(Coordination: Strong threshold problem due to pre-packaging).* In HPB the national bidder has an advantage over the regional bidders due to the limitation of package bids. In fully CAs regional bidders can often outbid the national bidder, although they do not manage to win all items.

Support. In Figure 8 we find the national bidder to win considerably more often a huge package of items in HPB than in CC and PDP, in both value models. We extend our analysis to the provisional winners in each round (Figure 9) and find that the national bidder managed to win all items in nearly 45% of the rounds in HPB in the local-SVM. In the results of the global-SVM, we see that bidders in PDP are able to outbid the

(a) All items of the local-SVM

(b) National circle of the global-SVM

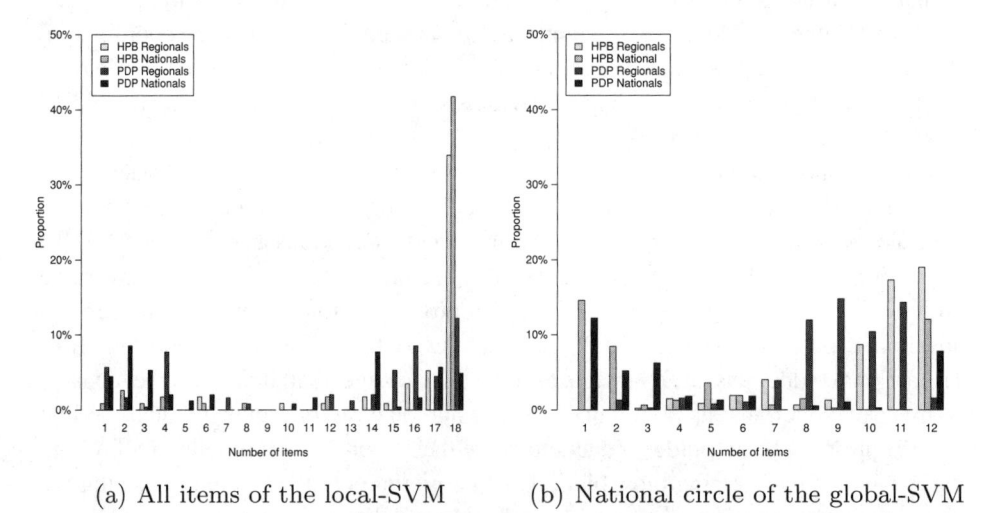

Figure 9. Number of items regional and national bidders provisionally win in a round.

national bidder by winning not all items, which partially explains the observed high number of unsold items. In Table 4 we find only ⌀ 0.1 of the items to remain unsold in the HPB auction, compared to ⌀ 1.0 in CC and ⌀ 2.5 in PDP (HPB ≺*** CC ≺*** PDP). □

In the PDP auction in the global-SVM, a large number of items remain unsold and the efficiency is low. The combination of activity and purchase limits (cf. Section 4.4.2), the XOR bidding language for the regional bidders, and a fully CA format leads to such a coordination problem. The CC auction, also a fully CA format, better solves this coordination problem due to the large number of bids submitted by the bidders. However, also in the CC auctions in the global-SVM one item remains unsold on average. In contrast, HPB, with the pre-packaging, solves the coordination problem by limiting the number of overlapping bids, resulting in the lowest number of unsold items. With the introduction of opening bids in the local-SVM, all items are sold in CC and PDP, while in HPB pre-packaging ensures complete sales.

We find HPB to favor the national bidder. One reason for this is that the national bidder's preferences are perfectly matched by the biggest package available in HPB. Regional bidders, in contrast, usually find their preferences not matched, so they might have to overbid their valuation on a smaller package or an individual item to be able to win a package not included in the pre-packaging. The risk for them is to win only part of that package and paying too much for it, which is the exposure problem. Figure 10 describes the ratio of final bids that regional bidders submitted in different auction formats. We classified bids under or at, and over the bidder's valuation.

Result 10 *(Exposure risk due to pre-packaging).* Pre-packaging in HPB forces the regional bidders to overbid their valuations. In the fully CA formats CC and PDP, overbidding is rare.

(a) Final bids in the local-SVM   (b) Final bids in the global-SVM

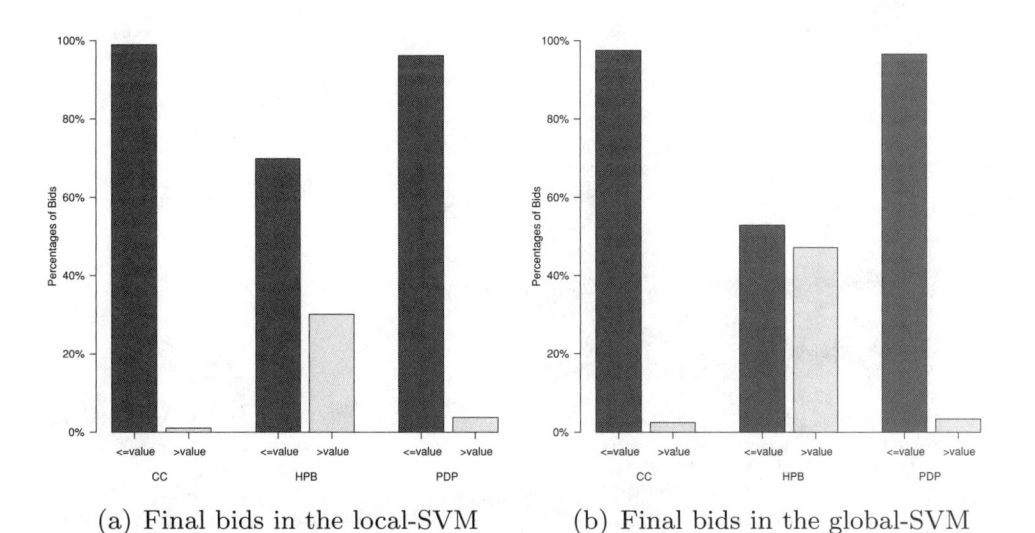

Figure 10. Percentage of the regional bidders' final bids over their respective valuation compared to the share of bids on or below their valuation.

Support. Figure 10 reveals that 47.2% of the regional bidders' final bids in HPB in the global-SVM are above their respective valuation, as well as 30.1% in the local-SVM. In CC and PDP, the number of final bids above the valuation is below 3.8%. □

4.4. Individual Bidder Behavior

Section 4.2 shows that bidders evaluate only a small number of packages. Since an equilibrium strategy is unknown in the three auction formats, the question is what strategy bidders follow and whether they bid on all evaluated packages.

4.4.1. Analysis of Individual-Item and Package Bids

Pre-packaging in HPB limits the number of packages a bidder is able to bid on. Figure 11 illustrates on how many distinct packages a bidder submits a bid in each auction format and value model.

Result 11 *(Package bids vs. individual-item bids).* Independent of the number of possible bids, bidders only submit bids on a small number of packages. Excluding the opening bids of the first round, bidders only submitted up to 15 distinct bids during an auction. Pre-packaging in HPB leads to small number of package bids submitted by regional bidders.

Support. Figure 11 shows the number of individual-item bids vs. the number of package bids. In HPB, the regional bidders submit ⌀ 0.2 distinct package bids during an auction in the global-SVM and ⌀ 1.57 in the local-SVM. In contrast, regional bidders in CC submitted ⌀ 7.7 package bids in the global-SVM and ⌀ 15.2 in the local-SVM. □

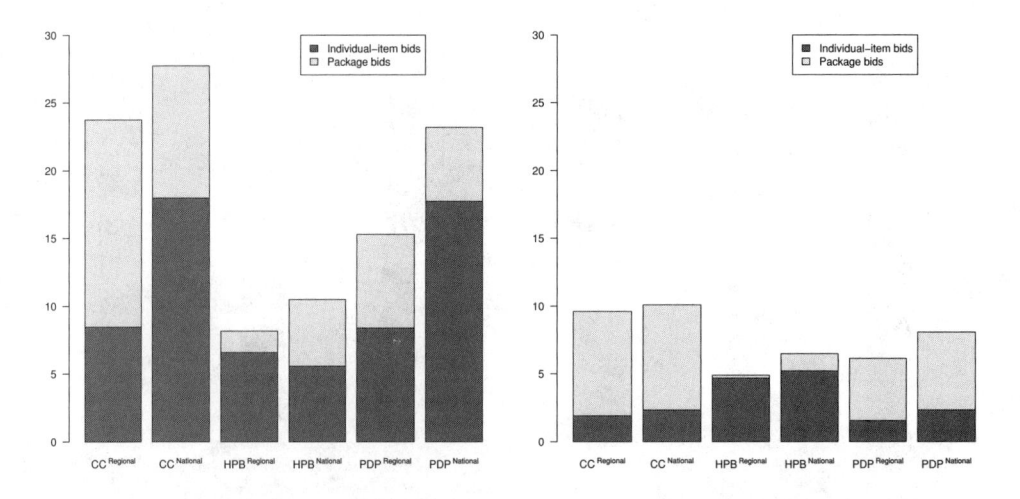

(a) Distinct package bids in the local-SVM

(b) Distinct package bids in the global-SVM

Figure 11. Number of distinct packages on which bidders submitted a bid.

On average, bidders only bid on six to ten different packages throughout the auction, independent of auction format or bidder type (regional or national) in the global-SVM, although they had 56 or 4095 packages with positive valuations, respectively. Note that since HPB is not fully combinatorial, regional bidders have only up to seven and national bidders 16 different packages to bid on in the global-SVM, including the six and twelve individual items respectively. CC and PDP collected only a few individual-item bids (fewer than two per bidder and auction) in the global-SVM. In contrast, the regional bidders in HPB hardly made use of any package bid.

Due to a significantly larger number of possible bids in the local-SVM, we observed that bidders bid on more distinct packages on average compared to the global-SVM. However, the 15 to 27 distinct package bids throughout the auctions in the local-SVM still comprise only a small proportion of the possible bids (63-2,047 for regional bidders). Because of the restricted bid possibility in HPB, only eight to ten different packages were bid on. Unlike in the global-SVM, bidders bid on packages in HPB more frequently. Clearly, with the opening bids we forced bidders to more bid activity in the first round, which bias the average number of submitted bids in the local-SVM. Taking this into account, bidders did not bid on many more different packages on average than in the global-SVM, which suggests cognitive barriers in the number of packages a bidder can focus on independent of the number of possible bids.

4.4.2. Impact of Activity and Purchase Limits

Regional bidders in the global-SVM have valuations for six distinct items, but have an activity and purchase limit of four items, i.e. they are only able to place bids on four different items in a round. Bidders in the local-SVM are not limited initially. We explore whether these limitations have an impact on bidder behavior by analyzing the distinct items the bidders placed a bid on. Figure 12 shows on how many distinct items

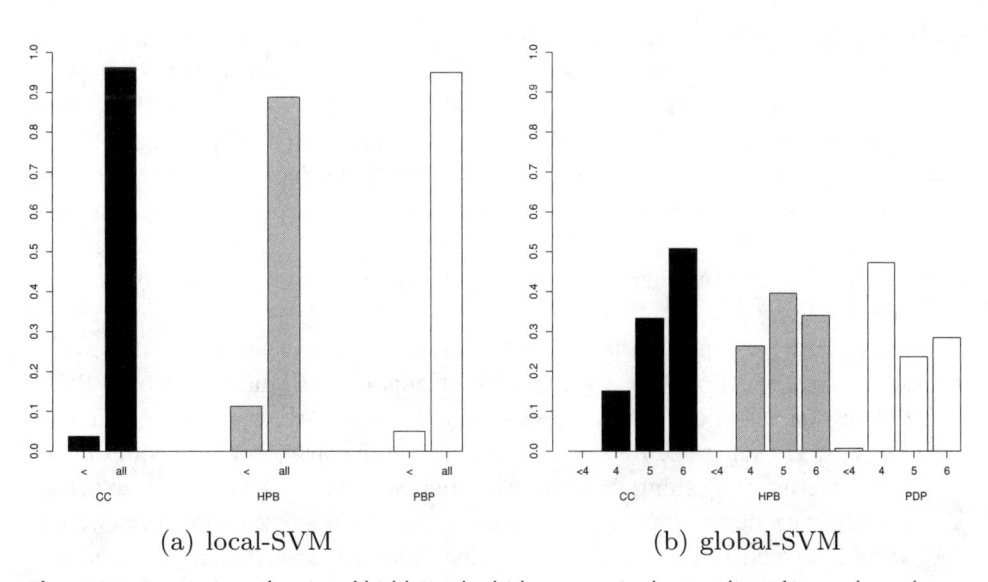

(a) local-SVM

(b) global-SVM

Figure 12. Proportion of regional bidders who bid on a particular number of items throughout the auction.

a regional bidder placed a bid throughout the auction process, either individually or in a package.

Result 12 *(Impact of tight activity limits).* In the global-SVM more than half of the bidders in the PDP auction and HPB did not bid on all items with positive valuation, due to jump bids and activity limits. This phenomenon vanished in the local-SVM.

Support. As shown in Figure 12, about 50% of the regional bidders in the PDP auction bid only on four items in the global-SVM. In HPB almost 30% of the regional bidders only bid on four items. This means that they never bid on two of the items for which they have positive valuations, neither individually nor in a package. Only very few bidders in the local-SVM did not bid on all items of interest (Figure 12). □

A reason for such behavior is jump bidding, which leads to the fact that many bidders win particular items for many rounds. If, due to a jump bid, they win a package of four items, they cannot submit bids on packages including any other items in these rounds in the PDP auction. Afterwards, the ask prices for other items often increase to a level such that a bid is not profitable. In the CC auction no jump bids are possible and no winning bids are announced in the auction process, so that bidders submit bids on many more items. In HPB, regional bidders face a similar problem as in the PDP auction; due to the activity limit and the announcement of the provisional winning bids, bidders are not always able to bid on all desired items. In the local-SVM, without activity limits and with opening bids, the situation is different. Most bidders submit bids on all items of interest.

5. Conclusions

The design of combinatorial auctions has been and remains a research challenge. Auction formats with strong game-theoretical solution concepts have shown to be impractical, or equilibrium strategies only hold for restricted valuations (Ausubel and Milgrom, 2006; Schneider et al., 2010). Linear-price combinatorial auctions have shown to be robust in simulations and lab experiments with small value models (Bichler et al., 2009; Kagel et al., 2009; Scheffel et al., 2010). The efficiency of combinatorial auctions in situations with more than 10 items of interest has not received much attention, although it is not clear whether the results of small combinatorial auctions also transfer to larger instances.

In this paper, we report on laboratory experiments comparing the Combinatorial Clock auction, an auction format with pseudo-dual prices, and the Hierarchical Package Bidding format in large value models with 18 items. All these auction formats have been used or suggested for selling spectrum, but they have not been compared in the same experimental setting. We use two different value models to analyze aggregate performance metrics such as efficiency and auctioneer's revenue share, as well as individual level bidder behavior. There are minor differences in efficiency and revenue across the auction formats, but the efficiency of all these linear-price formats is fairly high.

Interestingly, we find the limited number of packages that bidders evaluate to be the greatest barrier to efficiency, much more so than differences in the auction formats. On average, bidders evaluate about 14 packages besides the individual items, most of them in the first auction round, although they have a much larger number of profitable packages. They only submit six to 14 distinct package bids during an auction overall. We conjecture that this has to do with cognitive limits in terms of the number of items people can simultaneously concentrate on as have been analyzed in psychology for years (Miller, 1956). Practical auction formats need to take such cognitive barriers into account, either by limiting the number of items, by pre-packaging, or by advanced tooling, which allows a large number of packages to be easily explored. We also observe that changes in the auction rules matter. For example, activity rules can cause inefficiency in pseudo-dual price auctions with jump bids, while opening bids in the first round increase efficiency. While much fundamental theory in algorithmic mechanism design has been developed, the design of combinatorial auctions remains a formidable research challenge.

A. Auction Results

Table 8. *Auction results in the global-SVM*

Auction Format	Auction Number	Wave	$E(X)$	$R(X)$	Payoff Nationals	Payoff Regionals	Number of Bids	Unsold Items	Rounds
CC	1	W1	0.96	0.90	2.00	0.80	330.00	1.00	16.00
CC	2	W1	0.93	0.85	0.00	1.30	235.00	0.00	14.00
CC	3	W1	0.96	0.81	0.00	2.60	290.00	0.00	17.00
CC	4	W1	0.91	0.82	0.00	1.60	304.00	2.00	14.00
CC	1	W2	0.82	0.63	0.50	3.20	197.00	0.00	8.00
CC	2	W2	0.91	0.70	0.00	3.50	176.00	1.00	18.00
CC	3	W2	0.83	0.65	0.00	3.00	197.00	2.00	15.00
CC	4	W2	0.98	0.77	0.00	3.40	198.00	0.00	23.00
CC	5	W2	0.74	0.43	0.00	5.20	70.00	3.00	6.00
CC	6	W2	0.85	0.62	0.00	3.70	114.00	3.00	12.00
CC	1	W3	0.86	0.73	0.50	2.10	313.00	0.00	21.00
CC	2	W3	0.92	0.76	0.00	2.70	254.00	1.00	13.00
CC	3	W3	0.83	0.55	0.00	4.60	183.00	1.00	18.00
CC	4	W3	0.93	0.84	0.00	1.40	243.00	0.00	12.00
CC	5	W3	0.95	0.86	0.40	1.50	277.00	0.00	20.00
CC	1	W4	0.91	0.79	6.80	0.70	236.00	0.00	14.00
CC	2	W4	0.95	0.81	−1.40	2.60	287.00	0.00	15.00
CC	3	W4	0.86	0.81	0.00	0.90	199.00	3.00	18.00
CC	4	W4	0.94	0.81	0.20	2.10	189.00	0.00	13.00
CC	5	W4	0.84	0.74	−4.40	2.30	161.00	3.00	14.00
CC	6	W4	0.94	0.78	0.00	2.70	257.00	0.00	26.00
HPB	1	W1	0.94	0.58	−0.10	6.00	62.00	0.00	9.00
HPB	2	W1	0.98	0.83	0.00	2.60	64.00	0.00	10.00
HPB	3	W1	0.98	0.74	0.00	4.10	80.00	0.00	17.00
HPB	4	W1	0.85	0.90	−7.40	0.50	97.00	0.00	20.00
HPB	5	W1	0.96	0.79	0.00	2.80	93.00	0.00	15.00
HPB	6	W1	0.87	0.74	1.20	2.10	69.00	0.00	7.00
HPB	1	W2	0.99	0.50	1.40	7.90	66.00	0.00	13.00
HPB	2	W2	0.80	0.70	−9.30	3.30	93.00	0.00	10.00

(*cont.*)

Table 8. *(cont.)*

Auction Format	Auction Number	Wave	$E(X)$	$R(X)$	Payoff Nationals	Payoff Regionals	Number of Bids	Unsold Items	Rounds
HPB	3	W2	0.89	0.63	1.10	4.10	96.00	0.00	15.00
HPB	4	W2	0.86	0.69	0.00	2.90	88.00	0.00	9.00
HPB	5	W2	0.92	0.62	0.00	4.90	83.00	0.00	19.00
HPB	6	W2	0.89	0.80	3.10	1.00	113.00	0.00	14.00
HPB	1	W3	0.86	0.69	0.50	2.70	68.00	0.00	11.00
HPB	2	W3	0.88	0.64	0.40	4.00	76.00	0.00	16.00
HPB	3	W3	0.98	0.57	0.00	6.80	66.00	0.00	8.00
HPB	4	W3	0.86	0.76	4.60	1.00	69.00	0.00	12.00
HPB	5	W3	0.88	0.73	0.00	2.40	80.00	0.00	14.00
HPB	6	W3	0.82	0.67	7.30	1.40	75.00	1.00	15.00
HPB	1	W4	0.99	0.78	14.90	0.90	96.00	0.00	11.00
HPB	2	W4	0.90	0.68	1.40	3.30	72.00	1.00	12.00
HPB	3	W4	0.89	0.82	−0.80	1.20	99.00	0.00	20.00
HPB	4	W4	0.91	0.68	−0.20	3.90	113.00	0.00	17.00
HPB	5	W4	0.99	0.73	0.00	4.30	75.00	0.00	7.00
HPB	6	W4	0.99	0.65	0.00	5.70	77.00	0.00	18.00
PDP	1	W1	0.67	0.58	0.00	1.40	57.00	6.00	17.00
PDP	2	W1	0.82	0.52	−1.80	5.40	60.00	3.00	10.00
PDP	3	W1	0.75	0.65	7.50	0.40	73.00	3.00	10.00
PDP	4	W1	0.94	0.64	1.80	4.60	51.00	0.00	8.00
PDP	5	W1	0.75	0.43	0.00	5.50	53.00	4.00	10.00
PDP	6	W1	0.84	0.50	0.00	5.70	43.00	2.00	11.00
PDP	1	W2	0.92	0.64	2.00	4.30	66.00	1.00	10.00
PDP	2	W2	0.81	0.58	0.30	3.70	63.00	2.00	10.00
PDP	3	W2	0.82	0.71	0.40	1.80	94.00	2.00	11.00
PDP	4	W2	0.85	0.66	0.20	3.10	72.00	3.00	13.00
PDP	5	W2	0.87	0.63	0.00	4.00	75.00	3.00	19.00
PDP	6	W2	0.86	0.55	0.00	5.20	54.00	5.00	9.00
PDP	1	W3	0.89	0.76	0.00	2.20	79.00	2.00	11.00
PDP	2	W3	0.96	0.66	0.90	4.90	107.00	1.00	9.00
PDP	3	W3	0.95	0.72	0.00	4.00	125.00	1.00	18.00
PDP	4	W3	0.89	0.74	0.40	2.50	93.00	1.00	14.00
PDP	5	W3	0.87	0.76	−3.40	2.40	133.00	0.00	14.00
PDP	6	W3	0.78	0.73	0.00	0.90	79.00	5.00	11.00
PDP	1	W4	0.87	0.74	0.00	2.10	67.00	3.00	7.00
PDP	2	W4	0.90	0.67	1.40	3.50	57.00	4.00	11.00
PDP	3	W4	0.87	0.63	2.80	3.60	56.00	3.00	9.00
PDP	4	W4	0.80	0.68	0.00	2.10	51.00	4.00	9.00
PDP	5	W4	0.94	0.73	0.40	3.40	46.00	0.00	7.00
PDP	6	W4	0.86	0.65	0.00	3.60	40.00	1.00	10.00

Table 9. Auction results in the local-SVM

Auction Format	Auction Number	Wave	$E(X)$	$R(X)$ Payoff	Nationals	Payoff Regionals	Number of Bids	Unsold Items	Rounds
CC	1	W2	0.83	0.83	0.20	0.00	375.00	0.00	16.00
CC	2	W2	0.95	0.90	0.40	0.90	322.00	0.00	13.00
CC	3	W2	0.93	0.83	0.00	1.80	362.00	0.00	14.00
CC	4	W2	0.86	0.84	2.00	0.00	417.00	0.00	12.00
CC	1	W3	0.97	0.80	1.30	3.00	279.00	0.00	11.00
CC	2	W3	0.92	0.85	0.00	1.40	466.00	0.00	11.00
CC	3	W3	0.87	0.78	0.90	1.50	484.00	0.00	11.00
CC	4	W3	0.98	0.94	0.00	0.90	422.00	0.00	14.00
CC	1	W4	0.98	0.87	1.70	1.80	581.00	0.00	16.00
CC	2	W4	0.88	0.88	0.20	0.00	647.00	0.00	13.00
CC	3	W4	0.92	0.85	0.20	1.50	526.00	0.00	12.00
CC	4	W4	0.94	0.82	0.00	2.30	596.00	0.00	11.00
CC	1	W5	0.92	0.81	0.00	2.10	442.00	0.00	12.00
CC	2	W5	0.98	0.91	0.00	1.40	353.00	0.00	13.00
CC	3	W5	0.94	0.87	0.00	1.40	605.00	0.00	12.00
CC	4	W5	0.94	0.86	1.40	1.20	678.00	0.00	12.00
HPB	1	W2	0.83	0.77	5.80	0.00	84.00	0.00	5.00
HPB	2	W2	0.88	0.74	14.00	0.00	60.00	0.00	5.00
HPB	3	W2	0.97	0.91	0.00	1.30	91.00	0.00	4.00
HPB	4	W2	0.86	0.78	8.30	0.00	70.00	0.00	6.00
HPB	1	W3	0.55	0.62	0.00	−1.40	96.00	0.00	6.00
HPB	2	W3	0.91	0.83	0.00	1.70	86.00	0.00	7.00
HPB	3	W3	0.79	0.73	5.90	0.00	109.00	0.00	8.00
HPB	4	W3	0.94	0.82	0.00	2.40	124.00	0.00	8.00
HPB	1	W4	0.86	0.83	2.30	0.00	79.00	0.00	6.00
HPB	2	W4	0.91	0.86	0.00	0.90	74.00	0.00	6.00
HPB	3	W4	0.83	0.74	8.60	0.00	79.00	0.00	7.00
HPB	4	W4	0.90	0.96	0.00	−1.20	95.00	0.00	7.00
HPB	1	W5	0.83	0.82	2.80	−0.30	109.00	0.00	7.00
HPB	2	W5	0.87	0.84	3.20	0.00	72.00	0.00	8.00
HPB	3	W5	0.97	0.83	0.00	2.80	120.00	0.00	8.00
HPB	4	W5	0.97	0.85	0.80	2.30	109.00	0.00	6.00
PDP	1	W2	0.81	0.91	−9.70	0.10	175.00	0.00	9.00
PDP	2	W2	0.87	0.95	0.00	−1.60	190.00	0.00	11.00
PDP	3	W2	0.90	0.81	0.50	1.60	192.00	0.00	11.00
PDP	4	W2	0.90	0.87	6.40	−0.60	115.00	0.00	11.00
PDP	1	W3	0.89	0.74	7.90	1.40	146.00	0.00	7.00
PDP	2	W3	0.89	0.85	0.00	0.90	145.00	0.00	8.00
PDP	3	W3	0.87	0.83	0.00	0.90	156.00	0.00	9.00
PDP	4	W3	0.98	0.93	0.00	1.10	146.00	0.00	6.00
PDP	1	W4	0.94	0.79	0.60	2.80	199.00	0.00	15.00
PDP	2	W4	0.83	0.77	4.70	0.20	100.00	0.00	9.00
PDP	3	W4	0.92	0.77	0.20	2.90	111.00	0.00	8.00
PDP	4	W4	0.91	0.69	0.00	4.40	97.00	0.00	8.00
PDP	1	W5	0.98	0.86	0.00	2.30	116.00	0.00	10.00
PDP	2	W5	0.88	0.86	1.60	0.10	123.00	0.00	8.00
PDP	3	W5	0.88	0.84	4.50	−0.10	111.00	0.00	7.00
PDP	4	W5	0.94	0.71	1.40	4.30	97.00	0.00	7.00

Acknowledgements

The authors thank Jacob Goeree and Riko Jacob for their helpful comments and suggestions. The financial support from the Deutsche Forschungsgemeinschaft (DFG) (BI 1057/3-1) is gratefully acknowledged.

References

Ausubel L, Milgrom P (2006) The lovely but lonely Vickrey auction. In: Cramton P, Shoham Y, Steinberg R (eds) Combinatorial Auctions, MIT Press, Cambridge, MA

Ausubel L, Cramton P, Milgrom P (2006) The clock-proxy auction: A practical combinatorial auction design. In: Cramton P, Shoham Y, Steinberg R (eds) Combinatorial Auctions, MIT Press, Cambridge, MA

Ausubel LM, Cramton P, McAfee RP, McMillan J (1997) Synergies in wireless telephony: Evidence from the broadband PCS auctions. Journal of Economics and Management Strategy 6(3):497–527

Banks J, Ledyard J, Porter D (1989) Allocating uncertain and unresponsive resources: An experimental approach. RAND Journal of Economics 20:1–25

Banks J, Olson M, Porter D, Rassenti S, Smith V (2003) Theory, experiment and the FCC spectrum auctions. Journal of Economic Behavior and Organization 51:303–350

Bichler M, Shabalin P, Pikovsky A (2009) A computational analysis of linear-price iterative combinatorial auctions. Information Systems Research 20(1):33–59

Bikhchandani S, Ostroy JM (2002) The package assignment model. Journal of Economic Theory 107(2):377–406

Brunner C, Goeree JK, Hold C, Ledyard J (2009) An experimental test of flexible combinatorial spectrum auction formats. American Economic Journal: Micro-Economics forthcoming

Chen Y, Takeuchi K (2009) Multi-object auctions with package bidding: An experimental comparison of Vickrey and ibea. Games and Economic Behavior, DOI 10.1016/j.geb.2009.10.007

Cramton P (2009) Auctioning the Digital Dividend, Karlsruhe Institute of Technology

Cramton P (2013) Spectrum auction design. Review of Industrial Organization 42(2):161–190

Cramton P, Shoham Y, Steinberg R (eds) (2006a) Combinatorial Auctions. MIT Press, Cambridge, MA

Cramton P, Shoham Y, Steinberg R (2006b) Introduction to combinatorial auctions. In: Cramton P, Shoham Y, Steinberg R (eds) Combinatorial Auctions, MIT Press, Cambridge, MA

Day R, Raghavan S (2007) Fair payments for efficient allocations in public sector combinatorial auctions. Management Science 53:1389–1406

FCC (2007) Auction of 700 Mhz band licenses scheduled for January 24, 2008. Federal Communications Commition Public Notice (DA 07-4171), URL http://hraunfoss.fcc.gov/edocs_public/attachmatch/DA-07-4171A1.pdf

Goeree J, Lien Y (2016) On the impossibility of core-selecting auctions. Theoretical Economics 11(1)

Goeree JK, Holt CA (2010) Hierarchical package bidding: A paper & pencil combinatorial auction. Games and Economic Behavior 70(1): 146–169

Hastie T, Pregibon D (1992) Generalized linear models. In: Chambers J, Hastie T (eds) Statistical Models in S, Wasworth and Brooks/Cole, Pacific Grove, California

Hollander M, Wolfe DA (1973) Nonparametric statistical inference. John Wiley & Sons, New York, USA

Kagel JH, Lien Y, Milgrom P (2009) Ascending prices and package bids: An experimental analysis. In: AEA Conference

Kelso AS, Crawford VP (1982) Job matching, coalition formation, and gross substitute. Econometrica 50:1483–1504

Kwasnica T, Ledyard JO, Porter D, DeMartini C (2005) A new and improved design for multi-objective iterative auctions. Management Science 51(3):419–434

Lamy L (2009) Core-selecting package auctions: A comment on revenue-monotonicity. International Journal of Game Theory 37

Ledyard J, Porter D, Rangel A (1997) Experiments testing multiobject allocation mechanisms. Journal of Economics, Management, and Strategy 6:639–675

Meeus L, Verhaegen K, Belmans R (2009) Block order restrictions in combinatorial electric energy auctions. European Journal of Operational Research 196:1202–1206

Milgrom P (2000) Putting auction theory to work: The simultaneous ascending auction. Journal of Political Economy 108(21):245–272

Miller GA (1956) The magical number seven, plus or minus two: Some limits on our capacity for processing information. Psychological Review 63:81–97

Moreton PS, Spiller PT (1998) What's in the air: Interlicense synergies in the Federal Communications Commission's broadband personal communication service spectrum auctions. Journal of Law and Economics 41(2):677–716

Porter D, Smith V (2006) FCC license auction design: A 12-year experiment. Journal of Law Economics and Policy 3

Porter D, Rassenti S, Roopnarine A, Smith V (2003) Combinatorial auction design. Proceedings of the National Academy of Sciences of the United States of America (PNAS) 100:11,153–11,157

Rassenti S, Smith VL, Bulfin RL (1982) A combinatorial auction mechanism for airport time slot allocations. Bell Journal of Economics 13:402–417

Rothkopf MH, Pekec A, Harstad RM (1998) Computationally manageable combinatorial auctions. Management Science 44:1131–1147

Scheffel T, Pikovsky A, Bichler M, Guler K (2010) An experimental comparison of linear and non-linear price combinatorial auctions. Information Systems Research 22(2):346–368

Schneider S, Shabalin P, Bichler M (2010) On the robustness of non-linear personalized price combinatorial auctions. European Journal of Operational Research 206(1):248–259

CHAPTER 29

Do Core-Selecting Combinatorial Clock Auctions Lead to High Efficiency? An Experimental Analysis of Spectrum Auction Designs

Martin Bichler, Pasha Shabalin, and Jürgen Wolf

1. Introduction

There has been a long discussion on appropriate auction mechanisms for the sale of spectrum rights (Porter and Smith, 2006). Since 1994, the Simultaneous Multi-Round Auction (SMRA) has been used worldwide (Milgrom, 2000). The SMRA design was very successful, but also led to a number of strategic problems for bidders (Cramton, 2013). The *exposure problem* is central and refers to the risk for a bidder to make a loss due to winning only a fraction of the bundle of items (or blocks of spectrum) he has bid on at a price which exceeds his valuation of the won subset.

Combinatorial auctions (CAs) allow for bids on indivisible bundles avoiding the exposure problem. The design of such auctions, however, led to a number of fundamental problems, and many theoretical and experimental contributions during the past ten years (Cramton et al., 2006b). The existing experimental literature comparing SMRAs and CAs suggests that in the presence of significant complementarities in bidders' valuations and a setting with independent private and quasi-linear valuations, combinatorial auctions achieve higher efficiency than simultaneous auctions (Banks et al., 1989; Ledyard et al., 1997; Porter et al., 2003; Kwasnica et al., 2005; Brunner et al., 2010; Goeree and Holt, 2010). Since 2008 several countries such as the U.K., Ireland, the Netherlands, Denmark, Austria, Switzerland, and the U.S. have adopted combinatorial auctions for selling spectrum rights (Cramton, 2013). While the U.S. used an auction format called Hierarchical Package Bidding (HPB) (Goeree and Holt, 2010), which accounts for the large number of regional licenses, the other countries used a Combinatorial Clock Auction (CCA) (Maldoom, 2007; Cramton, 2009), a two-phase auction format with primary bid rounds (aka. clock phase) for price discovery, which is extended by a supplementary bids round (aka. supplementary phase). The CCA design used in those countries has a number of similarities to the Clock-Proxy auction, which was proposed by Ausubel et al. (2006). It was used for the sale of blocks in a single spectrum band (i.e., paired and unpaired blocks in the 2.6 GHz band) and for the sale of multiple bands in Switzerland.[1]

Although, spectrum auction design might appear specific, the application is a representative of a much broader class of multi-object markets as they can be found in logistics and industrial procurement. Spectrum auctions are very visible in public and successful designs are a likely role model for other domains as well. The main contributions of this paper are the following:

- To our knowledge, this is the first lab experiment on the CCA, which we compare to the SMRA. We used an implementation of the CCA and the SMRA, which mirrors the auction rules used in the field and derive a number of properties of these auction rules. While most experimental studies in this field focus on small markets with a few blocks only, we intentionally used an experimental design which resembles real market environments. This is an important complement to other studies, as results of small combinatorial auctions do not necessarily extend to larger ones (Scheffel et al., 2012).[2]
- We show that the efficiency in the CCA was not higher than that of SMRA and due to the low number of bundle bids actually significantly lower in the multiband setting. Auctioneer revenue was considerably lower than in SMRA wich can be explained by the CCA second-price payment rules. However, the auctioneer revenue in CCA was also significantly lower than in CCA simulations where we had artificial bidders submit bids on all possible combinations truthfully with the same value models.
- We also analyzed bidder behavior in the CCA. In particular, in the multiband value model bidders select only a small fraction of all possible bundles for practical, but also for strategic reasons. While restricted bundle selection has recently been discussed in the experimental literature on other combinatorial auction designs (Scheffel et al., 2012), the paper analyzes the specific effects it can have on the efficiency of the CCA with a core-selecting payment rule. Although bid shading in core-selecting auctions is a concern in the theoretical literature, we found most bundle bids in the supplementary phase of the CCA to be at the valuation and only limited bid shading.
- In complex environments such as spectrum auctions there is a danger that external validity of lab experiments is not given as bidders in the field are better prepared than in the lab. To address this point to some extent, we also conducted *competitions*, where bidders had additional information about equilibrium strategies, known auction tactics, and two weeks of time to prepare a bidding strategy in a team of two people. While the bidder payoff in SMRAs was significantly higher than in the lab, in the CCA bidding behavior was not much different to the lab.

In the next section, we revisit the existing experimental literature on spectrum auctions and combinatorial auctions. In Section 3, we discuss the auction formats and game-theoretical results relevant to our study. Section 4 describes the experimental design, while Section 5 summarizes the results of our experiments. Finally, in Section 6 we provide conclusions and a discussion of further research in this area.

---

[1] Note that Porter et al. (2003) have defined a combinatorial clock auction, which is different to the one described in this paper and in Maldoom (2007) and Cramton (2013), and only consists of a single clock phase.
[2] Also Goeree and Holt (2010) used realistic value models in an effort to provide guidance for regulators in the USA.

2. Related literature

There is a substantial experimental literature on spectrum auction design. One strand of literature on spectrum auctions tries to analyze and explain specific strategic situations, as they occured in particular auctions either game-theoretically, experimentally, or based on data from the field (Klemperer, 2002; Ewerhart and Moldovanu, 2003; Bajari and Yeo, 2009). Another strand analyzes the mechanisms used in spectrum auctions based on related settings in the lab (Abbink et al., 2005; Banks et al., 2003; Seifert and Ehrhart, 2005). For example, Abbink et al. (2005) found differences in results between experiments with experienced vs. inexperienced students. Also in the field, bidders typically work in teams of experts and they spend significant amounts of time to prepare for the auction. In our analysis, we introduced competitons to analyze the impact of experienced teams on bidding strategies.

Motivated by spectrum auctions in the U.S., a number of experimental studies compared different combinatorial auction formats (Cramton et al., 2006a) and analyzed the question under which conditions combinatorial auctions are superior to SMRA. In an early study, Ledyard et al. (1997) compared SMRA with a sequential auction and a combinatorial auction within various value models. They found that combinatorial auction are best suited for environments with value complementarities.

Experiments conducted by Banks et al. (1989), Banks et al. (2003), and Kwasnica et al. (2005) find a positive effect of bundle bidding on efficiency when complementarities are present. Brunner et al. (2010) compared a standard SMRA auction with a single-stage combinatorial clock auction and a FCC format that augmented an SMRA auction to allow for bundle bids. Here, the combinatorial clock auction provided the highest efficiencies and the highest seller revenues. More recently, the Hierarchical Package Bidding (HPB) format which has been developed for the spectrum auctions in the U.S. was compared to SMRA and Modified Package Bidding (a format with pseudo-dual linear prices) by Goeree and Holt (2010). HPB outperformed the other two auction formats in terms of efficiency and auction revenue. Scheffel et al. (2012) extended this work and showed that restricted bundle selection is the most important reason for inefficiencies in larger auctions with more than a few blocks only. We make a similar observation for the CCA in this paper. More specific literature on the SMRA and the CCA will be discussed in the next section.

3. The auctions

In the following, we describe the SMRA and the CCA and discuss equilibrium bidding strategies. Beyond this, we want to summarize important characteristics of the CCA, as the specific rules of the auction format have found little attention in the academic literature so far.

3.1. The Simultaneous Multi-Round Auction

The SMRA is a generalization of the English auction for more than one block. All the blocks are sold at the same time, each with a price associated with it, and the bidders can bid on any of the blocks. The bidding continues until no bidder is willing to raise the bid on any of the blocks. Then the auction ends with each bidder winning the blocks on which he has the high bid, and paying its bid for any blocks won (Milgrom, 2000). There are differences in the level of information revealed about other bidders' bids in different countries. Sometimes all information is revealed after each round, sometimes only prices of the currently winning bids are published. A detailed description of the activity rules and the SMRA auction format used in our experiments can be found in Section 4.

There has been limited theoretical research on SMRA. If bidders have substitute preferences and bid straightforwardly, then the SMRA terminates at a Walrasian equilibrium (Milgrom, 2000), i.e., an equilibrium with linear prices. *Straightforward bidding* means that a bidder bids on the bundle of blocks, which together maximizes her payoff at the current ask prices. Gul and Stacchetti (1999) showed that if goods are substitutes, then ascending and linear-price auctions cannot implement the VCG outcome. Milgrom (2000) has also shown that with at least three bidders and at least one non-substitutes valuation no Walrasian equilibrium exists. Bidder valuations in spectrum auctions typically include complementarities.

In an environment with substitutes and complements SMRA results in an exposure problem. A number of laboratory experiments document the negative impact of the exposure problem on the performance of the SMRA (Brunner et al., 2010; Goeree and Lien, 2012; Kwasnica et al., 2005; Kagel et al., 2010). Therefore, the exposure problem has become a central concern. Goeree and Lien (2010) provided a Bayes-Nash equilibrium analysis of SMRA considering complementary valuations and the exposure problem. They show that due to the exposure problem, the SMRA may result in non-core outcomes, where small bidders obtain blocks at very low prices and seller revenue can be decreasing in the number of bidders just like in the Vickrey-Clarke-Groves auction (VCG) (Ausubel and Milgrom, 2006). Regulators have tried to mitigate this problem via additional rules, such as the possibility to withdraw winning bids. However, such rules can also provide incentives for gaming behavior. SMRAs also allow various forms of signaling and tacit collusion, but such behavior is reduced if the identity of bidders is not revealed and the auctioneer only allows for pre-defined jump bids.

3.2. The Combinatorial Clock Auction

The Combinatorial Clock Auction (CCA) is a two-phase combinatorial auction format which was introduced by Cramton (2009) and in an earlier version by Ausubel et al. (2006). In contrast to SMRA, the auction avoids the exposure problem by allowing for bundle bids. Maldoom (2007) describes a version as it has been used in spectrum auctions across Europe. We will refer to this version as the CCA, as this name is used in applications for spectrum sales.

In a CCA, bids for bundles of blocks are made throughout a number of sequential, open rounds (the primary bid rounds or clock phase) and then a final sealed-bid round (the supplementary bids round). In the primary bid rounds the auctioneer announces prices and bidders state their demand at the current price levels. Prices of bands with excess demand are increased by a bid increment until there is no excess demand anymore. Jump bidding is not possible. In the primary bid rounds, bidders can only submit a bid on one bundle per round. This rule is different to earlier proposals by Ausubel et al. (2006). These primary bid rounds allow for price discovery. If bidders bid straightforward on their payoff maximizing bundle in each round and all goods get sold after the clock phase, allocation and prices would be in competitive equilibrium. It might well be that there is excess supply after the clock phase, however. The sealed-bid supplementary bids phase and a Vickrey-closest core-selecting payment rule induce truthful bidding and avoid incentives for demand reduction. This is because core payments in Day and Milgrom (2007) are computed such that a losing bid of a winner does not increase his payment for his winning bid. The winner determination after the supplementary bids round considers all bids, which have been submitted in the primary bid rounds and the supplementary bids round and selects the revenue maximizing allocation. The bids by a single bidder are mutually exclusive (i.e., the CCA uses an XOR bidding language).

Activity rules should provide incentives for bidders to reveal their preferences truthfully and bid straightforwardly already in the primary bid rounds. Bidders should not be able to shade their bids and then provide large jump bids in the supplementary bids round. An eligibility-points rule is used to determine activity and eligibility to bid in the primary bid rounds. Each block in a band requires a certain number of eligibility points, and a bidder cannot increase his activity across rounds. In the supplementary bids round, revealed preferences during the primary bid rounds are used to derive relative caps on the supplementary bids that impose consistency of preferences between the primary and supplementary bids submitted. The consequence of these rules is that all bids are constrained relative to the bid for the final primary package by a difference determined by the primary bids. This should set incentivizes for straightforward bidding in the primary bid rounds.

3.3. Properties of the CCA

Since the CCA is a relatively new auction format and there is not much literature available on the specific auction rules that we analyse in our experiments, we will first provide a discussion of relevant properties of these rules. These properties will not be tested in the lab, but they provide an understanding about this specific auction format. We will also provide a summary of relevant game-theoretical literature on core-selecting auctions. In what follows, we analyze the CCA with respect to straightforward bidding in the primary bid rounds, incentive compatibility, envy-freeness, and possibilities for spiteful bidding.

3.3.1. Straightforward Bidding

The CCA is designed to incentivize straightforward bidding in the primary bid rounds and truthful bidding in the supplementary bids round.

Proposition 1. *If a bidder follows a straightforward bidding strategy in the primary bid rounds of a CCA with an anchor activity rule, then the activity rule will not restrict him from bidding his maximum valuation on every bundle in the supplementary bids round.*

A detailed description of the activity rules and these propositions can be found in the Appendix 7.1. Unfortunately, straightforward bidding is not always possible if a simple eligibility-points rule is used in the primary bid rounds.

Proposition 2. *If valuations for at least two bundles A and B are full substitutes with $v(A \cup B) = max(v(A), v(B))$ and the bundle of higher valuation A requires less bid rights than the lower valued bundle B, straightforward bidding is not possible due to the activity rule in the primary bid rounds.*

Strategies become difficult in this situation. Let's assume, a bidder wants either two blocks in band I or four blocks in band II, which have a lower value to him. All blocks have the same number of bid rights. A bidder needs to bid on a bundle, which is at least as large as his largest bundle of interest (4 blocks in this example) in the early rounds in order to be able to bid on his most preferred bundle in later rounds. If the bidder bid only on band II, he could end up winning his second preferred option, and would not be able to reveal his true valuation for the bundle in band I. If he bids on two blocks in band I and two blocks in band II, in order to switch to 4 blocks in band II eventually, he could well end up winning all four blocks effectively making a loss.[3]

3.3.2. Incentive-Compatibility

In order to minimize incentives for bid shading, the CCA design implements a closest-to-Vickrey core-selecting payment rule (Day and Milgrom, 2007; Day and Raghavan, 2007). Such payments guarantee that there is no group of bidders who can suggest an alternative outcome preferable to both themselves and the seller, and are minimized given this condition. It is known that the VCG outcome is in the core, if goods are substitutes, but no ascending auction can always implement the VCG outcome even if this condition holds (Gul and Stacchetti, 2000). If goods are complements, a bidder in a CCA still has an incentive to shade his bids and not reveal his true valuations, as he can possibly increase his payoff. So, after the primary bid rounds, if a bidder has a standing bid[4] on his most preferred bundle, he does have an incentive to minimize his bids in the supplementary bids phase and not reveal his true valuation.

The following two propositions define "safe supplementary bids", which cannot become losing based on the CCA activity rules if the bidders have a standing bid after the primary bid rounds (proofs can be found in Appendix 7.2). Let π describe the vector of ask prices in the last primary bid round, $b_j^p(X_j)$ is the last round bid of bidder $j \in J$ on a bundle after the primary bid rounds, and b_j^s a supplementary bid.

Proposition 3. *If demand equals supply in the final primary bid round, any single supplementary bid $b_j^s(X_j) > b_j^p(X_j)$ cannot become losing.*

---

[3] Recent modifications proposed in countries such as Canada address this problem with a revealed preference rule which is used in addition to the eligibility-points rule in the primary bid rounds (www.ic.gc.ca/eic/site/smt-gst.nsf/eng/sf10363.html). Also, the supplementary bids must satisfy revealed preference with respect to each eligibility reducing clock round after the last round in which the bidder had sufficient eligibility to bid on the package, as well as with respect to the final clock round.

[4] A bid is *standing* if its bid price is equal to the ask price of the last primary bids round.

This is because the supplementary bids of competitors on their standing bundle bid from the final primary bid round does not impact the safe supplementary bid of a bidder $j \in J$. Any additional items added by competitors to their standing bundle bid cannot increase the supplementary bid price by more than the ask price in the last of the primary bid rounds. If the bidder submits additional supplementary bids on packages not containing X_j, his bid $b_j^s(X_j)$ can well become losing, as can easily be shown by examples. The anchor rule also applies to bundles which are smaller than the standing bid of the last primary bid round.

Proposition 4. *If there is zero demand on bundle M after the last primary bid round, a single supplementary bid of a standing bidder $b_j^s(X_j) > b_j^p(X_j) + \pi M$ cannot become losing.*

This can be shown by an example where a losing bidder on all blocks reduces his demand to null in the last primary bids round. Bidder j needs to make sure that he wins, even if this losing bidder submits a bid on all blocks at the ask prices of the last primary round.

There have been a number of recent game-theoretical papers on core-selecting auctions. Day and Milgrom (2007) characterize a full information Nash equilibrium and show that bidder-optimal core prices minimize the incentives for speculation. Goeree and Lien (2012), Sano (2012a), and Ausubel and Cramton (2011) analyze the Bayesian Nash equilibrium of sealed-bid core-selecting auctions with single-minded bidders. Goeree and Lien (2012) derive an equilibrium of the nearest-Vickrey core-selecting auction and show that in a private values model with rational bidders, auctions with a core-selecting payment rule are on average further from the core than auctions with a VCG outcome. They also show that no Bayesian incentive-compatible core-selecting auction exists, when the VCG outcome is not in the core. Ausubel and Cramton (2011) analyze different core-selecting auction rules and the case of correlated values.

Sano (2012b) recently provided a Bayesian analysis of an ascending core-selecting auction with independent private values and shows that such an auction can even lead to an inefficient non-bidding equilibrium with risk-neutral bidders. Guler et al. (2016) shows that risk aversion leads to a lower likelihood of a non-bidding equilibrium, but that this possibility still exists depending on the prior distributions. Complementary valuations and the threshold problems as described in these papers are practically relevant in many spectrum markets with national and regional bidders. There is a fundamental *free-rider problem* in such threshold problems, where one regional bidder can try to increase expected payoff at the expense of other regional bidders. Truth-telling is no equilibrium strategy in such situations.

3.3.3. Lack of Envy-Freeness and Spiteful Bidding

For a characterization of the auction format, it is also worthwhile mentioning that bidders in a CCA or a VCG mechanism do not necessarily pay the same price for identical blocks. Suppose there are two bidders and two homogeneous units of one item. Bidder 1 submits a bid of $5 on one unit, while bidder 2 submits a bid of $5 on one unit and a bid of $9 on two units. Each bidder wins one unit, but bidder 1 pays $4 and the bidder 2 pays zero. This difference is due to the asymmetry of bidders, and this asymmetry leads to a violation of the law of one price, a criterion, which is often seen desirable in market design (Cramton and Stoft, 2007). Although arbitrage is avoided as bidders typically cannot sell licenses among each other immediately after a spectrum auction, different prices for the same spectrum are difficult to justify in the public and violate the goal of envy-freeness of an allocation for general valuations (Papai, 2003).

Finally, spiteful bidding needs to be taken into account when designing a CCA for a particular application. Bidders in spectrum markets may spitefully prefer that their rivals earn a lower surplus. This is different from the expected utility maximizers typically assumed in the literature. Spiteful bidding has been analyzed by Morgan et al. (2003) and Brandt et al. (2007), who show that the expected revenue in second-price auctions is higher than the revenue in first-price auctions with spiteful bidders in a Bayes Nash equilibrium. While spiteful bidding is possible in any auction, the two-stage CCA provides possibilities to submit spiteful supplementary bids with a no risk of actually winning such a bid, if all goods are sold after the primary bid rounds and standing bidders try to win their standing bid in the supplementary bids round. We provide an example in the Appendix 7.3.

4. Experimental Design

In what follows, we characterize the economic setting and the two value models of our experiments. Then we provide further details on the auction rules used in our experiments, the treatment structure, and the organization of our experiments.

4.1. Value Models

We used two value models reflecting the characteristics of bidder valuations in the field. Four bidders competed in both value models.

4.1.1. The Base Value Model

In the *base value model*, we used a band plan with two bands of blocks as it can be found in several European countries.[5] There are 14 (paired) blocks in band A and 10 (unpaired) blocks in band B. Bidders have a positive valuation for up to 6 blocks in each band with free disposal for bundles greater than that. Each bidder receives a base valuation v_A and v_B for each of the bands. The base valuations represented the valuations of a single block within each band and were drawn randomly from a uniform distribution, v_A in the range of [120, 200] and v_B in the range of [90, 160]. We modeled ascending complementarities in the valuation of bundles of several A blocks. In the A band, a bundle of two blocks receives a value of $1.2 * 2 * v_A$, i.e., a complementarity

---

[5] The frequencies of the 2.6 GHz band are available for mobile services in all regions of Europe. It includes 190 MHz which are divided into blocks of 5 MHz which can be used to deliver wireless broadband services or mobile TV. In particular, there are two standards which will likely be used in the 2.6 GHz band, LTE and WiMAX. LTE uses paired spectrum (units of 2 blocks), while WiMAX uses unpaired spectrum (units of 1 block). While some European countries auctioned the 2.6 GHz band solely, others combined several spectrum bands in one auction.

bonus of 20% on top of the base valuations. The complementarity in the value model rises with the number of blocks in the bundle. A bundle of three blocks has a complementarity of 40% and a bundle of four blocks of 80%[6]. There was no additional bonus for the fifth and sixth A-block. The valuations in the band B were purely additive. The total valuation of blocks from both bands is the sum of valuations within the bands. In total, each bidder is interested in up to $7 * 7 - 1 = 48$ different bundles.

In other words, four blocks in band A have the highest per block valuation to all bidders. If all bidders aim for at least four A-blocks and with 14 blocks on sale, it is possible that either two or three bidders get this bundle, while the other bidders win only two or three blocks in the band A. We assume block valuations to be bidder-specific, but the synergy structure of bundles to be the same for all bidders. In some experiments, we also vary the synergies.

4.1.2. The Multiband Value Model

The *multiband value model* is inspired by a recent applications of the CCA for the sale of multiple bands. The multiband value model had also 24 blocks, four bands with six blocks each. Band A was of high value to all bidders, whereas bands B, C, and D had lower value. As in the base value model, each bidder received a base valuation for a block in each band. Base valuations are uniformly distributed: v_A was in the range of [100, 300] while v_B, v_C and v_D were in the range of [50, 200]. Again bidders had complementary valuations for bundles of blocks.

In all bands, bundles of two blocks resulted in a bonus of 60% on top of the base valuations, bundles of three blocks in a bonus of 50%. For example, if the base value was 200, then the valuation for two blocks is 640, for three blocks 900, and for four blocks 1.100. Similar to the base value model, more blocks were valued at the base valuation and did not add any extra bonus. Overall, bidders in the multiband setting could bid on $7^4 - 1 = 2,400$ different bundles, which is significantly more than the 48 bundles in the base value model.

The structure of the value model and the distribution of the block valuations of all bands were known to all bidders. Bidders used an artificial currency called Franc. Although the value models resemble characteristics of spectrum sales, this was not communicated to the subjects in the lab (neutral framing). Note that we used start prices of 100 Franc in the band A and 50 Franc in the bands B to D. The bid increments were 20 Franc in the band A and 10 Franc in bands B to D.

4.2. Detailed Auction Rules in the Experiments

Our experiments were conducted using a Web-based software tool, which implemented the SMRA and the CCA.[7]

---

[6] This reflects the valuation in the 2.6 GHz band. Four 5-MHz blocks allow for peak performance rates with LTE and provide maximum value to all bidders.

[7] Our implementation of the CCA followed the exact rules that were specified in the guidelines for the Austrian spectrum auction in 2010 at www.rtr.at/en/tk/FRQ_2600MHz_2010_VA. Our software implementation can be made available for tests or replication experiments.

4.2.1. Simultaneous Multi-Round Auction

In SMRA all blocks were sold at the same time with an individual price for each block. After each round the provisional winner of each block was determined by the highest bid. Ties were broken randomly. A bid on a block had to exceed the standing high bid by at least the minimum increment. Jump bids were restricted to predefined levels (click-box) to prevent signaling, and the identity of the bidders was unknown in the auction.

An activity rule restricted the number of blocks a bidder can bid on across all bands. Following recent SMRA designs, we implemented a stacked activity rule with two activity levels. At the beginning each bidder was eligible to bid on all blocks at sale. In the first three rounds bidders were required to use only 50% of their eligibility to maintain all eligibility points for the next round. From the fourth round on, 100% were required. At the beginning of each round all bids from the previous round (winning and losing) were revealed to all bidders. Finally, the auction terminated if no bidder submitted a bid within one round.

4.2.2. Combinatorial Clock Auction

As introduced in Section 3, the CCA is composed of the primary bid or clock rounds and the supplementary bids round. All blocks within one band had the same price. In the base value model, there was one price for the A band and one for the B band, in the multiband value model there were separate prices for all four bands. The auctioneer announced the new ask price for each band in each round of the clock phase and bidders decided on the quantitites of blocks they wanted to bid on within each band. The quantities specified in all bands formed one bundle bid. Each bidder could submit at most one bid in each round. If there was excess demand (i.e., if the combined demand of all bidders within one band exceeded the number of blocks) in at least one band, a new round is started with higher prices for the bands with excess demand. In the experiments bidders did not know about the level of excess demand, which is in line with the auction rules used for example in Austria.[8] Bidders did not learn about other bidders bids, only whether there was excess demand in each band or not in the previous round. In our experiments, each bidder started with eligibiliy for all blocks in the first round. The primary bids phase ended after there was no excess demand in any bands any more.

The supplementary bids round allowed for as many sealed bids as desired by every bidder. They were able to bid on any combination of blocks regardless of the bids of the first phase. Only the maximum bid price was limited by the anchor rule. At the end of the round the optimal allocation was calculated using all bids from both phases. Then the bidder-optimal core-selecting payments were calculated using a quadratic program following Day and Raghavan (2007). All optimizations were performed using the IBM/CPLEX optimizer (version 11).

---

[8] In the more recent Swiss auction, the level of excess demand was revealed to bidders during the auction. In such cases, where all blocks indeed get sold after the primary bid rounds, standing bidders would have no incentive to raise their bid more than a minimal increment, as otherwise they cannot become losing.

Table 1. *Treatment structure*

Treatment no.	Auction format	Value model	Bidder	Auctions
1	SMRA	Base	Lab	20
2	CCA	Base	Lab	20
3	SMRA	Base	Competition	9
4	CCA	Base	Competition	9
5	SMRA	Multiband	Lab	16
6	CCA	Multiband	Lab	16

4.3. Competitions

In addition to the lab experiments with unprepared subjects, we conducted experiments with experienced subjects. In these experiments subjects were recruited from a class on auction theory and market design and were grouped into teams of two persons. The subjects were invited to the lab two weeks prior to the experiment and received the same introduction as lab subjects. In addition, we provided them with information and literature on previous spectrum auctions describing known strategies and tactics of bidders in the field. During the two weeks they prepared their own strategy and wrote a paper to describe their strategy. We refer to these experiments as "competitions" in the following to highlight these differences. In order to defy collusion among the bid teams in a coalition, we told them that they would immediately be excluded if such collusion would be observed. We also analyzed bid data to understand whether they followed the bid strategies that they described in their paper and if there were signs of collusion.

4.4. Treatment Structure

We considered two major treatment factors, *auction format* and *value model*, with each having two levels (SMRA and CCA, base and multiband). In addition, we analyzed the base value model treatments with bidders in the lab and in the competition, which yields another treament factor *bidder*. Overall, we get six *treatments* in total (Table 1).

For each replication or "wave" we generated new sets of random values for all the bidders. We used the same sets of values across auction formats to reduce performance differences due to the random draws. For the base value model we drew valuations for five waves randomly. Each wave consisted of four different auctions which were conducted in the lab within one session. We ran between subjects experiments with four bidders in each session.

In addition to the auctions in Table 1, we organized 4 sessions with another multiband value model, where synergies were different across bidders. These experiments were only conducted to make sure that the synergies do not have a significant impact on our main results, which they did not. Overall, we organized 28 sessions with 106 auctions. In two of the sessions with competitions, we ran only 1 auction and not 4.

For each treatment combination a run with CCA and a run with SMRA was conducted in the lab. All auctions of waves A, B and the first auction of wave C were also used in the competition with both auction formats to enable a direct comparison to the lab. In the multiband value model we defined four waves with four different auctions each.

4.5. Procedures and Organization

112 students participated in all the experiments and competitions in 2010 and 2011. Subjects were recruited from the departments of mathematics and computer science. Each subject participated either in one CCA- or in one SMRA-session but never in both. One session comprised all four auctions of one wave and took on average four hours. In the competition, there was one session with five auctions.

To reduce differences between lab sessions, the introduction was delivered through a video. Each participant received a handout and was able to pause the video whenever necessary. An experimentator was present to answer questions. Subjects were then made familiar with the auction software through a demo auction. In addition, a software tool to analyze bundle valuations and payoffs was provided to all subjects. This tool showed a simple list of all available bundles which could be sorted by bundle size, bidder individual valuation, or the payoff based on current prices. In order to ensure a full understanding of the economic environment, the value model, the auction rules, and the financial reward scheme all subjects had to pass a web-based test.

At the beginning of each auction all subjects received the individual draw of valuations, the distribution of valuations, and the information about the complementarities. Each round in SMRA and the primary bid rounds of the CCA took 3 minutes. The supplementary bids phase of the CCA lasted around 10 minutes to provide enough time for bid submission. The subjects could ask for more time if required.

After each session subjects were compensated financially. The total compensation resulted from a 10 Euro show up fee and the auction reward. The auction reward was calculated by a 3 Euro participation reward plus the sum of all auction payoffs converted from Franc to Euro by a 12:1 ratio. Negative payoffs were deducted from the participation reward. Due to the different payment rules in both auction formats, payoffs in CCAs were higher than in SMRAs. Therefore we leveled the expected payoff per participant by financially compensating three random out of the four auctions of the SMRA sessions, while only two out of four auctions were compensated in CCA sessions. On average each subject received 93.52 EUR. The rewards for pairs of subjects in the competition were similar to the ones for single subjects in the lab.

5. Results

First, we present efficiency and revenue of the different auction formats on an aggregate level. Then, we discuss individual bidder behavior in both auction formats and differences between lab and competition.

5.1. Efficiency and Revenue

We use *allocative efficiency* $E(X)$ as a primary aggregate measure to compare different auction mechanisms.[9] Efficiency $E(X)$ cannot easily be compared between different

---

[9] We measure efficiency as $E(X) := \frac{\text{actual surplus}}{\text{optimal surplus}} \times 100\%$.

Table 2. *Aggregate measures of auction performance*

Value model	Auction	Bidder	$E(X)$	$E(X)^*$	$R(X)$	Unsold blocks
Base	SMRA	Lab	96.16%	63.27%	83.74%	0
Base	SMRA	Competition	98.57%	87.28%	75.06%	0
Base	CCA	Lab	96.04%	63.96%	64.82%	0
Base	CCA	Competition	94.15%	47.17%	55.38%	0.44 (1.9%)
Multiband	SMRA	Lab	98.46%	93.85%	80.71%	0
Multiband	CCA	Lab	89.28%	56.71%	33.83%	1.25 (5.2%)

value models. Therefore, we also calculate **relative efficiency**, $E(X)^*$.[10] In addition, we measure **revenue distribution**, which shows how the resulting total surplus is distributed between the auctioneer and bidders.[11] For the pairwise comparisons of these metrics we use the rank sum test for clustered data by Somnath05.[12]

Result 1: The efficiency of SMRA was not significantly different to the CCA in the base value model in both, the lab (SMRA $\sim$ CCA, $p = 0.247$) and the competition (SMRA $\sim$ CCA, $p = 0.781$). In contrast, the efficiency of the CCA was significantly lower than that of SMRA in the multiband value model in the lab (SMRA $\succ^*$ CCA, $p < 0.012$).

Support for result 1 is presented in Figure 1 and Table 2. A reason for the low efficiency of the CCA in the multiband value model was the number of unsold blocks. On average, 5.2% or 1.25 blocks remained unsold in this value model. The efficiency of CCAs where all blocks were sold was 93.61% (SMRA $\sim$ CCA, $p = 0.1563$). The CCA in the multiband value model was the only environment where a significant number of blocks remained unsold (CCA $\succ^{**}$ 0, $p = 0.0020$). In competitions, only two CCAs in the base value model terminated with blocks unsold, and the number of unsold blocks

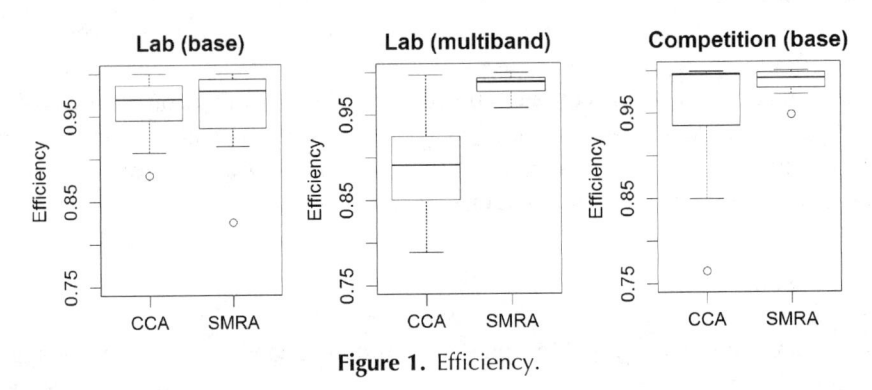

Figure 1. Efficiency.

---

[10] For this, we compute the mean of the social welfare over all possible allocations assuming that all goods are sold (Kagel et al., 2010). For this definition, the relative efficiency of an efficient allocation ist still 100% while the mean of random assignments of all blocks is 0%. Note that allocations below the mean have negative relative efficiency.

[11] We measure measure auction revenue share as $R(X) := \frac{\text{auctioneer's revenue}}{\text{optimal surplus}} \times 100\%$.

[12] $\sim$ indicates an insignificant order, $\succ^*$ indicates significance at the 5% level, and $\succ^{**}$ indicates significance at the 1% level.

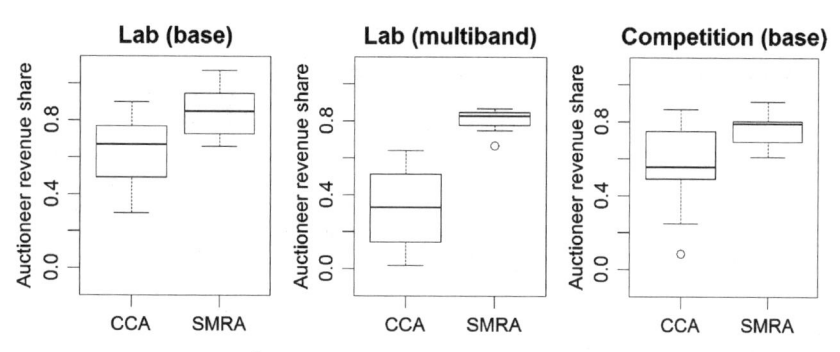

Figure 2. Auctioneer revenue share.

was small. In SMRA, bidders in competitions achieved higher efficiency than lab bidders while in the CCA lab bidders achieved higher efficiency. Both differences were not significant (SMRA: $p = 0.917$, CCA: $p = 0.297$).

Relative efficiency helps to compare the performance between value models. SMRAs led to significantly higher relative efficiency in the multiband than in the base value model (multiband $\succ^*$ base, $p = 0.0439$).[13] With more bands of blocks, bidders tended to focus on the bands for which they had high valuations and they were willing to take an exposure risk in these bands. The CCA had a higher relative efficiency in the base than the multiband value model even though the difference was not significant (SMRA $\sim$ CCA, $p = 0.774$). This insignificance is due to the large variance of relative efficiency values.

Result 2: The auctioneer revenue of SMRA was significantly higher than that of the CCA in both value models in the lab (SMRA $\succ^*$ CCA, base: $p = 0.034$; SMRA $\succ^{**}$ CCA, multiband: $p < 0.007$). The differences between lab and competitions were not significant for both auction formats (CCA: $p = 0.684$, SMRA: $p = 0.230$) in the base value model.

Support for result 2 can be found in Figure 2 and Table 2. Auctioneer revenue share of both auction formats was higher in the base than in the multiband value model. The difference was significant for the CCA (base $\succ^{**}$ multiband, $p < 0.001$). The payment rule of the CCA had a significant impact and led to low revenue given the discounts, the low number of bids and bidders. Another reason for the difference is the number of unsold blocks in CCA auctions.

The auctioneer revenue of CCAs was higher in the base value model than in the multiband value model. Again, the number of possible bundles serves as an explanation, since it causes difficulties when bidders try to coordinate and find the efficient solution. Given the low number of bundle bids, the second best allocation was often much lower and resulted in high discounts and low payments. 5 of 16 CCAs in the multiband value model terminated with an auctioneer revenue share of 30% or less. One auction yielded as little as 2% auctioneer revenue share. Low revenue was also a phenomenon in the field.

---

[13] Note that relative efficiency emphasizes results below the mean disproportionately.

Table 3. *Aggregate simulations results*

Value Model	Auction Format	Bidder	$E(X)$	$E(X)^*$	$R(X)$
Base	SMRA	Simblock	92.49%	30.63%	64.66%
Base	CCA	SimDirect	100.00%	100.00%	86.16%
Multiband	SMRA	Simblock	90.55%	60.90%	58.92%
Multiband	CCA	SimDirect	100.00%	100.00%	75.72%

In SMRA, some efficiency losses can occur due to the exposure problem and it is interesting to understand what the efficiency of the auctions would be with straightforward bidders, who always bid on their payoff maximizing bundle. We implemented simulations with artificial bidders bidding in the SMRA and the CCA. In SMRA, the bidders do not take an exposure risk, and bid up to their valuations per block (*Simblock* bidders). In the CCA bidders always bid on their payoff-maximizing bundle in the primary bid rounds, and they submitted a truth-revealing bid on all bundles with positive value in the supplementary bids round (*SimDirect* bidders). This helps understand the difference in efficiency and revenue, which is due to the bundle selections of bidders in the lab.

Table 3 presents aggregate results of these simulations. Efficiency and Revenue of the SMRA format is lower compared to the results from the lab. This is due to the fact that the bidders always avoid exposure. Efficiency in the CCA format is 100% with truth revealing bidders because there is no exposure risk. The difference in revenue to the lab results is substantial. This phenomenon is particularly strong in the multiband value model.

5.2. Bidder Behavior in the CCA

Result 3: On average, there was modest bid shading in the supplementary bids round in both value models in the lab and in the competition.

Figure 3 shows whether bidders bid below, at or above their valuation on a bundle in the supplementary bids round. The figure also plots a regression line in addition to the diagonal with a slope of one (the truthtelling strategy). The slope of this regression can serve as an estimator for bid shading. For the base value model the slope is 0.90 (adjusted $R^2 = 0.77$) for data from the lab and 0.896 (adjusted $R^2 = 0.84$) for the bid data from the competition. In the multiband value model the slope of the regression was 0.68 (adjusted $R^2 = 0.796$). This type of bid shading is in line with theoretical predictions.

In the base value model in the competition only 3.2% of the bids were above their valuation while in the lab there were 22.6% of the bids above the valuation. A single lab bidder in the base value model bid consistently above his valuations, which led to this effect. Without this bidder only 12.8% of the bids were above the valuations. In the multiband value model only 6.4% of the bids were above the valuation.

Next, we want to understand the selection of bundles in the supplementary bids round. With a VCG payment rule and independent private values, bidders would have a dominant strategy to bid on all bundles with a positive utility. The CCA does not have

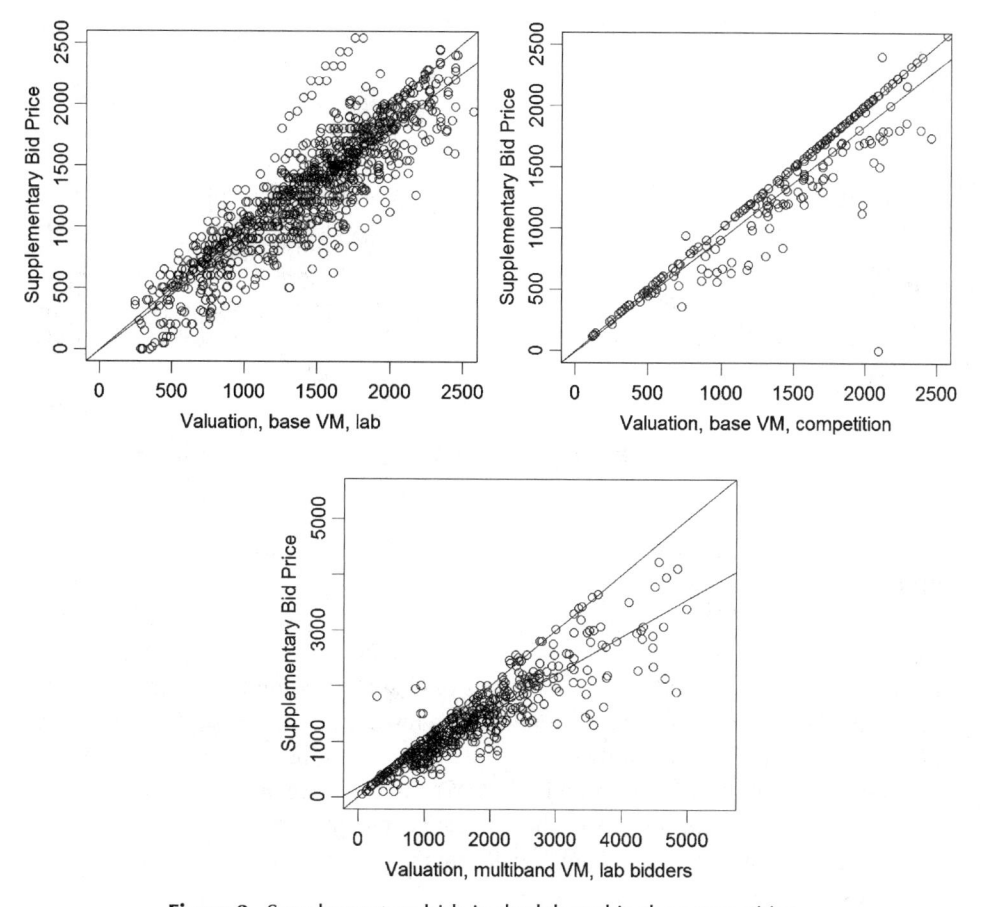

Figure 3. Supplementary bids in the lab and in the competition.

a dominant strategy, but it is also not obvious, how bidders would select their bundles strategically in such a setting to improve expected utility.

Result 4: Bidders bid only on a fraction of bundles with a positive value in the supplementary bid phase. Bidders in the lab bid only in 23.67% (11.36 bids) of 48 potential bundle bids they could bid on in the base value model and 0.06% (8.33 bids) of the 2,400 possible bundles in the multiband value model in the supplementary bids round. In the competition bidders submitted on average 12.6% (6.06 bids) in the base value model.

Lab bidders in the large multiband value model actually submitted less bundle bids in absolute numbers than in the smaller base value model. Some bidders in the small value models actually submitted bids on almost all bundles in the supplementary bids round. For example, there were bidders who submitted 36 of 48 possible bundle bids in the base value model. In contrast, in the multiband value model bidders submitted 22 bids in the supplementary bids round at a maximum. A focus on only subsets of all possible bundles was also found by Scheffel et al. (2012) for other combinatorial auction formats with a larger number of bundles and can be seen as a consequence of the communication complexity of combinatorial auctions (Nisan and Segal, 2006).

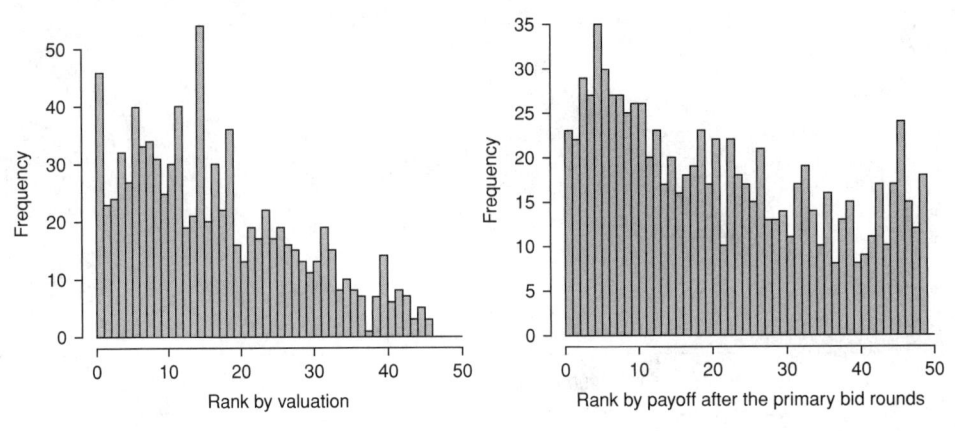

Figure 4. Rank of supplementary bids by valuation or payoff after the primary bid rounds, base value model, lab.

A low number of bids can also be observed in the field. For example, in the L-band auction in the U.K. in 2008, 17 specific blocks were sold resulting in 131,071 possible bundles, but the 8 bidders only submitted up to 15 bids in the supplementary bids round (Cramton, 2008). Similarly, in the 10-40 GHz auction in the U.K. in 2008 bidders could bid on 12,935 distinct bundles. Eight bidders only submitted up to 22 bundles, while one submitted 106 and another 544 bundle bids (Jewitt and Li, 2008). It might be that bidders were just unprepared and have not fully understood the consequences of particular strategies in the CCA (Jewitt and Li, 2008). Another explanation is that practical reasons keep bidders from submitting hundreds or even thousands of bids. It is interesting to understand, how bidders select bundles in the primary and in the supplementary bid rounds.

Result 5: Bidders selected bundles in the supplementary bids round based on synergies in the value model, their relative strength with respect to the prior distribution, and ask prices after the primary bid rounds.

Let us first look at the base value model. We have calculated the rank of each bid submitted based on the valuation of the bundle for a bidder, and based on the payoff the bidder had for the bundle at the end of the primary bid rounds. Figures 4 and 5 show that bidders exhibit a tendency to select bundles with a better rank based on payoff in both the lab and the competition. The histogram to the left shows the frequency of bids ordered by valuation for these bids, while the histogram to the right shows the frequency of bids ordered by payoff at prices of the last primary bid rounds. Bidders also submit a considerable number of bids when their payoff at prices of the last primary bid round is low or negative in the base value model. Figure 6 reveals that most bids in band A are on four blocks, where the synergies were highest. In contrast, bidders bid on up to six blocks in band *B*, where there was no synergy for larger bundles. There is also a significant difference between weak and strong bidders. Bidders with a higher base valuation in band A are classified as strong, with a base valuation lower than the second order statistic as weak. Figure 6 shows the frequency of bids on a certain number of blocks for strong bidders in the lab and in the competition in the top row and for weak bidders in the lab and in the competition in the bottom row. While only a few strong

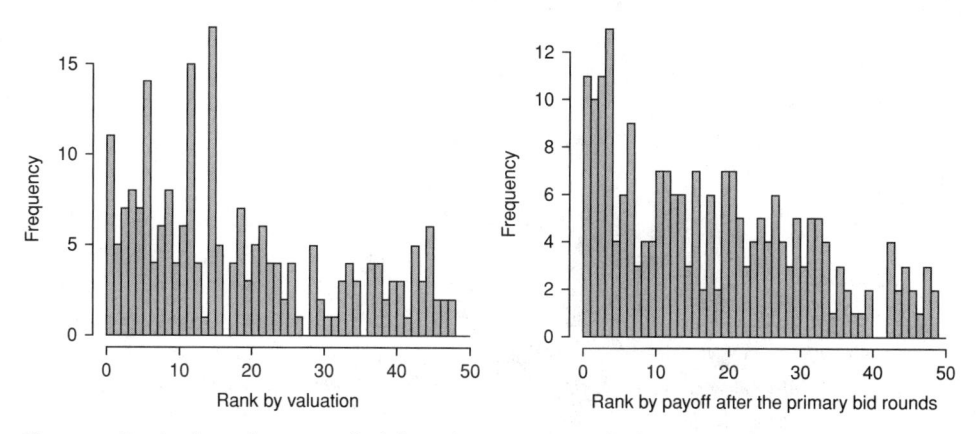

Figure 5. Rank of supplementary bids by valuation or payoff after the primary bid rounds, base value model, competition.

bidders submitted bids on less than four blocks in the A band, weak bidders typically did submit such bids. This is even more pronounced in the competition. The number of blocks in bids of weak or strong bidders, as well as those of bidders in the lab and in the competition are significantly different ($\alpha = 0.01$).

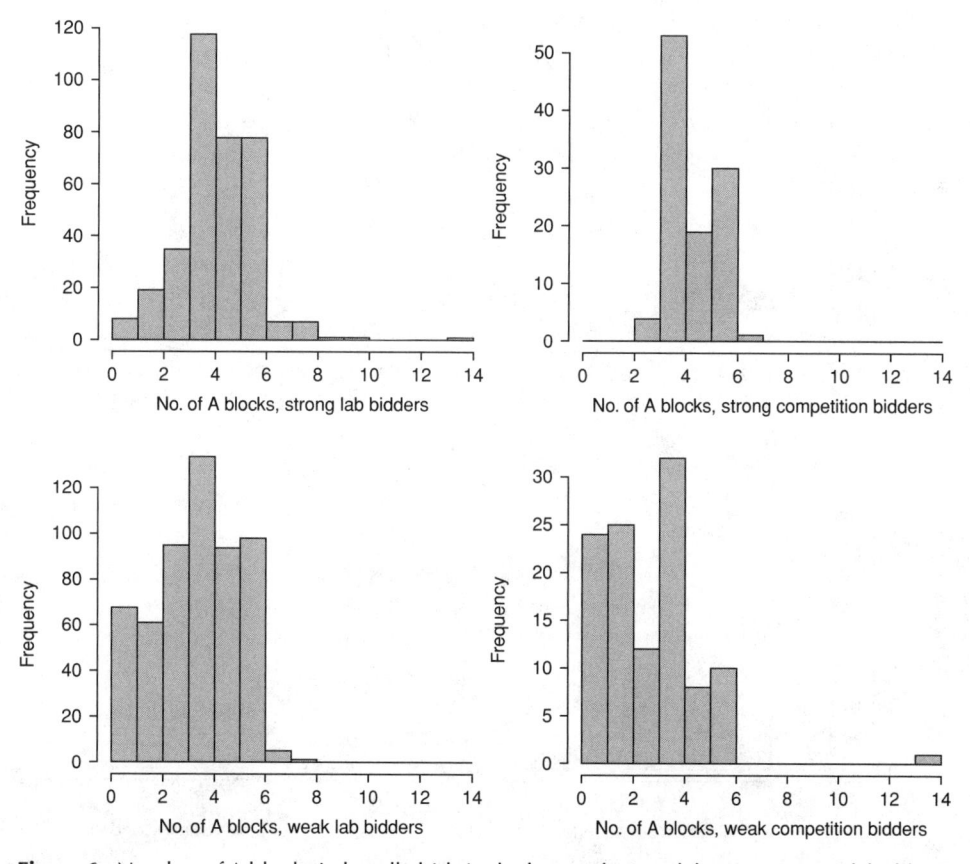

Figure 6. Number of A blocks in bundle bids in the base value model, strong vs. weak bidders, lab and competition.

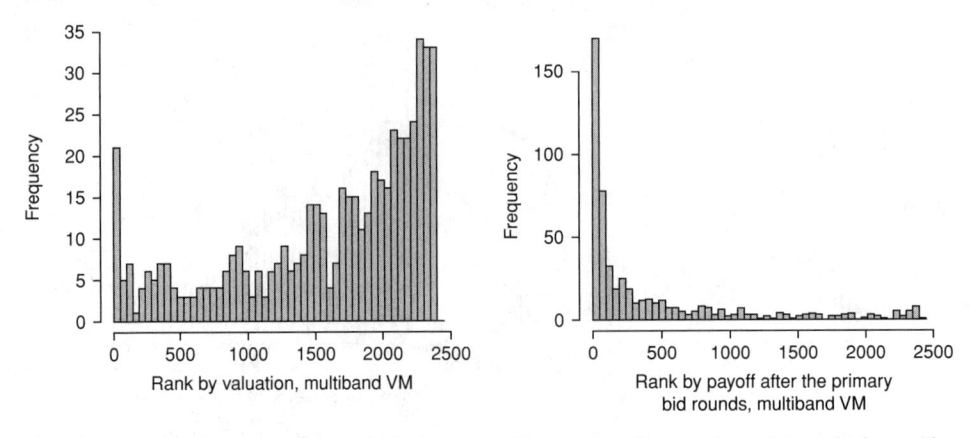

Figure 7. Rank of supplementary bids by valuation or payoff after the primary bid rounds, multiband value model, lab.

Next, we analyze the bids in the multiband value model, where bidders had 2,400 bundles to choose from. Figure 7 shows the frequency of bids ranked by valuation or by payoff after the primary bid rounds for the multiband value model. Apparently, bidders used the ranking based on payoff as a guidance to select bundles, whereas the ranking of valuations did not influence decisions strongly. Figure 8, however, shows that again

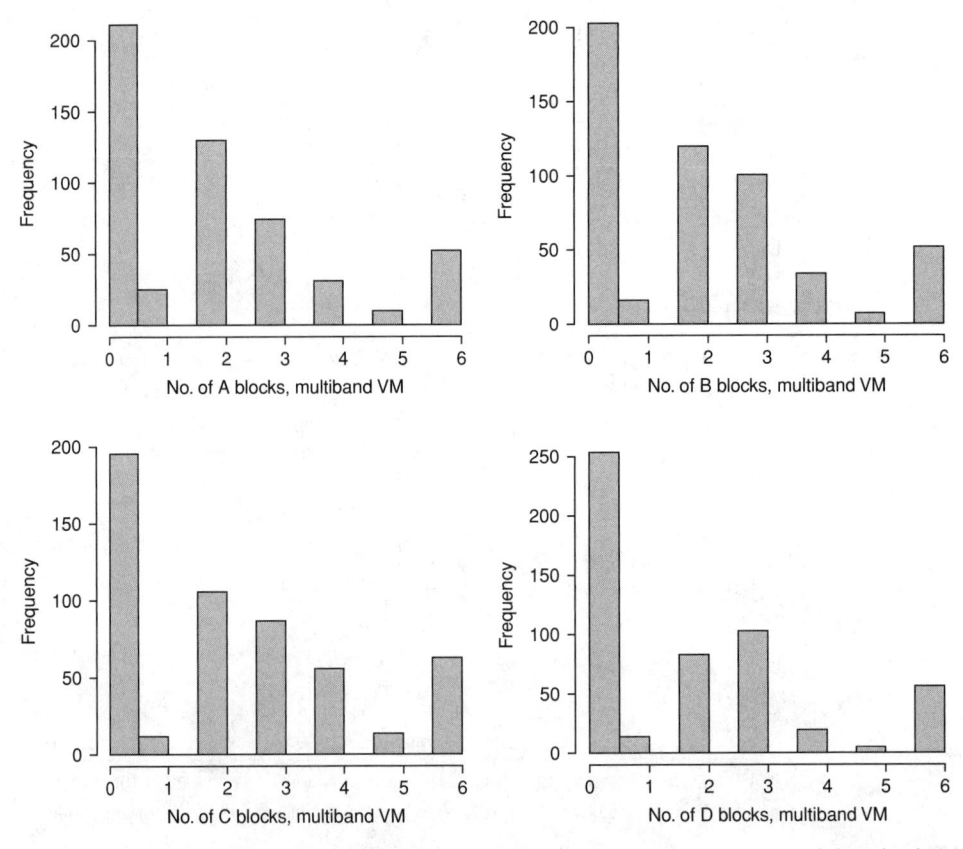

Figure 8. Number of blocks in bundle bids in the supplementary bids round, multiband value model.

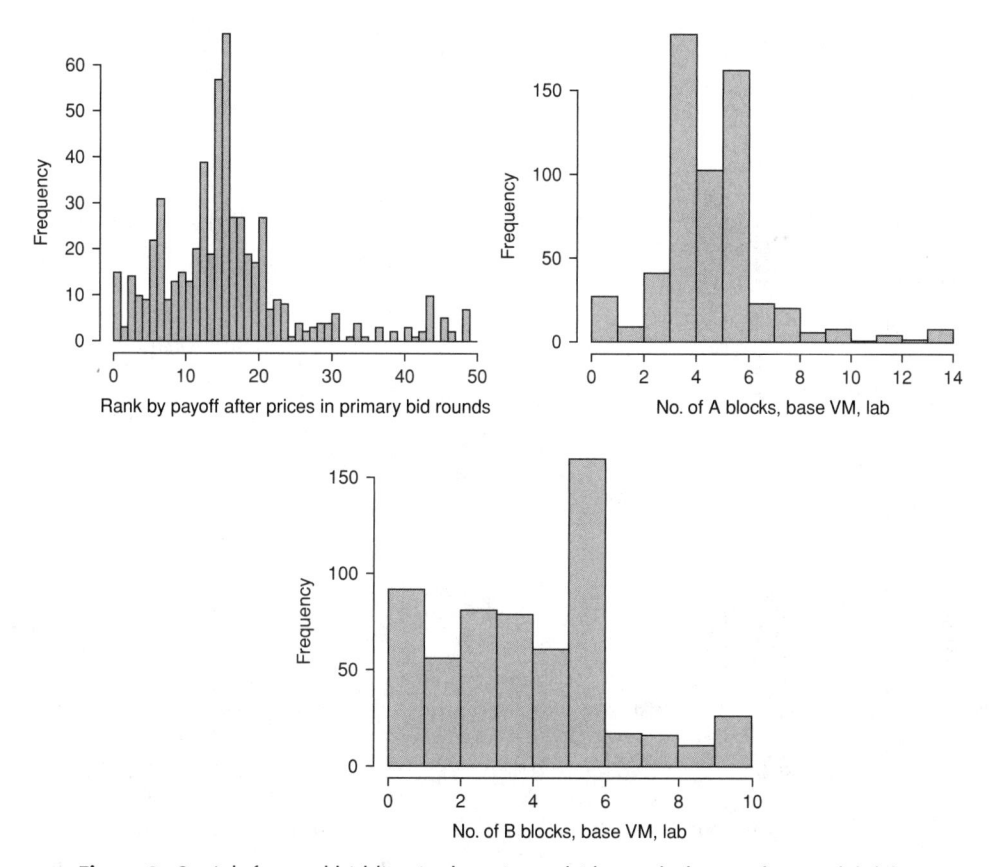

Figure 9. Straightforward bidding in the primary bid rounds, base value model, lab.

information about the synergies has influenced the bundle selection in the A, B, C, and D band. Most bundles included two or three blocks of a band only. Those bundles also had the highest synergy in the multiband value model with two blocks having higher synergies than three blocks.

In summary, in both value models bidders in the lab and in the competition used information about the synergies in the value models and tried to win those bundles with the highest synergies. In the multiband value model, they also selected the top-ranked bundles based on payoff. These observations are in line with a recent paper by Scheffel et al. (2012), which found that bidders often use simple heuristics for bundle selection in combinatorial auctions.

Finally, we also analyzed, whether bidders bid straightforwardly in the primary bid rounds. In other words, do they select the bundle with the highest payoff in each round?

Result 6: Bidders did not follow a straightforward bidding strategy in the primary bid rounds in both value models.

Figures 9 shows that in both, the competition and the lab, bidders did not bid straightforwardly, and the payoffly in each round was not the primary criterion for the bundle that bidders selected in the base value model. The histogram to the left shows the rank by payoff based on prices in the last round of the primary bids phase. The two

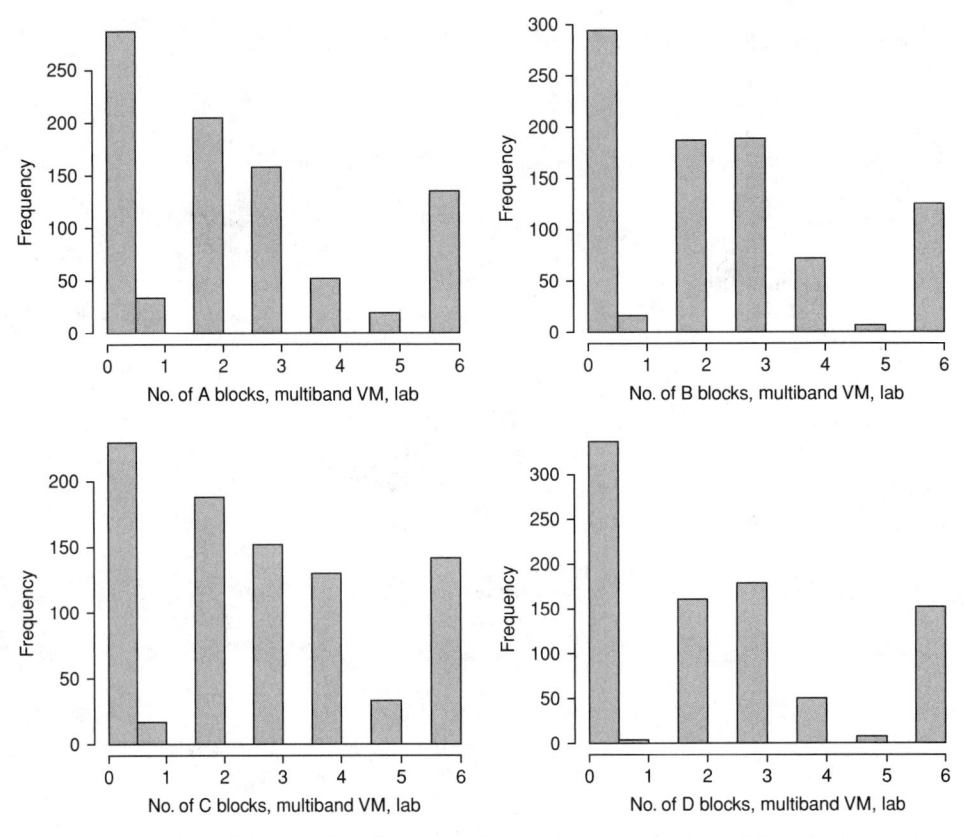

Figure 10. Number of blocks in bundle bids in the primary bid rounds, multiband value model.

histograms to the right show the frequency of bids in the A and B band respectively for lab bidders. The pattern in the competition was similar. The activity rule in the primary bid rounds is one explanation. Bidders were not allowed to increase the number of blocks in a bundle bid across rounds. Therefore, they often started out with large bundle bids in the initial rounds, rather than selecting their payoff maximizing bundle. In the multiband value model bidders selected bundles with two or three blocks in a band most frequently, which exhibited the highest synergies (see Figure 10).

5.3. Bidder Behavior in the SMRA

The focus of this paper is the CCA. However, in the following, we also provide a summary of the main findings about bidding in the SMRA. We will focus on the likelihood of taking an exposure risk and jump bidding.

Exposure risk is a central strategic challenge of the SMRA in the presence of complementary valuations. Strong bidders with a high valuation might want to take this risk, while weak bidders would decide to reduce demand in order to keep prices low. Alternatively, weak bidders could try to pretend to be strong and bid aggressively at the start hoping others believe the threat and reduce their demand.

Table 4. *Share of bidders who take different levels of exposure risk in the base value model*

Bidder	Band	Strength	No. of bidders	$bid > v_A$	$bid > 1.2 * v_A$	$bid > 1.4 * v_A$
Lab	A	All	80	88.75%	72.50%	56.25%
Lab	A	Strong	39	87.18%	64.10%	53.85%
Lab	A	Weak	41	90.24%	80.49%	58.54%
Comp	A	All	36	86.11%	55.56%	36.11%
Comp	A	Strong	18	83.33%	55.56%	22.22%
Comp	A	Weak	18	88.89%	55.56%	50.00%

Result 7: In the base value model, strong bidders took an exposure risk less often than weak bidders in the lab and the competition. In contrast, strong bidders took exposure risk more often than weak bidders in all four bands of the multiband value model. Strong bidders also took higher levels of exposure risk more often than weak bidders in the more valuable band A while weak bidders took it more often in the other bands. Bidders in competitions took exposure risk less often than bidders in the lab.

Both value models encompass different levels of synergies which define several stages of exposure risk for the bidders. Synergies in band A in the *base value model* rise with the bundle size for up to four blocks. First, we analyzed how many of the bidders submitted a bid in band A, which was higher than their base valuation $bid > v_A$, higher than their valuation on a bundle of two blocks ($bid > 1.2 * v_A$), and even higher than their valuation for a package of three blocks ($bid > 1.4 * v_A$). Bidders with a base valuation lower than the second order statistic were again classified as weak, the others as strong.

In the *multiband value model* we have decreasing synergies. Due to the higher synergy of two blocks, bidders can bid on three blocks and then reduce from three to two blocks without making losses. Only a reduction from two blocks to one block will lead to a loss if prices rise above the base valuation. The different degrees of exposure risk

Table 5. *Share of bidders who take different levels of exposure risk in the multiband value model*

Bidder	Band	Strength	No. of bidders	$bid > v_A$	$bid > 1.5 * v_A$
Lab	A	All	64	79.69%	21.88%
Lab	A	Strong	24	91.67%	25.00%
Lab	A	Weak	40	72.50%	20.00%
Lab	B	All	64	81.25%	17.19%
Lab	B	Strong	33	90.91%	12.12%
Lab	B	Weak	31	70.97%	22.58%
Lab	C	All	64	84.38%	25.00%
Lab	C	Strong	36	86.11%	11.11%
Lab	C	Weak	28	82.14%	42.86%
Lab	D	All	64	81.25%	20.31%
Lab	D	Strong	34	91.18%	14.71%
Lab	D	Weak	30	70.00%	26.67%

Table 6. *Bidders with negative payoff*

Value model	Bidder	Bidders with negative payoff	Total bidders	Share
Base	Lab	19	80	23.75%
Base	Competition	5	36	13.89%
Multiband	Lab	5	64	7.81%

in both value models explain the higher number of exposure problems in the base value model compared to the multiband value model (Table 6).

In the base value model the competitive situation makes it easier to win for strong bidders. With 14 blocks in band A the strong bidders expect to win four blocks each while the weaker bidders have to split the remaining six blocks. Weaker bidders cannot win four blocks without taking exposure risk, which explains why they take more exposure risk.

With only six blocks per band in the multiband value model, it is likely that either two bidders win three blocks each or three bidders win two blocks each. Weak bidders are less willing to take exposure risk of ending up with only one block. The strong bidders often face strong competitors within the same band which forces them to take exposure risk. Within all four bands of the multiband value model strong bidders take exposure risk more often than weak bidders.

In competitions, strong as well as weak bidders are more careful about exposure risk. In particular, this is the case for higher levels of exposure. Table 6 shows that a lower percentage of bidders in competitions experience losses in the base value model. While 23.75% of the bidders in the lab receive negative payoff, only 13.89% of the bidders in the competition make a loss due to taking an exposure risk. Since bidders have four bands with complementarities to coordinate in the multiband value model, their risk of a loss is lower and only 7.81% of bidders actually make a loss.

We did not find evidence for tacit collusion in SMRA, neither in the lab nor in the competition although the stacked activity rule gave bidders the possibility to bid on a lower number of items without losing the chance of bidding on larger bundles in later rounds. Bidders signaled their preferences, but none of the auctions resulted in agreements at low revenue.

Jump bids can be interpreted as signaling. By bidding more than the ask price a bidder can signal preferences and discourage other bidders from bidding on this block. Such a signal can also be given by raising the bid on an block, for which the bidder already holds the standing high bid.

Result 8: Jump bids were used by all bidders in all treatments. Bidders in the lab used jump bids more often than bidders in the competition (lab $\succ^*$ competition, $p = 0.0103$). Bidders in the competition submitted lower jump bids.

Table 7 shows that jump bids were heavily used across all bands.

Of course the level of jump bids varies. Table 8 shows the number of jump bids of different levels as percentage of the total number of bids. Low jump bids are those bids which exceed the ask price by 1 and 2 Franc (two lowest steps of the click-box),

Table 7. *Jump bids by band*

Value model	Bidder	Band	Avg. no. of jump bids	Avg. no. of bids	Share (%)
Base	Lab	A	12.91	23.23	55.60
Base	Lab	B	9.54	17.16	55.57
Base	Lab	All bands	22.45	40.39	55.59
Base	Comp	A	8.58	17.14	50.08
Base	Comp	B	7.83	19.08	41.05
Base	Comp	All bands	16.42	36.22	45.32
Multiband	Lab	A	8.09	15.33	52.80
Multiband	Lab	B	6.88	14.72	46.71
Multiband	Lab	C	9.22	15.72	58.65
Multiband	Lab	D	7.31	13.75	53.18
Multiband	Lab	All bands	31.5	59.52	52.93

medium jump bids exceed the ask price by 5 and 10 Franc (two steps in the middle) and high jump bids by 20 and 50 Franc (two top steps). Low jump bids can be used to avoid ties.

Bidders in both, the lab and competition, used low jumps, and there was no significant difference (lab $\sim$ competition, $p = 0.5270$). Medium and high jump bids are used to demonstrate strength and discourage other bidders from bidding on this very block. We found that bidders in competitions used medium and high jump bids less often than lab bidders (medium: lab $\succ^{**}$ competition, $p = 0.0004$; high: lab $\succ^{**}$ competition, $p = 0.0032$).

Result 9: Bidders of all treatments placed bids on blocks that they have provisionally won in the previous round (own blocks). In bands of higher valuation (band A in both value models) bidders used a higher number of bids on own blocks than in other bands.

Support for this result is presented in Table 9. Bidders in competition used almost 10 times more bids on own blocks in the more valuable A band (9.40%) than in band B (1.89%). Bidders in the lab submitted bids on own blocks in band A (6.73%) and band B (4.01%).

6. Conclusions

One result of this study is that the CCA did not yield higher efficiency in the small base value model and performed significantly worse than SMRA in the multiband value model. Revenue was significantly lower in all treatments and sometimes blocks

Table 8. *Jump bids by step size on all bands*

Value model	Bidder	Avg. no. of bids	All	Low	Medium	High
Base	Lab	40.39	55.59%	21.11%	21.02%	13.46%
Base	Comp	36.22	45.32%	24.77%	12.81%	7.75%
Multiband	Lab	59.52	52.93%	18.09%	14.12%	20.71%

Table 9. *Bids on own blocks*

Value model	Bidder	Band	Avg. no. of bids on own blocks	Avg. no. of bids	Share (%)
Base	Lab	A	1.56	23.23	6.73
Base	Lab	B	0.69	17.16	4.01
Base	Lab	All bands	2.25	40.39	5.57
Multiband	Lab	A	0.63	15.33	4.08
Base	Comp	A	1.61	17.14	9.40
Base	Comp	B	0.36	19.08	1.89
Base	Comp	All bands	1.97	36.22	5.44
Multiband	Lab	B	0.48	14.72	3.29
Multiband	Lab	C	0.64	15.72	4.08
Multiband	Lab	D	0.36	13.75	2.61
Multiband	Lab	All bands	2.11	59.52	3.54

remained unsold in spite of sufficient demand. This was due to the low number of bundle bids and the CCA payment rule. In the CCA, bidders only submitted a small subset of all possible bundle bids. Bidders used heuristics to select these bundles, mainly based on their strength and the synergies in the value model. In real-world applications bidders cannot be expected to submit hundreds or thousands of bids in an auction, even with decision support tools available. It might also be difficult to get an agreement among all stakeholders in a company for thousands of package valuations, even though all of these packages of licenses can have a value to a large bidder. As our experiments show, this can have a significant negative impact on efficiency and revenue in CCAs with many blocks.

It is interesting to note that if bidders submitted bids on all possible bundles truthfully, as was the case in our simulations, the revenue of the CCA was much higher and comparable to the revenue of SMRA in the lab. In comparison, the SMRA elicited the valuations of bidders on individual blocks sufficiently well to allow for high efficiency even in the multiband value model. In particular, strong bidders often took the exposure risk to win a bundle with high synergies, such that the negative effect on efficiency was mitigated.

Of course, our results need to be interpreted with the necessary care. First, our results do not necessarily generalize to very different value models. We also assumed all bidders to have the same synergies. This was motivated by our application domain and the fact that synergies often arise from a mobile operator's ability to achieve peak performance with a technological standard after winning a certain amount of bandwidth, and these synergies are the same for all operators. We ran 4 additional sessions with a multiband value model and different synergies across bidders, but saw no significant impact on the results of our study. Second, one can argue that with even more time to prepare, bidders might behave differently. We conjecture that for a sufficiently large number of blocks, bidders will not be able to elicit and submit bids on the exponential number of bundles with positive value. Simplification and compact bid languages, which allow expressing the main synergies of a value model with a few parameters only, might be a remedy and further research is needed in this area. Such approaches have been used for procurement (Bichler et al., 2011), but also for spectrum auctions in the field (Goeree and Holt, 2010).

7. Appendix

7.1. CCA Activity Rules

The primary bid rounds help reducing value uncertainty in the market. Activity rules should provide incentives for bidders to reveal their preferences truthfully and bid straightforwardly in the primary bid rounds. Bidders should not be able to shade their bids and then provide large jump bids in the supplementary bids round. We will describe the activity rules and derive some useful propositions.

An eligibility points rule is used to enforce activity in the primary bid rounds. The number of bidder's eligibility points is non-increasing between rounds, and it limits the number of blocks the bidder can bid on in subsequent rounds. In the supplementary bids round, the following rules apply:

- There is no limit on the supplementary bid that can be made for the bundle bid in the final primary bid round.
- The supplementary bid for any other bundle A is subject to a cap determined in the following way:
  1. First, we determine the last primary bid round in which the bidder would have been eligible to bid for bundle A. Call this round the anchor round n. This will either be the final round or some round in which the bidder dropped its eligibility to bid (by reducing the number of blocks bid for) and gave up the opportunity to bid for bundle A in later primary bid rounds.
  2. Suppose that the bidder bids for bundle B in round n; The supplementary bid for bundle A cannot exceed the bid for bundle B (i.e., the supplementary bid for this bundle, if one is made, or otherwise the primary round bid) plus the price difference between bundles A and B that applied in round n (i.e., the additional blocks are priced at the ask prices in round n).

Note that after bidding on a bundle B in the final primary bid round, a bidder can still submit a bid on a larger bundle A in the supplementary bids round. However, the bid price for A cannot be higher than the bid for bundle B plus the price difference between bundles A and B that applied in round n. Since he has the opportunity to choose between A and B back in round n, and to opt for B, the bidder reveals the relative value between these two bundles. In the supplementary bids round, the bidder cannot reverse this reported preference by submitting a high bid on A. If a bidder bids on a bundle C, which is smaller than B, the bundle price is also bounded by the supplementary bid for B minus the price difference between the bundles B and C in round n. In other words, this is a revealed preference constraint. However, this constraint is only applied to the supplementary bids phase and with respect to the last primary bid round, where the bidder had sufficient eligibility points to submit a bid on C. The proposed rules for new spectrum auctions to be organized in the future also apply the revealed preference constraint to the primary bid rounds. Also, the constraint needs to be satisfied with respect to any eligibility-reducing primary bid round after the one, where the bidder could submit a bid on bundle C for the last time.[14]

Proposition 1. *If a bidder follows a straightforward bidding strategy in the primary bid rounds of a CCA with an anchor activity rule, then the activity rule will not restrict him to bid his maximum valuation on every bundle in the supplementary bids round.*

Proof. Let's assume, a bidder bids straightforwardly, i.e., he submits a bid on his payoff maximizing bundle in every round. Throughout the primary bid rounds he might have switched from a bundle A to a bundle B in a round n, when $v(B) - p_n(B) > v(A) - p_n(A)$, where $p_n(A)$ is the price of bundle A in round n. For the bundle A the bidder did not necessarily bid up to his true valuation in the primary bids round. Based on the anchor rule, in the supplementary bids round s the bidder can submit a maximum bid of $p_s^{max}(A) = v(B) + p_n(A) - p_n(B)$, if he bid his true valuation $p_s(B) = v(B)$. As a result $p_s^{max}(A) > v(A)$ such that the bidder can bid up to his true valuation on A in the supplementary bids phase. Note that the same argument applies for bundle bids, which were submitted in the primary bid rounds before A, after the bidder has revealed his true valuation $v(A)$ in the supplementary bids round. The proof also applies to bundles C, on which the bidder has never submitted a bid in the primary bid rounds, as long as $v(B) - p_n(B) > v(C) - p_n(C)$ in a round n, where the bidder had sufficient eligibility points to bid on this bundle. □

Unfortunately, bidding straightforward is not always possible.

Proposition 2. *If valuations for at least two bundles A and B are full substitutes with $v(A \cup B) = max(v(A), v(B))$ and the bundle of higher valuation A requires less bid rights than the lower valued bundle B, straightforward bidding is not possible due to the activity rule in the primary bid rounds.*

Proof. The activity rule in the primary bid rounds does not allow to increase the number of bid rights in later rounds. Suppose there are two bundles A and B with $|A| < |B|$ and $v(A \cup B) = max(v(A), v(B))$. If the value of a bundle A $v(A) > v(B)$ and prices in A rise such that the payoff of A becomes smaller than that of B, a bidder would not be able to switch to bundle B, and would therefore not be able to bid straightforward on his payoff maximizing bundle at the prices. □

7.2. Safe Supplementary Bids

We first introduce some necessary notation. Let I denote the supply of blocks, and $b_j^p(X_j) \in B$ the standing bid of bidder $j \in J$ on bundle $X_j \subseteq I$ after the primary bid phase. In addition, let $r^s(I)$ denote the revenue of the optimal allocation after the supplementary bids phase including all bids B in both phases. $r^p(I)$ describes the value with only standing bids in the last round of the primary bids phase. $r_{-b_j^p}^s(I)$ denotes the auctioneer revenue in the optimal allocation without all bids of bidder $j \in J$ on bundle

---

[14] See the specification of the proposed auction rules in Canada at www.ic.gc.ca/eic/site/smt-gst.nsf/eng/sf10363.html.

X_j. $X_j^C = I \setminus X_j$ is the set of blocks complementary to X_j. We refer to π as the ask price vector in the last primary bid round.

Proposition 3. *If demand equals supply in the final primary bid round, any single supplementary bid $b_j^s(X_j) > b_j^p(X_j)$ cannot become losing.*

Proof. In the last primary bid round, there is a demand of exactly I blocks, if demand equals supply. Bidder $j \in J$ submits a bid $b_j^p(X_j)$ in the last primary bids round, his standing bid after the primary bid rounds. Let $b_j^s(X_j) > r_{-b_j^p}^s(I) - r_{-b_j^p}^s(X_j^C)$ be the bid price that bidder j needs to submit, in order to win X_j after the supplementary bid round. Due to the anchor rule, j's competitors $k \in J$ with $k \neq j$ can only increase their bids without limits on bundles $X_k \subseteq X_j^C$, which were submitted in the last primary bids round. Any high supplementary bid $b_k^s(X_k^p)$ on a bundle X_k^p from k's standing bid after the primary bid rounds, will increase $r_{-b_j^p}^s(I)$ as well as $r_{-b_j^p}^s(X_j^C)$ and cannot impact $b_j^s(X_j)$. Supplementary bids on packages different to the standing bid of bidder k are restricted by the anchor rule such that $b_k^s(X_k^s \cup Z) \leq b_k^s(X_k^s) + \pi Z$ and $b_k^s(X_k^s \setminus Z) \leq b_k^s(X_k^s) - \pi Z$. As a result, any supplementary bid $b_j^s(X_j) > b_j^p(X_j) = \pi X_j$ must be winning. □

If there is excess supply in the last round of the primary bid phase, a last primary round bid b_j^p can become losing, because even if no supplementary bids were submitted, the auctioneer conducts an optimization with all bids submitted at the end, which might displace b_j^p. This raises the question for the safe supplementary bid $b_j^s(X_j^p)$, which ensures that the bidder j wins the bundle X_j^p of his standing bid from the primary round after the supplementary bids phase.

Proposition 4. *If there is zero demand on bundle M after the last primary bid round, a single supplementary bid of a standing bidder $b_j^s(X_j) > b_j^p(X_j) + \pi M$ cannot become losing.*

Proof. Let's assume a bidder j bidding on a bundle X_j^p in the last primary bid round and two other competitors, who bid on a bundle with all blocks I in the previous to last round of the primary bid phase. In the last round the two competitors reduce demand to zero so that X_j^C blocks have zero demand after the last primary bid round. Now, at least one of the competitors submits a supplementary bid on the bundle I at the prices of the last primary round πI. Bidder j can only win, if he increases his bid to $b_j^s(X_j) > b_j^p(X_j) + \pi X_j^C$.

Actually, πX_j^C is the maximum markup that bidder j has to pay to become winning with certainty. To see this, note that the bid by any competitor $k \in J$ with $k \neq j$ on any subset of X_j^C in this example is limited by π due to the activity rule. Now, if there was another standing bid by a competitor k on any subset of X_j^C after the last primary bid round, the maximum markup of j cannot increase. Due to the anchor rule, j's competitors k can only increase their bids without limits on bundles $X_k \subseteq X_j^C$, which were submitted in the last primary bids round. Such bids will increase $r_{-b_j^p}^s(I)$ as well as $r_{-b_j^p}^s(X_j^C)$ and cannot impact $b_j^s(X_j)$. Supplementary bids on packages different to the standing bid of bidder k are restricted by the anchor rule such that $b_k^s(X_k^s \cup Z) \leq b_k^s(X_k^s) + \pi Z$ and $b_k^s(X_k^s \setminus Z) \leq b_k^s(X_k^s) - \pi Z$.

	Bidder N	Bidder R_{11}	Bidder R_{21}	Bidder R_{12}	Bidder R_{22}
Round 1	(AC) = $2	(A) = $1	(AB) = $2	(CD) = $2	(CD) = $2
...					
Round 15	(AC) = $30	(A) = $15	(AB) = $16	(CD) = $16	(CD) = $16
Round 16	(AC) = $32	(A) = $16	(2B) = $2	(CD) = $17	(CD) = $17
...					
Round 20	(AC) = $40	(A) = $20	(2B) = $2	(CD) = $21	(CD) = $21
Round 21		(A) = $21	(2B) = $2	(CD) = $22	(CD) = $22
...					
Round 40		(A) = $21	(2B) = $2	(CD) = $41	(CD) = $41
Round 41		(A) = $21	(2B) = $2	(CD) = $42	(D) = $1
			– Termination –		

Figure 11. Bids in the primary bid rounds.

Similarly, if the other bidders reduce their demand such that a package M with $|M| < |X_j^C|$ blocks remain unsold after the primary bid rounds, bidder j has to increase his standing bid by no more than πM to become winning after the supplementary bid round with certainty. □

7.3. Spiteful Bidding

In the following, we will provide a brief example of a CCA, in which a bidder can submit a spiteful bid, which increases the payments of other bidders with little risk of winning such a bid. If there is no excess supply after the primary bid rounds and bidders do not submit additional smaller bundle bids in the supplementary bid round, such bids would not stand a chance of winning.

Consider one region in which 3 blocks A (1 unit) and B (2 units), and one region in which 3 blocks C (1 unit) and D (2 unites) are up for auction among one national and several regional bidders. Start prices are $1 for all blocks and prices for overdemanded blocks are increased by $1 per round. Each block corresponds to one eligibility point.

The national bidder N is only interested in winning block A in each of the two regions for at most $40, i.e. he is not willing to switch to other packages. Regional bidder R_{11} is only interested in obtaining block A in his region. Regional bidder R_{21}

	Bid Price	VCG Payment	CCA Payment
National N (AC)	($40)		
Regional R11 (A)	$21	$14	$14
Regional R21 (2B)	$2	$0	$0
Regional R12 (AB)	$42	$40	$40
Regional R22 (B)	$1	$0	$0
	$66	$54	$54

Figure 12. Payments after the supplementary bids round without additional supplementary bids.

	Bid Price	VCG Payment	CCA Payment
National N (AC)	($40)		
Regional R11 (A)	$21	$20	$20
Regional R21 (2B)	$2	$0	$0
Regional R12 (AB)	$42	$40	$40
Regional R22 (B)	$1	$0	$0
	$66	$60	$60

Figure 13. Payments after the supplementary bids round with an additional supplementary bid by bidder R_{21} on AB for $22.

wants prefers AB over 2B. He is willing to switch from AB to 2B, if prices differ by at least $15. Regional bidders R_{12} and R_{22} would like to obtain CD. Bidder R_{2_2} is weaker and willing to bid on D after if he is overbid.

Figure 11 illustrates the primary bid rounds, while Figure 12 describes the payments if no supplementary round bid was submitted. Finally, Figure 13 illustrates the payments if bidder R_{2_1} submitted a spiteful bid on AB for $22, the package price in the final round for which he would still be eligible according to the activity rule. Consequently, the payment of the regional competitor increases by $6. Such bids are possible due to the initial eligibility points rule, which might not reflect the market value of different lots appropriately.

References

Abbink K, Irlenbusch B, Pezanis-Christou P, Rockenbach B, Sadrieh A, Selten R (2005) An experimental test of design alternatives for the british 3g - umts auction. European Economic Review 49:1197–1222

Ausubel L, Cramton P (2011) Activity rules for the combinatorial clock auction. Tech. rep., University of Maryland

Ausubel L, Milgrom P (2006) The lovely but lonely Vickrey auction. In: Cramton P, Shoham Y, Steinberg R (eds) Combinatorial Auctions, MIT Press, Cambridge, MA

Ausubel L, Cramton P, Milgrom P (2006) The clock-proxy auction: A practical combinatorial auction design. In: Cramton P, Shoham Y, Steinberg R (eds) Combinatorial Auctions, MIT Press, Cambridge, MA

Bajari P, Yeo J (2009) Auction design and tacit collusion in fcc spectrum auctions. Information Economics and Policy 21:90–100

Banks J, Ledyard J, Porter D (1989) Allocating uncertain and unresponsive resources: An experimental approach. RAND Journal of Economics 20:1–25

Banks J, Olson M, Porter D, Rassenti S, Smith V (2003) Theory, experiment and the FCC spectrum auctions. Journal of Economic Behavior & Organization 51:303–350

Bichler M, Schneider S, Guler K, Sayal M (2011) Compact bidding languages and supplier selection for markets with economies of scale and scope. European Journal on Operational Research 214:67–77

Brandt F, Sandholm T, Shoham Y (2007) Spiteful bidding in sealed-bid auctions. In: 20th International Joint Conference on Artificial Intelligence (IJCAI), pp 1207–1214

Brunner C, Goeree JK, Holt C, Ledyard J (2010) An experimental test of flexible combinatorial spectrum auction formats. American Economic Journal: Micro-Economics 2(1):39–57

Cramton P (2008) A review of the l-band auction. Tech. rep.
Cramton P (2009) Auctioning the Digital Dividend, Karlsruhe Institute of Technology
Cramton P (2013) Spectrum auction design. Review of Industrial Organization 42(2):161–190
Cramton P, Stoft P (2007) Why we need to stick with uniform-price auctions in electricity markets. Electricity Journal 26:26–37
Cramton P, Shoham Y, Steinberg R (eds) (2006a) Combinatorial Auctions. MIT Press, Cambridge, MA
Cramton P, Shoham Y, Steinberg R (2006b) Introduction to combinatorial auctions. In: Cramton P, Shoham Y, Steinberg R (eds) Combinatorial Auctions, MIT Press, Cambridge, MA
Day R, Milgrom P (2007) Core-selecting package auctions. International Journal of Game Theory 36:393–407
Day R, Raghavan S (2007) Fair payments for efficient allocations in public sector combinatorial auctions. Management Science 53:1389–1406
Ewerhart C, Moldovanu B (2003) The german umts design: Insights from multi-object auction theory. In: Illing G (ed) Spectrum Auction and Competition in Telecommunications, MIT Press
Goeree J, Holt C (2010) Hierarchical package bidding: A paper & pencil combinatorial auction. Games and Economic Behavior 70(1):146–169, DOI 10.1016/j.geb.2008.02.013
Goeree J, Lien Y (2010) An equilibrium analysis of the simultaneous ascending auction. Working Paper, University of Zurich
Goeree J, Lien Y (2012) On the impossibility of core-selecting auctions. Theoretical Economics
Gul F, Stacchetti E (1999) Walrasian equilibrium with gross substitutes. Journal of Economic Theory 87:95–124
Gul F, Stacchetti E (2000) The english auction with differentiated commodities. Journal of Economic Theory 92:66–95
Guler K, Petrakis J, Bichler M (2016) Ascending combinatorial auctions with risk averse bidders. INFORMS Group Decision and Negotiation 25(3):609–639
Jewitt I, Li Z (2008) Report on the 2008 uk 10-40 Ghz spectrum auction. Tech. rep., URL http://stakeholders.ofcom.org.uk/binaries/spectrum/spectrum-awards/completed-awards/jewitt.pdf
Kagel J, Lien Y, Milgrom P (2010) Ascending prices and package bids: An experimental analysis. American Economic Journal: Microeconomics 2(3)
Klemperer P (2002) How (not) to run auctions: the European 3G telecom auctions. European Economic Review 46(4-5):829–848
Kwasnica T, Ledyard JO, Porter D, DeMartini C (2005) A new and improved design for multi-objective iterative auctions. Management Science 51(3):419–434
Ledyard J, Porter D, Rangel A (1997) Experiments testing multiobject allocation mechanisms. Journal of Economics and Management Strategy 6:639–675
Maldoom D (2007) Winner determination and second pricing algorithms for combinatorial clock auctions. Discussion paper 07/01, dotEcon
Milgrom P (2000) Putting auction theory to work: The simultaneous ascending auction. Journal of Political Economy 108(21):245–272
Morgan J, Steiglitz K, Reis G (2003) The spite motive and equilibrium behavior in auctions. Contributions to Economic Analysis and Policy 2
Nisan N, Segal I (2006) The communcation requirements of efficient allocations and supporting prices. Journal of Economic Theory 129:192–224
Papai S (2003) Groves sealed bid auctions of heterogeneous objects with fair. Social Choice and Welfare 20:371–385
Porter D, Smith V (2006) FCC license auction design: A 12-year experiment. Journal of Law Economics and Policy 3

Porter D, Rassenti S, Roopnarine A, Smith V (2003) Combinatorial auction design. Proceedings of the National Academy of Sciences of the United States of America (PNAS) 100:11,153–11,157

Sano R (2012a) Incentives in core-selecting auctions with single-minded bidders. Games and Economic Behavior 72:602–606

Sano R (2012b) Non-bidding equilibrium in an ascending core-selecting auction. Games and Economic Behavior 74:637–650

Scheffel T, Ziegler A, Bichler M (2012) On the impact of package selection in combinatorial auctions: An experimental study in the context of spectrum auction design. Experimental Economics 15(4):667–692

Seifert S, Ehrhart KM (2005) Design of the 3G spectrum auctions in the UK and Germany: An experimental investigation. German Economic Review 6(2):229–248

CHAPTER 30

Spectrum Auction Design: Simple Auctions For Complex Sales

Martin Bichler, Jacob K. Goeree, Stefan Mayer, and Pasha Shabalin

1. Introduction

The 1994 sale of radio spectrum for "personal communication services" (PCS) marked a sharp change in policy by the US Federal Communications Commission (FCC). Before turning to auctions the FCC had allocated valuable spectrum on the basis of comparative hearings (also known as "beauty contests") and lotteries. Nobel laureate Ronald Coase long advocated that market-based mechanisms would improve the allocation of scarce spectrum resources, but his early insights were ignored for decades [1]. The PCS auction raised over six hundred million dollars for the US treasury and it was widely considered a success. Several authors discuss the advantages and disadvantages of auctions and beauty contests for allocating scarce spectrum [2, 3, 4]. For example, some argue that financially strong bidders might have advantages over weaker bidders in an auction, while others argue that with efficient capital markets such differences should be less of a concern. Nowadays, spectrum is predominantly assigned by auction, both in the US and elsewhere [5, 6], and in this paper we focus on questions of auction design.

The simultaneous multi-round auction (SMRA), which was designed for the US FCC in the early 90's has been the standard auction format for selling spectrum world wide for many years. It auctions multiple licenses for sale in parallel and uses simple activity rules which forces bidders to be active from the start. Despite the simplicity of its rules there can be considerable strategic complexity in the SMRA when there are synergies between licenses that cover adjacent geographic regions or between licenses in different frequency bands. Bidders who compete aggressively for a certain combination of licenses risk being exposed when they end up winning an inferior subset at high prices. When bidders rationally anticipate this *exposure problem*, competition will be suppressed with adverse consequences for the auction's performance. The exposure problem has led auction designers to consider combinatorial auctions, which enable bidders to express their preferences for an entire set of licenses directly. In fact, the design of spectrum auctions is seen as a pivotal problem im multi-object auction design and successful solutions are a likely role-model for other public or private sector auctions such as transportation or industrial procurement.

Since 2008, the combinatorial clock auction (CCA) has been used by regulators in various countries such as the Austria, Australia, Canada, Denmark, Ireland, the Netherlands, and Switzerland to sell spectrum.[1] The CCA combines an ascending auction where individual license prices rise over time (clock phase) in response to excess demand, with a sealed-bid supplementary phase. In addition, the auction uses a complex activity rule to set incentives for bidders to bid actively from the start [10]. Unlike the SMRA, bidders can demand combinations of licenses as well as individual licenses.

Combinatorial auctions can employ different types of bid languages, such as OR and XOR languages. Both allow bidders to submit indivisible bids on packages. For example, if a bidder bids on packages $\{A, B\}$ and $\{C, D\}$, he would only be assigned one of the packages at most with an XOR language. With an OR language he might win both packages. This way, the number of different bids is reduced substantially. However, if a bidder only wants to win one of the two packages and not both, he cannot express this in a pure OR language. Actually, the OR language can only express superadditive valuations.

The CCA has employed an XOR bid language so far, but this comes at the price of high communication complexity.[2] With thirty licenses the number of possible combinations already exceeds a billion, which are far too many for bidders to express their values for.[3] This can lead to inefficiencies because the winner-determination algorithm allocates the spectrum as if missing bids for certain combinations reflect zero values for the bidders. Often the number of possible bids per bidder even has to be capped to a few hundred in order to keep the winner-determination problem feasible. In the bid data that was recently released by Ofcom for the CCA that was conducted in the UK in 2013 bidders submitted bids on between 8 and 62 packages in the supplementary round from 750 possible package bids considering the spectrum caps.[4] It is unlikely that bidders had a zero value for all the other packages.

In spectrum auctions it is typically common knowledge what combinations of licenses generate the most synergies.[5] In this paper, we study how the introduction

---

[1] A single stage combinatorial clock auction has been proposed in [7]. Such a single-stage ascending clock auction format was used in Nigeria [8], for example. In contrast, we discuss the two-stage combinatorial clock auction that has been used in spectrum auctions throughout the world in the past five years [9].

[2] This is separate from the issue of computational complexity for the auction designer, i.e., how to determine which bids are winning, which is known to be an NP-hard computational problem. Nisan [11] point out that for fully efficient allocations and general valuations the communication requirements grow exponentially.

[3] Spectrum auctions with dozens of licenses have been conducted in Austria, Australia, Switzerland, the Netherlands, Ireland, and the UK. For example, in the 2012 auction in the Netherlands, 41 spectrum licenses in the 800 MHz, 900 MHz and 1800 MHz bands were sold. Switzerland auctioned 61 licenses distributed over 11 bands in 2012. Canada used a CCA for 98 licenses in 2014. Although not all packages will have a value for bidders in such auctions, large national bidders will not be able to submit bids for all packages with positive value in auctions with this many items.

[4] http://stakeholders.ofcom.org.uk/spectrum/spectrum-awards/awards-archive/completed-awards/800mhz-2.6ghz/auction-data/

[5] For example, there is high complementarity within the 800 MHz band in most European auctions, where a package of two licenses often has much higher value than two times the value of a single license. For the new LTE mobile communication standard, telecom companies typically aim for four adjacent blocks of spectrum (i.e., 20 Mhz) in higher bands to fully leverage the new standard.

of a simple bid language, tailored to capture the main synergies, affects the performance of multi-band spectrum auctions. Our bid language allows bidders to specify either-or bids on packages within a band (XOR) while bids for packages in different bands are considered additive (OR). This way, the number of possible bids is reduced substantially. Although elements of the bid language can be used in practice, we do not suggest there is a one-size-fits-all bid language. Rather, we want to understand the potential benefits of such an OR-of-XOR bid language over a fully expressive one. Interestingly, the design of compact bid languages has not been an issue in the design of spectrum auctions in different countries and a fully expressive XOR bid language has always been used for the CCA.

Besides the bid language, another defining feature of the CCA is the *core-selecting payment rule*. Theoretical considerations for this payment rule are based on the Vickrey-Clarke-Groves (VCG) mechanism, which has a simple dominant strategy for bidders to submit their valuations truthfully. The VCG mechanism, however, can lead to outcomes where the winners pay less then what losing bidders are willing to pay with their bids.

A simple example should illustrate the problem. Suppose three bidders 1, 2 and 3 bid on two items A and B. Bidder 1 is only interested in item A and the bundle $\{A, B\}$ for \$2 since we assume free disposal. Bidder 2 is only interested in B and the bundle of both items for \$2, and bidder 3 is willing to pay \$0 for each item and \$2 for the bundle. Bidders 1 and 2 are winners and in a VCG auction they get a discount equaling their marginal contribution to the overall revenue. Bidder 1 would win A and pay his bid price of \$2 minus (\$4-\$2), i.e., the difference in revenue with and without him. This means bidder 1 would pay zero and likewise also the second bidder would pay zero in a VCG auction resulting in zero revenue, although there was another losing bidder, who expressed a willingness to pay \$2.

To avoid such "non-core" outcomes with respect to the bids, the core-selecting payment rule has been used in the CCA. This payment rule is sufficiently complex that it generally does not allow for a game-theoretic analysis and its outcomes can appear non-transparent as small changes in the package bids selected by the bidders can lead to substantial variations in the payments.[6] Moreover, the payments are not known until after the auction, which precludes bidders from reporting to management about the progress of the auction and about expected payments. These issues do not arise with a simple pay-as-bid payment rule as used, for instance, in the Romanian spectrum auction in 2012.

Avoiding uncompetitively low revenue as is possible in a VCG mechanism was one of the original design goals of core-selecting payment rules [13]. Arguably, revenue is an important result in spectrum auctions even though it rarely is an official design goal. Note that efficiency cannot be analyzed in the field where the bidders' valuations stay private. Transparency of the auction process and the law of one price (for a license across all bidders) are additional design goals apart from efficiency and revenue that matter in spectrum auction design and there are trade-offs between these goals. Both, the CCA and the VCG mechanism do not satisfy the law of one price and for example in

---

[6] For a simplified setting, Goeree and Lien [12] show that the "core selecting" payment rule may result in prices that are further from the core than Vickrey prices.

Switzerland one of the bidders had to pay almost 482 million Swiss Francs and another bidder close to 360 million Francs although they won a similar set of licenses.[7] These problems have led to discussions among regulators and telecoms on pros and cons of different auction designs used for selling spectrum. In particular, stakeholders need to understand the impact of different bid languages and different payment rules on the overall efficiency and revenue of the auction.

We have implemented the two-stage CCA with all the activity, allocation, and core-selecting payment rules as it is used in the field, but also the alternative treatments and analyzed them in lab experiments. The different treatments of our experiment allow us to measure how auction revenue varies when using the pay-as-bid or core-selecting payment rule. We consider the treatment variations, simple versus complex bid language and simple versus complex payment rule, for both ascending and sealed-bid formats. We find that simplicity of the bid language has a substantial positive impact on the auction's efficiency and simplicity of the payment rule has as a substantial positive impact on the auction's revenue.

2. Experimental Design

In what follows, we characterize treatment variables, in particular the value model, the bidding language, and the auction formats, before we discuss details of the organization of our experiments.

2.1. The Value Model

In this paper we will draw on the multi-band value model used in earlier experiments [10], which has four bands with 6 licenses each. Within a band, each individual block has the same value for bidders so that there are essentially $7^4 - 1 = 2,400$ different packages. The structure of the value model and the distribution of the block valuations of all bands are known to all bidders. In particular, band A is of high value to all bidders and bands B, C, and D are less valuable. Bidders receive base valuations for items in each band. Base valuations are uniformly distributed: v_A was in the range of [100, 300] while v_B, v_C, and v_D were in the range of [50, 200]. Furthermore, bidders have complementary valuations for bundles of blocks within bands, but not across bands. In all bands, bundles of two blocks resulted in a bonus of 60% on top of the base valuations, while bundles of three or more blocks resulted in a bonus of 50% for the first three blocks. For example, if the base value was 100 experimental Francs, then the valuation for two blocks was 320, for three blocks 450, and for four blocks 550. Although the value models resemble characteristics of actual spectrum sales, this was not communicated to the subjects in the lab to maintain a neutral framing.

2.2. Bid Languages

The bidding language in a CA specifies the kinds of bids that can be placed by a bidder. Under the fully expressive XOR bid language, bids can be placed on any of the 2,400

---

[7] Results of the Swiss auction can be found at www.news.admin.ch/NSBSubscriber/message/attachments/26004.pdf.

```
    ┌─XOR─┐
   ┌─┬─┬─┐
   │2A│4A│6A│
┌OR┤2B│4B│6B│
   │2C│4C│6C│
   │2D│4D│6D│
   └─┴─┴─┘
```

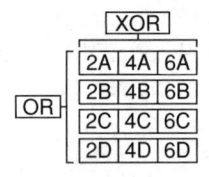

Figure 1. Compact bid language.

different packages with the understanding that at most one of the bids can become winning. This bid language has been used in all spectrum auctions so far, as it allows expressing all possible preferences including complements and substitutes. As already introduced, this expressivity comes at the price of an exponential number of possible packages. There have been attempts to design bid languages allowing bidders to express their preferences with a lower number of parameters. For example, the OR* language extends the OR language to allow a bidder to introduce some number of bidder-specific dummy items. The dummy items have no value, but they allow the bidder to constrain the sets of bids that can be selected by the auctioneer. This can substantially reduce the number of bids that a bidder has to submit [14]. Apart from such generic bid languages, domain-specific bid languages have been proposed, which leverage common knowledge about the utility functions of bidders in specific markets [15].

In our experiments, we use an OR-of-XOR bid language, which draws on the observation that typically there are high synergies among licenses within a band, but lower synergies across bands. In the experiments bids could be submitted on 2, 4, and 6 lots only in each of the bands and at most one of the bids within a band could become winning. However, a bidder could win multiple bids in different bands, i.e. we use an OR bid language across bands. Overall, bidders can submit $3 * 4 = 12$ bids in each round, and win a maximum of 4 bids (one bid per band) in this OR-of-XOR language (see Figure 1). In our value model there were no cross-band synergies. Even if there were synergies across bands, bidders can often handle the remaining exposure risk well. Overall, the bid language and the value model might differ for specific applications, but the experiments allow us to estimate the differences in efficiency of a compact bid language compared to an XOR bid language, which has been used in spectrum auctions so far.

2.3. Treatment Structure

We analyze two variations, simple (S) and complex (C), of the bid language and payment rule. In particular, we consider a compact bid language versus a fully expressive bid language, and a pay-as-bid versus a bidder-optimal core-selecting payment rule. We do so for both ascending (A) and sealed-bid (SB) auctions. The different treatments are denoted F_{LP} where $F = \{A, SB\}$ denotes the format and the subscripts $L = \{S, C\}$ and $P = \{S, C\}$ indicate the bid language and payment rule respectively (see Table 2). For example, the CCA is denoted A_{CC} while SB_{SS} denotes a sealed-bid auction with a compact bid language and a pay-as-bid payment rule. The only ascending auction format with a fully expressive bid language we consider is the A_{CC} (and not A_{CS}) since

Table 1. *Treatment structure of the experiments*

Treatment	Auction format	Bid language	Payment rule	Auctions
1 (*SMRA*)	ascending	single-item	simple	16
2 (A_{CC})	two-stage	complex	complex	16
3 (SB_{SC})	sealed-bid	simple	complex	16
4 (SB_{SS})	sealed-bid	simple	simple	16
5 (SB_{CS})	sealed-bid	complex	simple	16
6 (SB_{CC})	sealed-bid	complex	complex	16
7 (A_{SC})	ascending	simple	complex	16
8 (A_{SS})	ascending	simple	simple	16

it is the incumbent standard, this means the CCA.[8] Instead of the A_{CS} we include the SMRA, which used to be the standard and also has a simple pay-as-bid payment rule and a (super) compact bid language, i.e. OR bidding within and across bands.

The sealed-bid formats are straightforward in that bids can be submitted only once, after which the winner-determination problem is solved and prices are computed. In contrast, the ascending auctions consist of an unknown number of rounds and at the start of each round ask prices for all licenses are announced. Based on these ask prices, bidders report whether they are interested in 0, 2, 4, or 6 licenses in each of the four bands. If there is excess demand (i.e., if the combined demand of all bidders exceeds the number of licenses available) in at least one band, a new round starts with higher ask prices for the bands with excess demand. Prices in the first round are set to 100 for items in the A band and to 50 in the B, C, and D bands. The price increment in the A band is 20 while in the B, C and D bands it is set to 15. A bidder has to submit at least one bid in each round to bid again for bundles in the next round. When there is no more excess demand in any of the bands the winner determination problem is solved

Table 2. *Aggregate measures of auction performance*

Auction	E	R	Unsold licenses
SMRA	98.51%	81.96%	0
A_{SS}	95.92%	86.62%	0
A_{SC}	97.26%	78.96%	0
A_{CC}	89.33%	37.41%	1.25 (5.2%)
SB_{SS}	94.33%	91.05%	0
SB_{SC}	97.21%	77.28%	0
SB_{CS}	88.56%	89.62%	0.82 (3.4%)
SB_{CC}	91.76%	65.53%	0.31 (1.3%)

---

[8] Ascending auction formats with an XOR bid language, a pay-as-bid payment rule and non-linear and personalized ask prices have already been tested in the lab [16], but the number of auction rounds renders them impractical for larger auctions with more than 10 items.

considering *all* bids submitted during the entire auction. If the computed allocation does not displace an active bidder from the last round the auction terminates, otherwise the price is incremented in those bands where a bidder was displaced to give now losing bidders a chance to improve their bid.[9]

We put great emphasis on the external validity of the experiments. The value models were modeled after real-world auctions, and bidders were provided decision support in selecting their packages based on value or payoff after each round which reflects practices in the field. Still, there might be phenomena in the field that we do not observe in the lab. For example, in our experiments we provided strong incentives to maximize payoff. Bidders in the field might be spiteful and try to block other competitors from getting their preferred allocation or drive up their prices on items other bidders desire. In a combinatorial clock auction they might also try to increase payments of others by submitting bids with a high probability of losing [10]. Such issues are unlikely in the lab. Still, the lab results help understand many aspects bidder behavior such as restricted bundle selection and its consequences on auction efficiency and revenue.

2.4. Procedures and Organization

We used the same sets of value draws ("waves") across treatments to reduce performance differences due to the random draws. Each wave was used to run four different auctions, which combined define one session. We ran between subjects experiments with four bidders in each session. The experiments were conducted from June 2012 to March 2013 with subjects from computer science, mathematics, physics, and mechanical engineering. The subjects were recruited via e-mail. Each subject participated in a single session only.

The sessions with the ascending auction took around four hours and the sealed bid auctions between 1.5 and 2.5 hours. At the start of each session the environment, the auction rules and all other relevant information was explained to the participants. The instructions were read aloud and participants had to pass a test before they were admitted to start the experiment.

A spreadsheet tool was provided to subjects to analyze payoffs and valuations in each round. This tool showed a simple list of available bundles, which could be sorted by bundle size, bidder individual valuations, or payoffs based on current prices in the ascending auction formats. At the start of each auction, subjects received their individual value draws, information about the value distributions and their synergies for certain bundles. Each round in the ascending auction took 3 minutes. The time given to the subjects in the sealed bid formats varied between 20 and 25 minutes (although subjects could always ask for more time when needed).

After all four auctions were completed, subjects were paid. The total compensation consisted of a 10 Euro show up fee and an auction reward, which was calculated as a 3 Euro participation reward plus the auction payoff converted to Euros at a 12:1 ratio. Negative payoffs were deducted from the participation reward. To compensate for the different durations of the ascending and sealed-bid auction formats, and for

---

[9] This procedure is in line with the single stage combinatorial clock auction [7]. A theoretical analysis of this auction format can be found in [17].

the differences in earnings stemming from the payment rules, we paid two out of four randomly drawn auctions in A_{SC}, three out of four in A_{SS}, 1.5 out of four auctions in SB_{CS} and SB_{SS},[10] and one out of four auctions in SB_{CC} and SB_{SC}. On average, each subject earned 70.94 EUR in A_{SC} and 69.75 EUR in A_{SS}, 37.69 EUR in the sealed bid auction with compact bid language (SB_{SC}, SB_{SS}) and 42.16 EUR in the sealed bid expressive auction (SB_{CC}, SB_{CS}).

3. Results

We will first present aggregate results, i.e., efficiency and revenue of the different auction formats, and then discuss individual bidder behavior. For the pairwise comparisons of various metrics we use the rank sum test for clustered data to reflect that the auctions were conducted in sessions with the same set of subjects [18].

3.1. Efficiency and Revenue

We compare auction formats in terms of *allocative efficiency*

$$E = \frac{\text{actual surplus}}{\text{optimal surplus}} \times 100\%$$

and in terms of *revenue distribution*

$$R = \frac{\text{auctioneer's revenue}}{\text{optimal surplus}} \times 100\%$$

which shows how the resulting total surplus is distributed between the auctioneer and the bidders. Optimal surplus describes the resulting revenue of the winner-determination problem if all valuations of all bidders were available, while actual surplus considers the true valuations for those packages of bidders selected by the auction. In contrast, auctioneer's revenue used in the revenue distribution describes the sum of the bids selected by the auction, not their underlying valuations.

Regulators typically aim for competition and high efficiency in the downstream wireless telecommunications market after the auction. Note that this is different from allocative efficiency of the auction outcome, which has been a primary concern in auction theory. Proponents of using auctions as a means to allocate spectrum argue that allocative efficiency of an auction places spectrum with those who value it most and who are therefore likely to develop it most effectively. Of course, the result of an efficient auction could be a monopoly, which is why auction designs sometimes include set aside blocks for new entrants and license acquisition limits (aka. caps) for bidders [2, 19]. In this paper, we assume sufficient competition in a market of telecom operators and analyze allocative efficiency as a desirable goal.

Although, high revenue is rarely an official design goal for regulators, it is always an issue after spectrum auctions in the absence of bidders' true valuations. Whether high revenue is seen as another goal in addition to efficiency or low payments for the

---

[10] The first auction that was drawn was paid fully and for the second auction only half the payoff.

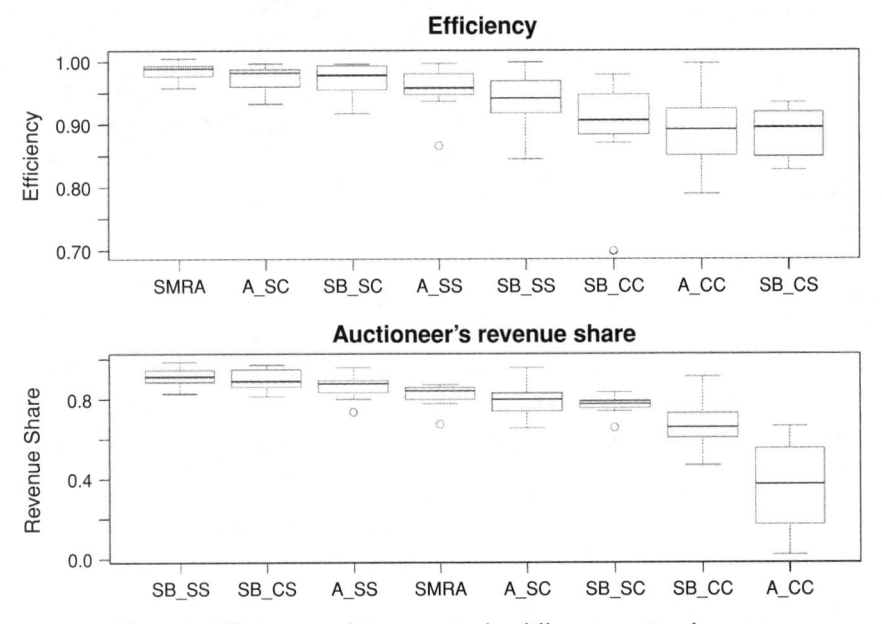

Figure 2. Efficiency and Revenue in the different auction formats.

bidders are seen as desirable depends on the overall telecommunications policy [2, 4]. In any case, it is important to understand the potential impact of payment rules on the revenue of an auction.

Result 1: (i) Formats with a compact bid language are more efficient than those with a fully expressive language. To some extent the efficiency loss with a fully expressive bid language is due to the fact that items remain unsold, which does not happen with a compact bid language. (ii) Among the formats with a fully expressive bid language there are no efficiency differences. (iii) Among the formats with a compact bid language only the SMRA yields significantly, albeit not substantially, higher efficiency.[11]

Result 1 is illustrated in Figure 2 and Table 2. The intuition behind the efficiency loss with fully expressive bid languages is that few bids among the 2,400 possible bids are selected (see Sections 2.2 and 2.3). The winner-determination algorithm assigns zero value to all packages not bid for, which distorts from the optimal allocation especially when the submitted bids create a fitting problem. Somewhat surprisingly, the SMRA comes out ahead despite the substantial complementarities within bands. Bidders did a good job in dealing with the resulting exposure risk, with high-value bidders taking more exposure risk and low-value bidders less.

A multiple linear regression confirms the impact of bid language (compact or fully expressive) on efficiency, while the payment rule (core-selecting or pay-as-bid) and the format (ascending or sealed-bid) have no significant effect (see Table 3).

---

[11] In more detail, SMRA $\succ^*$ A_{SC} $\sim$ SB_{SC} $\sim$ A_{SS} $\sim$ SB_{SS} $\succ^*$ SB_{CC} $\sim$ A_{CC} $\sim$ SB_{CS}, where $\sim$ indicates an insignificant order, $\succ$ indicates significance at the 10% level, $\succ^*$ indicates significance at the 5% level, and $\succ^{**}$ indicates significance at the 1% level.

Table 3. *Impact of bid language, payment rule, and auction format on efficiency (adjusted $R^2 = 0.4239$)*

Coefficients	Estimate	Pr(> $|t|$)
Intercept	0.9759	$< 2e-16$
XOR bid language	-0.0728	$1.36e-15$
Pay-as-bid payment rule	-0.0104	0.165
Auction format	-0.0081	0.279

Result 2: Formats with a pay-as-bid payment rule yield higher revenue than those with a core-selecting payment rule. Among the formats with a pay-as-bid payment rule only the SMRA yields significantly and substantially less revenue. Among the formats with a core-selecting payment rule those with a fully expressive bid language yield significantly and substantially less revenue.[12]

Support for result 2 can be found in Figure 2 and Table 2. The higher revenue for pay-as-bid sealed-bid auction formats might be explained by risk aversion. Auction format, bid language, and payment rule all have a significant impact on auctioneer revenue, see Table 4.

3.2. Bidder Behavior in Ascending Auctions

Result 3: Bidders in an ascending auction with a compact bid language select their bundles mainly based on payoff. Bidders did not only bid on their highest valued bundles, but on 72.9% of all bundles with a positive payoff. The payment rule did not have an impact on bundle selection. A fraction of 7.83% of all bids were above value in the A_{SC} auction compared to only 0.32% in the A_{SS} auction. In the supplementary phase of the two-stage CCA (A_{CC}) only a small fraction (0.06%) of the 2,400 possible bids were submitted.

Note that in the clock phase of the CCA bidders are only allowed to submit a single package bid per round. Figure 3 shows how many bids were submitted on the bundle with the highest payoff (dark grey), the second and third highest payoff, and on how many bundles with a positive payoff were not bid on (light grey). The three bars summarize the distribution of such bids in the first, middle, and final third of all auction

Table 4. *Impact of bid language, payment rule, and auction format on auctioneer's revenue (adjusted $R^2=0.5827$)*

Coefficients	Estimate	Pr(> $|t|$)
Intercept	0.6656	$< 2e-16$
XOR bid language	-0.1738	$3.93e-14$
Pay-as-bid payment rule	0.1794	$7.58e-13$
Auction type	0.1435	$2.89e-09$

---

[12] In more detail, $SB_{SS} \sim SB_{CS} \sim A_{SS} \succ^* \text{SMRA} \succ A_{SC} \sim SB_{SC} \succ^* SB_{CC} \succ^* A_{CC}$.

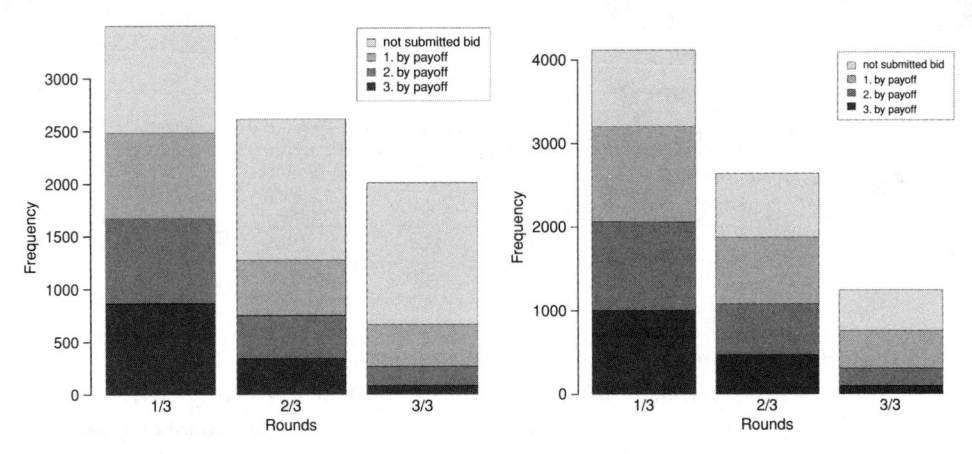

Figure 3. Distribution of bids by payoff in the A_{SS} (left) and A_{SC} (right) auction.

rounds (recall that the number of rounds varies across auctions). The two panels highlight that bidders did not only bid on the payoff maximizing bundle. Initially, they even submit more bids on bundles with the second or third highest payoff. We conjecture that bidders compare valuations rather than payoffs in the initial rounds.

Bids were frequently above value with the core-selecting payment rule, which might be due to the fact that the payment is lower than the submitted bid in this case.

3.3. Bidder Behavior in Sealed-Bid Auctions

Result 4: Bidders in core-selecting sealed-bid auctions with a compact bid language bid on all possible bundles. Bidders in sealed-bid auctions with a fully expressive bid language bid only on 2.42% of all 2,400 possible packages. There was more bid shading with the pay-as-bid payment rule compared to the core-selecting payment rule.

Figure 4 and Table 5 provide support for this result. We also estimated a linear regression with valuation as a covariate to explain bid prices (and bidder ID to control for unobserved heterogeneity among bidders). The intercept (α) and the slope (β) of the bidding function can be found in Table 6. The β coefficients are lower for pay-as-bid auctions, which indicates higher bid shading for higher valuations. The estimation results are shown by the dashed lines in Figure 4.

4. Discussion and Conclusions

The CCA is being increasingly used by regulators world-wide to sell spectrum licenses in multi-band auctions where bidders can submit bids on thousands or millions of different packages. The large number of possible bids introduces communication complexity into the auction, and it seems realistic to assume that bidders will typically submit bids only for a much smaller subset. Since the winner-determination algorithm assumes all other packages have zero value, the missing bids problem can have adverse effects for the auction's efficiency and revenue. This missing bid problem arises in

30 SIMPLE AUCTIONS FOR COMPLEX SALES

Table 5. *Truthful bidding in sealed-bid auctions*

Format	truthful	overbidding	underbidding
SB_{SS}	0%	0.99%	99.01%
SB_{CS}	0%	1.23%	98.77%
SB_{SC}	32.34%	22.05%	45.61%
SB_{CC}	18.11%	4.55%	77.34%

Table 6. *Estimated bid functions:* $b = \alpha + \beta v$

	α	β	p-value	adjusted R^2
SB_{SS}	0.5601	0.8834	0.0086	0.917
SB_{CS}	−0.0129	0.953	0.0033	0.986
SB_{SC}	−76.3868	0.9921	0.0056	0.975
SB_{CC}	−0.5637	0.9736	0.0029	0.986

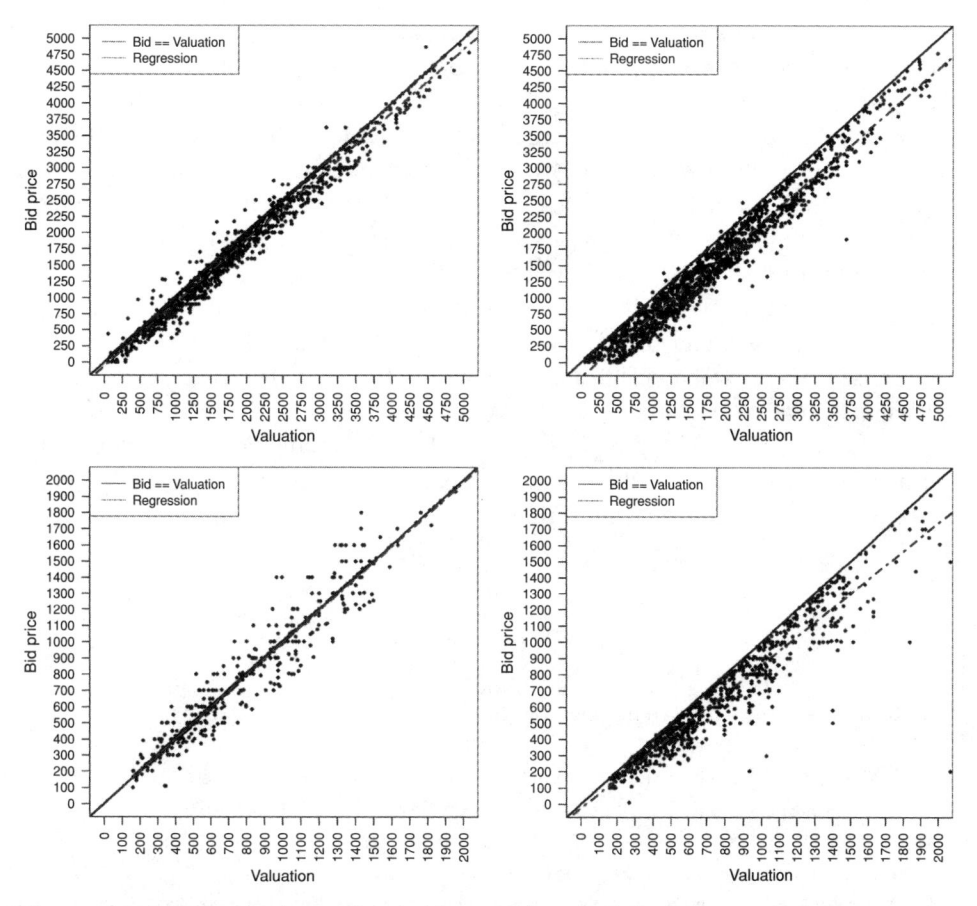

Figure 4. Bid shading in the auctions with core-selecting (left) and pay-as-bid auction (right) with a fully expressive bid language (top) and a compact language (bottom).

any combinatorial auction that uses a fully expressive (XOR) bid language unlike, for instance, in the SMRA that employs a simpler OR language. Regulators therefore face a trade-off between the SMRA's exposure problem and the CCA's communication complexity, both of which negatively impact the auction's efficacy.

In this paper we consider a middle-ground solution that aims to mitigate both the exposure problem and communication complexity. In particular, we analyze a bid language that drastically reduces the number of possible bids that can be submitted. First, the bid language assumes bids in different bands are additive, like in the SMRA, so that across bands multiple bids can be winning. In addition, we allow only for bids on packages of 2, 4, or 6 licenses within a band and at most one such bid can be winning. This reduces the number of possible bids from 2,400 to 12. Although, bid languages will be different depending on the application and there might be some complementarities across bands as well, the experimental results demonstrate that a simpler compact bid language yields significantly and substantially higher efficiency levels compared to a complex and fully expressive XOR bid language.

The results of our experiments do not suggest that SMRA always outperforms combinatorial auctions in markets with many licenses. Complementarities might be such that the exposure problem for large bidders creates a substantial strategic problem. Actually, in spectrum auctions with many regional licenses such as in Canada or the USA nationwide carriers will have preferences for the availability of package bids. The results of the experiments do suggest, however, that a fully expressive XOR bid language leads to substantial efficiency losses for larger combinatorial auctions. Even though it will take extra effort to design an appropriate bid language and find an agreement among the stakeholders in an auction, this design decision is essential and it shall not be ignored.

Besides complexity of the bid language we also studied how complexity of the payment rule affects auction performance. In particular, we compare a pay-as-bid rule to the core-selecting payment rule that underlies the CCA. We find that auction revenue is substantially higher with the simpler pay-as-bid rule. The pay-as-bid rule avoids uncertainty about how much a bidder has to pay for a bid at the end of an auction, if this bid becomes winning. Taken together our results underline the benefits of simplicity – both of the bid language and the payment rule.

References

[1] R. H. Coase, The federal communications commission, Journal of law and economics 2 (1959) 1–40.
[2] J. McMillan, Why auction the spectrum?, Telecommunications Policy 3 (19) 191–199.
[3] T. Valletti, Spectrum trading, Telecommunications Policy 25 (2001) 655–670.
[4] A. Morris, Spectrum auctions: Distortionary input tax or efficient revenue instrument?, Telecommunications Policy 29 (2005) 687–709.
[5] A. Gruenwald, Riding the us wave: spectrum auctions in the digital age, Telecommunications Policy 25 (2001) 719–728.
[6] R. S. Jain, Spectrum auctions in india: lessons from experience, Telecommunications Policy 25 (2001) 671–688.

[7] D. Porter, S. Rassenti, A. Roopnarine, V. Smith, Combinatorial auction design, Proceedings of the National Academy of Sciences of the United States of America (PNAS) 100 (2003) 11153–11157.

[8] C. Doyle, P. McShane, On the design and implementation of the GSM auction in Nigeria - the world's first ascending clock spectrum auction, Telecommunications Policy 27 (2003) 383–405.

[9] P. Cramton, Spectrum auction design, Review of Industrial Organization 42 (2013) 161–190.

[10] M. Bichler, P. Shabalin, J. Wolf, Do core-selecting combinatorial clock auctions always lead to high efficiency? An experimental analysis of spectrum auction designs, Experimental Economics (2013) 1–35.

[11] N. Nisan, I. Segal, The communication complexity of efficient allocation problems, in: DIMACS workshop on Computational Issues in Game Theory and Mechanism Design, Minneapolis, MI, 2001.

[12] J. Goeree, Y. Lien, On the impossibility of core-selecting auctions, Theoretical Economics.

[13] R. Day, P. Milgrom, Core-selecting package auctions, International Journal of Game Theory 36 (2008) 393–407.

[14] N. Nisan, Bidding languages, in: P. Cramton, Y. Shoham, R. Steinberg (Eds.), Combinatorial Auctions, MIT Press, Cambridge, MA, 2006.

[15] M. Bichler, S. Schneider, K. Guler, M. Sayal, Compact bidding languages and supplier selection for markets with economies of scale and scope, European Journal on Operational Research 214 (2011) 67–77.

[16] T. Scheffel, A. Pikovsky, M. Bichler, K. Guler, An experimental comparison of linear and non-linear price combinatorial auctions, Information Systems Research 22 (2011) 346–368.

[17] M. Bichler, P. Shabalin, G. Ziegler, Efficiency with linear prices? A theoretical and experimental analysis of the combinatorial clock auction, INFORMS Information Systems Research 24 (2013) 394–417.

[18] S. Datta, G. Satten, Rank-sum tests for clustered data, Journal of the American Statistical Association 100 (2005) 908–915.

[19] R. Earle, D. Sosa, Spectrum auctions around the world: An assessment of international experiences with auction restrictions, Tech. rep., AnalysisGroup (July 2013).

PART V
The Bidders' Perspective

CHAPTER 31

Winning Play in Spectrum Auctions*

Jeremy I. Bulow, Jonathan Levin, and Paul R. Milgrom

1. Introduction

Since being pioneered by the U.S. in 1994, simultaneous ascending auctions have become a common mechanism to allocate spectrum rights.[1] Spectrum auctions can involve billions of dollars and companies bidding in these auctions regularly create specialized bidding teams and hire experts in auction theory to develop bidding strategies. Nevertheless, the results can be surprising. In the FCC's auction of Advanced Wireless Service spectrum, price arbitrage failed so dramatically that one new entrant was able to purchase essentially nationwide coverage for about a third (more than a billion dollars) less than what incumbent carriers paid for equivalent spectrum in the same auction. At the same time, the other prospective nationwide entrant exited the auction early and filed a letter with the FCC claiming that the auction rules disadvantaged new entrants!

Results of this sort raise questions for economists. Does the apparent failure of the Law of One Price indicate a fundamental flaw in auction design? If not, why must such auctions be complicated? What are the issues that create strategic complexity for bidders? And to what extent can the tools of economic theory provide insights that facilitate effective bidding in highly complex environments?

We start by explaining some of the reasons why large spectrum auctions are necessarily complicated, and why the Law of One Price can fail so dramatically in a spectrum auction. We emphasize two difficulties facing bidders: *exposure problems*, which are essentially the problems of bidders wishing to acquire complementary licenses, and *budget constraints*, which we argue are ubiquitous. We explain why these difficulties make bidding in simultaneous ascending auctions complicated, and also why they would complicate bidding in other auction designs.

---

* The authors advised SpectrumCo in FCC Auction 66. We thank John Hegeman and Marcel Priebsch for excellent research assistance on this paper.
[1] Among the countries that have used such auctions to allocate spectrum are the US, Canada, Mexico, Australia, New Zealand, India, Germany, the Netherlands and the United Kingdom.

Exposure problems create fundamental difficulties for a new entrant seeking to compete head-to-head with incumbent nationwide wireless carriers in the US. Such an entrant needs to acquire adequate bandwidth in every major metropolitan area, but because licenses covering cities or regions are sold individually, the entrant could commit to spending billions of dollars winning spectrum licenses before discovering that the total price for the bundle of licenses it seeks makes the whole entry unaffordable or unprofitable. It could then be left to dispose of extensive holdings at fire-sale prices.[2]

The exposure problem, as well as the difficulties created by budget constraints, arise because there is uncertainty about the final auction prices. A bidder who knew what final prices would be would face no exposure problem and have no difficulty deciding how to allocate its limited budget. Information early in the auction about the final auction prices can therefore be extremely valuable to bidders.

Remarkably, it turns out that in large spectrum auctions, information sufficient to forecast final price levels is often available early in the auction. We document this previously unnoticed pattern using data from large FCC auctions. We also provide a simple theory that is broadly consistent with the facts. According to our theory, it is bidders' *budgets*, as opposed to their *license values*, that determines average prices in a spectrum auction.

We then explore the dynamics of simultaneous ascending auctions, in which prices of various licenses may follow a variety of increasing paths. We show that bidders facing exposure problems and budget constraints may wish to manipulate the price paths so larger licenses reach their final prices earlier in the auction than smaller ones. We describe the tactics available to bidders to accomplish that. And we explain the sometimes conflicting interests of new entrants regarding auction timing. Finally, we explain how bidders facing competitors who must deal with exposure and budget problems can disadvantage them by manipulating the price path in other ways.

We illustrate the practical application of these ideas using the experience of the U.S. Advanced Wireless Service auction mentioned above. In that auction, held in the late summer of 2006, the FCC auctioned 90 MHz of nationwide spectrum divided into 1122 licenses. The sale, in which winning bids ultimately totalled 13.9 billion dollars, attracted 168 bidders including two potential nationwide entrants: a consortium of cable television companies and a rival consortium of satellite television companies. These bidders, due to budget and exposure problems, faced by far the most difficult strategic problems.

During the auction, the satellite consortium exited earlier than any other major bidder, without buying a single license. The cable consortium, bidding under the name SpectrumCo, acquired licenses covering 91.2 percent of the U.S. population at prices that were much lower than those paid by the other large buyers. At the per unit prices paid by the major incumbent carriers, SpectrumCo's licenses would have cost more than 3.5 billion dollars — as it worked out, SpectrumCo paid less than 2.4 billion dollars.[3] Seneca famously remarked that "Luck is what happens when preparation meets opportunity." We describe in the final section of the paper how the elements of

---

[2] This "exposure problem" has been the driving force behind attempts to create new auctions in which bidders can bid for packages of licenses. Essays in the book by Cramton, Shoham, and Steinberg (2005) deal with various aspects of this mechanism design problem.

[3] The most common unit for measuring the size of a spectrum license or a collection of licenses is MHz-pop, which is calculated by multiplying the bandwidth in MHz by the population covered. Due to both spectrum

auction strategy analyzed here, specifically tactics to control the pace of the auction and decisions guided by budget-based price forecasts, put SpectrumCo in a position to be lucky.

2. Why Spectrum Auctions are Complicated

Spectrum auctions in the United States and other countries have typically been conducted using a simultaneous multiple round format. There are both practical and theoretical rationales for this choice. For instance, under certain conditions a simultaneous ascending auction results in competitive market-clearing prices. Indeed, suppose that bidders view licenses as substitutes, and that bidding is "straightforward" meaning that every round each bidder make offers on the set of licenses that give it the most surplus at current prices. Milgrom (2006) proved that prices would rise and ultimately stop at approximate competitive equilibrium levels (see also Gul and Stacchetti, 2000).

As an example, consider an auction like the British 3G auction held in 1999. Five national licenses were sold, with no bidder being allowed to buy more than one. Two licenses were bigger than the other three. In this auction, straightforward bidding, with each bidder making the qualifying offer that would give it the most surplus if the auction ended immediately, was a sensible strategy. The simple design helped allocate the spectrum to those willing to pay the most.

In auctions where bidders can buy multiple items, however, both bidders and auction designers face more serious challenges. For starters, to keep the auction moving forward the FCC employs an "activity rule" requiring each bidder to make offers in each round on at least a certain percentage of the "quantity" of spectrum which it is eligible to buy, the percentage increasing across two or three auction stages. If a bidder makes offers on a smaller amount of spectrum, its eligibility is reduced.

While the reasons for such an activity rule are compelling, the rule makes it more difficult for bidders to move back and forth between say a 30 MHZ license and a 10 MHZ license covering the same geographic area but only absorbing a third as much eligibility. Even moving back and forth between a 30 MHZ license and three 10 MHZ licenses can be tricky, as doing so would require being outbid on the three smaller licenses simultaneously. This barrier to arbitrage helps create the possibility of large price differences among nearly identical spectrum footprints.

Activity rules are one reason that in spectrum auctions bidding activity often starts on the larger licenses and then moves to the smaller licenses (though there are also strategic reasons for this, as we will explain below). Figure 1 illustrates the general pattern for the AWS auction. The two curves plot the round-by-round fraction of bids, by number, that are made on licenses that are larger or smaller than the median license, according to the FCC's quantity measure ("points"). In this auction, the larger licenses on average saw more bidding for the first 50 or so rounds, after which the pattern reversed.[4]

---

scarcity and differential build-out costs, prices of large urban areas are higher in MHz-pop, so our calculation may somewhat understate SpectrumCo's price differential.

[4] The figure ends after the Round 82, at which point there were 75 bids. In later rounds, there were never more than 50 bids and generally far fewer, and the fractions plotted in the chart fluctuate widely due to the small numbers.

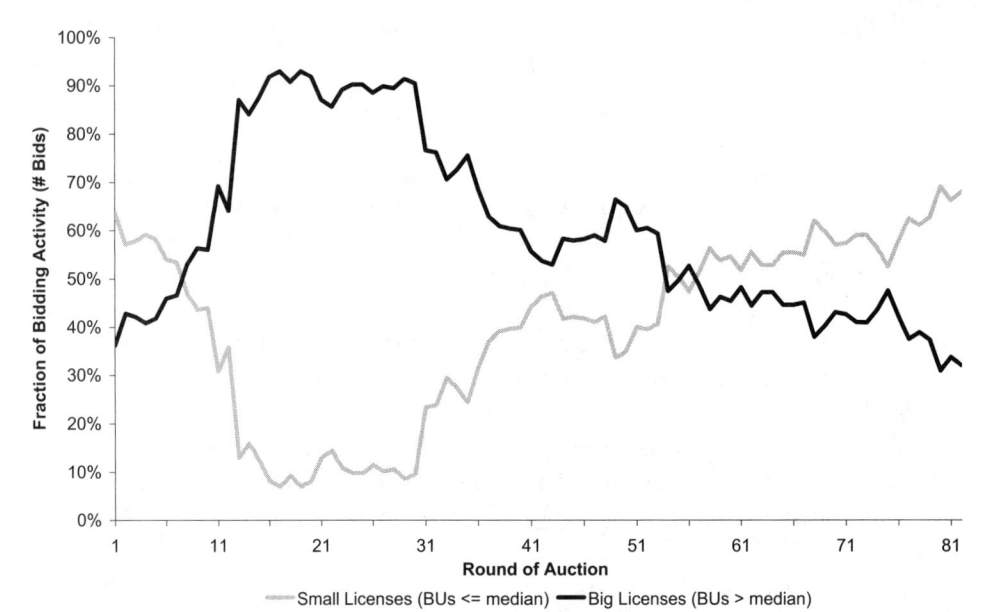

Figure 1. Pattern of Bidding Activity in Auction 66.

Of course, there is much more variation in the underlying data. Some small markets clear early, and some large markets can clear quite late. To provide a sense of the variability, Figure 2 plots the round of the last bid for each license against the size of the license for both the AWS auction and FCC Auction 35, another large auction. What is particularly important to note here is the wide range of rounds in which different licenses received their final bids. It is exactly this dispersion that creates problems for a bidder trying to assemble a package and worried about a failed aggregation. We will show below that it also gives these bidders an incentive to manage the pace of the auction, rather than bidding in straightforward fashion.

Our model will assume that a bidder who needs two licenses and acquires only one cannot resell its license. That assumption requires some justification. After all, the bidder could try after the auction to sell to the bidder that placed the final competing bid on the unwanted license, often only one increment below the sale price. The losing bidder, however, may have redirected its limited budget elsewhere, or met its needs some other way, or it may also have attempted and failed to assemble a collection of licenses, eliminating its interest in the single license. Furthermore, even if there is continued interest, the underbidder will know that it offered the highest alternative price putting it in a strong bargaining position. So while unwanted licenses do often have some salvage value, rational bidders anticipate incurring significant losses in trying to sell them.

A second important complication for bidders who cannot forecast final prices arises from limited budgets. Suppose a bidder is targeting two particular licenses that it believes would be profitable at prices up to 200 million and 100 million dollars, respectively, but has been able to raise only 150 million dollars in capital. If the rules require that the bidder remain active on both licenses or else lose the eligibility to purchase

31 WINNING PLAY IN SPECTRUM AUCTIONS

(a)

(b)

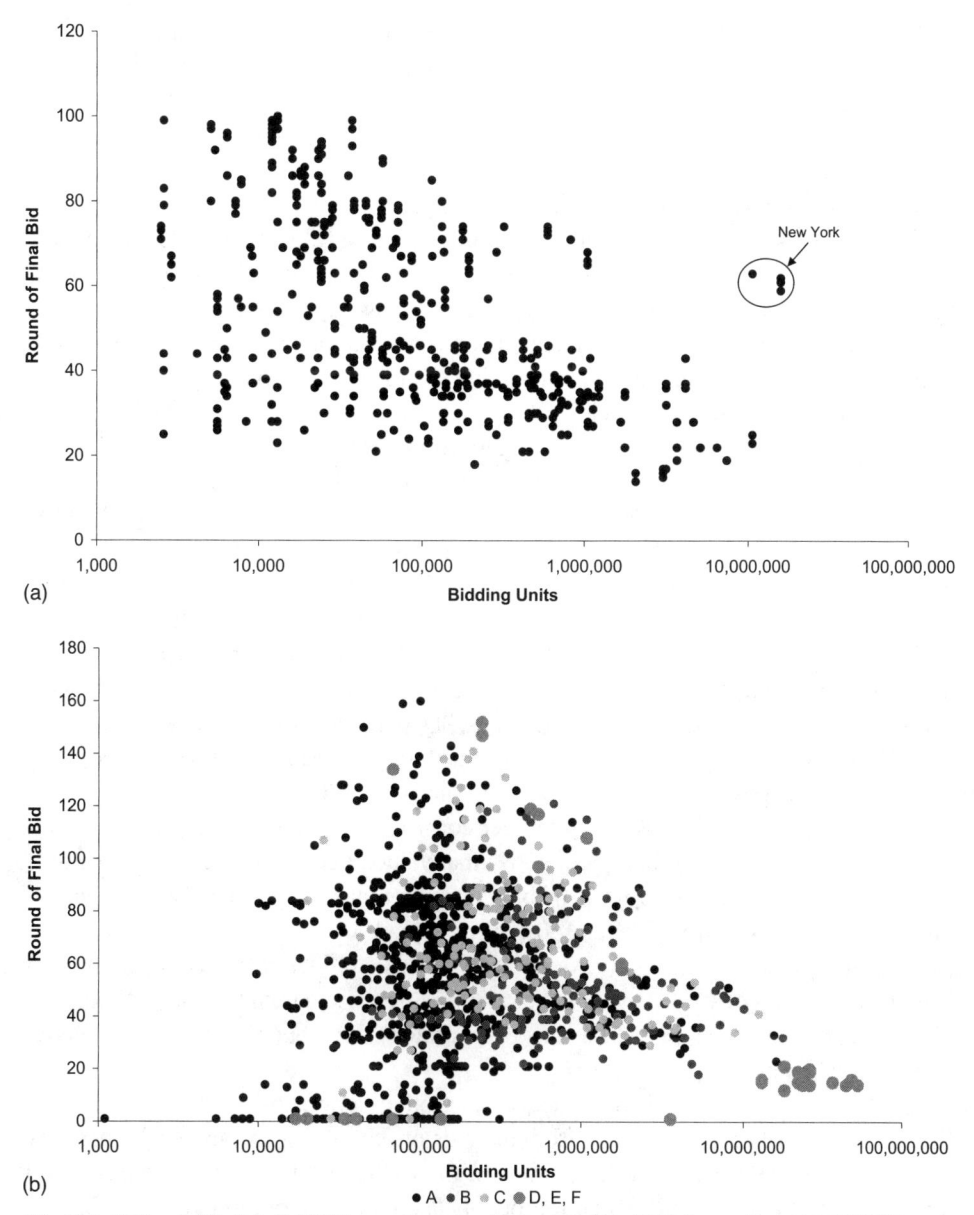

Note: Blocks D,E,F were divided into 12 REAG licenses; blocks B and C were divided into 176 EA licenses; block A into 734 CMA licenses.

Figure 2. (a) Round of Final Bid by License Size (Auction 35). (b) Round of Final Bid by License Size (Auction 66).

both, and the current price of each license if 50 million, what should the bidder do? If it continues bidding on only the more valuable license, it passes up the opportunity to win both. But if it bids for both, it might win the less valuable license in the current round and later find that its budget constraint blocks it from buying the other license, which may be a much better bargain.

Theoretically, some of the these difficulties could be addressed using alternative auction designs. For instance, a natural way to address the exposure problem is to permit package bids as well as bids on individual licenses.[5] Package bidding, however, comes with its own difficulties including complexity problems (there may be very many potential packages!), coordination problems (bidders need to make bids that fit together in reasonable ways) and strategic problems which depend on the particular auction rules.

To illustrate some of the issues, consider the Vickrey auction. In a Vickrey auction, each bidder can make offers on any license packages, and licenses are allocated to maximize the total bidder value assuming that bids reflect true value. Each bidder receives a surplus equal to the difference between the maximized value conditional on its participation and absent its participation. So the price a bidder pays equals its stated value for the licenses it receives, less its incremental contribution to auction surplus. Under well known assumptions bidders have an incentive to bid straightforwardly in the Vickrey auction and the allocation is efficient.

Unfortunately the Vickrey auction suffers from serious problems, many of which are described and analyzed in Ausubel and Milgrom (2007). Both the budget and the exposure problem have analogues in the Vickrey auction. The exposure problem is "solved" only at the cost of a potentially enormous reduction of seller revenues. For example, suppose that a package bidder offers 10 for licenses A and B, and faces one individual bidder on each license. Say these bidders make bids of 9 each. Each will be credited with creating a surplus of 8 (18 from the efficient allocation less 10 if either of the individual bidders did not exist and the licenses were sold to the package bidder) and so would pay a Vickrey price of 1. The seller will have total revenue of 2 — well outside the core, given that there was a package bidder willing to pay 10. The Vickrey solution to the exposure problem in this example is very costly for the seller.[6]

The second problematic issue — budget constraints — also create problems for a Vickrey auction. Consider the bidder above that valued license A at 200 and license B at 100, and had a budget of 150. If its bids report its maximum price for each license or bundle of licenses, it will offer 150 for A alone, 100 for B alone, and 150 for the pair. The mechanism will treat it exactly the same as a bidder who places *no value* on B in the event it receives A, so that the bidder will never win both licenses.[7] Evidently, bidding its maximum price can be very poor strategy for a budget-constrained bidder in a Vickrey auction.

Finally, a drawback of Vickrey auctions and other "one-shot" auctions is that they limit the ability of bidders to learn from the bids of others. For example, a bidder may think that it is the most efficient operator in market A and therefore "should" win a license there. It may also think that it probably "should" lose out in market B. Finally, it may believe that a license in market B will be worth exactly as much to it as a license in market A. However, it has an extremely noisy estimate of what that value is. How should this bidder proceed in a Vickrey auction? Should it bid the same amount for

---

[5] The FCC's recent auction of 700 MHz spectrum allowed a very limited set of package bids. The upcoming British auction of WiMax spectrum will use an even more ambitious package bidding design.

[6] This example was presented by Jeremy Bulow at the FCC's May 2000 conference on combinatorial bidding. More complete descriptions of the faults of the Vickrey auction can be found in Ausubel and Milgrom (2005) and Rothkopf (2007). This problem and others spurred subsequent research into core-selecting package auctions (Day and Milgrom, 2008).

[7] Hegeman (2008) proposes an efficient auction for budget constrained bidders, but in the case of indivisible goods his mechanism involves the use of lotteries, which likely would be problematic in high-stakes spectrum auctions.

Table 1. *Spectrum Price per MHz-Pop in Past Auctions*

Auction	Description	Date	Revenue (US $m)	Price per MHz-Pop
5	PCS C Block	May 1996	13,429	1.77
10	PCS C Block Re-Auction	July 1996	697	1.50
11	PCS D,E,F Block	Jan 1997	2,716	0.36
22	PCS	March 1999	533	0.20
34	800 MHz Auction	Sept 2000	337	0.18
35	PCS C&F Block	Jan 2001	17,596	4.37
58	Broadband PCS	Jan 2005	2,250	1.05
66	AWS Auction	Sept. 2006	13,879	0.54
73	700 MHz	March 2008	19,592	1.11

each license, believing they are equally valuable? What if its value estimates are much higher than most of its competitors? Will it overspend and buy licenses where it is not the efficient operator? On the other hand, what if its estimates are too low? Will it be shut out of licenses that it knows it should win?

These problems have led many economists concerned with auction design to favor multiple round auction formats when there is substantial price and value uncertainty.[8] Given this, we now take up how bidders in simultaneous multiple round auctions can deal with the strategic complexities, particularly the exposure and budget problems.

3. Price Forecasts and Bidder Budgets

3.1. The Price Forecasting Surprise

For a bidder facing a serious exposure problem, such as SpectrumCo in the AWS auction, the central strategic question is whether and when to exit the auction. An accurate early prediction of final prices can allow the bidder to avoid both kinds of exit mistakes, namely, the mistake of exiting too early, when final prices turn out to be low enough for successful entry, and the mistake of exiting too late, when final prices are found to be too high only after the bidder has won some licenses. Accurate price predictions are also valuable for bidders without an exposure problem but facing serious budget constraints, because it allows them to focus their spending on the licenses that will prove to be the best values.

Spectrum auction prices, however, can be hard to predict before the auction. The spectrum offered in each auction often has its own unique characteristics or restrictions imposed by the regulator, and even when an attempt is made to control for these differences, spectrum prices have fluctuated wildly over time. Table 1 shows the average prices per MHz-pop for 9 FCC spectrum auctions between 1995 and 2008. The variation is dramatic and much of it not easily explained by the nature of the spectrum

---

[8] There are also important considerations that can weigh against multiple round formats, including the need to avoid collusion and encouraging entry. As Klemperer (2002) convincingly argues, these considerations are the most important ones for many auction designs. Also, because our focus here is on spectrum auctions run by governments we assume that the seller has a great deal of power in setting the rules. In other situations, buyers may be able to take actions that disrupt or pre-empt an auction, or may refuse to participate unless they deem the rules sufficiently favorable to their interests.

for sale or industry conditions. Even forecasts made just prior to an auction by investment banks tend to have high variance. Prior to the AWS auction, analyst estimates of auction revenue ranged from $7 to $15 billion. For the recent 700 MHz auction, they varied over an enormous range – from $10 to $30 billion.[9]

Can information from early bidding usefully reduce the uncertainty? In an ascending auction for a single item, bidders who quit early provide some statistical information, but the only certain conclusion when the price reaches p is that the final price will be no lower. With multiple items being sold, more information may be available. High prices per MHz-pop for licenses that have seen early bidding can be a clue about the values bidders place on licenses that have not yet seen much activity.

For simultaneous multiple round auctions, we have found a simple and surprisingly powerful approach to making price forecasts that can be useful for bidding decisions. Our approach focuses attention not on bidder values, which are so emphasized in traditional auction theory, but on bidder budgets.

In major spectrum auctions, even large corporations need to raise or put aside money in advance to finance their spectrum purchases. Many of these companies also have a broad set of target licenses. If these licenses are substitutes and the budget constraint is binding, the bidder's optimal purchase will involve spending its whole budget or nearly so. Of course not every bidder falls into this category. For bidders with tight budgets and narrow interests, or for entrants with all-or-nothing goals, rising prices could lead them to spend zero once the prices of target licenses rise too high.

If bidders in the first category account for enough of the money in the auction, a previously unexplored pattern becomes identifiable in the data. Define a bidder's *exposure* to be the sum of all of its bids in a given round, including its standing high bids from the prior round and all of its new bids in the current round, whether provisionally winning or not. This is the largest amount that a bidder might have to pay if all of its bids were to become winning. If a bidder faces a binding budget constraint and has broad interests, then as prices increase from round to round, its total exposure will eventually level off at an amount approximating its budget. If all bidders were to fall in this category, then the total exposure of all bidders in the auction would rise to the level of the aggregate bidder budgets and level off, forecasting the final auction prices. As prices rise, bidders will narrow the set of licenses on which they bid, the identities of the provisionally winning bidders on various licenses will change, and total winning bids will continue to rise, but final total winning bids will be forecast early and well by total exposure.

Does such a pattern exist in the data? Figure 3 shows the pattern of total exposure and total prices in Auction 35, which was the largest US spectrum auction in the years before the AWS auction. At round 10, total prices in the auction were still less than one-third of their eventual level, but total exposure had approached its final level. Forecasting that total prices would be equal to total exposure from that point forward would lead to errors that are mostly less than 10% – close enough to guide some of the most critical strategic calculations. A potential new entrant who decided, based on that forecast, that prices would become too high could stop bidding while prices were still far

---

[9] In both of these cases, the final auction revenue fell between the extremes. In other cases, such as in some of the European 3G auctions, revenues have greatly exceeded or fallen short of analyst expectations.

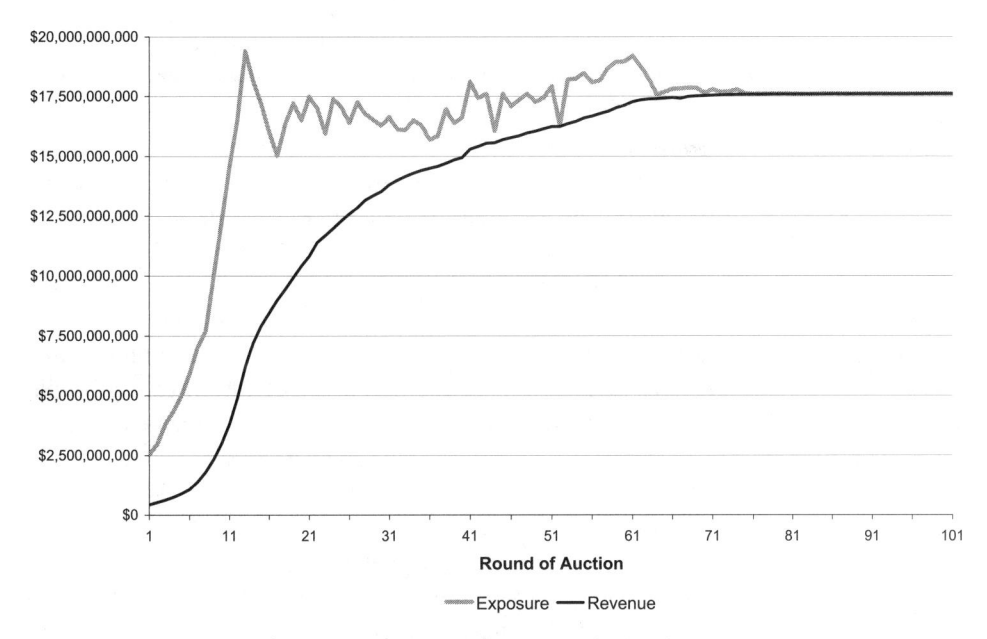

Figure 3. Revenue and Exposure in Auction 35.

below their final levels. The entrant would likely be topped on most or all of its provisionally winning bids and, even if it were not, its early withdrawal would mean that it acquired licenses at only a small fraction of the average auction price, greatly reducing any expected loss on resale.

Figure 4 shows the similar pattern of exposure and revenue for the AWS auction. Again, total exposure provides a remarkably accurate early forecast for total prices in the auction. Exposure peaked at $14.2 billion in round 11 and final auction revenue

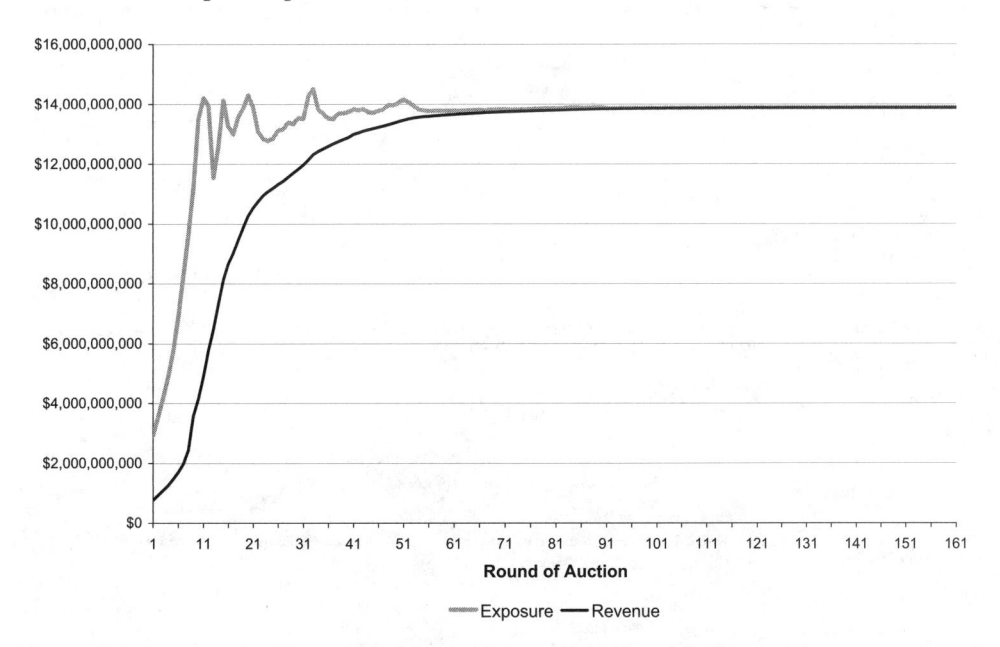

Figure 4. Revenue and Exposure in Auction 66.

was $13.9 billion. The large drop in exposure in round 13 is largely due to the exit of Wireless DBS, a joint venture of the two satellite TV companies Echostar and DirecTV. In Round 12, Wireless DBS's exposure was $2.025 billion; it dropped to $196 million in round 13 and subsequently to zero. From round 15 onwards, however, a bidder who estimated final total prices to be equal to current total exposure would never have made an error larger than 10%, despite the fact that the total price was still 40% below its final value.

In the AWS auction, in fact, the ability to forecast prices early in the auction had another key implication. Early bidding in that auction focused almost entirely on the 40 MHz of spectrum that was divided into large REAG licenses, before turning to the 50 MHz of spectrum that was divided into smaller EA and CMA licenses. By round 15, it was possible to forecast that cumulative high bids on the REAG licenses were so high relative to the total budgets in the auction that the smaller licenses would sell for a steep discount. This allowed SpectrumCo — alone among the major bidders — to make an early switch to the smaller licenses.

Does the budget constraint theory actually account for this observed pattern? Figure 5 provides additional detail, plotting the exposure of the largest individual bidders in Auctions 35 and 66. These figures suggest more than one pattern of bidding. In Auction 35, AT&T, Cingular, and many of the smaller bidders exhibit bidding patterns that suggest a binding budget, and one that could have been inferred early on. Verizon eventually may have hit a budget constraint, but if so not until relatively late in the auction.

In the AWS auction, all of the major winners — T-Mobile, Verizon, SpectrumCo, MetroPCS, Cingular, Leap/Denali and Barat — exhibit budget-constrained patterns. Yet there are also clear and interesting exceptions. Two large bidders, the Dolans and Wireless DBS, stand out. Wireless DBS was presumably an all-or-nothing entrant. The Dolans' bidding reveals a single objective – to acquire a New York license, presumably

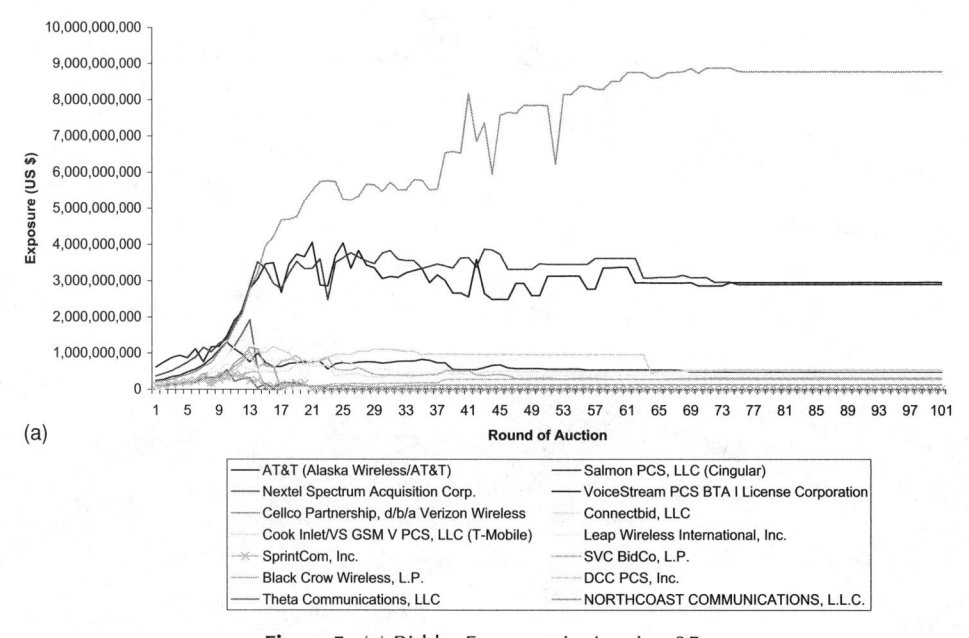

Figure 5. (a) Bidder Exposure in Auction 35.

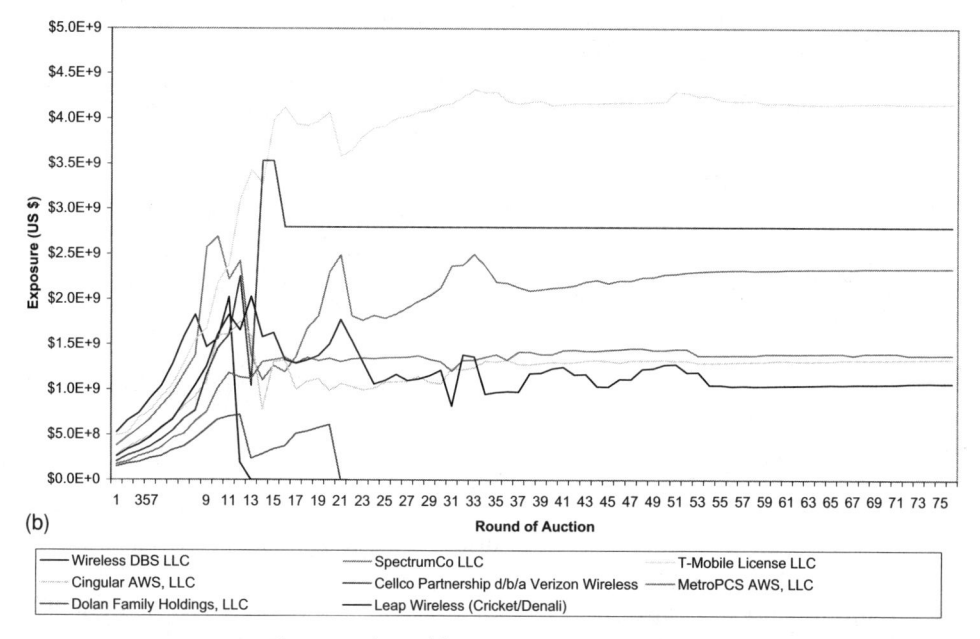

Figure 5. (b) Bidder Exposure in Auction 66.

to complement Cablevision, their New York cable franchise. When the prices became prohibitive, these bidders found no desirable substitutes upon which to spend their budgets, so they exited from the auction.

Using exposure to forecast prices would not have worked perfectly in every FCC auction. Exposure in smaller auctions sometimes has peaked well above final revenue, as one might expect with independent ascending auctions not tied together by a budget constraint. Figure 6 shows the ratio of maximum auction exposure to final auction revenue for 10 previous FCC auctions. For the larger auctions, exposure does not rise much above ten percent over final auction revenue, but there is greater variance for the

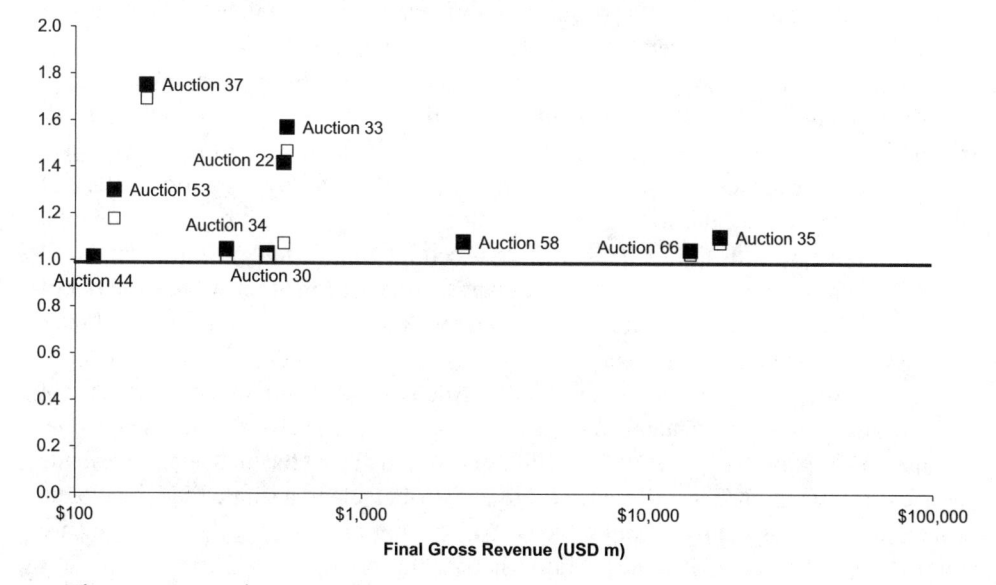

Figure 6. Ratio of Maximum Exposure to Final Gross Revenue in Major FCC Auctions.

Table 2. *Exposure and Revenue Rise in Major Auctions*

Auction	Rounds in Auction	First Round: PWB ≥ 90% of Final Revenue	First round: Exp. ≥ 90% of Final Revenue	PWB/Final Rev. when Exp. ≥ 90% of Final Revenue
22	78	21	6	0.49
30	73	32	14	0.52
33	66	29	11	0.16
34	76	36	20	0.41
35	101	48	12	0.21
37	62	27	6	0.16
44	84	33	21	0.82
53	49	19	11	0.33
58	91	37	4	0.38
66	161	36	10	0.25

smaller auctions. We note, however, that in some of these auctions exposure peaked for just one or two rounds. Figure 6 also displays, using smaller hollow squares, the same exposure to final revenue plot only using third highest round of auction exposure. Haile, Meidan and Orszag (2013) use the insight that exposure sometimes briefly spikes above final auction revenue to derive even more robust relationships between exposure during the auction and final revenue. Their analysis also includes Auction 73, a large auction that occurred after the AWS auction.

What Figure 6 does not show is that in these auctions, as in Auctions 35 and 66, auction exposure also climbed much faster than auction revenue and so provided a usefully *early* forecast of final auction revenues. We document this in Table 2 which, for each auction, reports (1) the round in which auction revenue reached 90% of its final level, (2) the round in which auction exposure reached 90% of final auction revenue, and (3) the ratio of auction revenue in that round to final auction revenue. The choice of 90% is, of course, arbitrary, but it is a reasonable choice because 10% is the smallest bid increment that the FCC used for individual licenses in the AWS auction. What Table 2 shows is that in nine of ten cases auction exposure reached 90% of final auction revenue at a point where auction revenue was under half its final level.

To see how useful such a forecast is for a new entrant planning nationwide entry, consider an entrant whose actual total budget differs from what is required for entry by more than one bid increment, that is, by more than 10%. Imagine that the entrant's decision about whether to commit or exit is guided by the following simple rule: exit the auction only when the forecasted price for a national footprint based on the total exposure price exceeds the available budget. According to the tables, in past auctions, such a rule would have recommended an inappropriate exit in at most one case out of ten and, at the time of recommended exit, prices would have been at less than half their final levels in nine cases out of ten.

The results of the recent 700 MHz auction – Auction 73 – highlight the importance of avoiding an entirely mechanical analysis of this – or any – price-forecasting technique. In auction 73, exposure peaked at $25.6 billion in round 27, but the final auction revenue was just $19.6 billion. The difference is largely attributable to a single bidder, Google, which had a provisionally winning bid of $4.7 billion on a national package license through Round 27, but exited the auction when its bid was topped by Verizon. Google's behavior had been widely discussed as a possibility even before the auction, because of its unusual role and objectives.[10]

3.2. Why Bidder Budgets?

Why do the teams representing large bidders in spectrum auctions face budget constraints? Superficially, the answer appears simple. Bidding in a spectrum auction requires a substantial amount of cash-on-hand, and raising this money from external capital markets takes time. In turn the capital markets may want to deliver money against a promised acquisition. If this pattern is optimal, then it is hardly surprising that the same pattern of capital budgeting could emerge in companies funding a division bidding in a spectrum auction.

The harder questions concern why this pattern of capital budgeting prevails and why prices vary so widely over time. In practice, incumbent firms can often substitute for additional spectrum by using existing spectrum more intensively, by building more cell sites or using other spectrum enhancing technologies. It seems unlikely that the shadow cost of spectrum fluctuates so substantially over time. Nevertheless, evidence from behavior in spectrum auctions suggests that bidding teams often face budget constraints and yet have considerable freedom in deciding which licenses to buy within their fixed budgets. Such a pattern might be rationalized if the bidding team has better information about the relative values of different licenses but also has either different incentives or different beliefs about factors like demand growth that affect the value of the entire business.[11]

4. Controlling the Pace of the Auction

By far the most dramatic moment in the AWS auction occurred in round 9, when SpectrumCo made a jump bid doubling the prices of large REAG licenses covering the Northeast and Western United States. This move, which we referred to at the time as the "shake-out tactic", was intended to resolve competitive uncertainty and favorably align relative prices in the auction. In particular, it aimed to alleviate the risk that SpectrumCo might end up purchasing the licenses across the interior U.S. but fail to purchase licenses covering the large cities on the coasts.

In this section, we explain why bidders facing exposure or budget problems almost always have an incentive to control the relative pace of price increases of different

---

[10] Google had lobbied the FCC to include an "open access" band – band C – in the auction. Under the auction rules, if the FCC-set $4.6 billion reserve were met, the winner of that band would be required to allow the operation of devices and software from independent providers (such as Google). If the reserve were not met, then the open-access provision would be removed and the licenses made available for re-auction. Google participated in the auction until the reserve was met and the open-access provision was triggered. It then immediately ceased bidding. Google's behavior highlights the important role of competitor analysis for forecasting auction prices.

[11] Our own experience in these auctions indicates that one difficulty in relying solely on net present value estimates of license values is that these kinds of estimates are extremely sensitive to assumptions about interest rates, demand growth, market share, and so on. For example, we know of a successful bidder that prior to a major FCC auction estimated the value of a Chicago area license at $30 per covered person, plus or minus $60. With such a wide range of values, a binding budget constraint may be a sensible way to focus a bidding team on relative values.

items in a simultaneous ascending auction. In the process, we identify optimal bidding patterns in a stylized model of the auction, characterize the welfare effects, and then explain the practical implications.

4.1. An Illustrative Example

Consider a bidder who is interested in acquiring two licenses, A and B. It is willing to pay 10 for the package but regards each individual license separately as worthless. It does not know the values that others place on the licenses, but thinks that the amount it will have to pay to win license A is uniformly distributed between 0 and 10, and the amount to win B is independent and uniformly distributed between 0 and 6. No package bidding is possible.

Participating in the auction is profitable in expectation — even a brute force strategy of buying both licenses regardless of the price earns an expected profit of 2 — but exactly how profitable depends on the bidding dynamics.

Suppose for example that license B sells first. If the buyer purchases B for p_B, she will certainly want to buy A, but could lose money overall if the price of A goes above $10 - p_B$. Her best strategy is to bid for B only until the price reaches 5, at which point she would make zero in expectation from winning. This strategy gives expected profit of 2.0833. It is better for the bidder if A sells first. Her optimal strategy is then to bid for A until its price reaches 7, allowing her an ex ante expected profit of 2.45.

In a simultaneous ascending auction the buyer ideally would like to see the price of A rise faster than B until she either wins a license or decides to exit. If no license clears earlier, the buyer would have the price of A reach 6 and the price of B reach 2 at exactly the same time, and then quit. This raises her expected profit to 2.533.

Intuitively, the buyer prefers that prices rise in a way that conveys as much information as possible before it must commit to buy or not. That tends to make the buyer prefer a faster increase in the price of the more uncertain license. In this case, there is initially more uncertainty about the price of license A and, so long as the other bidders remain active, the buyer's best policy is to raise the prices until they reach (6,2). At that price vector, the remaining uncertainty about both license prices is the same and winning either license A at price 6 or license B at price 2 leads to expected profits of zero.

This illustrates a general principle: on the buyer's most preferred price path, she exits at a point where she would get exactly zero expected profit from winning *either* license at its current price, given her conditional expectation about the other license price. This principle, which can be inferred from reasoning about the first-order optimality condition, also applies if the bidder needs to assemble multiple licenses, or if the licenses have some stand-alone value.[12]

---

[12] While we will focus on controlling prices to manage the exposure problem, we note that a bidder with additive license values may want to do the same if it has a budget constraint. To illustrate, consider a bidder that values license 1 at $2v$ and license 2 at v, but has a budget b with $v < b < 3v$. Ideally, this bidder would like the price of license 1 to rise to $(b + v)/2$ by the time the price on the second license reaches $(b - v)/2$. It would be happy to buy either license at a lower price, regardless of what would be required to buy the other. Once the target prices are reached, however, the buyer cannot afford both licenses, and simply wants to maintain a constant

4.2. Managing the Exposure Problem

We now consider a more general case where an entrant is willing to pay a premium for two licenses over the sum of its individual license valuations. Let the entrant's value for license i individually be v_i while the value of the package is $v_{12} > v_1 + v_2$. Suppose the entrant has a budget of b, and for simplicity that $b \geq v_{12}$. There is one competitor on each license (we will generalize this later), and they have independent unknown values c_1 and c_2, drawn from distributions $F_i(\cdot)$ with densities $f_i(\cdot)$. We will assume that these competitors bid "straightforwardly," that is, each remains active on its license until the license price exceeds its valuation.

We consider a hypothetical setting where the license prices, denoted p_1 and p_2, rise continuously and the package bidder can choose the price path as a function of the activity of the other bidders. At any price p, the package bidder knows both p and the prior bidding by the competitors; observing the latter is the same as observing $\min(c_i, p_i)$ for $i = 1, 2$. Denote this historical information by $h(p)$. A strategy for the entrant consists of a pair (P, d) where P is history-dependent price path d is a history-dependent decision rule specifying whether to exit or continue bidding on each license.

An expected-profit maximizing entrant will continue bidding on each individual license i at least until the price reaches the entrant's stand-alone value v_i and will never buy i at a price exceeding $v_{12} - v_j$. A graphical approach will help explain the decision rule d in more detail. First, suppose that the price pair (p_1, p_2) is reached and the individual bidder on license i then exits the auction. Conditional on that event, the entrant expects a profit of

$$\pi_i(p_1, p_2) = v_i - p_i + Q_j(p_1, p_2). \tag{1}$$

The final term represents the "option value" of continued bidding on license j, having purchased license i:[13]

$$Q_j(p_1, p_2) = \int_{p_j}^{\max\{b-p_i, p_j\}} \max\{0, (v_{12} - v_i - c_j)\} dF_j(c_j | c_j \geq p_j). \tag{2}$$

Suppose that the entrant selects a path of prices passing through the point (p_1, p_2) and plans to stop bidding on both licenses at that point if both competitors are still active. Increasing p_i slightly will increase the entrant's profit if $\pi_i(p_1, p_2) > 0$ and decrease it if $\pi_i(p_1, p_2) < 0$, so optimality requires that

$$\pi_1(p_1, p_2) = 0 \quad \text{and} \quad \pi_2(p_1, p_2) = 0. \tag{3}$$

Figure 7 depicts the two curves satisfying equations (3). To understand the picture, observe first that if $p_i < v_i$, then since $Q_j(p_i, p_j) \geq 0$, $\pi_i > 0$. This says that continuing to bid on license i is always profitable in this case. If $p_i = v_i$, there is no immediate payoff to winning license i, but nevertheless $\pi_i > 0$ provided that there is a chance of profiting from the addition of license j, i.e. if $p_j < v_{12} - v_i$. Similarly, it may be

---

price difference of v between the two licenses. That is, the bidder wants to keep the relative price higher on the more valuable (or larger) license.

[13] This "option value" analogy can be made precise. If the entrant's budget b is sufficiently large, then $Q_j(p_i, p_j) = \mathbb{E}[\max\{0, v_{12} - v_i - c_j\} | c_j > p_j]$, which is the value of a put option on c_j with exercise price $v_{12} - v_i$, conditional on $c_j \geq p_j$.

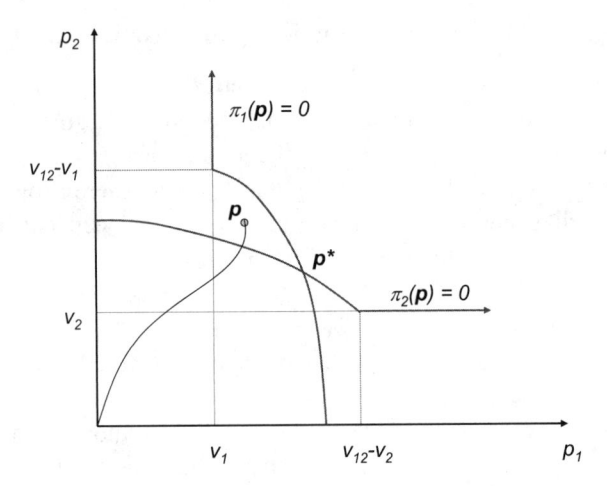

Figure 7. Managing Prices with a Single Entrant.

desirable to purchase license i even if $p_i > v_i$ provided p_j is sufficiently low. In this region, the curve satisfying $\pi_i = 0$ slopes down because π_i is strictly decreasing in both license prices.

The two curves satisfying (3) must cross at some p^* satisfying $v_i < p_i^* < v_{12} - v_j$. The reason is that if $p_i \geq v_{12} - v_j$, then $Q_j(p_i, p_j) = 0$ and so $\pi_i < 0$: buying license i could not possibly lead to positive expected profits. Also, $\pi_j > 0$ for sufficiently small p_j. In Figure 7, the crossing point is unique, which must be the case given the following condition.

(U) If (3) holds at (p_1, p_2), then $\left| \frac{\partial Q_2/\partial p_1}{\partial Q_2/\partial p_2 - 1} \right| > \left| \frac{\partial Q_1/\partial p_1 - 1}{\partial Q_1/\partial p_2} \right|$.

Condition (U) states that any crossing point, the curve defined by $\pi_1 = 0$ is strictly steeper than the curve defined by $\pi_2 = 0$. As both curves slope down, this means at most one intersection.

We can now state our first result.

Proposition 1. *Assume that (U) holds. The optimal strategies for the entrant are characterized as follows: Raise prices (along any path) to the unique price pair $p^* = (p_1^*, p_2^*)$ that solves (3); drop out at p^* if both competitors are still active at the point; otherwise, if the individual bidder for license i is the first to drop out, continue bidding on license j until $p_j = \min\{b - p_i, v_{12} - v_i\}$.*

Proof. Recall that any strategy in which the entrant plans its initial exit at a price p such that $p_j > v_{12} - v_i$ cannot be optimal. So we can restrict ourselves to strategies that involve an initial exit price vector $\hat{p}$ satisfying $p_j \leq v_{12} - v_i$ for $j = 1, 2$.

Conditional on the initial exit being planned for price vector $\hat{p}$, the path of prices leading to $\hat{p}$ is irrelevant. Why? If both individual bidders have values above $\hat{p}$, they will not exit along any path to $\hat{p}$, so the path does not affect payoffs. If one individual bidder, say i, has a value below $\hat{p}_i$, it will exit along any path to $\hat{p}$ and its exit will make it optimal for the entrant to remain bidding on j at least to $v_{12} - v_i \geq \hat{p}_j$. By the same logic, if both individual bidders have values below $\hat{p}$, the entrant will win both licenses regardless of the path toward $\hat{p}$.

Finally, we argue that a price path leading to initial exit at p^* is best. Consider a strategy with initial exit at $\hat{p} \neq p^*$, where $\pi_i < 0$ for some i. As all paths to $\hat{p}$ yield equivalent payoff, consider the path where just prior to reaching $\hat{p}$, only the price of license i is rising (such a path must exist because $\pi_i < 0$ at $\hat{p}$, then $\hat{p}_i > 0$). The entrant does better to follow this path and drop out a bit before $\hat{p}$. Next, consider a strategy with initial exit at prices $\hat{p}$ satisfying $\pi_i > 0$ and $\pi_j \geq 0$. Rather than exit at $\hat{p}$, the entrant does better to follow a continuation path in which the price rises just on i, which makes strictly higher expected profits than exiting on one or both licenses at $\hat{p}$. So $\pi_1(\hat{p}) = \pi_2(\hat{p}) = 0$ at any optimum. Q.E.D.

What does the result imply about the direction in which the entrant should push prices? A simple case is where the individual bidder valuations have a constant hazard rate, the same on both licenses. In this case, the entrant's optimal exit point satisfies: $p_1^* - v_1 = p_2^* - v_2$. That is, the bidder will manage prices to equalize the exposure risk across the two licenses.

More generally, the entrant tries to limit exposure risk in the following sense. Along an optimal price path, the entrant may purchase a license at an immediate loss (i.e. at a price $p_i > v_i$), but it never risks a purchase that leads to negative conditional expected profit. That is, along an optimal path, we always have $\pi_1, \pi_2 \geq 0$. In fact, this characterizes the optimal choice p^*. Indeed, for any strictly increasing price path that does not pass through p^*, the entrant would always want to continue bidding beyond the point where $\pi_i = 0$ for some i, and $\pi_j > 0$ for some j, risking a small expected loss if bidder i drops first in favor of a larger expected gain if bidder j drops first.

The case with multiple individual bidders on both licenses can be easily deduced from the preceding analysis. Generally, there are multiple optimal strategies for the entrant. The following proposition reports one.

Proposition 2. *Assume that (U) holds and that there is one or more individual bidders for each license and at least two for some license, with all individual values independently and continuously distributed. Then, one optimal strategy for the entrant is the following: Increase the prices for licenses 1 and 2 at any rate until a price $\hat{p}$ is reached at which there is just one other remaining bidder for each license. (1) If $\pi(\hat{p}) = (\pi_1(\hat{p}), \pi_2(\hat{p})) \geq 0$, the problem is the one analyzed in Proposition 1 and the optimal continuation is the same. (2) If $\pi(\hat{p}) \ll 0$ (negative in both components), exit. (3) If $\pi_i(\hat{p}) \leq 0$, $\pi_j(\hat{p}) > 0$, increase price p_j either until $\pi_j(\hat{p}_i, p_j) = 0$, in which case exit, or until the single remaining j competitor drops out. In the second case, raise the price of i until it reaches $\min(b - p_i, v_{12} - v_j)$.*

Proof. Fix any strategy $\sigma = (P, d)$ for the entrant. It is convenient to introduce a time structure, so let $P(t)$ be the corresponding time-and-history-dependent price path and $d(t)$ the decision path, where $d_i(t) = 0$ or 1 depending on whether the package bidder is still active or has exited by time t. We construct a new strategy $\hat{\sigma} = (\widehat{P}(t), \widehat{d}(t))$ as follows. First, the strategy increases prices through the point $\hat{p}$ while the package bidder remains active on both licenses: $\widehat{d_i}(\widehat{t}) = 0$ for $i = 1, 2$. Because values are continuously distributed, with probability one, all licenses still have active individual bidders when $\hat{p}$ is reached at some time $\widehat{t}$. The strategy $\hat{\sigma}$ continues after $\widehat{t}$ by setting $\widehat{P_i}(\widehat{t} + t) = \max(P_i(t), \hat{p}_i)$ and $\widehat{d}(\widehat{t} + t) = d(t)$. By construction, the licenses with active individual bidders are the same at prices $P(t)$ and $\widehat{P}(\widehat{t}+t)$. Also, since $\widehat{P}(\widehat{t}+t) \geq P(t)$, the decision rule $\widehat{d}$ is feasible. By construction, the entrant acquires a license at time t using σ if and only if it acquires the same license at time $\widehat{t}+t$ using $\widehat{\sigma}$. Consequently, without loss of optimality, we may limit attention to strategies like $\widehat{\sigma}$ that begin by raising take the initial prices to $\widehat{p}$.

For cases (1) and (2) of Proposition 2, the optimal continuation from $\widehat{p}$ follows from the proof of Proposition 1 (just as if $\widehat{p} = (0, 0)$). Also, optimal behavior with just one remaining competitor was analyzed in the proof of Proposition 1.

The new case is the one with $\pi_i(\widehat{p}) \leq 0, \pi_j(\widehat{p}) > 0$. Again, mimicking the proof of Proposition 1 establishes that there is some $\widetilde{p} > \widehat{p}$ such that any path from $\widehat{p}$ through $\widetilde{p}$ is optimal and all such paths yield the same expected payoff. If $\widetilde{p}_i > \widehat{p}_i$, then one such path passes through $(\widehat{p}_i, \widetilde{p}_j)$ and then continues to $\widetilde{p}$, but that path has a lower payoff than the same path stopped at $(\widehat{p}_i, \widetilde{p}_j)$. It follows that at the optimum, $\widetilde{p}_i = \widehat{p}_i$. *Q.E.D.*

Following auction 66, responding to the tactics that SpectrumCo had employed in that auction, the FCC moved to limit the ability of bidders to engage in jump bidding, which can be a key tool for controlling the pace of the auction. Was the FCC right to do that? Does bidder control of the price path as described damage efficiency?

In the preceding model, regardless of the price path, the entrant's total payment for any licenses it acquires is the sum of its competitors' values, that is, the entrant pays the social opportunity cost of any licenses it acquires. Consequently, an entrant that maximizes its own net profits necessarily maximizes the net auction surplus.

Proposition 3. *If the entrant controls the feasible path of prices to maximize its expected profits, then any effective restriction on the entrant's control reduces expected total social surplus from the auction.*

We offer several caveats to the application of Proposition 3. First, the auctioneer typically has criteria other than total social surplus, including revenues, market structure concerns, and more. Even in the context of efficiency analysis, the model itself omits two potentially important considerations.

One is illustrated by auction 66, in which there were *two* potential national entrant, SpectrumCo and Wirelss DBS, not just one as specified in our model. In general, two entrants with different values would prefer different price paths, and they cannot both maximize social surplus. Moreover, an entrant who could choose the price path to maximize its own net profit would not internalize the effect of its choice on the profits and entry decisions of the second entrant, so its choice would not generally be efficient.

Even in the model with just one entrant, we have treated individual bidders as passive automata and not as strategic players. A strategic individual bidder for license j might seek a path of prices to disadvantage the entrant. A simple example illustrates this possibility. Suppose that the entrant has a value of 2 for a pair of licenses and zero for an individual license, that the individual bidder for license 1 has a value of 1, and that the individual bidder for license 2 has an uncertain value c which is uniformly distributed on $[0, 3]$. If the price of license 2 climbs first, the entrant will bid up to 1 and will win both licenses if that is efficient. Bidder 2 will win a license only when $c > 1$ and its profit will be $c - 1$. If bidder 2 could force the price of license 1 to rise first, and force its price to rise first to 1, the entrant would find it unprofitable to bid up to the clearing price for license 1, so bidder 2 would win more often, in this example acquiring a license for all values of c and at a price of zero. Similar examples can be constructed even when there is no value complementarity between the licenses, but bidding on the two is connected by a common budget. In that way, bidding up the price of license 1 and reduce the entrant's ability to compete for license 2.

4.3. Managing Prices in Practice

A difference between our stylized model and a real auction is that in reality prices do not rise continuously, and a buyer cannot perfectly control the pace of the auction. Nevertheless, the FCC rules do give bidders at least two ways to influence pacing.

(1) *Holding back demand*: In the early rounds of FCC auctions, bidders need not bid on all the spectrum they are eligible to win. For example, early in the $7 billion "AB" auction of 1995, bidders could maintain their eligibility by making offers on just one third as much spectrum. So bidders could simply defer bidding on many target properties.

(2) *Parking*. To the extent that activity rules do require a bidder to place bids, it can "park" eligibility by bidding on non-target licenses, planning to switch later to the licenses of main interest. This tactic, too, can affect the relative rate of price increase among licenses.

(3) *Jump bidding*: Though seldom used as a strategic tool prior to the AWS auction, jump bids can similarly raise the price on one license ahead of others. A jump bid made early in the auction can be ineffective in altering the relative rates of price increase, because any changes can be undone by competitors' responses. A jump bid late in the auction entails some risk of overpaying. This makes the timing and analysis of jump bids subtle, but far from impossible.

5. The AWS Auction

The AWS auction provides a dramatic illustration of how jump bidding and the strategic considerations analyzed above can play out in practice. The auction was held in the late summer of 2006. The FCC offered for sale 90 MHz of spectrum covering all of the United States and its territories, divided into 1122 licenses. One 20 MHz layer of spectrum (the "A" band) was divided into 734 Cellular Market Area (CMA) licenses. Two other layers, one 20 MHz and the other 10 MHz (the "B" and "C" bands), were carved into 176 Economic Area (EA) licenses. Finally, three spectrum layers of 10, 10 and 20 MHz respectively (the "D", "E" and "F" bands), were divided into 12 Regional (REAG) licenses.

Traditional thinking about the exposure problem highlights large licenses as offering the most protection from exposure, so the large REAG licenses appeared to offer the easiest route to a nationwide aggregation. Of the 12 licenses covering the US and its territories, a footprint covering the contiguous U.S. required just 6 — the Northeast, Southeast, Great Lakes, Mississippi Valley, Central and West regions. The other 6 licenses in each REAG band covered Alaska, Hawaii and various U.S. territories (e.g. American Samoa), and had much lower values. Based on historical prices, the REAG licenses in different regions were not expected to settle at equal prices, either in absolute terms or on a per MHz-pop basis. For example, the price of a license covering the densely populated Northeast would normally be much higher even on a per MHz-pop basis than, say, the price of the Mississippi Valley license.

The auction attracted 168 bidders, including incumbent carriers Verizon, Cingular, T-Mobile, MetroPCS, and Leap Wireless, and the two potential national entrants: SpectrumCo (the cable consortium) and Wireless DBS (the satellite TV consortium). Bidding was expected to be fierce. Prior to the auction, SpectrumCo's economic advisors used rough budget estimates to develop three possible scenarios for final auction prices. The analysis strongly suggested that at most one of the national entrants would be able to complete a successful aggregation, and also that the entrants very possibly had quite similar resources at their disposal.

As a further complication, SpectrumCo's advisors were concerned that incumbent carriers, naturally concerned about the prospect of a nationwide entrant, might seek to purchase large amounts of the spectrum covering key markets such as New York and Los Angeles. Successful national entry without these markets was thought to be impossible, partly because the scarcity of spectrum in these markets relative to their large populations makes it too difficult to buy after-market spectrum or negotiate a roaming agreement.

Historically, most early bidding in FCC auctions focuses on the largest, most valuable licenses, and auction 66 was no exception: initial bidding centered on the REAG licenses. Within the set of these licenses, however, an unusual pattern emerged. The FCC had set essentially uniform starting prices for the licenses (with prices measured on a per MHZ-pop basis) and had set the same minimum percentage price increments for all licenses that attract high activity. With almost all bidding taking place at the minimum price in each round, prices across the REAG licenses remained uniform even as they climbed far from their starting values.[14]

As we have explained, the final market clearing prices were not likely to be nearly so uniform. If the early pattern of bidding persisted, bidding would likely close on the less valuable REAG licenses like those covering the Mississippi valley and the Mountain states long before the final prices were determined for licenses covering the northeast and west coast.

From the perspective of SpectrumCo, this timing posed a serious danger: SpectrumCo could wind up winning licenses covering the interior U.S., only to find that price of spectrum covering the coasts had become prohibitive. A further concern was that SpectrumCo and Wireless DBS could each win licenses in the interior U.S., virtually guaranteeing that at least one of them would was left with an economically untenable partial footprint. And, with minimum prices rising at 20% per round and rounds occurring every two hours, these dangers were increasingly imminent.

At this point, SpectrumCo decided to execute its pre-planned "shake-out" tactic, submitting a maximal set of jump bids on all of the Northeast and West REAG licenses[15].

---

[14] Small rural licenses had starting prices that were somewhat discounted ($0.03 per MHz-pop, as opposed to $0.05 for all the other licenses). The minimum increment rules also meant that licenses receiving different number of bids did not rise in lockstep.

[15] Because the auction rules define a mechanism with so many information sets, SpectrumCo could not pre-plan an entire strategy. Instead, we created a playbook of tactics that could be employed in certain qualitative situations. The "shake-out" tactic was one of those.

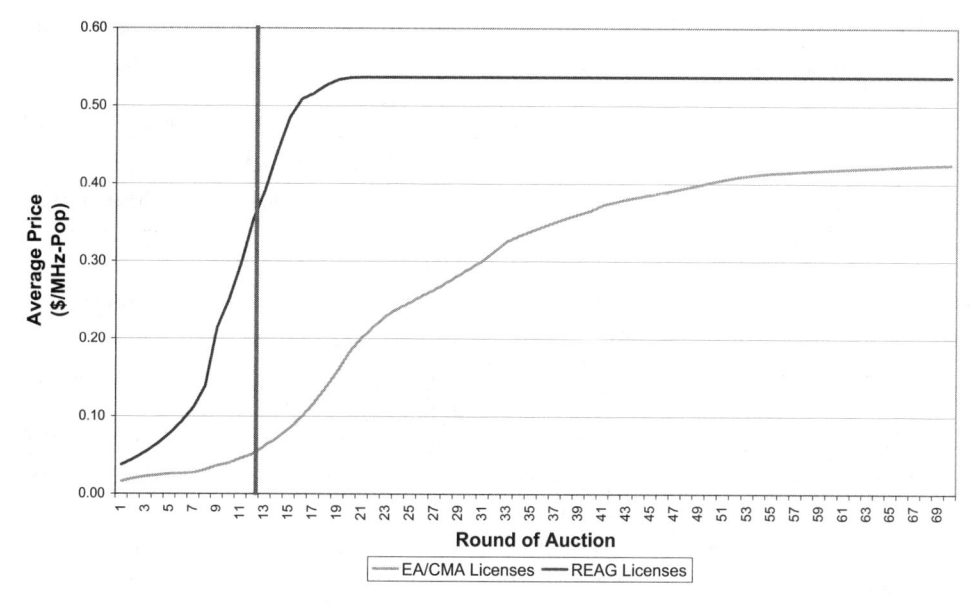

Figure 8. Price across AWS Bands.

It submitted these bids on the first round of the Monday morning after the auction started, timed to give competitors only a few hours to respond. The jump bids doubled the prices on the Northeast and West licenses (from roughly $0.20 per MHz-pop to $0.40), but SpectrumCo did not assign a large probability to these bids becoming winning. While it would have been satisfied with that outcome, the primary goal was to resolve as much uncertainty as possible and align relative prices.

The jump bids left Wireless DBS in a position of having to raise their bids and loss exposure by hundreds of millions of dollars with just two hours notice and in unexpected circumstances. Moreover, the shift in relative prices may not have favored their business model, as the satellite television subscriber base skews more toward rural areas than that of cable operators. Wireless DBS responded at first by buying time, taking its first waiver, and then by exiting the auction.[16] Meanwhile SpectrumCo's bids were topped and the REAG prices climbed still higher.

SpectrumCo's attention turned next to its own key decisions about whether to remain active and how to acquire its desired footprint at the lowest possible price. Forecasting based on budget exposure played a central role. At the end of the day on Monday (following Round 12), auction revenue was still just $6 billion but auction exposure had peaked at $14.2 billion. Meanwhile prices across the spectrum bands had diverged so remarkably that the high bids on the 40 MHz of REAG licenses, just 44% of the total spectrum in each area, accounted for over 85% of the total high bids (see Figure 8). According to the budget model, even if the pattern of bidding were to reverse immediately, the REAG licenses would wind up over-priced relative to the smaller EA and

---

[16] This sequence of events does not imply that Wireless DBS erred. It faced, on very short notice, the need to make a multibillion dollar investment decision in an unprecedented auction environment with the timing pattern of bids reversed from what it likely expected. The risks may have been magnified because the expensive licenses were not necessarily in the consortium's strongest markets. And, even if its bidding team could sort out the situation in real time, continuing to bid would have required educating and then getting approval from the senior management of the two consortium companies within just hours, certainly a daunting task.

CMA licenses. In response, SpectrumCo switched its bidding to the smaller licenses in the first round on Tuesday morning.[17]

By mid-day Wednesday, the budget model indicated that SpectrumCo's switch might be rewarded handsomely. The cumulative high bids on the 40 MHz of REAG licenses reached $7.6 billion, versus just $2.3 billion for the 50 MHz of EA and CMA licenses. Given the earlier exposure peak, the budget forecast implied that the smaller licenses were unlikely to sell for more than $6.7 billion in total, or $0.47 per MHz-pop compared to the current and still-rising price of $0.67 per MHz-pop for the REAG licenses.

It is tempting to ask why other bidders, who had access to precisely the same information, did not identify the same opportunity. Is this evidence of "irrationality" in the bidding? Only in the sense that chess grandmasters are irrational because they change their play over time. What the pattern of bidding highlights is that the auction game is similarly complicated, and that the incumbent bidders who dominated the auction, not facing the same challenging entry decision as SpectrumCo, may have devoted less resources to forecasting final prices early in the auction.

While SpectrumCo correctly anticipated lower average prices on the smaller EA licenses, it faced another problem in completing a successful aggregation. While it was absolutely essential to win the major markets, it was less crucial to win smaller markets or sparsely populated areas. Again, SpectrumCo faced an exposure problem, this time the problem of putting together, within its budget, enough of the 176 EA licenses to have a meaningful national footprint.

To minimize risk, and again control the pace of price changes, SpectrumCo initially bid for 30 MHz in the major areas, eschewing less valuable licenses. A simple calculation at the time showed that if the EA and CMA licenses sold for a total of $6.5 billion, and SpectrumCo could likely acquire 20 MHz nationwide, or 40% of the smaller licenses, for roughly $2.6 billion.[18] Anticipating this eventuality, SpectrumCo adopted a "steadfast" strategy, strongly defending the 20 MHz B band licenses, while slowly allowing itself to be bid off the 10 MHz C band licenses and meanwhile using the freed up eligibility to fill in the non-major market B band licenses.[19]

This strategy led to some interesting late auction decisions as the budget forecast was borne out. One implication of the budget hypothesis, in its strictest form, is that a bidder that reduces its bidding by a dollar will reduce total prices by a dollar. SpectrumCo was looking to purchase roughly 40% of the EA and CMA spectrum, so when its high bid on a license such as Lexington, KY was topped, conceding the license promised to save roughly 40% of the bid amount in spending on other licenses. At the same time, there was a strategic reputational issue. A show of weakness might encourage another round of bidding on a SpectrumCo license such as New York. With the large bid increments, such a development could easily end up costing $50 million dollars.

---

[17] The switch was not totally committing, in the sense that it would have been possible, and at least early on not terribly difficult to switch back into the REAG band. Nevertheless, there was a strong view at the time that the EA licenses offered the best route to success.

[18] While the exact size of SpectrumCo's budget cannot be disclosed directly (and indeed was not known precisely by its outside advisors at this point in the auction), the exposure plot in Figure 5(b) shows its bidding topping out several times in the vicinity of $2.5 billion.

[19] SpectrumCo ultimately purchased just one C block license, covering Dallas.

Balancing these considerations, SpectrumCo maintained its steadfast strategy, almost without exception, to the end of the auction.[20]

When the auction finally ended, the REAG licenses had sold for an average of $.705 per MHz-pop. The C licenses sold for $0.548. The 734 A licenses sold for $0.417.[21] The B licenses, on which Spectrumco's bids represented over 95 percent of the money spent, went for $0.451. SpectrumCo acquired 20 MHz of spectrum covering virtually the entire country for $2.378 billion. By contrast, the two largest bidders in the auction, Verizon and T-Mobile, between them acquired 40 MHz for $6.99 billion.[22] By that standard, SpectrumCo saved more than $1.1 billion.

6. Conclusion

Keynes (1936) famously concluded in the *General Theory*, "Practical men, who believe themselves to be quite exempt from any intellectual influence, are usually the slaves of some defunct economist." But in the modern economy, particularly in fields such as portfolio theory and auction theory, the time to implementation has shrunk. Ideas developed by economists have not only played a role in designing the market games that allocate many important resources, but they also provide the insights necessary to play these games at the highest level.

In the AWS auction, SpectrumCo's ability to alter the relative pace of price increases of the large licenses, combined with its ability to forecast final total prices, enabled it to take two calculated risks. Its "bookends" jump-bid strategy enabled it to discover that the cost of assembling a national footprint using major REAG licenses would likely become more than it was willing to pay. The strategy also forced Spectrumco's most direct competitor, Wireless DBS, into making billion dollar decisions with just hours of notice. SpectrumCo's ability to forecast total auction revenue gave it the confidence that it could assemble a large number of smaller licenses into a national footprint within its available budget, making bidding on these licenses a good calculated risk in spite of the exposure problem.

In the later part of the auction, SpectrumCo's strategy of steadfast bidding on its network of B licenses may have encouraged other bidders to devote most of their attention to other blocks. While tactically the other large bidders (Verizon in particular) could have taken actions to raise SpectrumCo's costs, it seemed unlikely that such tactics would be employed and they were not. The net result was a national wireless footprint at a savings of more than a billion dollar relative to other large competitors' prices. While opportunities to achieve such successes are hard to come by and the fine details

---

[20] There were only four markets priced at over $2 million in which Spectrumco did not acquire licenses — St. Louis ($23.5 million), Cincinnatti ($21.9 million), Greenville, S.C. ($5.2 million), and Lake Charles, LA ($3.6 million).

[21] The low prices overall for the A licenses mask significant differences between the large and small markets. For example, in the five top markets (New York, Chicago, Los Angeles, Washington, and Philadelphia) the B licenses were 21 percent cheaper than the A licenses (9 percent greater cost for areas covering 38 percent more people) but in the remainder of the country the A licenses were roughly 15 percent cheaper.

[22] Verizon spent $2.81 billion almost exclusively on the REAG bands, paying an average price of $.731. T-Mobile spent $4.18 billion, 70 percent on REAG licenses and half the rest on the two most valuable A licenses (New York and Chicago). A collection of cheaper small licenses reduced its average cost to $.630.

of every auction are different, this experience suggests a high value to careful economic and game theoretic analysis in complex high-stakes auctions.

References

Albano, Gian Luigi, Fabrizio Germano and Stefano Lovo, "A Comparison of Standard Multi-Unit Auctions with Synergies," *Economics Letters*, 2001, 71, pp. 55–60.

Ausubel, Lawrence and Paul Milgrom, "The Lovely But Lonely Vickrey Auction," in Cramton, Shoham and Steinberg, *Combinatorial Auctions*, 2006.

Avery, Christopher, "Strategic Jump Bidding in English Auctions," *Review of Economic Studies*, April 1998, 65(2), pp. 185–210

Brusco, Sandro and Guiseppe Lopomo. "Collusion via Signalling in Simultaneous Ascending Bid Auctions with Heterogeneous Objects, with and without Complementarities," *Review of Economic Studies*, 2002, 69(1), pp. 1–30.

Cramton, Peter, Yoav Shoham, and Richard Steinberg, *Combinatorial Auctions*. Cambridge, MA: MIT Press, 2006.

Day, Bob and Paul Milgrom, "Core-Selecting Auctions," *International Journal of Game Theory*, 36, 2008, 393–407.

Gul, Faruk and Ennio Stacchetti, "The English Auction with Differentiated Commodities," *Journal of Economic Theory*, 2001.

Haile, Philip, Maya Meidan and Jonathan Orszag, "The Impact on Federal Revenues from Limiting Participation in the FCC 600 MHz Spectrum Auction," Compass Lexecon Working Paper, 2013.

Keynes, John Maynard, *The General Theory of Employment, Interest and Money*, New York: Harcourt, Brace and Company, 1936.

Klemperer, Paul, "What Really Matters in Auction Design," *Journal of Economic Perspectives,* 2002.

Milgrom, Paul, "Putting Auction Theory to Work: The Simultaneous Ascending Auction." *Journal of Political Economy*, 2006, 108(2), pp. 245–72.

Milgrom, Paul, *Putting Auction Theory to Work*, Cambridge: Cambridge University Press, 2004.

Milgrom, Paul, "Package Auctions and Package Exchanges," *Econometrica*, 2007, 75(4), pp. 935–966.

Salant, David, "Up in the Air: GTE's Experience in the MTA Auction for Personal Communication Services Licenses," *Journal of Economics and Management Strategy,* 1997, 6(3), pp. 549–572.

Zheng, Charles. "Jump Bidding and Overconcentration in Decentralized Simultaneous Ascending Auctions," Iowa State Working Paper, 2006.

CHAPTER 32

Up in the Air: GTE's Experience in the MTA Auction for Personal Communication Services Licenses

David J. Salant[*]

1. Introduction

In a series of auctions starting in 1994, the Federal Communications Commission (FCC) sold the rights to provide personal communications services (PCS) using the electromagnetic spectrum. The rights are defined by both wavelength and geographic coverage. The largest of these auctions, for ninety-nine 30-MHz licenses in 51 major trading areas (MTAs) in the United States and its territories (two per MTA, except one each in New York, Los Angeles, and Washington, DC) began on December 5, 1994. Thirty bidders registered and qualified for the auction and spent over $7 billion acquiring licenses. This report describes how our team (for GTE) answered the following question: Given the information we had about GTE's valuations, budgets and objectives, the valuations, budgets and objectives of rival bidders, and the rules, how should we bid to achieve the best attainable outcome?

In developing our auction strategy we relied on (1) our knowledge of the other bidders and market opportunities, (2) our understanding of GTE's senior management's objectives, and (3) the ability to adapt game theory to model the behavior of bidders in the auction. Going into the auction, GTE's senior management gave our bidding team an outline of the company's objectives. Because we had limited information about our rivals, we needed to make some guesses about their interests, valuations, and budgets, as well as about the auction's duration and the price threshold.

The FCC used a novel simultaneous, multiple-round ascending-bid SMR format for this auction (see Cramton (1995a) or McAfee and McMillan (1995) for a more detailed description). In summary, 30 firms simultaneously bid on 99 PCS licenses in a sequence of rounds. At the end of each round, the FCC posted all the bids from that round, and the bidders then had the chance to place new bids in the next round (subject to constraints

---

[*] During the auction, I was a Principal Member, Technical Staff, at GTE Laboratories and was a member of GTE's bidding team. Their support, especially that of Ed Horner and Bill Pallone, is greatly appreciated. I would also like to thank Bob Rosenthal and Rob Porter for comments on an earlier draft. I also wish to acknowledge helpful input from Carol Bjelland, Terry O'Connor, Mike Weintraub, Ron Williams, and Israel Zibman.

imposed by the activity rules). The auction rules specified that the auction would close on all 99 properties at the same time.

This was one of the few times this auction format had ever been used. Although filings at the FCC by the leading auction experts preceded the FCC's adoption of the SMR format, no one had fully analyzed the equilibria. Most of our rivals had auction experts helping them determine their strategies.[1] The expert advice should have ensured that our rivals had well-formulated plans. However, we still needed to form guesses as to what strategies our rivals would use. Our ideas were speculative, as (1) there was, and is, no reliable theory on which to predict our rivals' bidding and (2) we had limited knowledge about their objectives and resources. Moreover, there was no reason to believe that each of the other bidding teams would adhere to a single well-formulated strategy throughout the auction.

In what follows I explain how our team devised its bidding strategy using game theory. I also explain why GTE's senior management concluded that our bidding team did well in the auction. Section 2 describes GTE and the bidding problem we faced going into the auction. Section 3 outlines how we formulated our bidding strategy. In Section 4, I describe a computerized auction simulation model we developed to assist us in formulating our bidding strategy. Section 5 describes some of the key decision points we faced and how we fared. Section 6 concludes with a discussion of the value and limitations of using game theory to formulate our bidding strategy.

2. How GTE Prepared for the Auction

This section provides relevant facts about GTE, its objectives, and the information we had about rival bidders going into the auction.

2.1. GTE and Its Objectives

To understand the problem facing our team going into the auction, it is necessary to first describe GTE, its auction competitors, and how PCS licenses would complement GTE's other lines of business. The spectrum licenses for sale would allow GTE to offer PCS. Going into the auction, the concept of PCS was unclear because every bidder had its own definition. Every definition of PCS, however, included wireless voice telephone service. PCS is thus best described, although somewhat imprecisely, as "advanced cellular services." Depending on the price charged, PCS can substitute for cellular service and/or regular plain old telephone service (commonly referred to as POTS). PCS can also be provided in conjunction with POTS.

A local exchange (telephone) carrier (LEC) might want to acquire PCS licenses within its telephone service areas for the defensive purpose of protecting its franchise area as well as to expand its service offerings. In addition, many parts of a LEC's facilities can be used to reduce the cost and speed deployment of PCS. LECs were allowed to acquire PCS licenses within their franchise areas. However, cellular carriers,

---

[1] Among the experts working for various bidders were Jeremy Bulow and Barry Nalebuff for Bell Atlantic, Rob Gertner for WirelessCo, Ron Harstead for Southwestern Bell Corporation, R. Preston McAfee for Airtouch and PCS PrimeCo, Paul Milgrom and Bob Wilson for PacTel, John Riley for BellSouth, and Michael Rothkopf for USWest. GTE had Rob Porter and Bob Rosenthal advising.

being viewed as providers of virtually perfect PCS substitutes, were prohibited from bidding on PCS in areas where they had a significant cellular presence (meaning that the cellular carrier's franchise area covered more than 20 percent of the population in the MTA). However, cellular carriers could acquire MTAs in areas adjacent to their cellular holdings.

GTE is one of the largest telecommunications firms in the world, with a market value of over $30 billion. In terms of telephone subscriber numbers, GTE was the largest LEC in North America, larger than any of the seven Regional Bell Operating Companies (RBOCs) at the time of the auction. GTE had over 15 million access lines in the US. GTE is also one of the five largest cellular carriers in the US.

Another significant fact about GTE is that although it is basically as large or larger than any of the RBOCs, it does not enjoy their regional concentration. The RBOCs' telephone service areas are each concentrated in one part of the country.[2] In the MTAs where GTE provides POTS, the local RBOC usually has a significantly higher percentage of both the cellular and the local telephone businesses than does GTE. The major exceptions are the Tampa and Honolulu MTAs, where, because of its cellular holdings, GTE was not eligible to bid.

GTE's position in the Los Angeles MTA provides an example of the types of situations facing the company. GTE provides telephone service to over 20 percent of the people in this MTA, which includes Los Angeles, Santa Barbara, Orange County, and San Diego. However, PacTel, the local RBOC, serves over twice as many households in the Los Angeles MTA as does GTE. (GTE and PacTel do not compete directly for telephone service, as they serve geographically distinct markets.)

Because its cellular and telephone franchise areas do not overlap very much (approximately 20 percent of its cellular markets coincide with its LEC franchises and vice versa), GTE could bid on licenses to serve many markets where it had a significant number of telephone lines.[3] The most notable case where GTE had a significant wireline presence and was eligible to bid for a PCS license was Seattle. GTE was eligible to bid in 34 MTAs. The markets in which GTE was eligible and also had a large number of telephone access lines also included Chicago, Detroit, Cincinnati, and St. Louis. In total, GTE had at least 100,000 access lines in 10 of the MTAs in which it was eligible to bid and at least 50,000 access lines in 18 of those MTAs. In only a few MTAs, such as Tampa and Hawaii, was there a significant overlap of GTE's cellular and telephone coverage. GTE would have been eligible to bid in Dallas and Los Angeles had it not signed deals before and during the auction.[4] GTE had a strong presence in Atlanta, where its headquarters for cellular operations are located.

Thus, there were a large number of MTAs in which GTE had some interest because of a desire to obtain wireless coverage for its wireline service areas. In addition, GTE's

---

[2] NYNEX serves New York and the New England states, Bell Atlantic the mid-Atlantic states, BellSouth the southeast, SBC, south central and southwestern US, Ameritech the upper Midwest, PacTel California and Nevada, and US West the northwest. GTE has some market presence in each of these regions, although it is minimal in the area served by NYNEX.

[3] The auction rules did not permit a firm to bid on an MTA in which it also owned cellular franchises covering over 20 percent of the population.

[4] GTE swapped its Portland cellular franchise for San Diego during the auction. San Diego is in the Los Angeles MTA. Because GTE entered into a resale agreement with SBC corporation in Dallas and Houston shortly prior to the auction, GTE was ineligible to bid on the Dallas MTA.

cellular serving area covered much of the southeast. However, there were a number of holes that an additional MTA, such Atlanta, Jacksonville, or Washington, DC, could help fill in.

Prior to the auction, our team assessed the value of each license for which GTE was eligible to bid. These valuations, in part, reflected synergies between existing operations. The budget parameters determined by GTE's management were in part based on these valuations. The team felt that a few of the MTAs were unlikely to be very profitable regardless of the license cost. GTE did have some interest in acquiring licenses in virtually every major market as well as almost every MTA in which it had a significant fraction of the local telephone business. GTE's senior management had a particular interest in acquiring licenses in Seattle and Atlanta.[5]

2.2. Competitive Assessments

Prior to the auction, our team assessed the competition on a market-by-market basis. One aspect of the auction, which had a significant bearing on our bidding strategy, was the overall amount of competition and how the level of competition varied across markets. Probably the best measure of the overall level of competition is the aggregate population eligibility of all the bidders. In the PCS A- and B-block auction, there were 99 licenses whose aggregate population (pops) totaled approximately 450 million. The ability of bidders to make bids and win licenses (that is, their eligibility) as well as their bidding activity was measured in terms of pops. Initially, aggregate eligibility was approximately 1.9. This means that there were, on average, no more than two bidders per pop, or per license. Although an auction with 10 objects and 11 bidders each seeking one object could prove to be highly competitive, this did not seem to be the case in most of the markets. Only a few of the markets were strongly contested. The largest three bidders were WirelessCo (a consortium of Sprint and three of the largest cable operators: TCI, Cox, and Comcast), AT&T, and PCS PrimeCo (a consortium consisting of three Regional Bell Operating Companies: NYNEX, Bell Atlantic, and US West, plus Airtouch, the cellular carrier that was formerly part of PacTel). They were competing against each other for licenses only in Chicago, New Orleans, St. Louis, Milwaukee, Memphis, and Birmingham. Moreover, Memphis went to Powertel and Southwestern Bell, and one side of Birmingham went to Powertel. At the outset, the industry's expectation was that these three bidders had the largest budgets and the highest private-value components for their valuations. Some of the other local exchange carriers, viz., GTE, Ameritech, BellSouth, PacBell, and Southwestern Bell, could also be expected to have fairly high private-value components and significant amounts to spend. However, our team felt that the three bidders seeking national networks would include an additional component in their valuation that reflected the superadditivity from acquiring a national network.

Markets such as New York, Los Angeles, Charlotte, Dallas, Philadelphia, Boston, and Minneapolis had relatively little effective competition from the major bidders. The lack of competition in these markets was due to several factors:

---

[5] Since the close of the auction, GTE has sold off its Atlanta and Denver licenses. I am no longer with GTE. I can only guess that there has been some change in managerial direction corresponding to the turnover that has occurred in the ranks of senior management.

1. At least one of the three largest bidders was not eligible to win a license in most markets, due to its existing cellular holdings.
2. The preauction agreements, especially that of the PCS PrimeCo coalition involving Airtouch, Bell Atlantic, NYNEX, and USWest, reduced the number of bidders and limited their eligibility. The PCS PrimeCo agreement essentially reduced four bidders to one, and also reduced the number of different markets in which those firms could bid. For instance, had it not entered the coalition, NYNEX could have competed for the Los Angeles and Washington DC licenses. Other significant agreements included the one between Sprint and its three cable-television partners (TCI, Comcast, and Cox) to form WirelessCo, and the resale agreement between GTE and SBC. Other confidential negotiations during the auction probably reduced competition even further, as evidenced by the WirelessCo agreement with the Pioneer's Preference winner in Washington during the auction.
3. Most of the LECs were reluctant to bid outside their existing local exchange franchise areas. For instance, BellSouth has extensive cellular holdings, but apparently was not interested in filling in gaps or extending the footprint of its wireless network with new MTA licenses.
4. Only a few firms had the resources to develop large markets. Buildout costs could easily be two or three times the cost of acquiring licenses. Additionally, the technical and marketing personnel needed to build and roll out services would likely be scarce in the next few years. A startup, with limited staff—even a Craig McCaw—could find it difficult and very expensive to get services to market.
5. MCI, potentially one of the largest bidders, decided not to participate in the auction. A possible explanation for its absence is that the auction rules did not permit it to bid for a national license.
6. The other options available certainly affected bidders' valuations. Such options included acquiring coverage via cellular deals; participating in the upcoming auction for the smaller basic trading areas; postauction agreements across different areas, such as the resale or roaming agreements that exist for cellular; and other related activities, such as investing in video and/or broadband networks, or investing outside the United States.

Beyond this overall assessment of competition intensity, we also made assessments regarding the amount of competition in each MTA. To do so, we informally divided the eligible bidders in each MTA into larger (those likely to have relatively high valuations) and smaller (those with relatively low valuations). Thus, we thought GTE would be able to outbid most small bidders in a market but we were very uncertain about GTE's ability to outbid any large bidder in any market.

In the remainder of this section I provide a brief overview of the stronger bidders—those we thought might have valuations higher than GTE's for licenses in which GTE had a significant interest. Other bidders, such as Poka Lambro, CCI Data, and Century Communications, are not discussed in detail, because we considered them to be smaller bidders.[6]

---

[6] Among the "small" bidders, Western PCS surprised us by being willing to pay prices, in some western markets, that were close to what GTE was willing to pay. We were not greatly concerned about the other cable companies, such as Continental or Cox, where they were bidding on their own and not as part of the WirelessCo consortium. Although these cable operators were large firms, it was not clear that they would be willing to invest a large

As noted above, AT&T, WirelessCo, and PCS PrimeCo were the three biggest bidders in the auction. We expected each would be willing to outbid almost anyone else almost anywhere. This was true wherever they were eligible to bid. AT&T is the largest long-distance carrier in the US and one of the largest firms in the world. It also has some of the best technological capabilities. One strong rationale for AT&T, and also for Sprint (WirelessCo), entering the PCS market was to limit access payments to the LECs for originating and terminating long-distance calls. These fees comprise close to half of AT&T's long-distance revenues.

PCS PrimeCo was the largest consortium of local telephone companies and cellular carriers in the auction. Going into the auction, they had cellular franchises in approximately half the MTAs and telephone service in close to half the country. Sprint, the lead partner in the WirelessCo consortium, is the third largest long-distance carrier in the US. The other firms in WirelessCo—TCI, Comcast, and Cox—are three of the largest cable operators in the US. As a long-distance carrier, Sprint had a strong incentive to acquire PCS licenses to avoid access charges. Its cable television partners had substantial facilities that could aid deployment of service.

We also considered any RBOC to be a major bidder in any MTA in which it also provided local telephone service. For example, we viewed PacTel as a very strong bidder in its home MTAs, Los Angeles and San Francisco. We thought PacTel would be a much weaker bidder elsewhere, especially east of the Rockies. Similarly, we thought Ameritech would be a strong bidder in its home markets of Cleveland and Indianapolis.

In relatively few MTAs was the local RBOC eligible to bid. This is because in most MTAs the local RBOC owns one of the two cellular franchises. The most prominent exception was in California. Prior to the auction, PacTel had spun off its cellular division into a separate entity called Airtouch. In virtually every MTA in which the RBOC that provided local telephone service was eligible to bid, that RBOC won the license.[7]

Los Angeles was the only significant MTA in which GTE could compete with the local RBOC, namely PacTel. We thought it likely that PacTel would outbid us for the one available Los Angeles license.[8] Both PacTel and GTE would have an interest in acquiring this license, for many of the same reasons. One was the protection that a PCS license could provide to its wireline business. Another was the synergies between the two networks. Some of the same facilities and distribution channels could be used for both. If these synergies were roughly proportional to the population sizes of the telephone franchise areas, then PacTel would be willing to pay a lot more than GTE for the license, having more than twice as large a telephone market in the Los Angeles MTA. Therefore, despite its potential attractiveness, the Los Angeles MTA was not one of GTE's main objectives in the auction. Through a swap of its Portland cellular franchise for the San Diego franchise during the auction, GTE gave up its eligibility to win this license during the auction.

---

amount in PCS with the possible exception of one or two MTAs in which a cable operator served a significant fraction of the population, e.g., Continental in Minneapolis.

[7] Some examples include PacTel winning licenses in the Los Angeles and San Francisco MTAs, BellSouth in the Charlotte and Knoxville MTAs, Ameritech in the Cleveland and Indianapolis MTAs, and Southwestern Bell in the Memphis, Little Rock, and Tulsa MTAs.

[8] The FCC had awarded the other license as a Pioneer's Preference.

We had some concerns about a number of other bidders, especially Powertel, APT, and Alaacr. We thought American Personal Telecommunications (APT) could be aggressive and was certainly going to be opportunistic. APT was the PCS Auction name for TDS, a sizable telephone and cellular provider.[9] We thought that Powertel, an electric utility, could be a strong bidder, especially in the South. Like many others, we were uncertain about how Craig McCaw's organization, Alaacr, would bid. Craig McCaw's personal wealth meant that Alaacr could be considered a serious competitor in any market.

In short, in each market we tried to assess how strong a competitor each bidder would be. Then we made market-by-market assessments of the competition. For example, in Chicago, AT&T, PCS PrimeCo, WirelessCo, GTE, and APT were all eligible to bid. Thus, we thought this would be a very costly license to acquire. In New York, on the other hand, PCS PrimeCo and AT&T were not eligible. We thought that New York would likely sell at a lower price per pop than would Chicago. (New York sold for $17/pop, that is, the price of the license divided by the population of the MTA was $17; Chicago sold for approximately $31.50/pop for each of the two licenses). In Charlotte (the sixth-largest MTA), the only significant bidders seemed to be AT&T and BellSouth. And the license for it sold relatively cheaply (approximately $7.00/pop).

We looked at the competition for each MTA for which we might want to win a license. In many markets GTE's main competition was AT&T and WirelessCo. In addition, APT and Alaacr were perceived as potentially significant competitors, especially in Seattle.

3. Devising a Bidding Strategy

We entered the auction after forming projections about potential net present values of the different licenses, other GTE corporate objectives, a budget constraint, and rival bidders' interests. However, we really had no way of accurately assessing rival valuations and budgets. Since there is no theory about determining the optimal bid in such a situation, this was our most significant question.

GTE spent over six months prior to the auction preparing its valuations. These efforts provided a good basis for evaluating how high prices would be at any point in the auction, since these numbers provided an idea about how much other bidders might be willing to pay. Shortly before the auction started, our team made some assessment of competitors' valuations. We did this to determine where competition was likely to be light and prices correspondingly low. This preauction assessment of the likely competition for different licenses helped the team select its target markets in the early phases of the auction.

We could conjecture that our rivals would use similar approaches to assess valuation to what GTE did. Each firm's valuation consists of a common value component and a private value component. We surmised that the common value component of GTE's rivals' valuations would be approximately the same as GTE's, net the difference in

---

[9] Robert Weber was advising TDS. We thought that Weber would be able to assist TDS to develop an aggressive, well-formulated, and opportunistic bidding strategy.

forecast errors. We estimated the private values for the more significant bidders in each market, such as the IXCs, who might be seeking to avoid access charges, and other LECs.

Our next step was to examine each of the roughly thirty markets in which we could bid and assess the competition in each of the markets. As noted above, GTE's senior management gave our bidding team strict orders to win licenses in Atlanta and Seattle. We were also told how much we could spend. We were not worried that we would be unable to win one of the two Seattle licenses. The most significant other bidder in Seattle was WirelessCo. PacTel, Alaacr (Craig McCaw), and APT were also eligible to bid for the Seattle licenses. PCS PrimeCo and AT&T were not. We did not feel that any of the bidders other than WirelessCo would be willing to outbid us there. Our only concern was that PacTel or Alaacr might drive up the price beyond what we were authorized to spend. We also thought that APT would be opportunistic. WirelessCo and GTE won the two Seattle licenses. Our preauction assessment of the competition for the Seattle licenses proved to be accurate.

Our view of the situation in Atlanta was much less sanguine. AT&T and WirelessCo were both eligible. In addition, there was another potentially strong competitor, Powertel. Powertel, a well-funded electric utility, could have been willing and able to outbid us for an Atlanta license. Powertel did eventually outbid GTE in Jacksonville. It also outbid AT&T, PCS PrimeCo, and WirelessCo in Memphis, and AT&T and PCS PrimeCo in Birmingham. Thus, the situation we faced in Atlanta involved two licenses and four potentially strong competitors. GTE's senior management would not be pleased if we failed to win one of the Atlanta licenses. Senior management also wanted to limit the amount GTE spent in the auction or on any one MTA.

We made similarly detailed analyses of each of the markets in which we might bid. We divided the 51 MTAs into three broad categories: (1) markets we thought would not be very competitive, such as Seattle and Minneapolis (where we did not think there would be more strong competitors than there were licenses); (2) other markets that would be competitive only if GTE were trying to acquire a license, such as New York, Cincinnati, and Detroit (in these markets there were an equal number of licenses and strong competitors not counting GTE); (3) The most competitive markets, such as Chicago, Milwaukee, and St. Louis (the Big Three were eligible, and we thought competition would be strong in those markets even without GTE).

Since the format of the auction was novel, there was little in the literature to provide guidance as to the outcome. The possible synergies and the budget constraints facing many of the bidders made it even more difficult to predict the outcome.[10] An efficient auction format would be expected to result in an allocation where licenses in each market would be won by the firms having the highest valuations.[11] The expected price would be approximately the highest losing bidder's valuation. However, because of the possible synergies and the fact that bidders faced budget constraints, the SMR format would not necessarily result in such an outcome. The budget constraints seemed

---

[10] Pitchik and Schotter (1988) and Gale and Stegeman (1995) have studied auctions in which bidders can be budget-constrained. However, the format of those auctions bears little resemblance to the PCS auction.

[11] As Cramton (1995a) pointed out, bidders could withhold demand to depress price. In such a case it would not be expected that each license would be won by the bidders having the highest valuations, and at a price determined by the highest valuation among the losing bidders.

to be significant during the auction. Additionally, we could not ignore the fact that firms had private information and could possibly enter bids as a means of sending signals.

Going into the auction, one synergy that we felt might be important was the advantage a firm could gain by obtaining a national license. Three firms were seeking to acquire national footprints, AT&T, PCS PrimeCo, and WirelessCo. Given their preauction holdings of cellular licenses, it was logically possible that all three could assemble an almost complete national license; there were only a few MTAs in which at least one of the three would necessarily have a hole, most notably Chicago, St. Louis, and Milwaukee. Aside from the efforts of the Big Three to put together national licenses, most markets were large enough so that there were few synergies across PCS licenses about which the team needed to be concerned.[12]

On the other hand, synergies between PCS and other existing operations were significant. GTE, in part, adjusted its valuations to account for possible synergies with its existing LEC facilities and market presence. Given the activity of RBOCs in their own franchise areas, it seems likely that they too felt there were potential synergies between PCS and their traditional service offerings. In addition, the two longdistance carriers in the auction, AT&T and WirelessCo (Sprint), had potential synergies between their long-distance businesses and PCS.

Synergies across MTAs can create an exposure problem. A bidder might end up having to bid beyond its standalone value on two or more licenses in order to stand a chance at winning the combination; this exposes the bidder to the risk that it will end up only winning some, and not all, of the licenses in the combination.[13] The combination of budget constraints and the activity rules can create a similar exposure problem. A firm might maintain activity on more licenses than it wants to win. Failing to do so might deprive it of the flexibility it might need late in the auction (should it wish, for instance, to switch to a larger market because the price in a smaller market had increased).

Two additional aspects of the auction made devising a bidding strategy difficult:

1. Even in an auction in which every bidder knows every other bidder's valuations and budgets, there is a coordination problem. Given the multiround format of the auction and the budget constraints, the outcome can be very path-dependent.
2. The fact that there is imperfect information meant that bidders might wish to influence rivals' beliefs by signaling. The auction format provided bidders with ample opportunity to send signals, but did not provide a means by which any signal could have an unambiguous interpretation. For instance, a large hike in a bid could indicate a strong interest in a license or an attempt to scare away rivals.

---

[12] One prominent exception was GTE's valuation of Jacksonville. Late in the auction, when it appeared likely that GTE had won Atlanta, the valuation of the adjacent Jacksonville license was revised upward to reflect some synergies.

[13] Bykowsky et al. (1995) and Bykowsky and Cull (1995) discussed the exposure and threshold problems. This was also discussed by Milgrom and Wilson and McAfee in their filings on behalf of PacTel and Airtouch, respectively. An example of the problem is as follows. Suppose a bidder has standalone valuations on two MTAs, say New York and Boston, of $300 million and $150 million, respectively. That bidder might value the pair at, say, $600 million. If the bidder were to win one, say New York at $400 million, and not Boston, it would end up paying more than the standalone valuation for New York.

As I have mentioned above, GTE had a strong presence in well over half of the 34 MTAs in which it was eligible to bid. We felt that other bidders would have an extremely difficult time guessing where our interests might be strongest. We thought that Los Angeles and Seattle were the two MTAs others might think we were most interested in. Those were the two MTAs in which GTE had the largest number of access lines. Also, GTE has a strong cellular presence on the West Coast. However, GTE never entered a bid in the Los Angeles MTA.

Partly because we did not think we would intimidate any other major bidders, we thought that it would be unwise to signal our intentions. We were most concerned that other bidders might ascertain our true interests in Atlanta. They could possibly take advantage of this interest to maintain their own eligibility at no cost. However, we were also concerned that Sprint or Powertel might set aside more of what might be limited budgets for Atlanta if they knew our true interests. Thus, we thought that it would be wise to defer bidding in Atlanta until late in the auction. We hoped to wait until our competitors, mainly WirelessCo and Powertel, had committed a great deal of money to other markets, before entering bids for the Atlanta MTA.

The auction is a bit like a poker game in that bidders might be able to take advantage of information about rivals' hole cards, i.e., their objectives, valuations, and budget constraints. It is probably unwise to bid in a manner that reveals too much about one's intentions during the auction. A bidder who, for instance, knows one rival's valuation in one market can take advantage of that information by continuing to bid there, maintaining eligibility and raising the rival's costs. Thus, a bidder might wish to remain active on markets that are of secondary importance until late in the auction, and only near the end switch to bidding on primary targets. (This carries the risk that the close of the auction can be misjudged.) Additionally, where rivals' budget constraints are likely to impinge, there can be strategic advantages from doing so, in that a bidder can make rivals spend more on some markets, leaving them with less to spend in other markets. Timing is critical. A crucial issue is predicting what other firms will do. A coy bidder can end up using all of its eligibility on licenses that are not as valuable as others that might have been available. It becomes difficult to enter new markets in stage III, unless a bidder standing high bids in one round gets topped in subsequent rounds. On the other hand, switching too early could mean jeopardizing chances at winning primary targets, overpaying, or giving up prematurely on licenses that might have been the best deals. Moreover, maintaining maximum flexibility requires that bidding on groups of properties be in a sequence dependent on market size and valuation. In particular, the auction rules meant that during stage III, backup options always had to result in smaller-sized blocks than the primary options. Our bidding team spent much time devising plans for managing eligibility in order to maintain maximum flexibility.

We followed the above principles in devising our bidding strategy. In the earlier stages of the auction, most of our bidding activity was in markets of secondary interest to GTE. Our preauction assessment and computer simulations of the auction (described below) were very useful guides in timing when GTE would stop bidding in some markets and start bidding in others. Given our strategy of deferring bidding on high-priority markets, especially Atlanta, until the later stages of the auction, it was crucial that we carefully manage the timing of when we would switch. If the auction closed before we started bidding in Atlanta and Seattle, we could end up winning lower-priority markets; we might not have the flexibility, unless we risked a substantial penalty from a bid withdrawal to win those markets. As noted above, we did not want to switch too early, fearing that we would reveal our interests and drive prices higher than they would otherwise go, and possibly beyond GTE's willingness to pay.

In devising our bidding strategy, we needed to make assessments about how high to bid and how long to bid on less attractive (to GTE) markets, before moving to more attractive markets. I cannot prove that, by not bidding in Atlanta and Seattle and concealing our preferred choices until late in the auction, we had any significant effect on the outcome. However, had we bid aggressively on those markets from the outset, others might have taken advantage of this interest to reduce our budget or maintain their own flexibility. Because of this possibility, we did extensive calculations and computer simulations that made use of information about valuations and budgets that could be assessed from auction bidding behavior. It seems that Craig McCaw (Alaacr) took advantage of PacTel's interest in San Francisco and Los Angeles by jacking up the prices in those markets and preserving his own options. We did not want this to happen to us.

We knew that our early-round bids would represent no commitment. For example, GTE's opening bid of $50 million on New York represented 78 million pops in eligibility (= 26 million × 3, where being active on 1 pop in stage I gave the bidders 3 pops of eligibility). This bid put no money at risk, as it was inconceivable that a $50 million bid for New York would stand. However, GTE's high bid in Chicago for $270 million near the end of stage II represented much more of a commitment. We still felt that $270 million was not enough to win a Chicago license. But at that point, we could not be sure. (The winning bids in Chicago in fact exceeded $370 million.)

We were also concerned with how to manage our eligibility. Eligibility management was one of the trickier aspects of the auction. Managing eligibility is especially difficult in stage III of the auction, since the set of properties that a bidder wishes to buy depends on the prices, and some options are mutually exclusive.

For instance, suppose a bidder is interested in obtaining licenses in as many markets from a list as possible, subject to an overall budget constraint and valuation limits on each of the markets. If revealing interests to rival bidders is not a concern, the optimal strategy is to first bid on all the markets on the list as long as the sum of the bids remains below the overall budget. However, once the prices rise above a level where it is possible to purchase all the licenses on the list within the overall budget, then there are tactical choices to be made. Among the options is to continue bidding on all the markets for a while, which means there is some, albeit small, chance of winning all the markets and being overcommitted. It is clearly reasonable to take the risk of being just a little bit overcommitted, but there is a question of how far to go; the answer depends largely on the degree of the firm's risk aversion and budget flexibility. Furthermore, as prices rise, eligibility reductions need to be managed to maintain the greatest flexibility. Some of the tactics that can accomplish this include bidding on highly contested markets, double bidding (that is, submitting bids on both licenses in a market at the same time), and bidding in a way that minimizes chances of having a high bid in any one round. However, senior management prefers that the bidding team appear, at least to outsiders such as financial analysts and shareholders, to have a well-thought-out strategy. This may limit the extent to which these tactics can be used.

4. Simulations

We ran computer simulations of the auction, in large part, to assess how long the auction would last and to help us decide the timing of tactical moves during the auction. Our simulation software followed the procedures specified by the FCC for the MTA auction. We ran simulations in which the number of bidders, licenses, valuations, and bidder strategies came as close as possible to what we thought would happen in the actual auction. We did this by developing programs that could simulate actual bidder behavior and calculate the outcome of a multiple-round auction based upon the simulated bidder programs and the values we entered for each bidder's budget, initial eligibility, and valuations. The simulation programs specified the licenses and the amount bidders would bid for the properties during each round. In addition, we added small random jumps to the amounts bid.[14]

We imposed a number of restrictions on simulated bidders that did not exist in the actual auction. We did not allow bidders to bid simultaneously on both the A and B licenses in an MTA. Also, the simulations did not allow bidders to withdraw bids. Despite these restrictions, the simulation fairly accurately captured the bidding behavior in the real auction. Our simulations employed a number of different bidding strategies, that are described as follows:

- *as is.* The list is used in the order given by the bidder's specification.
- *percent return.* Consider licenses with the largest percent return first:

$$\text{percent return} = \frac{\text{valuation} - \text{bid price}}{\text{valuation}}.$$

- *net return.* Consider license with the largest net return first:

$$\text{net return} = \text{valuation} - \text{bid price}.$$

- *property price.* Consider the cheapest properties first.
- *importance.* Consider the most important properties first. The importance of a license is specified before the auction. This rating can define classes of importance or individual rank orderings.
- *price per pop.* Consider the properties with the least price per pop first.
- *biggest pop.* Consider the properties with the most MHz-pops first.

Each rule refers to a sorting rule that the algorithm used to generate bids. The *property price, price per pop*, and *biggest pop* rules created bids for the highest-ranking properties until the bidder fulfilled the required activity to maintain its eligibility. The other algorithms generated bids until the bidders' budgets were exhausted. The budget parameters were inputs to the simulations.

---

[14] The random component was drawn from a uniform distribution over a user-defined interval.

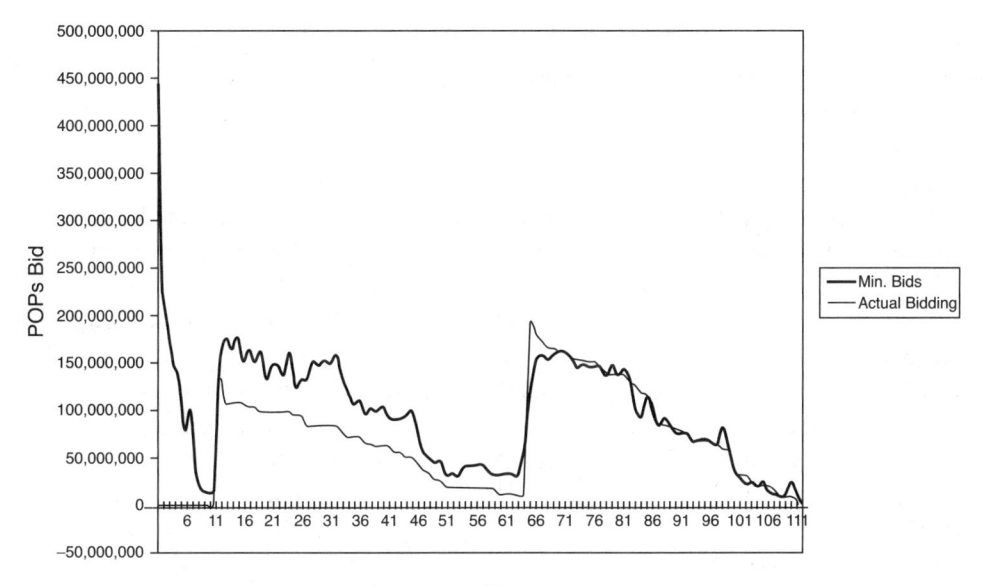

Figure 1. Bidding Activity.

In addition, we allowed simulated bidders to combine strategies. For example, we considered strategies that combined *importance* and *price per pop*; in this case, the licenses are first ranked by *importance* and then by *price per pop*. We also allowed the simulated bidders to use one set of strategies during one part of the auction and another set during another part of the auction. This allowed for the possibility that bidders might play a *snake-in-the-grass* strategy: wait until (they believed) the auction was about to end and then switch to a sincere bidding rule. The criterion the simulation program used for judging the proximity of the auction's close was based on the total license prices as a percentage of their expected valuations. Bidders using such two-phased strategies were parametrized by a cutover value, at which point the simulated bidder will begin to use its final bidding strategy.

Figure 1 plots the bids that were needed for all bidders to maintain their eligibilities versus actual bids. It demonstrates that bidding activity rarely differed much from what was required for bidders to maintain their eligibility. It also suggests that the snake-in-the-grass algorithms were a useful model of bidder behavior.

We adjusted our simulations throughout the first two stages of the auction to account for information revealed during the actual auction. License valuations, importance, and budgets of the various participants were adjusted as information presented itself. Likewise, new bidding strategies were considered as the auction progressed.

5. Revising the GTE Bidding Strategy during the Auction

In Section 3, I described GTE's basic bidding strategy: Bid to maintain eligibility, avoid sending signals, and defer bidding in Atlanta and Seattle until late in the auction. In this section, I describe the game-theoretic issues we considered prior to and during the auction, as well as the adjustments we needed to make during the auction.

5.1. Preauction Strategic Concerns

We viewed bidder valuations as consisting of private and common value components. Going into the auction, we felt that the long-distance carriers, AT&T and WirelessCo (Sprint), would have large private value components in their valuations. We also thought that the RBOCs would also have large private value components to their valuations in their home markets. Further, we felt that the private values would dwarf the forecast error of the common values. The results of the auction are consistent with this view.

The unimportance of the *winner's curse* was ironic. The winner's curse was a significant factor that led the FCC to adopt the SMR format, and it was one of the main topics discussed in the filings by leading game theorists who recommended various auction formats. A winner's curse occurs in common-value auctions for a single object when the most optimistic bidder wins but pays too much. However, the winner's curse is not nearly as relevant in an auction in which bidders have independent private values as in one in which there are common values.

The auction format did compel bidders to reveal significant information about their valuations. So, indirectly, the logic of the winner's curse was probably relevant. For instance, our observations about what other bidders were offering in other markets provided some reassurance about our own valuations. This would be the case if the factors that determine the value of one license, such as potential demand, equipment costs, and access charges paid to the LEC, were common across bidders and licenses. The lack of close competition in many markets meant that the winners need not worry a great deal about overpaying.[15]

There were a number of other concerns that the simultaneous ascending bid design helped alleviate. One potential problem was *exposure*. An exposure problem arises when a bidder enters bids on multiple markets that exceed the standalone values of those markets.[16] The bidder might do this where, owing to synergies, the value of the package exceeds the sum of the parts. If the bidder then is outbid on some pieces, but not all, then that bidder will pay more than was intended for the markets won. We were concerned that GTE, or its rivals, could encounter this problem during the auction.

The auction format did allow bidders to first bid on the larger markets for which synergies existed, and then fill in the gaps. For example, a bidder seeking to win New York and Boston licenses could remain active in New York and bid later in Boston. Although there was also some risk that a high bid on the larger MTAs would be upset very late in the auction, the eligibility rules meant this risk was small. Our team based its bidding strategy on the assumption that the larger markets would effectively close before the smaller ones.

We were very concerned about how budget constraints could affect bidding. Most of the theoretical literature ignores budget constraints.[17] In the MTA auction, budget

---

[15] For example, in a common-value auction with two identical licenses and three bidders who have unbiased estimates of the value of the license, the expected value of the lowest valuation is less than the true value of the license. More specifically, suppose each bidder's forecast error is uniformly distributed on an interval $[-a, +a]$. On average the forecast of the least optimistic bidder will be the true value less $a/2$.

[16] See Bykowsky and Cull (1995).

[17] Exceptions include Gale and Stegeman (1995) and Pitchik and Schotter (1988).

constraints appeared to limit bids. Where there were a number of MTAs among which we were indifferent, and we were merely attempting to acquire those with the best ratio of value to price, the activity rules in stage III made it very difficult to manage eligibility and preserve our options. For example, if we were to stop bidding on Philadelphia and Boston and were to begin bidding on the smaller Washington and Atlanta MTAs, we would not have the eligibility to return to Philadelphia and Boston should prices make those markets more attractive. Thus, to preserve our options, we felt that it would be prudent to bid, at times, for more licenses than we really wanted to acquire and to have at stake more money than we were authorized to spend.

5.2. A Chronology of the MTA Auction as it Affected GTE

One of our main concerns was to determine how long it would be desirable to lie low by placing bids just to maintain eligibility, and when it would be in GTE's interests to be bidding on targeted markets. In the simulations we ran to help determine optimal timing, we assumed most bids would be made to maintain eligibility. The events of the first phase of the auction helped confirm these beliefs. In particular, there were essentially no bids made in stage I of the auction that could seriously be considered winning bids.

Almost all of the bids made in the first 11 rounds were close to the minimum bid levels set by the FCC. The few bids substantially above the minima, such as GTE's opening bid of $50 million in New York and subsequent jumps in New York by Alaacr and WirelessCo, could not be considered anything other than bids designed to maintain eligibility for a few rounds or to push the pace at which prices rose in those markets. This is not surprising—there was little incentive to make more than the minimal bid, other than perhaps to send a signal or to move the auction along.[18] Although there were a number of anomalous bids that could be interpreted as signals, there seemed to be little effective signaling, and there were few bids that exceeded the minimum by more than a few percent. This cautious bidding continued throughout the auction.

We conjectured that there would be an increase in bidding activity when the auction entered the second stage. At that point bidders would either have to increase their bidding activity or reduce their eligibility. At the beginning of stage II, prices were sufficiently low in enough markets that there was little incentive for bidders to reduce eligibility. This is what happened when the auction entered stage II in round 12.

We also thought that bidding activity would only gradually decrease as prices rose and bidders decided to reduce their eligibility. This is what actually happened in stages II and III. By the end of stage II, we felt that any bid could conceivably be a final one, and thus, bids would likely reveal information about each firm's true intentions. This was not the case in stage I or much of stage II, where bidders had no need to reveal much about what markets they wanted to acquire. So, for instance, throughout

---

[18] With two exceptions, there seems to have been no effective signaling during the auction. And the exceptions did not seem to deter competition so much as to facilitate coordination. The cases where signaling seemed to be effective were (1) PCS PrimeCo and AT&T's apparent coordinated bidding and (2) the bid withdrawal by WirelessCo from Tampa and Houston, which apparently invited American Portable communications to bid there.

most of stage II we were still somewhat unsure about whether Alaacr or PacTel might want to acquire a Seattle license.

Once the auction entered stage III we felt more comfortable about our prospects in Seattle. We thought it was best to begin bidding in Atlanta once the auction entered stage III. At that time, two large questions remained: (1) would GTE be able to win an Atlanta license without violating the budget limitations imposed by senior management, and (2) what licenses, besides Seattle and Atlanta, would we be able to win? We had a wish list of about a dozen licenses, and a discretionary budget. We hoped to win as many of those as possible given our budget constraints.

It became difficult for us to manage eligibility in stage III. In stage III, the auction essentially reduces to a set of sequential auctions for each bidder, who has a list of mutually exclusive options. Such bidders have to bid first on the option with the largest population, and then work down to options with smaller aggregate populations. Our team spent a great deal of effort working out possible sequences of strategies we might want to pursue as prices rose.

For instance, suppose we wanted to win one license having between 5 and 6 million pops. The four MTAs that fall in this range are Minneapolis, Tampa, Houston, and Miami. In order to maximize our chances at winning one of these MTAs, and ignoring our valuations, we would want to first bid on Minneapolis (5.9 million pops), then on Tampa (5.4 million), then on Houston (5.2 million), and last on Miami (5.1 million). It could turn out that Minneapolis sells at a better price than do any of the other MTAs, and that Miami sells at a worse price. However, the activity rules might not allow us to withdraw from Minneapolis in order to bid on Miami. It could also turn out that the firm that won Miami would prefer Minneapolis and vice versa. This would be an example of an inefficient outcome. However, we might not want or have the budget and eligibility to switch to Miami and risk losing both Miami and Minneapolis.

This was the type of situation we faced in round 96, when we were topped in Denver, Phoenix, Salt Lake City, and Jacksonville. We wanted to enter new bids in Denver and Seattle without losing the option to reenter Phoenix and Jacksonville. However, if we stopped bidding on Phoenix and were later topped in Salt Lake City, we would be unable to resume bidding on Phoenix.

5.3. The End of the Auction

GTE had three objectives during the early rounds: (1) avoid bidding on the target markets, partly to keep rivals from guessing that GTE's target markets were Atlanta and Seattle; (2) maintain eligibility; and (3) push the price up in nontarget markets, partly to avoid escalating the price in target markets and partly to induce rivals to spend more money on those markets. These efforts might not have made a difference; I cannot demonstrate that GTE would not have won the Atlanta license if we had bid more naively. Our bidding strategy was designed to improve our prospects for success as defined by criteria established by GTE's senior management.

Going into the auction, we thought that Atlanta would be heavily contested. GTE would be competing against AT&T, Powertel, and WirelessCo for the two licenses. WirelessCo spent over $2 billion. However, it apparently was concerned about having to commit to spending an additional $200 million in order to acquire an Atlanta license. One possible reason that WirelessCo ran out of money, or at least stopped bidding in

Atlanta, was that they had already committed sizable sums in other markets by the time bidding intensified in Atlanta—although the unlikely possibility that WirelessCo was not that interested in Atlanta remains. The nature of the auction meant that a bidder who could effectively make the last bids would have a strategic advantage. The better values would only become apparent at the end of the auction, and it would be advantageous to be able to only bid on the better-priced markets and not commit to anything in the higher-priced markets. To preserve the flexibility needed to enter the late bids, a bidder had to maintain both eligibility and uncommitted pops (a bidder would have *uncommitted pops* if that bidder's eligibility at the end of a round exceeded its high bids; this could only occur when its bids were getting outbid). One of our team members, Ray Zibman, coined the phrase "losing is leading" to describe this phenomenon.

The difficulty a bidder faced in trying to use a "losing is leading" strategy is that it was hard to be certain which bids will be topped in subsequent rounds. Our efforts to preserve flexibility were inherently risky. Management and shareholders had definite limits on how much risk they felt would be consistent with GTE's policy. Due, in part, to limits on how far the team felt it could go in assuming the risk of winning licenses for nontargeted markets, GTE was limited in its ability to increase WirelessCo's costs in markets where GTE was the third major competitor, such as Detroit. However, APT also significantly raised WirelessCo's cost of acquiring a number of licenses, most notably Washington and San Francisco. The competition from both APT and GTE probably kept WirelessCo from winning a few more licenses, such as Atlanta and Chicago. GTE was interested in a limited number of other markets. We had no incentive to bid on the more attractive markets until late in the auction, when prices were closer to their final values. Our aim was to bid conservatively in markets of relatively little interest in order to maintain eligibility, and gradually shift to markets in which GTE had greater interest. We had relatively little interest in many of the East Coast markets that we bid on during the early rounds of the auction. We had stronger interests in the midwest markets, such as Cincinnati and St. Louis; we started bidding in those markets relatively early in the auction. As stated above, GTE's strongest interest was to acquire Atlanta and Seattle licenses. We waited until late in the auction to begin bidding for those licenses. The tricky part of the timing was to not stay too long in secondary markets, nor to stop bidding in them too quickly.

6. A Final Assessment of How GTE Fared

In the auction, GTE paid $400 million for Atlanta, Cincinnati, Denver, and Seattle. Given the constraints it faced, GTE did about as well as could have been hoped. Of the four markets GTE acquired, Seattle and Cincinnati fit in nicely with GTE's cellular and wireline networks. At the time, GTE's senior management felt Atlanta and Denver were very attractive and high-growth markets that fit in with a regional strategy. It is hard to imagine an alternative strategy for GTE that would have resulted in the company winning many more markets, or a more attractive combination of markets, given the limits on the amount of capital that senior management felt it prudent to invest in PCS.[19]

---

[19] Postauction transactions could be, and were, used to reduce GTE's stake in the PCS industry. However, there is significant uncertainty and large transactions costs associated with such transactions.

Overall, GTE managed the eligibility and timing fairly well. Our team won the two licenses we set out to win at prices well within our limits. This is not to say that GTE got every prime market it might have wanted. However, given all the circumstances, GTE got very good deals in each of the markets, especially Atlanta, for which GTE paid $14 million less than AT&T did for the other Atlanta license.

At the end, there were probably no bids that GTE would have wished to make that had been prevented by eligibility constraints. The rules made it difficult, especially in stage III of the auction, for anyone to flip back and forth between various combinations of MTAs. GTE proceeded in a fairly logical sequence, which resulted in close to the best attainable set of MTAs at near-minimal prices. We heavily relied on *basic* game-theoretic principles in formulating our bidding strategy. However, we could not rely on standard textbook models, such as those of the winner's curse or of the signaling literature. I and GTE's bidding team differed from Cramton (1995b) view about the usefulness of signals. We could not rule out the possibility that bid withdrawals, jump bids, and bidding both sides of a market were any more than cheap talk. For example, we could not rule out the possibility that a jump bid was any more than one bidder attempting to obtain a license cheaply by scaring off rivals.

We used a simple model of the PCS auction in which bidders faced budget constraints. The most rudimentary analysis suggested that bidders had strong incentives to only submit those bids needed to maintain their eligibility until prices reached a level which reduced demand. This simple observation helped us more accurately estimate how long the auction would run. A simple model of budget-constrained bidders suggested that the outcome of the auction could be very path-dependent, as bidders ran out of money. This simple observation was behind much of our bidding strategy.

References

Bykowsky, M. and R. Cull, 1995, "Broadband PCS (MTA) Auction: An Empirical Examination," Working Paper, NTIA, October.

Bykowsky, M., R. Cull, and J. Ledyard, 1995, "Mutually Destructive Bidding: The FCC Auction Design Problem," Working Paper 916, California Institute of Technology, January.

Cramton, P., 1995a, "Money Out of Thin Air: The Nationwide Narrowband PCS Auction," *Journal of Economics and Management Strategy*.

Cramton, P., 1995b, "The PCS Spectrum Auctions: An Early Assessment," Working Paper, Department of Economics, University of Maryland.

Gale, I.L. and M. Stegeman, 1995, "Sequential Bidding for Endogenously Valued Objects," Working Paper, Antitrust Division, Department of Justice, June.

McAfee, R.P. and J. McMillan, 1995, "Analyzing the Airwaves Auction," University of Texas at Austin and University of California, San Diego, Mimeo, July.

Pitchik, C. and A. Schotter, 1988, "Perfect Equilibria in Budget-Constrained Sequential Auctions: An Experimental Study," *The Rand Journal of Economics*, 19 (3, Autumn), 363–388.

CHAPTER 33

Bidding Complexities in the Combinatorial Clock Auction*

Vitali Gretschko, Stephan Knapek, and Achim Wambach

1. Introduction

The Combinatorial Clock Auction (CCA) is an innovative auction design that has been used in many recent auctions of spectrum for telecommunication use.[1] The CCA is based on ideas from modern microeconomic theory and combines package bidding with dynamic price discovery in a two-stage design.

In the *clock phase*, bidders express their demands at increasing prices in each of the auctioned categories of spectrum lots until the indicated demand matches the available supply. In the *supplementary phase,* bidders can improve their bids from the clock phase and submit additional bids for other desired combinations of lots. To induce truthful bidding in the clock phase, bids in the supplementary phase are constrained by a cap that is based on the clock bids. To determine winnings and prices all bids of a particular bidder are treated as mutually exclusive package bids and the combination of packages that maximizes the value as expressed by the bids is the winning allocation. The prices are determined through (a variant of) the *Vickrey* (second-price) rule by calculating the opportunity cost imposed by each bidder on her competitors. In general, there exist many more additional details like reserve prices, caps, or activity rules that have to be considered when designing a CCA.[2]

While the design is quite complex, the promise of the CCA is that bidding is simple. Regulators argue that in a CCA *truthful bidding* is close to optimal independent of the bidding strategy of the competitors. Regulators claim that the CCA "allows bidders to use a simple strategy" and "allows the participants to evaluate the spectrum without [...] shadow bids."[3] If truthful bidding is indeed close to optimal independent of the

---

* We thank Nicolas Fugger for helpful comments and suggestions.
[1] The CCA was used among other auctions in the United Kingdom 2008 and 2013, Austria 2010 and 2013, Denmark 2010, Netherlands 2010, Ireland 2012, Switzerland 2012, Czech Republic 2012, Australia 2013, and Canada 2014.
[2] See Ausubel and Baranov (2014) for an excellent overview. A detailed description of the rules used in this paper can be found in Section 2.1.
[3] See, for example, Cramton (2013) and ComReg (2010). The quotes are from CTU (2011).

competitors' behavior, this would be useful for the participants. In this case there is no need for strategizing and bidders could simply quote on the packages that lead to largest profit in the clock phase. In particular, bidders could focus their resources on determining the correct valuations and would not need to worry about the preferences of the competitors or their potentially erratic or spiteful behavior.

The goal of this article is to demonstrate with clear and simple examples that truthful bidding is neither a dominant strategy nor close to optimal. Thus, bidding truthfully can significantly hurt bidders. We will provide two examples in which a bidder can either gain by overstating or understating her demand during the clock phase. In both cases, whether understating or overstating demand is preferable to truthful bidding or not crucially depends on the behavior of the competitors and thereby their preferences. More precisely, whenever her competitor incurs a (small) cost of submitting bids in the supplementary round, the bidder can gain by expanding her demand in the clock phase and deter supplementary bids of her competitor. Whenever her competitor has a preference for raising the payments of his rivals, she can gain by understating her demand in the clock phase and limit the opportunity of riskless bids aimed at increasing the price. The two examples demonstrate that the success of a bidding strategy in the CCA may crucially depend on the bidding behavior of the competitors. Moreover, there is no straightforward way to adjust the bidding strategy. Thus, the CCA cannot deliver on the promise that bidding is simple. Bidders face complex strategic decisions when preparing for the CCA.[4]

The idea that truthful bidding may be a dominant strategy is based on the opportunity cost pricing rule. This pricing rule is adapted from static Vickrey auctions in which truthful bidding is a dominant strategy. However, as our examples will demonstrate, the dynamic nature of the CCA can make bidding truthfully suboptimal. Both examples exploit the fact that the bidders can observe the bidding and outcome of the clock phase and condition their strategies on this outcome. Such a strategy is clearly not possible in a static Vickrey auction. In the first example, the competitor is deterred from submitting bids in the supplementary round after observing in the clock phase that he is not able to win a profitable package. In the second example, the raising of the cost of a particular bidder is only riskless if the spiteful bidders can observe the allocation and prices in the clock phase.

Another advertised feature of the CCA is that once the clock phase is over and the Final Cap rule is in place, it is optimal for bidders to submit bids only on a handful of packages (*truncated bidding*).[5] More precisely, if demand equals supply at the end of the clock phase the final allocation cannot change and thus no further bids are required.[6] On one hand, this is a very desirable feature for mostly three reasons. First, there is no need to calculate a precise value for all possible packages. Second, if bidders need to secure a minimal package to ensure business continuity, they value greatly to be in full control of the final allocation. Third, there is no need for bidders to reveal

---

[4] To keep the examples as simple as possible, we restrict ourselves to an environment in which bidders are fully able to observe each other bids in the clock phase. A usual implementation of the CCA does not provide such information which makes manipulation more difficult but not impossible.

[5] For example, see ComReg (2010).

[6] If demand is less then supply at the end of the clock phase, bidders can secure the final allocation by raising their bids on the final package by the value of the unsold blocks at the final round prices.

their full valuations to the regulator, as this information might be used against them in the future. On the other hand, bidders might be concerned that if the final allocation cannot change after the clock phase, they will lose flexibility. Furthermore, their rivals' strategies aimed at raising the costs become riskless if such bids have no effect on the final allocation. Thus, in some implementations of the CCA the Final Cap is replaced by a Relative Cap.[7] We provide a simple example in which if only a Relative Cap is imposed, the final allocation can change dramatically if the bidders submit truncated bids as described above. We then argue that with a Relative Cap the bidders lack the relevant information to calculate the sufficient bids to secure the final clock-round allocation. Moreover, as was demonstrated by Ausubel and Baranov (2014), the bidders would need to place bids that exceed the final clock phase bids by up to several orders of magnitude.

Overall, preparing a bidding strategy for a CCA is a difficult task. It is not sufficient to focus on the derivation of the own valuations and rely on truthful bidding. Depending on the preferences and valuations of the competing bidders different strategies become optimal. Thus, when preparing for a CCA a strong focus on the values and preferences of the competition is needed. However, such a focus implies a level of complexity that is hard to handle. A good way to address this complexity is to use simulation software that allows to quickly generate a vast variety of bidding scenarios. These simulations should not be viewed as a way to forecast outcomes but rather as a tool to facilitate the understanding of how bidding behavior could be influenced by preferences. After the derivation of bidding strategies, the overall complexity makes it difficult to educate the top-level management and the supervisory board on the insights of the project team. Mock auctions, time-lapse versions of the actual auction, can in this case be used to familiarize the decision makers with different scenarios and the potential impact of their decisions.

The paper proceeds as follows. In Section 2 we outline a simplified version of the rules and summarize the recommendations of the regulators with regard to the bidding behavior. In Section 3 we show that truthful bidding is not a dominant strategy and argue that depending on the preference of the competitors a bidder may gain from either overstating or understating her demand. In Section 4 we demonstrate how the final allocation may change if merely a Relative Cap is imposed. In Section 5 we discuss further issues in the CCA that complicate the development of a bidding strategy. In Section 6 we conclude by briefly discussing the implications of the findings for the preparation of a bidding strategy in the CCA. Throughout the article we will use simple examples and verbal explanations to illustrate our points and refer the readers to other articles for a more formal analysis of the discussed issues.

2. Auction Rules and Recommended Behavior

In what follows, we describe a simplified version of the rules of a Combinatorial Clock Auction and summarize the recommendations for bidding behavior given by regulators. Our goal in this article is to demonstrate with simple examples that deriving a bidding

---

[7] The precise description of the cap rules can be found in Section 2.1.

strategy in a CCA is not a simple task. Thus, we focus on rules that are essential for our examples and comment on further rules and resulting complexities in Section 5. Moreover, we restrict our attention to a simplified setting with at most three bidders and two categories of spectrum licenses. This approach is without loss with respect to the goals of this paper as the omitted rules do not simplify the bidder's decision but rather add another layer of complexity.

2.1. Rules of the Combinatorial Clock Auction

We consider the sale of spectrum licenses to up to three bidders: bidder A, bidder B and bidder C. The licenses are organized in up to two categories: category α and category β. Category α contains n_α identical lots and category β contains n_β identical lots.[8] The auction consist of two phases, the clock phase and the supplementary phase.

Clock Phase

The clock phase consists of multiple rounds. In each round t the auctioneer announces prices for each of the categories. That is, she announces $p^t = (p_\alpha, p_\beta)$ with p_α (p_β) denoting the price of one lot in category α (category β). After the announcement of the auctioneer, the bidders decide how many lots they desire at the announced prices in each of the categories. That is, each bidder decides on $q^t = (n_\alpha^t, n_\beta^t)$ with n_α^t (n_β^t) denoting the number of lots she desires in category α (category β).[9] Bidders are constrained by the *eligibility rule*: the overall demand in round t needs to be below or equal to the overall demand in round $t-1$ ($t-2, \ldots, 1$). That is, $n_\alpha^t + n_\beta^t \leq n_\alpha^{t-1} + n_\beta^{t-1}$. Let $N^t := n_\alpha^{t-1} + n_\beta^{t-1}$ denote the *eligibility* of the bidder in round t. The initial eligibility is the sum of all available lots, i.e., $N^1 = n_\alpha + n_\beta$. If in round t the overall demand (i.e., the sum of the demands of the three bidders) in some category exceeds the available supply in this category, the auction moves on to round $t+1$ and the auctioneer increases the price in this category. The clock phase closes as soon as the overall demand in each category is below (or equal to) the available supply in the category. Denote the final round as round f and the desired demand of a bidder in round f as q^f.

Supplementary Phase

The supplementary phase commences after the closing of the clock phase. In the supplementary phase, each bidder can make an arbitrary number of additional bids on any possible combination of lots. That is, a bid in the supplementary phase is a package $q^s = (n_\alpha^s, n_\beta^s)$ and a (maximum) price $b(q^s)$ that a bidder would pay for this package. Moreover, all packages that the bidder bids on during the clock phase are considered as supplementary bids if the bidder does not increase her bid for such a package. That is, if in round t the bidder indicated demand $q^t = (n_\alpha^t, n_\beta^t)$, her bid on this package is

---

[8] For example, in most European auctions the spectrum was organized in nationwide licenses of 2×5 MHz blocks in predefined bands. In this case, the categories correspond to spectrum bands (e.g., 800 MHz band) and the lots correspond to blocks of spectrum in this band (e.g., six 2×5 MHz blocks in the 800 MHz band).

[9] To economize on notation, we forgo to index q^t with respect to the bidder whenever it does not cause confusion.

$b(q^t) = q^t \times p^t = n_\alpha^t p_\alpha^t + n_\beta^t p_\beta^t$. The bid on the final round package $b(q^f)$ is not constrained. The bid on any other package is constrained by a *cap*. In this article we will consider two different caps:[10]

Relative Cap

If the Relative Cap is imposed, only packages with a smaller eligibility than the final clock package are constrained by inequality (2). If a bidder wants to bid on a package q^x which requires a higher eligibility than the final package, say $N_x > N_f$, then her bid is constrained by the bid she placed in the last clock round where her eligibility was sufficient to bid on q^x. To illustrate, suppose that during the clock phase in round t the bidder reduced her demand from N_x to $N_t < N_x \leq N_{t-1}$. Her bid on package q^x is then constrained by

$$b(q^x) \leq b(q^t) + (q^x - q^t) \times p^t. \quad (1)$$

Inequality (1) implies that the bid for package q^x is constrained by the bid for the package q^t plus the difference in package prices at the prices in round t. The idea of the Relative Cap is that as the bidder was not willing to bid on the larger bundle in round t, when she was still allowed to do so, she "revealed" that she was at most willing to pay this differences in prices to obtain a bundle of size N_x compared to the smaller bundle of size N_t. Note that if $N_t > N_f$, the price the bidder can bid on bundles of size N_t is constrained in a similar manner. Recursively, the maximum bid on q^x is determined by the maximum bid on the final clock-round bundle.

Relative Cap plus Final Cap

If additionally to a Relative Cap a Final Cap is imposed, the maximum bid for a package q is constrained by the bid on the final clock package:

$$b(q) \leq b(q^f) + (q - q^f) \times p^f. \quad (2)$$

Inequality (2) implies that the bid for a package q is constrained by the bid for the final clock round package plus the difference in package prices at the final round of the clock phase. The idea of the Final Cap is that it is unreasonable for a bidder to claim a higher value for a smaller package than the package she bid on in the final round, considered she could have bid on this smaller package in the final round. Moreover, for a larger package, it is unreasonable to claim a value higher than the price difference to the final package at final prices, considered she had decreased her demand at lower prices.

Winner Determination and Payments

Once all bids are collected, the auctioneer determines the combination of bids that maximizes the overall values as reflected by the bids. The sum of demands in each

---

[10] See Ausubel and Baranov (2014) for more possible cap rules.

category reflected in the winning bids has to be below or equal to the available supply. We call this allocation the winning allocation.

In the logic of a second-price auction (Vickrey auction), the price a winning bidder has to pay is equal to the lowest bid that she would have needed to bid in order to make the winning allocation the winning allocation. That is, the auctioneer determines the value of a hypothetical second-best allocation, where all bids of this particular bidder are excluded. This bidder has to pay the difference of the value of this hypothetical second-best allocation minus the value of the bids of all the other winning bidders in the final allocation, as long as this expression is larger than zero. Otherwise, she has to pay zero. Thus, each winner pays her opportunity cost. To illustrate, denote by q_A^w, q_B^w, and q_C^w the packages bidders A, B, and C receive in the winning allocation with their respective bids $b(q_A^w)$, $b(q_B^w)$, and $b(q_C^w)$. Moreover, denote by q_B^{sb} and q_C^{sb} the packages bidders B and C receive in the second-best allocation if A was excluded with their respective bids $b(q_B^{sb})$, and $b(q_C^{sb})$. The price bidder A has to pay is then

$$P_A = b(q_B^{sb}) - b(q_B^w) + b(q_C^{sb}) - b(q_C^w)$$

if P_A is larger than zero. Otherwise, she does not have to pay anything.

2.2. Recommended Behavior

Truthful Bidding

Even though the rules of the CCA are complex, even in the simplified version presented above, the promise is that bidding in the CCA becomes simple. As the rules entail many features of the Vickrey auction, a straightforward recommendation to a particular bidder would actually be to bid truthfully independent of the bidding behavior of the competitors. Bidding truthfully implies that during the clock phase the bidder should bid on an eligible combination of frequencies which maximizes her (hypothetical) profit at the given clock prices. In the supplementary phase, truthful bidding is also a sensible recommendation as the caps on the supplementary bids are such that truthful bids are possible given truthful bidding in the clock phase.

The optimality of truthful bidding would be a very desirable feature from the bidders perspective as in preparing the auctions the bidders could fully concentrate on determining the right valuations and would not need to worry about their competitors' preferences or potentially erratic behavior. Thus, we would like to stress that truthful bidding being a dominant strategy is of great importance as compared to truthful bidding merely being an equilibrium.

Truncated Bidding

Even if bidders do not have to worry about their competitors' behavior, bidding on all possible combinations of lots in the supplementary phase might still be very demanding. Thus, if a Final Cap is imposed, the following recommendation would simplify bidding significantly. If at the end of the clock phase, all lots have been allocated, i.e., in all categories demand exactly equals supply, no additional bid in the supplementary bid is required as this allocation cannot be changed by any combination of bids in the supplementary phase. If, however, at the end of the clock phase in some category the demand is strictly below the supply, the bidders who are bidding on lots in this category have to increase their bids by the values of the unsold lots at the final clock prices.[11] That the allocation cannot change in the supplementary phase not only reduces the complexity, it is also important to bidders who need to secure a minimal package to ensure business continuity after the auction.[12]

3. Truthful Bidding in the Clock Round is Not a Dominant Strategy

As stated above, the optimality of truthful bidding irrespective of the strategy of competitors' would be a desirable feature for bidders in a CCA. In this case, bidders would not have to worry about the behavior and preferences of their competitors. Unfortunately, this property of the Vickrey auction does not extend to the CCA. Janssen and Karamychev (2013) and Levin and Skrzypacz (2014) showed that truthful bidding is not dominant but merely an equilibrium (one of many possible equilibria). That is, the optimal bidding behavior depends on the bids of the other bidders.

One of the crucial differences between the static Vickrey auction and the CCA is that in the CCA bidders can condition their strategies on the outcomes of the clock phase. Whenever the allocation cannot change after the clock phase, the second-price rule implies that bidders can only change the prices paid by their competitors but do not influence their own payoffs. Thus, if bidders only care about their profits as measured by willingness to pay minus the price paid, they are indifferent among all bids in the supplementary phase. The optimal bidding in the clock phase of a profit-maximizing bidder then crucially depends on how her competitors resolve this indifference. This is an important observation as it demonstrates that even if a bidder has straightforward profit-maximizing preferences she still has to worry about the preferences of her competitors.

In what follows we discuss two examples in which either overstating or understating the demand is optimal depending on the preferences of the competitors. In each subsection we will start with the description of the setting and the outcome of truthful bidding. We then introduce a particular preference on how to resolve the indifference about bids in the supplementary phase for one of the bidders and show how the other bidders should optimally change their bids as a reaction to this preference. To keep the examples as simple as possible, we restrict ourselves to an environment in which bidders are fully able to observe each other bids in the clock phase.

---

[11] The optimality of truncated bidding with a final cap was first observed by Bichler et al. (2013).

[12] DotEcon writes on this issue on behalf of the Irish regulator ComReg (ComReg, 2010): "The problem of business continuity risks and the difficulty for an incumbent bidder in valuing the retention of spectrum to serve existing GSM customers would be better addressed by a CCA than a SBCA [Sealed Bid Combinatorial Auction] (provided the appropriate activity rule is used for the CCA). This issue was not considered in our previous report as we judged that there the probability of an incumbent not bidding to reflect a high value on business continuity was insignificant. However, incumbents have taken a contrary view about this probability in their consultation responses."

Table 1. *Bids and outcomes if both bidders bid truthfully*

Round	Price of α	Demand of A	Demand of B
1	0	2	2
...	...	2	2
5	4	2	2
6	5	2	1
7	6	2	1
8	7	2	1
9	8	0	1

Clock phase

Package of	Bid of A	Bid of B
1 lot	–	11
2 lots	–	15

Supplementary phase

	Bidder A	Bidder B
Allocation	0	2
Price	0	14

Outcome

3.1. Overstating Demand in the Clock Phase Can be Optimal if Bidding in the Supplementary Phase is Costly

In this section we provide an example in which one bidder may profit from overstating her values in the clock phase depending on the bidding behavior of the other bidder. Suppose two bidders A and B bid on two lots in one category α. The willingness to pay of bidder A for two lots is 14 and zero for a single lot. Bidder B is willing to pay up to 15 for two lots and up to 11 for one lot.

Truthful Bidding

If both bidders bid truthfully in the clock phase, bidder B reduces her demand from 2 to 1 when the price of a lot of α reaches 5. Bidder A reduces her demand from 2 to zero as soon as the clock price of α reaches 8. In the supplementary phase, bidder B increases her bid for a single lot to 11. Moreover, she raises her bid for two lots to 15. As a result, bidder B wins both lots and pays a price of 14. This is summarized in Table 1.

Bidding Cost

To illustrate that bidding truthfully is not always optimal for bidder A, we induce bidder B with a slightly different bidding behavior by assuming that for bidder B it is costly to submit bids in the supplementary phase. More precisely, the cost of submitting a bid in the supplementary phase for bidder B is arbitrary small but strictly positive. That is, whenever bidder B could strictly gain from submitting a bid in the supplementary phase, she will do so. That bidding is (slightly) costly in the supplementary phase is not an unreasonable assumption in real-life auctions. The preparation of the bid documents for the supplementary phase requires at the very least some time by an employee who could have been productively performing a different task. Moreover, all bids have to be approved by the board, which typically requires the board to meet and to go through the determination of the bid. Another way to think about the cost of submitting a bid is to consider the drawbacks from exposing the true valuation on a combination of spectrum lots. This information might at a later stage be used by the regulator or the competitors to extract rents from the bidder in question.

Table 2. *Bids and outcomes if bidder B has costs of submitting supplementary bids*

Round	Price of α	Demand of A	Demand of B
1	0	2	2
...	...	2	2
5	4	2	2
6	5	2	1
7	6	2	1
8	7	2	1
9	8	2	1
10	9	0	1

Clock phase

Package	Bid of A	Bid of B
1	–	–
2	–	–

Supplementary phase

	Bidder A	Bidder B
Allocation	2	0
Price	8	0

Outcome

Now suppose that bidder B bids truthfully in the clock phase and reduces his demand from 2 to 1 when the price of a lot of α reaches 5. Bidder A, however, deviates from truthful bidding and reduces her demand from 2 to zero when the clock price of α reaches 9 instead of 8. In this case, there exists no profitable bid in the supplementary phase for bidder B that would make him win one of the lots. As A has placed a bid of 16 on two lots in the clock phase, neither bidding 15 for two lots or bidding 11 for one lot will change the allocation in favor of B. Thus, as bidding in the supplementary phase is costly, bidder B would refrain from placing her truthful bids. It follows that bidder A wins both objects at a price of 8. This is summarized in Table 2.

Summing up, compared to truthful bidding, bidder A has profited from overstating her demand in the clock phase, thus deterring bidder B from placing additional bids in the supplementary phase. As stated above, this is due to the two-stage nature of the CCA; bidder B could only be deterred from submitting truthful bids by giving him the opportunity to observe the outcome of the clock phase.

3.2. Understating Demand in the Clock Phase Can be Optimal if Competitors Try to Raise Costs

In this section we provide an example in which a bidder may significantly profit from understating her values in the clock phase depending on the bidding behavior of the other bidders. Suppose three bidders, A, B, and C, bid on two lots in each of two categories α and β. The willingness to pay of bidder A is 15 for two lots of α, 8 for one lot of α, and zero for any number of lots of β. Bidder B values one lot of α at 8 and has a willingness to pay of zero for any other package. The willingness to pay of bidder C is 15 for two lots of β and zero for any other package.

Truthful Bidding

If all bidders bid truthfully, bidder A reduces her demand in the clock phase from 2 lots of α to 1 lot at the price of 8. Bidders B and C do not change their demand throughout the clock phase. The clock phase ends as soon as bidder A reduces her demand. In the supplementary phase bidder A submits a bid of 15 for two lots of α and bidder C submits

Table 3. *Bids and outcomes with truthful bidding*

Round	Price of α	Price of β	Demand of A	Demand of B	Demand of C
1	0	0	(2,0)	(1,0)	(0,2)
...	...	0	(2,0)	(1,0)	(0,2)
7	7	0	(2,0)	(1,0)	(0,2)
8	8	0	(1,0)	(1,0)	(0,2)

Clock phase

Package	Bid of A	Bid of B	Bid of C
(1,0)	–	–	–
(2,0)	15	–	–
(0,2)	–	–	15

Supplementary phase

	Bidder A	Bidder B	Bidder C
Allocation	(1,0)	(1,0)	(0,2)
Price	0	6	0

Outcome

a bid of 15 on two blocks of β. Bidder B does not need to submit any additional bids as her values are reflected in her clock phase bid. As a result, bidders A and B receive one lot of α each and bidder C receives both lots of β. Bidders A and C pay 0, bidder B pays 6.[13] The results are summarized in Table 3.

Raising Rivals' Costs

To illustrate that reducing demand in the clock phase may be beneficial for bidder A, we induce bidder C with a different bidding behavior by assuming that bidder C weakly prefers if bidder A has to pay more in the auction. More precisely, if bidder C has the choice between two bids that give her the same profit, she will choose the bid that increases the payment of A. That is, bidder C is not willing to sacrifice some of her profits to make bidder A pay more. However, whenever increasing the payment of A has no impact on bidder C's profits, she will increase the payment of bidder A.[14]

Suppose first, bidder A does not change her bidding behavior in the clock phase and demands two lots of α up to a price of 7 and reduces her demand to one lot of α at a price of 8. In this case, bidder C can increase the price bidder A has to pay without any risk for himself by submitting a supplementary bid of 16 for (2, 0). Thus, bidder C changes the next best alternative for the calculation of the payment of bidder A. The allocation remains unchanged as compared to truthful bidding. However, bidder A has to pay 8 instead of 0 due to the malicious bid of bidder C. By the same token, the amount that

---

[13] The second-best allocation for the other bidders without either bidder A or bidder C is the same as the resulting allocation. Thus, by the Vickrey pricing rule the prices for both bidders must be 0. The second-best allocation without bidder B would be that bidder A receives 2 lots of α at a price of 14. Thus the opportunity cost imposed by bidder B is $14 - 8 = 6$.

[14] Janssen and Karamychev (2013) provide an equilibrium analysis of bidding in the CCA with such preferences. The point we make here is much simpler: even if a bidder is only concerned with profit maximization she needs to take into account the preferences of their rivals. Marsden and Sorensen (2017) provide further examples on how a preference for good relative outcomes changes bidding in a CCA.

Table 4. *Bids and outcomes if Bidder C prefers to raise rivals' costs*

Round	Price of α	Price of β	Demand of A	Demand of B	Demand of C
1	0	0	(2,0)	(1,0)	(0,2)
...	...	0	(2,0)	(1,0)	(0,2)
7	7	0	(2,0)	(1,0)	(0,2)
8	8	0	(1,0)	(1,0)	(0,2)

Clock phase

Package	Bid of A	Bid of B	Bid of C
(1,0)	–	–	–
(2,0)	15	–	16
(0,2)	–	–	15

Supplementary phase

	Bidder A	Bidder B	Bidder C
Allocation	(1,0)	(1,0)	(0,2)
Price	8	8	0

Outcome

bidder B has to pay increases from 6 to 8. That is, bidders A and B bid the full amount of their bids. The situation is depicted in Table 4.

Given the preference of bidder C to raise her competitors' costs, bidder A benefits from reducing her demand in the first round of the clock phase followed by placing a truthful bid on one lot of α. In this case, bidder C has no maneuver to increase the bid of A without bearing the risk of winning a worthless package. Thus, bidder C refrains from a malicious bid and bids truthfully in the supplementary phase. This is depicted in Table 5.

Note that reducing the demand in the clock phase comes at a cost for bidder A: she is not able to submit a truthful bid on two lots of α. Thus, bidder A is giving up on the opportunity to outbid bidder B on the second lot of α. Overall, bidder A faces a complex trade-off that in general depends on her belief about her competitors' preferences.

Table 5. *Bids and outcomes if Bidder A anticipates the preference of Bidder C for raising rivals' costs*

Round	Price of α	Price of β	Demand of A	Demand of B	Demand of C
1	0	0	(1,0)	(1,0)	(0,2)

Clock phase

Package	Bid of A	Bid of B	Bid of C
(1,0)	8	8	–
(2,0)	8	–	–
(0,2)	–	–	15

Supplementary phase

	Bidder A	Bidder B	Bidder C
Allocation	(1,0)	(1,0)	(0,2)
Price	0	0	0

Outcome

That bidders may benefit from raising the payments of their rivals is a reasonable assumption for real-life spectrum auctions given that all bidders will compete in the aftermarket. Moreover, the management of companies as well as operative project teams involved in the preparation of the bidding are very much interested in not paying more for comparable frequencies than their competitors. One of the reasons is that management tries to limit the available budget of their competitors for necessary infrastructure investments following the auction. Another reason is that all involved management and project members carry a significant personal risk for their careers, if their own company seemingly overpaid in the auction. In fact, Ofcom (2012) decided against a Final Cap in the UK 4G auction because "[A Final Cap] allows bidders to place bids in the Supplementary Bids Round that they know cannot win but that might raise the prices other bidders pay for spectrum." Unfortunately, as we have demonstrated above, the ability to risklessly raise the cost of the competitors is not only driven by the cap regulation but rather by the dynamic nature of the CCA. Bidders could observe their competitors' demands in the clock phase and tailor their bids to the sole purpose of raising rivals' costs without running the risk of winning undesirable packages.

4. If a Relative Cap is Imposed, Truncated Bids are Not Tractable

In this section we focus on the feasibility of truncated bidding. That is, the property that to secure the final allocation in the supplementary phase it is sufficient to bid only on a handful of packages. As stated above, truncated bidding is desirable for bidders for various reasons. For example, to reduce complexity, to limit the exposure of the true valuation or to secure business continuity.[15] However, truncated bidding is only optimal if a Final Cap is imposed on the bids in the supplementary phase. Nevertheless, imposing a Final Cap has also drawbacks. For example, bidders worry that if the allocation cannot change in the supplementary phase, their rivals would be able to risklessly raise their payments.[16] However, if instead a Relative Cap is imposed, truncated bidding may not be sufficient to secure the final package. In this section we provide a simple example on how the final allocation can change in the supplementary phase if a Relative Cap is imposed. Moreover, we comment on the difficulties that bidders face if they want to secure the final allocation in the final round. A more general discussion on different cap rules beyond the final and the Relative Cap can be found in Ausubel and Baranov (2014).

Suppose three bidders A, B, and C bid on two lots in each of two categories α and β. Suppose furthermore that the clock phase consists of two rounds of bidding. Bidder A demands a package of $(2, 1)$ in either round. Bidder B demands a package of $(0, 1)$

---

[15] There are several further reasons why bidders would like to be able to reduce uncertainty about the final allocation. For example, bidders' valuations are usually based on the revenue that they can achieve in the telecommunications market after the auction. Thus, these values crucially depend on the packages that are won by the competitors. If the individual demands of the bidders are transparent in the clock phase, bidders can condition their bids on the demands of their competitors. This is clearly not possible in the supplementary phase.

[16] See Ofcom (2012).

Table 6. *The allocation can change in the supplementary phase*

Round	Price of α	Price of β	Demand of A	Demand of B	Demand of C
1	0	0	(2,1)	(0,1)	(0,2)
2	0	10	(2,1)	(0,1)	(0,0)

Clock phase

Package	Bid of A	Bid of B	Bid of C
(0,2)	–	–	20
(2,1)	–	–	20

Supplementary phase

	Bidder A	Bidder B	Bidder C
Allocation	(0,0)	(0,1)	(2,1)
Price	0	0	10

Outcome

in each round. Bidder C demands a package of (0,2) in round 1 and after the price of β rises to 10 in round 2 drops her demand to (0,0).[17]

In the supplementary phase, bidders A and B engage in truncated bidding and make no additional bid, i.e., the final bid of A is 10 for the package (2,1). B bids 10 for the package (0,1). Bidder C, however, bids 20 on the package (0,2). This is feasible as the last round when she was able to bid on this package is the final round and 20 is the price for this package at the final round prices. Moreover, bidder C hands in a bid of 20 on the package (2,1). This is also feasible as the first round when she was bidding on a bundle with eligibility smaller than three was round one. Thus, the bid she can place on (2,1) is 20, which is the bid for the package (0,2) of round one plus the difference in the value of the packages (2,1) and (0,2) at the prices of round one. This difference, however, is equal to zero. With this bids, bidder C outbids bidder A and changes the allocation of the final round. In the end, bidder C receives the package (2,1) at a price of 10, bidder B receives the package (0,1) at the price of 0 and bidder A does not receive anything. This situation is depicted in Table 6.

The example demonstrates that truncated bidding as described in Section 2.2 is not sufficient to secure the final allocation of the clock phase. This is always the case if one of the bidders in some round reduces her overall eligibility but increases her demand in at least one category and the overall price of the package. However, this does not mean that securing the final allocation from the clock phase is generally impossible. In principle, bidders could use the Relative Cap constraints as described in Section 1 to calculate the bids that are sufficient to secure the final clock phase package. However, there are two complications with this. First, this would require the exact knowledge on the eligibility of all the other bidders at the start of the auction and full transparency of all demands during the clock phase, which is not given in most implementations of the CCA. Second, the bids needed may be well above the final clock phase prices. Ausubel and Baranov (2014) calculated the sufficient bids on the final clock round package in the UK 4G auction as a percentage of the final clock phase prices. Their results are reported in Table 7.

---

[17] As in this section we are only concerned with the feasibility of truncated bidding, we will abstract from true valuations to simplify exposition.

Table 7. *Required bids in the supplementary phase to secure the final clock phase package as a percentage of the final prices in the UK 4G auction 2013*

Bidder	Sufficient bid [in % of final prices]
Vodafone	245%
Telefonica	282%
EE	697%
Niche	1092%

5. Discussion of Further Complexities

In the previous sections we have demonstrated that deciding on a bidding strategy is not as straightforward as the usual recommendations suggest. Truthful and truncated bidding may be harmful for bidders. Moreover, the optimal bidding strategy may crucially depend on the bidding strategies and thereby on the preferences of their competitors. In this section we will comment upon further complications that may arise during the preparation of a bidding strategy. These complications arise due to particular design features or due to particularities of decision making inside firms.

Threats and Coercion

In the context of spectrum auctions, it is conceivable that there are lots that a specific telecommunication company needs to acquire. For example, if the regulator is not able to freely rearrange spectrum after the auction, some of the spectrum will be organized in specific lots rather than generic lots.[18] In this case, companies owning lots that are adjacent to the lots at auction may have a significantly higher valuation for such lots than other lots in the same band and this is potentially known to other bidders. This knowledge can be leveraged by their competitors in the clock phase. Punitive bidding on such lots could be used to coerce a bidder to reduce her demand on lots in other categories that are desired by her competitors. While this is true for any dynamic format, the Vickrey pricing of the CCA may make such a strategy almost costless as it is possible to increase the prices of a competitor without influencing own payments.

Close-to-Vickrey Core Pricing

In most implementations of the CCA payments are not determined purely according to the Vickrey pricing rule. It is further required that the sum of the final payments should not be smaller than the sum of bids of an alternative set of bidders. That is, it is required that the outcome is in the core. This potentially leads to an increase in the payments of one or more winning bidders. Thus, as the own bid is an upper limit for the own payment, a bidder can now influence her own payments by her own bids. This in turn destroys the logic behind truthful bidding.[19] Even worse, Goeree and Lien (2016)

---

[18] In a CCA this would imply having more categories of lots within the same band.
[19] See Marsden and Sorensen (2017) for comprehensive examples of this effect.

demonstrate that if the outcome of the Vickrey auction is not in the core, there exists no auction with a dominant strategy equilibrium that is in the core.

CFOs Want to Have Control Over Their Expenses

Typically there is a lot at stake for bidders participating in CCAs and the corresponding valuations amount to significant investments for companies if they would have to be spent. Therefore, during the auction, bidders usually have to report to their supervisory board on the development of the auction and especially on the momentary expenses at that stage of the auction. Additionally, CFOs usually want to keep close control over their expenses and the value at stake. However, in the CCA, the Vickrey pricing rule (or even more complex the "Close-to-Vickrey core prices") makes this requirement hard to fulfill. Any computation of the actual payments would require knowledge of individual bids of competitors which is not available. Thus, in most cases, only estimates of varying reliability (and hence often very limited usefulness) can be provided. Despite the limited relevance of auction expenses for optimal bidding behavior (apart from budget limitations), bidders nevertheless tend to put huge effort in determining actual expenses or even act according to actual expenses and not clock prices.

Budget Constraints

In many cases, bidding according to valuations is limited by available resources. From a bid strategy point of view, budget limitations force bidders to change to "cheaper" packages if the absolute budget limit is reached during the auction, although another package would still be more profitable. Additional problems might occur due to a change of a budget limit during the auction, which might lead to ex-ante suboptimal bids. Note that these problems are not specific to the CCA. However, there are two problems that are CCA specific. First, bidders do not like to gamble in spectrum auctions. Thus, the budget constraint is usually interpreted such that no bid can be placed above the budget. However, as Vickrey pricing implies that bidders hardly ever need to pay their bids, bidders do not utilize significant parts of their budget. Thus, the budget constraint is more restricting than in other formats with a more direct payment rule. Second, even with a Final Cap, there might be not enough budget to secure the end-of-clock-phase allocation. Thus, to gain certainty, it may be beneficial for a bidder to reduce her demand in the clock phase even if her current preferred package is well within the budget constraints.[20]

6. Conclusion

We have demonstrated with simple examples that optimal bidding in a CCA crucially depends on the behavior and thereby preferences of the competitors. Thus, when preparing for a CCA it is not sufficient to focus on the determination of the own valuations and rely on truthful bidding. There is no straightforward way to adjust the

---

[20] See Janssen et al. (2017) and Fookes and McKenzie (2017) for a thorough analysis of the impact of budget constraints on the CAA. Moreover, Marsden and Sorensen (2017) provide some comprehensive examples on how budget constraints may affect bidding.

bidding strategy. Depending on the preferences and valuations of the competitors different strategies become optimal. For example, we demonstrated in Section 3 that both increasing the demand and decreasing the demand in the clock phase might be optimal.

Our results imply that when preparing for a CCA the project team needs to focus on the competing bidders. It should have a model in place to estimate the valuations of the competitors and use it to understand the potential final outcomes of the auction. Moreover, different scenarios should be considered with respect to the competitors' preferences that are not captured in the valuations. At this, the preference for raising rivals' costs should take a prominent role. Once there is a basic understanding of the competition, the next step would be to understand how the derived preferences might translate into bidding behavior in the auction at hand. For this it is important to understand which degrees of freedom the specific set-up gives to the competition to act according to their potential preferences.

Conditioning the strategy on the behavior of competitors is in general a very delicate task, since a CCA with many packages quickly reaches a complexity level that is difficult to handle. One way to address this challenge is to use simulations with automated bidders. Such simulations allow to generate many different bidding scenarios and to test the sensitivities of the made assumptions on the valuations and preferences.

Once the project team has understood the different scenarios they usually need to explain the insights to the top-level management and the supervisory board. The complexity of the CCA makes it difficult to achieve a sufficient understanding in the limited amount of time that is available when dealing with the top-level decision makers of a company. Mock auctions can help to facilitate this process. Mock auctions are a time-lapse version of the actual auction and can be used to generate different decision scenarios. This gives the management the opportunity to understand the workings of the CCA and the potential impact of their decisions concerning budgets and bidding strategy.

Summing up, in the CCA the optimal bidding strategy crucially depends on the behavior of the competitors and the derivation of the optimal bid strategy is not straightforward. Thus, the derivation of an optimal bidding strategy is a complex task that requires careful preparations and analysis.

References

Ausubel, L. M. and Baranov, O. V. (2014) "A practical guide to the combinatorial clock auction," *mimeo.*

Bichler, M., Shabalin, P., and Wolf, J. (2013) "Do core-selecting combinatorial clock auctions always lead to high efficiency? An experimental analysis of spectrum auction designs," *Experimental Economics*, 16, 511–545.

ComReg (2010) "Award of liberalised spectrum in the 900MHz and other bands," *ComReg Document 10/71a.*

Cramton, P. (2013) "Spectrum auction design," *Review of Industrial Organization*, 42, 161–190.

CTU (2011) "Basic principles of tender/auction for the assignment of rights to use radio frequencies in the 800 MHz, 1800 MHz and 2600 MHz bands," Ref: CTU-80 070/2011–20.

Fookes, N. and McKenzie, S. (2017) "Impact of budget constraints on the efficiency of combinatorial auctions," in *Handbook of Spectrum Auction Design*, Cambridge University Press.

Goeree, J. K. and Lien, Y. (2016) "On the impossibility of core-selecting auctions," *Theoretical Economics,* 11, 41–52.

Janssen, M. and Karamychev, V. (2013) "Gaming in combinatorial clock auctions," *mimeo*.

Janssen, M., Karamychev, V., and Kasberger, B. (2017) "Budget constraints in VCG and the CCA," in *Handbook of Spectrum Auction Design*, Cambridge University Press.

Levin, J. and Skrzypacz, A. (2014) "Are dynamic Vickrey auctions practical? Properties of the combinatorial clock auction," *mimeo*.

Marsden, R. and Sorensen, S. (2017) "The combinatorial clock auction from a bidder perspective," in *Handbook of Spectrum Auction Design*, Cambridge University Press.

Ofcom (2012) "Assesment of future mobile competition and award of 800 MHz and 2.6 GHz," http://stakeholders.ofcom.org.uk/consultations/award-800mhz-2.6ghz/statement/.

CHAPTER 34

Strategic Bidding in Combinatorial Clock Auctions – A Bidder Perspective

Richard Marsden and Soren T. Sorensen

1. Introduction

Proponents of the Combinatorial Clock Auction (CCA) for spectrum awards often argue that the format offers two important advantages over the Simultaneous Multiple Round Auction (SMRA): it eliminates aggregation risk by allowing for package bids; and it facilitates efficient outcomes through incentives for straightforward value-based bidding.[1] While the first argument is irrefutable, the second argument is often questioned by practitioners, and recently by the academic literature on auctions as well.[2]

In this paper, we address the question of whether the CCA provides incentives for straightforward bidding in the context of spectrum awards. Straightforward bidding in a CCA describes a strategy where a bidder: (a) bids in each clock round for the package with the highest surplus (intrinsic value of package less clock price of package); and (b) submits bids at full valuation for all desired packages in the supplementary round.[3] The paper builds on the authors' practical experience with designing spectrum auctions and advising bidders participating in them, spanning more than 15 years.[4]

---

[1] For example, in the abstract of his paper on the CCA, Cramton (2013) says that "the pricing rule and information policy are carefully tailored to mitigate gaming behavior." Similarly, the Irish regulator ComReg (2012a, p. 70) states that their consultancy firm DotEcon notes, "the second price rule is utilized to disincentivise gaming behavior and encourage straightforward bidding ... the second price rule largely removes any need for a bidder to consider bidding strategies of other bidders...." And Ofcom (2015, A8.122) states that "the fundamental rationale for the CCA as an auction format is that it provides incentives for straightforward bidding by bidders."

[2] See e.g. Janssen (2015); and Levin and Skrzypaczy (2014); and Chapter 33, this volume. In some other literature, the term "truthful bidding" is used as a synonym for "straightforward bidding."

[3] For a description of the structure of the CCA, including the role of the clock rounds and sealed bid supplementary round, see Ausubel et al. (2006).

[4] At NERA Economic Consulting, the authors have advised mobile operators on bid strategy in CCAs in countries around the world, including Montenegro (2016), Canada (2014 and 2015), Slovenia (2014), Slovakia (2013), Australia (2013), UK (2013), Denmark (2012), Ireland (2012 and 2017), and Switzerland (2012). In addition, Richard Marsden, in his previous role, was a lead member of the team that designed and implemented CCAs in the UK (2005–2010), Denmark (2010), and the Netherlands (2010). The authors also have extensive experience in design, implementation and strategy advice concerning other auction formats, such as the SMRA, Clock Auction, and sealed bids.

Evidence from recent spectrum awards using the CCA format suggests that bidders often do not bid straightforwardly in practice, and there are multiple explanations for this. In Section 2, we identify reasons why bidders may not follow the advice of straightforward bidding. Our list includes complexity, the impact of the core pricing principle, budget constraints, and preference for good relative outcomes. Each of these factors provides a rationale for deviating from straightforward bidding in a CCA, under certain circumstances.

Spectrum holdings play a role in determining the level and type of services that mobile operators can offer their customers; spectrum is particularly important for roll-out of next-generation LTE services, and associated service attributes such as speed, network capacity and in-building coverage. Under certain conditions, this may create so-called "strategic investment" value for operators, that is value in excess of the intrinsic value to operators which is derived from the competitive benefits of denying spectrum to rival operators. In Section 3, we discuss whether the CCA format may create stronger incentives for bidding based on strategic value relative to other auction formats.

In Section 4, we make some concluding remarks and discuss how strategic bidding incentives may be affected by possible future modifications to the CCA format. We conclude that it is typical for some or all bidders in a CCA to have strategic incentives to deviate from straightforward bidding. Such incentives may be specific to local circumstances and vary greatly from bidder to bidder. In this respect, the CCA is no different from other formats, although the incentives that apply to bidders may vary significantly from those they would face under other common spectrum auction formats, such as the SMRA or clock auction.

2. Reasons for Deviating from Straightforward Bidding in a CCA

In a number of CCAs, regulators have gone so far as to advise bidders to consider adopting a straightforward bidding approach, de facto arguing that it is a (weakly) dominant strategy and provides bidders with a simple way to overcome the complexity of the CCA.[5] The theoretical basis for such advice is the close similarity between the winner and price determination mechanisms in the CCA and the VCG mechanism (or Vickrey auction), which has the property that in simple cases bidders can do no better than submitting bids that correspond to their intrinsic valuations. However, the CCA may differ from a simple Vickrey auction in two important aspects. First, multi-band CCAs may feature large numbers of potential bid options (packages), so there may be practical constraints on a bidder's ability to value and bid straightforwardly on available packages. Second, the pricing rule in a CCA is a modified Vickrey pricing rule, which does not have the property that bidders can do no better than submitting bids that correspond to their intrinsic valuations. For these two reasons alone, straightforward bidding can

---

[5] For example, Irish regulator ComReg states that "using a CCA makes a 'strategic demand reduction' strategy redundant and that ComReg's auction format proposal encourages straightforward bidding." See Paragraph 3.48 of ComReg (2012c).

Table 1. *Overview of reasons for deviating from straightforward bidding in a CCA*

#	Reason for deviation	Effect on price level (direction)	Importance in practice
1	Complexity	Down	Usually small
2	Pricing rule	Ambiguous	Small
3	Budget constraint	Ambiguous	Can be large
4	Relative outcome	Up	Can be large

be dismissed outright as a dominant strategy in a CCA. Moreover, as we will argue below, it may also not be a good strategy if bidders face budget constraints or care about relative outcomes.

When advising bidders on how to approach bidding in a CCA, we typically start by introducing the notion of straightforward bidding as a default strategy for all bidders. As with any spectrum auction, the challenge is then to identify circumstances that may lead bidders to deviate from straightforward bidding.

Table 1 lists four reasons why bidders may deviate; in each case, we characterise them with respect to how they might affect prices relative to the market level based on intrinsic valuations (up, down or ambiguous) and how big an impact they may have on prices (large, small or ambiguous). This assessment is based on our practical experience with advising bidders in CCAs and our reasoning is explained in detail in the following sub-sections.

We deliberately exclude strategic demand reduction from this list. Although there are a number of examples of multi-band CCAs where bidders could, in theory have benefited from tacit agreement to reduce demand, our experience is that the auction format is highly effective in deterring such behaviour. We return to this point in our discussion of price driving activity associated with securing good relative outcomes in Section 2.4 below.

In the following subsections, we discuss how bidders in practice respond to each of these four reasons for deviating from straightforward bidding.

2.1. Complexity

Multi-band CCAs with many lots are complex. Depending on the lot structure, the number of bid options (packages that bidders can bid on) may vary from many hundreds up to billions.[6] This may be a challenge for bidders from a valuation perspective and for the auctioneer in terms of the speed of processing the winner and price determination algorithms. As a result, the number of bids each bidder can submit in the supplementary round is often limited to say 5000 bids or less.

In many auctions this is not a big concern. As an example, no bidder in the UK 4G auction (2013) could bid on more than 3000 packages owing to spectrum caps and competition constraints, which implied that bid options were not artificially limited in the supplementary round. On the other hand, for CCAs with regional lots, such as

---

[6] For a discussion of the complexity associated with having large numbers of bid options, see: Bichler et al. (2014).

Australia (2013), Canada (2014, 2015), and Ireland (2017), the limitation of bid options implied that bidders would in fact not have been able to bid straightforwardly. Bidders have to select packages based on some criteria other than simply bidding on everything at value; those packages for which they do not place a bid de facto receive a zero bid, which differs from actual valuations.

In theory, the possibility that bidders fail to bid on some packages owing to limitations on their number of bids means that price levels may be depressed. If bids are missing, prices can only decrease (or stay unchanged), but never increase. This issue has clearly been a concern for some regulators. For example, for the 2015 Canadian 2500 MHz auction, where the number of package bid options was exceptionally large owing to the award of lots on a regional basis, Industry Canada limited bidders to 500 supplementary bids. However, in addition to this cap, it also allowed bidders to submit so-called "OR" bids, which were bids for any combination of specified lots added to their final clock round package.[7] This measure provided a simplified language for submitting many hundreds of additional bids, although it also imposed a new level of complexity on bidders.[8]

In practice, our experience is that the limitation of bid options is typically a minor issue. The fact that the supplementary round happens after the clock phase means that bidders always have some ability to narrow their attention to a smaller set of packages which are relevant either as potential winning bids or as price-setting bids. Other potential bids that will clearly have no impact on outcome can be discarded. A well-prepared bidder is typically able to identify the supplementary bids that matter, provided aggregate demand data is available (and Auctioneers typically give bidders two or more days to prepare these bids).

2.2. Modified Vickrey Pricing Rule

The pricing rule used in all CCAs to date is a modified Vickrey pricing rule,[9] which is based on the principle of "happy winners, happy losers," i.e. the so-called core pricing principle. Under this principle, prices have the property that no bidder (or group of bidders) has offered to pay more than the prices paid by any winner (or group of winners).[10]

The following simple example illustrates how bidders may gain from deviating from straightforward bidding under such a pricing rule. Suppose there are two lots available in the auction and there are three bidders; each bidder has a valuation for winning two lots, but only two of the bidders have a valuation for winning one lot, as outlined in Table 2 below.

---

[7] Industry Canada (2014) states that "The limit on the number of different supplementary round packages that a bidder will be allowed to place will be announced after the bidder qualification has occurred, but will be no less than 500 different packages." Industry Canada later confirmed the limit of 500 bids plus up to three packages of "OR" bids.

[8] See e.g. Nisan (2006) for a discussion of bidding language in combinatorial auctions, and the issue of expressiveness vs. simplicity.

[9] Examples where a pure Vickrey pricing rule was used are assignment rounds in the Swedish 1800 MHz auction (2011) and the New Zealand 700 MHz auction (2013).

[10] The core pricing principle in the context of the CCA was first proposed by Ausubel et al. (2006).

Table 2. *Example of valuations*

	Valuation for 1 lot	Valuation for 2 lots
Bidder A	100	150
Bidder B	100	150
Bidder C	–	150

First, suppose all bidders bid their value. In the winning allocation, bidders A and B receive one lot each. This allocation generates a value of 200, while any alternative allocation (when taking out either or both of the winning bidders) has a value of 150. The Vickrey price for both winning bidders is 50, but they jointly have to pay 150 to make bidder C a happy loser, hence a "core adjustment" of 25 is added to the Vickrey price for each winning bidder.[11]

Table 3 provides a summary of the winning allocation, prices and surplus for all bidders, when they all bid straightforwardly.

Essentially, the opportunity cost imposed by bidder C is shared between the winning bidders, A and B. Our continued example below shows how the ability to shift the shared surplus from one bidder to another creates an incentive to deviate from straightforward bidding.

Suppose bidder A inflates his bid for 2 lots from 150 to 160, i.e. a deviation from straightforward bidding. The winning allocation is unchanged, but the Vickrey price for bidder B is now 60, while it remains at 50 for bidder A. As a result, the two winning bidders only have to share an additional opportunity cost of 40, imposed by the losing bidder C. As this is shared equally, bidder B enjoys a reduced payment and higher surplus as a result of her strategic bidding behaviour, as illustrated in Table 4. Of course, bidder A has similar incentives.

The effect of such behaviour on the overall price level is ambiguous. In this particular example, the auction revenue is unchanged as the winning allocation does not change, and the winning bidders jointly pay the same amount. Aggregate revenues will only be affected when bidders "get it wrong" and the attempt to shift pricing onto rival bidders fails (e.g. bidder A inflates his bid so much that he actually wins both lots). It should also be noted that shifting pricing onto rival bidders does not necessarily require inflation of bids. In the example above, bidder A would have created the same effect if he had decreased his bid for 1 lot to 90 and left his 2 lot bid unchanged. The key is that

Table 3. *All bidders bid straightforwardly*

	Winning package	Vickrey price	Core adjustment	Core price	Surplus
Bidder A	1 lot	50	25	75	25
Bidder B	1 lot	50	25	75	25
Bidder C	–	–	–	–	0
Total	2 lots	100	50	150	50

---

[11] Here we assume the core adjustment is calculated by minimising the sum of squared differences between Vickrey prices and base prices, i.e. so-called Vickrey nearest core prices.

Table 4. *Bidder A inflates bid for 2 lots*

	Winning package	Vickrey price	Core adjustment	Core price	Surplus
Bidder A	1 lot	50	20	70	30
Bidder B	1 lot	60	20	80	20
Bidder C	–	–	–	–	0
Total	2 lots	110	40	150	50

bidder A inflates the difference between his bids for 1 and 2 lots, such that bidder B faces a larger opportunity cost.

It is widely recognised that the pricing rule in CCAs creates an incentive to deviate from straightforward bidding.[12] However, in practice, we have only come across this issue as a relevant factor in bid determination in the context of a sealed-bid auction with a particularly simple lot structure that used the CCA pricing and winner determination rules. With multi-band auctions, our experience is that there are too many package options for each bidder to predict reliably how its bids may affect price outcomes where core adjustments are involved. Moreover, in large auctions, such considerations are typically second order compared to issues such as budget constraints and relative outcomes, which we turn to next.

2.3. Budget Constraints

In high-value spectrum auctions, bidders often face hard budget constraints, with the implication that they may be unable to express their true value differences between alternative packages; hence straightforward bidding for all target packages may be impossible.[13] This is often a key issue for smaller bidders, but in large multi-band auctions it may affect even large well-financed bidders.

To illustrate the point, consider a simple example where bidder A is competing against bidder B in a one-band CCA for two lots. Suppose that bidder A's valuation is 100 for 1 lot and 150 for 2 lots but she has a budget constraint of 125. By implication, she enters the process knowing that she will not be able to express her full value for the larger package, so strictly straightforward bidding is not an option.

In general, Bidder A has two different strategy options (as well as many variations within these two extremes):

- **Delta preserving strategy**: She could reduce the overall level of her bids and preserve the delta between them: bid 125 for 2 lots and 75 for 1 lot. This strategy carries the risk of winning nothing when she could have won 1 lot. However, if she does win 1 lot, this approach ensures that bidder B pays her full opportunity cost.
- **Delta reducing strategy**: She could cap her bids at the budget level, thereby reducing the delta between bids: bid 100 for 1 lot and 125 for 2 lots. This strategy maximizes her likelihood of winning 1 lot, but may mean foregoing a winnable second lot and/or having bidder B secure the same spectrum at a lower price.

---

[12] See e.g. Goeree and Lien (2016).
[13] See e.g. Jansen et al. (2015).

This dilemma is most acute in the context of a sealed-bid combinatorial auction. To an extent, the CCA can help ease the budget constraint dilemma for the bidder, owing to price discovery in the clock rounds. In our example with a CCA, the bidder could adopt a straightforward bidding strategy for 2 lots up to a price of 50, then drop back to 1 lot and bid on this package up to its valuation of 100. Suppose the clock rounds end at a price of 74 per lot. At this price, bidder A will face one of two possible situations:

- First, in case there were no unsold lots (which by implication must mean that bidder B reduced demand from 2 lots to 1 lot), bidder A has a so-called knockout bid for 1 lot at a price of 75.[14] If she submits a bid of 75 for 1 lot, she is guaranteed to win at least this lot at a price that does not exceed 75. In this case, the bidder can also express an incremental value of 50 for a second lot, i.e. bid 125 for 2 lots, which is just within her budget limit. In this case, bidder B will pay bidder A's full opportunity cost for 1 lot (of course, if the closing clock price had been higher, this would not have been possible).
- Second, in case there is one unsold lot at the end of the clock rounds (which by implication must mean that bidder B reduced demand from 2 lots to zero), the bidder faces a knockout bid of 149 for 1 lot (by adding the value of the unsold lot). Such a knockout bid exceeds the bidder's budget constraint. In this case, the bidder has little option but to submit a bid of 100 for 1 lot, so as to maximize her chances of getting at least 1 lot, and a bid of 125 for 2 lots. In this case, if bidder B does bid for and win 1 lot, she will only pay 25, much less than bidder A's true valuation differential.

The budget constraint issue is generic to second-price auctions where bidders typically pay less than their winning bid.[15] In our simple example, bidder A can reliably predict that she will not win 2 lots, so at least the clock rounds provide guidance on how best to deviate from straightforward bidding, i.e. focus on a delta-reducing strategy. However, with more complex lot structures, especially ones with multiple bands and substitutable lots across bands, bidders may not face such a clear choice. In our work for bidders on European multi-band auctions, we have often identified situations where an operator with a budget constraint faced a non-obvious choice between the delta-preserving and delta-reducing strategies.

The dilemma is particularly acute if a bidder ends the clock rounds on a large target package but cannot afford the knockout bid to guarantee that package. In this case, the bidder faces a choice between bidding as high as possible for that package and not making any back-up bids for smaller packages, or submitting back-up bids that may outcompete her preferred package. The best course of action is non-obvious, as it depends on predicted competitor bids. In either case, it involves deviation from straightforward bidding.

One way that bidders can attempt to ease uncertainty is by tracking the activity of rival bidders during the clock rounds, so as to predict their supplementary round constraints. Taking into account spectrum caps and activity rules, it may be possible to

---

[14] A knockout bid is a bid in the supplementary round that guarantees a bidder wins his/her final clock round package. A widely recognised rule of thumb for a knockout bid is to simply add to your final round bid the value of unsold lots at final round clock prices plus a minimum unit (e.g. $1,000) to avoid ties. However, the validity of this rule depends on the details of activity rules etc. In the context of the Irish CCA (2012), ComReg provided an exact formula for the knockout bid, see Annex 9 of ComReg (2012b).

[15] See e.g. Ausubel and Milgrom (2006).

identify a lower knockout bid level than simply adding the value of unsold lots, which may help to avoid reaching a budget limit. However, this is not straightforward, as Auctioneers typically only provide aggregate demand data, meaning bidders have to form estimates based on the predictable demand of rival bidders, i.e. using industry knowledge.

As bidders with budget constraints typically underbid for larger packages, it may be expected that this issue will normally be associated with lower overall prices. This is because budget-constrained bidders may express an artificially low delta for large packages that set the opportunity cost for rival bidders. However, if bidders anticipate that their rivals have budget constraints, this may also create incentives for over-bidding. Consider our simple example above. Suppose bidder B correctly predicts that she has similar values to bidder A but she has no budget constraint. If she bids straightforwardly, both bidders will win 1 lot at a price of 50. However, if she overbids for 2 lots, then reduces to zero lots at say a price of 60 per lot, she will force bidder A to confront her budget constraint. Bidder A will bid her maximum 100 for one lot and 125 for 2 lots, so bidder B will only pay 25 to win 1 lot, a gain of 25. Depending on the supplementary bids made by bidder B, she could force bidder A to pay anything up to 95 without the risk of winning 2 lots. Therefore, in situations where one bidder can anticipate a rival's valuations and budget constraint, it is possible that prices could rise for one or more bidders owing to the impact of strategic over-bidding.

Based on observations in our work supporting bidders, we believe that this type of behaviour has affected prices in a number of multi-band CCAs. A typical situation is as follows. Bidders are aware that the knockout bid for a final round package is roughly equal to the final round bid for that package plus the value of unsold lots. One or more bidders identifies an opportunity to close the clock rounds by reducing demand to an extent that creates unsold lots, which in turn inflates the required knockout bids for rival bidders anxious to secure their final round packages. If the bidder is fortunate, its rivals will spend more of their limited budget on securing a preferred package (or even a smaller back-up package) and be unable to express full value for larger packages that will set prices. The presence of spectrum caps or set asides may in some cases even make this a risk-free strategy.

2.4. Preference for Good Relative Outcomes

Relative outcomes matter in spectrum auctions. An obvious way to evaluate the success of a bidder's strategy after the auction is to compare the prices that different firms have paid for similar spectrum. A bid strategy is often considered to be unsuccessful if another bidder paid a substantially lower "per MHz" price for similar spectrum. Good relative outcomes may be particularly important for bidders acquiring less spectrum than others in an auction, as such bidders must rationalise to shareholders why they won less.

In a large multi-band spectrum auction, the amounts spent may also be sufficiently large as to impact on their ability to spend after the auction on marketing and network upgrades. Bidders that spend more in large auctions may find it harder to gain budget from shareholders for other activities or they may find it more expensive to secure external finance for future investments. Thus, increasing the price rivals pay for spectrum in an auction may delay or otherwise obstruct investments by competitors in increasing the quality of their networks. This may give the bidder who gets a good relative outcome a competitive advantage in the market after the auction vis-à-vis its rivals.

As bidders set each other's prices in a CCA, it is common for bidders to identify opportunities for deviations from straightforward bidding that can increase a rival's costs without affecting the bidder's own price. Typically these involve over-bidding on lots where rivals have predictable demand. Bidders can and do act on these opportunities. In the academic literature, this has been described as "spiteful" bidding,[16] but we prefer to avoid this term because it implies that such behaviour is purely offensive.

In practice, our experience is that bidders often engage in price-setting behaviour for two primarily defensive reasons:

- If they don't engage in such tactics, they fear that others will and they will pay much more for similar spectrum. In this context, the example of Sunrise, which paid more for a strictly smaller package of spectrum than Swisscom in the 2012 Swiss 4G auction, is often held up as an example of an outcome that no one wants to replicate.
- By over-bidding, it may be possible to exert price and budget pressure on a rival that would otherwise not exist, thereby increasing the chance that they surrender other target lots. This may be particularly relevant in a multi-band auction, when bidders are bidding on large packages of lots that are both substitutes and complements, and the value of the spectrum may account for a substantial proportion of enterprise value.

A simple example of a 900 MHz band with 7 lots, which is loosely based on a real situation from a European multi-band auction, illustrates the point. Suppose there are three bidders and a spectrum cap of three lots per bidder. Bidder A is a strong incumbent which predictably will demand 3 lots and has valuations and budget that cannot be beaten. Bidder B is a smaller incumbent with a legacy 900 MHz network; it has a substantial value on 3 lots and a minimum demand of 2 lots. Bidder C is the smallest incumbent and its legacy network is less dependent on 900 MHz; it has a substantial value for 2 lots and a minimum demand of 1 lot. Suppose that for all bidders the value of incremental lots is strictly descending and, for simplicity, that the reserve price is zero.

If the bidders bid straightforwardly, the clock rounds will end in one of two ways:

- Bidder C reduces demand to 1 lot. The other two bidders each win 3 lots and pay an identical price, based on the opportunity cost of bidder C not winning a second lot. Effectively the two larger bidders are paying opportunity cost for 1 lot while winning 3 lots.
- Bidder B reduces demand to 2 lots. In this case, bidder A wins 3 lots and bidders B and C each win 2 lots. Observe that bidder C did not reduce demand, so he imposes no opportunity cost on his rivals. As a result, bidder B takes 2 lots at zero price. However, bidder C has to pay the opportunity cost of denying 1 lot to bidder B. Meanwhile, bidder A pays the same price as Bidder C, even though it wins one more lot.

Neither outcome is very attractive for our smallest bidder, bidder C. In the second case, he faces paying a substantially higher price per MHz than his two stronger rivals. Is

---

[16] See e.g. Morgal et al. (2003) and Jansen et al. (2015).

there anything he can do to avoid this? Yes, he could start the auction by bidding on 3 lots and keep bidding on 3 lots until either (a) bidder B reduces to 2 lots, thereby accommodating his real demand for 2 lots; or (b) he becomes worried about the risk of winning 3 lots, in which case he reduces demand to zero lots, knowing he can pick up 1 lot in the supplementary round owing to the spectrum cap. This is a better strategy than straightforward bidding: in case (a) he pays the same price for 2 lots as in the straightforward case, but he can impose the same cost on bidder B, so he does not look like he overpaid; and in case (b) he only wins 1 lot, but pays zero and imposes much higher prices on his rivals, so his decision to reduce demand can be portrayed as prudent. Moreover, by imposing greater price pressure on bidder B through the clock rounds, he increases the likelihood that bidder B reduces demand without any impact on his own price.

This simple example illustrates that the key to securing a good relative outcome in a CCA is exploiting the predictable demand of rival bidders. Without such predictability, a bidder who is bidding above valuation is exposed to the risk of winning lots at prices he does not want to pay. For this reason, this kind of over-bidding is particularly likely to be observed in legacy bands, such as 900 MHz, where some bidders have predictable irreducible demand levels owing to need to provide service continuity to their legacy GSM and 3G customer bases. In the context of a multi-band auction, this has the potential to lead to substantial asymmetries in relative prices during the clock rounds, as bidders deliberately drive up prices in some bands (e.g. legacy bands) where rivals have predictable demand, while bidding more straightforwardly in other bands.

The general importance of good relative outcomes is of course independent of the auction format that is used. However, owing to the pricing rule, this motive can lead to very different behaviour in a CCA as compared to, say, an SMRA. Specifically, identical situations can lead to incentives for over-bidding in a CCA and under-bidding in an SMRA. Consider our 900 MHz example above. Had an SMRA been used, whichever of bidder B and C was considered to have the weaker value for a marginal lot would have had a very strong incentive to reduce demand and end the auction immediately, with all bidders paying the zero reserve price. Even if a bidder cannot unilaterally close an SMRA through demand reduction, he may anticipate that a unilateral demand reduction may trigger a reciprocal demand reduction that does close the auction, if he anticipates that another bidder also prefers a low-price outcome.

Of course, bidders in CCAs might also benefit from demand reduction. However, the pricing rule tends to undermine any attempt to broker tacit coordination. Specifically, bidders may be reluctant to surrender excess demand for fear that rivals may not reciprocate. With auctions with large numbers of lots, it is possible that bidders might offer small reciprocal reductions over many rounds. We have modelled this in mock auctions and have seen some evidence of such behaviour in actual CCAs (it is possible if aggregate demand data is revealed). However, in practice, we have observed that such reductions often only happen as bidders reach or surpass actual incremental valuations, and are difficult to distinguish from straightforward bidding. In particular, in a CCA, it may be difficult to coordinate the final round of reciprocal demand reduction that closes an auction, as each bidder may fear that the other will not reciprocate with a demand reduction once the threat of further price increases through opportunity cost is removed. This contrasts with an SMRA, where the last bidder to reduce demand still has a unilateral incentive to reduce demand to secure a lower price.

3. Strategic Investment Incentive

We next turn to the question of intrinsic value versus strategic investment value in the context of spectrum auctions. A basic premise behind straightforward bidding in a CCA is that the valuations of a particular bidder are based on intrinsic value. In a spectrum auction context, this means that valuations are independent of the amount of spectrum that other bidders win, and the distribution of spectrum between rival bidders. This is a bold assumption. In practice, especially in large multi-band auctions where a substantial amount of spectrum by value is available, bidders may have a strategic investment incentive to foreclose downstream competition by bidding above intrinsic value.

The UK communications regulator Ofcom provides a definition of the strategic investment incentive:[17]

> Strategic investment, where a bidder, with the aim of foreclosing downstream competition, bids above its intrinsic value of spectrum to prevent it being acquired by the bidder's downstream competitors (We distinguish here between intrinsic value and strategic investment value to a bidder. Intrinsic value is the bidder's value of the spectrum in the absence of strategic considerations). Such a bidding strategy (whether or not it achieves its aim) by one or more bidders could result in auction prices that overstate market value.

It should be noted that bidders themselves may have little or no incentive to distinguish between their own intrinsic value and strategic investment value, as they accrue benefits equally from both. However, this difference may matter to a regulator (and to rival bidders), as the assumption that an auction will produce an efficient outcome rests on the premise that bids are based on intrinsic value alone.

Some spectrum auctions are more vulnerable to foreclosure than others. Our view is that the CCA is more vulnerable to foreclosure attempts than the SMRA because a failed foreclosure attempt may be less costly in a CCA. A failed attempt to foreclose rival bidders in a CCA may simply have the effect of setting higher prices for rival bidders, whereas a similar attempt in an SMRA will drive up prices for all bidders, *including the forecloser*.

The possibility of there being a viable foreclosure strategy increases if:

1. a large proportion of all available mobile spectrum in a country (measured by value) is included in the auction; and
2. spectrum caps are relatively lax so that bidders are allowed to bid for a large proportion of the available spectrum (measured by value).

We argued, in Section 2, that the type of strategic bidding associated with the CCA format is more likely to be associated with over-bidding than under-bidding. This contrasts with the SMRA or simple clock auction, where such incentives may be offset by

---

[17] See Ofcom (2014), §A7.87.

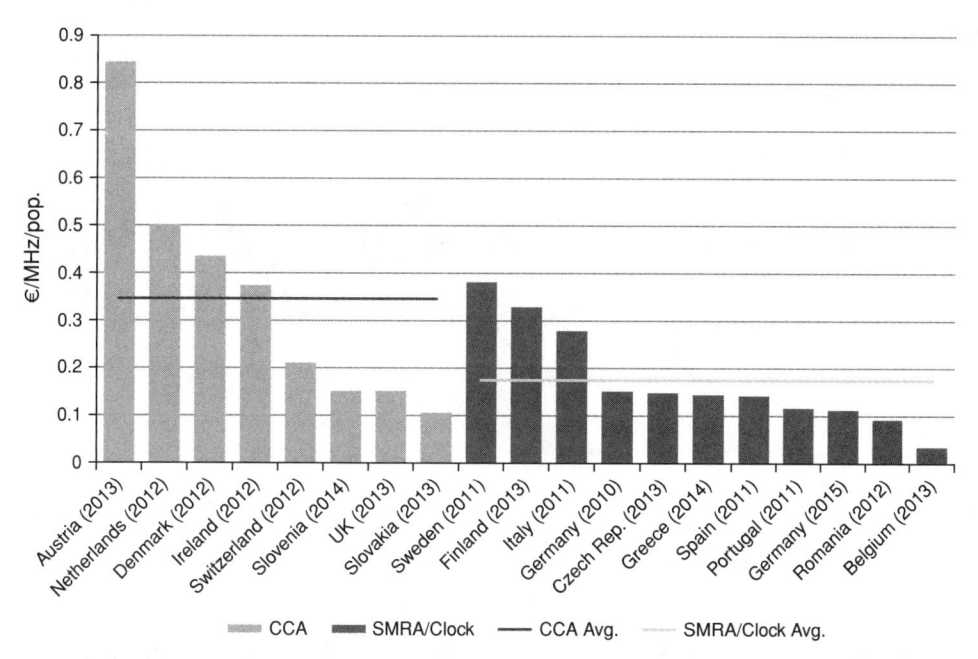

Figure 1. Per MHz per pop prices for European 4G auctions, by auction format.
Source: NERA database of spectrum awards.

incentives for demand reduction. If this analysis is correct, one might expect that across a broad sample of auctions, CCAs will tend to realise higher prices than SMRAs. In this section, we have further argued that incentives for strategic over-bidding may be intensified if there is potential to foreclose the market. Again, this contrasts with other formats where there may be off-setting incentives for demand reduction. If our supposition is correct then we would expect to see the highest prices in countries that ran large, multi-band CCAs with "lax" spectrum caps.

As an initial test of our hypothesis, we carried out a simple analysis of price per MHz pop data[18] of recent European 4G auctions, comparing outcomes of awards using a CCA format to those using either a SMRA or clock auction. The results are illustrated in Figure 1. As can be seen, CCAs (€0.35/MHz/pop) have on average been associated with higher-price outcomes than SMRAs and clock auctions (€0.18/MHz/pop). It should be noted that we have not weighted our sample to take account of the different "quality" of spectrum by band available in each award,[19] or considered any of the many local factors that may have affected prices, so caution should be exercised in interpreting this data. Nevertheless, the results support the possibility that incentives for over-bidding in CCAs and/or under-bidding in SMRAs and clock auctions may have an effect on outcomes.

---

[18] Price divided by the amount of spectrum in MHz divided by the population. This is standard measure for comparing spectrum prices across countries.

[19] These awards vary significantly with respect to the amount of low- and high-frequency mobile bands. Prices for awards with larger proportions of high-frequency spectrum (e.g. Belgium, Denmark and Sweden, which were 800 MHz only) may be overstated. In addition, some awards (e.g. Switzerland, UK and both German auctions) included unpaired frequencies which typically have lower value than paired frequencies, so prices for these awards may be understated.

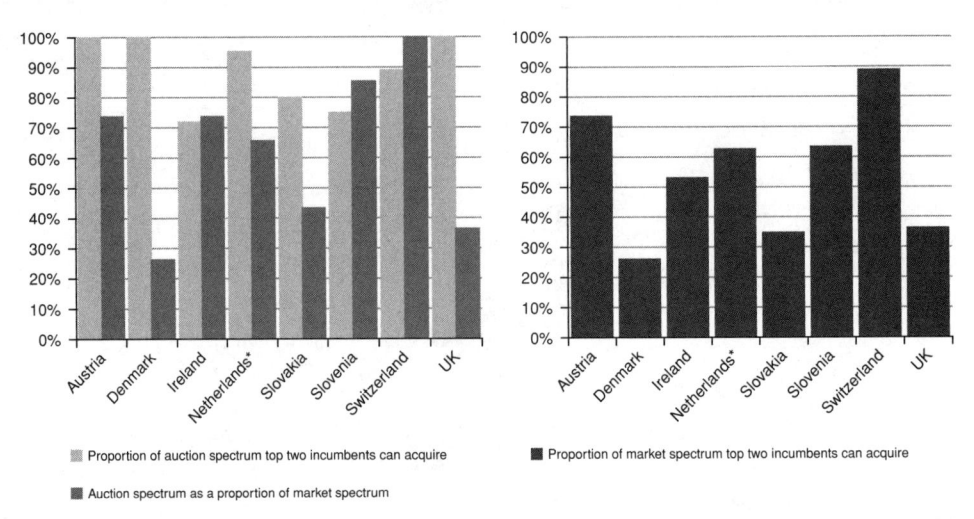

Figure 2. Potential to foreclose market in European 4G auctions.
Source: NERA database of spectrum awards.

For the subset of CCA auctions, we further considered the scope for foreclosure of local spectrum markets. The left-hand side of Figure 2 illustrates the level of the two foreclosure factors we identified above (high availability of spectrum and lax spectrum caps) in eight recent European 4G auctions that used the CCA format while the right-hand side illustrates the combined effect of the two factors.[20]

Again, the data is consistent with our supposition. Notably, the two auctions that produced the highest per MHz per pop prices – Austria (€0.84) and Netherlands (€0.50) – are near the top of the list for potential for foreclosure. Both these awards featured intense bidding during the clock rounds driven either exclusively or primarily by competition between three incumbent bidders, a market situation that elsewhere has often been associated with demand reduction and price outcomes. It seems possible that these auctions were affected by bidding based on strategic investment incentives in the context of a CCA. However, a robust conclusion is not justified given the limited sample size and challenges in comparing data across countries.

4. Concluding Remarks

Our experience as advisors to participants in recent CCAs is that bidders often do not bid straightforwardly in practice. There are compelling explanations why bidders rationally deviate from straightforward bidding, most notably the impact of budget constraints, preferences for good relative outcomes and opportunities to realise strategic investment value.

---

[20] For the purpose of this analysis we have used the values of different spectrum bands set forth by Ofcom in its February 2015 consultation of UK annual licence fees: 800 MHz: 33; 900 MHz: 23; 1800 MHz: 13; and 2600 MHz: 6. For 2.1 GHz, we interpolate a ratio of 10. Note our use of these values is for illustrative purposes only and does not represent any endorsement of the values, which are subject to consultation. See Ofcom (2015, page 2).

The strategic incentives to deviate from straightforward bidding that matter most in a CAA context typically result in overbidding relative to intrinsic value, potentially by a large amount. A crucial difference between the CCA and its main alternative multi-round formats, the SMRA and the standard clock auction, is the absence of any meaningful incentives for demand reduction. All auction formats are vulnerable to low participation, but in situations where participation is healthy, the CCA is likely less vulnerable to under-bidding (relative to intrinsic value) and more vulnerable to over-bidding. Our supposition is supported by anecdotal evidence from recent European 4G auctions, where awards using CCAs have tended to realise higher prices on a per MHz per pop basis than those using SMRAs or clock auction formats (see Figure 1). Clearly, the impact of auction format on strategic bidding is only one of many explanations for differences in prices across auctions, but – in our view – it is an important one.

In a spectrum auction, the incentives for strategic bidding are asymmetric, potentially varying both across bidders and across spectrum bands. Predictable differences in demand and financial commitment, for example driven by market share and local status or differences in current spectrum holdings, create strategic bidding opportunities. In turn, such opportunities are closely linked to the availability of spectrum by band and activity rules, especially spectrum caps. In the CCA context, this means that the scope for one bidder to overbid may be much greater than another, and the potential for such behaviour may vary greatly across bands. Even if the same over-bidding opportunities exist with an SMRA or clock auction, they may be constrained by countervailing incentives to moderate demand. As a general observation, legacy bands, such as 900 MHz or 1800 MHz, may be particularly vulnerable to over-bidding because some incumbents typically have irreducible and predictable minimum demands associated with servicing their legacy 2G and 3G customer bases.

The CCA auction was developed to address efficiency concerns with the SMRA format, such as aggregation risk. However, its recent popularity with regulators in Europe and beyond may have as much to do with its perceived effectiveness in undermining incentives for demand reduction as it has to do with efficiency concerns. Given the information now available about past awards, we think it would be disingenuous for any regulator to claim that the CCA format diminishes strategic behaviour. Indeed, when deciding whether or not to use the CCA for a particular spectrum award, regulators should weigh carefully the impact of the format on strategic bidding incentives, and how these will be affected by decisions on specific rules, such as spectrum caps.

This raises the question whether the CCA format could be refined to reduce its exposure to strategic bidding. The concerns we have raised in this paper regarding budget constraints and preferences for good relative outcomes are both linked to the fact that the CCA uses a second-price rule. Opportunities to exploit the rules are also linked to the activity rules linking the clock and supplementary rounds.

Ausubel (2013) has suggested changing the activity rules of the CCA in a way that turns the CCA into an auction that is "effectively" first price. Most CCAs to date have used a pure eligibility points-based activity rule for clock rounds (bidders can bid on any package subject to never increasing eligibility points of the package they are bidding on). A smaller number of CCAs have in addition introduced revealed preference considerations, allowing bidders to violate the pure eligibility points-based activity rule, subject to not violating revealed preferences (as defined by the WARP condition).[21] A possible next step, explored by Ausubel (2015), would be to abandon the eligibility points-based activity rule altogether and introduce a GARP-based activity rule,[22] which is a stronger (i.e. more restrictive) condition than WARP. The effect of such a change to the activity rule would be that bidders always risk paying their full bid amount, thereby effectively turning the CCA into a first price auction. This would also ensure a stronger link between the allocation at the end of clock rounds and the final auction outcome.

While the details of such modifications to the CCA have yet to be worked out, it is worth noting that from a bidder perspective a GARP-based activity rule would place greater emphasis on the quality of bidder valuation models. Under activity rules used in all CCAs to date, bidders have had some flexibility to modify assumptions and adjust valuations during clock rounds, in response to observed bidding behaviour. In contrast, a GARP-based activity rule would be more restrictive and bidders may find themselves very constrained in later clock rounds owing to bidding patterns in earlier clock rounds. This is problematic, as in our experience bidder valuations are often elastic. Valuations are based on multiple assumptions, each of which may be quite uncertain, and are often adjusted during an auction if there is new information about competitors based on bids in the auction. This is particularly relevant for large multi-band auctions, where it may be hard for operators to differentiate between the incremental intrinsic value of the spectrum and the broader value of a mobile network operator's entire business.

This debate regarding the future format of the CCA reveals a tension between using an open multiple-round format, with the purpose of allowing bidders to update valuations in response to price discovery, and restricting bidding behaviour to pre-determined valuation profiles. On the one hand, introducing a GARP-based activity rule might ease some of the concerns regarding strategic bidding we have discussed in this paper, and on the other hand it might be unduly restrictive for bidders in the context of spectrum auctions. Thus, even if a new CCA format diminishes scope for strategic bidding, it will probably raise other concerns. The debate over strategic behaviour in spectrum auctions, and associated risks, looks set to continue for the foreseeable future.

References

Ausubel, L. M. (2013). "An enhanced combinatorial clock auction," presentation available at www.ausubel.com/auction-papers/ausubel-baranov-enhanced-cca-slides-26Feb2013.pdf

Ausubel, L. M., Cramton, P., and Milgrom, P. (2006). "The clock-proxy auction: a practical combinatorial auction design," in *Combinatorial Auctions*, edited by Peter Cramton, Yoav Shoham, and Richard Steinberg, pp. 115–38. Cambridge, MA: MIT Press.

---

[21] Examples include CCA auctions in Ireland (2012 and 2017), Australia (2013) and Canada (2014, 2015).

[22] GARP and WARP, in the context of auctions, are two different tests of rationality/consistency in observed bidding behaviour. To simplify, an activity rule that imposes GARP/WARP does not allow bidders to submit bids that violate preferences they have expressed in earlier bidding rounds. The main difference between WARP and GARP is that WARP rules out bids that are in *direct* conflict with earlier bids, while GARP also rules out bids that are *indirectly* in conflict with earlier bids via a chain of revealed preferences/bids. Hence, GARP is a more restrictive consistency requirement.

Ausubel, L. M. and Milgrom, P. (2006) "The lovely but lonely Vickrey auction," in *Combinatorial Auctions*, edited by Peter Cramton, Yoav Shoham, and Richard Steinberg, pp. 17–40. Cambridge, MA: MIT Press.

Bichler, M., Goeree, J., Mayer, S., and Shabalin, P. (2014). "Spectrum auction design: simple auctions for complex sales," *Telecommunications Policy*, 38(7): 613–622.

ComReg (2012a). "Multi-band spectrum release – release of the 800 MHz, 900 MHz and 1800 MHz Radio Spectrum Bands," ComReg 12/25, 16 March 2012.

ComReg (2012b). "Multi-band spectrum release, information memorandum," ComReg 12/52, 25 May 2012.

ComReg (2012c). "Multi-band spectrum release, response to consultation on the draft information memorandum," ComReg 12/50, 25 May 2012.

Cramton, P. (2013). "Spectrum auction design" *Review of Industrial Organization*, 42(2), 161–190.

Goeree J. K. and Lien, Y. (2016). "On the impossibility of core-selecting auctions," *Theoretical Economics*, 11:41–52.

Industry Canada (2014). "Licensing framework for broadband radio service (BRS) – 2500 MHz Band." Industry Canada SLPB-001-14, January 2014.

Janssen, M. (2015). "Price distortions in combinatorial clock auctions, a theoretical perspective," available at http://stakeholders.ofcom.org.uk/binaries/consultations/annual-licence-fees-further-consultation/responses/Telefonica_-_annex_2.pdf.

Levin, J. and Skrzypacz, A. (2014). "Are dynamic Vickrey auctions practical? Properties of the combinatorial clock auction," unpublished working paper, September 2014.

Morgan, J., Steiglitz, K., and Reis, G. (2003). "The spite motive and equilibrium behavior in auctions," *Contributions to Economic Analysis & Policy*, 2(1):1102–1127.

Nisan, N. (2006). "Bidding languages for combinatorial auctions," in *Combinatorial Auctions*, edited by Peter Cramton, Yoav Shoham, and Richard Steinberg, pp. 215–231. Cambridge, MA: MIT Press.

Ofcom (2014). "Annual licence fees for 900 MHz and 1800 MHz spectrum – further consultation," Ofcom, 1 August 2014.

Ofcom (2015). "Annual licence fees for 900 MHz and 1800 MHz spectrum: provisional decision and further consultation," Ofcom, 19 February 2015.

CHAPTER 35

Impact of Budget Constraints on the Efficiency of Multi-lot Spectrum Auctions

Nicholas Fookes and Scott McKenzie

1. Introduction

Having advised mobile operators in a number of high-stakes spectrum awards, we offer a bidder's perspective on the challenges introduced by budget constraints in multi-lot auctions, and examine their impact on outcomes under the most prevalent formats. We consider in particular Combinatorial Auctions (CA), of which Combinatorial Clock Auctions (CCA) are a specific type, and Simultaneous Multiple Round Ascending Auctions (SMRA).

In contrast with SMRA, CAs involve a final sealed-bid round during which participants submit concurrent bids for different bundles or 'packages' of lots. The winning bids are selected such as to maximise the sum of corresponding bid values, under the proviso that each bidder wins at most one of its bids and that each lot is sold no more than once. Since each bidder either secures an entire package or nothing at all, there is no risk of winning and paying for an unwanted subset of a targeted allocation. This is a particularly attractive property, to which CAs must owe much of their popularity.

Furthermore, a generalised second-price rule is often applied, under which successful bidders pay no more than the minimum required to justify their allocation and to avoid 'unhappy losers'. The latter entails prices such that given their bids, losers may be deemed 'happier' to forego goods than pay the higher amount needed to trump winning rivals. Ostensibly, this incentivises truthful bidding, which, in turn, enhances the prospects of directing the available goods to those who value them the highest.[1]

In theory, therefore, second-price CAs simplify bid strategy while promoting efficiency. However, both of these advantages may be compromised if any of the participants are constrained by budgets rather than by their valuations for the goods on offer. Strict budget constraints may prevent bidders from expressing their relative preferences

---

[1] See Cramton et al. (2006). An allocation is generally considered efficient if the aggregate value attributed by the winners to their winning lots is maximised, subject to any restrictions (such as acquisition caps) that may be imposed by the seller. While this criterion needs to be qualified when budget constraints apply, if the amounts bid reflect true valuations for all eligible packages, a CA will always yield efficiency in this sense.

across the full range of packages of interest. This introduces a strategic dilemma. If they bid full value on smaller packages, such bidders will need to suppress marginal bid prices at the top end of the range to respect their budget limits. Doing so reduces the chances of winning a superior package. A constrained bidder may choose instead to create extra headroom by cutting bid prices at the lower end. But this increases the risk of winning nothing at all. In such circumstances therefore, bidders are liable to make suboptimal choices that lead them to forego goods when securing these might have been efficient.

This invites the following questions in particular:

- Does SMRA provide a more robust mechanism, and should it be preferred in a budget-constrained environment?
- Is it possible to accommodate budget constraints within a second-price CA framework, and if so, what impact would this have on bid incentives?
- What further options might policy-makers consider?

In brief, our view is that it is almost always easier for bidders to manage budget limits in an SMRA than in a second-price CA. Constrained bidders are likely to experience the sealed-bid element of a second-price CA as a lottery, because there is no opportunity to adjust bids in light of actual prices (note that unconstrained bidders should not need to). Notwithstanding concerns around strategic demand moderation, SMRA might therefore be preferred over second-price CA in a budget-constrained environment. However, the relative pros and cons are less clear-cut if there are complementarities in the valuations:[2] pursuing bundles is risky in an SMRA when lot prices exceed individual lots values, as one may be stuck with an unprofitable subset of the bundle.

We are sceptical about the scope to reflect budget constraints directly within the framework of a second-price CA, because incentive-compliance breaks down when bidders are dispensed from putting their money where their mouths are: identifying the optimal outcome is impossible when the underlying metric is vulnerable to manipulation. If the format cannot be fixed to accommodate hard cash limits, then the second-price CA loses one of its key selling points: the promotion of efficiency.

One remaining question is whether the drawbacks of SMRA can be sufficiently mitigated to turn it into an unequivocal format of choice, whenever budget constraints are likely to appear. Judicious packaging of lots or allowing spectrum pooling in certain circumstances may help alleviate possible exposure to unprofitable allocations. A proxy SMRA process might also be considered, in which the participants supply comprehensive package valuations as well as any budget limit. Virtual proxy agents would bid accordingly. While this might rule out coordinated demand reduction, which would be a clear benefit from a policy standpoint, the sealed-bid nature of the process would also exclude interactive price discovery. Finally, in markets where new entrants and speculators can be excluded, one might consider redistributing a proportion of the proceeds[3] equally among the participants. If 100% of the proceeds were redistributed, this

---

[2] Two objects A and B are referred to as complements if the value of the bundle (A + B) is greater than the value of (A) on its own plus that of (B) on its own. They are substitutes if the value of (A + B) is less than that of (A) plus that of (B).

[3] By proceeds, we mean the revenues that would accrue to the seller in the absence of any redistribution.

would effectively result in a mutual trading process, in which bidders exchange claims to resources for cash. The net price exposure would be reduced, allowing participants to bid closer to their true valuations. While such a scheme could also defuse collective demand-moderation incentives, raising the prices paid by rivals would be immediately rewarding. To dissuade this, the percentage of the proceeds that is redistributed would need to be chosen such that overbidding is more likely a priori to generate losses than gains.

This chapter is structured as follows. In Section 2, we discuss possible reasons for the emergence of budget constraints in the mobile communications industry. Section 3 examines the implications of hard budget limits in second-price CAs, in light of a case study on the 2013 UK Digital Dividend auction. Budget-constrained bidding in SMRAs is covered in Section 4. In Section 5, we consider the meaning of efficiency when hard budget limits apply, and introduce a tentative outline of a proxy SMRA. A mechanism based on the partial redistribution of auction proceeds is explored in Section 6. Within the rest of this chapter, the acronym CA will be used to refer to the commonly used 'second-price Combinatorial Auction' formats unless otherwise stated.

2. The Reality of Budget Constraints in the Mobile Industry

If markets were perfect, the supply of funds available to companies from investors and lenders would match demand. All players would be equally well informed, and there would be no budget constraints preventing companies from acquiring value-enhancing assets. However, our own experience is at odds with this theoretical ideal: we have often observed significant differences between spectrum valuations and available budgets.

In some cases, budget constraints may be manifestations of 'Principal–Agent' problems,[4] potentially exacerbated by a lack of confidence in the valuations of the assets on offer. Such lack of confidence is not entirely surprising: estimating spectrum value is an exacting task, ridden with uncertainty. Cramton observes that, in practice, 'bidders almost never have a completely specified valuation model'.[5] We have also found that distinct methodologies are often used by separate operators.[6] These may yield widely differing results, even under identical inputs. Coupled with this, spectrum valuations typically rely on long-term demand forecasts which, in truth, may be anyone's guess. Assumptions also need to be made about the firm's long-term commercial and technical strategy, as well as that of the competition.

---

[4] See Ausubel et al. (2013, p. 2): 'In Burkett's work, the budget constraint is a control mechanism that a principal (e.g., the corporate board) imposes on an agent (e.g., the manager delegated to bid for an asset) in order to curb managerial discretion such as empire building'.

[5] See Cramton (2008, p. 3).

[6] A commonly used methodology takes market demand and market shares as fixed inputs. The value of additional spectrum corresponds with the avoided network costs needed to serve forecast usage volumes (see for example Doyle, 2010). An alternative approach is to compute the market-share impact of relative capacity constraints under different spectrum scenarios, recognising that there are practical limits to network densification.

However, financial constraints may also have external causes:

- With the increasing maturity of the sector over the past decade, investors have tended to run telecoms businesses for cash in order to maximise dividends and share buy-backs;
- Since the global financial crisis, banks have been wary of increasing lending to telecom operators, not least due to high fixed-cost nature of the business and the lack of revenue growth;
- Much of the value of spectrum is 'defensive' rather than incremental: operators need to secure spectrum to maintain their relative market position, and it is often hard to galvanise investors with such a narrative.[7]

A further issue is that access to capital is often highly skewed. At one extreme, a state-backed former monopoly can mobilise a disproportionate war chest to bid for spectrum, whist enjoying close to sovereign interest rates on its debt. The financial advantages that accrue from state patronage and near-monopoly rents are not necessarily connected with operational performance or even efficiency of spectrum use. A highly leveraged private-equity backed challenger, in contrast, may have limited funding headroom and correspondingly tight budget constraints. As discussed in the following section, such disparities are especially problematic in a CA.

3. Budget-constrained Bidding in Combinatorial Auctions

In its report on the 2013 auction of 4G spectrum in the United Kingdom, the National Audit office (NAO) observes that 'at least two bidders appeared to be subject to budget constraints [...] potentially limiting the efficiency of the auction'.[8] Accordingly, the NAO urges the UK Office for Communications (Ofcom) to consider the financial position of the bidders involved when designing future auctions. This call to action is bound to resonate beyond the walls of Ofcom, and it seems appropriate to consider what occurred during the auction itself.[9]

Of particular interest is the Principal Stage of the award, which dealt with the allocation and pricing of generic spectrum lots.[10] A second-price Combinatorial Clock Auction (CCA) was used: a rising clock phase was held ahead of the final, sealed 'supplementary-bid' round. The required consistency of final supplementary bids with revealed preferences in the clock phase was designed to incentivise truthful bidding and allow for price discovery. The table below shows the quantities of spectrum acquired

---

[7] In our experience, the sum of private valuations for incremental spectrum often significantly exceeds the collective value of the resources to the industry, precisely because of the defensive element.

[8] Comptroller and Auditor General (2014, pp. 6 and 10).

[9] Our simplified analysis is based on the data made public by Ofcom after the auction, which includes the entire bid histories of each of the participants.

[10] Spectrum lots were offered on a generic basis during the Principal Stage of the auction, with the exception of a specific 2 × 10 MHz block of 800 MHz spectrum with a coverage obligation. This block was secured by O2 (Telefonica); however, there was no price differential with the generic 800 MHz spectrum at the end of the clock phase. The positioning of winning allocations in each band were determined during the subsequent Assignment Stage. The spectrum on offer included 2 × 30 MHz in the 800 MHz band, and 2 × 70 MHz of paired Frequency Division Duplex (FDD) plus 35 MHz of Time Division Duplex (TDD) in the 2.6 GHz band.

by the winning bidders at the end of the Principal Stage, and respectively the winning bid prices and prices paid:

Principal Stage results	800 MHz	2.6 GHz FDD	2.6 GHz TDD	Bid price	Price paid
EE	2 × 5 MHz	2 × 35 MHz	–	£1,050m	£589m
Vodafone	2 × 10 MHz	2 × 20 MHz	20 MHz	£2,075m	£791m
O2 (Telefonica)	2 × 10 MHz	–	–	£1,219m	£550m
H3G	2 × 5 MHz	–	–	£566m	£225m
BT	–	2 × 15 MHz	15 MHz	£340m	£186m
Total	2 × 30 MHz	2 × 70 MHz	35 MHz	£5,249m	£2,321m

The final clock round prices per lot and the aggregate clock-prices of allocated bandwidth are provided below:

Final clock prices	800 MHz	2.6 GHz FDD	2.6 GHz TDD	Total
Price per 5 MHz block	£423m	£92m	£24m	–
Price of allocated lots	£2,538m	£1,288m	£171m	£3,997m

Final prices paid for the generic lots were 44% of the winning bid values and 59% of the final clock prices. While such disparities may not be unusual in second-price auctions, they do rather seem to defeat the object of price discovery.

What is striking about this auction, however, is the bid behaviour of O2 (Telefonica), both during the clock phase and in the supplementary round. Based on the data, we observe the following in particular:

- While O2's clock bids suggest a marginal value of at least £6m for 15 MHz of TDD spectrum, its marginal bid price for this resource was reduced to zero in the supplementary round (the first sign of a biting cash constraint).
- O2 opened with a bid for a package containing 2 × 70 MHz of bandwidth, yet its highest supplementary bid was for only 2 × 10 MHz at 800 MHz plus 2 × 10 MHz at 2.6 GHz, with a bid price of £1,347m.
- This yields a strong suggestion that O2 had a hard budget limit of £1,347m.
- O2 topped up the amount of its final clock bid by £373m in the supplementary round, significantly limiting the headroom for bids on larger packages.
- In the event, O2 won its supplementary bid of £1,219m for just 2 × 10 MHz at 800 MHz, and paid £550m (45% of its bid price and 65% of the package price in the final clock round).
- Had O2 known that it would pay so little for its core 800 MHz target, it might not have added £373m to its final clock bid in the supplementary round (it would have been among the winners in any case).
- This would have given it extra headroom to place bids on larger packages within a £1,347m envelope.

The extent to which O2's supplementary bid strategy appears to have reduced its headroom for larger bids is made plain in Figure 1. O2's £128m marginal bid price for an

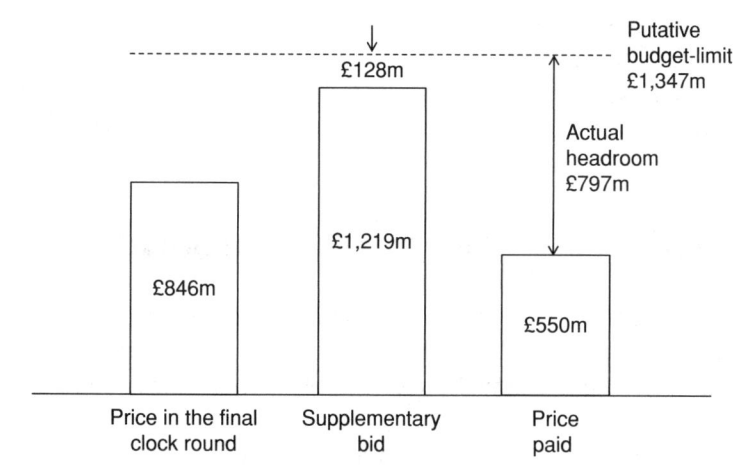

Figure 1. Prices for 2 × 10 MHz.

additional 2 × 10 MHz at 2.6 GHz may plausibly have understated its full marginal value for this extra bandwidth.

Whether or not this influenced the final allocation remains for us a matter of conjecture.[11] The key point here is that gauging price exposure in a multiband CCA is notoriously difficult.[12] Moreover, the clock rounds in the UK auction concluded with spare capacity of 2 × 10 MHz at 800 MHz plus 2 × 5 MHz at 2.6 GHz FDD. From O2's perspective, this would have substantially increased the amount required to be confident of securing its minimum 800 MHz target. Managing hard budget limits in such situations is especially challenging.

The uncertainty introduced by the one-shot, sealed-bid element of a CA is liable to induce constrained bidders to adopt unduly conservative strategies: spectrum is a key input, and such bidders may reckon that it is better to maximise the chances of winning small than risk walking away with nothing. Rather than aiding price discovery, confusing price signals in the clock phase of a CCA may exacerbate this problem: bidders can easily be wrong-footed if these suggest significantly higher costs than those to which they are actually exposed.

Conservative bidding tends to benefit competitors: lower marginal bid prices limit the prospects of securing additional bandwidth and may lead to lower prices paid by rival bidders.[13] This opens a potential avenue for strategic bidding in a CCA. For example, a state-backed operator with access to vast cash reserves might seek to 'spook'

---

[11] However, the NAO's view that efficiency may have been compromised is supported by a report in the *Guardian* newspaper from 14 April 2013, which quotes one of the bidders: '[Ofcom] neither raised the amount that the government was looking for, nor did it ensure that spectrum found its way into the hands of everybody who wanted it' (Garside, 2013). This suggests there was at least one 'unhappy' loser (or partial loser).

[12] Computing this exposure from aggregate bid data is not intractable: see for example Janssen et al., 'Budget Constraints in Combinatorial Clock Auctions' (Chapter 16 in this volume). However, the difference between the expected and maximum theoretical prices can be very large.

[13] In the event, prices paid were reasonably uniform in the UK auction, with the exception of H3G, which benefited from a guaranteed 'minimum portfolio package' at the reserve price. However, bidders will have been acutely aware of the possibility of significant price differentials such as those observed in the previous multiband CCA in Switzerland.

constrained challengers into bidding full value on their final clock package. Overstating earlier clock bids and creating artificial gaps (i.e. spare supply) in the final clock-round configuration are possible means of achieving this. The mere existence of predatory bidding opportunities of this kind should give rise to significant concern among bidders and policy-makers alike.

4. Is SMRA More Robust and Should It Be Preferred?

The price transparency in an SMRA offers far greater scope for bidders to manage budget limits. There is no sealed-bid round, nor any need to second-guess one's cash exposure: this is visible throughout. Winners pay the amount of their winning bids, and participants can react to events at every stage of the process.

SMRAs can be construed as tacit negotiations between rival operators: the faster they all agree on the final distribution of lots, the less everyone pays. Whereas CAs may allow bidders to pursue unrealistically large packages, the feedback between what one bids and what one pays in an SMRA incentivises an ultimate focus on winnable resources: the contrary is liable to drive unnecessary unit-price inflation. Pushing rivals to their limits is costly. Other than strategic demand reduction, which potentially rewards *all* participants, there is no obvious means of exploiting the cash constraints of competitors to obtain lots at a discount.[14] An added benefit from a bidder perspective is that SMRA excludes significant price disparities that may skew the financial playing field and cause embarrassment.[15]

In the absence of complements, bidders are thus likely to feel less exposed to strategic risks caused by hard budget limits in an SMRA than in a CA. SMRA would then appear to offer a more robust mechanism from a bidder's perspective, especially in auctions involving large quantities of incremental as well as expiring spectrum licences, where such constraints are most prone to arise.

But when the value of some packages is greater than the sum of the values of their parts, SMRA may lead to exposure problems. Pursuing bundles is risky once lot prices exceed individual lot values: failure to secure the bundle may result in a loss-making allocation, if bidders are unable to extricate themselves from their 'standing high bids'. Paradoxically, budget constraints could exacerbate this problem in certain instances.[16] The emergence of LTE technology increases the likelihood of this kind of scenario: while the marginal benefits of extra capacity tend to decline with each additional block of spectrum, the cost of adding bandwidth to an existing channel (within a 20 MHz

---

[14] In full-disclosure SMRAs with specific, regional lots, stronger bidders may use retaliation bids to persuade weaker rivals to stay out of certain areas (see for example Klemperer, 2002, p. 171). However, it seems to us that the success of such strategies rests more heavily on the desire of bidders to maximise their value-surplus than on a need to respect bid limits.

[15] In a second-price CA, the amounts paid by winners are a function of the values of rivals' unsuccessful bids. As a result, it is possible to pay more in absolute terms than rival winners, yet secure an inferior package. Such outcomes are especially difficult to explain to stakeholders. As Janssen et al. demonstrate in Chapter 16, conservative bidders may be more exposed to adverse pricing differentials than unconstrained or risk-taking rivals. Moreover, bidders with a spite motive may overstate their valuations for packages that are guaranteed not to win, with a view to raise prices paid by rivals.

[16] See Brusco and Lopomo (2005, p. 9). This paper gives a surprising example of an auction for two identical lots in which exposure risk forces a bidder with a higher budget to drop out in favour of a weaker rival.

limit) is zero. Moreover, the performance in terms of headline data rates and latency improve with channel size.[17] Where the latter effects dominate, complementarities will inevitably appear. The same may occur when spectrum is sold on a regional basis: a national footprint may be worth more than the sum of the values of regional licences.

A further policy concern is that SMRAs may lead to inefficient allocations. Given the demand-moderation strategies that this format invites, scarce goods may fail to reach those who value them the highest. This frustrates the central objective of spectrum auction design: to achieve the value-maximising distribution of resources, subject to applicable competition safeguards.

In light of the SMRA's own drawbacks, policy-makers may be reluctant to abandon CAs altogether. One of the remaining questions is whether budget constraints could be accommodated directly within the framework of a CA. This hinges on the prospects of determining the most efficient outcome with a reasonable degree of confidence. As illustrated through simple examples in the next section, however, such prospects seem remote.

5. Efficiency in a Budget-constrained Setting

Characterisation of Strategic Alternatives for Budget-constrained Bidders

Figure 2 depicts two extreme choices that a (hard) budget-constrained bidder could rationally make in defining marginal bid prices across N packages of interest, consistent with its relative preferences.[18] (Note that this diagram relates to packages and not necessarily to individual lots.) The packages $1, \ldots, N$ are taken to be ordered by ascending bidder valuations, such that the value V_j ascribed by the bidder to any package j is the sum $(MV_1 + MV_2 + \ldots + MV_j)$ of marginal valuations, and the bid price B_j is the sum $(MB_1 + MB_2 + \ldots + MB_j)$ of marginal bid prices. The budget limit L is taken here to be the same for all packages. This may be visualised as a 'bid window' that one can shift along the range of packages of interest. The final position of the bid window, reflecting a 'shift factor' h between 0 and $V_N - L$, dictates the construction of a corresponding matrix of bids that is representative of marginal values between the packages within this window.

Figure 2(a) illustrates a conservative bid strategy, with $h = 0$. In this case, the bids for packages 1 and 2 reflect full value, while B_3 is curtailed and $B_j = B_3$ for all $j > 3$. All other marginal bid prices are zero. Figure 2(b) shows the other extreme. Here, $B_1 = 0$ and B_2 is curtailed, while the marginal bid prices for all remaining packages are equal to the marginal valuations.

A bid matrix reflecting a shift value $h = 0$ minimises the risk of winning nothing. Note further that, given a set of rival bids, there is no shift value $h > 0$ that would enable a bidder to win a non-zero package when nothing would be secured with

---

[17] However, carrier aggregation with 'LTE advanced' might attenuate complementarities between spectrum quantities in individual bands, by allowing smaller blocks to be combined with LTE spectrum in other bands.
[18] While other approaches may be conceived, such as compressing the values of all N packages to fit within the bid limit, we feel that ours lends itself better to a description of conservative versus aggressive bid strategies.

(a) Suppression of marginal bid prices for highest packages.

(b) Suppression of marginal bid prices for lowest packages.

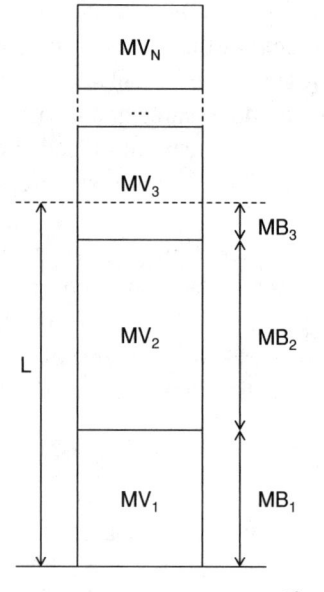

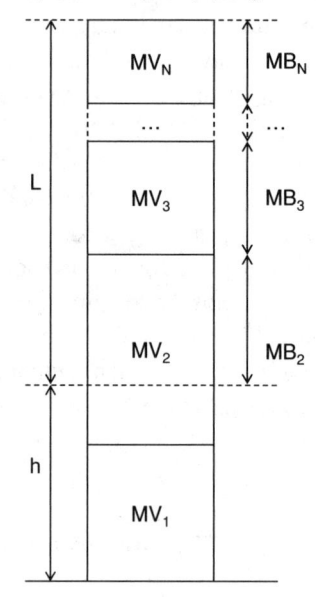

Figure 2.

$h = 0$. Conversely, if a non-zero package can be won with a given shift factor $h > 0$, then a smaller value of h could never result in a more valuable package being acquired.

Impact of Budget Constraints on CA Outcomes

To illustrate the impact of the value of the shift-factor h on outcomes, consider a second-price CA for two identical lots, in which two bidders participate. Let the valuations for the two feasible packages and budget limits be as in the table below. For ease, we assume that the reserve price per lot is zero in all the illustrative examples within the rest of this chapter.

Example 1	Value for 1 lot	Value for 2 lots	Budget limit
Bidder A	100	200	100
Bidder B	30	90	90

Note that Bidder B is not constrained by its budget in this example, and can therefore always bid full value on both packages. Suppose that Bidder A pursues a conservative bid strategy, reflecting a shift-value $h = 0$. The bids and the corresponding outcome are then as follows:

Example 1a	Bid for 1 lot	Bid for 2 lots	Lots secured	Price paid
Bidder A	100	100	1	60
Bidder B	30	90	1	0

The aggregate bid price is maximised by allocating one lot to each of the bidders (total = 130). The winning bid prices also correspond with the unconstrained valuations ascribed by each bidder to their allocated lot.

Prices paid by each bidder are computed by considering the amount that the auctioneer *could* have obtained by selling the winning lot to the other bidder, on the basis of the latter's bid prices. Hence Bidder A's second-price is Bidder B's marginal bid price of 60 for a second lot. Bidder B's price is zero, because on the basis of its bids, Bidder A was not prepared to pay a higher price to secure a second lot. Notice that Bidder A is effectively an unhappy loser in this example, since it was in fact willing to pay more than Bidder B's price to add Bidder B's lot to its own holdings. Bidder A's choice of bids is suboptimal, and Bidder B only secures its lot by virtue of the latter.

If Bidder A were instead to maximise its marginal bid price for the second lot within its budget limit, corresponding with a shift value $h = 100$, then the bids and outcome would be as follows:

Example 1b	Bid for 1 lot	Bid for 2 lots	Lots secured	Price paid
Bidder A	0	100	2	90
Bidder B	30	90	0	0

In this case, Bidder A wins both lots and pays Bidder B's bid price for 2 lots. The aggregate winning bid price is now reduced from 130 to 100. Viewed in isolation, this reduction would suggest that the outcome in Example 1b is less efficient than that in Example 1a. But this would lead to the absurd conclusion that one might need to rely on suboptimal bids by one or more of the participants to achieve economic efficiency.

Furthermore, the unconstrained value associated with this allocation is now 200 (Bidder A's valuation, ignoring the budget limit), versus 130 in Example 1a. This points towards Example 1b being the preferable outcome. However, if one were to allocate goods on the basis of unconstrained valuations while limiting prices to reported budgets limits, the participants could then overstate their valuations and understate their budgets with impunity.

Outcomes with Mutually Optimised Bid Matrices

The optimal bid matrix for an individual bidder could be taken to be that which secures the most valuable package within reach, given all rival bids. Such a matrix corresponds with the highest shift factor $h \in [0, V_N - L]$ that is non-losing.[19] If there are no values of h that can prevent the bidder being knocked out, $h = 0$ should be considered optimal. This is because $h = 0$ always minimises the risk of winning nothing, and even when nothing is won, it is nearest to succeeding. Note further that for unconstrained bidders, $h = 0$ by definition.

The mutual optimum, then, reflects the simultaneous attempt by all constrained bidders to maximise their own shift values without knocking themselves out. On aggregate, this translates into the maximisation of the sum of shift values, subject to there being no bidder with $h > 0$ that wins no lots. This may yield multiple equilibria, as the following

---

[19] For simplicity, we assume here that L is greater than all marginal values MV_j.

examples illustrate. Note that both of these produce tied outcomes. To disambiguate, we select the outcome that maximises the number of winners in each case.

Example 2	Value for 1 lot	Value for 2 lots	Budget limit
Bidder A	80	130	80
Bidder B	100	150	130

Our own simulations indicate that the highest sum of shift values (subject to $h = 0$ for any bidder winning no lots) is 50. With shift values $h(A) = 50$ and $h(B) = 0$, we obtain:

Example 2a	Bid for 1 lot	Bid for 2 lots	Lots secured	Price paid	Value surplus
Bidder A	30	80	1	30	50
Bidder B	100	130	1	50	50

And with $h(A) = 30$ and $h(B) = 20$:

Example 2b	Bid for 1 lot	Bid for 2 lots	Lots secured	Price paid	Value surplus
Bidder A	50	80	1	50	30
Bidder B	80	130	1	30	70

The aggregate price paid and the aggregate value surpluses are the same in both Example 2a and 2b. To disambiguate between these, one might seek to minimise differences in value surpluses achieved by all bidders. Example 2a may thus be taken here to reflect the preferred outcome.

Unfortunately, an auction mechanism that automatically determined the mutually optimised bid set would still be vulnerable to strategic bidding. Suppose for example that Bidder B overstates its marginal valuation for the second lot by up to 30, while maintaining its true budget limit of 130. This would yield the following equilibrium:

Example 2c	Bid for 1 lot	Bid for 2 lots	Lots secured	Price paid	Value surplus
Bidder A	80	80	1	80	0
Bidder B	50	130	1	0	100

Here, Bidder B's insincerity is rewarded with a lower final price, due to the downward pressure of its bid-input on Bidder A's equilibrium shift-value. Bidder A's price has also increased by 50 to 80, which would serve possible spite motives held by Bidder B. Moreover, there appears to be little risk associated with this strategy: if Bidder B overplays its hand by reporting a marginal value greater than 80 for a second 2 lot, it wins both lots and pays 80 (Bidder A's budget limit), yielding a value-surplus of 70. This is the same as that in Example 2b, but now Bidder A is knocked out as a bonus.[20]

---

[20] If the benefit of knocking out Bidder A is already reflected in its valuation, Bidder B should have no economic preference between the outcomes in Examples 2b and 2c. However, it might derive a degree of *schadenfreude* from the latter.

Implications for Auction Design

In short, it does not seem to be feasible to accommodate budget limits directly within a second-price CA framework. Accordingly, the quest for perfect efficiency and incentive compliance might be futile in a cash-constrained environment, in which case it may no longer be worth insisting on the mathematical purity of a CA.

The drawbacks of SMRA-based schemes may seem more acceptable in this light. Nevertheless, these should not be underplayed. While policy-makers may be concerned with joint demand-moderation strategies, the greatest issue from a bidder perspective is undoubtedly exposure risk. Fortunately, mitigation may exist for both.

In our experience, complementary spectrum values tend to be most pronounced between blocks of 2×5 MHz and 2×10 MHz of LTE spectrum. This is because the marginal benefits of extra capacity and performance tend to decline more rapidly between 2×10 MHz and 2×20 MHz. Packaging the lots in such a way as to exclude isolated allocations of 5 MHz blocks to individual operators might thus help alleviate exposure risk. Allowing bidders to specify a minimum package size, as proposed by Ofcom for the forthcoming 2.3 GHz and 3.4 GHz award in the UK, would also achieve this.[21] Alternatively, one might allow spectrum pooling between operators under certain circumstances, such that a winner of a 5 MHz lot might be able to benefit from the cost and performance synergies associated with wider channel deployments.[22] This would increase the value of a single 5 MHz lot relative to that of 10 MHz. In our view, the balance of pros and cons would shift decisively in favour of SMRA, if exposure issues can indeed be sufficiently mitigated.

Policy concerns related to demand-moderation strategies might be abated by using a proxy SMRA mechanism, in which bidders supply comprehensive valuation sets and budget limits for the resources on offer.[23] Virtual proxy agents would then bid accordingly, on behalf of the participants.[24] This would preclude coordinated demand reduction, albeit the incentive to jointly moderate demand would remain.[25]

Budget constraints are easier to manage in an SMRA, although the possible threats to efficiency would remain. However, if new market entry can be discounted and

---

[21] See 'Notice of Ofcom's proposal to make regulations in connection with the award of 2.3 GHz and 3.4 GHz spectrum', consultation document published on 26 October 2015. The proposal is to allow bidders to specify a floor of between 5 and 20 MHz in the 3.4 GHz band, such that these bidders either win a package of that size (or greater) or no spectrum in the 3.4 MHz band at all.

[22] Some degree of coordination might be necessary between infrastructure-sharing partners to enable feasible spectrum-sharing outcomes. This might be achieved by aggregating bids though joint bidding vehicles. A proxy SMRA process (as sketched out further below) might also, conceivably, accommodate conditional bids.

[23] Note that if there are many lots and categories, the number of possible packages may be extremely large. It may not be feasible to value every single one of these. If necessary, the proxy auction could be put on hold to allow participants to fill in relevant gaps in their bid input.

[24] In any round, the proxy agents would bid the lowest amount necessary to pursue the most profitable package whose cost does not exceed the budget limit. Whereas accommodating budget limits directly in a CA framework is impractical, there is no issue with a proxy SMRA because there is no mutual optimisation taking place: the inputs to the proxy process are static.

[25] In a regular SMRA for generic objects, strong bidders may be tempted to drop lots prematurely if their weaker rivals do so, with a view to reach a rapid conclusion and avoid price inflation. A proxy process would deprive bidders of an opportunity to adjust their strategy on the basis of that of their rivals. This does not prevent unilateral demand moderation, reflected directly in the bid inputs. But unilateral demand moderation by a strong bidder is risky, because this may lead to a less profitable (and efficient) outcome if its weaker rivals continue to bid robustly. Note that a proxy process would also preclude interactive price discovery.

speculative bidders can be kept at bay, one might envisage redistributing a proportion of the auction proceeds equally between the participants. This may alleviate the impact of budget constraints on efficiency, by allowing participants to bid closer to their true valuations. In addition, such an approach may help defuse demand-moderation incentives, by reducing the potential gains that strategic demand reduction may yield. We explore this idea in greater detail in the following section.

6. Auctions with Partial Redistribution of Proceeds

Consider a rising clock auction with small increments, in which N identical lots are allocated. The small increments ensure that the winning bid prices per lot are approximately uniform. Our suggested redistribution mechanism would yield the following:

$$P(n) = n \cdot b - a \cdot (b - r) \cdot N/M$$

in which:

$P(n)$ is the net price paid by a winner of n lots;
b is the winning bid price per lot;
r is the reserve price per lot;
a is the proportion of the proceeds (after deducting the reserve price) that is redistributed;
M is the number of bidders.

A bidder that wins zero lots thus achieves a surplus amounting to its equal share of the proceeds, and the net price paid by each winner is reduced by the same amount. If the proportion a that is redistributed is 100%, the seller receives the reserve price of allocated lots, while any additional proceeds (that would otherwise accrue to the auctioneer) is returned to the bidders. In a spectrum award, this would mean that the industry collectively pays the exact amount that the state is willing to accept for the bandwidth on offer. This might be the cost of clearing spectrum, for example, or the opportunity cost associated with the deprival of alternative industrial uses of the bandwidth.

Key Policy Considerations

Forgoing potential surpluses is unlikely to be popular with finance ministers. Partial redistribution of spectrum auction proceeds may nevertheless reflect sound policy, for the following reasons.

First, and subject to the applicable reserve prices, this approach mitigates the risk that operators overpay collectively for spectrum. Such overpayments, if significant, could deprive the industry of capital, which could bear adversely on retail prices and investment. In today's money, the £22.5 billion raised by the UK government during the 3G auction in April 2000 is over 90% of current industry value.[26] We also estimate that

---

[26] According to the Bank of America Merrill Lynch Global Wireless Matrix 1Q15, Vodafone Group's Enterprise Value (i.e. the value of its equity plus debt) is 7.3× projected 2015 EBITDA. Applying this multiple to projected UK industry EBITDA would yield an aggregate Enterprise Value of GBP 33.5 billion for the UK mobile

operators spent between a quarter and a third of aggregate enterprise value in the multi-band auctions in the Netherlands (2012) and Austria (2013), both of which used a CCA format.[27] The risk of overpayment arises because the value of spectrum to individual operators includes the value of depriving rivals of a key input. On aggregate, these private valuations are likely to exceed the value added to the industry as a whole. If such transfers to the state were replicated each time spectrum usage-rights were renewed, the capital-base of the industry would soon erode, with potentially significant negative consequences for consumers and the wider economy.

Second, partial redistribution of proceeds may yield efficiency benefits that outweigh the impact on direct gains to the public purse. Hazlett et al. make a cogent case against the use of spectrum policy as an instrument to finance the state. One of their main points is that the social gains generated by the mobile industry far outstrip direct auction revenues, and that measures that jeopardise the former in favour of the latter are 'penny wise, pound foolish'. Based on their analysis of US market data for 2001–2008 versus FCC auction receipts over the same period, they obtain a ratio of 240× in favour of retail market efficiencies.[28] Allowing bidders to trade their claims to spectrum for cash could also yield long-term benefits for competition. An operator that relinquishes spectrum in favour of a more efficient user at least receives financial compensation for doing so. The extra capital may be used to mitigate the loss of spectrum, *inter alia*, by enhancing network capacity and quality through further infrastructure investment.

An obvious concern is that redistributive auctions might attract speculative bidders. New market entry might be discounted in mature markets, however, and stringent pre-qualification criteria could be applied to exclude prospective free-riders. Redistributions could also take the form of fiscal incentives, applied to certain allowable investments and operating costs.[29] This might counter the perception that undue windfalls were being offered *gratis* to private enterprises.

Redistributive Auction for a Single Object

Consider an auction for a single object, in which the seller is solely interested in identifying the value-maximising allocation. Let there be two bidders with the values and budget limits as shown in the table below. Again, we assume that the reserve price is zero.

Example 3	Lot value	Budget limit
Bidder A	70	50
Bidder B	60	60

---

network sector. Adjusting for inflation (based on World Bank data), GBP 22.5 billion is GBP 30.9 billion in today's money.

[27] Applying the same 7.3× multiple to Dutch and Austrian market EBITDA over the period 2012–2015 yields spectrum costs between 23% and 27% of Enterprise Value in the Netherlands and 31% to 37% in Austria (based on Bank of America Merrill Lynch Global Wireless Matrix 1Q13 data and projections). The auctions raised €3.8 billion and €2 billion respectively.

[28] Hazlett et al. (2012).

[29] In this case, the impact on budget constraints would be indirect: the constrained bidder might be able to raise additional funds on the basis of the deferred fiscal rebate.

Suppose in the first instance that a sealed, second-price auction is applied, without any redistribution of proceeds. If both bidders bid their actual budget limits, the object would accrue to Bidder B at a price of 50 (the amount of Bidder A's bid). Although Bidder A has a higher valuation, its budget constraint causes it to lose. Suppose now, instead, that 100% of the proceeds are redistributed. Bidder A is then in a position to bid its full value of 70: its maximum cash exposure, after the equal redistribution of proceeds, would be 70 – 50% × 70 = 35, which is well within its budget limit.[30] Given that Bidder B's valuation is in fact 60, Bidder A wins at a bid price of 60, paying a net price of 30 after redistribution. Bidder A thus obtains a net surplus of 40 (its valuation of 70 minus the payment of 30 to Bidder B), while Bidder B achieves a profit of 30. A capped-revenue auction may thus promote efficiency, by allowing the bidder with the highest valuation (if not the most cash) to win.

But here's the rub. By overstating its value to fractionally less than 70, Bidder B could increase its profit by 5 from 30 to 35 (50% of the increase in Bidder A's winning bid price). Yet if Bidder B overplays its hand, it wins the lot and pays a net price of 35, reducing its surplus to 25 (its value of 60 minus the payment of 35 to Bidder A). Assuming it knows nothing about Bidder A's bid limit, therefore, Bidder B stands to gain as much as it may lose by overbidding. In these circumstances, Bidder B might still be tempted to overbid on the basis of its expectations of Bidder A's bid limit.

By reducing the proportion a that is redistributed to less than 100%, however, overbidding becomes more likely a priori to generate losses than gains. If $a = 80\%$, for example, overbidding delivers at most 40% of any increase in the bid price. But if the overstated bid is winning, Bidder B pays 100% of the winning bid price minus its 40% share of the proceeds, reducing its surplus by up to 60% of the delta. Accordingly, the ratio of potential losses to potential gains from overbidding is now 3:2. If we introduce two extra bidders, this ratio increases to 4:1, because proceeds are now split four ways.[31] Such asymmetries in risk versus opportunity may well be sufficient to keep bidders honest.[32] A balance does need to be struck: the greater this asymmetry, the lower the mitigation for budget constraints that is offered by the redistribution mechanism. Taking Example 3 with four bidders, $a = 80\%$ would enable Bidder A to bid up to 62.5, which would still be enough to beat Bidder B. But if a drops to 66.6% or below, Bidder A can no longer win, even though it might be efficient for it to do so.[33]

---

[30] In this example, the redistribution of proceeds solves the budget-constraint problem entirely. If Bidder A's budget limit is less than 35, however, its bid will still be curtailed. Nevertheless, the redistribution of proceeds will at least reduce the risk of inefficient allocations caused by budget constraints.

[31] The potential gain is 80%/4 = 20%, while the potential loss (i.e. reduction in surplus) is 100% – (80%/4) = 80% of the bid delta. In this calculation, we ignore the possibility that losing rivals might in any case have carried on bidding beyond the strategic bidder's own natural bid ceiling. Foregone 'free-riding' opportunities would push the net losses from overbidding even higher.

[32] A single sealed-bid approach also reduces the risk of collusive risk-taking. If a rising multi-round auction were applied with full visibility of aggregate demand, weaker participants might be tempted mutually to overbid while three or more players remain: the risk of an abrupt end to the auction could be deemed sufficiently small, as this would require two bidders to drop out at the same price point. If, in addition, the clock price is lower than their expectation of the strongest bidder's bid limit, they might calculate that the benefits of driving up the winning bid price outweigh the risks.

[33] With $a = 80\%$, each of the four bidders receives 20% of the proceeds such the winner's net payment is 80% of the winning bid price. Bidder A's budget limit is 50, such that it can afford to bid up to 50/80% = 62.5. With

Redistributive Auction for Multiple Lots

In principle, this approach can be extended to any multi-lot auction format. However, a second-price CA may lead to outcomes in which some bidders pay higher total amounts yet obtain inferior packages.[34] Redistribution of a share of proceeds could then lead to incongruous situations in which winners of large packages receive net payments from winners of smaller packages. This could yield the impression that money was taken from the poor to give to the rich. For this reason, redistribution may more suitably be implemented within an SMRA-based framework.

Consider partial redistribution of proceeds within a (proxy) SMRA for N identical lots, with M participants. Let the bid prices rise by small increments in rounds. Suppose further that the smallest admissible non-zero package contains m lots, where $m \geq 1$. By overbidding on m lots, a bidder may thus raise the price of all N lots, which bears on the value of its own share of redistributed proceeds. The redistribution parameter a would, again, need to be chosen such as to discourage this. The ratio c of potential losses to potential gains from overbidding is given by the following formula:[35]

$$c = m \cdot M/(a \cdot N) - 1$$

Solving for a yields:

$$a = m \cdot M/((1+c) \cdot N)$$

We define the 'budget gain' $g(n)$ as the maximum extra % that redistribution allows a constrained bidder to bid on n lots without risk of net above-budget payments. Assuming zero reserves, $g(n)$ is given by the following:[36]

$$g(n) = a \cdot N/(n \cdot M - a \cdot N)$$

Note that the budget gain is reduced when reserve prices are introduced into the equation. The higher the reserves, therefore, the lower the scope to mitigate the distortive impact of budget constraints. Suppose now that the auctioneer seeks a ratio $c = 3:2$

$a = 66.6\%$, each bidder receives 16.65% of the proceeds and the winner pays 83.35% of the winning bid price. Bidder A can then afford to bid up to 50/83.35% < 60, which is less than Bidder B's bid limit.

[34] In the Swiss multiband CCA in 2012, for example, Sunrise paid over CHF 120 million more for a smaller package than the state-backed incumbent Swisscom (which enjoyed close to 60% market share). This represents an adverse price differential for Sunrise of over 33%.

[35] Derivation: let a losing bidder's value for m lots be $V = m \cdot v$, let e be the extra amount that it considers bidding per lot (over and above v) and, for completion, let r be the reserve price per lot. If our bidder drops out at a lot price of v, it is certain to achieve a surplus $S \geq a \cdot (v-r) \cdot N/M$, which is its share of the proceeds. If its overstated bid is winning at a price v', however, it achieves a surplus $S' = V - (v' \cdot m - a \cdot (v'-r) \cdot N/M)$.

At the limit $v' - v = e$, and by substituting $m \cdot v$ for V, we may express the potential loss from overbidding as $\Delta S^- = S' - S = (a \cdot N/M - m) \cdot e$.

If the overstated bid is losing, our bidder's strategy nets it a maximal increase in surplus $\Delta S^+ = a \cdot N/M \cdot e$.

The ratio of potential losses to gains can thus be expressed as $c = -\Delta S^-/\Delta S^+ = m \cdot M/(a \cdot N) - 1$. Note that this is a lower bound: the calculation does not take account of forgone free-riding opportunities (as a result of other bidders causing price inflation).

[36] Derivation: let L be the budget limit, and b_{max} the maximum bid price per lot that still respects the limit (after partial redistribution of the proceeds). The net price paid for n lots is $P(n) = n \cdot b - a \cdot (b-r) \cdot N/M$.

At the limit, with $r = 0$, we obtain $L = b_{max} \cdot (n - a \cdot N/M)$. The maximum bid price per lot is therefore given by: $b_{max} = L/(n - a \cdot N/M)$. Now let b_0 be the maximum budget-respecting bid price per lot for n lots if there were no redistribution, i.e. if $a = 0$. This is given by $b_0 = L/n$. Rearranging yields $b_{max}/b_0 = n \cdot M/(n \cdot M - a \cdot N)$ such that $g(n) = b_{max}/b_0 - 1 = a \cdot N/(n \cdot M - a \cdot N)$.

to dissuade overbidding. For different illustrative values of N, M, m and n, we then obtain the following values for the redistribution parameter a and the budget gain $g(n)$, assuming zero reserves:

Example 4	N	M	m	c	a	n	$g(n)$
Example 4a	3	4	1	3:2	53.3%	2	25%
Example 4b	6	4	1	3:2	26.7%	2	25%
Example 4c	14	4	2	3:2	22.9%	4	25%

With these values, a bidder can bid up to a quarter above its budget limit for a package of 2 lots in the 3 and 6-lot scenario or for a package of 4 lots in the 14-lot scenario. This is potentially significant in a cash-constrained environment: such budget gains could help restore unconstrained value maximisation as a truthful efficiency metric, by allowing participants to bid (closer to) their true values.

In multiband spectrum auctions, distinct redistribution parameters could apply to each band. In a combined auction for 6 lots of 800 MHz spectrum plus 14 lots of 2.6 GHz spectrum, for example, one could choose to redistribute 26.7% of the proceeds from the 800 MHz band plus 22.9% of those from 2.6 GHz (assuming we exclude single allocations of 2×5 MHz at 2.6 GHz). Overbidding in either band is then, a priori, equally risky.[37]

Finally, partial redistribution of proceeds diminishes the potential gains from strategic demand reduction. This should help defuse demand suppression incentives, as unilateral demand moderation becomes *relatively* more risky. A redistributive mechanism thus has the potential to address multiple policy concerns relating to allocation efficiency.

7. Concluding Remarks

We have sought to provide a bidder perspective on the prevailing multi-lot spectrum auction formats. Our conclusion is that SMRA and CA as currently applied may both give rise to significant problems for bid-strategy development.

Taking budget constraints directly into account within a CA seems impractical because it is likely to yield perverse bid incentives. However, the exposure risk in an SMRA might be mitigated by packaging lots such as to preclude LTE deployments in 2×5 MHz blocks. If the exposure problem can indeed be sufficiently alleviated, our view is that SMRA should unequivocally be preferred over CA whenever budget constraints are likely to arise. A proxy SMRA would rule out coordinated demand reduction, but at the plausibly acceptable cost of excluding interactive price discovery.

We suggested a further avenue for auction design, involving partial redistribution of auction proceeds among participants. This might further alleviate the budget-constraint

---

[37] We note that asymmetry in the redistribution percentage could lead to bias the bid prices. For example, if lots in two separate categories were perfect substitutes, bidders would be incentivised to bid higher relative amounts on the lots yielding the greater redistribution percentage, while discounting the others. This would maximise net surplus. However, because such a bias would apply equally to all bidders, we do not anticipate that this would interfere with the efficiency of the process.

problem, provided that speculative bidders can be kept at bay. This would also require public policy to be genuinely focused on optimising the allocation of scarce resources, rather than on raising funds for the state. A redistributive mechanism would also help defuse demand-moderation incentives.

Second-price Combinatorial Auctions gave us a fleeting glimpse of almost perfect incentive-compliance and efficiency, and anything short of this might now seem disappointing. Unfortunately, budget constraints appear to defeat this ideal. The existence of such constraints risks turning auction design into a game of spot the lesser evil.

References

Ausubel, L. M., Burkett, J. E. and Filiz-Ozbay, E. (2013) 'An experiment on auctions with endogenous budget constraints', working paper, December. Available at http://econweb.umd.edu/~filizozbay/Endo_budget.pdf

Brusco, S. and Lopomo, G. (2005) 'Simultaneous ascending auctions with complementarities and known budget constraints', working paper, August. Available at www.stonybrook.edu/commcms/economics/research/papers/2005/budcomp.pdf

Comptroller and Auditor General (2014) '4G radio spectrum auction: lessons learned', report by the Comptroller and Auditor General to the House of Commons, 6 March 2014, National Audit Office. Available at www.nao.org.uk/wp-content/uploads/2015/03/4G-radio-spectrum-auction-lessons-learned.pdf

Cramton, P. (2008) 'Auctioning the digital dividend', International Workshop on Communication Regulation in the Age of Digital Convergence, Karlsruhe Institute of Technology, December. Available at www.cramton.umd.edu/papers2005-2009/cramton-auctioning-the-digital-dividend.pdf

Cramton, P., Shoham, Y. and Steinberg, R. (2006) 'Introduction to combinatorial auctions', in P. Cramton, Y. Shoham and R. Steinberg (eds), *Combinatorial Auctions,* MIT Press. p. 32. Available at ftp://cramton.umd.edu/ca-book/cramton-shoham-steinberg-combinatorial-auctions.pdf

Doyle, C. (2010) 'The need for a conservative approach to the pricing of radio spectrum and the renewal of radio spectrum licences', advocacy paper prepared for the Australian Mobile Telecommunications Association, December, Section 5.1, pp. 21–22. Available at www.amta.org.au

Garside, J. (2013) '4G auction: treasury could have raised £3bn more', *Guardian*, 15 March. Available at www.theguardian.com/technology/2013/mar/15/4g-auction-chancellor-loses-out

Hazlett, T. W., Muñoz, R. E. and Avanzini, D. R. (2012) 'What really matters in spectrum allocation design', *Northwestern Journal of Technology and Intellectual Property*, 10(3), 102–103. Available at http://scholarlycommons.law.northwestern.edu/cgi/viewcontent.cgi?article=1159&context=njtip

Klemperer, P. (2002) 'What really matters in auction design', *Journal of Economic Perspectives* 16(1), 169–189. Available at: http://cramton.umd.edu/econ415/klemperer-what-really-matters-in-auction-design-jep-2002.pdf

PART VI
Secondary Markets and Exchanges

CHAPTER 36

Spectrum Markets: Motivation, Challenges, and Implications

Randall Berry, Michael L. Honig, and Rakesh V. Vohra

I. Introduction

The continued growth of wireless networks and services depends on the availability of adequate spectrum resources. Accelerating demand for those resources, due to the popularity of portable data-intensive wireless devices, are testing the limits of current commercial wireless networks, underscoring the need for changes in current spectrum allocations. This has prompted the Federal Communications Commission (FCC) in the United States to consider ways to increase the supply of spectrum allocated to broadband access and to introduce techniques for improving the utilization of existing allocations [1, Ch. 5].

Spectrum allocations generally fall into one of two categories: a *licensed* allocation gives exclusive use rights to the licensee, whereas an *unlicensed* allocation corresponds to the commons model in which the band can be shared by different applications and service providers [2]. Licensed spectrum typically carries restrictions on how it can be used, and is generally not transferable. Although these restrictions have been alleviated to some extent by the introduction of secondary spectrum markets [3], existing rules still inhibit the reallocation of spectrum to more efficient uses.

In contrast to the current "command and control" method for licensing spectrum, a spectrum *market* is based on a notion of spectrum property rights, which can be traded among buyers and sellers. The potential benefits of spectrum markets for increasing the efficiency of spectrum allocations is widely acknowledged. Thus far related discussions have focused on secondary markets, which allow service providers with licensed spectrum to lease their spectrum to other service providers. Transactions must be filed with the FCC for approval (which are automatic in some scenarios), introducing delays that increase transaction costs [3].

Here we reconsider the spectrum allocation problem without existing regulatory constraints. We start by providing general motivations for introducing spectrum markets. That is, a basic policy choice is whether to define and enforce spectrum property rights. From a social welfare point of view, this choice ultimately depends on whether spectrum is scarce, that is, if demand for it exceeds supply when it is free. If spectrum is abundant, then it can be made freely available (subject to appropriate power constraints), as in the commons model. If spectrum is scarce, then an allocation mechanism becomes necessary to mitigate interference and avoid a "tragedy of the commons."

To determine whether spectrum should be viewed as a scarce resource, we estimate in Section III the achievable rate per user assuming that spectrum assigned to non-government services between 150 MHz and 3 GHz is available for mobile broadband access. The calculation assumes a cellular infrastructure with a fixed density of Access Points (APs), and accounts for interference between adjacent cells using standard large-scale propagation models. Although the answer depends on assumptions concerning frequency reuse in different parts of the band, the power constraint, and the distance of the user from the cell boundary, we conclude that extensive spectrum sharing in the range considered (with a managed infrastructure) could provide a few Mbits/sec per user. While this is a relatively large number for many types of services, it is small enough that some distributed spectrum management is likely to be necessary to provide for a wide range of future services.

Spectrum markets are subsequently described along with implications for wireless services and networks. A challenge in creating such markets is how to define the spectrum assets being traded. While there has been a great deal of discussion about this in the legal literature (e.g., see [4]–[7]), here, our emphasis is on how spectrum markets may affect the provision of wireless services. Specifically, we speculate that the distinction between owned and leased spectrum assets would give rise to a two-tier market: in the upper tier spectrum property rights at particular locations (APs) are traded among spectrum owners (as in a commodities market); in the lower tier spectrum owners rent or lease their spectrum to service providers at particular APs via a spot market run by spectrum brokers. (Spectrum spot markets have been previously proposed assuming a given supply of spectrum, e.g., assigned by a regulator [8]. A key difference here is that the supply of spectrum at the lower tier is determined by the upper-tier spectrum market.)

An important feature of two-tier spectrum markets is that the market for spectrum is separated from the market for wireless services. That would allow efficient and flexible allocation of spectrum, while lowering entry barriers for wireless service providers. We conclude with a discussion of related interference management issues and implications for wireless system design.

II. The Motivation for Spectrum Markets

From an economic perspective a common objective of any resource allocation is to maximize efficiency, meaning the total utility (summed over all agents requesting the resource) derived from the allocation. Determining an efficient spectrum allocation is complicated by propagation characteristics, which can vary substantially across frequency and locations, and variations in application requirements (e.g., voice, internet, broadcast, emergency, etc). Hence the portion of the spectrum most suitable for a particular application can change over location and time. The relative value of the applications to consumers can also change across locations, times, and user groups. Moreover, the mapping of application requirements to spectrum depends on available technologies and their costs, which change over time. It is well known that properly designed markets are an effective approach for solving these types of problems. After highlighting the main problems with centralized (command and control) allocations, we compare market solutions with other proposed methods for making more efficient use of spectrum.

A. Problems with Centralized Allocations

Allocations of spectrum to different applications by government agencies, such as the FCC, are typically static, i.e., they apply for many years. Hence changes in traffic demands, potential applications, user preferences, and available technologies over time and locations have led to inefficient use of spectrum resources. *Dynamic* spectrum allocation that adapts to these variations over time- and geographic-scales of interest is, of course, extremely difficult to accomplish via a centralized allocation scheme due to the overwhelming amount of information and computation required.

These problems, based primarily on economic considerations, are not unique to spectrum. (They apply to land assets as well.) That prompted an early critique of the command and control model for spectrum allocation, and a proposal for spectrum property rights by the economist R. Coase [9]. More recently, spectrum markets have been proposed and discussed in [2], [4], [5], [10].

Inefficient spectrum use is one consequence of the current policy. A second is that it erects formidable entry barriers to the market for wireless services. This is due in part to the high degree of complementarity among spectrum licenses. To offer a wireless service over a broad coverage area, a potential entrant must acquire a package of associated spectrum licenses. If the service is tied to a given spectrum block, then the entrant must bid for that block across different geographic regions. Therefore, the value a provider obtains from a license is contingent on the bundle of licenses already owned. The resulting high cost of spectrum combined with the high infrastructure investment makes it difficult to enter the market on a small scale (e.g., within a small geographic region). Hence the current cellular market is confined to service providers that can make a huge initial investment (i.e., several billion dollars). The limited amount of competition means that service providers can potentially exert considerable influence over related markets (e.g., third-party hardware and software). (This also creates incentives for "rent-seeking" behavior [4].)

We consider instead a scenario in which spectrum is made available for sharing among many different applications across a large geographic region. Our main assumption is that the spectrum is partitioned into a set of spectrum assets that can be allocated among agents (service providers) at different locations. Spectrum markets might therefore be associated with a network of APs, which includes the current cellular infrastructure of base station towers. Each AP would have a set of particular spectrum assets, which are allocated among agents by a spectrum broker. Our main focus is on the scenario where spectrum is used to provide network access via these APs. Such connections could provide the commercial wireless services available today including voice, internet access, and broadcast radio/television.

B. Spectrum Sensing and Harvesting

In addition to the preceding criticisms of command and control allocations concerning *economic* efficiency, there are also the following engineering criticisms that pertain to *spectral* efficiency (i.e., bits per second per Hz):

1) Static assignments cannot exploit statistical multiplexing of traffic across different applications over shorter time scales. Even if the spectrum assignments are able to match *average* long-term demand, there are typically large fluctuations in demand, which lead to inefficient allocations over shorter time periods.
2) Centralized allocations often hinder the introduction of new technologies and services.

The first criticism can be addressed through the introduction of cognitive radios that seek out and exploit (or "harvest") idle bands in real-time [11]. This is the basis for the primary/secondary model for sharing vacant broadcast television bands (IEEE 802.22 standard). While schemes for harvesting idle spectrum would help to increase the spectral efficiency associated with particular bands, they do not address the previous issues concerned with static allocations of large blocks of spectrum to particular applications. Also, applications supported by spectrum harvesting are limited by the interference constraints with primary users of the band. Hence without the flexibility of reallocating this spectrum to higher-utility applications via a market, social costs due to inefficient allocations are still incurred.

Ideally, spectrum sensing and harvesting could be *combined* with spectrum markets. For example, the spectrum owner or licensee could negotiate a fixed usage fee for secondary users subject to acceptable interference constraints. (In contrast to the development of the IEEE 802.22 standard, interference constraints imposed on both secondary and primary users could vary substantially depending on the usage fee.) That would increase spectral efficiency while allowing markets to determine economically efficient allocations of spectrum to applications along with interference levels between primary and secondary users.

C. Markets and the Spectrum Commons

The motivation for spectrum markets is predicated on the assumption that spectrum is a scarce resource. That in turn depends on propagation characteristics, the transmitted power, which determines range and interference levels, and the nature of the traffic demands. For applications requiring *short-range* communications over links spanning no more than a few meters, there is an abundance of spectrum above 3 GHz that can be used. Interference is unlikely to be a major concern in these scenarios unless the density of wireless devices becomes very large. Hence the demand for short-range applications can be satisfied with a spectrum commons at high frequencies.

Longer-range communications (e.g., over 50 meters) requires lower frequencies, where interference becomes problematic. There are two primary concerns with using a spectrum commons for these types of applications. First, the propagation range becomes more difficult to predict at lower frequencies, since depending on the environment, signals may propagate much farther in certain directions than in other directions. This complicates interference management, especially without restrictions on where

APs can be deployed. (To alleviate this concern, at lower frequencies the commons model could be combined with a cellular infrastructure, which restricts AP locations, or restricts the use of particular frequencies at certain locations. Subject to those constraints, spectrum could otherwise be freely available.)

The second more fundamental concern is that as the demand for wireless services grows, demand may eventually exceed supply, creating excessive interference and lowering overall utility. Although the rate calculation in the next section gives a rough indication of whether spectrum is scarce or abundant, ultimately the value of a particular spectrum asset can be determined only through a market. If spectrum is truly abundant, then the prices of all spectrum assets will fall to zero, in which case the spectrum market reduces to the commons model [5]. (In practice, the price may not be zero, but rather large enough to cover any costs required to police the spectrum for violations of power constraints.) Of course, in general prices of spectrum assets should vary across frequencies and locations according to variations in demand and interference levels.

The preceding discussion implies that the frequencies at which the spectrum market transitions to a commons model can be automatically determined by the spectrum market (see also [12]). Namely, at high enough frequencies the price of the spectrum assets should be zero, since the propagation range is highly confined and therefore useful only to a small number of devices.

D. Spectral Efficiency, Cost, and the Supply Curve

The cost of providing a wireless service includes the cost of spectrum plus the cost of the associated devices and systems. Efficient use of spectrum should balance these two costs. This is reflected in the current design of wireless equipment and standards, which have been developed under the assumption that spectrum scarcity poses a major limitation on system capacity and revenues. The high cost of spectrum has led to the development of cellular standards with sophisticated (expensive) air interfaces, which attempt to maximize spectral efficiency.

Inexpensive spectrum, obtained through a spectrum market, would likely motivate the deployment of inexpensive devices and systems, which operate at much lower spectral efficiencies, compared with current cellular standards. (In addition, less expensive devices may not be able to support as wide a range of frequencies, further limiting the amount of spectrum that can be effectively shared.) In economic terms this implies that if a market is used to allocate spectrum, then the "supply" of spectrum is not inelastic. (Here "spectrum supply" loosely refers to the amount of services that can be provided with a particular spectrum resource.) Even if the physical amount of spectrum available for a particular service is fixed, the spectrum supply effectively increases with the price since more expensive spectrum justifies the use of more expensive equipment that achieve higher spectral efficiencies. (Hence the same amount of physical spectrum can provide more services.)

This discussion is illustrated in Fig. 1, which shows supply and demand for a particular (fictitious) spectrum resource versus price. (The meaning of "spectrum resource" is discussed in Section IV.) The supply curve has a positive slope, and an equilibrium price p^* is shown, which determines whether or not the spectrum should be used as a commons. If p^* is sufficiently small, then the transaction costs incurred from running a

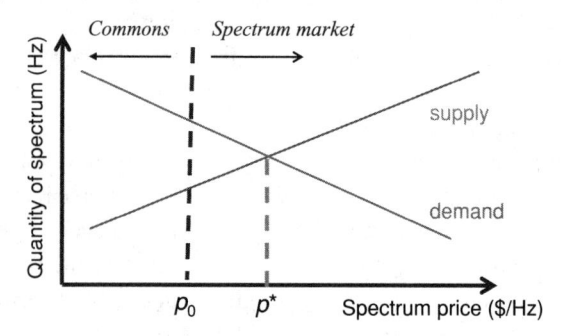

Figure 1. Comparison of equilibrium spectrum price p^* with the transition price p_0. For the case shown $p^* > p_0$ implies that a spectrum market is more efficient than a commons model.

market exceed the cost of interference (or loss in utility) with a commons model. This is represented by the boundary $p = p_0$ in the figure. Since p^* is shown to the right of the boundary, a spectrum market is needed to coordinate spectrum usage, whereas if p^* were instead to the left of the boundary, the commons model would be more efficient.

III. Is Spectrum Scarce?

We now examine the assumption that spectrum is a scarce resource. This depends in part on regulatory and economic considerations, but is ultimately a technical problem of determining if network architectures can scale effectively as their applications grow [13]. Here we address this question for a specific network architecture, namely an infrastructure of APs, which represents the primary network architecture deployed today for commercial wireless services. (Moreover, this architecture has better scaling properties than alternatives, such as mesh networks, given a sufficient number of APs.)

Our objective is to give a rough estimate of what rates could be provided with more extensive spectrum sharing. We therefore assume that all spectrum between 150 MHz and 3 GHz is pooled for commercial services, excluding spectrum currently assigned for military and government use. The particular bands used in the calculation are shown in Table 1. Note that broadcast television bands are included in this list. Demand for those services might be satisfied by a combining wire-line cable services with the

Table 1. *Frequencies used to calculate achievable rates*

Broadcasting TV (total: 348 MHz)	174–216 MHz, 470–608 MHz, 614–764 MHz, 776–794 MHz
Fixed, Mobile, Satellite, Amateur	150.8–157.0375 MHz, 157.1875–162.0125 MHz, 173.2–173.4 MHz, 450–460 MHz,
(total: 669.7625 MHz)	764–776 MHz, 794–902 MHz, 928–932 MHz, 935–941 MHz, 944–960 MHz, 1390–1395 MHz, 1427–1429 MHz, 1850–2025 MHz, break 2110–2200 MHz, 2300–2310 MHz, 2385–2417 MHz, 2450–2483.5 MHz, 2500–2655 MHz
The table is based on U.S. Frequency Allocation Table as of October 2003, and includes all non-Federal Government exclusive spectrum between 150 MHz and 3 GHz. The total bandwidth shown in the table is 1.018 GHz.	

wireless infrastructure assumed here. (Current broadcast services, such as television, could be offered as multicast services over this type of cellular architecture. That would make more efficient use of spectrum since the signal would be transmitted in a given cell only when someone in the cell requests it.)

We also assume that the APs are deployed with particular frequency reuse patterns for interference mitigation. This should give an optimistic indication of what rates are achievable with full coordination among service providers; a relatively low rate per user indicates that spectrum is scarce, and needs to be carefully managed, whereas a very high rate indicates that simple spectrum management schemes (e.g., the commons model) are likely to be adequate.

A. Cellular Model

To compute an achievable rate per user, a cellular topology with hexagonal cells is assumed over which both APs and users are uniformly spread with densities ρ_{ap} and ρ_u, respectively. Each AP serves the same number of users. Here we focus on the achievable rate for the *downlink*, i.e. communication from the AP to each user. We expect similar results for the uplink.

We make the following assumptions:

1) The entire set of available frequencies in Table 1 is quantized into 1 MHz pieces, which are allocated across the APs according to a standard frequency reuse pattern.
2) Each AP transmits with uniform power spectral density over the set of assigned channels.
3) Each AP applies Time-Division Multiple Access (TDMA) to multiplex users within the cell.
4) Interference from neighboring cells only is taken into account. Also, we assume that the signal attenuation is determined according to large-scale propagation models, and do not account for random fluctuations (fading).

For a particular 1 MHz channel at frequency f, the rate for a particular user at distance d from the AP is assumed to be the Shannon rate

$$R(d, f) = \frac{1}{n} \log\left(1 + \text{SINR}(d, f)\right), \tag{1}$$

where $n = \rho_u/\rho_{ap}$ is the number of users per cell, and SINR is the Signal-to-Interference-Plus-Noise Ratio given by

$$\text{SINR} = \frac{P_r(d, f)}{N_0 + \sum_{i \in \mathcal{I}} P_r(d_i, f)}. \tag{2}$$

Here, $P_r(d, f)$ is the received power for a user at distance d at frequency f, $\mathcal{I}$ is the set of interfering APs, and $P_r(d_i, f)$ is the received interference power from the i-th interfering AP.

This rate assumes optimal coding and delay-tolerant applications and so gives an optimistic estimate of the rate that can be obtained. (However, we ignore the possibility of using multiple antennas and cooperative techniques to increase the achievable rate per mobile.) In principle, we can account for channel variations (fading), practical

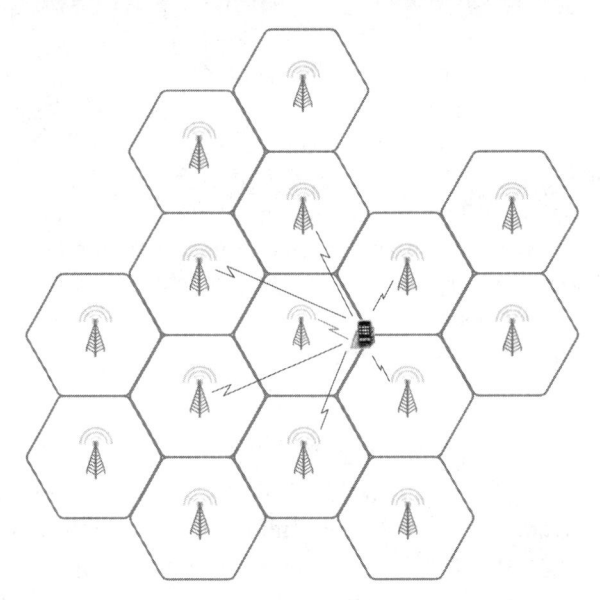

Figure 2. Cellular system used to calculate achievable rates. The location of the worst-case user with the lowest rate is shown. Only interference from the neighboring cells is taken into account.

coding schemes with delay constraints, interference from more distant cells, etc. by adding an appropriate margin to the SINR.

The achievable rate for a particular user depends upon where they are in the cell. The lowest rate corresponds to a user at the corner of a cell, as illustrated in Fig. 2. It is then straightforward to compute the SINR in (2) based on the hexagonal geometry. To determine the received power $P_r(d, f)$ we use Hata's outdoor propagation model for the frequency range 150 MHz to 1.5 GHz, and its extension to PCS for $f > 1.5$ GHz (see [14, Ch. 4]). The rate is then obtained by quantizing the spectrum bands shown in Table 1 into 1 MHz pieces, and summing the rate function over those bands. To account for losses expected in practice (e.g., due to other channel impairments) a 6 dB margin is subtracted from the SINR. Also, we assume a fixed power per unit area, i.e., the total power across the area covered does not scale with the density of APs. That also constrains the background interference level.

B. Achievable Rate Results

Results from the preceding calculation are shown in Fig. 3. For these plots the transmit power density is $P_t = -40$ dBm/Hz for all APs, the noise power spectral density is $N_0 = -174$ dBm/Hz, the base station antenna height is 30 m, and the receiver antenna height is 1 m.

Figure 3(a) shows achievable rates for the worst-case user at a corner point of the cell versus user density. Different curves are shown for different values of the cell radius r. For these plots the frequency reuse factor N is chosen such that the rate at each frequency is maximized. (Possible values are 1, 3, 4, or 7.) Hence the rate per user decreases as the cell radius increases. (This is not necessarily true if N is fixed, since the

(a) rate/user vs ρ_u

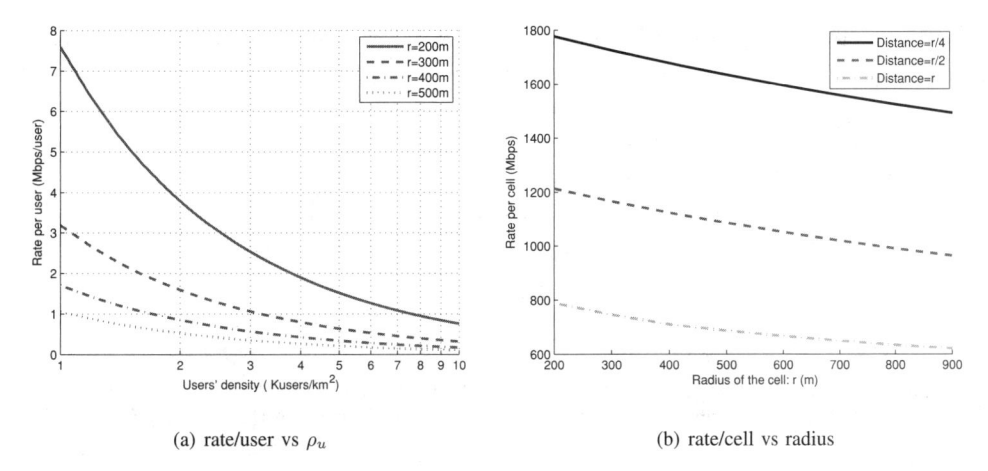

(b) rate/cell vs radius

Figure 3. Achievable rates with a cellular infrastructure assuming all frequencies shown in Table 1 are available for sharing: (a) worst-case rate per user versus user density; (b) rate per cell versus cell radius.

interference decreases with the cell radius.) As a specific example, for a large city like Chicago, which has a population density of approximately 4000 people per square km, the worst-case rate per user increases from about 0.3 to 2 Mbps as the cell radius shrinks from 500 m to 200 m. (For this example, the spectrum efficiency is about 0.7 bps/Hz per cell, which is less than that expected for Long-Term Evolution cellular systems. This is because the user is located on a cell boundary and we assume omni-directional antennas without cell sectorization.)

Figure 3(b) shows how the achievable rate varies with the distance from the AP. The achievable rate per cell at different distances from the AP is shown versus the cell radius. (The rate per user is then obtained by dividing this rate by the user density.) These results indicate that the rate increases by about 50% when moving from the edge of the cell to distance $r/2$, and more than doubles if the distance decreases to $r/4$.

C. Interpretation

The preceding results indicate that if a cellular infrastructure with cell radii less than 200 m has access to all of the bandwidth in Table 1, then rates well above 1 Mbps could, in principle, be made continuously available to every member of a dense urban population. Furthermore, the achievable rate increases substantially with the density of APs, and with the fraction of inactive users (as opposed to assuming all users are active).

Since the range of rates indicated here are sufficient to support a wide range of near-term mobile services, one might conclude that simple allocation schemes, such as those based on a commons model, may be adequate for spectrum allocation. However, the rates shown in the preceding section are optimistic in that the availability of a managed infrastructure with coordinated frequency reuse has been assumed. Also, we have assumed an extreme case in which a large amount of spectrum currently assigned to many different applications is pooled for shared use. If less spectrum is available, the rates decrease accordingly.

As discussed in Section II-D, another reason why spectrum may still be scarce even with extensive sharing is that the additional spectrum would encourage the use of simpler systems having lower spectral efficiency, e.g., by using modulation and coding schemes that operate at rates well below the Shannon limit. Additional bandwidth also enables a reduction in transmit power and associated interference. Finally, although the rates reported here may seem large (especially when divided among a relatively small set of active users), it is possible that in the long-term new applications may arise that require rates on the order of (or beyond) what are indicated here.

Hence, we conclude that even with extensive spectrum sharing and with coordination of spectrum resources across APs, the demand for spectrum may exceed supply as users, applications, and systems proliferate. We therefore discuss how spectrum markets might be defined that achieve efficient allocations and benefit consumers of wireless services.

IV. The Challenge: Spectrum Property Rights

A basic requirement for any market is to define clearly the asset being traded. The purpose of property rights in spectrum is to limit the amount of interference an owner (or licensee) receives from transmitters operated by other owners (or licensees). The definition of these assets influences the technical constraints under which wireless networks operate, the valuation of spectrum by market participants, and the resulting market mechanism. This has prompted extensive discussions of how such rights should be defined in the legal literature [4]–[7], [9]. We briefly highlight some of the key issues, and subsequently propose a definition based on transmitted power mask that will be used as a basis for the subsequent discussion of market features.

From an economic point of view, the definition of a spectrum property right (or asset) should satisfy the following criteria:

1) It should be *clear* and *easily enforced*.
2) It should be *transparent*, meaning that it is straightforward to determine the benefit of ownership (or rental).
3) It should facilitate efficient allocations (e.g., avoid "natural" monopolies).
4) It should be *flexible*, so that it can be applied to different radio environments with varying traffic and propagation characteristics.

Satisfying all of these criteria is difficult due to the fundamental problem of interference management in wireless networks. For example, to satisfy the second criterion a property right would ideally guarantee the owner of a spectrum asset that received interference power from transmitters using other spectrum assets will not exceed a given value. Due to random propagation characteristics, such a right would be difficult if not impossible to enforce, failing the first criteria. Moreover, for many applications, attempting to approximate such a right (e.g., by enforcing large frequency re-use distances) would be overly conservative and lead to inefficient use of spectrum, failing the third property. (Such a conservative approach has been common practice and has contributed to the current inefficiencies in spectrum use.) Hence in practice the preceding criteria must be relaxed when defining spectrum property rights.

A. Power Mask

In its most general form a spectrum property right can be defined in terms of constraints on transmitted and/or received power over frequencies, time, and space. This definition should depend on the propagation environment, traffic characteristics, and application requirements, which determine "acceptable" interference levels. For example, power and interference constraints for a community broadcast type of service in a rural area are clearly different from those for a high-speed data service in a city center.

Attempting to divide regions of frequency/time/space into a set of spectrum property assets, which *a priori* accounts for all of the previous factors is clearly impractical. Rather, an initial (perhaps coarse) definition of spectrum asset can be provided, which is subsequently refined through negotiations among spectrum owners [7]. In general, there is a tradeoff between the *front-end* cost of defining an initial set of spectrum property rights to minimize potential interference, and the *back-end* cost of negotiating subsequent changes to those rights once a market is introduced [6]. A challenge is to provide a set of definitions for spectrum property rights that balance those costs.

Because of the difficulties associated with constraining received power, we will assume that a spectrum asset is defined in terms of a *transmitted* power mask, which limits radiated power in a particular band (and outside the band) in a particular geographic region for a given time duration. This can serve as the basis for the spectrum markets described in the next section, provided that the power masks can vary over locations, times, and frequencies. Specifically, the power mask may depend on factors such as the antenna height and the distance of the AP or mobile to the boundary of the given geographic area. (Such restrictions currently exist in some bands.) Similarly, the power limit may increase with frequency (due to higher attenuation) and also increase when a neighboring system is expected to be lightly loaded.

In practice, the variations of spectrum power masks over time/frequency/space should be negotiated by neighboring service providers and spectrum owners, depending on their intended applications. (Rather than change the definition of power mask, the negotiations could instead settle upon monetary compensation for high interference levels.) Defining spectrum property rights in this way provides great flexibility, since the spectrum assets can be adapted to the environment and applications, and can evolve according to user demands and changes in technology. However, it does not give hard guarantees about received interference. Instead constraining transmit power would provide reasonable (statistical) expectations about neighboring interference levels. Moreover, if a spectrum owner or service provider requires a stronger guarantee about the level of interference, it would have the option of acquiring the neighboring spectrum assets to prevent other transmitters from using them.

B. Owning Versus Leasing

For the spectrum markets to be described there is an important distinction between owning and renting (or leasing) a spectrum asset. The definition of the spectrum asset depends on this distinction. Namely, an *owned* spectrum asset has a long (perhaps unlimited) duration, and is traded as a land right. Spectrum assets can be *rented* or leased by the spectrum owner. The duration of the spectrum asset being rented can vary across frequencies, agents, and locations, and determines market dynamics. A short duration (say, less than a day) may be associated with a spot market for short-term commercial use (analogous to electricity markets [15]), whereas a long duration (e.g., years) may be associated with broadcast services that require continual use of spectrum. Note, in particular, that a spectrum owner could conceivably decide to switch applications (e.g., migrate from broadcast to cellular), or sell spectrum rights to another owner once a rental agreement has expired.

V. Two-Tier Spectrum Market

Allowing spectrum property rights to be flexibly defined and traded would produce major changes in markets for wireless services. The most visible of those changes would be the separation of spectrum ownership from the provision of wireless services. One reason a service provider may prefer to rent or lease spectrum, rather than trade it as an owner, is that leasing carries relatively low transaction costs compared to buying and selling the spectrum asset. This becomes especially important for services which require intermittent use of spectrum. (Examples may include emergency or monitoring, and delay-tolerant applications, such as video downloads, that can exploit periods of light usage.) A second reason is that leasing avoids "maintenance" costs, such as negotiating power levels with neighboring spectrum owners along with associated policing functions. Hence leasing or renting spectrum on a short-term basis allows for flexible and efficient allocations that vary over time and locations.

The distinction between owned and rented spectrum assets would lead to the two-tier spectrum market shown in Fig. 4. The upper tier consists of spectrum owners that

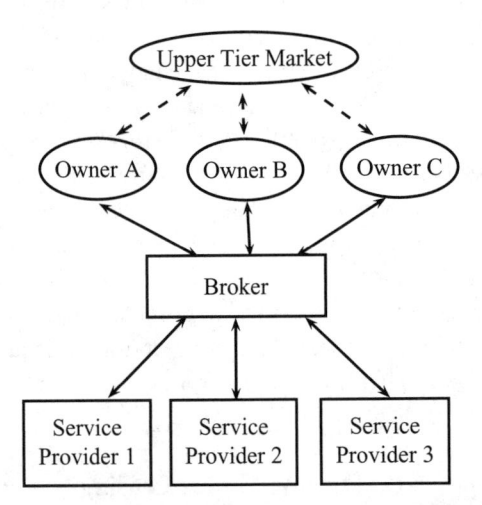

Figure 4. Two-tier spectrum market: Spectrum assets corresponding to particular locations are traded by owners at the upper tier; those assets are then rented or leased to service providers via lower-tier spot markets at the APs. The spot market at a particular AP (or set of APs) is managed by a spectrum broker.

buy and sell spectrum rights with unlimited duration. Owners could choose to rent or lease their spectrum to service providers at the lower tier market through a spectrum broker, which manages spectrum assets at particular locations. The upper- and lower-tier markets would likely operate at two different time-scales: owned spectrum assets at the upper tier might be traded on a relatively slow time scale (months or years), whereas rented spectrum assets at the lower tier could be negotiated over short time scales (e.g., hours or even minutes, depending on the application).

We next describe additional features of two-tier spectrum markets. They are organized as properties associated with the spectrum owners, service providers and the spectrum broker.

1) Spectrum Owners: Spectrum assets at a particular AP, or geographic region, would be traded according to a conventional market mechanism, as in a commodity market. An issue, which arises with a spectrum asset, is that its value depends on the interference generated by nearby mobiles and APs, which can change over time. This may encourage aggregation of spectrum assets across neighboring locations. Strong interference between nearby APs with different owners may have to be resolved through additional negotiations.

Spectrum owners would have an incentive to rent their assets to service providers with applications that generate the most revenue. (That could, of course, vary with the time of day.) Allowing owners to trade spectrum assets implies that each asset could be reassigned to applications that generate higher revenue, or alternatively, to groups that want to purchase spectrum for non-commercial purposes (e.g., community broadcast). Furthermore, with extensive spectrum sharing, as assumed in Section III, many spectrum assets would be available at each AP and the transaction costs for trading spectrum would presumably be low. Hence the way spectrum is used would be determined by market supply and demand, and the price of a particular spectrum asset would be tied to the long-term expected revenue it is expected to generate.

2) Service Providers: A service provider offers a set of wireless services to end customers through a particular pricing scheme. With spectrum markets a service provider could purchase (rent) spectrum on a short-term basis. As a consequence, a service provider need not build out a national footprint of APs, which use the same spectrum. The spectrum could be rented via the spectrum spot market at desired locations according to customer demand. This, of course, assumes that the customers have frequency-agile radios that are able to switch to the assigned band and use the appropriate modulation and coding format. (Notifying the end-user what particular band to use would also require some signaling overhead, although that is likely to incur a relatively small cost.)

The service provider could also conceivably rent the necessary equipment at an AP from an equipment manufacturer. (That cost would also account for the cost of the tower on which it is mounted.) Hence the combination of spectrum markets and equipment rentals could dramatically lower the entry (sunk) costs for a service provider, potentially increasing competition along with service options.

Service providers may also provide an arbitrage function for customers. Namely, large fluctuations in the demand for particular wireless services may cause large price fluctuations for spectrum rental. Since end customers typically prefer predictable (e.g., flat-rate) pricing plans, a service provider may provide such an option, but with a premium, which accounts for statistical fluctuations in the price of spectrum. (Alternatively, the arbitrage function may be performed by third-party resellers.) A service provider may also choose to negotiate longer-term contracts for spectrum with the spectrum owner to provide more reliable Quality of Service. (Another possibility is to create a spectrum "futures" market, analogous to current electricity futures markets [15], in which rights to use particular spectrum assets at future times are traded.)

3) Spectrum Broker: The lower-tier spot market for spectrum assets at each AP could be managed by a spectrum broker, which determines how spectrum assets are allocated among service providers, and how much each service provider pays for each spectrum asset. The allocation method, or mechanism, must balance efficiency with complexity, a topic discussed in the literature on *algorithmic mechanism design* [16].

For example, the allocation could be determined through an auction mechanism in which the broker collects bids to buy from the service providers, bids to sell from the spectrum owners, and subsequently determines the allocation along with the price for each spectrum asset. The auction would then be repeated as spectrum assets become available (i.e., as they are released by service providers).

Alternatively, the spectrum broker could announce a set of prices for the available spectrum assets, and adjust the prices over time to maximize expected revenue or to clear the market periodically. This approach is generally simpler, and requires less overhead (information exchange) than an auction mechanism. However, a well-designed auction mechanism can achieve either a higher efficiency or more revenue (whichever is the objective). The choice between these two approaches should depend on the "thickness" of the market; with relatively few buyers and sellers (a "thin" market) an auction mechanism becomes simple to implement, so may be preferred. With many buyers and sellers the loss in efficiency (or revenue) with the pricing scheme becomes small, so that the pricing scheme may be preferred.

With either approach the protocol for information exchange (bids for assets or price adjustments) could be automated and run on a spectrum server. (See also [8], which proposes a related type of spectrum server.) Hence the lower-tier market for renting spectrum assets could operate on a very fast time scale with small transaction costs.

VI. Interference Management

For the spectrum markets considered here interference management at a basic level is accomplished by the power masks corresponding to spectrum property rights. However, fixed power masks (as defined today) would lead to inefficiencies due to changes in the deployment of wireless networks (e.g., density of APs), demand for spectrum across time and locations, and propagation characteristics. Hence depending on the location and time, some power masks may be too stringent (lowering the value of the services provided), and some may permit excessive interference (lowering the value of services at neighboring locations). We next describe ways in which spectrum markets may address this issue. Much of the following discussion applies to both owners and service providers, which we will refer to as agents.

A. Local Cooperation

Rather than fixing power masks across time and locations, power masks could be *adapted* through negotiations, or by means of particular protocols. For example, adjacent owners may negotiate cross-rental agreements for the same spectrum, or alternatively, agree to *cross-payments* for reducing or increasing interference. Specifically, an agent A at a particular AP could offer to pay agent B at a neighboring AP to reduce interference by imposing a smaller power mask on its customers. Alternatively, agent A could pay B to accept more interference, allowing A to increase its power mask.

To maximize efficiency (i.e., total utility for both service providers), the cross-payment for interference should match the loss in utility (or externality) incurred by the neighboring agent that agrees to reduce its power, or accept more interference. To find this cross-payment, the agents would need to exchange information about their utilities and cross-channel gains. This may be worthwhile, provided that the signaling overhead is manageable (i.e., occurs on a sufficiently slow time scale), and the agents are willing to exchange this information. For example, that might be the case if the same agent is managing spectrum assets across neighboring APs, as in current cellular systems. (The APs may then exchange "interference prices" to adjust powers, taking into account externalities due to interference [17].)

Competing agents at neighboring APs may not be willing to exchange information about actual utilities (e.g., expected revenue), but could negotiate cross-payments through a bargaining procedure. Moreover, there would be the possibility of negotiating a *protocol* for adapting powers that specifies how power limits and cross-payments are determined as a function of measured interference at particular locations. In this way, power masks can adapt to changing application requirements, network topology (e.g., if an agent wishes to add or remove an AP), and anticipated traffic.

We also point out that local cooperation can be applied to the commons model (e.g., 802.11 networks). An example of this, which requires minimal signaling overhead, is distributed dynamic channel allocation. As the traffic and network load increase, more extensive local cooperation, such as interference pricing and cross-payments, might be introduced. For example, an AP may offer to carry traffic from neighboring locations, rather than accept the additional interference from another AP. (The effect of such negotiations on the density of APs is considered in [18].) Referring to Fig. 1, by mitigating interference through local cooperation, the threshold price p_0 increases, thereby extending the range of frequencies over which the commons model is more efficient than a spectrum market.

B. Asset Aggregation

To avoid negotiating interference levels with neighboring agents, an agent may try to purchase or rent similar spectrum assets at neighboring locations. In that way, the agent would have more control over interference within a particular region. (It may then be willing to allow more interference at the boundaries from different APs.) Also, the agent may wish to purchase or rent similar spectrum assets at neighboring APs to ensure adequate coverage. An agent's valuation for "bundles" of spectrum would then exhibit *complementarities*, i.e., the value an agent places on a particular asset may be greater if the agent owns (or rents) neighboring assets. (More generally, the value of a bundle of assets may be greater than the sum of the values of each asset alone.)

The existence of complementarities implies that spectrum markets should allow spectrum assets to be *aggregated*. Agents could then bid for bundles of assets across different locations. (The power constraints within the region could be relaxed provided that the total received power at the boundary stays the same.) This complicates the design of spectrum markets, since a particular asset may appear in many bundles, each containing a different combination of assets. Finding an efficient allocation in general can be computationally difficult and may require excessive information exchange (bids and asks) [19]. These costs could be reduced by adopting a less efficient mechanism.

Finally, complementarities for spectrum assets may also exist over adjacent frequency bands. For example, an agent may wish to own or rent those bands to manage adjacent-channel interference, e.g. to allow for simpler receiver filters. Hence spectrum assets might be bundled across both frequency and spatial locations.

VII. Implications for Wireless System Design

Because spectrum has been viewed as a scarce resource, wireless systems engineering has put a premium on spectral efficiency (bits/sec/Hz). As spectrum becomes more abundant through extensive sharing, the importance of spectral efficiency may diminish. Instead, the design objective with extensive sharing may shift toward lowering equipment cost and/or increasing power efficiency. That is, power could be reduced to provide longer battery life and reduce interference. Hence an abundance of spectrum may encourage the use of less expensive, low-power, wideband (spread spectrum) systems.

Another consequence of more abundant spectrum is that the economic benefit from spectrum use becomes limited by transaction costs for repeated spectrum (re)allocations. Hence an objective should be to minimize those costs. This suggests developing standards for broker mechanisms for the lower-tier spot market that can be included as a core part of cognitive radios (in addition to standard air interfaces). Service providers could then build applications, which exploit standard dynamic protocols for acquiring spectrum.

Alternatively, the commons model might be used in some bands with distributed interference management schemes, such as interference pricing and local negotiations. Those schemes effectively *introduce* transaction costs to mitigate interference, which may be needed as the density of nodes increases. The choice between these two approaches (market or commons) would again depend on the price of spectrum and associated transaction costs.

VIII. Conclusions

Allowing large parts of the radio spectrum to be traded and rented across geographic locations and time would provide incentives for more efficient use, and encourage alternative models for dynamically sharing spectrum. A key consequence is that spectrum ownership could be separated from the provision of wireless services. That would lower entry barriers, and thereby facilitate the introduction of more diverse sets of services. More abundant spectrum may also motivate the deployment of different types of radio systems, which operate at lower spectral efficiencies.

We have highlighted some of the main technical features and issues associated with spectrum markets. Additional issues not discussed include networking functions, such as handoff, and spectrum policing to enforce property rights. Perhaps more difficult to resolve are the policy issues associated with the transition away from current allocations. It remains to be seen whether the potential benefits of spectrum markets can overcome the obstacles to the necessary spectrum policy reforms.

Acknowledgment

The authors thank Hang Zhou for generating the plots in Figs. 3(a)–3(b), and Junjik Bae, Eyal Beigman, Hongxia Shen, and Hang Zhou for their contributions to an earlier version of this paper presented at DySPAN 2008. This work was supported by the U.S. National Science Foundation under grants CNS-0519935 and CNS-0905407.

References

[1] Federal Communications Commission, "The national broadband plan," March 2010. [Online]. Available: www.broadband.gov/plan/5-spectrum.

[2] J. M. Peha, "Approaches to spectrum sharing," *IEEE Communications Magazine*, vol. 43, no. 2, pp. 10–12, 2005.

[3] FCC, "The development of secondary markets - report and order and further notice of proposed rule making," Federal Communications Commission Report 03-113, 2003.

[4] T. W. Hazlett, "The wireless craze, the unlimited bandwidth myth, the spectrum auction faux pas, and the punchline to Ronald Coase's 'big joke': An essay on airwave allocation policy," *Harvard Journal of Law & Technology*, vol. 14, no. 2, pp. 335–567, 2001.

[5] G. R. Faulhaber and D. J. Farber, "Spectrum management: Property rights, markets, and the commons," *Rethinking rights and regulations: institutional responses to new communication technologies*, pp. 193–226, 2003.

[6] P. J. Weiser and D. Hatfield, "Spectrum policy reform and the next frontier of property rights," *George Mason Law Review*, vol. 15, p. 549, 2007.

[7] T. W. Hazlett, "A law & economics approach to spectrum property rights: A response to Weiser and Hatfield," *George Mason Law Review*, vol. 15, no. 4, 2008.

[8] O. Ileri and N. Mandayam, "Dynamic spectrum access models: toward an engineering perspective in the spectrum debate," *IEEE Communications Magazine*, vol. 46, no. 1, pp. 153–160, January 2008.

[9] R. H. Coase, "The Federal Communications Commission," *The Journal of Law and Economics*, vol. 2, no. 1, p. 1, 1959.

[10] E. R. Kwerel and J. R. Williams, "Moving toward a market for spectrum," *Regulation: The Review of Business and Government*, 1993.

[11] Q. Zhao and B. M. Sadler, "A survey of dynamic spectrum access," *IEEE Signal Processing Magazine*, vol. 24, no. 3, pp. 79–89, May 2007.

[12] M. Cave and W. Webb, "Spectrum licensing and spectrum commons: where to draw the line," in *International Workshop on Wireless Communication Policies and Prospects: A Global Perspective*, 2004.

[13] D. P. Reed, "How wireless networks scale: the illusion of spectrum scarcity," in *International Symposium On Advanced Radio Technologies (ISART)*, 2002.

[14] T. S. Rappaport, *Wireless Communications: Principles and Practice*, 2nd ed. Prentice Hall, 1996.

[15] S. Hunt and G. Shuttleworth, *Competition and Choice in Electricity*. John Wiley & Sons, May 1996.

[16] N. Nisan, T. Roughgarden, E. Tardos, and V. V. Vazirani, Eds., *Algorithmic Game Theory*. Cambridge Univ. Pr., 2007.

[17] J. Huang, R. Berry, and M. Honig, "Distributed interference compensation for wireless networks," *IEEE Journal on Selected Areas in Communications*, vol. 24, no. 5, pp. 1074–1084, May 2006.

[18] J. Bae, E. Beigman, R. Berry, M. Honig, and R. Vohra, "Incentives and resource sharing in spectrum commons," *IEEE Symposium on New Frontiers in Dynamic Spectrum Access Networks (DySPAN 2008)*, pp. 1–10, Oct. 2008.

[19] H. Zhou, R. Berry, M. Honig, and R. Vohra, "Complementarities in spectrum markets," in *Allerton Conference on Communications, Control and Computing*, Monticello, IL, 2009.

CHAPTER 37

Designing the US Incentive Auction[*]

Paul R. Milgrom and Ilya R. Segal

The US government recently completed a new kind of double auction, which aquired television broadcast rights from current TV broadcasters and sold mobile broadband licenses. Known as the "incentive auction" because it was intended to create an incentive for TV broadcasters to relinquish their licenses, this auction involved tens of billions of dollars of payments and determined an allocation that satisfies millions of interference constraints. In this paper, we describe why the problem is so complex, outline the auction's final design adopted by the Federal Communication Commission, and describe how this design overcame the novel challenges that the FCC faced.

1. Background and Challenges

The incentive auction was mandated by an act of Congress in February 2012. In the years since the introduction of the iPhone, iPad, and similar devices, there has been explosive growth in the demand for broadband services and the spectrum it uses. The US government responded with its National Broadband Plan, which aims to make more spectrum available for broadband services partly by clearing some bands currently in other uses. Some spectrum used for UHF television broadcasts was very well suited to broadband. Since 90% of US households had access to cable or satellite television, over-the-air broadcasting appeared to be less important than in an earlier era. Also, the switch to digital TV signals made it possible to transmit several standard-definition signals using a single TV channel.

---

[*] This paper, which was initially written in 2013, has been changed to bring citations up to date, to mention treaty developments involving Mexico and Canada, and to acknowledge that the proposed auction rules were eventually adopted and used. We thank our Auctionomics co-consultants Jonathan Levin and Kevin Leyton-Brown, who participated in creating the design described here. Any opinions expressed here are those of the authors and not the FCC.

1.1. The Roles of Government and Auction

Economists naturally ask: Why not just rely on private transactions to shift spectrum resources to their most valuable uses? Why is value added by a government intervention?

The answer is that spectrum reallocation has the characteristics of a collective action problem. In order to use spectrum most effectively, the frequencies used for TV broadcasting need to be nearly the same across the whole country and the wireless broadband uses must be separated from TV frequencies and coordinated to avoid unacceptable interference between uplink and downlink uses.

Without government mandates, there could have been an important broadcaster holdout problem. A single broadcaster on, say, channel 48 in some city could have effectively blocked the use of spectrum on that channel and adjacent ones over a wide area. To resolve such problems, the law allowed the FCC to require broadcasters that do not relinquish their licenses to move to a different broadcast channel, for example, from channel 48 to channel 26, provided the government makes "all reasonable effort" to ensure that the coverage area of the station is not substantially reduced by interference from other broadcasters. With the authority to retune broadcasters who do not relinquish their spectrum, the FCC could make all broadcasters compete for the right to relinquish their licenses for compensation, regardless of the channels they are initially assigned (including channels that the FCC does not intend to clear, but into which other broadcasters could be retuned). And some authority was needed to interpret and implement the words of the mandate of substantially preserving the post-repacking coverage of the remaining broadcasters.

To ensure that any transactions benefit the broadcasters, the law required that participation in the incentive auction be voluntary. Even broadcasters that did not sell their rights would benefit, because increased scarcity would make their remaining broadcast licenses more valuable. An auction would preserve high-value programming for consumers by encouraging only the lowest-value broadcasters to become sellers.

1.2. Special Challenges

Three technical characteristics made the incentive auction especially challenging compared to any previous two-sided auction.

First, the items being bought and sold – wireless broadband licenses and TV broadcast licenses – were not the same, so it was not sufficient to transfer a set of rights from seller to buyer. Compared to broadband licenses, television broadcast licenses involve different power restrictions, different coverage areas, different amounts of bandwidth, and different protections from interference. This made both pricing and determining the quantities of each kind of license to buy and sell unusually challenging.

Second, the problem of moving television stations that do not sell to new channels was very complex. There were about 3,000 broadcast stations in the United States and Canada that might need to be repacked[1] and millions of pairwise constraints among

---

[1] Canada had agreed to a joint repacking in order to also clear a swath of spectrum for mobile use and coordinate cross-border interference.

them, mostly of the form: it is not allowed to assign station X to channel Y and station X' to channel Y', because that would cause unacceptable interference. Even if the bidders' values were known, the combinatorial problem of computing an optimal repacking would have been computationally intractable.

But bidders' values were not known, and the auction needed to elicit information about them, and the practical impossibility of optimization implied that Vickrey prices, too, could not be computed.[2] Paid-as-bid pricing would impose such intractable challenges for small bidders, whose participation is essential, that they might prefer to sit out the auction.

The FCC also wanted to allow UHF broadcasters to continue broadcasting in the VHF band and VHF broadcasters to go off air, both in exchange for suitable compensation. Also, since the interference that remains after the auction depends on which stations are eventually cleared, some of the new mobile broadband licenses may be impaired to a degree that cannot be known before the auction. Uncertainty about the product is an enemy of a successful auction and needs to be minimized.

2. Three Part Plan and Desired Properties

The final auction design, based on the initial proposal of Milgrom et al. (2012) and finalized by the Federal Communication Commission (2015), had three conceptual components: (i) a forward auction in which buyers would acquire broadcast licenses, (ii) a reverse auction in which sellers would offer their broadcast licenses, and (iii) a coordination/clearing rule to determine what quantities of licenses of each kind should be traded.

The design for the forward auction had a goal of reducing substantially the time to completion compared to the FCC's traditional simultaneous multiple round auction. It also aimed to build the auction so that information elicited for one quantity target could be reused if the quantity target changes, so that multiple complete forward auctions would not be required.

The goal for the reverse auction was to acquire enough licenses to clear a predetermined amount of spectrum at a reasonable cost, and to encourage participation by small broadcasters by making the auction nearly strategy-proof for them.

The clearing/coordination rule needed to ensure compatibility of the outcomes of the forward and reverse auctions and to account for the FCC's revenue objectives, which included raising funds to support a new public safety communications network.

3. The Economic Design

3.1. Forward Clock Auction

To achieve the speed-up required by the incentive auction, the forward auction was designed as a *clock auction* with an extra feature called *intra-round bidding* and with

---

[2] Approximate optimization was possible but would not be a suitable basis for strategy-proof pricing. This issue has been examined in the field known as algorithmic mechanism design. See Milgrom and Segal (2017) and the references therein.

a separate *assignment round*. Clock auctions are conceptually similar to the FCC's familiar simultaneous multiple round auction, but with prices increasing on all licenses until demand falls to match the target supply.

To illustrate the difference, imagine a simultaneous multiple round auction for four identical licenses with five bidders and a limit of one license per bidder. At each round, there are four provisionally winning bidders and the fifth bidder raises the price of one license. It takes four rounds for the price of all four licenses to rise by one increment. A clock auction raises the price of all four licenses in every round so long as there is excess demand. In the incentive auction, it seems possible that this feature, alone, could reduce the auction time by about half.

Intra-round bidding is a feature that allows any bidder to indicate that it wishes to change its demand at prices between the proposed beginning and ending prices for a round, and prices generally stop rising when there is no excess demand for a product. In the preceding example, intra-round bidding allows the clock auction to use very large increments without overshooting the clearing price. Doubling the increment compared to the FCC's typical SMRA could cut the auction time in half again, so the combination of factors could shorten the auction duration by about 75%.

Because the licenses in a category in the clock auction are not actually identical, there needs to be a further procedure to determine which winner gets which licenses. In Europe, it has become common to use an assignment round to achieve that, and our design specified a similar procedure for the US.

For a multi-product auction, intra-round bidding is more subtle, but we omit the details here for reasons of length. Overall, a 75% reduction in time-to-completion appears to be a reasonable target for such auctions.

3.2. Reverse Clock Auction

In the reverse auction, each bidder was offered tentative prices for several options that it could exercise, including to go off-air or, in exchange for compensation, to shift to broadcast in a lower band. The prices were initialized at high levels and were reduced over time, and during the auction each bidder indicated its currently preferred option. The initial clearing target was set to be consistent with the broadcaster's initial choices. To ensure that all transactions were voluntary, whenever a bidder's price for its preferred option was reduced, it was given an opportunity to exit and so continue to broadcast in its original "home" band.

The clock auction maintained a tentative feasible assignment of stations to bands and channels that avoided unacceptable interference. In every round, the auction ran a "feasibility checker" to determine whether there would still be a feasible channel assignment if any given station was moved into any given band. If it was infeasible for a station to return to its home band, the auction could not reduce the station's current price offer. Feasibility checking is an NP-hard problem similar to "graph coloring"; however, unlike the value-maximization problem, it can usually be solved in reasonable time. Since the constraints were known well in advance of the auction, modern computer science techniques were used to develop a specialized algorithm that was expected to solve as many as 99% of potential feasibility checking problems within seconds (see Frechette et al., 2015). Also, feasibility checking for different stations could be done in parallel using hundreds of processors, as needed. When a checker did not find a feasible way to assign a station into a given band within the available time, the auction treated the move as infeasible.

Among the questions an economist might ask about this auction design are the following: (1) Can the auction achieve reasonably high efficiency and low cost of spectrum clearing? (2) How should clock price reductions be computed to advance those goals? (3) Does the auction avoid incentives for gaming? (4) How can it be integrated with the forward auction (i.e., how will the amount of spectrum to clear be determined)? These questions were examined both theoretically and in simulations.

Theoretical Analysis

Equivalence Results – Milgrom and Segal (2017) (hereafter MS) offer a theoretical analysis which assumes that all bidders are single-station owners who know their station values and are "single-minded," that is, willing to bid only for a single option. This assumption is reasonable for commercial UHF broadcasters that view VHF bands as ill-suited for their operations and for non-profit broadcasters that are willing to move for compensation to a particular VHF band but that view going off-air as incompatible with their mission. A single-minded bidder may reasonably adopt a "cutoff strategy," in which it bids for its acceptable option until the compensation offered for that option reaches its minimum acceptable price, at which point it exits. MS find that when bidders use such strategies, the class of clock auctions described above is equivalent to the class of sealed-bid *deferred-acceptance threshold auctions*, which use one of a class of deferred acceptance algorithms that prioritize bids for rejection by assigning a score to each bid at each round and rejecting the bid with the highest positive score until no such bids remain. The auction then pays each accepted bidder its "threshold price," which is the maximum price that bidder could bid that would have still been accepted, given the bids by the others.

Incentives – What are the individual and group incentives for bidders in the reverse auction? MS show that for "single-minded" bidders, any clock auction with any information disclosure policy, or the equivalent sealed-bid DA heuristic auction with threshold prices, is *weakly group strategy-proof:* for every coalition of bidders and every possible strategy profile of bidders outside the coalition, there is no coalitional deviation from truthful bidding that strictly benefits all of its members.[3]

Advantages of Clock Auctions

In spite of the equivalence described above, clock auctions do offer some distinct advantages over the equivalent sealed-bid auctions:

1. In contrast to sealed-bid auctions with threshold pricing, the strategy-proofness of a clock auction is obvious even to bidders who do not understand the auction algorithm

---

[3] The idea of the proof is simple: consider the first round of the clock auction in which a deviation occurs. If an agent deviates by exiting while his price offer exceeds his value, he receives a zero net payoff, while if he deviates by not exiting when his price offer falls below his value, he will ultimately receive a nonnegative net payoff; in both cases, he does not strictly benefit from the deviation.

or trust the auctioneer's computations: as long as a bidder believes that his price offer is only going to be reduced, it is clear that there is no gain to exiting while the price offer is above his value or to staying in the auction when your price offer is below your value. (A formal notion of "obvious strategy-proofness" that captures some of this intuition is offered by Li, 2017.)

2. Unlike sealed-bid auctions, clock auctions do not require bidders to report or even know their exact values. For one thing, the winners in a clock auction will not have to evaluate their exact values: they only need to verify that they are interested in selling at their final clock prices. In fact, MS show that for single-minded private-value bidders, clock auctions can be characterized by the property of "winners' unconditional privacy": they elicit minimal information about winners' values that is required to establish that they should be winning. Also, in the case of a common-value component (for example, this being the resale values of TV spectrum, about which different bidders could have different information), information feedback during the clock auction may help aggregate the common-value information among bidders, resulting in lower cost (cf. Milgrom and Weber, 1982).

3. Clock auctions can accommodate bidders who are interested in more than one option, by permitting them to switch between options as their prices are reduced. Such auctions for multi-minded bidders are generally no longer strategy-proof. For example, multi-station owners might profitably engage in "supply reduction," while bidders who are willing to switch between bidding to go off-air and bidding to move to a lower band might be able to influence its prices by choosing when to switch.[4] However, potential gains from such gaming would be small in a "large-market" setting in which there are many participants competing in each region. (The FCC's restrictions on station cross-ownership in each region could help ensure such competitiveness.) Furthermore, a clock auction can provide some information feedback that could be useful for bidders to account for value complementarities or substitutabilities across their stations in their bidding strategies, but not detailed enough to facilitate effective price-manipulation strategies.

Price Reduction Algorithm for Efficiency and Cost Reduction

Given the equivalence result, a price reduction algorithm for single-minded bidders is equivalent to a scoring rule, and one could evaluate such scoring rules according to the efficiency of the resulting outcome and/or the sum of the threshold prices paid. If stations were always substitutes and optimization in reasonable time were possible, then as MS show, one can adapt the results of Ausubel (2004) and Bikhchandani et al. (2011) to show that a simple clock auction can implement the efficient allocation and determine Vickrey prices. In the actual problem, stations were not always substitutes, because it would sometimes be possible to buy out two small stations instead of one large one, or to pay a UHF station to move to VHF and a VHF station to go off-air instead of paying a UHF station to go off-air. Also, given reasonable time

---

[4] There do exist clock auctions for multi-parameter bidders which sustain truthful bidding as an ex post Nash equilibrium, by implementing the Vickrey outcome and thereby eliminating the incentive for price manipulation (see, e.g., Ausubel and Milgrom, 2002). However, this only works when bidders are substitutes in the total value and when price adjustments are guided by exact optimization, and neither assumption holds for FCC's reverse-auction problem.

limits, no algorithm could always find efficient allocations, or compute Vickrey prices. Nevertheless, as MS show, it may sometimes still be possible for some deferred acceptance threshold auction and its corresponding clock auction to be strategy-proof and guarantee near-efficiency.

One case in which a simple heuristic approach achieves great results is when all participating UHF stations interested only in off-air and not in going to VHF. The simplest problem arises when the interference graph is a union of regional "cliques" (a "clique" is a set of stations that all have same-channel interference constraints with each other) with no cross-clique constraints and no cross-channel constraints. Thus, assuming that there are k channels available in each region, the interference constraints allow us to assign any set of k stations, but no more than k, to broadcast in each region. In that case, a clock auction that offers the same descending price to all stations (i.e., no scoring) results in assigning the k most valuable stations in each region, which is efficient, and in paying each winner the value of the lowest assigned station in its region – the Vickrey price.

MS study an example in which the interference graph is more complicated than that. For example, some stations in Philadelphia may have interference constraints with New York stations, and others have constraints with Washington, DC stations. Nevertheless, MS show that if there are "not too many" cross-regional constraints then some simple clock auction would still be exactly or approximately efficient, assigning channels to the m highest-value stations in each region, with m less than k.

The actual interference constraints were even more complicated than that, and simulations suggested that it was possible to improve efficiency by reducing the scores of stations that had many interference constraints. In terms of the clock auction, this meant offering higher prices to stations, which, if assigned, would cause more interference.

The problem of pricing VHF options and prices for VHF stations was substantially more complicated than the problem of pricing offers for UHF stations to go off air, as the problem of efficiently repacking stations across bands was less amenable to heuristic solutions. The price reduction algorithm designed for the auction, while not guaranteed to achieve a nearly-efficient repacking, extracted substantial additional value from VHF bands. The algorithm has the following properties:

- The price offers to UHF stations to go off-air was set proportionately to their "volumes" (determined by their population and interference), and reduced proportionately in each round of the auction. This proportionate reduction ensured a speedy auction (given the price decrements used, each stage of the reverse auction finished in 54 rounds).
- The compensation offered to a station for going to a given band never exceeded the compensation offered to the same station for going to a lower band or for going off air. A station with a lower home band was never offered a higher compensation for a given option than an otherwise identical station with a higher home band. This reflected the consensus view that higher-frequency bands are more desirable for digital TV broadcasting.
- For stations from different bands that had identical location, covered population, and interference, the auction always equalized the total compensation for any combination of their moves that would clear a single UHF channel without changing the

availability of channels in either VHF band.[5] This equalization was desirable for efficient utilization of VHF bands. In particular, it linked the price offers to a VHF station to those to an otherwise identical UHF stations.

- For any station, the difference between price offers for any two options could only decrease during the auction. This ensured that stations without wealth effects (i.e., those whose objective was to maximize the sum of broadcast value and the compensation received in the auction) could only want to switch to a higher option as the auction progressed. The auction also enforced an "activity rule" preventing stations from switching to a lower option than the one currently held, so as to eliminate the possibility of price manipulation by switching back and forth between options.[6]

- The reduction in a station's price offer to a particular band was determined by the "vacancy" (the number of still-available channels in the tentative assignment) in the station's area in that band relative to the vacancy in the same area in other bands. The goal of this approach was to incentivize stations to fill different bands in each area at the same rate, in order to balance the possibility of the following two kinds of inefficiencies: (i) UHF stations "freezing" off-air before VHF is filled, leaving unused VHF channels to which UHF stations can no longer move, and (ii) VHF getting too congested while a lot of space is still left in UHF, which may induce low-value stations to move to VHF just because they fit well with the stations that are currently assigned there but are bound to eventually move to UHF.

The reverse auction was a component of a larger incentive auction. When clearing proved too costly relative to forward auction revenue, the clearing target needed to be reduced (as described in Section 3.3 below), which could have entailed substantial losses in overall efficiency. Therefore, cost-reduction measures may have enhanced overall efficiency even if they risked some efficiency distortions in deciding which stations to assign. One such cost-reduction measure that was implemented scored bids based not only on the interference constraints they created but also on their covered population, which was an important determinant of the market value of TV spectrum.

Paid-as-Bid Equivalence

One might wonder whether a reverse clock auction, or the equivalent deferred acceptance threshold auction, could be criticized on the grounds that it "overpays" bidders relative to their willingness to accept. Of course, if a paid-as-bid auction is used instead, then reverse-auction bidders will bid above their values. In such a complicated setting, their exact bidding strategy and the resulting outcome is hard to predict. However, for a simple theoretical benchmark, suppose that bidders are single-minded, have full information about each other's values, and play a Nash equilibrium of the paid-as-bid auction. For this case, MS show that the paid-as-bid auction with a DA heuristic

---

[5] Such combinations included: (i) a UHF station going off the air, (ii) a UHF station going to upper VHF and an upper VHF station going off the air, (iii) a UHF station going to lower VHF and a lower VHF station going off the air, and (iv) a UHF station going to upper VHF, an upper VHF station going to lower VHF, and a lower VHF station going off the air.

[6] However, this restriction makes it more difficult to bid for bidders with strong wealth effects, e.g., those who have a "revenue target" and so might wish to switch to a lower, better-compensated option as prices are reduced.

assignment rule has a full-information Nash equilibrium outcome that is the same as the truthful-bidding outcome of the threshold-price auction with the same heuristic assignment rule. In this sense, using threshold pricing "does not cost extra" given the heuristic assignment rule we are using.

The paid-as-bid auction typically has many Nash equilibria, but in certain cases, a unique outcome is selected by iterated deletions of weakly dominated strategies. MS show that this happens for deferred acceptance auctions that use "non-bossy" assignment rules – ones in which a losing bidder that switches to a different losing bid does not alter the set of winners.[7]

3.3. Coordination Clearing Rule

Having described the forward and reverse auctions, there remains the task of coordinating them to buy and sell compatible sets of licenses. The FCC design coordinated these using a series of declining clearing targets, as follows.

The process began by setting reserve prices for broadcasters in the reverse auction. Broadcasters who register to bid commited to sell at the reserve prices, which also served as the reverse auction starting prices. A rule determined the number of broadcast channels that could be cleared across the country using the initial offers and a corresponding set of broadband licenses.

The forward and reverse auctions followed, determining prices for both buyers and sellers. The auction then checked a stopping criterion: if net revenue (forward auction revenue minus reverse auction payments) was sufficiently high and if another FCC-determined criterion was satisfied, then the auction would end. The extra criterion depended on the amount of spectrum cleared and on the average prices in the forward auction. Otherwise, the clearing target was reduced, the products in the forward auction were adjusted, and prices continued to descend in the reverse auction and ascend in the forward auction.

Another novel feature of the forward auction was its use of "extended rounds," which were to be triggered if the auction revenue was still too low when demand has fallen to be equal to target supply. In that case, provided that the revenues are within 25% of the target, prices in the forward auction would continue to rise, in order to give bidders in the forward auction an opportunity to raise their bids to meet the prices set in the reverse auction. Details are not reported here.

A final novel feature of the design was the "conditional reserve." The FCC faced a novel challenge between promoting competition among wireless carriers and raising sufficient revenues to satisfy the stopping criterion. On one hand, the FCC wanted to ensure that the low-band spectrum, including that sold in this auction, was not all held by just one or two companies, who might acquire too much market power in that way. On the other hand, if the FCC were to limit competition in the auction by setting aside spectrum for bidders without existing holdings of low-band spectrum, the forward auction revenues might be reduced, and it might fail to clear a large number of

---

[7] As an additional theoretical point, MS show that dominance-solvability of a paid-as-bid auctions with a non-bossy monotonic assignment rule is *equivalent* to the assignment rule being implementable with a DA heuristic (or a clock auction).

TV channels. The solution the FCC adopted was the conditional reserve, according to which spectrum was set aside only when the net revenue condition from the stopping criterion was met. Moreover, the number of licenses to be set aside was the smaller of a pre-determined number or the number of licenses being demanded by qualified bidders at the time the criterion was satisfied. In that way, the set aside did not threaten clearing, but could have been effective at encouraging competition once the clearing condition had been met.

4. Conclusion

The FCC incentive auction posed unusual challenges but also promised a rare opportunity to contribute to the wireless broadband infrastructure of the United States. We are excited to have been part of the team that designed this new auction.

References

Ausubel, L. (2004). "An Efficient Ascending Auction for Multiple Objects," *American Economic Review* 94(5), 1452–1475.

Ausubel, L. and Milgrom, P. (2002). "Ascending Auctions with Package Bidding," *Frontiers of Theoretical Economics* 1(1).

Bikhchandani, S., de Vries, S., Schummer, J., and Vohra, R. (2011). "An Ascending Vickrey Auctions for Selling Bases of a Matroid," *Operations Research* 59(2), 400–413.

Frechette, A., Newman, N., and Leyton-Brown, K. (2015). "Solving the Station Repacking Problem," *International Joint Conference on Artificial Intelligence (IJCAI)*.

Federal Communication Commission (2015). "Public Notice 1578: Broadcast Incentive Auction Scheduled To Begin on March 29, 2016: Procedures for Competitive Bidding In Auction 1000, Including Initial Clearing Target Determination, Qualifying To Bid, And Bidding in Auctions 1001 (Reverse) and 1002 (Forward)," AU Docket No. 14-252, GN Docket No. 12-268, WT Docket No. 12-269, MB Docket No. 15-146.

Li, S. (2017). "Obviously Strategy-Proof Mechanisms," SSRN working paper.

Milgrom, P. and Weber, R. (1982). "A Theory of Auctions and Competitive Bidding," *Econometrica* 50(5), 1089–1122.

Milgrom, P. and Segal, I. (2017). "Deferred Acceptance Auctions and Radio Spectrum Reallocation," Stanford University working paper.

Milgrom, P., Ausubel, L., Levin, J., and Segal, I. (2012). "Appendix C: Incentive Auction Rules Option and Discussion," http://hraunfoss.fcc.gov/edocs_public/attachmatch/FCC-12-118A2.pdf.

CHAPTER 38

Solving the Station Repacking Problem[*]

Alexandre Fréchette, Neil Newman, and
Kevin Leyton-Brown

1. Introduction

Over 13 months in 2016–17, the US government held an innovative "incentive auction" for radio spectrum, in which television broadcasters were paid to relinquish broadcast rights via a "reverse auction", remaining broadcasters were repacked into a narrower band of spectrum, and the cleared spectrum was sold to telecommunications companies. The stakes were enormous: the auction was forecast to net the government tens of billions of dollars, as well as creating massive economic value by reallocating spectrum to more socially beneficial uses (Congressional Budget Office 2015). As a result of both its economic importance and its conceptual novelty, the auction has been the subject of considerable recent study by the research community, mostly focusing on elements of the auction design (Bazelon, Jackson, and McHenry 2011; Kwerel, LaFontaine, and Schwartz 2012; Milgrom et al. 2012; Calamari et al. 2012; Marcus 2013; Milgrom and Segal 2014; Dütting, Gkatzelis, and Roughgarden 2014; Vohra 2014; Nguyen and Sandholm 2014; Kazumori 2014). After considerable study and discussion, the FCC has selected an auction design based on a descending clock (FCC 2014c; 2014a). Such an auction offers each participating station a price for relinquishing its broadcast rights, with this price offer falling for a given station as long as it remains repackable. A consequence of this design is that the auction must (sequentially!) solve hundreds of thousands of such repacking problems. This is challenging, because the repacking problem is NP-complete. It also makes the performance of the repacking algorithm extremely important, as every failure to solve a single, feasible repacking problem corresponds to a lost opportunity to lower a price offer. Given the scale of the auction, individual unsolved problems can cost the government millions of dollars each.

This chapter shows how the station repacking problem can be solved exactly and reliably at the national scale. It describes the results of an extensive, multi-year investigation into the problem, which culminated in a solver that we call SATFC. This solver

---

[*] A more recent description of SATFC, along with more thorough analysis, can be found in a forthcoming CACM article Newman, Fréchette and Leyton-Brown (2017).

combines a wide variety of techniques: SAT encoding; algorithm configuration; algorithm portfolios; and a bevy of problem-specific speedups, including a powerful and novel caching scheme that generalizes from different but related problem instances. Overall, SATFC solves virtually all problems in a previously-unseen test set—99.6%—within one minute. It was adopted by the FCC for use in the incentive auction (FCC 2014b). SATFC is open-source; pointers both to the solver and to data used in this chapter are available at www.cs.ubc.ca/labs/beta/Projects/SATFC.

In what follows, we begin by defining the station repacking problem and explaining some of its salient properties (Section 2). We then discuss the encodings we considered and the tools we leverage (Section 3) before detailing the problem-specific speedups that make our approach effective (Section 4). We report experimental results throughout, establishing a baseline in Section 2 and then showing the extent to which each of our extensions strengthens our solver.

2. The Station Repacking Problem

Prior to the auction, each US television station $s \in S$ was assigned a channel $c_s \in C \subseteq \mathbb{N}$ that ensured that it did not excessively interfere with other, nearby stations. The FCC reasons about what interference would be harmful via a complex, grid-based physical simulation ("OET-69" (FCC 2013)), but has also processed the results of this simulation to obtain a CSP-style formulation listing forbidden pairs of stations and channels, which it has publicly released (FCC 2014e). Let $I \subseteq (S \times C)^2$ denote a set of *forbidden station–channel pairs* $\{(s, c), (s', c')\}$, each representing the proposition that stations s and s' may not concurrently be assigned to channels c and c', respectively. The effect of the auction will be to remove some broadcasters from the airwaves completely, and to reassign channels to the remaining stations from a reduced set. This reduced set will be defined by a *clearing target*: some channel $\bar{c} \in C$ such that all stations are only eligible to be assigned channels from $\overline{C} = \{c \in C : c < \bar{c}\}$. The sets of channels *a priori* available to each station are given by a *domain* function $D : S \to 2^{\overline{C}}$ that maps from stations to these reduced sets. The *station repacking problem* is then the task of finding a repacking $\gamma : S \to \overline{C}$ that assigns each station a channel from its domain that satisfies the interference constraints: i.e., for which $\gamma(s) \in D(s)$ for all $s \in S$, and $\gamma(s) = c \Rightarrow \gamma(s') \neq c'$ for all $\{(s, c), (s', c')\} \in I$. It is easy to see that the station repacking problem is NP-complete; e.g., it generalizes graph coloring (see below). It also falls under the umbrella of *frequency assignment problems* (see Aardal et al. (2007) for a survey and a discussion of applications to mobile telephony, radio and TV broadcasting, satellite communication, wireless LANs, and military operations).

Luckily, there is reason to hope that this problem could nevertheless be solved effectively in practice. First, we only need to be concerned with problems involving subsets of a fixed set of stations and a fixed set of interference constraints: those describing the television stations currently broadcasting in the United States. Channels can be partitioned into three equivalence classes: LVHF (channels 1–6), HVHF (channels 7–13), and UHF (channels 14–$\bar{c}$, excepting 37), where $\bar{c} \leq 51$ is the largest available UHF channel set by the auction's clearing target, and 37 is never available. No interference constraints span the three equivalence classes of channels, giving us a straightforward way of decomposing the problem. A problem instance thus corresponds to a choice

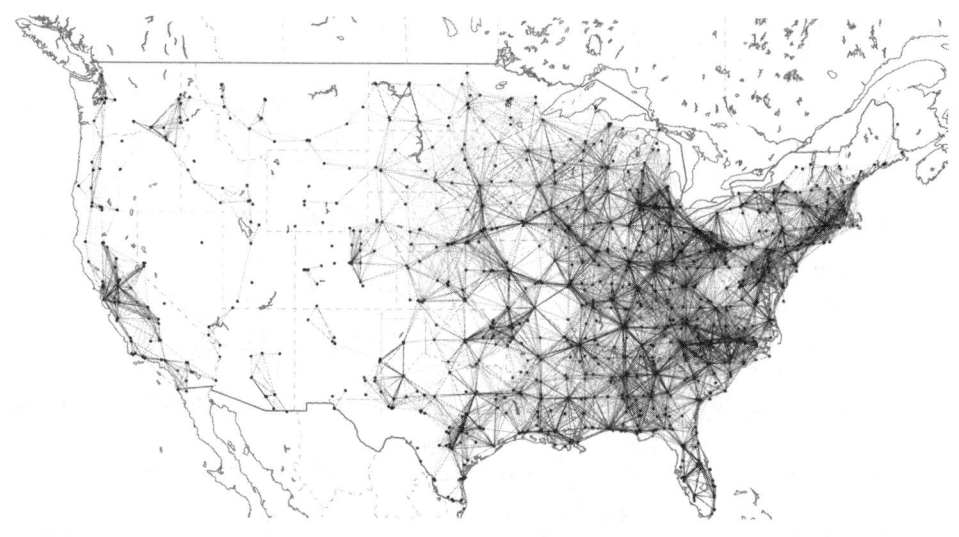

Figure 1. Interference graph derived from the FCC's May 2014 constraint data (FCC 2014e).

of stations $S \subseteq \mathcal{S}$ and channels $C \subseteq \mathcal{C}$ to pack into, with domains D and interference constraints I implicitly being restricted to S and C; we call the resulting restrictions $\mathcal{D}$ and $\mathcal{I}$.

Let us define the *interference graph* as an undirected graph in which there is one vertex per station and an edge exists between two vertices s and s' if the corresponding stations participate together in any interference constraint: i.e., if there exist $c, c' \in C$ such that $\{(s, c), (s', c')\} \in I$. Figure 1 shows the US interference graph. We know that every repacking problem we will encounter will be derived from the restriction of this interference graph to some subset of S. This suggests the possibility of doing offline work to capitalize on useful structure present in the interference graph. However, this graph involves a total of $|\mathcal{S}| = 2173$ stations, and the number of possible subsets is exponential in this number. Thus, it is not possible to exhaustively range over all possible subsets. Nevertheless, meaningful structure exists, whether leveraged explicitly or implicitly (see e.g., a computational analysis of unavoidable constraints by Kearns and Dworkin (2014)).

Interference constraints are more structured than in the general formulation: they come in only two kinds. *Co-channel constraints* specify that two stations may not be assigned to the same channel; *adjacent-channel constraints* specify that two stations may not be assigned to two adjacent channels. Hence, any forbidden station–channel pairs are of the form $\{(s, c), (s', c)\}$ or $\{(s, c), (s', c + 1)\}$ for some stations $s, s' \in \mathcal{S}$ and channel $c \in \mathcal{C}$. Note that if we were dealing exclusively with co-channel constraints, we would face a graph coloring problem.

Interestingly, our cost function is asymmetric: it is terrible to incorrectly conclude that a repacking problem is feasible (this could make the auction outcome infeasible) whereas the consequences are fairly mild for wrongly claiming that a given repacking is infeasible (this prevents a price offer from being lowered, costing the government money, but does not pose a fundamental problem for the auction itself). Thus, whenever a problem cannot be solved within the given amount of time, we can safely treat it as though it was proven infeasible—albeit at some financial cost.

The last key property that gives us reason to hope that this large, NP-complete problem can be tamed in practice is that we are not interested in worst-case performance, but rather in good performance on the sort of instances generated by actual reverse auctions. The question of which stations will need to be repacked in which order depends on the stations' valuations, which depend in turn (among many other factors) on the size and character of the population reached by their broadcasts. The distribution over repacking orders is hence far from uniform. Second, descending clock auctions repeatedly generate station repacking problems by adding a single station s^+ to a set S^- of provably repackable stations. This means that every station repacking problem $(S^- \cup \{s^+\}, C)$ comes along with a partial assignment $\gamma^- : S^- \to C$ which we know is feasible on restricted station set S^-.

In order to validate the auction design, the FCC has run extensive simulations of the auction, based on a wide variety of assumptions about station participation and bidding behavior. We obtained anonymized versions of some repacking problems from five such simulations,[1] and randomly partitioned them into a training set of 100,572 examples, a validation set of 1,000 examples, and a test set of 10,000 examples. All of these problems were nontrivial in the sense that the auction's previous solution could not be directly augmented with an assignment to the newly introduced station; such trivial problems (80–90% of the total encountered in a typical auction simulation) are solved directly in the FCC's software without calling an external solver. Preliminary experiments showed that repacking problems in the VHF bands are very easy, because these bands contain at most 7 channels. We thus constrained ourselves to the much harder UHF band, fixing the interference model and setting the clearing target such that $|C| = 16$. The encoded test problems contained between 453 and 16,299 variables (averaging 8,654) and between 3,197 and 342,450 clauses (averaging 146,871) of which between 210 and 228,042 (averaging 86,849) were interference clauses. Station domains ranged from 1–16 channels (averaging 14). We used test instances solely to perform the benchmarking reported here, having performed preliminary experimentation and optimization using the training and validation sets. We chose a cutoff time of 60 seconds, reflecting the constraints involved in solving up to hundreds of thousands of problems sequentially in a real auction.

3. Encoding and Tools

3.1. Initial Efforts

The FCC's initial investigations included modeling the station repacking problem as a mixed-integer program (MIP) and using off-the-shelf solvers paired with problem-specific speedups (FCC 2014d). Unfortunately, the problem-specific elements of this

---

[1] This small set of simulations explores a very narrow set of answers to the questions of which stations would participate and how bidders would interact with the auction mechanism; it does not represent a statement either by us or by the FCC about how these questions are likely to be resolved in the real auction. At the time of writing, this data represented the best proxy available for the sorts of computational problems that would arise in practice. It is of course impossible to guarantee that variations in the assumptions would not yield computationally different problems.

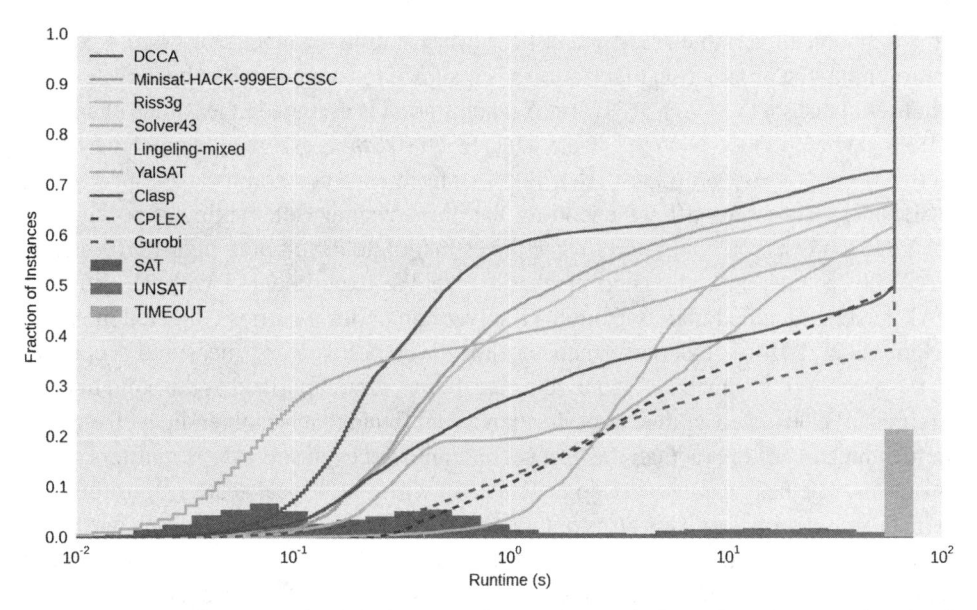

Figure 2. ECDF of runtimes for default MIP and SAT solvers. The bars show fraction of SAT and UNSAT instances binned by their (fastest) runtime. Although present, unsatisfiable instances form an insignificant portion of instances solved.

solution were not publicly released, so we do not discuss them further in this chapter. Instead, to assess the feasibility of a MIP approach, we ran what are arguably the two best-performing MIP solvers—CPLEX and Gurobi—on the test set described above; the results are summarized as the dashed lines in Figure 2. To encode the station repacking problem as a MIP, we create a variable $x_{s,c} \in \{0, 1\}$ for every station–channel pair, representing the proposition that station s is assigned to channel c. We then add the constraints $\sum_{c \in D(s)} x_{s,c} = 1 \forall s \in S$, and $x_{s,c} + x_{s',c'} \leq 1 \forall \{(s, c), (s', c')\} \in I$. These constraints ensure that each station is assigned to exactly one channel, and that interference constraints are not violated.

On our experimental data, both solvers solved under half of the instances within our cutoff time of one minute. Such performance would likely be insufficient for deployment in practice, since it means that most stations would be paid unnecessarily high amounts due to computational constraints (recall that each station gives rise to many feasibility checking problems over the course of a single auction).

3.2. SAT Encoding

We propose instead that the station repacking problem should be encoded as a propositional satisfiability (SAT) problem. This formalism is well suited to station repacking, which is a pure feasibility problem with only combinatorial constraints.[2] The SAT reduction is straightforward (and similar to the MIP reduction just described): given a station repacking problem (S, C) with domains D and interference constraints I, we

---

[2] Of course, it may nevertheless be possible to achieve good performance with MIP or other techniques; we did not investigate such alternatives in depth.

create a boolean variable $x_{s,c} \in \{\top, \bot\}$ for every station–channel pair $(s, c) \in S \times C$, representing the proposition that station s is assigned to channel c. We then create three kinds of clauses: (1) $\bigvee_{d \in D(s)} x_{s,d}$ $\forall s \in S$ (each station is assigned at least one channel); (2) $\neg x_{s,c} \vee \neg x_{s,c'}$ $\forall s \in S$, $\forall c, c' \neq c \in D(s)$ (each station is assigned at most one channel); (3) $\neg x_{s,c} \vee \neg x_{s',c'}$ $\forall \{(s, c), (s', c')\} \in I$ (interference constraints are respected).

Besides parsimony, a SAT encoding has the advantage of making it possible to leverage the research community's vast investment into developing high-performance SAT solvers (see e.g., Järvisalo et al. (2012)). We experimented with 18 different SAT solvers, obtained mainly from SAT solver competition entries collected in AClib (Hutter et al. 2014b). The performance of the seven best solvers, measured according to the number of instances solved by the cutoff time, is summarized as the solid lines in Figure 2. We observed a range of performance, but found that no solver did well enough to recommend use in practice: the best solver could not even solve three quarters of the instances. The best was DCCA (Luo et al. 2014), which as a local search algorithm can only prove satisfiability. This turns out not to be a major impediment because, due to the way a descending clock auction works, the instances we encounter are predominantly satisfiable. More specifically, out of the 9963 test instances that we were able to solve by any means, including the solvers introduced later in this chapter, 9871 (99.07%) were satisfiable and only 92 (0.93%) unsatisfiable. We illustrate the distribution of instances labeled by feasibility at the bottom of Figure 2.

3.3. Meta-Algorithmic Techniques

In recent years, there has been increasing development of artificial intelligence techniques that reason about how existing heuristic algorithms can be modified or combined together to yield improved performance on specific problem domains of interest. These techniques are called *meta-algorithmic* because they consist of algorithms that take other algorithms as part of their input. For example, *algorithm configuration* consists of setting design decisions exposed as parameters to optimize an algorithm's average performance across an instance distribution. This approach has proven powerful in the SAT domain, as many SAT solvers expose parameters that can drastically modify their behavior, from probability of random restarts to choice of search heuristics or data structures (Hutter et al. 2014a). We performed configuration using the Sequential Model-based Algorithm Configuration algorithm, or SMAC (Hutter, Hoos, and Leyton-Brown 2011).

Unfortunately, even after performing algorithm configuration, it is rare to find a single algorithm that outperforms all others on instances of an NP-hard problem such as SAT. This inherent variability across solvers can be exploited by *algorithm portfolios* (Gomes and Selman 2001; Nudelman et al. 2003). Most straightforwardly, one selects a small set of algorithms with complementary performance on problems of interest and, when asked to solve a new instance, executes them in parallel.

Finally, algorithm configuration and portfolios can be combined. Hydra (Xu, Hoos, and Leyton-Brown 2010) is a technique for identifying sets of complementary solvers from highly parameterized design spaces via algorithm configuration, by greedily adding configurations that make the greatest possible marginal contribution to an existing portfolio. Specifically, we create a (parallel) portfolio by greedily selecting the algorithm that most improves its performance. In our experiments we measured improvement by percentage of instances solved within a one-minute cutoff. Therefore, we started off by picking the best solver, then used algorithm configuration to construct many new solvers that complement it well, then identified the next best solver given that solver, and so on.

Performing this procedure on our experimental data, we committed to the best default SAT solver, DCCA (which unfortunately exposes no parameters), and configured the remaining 17 solvers with the objective of maximizing marginal contribution. The solver clasp (Gebser et al. 2007) was the best in this regard, improving the number of instances solved in our validation set by 4% when executed in parallel with DCCA. We could then have performed further Hydra iterations; however, our first two enhancements from Section 4 end up altering our configuration scenario. Hence, we revisit the impact of Hydra at the end of Section 4.2.

4. Problem-Specific Enhancements

We now describe the novel methods we developed to bring our SAT-based station repacking solver to the point where it could deliver high performance in practice.[3] In what follows, we assume that the sets S and C consist of all stations and channels, domains $\mathcal{D}$ and interference constraints $\mathcal{I}$ are fixed, and that we are given a station repacking instance $(S = S^- \cup \{s^+\}, C)$ along with a feasible, partial assignment γ^-.

4.1. Incremental Station Repacking

Local Augmenting. On a majority of problem instances, a simple transformation of γ^- is enough to yield a satisfiable repacking: we consider whether it is possible to assign s^+ to a channel and update the channel assignments of the stations in s^+'s neighborhood, holding the rest of γ^- fixed. Specifically, we find the set of stations $\Gamma(s^+) \subseteq S$ that neighbor s^+ in the interference graph, then solve the reduced repacking problem in which all non-neighbors $S \setminus \Gamma(s^+)$ are fixed to their assignments in γ^-. Observe that a feasible repacking for this reduced problem is also feasible on the full set; on the other hand, if we prove that the reduced problem is infeasible, we cannot conclude anything. The value of this approach is in its speed: the sparseness of the interference graph often yields very small neighborhoods, hence small reduced problems.

To evaluate this technique experimentally, we performed local augmentation on all our instances and used the training set to configure each SAT solver for best performance on this new instance distribution. DCCA was once again the best-performing solver. The performance of this altered version of DCCA, dubbed DCCA-preSAT, is shown as one of the dotted lines in Figure 5: it solved 78.5% of the instances and then stagnated within 0.1 seconds.

---

[3] While we only discuss positive results, we unsuccessfully explored many other avenues, notably including incremental SAT solvers, various heuristics, and local consistency techniques.

Starting Assignment for Local Search Solvers. Local search solvers such as DCCA work by searching a space of complete assignments and seeking a feasible point, typically following gradients to minimize an objective function that counts violated constraints, and periodically randomizing. When working with such solvers we can leverage γ^- in a second way, by assigning the stations in γ^- to their channels in γ^- and randomly assigning a channel for s^+. If a solution does indeed exist near this starting point, such an initialization can help us to find it much more quickly (although there is no guarantee that the solver will not immediately randomize away to another part of the space). We observe that this approach does not generalize the "local augmenting" approach, as we do not constrain the local search algorithm to consider only s^+'s (extended) neighborhood.

The performance of DCCA starting from the previously feasible assignment, which we dub DCCA+, is shown in Figure 5. It solved 85.4% of the instances, clearly dominating the original, randomly initialized DCCA.

4.2. Problem Simplification

We now describe two ways of simplifying problem instances.

Graph Decomposition. First, the subgraph of the interference graph induced by the set of stations considered in a particular problem instance is usually disconnected. It can therefore help to identify disconnected components and solve each separately, giving smaller instances to our SAT solvers. An additional benefit is that if we identify a single component as infeasible, we can immediately declare the entire problem infeasible without looking at all of its components. In practice, we decompose each problem into its connected components and solve each component sequentially, starting with the smallest component. While we could have instead solved each component in parallel, we found that in practice runtimes were almost always dominated by the cost of solving the largest component, so that it did not make much of a difference whether or not the components were solved simultaneously. This did not mean that decomposition was not worth doing—we also found that the largest component was often considerably smaller than the full problem (e.g., first, second and third quartiles over number of stations of 346, 458, and 559 after simplification as compared to 494, 629, and 764 before).

Underconstrained Station Removal. Second, in some cases we can delete stations completely from a repacking problem and thereby reduce its size. This occurs when there exist stations for which, regardless of how every other station is assigned, there always exists some channel into which they can be packed. Verifying this property exactly costs more time than it saves; instead, we check it via the sound but incomplete heuristic of comparing a station's available channels to its number of neighboring stations. This problem simplification complements graph decomposition: we perform it first in order to increase the number of components into which we will be able to decompose the graph. We observe that a few important stations of high degree are often underconstrained; these are the stations whose removal makes the biggest difference to graph decomposition.

4.3. Hydra Revisited

The problem-specific enhancements we have discussed so far impact the Hydra procedure: incremental solvers solve many instances extremely quickly, allowing the remaining solvers in the portfolio to concentrate their efforts elsewhere; our problem simplifications change instances enough to reduce correlation between solvers. We thus augmented the set of solvers available to Hydra to include `DCCA-presat`, `DCCA+`, and all base SAT solvers given simplified problem instances. Our first two rounds of Hydra, already described, identified our base `DCCA-preSAT` and `DCCA+` solvers. Our third round selected a configured version of `clasp`; we dub this new contributor `clasp-h1`. The fourth iteration found a second `clasp` configuration that operates on simplified instances; we dub that `clasp-h2`. The (test-set) performance of these two `clasp` configurations is shown in Figure 5. `clasp-h1` solves 2.8% of the (test-set) instances previously unsolved by the (`DCCA-preSAT`; `DCCA+`) portfolio, and `clasp-h2` solves an additional 0.9% that were unsolved by the 3-solver portfolio. The next Hydra step yielded only a 0.2% (validation-set) improvement, and we found it desirable to obtain a portfolio that could be run on a 4-core workstation, so we stopped with this set of 4 solvers.

4.4. Caching Instances

So far, we have concentrated on building the best station repacking solver possible, based on no contextual information except a previous assignment. However, our advance knowledge of the constraint graph means that we have considerably more context. Furthermore, it is feasible to invest an enormous amount of offline computational time before the actual auction, in order to ensure that the repacking problem can be solved quickly online. We investigated a wide range of strategies for leveraging such offline computation (including incremental SAT solving), but had the most success with a novel caching scheme we call *containment caching*.

Containment Caching. "Caching" means storing the result of every repacking problem solved on our training set, for reference at test time. Unfortunately, in experiments on our validation set, we observed that it was extremely rare to encounter previously-seen problems, even given our training set of over 100,000 problems. However, observe that if we know whether or not it is possible to repack a particular set of stations S, we can also answer many different but related questions. Specifically, if we know that S was packable then we know the same for every $S' \subseteq S$ (and indeed, we know the packing itself—the packing for S restricted to the stations in S'). Similarly, if we know that S was unpackable then we know the same for every $S' \supseteq S$. This observation dramatically magnifies the usefulness of each cached entry S, as S can be used to answer queries about an exponential number of subsets or supersets (depending on the feasibility of repacking S).

We call a cache meant to be used in this way a *containment cache*, because it is queried to determine whether one set contains another (i.e., whether the query contains the cache item or vice versa). To the best of our knowledge, containment caching is a novel idea. A likely reason why this scheme is not already common is that querying a containment cache is nontrivial; see below. We observe that containment caching is applicable to any family of feasibility testing problems generated as subsets of a master set of constraints, not just to spectrum repacking.

In more detail, containment caching works as follows. We maintain two caches, a *feasible cache* and an *infeasible cache*, and store each problem we solve (including both full instances and their components resulting from problem simplifications) in the appropriate cache. When asked whether it is possible to repack station set S, we proceed as follows. First, we check whether the feasible cache contains a superset of S, in which case the original problem is feasible. If we find no matches, we check to see whether a subset of S belongs to the infeasible cache, in which case the original problem is infeasible. If both queries fail, we simplify and decompose the given instance and query each component in the feasible cache. We do not check the infeasible cache again after problem simplification, because a subset of a component of the original instance is also a subset of the original instance.

Querying the Containment Cache. Containment caching is less straightforward than traditional caching, because we can not simply index entries with a hash function. Instead, an exponential number of keys could potentially match a given query. We were nevertheless able to construct an algorithm that solved this problem extremely quickly: within an average time of 30 ms on a cache of nearly 200,000 entries.[4]

Specifically, our approach proceeds as follows. Offline, we build (1) a traditional cache $\mathfrak{C}$ indexed by a hash function and—in the case of feasible problems—storing solutions along with each problem; and (2) a secondary cache $\mathfrak{C}_o$ containing only a list of station sets that appear in $\mathfrak{C}$. This secondary cache is defined by an ordering o over the stations, which we choose uniformly at random. We represent each station set stored in $\mathfrak{C}_o$ as a bit string, with the bit in position k set to 1 if and only if the k-th station in ordering o belongs to the given station set. We say that one station set is larger or smaller than another by interpreting both station set bit strings as integers under the ordering o and then comparing the integers. Appealing to this ordering, we sort the entries of $\mathfrak{C}_o$ in descending order. We give an example in Figure 3: the left and center diagrams respectively illustrate a set of six subsets of the power set $2^{\{a,b,c,d,e\}}$ and a secondary cache constructed based on these sets along with a random ordering over their elements. As the figure suggests, secondary caches are very compact: a cache of 200,000 entries, each consisting of 2,000 stations/bits, occupies only 50 MB. We can thus afford to build multiple secondary caches $\mathfrak{C}_{o_1}, \ldots, \mathfrak{C}_{o_\ell}$, based on the same set of station sets but ℓ different random orderings.

We now explain how to query for a superset, as we do to test for feasible solutions; the algorithm for subsets is analogous. (A sample execution of this algorithm is illustrated in Figure 3 (right) and explained in the caption.) Given a query S, we perform binary search on each of the ℓ secondary caches to find the index corresponding to S itself (if it is actually stored in the cache) or of the smallest entry larger than S (if not); denote the index returned for cache $\mathfrak{C}_{o_k}$ as i_k. If we find S, we are done: we retrieve its corresponding solution from the main cache. Otherwise, the first i_1 entries in cache $\mathfrak{C}_{o_1}$ contain a mix of supersets of S (if any exist) and non-supersets that contain one or

---

[4] There is a literature on efficiently finding subsets and supersets of a query set (Hoffmann and Koehler 1999; Savnik 2013; Charikar, Indyk, and Panigrahy 2002; Patrascu 2011). However, our algorithm was so fast in our setting that we did not explore alternatives; indeed, our approach may be of independent interest.

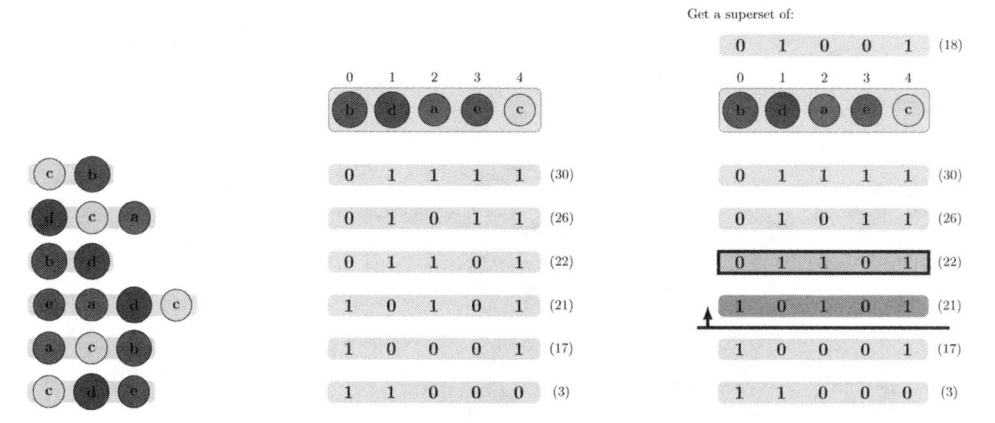

Figure 3. Containment caching example. Left: six elements of the power set $2^{\{a,b,c,d,e\}}$. Center: a secondary cache defined by a random ordering over the five elements, with each of the sets interpreted as a bit string and sorted in descending order. Right: the result of querying the containment cache for supersets of $\{c, d\}$. The query (18) does not exist in the cache directly; the next largest entry (21) is not a superset (i.e., 01001 does not bitwise logically imply 10101); the cache returns $\{a, c, d\}$ (22).

more stations not in S that appear early in the ordering o_1. Likewise, the first i_2 entries in $\mathfrak{C}_{o_2}$ contain the same supersets of S (because $\mathfrak{C}_{o_1}$ and $\mathfrak{C}_{o_2}$ contain exactly the same elements) and a different set of non-supersets based on the ordering o_2, and so on. We have to search through the first i_k entries of some cache $\mathfrak{C}_{o_k}$; but it does not matter which $\mathfrak{C}_{o_k}$ we search. We thus choose the shortest list: $k = \arg\min_j i_j$. This protects us against unlucky situations where the secondary cache's ordering yields a large i_k: this is very unlikely to happen under all ℓ random orderings. The superset search itself can be performed efficiently by testing whether the cached bit string is bitwise logically implied by the query bit string. If we find a superset, we query the main cache to retrieve its solution.

Evaluation. To build the cache we used to evaluate our approach, we ran our 4-solver parallel portfolio with a 24-hour cutoff time on all instances from both our training and validation sets,[5] along with all of their simplified versions, terminating runs when one solver completed. To speed up this process, we constructed the cache in a bootstrapped fashion: we made use of a partial cache even as we were working through the set of instances designed to populate the cache, thereby solving subsequent instances more quickly. In the end, we obtained a cache of 185,750 entries at a cost of roughly one CPU month. The largest feasible problem in our cache contained 1170 stations, and

---

[5] A word of warning: Our experiments may exaggerate cache performance because of the way we partitioned our training and test data: related instances from the same auction run can appear in different sets. We would have preferred to use a test set consisting of entirely distinct auction simulations; however, the data provided to us by the FCC did not distinguish the auction runs from which each instance was taken. We do note that SATFC achieved very strong performance even without the cache—solving 98% of the instances in under a minute—meaning that even if we do overestimate cache performance here, our overall qualitative findings would not change. Subsequent, informal experimentation on new data (Newman, Fréchette and Leyton-Brown (2017)) assures us that containment caching continues to "solve" a large fraction of instances even when entire auction runs are excluded from training, albeit usually fewer than the 98% reported below.

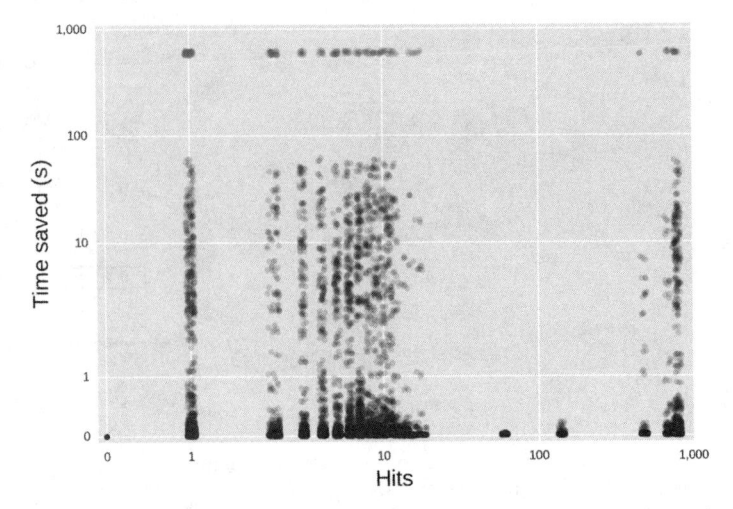

Figure 4. Time saved per cache hit. Each point represents a cache hit on a particular key: the x-axis represents the number of times the corresponding key was hit, while the y-axis represents the amount of time each individual cache hit saved. For visualization purposes we count the runtime of unsolved instances as 10 times the cutoff time and color such points in red.

the smallest infeasible problem contained 2 stations. We built $\ell = 5$ secondary caches based on different random station orderings.

We interpret the containment cache as a standalone solver that behaves as follows: (1) checking whether a full instance has a superset in the feasible cache; (2) if not, checking whether a subset of the instance belongs to the infeasible cache; (3) simplifying the instance and then asking whether all of its components have supersets in the feasible cache. This solver's runtime is equal to the appropriate cache lookup time(s) plus the time to perform problem simplification, if initial cache lookup fails; we report its performance in Figure 5. It far outperformed all other algorithms, solving 98.2% of the instances on its own; however, this "solver" obviously works only because we were able to obtain a database of solved instances via our other algorithms. We also note that its performance would continue to improve if we obtained an even larger training set and performed more offline computation, as we intend to do in preparation for the incentive auction.

To investigate more deeply how the cache functioned, Figure 4 shows a scatter plot relating the frequency with which different keys were "hit" in the cache and the amount of time the best set of remaining solvers would have taken to solve the instances if the cache had not been used. This analysis shows that only a handful of keys were hit more than ten times, but that infrequently hit keys contributed significantly to the total time saved by the cache. Furthermore, many hits saved more than our one-minute cutoff time, justifying our investment in very long runs while populating the cache.

5. Conclusions: Putting It All Together

Station repacking is an economically important problem that initially seemed impractical to solve exactly. We have shown how to combine state-of-the-art SAT solvers, recent

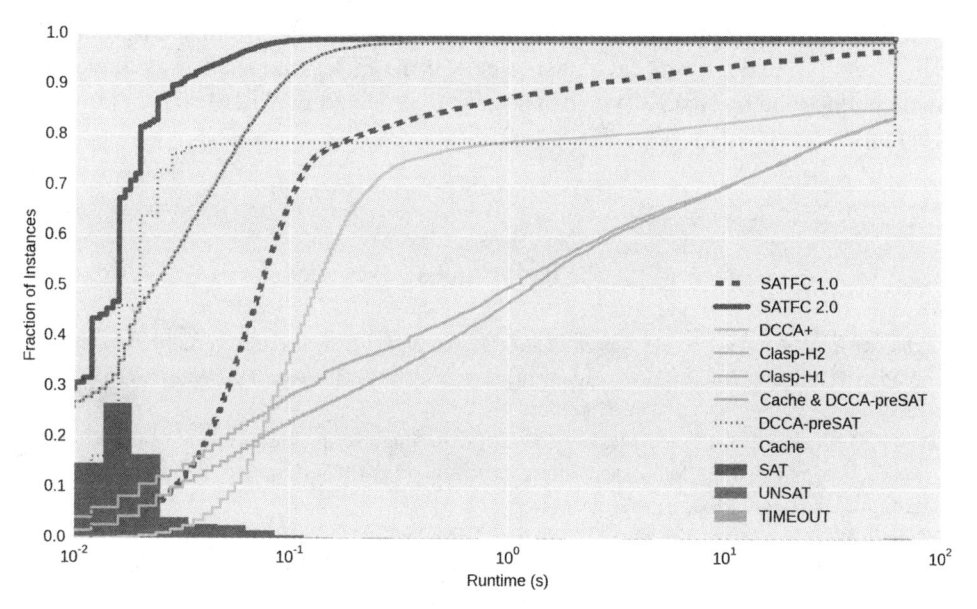

Figure 5. ECDF of runtimes of the SAT solvers we include in our final portfolio. The bars show fraction of SAT and UNSAT instances binned by their (fastest) runtime.

meta-algorithmic techniques, and further speedups based on domain-specific insights to yield a solver that meets the performance needs of the real incentive auction. Specifically, we identified a powerful parallel portfolio[6] of four solvers: a containment cache followed by DCCA-preSAT; DCCA+; clasp-h1; and clasp-h2. This portfolio, which we named *SATFC 2.0* (for SAT-based Feasibility Checker) achieved impressive performance (shown in Figure 5) as the SATFC 2.0 line), solving 99.0% of test instances in under 0.2 seconds, and 99.6% in under a minute. Moreover, the contribution of one of its main components, the containment cache, will continue to increase at negligible (online) CPU cost as we base it on more data.

Finally, we note SATFC 2.0's significant improvement over our previous solver SATFC, which was adopted and officially released by the FCC in November 2014 (FCC 2014b), albeit never previously discussed in an academic publication. Most of the new ideas presented in this chapter go beyond SATFC, which is a single-processor sequential portfolio: it works by carrying out a SAT encoding, performing the "local augmenting" idea described in Section 4, and then executing a version of clasp that we identified via algorithm configuration. Although SATFC achieved quite good performance within long cutoff times, it is both dramatically less able to solve instances within very short timescales and able to solve considerably fewer instances overall.

Acknowledgments. We gratefully acknowledge support from Auctionomics and the FCC; helpful conversations with Paul Milgrom, Ilya Segal, and James Wright; assistance from past research assistants Nick Arnosti, Guillaume Saulnier-Comte, Ricky Chen, Alim Virani, and Chris Cameron; experimental infrastructure assistance from

---

[6] Our intention is for SATFC to be run in parallel. However, since our portfolio achieves excellent performance in under a second, sequential execution would therefore achieve the same performance in under four seconds.

Steve Ramage; and help gathering data from Ulrich Gall, Rory Molinari, Karla Hoffman, Brett Tarnutzer, Sasha Javid, and others at the FCC. This work was funded by Auctionomics and by NSERC via the Discovery Grant and E.W.R. Steacie Fellowship programs.

References

Aardal, K. I.; Van Hoesel, S. P.; Koster, A. M.; Mannino, C.; and Sassano, A. 2007. Models and solution techniques for frequency assignment problems. *Annals of Operations Research* 153(1):79–129.

Bazelon, C.; Jackson, C. L.; and McHenry, G. 2011. An engineering and economic analysis of the prospects of reallocating radio spectrum from the broadcast band through the use of voluntary incentive auctions. TPRC.

Calamari, M.; Kharkar, O.; Kochard, C.; Lindsay, J.; Mulamba, B.; and Scherer, C. B. 2012. Experimental evaluation of auction designs for spectrum allocation under interference constraints. In *Systems and Information Design Symposium*, 7–12. IEEE.

Charikar, M.; Indyk, P.; and Panigrahy, R. 2002. New algorithms for subset query, partial match, orthogonal range searching, and related problems. In *Automata, Languages and Programming*. Springer. 451–462.

Congressional Budget Office. 2015. Proceeds from auctions held by the Federal Communications Commission. www.cbo.gov/publication/50128.

Dütting, P.; Gkatzelis, V.; and Roughgarden, T. 2014. The performance of deferred-acceptance auctions. In *Proc. of EC*, EC '14, 187–204. New York, NY, USA: ACM.

FCC. 2013. Office of engineering and technology releases and seeks comment on updated OET-69 software. *FCC Public Notice* DA 13–138.

FCC. 2014a. Comment sought on competitive bidding procedures for broadcast incentive auction 1000, including auctions 1001 and 1002. *FCC Public Notice* 14–191.

FCC. 2014b. FCC feasibility checker. http://wireless.fcc.gov/incentiveauctions/learn-program/repacking.html.

FCC. 2014c. In the matter of expanding the economic and innovation opportunities of spectrum through incentive auctions. *FCC Report & Order* FCC 14–50. particularly section IIIB.

FCC. 2014d. Information related to incentive auction repacking feasibility checker. www.fcc.gov/document/information-related-incentive-auction-repacking-feasability-checker.

FCC. 2014e. Repacking constraint files. http://data.fcc.gov/download/incentive-auctions/Constraint_Files/.

Gebser, M.; Kaufmann, B.; Neumann, A.; and Schaub, T. 2007. clasp: A conflict-driven answer set solver. In *Logic Programming and Nonmonotonic Reasoning*. Springer. 260–265.

Gomes, C. P., and Selman, B. 2001. Algorithm portfolios. *AIJ* 126(1):43–62.

Hoffmann, J., and Koehler, J. 1999. A new method to index and query sets.

Hutter, F.; Lindauer, M.; Bayless, S.; Hoos, H.; and Leyton-Brown, K. 2014a. Configurable SAT solver challenge (CSSC) (2014). http://aclib.net/cssc2014/index.html.

Hutter, F.; López-Ibáñez, M.; Fawcett, C.; Lindauer, M.; Hoos, H. H.; Leyton-Brown, K.; and Stützle, T. 2014b. AClib: A benchmark library for algorithm configuration. In *LION*. Springer. 36–40.

Hutter, F.; Hoos, H. H.; and Leyton-Brown, K. 2011. Sequential model-based optimization for general algorithm configuration. In *Proc. of LION*, 507–523.

Järvisalo, M.; Le Berre, D.; Roussel, O.; and Simon, L. 2012. The international SAT solver competitions. *AI Magazine* 33(1):89–92.

Kazumori, E. 2014. Generalizing deferred acceptance auctions to allow multiple relinquishment options. *SIGMETRICS Performance Evaluation Review* 42(3):41–41.

Kearns, M., and Dworkin, L. 2014. A computational study of feasible repackings in the FCC incentive auctions. *CoRR* abs/1406.4837.

Kwerel, E.; LaFontaine, P.; and Schwartz, M. 2012. Economics at the FCC, 2011–2012: Spectrum incentive auctions, universal service and intercarrier compensation reform, and mergers. *Review of Industrial Organization* 41(4):271–302.

Luo, C.; Cai, S.; Wu, W.; and Su, K. 2014. Double configuration checking in stochastic local search for satisfiability. In *AAAI*.

Marcus, M. J. 2013. Incentive auction: a proposed mechanism to rebalance spectrum between broadcast television and mobile broadband [spectrum policy and regulatory issues]. *Wireless Communications* 20(2):4–5.

Milgrom, P., and Segal, I. 2014. Deferred-acceptance auctions and radio spectrum reallocation. In *Proc. of EC*. ACM.

Milgrom, P.; Ausubel, L.; Levin, J.; and Segal, I. 2012. Incentive auction rules option and discussion. *Report for Federal Communications Commission. September* 12.

Newman, N.; Fréchette A.; and Leyton-Brown, K. 2017. Deep optimization for spectrum repacking. Communications of the ACM, in press.

Nguyen, T.-D., and Sandholm, T. 2014. Optimizing prices in descending clock auctions. In *Proc. of EC*, 93–110. ACM.

Nudelman, E.; Leyton-Brown, K.; Andrew, G.; Gomes, C.; McFadden, J.; Selman, B.; and Shoham, Y. 2003. Satzilla 0.9. Solver description, International SAT Competition.

Patrascu, M. 2011. Unifying the landscape of cell-probe lower bounds. *SIAM Journal on Computing* 40(3):827–847.

Savnik, I. 2013. Index data structure for fast subset and superset queries. In *Availability, Reliability, and Security in Information Systems and HCI*, 134–148. Springer.

Vohra, A. 2014. On the near-optimality of the reverse deferred acceptance algorithm.

Xu, L.; Hoos, H. H.; and Leyton-Brown, K. 2010. Hydra: Automatically configuring algorithms for portfolio-based selection. In *AAAI*, 210–216.

CHAPTER 39

ICE: An Expressive Iterative Combinatorial Exchange

Benjamin Lubin, Adam I. Juda, Ruggiero Cavallo, Sébastien Lahaie, Jeffrey Shneidman, and David C. Parkes

1. Introduction

Combinatorial exchanges combine and generalize two different mechanisms: double auctions and combinatorial auctions. In a double auction (DA), multiple buyers and sellers trade units of an identical good (McAfee, 1992). In a combinatorial auction (CA), a single seller has multiple heterogeneous items up for sale (de Vries & Vohra, 2003; Cramton, Shoham, & Steinberg, 2006). Buyer valuations can exhibit complements ("I want *A* and *B*") and substitutes ("I want *A* or *B*") properties. CAs provide an expressive bidding language to describe buyer valuations. A common design goal in DAs and CAs is to implement the *efficient allocation*, which is the allocation that maximizes social welfare, i.e. the total value.

A *combinatorial exchange* (CE) (Parkes, Kalagnanam, & Eso, 2001) is a combinatorial double auction and allows multiple buyers and sellers to trade on multiple, heterogeneous goods. A motivating application is to the reallocation of U.S. wireless spectrum from low-volume television stations to digital cell phone services (Cramton, Kwerel, & Williams, 1998; Cramton, Lopez, Malec, & Sujarittanonta, 2015). An *incentive auction* has been proposed for this application. This auction design uses a reverse auction to buy back existing spectrum rights followed by a forward auction to sell these rights to new owners.[1] (See "Designing the US Incentive Auction" by Mulgrom and Segal.) One advantage of the incentive auction design is that it enables the use of existing CA technology for both the forward and reverse stages. In addition, the proposed design uses optimization to solve the complex repacking problem of shifting incumbent users' allocation to new bands, which can free up a significant amount of bandwidth (Frechette, Newman, & Leyton-Brown, 2015).

CEs present an alternative design where both demand-side and supply-side price discovery happen in tandem, leading to coordinated information revelation and potential

---

[1] Voucher-based schemes (Kwerel & Williams, 2002) are a predecessor to the incentive auction design. The idea is to collect all goods from sellers, and then run a one-sided auction in which sellers can buy-back their own goods with vouchers that provide a seller with a share of the revenue collected on their own goods.

efficiency gains. The separation between stages in an incentive auction opens the possibility of the government purchasing either too little or too much spectrum in the reverse stage for subsequent sale in the forward stage. CEs avoid this problem by enabling the simultaneous purchase and sale of bandwidth by incumbents and new entrants alike. The (approximate) clearing prices obtained in CEs also provide meaningful value estimates for various spectrum combinations. These prices are useful in guiding demand and supply revelation (or *preference elicitation*, in the language of computer science). In addition, these prices can find potential use in secondary resale markets or for guiding the operation of subsequent instantiations of the exchange.

CEs are also applicable to airport takeoff and landing slot allocation (Ball, Donohue, & Hoffman, 2006; Vossen & Ball, 2006), in financial markets (Saatcioglu, Stallaert, & Whinston, 2001; Bossaerts, Fine, & Ledyard, 2002; Fan, Stallaert, & Whinston, 1999), and to allocate computational resources (Fu, Chase, Chun, Schwab, & Vahdat, 2003).[2] CEs can be applied to trade tasks within multi-robot systems, providing a new tool for the design of multi-agent systems (Gerkey & Mataric, 2002; Bererton, Gordon, & Thrun, 2003; Dias, Zlot, Kalra, & Stentz, 2006). CEs can also be used to allow multiple buyers to collaborate in a single sourcing event, each buyer perhaps representing a different profit center within an organization; see the associated work on expressive sourcing in CAs (Sandholm, 2007).

The main contribution of this paper is the design of the first *fully expressive, iterative* combinatorial exchange (ICE). We share the motivation of earlier work on iterative CAs: we wish to mitigate elicitation costs by guiding them to parts of the allocation space where they may enjoy a competitive advantage. The ICE design does this by quoting item prices and using activity rules to promote credible early bids. Market designs that are iterative rather than sealed-bid, one-shot are important because determining the value for even a single trade can be a challenging problem in domains of interest (Sandholm & Boutilier, 2006; Compte & Jehiel, 2007). Moreover, bidders often wish to reveal as little information as possible about their valuations to competitors.

At a high level, the design of ICE has the following components. A bidder represents the trades of interest by placing a bid in the *tree-based bidding language* (*TBBL*), which can provides a succinct way of representing values for trades in a CE. A bidder must also annotate the tree with lower and upper bounds, which have the effect of describing a set of possible valuation functions. In each round, the exchange uses each bidder's set of possible valuation functions (according to current lower and upper bounds) to select a provisional trade and provisional payments, and also to quote prices on items. Each bidder is required to refine the bounds (and thus the uncertainty about his valuation function) until there is a well defined trade that would be optimal for the bidder at current prices, whatever his actual valuation function. In this way, the design provides a hybrid between a demand-revealing auction process and a direct-revelation mechanism. Item prices guide demand revelation in each round, but bids are not statements about packages (or trades) demanded at current prices but rather placed through a reported valuation in the *TBBL* language and modified across rounds by changing upper and lower bounds. These reported valuation functions are eventually used to clear the exchange, determining the trade and payments.

---

[2] Since the journal version of this paper was published Google experimented with an application of a one-shot version of ICE for their internal compute cluster allocations (Stokely, Winget, Keyes, Grimes, & Yolken, 2009).

We highlight the following technical contributions:

- *TBBL* extends earlier bidding languages to support participants who are simultaneously buying some goods and selling others, to allow for valuation bounds, and to introduce generalized *choose operators* to provide more succinct representations than the OR* and $\mathcal{L}_{GB}$ languages for CAs (Boutilier & Hoos, 2001; Nisan, 2006). *TBBL* can be directly encoded within a mixed-integer programming (MIP) formulation of the winner determination problem.
- Despite quoting prices on items (and not packages of items), ICE is able to converge to the efficient trade if bidders are straightforward and keep their true valuation function within their bounds that represent their bids in each round. The proof of efficiency is via duality theory when prices are sufficiently accurate and otherwise achieved by analysis of the set of valuations prescribed by each agent.
- We introduce two new activity rules to require properties of bids across rounds. The first activity rule is the *modified revealed-preference activity rule* (MRPAR), which requires each bidder to make precise which trade is most preferred in each round given current prices. The second is the *delta improvement activity rule* (DIAR), which requires each bidder to refine his bid either to improve price accuracy or to prove that no improvement is possible. Together, there two rules ensure that useful progress towards identifying the efficient trade is made in each round.

The main design innovation is the use of bounds from bidders on their valuation functions for various trades, defining a set of possible valuations. A similar idea can be seen in the eBay proxy agent design, where a bidder's current bid can be interpreted as a lower-bound on her value, this bound refined (upwards) over time. It is crucial that we have both lower and upper bounds in the present setting, since this is an exchange and without this information we would not have enough guidance to determine a provisional allocation in early rounds. Moreover, by eliciting valuation functions ICE is able to implement the efficient allocation even though it uses item prices and even though item prices need not support the efficient allocation in competitive equilibrium. A quantitative bound on the potential inefficiency of the provisional allocation in each round can be computed by reasoning directly about the bounds on valuation functions.

In regard to the final payments charged of bidders, ICE is flexible and can be configured to be used with different payment rules. For a given payment rule, the prices that are quoted in each round are designed to approximate these payments. But the final payments are determined using the payment rule rather than on the basis of these item prices. In this sense, the design of payment rules is orthogonal to the main challenges addressed in this paper, and is not the main focus of our work. For concreteness in describing the exchange design, we adopt the *threshold payment rule* (Parkes et al., 2001). This rule minimizes the *ex post* regret to a bidder for truthful bidding, when holding the bids from other participants fixed, across all budget-balanced payment rules (for exchanges that clear to maximize total revealed value). Alternative payment rules for CEs and their equilibria have been studied in the work of Lubin and Parkes (2009) and the work of Lubin (2015); see also the work of Milgrom (2007).[3]

---

[3] In domains where a *truthful* sealed bid mechanism is available, the payment rule from such a mechanism can be adopted within the ICE framework. The effect of this is for ICE to support *straightforward bidding* in an

ICE is fully implemented in Java (with a C-based MIP solver). We present empirical results across a wide range of simulated domains, both to demonstrate its scalability and also to provide a qualitative understanding of the characteristics of our mechanism. Our empirical results (with straightforward bidders) show that the exchange quickly converges to the efficient trade. For example, this requires an average of seven rounds in a domain with 100 goods of 20 different types and eight bidders, with *TBBL* valuations that contain an average of more than 100 nodes. For a particular measure of residual uncertainty about bidder valuations, bidders leave more than 50% of their valuation function undefined in ICE in this same domain. ICE terminates in less than ten minutes when simulated on a 3.2GHz dual-processor dual-core workstation with 8GB of memory. This includes the time for all winner determination, pricing, and activity rules, as well as the time to simulate agent bidding strategies.

1.1. Related Work

Many ascending-price one-sided CAs are known in the literature (Parkes & Ungar, 2000a; Wurman & Wellman, 2000; Ausubel & Milgrom, 2002; de Vries, Schummer, & Vohra, 2007; Mishra & Parkes, 2007). Direct elicitation approaches, in which bidders respond to explicit queries about their valuations, have also been proposed for one-sided CAs (Conen & Sandholm, 2001; Hudson & Sandholm, 2004; Lahaie & Parkes, 2004; Lahaie, Constantin, & Parkes, 2005). Of particular relevance are ascending-price designs with item prices (Dunford, Hoffman, Menon, Sultana, & Wilson, 2003; Kwasnica, Ledyard, Porter, & DeMartini, 2005). In computing (approximately competitive) item prices, we generalize and extend these methods. Building on the work of Rassenti, Smith, and Bulfin, 1982, these earlier papers consider bids on bundles individually, and find prices that are exact on winning bids and minimize the pricing error to losing bids. Generalizing to the *TBBL* expressive language, we compute prices that minimize the worst-case pricing error over all *bidders* (rather than bids on individual trades), considering the most preferred trade consistent with the *TBBL* bid of each bidder. As in the work of Dunford et al. (2003) and Kwasnica et al. (2005) we incorporate additional tie-breaking stages, in our case to minimize the pricing error followed by the disagreement between prices and provisional payments.

Item prices rather than (non-linear) package prices are important in practical applications. Such prices are adopted in the U.S. FCC's wireless spectrum auctions (Cramton, 2006), within clock auctions for the procurement of electricity generation (Cramton, 2003), and in an auction design for airport landing rights at Laguardia airport (Ball et al. 2007). Item prices supporting competitive equilibrium exist in two-sided markets with indivisible goods in which each agent will buy or sell a single item (but may be interested in multiple different items) (Shapley & Shubik, 1972). But these kind of simple, competitive equilibrium prices do not exist in general CEs; see the work of Kelso and Crawford (1982), Bikhchandani and Mamer (1997), Bikhchandani and Ostroy (2002), and O'Neill, Sotkiewicz, Hobbs, Rothkopf, and Stewart (2005) for related discussions.

---

ex post Nash equilibrium; see Mishra and Parkes (Mishra & Parkes, 2007) for related observations in the context of iterative CAs. Here, a truthful mechanism is one in which truthful bidding is a dominant-strategy equilibrium. Straightforward bidding is any strategy in ICE in which bids are refined in a way that keeps an agent's true valuation function within the set of valuations prescribed by its bounds.

ICE has a *proxy-based architecture*, in the sense that bidders submit and refine bounds on *TBBL* bids directly to the exchange, with this information used to drive price dynamics and clear the exchange. Earlier work considered proxied approaches in ascending-price CAs (Parkes & Ungar, 2000b; Ausubel & Milgrom, 2002); see also Ausubel et al., 2006.

Activity rules are known to be an important design feature of iterative auctions. For example, the Milgrom-Wilson activity rule, which requires a bidder to be active on a minimum percentage of the quantity of the spectrum for which he is eligible to bid, is a critical component of the FCC's auction rules (Milgrom, 2004). ICE adopts a variation on the *revealed-preference activity rule* of the clock-proxy auction.

Efficiency and budget-balance is not possible in CEs because of the Myerson-Satterthwaite impossibility result (Myerson & Satterthwaite, 1983). Given this, Parkes et al. study sealed-bid combinatorial exchanges and introduced the Threshold payment rule (Parkes et al., 2001); see the work of Milgrom (2007) and Day and Raghavan (2007) for a related discussion. Double auctions in which truthful bidding is in a dominant strategy equilibrium are known for unit demand settings (McAfee, 1992) as well as slightly more expressive domains (Babaioff & Walsh, 2005; Chu & Shen, 2008). However, no truthful, budget-balanced mechanisms with approximate efficiency are known for the general CE problem.

Smith, Sandholm, and Simmons previously studied iterative CEs, but handle only limited expressiveness and adopt a direct-query based approach that does not scale (Smith et al., 2002). A novel feature in this earlier design (not supported here) is *item discovery*, where the items available to trade may not be known in advance. Earlier work has also considered sealed-bid combinatorial exchanges for basket trades in financial markets, including aspects of expressiveness and winner determination (Saatcioglu et al., 2001).

Several bidding languages have been proposed for CAs, a number of which are *logical bidding languages* and allow bidders to represent the logical structure of their valuation over goods (Nisan, 2006). Closest to *TBBL* is the $\mathcal{L}_{GB}$ language (Boutilier & Hoos, 2001), which allows for arbitrarily nested logical expressions, supports standard propositional logic operators, and also provides a *k-of* operator, used to represent a willingness to pay for any k of some set of trades; see also the work of Rothkopf, Pekeč, and Harstad (1998) for a less general, tree-based bidding language. Boutilier achieves good scalability of winner determination by coupling the $\mathcal{L}_{GB}$ language with a MIP solver (Boutilier, 2002). *TBBL* shares some structural elements with the $\mathcal{L}_{GB}$ language but has important differences in its semantics. In $\mathcal{L}_{GB}$, the semantics are those of propositional logic, with the same items in an allocation able to satisfy a tree in multiple places. Although this can make bids in $\mathcal{L}_{GB}$ especially succinct in some settings, the semantics of *TBBL* are modular, so that the value of each component of a bid tree can be understood in isolation of the rest of the tree.

1.2. Outline

Section 2 provides preliminaries, including a definition of what we mean by an efficient trade and competitive equilibrium prices. Section 3 defines a sealed-bid CE, introduces the *TBBL* language, and provides the MIP formulation that we use for winner determination. Section 4 extends *TBBL* to allow for valuation bounds, defines the MRPAR and DIAR activity rules and the method to determine item prices, and presents the main theoretical results. Section 5 gives a number of illustrative examples. Section 6 presents the main experimental results. We conclude in Section 7. The Appendix provides an algorithm for each of the two activity rules and details on the straightforward bidding strategy adopted by bidders for the purpose of our simulations.

2. Preliminaries

The basic environment considers a set of bidders, $N = \{1, \ldots, n\}$, who are interested in trading multiple units of distinct, indivisible goods, where the set of different types of goods is denoted $G = \{1, \ldots, m\}$. Each bidder has an initial endowment of goods and a valuation for different trades. Let $x^0 = (x_1^0, \ldots, x_n^0)$ denote the initial endowment of goods, with $x_i^0 = (x_{i1}^0, \ldots, x_{im}^0)$ and $x_{ij}^0 \in \mathbb{Z}_+$ to indicate the number of units of good type $j \in G$ initially held by bidder $i \in N$. A *trade* $\lambda = (\lambda_1, \ldots, \lambda_n)$ denotes the *change* in allocation, with $\lambda_i = (\lambda_{i1}, \ldots, \lambda_{im})$ and $\lambda_{ij} \in \mathbb{Z}$ denoting the change in the number of units of item j to bidder i. Let $M = \sum_{i \in N} \sum_{j \in G} x_{ij}^0$ denote the total supply in the exchange. We write $i \in \lambda$ to denote that bidder i is *active* in the trade, i.e., buys or sells at least one item.

2.1. The Efficient Trade

Each bidder has a value $v_i(\lambda_i) \in \mathbb{R}$ for his component of trade λ. This value can be positive or negative, and represents the *change in value* between the final allocation $x_i^0 + \lambda_i$ and the initial allocation x_i^0. The valuation and initial allocation information is private to each bidder, and we assume that there are no externalities, so that each bidder's value depends only on his individual trade. We assume *free disposal*, so that $v_i(\lambda_i') \geq v_i(\lambda_i)$ for trade $\lambda_i' \geq \lambda_i$, i.e., for which $\lambda_{ij}' \geq \lambda_{ij}$ for all j. Let $v(\lambda) = \sum_i v_i(\lambda_i)$.

Utility is modeled as quasi-linear, with $u_i(\lambda_i, p) = v_i(\lambda_i) - p$ for trade λ_i and payment $p \in \mathbb{R}$. This implies that bidders are modeled as being risk neutral and assumes that there are no budget constraints. The payment, p, can be negative, indicating the bidder may receive a payment for the trade. We use the term *payoff* interchangeably with *utility*. Because of quasi-linearity, any Pareto optimal (i.e., efficient) trade will maximize the social welfare, which is equivalent to the total increase in value to all bidders due to the trade. Given an instance of the CE problem, defined by tuple (v, x^0), i.e., a valuation profile $v = (v_1, \ldots, v_n)$ and an initial allocation $x^0 = (x_1^0, \ldots, x_n^0)$, the efficient trade λ^*, is defined as follows:

Definition 1. Given CE instance (v, x^0), the efficient trade λ^* solves

$$\max_{(\lambda_1, \ldots, \lambda_n)} \sum_i v_i(\lambda_i) \tag{1}$$

$$\text{s.t.} \quad \lambda_{ij} + x_{ij}^0 \geq 0, \quad \forall i, \forall j \tag{2}$$

$$\sum_i \lambda_{ij} = 0, \quad \forall j \tag{3}$$

$$\lambda_{ij} \in \mathbb{Z}$$

Constraints (2) ensure that no bidder sells more items than he has in his initial allocation. By free disposal, we can impose strict balance in the supply and demand of goods at the solution in constraints (3), i.e., we can allocate unwanted items to any bidder. We adopt $\mathcal{F}(x^0)$ to denote the set of *feasible trades*, given these constraints and given an initial allocation x^0, and $\mathcal{F}_i(x^0)$ for the set of feasible trades to bidder i. Note that valuation function v_i cannot be explicitly represented as a value for each possible trade to bidder i, because the number of such trades scales as $O(s^m)$, where s is the maximal number of units of any item in the market and there are m different items. The *TBBL* language (introduced in Section 3) leads to a succinct formulation of the efficient trade problem as a mixed-integer program.

The initial allocation x_i^0 may be private to agent i. We assume throughout that bidders are truthful in revealing this information, which we motivate by supposing that participants cannot sell items that they do not actually own (or pay a suitably high penalty if they do).

2.2. Competitive Equilibrium Prices

Item prices, $\pi = (\pi_1, \ldots, \pi_m)$, define a price π_j on each good so that the price to bidder i on a trade λ is defined as $p^\pi(\lambda_i) = \sum_j \lambda_{ij} \pi_j = \lambda_i \cdot \pi$. Such prices play an important role in ICE. Of particular interest is the set of competitive equilibrium prices:

Definition 2. Item prices π are competitive equilibrium (EQ) prices for CE problem (v, x^0) if there is some feasible trade $\lambda \in \mathcal{F}(x^0)$ such that:

$$v_i(\lambda_i) - p^\pi(\lambda_i) \geq v_i(\lambda_i') - p^\pi(\lambda_i'), \quad \forall \lambda_i' \in \mathcal{F}_i(x^0), \tag{4}$$

for every bidder i. We say that such a trade, λ, is **supported** by prices π.

Theorem 1 (*Bikhchandani & Ostroy, 2002*). *Any trade λ supported by competitive equilibrium prices π is an efficient trade.*

In practice, exact EQ prices are unlikely to exist. Instead, it is useful to define the concept of *approximate* EQ prices and an approximately efficient trade:

Definition 3. Item prices π are δ-approximate competitive equilibrium (EQ) prices for CE problem (v, x^0) and $\delta \in \mathbb{R}_{\geq 0}$, if there is some feasible trade $\lambda \in \mathcal{F}(x^0)$ such that:

$$v_i(\lambda_i) - p^\pi(\lambda_i) + \delta \geq v_i(\lambda_i') - p^\pi(\lambda_i'), \quad \forall \lambda_i' \in \mathcal{F}_i(x^0), \tag{5}$$

for every bidder i.

At δ-approximate EQ prices, there is some trade for which every bidder i is within $\delta \geq 0$ of maximizing his utility. Furthermore, we say that trade λ is *z-approximate* if the total value of the trade is within z of the total value of the efficient trade.

Theorem 2. *Any trade λ supported by δ-approximate EQ prices π is a $2\min(M, \frac{n}{2})\delta$-approximate efficient trade.*

Proof. Fix instance (v, x^0) and consider (λ, π). For any trade $\lambda' \neq \lambda$ we have

$$\sum_{i \in \lambda \cup \lambda'} [v_i(\lambda_i) - p^\pi(\lambda_i) + \delta] \geq \sum_{i \in \lambda \cup \lambda'} [v_i(\lambda_i') - p^\pi(\lambda_i')], \tag{6}$$

by δ-EQ prices and because values and prices are zero for bidders that do not participate in a trade. We have $\sum_{i \in \lambda \cup \lambda'} p^\pi(\lambda_i) = \sum_{i \in \lambda \cup \lambda'} p^\pi(\lambda'_i) = 0$ (since $\sum_i p^\pi \lambda''_i = \sum_i \lambda''_i \cdot \pi = \sum_i \sum_j \lambda''_{ij} \pi_j = \sum_j \pi_j \sum_i \lambda''_{ij} = 0$, with $\sum_i \lambda''_{ij} = 0$ for all j, for all $\lambda''_i \in \mathcal{F}(x^0)$). Then, $\sum_i v_i(\lambda_i) + \sum_{i \in \lambda \cup \lambda'} \delta_i \geq \sum_i v_i(\lambda'_i)$. Fix $\lambda' := \lambda^*$, for efficient trade λ^*. Then, $\sum_i v_i(\lambda_i) + \Delta \geq \sum_i v_i(\lambda^*_i)$, where

$$\Delta = \sum_{i \in \lambda \cup \lambda'} \delta_i \leq \min(2A^\#(x^0), n)\delta \leq \min(2\min(M, n), n)\delta = 2\min\left(M, \frac{n}{2}\right)\delta \quad (7)$$

Here $A^\#(x^0)$ is the maximal number of bidders that can trade in a feasible trade given x^0. The second inequality follows because no more bidders can trade than there are number of goods to trade or bidders in the market and thus $A^\#(x^0) \leq \min(M, n)$. □

3. Step One: A *TBBL*-Based Sealed-Bid Combinatorial Exchange

We first flesh out the details for a non-iterative, *TBBL*-based CE in which each bidder submits a sealed bid in the *TBBL* language.

Bidding language. The tree-based bidding language (*TBBL*) is designed to be expressive and succinct, to be entirely symmetric with respect to buyers and sellers, and to easily provide for bidders that are both buying and selling goods; i.e., ranging from simple swaps to highly complex trades. Bids are expressed as annotated *bid trees*, and define a bidder's *change in value* for all possible trades. The main feature of *TBBL* is that it has a general *interval-choose logical operator* on internal nodes coupled with a rich semantics for propagating values within the tree. Leaves of the tree are annotated with traded items and all nodes are annotated with changes in values (either positive or negative). *TBBL* is designed such that these changes in value are expressed on *trades* rather than the total value of allocations. Examples are provided in Figures 1–3, described in detail below.

Consider bid tree T_i from bidder i. Let $\beta \in T_i$ denote a node in the tree, and let $v_i(\beta) \in \mathbb{R}$ denote the value specified at node β (perhaps negative). Let $Leaf(T_i) \subseteq T_i$ be the subset of nodes representing the leaves of T_i and let $Child(\beta) \subseteq T_i$ denote the children of node β. All nodes except leaves are labeled with the *interval-choose* operator $IC^y_x(\beta)$. Each leaf β is labeled as a *buy* or *sell*, with units $q_i(\beta, j) \in \mathbb{Z}$ for the good j associated with leaf β, and $q_i(\beta, j') = 0$ otherwise. The same good j may simultaneously occur in multiple leaves of the tree, given the semantics of the tree described below.

The IC operator defines a range on the number of children that can be, and must be, satisfied for node β to be satisfied: an $IC^y_x(\beta)$ node (where x and y are non-negative integers) indicates that the bidder is willing to pay for the satisfaction of at least x and at most y of his children. With suitable values for x and y the operator can include many logical connectors. For instance: $IC^n_n(\beta)$ on node β with n children is equivalent to an AND operator; $IC^n_1(\beta)$ is equivalent to an OR operator; and $IC^1_1(\beta)$ is equivalent to an XOR operator.[4]

---
[4] This equivalence implies that *TBBL* can directly express the XOR, OR and XOR/OR languages (Nisan, 2006).

We say that the *satisfaction* of an $\text{IC}_x^y(\beta)$ node is defined by the following two rules:

R1 Node β with $\text{IC}_x^y(\beta)$ may be *satisfied* only if at least x and at most y of its children are *satisfied*.

R2 If some node β is *not satisfied*, then none of its children may be *satisfied*.

One can consider **R1** as a first pass, that defines a set of candidates for satisfaction. This candidate set is then refined by **R2**. Besides defining how value is propagated, by virtue of **R2** our logical operators act as *constraints* on what trades are *acceptable* and provide necessary and sufficient conditions.[5]

Given a tree T_i, the (change in) value of a trade λ is defined as the sum of the values on all satisfied nodes, where the set of satisfied nodes is chosen to provide the *maximal* total value. Let $sat_i(\beta) \in \{0, 1\}$ denote whether node β in tree T_i of bidder i is satisfied, with $sat_i = \{sat_i(\beta), \forall \beta \in T_i\}$. For solution sat_i to be valid for tree T_i and trade λ_i, written $sat_i \in valid(T_i, \lambda_i)$, then rules **R1** and **R2** must hold for all internal nodes $\beta \in \{T_i \setminus Leaf(T_i)\}$ with $\text{IC}_x^y(\beta)$:

$$x \, sat_i(\beta) \leq \sum_{\beta' \in Child(\beta)} sat_i(\beta') \leq y \, sat_i(\beta) \qquad (8)$$

Equation (8) enforces the *interval-choose* constraints, by ensuring that no more and no less than the appropriate number of children are *satisfied* for any node that is *satisfied*. The constraint also ensures that any time a node other than the root is satisfied, its parent is also satisfied. We further require, for $sat_i \in valid(T_i, \lambda_i)$, that the total increase in quantity of each item across all satisfied leaves is no greater than the total number of units awarded in the trade:

$$\sum_{\beta \in Leaf(T_i)} q_i(\beta, j) sat_i(\beta) \leq \lambda_{ij}, \quad \forall j \in G \qquad (9)$$

By free disposal, we allow here for a trade to assign additional units of an item over-and-above that required in order to activate leaves in the bid tree. This works for sellers as well as buyers: for sellers a trade is negative and this requires that the total number of items indicated sold in the tree is at least the total number of items that are traded away from the bidder in the trade.

Given these constraints, the total value of trade λ_i, given bid tree T_i from bidder i, is defined as the solution to an optimization problem:

$$v_i(T_i, \lambda_i) = \max_{sat_i} \sum_{\beta \in T_i} v_i(\beta) sat_i(\beta) \qquad (10)$$

s.t. (8), (9)

Example 1. Consider an airline operating out of a slot-controlled airport that already owns several morning landing slots, but none in the evening. In order to expand its business the airline wishes to acquire at least two and possibly three of the evening

---

[5] **R1** naturally generalizes the approach taken in $\mathcal{L}_{GB}$, where an internal node is satisfied according to its operator and the subset of its children that are satisfied. The semantics of $\mathcal{L}_{GB}$, however, treat logical operators only as a way of specifying when added value (positive or negative) results from attaining combinations of goods. Our use of **R2** also imposes constraints on acceptable trades.

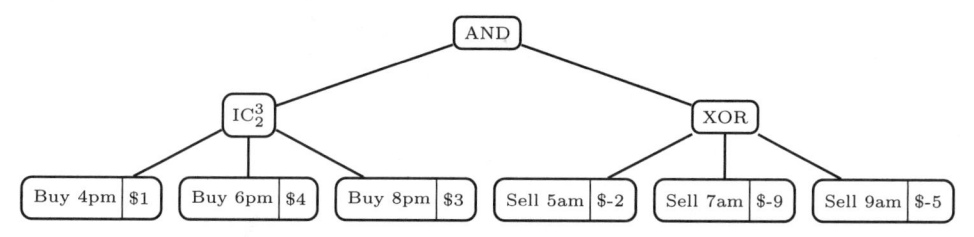

Figure 1. A *TBBL* tree for an airline interested in trading landing slots.

slots. However, it needs to offset the cost of this purchase by selling one of its morning slots. Figure 1 shows a *TBBL* valuation tree for expressing this kind of swap.

Example 2. Consider a mobile operator that wishes to acquire spectrum blocks in the 'partial economic areas' corresponding to the regions of New York, New Jersey, and Massachusetts (PEA1, PEA6, PEA7). The particular blocks are G (10 MHz) and H (5 MHz). Due to synergies, the acquisition must cover at least two areas to be worthwhile. Within areas, blocks are additive, though the larger G block is worth more per MHzPop than the smaller H block. A buyer bid representing these values can be succinctly represented in the *TBBL*, as illustrated in Figure 2. By replacing the root node with an OR-of-AND, the operator could also enforce the areas to be contiguous (i.e., always include New York).

A crucial aspect of *TBBL* is that it represents values for trades not allocations.[6] In earlier work, we demonstrate natural instances for which *TBBL* is *exponentially more succinct* than OR* and $\mathcal{L}_{GB}$ (Cavallo et al., 2005). In fact, *TBBL*'s succinctness is incomparable to OR* and $\mathcal{L}_{GB}$ but can be extended in simple ways to strictly dominate these earlier languages.

Winner Determination. The problem of determining an efficient trade given bids is called the *winner determination* (WD) problem. The WD problem in CAs (and thus also in CEs) is NP-hard (Rothkopf et al., 1998). The approach we adopt here is to formulate the problem as a mixed-integer program (MIP), and solve with branch-and-cut algorithms (Nemhauser & Wolsey, 1999). A similar approach has proved successful for solving the WD problem in CAs (de Vries & Vohra, 2003; Boutilier, 2002; Sandholm, 2006).

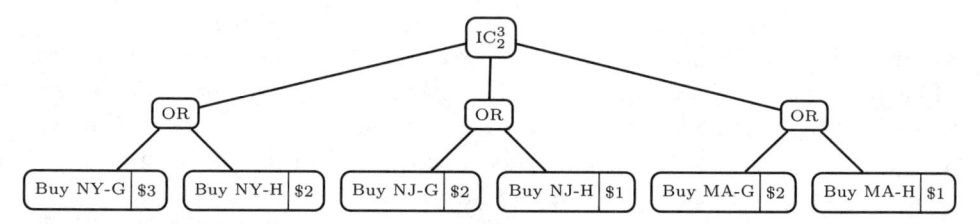

Figure 2. A *TBBL* tree for a bidder interested in a two blocks of spectrum in New York, New Jersey and Massachusetts.

---

[6] Working with numerous examples quickly reveals that it is difficult to succinctly capture even simple trades in languages that specify values on allocations rather than trades, as is the case with all existing languages.

Given some tree T_i, it is useful to adopt notation $\beta \in \lambda_i$ to denote a node $\beta \in T_i$ that is satisfied by trade λ_i. We can now formulate the WD problem for bid trees $T = (T_1, \ldots, T_n)$ and initial allocation x^0:

$$WD(T, x^0) : \max_{\lambda, sat} \sum_i \sum_{\beta \in T_i} v_i(\beta) sat_i(\beta)$$

s.t. (2), (3)

$$sat_i \in valid(T_i, \lambda_i), \qquad \forall i$$

$$sat_i(\beta) \in \{0, 1\}, \lambda_{ij} \in \mathbb{Z},$$

where $sat = (sat_1, \ldots, sat_n)$. The tree structure is made explicit in this MIP formulation: we have decision variables to represent the satisfaction of nodes and capture the logic of the *TBBL* language through linear constraints; a related approach approach has been considered in application to $\mathcal{L}_{GB}$ (Boutilier, 2002). By doing this, there are $O(nB + mn)$ variables and constraints, where B is the maximal number of nodes in any bid tree. The formulation determines the trade λ while simultaneously determining the value to all bidders by activating nodes in the bid trees.

Payments. Given reported valuation functions $\hat{v} = (\hat{v}_1, \ldots, \hat{v}_n)$ from each bidder, the Vickrey-Clarke-Groves (VCG) (e.g. Krishna, 2002) mechanism collects the following payments from each bidder:

$$p_{vcg,i} = \hat{v}_i(\lambda_i^*) - (V(\hat{v}) - V_{-i}(\hat{v})), \qquad (11)$$

where λ^* is the efficient trade, $V(\hat{v})$ is the reported value of this trade and $V_{-i}(\hat{v})$ is the reported value of the efficient trade in the economy without bidder i, where $v_{-i} = (v_1, \ldots, v_{i-1}, v_{i+1}, \ldots, v_n)$. Let us refer to $\Delta_{vcg,i} = V(\hat{v}) - V_{-i}(\hat{v})$ as the *VCG discount*. The problem with the VCG mechanism in the context of a CE is that it may run at a budget deficit with negative total payments. An alternative payment method is provided by the Threshold rule (Parkes et al., 2001):

$$p_{thresh,i} = \hat{v}_i(\lambda_i^*) - \Delta_{thresh,i}, \qquad (12)$$

where the discounts $\Delta_{thresh,i}$ are picked to minimize $\max_i(\Delta_{vcg,i} - \Delta_{thresh,i})$ subject to $\Delta_{thresh,i} \leq \Delta_{vcg,i}$ for all i and $\sum_i \Delta_{thresh,i} \leq V(\hat{v})$. Threshold payments are exactly budget balanced and minimize the maximal deviation from the VCG outcome across all balanced rules.

Example 3. Consider the two bidders in Figure 3. Bidder 1 will potentially sell one of his items (*A* or *B*) if he can get Bidder 2's item, *C*, at the right price. Bidder 2 is interested in buying one or both of Bidder 1's items and also selling his own item. We consider each of the possible trades: If Bidder 1 trades *A* for *C* he gets $2 of value and Bidder 2 gets $7. If Bidder 1 trades *B* for *C* he gets $-2 of value and Bidder 2 gets $2. And if no trade occurs both bidders get $0 value. Therefore the efficient trade is to swap *A* for *C*.

Because the efficient trade creates a surplus of $9 and removing either bidder results in the null trade, both bidders have a Vickrey discount of $9. Thus if we use VCG payments, Bidder 1 pays $2 - $9 = $-7 and Bidder 2 pays $7 - $9 = $-2 and the

Bidder 1

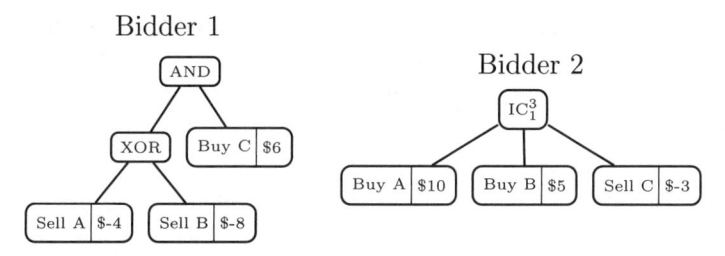

```
         AND
        /    \
      XOR    Buy C | $6
     /    \
Sell A|$-4  Sell B|$-8
```

Bidder 2

```
         IC₁³
       /   |   \
  Buy A   Buy B   Sell C
  $10      $5      $-3
```

Figure 3. Two bidders and three items {A, B, C}. The efficient trade is for bidder 1 to sell A and buy C.

exchange runs at a deficit. The Threshold payment rule chooses payments that minimally deviate from VCG while maintaining budget balance. This minimization reduces the discounts to $4.50, and thus Bidder 1 pays $2 − $4.50 = $ − 2.50 and Bidder 2 pays $7 − $4.50 = $2.50.

4. Step Two: Making the Exchange Iterative

Having defined a sealed-bid, *TBBL*-based exchange we can now modify the design to make it iterative. Rather than provide an exact valuation for all interesting trades, a bidder annotates a single *TBBL* tree with upper and lower bounds on his valuation. The ICE mechanism then proceeds in rounds, as illustrated in Figure 4.

ICE is a *proxied design* in which each bidder has a proxy to facilitate his valuation refinement. In each round, a bidder responds to prices by interacting with his proxy agent in order to tighten the bounds on his *TBBL* tree and meet the activity rules. The exchange chooses a *provisional valuation profile* (denoted $v^\alpha = (v_1^\alpha, \ldots, v_n^\alpha)$ in the figure), with the valuation v_i^α for each bidder picked to fall within the bidder's current valuation bounds (and to tend towards the lower valuation bound as progress is made towards determining the final trade). Then, the exchange computes a *provisional trade* λ^α and checks whether the conditions for moving to a last-and-final round are satisfied. Approximate equilibrium prices are then computed based on valuation profile v^α and trade λ^α and a new round begins. In the last-and-final round, the final payments and the trade are computed in terms of *lower* valuations; the semantics are that these

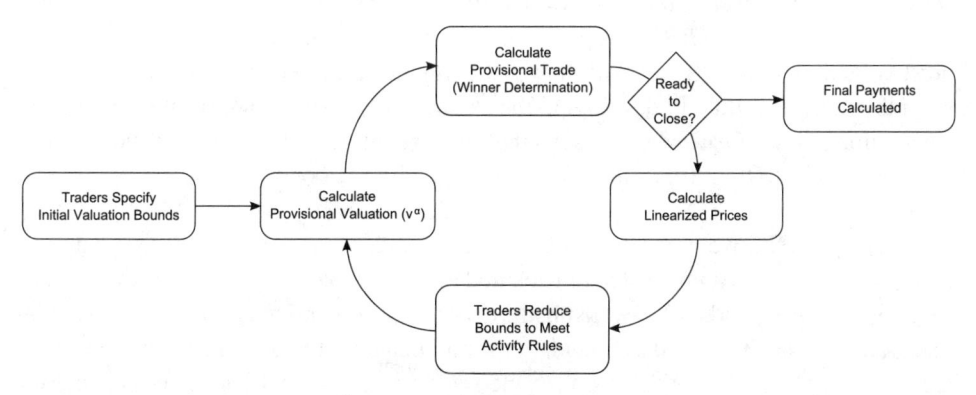

Figure 4. ICE system overview.

lower bounds guarantee that a bidder is willing to pay at least this amount (or receive a payment of this amount) in order to complete the trade.

Let $\underline{v}_i$ and $\overline{v}_i$ denote the lower and upper valuation functions reported by bidder i in a particular round of ICE, and adopt WD(v) to denote the WD problem for valuation profile $v = (v_1, \ldots, v_n)$. ICE is parameterized by a target approximation error $\Delta^* \in (0, 1]$, which requires that the total value from the optimal trade $\underline{\lambda}$ given the current lower-bound valuation profile (i.e., $\underline{\lambda}$ solves WD($\underline{v}$)) is close to the total value from the efficient trade λ_i^*:

$$\text{EFF}(\underline{\lambda}) = \frac{\sum_i v_i(\underline{\lambda}_i)}{\sum_i v_i(\lambda_i^*)} = \frac{v(\underline{\lambda})}{v(\lambda^*)} \geq \Delta^* \tag{13}$$

However, the true valuation v and thus the trade λ^* are unknown to the ICE, and thus we will later introduce techniques to establish this bound.

In each round, ICE goes through the following steps:

1. If this is the last-and-final round, then implement the trade that solves WD($\underline{v}$) and collect Threshold payments defined on valuations $\underline{v}$. STOP.

ELSE,

2. Solve WD($\underline{v}$) to obtain $\underline{\lambda}$. Use valuation bounds and prices to determine a lower-bound, ω^{eff}, on the allocative efficiency EFF($\underline{\lambda}$) of $\underline{\lambda}$. If $\omega^{\text{eff}} \geq \Delta^*$ then the next round will be designated the last-and-final round.
3. Set $\alpha \in [0, 1]$, with α tending to 1 as ω^{eff} tends to 1, and provisional valuation profile $v^\alpha = (v_1^\alpha, \ldots, v_n^\alpha)$, where $v_i^\alpha(\lambda_i) = \alpha \underline{v}_i(\lambda_i) + (1 - \alpha) \overline{v}_i(\lambda_i)$, expressed with a *TBBL* tree in which the value on node $\beta \in T_i$ is $v_i^\alpha(\beta) = \alpha \underline{v}_i(\beta) + (1 - \alpha) \overline{v}_i(\beta)$.
4. Solve WD(v^α) to find *provisional trade* λ^α, and determine Threshold payments for provisional valuation profile, v^α.
5. Compute item prices, $\pi \in \mathbb{R}_{\geq 0}^m$, that are approximate CE prices given valuations v^α and trade λ^α, breaking ties to best approximate the provisional Threshold payments and finally to minimize the difference in price between items.
6. Report (λ_i^α, π) to each bidder $i \in N$, and whether or not the next round is last-and-final.

In transitioning to the next round, the proxy agents are responsible for guiding bidders to make refinements to their lower- and upper-bound valuations in order to meet activity rules that ensure progress towards the efficient trade across rounds. In what follows, we (a) extend *TBBL* to capture lower and upper valuation bounds, (b) describe our two activity rules, (c) explain how we compute price feedback, (d) provide our main theoretical results. In developing theoretical and experimental results about ICE we assume *straightforward* bidders, so that bidders refine upper and lower bounds on valuations to keep their true valuation consistent with the bounds.

Extending *TBBL*. We first extend *TBBL* to allow bidder i to report a lower and upper bound $(\underline{v}_i(\beta), \overline{v}_i(\beta))$ on the value of each node $\beta \in T_i$, which in turn induces valuation functions $\underline{v}_i(T_i, \lambda_i)$ and $\overline{v}_i(T_i, \lambda_i)$, using the exact same semantics as in (10). The bounds on a trade can be interpreted as bounding the payment that the bidder considers acceptable. The bidder commits to complete the trade for a payment less than or equal to the lower-bound and to refuse to complete a trade for any payment greater than the

Bidder 1

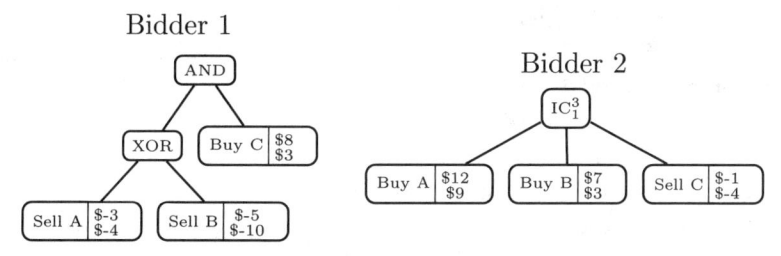

```
         AND
        /   \
      XOR   Buy C | $8
     /   \         $3
Sell A   Sell B
 $-3      $-5
 $-4      $-10
```

Bidder 2

```
          IC³₁
        /  |  \
   Buy A  Buy B  Sell C
   $12    $7     $-1
   $9     $3     $-4
```

Figure 5. Two bidders, each with partial value information defined on their bid tree. One can already prove that the efficient trade is for bidder 1 to sell A and buy C.

upper-bound. The exact value, and thus true willingness-to-pay, remains unknown except when $\underline{v}_i(\beta) = \overline{v}_i(\beta)$ on all nodes. We say that bid tree T_i for bidder i is *well-formed* if $\underline{v}_i(\beta) \leq \overline{v}_i(\beta)$ for all nodes $\beta \in T_i$. In this case we also have $\underline{v}_i(T_i, \lambda_i) \leq \overline{v}_i(T_i, \lambda_i)$ for all trades λ_i. We refer to the difference $\overline{v}_i(\beta) - \underline{v}_i(\beta)$ as the *value uncertainty* on node β. The efficient trade can often be determined with only partial information about bidder valuations. Consider the following simple variant on Example 3:

Example 4. The structure of the bidders' trees in Figure 5 is the same as in Example 3 but the nodes are annotated with bounds. Let $x \in [3, 8]$ denote Bidder 1's true value for "buy C" and $y \in [-4, -1]$ denote Bidder 2's true value for "sell C." The three feasible trades are: (1) trade A and C, (2) trade B and C, (3) no trade. The first trade is already provably efficient. Fixing x and y, its minimal value is $-4 + 9 + x - y$ and this is at least $-5 + 7 + x - y$, the value of the second trade. Moreover, its worst-case value is $-4 + 9 + 3 - 4 \geq 0$, the value of the null trade.

4.1. Activity Rules

Activity rules are used to guide the preference elicitation process in each round of ICE. Without an activity rule, a rational bidder would likely wait until the last moment to revise his valuation information, free-riding on the price discovery enabled by the bids of other participants. If every bidder were to behave this way then the exchange would reduce to a sealed-bid mechanism and lose its desirable properties.[7] Thus, activity rules are critical in mitigating opportunities for strategic behavior.[8] ICE employs two activity rules. In presenting our activity rules, we will not specify the explicit consequences of failing to meet an activity rule. One simple possibility is that the default action is to automatically set the upper valuation bound on every node in a bid tree to the maximum of the *provisional price on a node*[9] and the lower-bound value on that node. This is entirely analogous to when a bidder in an ascending-clock auction stops bidding at a price: he is not permitted to bid at a higher price again in future rounds.

---

[7] This problem has been described as the "snake in the grass" problem. See Kwerel's forward in Milgrom's book (2004).

[8] There is no conflict here with our assumption about straightforward bidding: we design for the strategic case despite assuming straightforward bidding to provide for tractable theoretical and experimental analysis; moreover, the presence of activity rules helps to motivate straightforward bidding.

[9] The provisional price on a node is defined as the minimal total price across all feasible trades for which the subtree rooted at the node is satisfied.

Modified Revealed-Preference Activity Rule (MRPAR). The first rule, MRPAR, is based on a simple idea. We require bidders to refine their valuation bounds in each round, so that there is some trade that is optimal (i.e., maximizes surplus) for the bidder given the current prices and for all possible valuations consistent with the bounds. MRPAR is loosely based around the revealed-preference based activity rule, advocated for the clock-proxy auction in a one-sided CA (Ausubel et al., 2006).

Let $v'_i \in T_i$ for *TBBL* tree T_i denote that valuation v'_i is consistent with the value bounds in the tree. If the bounds are tight everywhere, then v'_i is exactly the valuation function defined by tree T_i. A variant (RPAR), requires that there is enough information in valuation bounds to establish that one trade is weakly preferred to all other trades at the prices, i.e.

$$\exists \check{\lambda}_i \in \mathcal{F}_i(x^0) \text{ s.t. } v'_i(\check{\lambda}_i) - p^\pi(\check{\lambda}_i) \geq v'_i(\lambda'_i) - p^\pi(\lambda'_i), \forall v'_i \in T_i, \forall \lambda'_i \in \mathcal{F}_i(x^0) \quad \text{(RPAR)}$$

Note that a bidder can always meet this rule by defining an *exact* valuation $\hat{v}_i$ and tight value bounds on every node in his bid tree; in this case, trade $\check{\lambda}_i \in \arg\max_{\lambda_i \in \mathcal{F}_i(x^0)}[\hat{v}_i(\lambda_i) - p^\pi(\lambda_i)]$ satisfies RPAR. We say that prices π are *strict EQ prices* for $(v^\alpha, \lambda^\alpha)$ when:

$$v_i^\alpha(\lambda_i^\alpha) - p^\pi(\lambda_i^\alpha) > v_i^\alpha(\lambda'_i) - p^\pi(\lambda'_i), \quad \forall \lambda'_i \in \mathcal{F}_i(x^0) \setminus \{\lambda_i^\alpha\}, \quad (14)$$

for every bidder $i \in N$.

Theorem 3. *If prices π are strict EQ prices for provisional valuation profile v^α and trade λ^α, and every bidder i retains v_i^α in his bid tree after meeting RPAR, then trade λ^α is efficient when all bidders are straightforward.*

Proof. Fix bidder *i*. Let $\check{\lambda}_i$ denote the trade that satisfies RPAR. Because v_i^α is consistent with the revised bid tree of bidder *i*, we have:

$$v_i^\alpha(\check{\lambda}_i) - p^\pi(\check{\lambda}_i) \geq v_i^\alpha(\lambda'_i) - p^\pi(\lambda'_i), \quad \forall \lambda'_i \in \mathcal{F}_i(x^0). \quad (15)$$

Moreover, we must have $\check{\lambda}_i = \lambda_i^\alpha$, because $v_i^\alpha(\lambda_i^\alpha) - p^\pi(\lambda_i^\alpha) > v_i^\alpha(\lambda'_i) - p^\pi(\lambda'_i)$ by the strictness of prices. Instantiating RPAR with this trade, and with true valuations $v_i \in T_i$ (since bidders are straightforward), we have:

$$v_i(\lambda_i^\alpha) - p^\pi(\lambda_i^\alpha) \geq v_i(\lambda'_i) - p^\pi(\lambda'_i), \quad \forall \lambda'_i \in \mathcal{F}_i(x^0), \quad (16)$$

from which prices p^π are EQ prices with respect to true valuations. The efficiency claim then follows from the welfare theorem, Theorem 1. □

In particular, the provisional trade is efficient given strict EQ prices when every bidder meets the rule without modifying his bounds in any way. Strict EQ prices are required to prevent problems involving ties:

Example 5. The *TBBL* trees shown in Figure 6 will have no trade occur at the truthful valuation (which is indicated in bold between the value bounds). However, suppose $\alpha = 0$ so that at the provisional valuations it is efficient for *A* to be traded. Prices $\pi = (6, 2)$ are EQ (but not strict EQ) prices given v^α and λ^α, with the buyer indifferent between buying *A* and buying *B* and the seller indifferent between selling *A*, selling *A* and *B*, or making no sale. The buyer passes RPAR without changing his bounds because the

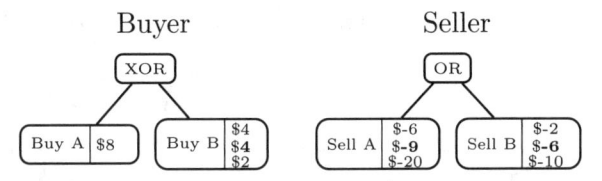

```
         Buyer              Seller
          XOR                 OR
         /   \              /    \
   ┌─────┐ ┌─────┬──┐  ┌─────┬───┐ ┌─────┬────┐
   │     │ │     │$4│  │     │$-6│ │     │$-2 │
   │Buy A│$8│Buy B│$4│  │Sell A│$-9│ │Sell B│$-6 │
   │     │ │     │$2│  │     │$-20││     │$-10│
   └─────┘ └─────┴──┘  └─────┴───┘ └─────┴────┘
```

Figure 6. An example to illustrate the failure of the simple RPAR rule without strict EQ prices. True values are shown in bold and are such that the efficient outcome is no trade.

bounds already establish that he (weakly) prefers A to B, and prefers A to no trade, at all possible valuations. Similarly, the seller passes RPAR without changing his bounds because the bounds establish that he weakly prefers no trade to selling any combination of A and B given the current prices. Thus, we have no activity even though the current provisional trade is inefficient.

In order to better handle these sorts of ties, we slightly strengthen RPAR to *modified RPAR* (MRPAR), which requires that there exists some $\check{\lambda}_i \in \mathcal{F}_i(x^0)$ such that

$$\theta_i^\pi(\check{\lambda}_i, \lambda_i', v_i') \geq 0, \quad \forall v_i' \in T_i, \forall \lambda_i' \in \mathcal{F}_i(x^0) \qquad (17)$$

and either $\check{\lambda}_i = \lambda_i^\alpha$ or $\theta_i^\pi(\check{\lambda}_i, \lambda_i^\alpha, v_i') > 0, \quad \forall v_i' \in T_i.$ (18)

where $\theta_i^\pi(\lambda_i, \lambda_i', v_i') = v_i'(\lambda_i) - p^\pi(\lambda_i) - (v_i'(\lambda_i') - p^\pi(\lambda_i'))$ denotes the profit to bidder i for trade λ_i over λ_i' given v_i' and prices π. (17) is RPAR and the additional requirements enforce that the satisfying trade $\check{\lambda}_i$ is either λ_i^α or strictly preferred to λ_i^α. This need to show a strict preference over λ_i^α prevents the deadlock shown in Example 5. The seller has shown only a *weak* preference for not trading over selling A. With MRPAR, the seller must also show that he strictly prefers $\check{\lambda}_i$, in this case by reducing the upper-bounds on both A and B, thus ensuring progress.

The actual rule adopted in ICE is δ-MRPAR, parameterized with *accuracy parameter* $\delta \geq 0$, and providing a relaxation of MRPAR which is useful even when there are no exact EQ prices defined with respect to $(\lambda^\alpha, v^\alpha)$ in some round.

Definition 4. Given provisional trade λ^α, item prices π, and accuracy parameter $\delta \geq 0$, δ-MRPAR requires that every bidder i refines his value bounds so that his *TBBL* tree T_i satisfies:

$$\theta_i^\pi(\lambda_i^\alpha, \lambda_i', v_i') \geq -\delta, \quad \forall v_i' \in T_i, \forall \lambda_i' \in \mathcal{F}_i \qquad (19)$$

or, that there is some $\check{\lambda}_i \in \mathcal{F}_i(x^0)$ such that

$$\theta_i^\pi(\check{\lambda}_i, \lambda_i', v_i') \geq 0, \quad \forall v_i' \in T_i, \forall \lambda_i' \in \mathcal{F}_i(x^0) \qquad (20)$$

$$\theta_i^\pi(\check{\lambda}_i, \lambda_i^\alpha, v_i') > \delta, \quad \forall v_i' \in T_i \qquad (21)$$

We can check that δ-MRPAR reduces to MRPAR for $\delta = 0$. Phrasing the description to allow for the rule to be interpreted with and without the δ relaxation, δ-MRAPR requires that *each bidder must adjust his valuation bounds to establish that the provisional trade is [within δ of] maximizing profit for all possible valuations* (19), *or some other trade satisfies RPAR* (20) *and is strictly preferred [by at least δ] to the provisional* *trade* (21). Just as for RPAR, one can show that a bidder can always meet δ-MRPAR (for any δ) by defining an exact valuation.[10]

Lemma 1. *If every bidder i meets δ-MRPAR without precluding v_i^α from his updated bid tree, and prices π are δ-approximate EQ prices with respect to provisional valuation profile v^α and trade λ^α, and bidders are straightforward, then the provisional trade is a $2\min(M, \frac{n}{2})\delta$-approximate efficient trade.*

Proof. Fix bidder i. By δ-EQ, we have $\theta_i^\pi(\lambda_i^\alpha, \lambda_i', v_i^\alpha) \geq -\delta$ for all $\lambda_i' \in \mathcal{F}_i(x^0)$. Consider any $\check{\lambda}_i \neq \lambda_i^\alpha$. Because v_i^α remains in the bid tree, we must have $\theta_i^\pi(\check{\lambda}_i, \lambda_i^\alpha, v_i^\alpha) \leq \delta$ and δ-MRPAR cannot be satisfied via (20) and (21). Therefore, δ-MRPAR is satisfied for every bidder via (19) and with provisional trade λ^α the satisfying trade. Therefore we prove that prices, π, are δ-approximate EQ prices for all valuations, and including the true valuation since bidders are straightforward and this is within their bounds. The efficiency of the trade follows from Theorem 2. □

This in turn provides a proof for the efficiency of ICE when approximate CE prices exist upon termination. Suppose that ICE is defined to terminate as soon as prices are δ-accurate and v^α is retained in the bid tree by all bidders in meeting the activity rule, or when quiescence is reached and no bidder refines his bounds in meeting the rule. In this variation, the provisional trade λ^α is the trade finally implemented.

Theorem 4. *ICE with δ-MRPAR is $2\min(M, \frac{n}{2})\delta$-efficient when prices are δ-accurate with respect to $(v^\alpha, \lambda^\alpha)$ upon termination and bidders are straightforward.*

Proof. When ICE terminates either (a) prices are δ-accurate and v^α is retained in the bid tree by all bidders and we can appeal directly to Lemma 1, or (b) no bidder refines his bounds in meeting δ-MRPAR, in which case v_i^α remains in the space of valuations consistent with the bid tree for each bidder. □

We also have the following corollary, which considers the property of ICE for a domain in which approximately accurate EQ prices exist:

Corollary 1. *ICE with δ-MRPAR is $2\min(M, \frac{n}{2})\delta$-efficient when δ-accurate competitive equilibrium prices exist for all valuations in the valuation domain and when all bidders are straightforward.*

Specializing to domains in which exact EQ prices exist (e.g., for unit-demand preferences as in the assignment model of Shapley and Shubik, 1972; see also the work of Bikhchandani and Mamer, 1997) then ICE with MRPAR is efficient for straightforward bidders.

Example 6. To illustrate the δ-MRPAR rule consider a single bidder with a valuation tree as in Figure 7(a). Suppose the provisional trade λ_i^α allocates A to the bidder, and with prices $\pi_A = 3$, $\pi_B = 4$ and $\delta = 2$. Here the bidder has satisfied δ-MRPAR because the guaranteed $\$2 - \$3 = \$-1$ payoff from A is within δ of the possible $\$5 - \$4 = \$1$

---

[10] Let v_i denote this valuation. If δ-MRPAR is not satisfied via (19) then $\check{\lambda}_i \in \arg\max_{\lambda_i \in \mathcal{F}_i(x^0)}[v_i(\lambda_i) - p^\pi(\lambda)]$ will satisfy δ-MRPAR. This satisfies (20) by construction. Now, let λ_i' denote the trade with $v_i(\lambda_i') - p^\pi(\lambda_i') > v_i(\lambda_i^\alpha) - p^\pi(\lambda_i^\alpha) + \delta$. We have $v_i(\check{\lambda}_i) - p^\pi(\check{\lambda}_i) \geq v_i(\lambda_i') - p^\pi(\lambda_i') > v_i(\lambda_i^\alpha) - p^\pi(\lambda_i^\alpha) + \delta$, and (21).

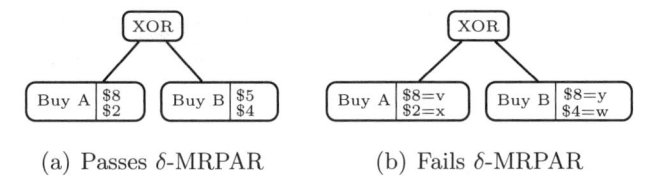

```
        XOR                              XOR
       /   \                            /   \
  Buy A $8  Buy B $5              Buy A $8=v  Buy B $8=y
       $2       $4                     $2=x       $4=w

    (a) Passes δ-MRPAR              (b) Fails δ-MRPAR
```

Figure 7. δ-MRPAR where the provisional trade is "Buy A", $\pi_A = 3$, $\pi_B = 4$ and $\delta = 2$.

payoff from B. Now consider Figure 7(b), with a relaxed upper-bound on "buy B" of $\$8$. Now the bidder fails δ-MRPAR because the guaranteed $\$-1$ payoff from A is not within δ of the possible payoff from B of $\$8 - \$4 = \$4$. Let $[x, v]$ and $[w, y]$ denote the lower and upper bounds, on "buy A" and "buy B" respectively, as revised in meeting the rule. To pass the rule, the bidder has two choices:

- Demonstrate λ_i^α is the best response. To do so the bidder will need to adjust x and y to make $x - 3 \geq y - 4 - 2 \Rightarrow y - x \leq 3$; e.g., values $x = \$2, y = \$5$ solve this, as in Figure 7(a), as do many other possibilities.
- OR Demonstrate that another trade (e.g., "buy B") is more than $\$2$ better than λ_i^α, i.e., $w - 4 > v - 3 + 2 \Rightarrow w - v > 3$, and "buy B" is weakly better than the null trade, i.e., $w - 4 \geq 0$. For instance, if the bidder's true values are $v_A = \$3, v_B = \$8$ then $x \leq 3 \leq v$ and $w \leq 8 \leq y$ and the rule cannot be satisfied in the first case. But, the buyer can establish that "buy B" is his best-response, e.g., by setting $v = \$4, w = \$7$, or $v = \$3, w = \$6$.

Remark: Computation and Bidder Feedback. The definition of MRPAR naively suggests that checking for compliance requires explicitly considering all valuations $v_i' \in T_i$ and all trades $\lambda_i' \in \mathcal{F}_i(x^0)$. Fortunately, this is not necessary. We present in the Appendix a method to check MRPAR given prices π, provisional trade λ_i^α and bid tree T_i by solving three MIPs. Moreover, we explain that the solution to these MIPs also provides nice feedback for bidders. ICE can automatically identify a set of nodes at which a bidder needs to increase his lower bound and a set of nodes at which a bidder needs to decrease his upper bound in meeting MRPAR.

Delta Improvement Activity Rule (DIAR). With only δ-MRPAR, it is quite possible for ICE to get stuck, with all bidders satisfying the activity rule without changing their bounds, but with the prices less than δ accurate (with respect to $(\lambda^\alpha, v^\alpha)$). Therefore, we need an activity rule that will continue to drive a reduction in value uncertainty, i.e., the gap between upper bound values and lower bound values, even in the face of inaccurate prices, and ideally in a way that remains price-directed, in the sense of using prices to determine which trades (and in turn which nodes in *TBBL* trees) each bidder should be focused on.

We introduce for this purpose a second (and novel) activity rule (DIAR), which fills this role by requiring bidders to reveal information so as to improve price accuracy and, in the limit, full information on the nodes that matter. Defined this way, the DIAR rule very nicely complements the δ-MRPAR rule. Because we can establish the efficiency of the provisional trade directly via the valuation bounds, as we will see in Section 4.3, we do not actually need fully accurate prices in order to close the exchange. Thus, the DIAR rule does not imply that bidders will reveal full information. Rather, the presence of DIAR ensures both good performance in practice as well as good theoretical properties. In our experiments we enable DIAR in all rounds of ICE, and it fires in parallel with δ-MRPAR. In practice, we see that most of the progress in refining valuation information occurs due to δ-MRPAR, and that *all* the progress in early rounds occurs due to δ-MRPAR. Experimental support for this is provided in Section 6.[11]

Before providing the specifics of DIAR, we identify a node $\beta \in T_i$ in the bid tree of bidder i as *interesting* for some fixed instance (v, x^0), when the node is satisfied in *some* feasible trade. We have the following lemma:

Lemma 2. *If there is no value uncertainty on any interesting nodes in the bid trees of any biders, and bidders are straightforward, then λ^α is efficient.*

Proof. No value uncertainty and thus exact information about the value on all *interesting* nodes implies that the difference in value is exactly known between all pairs of feasible trades because for all uninteresting nodes, either the node is never satisfied in any trade (and thus its value does not matter) or the node is satisfied in every trade and thus its actual value does not matter in defining the difference in value between pairs of trades. Only the difference in value between pairs of trades is important in determining the efficient trade. □

DIAR focuses a bidder in particular on interesting nodes that correspond to trades for which the pricing error is large, and where this error could still be reduced by refining the valuation bounds on the node. Given prices π and provisional trade λ_i^α, the main focus of DIAR is the following upper-bound $\overline{\delta}_i^k$, on the amount by which prices π might misprice some trade $\lambda_i^k \in \mathcal{F}_i(x^0)$ with respect to bidder i's true valuation:

$$\overline{\delta}_i^k = \max_{v_i' \in T_i}[v_i'(\lambda_i^k) - p^\pi(\lambda_i^k) - (v_i'(\lambda_i^\alpha) - p^\pi(\lambda_i^\alpha))] \qquad (22)$$

We call this the *DIAR error on trade* λ_i^k, and note that it depends on the current prices as well as the current bid tree and provisional trade, but not the true valuation which is unknown to the center. The DIAR error provides an upper bound on the additional payoff that the bidder could achieve from trade λ_i^k over trade λ_i^α. If we order trades, $\lambda_i^1, \lambda_i^2, \ldots$, so that λ_i^1 has maximal DIAR error, then $\overline{\delta}_i^1 \geq \delta_i$, where $\delta_i = \max_{\lambda_i' \in \mathcal{F}_i(x^0)}[v_i^\alpha(\lambda_i') - p^\pi(\lambda_i') - (v_i^\alpha(\lambda_i^\alpha) - p^\pi(\lambda_i^\alpha))]$ is the pricing error with respect to the provisional trade and provisional valuation profile. This is the error that the pricing algorithm is designed to minimize in each round, and the same error that is used in Theorem 2 in reference to δ-accurate prices. Thus, we see that *the maximal DIAR error also bounds the amount by which prices are approximate EQ prices*, and that if $\overline{\delta}_i^1 \leq 0$ for all bidders i then the current prices π are exact EQ prices with respect to $(\lambda^\alpha, v^\alpha)$.

To satisfy DIAR a bidder must reduce the DIAR error on the trade with the largest error for which the error can be reduced (some error may just be intrinsic given the current prices and not because of uncertainty about the bidder's valuation), or establish by providing exact value information throughout the tree that none of the DIAR error on any trades is due to value uncertainty. Figure 8 illustrates the difference between

---

[11] In a variation on the way ICE is defined, DIAR could be used only in rounds in which the price error for the provisional valuation and trade is greater than the error associated with δ-MRPAR. This is because δ-MRPAR is sufficient for approximate efficiency when prices are accurate enough.

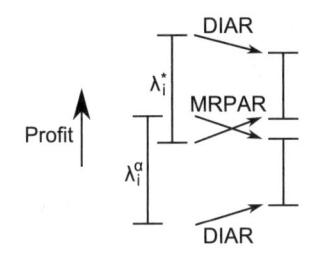

Figure 8. Stylized effect of MRPAR and DIAR on the bounds of the λ_i^α and λ_i^* trades.

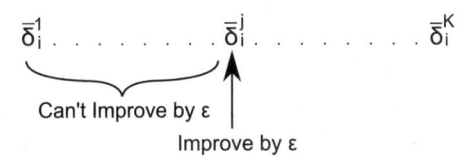

Figure 9. Trades for bidder i, ordered with DIAR error reducing from left to right. The bidder must reduce, by at least ϵ, the DIAR error on the trade with the greatest error for which this is possible and prove (via valuation bounds) that it is impossible to improve by ϵ any trades with larger error.

MRPAR and DIAR. A bidder can satisfy MRPAR by making it clear that the lower bound on payoff from some trade is greater than the upper bound on all other trades, but still leave large uncertainty about value. DIAR requires that a bidder also refine this upper bound if it is on a node that corresponds to a trade for which the DIAR error (and thus potentially the actual approximation in prices) is large. The rule is illustrated in Figure 9.

DIAR is parameterized by some $\epsilon \geq 0$. We refer to the formal rule as ϵ-DIAR:

Definition 5. To satisfy ϵ-DIAR given provisional trade λ_i^α and prices π, the bidder must modify his valuation bounds to:
(a) reduce the DIAR error on some trade, $\lambda_i^j \in \mathcal{F}_i(x^0)$, by at least ϵ **and**
(b) prove that error $\overline{\delta}_i^k$ **cannot** be improved by ϵ for all trades $\lambda_i^k \in \mathcal{F}_i(x^0)$ for $1 \leq k < j$,
or (c) establish that $\overline{\delta}_i^k$ cannot be improved by ϵ on **any** trade $\lambda_i^k \in \mathcal{F}_i(x^0)$.

In particular, even if the bidder is in case (c) above, he will still be forced to narrow his bounds and progress will be made towards bounding efficiency. In practice, we define the ϵ parameter to be large at the start and smaller in later rounds.

Example 7. Consider the tree in Figure 10(a) when the provisional trade is "buy A", prices $\pi = (\$4, \$5, \$6)$ and DIAR parameter $\epsilon = 1$. The DIAR error on each trade,

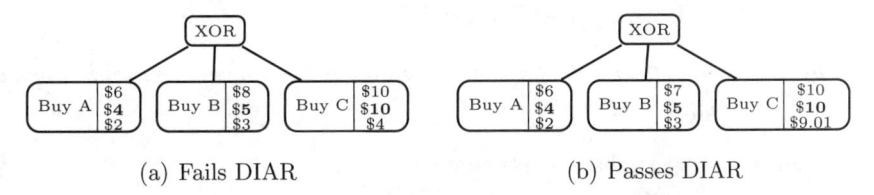

Figure 10. Respecting DIAR where the provisional trade is "Buy A", $\pi_A = 4, \pi_B = 5, \pi_C = 6$ and $\epsilon = 1$.

defined via (22), and listed in decreasing order, are:

$$C \to \overline{\delta}^1 = (\$10 - \$6) - (-\$2) = \$6$$
$$B \to \overline{\delta}^2 = (\$8 - \$5) - (-\$2) = \$5$$
$$\emptyset \to \overline{\delta}^3 = (\$0 - \$0) - (-\$2) = \$2$$
$$A \to \overline{\delta}^4 = (\$2 - \$4) - (-\$2) = \$0,$$

where $-\$2 = \$2 - \$4$ is the worst-case profit from the provisional trade. Now, we see that $\overline{\delta}^1$ cannot be made smaller by lowering the upper-bound on leaf "buy C" because this bound is already tight against the truthful value of \$10. Instead the bidder must demonstrate that a decrease of $\epsilon = 1$ is impossible by raising the lower bound on "buy C" to 9.01. However $\overline{\delta}^2$ can be decreased by $\epsilon = 1$, by reducing the upper-bound on "buy B" from 8 to 7, giving us the tree in Figure 10(b).

Lemma 3. *When ICE incorporates DIAR, a straightforward bidder must eventually reveal complete value information on all interesting nodes in his bid tree as $\epsilon \to 0$.*

Proof. Fix provisional trade λ_i^α and consider trade, $\lambda_i^1 \in \mathcal{F}_i(x^0) \neq \lambda_i^\alpha$, with the maximal DIAR error. Continue to assume straightforward bidders. Recall that $v_i(\beta)$ denotes a bidder's true value on node β in his *TBBL* tree. By case analysis on nodes $\beta \in T_i$, meeting the DIAR rule on this trade as $\epsilon \to 0$ requires:

(i) Nodes $\beta \in \lambda_i^1 \setminus \lambda_i^\alpha$. Decrease the upper-bound to $v_i(\beta)$, the true value, to reduce the error. Increase the lower-bound to $v_i(\beta)$ to prove that further progress is not possible.
(ii) Nodes $\beta \in \lambda_i^\alpha \setminus \lambda_i^1$. Increase the lower-bound to $v_i(\beta)$, the true value, to reduce the error. Decrease the upper-bound to $v_i(\beta)$ to prove that further progress is not possible.
(iii) Nodes $\beta \in \lambda_i^\alpha \cap \lambda_i^1$. No change is required.
(iv) Nodes $\beta \notin \lambda_i^1 \cup \lambda_i^\alpha$. No change is required.

Continue to fix some λ_i^α, and consider now the impact of DIAR as $\epsilon \to 0$ and as the rule is met for successive trades, moving from λ_i^1 to λ_i^2 and onwards. Eventually, the value bounds on all nodes $\beta \notin \lambda_i^\alpha$ but *in at least one other feasible trade* are driven to truth by (i), and the value bounds on all nodes $\beta \in \lambda_i^\alpha$ but *not in at least one other feasible trade* are driven to truth by (ii). Noting that the null trade is always feasible, the bidder will ultimately reveal complete value information except on nodes that are not satisfied in any feasible trade. □

Putting this together we have the following theorem, which considers the convergence property of ICE when DIAR is the only activity rule.

Theorem 5. *ICE with the ϵ-DIAR rule will terminate with the efficient trade when all bidders are straightforward and as $\epsilon \to 0$.*

Proof. Immediate by Lemma 2 and Lemma 3. □

In practice, we use both δ-MRPAR and DIAR and the role of DIAR is to ensure convergence in instances for which there do not exist good, supporting EQ prices. The use of DIAR does *not* lead, in any case, to full revelation of bidder valuations because we can prove efficiency directly in terms of valuation bounds on different trades (see Section 4.3).

Remark: Computation and Bidder Feedback. We present in the Appendix a method to check ϵ-DIAR given prices π, provisional trade λ_i^α, the bidder's bid tree from the past round and proposed new bid tree by solving two MIPs. Moreover, the solution to these MIPs also provides nice feedback for bidders. ICE can automatically identify the trade, and in turn the corresponding nodes in the bid tree, for which the bidder must provide more information.

4.2. Computing Item Prices

Given the provisional trade λ^α, provisional valuations v^α, and given that provisional payments have also been determined (according to the payment rule, such as Threshold, adopted in the exchange), approximate clearing prices are computed in each round according to the following rules:

I: **Accuracy (ACC).** First, we compute prices that minimize the maximal error in the best-response constraints across all bidders.
II: **Fairness (FAIR).** Second, we break ties to prefer prices that minimize the maximal deviation from provisional payments across all bidders.
III: **Balance (BAL).** Third, we break ties to prefer prices that minimize the maximal difference in price across all items.

Taken together, these steps are designed to promote the informativeness of prices in driving progress across rounds. Balance is well motivated in domains where items are more likely to be similar in value than dissimilar, preferring prices to be similar across items and rejecting extremal prices. Note that these prices may ascend or descend from round to round – but that they will in general tend towards increasing accuracy, as we shall see experimentally in Section 6.

Example 8. Consider the example in Figure 11 with one buyer interested in buying AB and one seller interested in selling AB. Here the buyer's and seller's values for each item are 8 and –6 respectively. The efficient outcome given these values is for the trade to complete. ACC requires $12 \leq \pi_A + \pi_B \leq 16$, and thus allows a range of prices. The Threshold payment splits the difference, so that the buyer pays 14 to the seller and so FAIR adds the constraint $\pi_A + \pi_B = 14$. Finally, BAL requires $\pi_A = \pi_B = 7$.

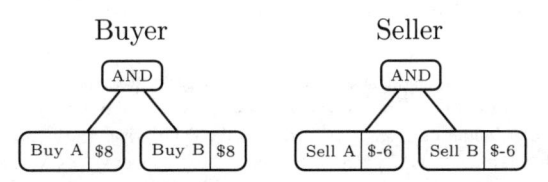

Figure 11. An example to illustrate pricing. ACC prices AB between $12 and $16, FAIR narrows this to $14 and BAL requires $A = \$7, B = \$7$.

Each of the three stages occur in turn. In the interest of space, here we only present the basic formulation of the Accuracy stage: We define maximally accurate EQ prices by first considering the following LP:

$$\delta^*_{\text{acc}} = \min_{\pi, \delta_{\text{acc}}} \delta_{\text{acc}}$$

$$\text{s.t.} \quad v_i^\alpha(\lambda_i') - \sum_j \pi_j \lambda_{ij}' \leq v_i^\alpha(\lambda_i^\alpha) - \sum_j \pi_j \lambda_{ij}^\alpha + \delta_{\text{acc}}, \quad \forall i, \forall \lambda_i' \in \mathcal{F}_i(x^0) \quad (23)$$

$$\delta_{\text{acc}} \geq 0,$$

$$\pi_j \geq 0, \quad \forall j \in G$$

These prices minimize the maximal loss in payoff across all bidders for trade λ^α compared to the trade that a bidder would most prefer given provisional valuation v^α, i.e., minimize the maximal value of $\theta_i^\pi(\lambda_i^*, \lambda_i^\alpha, v_i^\alpha)$, where $\lambda_i^* = \arg\max_{\lambda_i \in \mathcal{F}_i(x^0)}[v_i^\alpha(\lambda_i) - p^\pi(\lambda_i)]$. Prices that solve this LP are then refined lexicographically, fixing the worst-case pricing error (ACC) and then working down to try to additionally minimize the next largest pricing error and so on. Given maximally accurate prices, this then triggers a series of lexicographical refinements to best approximate the payments (FAIR) without reducing the pricing accuracy, and eventually a series of lexicographical refinements to try to maximally balance prices across distinct items (BAL). In addition to further improving the quality of the prices, this process also ensures uniqueness of prices.

Each of the Accuracy, Fairness and Balance problems have an exponential number of constraints because the price accuracy constraints (23) (which are carried forward into the subsequent stages) are defined over all trades $\lambda_i' \in \mathcal{F}_i(x^0)$ and all bidders i. It is therefore infeasible to even write these problems down. Rather than solve them explicitly, we use *constraint generation* (e.g. Bertsimas & Tsitsiklis, 1997) and dynamically generate a sufficient subset of constraints. Constraint generation (CG) considers a relaxed program that only contains a manageable subset of the constraints, and solves this to optimality. Given a solution to this relaxed program, a subproblem is used to either prove that the solution is optimal to the full program, or find a violated constraint in the full problem that is introduced, with the modified program re-solved. In this case the subproblem is a variation of the winner determination IP from Section 3, and can be succinctly formulated and solved via branch-and-cut.[12]

4.3. Establishing Bounds on Efficiency

Consider some round t in ICE. The round starts with the announcement of prices, denote them π^t, and the provisional trade. The round ends with every bidder having

---

[12] The pricing step is the most computationally intensive of all steps in ICE and therefore heavily optimized. In practice, we have found it useful to employ heuristics to seed the set of constraints used in CG. We have also developed algorithmic techniques to speed the search for the appropriate set of constraints in the context of lexicographic refinement: *provisional locking* of multiple lexicographic values for each CG check, and *lazy constraint checks* in which only a subset of the conditions for CG are routinely checked, even though the complete set is eventually enforced. Please see the technical report at www.eecs.harvard.edu/~blubin/ice for complete details of the pricing method.

met the δ-MRPAR and ϵ-DIAR activity rules. The question to address is: *what can be established about the efficiency of the trade defined on lower-bound valuations at the end of the round?* It is perhaps unsurprising that MRPAR by itself is sufficient to provide efficiency claims when prices are suitably accurate. What is interesting is that the coupling of MRPAR with DIAR ensures that ICE converges to a provably efficient trade in all cases, with efficiency often established independently of prices by reasoning directly about lower and upper valuation bounds. For the theoretical analysis of convergence to efficiency, we assume *straightforward bidders*, by which we mean *a bidder that always retains his true valuation within the valuation bounds*. (All results could equivalently be phrased in terms of efficiency claims with respect to reported valuations.)

At the closing of each round, ICE makes a determination about whether to move to the last-and-final round. Bidders are notified when this occurs. This last-and-final round provides a final opportunity for bidders to update their lower valuation bound information (without exceeding their upper bounds). The exchange finally terminates with the efficient trade and payments determined with respect to the lower valuation bounds: it is these lower bounds that can be considered to be the ultimate bid submitted by each bidder when ICE terminates. Let $\underline{\lambda} \in \arg\max_{\lambda \in \mathcal{F}(x^0)} \sum_i \underline{v}_i(\lambda_i)$ denote the trade that is optimal given the lower bound valuations. As explained in Section 4, ICE is parameterized by target approximation error, Δ^*, providing a lower-bound on the relative efficiency of $\underline{\lambda}$ to the efficient trade λ^* for true valuations. The challenge is to obtain useful bounds on the relative efficiency $\text{EFF}(\underline{\lambda})$ of trade $\underline{\lambda}$. We provide two methods, one of which is price-based and uses duality theory and the second of which directly reasons about the bounds on bidder valuations. We now consider each in turn.

A Price-based Proof of Efficiency. We have already seen in Section 2.2 that a bound on the efficiency of provisional trade λ^α can sometimes be established via prices. This provides a method to establish a bound on the efficiency of trade $\underline{\lambda}$. Fix some $\delta \geq 0$. For v^α denoting the provisional valuation profile at the start of round t, and λ^α the corresponding provisional trade, we know that if

(a) bidders meet δ-MRPAR while leaving v^α within their bounds,
(b) prices π^t were δ-approximate EQ prices for v^α and λ^α, and
(c) λ^α is equal to $\underline{\lambda}$, i.e., the efficient trade given the refined lower bound valuations,

then trade $\underline{\lambda}$ is a $2\min(M, \frac{n}{2})\delta$-approximation to the efficient trade λ^* by Theorem 2. We have $\sum_i v_i(\underline{\lambda}_i) + 2\min(M, \frac{n}{2})\delta \geq \sum_i v_i(\lambda_i^*)$, and then,

$$\text{EFF}(\underline{\lambda}) = \frac{\sum_i v_i(\underline{\lambda}_i)}{\sum_i v_i(\lambda^*)} \geq 1 - \frac{2\min(M, \frac{n}{2})}{\sum_i v_i(\lambda_i^*)}\delta \geq 1 - \frac{2\min(M, \frac{n}{2})}{\max_{\lambda \in \mathcal{F}(x^0)} \sum_i \overline{v}_i(\lambda)}\delta, \quad (24)$$

which we define as ω^{price}. Conditioned on (a–c) being met, so that the bound is available, it will satisfy $\omega^{\text{price}} \geq \Delta^*$ for a small enough δ parameter. When the bound is not available we set $\omega^{\text{price}} := 0$.

A Direct Proof of Efficiency. We also provide a complementary, direct, method to establish the relative efficiency of $\underline{\lambda}$ by working with the refined valuation bounds at

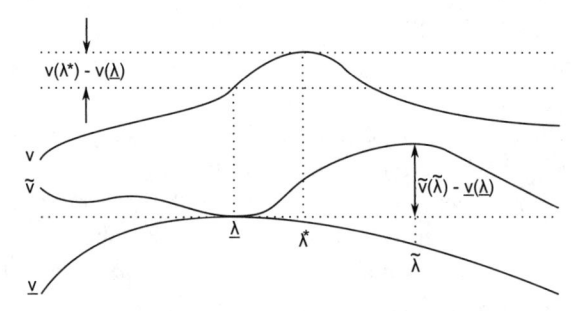

Figure 12. Determining an efficiency bound based on lower and upper valuations.

the end of round t. First, given a bid tree T_i, it is useful to define the ***perturbed valuation with respect to a trade*** λ_i, by assigning the following values to each node β:

$$\tilde{v}_i(\beta) = \begin{cases} \underline{v}_i(\beta), & \text{if } \beta \in sat_i(\lambda_i) \\ \overline{v}_i(\beta), & \text{otherwise,} \end{cases} \quad (25)$$

where $\beta \in sat_i(\lambda_i)$ if and only if node β is satisfied given tree T_i and lower bound valuations $\underline{v}_i$ on nodes, and given trade λ_i. The valuation function $\tilde{v}_i$ associated with TBBL tree T_i is defined to minimize the value on nodes satisfied by trade λ_i and maximize the value on other nodes. With this concept we can now establish the following bound,

$$\text{EFF}(\underline{\lambda}) = \frac{v(\underline{\lambda})}{v(\lambda^*)} \geq \min_{v' \in T, \lambda' \in \mathcal{F}(x^0)} \left[\frac{v'(\underline{\lambda})}{v'(\lambda')} \right] = \min_{\lambda' \in \mathcal{F}(x^0)} \left[\frac{\tilde{v}(\underline{\lambda})}{\tilde{v}(\lambda')} \right] = \frac{v(\underline{\lambda})}{\tilde{v}(\tilde{\lambda})}, \quad (26)$$

which we define as ω^{direct}. Here we use the notation $\tilde{v} = (\tilde{v}_1, \ldots, \tilde{v}_n)$, and let $\tilde{\lambda}$ be the trade that maximizes $\sum_i \tilde{v}_i(\lambda_i)$ across all feasible trades. The first inequality holds because the domain of the minimization includes $v \in T$ and trade $\lambda' = \lambda^*$. The first equality holds because for any $\lambda' \neq \underline{\lambda}$, the worst-case efficiency for $\underline{\lambda}$ occurs when the value $v' \in T$ is selected to minimize the value on nodes $\underline{\lambda} \setminus \lambda'$, maximize the value on nodes $\lambda' \setminus \underline{\lambda}$, and minimize the value on shared nodes, $\lambda' \cap \underline{\lambda}$. Whatever the choice of λ', this valuation is provided through perturbed valuation $\tilde{v}$. For the final equality, $\tilde{v}(\underline{\lambda}) = \underline{v}(\underline{\lambda})$ by definition, and the optimal trade λ' is that which maximizes the value of the denominator, i.e., trade $\tilde{\lambda}$. Figure 12 schematically illustrates the various trades and values used in this bound, and in particular provides some graphical intuition for why $\tilde{v}(\tilde{\lambda}) - \underline{v}(\underline{\lambda}) \geq \tilde{v}(\lambda^*) - \underline{v}(\underline{\lambda}) = \max_{v' \in T}[v'(\lambda^*) - v'(\underline{\lambda})] \geq v(\lambda^*) - v(\underline{\lambda})$.

Combining Together. Given the above methods we can establish lower-bound $\omega^{\text{eff}} = \max(\omega^{\text{price}}, \omega^{\text{direct}})$ on the relative efficiency of trade $\underline{\lambda}$. ICE is defined to move to the last-and-final round when either of the following hold:

(a) the error bound $\omega^{\text{eff}} \geq \Delta^*$
(b) there is no trade even at optimistic (i.e., upper-bound) valuations.

Combining this with Theorem 5, we immediately get our main result.

Theorem 6. *When ICE incorporates δ-MRPAR and ϵ-DIAR and when all bidders are straightforward, then the exchange terminates with a trade that is within target approximation error Δ^*, for any $\Delta^* \geq 0$ as $\epsilon \to 0$.*

The use of ϵ-DIAR by itself is sufficient to establish this result. However, it is the use of prices and MRPAR that drives most elicitation in practice, particularly as we fix δ in δ-MRPAR to a tiny constant in actual use. Empirical support for this, along with the quality of the price-based bound and the direct efficiency bounds, is provided in Section 6. For the ϵ parameter in ϵ-DIAR, we find that a simple rule:

$$\epsilon := \frac{1}{2n} \sum_i \sum_{\beta \in T_i} \frac{\overline{v}_{i \in N}(\beta) - \underline{v}_i(\beta)}{|T_i|}, \quad (27)$$

works well. This tends towards zero as more value information is revealed by participants.

One last element of the design of ICE is the precise method by which the provisional valuation profile $v^\alpha = \alpha \underline{v} + (1-\alpha)\overline{v}$ is constructed. This is important because it is then used to determine the provisional trade and price feedback. An approach that works well is to define $\alpha := \max(0.5, \omega^{\text{eff}})$. We find that the lower bound of 0.5 is a useful heuristic for early rounds when ω^{eff} is likely to be small, making ICE adopt a provisional valuation in the middle of the valuation bounds when not much is known. The effect is then to push α towards 1 and thus v^α towards $\underline{v}$ as the efficiency bound ω^{eff} improves.[13]

5. Illustrative Examples

In this section we illustrate the behavior of the exchange on two simple examples. These examples are provided to give a qualitative feel for its behavior. To construct the examples we populate ICE with very simple, automated bidding agents. These agents use MIP-guided heuristics to minimize the amount of information revealed in the course of passing the activity rules, while maintaining their true value within their lower- and upper-bounds (i.e., they act in a 'straightforward' way). Their reluctance to reveal information models a basic tenet of our design, that it is costly for participants to refine and then reveal information about their values for different trades.[14]

In this section, and also in presenting our main experimental results, we do *not* move to a last-and-final round. Rather, the bidding agents are programmed to continue to improve their bids past the round at which efficiency is already proved (and when a last-and-final round would ordinarily be declared), and until payments are within some desired accuracy tolerance. We do this to avoid the need to program agents with a strategy for how to bid in the last-and-final round.

Example 9. Consider a market with a no-reserve seller of two items **A** and **B**, and three buyers. AgentA demands **A** with a value of $8, AgentB demands **B** with a value of $8, and AgentAB demands **A** AND **B** with a value of $10. Figure 13(a) shows that

---

[13] In some domains, it may also be important to require that *payments* (rather than just the efficiency of trade λ) be accurate enough before moving to the last-and-final round. A bound on payments can be computed in an analogous way to that on efficiency. Whether this is required in practice is likely domain-specific and to depend, for instance, on whether the payments tend to be accurate anyway by the time the trade is approximately accurate, and also on the impact on strategic behavior.

[14] A detailed explanation of the operation of these bidding agents is provided in the Appendix.

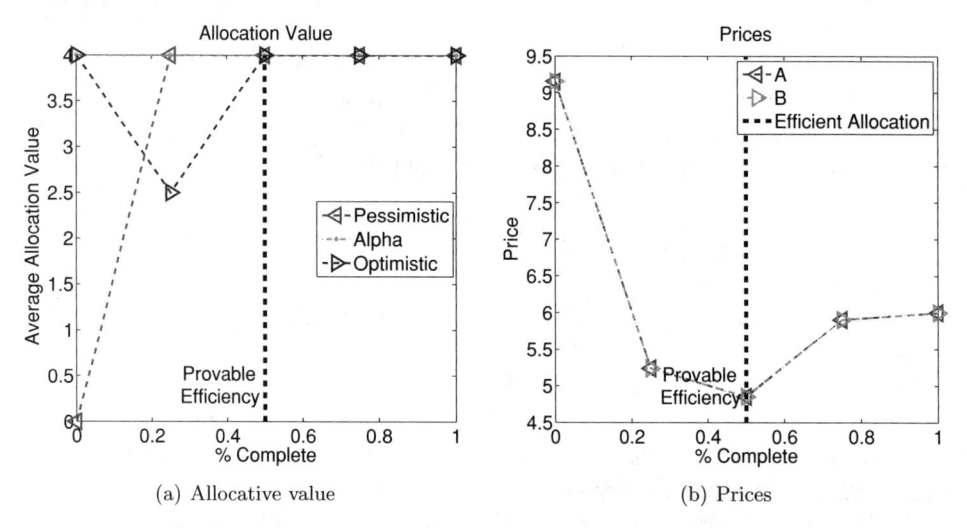

Figure 13. AgentA: **A** $8, AgentB: **B** $8, AgentAB: **A** AND **B** $10.

the exchange quickly discovers the correct trade. A price between $5 and $8 will be accurate in this situation, and we can see that the prices in Figure 13(b) quickly meet this condition. Fairness drives the prices towards $6, which will be the eventual Threshold payments to AgentA and AgentB. Balance ensures that the prices remain the same for the two items.

Example 10. Consider an example with a seller offering **A** for a reserve of $10, an agent who looks to swap goods and is willing to pay $8 if he can swap his **B** for **A**, and a buyer willing to pay $4 for **B**. In this more complex example it takes 4 rounds, as illustrated in Figure 14(a), for a trade to be found at the lower values. Revelation drives progress towards a completed trade, and as we can see in Figure 14(b), this

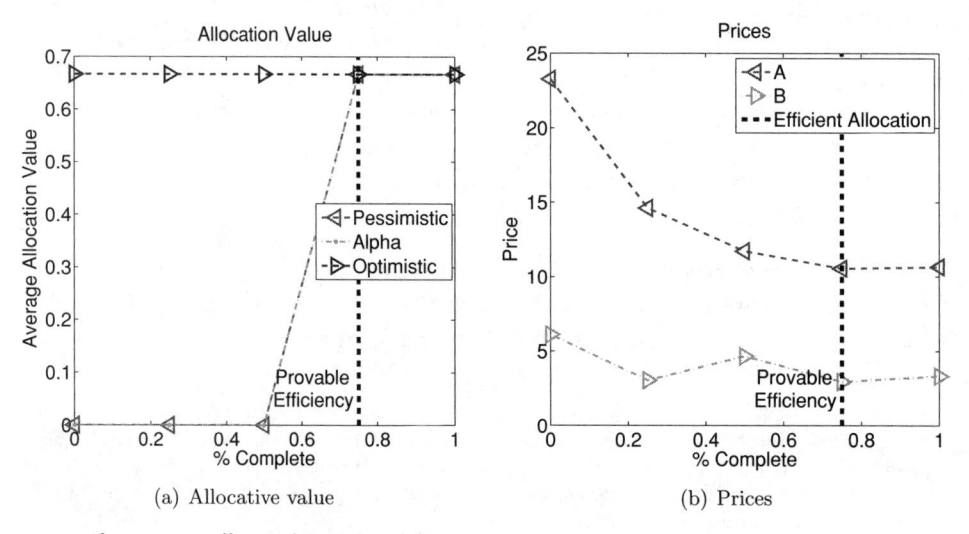

Figure 14. Seller **A** -$10, agent who wants to: swap **B** for **A** $8, and buyer **B** $4.

Table 1. *Exchange components and code breakdown*

Component	Purpose	Lines
Agent	Strategic behavior and information revelation decisions	2001
Model	XML support to load goods and true valuations	1353
Bidding Language	Implements *TBBL*	2497
Exchange Driver & Communication	Controls exchange, and coordinates agent behavior	1322
Activity/Closing Rule Engines	MRPAR, DIAR and Closing Rules	1830
WD Engine	Logic for WD	685
Pricing Engine	Logic for three pricing stages	1317
MIP Builders	Translates from engines into our optimization APIs	2206
Framework & Instrumentation	Wire components together & Gather data	2642
JOpt	Our Optimization API wrapping CPLEX	2178
Instance Generator	Random Problem Generator	497

is reflected in falling prices on the goods. Thus we can see that the price feedback is providing accurate information to the participants: only when the price eventually becomes low enough do the buying bidders actually want a trade to occur– and that is also when the exchange's provisional trade switches. It is also worth noting that the greater value the agents place on good **A** result in it having a net higher price than for good **B**.

6. Experimental Analysis

In this section we report the results of a set of experiments that are designed to provide a proof-of-concept for ICE. The results illustrate the scalability of ICE to realistic problem sizes and provide evidence of the effectiveness of the elicitation process and the techniques to bound the efficiency of the provisional trade.

Implementation. First, a brief aside on our experimental implementation. ICE is approximately 20,000 lines of extremely tight Java code, broken up into the functional packages described in Table 1.[15] The prototype is modular so that researchers may easily replace components for experimentation.[16] Because of ICE's complexity, it is essential that the code be constructed in a rigid hierarchy that avoids obscuring the high level logic behind the details of generating, running and integrating the results of MIPs. To this end, the system is written in a series of progressively more abstract mini-languages, each of which defines a clean, understandable API to the next higher level of logic. Our hierarchy provides a way to hide the extremely delicate steps needed to

---

[15] Code size is measured in physical source line of code (SLOC).
[16] Please contact the authors for access to the source code.

handle the numerical issues that come out of trying to repeatedly solve coupled optimization problems, where the constraints in one problem may be defined in terms of slightly inaccurate results from an earlier problem. Most of the constraints presented in this paper must be carefully relaxed and monitored in order to handle these numerical precision issues. At the bottom of this hierarchy the MIP specification is fed into our generalized back-end optimization solver interface[17] (we currently support CPLEX and the LGPL-licensed LPSolve), that handles machine load-balancing and parallel MIP/LP solving. This concurrent solving capability is essential, as we need to handle tens of thousands of comparatively simple MIPs/LPs.

Experimental Set-up. In the experiments, the δ-parameter in MRPAR is set to near zero and both the MRPAR and DIAR activity rule fire in every round. The rule used to define the ϵ-parameter in DIAR is exactly as described in Section 4.1. We adopt the same straightforward bidding agents that were employed in Section 5 (see the Appendix for details). In simulation, we adopt the Threshold payment rule and terminate ICE when the per-agent error in payment relative to the correct payment is within 5% of the average per-agent value for the efficient trade. On typical instances, this incurs an additional 4 rounds beyond those that would be required if we had a last-and-final round. All timing is wall clock time, and does not separately count the large number of parallel threads of execution in the system. The experiments were run on a dual-processor dual-core Pentium IV 3.2GHz with 8GB of memory and CPLEX 10.1. All results are averaged over 10 trials. The problem instances are available at www.eecs.harvard.edu/~blubin/ice.

Our instance generator begins by generating a set G of good *types*. Next, for each $j \in G$ it creates $s \geq 1$ copies of each good type, forming a total potential supply in the market of $s|G|$ goods (exactly how many units are in supply depends on the precise structure of bid trees). Each unit is assigned to one of the bidders uniformly at random. The generator creates a bid tree T_i for each bidder by recursively growing it, starting from the root and adopting two phases. For the tree above *depthLow*, each node receives a number of children drawn uniform between *outDegreeLow* and *outDegreeHigh* (a percentage of which are designated as leaves), resulting in an exponential growth in the number of nodes during this phase. By the *width* at some depth we refer to the number of nodes at that depth. Below this point, we carefully control the expected number of children at each node in order to make the expected width conform to a triangle distribution over depth from *depthLow* to *depthMid* to *depthHigh*: we linearly increase the expected width at each depth between *depthLow* and *depthMid* to a fixed multiple (ξ) of the width at *depthLow*, and then linearly decrease the expected width back to zero by *depthHigh*.[18] This provides complex and deep trees without inherently introducing an exponential number of nodes.

Each internal node must be assigned the parameters for its interval choose operator. We typically choose y with a high-triangle distribution between 1 and the number of children and x with a low-triangle distribution between 1 and y. This will bias towards

---

[17] www.eecs.harvard.edu/econcs/jopt.
[18] Note that by setting *depthLow=depthMid=depthHigh* one can still grow a full tree of a given depth by eliminating phase 2.

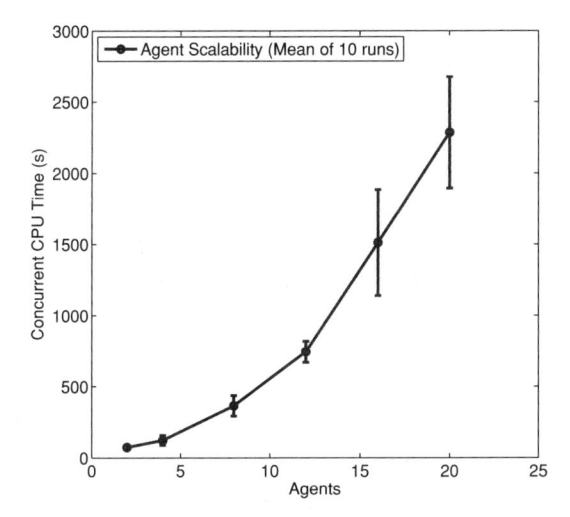

Figure 15. Effect of the number of bidders on the run-time of ICE.

the introduction of IC operators that permit a wide choice in the number of children. Each internal node is also assigned a bonus drawn according to a uniform distribution. Each leaf node is assigned as a "buy" node with a probability $\psi \in [0, 1]$, and then a specific good type for that node is chosen from among those good types for sale in the market. The node is assigned a quantity by drawing from a low-triangle distribution between 1 and the total number in existence.[19] A unit value for the node is then drawn from a specific "buy" distribution, typically uniform, which is multiplied by the quantity and assigned as the node's bonus. The leaf nodes assigned as "sell" nodes have their goods and bonuses determined similarly, this time with goods selected from among those previously assigned to the bidder.[20]

6.1. Experimental Results: Scalability

The first set of results that we present focuses on the computational properties of ICE. Figure 15 shows the runtime performance of the system as we increase the number of bidders while holding all other parameters constant. In this example, 100 goods in 20 types are being traded by bidders with an average of 104 node trees. The graph shows the total wall clock time for all parts of the system. While we see super-linear growth in solve time with the number of bidders, the constants of this growth are such that markets with large numbers of bidders can be efficiently solved (solving for

---

[19] The total number of goods of a given type in existence may not actually be available for purchase at any price given the structure of seller trees. Thus a bias towards small quantities in "buy" nodes and large quantities in "sell" nodes produces more interesting problem instances.

[20] In our experiments, we vary $2 \leq |G| \leq 128$, $1 \leq d \leq 128$, $2 \leq |N| \leq 20$, $2 \leq outDegreeLow \leq 8$, $2 \leq outDegreeHigh \leq 8$, $2 \leq depthLow \leq 6$, $2 \leq depthMid \leq 6$, $2 \leq depthHigh \leq 8$, set a balanced buy probability $\psi = 0.5$, and set width multiplier during the second phase to $\xi = 2$. In these examples, buy node bonuses were drawn uniformly from $[10, 100]$, sell nodes bonuses were drawn uniformly from $[-100, -10]$ and internal nodes bonuses uniformly from $[-25, 25]$.

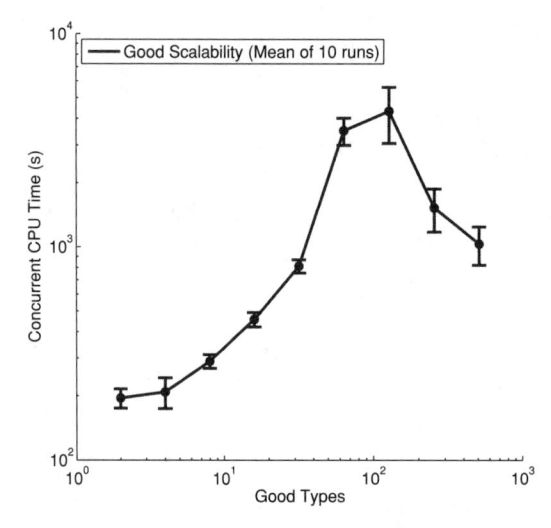

Figure 16. Effect of the number of good types on the run-time of ICE.

20 bidders in around 40 minutes). The error bars in all plots are for the standard error of the statistic.

In Figure 16 we can see the effect of varying the number of types of goods (retaining 5 units of each good in the supply) on computation time. For this example we adopt 10 bidders, and the same tree generation parameters. A likely explanation for eventual concavity of the run-time performance is suggested by the decrease in the average (item) price upon termination of ICE as the number of types of goods are increased (see Figure 17). The average price provides a good proxy for the competitiveness of the market. Adding goods to the problem will initially make the winner determination problem more difficult, but only until there is a large over-supply, at which point the outcome is easier to determine.

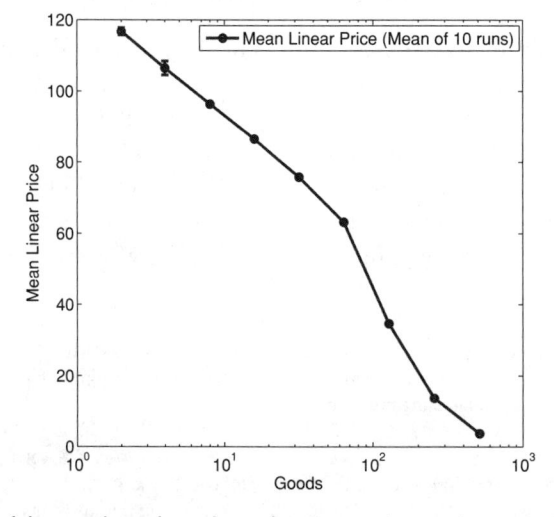

Figure 17. Effect of the number of goods on the average item price upon termination of ICE.

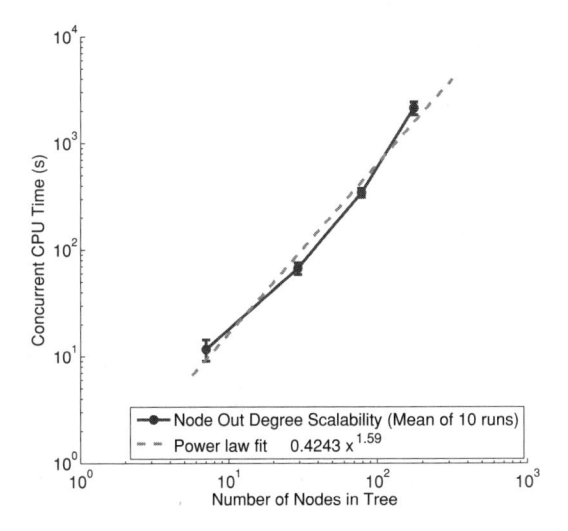

Figure 18. Effect of bid tree size on run-time of ICE: Varying the node-out degree.

Figures 18 and 19 illustrate the change in run time with the size of bid trees. Here we use only the first phase of our tree-generator to avoid confounding the effects of size with structural complexity. In both experiments, 100 goods in 20 types were being traded by 10 bidders. In Figure 18 we vary the number of children of any given node while in Figure 19 we vary the depth of the tree. Increasing the branching factor and/or tree depth results in an exponential growth in tree size, which necessarily corresponds to an exponential growth in runtime. However, if we account for this by instead plotting against the number of nodes in the trees, we see that both graphs indicate a *near-polynomial increase in runtime with tree size*. We fit a polynomial function to this data of the form $y = Ax^b$, indicating that this growth is approximately of degree 1.5 in the range of tree sizes considered in these experiments.

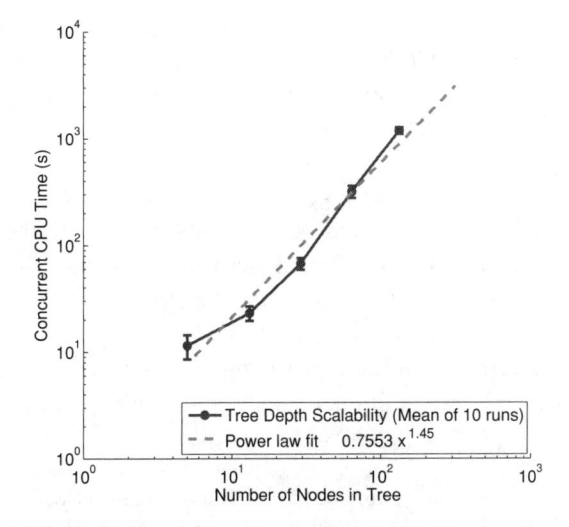

Figure 19. Effect of bid tree size on run-time of ICE: Varying the tree depth.

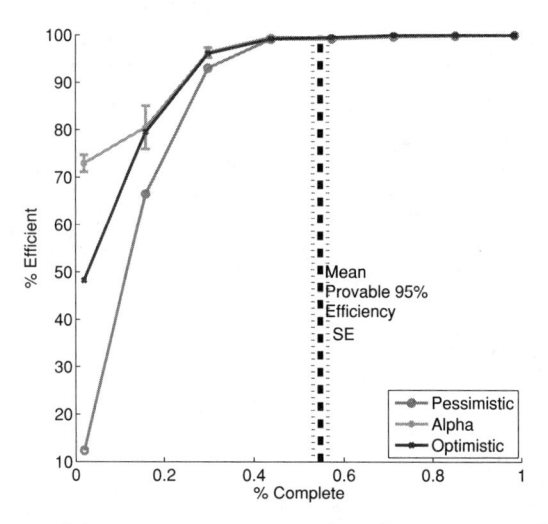

Figure 20. Efficiency of the optimistic, provisional, and pessimistic trades across rounds.

6.2. Empirical Results: Economic Properties

The second set of results that we present focus on the economic properties of ICE: the efficiency of trade across rounds, the effectiveness of preference elicitation, and the accuracy and stability of prices. For this set of experiments we average over 10 problem instances, each with 8 bidders, a potential supply of 100 goods in 20 types, and bid trees with an average of 104 nodes.

Figure 20 plots the *true* efficiency of the trades computed at pessimistic (lower bounds $\underline{v}$), provisional (α-valuation v^α) and optimistic (upper bounds $\bar{v}$) valuations across rounds. In this graph and those that follow, the x-axis indicates the number of rounds completed as a percentage of the total number of rounds until termination which enables results to be aggregated across multiple instances, each of which can have a different number of total rounds.[21] The vertical (dashed) line indicates the average percentage complete when the trade is provably 95% efficient. The exchange remains open past this point while payments converge (and because we simulate the outcome of the last-and-final round by continuing progress with our straightforward bidding agents). The two lines on either side represent one standard error of this statistic.

In Figure 20, we see that the exchange quickly converges to highly efficient trades, taking an average of 6.8 rounds to achieve efficiency. In general, the optimistic trade (i.e., computed from upper bounds $\bar{v}$) has higher (true) efficiency than the pessimistic one (i.e., computed from lower bounds $\underline{v}$), while the efficiency of the provisional trade λ^α is typically better than both. This justifies the design decision to adopt the provisional valuations and provisional trade in driving the exchange dynamics. It also suggests that exchanges with the traditional paradigm of improving bids (i.e., increasing

---

[21] Each data point represents the average across the 10 instances, and is determined by averaging the underlying points in its neighborhood. Error-bars indicate the standard error (SE) of this mean. Thus, the figures are essentially a histogram rendered as a line graph.

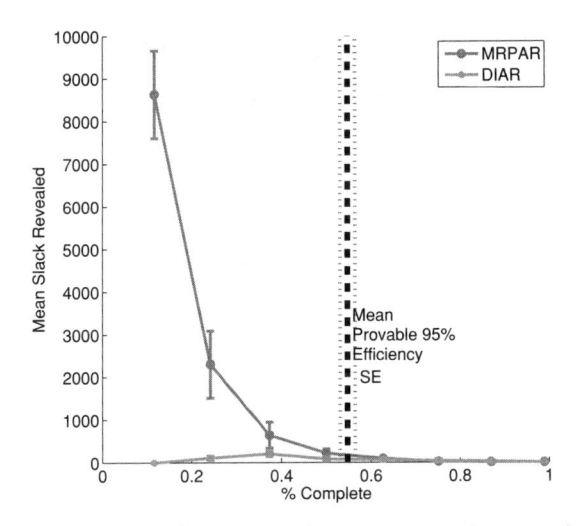

Figure 21. Average reduction in value uncertainty due to each rule.

lower bound claims on valuations) would allow little useful feedback in early rounds: the efficiency of the pessimistic trade—all that would be available without information about the upper-bounds of bidder valuations—is initially poor.

Figure 21 shows the average amount of revelation caused by MRPAR and DIAR in each round of ICE. Revelation is measured here in terms of the *absolute tightening of upper and lower bounds, summed across the bid trees*. The MRPAR activity rule is the main driving force behind the revelation of information and the vast majority of revelation (in absolute terms) occurs within the first 25% of rounds. DIAR plays a role in making progress towards identifying the efficient trade but only once MRPAR has substantially reduced the value uncertainty. One can think of MRPAR as our rocket's main engine, and DIAR as a thruster for mid-course correction. ICE determines the efficient trade when the average node in a *TBBL* tree still retains a gap between the upper and lower bounds on value at the node equal to around 62% of the maximum (true) value that a node could contribute to a bidder's value, roughly the maximum marginal value contributed by a node over all feasible trades. We see that ICE is successful in directing preference elicitation to information that is relevant to determining the efficient trade.

We now provide two different views on the effectiveness of prices. Figure 22 shows the mean percentage absolute difference between the prices computed in each round and the prices computed in the final round. Prices quickly converge. In our experiments we have driven the exchange beyond the efficient solution in order to converge to the Threshold payments, but we see that most of the price information is already available at the point of efficiency. Figure 23 provides information about the quality of the price feedback. We plot the 'regret', averaged across bidders and runs, from the best-response trade as determined from intermediate prices in comparison to the best-response to final prices, where the regret is defined in terms of lost payoff at those final prices. Define the regret to bidder i for his best response $\lambda'_i = \arg\max_{\lambda_i \in \mathcal{F}_i(x^0)} [v_i(\lambda_i) -$

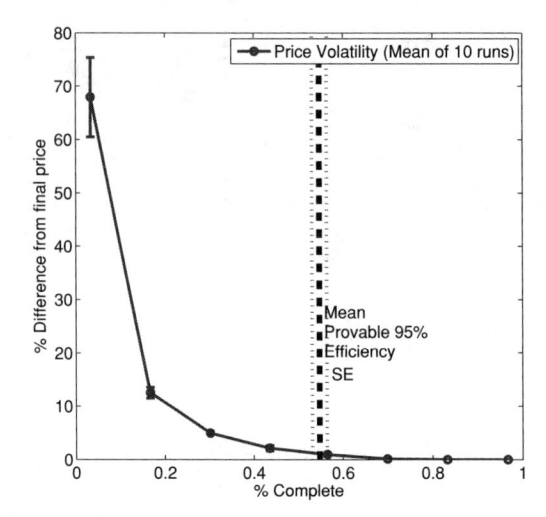

Figure 22. Price trajectory: Closeness of prices in each round to the final prices.

$p^{\hat{\pi}}(\lambda_i)$], to prices $\hat{\pi}$, given that the final prices are π, as:

$$\text{Regret}_i(\lambda'_i, \pi) = \left(1 - \frac{v_i(\lambda'_i) - p^\pi(\lambda'_i)}{\max_{\lambda_i \in \mathcal{F}_i(x^0)} v_i(\lambda_i) - p^\pi(\lambda_i)}\right) \times 100\%. \quad (28)$$

As the payoff from trade λ'_i, when evaluated at prices π, approaches that from the best-response trade at prices π, then $\text{Regret}_i(\lambda'_i, \pi)$ approaches 0%. Figure 23 plots the *average regret across all bidders* as a function of the number of rounds completed in ICE. The regret is low: 11.2% when averaged across all rounds before the efficient trade is determined and 7.0% when averaged across all rounds. That regret falls across rounds also shows that prices become more and more informative as the rounds proceed.

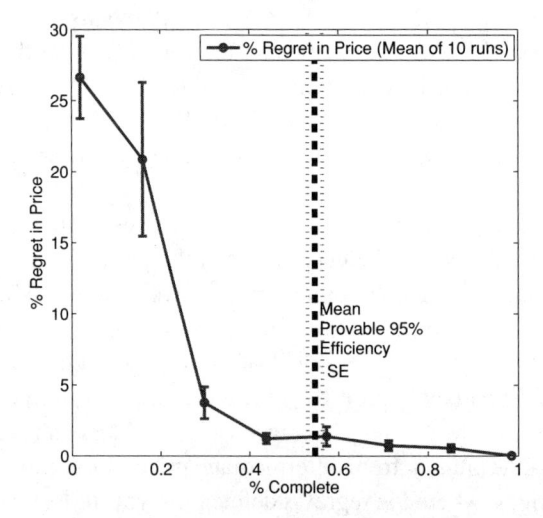

Figure 23. Regret in best-response by bidders due to price inaccuracy relative to final prices.

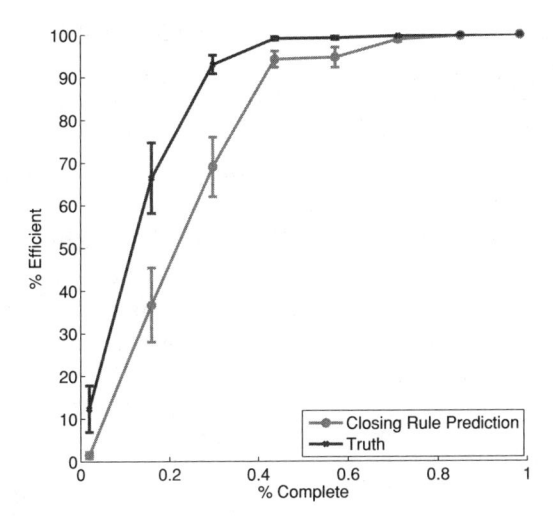

Figure 24. Comparison between the actual efficiency of the pessimistic trade and the ω^{direct} bound.

Finally, we present experimental results that relate to the two methods that ICE employs to bound the final efficiency of the pessimistic trade. The total pricing error across all bidders in each round as determined within pricing in terms of $(\lambda^\alpha, v^\alpha)$, and normalized here by the total *true* value of the efficient trade, is already small (at 8.5%) in initial rounds and falls to around 3% by final rounds of ICE. This suggests that a price-based bound is quite informative, although note that this is defined in terms of the error given $(\lambda^\alpha, v^\alpha)$ and does not immediately map to a price-based accuracy claim for true valuations and for the current trade $\underline{\lambda}$ defined on lower bound valuations. Figure 24 compares the actual efficiency of the pessimistic trade $\underline{\lambda}$ in each round with that estimated by the ω^{direct} bound on efficiency that is available to the exchange. This confirms that the direct bound is reasonably tight, and effective in bounding the true efficiency regardless of the accuracy of the prices.

7. Conclusions

We designed and implemented a scalable and highly expressive iterative combinatorial exchange. The design incorporates a tree-based bidding language for combinatorial exchanges, a new method to construct approximate item prices, a proxy-based architecture with optimistic and pessimistic valuations coupled with price-based activity rules to drive price discovery and preference elicitation, and a method to bound the inefficiency of a provisional trade. By asking bidders to refine valuation functions across rounds, we are able to prove efficiency properties even thought all elicitation is guided by item prices. Empirical results with simulated, straightforward bidding agents confirm good economic and computational properties.

There are a number of opportunities for future work. For example, it would be interesting to instantiate the design to narrow domains. For example, an ICE design could be specified with application to two-sided spectrum markets. It would also be interesting to instantiate ICE in domains with truthful CEs, since this would bring straightforward bidding strategies into an *ex post* Nash equilibrium. For example, it might be possible to integrate methods such as trade-reduction (McAfee, 1992) and its generalizations (Babaioff & Walsh, 2005; Chu & Shen, 2008). A technical extension would be to modify the design to allow bidders to refine the structure and not just the valuation bounds on their *TBBL* bids. It would also be interesting to design iterative CEs for dynamic populations; see (Blum, Sandholm, & Zinkevich, 2006; Bredin, Parkes, & Duong, 2007) for results on the design of dynamic double auctions, and the work of Parkes (2007) for a survey on online mechanism design. Lastly, the incentive properties of ICE depend on the payment rule used, which argues for further analysis of the threshold rule and its alternatives.

Acknowledgments

This work was supported in part by NSF grant IIS-0238147. An earlier version of this paper appeared in *J. Art. Intell. Res.* **33** (2008) and the *Proc. 6th ACM Conference on Electronic Commerce* (2005). The *TBBL* language is also described in a workshop paper (Cavallo et al., 2005). Nick Elprin, Loizos Michael and Hassan Sultan contributed to earlier versions of this work. Our thanks to the students in Al Roth's Econ 2056 at Havard who participated in a trial of this system, and to Cynthia Barnhart for providing airline domain expertise. We would also like to thank Evan Kwerel and George Donohue for early motivation and encouragement. The paper's genesis is from the class CS 286r "Topics at the Interface of Computer Science and Economics" taught at Harvard in the Spring of 2004.

Appendix A. Computation for MRPAR

In this section, we show that MRPAR can be computed by solving a sequence of three MIPs. We begin by considering the special case of $\delta = 0$. The general case follows almost immediately. Define a *candidate passing trade*, λ_i^L, as:

$$\lambda_i^L \in \arg\max_{\lambda_i \in \mathcal{F}_i(x^0)} \underline{v}_i(\lambda_i) - p^\pi(\lambda_i) \tag{29}$$

breaking ties

(i) to maximize $\overline{v}_i(\lambda_i) - \underline{v}_i(\lambda_i)$

(ii) in favor of λ_i^α

This can be computed by solving one MIP to maximize $\underline{v}_i(\lambda_i) - p^\pi(\lambda_i)$, followed by a second MIP in which this objective is incorporated as a constraint and $\overline{v}_i(\lambda_i) - \underline{v}_i(\lambda_i)$ becomes the objective. Given perturbed valuation $\tilde{v}_i$, defined with respect to trade λ_i^L (as in Section 4), we can define a *witness trade*, λ_i^U, as:

$$\lambda_i^U \in \arg\max_{\lambda_i \in \mathcal{F}_i(x^0)} \tilde{v}_i(\lambda_i) - p^\pi(\lambda_i). \tag{30}$$

This can be found by solving a third MIP. Given prices π, provisional trade λ_i^α and bid tree T_i, the *computational* MRPAR rule (C-MRPAR) for the case of $\delta = 0$ can now be defined as:

(1) $\underline{v}_i(\lambda_i^L) - p^\pi(\lambda_i^L) \geq \tilde{v}_i(\lambda_i^U) - p^\pi(\lambda_i^U)$, **and**
(2) $\lambda_i^L = \lambda_i^\alpha$, **or** $\underline{v}_i(\lambda_i^L) - p^\pi(\lambda_i^L) > \tilde{v}_i(\lambda_i^\alpha) - p^\pi(\lambda_i^\alpha)$

We now establish that C-MRPAR is equivalent to MRPAR, as defined by (19)–(21).

Lemma 4. *Given trades λ_i and λ_i', prices π, and tree T_i, we have $\theta_i^\pi(\lambda_i, \lambda_i', v_i') \geq 0$, $\forall v_i' \in T_i$ if and only if $\underline{v}_i(\lambda_i) - p^\pi(\lambda_i) \geq \tilde{v}_i(\lambda_i') - p^\pi(\lambda_i')$, where $\tilde{v}_i$ is defined with respect to trade λ_i.*

Proof. Direction ($\Rightarrow$) is immediate since $\tilde{v}_i \in T_i$. Consider direction ($\Leftarrow$) and suppose, for contradiction, that $\underline{v}_i(\lambda_i) - p^\pi(\lambda_i) \geq \tilde{v}_i(\lambda_i') - p^\pi(\lambda_i')$ but there exists some $v_i' \in T_i$ such that $v_i'(\lambda_i) - p^\pi(\lambda_i) < v_i'(\lambda_i') - p^\pi(\lambda_i')$. Subtract $\sum_{\beta \in \lambda_i \cap \lambda_i'}[v_i'(\beta) - \underline{v}_i(\beta)]$ from both sides, where $\beta \in \lambda_i$ indicates that node β is satisfied by trade λ_i, to get

$$\sum_{\beta \in \lambda_i \setminus \lambda_i'} v_i'(\beta) + \sum_{\beta \in \lambda_i \cap \lambda_i'} v_i'(\beta) - \sum_{\beta \in \lambda_i \cap \lambda_i'} v_i'(\beta) + \sum_{\beta \in \lambda_i \cap \lambda_i'} \underline{v}_i(\beta) - p^\pi(\lambda_i)$$
$$< \sum_{\beta \in \lambda_i' \setminus \lambda_i} v_i'(\beta) + \sum_{\beta \in \lambda_i \cap \lambda_i'} v_i'(\beta) - \sum_{\beta \in \lambda_i \cap \lambda_i'} v_i'(\beta) + \sum_{\beta \in \lambda_i \cap \lambda_i'} \underline{v}_i(\beta) - p^\pi(\lambda_i') \quad (31)$$

$$\Rightarrow \sum_{\beta \in \lambda_i \setminus \lambda_i'} \underline{v}_i(\beta) + \sum_{\beta \in \lambda_i \cap \lambda_i'} \underline{v}_i(\beta) - p^\pi(\lambda_i) < \sum_{\beta \in \lambda_i' \setminus \lambda_i} \overline{v}_i(\beta) + \sum_{\beta \in \lambda_i \cap \lambda_i'} \underline{v}_i(\beta) - p^\pi(\lambda_i') \quad (32)$$

$$\Rightarrow \underline{v}_i(\lambda_i) - p^\pi(\lambda_i) < \tilde{v}_i(\lambda_i') - p^\pi(\lambda_i'), \quad (33)$$

which is a contradiction. □

Lemma 5. *Given trade λ_i, prices π, and tree T_i then $\theta_i^\pi(\lambda_i, \lambda_i', v_i') \geq 0$, $\forall v_i' \in T_i$, $\forall \lambda_i' \in \mathcal{F}_i(x^0)$, if and only if $\underline{v}_i(\lambda_i) - p^\pi(\lambda_i) \geq \tilde{v}_i(\lambda_i^U) - p^\pi(\lambda_i^U)$, where $\tilde{v}_i$ is defined with respect to trade λ_i and λ_i^U is the witness trade.*

Proof. Direction ($\Rightarrow$) is immediate since $\tilde{v}_i \in T_i$ and $\lambda_i^U \in \mathcal{F}_i(x^0)$. Consider direction ($\Leftarrow$) and suppose, for contradiction, that $\underline{v}_i(\lambda_i) - p^\pi(\lambda_i) \geq \tilde{v}_i(\lambda_i^U) - p^\pi(\lambda_i^U)$ but there exists some $\lambda_i' \in \mathcal{F}_i(x^0)$ and $v_i' \in T_i$ such that $\theta_i^\pi(\lambda_i, \lambda_i', v_i') < 0$. By Lemma 4, this means $\underline{v}_i(\lambda_i) - p^\pi(\lambda_i) < \tilde{v}_i(\lambda_i') - p^\pi(\lambda_i')$. But, we have a contradiction because

$$\underline{v}_i(\lambda_i) - p^\pi(\lambda_i) \geq \tilde{v}_i(\lambda_i^U) - p^\pi(\lambda_i^U) \quad (34)$$

$$= \max_{\lambda_i'' \in \mathcal{F}_i(x^0)} \tilde{v}_i(\lambda_i'') - p^\pi(\lambda_i'') \geq \tilde{v}_i(\lambda_i') - p^\pi(\lambda_i') \quad (35)$$

□

Theorem 7. *C-MRPAR is equivalent to δ-MRPAR for $\delta = 0$.*

Proof. Comparing (17) and (18) with C-MRPAR, and given Lemmas 4 and 5, all that is left to show is that it is sufficient to check λ_i^L, as the only candidate to pass MRPAR.

That is, we need to show that if there is some $\check{\lambda}_i \in \mathcal{F}_i(x^0)$ that satisfies MRPAR then candidate λ_i^L satisfies MRPAR. We argue as follows:

1. Trade $\check{\lambda}_i$ must solve $\max_{\lambda_i \in \mathcal{F}_i(x^0)}[\underline{v}_i(\lambda_i) - p^\pi(\lambda_i)]$. Otherwise, there is some λ_i' with $\underline{v}_i(\lambda_i') - p^\pi(\lambda_i') > \underline{v}_i(\check{\lambda}_i) - p^\pi(\check{\lambda}_i)$. A contradiction with (17).
2. Trade $\check{\lambda}_i$ must also break ties in favor of maximizing $\overline{v}_i(\lambda_i) - \underline{v}_i(\lambda_i)$. Otherwise, there is some λ_i' with the same profit as $\check{\lambda}_i$ at $\underline{v}_i$, with $\overline{v}_i(\lambda_i') - \underline{v}_i(\lambda_i') > \overline{v}_i(\check{\lambda}_i) - \underline{v}_i(\check{\lambda}_i)$. This implies $\overline{v}_i(\lambda_i') - \overline{v}_i(\check{\lambda}_i) > \underline{v}_i(\lambda_i') - \underline{v}_i(\check{\lambda}_i)$, and $\theta_i^\pi(\lambda_i', \check{\lambda}_i, \overline{v}_i) > \theta_i^\pi(\lambda_i', \check{\lambda}_i, \underline{v}_i)$. But, since λ_i' has the same profit as $\check{\lambda}_i$ at $\underline{v}_i$ we have $\theta_i^\pi(\lambda_i', \check{\lambda}_i, \underline{v}_i) = 0$ and so $\theta_i^\pi(\lambda_i', \check{\lambda}_i, \overline{v}_i) > 0$. This is a contradiction with (17).
3. Proceed now by case analysis. Either $\check{\lambda}_i = \lambda_i^\alpha$, in which case we are done because this will be explicitly selected as candidate passing trade λ_i^L. For the other case, let Λ_i^L denote all feasible solutions to (29) and consider the difficult case when $|\Lambda_i^L| > 1$. We argue that if $\check{\lambda}_i \in \Lambda_i^L$ satisfies MRPAR, then so does any other trade $\lambda_i' \in \Lambda_i^L$, with $\lambda_i' \neq \check{\lambda}_i$. By MRPAR, we have $\theta_i^\pi(\check{\lambda}_i, \lambda_i', v_i') \geq 0, \forall v_i' \in T_i$. In particular, $\tilde{v}_i(\check{\lambda}_i) - p^\pi(\check{\lambda}_i) \geq \tilde{v}_i(\lambda_i') - p^\pi(\lambda_i')$, where $\tilde{v}_i$ is defined with respect to $\check{\lambda}_i$, and equivalently,

$$\underline{v}_i(\check{\lambda}_i) - p^\pi(\check{\lambda}_i) \geq \tilde{v}_i(\lambda_i') - p^\pi(\lambda_i'). \tag{36}$$

On the other hand,

$$\underline{v}_i(\check{\lambda}_i) - p^\pi(\check{\lambda}_i) = \underline{v}_i(\lambda_i') - p^\pi(\lambda_i'), \tag{37}$$

since both are in Λ_i^L. Taking (36) together with (37), we must have that λ_i' satisfies no uncertain value nodes in T_i not also satisfied in $\check{\lambda}_i$. Moreover, since $\overline{v}_i(\check{\lambda}_i) - \underline{v}_i(\check{\lambda}_i) = \overline{v}_i(\lambda_i') - \underline{v}_i(\lambda_i')$, both trades must satisfy *exactly* the same uncertain value nodes. Finally, by (37) the profit from all fixed value nodes in T_i must be the same in both trades. We conclude that the profit is the same for *all* $v_i' \in T_i$ for $\check{\lambda}_i$ and λ_i' at the current prices and MRPAR is satisfied by either trade. □

To understand the importance of the tie-breaking rule (i) in selecting the candidate passing trade, λ_i^L, in C-MRPAR, consider the following example for MRPAR with $\delta = 0$:

Example 11. A bidder has XOR(+A, +B) and a value of 5 on the leaf +A and a value range of [5,10] on leaf +B. Suppose prices are currently 3 for each of A and B and $\lambda_i^\alpha = +B$. The MRPAR rule is satisfied because the market knows that however the remaining value uncertainty on +B is resolved the bidder will always (weakly) prefer +B to +A and +B is λ_i^α. Notice that both +A and +B have the same pessimistic utility, but only +B can satisfy MRPAR. But +B has maximal value uncertainty, and therefore this is selected over +A by C-MRPAR.

To understand the importance of selecting, and evaluating, λ_i^U with respect to $\tilde{v}_i$ rather than $\overline{v}_i$, consider the following example (again for $\delta = 0$). It illustrates the role of shared uncertainty in the tree, which occurs when multiple trades share a node with uncertain value and the value, although uncertain, will be resolved in the same way for both trades.

Example 12. A bidder has XOR(+A, +B) and value bounds [5, 10] on the *root* node and a value of 1 on leaf +A. Suppose prices are currently 3 for each of A and B and

$\lambda_i^\alpha = +B$. The MRPAR rule is satisfied because the bidder strictly prefers $+A$ to $+B$, whichever way the uncertain value on the root node is ultimately resolved. C-MRPAR selects λ_i^L as "buy A", with payoff $\underline{v}_i(\lambda_i^L) - p^\pi(\lambda_i^L) = 5 + 1 - 3 = 3$. At valuation $\bar{v}_i$, the witness trade "buy B" would be selected and have payoff $10 - 3 = 7$ and seem to violate MRPAR. But, whichever way the uncertain value at the root is resolved it will affect $+A$ and $+B$ in the same way. This is addressed by setting $\tilde{v}_i(\beta) = \underline{v}_i(\beta) = 5$ on the root node, the same value adopted in determining the payoff from λ_i^L. Evaluated at $\tilde{v}_i$, the witness is "buy A" and (1) of C-MRPAR is trivially satisfied while (2) is satisfied since $3 > 5 - 3 = 2$.

For δ-MRPAR with $\delta > 0$, we adopt a slight variation, with a δ-C-MRPAR procedure defined as:

(1) Check $\theta_i^\pi(\lambda_i^\alpha, \lambda_i', v_i') \geq -\delta$ for all $v_i' \in T_i$, all $\lambda_i' \in \mathcal{F}_i(x^0)$ directly, by application of Lemma 5 with valuation $\tilde{v}_i$ defined with respect to trade λ_i^α, and test

$$\underline{v}_i(\lambda_i^\alpha) - p^\pi(\lambda_i^\alpha) \geq \tilde{v}_i(\lambda_i^U) - p^\pi(\lambda_i^U) - \delta \tag{38}$$

(2) If this is not satisfied then fall back on C-MRPAR to verify (20) and (21), with candidate passing trade λ_i^L modified from (29) to drop tie-breaking in favor of λ_i^α and with the second step of C-MRPAR modified to require $\underline{v}_i(\lambda_i^L) - p^\pi(\lambda_i^L) > \tilde{v}_i(\lambda_i^\alpha) - p^\pi(\lambda_i^\alpha) + \delta$, again with $\tilde{v}_i$ defined with respect to λ_i^L.

The argument adopted in the proof of Theorem 7 remains valid in establishing that it is sufficient to consider λ_i^L, as defined in δ-C-MRPAR, in the case that λ_i^α does not pass the activity rule.

Appendix B. Computation for DIAR

The ϵ-DIAR rule can be verified by solving two MIPs. The first optimization problem identifies the trade with maximal DIAR error for which the current bounds refinement has improved this error by at least ϵ:

$$\Delta_i^P = \max_{\lambda_i \in \mathcal{F}_i(x^0)} [\tilde{v}_i^0(\lambda_i) - p^\pi(\lambda_i) - (\underline{v}_i^0(\lambda_i^\alpha) - p^\pi(\lambda_i^\alpha))] \tag{39}$$

$$\text{s.t.} \quad (\tilde{v}_i^0(\lambda_i) - p^\pi(\lambda_i) - (\underline{v}_i^0(\lambda_i^\alpha) - p^\pi(\lambda_i^\alpha)))$$
$$- (\tilde{v}_i^1(\lambda_i) - p^\pi(\lambda_i) - (\underline{v}_i^1(\lambda_i^\alpha) - p^\pi(\lambda_i^\alpha))) \geq \epsilon \tag{40}$$

$$= -C + \max_{\lambda_i \in \mathcal{F}_i(x^0)} \tilde{v}_i^0(\lambda_i) - p^\pi(\lambda_i) \tag{41}$$

$$\text{s.t.} \quad \tilde{v}_i^0(\lambda_i) - \underline{v}_i^0(\lambda_i^\alpha) - \tilde{v}_i^1(\lambda_i) + \underline{v}_i^1(\lambda_i^\alpha) \geq \epsilon, \tag{42}$$

where $\tilde{v}_i^0$ and $\tilde{v}_i^1$ are defined with respect to λ_i^α, v^0 and v^1 represent valuations defined before and after the bidder's refinement respectively, and $C = \underline{v}_i^0(\lambda_i^\alpha) - p^\pi(\lambda_i^\alpha)$. Note that the problem could be infeasible, in which case we define $\Delta_i^P := -\infty$.

The second optimization identifies the trade with maximal DIAR error for which v^1 still allows for the possibility of valuation bounds that provide an ϵ error reduction over v^0:

$$\Delta_i^F = \max_{\lambda_i \in \mathcal{F}_i(x^0)} [\tilde{v}_i^0(\lambda_i) - p^\pi(\lambda_i) - (\underline{v}_i^0(\lambda_i^\alpha) - p^\pi(\lambda_i^\alpha))] \tag{43}$$

$$\text{s.t.} \quad (\tilde{v}_i^0(\lambda_i) - p^\pi(\lambda_i) - (\underline{v}_i^0(\lambda_i^\alpha) - p^\pi(\lambda_i^\alpha)))$$
$$- (\underline{v}_i^1(\lambda_i) - p^\pi(\lambda_i) - (\check{v}_i^1(\lambda_i^\alpha) - p^\pi(\lambda_i^\alpha))) \geq \epsilon \tag{44}$$

$$= -C + \max_{\lambda_i \in \mathcal{F}_i(x^0)} \tilde{v}_i^0(\lambda_i) - p^\pi(\lambda_i) \tag{45}$$

$$\text{s.t.} \quad \tilde{v}_i^0(\lambda_i) - \underline{v}_i^0(\lambda_i^\alpha) - \underline{v}_i^1(\lambda_i) + \check{v}_i^1(\lambda_i^\alpha) \geq \epsilon, \tag{46}$$

where $\tilde{v}_i$ is defined with respect to λ_i^α, and $\check{v}_i$ is similarly defined with respect to λ_i. The second term in (44) recognizes that it remains possible to decrease the value on λ_i to the new lower-bound $\underline{v}_i^1(\lambda_i)$, while increasing the value on λ_i^α to the new upper-bound $\overline{v}_i^1(\lambda_i^\alpha)$ except on those nodes that are shared with λ_i, giving $\check{v}_i^1(\lambda_i^\alpha)$. We see that (46) is equivalent to:

$$\sum_{\beta \in \lambda_i \setminus \lambda_i^\alpha} [\overline{v}_i^0(\beta) - \underline{v}_i^1(\beta)] + \sum_{\beta \in \lambda_i^\alpha \setminus \lambda_i} [\overline{v}_i^1(\beta) - \underline{v}_i^0(\beta)] \geq \epsilon,$$

which calculates the amount of refinement that is still possible in service of reducing the DIAR error. Note the problem could be infeasible, in which case we define $\Delta_i^F := -\infty$. We ultimately compare the two solutions, and the bidder passes DIAR if and only if $\Delta_i^P \geq \Delta_i^F$.

Appendix C. The Automated Bidding Agents and Bidder Feedback

The bidding agents that are used for the simulation experiments are designed to minimize the amount of information revealed in order to pass the activity rules while remaining straightforward so that the true valuation is consistent with lower and upper valuations. In summarizing the behavior of the bidding agents, there are three things to explain: (a) the method that we adopt in place of the last-and-final round; (b) the feedback that is provided by ICE to bidders in meeting MRPAR and DIAR; and (c) the logic that is followed by the bidding agents. Rather than define a method for bidding agents to adjust their bounds in a last-and-final round, we keep ICE open in simulation past the point in which it would ordinarily go to the last-and-final round. Past this point, the bidding agents continue to refine their bounds and ICE terminates when the payments are within some desired accuracy. Each bidding agent in this phase reduces its uncertainty by some multiplicative factor on all nodes that are active in the current provisional trade or in any of the provisional trades for the economies with bidder i removed. This is adopted for simulation purposes only.

Our bidding agents operate in a loop, heuristically modifying their valuation bounds in trying to meet MRPAR and DIAR and querying the proxy for advice. The proxy provides guidance to help the bidding agent further refine its valuation so it can meet the activity rule. For both MRPAR and DIAR, the optimization problems that are solved in checking whether a bidder has satisfied the activity rule also provide information that can guide the bidder. First consider MRPAR and recall that λ_i^L is the candidate passing trade and λ_i^U is the witness trade. The following lemma is easy, and stated without proof:

Lemma 6. *When MRPAR is not satisfied for the current valuation bounds, a bidder must increase a lower bound on at least one node in $\{\lambda_i^L \setminus \lambda_i^U\}$, or decrease an upper bound on at least one node in $\{\lambda_i^U \setminus \lambda_i^L\}$, in order to meet the activity rule.*

Once a bidder makes some changes on some subset of these nodes, the bidder can inquire if he has passed the activity rule. The proxy can then respond "yes" or can revise the set of nodes on which the bidding agent should refine its valuation bounds. A similar functionality is provided for DIAR. This time the trade that solves the second MIP (with DIAR error Δ_i^F) is provided as feedback, together with information about how much the bidder must either further reduce the error, or further constrain the possibilities on this trade, to satisfy DIAR. The bidding agent can determine from this information which nodes it must modify, and by how much in total, and is free to decide how much to modify each node to satisfy the rule. The key to our agent design is the following lemma:

Lemma 7. *The trade with which a straightforward bidder passes MRPAR (for $\delta = 0$) must be a trade that is weakly preferred by the bidder to all other trades for his true valuation.*

Proof. By contradiction. Suppose true valuation $v_i \in T_i$ and trade $\check{\lambda}_i$ meets MRPAR but is not a weakly preferred trade at the true valuation and prices π. Then, there exists a trade $\lambda_i^* \in \mathcal{F}_i(x^0)$ such that $\theta_i^\pi(\lambda_i^*, \check{\lambda}_i, v_i) > 0$. But, this is a contradiction with MRPAR since $\theta_i^\pi(\check{\lambda}_i, \lambda_i', v_i') \geq 0$ for all $v_i' \in T_i$ and all $\lambda_i' \in \mathcal{F}_i(x^0)$, including $v_i' = v_i$ and $\lambda_i' = \lambda_i^*$. □

We use this observation to define a procedure UPDATEMRPAR by which a bidder can intelligently refine its valuation bounds to meet MRPAR. Let $\check{\lambda}_i$ be the trade with which we hope to pass MRPAR, and define $u_i(\lambda_i, \pi) = v_i(\lambda_i) - p^\pi(\lambda_i)$, $\underline{u}_i(\lambda_i, \pi) = \underline{v}_i(\lambda_i) - p^\pi(\lambda_i)$, $\tilde{u}_i(\lambda_i, \pi) = \tilde{v}_i(\lambda_i) - p^\pi(\lambda_i)$, where $\tilde{v}_i$ is defined with respect to candidate passing trade $\check{\lambda}_i$. The high-level approach is as follows:

function UPDATEMRPAR
$\quad \check{\lambda}_i \in \arg\max_{\lambda_i \in \mathcal{F}_i(x^0)} \underline{u}_i(\lambda_i, \pi)$
$\quad$ **if** $\underline{u}_i(\check{\lambda}_i, \pi) < 0$ **then**
$\quad\quad$ reduce slack on $\check{\lambda}_i$ by $\underline{u}_i(\check{\lambda}_i, \pi)$
$\quad$ **end if**
$\quad \lambda_i^U \in \arg\max_{\lambda_i \in \mathcal{F}_i(x^0)} \tilde{u}_i(\lambda_i, \pi)$
$\quad$ **while** $\underline{u}_i(\check{\lambda}_i, \pi) < \tilde{u}_i(\lambda_i^U, \pi)$ **do**
$\quad\quad$ Heuristically reduce upper bounds on $\lambda_i^U \setminus \check{\lambda}_i$ by $\tilde{u}_i(\lambda_i^U, \pi) - \underline{u}_i(\check{\lambda}_i, \pi)$
$\quad\quad$ If remaining slack heuristically reduce lower bounds on $\check{\lambda}_i \setminus \lambda_i^U$
$\quad\quad \lambda_i^U \in \arg\max_{\lambda_i \in \mathcal{F}_i(x^0)} \tilde{u}_i(\lambda_i, \pi)$
$\quad$ **end while**

```
        if λ̌_i ≠ λ_i^α then
            while u_i(λ̌_i, π) ≤ ũ_i(λ_i^α, π) do
                Heuristically reduce upper bounds on λ_i^α \ λ̌_i by ũ_i(λ_i^α, π) − u_i(λ̌_i, π)
                If remaining slack heuristically reduce lower bounds on λ̌_i \ λ_i^α
            end while
        end if
        return λ̌_i
end function
```

The bidding agent makes use of a couple of optimization modalities that are exposed by the proxy to the bidder. The procedure first chooses the most preferred trade at truth as the trade to pass MRPAR with $\check{\lambda}_i$; the bidding agent requests that the proxy finds this trade by solving a MIP. If the trade has negative profit, then the bidding agent attempts to demonstrate positive profit for this trade. Next, the bidding agent enters a loop, wherein it repeatedly requests the proxy to run a MIP that calculates a witness trade λ_i^U with respect to $\check{\lambda}_i$. As long as this witness has more profit than that of what should be the most preferred trade, the bidding agent adjust bounds so as to reverse this mis-ordering. Lastly, because the bidding agent must pass MRPAR, not merely RPAR, the bidding agent attempts to show a strict preference for $\check{\lambda}_i$ over λ_i^α when they are not identical.

In meeting DIAR, the bidding agent responds to the $\Delta^F \geq 0$ and $\epsilon \geq 0$ parameter provided by the proxy as follows. Let λ^F be the trade chosen in the maximization that calculates Δ^F. The high-level approach is as follows:

```
function UPDATEDIAR
    while Proxy says we still have not passed DIAR do
        if λ^F or λ^α can be modified to reduce DIAR error by ϵ over last round then
            Heuristically reduce the upper-bound slack in λ^F \ λ^α
            Heuristically reduce the lower-bound slack in λ^α \ λ^F
        else
            Heuristically reduce the upper-bound slack in λ^α \ λ^F
            Heuristically reduce the lower-bound slack in λ^F \ λ^α
        end if
    end while
end function
```

The bidding agent attempts to make the current failing trade pass DIAR if possible by reducing the error with respect to that trade. Otherwise, it reduces bounds to prove that DIAR could not be made to pass on that trade and loops on to the next trade.

References

Ausubel, L., Cramton, P., & Milgrom, P. (2006). The clock-proxy auction: A practical combinatorial auction design. In Cramton, P., Shoham, Y., & Steinberg, R. (Eds.). *Combinatorial Auctions*. MIT Press, chap. 5.

Ausubel, L. M., & Milgrom, P. (2002). Ascending auctions with package bidding. *Frontiers of Theoretical Economics*, *1*, 1–42.

Babaioff, M., & Walsh, W. E. (2005). Incentive-compatible, budget-balanced, yet highly efficient auctions for supply chain formation. *Decision Support Systems*, *39*, 123–149.

Ball, M., Donohue, G., & Hoffman, K. (2006). Auctions for the safe, efficient, and equitable allocation of airspace system resources. In Cramton, P., Shoham, Y., & Steinberg, R. (Eds.). *Combinatorial Auctions*. MIT Press, chap. 20.

Ball, M. O., Ausubel, L. M., Berardino, F., Cramton, P., Donohue, G., Hansen, M., & Hoffman, K. (2007). Market-based alternatives for managing congestion at new york's laguardia airport. In *Proceedings of the AirNeth Annual Conference*.

Bererton, C., Gordon, G., & Thrun, S. (2003). Auction mechanism design for multi-robot coordination. In *Proceedings of the 17th Annual Conference on Neural Information Processing Systems (NIPS-03)*.

Bertsimas, D., & Tsitsiklis, J. (1997). *Introduction to Linear Optimization*. Athena Scientific.

Bikhchandani, S., & Mamer, J. W. (1997). Competitive equilibrium in an exchange economy with indivisibilities. *Journal of Economic Theory*, *74*, 385–413.

Bikhchandani, S., & Ostroy, J. M. (2002). The package assignment model. *Journal of Economic Theory*, *107*(2), 377–406.

Blum, A., Sandholm, T., & Zinkevich, M. (2006). Online algorithms for market clearing. *Journal of the ACM*, *53*, 845–879.

Bossaerts, P., Fine, L., & Ledyard, J. (2002). Inducing liquidity in thin financial markets through combined-value trading mechanisms. *European Economic Review*, *46*(9), 1671–1695.

Boutilier, C. (2002). Solving concisely expressed combinatorial auction problems. In *Proceedings of the 18th Conference of the Association for the Advancement of Artificial Intelligence (AAAI-02)*, pp. 359–366.

Boutilier, C., & Hoos, H. (2001). Bidding languages for combinatorial auctions. In *Proceedings of the 17th International Joint Conference on Artificial Intelligence (IJCAI-01)*, pp. 1121–1217.

Bredin, J., Parkes, D. C., & Duong, Q. (2007). Chain: A dynamic double auction framework for matching patient agents. *Journal of Artificial Intelligence Research*, *30*, 133–179.

Cavallo, R., Parkes, D. C., Juda, A. I., Kirsch, A., Kulesza, A., Lahaie, S., Lubin, B., Michael, L., & Shneidman, J. (2005). TBBL: A tree-based bidding language for iterative combinatorial exchanges. In *Multi-Disciplinary Workshop on Advances in Preference Handling (IJCAI-05)*.

Chu, L. Y., & Shen, Z.-J. M. (2008). Truthful double auction mechanisms. *Operations Research*, *56*(1), 102–120.

Compte, O., & Jehiel, P. (2007). Auctions and information acquisition: Sealed-bid or Dynamic Formats?. *RAND Journal of Economics*, *38*(2), 355–372.

Conen, W., & Sandholm, T. (2001). Preference elicitation in combinatorial auctions. In *Proceedings of the 3rd ACM Conference on Electronic Commerce (EC-01)*, pp. 256–259. ACM.

Cramton, P. (2003). Electricity Market Design: The Good, the Bad, and the Ugly. In *Proceedings of the Hawaii International Conference on System Sciences*.

Cramton, P. (2006). Simultaneous ascending auctions. In Cramton, P., Shoham, Y., & Steinberg, R. (Eds.). *Combinatorial Auctions*. MIT Press, chap. 3.

Cramton, P., Kwerel, E., & Williams, J. (1998). Efficient relocation of spectrum incumbents. *Journal of Law and Economics*, *41*, 647–675.

Cramton, P., Lopez, H., Malec, D., & Sujarittanonta, P. (2015). Design of the reverse auction in the fcc incentive auction. Available at: www.cramton.umd.edu/papers2015–2019/cramton-reverse-auction-design-fcc-comment-pn.pdf.

Cramton, P., Shoham, Y., & Steinberg, R. (Eds.). (2006). *Combinatorial Auctions*. MIT Press.

Day, R., & Raghavan, S. (2007). Fair payments for efficient allocations in public sector combinatorial auctions. *Management Science, 53*(9), 1389.

de Vries, S., Schummer, J., & Vohra, R. V. (2007). On ascending Vickrey auctions for heterogeneous objects. *Journal of Economic Theory, 132*, 95–118.

de Vries, S., & Vohra, R. V. (2003). Combinatorial auctions: A survey. *INFORMS Journal on Computing, 15*(3), 284–309.

Dias, M., Zlot, R., Kalra, N., & Stentz, A. (2006). Market-based multirobot coordination: A survey and analysis. *Proceedings of the IEEE, 94*, 1257–1270.

Dunford, M., Hoffman, K., Menon, D., Sultana, R., & Wilson, T. (2003). Testing linear pricing algorithms for use in ascending combinatorial auctions. Tech. rep., SEOR, George Mason University.

Fan, M., Stallaert, J., & Whinston, A. B. (1999). A web-based financial trading system. *Computer, 32*(4), 64–70.

Frechette, A., Newman, N., & Leyton-Brown, K. (2015). Solving the station repacking problem. In *Proceedings of the 24th International Joint Conference on Artificial Intelligence (IJCAI-15)*.

Fu, Y., Chase, J., Chun, B., Schwab, S., & Vahdat, A. (2003). SHARP: an architecture for secure resource peering. In *Proceedings of the 19th ACM Symposium on Operating Systems Principles (SOSP-03)*, pp. 133–148. ACM.

Gerkey, B. P., & Mataric, M. J. (2002). Sold!: Auction methods for multi-robot coordination. *IEEE Transactions on Robotics and Automation, Special Issue on Multi-robot Systems, 18*, 758–768.

Hudson, B., & Sandholm, T. (2004). Effectiveness of query types and policies for preference elicitation in combinatorial auctions. In *Proceedings of the 3rd International Joint Conference on Autonomous Agents and Multi-Agent Systems (AAMAS-04)*, pp. 386–393.

Kelso, A. S., & Crawford, V. P. (1982). Job matching, coalition formation, and gross substitutes. *Econometrica, 50*, 1483–1504.

Krishna, V. (2002). *Auction Theory*. Academic Press.

Kwasnica, A. M., Ledyard, J. O., Porter, D., & DeMartini, C. (2005). A new and improved design for multi-object iterative auctions. *Management Science, 51*, 419–434.

Kwerel, E., & Williams, J. (2002). A proposal for a rapid transition to market allocation of spectrum. Tech. rep., FCC Office of Plans and Policy.

Lahaie, S., Constantin, F., & Parkes, D. C. (2005). More on the power of demand queries in combinatorial auctions: Learning atomic languages and handling incentives. In *Proceedings of the 19th International Joint Conference on Artificial Intelligence (IJCAI-05)*.

Lahaie, S., & Parkes, D. C. (2004). Applying learning algorithms to preference elicitation. In *Proceedings of the 5th ACM Conference on Electronic Commerce (EC-04)*, pp. 180–188.

Lubin, B. (2015). Games and metagames: Pricing rules for combinatorial mechanisms. Working Paper, Boston University.

Lubin, B., & Parkes, D. C. (2009). Quantifying the strategyproofness of mechanisms via metrics on payoff distributions. In *Proceedings of the 25th Conference on Uncertainty in Artificial Intelligence*, pp. 349–358. AUAI Press.

McAfee, R. P. (1992). A dominant strategy double auction. *Journal of Economic Theory, 56*, 434–450.

Milgrom, P. (2004). *Putting Auction Theory to Work*. Cambridge University Press.

Milgrom, P. (2007). Package auctions and package exchanges (2004 Fisher-Schultz lecture). *Econometrica, 75*, 935–966.

Mishra, D., & Parkes, D. C. (2007). Ascending price Vickrey auctions for general valuations. *Journal of Economic Theory, 132*, 335–366.

Myerson, R. B., & Satterthwaite, M. A. (1983). Efficient mechanisms for bilateral trading. *Journal of Economic Theory, 28*, 265–281.

Nemhauser, G., & Wolsey, L. (1999). *Integer and Combinatorial Optimization*. Wiley-Interscience.

Nisan, N. (2006). Bidding languages for combinatorial auctions. In Cramton, P., Shoham, Y., & Steinberg, R. (Eds.). *Combinatorial Auctions*. MIT Press, chap. 9.

O'Neill, R. P., Sotkiewicz, P. M., Hobbs, B. F., Rothkopf, M. H., & Stewart, Jr., W. R. (2005). Efficient market-clearing prices in markets with nonconvexities. *European Journal of Operations Research*, *164*, 269–285.

Parkes, D. C. (2007). On-line mechanisms. In Nisan, N., Roughgarden, T., Tardos, E., & Vazirani, V. (Eds.), *Algorithmic Game Theory*, chap. 16. Cambridge University Press.

Parkes, D. C., Kalagnanam, J. R., & Eso, M. (2001). Achieving budget-balance with Vickrey-based payment schemes in exchanges. In *Proceedings of the 17th International Joint Conference on Artificial Intelligence (IJCAI-01)*, pp. 1161–1168.

Parkes, D. C., & Ungar, L. H. (2000a). Iterative combinatorial auctions: Theory and practice. In *Proceedings of the 17th Conference of the Association for the Advancement of Artificial Intelligence (AAAI-00)*, pp. 74–81.

Parkes, D. C., & Ungar, L. H. (2000b). Preventing strategic manipulation in iterative auctions: Proxy agents and price-adjustment. In *Proceedings of the 17th Conference of the Association for the Advancement of Artificial Intelligence (AAAI-00)*, pp. 82–89.

Rassenti, S. J., Smith, V. L., & Bulfin, R. L. (1982). A combinatorial mechanism for airport time slot allocation. *Bell Journal of Economics*, *13*, 402–417.

Rothkopf, M. H., Pekeč, A., & Harstad, R. M. (1998). Computationally manageable combinatorial auctions. *Management Science*, *44*(8), 1131–1147.

Saatcioglu, K., Stallaert, J., & Whinston, A. B. (2001). Design of a financial portal. *Communications of the ACM*, *44*, 33–38.

Sandholm, T. (2006). Optimal winner determination algorithms. In Cramton, P., Shoham, Y., & Steinberg, R. (Eds.). *Combinatorial Auctions*. MIT Press, chap. 14.

Sandholm, T. (2007). Expressive Commerce and Its Application to Sourcing: How We Conducted $35 Billion of Generalized Combinatorial Auctions. *AI Magazine*, *28*(3), 45.

Sandholm, T., & Boutilier, C. (2006). Preference elicitation in combinatorial auctions. In Cramton et al. (Cramton, Shoham, & Steinberg, 2006), chap. 10.

Shapley, L. S., & Shubik, M. (1972). The assignment game I: The core. *International Journal of Game Theory*, *1*, 111–130.

Smith, T., Sandholm, T., & Simmons, R. (2002). Constructing and clearing combinatorial exchanges using preference elicitation. In *AAAI-02 workshop on Preferences in AI and CP: Symbolic Approaches*.

Stokely, M., Winget, J., Keyes, E., Grimes, C., & Yolken, B. (2009). Using a market economy to provision compute resources across planet-wide clusters. In *Proceedings for the International Parallel and Distributed Processing Symposium (IPPS-09)*, pp. 1–8.

Vossen, T. W. M., & Ball, M. O. (2006). Slot trading opportunities in collaborative ground delay programs. *Transportation Science*, *40*, 15–28.

Wurman, P. R., & Wellman, M. P. (2000). A*k*BA: A progressive, anonymous-price combinatorial auction. In *Proceedings of the 2nd ACM Conference on Electronic Commerce (EC-00)*, pp. 21–29.

CHAPTER 40

ACE: A Combinatorial Market Mechanism

Leslie Fine, Jacob K. Goeree, Tak Ishikida,
and John O. Ledyard

1. Introduction

At the time of the Clean Air Act of 1990, the Los Angeles basin was the only region in the country classified as an extreme non-attainment area for exceeding the National Ambient Air Quality Standards for ozone. Under pressure at the national level from the Environmental Protection Agency (EPA), and at the state level from the California Air Resources Board (CARB) to significantly reduce emissions, the South Coast Air Quality Management District (SCAQMD) launched a program for trading permits in nitrogen and sulfur oxides (NOx and SOx) in the Los Angeles basin. That program was strongly supported by environmental groups and large firms in the LA basin as the most cost effective way of attaining the desired reductions. This program, the REgional CLean Air Incentives Market (RECLAIM), was initialized in October of 1993 and has been operating since early 1994.[1]

The design of the tradable instruments was inevitably a compromise between regulatory interests and market efficiency. Regulators wanted to be able to control the timing and distribution of emissions as finely as possible with caps at many different locations and for many different time periods. To create liquidity, market designers wanted as few instruments as possible – a single aggregate cap would have been their preference. In the end, 136 different types of permits were created that could be used by a company to cover their emissions and, thereby, avoid the costs of abatement. This meant there were both substitutes and complements among the permits. Further, the markets for each of these permits would be illiquid. That created a big problem for the environmental engineers trying to choose between installing expensive abatement equipment or buying a portfolio of permits.

At the request of a number of firms, a new combinatoric market design was built and operated to help then and others deal with the complexities that the RECLAIM program created. In this chapter, we provide a description and analysis of that market. There are

---
[1] Our description of RECLAIM in this subsection is taken liberally from Cason and Gangadharan (1998). See also Fine (2001) and Carlson et al. (1993) for more information.

three parts. In the rest of this section, we provide more of the details of the RECLAIM program and describe the problems that created for environmental engineers. In Section 2, we describe the ACE-RECLAIM market design with special emphasis in Section 3 on the algorithms that drove the market. Finally, in Section 4 we look at the performance of that market.

1.1. The RECLAIM Program

RECLAIM is targeted at two major pollutants emitted from stationary sources: nitrogen oxides (NOx) and sulfur oxides (SOx). All facilities emitting 4 tons or more (per year) of NOx or SOx from permitted equipment were included in RECLAIM. Approximately 390 facilities were in the early NOx market, which collectively represented about 65 percent of the reported NOx emissions from all permitted stationary sources in the Basin. The SOx market consisted of 41 facilities, which represented about 85 percent of the reported SOx emissions from all permitted stationary sources.

Each facility in RECLAIM was allocated a certain number of RECLAIM Trading Credits (henceforth RTCs or simply *permits*) for equipment or processes that emit NOx or SOx. This allocation depended on the peak activity levels for each type of permitted equipment between 1989 and 1992. Each facility received an allocation for each year[2] between 1994 and 2000 based on a straight line rate of reduction calculated from the starting allocation to the allocation in the year 2000. For the years 2001 to 2003, the allocation levels were decreased further. Allocations for each year from 2004 to 2010 were to be equal to the 2003 allocation unless the AQMD decided that further reductions would be required. Average annual percentage reduction rates for facilities ranged between 7.1 and 8.7 percent in the NOx market and between 4.1 and 9.2 percent in the SOx market (SCAQMD, 1993). There is no banking allowed. At this point in the design there were $2 \times 17 = 34$ different types of permits to deal with. But more were yet to come.

Based on solid experimental evidence, the SCAQMD realized that this structure of permits would lead to extreme price volatility towards the end of each year. In a good business year, the need for permits by all firms would be high and at the end of the year prices could climb to as high as the fine for emitting too much. In a bad business year, the need for permits by all firms could be low and at the end of the year prices would drop, perhaps as low as zero. This price volatility would wreak havoc with rational planning efforts. To eliminate this problem, the SCAQMD adopted "cycles." Permits were identified by year and by cycle. There were two cycles; one beginning January 1 and one beginning July 1. As an example, 1994 cycle 1 permits could only be used to cover pollution emitted from January 1, 1994 to December 31, 1994. The 1994 cycle 2 permits could only be used to cover pollution emitted from July 1, 1994 to June 30, 1995. It was shown experimentally that such an overlapping structure of permits would serve to mitigate the extreme price volatility. In the end it was decided to accept a little more complexity in the number of types of permits for the reduced complexity due to excessive volatility. There were now $2 \times 34 = 68$ different types of permits. And yet more to come.

---

[2] A permit for 1998, for example, could only be used to cover emissions in 1998.

Due to regulatory worries about maintaining tight control over the distribution of pollution by the prevailing on-shore winds, the SCAQMD also identified permits by zones. The idea was that upwind zones could sell their permits to downwind zones but not vice versa. The SCAQMD originally wanted 37 zones. But, because the economists warned about serious complexities and thinness in trading, they settled for two: an inland zone and a coastal zone. Firms located in a coastal zone could only use permits identified as coastal. Inland firms could use either coastal or inland permits. In the end, therefore, for perfectly good regulatory considerations, there were a total of 136 different assets (68 per pollutant) that could be used to cover a firm's pollution: two pollutants (NOx and SOx), two zones (inland and coastal), two cycles, and 17 years (1994–2010).

Put yourself in the position of an environmental engineer dealing with the complexities of the RECLAIM program. A typical exercise would have the engineer deciding between installing abatement equipment or buying permits. Suppose abatement equipment costs $B to install and has a lifetime of T periods.[3] The equipment abates e_t units of emissions in period t. As an alternative, the engineer can cover the same emissions x by using some collection of the 136 permits. We let a_{kt} be the amount of permit of type k that can be used to cover one unit of emissions in t. For RECLAIM, $a_{kt} \in \{0, 1\}$. Suppose the engineer already has a portfolio of permits, $w \in \Re_+^K$. To avoid installing the abatement equipment she will have to buy a vector of permits $y \in \Re_+^K$ such that $y_k + w_k = \sum_t z_{kt}$ and $\sum_k a_{kt} z_{kt} \geq e_t$, where z_{kt} is the amount of permit k she will use to cover x_t.

If there is an active market for each of the 136 permits and a stable price of p_k for each, she has a fairly easy decision. Let

$$P = \min_{y,z} \sum_k p_k y_k,$$

subject to

$$y_k + w_k = \sum_t z_{kt}, \forall k, \quad (1)$$

$$\sum_k a_{kt} z_{kt} \geq x_t, \forall t,$$

$$y_k + w_k \geq 0, \forall k.$$

The last constraint eliminates short sales. Equation (1) is a straightforward linear programming problem which is easy to solve. She installs the abatement equipment if and only if $B - pw \leq P$. If $P < \$B - pw$, she buys the vector of permits y^* that solves (1).

But such thick markets never existed. As noted by Cason and Gangadharan (1998), the program placed minimal restrictions on how permits could be traded. Although the SCAQMD originally planned on putting out contracts for the design and management of a market system, brokers lobbied hard and succeeded in preventing that. Brokers

---

[3] A period can be a year, a quarter, a half-year, etc. For purposes of this example, it doesn't really matter.

suggested, without much real evidence, that traders could rely on brokers to help bring buyers and sellers together. RECLAIM brokers would thus bear the search costs for potential traders, of course for a fee.

Instead of leaving the market entirely to brokers, however, SCAQMD planners implemented an electronic bulletin board system (BBS) to help RECLAIM participants find trading partners and reduce search costs. The BBS was operated by SCAQMD, and anyone could obtain a password to access their computerized network. The BBS allowed firms to indicate trading interest by electronically posting offers to buy or sell RTCs. Other firms could then scroll through these offers and contact the offering firm to negotiate a transaction. Prices were not posted.

It turned out, however, this was not enough to enable firms to manage their environmental requirements.

1.2. The Environmental Engineer's Problem

The engineer must identify and pursue a sequence of deals from the SCAQMD bulletin board, buying permits of type k without knowing for sure what price she will be able to negotiate for permits of type, say, k'. And she has to make the decision now. A simple example illustrates the complications and risks this type of thin market causes.

Example 1 (Permits as Complements: The Need for AND Bids). There are two markets: one for A and one for B. Bidder 1 is our engineer who needs to buy both A and B. Bidder 2 is only in market A. Bidders 3 and 4 are only in market B. The amounts they want to buy or sell and their true values for that transaction are in the following table:

Bidder	$	Amount of A	Amount of B
1	400	50	50
2	−50	−50	
3	−180		−30
4	−120		−20

It is important to remember, for this and other examples, that *each bidder knows only her own $ value* and does not know those of the others.

If Bidder 1 has to participate separately in market A and market B, she has to decide what her separate bids in A and B will be. Suppose she assumes prices will be similar in the two markets and decides to bid 200 for 50 units in each (or 4 per unit). Having done that, suppose she buys her 50 units of A for 3 per unit or 150 total. So far so good. Now she must go into the market for B. She has two options. If she is unable to complete a trade for B, resell the A she bought and pay 400 to abate. Suppose she takes a loss from that of L. The cost of this outcome is $5 + \frac{L}{50}$ per unit. That is how much she is now willing to bid for B. But 3 and 4 are only willing to sell at a price greater than 6. If $L > 50$ she will pay something more than 6 for B. She will not have to abate (the socially desirable outcome) but will have paid more than 400 for the permits. This is not a good outcome. If $L < 50$ she will sell the 50 A and abate (the socially undesired outcome). She will also take a loss of L and therefore will have paid more than 400. This is not a good outcome. So, no matter what she does at this point she loses. Whatever the ultimate price of B, she paid too much for A.

She could have tried to negotiate harder when she bought A but if seller 2's willingness-to-accept had been 5 and seller 4 and 5 had a willingness-to-accept of 2, this would lead her to not buying A initially. After buying B she could go back to seller 2, but he might have completed another deal with someone else. No matter how she plays it she may end up not buying permits, which is the undesired outcome.

This is certainly a contrived example but imagine trying to string together trades for 15 different assets in 15 different markets. The example illustrates *the exposure problem*[4] that the environmental engineer faces when she must deal in *separate* markets for each type of permit or engage in a sequence of *separate*, complex multi-lateral negotiations. A permits and B permits are complements, like left and right shoes. Getting one without the other is no help. A buyer who has to deal in separate markets for A and B is *exposed* to the risk of incomplete transactions. To protect for this she may either bid lower than she really would be willing to or she may not participate at all. This destroys potential gains from trade and reduces trade below that which would be optimal.

Example 2 (Swaps and Endowments: The Need for OR Bids). Continuing Example 1, suppose Bidder 1's company had an initial allocation of permits from the SCAQMD and is willing to commit some of them to this project. Suppose she has 60 of A and 40 of B.[5] If she is going to use permits instead of abatement equipment, she will have to buy the following stream of permits: $(-10, 10)$. As in Example 1, she has a problem in deciding how much to offer in market A and how much to bid in market B. This is a swap she is willing to enter into if and only if $-10p_A + 10p_B \leq 400 - 60p_A - 40p_B$ or $50p_A + 50p_B \leq 400$. What she would like to do is to enter a deal to "sell 10 A and buy 10 B if $50p_A + 50p_B \leq 400$" OR "sell 60 A and 40 B if $50p_A + 50p_B > 400$" but not both. But the thin market won't allow such deals. Instead she must negotiate separately for A and B. Suppose when she goes to the A market, the price for A is \$4.50 and, believing that the price of B will be equal or greater than that, she sells her 60 permits. Then suppose that when she gets to the B market she learns the price for B is \$1.50. She will have to complete her sale of the 40 B at that price and install abatement equipment, paying out $400 - 270 - 60 = 70$. If, instead, she had only sold 10 A at the price of \$5 and bought 10 B at the price of \$1.50, she would have made \$30. The difference is \$100. And she may be the least cost abater, which makes this bad not only for her but also for society.

When markets are thick and prices are stable, buyers and sellers with complex needs are no more at risk than buyers and sellers with simple needs. But when markets are thin and prices are unpredictable, buyers and sellers with complex needs are at risk. Gangadharan (2002) has estimated that these problems were particularly bad in the early years of the RECLAIM program. The program consists of firms with very

---

[4] The exposure problem was identified and discussed at some length by Bykowsky et al. (2000).

[5] This is as if A are permits for 1995 and B are permits for 1996 and the firm is given a declining initial allocation.

different industrial structures, which use different technologies to produce very heterogeneous products, which means that matching a single buyer with a single seller with coincident wants is extremely unlikely. She found that these problems reduced the probability of trading by about 32%.

Individuals faced with this risk of incomplete trading often turn to brokers who generally claim to be able to find the collection of trades that will satisfy the bidder's price point. That is what happened with RECLAIM. The brokers were more than willing to help, but they charged 40% on each side of a trade, did not publish their prices, and did not let firms know what the alternatives in the market really were. They operated dark markets.[6]

After their experience with brokers, a collection of the firms that had been deeply involved in the design of RECLAIM came to us and asked that we consider creating a marketplace that would help them. Without knowing the language or even the exact concept, what they described as desirable was a *combinatorial exchange*. They wanted a transparent market place into which they could submit bids that expressed their complex needs and willingness-to-pay and from which they would get fair trades at understandable prices. The rest of the paper is about the market we built for them.

2. ACE-RECLAIM

The market, designed for RECLAIM, was initially operated under the name Automated Credit Exchange (ACE). The design was based on research conducted at Caltech in 1993.[7] The actual internet-based implementation was built by a team from Net Exchange that included Takashi Ishikida, Charles Polk, and Lance Clifner. It utilized an iterated combined-value call market to trade in quarterly trading sessions. We believe that it was the first internet-based commercial exchange. ACE-RECLAIM opened in April of 1995 and, by 1999, it had accounted for approximately 90% of all priced trades in the RECLAIM market.[8]

A functioning market has two main pieces: a market mechanism (the inside piece) and a market process (the outside piece). The inside piece includes the bid forms, the winner determination algorithm and the pricing algorithms. The outside piece includes the participation rules, bidding rules, what is displayed to bidders and when, stopping rules, enforcement rules, etc. We will first describe the outside piece of the ACE-RECLAIM market.

---

[6] A quote from Duffie (2012) shows how pervasive the problem is: "Over-the-counter (OTC) markets for derivatives, collateralized debt obligations, and repurchase agreements played a significant role in the global financial crisis. Rather than being traded through a centralized institution such as a stock exchange, OTC trades are negotiated privately between market participants who may be unaware of prices that are currently available elsewhere in the market. In these relatively opaque markets, investors can be in the dark about the most attractive available terms and who might be offering them. This opaqueness exacerbated the financial crisis, as regulators and market participants were unable to quickly assess the risks and pricing of these instruments." This also happened in the early years of RECLAIM.

[7] For a detailed description of the design and experimental test-bedding of the ACE-RECLAIM mechanism, see Ishikida et al. (2001).

[8] For more on participation in ACE-RECLAIM, see Section 4.1.

The market flowed as follows:

- *Qualification and escrow*: The first step in any market is the qualification of the market participants. Since the SCAQMD rules allowed anyone to hold permits and therefore to trade them, anyone was allowed to participate in the market. But combinatorial markets create a problem that separated markets do not. Trades are likely to be multi-lateral and complex. This means that if anyone defaults by not delivering cash or permits at the end of the process, then many winners may find their trades invalidated. To avoid this unfortunate outcome, ACE-RECLAIM required any participant to provide in escrow, before the market opened, the cash they would be willing to spend and the permits they were willing to sell. ACE-RECLAIM then used these data to provide constraints on the bids the bidders were allowed to make. In particular, no bidder could bid in a way that would violate the cash and permit bounds, no matter what bids were accepted. This removed the possibility of default.
- *Bid submission*: Bids were submitted in two ways: either online, through a custom Windows bidding interface on a computer connected via a modem to the ACE-RECLAIM computer, or via fax to the market manager who would then input the bids herself.
- *Provisional winners, prices, and payments*: After bids are submitted, the ACE-RECLAIM computer calculated provisional winners, prices, and payments using the market mechanism described in Section 3.
- *Stopping rule, resubmissions, and the improvement rule*: ACE-RECLAIM was run as an iterative process. It proceeded in rounds. At the end of each round (once provisional winners, prices, and payments were announced) a stopping rule is calculated. The stopping rule for ACE-RECLAIM was very simple: (a) there must be at least 3 rounds, (b) after round 2, if surplus and volume do not increase by at least 5% from the previous round, the market will end, and (c) the market will end after round 5 if (b) has not taken place. If the rule is satisfied, the process goes to Clearance and Settlement to finish up.

   If the rule is not satisfied, the process returns to allow more bids. There are two features of ACE-RECLAIM that come into play at this point. All winning bids must be re-submitted as they are. They can then be improved on. Improvement can happen in several ways. Without going into too much detail, improvement usually means (a) for a bid to buy, a higher willingness to pay is entered or (b) for a bid to sell, a lower willingness to accept is entered.

Remark 1. ACE can also be run as a one-shot sealed bid or a continuous mechanism. Each such version can be used with a variety of stopping rules, e.g. based on eligibility and activity similar to the FCC's SMR rules or based on the increase in surplus over the last round (e.g. stop if less than 5%).

- *Sunshine*: An important aspect of combinatorial trading is that often bids of several individuals have to fit together, like pieces of a jigsaw puzzle, in order for them to trade. Thus it is often important to know what others are willing to trade or to show others what you are willing to trade. ACE-RECLAIM allowed individual bidders to choose how much of their own bids would be made public. They could choose (a) nothing, (b) quantities of permits only, or (c) quantities of permits and dollar amounts bid. This was called, respectively, (a) no sunshine, (b) partial sunshine, and (c) sunshine. ACE-RECLAIM is

an iterative process and many bidders chose (a) in early rounds and (b) or (c) in later rounds as they became more desperate to find a match of some kind.
- *Clearance and Settlement*: Once the stopping rule comes into play, ACE-RECLAIM takes the results of the last round and implements these: sending orders to the RECLAIM database for the appropriate reallocation of permits and sending orders to the escrow holder for the appropriate reallocation of cash between accounts.

3. ACE

ACE is the market mechanism (the inside piece) for ACE-RECLAIM. ACE is an acronym for Approximate Competitive Equilibrium, the philosophy behind the market mechanism design. It is a generalization of the well-known Uniform Price Double Auction (UPDA), which produces high efficiencies in simple markets.[9] UPDA is usually run like a sealed bid auction. Subjects submit bids (P, Q). If $Q > 0$, it is a buy order to be read as "I will buy up to Q units of the good for any price less than or equal to P." If $Q < 0$ then it is a sell order to be read as "I am willing to sell up to Q units of the good for any price greater than or equal to P." After all bids are submitted, UPDA picks winners by choosing that set of trades that would maximize the reported surplus. The price at which all transactions take place is the midpoint between the marginal units (both accepted and rejected). All winning buyers pay less than (sellers receive more than) or equal to their bid. It is the same price for all, there is no subsidy or tax for the market, and there is a strong incentive to reveal one's true willingness-to-pay (or accept) except at margin. Since the probability that any bid is marginal can be very low, this gives ACE a serious shot at being virtually incentive compatible.

Since the ACE market mechanism is to operate in a world that is more complex than UPDA was designed for, the bid structure, winners determination, and price determination pieces all had to be modified. The mechanism allows bidders to submit contingent orders. The winners determination algorithm then maximizes reported surplus, as in UPDA. The pricing algorithm approximates the UPDA price rule but still leaves bidders as well off after as if they had not participated.

3.1. Bids

The mechanism allows bidders to submit contingent orders. These can be ANDs (I want A if and only if I can also secure B) or ORs (I want either A or B).

Participants submit bids that are numbered $i = 1, \ldots, N$.

3.1.1. AND Bids

The basic bid of ACE is an AND bid. It allows a bidder to express interest in a collection of complements, all of which are needed. An AND bid is intended to protect the bidder from exposure in situations like Example 1.

---

[9] See Smith et al. (1982), Friedman (1993), and McCabe et al. (1993).

Definition 1. A simple AND bid numbered i is (b^i, x^i, F^i), where $b^i \in \Re$, $x^i \in \Re^K$, and $F^i \in [0, 1]$. K is the number of commodities.

The bid is read as "I will pay up to $b^i f^i$ for the vector $x^i f^i$ as long as $F^i \leq f^i \leq 1$."

The vector $x^i \in \Re^K$ contains the quantities offered or demanded, where K is the number of commodities. If $x_k^i > 0$, then x_k^i units are demanded. If $x_k^i < 0$, then x_k^i units are offered for sale. It is not required that all x_k^i be positive or negative, i.e. swaps are allowed. The number $b^i \in \Re$ is the maximum amount the bidder is willing to pay for x^i. ($|b^i|$ is the minimum amount a seller is willing to sell for if $b^i < 0$.) The number $F^i \in [0, 1]$ is the minimum fill number.

When $F^i = 0$, the bid is similar to a (P, Q) bid in a uniform price auction where bidders are willing to accept any amount $q \leq Q$ at a payment $t \leq Pq$. $F^i > 0$, indicates a minimum fill requirement which introduces non-convexities and discontinuities into the design problem. This is the most challenging part of the ACE design. When $F^i > 0$, the simple UPDA approach won't work, particularly the pricing.

Example 3 (Minimum Fill Required #1). Consider a market with a single commodity labeled A, with one buyer and three sellers. Buyer 1 is willing to pay 20 for 3 units but does not want to pay anything if he gets less than 3 units. That is, he requires his order be filled at the 100% level or not at all.

Bidder	$	Amount of A	% fill required
1	24	3	100
2	−2	−1	0
3	−4	−1	0
4	−10	−1	0

Gains from trade are maximized if all four participate in the deal. Even though bidder 4 requires $10 per unit and bidder 1 is only willing to pay $8 per unit, bidder 4 makes the deal possible for bidders 2 and 3 and should be included.

The minimum fill requirement creates several problems here. First, if 1 bids $8 per unit, 4 will never trade and 1 will not get the 3 units he wants. Second, if 1 agrees to pay 4 any amount greater than $10, both 2 and 3 would want the same deal. But then 1 cannot afford to pay every one at the rate of $10/unit. Often this configuration of values and desires will lead to no trade. The $8 in potential gains from trade is lost.

A standard thin market has difficulty dealing with minimum fill requirements. In the above example, the standard market makes it difficult to include Bid 4 in the final allocation even though it should be. In our next example, the standard market makes it difficult to exclude someone from the final allocation even though they should be.

Example 4 (Minimum Fill Required #2). In this example, there is an additional buyer and seller 4 has a lower reservation price.

Bidder	$	Amount of A	% fill required
1	24	3	100
2	−2	−1	0
3	−4	−1	0
4	−6	−1	0
5	10	1	0

Here the allocation that maximizes the gains from trade would include 1,2,3, and 4, and exclude 5. The price would probably be somewhere between 6 and 8. But 5 is more than willing to pay up to $10 and so might work hard to be included in the allocation. In fact, at that proposed price, 5 could offer 2 a deal. 5 can offer to buy 2's unit for $9. There is no price at which 5 is not willing to buy and 1 is willing to buy. Often this configuration of values and desires will lead to a single trade between 1 and 5 yielding a surplus of $8. Whereas the maximum surplus possible is $12. Potential gains from trade are lost.

3.1.2. OR Bids

An OR bid allows a bidder to express interest in a collection of substitutable possibilities. An OR bid is intended to protect a bidder from exposure in situations like Example 2.

Definition 2. An OR bid numbered i is $\{(b^{ij}, x^{ij}, F^{ij})\}_{j=1}^{J_i}$.

The bid is read as "I will accept one and only one $j \in J_i$ in which case I will pay up to $b^{ij} f^{ij}$ as long as $f^{ij} \in [0, 1]$." If $J_i = 1$, then this is exactly the same as the AND in Section 3.1.1.

Each (b^{ij}, x^{ij}, F^{ij}) is a simple AND bid. An OR bid requires that, at most, only one of these be accepted. That means there are $\delta^{ij}, \forall j \in J_i$ such that $f^{ij} \in [\delta^{ij} F^{ij}, \delta^{ij}]$, $\sum_j \delta^{ij} \leq 1$, and $\delta^{ij} \in \{0, 1\}$.

3.1.3. Characteristic Bids

A special OR was designed for ACE-RECLAIM to give environmental engineers a user-friendly way of expressing interest and value over a wide range of substitutable possibilities. Here we provide a slightly more general version. As we described in Section 1.2, the environmental engineer has a vector of permits $w \in \Re_+^K$ and wants to cover a stream of pollution $e \in \Re_+^T$. The engineer needs to buy a vector of assets x such that $x_k + w_k = \sum_t z_{kt}, \forall k$, and $\sum_k a_{kt} z_{kt} \geq e_t, \forall t$. And, if they want to avoid short sales, they would add the constraint that $x^i + w^i \geq 0$.

The characteristic bid generalizes the OR bid to allow an infinite variety of possible satisfactory trades to be considered without exposing the bidder to the risk of having more than one of those options active at a time.

Definition 3. A characteristic bid numbered i is $(b^i, A^i, e^i, w^i, F^i)$.

A characteristic bid is to be read as "I will accept one trade x^i as long as $A_t^i z_t \geq f^i e_t^i$, $x^i + w^i = \sum_t z_t$, $x^i + w^i \geq 0$, and $f^i \in [F^i, 1]$ in which case I will pay up to $f^i b^i$ for it."

If $F^i = 1$, then the characteristic bid lets the market know that the engineer will buy any combination of assets x as long (i) as she can add them to her endowment w in a way that covers e and (ii) it costs her no more than her abatement costs, \$$B$. She leaves it to the market to decide what works.

If $F^i = 0$, then the engineer is also allowing the market to consider buying her endowment w as long as she is paid at least \$0. If there are prices p then she would want to agree to that trade if and only if $(0, -w^i)$ solved $\max_{(f^i, x^i)} f^i b^i - p x^i$ subject to $A_t^i z_t \geq f^i e_t^i$, $x^i + w^i = \sum_t z_t$, $x^i + w^i \geq 0$, and $f^i \in [0, 1]$. In particular, letting $f^i = 1$, this means that she would agree to the trade if and only if $pw \geq \$B - P$, where P solves (1). As we will see below, the ACE market mechanism algorithms are designed with exactly this in mind.

Remark 2. A characteristic bid is a generalization of a simple bid. Suppose (b, x, F) is a simple bid. Consider the characteristic bid $(B, A, e, w, G) := (b, I, x, 0, F)$. This bid agrees to pay up to fb for any y such that $y \geq fx$. Since $y = fx$ is the dominant possibility this is just another form of a simple bid.

3.2. Winners Determination

Once all bids are submitted, a winners determination algorithm determines what trades will be matched and implemented. For ACE, winners are determined by maximizing the "reported surplus" of the trades subject to the restrictions imposed by the bidders and subject to no excess demand.

Definition 4 (Winners Determination). Let $\mathcal{S} = \{\text{AND bids}\}$. Let $\mathcal{O} = \{\text{OR bids}\}$. Let $\mathcal{Z} = \{\text{characteristic bids}\}$. The *winner determination problem* is:

$$S = \max_{(f, y)} \sum_{i \in \mathcal{S} \cup \mathcal{Z}} b^i f^i + \sum_{i \in \mathcal{O}} \sum_{j \in J_i} b^{ij} f^{ij}, \tag{2}$$

subject to

$$f^i \in \{0\} \cup [F^i, 1] \text{ for all } i \in \mathcal{S} \cup \mathcal{Z}, \tag{3}$$

$$A_t^i z_t^i \geq f^i e_t^i, \; x^i + w^i = \sum_t z_t^i, \; x^i + w^i \geq 0, \text{ for all } i \in \mathcal{Z}, \tag{4}$$

$$f^{ij} \in [\delta^{ij} F^{ij}, \delta^{ij}], \delta^{ij} \in \{0, 1\}, \sum_{j \in J_i} \delta^{ij} \leq 1 \text{ all } j \in J_i, \text{ all } i \in \mathcal{O}, \tag{5}$$

$$\sum_{i \in \mathcal{S}} x_k^i f^i + \sum_{i \in \mathcal{Z}} x_k^i + \sum_{i \in \mathcal{O}} \sum_{j \in J_i} x_k^{ij} f^{ij} \leq 0, k \in K. \tag{6}$$

Equation (3) is the minimum fill requirement for all bids except ORs; (4) is the restriction placed by characteristic bids; (5) is the restriction placed by OR bids; and (6) is the requirement that there be no excess demand. In the RECLAIM ACE market, if

$\sum x^i f^i < 0$ at the optimum, the transactions were made and the un-transacted credits were retired. That is, free disposal was possible for the market. We retain that feature here.

Remark 3. For ease of computation, the algorithm actually split the market into disjoint segments (no overlapping bids). This also aided the price computations which we pick up in the next section.

Remark 4. Maximizing surplus is not the only rule one might use. For example, if one is interested in extracting revenue from the mechanism, then maximizing surplus is rarely the best strategy. But for now we stay with surplus maximization.

We will need to identify winners and losers as we proceed.

Definition 5. Let (f^*, y^*) be the values that solve (2) and let $f^{*i} := \sum_{j \in J_i} f^{*ij}$ for each $i \in \mathcal{O}$. The set of *winners* is denoted by $W = \{i | f^{*i} > 0\}$. The set of *losers* is denoted by $L = \{i | f^{*i} = 0\}$. Those winners for whom $0 < f^{*i} < 1$ will be referred to as *marginal winners*.

3.3. Pricing

ACE pricing is about (a) providing useful signals to the market that reflect aggregate demand and supply, (b) not extracting any revenue, (c) achieving some measure of incentive compatibility, and (d) not rewarding inflexible bidders (i.e. those for whom allowing $f^i = 0$ in (2) would increase the surplus). Ideally, all of this can be accomplished if we can find competitive equilibrium prices that support the allocation found by the winners determination problem.

3.3.1. Competitive Equilibrium Prices

Competitive equilibrium allocations and prices satisfy two conditions: (i) each individual's allocation is just what they would want to buy at those prices and (ii) there is no excess demand. For ACE, the bids are the individuals. We want prices that support the winners determination allocation as a competitive equilibrium. They satisfy *ex-post self-selection*, or incentive compatibility, for the bidders.[10]

For simple bids, competitive equilibrium prices π would satisfy $b^i - \pi x^i \geq 0$ if $i \in W$ and $b^i - \pi x^i \leq 0$ if $i \in L$. For other bids, things are a little more complex.

For OR bids, let $f^{*ij} > 0$ in the winners determination solution; that is, j is the winning part of the OR. $f^{*ik} = 0$, for all $k \neq j$. Prices π would be regret-free if (a) $b^{ij} - \pi x^{ij} \geq 0$ and $b^{ij} - \pi x^{ij} = 0$ if $0 < f^{*ij} < 1$ and (b) $b^{ik} - \pi x^{ik} \leq b^{ij} f^{*ij} - \pi x^{ij} f^{*ij}$ for all $k \neq j$. Note that if ij is a winner with $f^{*ij} = 1$, this does not require ik to be a loser (i.e. $b^{ik} - \pi x^{ik} < 0$) only that it not be as good as ij at those prices. It does imply that if $\sum_{j \in J_i} f^{*ij} = 0$, then all the J_i bids should be losers at the prices π.

---

[10] These are also sometimes called *no arbitrage* or *no re-contracting* constraints.

For characteristic bids, let the solution to the winners determination problem (2) be (f^{*i}, x^{*i}). Prices would be regret-free if $b^i f^{*i} - \pi x^{*i} \geq b^i f^i - \pi x^i$ for all x^i such that $A^i_t z_t \geq f^i e^i_t, x^i + w^i = \sum_t z_t, x^i + w^i \geq 0, f^i \in [F^i, 1]$.

Remark 5. For RECLAIM Zone free bids, the restriction to regret-free prices was implemented with the constraint that the inland price be less than or equal to the coastal price. Because inland credits could not be used to cover coastal emissions, it imposed a simple requirement on the prices of the two zones no matter what year-cycle is involved. Only the inland firms care about the relative prices. If the coastal price is less than the inland price, inland firms would want to only buy coastal credits, thus driving the price up. If the inland price is less than the coastal price, the coastal firms could not drive the inland price up. No-regret prices are also arbitrage-free prices.

The no-arbitrage conditions are:

- If there are k, k', l, i such that $a^i_{kl} a^i_{k'l} > 0, \overline{X}^i_k > 0$, and $\overline{X}^i_{k'} > 0$, then $p^b_k = p^b_{k'}$.
- If there are k, k', l, i such that $a^i_{kl} a^i_{k'l} > 0, \overline{X}^i_k > 0$, and $\overline{X}^i_{k'} = 0$, then $p^b_k \leq p^b_{k'}$.
- If there are k, k', l, i such that $a^i_{kl} a^i_{k'l} > 0, \overline{X}^i_k < 0$, and $\overline{X}^i_{k'} < 0$, then $p^s_k = p^s_{k'}$.
- If there are k, k', l, i such that $a^i_{kl} a^i_{k'l} > 0, \overline{X}^i_k < 0$, and $\overline{X}^i_{k'} = 0$, then $p^s_k \geq p^s_{k'}$.

If the p^b and p^s did not satisfy these then $\overline{X}^i$ would not be a cost-minimizing coverage of $\overline{e}^i$ at those prices.

Definition 6 (competitive equilibrium prices). Let (f^*, x^*, δ^*) be the values of f that solve the winner determination problem (2).[11] Competitive equilibrium prices, π, satisfy the following:

$$f^{*i}(b^i - \pi \cdot x^i) \geq f^i(b^i - \pi \cdot x^i) \text{ for all } f^i \in [F^i, 1], i \in \mathcal{S}, \tag{7}$$

$$\sum_{j \in J_i} f^{*ij}(b^{ij} - \pi x^{ij}) \geq \sum_{j \in J_i} f^{ij}(b^{ij} - \pi x^{ij}) \text{ for all } f^i, \delta^i \text{ such that} \tag{8}$$

$$\sum_{j \in J_i} \delta^{ij} \leq 1, f^{ij} \in [\delta^{ij} F^{ij}, \delta^{ij}], \delta^{ij} \in \{0, 1\}, i \in \mathcal{O},$$

$$f^{*i} b^i - \pi x^{*i} \geq f^i b^i - \pi x^i \text{ for all } (f^i, x^i) \text{ such that} \tag{9}$$

$$f^i \in [F^i, 1], A^i_t z^i_t \geq f^i e^i_t, x^i + w^i = \sum_t z^i_t, x^i + w^i \geq 0, \text{ for all } i, \quad i \in \mathcal{Z},$$

$$\pi \cdot \left[\sum_{i \in \mathcal{S}} x^i f^{*i} + \sum_{i \in \mathcal{Z}} x^{*i} + \sum_{i \in \mathcal{O}} \sum_{j \in J_i} x^{ij} f^{*ij} \right] = 0, \tag{10}$$

$$\pi_k \geq 0 \text{ for all } k. \tag{11}$$

Equation (7)–(9) are the regret-free conditions on bidders. They require that *ex post* each bid is allocated in a way that maximizes the bid's surplus at those prices

---

[11] x^* is relevant for characteristic bids. δ^* is relevant for OR bids. f^* is relevant for all types of bids.

subject to the bid's restrictions. Equation (10) is Walras' Law, which requires that $\pi_k = 0$ when there is an excess supply of k. These conditions, along with *no excess demand* (6) from the winners determination problem, imply that the winners determination allocation and the prices π are a competitive equilibrium for the economy described by the bids.

There are two possible problems at this point. Competitive equilibrium prices may not exist. And, even if they do exist, they may not be unique.

Non-uniqueness. If (2) is convex, prices satisfying (7)–(11) exist.[12] The dual variables of the linear-programming problem will serve as these prices. But they may not be unique. There are two reasons.

One, in a thin market when a package matches another package, the range of prices satisfying no-regret can be wide. Consider the following example.

Example 5 (Opposing Swaps). There are two AND bids:

Bid	$	Amount of A	Amount of B
1	3	1	−1
2	−3	−1	1

Bid 1 indicates, for example, that some one is willing to pay 3 to swap 1 unit of B for 1 unit of A. In this case, p_A and p_B are not separately identified and all we can conclude is that $p_A - p_B = 3$. This is not a problem, since no bidder cares which particular price vector is selected from those.

Two, the bounds in Definition 6 may not be tight. This occurs for example if there is no marginal winner.

Example 6 (Single Asset – No Marginal Winner).

Bid	$	Amounts	$/unit
1	500	500	1
2	−400	−500	0.8

For this example, any price between $0.80 and $1 will be a competitive equilibrium price. This is a problem, because the price selected determines the distribution of the surplus among buyers and sellers.

To deal with non-uniqueness in prices, ACE chooses the prices that maximize the winning bidders' minimum per-unit surplus. This is the intuitive equivalent to

---

[12] Since (2) maximizes surplus, and "preferences" are quasi-linear, the winners allocation is Pareto-optimal. We can therefore apply the second welfare theorem to establish the existence of prices supporting that allocation.

Myerson–Satterthwaite's (1983) k-double auction when $k = 1/2$. It equalizes the per-unit surplus between buyers and sellers on the margin. In Example 6, ACE will set the price to $0.9.

Non-existence. Non-existence is way more serious than non-uniqueness. The prime cause of the non-existence of equilibrium prices is inflexibility due to a minimum fill requirement. In this section, we show how the inflexibility causes the non-existence through examples and discuss how it is resolved. For simplicity, the examples are set in a single asset market.

Example 7 (Excess Supply at Winners Determination). In this example, it is possible to find a regret-free price at the winners determination allocation but payments will not balance.

Bid	$	Amount	Min. scale(%)	$/unit
1	2500	2000	0	1.25
2	500	500	0	1.00
3	−1500	−3000	100	0.50

If all orders were flexible, a competitive equilibrium would exist at the price of $0.5 per unit and 500 units of order 3 would not trade. Because of the sell order's inflexibility, the market must absorbs 500 units of excess supply. But, because of this imbalance between the amounts bought and the amounts sold, no single price will allow balance of all payments and charges. With inflexibility someone must pay for this extra 500 units, or no trade will occur.

In Example 7, if the extra 500 units were not taken by the market maker, no trade would occur. So it is not unreasonable to have bids 1 and 2 pay for it. A natural way to do that, while preserving some measure of incentive compatibility, is to charge buyers a higher price than is paid to sellers. To implement that, we would charge each buy unit at $0.8181 and pay each sell unit at $0.68182. But this rewards the inflexibility of bid 3. If they were perfectly flexible, they would gain a surplus of 0 and get only 2500 units. If they are inflexible and we used this pricing scheme, they would get $3000 \cdot \$0.18182 = \$545.476$. An alternative would be to pay B3 at what she bid, $0.50/unit, and charge the buyers $1.20. But this punishes not only the inflexible part of bid 1 but also the flexible part. A better alternative is to pay the inflexible part (500 units) at $0.50 and then split the prices in a way that equalizes the per-unit surplus at the margin. In Example 7, this would mean each buyer would be charged $0.80 and the seller would be paid $0.70 for the 2500 flexible units and $0.50 for the 500 inflexible units. ACE does something similar to this.

Example 8 (Excess Supply from AND Bid). In this example, we illustrate another way in which a single price can be found to satisfy no-regret but will not balance payments. There are no minimum fill requirements here.

Bid	$	Amount of A	Amount of B
1	−100	50	−30
2	300	25	10
3	−100	−75	
4	100		10

All these bids will win in the winners determination problem. But there will be an excess supply of 10 units of of B. It is easy to find prices that satisfy no-regret for all bids. For example, $p_A = 2$, $p_B = 9$ will do the job. But because there is an excess supply, the market will absorb the 10 units and, as in Example 7, someone will have to pay the $9 \cdot 10 = \$90$ to Bid 1.

ACE will deal with this by splitting the buy and sell prices and then choosing them so as to equalize the minimum per unit surplus across orders.

Example 9 (Incompatible Surplus Requirements). In this example it is not even possible to find a regret-free price at the winners determination allocation.

Order	$	Amount	Min. scale(%)	$/unit
1	2000	2000	100	1.00
2	−425	−500	0	0.85
3	−980	−1000	0	0.98
4	−525	−500	0	1.05

Sell order 4 is allocated because buy order 1 is an all-or-none order and there is enough surplus from the part of trade made among orders 1, 2, and 3 to compensate for the surplus loss resulting from matching a part of order 1's demand and sell order 4. Without order 4 no trades will occur so 4 is included in the winners determination allocation.

Since sell order 4 asks more per unit than buy order 1 bids per unit, no single price can satisfy no-regret for both 4 and 1. For allocations in these examples, individual prices will have to be charged to some bids in order to satisfy no-regret. One option is to pay 4 what they bid, $525, and then split the buy and sell prices for the others to pay that. Since order 1 is inflexible, they would pay $1 per unit (as in Example 7 we don't want to reward inflexibility) while orders 2 and 3 would be paid $0.9833 per unit.

3.3.2. The ACE Pricing Algorithm

As we found out in the previous section, allowing bids that are able to express bidders' preferences opens up the possibility that competitive equilibrium prices will not exist. So one must settle for something less. In this section, we describe how ACE handles this.

We want prices that (a) provide useful signals to the market that reflect aggregate demand and supply, (b) do not extract or inject any revenue into the process, (c) achieve some measure of incentive compatibility, and (d) do not reward inflexible bidders. ACE accomplishes (a) by finding prices that are "as close as possible" to competitive equilibrium prices, accomplishes (b) by requiring payments and receipts to add up to zero, accomplishes (c) by ignoring losing bids[13] and choosing prices between all marginal winning bids (a double auction approach), and accomplishes (d) by paying inflexible bids at exactly what they bid.

ACE first identifies those inflexible bids or parts of those bids that prevent regret free prices from existing, as for example bid 4 in Example 9. Those inflexible bids are set aside and will be paid or pay at what they bid. There is now a price that is regret-free for the remaining bids. Unfortunately, payments may not balance at that price, as happens in Example 7. So ACE then splits the prices into a price vector for buyers, p^b, and a price vector for sellers, p^s, and searches for a pair that (i) satisfy no-regret for the remaining bids (or parts of bids), (ii) maximize the minimum per unit surplus for all of the remaining bids, and (iii) balance payments and receipts of all bids (including those set aside).

Step 1: Ignore losing bids and the losing parts of OR bids.
Begin with the allocation (f^*, y^*) from the winners determination problem (2). Let $\bar{I} \subset I$ be the set of allocated orders, (those $i \in \mathcal{S} \cup \mathcal{Z}$ with $f^{*i} > 0$ and those $i \in \mathcal{O}$ with $f^{*ij} > 0$). Let $\overline{\mathcal{S}} = \mathcal{S} \cap \bar{I}, \overline{\mathcal{Z}} = \mathcal{Z} \cap \bar{I}$, let $\overline{\mathcal{O}} = \mathcal{O} \cap \bar{I}$. For $i \in \mathcal{S} \cup \mathcal{Z}, \overline{B}^i = f^{*i}b^i$. For $i \in \mathcal{O}, \overline{B}^i = b^i \sum_{j \in J_i} f^{*ij} b^{ij}$. For $i \in \mathcal{S}, \overline{X}^i = f^{*i}x^i$. For $i \in \mathcal{O}, \overline{X}^i = \sum_{j \in J_i} f^{*ij} x^{ij}$. For $i \in \mathcal{Z}, \overline{X}^i = y^{*i}$, and $\overline{e}^i = f^{*i} e^i$.

Step 2: Determine which units are extra-marginal.
To determine those parts of the winning bids that are inflexible, ACE solves the following *fully flexible winners determination* problem:

$$\max_{(g,x)} \quad \sum_{i \in \bar{I}} \overline{B}^i g^i$$

$$\text{subject to} \quad |x^i| \leq |\overline{X}^i|, i \in \overline{\mathcal{Z}}$$

$$A^i_t z^i_t \geq g^i \overline{e}^i, x^i + w^i = \sum_t z^i_t, x^i + w^i \geq 0, i \in \overline{\mathcal{O}}$$

$$\sum_{i \in \overline{\mathcal{S}} \cup \overline{\mathcal{O}}} g^i \overline{X}^i_k + \sum_{i \in \overline{\mathcal{Z}}} x^i_k \leq 0, k \in K \tag{12}$$

$$0 \leq g^i \leq 1, i \in \bar{I}$$

Let $(\bar{g}, \bar{y})$ be the solution of (12). The inflexible part of bid i is $(1 - \bar{g}^i)\overline{X}^i$ for $i \in \overline{\mathcal{S}} \cup \overline{\mathcal{O}}$, and is $\overline{X}^i - \bar{y}^i$ for $i \in \overline{\mathcal{Z}}$.[14] ACE sets this part of these orders aside. They will be filled and they will pay $(1 - \bar{g}^i)\overline{B}^i$. Prices on the rest of the orders, called the fully

---

[13] In a thick market, losing bids can be used to drive incentive compatibility by leading winning bids to reveal their true values. In thin markets, losing bids can be used to manipulate prices and detract from incentive compatibility. Since ACE is designed for thin markets (it is not needed in thick markets), ACE ignores the losing bids from the winners determination problem (2) when determining prices. ACE ignores both losing simple bids (those with $f^i = 0$) and the losing parts of OR bids (those with $f^{ij} = 0$).

[14] A characteristic bid may not be itself inflexible but a part of it may be matching an inflexible part of another bid.

flexible orders, will have to "pay" for $D = \sum_{i \in \bar{I}}(1-\bar{g}^i)\bar{B}^i$. This usually requires that prices be split into buy and sell prices.

Step 3: Find a vector of buy prices and a vector of sell prices that is regret-free for as many of the fully flexible orders as possible, that balances all payments, and maximizes the minimum surplus per order.

Let $\bar{G} := \{i \in \bar{I} | \bar{g}^i > 0\}$ and $D := \sum_{i \in \bar{I}}(1-\bar{g}^i)\bar{B}^i$. $\bar{G}$ is the set of bids with a fully-flexible component. D is the contribution to the total payment from the units paying at their bid/ask prices.

For any vector y, let $y^+ := (\max\{0, y_1\}, \max\{0, y_2\}, \ldots, \max\{0, y_K\})$ and $y^- := (\min\{0, y_1\}, \min\{0, y_2\}, \ldots, \min\{0, y_K\})$. $\bar{X}_k^{i+}$ is the amount of asset k potentially bought at the market (buy) price, and $|\bar{X}_k^{i-}|$ is the amount of asset k potentially sold at the market (sell) price.

Solve the following pricing problem:

$$\max_{(m, p^b, p^s)} m, \tag{13}$$

subject to

$$p^b \bar{X}^{i+} + p^s \bar{X}^{i-} + m^i \sum_{k \in K} |\bar{X}_k^i| = \bar{B}^i, \, i \in \bar{G}, \tag{14}$$

$$p^b \bar{g}^i \bar{X}^{i+} + p^s \bar{g}^i \bar{X}^{i-} \leq p^b y^{i+} + p^s y^{i-}, \text{ for all } y^i \text{ such that}$$
$$A_t^i z_t^i \geq g^i \bar{e}_t^i, \, y^i + w^i = \sum_t z_t^i \geq 0, \, i \in \bar{G} \cap Z, \tag{15}$$

$$\sum_{i \in \bar{G} \setminus (\bar{G} \cap Z)} (p^b \bar{g}^i \bar{X}^{i+} + p^s \bar{g}^i \bar{X}^{i-}) + \sum_{i \in \bar{G} \cap Z} (p^b x^{i+} + p^s x^{i-}) + D = 0, \tag{16}$$

$$m^i \geq m, \, i \in \bar{G},$$

$$p_k^b \geq p_k^s \geq 0, \, k \in K.$$

In (14), m^i is the per-unit surplus that this bid will receive if the prices are (p^b, p^s). To be precise, each of the terms in (14) should be multiplied by $\bar{g}^i$, but, since they all just cancel, we are leaving them out. The regret-free condition for all bids is contained in (14) when $m^i \geq 0$. Problem (15) is the regret-free condition for $i \in Z$. It should be noted that the complete statement is $\bar{g}^i \bar{B}^i - \bar{g}^i (p^b, p^s) \cdot (\bar{X}^{i+}, \bar{X}^{i-}) \geq \bar{g}^i \bar{B}^i - (p^b, p^s) \cdot (y^{i+}, y^{i-})$. But the $\bar{g}^i \bar{B}^i$ cancels out. Finally, (16) is the requirement that all payments and receipts balance.

Remark 6. Note that (15), as well as (19) below, really involves an infinite number of constraints that define a convex set. As pointed out in Remark 5, for the RECLAIM implementation we were able to convert (15) to a finite set of constraints on the prices. Each application requires its own conversion. We do not have a general approach to accomplishing that.

If there is no inflexibility, then $D = 0$, and $p^b = p^s$ in the solution to (13). That is, a competitive equilibrium price will be found.

Step 4: *Check to see if appropriate prices have been found. If not, set aside more bids and repeat.*

(A) If the value of the objective function of (13) is nonnegative, ACE is done with this phase. ACE then enters into a clearance and settlement phase, determining what is allocated to each bid and how much they pay or are paid. We describe that process in Section 3.3.3.

(B) If the value of the objective function of (13) is negative, then an appropriate market price system does not exist because there is not enough surplus at the margin to pay for the inflexible parts of the winning bids.[15] In this case, ACE chooses some more units/orders to set aside and to pay at their bid/ask prices and then recomputes. This iterates until ACE finds an appropriate market price system for those bids that have not been set aside.

After the t-th iteration, there are two sets of orders. G_t is the set of the orders which have been set aside and which will pay at their bid. An order in G_t is (B^i, Y^i). H_t is the set of orders, or parts of orders, which have not yet been set aside. There are two types of orders in H_t: simple orders and cycle-zone free orders (B^i, Y^i) and characteristic orders (B^i, A^i, E^i, w^i). H_0 is set to be $\overline{G}$.

At the $t + 1$-st iteration, we solve

$$\max_{(m, p^b, p^s)} m,$$

subject to

$$p^b Y^{i+} + p^s Y^{i-} + m^i \sum_{k \in K} |Y_k^i| = B^i, i \in H_t, \tag{17}$$

$$p^b Y^{i+} + p^s Y^{i-} = B^i, i \in G_t, \tag{18}$$

$$p^b Y^{i+} + p^s Y^{i-} \geq p^b y^{i+} + p^s y^{i-}, \text{ for all } y^i \text{ such that}$$
$$A_t^i z_t^i \geq g^i \bar{e}_t^i, y^i + w^i = \sum_t z_t^i \geq 0, i \in H_t \cap Z, \tag{19}$$

$$\sum_{i \in H_t \setminus (H_t \cap Z)} (p^b Y^{i+} + p^s Y^{i-}) + \sum_{i \in H_t \cap Z} (p^b y^{i+} + p^s y^{i-}) + \sum_{i \in G_t} B^i = 0, \tag{20}$$

$$m^i \geq m, i \in H_t,$$

$$p_k^b \geq p_k^s \geq 0, k \in K.$$

If the value of the objective function is nonnegative, then ACE is done with the pricing computation and goes to clearance and settlement (see Section 3.3.3). If the value of the objective function is negative, then the orders that attain the minimum per unit surplus among the orders in H_t are now removed from H_t and added to G_t to get H_{t+1} and G_{t+1}.[16]

---

[15] This would be the case, for example, if bid 2 in Example 7 were $275 for 500 units or $0.55 per unit.

[16] We perform an extra step to determine the orders that are *truly* attaining the minimum per unit surplus. The step is necessary because of the non-uniqueness of prices; at some prices more orders can attain the minimum surplus than at other prices.

If H_{t+1} is empty, the surplus distribution is determined and ACE goes to clearance and settlement.[17] Otherwise ACE repeats Step 4.

3.3.3. Payments

When the ACE pricing algorithm is finished, there are two sets of orders, G_t and H_t, and two market price vectors, p^b and p^s. The orders in G_t will pay what they bid. The orders in H_t will pay at the market prices.

3.3.4. A Computational Note

The pricing algorithm took the split of the markets[18] further after removing losing bids (losing bids can overlap with winning ones and the removal can allow further division). This occurred between Steps 1 and 2. This was done to limit the number of bids that have to pay the deficit from the "pay-as-you-bid" portion of allocation and as a consequence of the decision not to impose arbitrage-free conditions on losing bids.

4. The Performance of ACE in RECLAIM

Although there have been many papers assessing the performance of RECLAIM,[19] ACE has been mostly invisible[20] because those assessments were aimed at the performance of RECLAIM and not at the performance of the markets themselves.

In this section, we will examine the performance of the ACE market in the RECLAIM program for the period from 1996 to 2000.[21] The ACE data explored in

---

Let W_t be the value of the objective function of LP (17). Let $H_t^{\min}$ be the set of orders whose per unit surplus w^js are found equal to W_t. Then solve the following LP for each j in $H_t^{\min}$:

$$\max \quad w^j$$
$$\text{subject to} \quad w^i \geq W_t, i \in H_t^{\min},$$
$$\text{all the constraints in (17).}$$

If the value of the objective function of the above is W_t, order j is considered attaining the minimum per unit surplus at the t-th iteration.

[17] There still is a chance that the price is not uniquely determined after the surplus distribution is determined because bids are as in Example 5. In such a case we choose prices that support the surplus distribution and are averaged. Suppose a buy order of $1000 for 500 units each of items A and B matches a sell order of $900 for 500 units each of items A and B. The per unit surplus of each order is $0.5. This surplus distribution can be achieved, for example, ($p_A = \$19$, $p_B = \$0$). We choose ($p_A = p_B = \$9.5$). This "averaging" is often consistent to the reference price information made available to traders in early round when there was no trade because asking prices by sellers are higher than bid prices by buyers. Suppose the buyer's bid is $900 and the seller's ask is $1000 in the example in this footnote. If there are no other orders involving assets A and B in the market, $0.9 is returned as the highest average bid price for both assets A and B and $1.0 is returned as the lowest average ask price for A and B.
[18] See Remark 3.
[19] See, for example, Fowlie et al. (2012), Fromm and Hansjurgens (1996), McCann (1996), and Thompson (2000).
[20] The exceptions are Klier et al. (1997), written before the first ACE market, and Fine (2001).
[21] This section draws heavily on Chapter 4 of Fine (2001) who had access to the ACE data from Net Exchange.

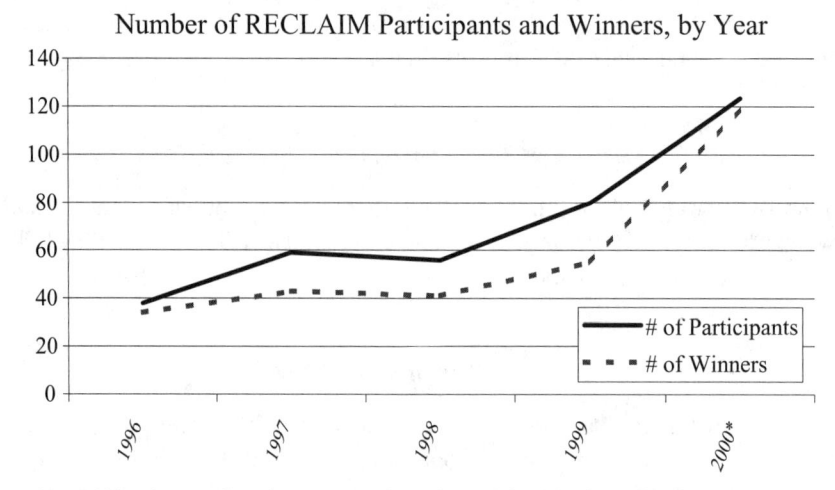

Number of RECLAIM Participants and Winners, by Year

Note: 2000 values are based on January 2000 data, divided by the proportion of participants and winners represented relative to the entire year in 1997, 1998, and 1999.

Figure 1. Participation in the ACE market, by session and year.

this section cover the markets from April 1996 to January of 2000.[22] There are two main reasons for choosing this time period. First, this is the period covered in Fine (2001). Second, it is the historically relevant period. RECLAIM began in late 1993. In the beginning, there were several avenues for trading but the main ones were a bulletin board run by the SCAQMD and semi-annual auctions run by the brokerage firm Cantor-Fitzgerald. ACE began in April 1996. In 2001, the California electricity market crisis caused serious disruptions in the operations of RECLAIM and by 2000 ACE had been sold by Net Exchange to Aeon. At that point the ACE data become suspect.[23]

To assess the performance of ACE in RECLAIM, we will examine the participation, bidding and pricing behavior exhibited by the participants in ACE market.

4.1. Participation

Over time more and more of the members of RECLAIM universe used ACE markets. In Figure 1, we show the number of bidders and winners participating in ACE markets from 1996 to 2000. Even though the number of RECLAIM facilities did not dramatically changed over this time period, the number of bidders and winners in the ACE markets increased yearly. In 1996 only 10% of RECLAIM members conducted trades through ACE. By 1999, 20% did. By 2000, the number increased to about 30%.

Even more dramatically, by 2000 the ACE market had accounted for approximately 90% of all trades-for-a-price[24] in the RECLAIM market. In 1999, RECLAIM recorded

---

[22] Markets were held at-least quarterly, with some years having five or six markets. Specifically, the data are from the following auctions: April 1996, July 1996, August 1996, October 1996, February 1997, April 1997, July 1997, October 1997, January 1998, April 1998, July 1998, October 1998, January 1999, April 1999, July 1999, August 1999, October 1999, and January 2000.

[23] For more on this, see Section 4.4 below.

[24] We are only interested in trades for a price, since zero price trades are either inter-facility within a firm or transfers to or from a permit broker, and are therefore not truly part of the competitive market.

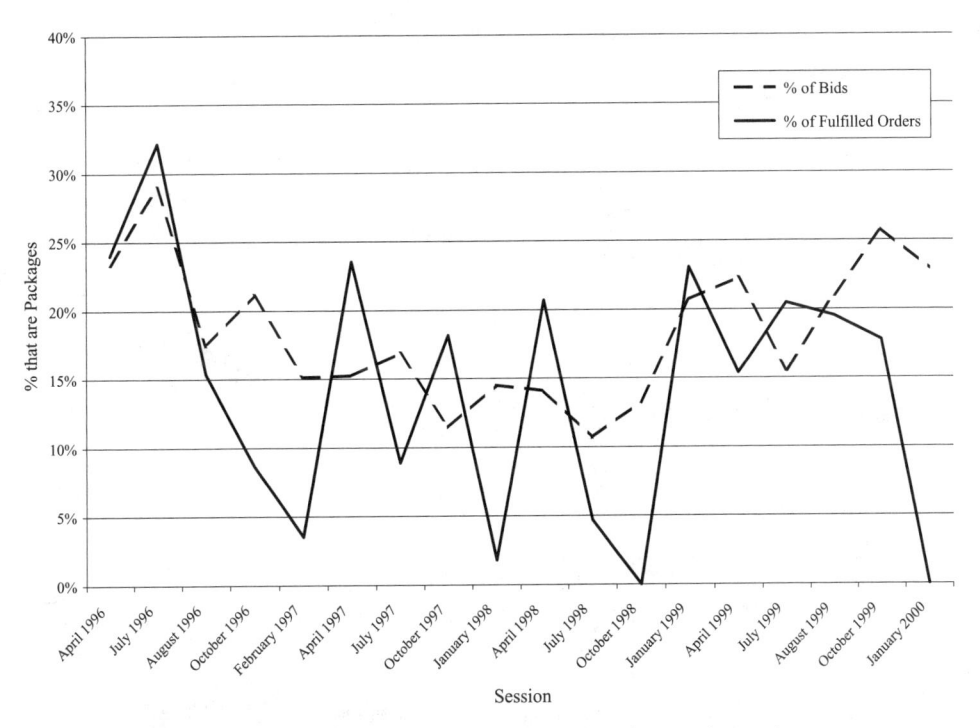

Figure 2. Percentage of bids and fulfilled orders that are for packages.

239 trades-for-a-price (219 in the NOx market and 20 in the SOx market). ACE conducted 213 of these trades, 208 of which were for NOx permits. That means 95% of all serious NOx trades went through the ACE markets.

As the RECLAIM program progressed, environmental engineers learned that the non-ACE trading mechanisms involved extreme transactions costs relative to the ease and efficiency of the ACE market. They were choosing the more efficient option.

4.2. Bidding Patterns

In each ACE market session, there are over 100 assets available for purchase. There is a significant stream of futures available to a RECLAIM trader. The ACE market is a powerful institution, allowing for bids that describe very complex preferences. Are the bidders using these options, or are the complex bidding features of the ACE mechanism simply window-trimming? In Figure 2 we display the percentages of package bids submitted and transacted. 18% of the bids submitted and 14% of the transactions are for packages. This seems to be a relatively small use of the combinatorial capability of the ACE algorithm. But taking a slightly different look at the data gives a deeper insight.

If we consider the RTCs as different assets because the date of effectiveness is different, we have a very thin data set from which it is difficult to perceive bidding and pricing patterns. So instead, we consider the RTCs not as dated objects (in terms of their year of validity) but as a stream of future contracts dated relative to the current

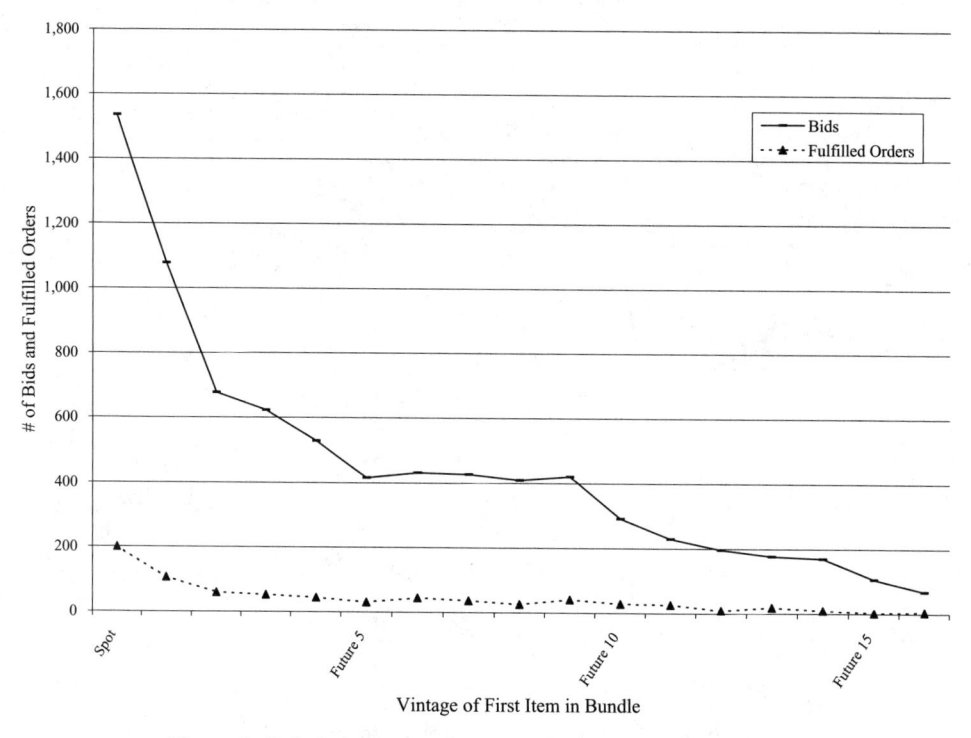

Figure 3. Relative vintage of assets in bids and fulfilled orders.

market. In Figure 3, we show the number of bids and fulfilled orders in the market as a function of the vintage of the earliest item in the bid.[25]

Once we make this adjustment and consider the permits in terms of their relative vintage, a pattern emerges. The number of bids entered is inversely proportional to the time horizon. Further, looking at Figure 4, it can be seen that there are really two markets: a spot or short-term and a planning or long-term market. The bidding patterns in the two markets are in stark contrast to each other. In the spot market (the earliest vintage currently available), approximately 91% of the bids are for a single asset. Additionally, there are no swappers at all in the spot market. That is, of those 9% of bids that are for more than one asset, the bids are either pure buys or pure sells. This is not unexpected. The spot market is a time for rebalancing planned with actual emissions. This can be easily done with single asset bids.

In the longer-term market the bidding is far more frequently for packages. 32% of bids are combined value (for the 72 permits more than 3 years in the future). This is also not unexpected. The futures market is a place to evaluate trade-offs between investment in abatement equipment and RTC acquisition. As we showed in Section 1.2, this involves swaps and packages. The frequency of package bidding in the long-term market is quite similar to that observed in the experiments discussed in Bossaerts et al. (2002). There it was shown that 20–30% package bidding was more than enough to

---

[25] We only consider the spot and futures markets up to 17 permits in the future (for a total of 9 years of forward contracts in any given trading session).

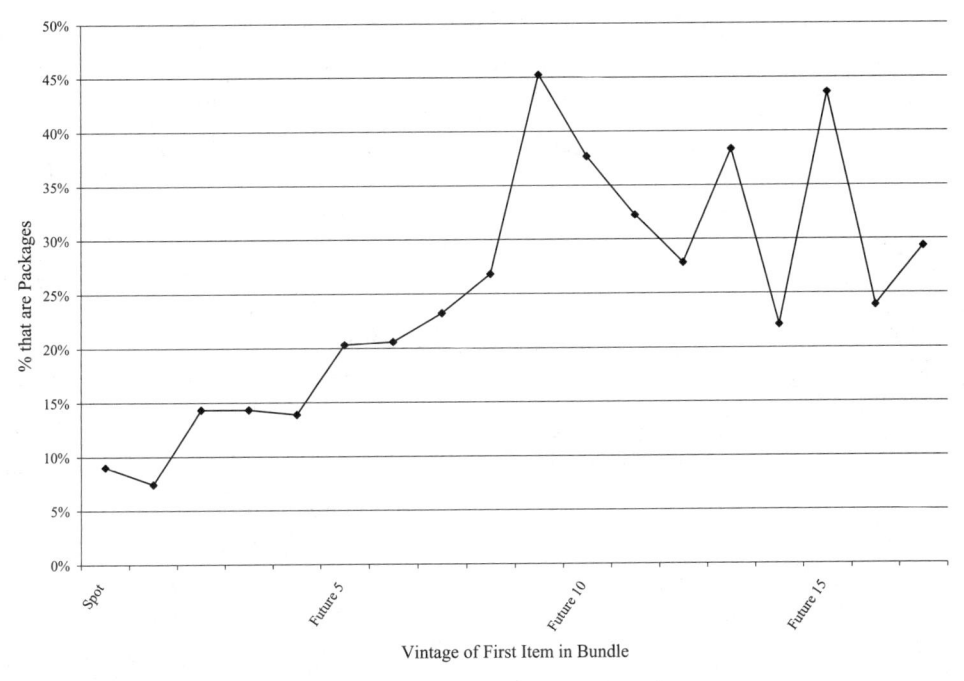

Figure 4. Percent of bids that are packages, by relative vintage.

create the liquidity necessary to allow efficient trading when there are strong complementarities as there are in RECLAIM.

4.3. Pricing Patterns

It is important to look separately at two different time trends: how prices move across ACE markets and how prices move across permit vintages.

In Figure 5, we detail the evolution of the NOx and SOx spot ACE price across ACE markets from April 1996 to January 2000. There is almost no trade in any of the markets except for NOx Zone 1, so henceforth we will use the data from this asset to demonstrate the patterns that emerge. The initial supply of RTCs was in excess of the reported emissions at the market's inception in 1994. The supply of RTCs was then reduced each year and in 1997–98 dipped below the total 1994 emissions level. It is at this point, that prices began an uphill climb.[26] From July 1997, the spot market price strictly increased over time. This is because facilities looked down the road, recognized the impending shortage of RTCs, and began rolling permits forward. In late 1999, the price of RTCs dramatically increases shortly after the expected cross of RTC

---

[26] Why do we observe non-zero prices in the spot market? While the number of RTCs available across both cycles has historically been in excess of reported emissions, this does not mean that in any given market there is an excess supply. Firms were either assigned to cycle 1 or cycle 2, and then allocated RTCs. Therefore, it is quite possible that there may not be enough permits of a given cycle to cover that entire year's emissions without benefit of permits from another cycle.

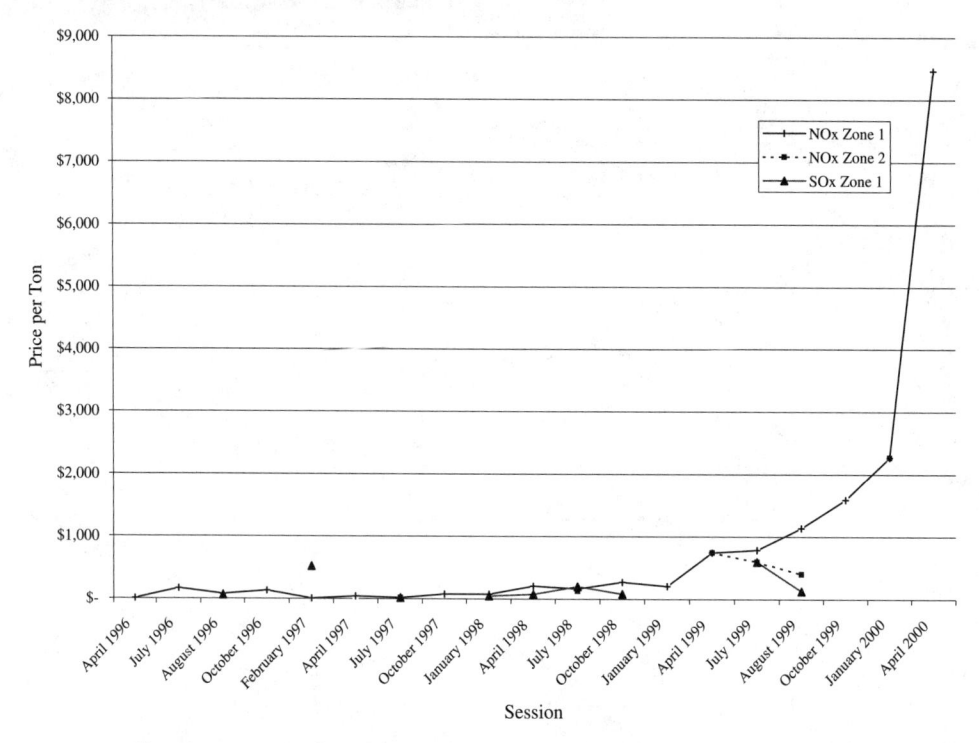

Figure 5. Spot market pricing in ACE markets, April 1996 to April 2000.

availability and reported emissions. However, even this highest price falls quite short of the AQMD's anticipated price of $11,257 per ton.[27]

In fact, prices have stayed quite low. Assuming these prices accurately reflect the marginal abatement costs, it appears that the planning flexibility offered by the RECLAIM program resulted in lower than expected marginal control costs, perhaps from the shift to a facility-wide performance standard.[28] Additionally, prior to the start of RECLAIM, facilities had incentive to misrepresent their emissions and true costs of abatement. Indeed the incentive historically has been to overstate the control costs during public hearings in order to deflect proposed command-and-control type regulations. This strategic reporting further accounts for the gap between predicted and actual RTC prices.

The April 2000 data indicate that prices were now approaching the predicted levels. The marginal cost of Best Available Control Technology (or BACT) for NOx was believed to be in the range of $3.50 to $4.50 per pound. The spot market price for permits in the April 2000 market was $4.23. In 2000, the RECLAIM program was just reaching the market transition point that Johnson and Pekelney (1996) believed would occur.

---

[27] Johnson and Pekelney (1996) built the Emissions Trading Model (ETM) to assess the potential economics and environmental impacts of RECLAIM's emissions trading program. It estimated trades that were likely to occur under the program and linked them to a general equilibrium model of the regional economy.

[28] See Bohi and Butraw (1997) for an excellent discussion of the impacts on control costs.

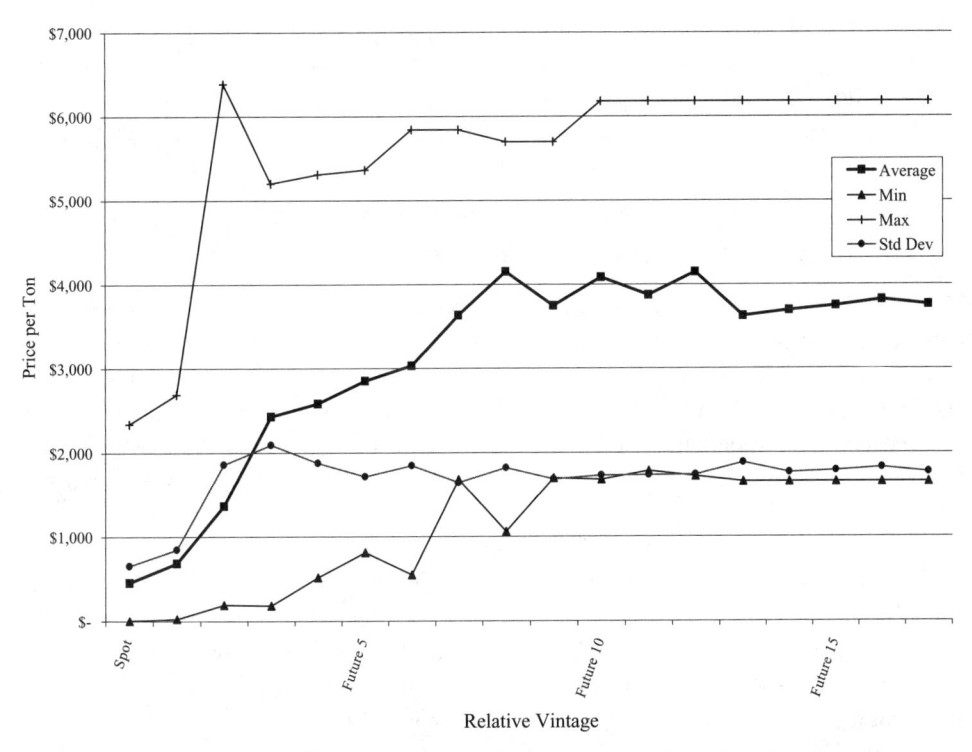

Figure 6. Price of NOx Zone 1 RTCs, as a function of relative vintage.

Now let us turn to how prices change across vintages. Again we can see differences between the short and long-term markets. As Figure 6 shows, from the spot market to about seven permits or 3.5 years into the future, the price of permits increases. Once we look to 4 years in the future and beyond, the pricing is remarkably stable. The combined-value mechanism, including its pricing algorithm, provides the liquidity necessary for stable, meaningful prices in the long-term market. This allows the long-term market to exhibit the stable pricing that many economists feared impossible.[29]

4.4. But, Who Will Guard the Guardians?

We cannot leave off our evaluation of ACE-RECLAIM without acknowledging a serious, practical problem that market designers generally just wave their hands at.[30] How can we keep the operators of the algorithms honest? The post-2000 history of ACE-RECLAIM illustrates the problem. In 1999/2000, Net Exchange, who had developed and run the ACE algorithms, sold ACE-RECLAIM to Aeon, the company that had handled most of the back-end details of the auctions, such as marketing, escrowing, etc. Between 2000 and 2004, something went very wrong. In June of 2004, Anne Sholtz,

---

[29] See Hausker (1992).
[30] One exception can be found in the Nobel Lecture of Leonid Hurwicz, "But, Who Will Guard the Guardians?" see Hurwicz (2007). We have adopted his title for this section.

the president of Aeon was arrested for being in a "scheme to defraud" numerous companies.[31] This included trading in front of the bidders and selling RTCs that had no counter side. She was sentenced in 2008.

The algorithms were fine, trustworthy, and reliable; the operator was not. How to design to protect against such misbehavior remains an open research question.

4.5. Conclusion

Karl Hausker (1992) eloquently voiced the concerns of many economists of the time about tradable permit programs such as RECLAIM. He was concerned that the long-term market would suffer from extreme thinness due to uncertainty, transactions costs, and other sources of market inefficiency. Although the long-term market does seem to have been thinly traded, the combined-value market mechanism used in the ACE market overcame this illiquidity, as predicted in Bossaerts et al. (2002). The ACE market mechanism became, over the first four years of the RECLAIM emissions credit trading program, the market venue of choice for polluting facilities.

The short-term and long-term markets exhibited dramatically different bidding and pricing patterns. The short-term trader places single-asset bids, making one-time adjustments to her predicted emissions levels. For the long-term planner, the combined value market provided clear, stable pricing and the ability to plan a pollution stream without the risk of attaining only part of that stream.

Since a single-asset bid is simply a degenerate form of a combined-value bid, imposing the combined-value structure on the short-term market certainly did no harm. Indeed, the additional liquidity provided by those few traders who traded in bundles including both short-term and long-term RTCs provided an important bridge between the two markets, improving liquidity in both.

References

Bohi, D. and Butraw, D. (1997) "SO_2 Allowance Trading: How do Expectations and Experience Measure Up," *The Electricity Journal*, 10(7): 67–75.

Bossaerts, P., Fine, L., and Ledyard, J. (2002) "Inducing Liquidity In Thin Financial Markets Through Combined-Value Trading Mechanisms," *European Economics Review*, 46(9): 1671–1695.

Bykowsky, M., Cull, R., and Ledyard, J. (2000) "Mutually Destructive Bidding: The FCC Auction Design Problem," *Journal of Regulatory Economics*, 17(3): 205–228.

Carlson, D., Forman, C., Ledyard, J., Olmstead, N., Plott, C., Porter, D., and Sholtz, A. (1993) *An Analysis and Recommendation for the Terms of the RECLAIM Trading Credit*, Pacific Stock Exchange Technical Report, April 27, 1993.

Cason, T. and Gangadharan, L. (1998) "An Experimental Study of Electronic Bulletin Board Trading for Emission Permits," *Journal of Regulatory Economics*, 14: 55–73.

Duffie, D. (2012) *Dark Markets: Asset Pricing and Information Transmission in Over-the-Counter Markets*, Princeton Lectures in Finance.

Fine, L. (2001) "Cooperative and Market-based Solutions to Pollution Abatement Problems," PhD Dissertation, California Institute of Technology, Pasadena, CA, USA.

---

[31] See *LA Times* (2004) and *Pasadena Weekly* (2010).

Fowlie, M., Holland, S., and Mansur, E. T. (2012) "What do Emissions Markets Deliver and to Whom? Evidence from Southern California's Nox Trading Program," *American Economic Review*, 102(2): 965–993.

Friedman, D. (1993) "How Trading Institutions Affect Financial Market Performance: Some Laboratory Evidence," *Economic Inquiry*, 31(3): 410–435.

Fromm, O. and Hansjurgens, B. (1996) "Emission Trading in Theory and Practice: An Analysis of RECLAIM in Southern California," *Environment and Planning*, 14: 367–384.

Gangadharan, L. (2000) "Transactions Costs in Pollution Markets: An Empirical Study," *Land Economics*, 76: 601–614.

Hausker, K. (1992) "The Politics and Economics of Auction Design in the Market for Sulfur Dioxide Pollution," *Journal of Public Analysis and Management*, 11: 553–572.

Hurwicz, L. (2007) "But, Who Will Guard the Guardians?," Nobel Lecture, www.nobelprize.org/nobel_prizes/economic-sciences/laureates/2007/hurwicz-lecture.html.

Ishikida, T., Ledyard, J., Olson, M., and Porter, D. (2001) "Experimental Testbedding of a Pollution Trading System: Southern California's *RECLAIM* Market," *Research in Experimental Economics*, 8: 185–220.

Johnson, S. L. and Pekelney, D. M. (1996) "Economics Assessment of the Regional Clean Air Incentives Market: A New Emissions Trading Program for Los Angeles," *Land Economics*, 72(3): 277–297.

Klier, T. H., Mattoon, R. H., and Prager, M. A. (1997) "A Mixed Bag: Assessment of Market Performance and Firm Trading Behavior in the NOx RECLAIM Programme," *Journal of Environmental Planning and Management*, 40(6): 741–774.

Los Angeles Times (2004) "Smog Credit Trader Held in Fraud Case," June 17, http://articles.latimes.com/2004/jun/17/local/me-smog17.

McCabe, K., Rassenti, S., and Smith, V. (1993) "Designing a Uniform-price Double-auction: An Experimental Evaluation," in: D. Friedman and J. Rust (eds.), *The Double Auction Market, SFI Studies in the Sciences of Complexity, Proc.* vol. XIV, 307–332, Addison-Wesley.

McCann, R. (1996) "Environmental Commodities Markets: 'Messy' versus 'Ideal' Worlds," *Contemporary Economic Policy*, 14: 85–97.

Myerson, R. and Satterthwaite, M. (1983) "Efficient Mechanism for Bilateral Trading," *Journal of Economic Theory*, 29: 265–281.

Pasadena Weekly (2009) www.pasadenaweekly.com/cms/story/detail/an_air_of_deceit/7616/.

SCAQMD (1993) *Annual Report*.

Smith, V. L., Williams, A. W., Bratton, W. K., and Vannoni, M. G. (1982) "Competitive Market Institutions: Double Auctions vs. Sealed bid-offer auctions," *American Economic Review*, 72: 58–77.

Thompson, D. (2000) "Political Obstacles to the Implementation of Emissions Markets: Lessons from RECLAIM," *Natural Resources Journal*, 40: 645–697.

Outlook

Martin Bichler and Jacob K. Goeree

Make things as simple as possible, but not simpler (Albert Einstein)

1. Introduction

From 1994 to 2008, spectrum was sold almost exclusively using the Simultaneous Multiple Round Auction (SMRA). The SMRA is based on simple rules, which make it easy to explain and implement, yet they create considerable *strategic complexity*. Since items have to be won one-by-one, bidders who compete aggressively for combinations of items risk paying too much if they ultimately win an inferior subset. This *exposure risk* suppresses bidding with adverse consequences for the auction's efficiency and revenue.

Since 2008, regulators worldwide have adopted the Combinatorial Clock Auction (CCA) to avoid exposure problems. The CCA is based on very complex rules, but the premise was that bidding would be "straightforward," i.e. bids would truthfully reflect valuations. Unfortunately, it is now well known that the CCA admits many other behaviors, including demand reduction, demand expansion, and predatory bidding (see Chapters 15–17). In particular, the CCA's supplementary stage may provide bidders with an opportunity to raise rivals' costs, which has led to some hard-to-defend outcomes.

In light of recent experiences with the CCA, regulators should be reassured about the advantages of combinatorial formats when synergies for adjacent geographic regions or contiguous blocks of spectrum are important. Market designers should beware of Einstein's advice and not regress to offering solutions that are too simplistic. Instead, they should take stock of two decades of field experience to pinpoint features essential to participating bidders and regulators. After all, spectrum auction design will only be truly successful if we are able to model their preferences correctly.

The standard paradigm in mechanism design assumes bidders with independent and private valuations, quasi-linear utility functions, and unlimited budgets, and regulators who aim to maximize efficiency or revenue of an auction in isolation, i.e. ignoring its effect on the downstream market. While these assumptions result in models that are elegant, they are not necessarily relevant. In what follows, we discuss objectives of regulators and bidders in spectrum auctions and how they differ from these "textbook" assumptions. These differences have an impact on the choice of the auction format. Furthermore, we discuss challenges for future auction designs and requirements for new models.

2. The Regulator's Perspective

Even in the idealized textbook environment where bidders' values are independent and private, the preferred choice of auction is non-obvious as bidders' values for combinations may be sub-additive or super-additive. The former possibility implies that bidders treat the different items as substitutes, in which case the SMRA is predicted to perform perfectly (Milgrom, 2004). The latter possibility turns out to be more realistic, however, which has stirred interest in combinatorial formats. These formats pose new design problems, as we discuss next.

2.1. Computational Complexity and Approximation

When bidders can express valuations for arbitrary combinations of items, optimal assignment becomes a computationally hard problem. Modern day optimization software typically allows for (near) optimal solutions, although very large auctions can still pose a problem. For some auctions, such as the US Incentive Auction (see Chapter 37), the regulator may not be able to guarantee full optimality and may need to approximate the welfare-maximizing allocation. This has led to fruitful research in computer science on *approximation mechanisms* that maintain truthfulness, but relax the goal of maximizing welfare (Nisan and Ronen, 2001). By now, worst-case bounds of approximation algorithms that satisfy strong game-theoretical equilibrium solution concepts are known for a number of problem types (Vazirani et al., 2007). Sometimes, these worst-case bounds can be low, however. It is important to go beyond general worst-case analyses by taking prior knowledge about specific markets into account.

2.2. Communication Complexity and Compact Bid Languages

Communication complexity is another fundamental problem. The term refers to the amount of information bidders need to communicate to be able to compute the efficient allocation. For some spectrum auctions, such as in the Canadian auction in 2014, there were around 100 licenses for sale. Bidders cannot possibly enumerate all packages ($\sim 2^{100}$ ignoring caps and floors) with the fully enumerative bid languages that have been used so far. Only a very small subset of all possible bids can be submitted and the vast majority will be missing. These "missing bids" are interpreted as expressing zero value by the winner-determination algorithm, causing inefficiencies and considerable randomness in allocations and prices.

Regulators need to be aware that higher expressiveness of the bid language does *not* necessarily lead to higher efficiency. Simplification has been introduced as a guiding principle in market design (Goeree and Holt, 2010; Milgrom, 2010) and the experimental results of Chapters 28 and 30 provide evidence that compact bid languages yield improved outcomes when there are many licenses. This result has largely been ignored in spectrum auction design in the field.

Compact bid languages leverage prior information about the structure of the bidders' preferences and elicit these with a small number of parameters. Examples are hierarchical package bidding (Goeree and Holt, 2010), which reduces the packages allowed in the auction to a hierarchy, or domain-specific languages as they are used in procurement auctions (Bichler et al., 2011; Goetzendorff et al., 2015).

2.3. Policy Goals and Allocation Constraints

Winner-determination algorithms pick the combination of bids that maximize seller revenue. The justification is that those with higher values can bid higher, so the revenue-maximizing allocation also maximizes the sum of bidders' values. In other words, by forcing bidders to "put their money where their mouths are," auctions result in efficient outcomes.

This argument, however, is persuasive only if the auction is being considered in isolation, i.e. without reference to the downstream market. Industry profits are highest when there is a monopoly, but it would be tenuous to praise the auction's efficiency when all spectrum is awarded to a single bidder. Indeed, regulators are more concerned with the well-functioning of the downstream market than revenue maximization in the auction *per se*. They need to strike a balance between incentives for investments and efficiency, and ensure enough competition in the end market to stimulate low consumer prices and quality of service.

To this end, regulators frequently use caps and set-aside licenses to avoid undesired allocations or to encourage participation by entrants. It is important that regulators are able to implement such policy decisions in the mechanism. While it is simple to consider *allocation constraints* in an optimization model computing the optimal allocation, such constraints have received little attention in the auction design literature, in particular with ascending auction designs (Petrakis et al., 2013).

3. Bidders' Preferences

Standard (spectrum) auction models assume that bidders' valuations are private and independent and that bidders have quasi-linear utility functions which aim to maximizing payoff. While these assumptions are convenient for doing theory, models based on these idealized assumptions may lead to wrong advice for both bidders and regulators.

3.1. Value Uncertainty, Value Interdependencies, and Value Endogeneity

Bidders spend substantial resources estimating the net present value of different spectrum packages prior to an auction. Such estimates, however, are highly uncertain. As a consequence, revenues in spectrum auctions are hard to predict. Even forecasts made just prior to an auction by investment banks tend to have high variance. For example, prior to the AWS auction in the US, analyst estimates of auction revenue ranged from $7 billion to $15 billion (see Chapter 31). Calculating the value of spectrum requires consideration of total market population, market penetration rates, market share, average revenue per unit, customer acquisition and activation costs, customer deactivations, etc. There are many other factors that make it hard to determine the value of spectrum (Korczyk, 2008). For example, the advent of media streaming and smart phones has probably led to a substantial change in valuations, compared to those that companies had twenty years ago. Such technical developments were probably not adequately considered in the valuations of the early spectrum auctions.

Note that value uncertainty can be specific to a bidder (Goeree and Offerman, 2003b), e.g. costs of roll out, or it can be common to all bidders, e.g. the adoption of new technologies such as media streaming. When both private and common value elements play role, auctions can no longer be fully efficient (Dasgupta and Maskin, 2000; Jehiel and Moldovanu, 2001; Goeree and Offerman, 2003a). The intuition is that a bidder with more optimistic expectations about the rate of technology adoption may outbid a more pessimistic rival with lower costs. Second-best mechanisms, i.e. those achieving the highest possible (albeit less-than-full) efficiency, have not sufficiently been explored despite their obvious importance for (spectrum) auction design.

Finally, it should be noted that the design of the assignment mechanism, auction or otherwise, will affect bidders' valuations. If bidders know the next award will done via an efficient auction, they may have more incentives to invest (to generate higher values) than if spectrum will be awarded by lottery. So the valuations are endogenous to the choice of assignment mechanism. Of course, if auctions entail large transfers from the private to the public sector, investment incentives may be suppressed. Regulators concerned with quality of service might well prefer less competitive mechanisms that leave more rents for the bidders. In any case, value endogeneity is yet another reason why regulators' objectives are more complex than simply "welfare maximization" or "revenue maximization" within the auction.

3.2. Allocative Externalities and Anonymous Pricing

Any spectrum auction is a unique event with important consequences for the competitive landscape that ensues afterward. The auction determines telecoms' positions in the aftermarket, and anticipating certain (dis)advantages associated with different outcomes, telecoms will adapt their bidding behavior in the auction accordingly (Goeree, 2003). For starters, telecoms will be interested in the entire allocation, not only in the packages they win and the prices they pay themselves. For example, the number of competitors and also their allocations can have a substantial impact on the revenues in the downstream market. End consumers pay a premium for the provider with the best network, and this is relative to rivals' spectrum holdings. In other words, the net present value of winning a set of spectrum licenses also depends on the allocation to competitors.

In the German spectrum auction in 2000, for instance, six bidders could have closed the auction if they all reduced demand to two units at a total auction price of 30 billion euros, but two bidders eventually drove up the revenue to 50 billion euros (and then gave up). This was described as an attempt to drive out another bidder from the downstream market, and it shows that externalities can be substantial. Bichler et al. (2017) discuss the impact of allocative externalities in the 2015 German spectrum auction and the role that the high transparency in the auction design played.

Being the "bandwidth leader" with the best connectivity can be a significant advantage in the end consumer market, and allocative externalities often play a role. However, the phenomenon has received relatively little attention in the literature (Jehiel and Moldovanu, 2005). The VCG mechanism would still determine the efficient allocation in dominant strategies, if bidders could express their preferences for all possible allocations. This, however, is unreasonable to assume in realistic markets due to the combinatorial explosion of possible allocations. Therefore, it is interesting to understand how bidders would bid in standard auction formats in the presence of allocative externalities, and how auction designs can address such externalities to avoid inefficiencies.

Telecoms care not only about what spectrum their rivals win but also how much they pay. The Vickrey-Clarke-Groves (VCG) mechanism and the two-stage CCA use non-anonymous payments, which can lead to undesirable outcomes. For example, in the 2012 Swiss spectrum auction that used the CCA, a small bidder (Sunrise) paid substantially more than a larger bidder (Swisscom) even though they won virtually the same amount of spectrum (see Chapter 17). In high-stakes spectrum auctions, payments are in the billions of dollars and a much higher payment in the auction can be a significant disadvantage in the downstream market. Predatory strategies to raise rivals' costs have been observed in a number of applications of the two-stage CCA and have also been analyzed theoretically (Bichler et al., 2013a; Janssen and Karamychev, 2017; Levin and Skrzypacz, 2016).

3.3. Principal–Agent Relationships and Budget Constraints

If financial markets were perfect, bidders would not face any budget constraints when acquiring valuable licenses. In reality, budget constraints are almost always an issue, which challenges the quasi-linear utility assumption usually made in mechanism design. Furthermore, private budget constraints defy strategy-proof mechanisms, even if bidders' values are private and independent and they aim to maximize payoff (Dobzinski et al., 2008).

Budget constraints are often a result of principal–agent relationships in bidding teams, with the management taking the role of the agent and the board of directors that of the principal. While the agent may have a good estimate of the value of a particular package, the principal does not. The principal wants to maximize profit, i.e. value minus cost, but the agent might prefer more valuable packages to less valuable ones, irrespective of cost. In other words, agents try to win their most preferred package as long as it fits within the budget, while the principal pays the bill (usually billions of dollars). Bichler and Paulsen (2015) explore environments where agents bid more aggressive than a principal would (if she possessed the agent's information), resulting in inefficient outcomes. Even though the principal controls the agent's budget, this may not be sufficient to incentivize the agent to bid optimally from the principal's viewpoint. In practice, the agent's hidden information makes the design of optimal contracts between principal and agent very difficult.

4. Discussion

Spectrum auction design has seen considerable progress, but the journey has just begun. Two decades of implementation in the field has made clear that regulators' objectives and bidders' preferences differ from standard "textbook" assumptions. These differences require us to rethink all aspects of the auction's design, i.e. the process, the bid language, and the payment rule.

Auction Process: Sealed-bid vs. Iterative

Iterative auctions provide valuable price feedback to bidders over a series of rounds. This price feedback can reduce winner's curse concerns when value uncertainty plays a role (Milgrom and Weber, 1982). It can also mitigate coordination problems, informing bidders about the intensity of their rivals' interests so that an informed trade-off between value and cost can be made.

Iterative auctions also make it easier for a board of directors to steer the bidding team during the course of an auction. In particular, iterative auctions may alleviate hidden-information problems and make it easier for the principal to implement his optimal strategy (Bichler and Paulsen, 2015).

In contrast, in sealed-bid auctions, value uncertainty can lead to surprising, and possibly problematic, outcomes. For example, in a first-price sealed-bid combinatorial auction conducted in Norway in 2013, one of the incumbents bid too low and did not win any spectrum. This incumbent was later forced to leave the market, and many argued this would not have happened if an iterative auction had been used.

Of course, the flip side is that entrants have better chances in sealed-bid formats, a point that has been advocated by Klemperer (2002). In an iterative auction, deep-pocketed incumbents have the opportunity to veto outcomes where entrants win licenses by topping their bids in a next round (recall, e.g., the German 2000 spectrum auction). Such pre-emptive behavior is more difficult in sealed-bid formats, which have been adopted by some countries for this reason.

And the flip side to the ability to solve coordination problems is that this resolution can, in principle, occur at any price levels. An iterative auction is more vulnerable to tacit collusion, e.g. when bidders decide on a strategy of mutual forbearance to divide the market as happened in the 1999 German spectrum auction (Grimm et al., 2003).

Bid Language: Expressiveness vs. Compactness

Combinatorial auctions provide bidders with a more flexible language to express their preferences for combinations of licenses, which is important when value complementarities (synergies) exist. However, a fully enumerative bid language, which allows bidders to submit bids on every possible package suffers from the *missing bids problem*, i.e. bidders can only specify bids for a small subset of the exponentially many packages. The missing bids problem can lead to substantial inefficiencies.

A compact bid language is less demanding in that it lets bidders specify packages of licenses with high synergies, but does not require an exponentially large set of bids.

Hierarchical package bidding (Goeree and Holt, 2010) is one example for a compact bidding language with regional licenses (see Chapter 20). Compact bid languages can also be developed for the award of national licenses (see Chapter 30) with some prior knowledge about the main synergies for bidders.

Regulators also need to be able to express their preferences and constraints. For example, allocation constraints can be used in the winner determination to avoid very unequal distributions of spectrum, when the policy goal is to achieve a competitive end-consumer market. Such constraints are important and they need careful design.

Payment Rules: Non-anonymous vs. Anonymous

The Vickrey-Clarke-Groves (VCG) mechanism plays a central role in mechanism design theory. Externality pricing to support efficient assignment is persuasive in the narrow context of a single auction where bidders' values are exogenously given, but less so when bidders compete in a downstream market afterwards. If one bidder has to pay considerably more than another for the same spectrum, as can be the case with VCG or the two-stage CCA, then this will have consequences for the downstream market. From a regulator's perspective, externality-based pricing may be *undesirable* as it typically means that smaller bidders (entrants) end up paying more than larger bidders (incumbents), because the externality they impose is larger. Such an outcome has adverse consequences for the well functioning of the downstream market. Anonymous and linear prices, such as used in the SMRA, HPB, or single-stage CCA, are preferable even if they do not necessarily lead to fully efficient outcomes (Bichler et al., 2013b).

5. Conclusion

Every theoretical model is built on assumptions and it is important to have them in mind when providing policy advice. Assuming that bidders have private and independent values and that regulators simply maximize efficiency (or revenue) of the auction in isolation, leads to elegant but not necessarily relevant modeling. Bidders in spectrum auctions face substantial value uncertainty, allocative and informational externalities, long-term investment concerns, budget constraints, etc. Regulators are mainly concerned about the well functioning of the downstream market, which requires a careful choice not only of the auction rules, but also of anti-collusion rules, spectrum caps, set asides, etc.

Taking account of these details and constraints is crucially important when implementing spectrum auction designs in the field. As argued above, they necessitate careful reconsideration of the auction process (iterative vs. sealed-bid), the bid language (compact vs. fully enumerative), and the payment rule (anonymous vs. non-anonymous).

As is typical in engineering disciplines, market designers often have to deal with changing, and sometimes conflicting, objectives. As such, it is unlikely that a single format will emerge that is preferable for all spectrum sales. For instance, large markets with many bidders and regional licenses (such as in the USA and Canada) require a different bid language than small national markets with only a few bidders.

While there is unlikely to be a "one-size-fits-all" design, two decades of spectrum sales in the field have affirmed Einstein's intuition that it is best to *"make things as simple as possible, but not simpler."* Simple pricing rules and simple bid languages can lead to high efficiency compared to complex ones (such as the CCA's "core pricing" and "fully expressive XOR bids"), see Chapter 30. That said, a bid language that is too simple, e.g. one that allows only for bids on individual licenses (as in the SMRA), can lead to low efficiency in the presence of significant synergies.

Fortunately, spectrum auction designs that strike a balance between simplicity and flexibility exist and have been successfully employed in high-stakes applications since 2008. Intuitive and transparent package auction designs, resulting from careful theoretical, laboratory, and simulation analyses, hold great promise for spectrum and other applications.

References

Bichler, M., Gretschko, V., Janssen, M., 2017. Bargaining in spectrum auctions: A review of the German auction in 2015. *Telecommunications Policy*.

Bichler, M., Paulsen, P., 2015. First-price package auctions in a principal-agent environment. In: *International Conference on Group Decision & Negotiation*.

Bichler, M., Schneider, S., Guler, K., Sayal, M., 2011. Compact bidding languages and supplier selection for markets with economies of scale and scope. *European Journal on Operational Research* 214, 67–77.

Bichler, M., Shabalin, P., Wolf, J., 2013a. Do core-selecting combinatorial clock auctions always lead to high efficiency? An experimental analysis of spectrum auction designs. *Experimental Economics* 1–35.

Bichler, M., Shabalin, P., Ziegler, G., 2013b. Efficiency with linear prices? A theoretical and experimental analysis of the combinatorial clock auction. *INFORMS Information Systems Research* 24, 394–417.

Dasgupta, P., Maskin, E., 2000. Efficient auctions. *Quarterly Journal of Economics* 115, 341–388.

Dobzinski, S., Lavi, R., Nisan, N., 2008. Multi-unit auctions with budget limits. In: *Foundations of Computer Science*. Philadelphia, PA, pp. 260–269.

Goeree, J. K., 2003. Bidding for the future: Signaling in auctions with an aftermarket. *Journal of Economic Theory* 108 (2), 345–364.

Goeree, J. K., Holt, C., 2010. Hierarchical package bidding: A paper & pencil combinatorial auction. *Games and Economic Behavior* 70 (1), 146–169.

Goeree, J. K., Offerman, T., 2003a. Competitive bidding in auctions with private and common values. *The Economic Journal* 113 (489), 598–613.

Goeree, J. K., Offerman, T., 2003b. Winner's curse without overbidding. *European Economic Journal* 47 (1), 625–644.

Goetzendorff, A., Bichler, M., Shabalin, P., Day, R., 2015. Compact bid languages and core pricing in large multi-item auctions. *Management Science* 61 (7), 1684–1703.

Grimm, V., Riedel, F., Wolfstetter, E., 2003. The third generation (UMTS) spectrum auction in Germany. In: Illing, G., Kuehl, U. (eds.), *Spectrum Auctions and Competition in Telecommunications*. CESifo Seminar Series.

Janssen, M., Karamychev, V., 2017. Spiteful bidding and gaming in combinatorial clock auctions. *Games and Economic Behavior* 100, 186–207.

Jehiel, P., Moldovanu, B., 2001. Efficient design with interdependent valuations. *Econometrica* 69 (5), 1237–1259.

Jehiel, P., Moldovanu, B., 2005. Allocative and informational externalities in auctions and related mechanisms. Tech. rep., SFB/TR 15 Discussion Paper.

Klemperer, P., May 2002. How (not) to run auctions: The European 3G telecom auctions. *European Economic Review* 46 (4–5), 829–845. http://linkinghub.elsevier.com/retrieve/pii/S0014292101002185

Korczyk, D. J., 2008. Considerations in valuing wireless spectrum. Discussion paper, American Appraisal Associates, Inc.

Levin, J., Skrzypacz, A., 2016. Properties of the combinatorial clock auction. *American Economic Review* 106, 2528–2551.

Milgrom, P., 2004. *Putting Auction Theory to Work*. Cambridge University Press.

Milgrom, P., 2010. Simplified mechanisms with an application to sponsored-search auctions. *Games and Economic Behavior* 70 (1), 62–70.

Milgrom, P. R., Weber, R. J., 1982. A theory of auctions and competitive bidding. *Econometrica* 50 (5), 1089–1122.

Nisan, N., Ronen, A., 2001. Algorithmic mechanism design. *Games and Economic Behavior* 35, 166–196.

Petrakis, I., Ziegler, G., Bichler, M., 2013. Ascending combinatorial auctions with allocation constraints: On game theoretical and computational properties of generic pricing rules. *Information Systems Research* 24 (3), 768–786.

Vazirani, V. V., Nisan, N., Roughgarden, T., Tardos, E., 2007. *Algorithmic Game Theory*. Cambridge University Press.